Handbook of Blind Source Separation

Independent Component Analysis and Applications

Handbook of Blind Source Separation

Independent Component Analysis and Applications

Edited by P. Comon and C. Jutten

ELSEVIER

AMSTERDAM • BOSTON • HEIDELBERG • LONDON • NEW YORK • OXFORD
PARIS • SAN DIEGO • SAN FRANCISCO • SINGAPORE • SYDNEY • TOKYO

Academic Press is an imprint of Elsevier

Academic Press is an imprint of Elsevier
The Boulevard, Langford Lane, Kidlington, Oxford, OX5 1GB, UK
30 Corporate Drive, Suite 400, Burlington, MA 01803, USA

First edition 2010

British Library Cataloguing in Publication Data
A catalogue record for this book is available from the British Library

Library of Congress Control Number: 2009941417

ISBN: 978-0-12-374726-6

For information on all Academic Press publications visit our website at elsevierdirect.com

Printed and bound in the United States
10 11 12 11 10 9 8 7 6 5 4 3 2 1

Contents

About the editors

The two editors are pioneering contributors of ICA. They wrote together the first journal paper on ICA, which appeared in *Signal Processing*, published by Elsevier in 1991, and received a best paper award in 1992, together with J. Hérault.

Pierre Comon is Research Director with CNRS, Lab. I3S, University of Nice, France. He has been Associate Editor of the *IEEE Transactions on Signal Processing*, and the *IEEE Transactions on Circuits of Systems I*, in the area of blind techniques. He is now Associate Editor of the *Signal Processing* journal, published by Elsevier. He has been the coordinator of the European network "ATHOS" on High-Order Statistics. He received the Monpetit prize from the French Academy of Sciences in 2005 (rewarding works with industrial applications), and the Individual Technical Achievement Award from Eurasip in 2006. He is Fellow of the IEEE, Emeritus Member of the SEE, and member of SIAM. He authored a paper in 1994 on the theoretical foundations of ICA; this paper still remains among the most cited both on the subject of ICA and blind techniques, in the whole signal processing community.

Christian Jutten is Professor at the University Joseph Fourier of Grenoble, France. He is currently associate-director of GIPSA-lab, a 300-people laboratory focused on automatic control, signal, images and speech processing. He has been Associate Editor of the *IEEE Transactions on Circuits and Systems I*, in the area of Neural Networks and Signal Processing techniques. He is currently Associate Editor of *Neural Processing Letters*, published by Kluwer. He was the co-organizer of the first international conference on Blind Source Separation and Independent Component Analysis, in 1999 (ICA 99). He was the coordinator of two European projects, one of them (BLISS) focused on Blind Source Separation and Applications. He received the Blondel Medal of the SEE in 1997 for his contributions in blind source separation. He is Fellow of the IEEE and Senior Member of Institut Universitaire de France. He co-authored a set of two papers in 1991, with J. Hérault and P. Comon, on the first algorithm for blind source separation and on theoretical foundation, which still remains in the top five papers cited on the subject of ICA and/or blind techniques, and in the whole signal processing community.

Preface

In signal processing, a generic problem consists in separating a useful signal from noise and interferences. Classical approaches of the twentieth century are based on *a priori* hypotheses, leading to parameterized probabilistic models. Blind Source Separation (BSS) attempts to reduce these assumptions to the weakest possible.

As shown in this handbook, there are various approaches to the BSS problem, depending on the weak *a priori* hypotheses one assumes. The latter include either statistical independence of source signals or their sparsity, among others.

In order to prepare this book, among the best worldwide specialists were contacted to contribute (cf. page xviii). One of them, Serge Degerine, passed away unexpectedly during the writing of Chapter 7. We would like to dedicate this book to his memory.

This handbook is an extension of another book which appeared in 2007 in French, and published by Hermes. The present version contains more chapters and many additions, provided by contributors with international recognition. It is organized into 19 chapters, covering all the current theoretical approaches, especially Independent Component Analysis, and applications. Although these chapters can be read almost independently, they share the same notations and the same subject index. Moreover, numerous cross-references link the chapters to each other.

Pierre Comon and Christian Jutten

Contributors

Laurent Albera
Rennes, France

Moeness Amin
Villanova, PA, USA

Massoud Babaie-Zadeh
Teheran, Iran

Rasmus Bro
Copenhagen, Denmark

Jean-François Cardoso
Paris, France

Marc Castella
Evry, France

Pascal Chevalier
Colombes, France

Antoine Chevreuil
Marne-la-Vallée, France

Andrzej Cichocki
Tokyo, Japan

Pierre Comon
Sophia-Antipolis, France

Lieven De Lathauwer
Leuven, Belgium

Ali Mohammad-Djafari
Orsay, France

Yannick Deville
Toulouse, France

Rémi Gribonval
Rennes, France

Aapo Hyvärinen
Helsinki, Finland

Christian Jutten
Grenoble, France

Amar Kachenoura
Rennes, France

Ahmad Karfoul
Rennes, France

Juha Karhunen
Helsinki, Finland

Kevin H. Knuth
Albany, NY, USA

Eric Moreau
Toulon, France

Lucas C. Parra
New York, USA

Jean-Christophe Pesquet
Marne-la-Vallée, France

Mark D. Plumbley
London, United Kingdom

Dinh-Tuan Pham
Grenoble, France

Lotfi Senhadji
Rennes, France

Dirk Slock
Sophia-Antipolis, France

Nadège Thirion-Moreau
Toulon, France

Ricardo Vigario
Helsinki, Finland

Emmanuel Vincent
Rennes, France

Arie Yeredor
Tel-Aviv, Israel

Vicente Zarzoso
Sophia-Antipolis, France

Michael Zibulevsky
Technion, Israel

CHAPTER

Introduction

1

C. Jutten and P. Comon

Blind techniques were born in the 1980s, when the first adaptive equalizers were designed for digital communications [67,33,10,28]. The problem was to compensate for the effects of an unknown linear single input single output (SISO) stationary channel, without knowing the input.

The scientific community used the word "blind" for denoting all identification or inversion methods based on output observations only. In fact, *blind techniques* in digital communications aimed at working when the "eye[1] was closed"; hence the terminology.

At the beginning, the word "unsupervised" was sometimes used (for instance in French the wording *autodidacte*), but it seems now better to be consistent with the worldwide terminology, even if this is not ideal, since comprehensible only in the context of digital communications.

The problem of blind source separation (BSS) differs from blind equalization, addressed previously by Sato, Godard and Benveniste, by the fact that the unknown linear system consists of several inputs and outputs: such a system is referred to as *multiple inputs multiple outputs* (MIMO). Initially restricted to memoryless channels, the BSS problem now encompasses all linear or nonlinear MIMO mixtures, with or without memory.

The BSS problem was first formulated in 1984, although theoretical principles, which drive source separation methods, were understood later. In this chapter, we briefly introduce the principles and main notations used in this book. A few ideas which contributed to the development of this research domain from its birth are reviewed. The present chapter ends with a short description of each of the 18 subsequent chapters.

1.1 GENESIS OF BLIND SOURCE SEPARATION

The source separation problem was formulated around 1982 by Bernard Ans, Jeanny Hérault and Christian Jutten [36,38,4,37], in the framework of neural modeling, for motion decoding in vertebrates [66]. It seems that the problem has also been sketched independently in the framework of communications [6]. First related contributions to Signal Processing conferences [38] and to Neural Networks conferences [4,37] appeared

[1] The "eye" is a diagram obtained on the oscilloscope when looking at a synchronized discrete (e.g. BPSK) signal.

around 1985. Immediately, these papers drew the attention of signal processing researchers, mainly in France, and later in Europe. In the neural networks community, interest came much later, in 1995, but very massively.

Since the middle of the 1990s, the BSS problem has been addressed by many researchers, with expertise in various domains: signal processing, statistics, neural networks, etc. Numerous special sessions have been organized on these topics in international conferences, for instance in GRETSI since 1993 (France), NOLTA'95 (Las Vegas, USA), ISCAS (Atlanta, USA), EUSIPCO since 1996, NIPS'97 post workshop (Denver, USA), ESANN'97 (Bruges, Belgique), IWANN'99 (Alicante, Spain), MaxEnt2006 (Paris, France).

The first international workshop fully devoted to this topic, organized in Aussois in the French Alps in January 1999, attracted 130 researchers world-wide. After the first international papers, published in 1991 in the journal *Signal Processing* [44,26,70], various international journals contributed to the dissemination of BSS: *Traitement du Signal* (in French), *Signal Processing, IEEE Transactions on Signal Processing, IEEE Transactions on Circuits and Systems, Neural Computation, Neural Networks*, etc. In addition, a Technical Committee devoted to blind techniques was created in July 2001 in the IEEE Circuits and Systems Society, and BSS is a current "EDICS" in *IEEE Transactions on Signal Processing*, and in many conferences.

Initially, source separation was investigated for instantaneous (memoryless) linear mixtures [38]. The generalization to convolutive mixtures was considered at the beginning of the 1990s [21]. Finally, nonlinear mixtures, except a few isolated works, were addressed at the end of 1990s [12,40,73]. In addition, independent component analysis (ICA), which corresponds to a general framework for solving BSS problems based on statistical independence of the unknown sources, was introduced in 1987 [42], and formalized for linear mixtures by Comon in 1991 [22,23]. Beyond source separation, ICA can also be used for decomposition of complex data (signals, images, etc.) in sparse bases whose components have the mutual independence property [29]. ICA also relates to works on sparse coding in theoretical biology presented by Barlow in 1961 [5,57], and other works on factor analysis in statistics [30,31,27,46].

The number of papers published on the subject of BSS or ICA is enormous: in June 2009, 22,000 scientific papers are recorded by Google Scholar in Engineering, Computer Science, and Mathematics. On the other hand, few books present the BSS problem and the main principles for solving it. One can mention a book written by specialists of Neural Networks [39], containing only algorithms developed within the Machine Learning community. The present book aims at reporting the state of the art more objectively. A book with a wider scope is now certainly needed. Another rather complete book appeared slightly later [20]. However, some ways of addressing the problem were still missing (semi-blind approaches, Bayesian approaches, Sparse Components Analysis, etc.), and we hope the present book will complement it efficiently. More specific problems, i.e. separation of audio sources [51], or separation in nonlinear mixtures [2], have been the subject of other contributions.

The present book is hopefully a reference for all aspects of blind source separation: problem statements, principles, algorithms and applications. The problem can be stated in various contexts including fully blind static (ICA), and convolutive or nonlinear mixtures. It can be addressed in blind or semi-blind contexts, using second-order statistics if sources are assumed colored or nonstationary, or using higher order statistics, or else using time-frequency representations. There is a wide variety of mathematical problems, depending on the hypotheses assumed. For instance, the case of underdetermined mixtures is posed in quite different terms for sparse sources; the Bayesian approach is quite different from approaches based on characteristic functions or cumulants, etc.

In the next section, we first present the biological problem which was at the origin of blind source separation, and locate it in the scientific context of the 1980s. Then, we explain how and why a few researchers became interested in this problem: the answers have been given by the researchers themselves, and this section is partly extracted from [45].

1.1.1 A biological problem

Blind source separation was first considered in 1982 from a simple discussion between Bernard Ans, Jeanny Hérault and Christian Jutten with Jean-Pierre Roll, a neuroscientist, about *motion decoding* in vertebrates. Joint motion is due to muscle contraction, each muscle fiber being controlled by the brain, through a motoneuron. In addition, on each fiber, the muscle contraction is measured, and transmitted to the central nervous system by two types of sensorial endings, located in tendon, and called primary and secondary endings. The proprioceptive responses of the two types of endings are presented in Figs 1.1 and 1.2 respectively, for simple joint motion, at constant angular speed. For reliability reasons, results are obtained by averaging unit sensory responses coming from a large number of fibers, related to the repetition of the same forced motion. Figures 1.1 and 1.2 present frequencygrams, i.e. a representation where a spike at time t is represented by a point at t on the x axis and with a value on y axis equal to the inverse of the interval between the spike and the previous one, i.e. corresponding to the instantaneous frequency. Following Roll, here are the main comments concerning the frequencygrams:

- For a constant joint location, responses of the two endings are constant, i.e. the spike instantaneous frequency is constant. The instantaneous frequency is increasing with muscle stretching. The frequency/stretching ratio is similar, on the average, on the two types of endings.
- During a joint motion at constant (stretching) speed, the instantaneous frequency appears as the superimposition of the constant signal (for the speed) on the signal related to the muscle stretching. This is true for the two types of endings, with some differences.
- The response of primary endings is characterized by an initial burst (derivative effect), at the beginning of the motion, while the typical response of secondary endings is low-pass.

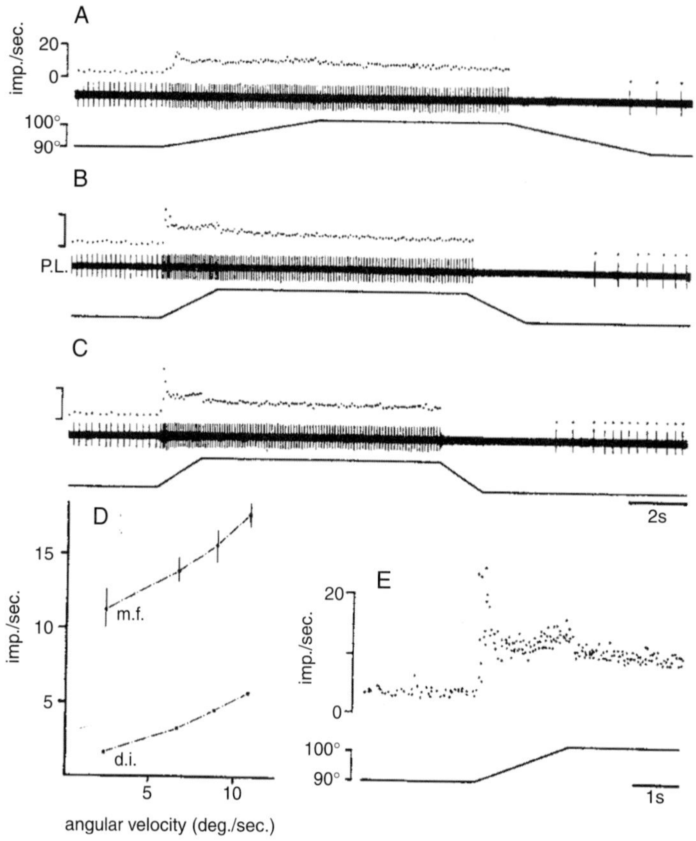

FIGURE 1.1

(A), (B) and (C) Responses of primary ending to a forced motion of a joint, with three constant speeds, (D) Frequency versus angular speed of the joint (E) Superimposition of many responses obtained for the same angular speed.
Source: From Roll [66].

- On average, the ratio frequency/speed is larger for primary endings than for secondary ones.

Surprisingly, while we could imagine that each type of ending only transmits one type of information, either stretching or speed, the proprioceptive information transmitted by endings is a mixture of stretching and speed information.

Neglecting transient phenomena (high and low-pass effects), and denoting $p(t)$ the angular position (related to muscle stretching), $v(t) = dp(t)/dt$ the angular speed, and $f_1(t)$ and $f_2(t)$ the instantaneous frequency of primay and secondary endings, respectively, one can propose the following model:

$$f_1(t) = a_{11}v(t) + a_{12}p(t)$$
$$f_2(t) = a_{21}v(t) + a_{22}p(t) \tag{1.1}$$

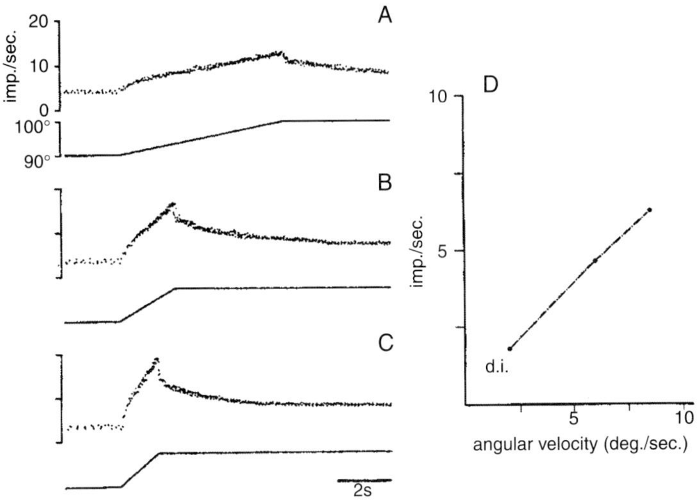

FIGURE 1.2

(A), (B) and (C) Responses of secondary ending for a forced motion of a joint, with three constant speeds, (D) Frequency versus angular speed of the joint.
Source: From Roll [66].

in which $v(t)$, $p(t)$ et a_{ij} are unknown, with the assumptions $a_{11} > a_{12}$ and $a_{22} > a_{21}$.

Estimating $v(t)$ and $p(t)$ from $f_1(t)$ and $f_2(t)$ seems impossible. However, even during forced motion, even with closed eyes, the central nervous system is able to separate joint speed and location while they are arriving as mixtures. As said by Mc Closkey [53] in 1978: "*Clearly, if spindle discharges are to be useful for kinesthetic sensations, the central nervous system must be able to distinguish which part of the activity is attributable to muscle stretch and which part is caused by fusimotor activity*".

Denoting $\mathbf{x}(t) = (f_1(t), f_2(t))^T$ and $\mathbf{s}(t) = (v(t), p(t))^T$, where T denotes transpose, and $\mathbf{A}$ the (mixing) matrix with general entry a_{ij}, this model is an instantaneous linear mixture:

$$\mathbf{x}(t) = \mathbf{A}\mathbf{s}(t), \tag{1.2}$$

in which we assume that the components ($v(t)$ and $p(t)$) of the source vector $\mathbf{s}(t)$ are statistically independent. Source separation is then based on estimating a separating matrix $\mathbf{B}$ such that the vector $\mathbf{y}(t) = \mathbf{B}\mathbf{x}(t)$ has mutually independent components. The first source separation algorithm was an adaptive algorithm based on a set of estimating equations [38], which cancelled higher (than 2) cross-moments of all the component pairs of the vector $\mathbf{y}(t)$, which is (under mild conditions) a simple independence criterion, as shown later [26].

1.1.2 Contextual difficulties

1.1.2.1 Independence

The first difficulty encountered for explaining the problem in the 1980s was related to statistical independence.

At first glance, this assumption seems very strong. In fact, we may think it rarely occurs, and it is more difficult to satisfy than second order independence i.e. non-correlation. In the above biological problem, it was indeed tricky to explain statistical independence between $p(t)$ and $v(t)$. In fact, one can remark that "*speed* $v(t)$ *is related to location* $p(t)$ *through* $v(t) = dp(t)/dt$. *The two variables then cannot be independent*". However, this functional dependence is not a statistical dependence. The knowledge of $p(t)$ at a given instant t does not provide any information on $v(t)$, and *vice versa*. In other words, while the random vectors of speed and position, $[v(t_1), v(t_2), \ldots, v(t_k)]^T$ and $[p(t_1), p(t_2), \ldots, p(t_k)]^T$ are generally not independent, the (marginal) random variables, $v(t)$ and $p(t)$, at any given time t are independent. Therefore, for instantaneous ICA algorithms, the dependence is irrelevant.

1.1.2.2 Second-order or higher-order statistics

Relationships between statistical independence and non-correlation are presented in any textbook on probability. However, in the 1980s, most signal models were assumed to be Gaussian, so that concepts of independence and non-correlation are the same. Fortunately, about at the same time, an increasing interest in methods based on higher order statistics (HOS) manifested itself. As a by-product, differences between statistical independence and lack of second order correlation were recognized and exploited. The first international workshop on HOS was held in Vail (Colorado, USA) in 1989, and contributed to the expansion of blind techniques.

1.1.2.3 Separable or not?

In 1983, Bienvenu and Kopp [11] showed that dominant eigenvectors of the spectral covariance matrix span the signal subspace, but cannot provide the sources. This result is based on algebraic arguments. Considering the extension of the linear model (1.2) from 2 mixtures of 2 sources ($v(t)$ and $p(t)$ in the above neurophysiological problem) to a larger (and equal) number of mixtures and sources, say $N = P$, the number of equations ($(N^2 - N)/2$) is smaller than the number of parameters ($(N^2 - N)$). Consequently, source separation has been considered as impossible to solve by most researchers for several years. During the poster session in GRETSI 85, comments on [38] raised surprise or skepticism. Two years later, in 1987, Lacoume, although he agreed with the results of Bienvenu and Kopp, thought that HOS (e.g. 4th order cumulants) could solve the problem [47], by introducing supplementary equations (like Nikias and others did in other problems). Some years later, he stated relationships governing source separation under the assumption of statistical independence. In particular, he designed (with Gaeta) a method based on maximum likelihood, in which source distributions were approximated by Gram-Charlier expansions [32]. During the same period, others also achieved this with other approaches (see Vail workshop in 1989). However, it can be observed that it took the scientific community nearly three years to realize that the separation problem could indeed be solved.

1.1.2.4 Source separation and neural networks

In 1985 and 1986, the first studies on source separation were presented in neural network conferences, Cognitiva'85 (Paris, France) and Snowbird'86 (Utah, USA). These communications attracted the interest of a few reseachers, but have been outshined by new interesting studies. For instance, at Cognitiva'85, Le Cun [49] published a new learning algorithm for multi-layer perceptrons (MLP) which became famous as the backpropagation algorithm. During Snowbird'86, most researchers were very excited by Hopfield models, Kohonen's self-organizing maps, MLP and backpropagation algorithms. For instance, Terry Sejnowski presented a nice exhibition of NetTalk, showing MLP applications. But during this conference, he began to be interested in source separation as he explained in 2000: *"Because I did not understand why your network model could get the results that it did, new students in my lab often were offered that question as a research problem. Shaolin Li, a Chinese postdoc, made some progress by combining beamforming with your algorithm. This project was started around 1991."*

1.1.2.5 Terminology

Keywords *source separation* and *independent component analysis (ICA)* have not been used from the beginning. For instance, first papers [36,38] had very long and intricate titles as in GRETSI'85, whose English translation is: *Detection of primary signals in a composite message using a neuromimetic architecture with unsupervised learning*. In 1986, the word *source discrimination* was used, but the keyword *source separation* became accepted only after 1987. Blind techniques also received a strange wording by P. Duvaut in 1990 [1], whose English translation corresponds to: "clairvoyant methods". Concerning ICA, the word was first introduced in 1987 [42] but the concept was formalized later by Comon in 1991 [22,23].

1.1.3 A few historical notes

1.1.3.1 A few pioneers

Although the GRETSI'85 and Snowbird'86 communications did draw some attention from a few researchers into this problem, they actually raised great interest from L. Kopp (Thomson-Sintra company) who engaged P. Comon for working on this problem in 1988. In this company, nobody wanted to work on this mysterious problem, for which it was difficult to guarantee any outcome within less than a year. During a workshop in Grenoble in September 1987, J.-F. Cardoso visited J. Hérault and C. Jutten, who explained to him the principles of source separation and showed him a real-time exhibition based on a source separation hardware demonstrator: a purely analog device based on operational amplifiers, transistors and audio amplifier that Jutten built in 1985, and which is able to separate, in real-time, two audio sources in a mixture controlled by potentiometers [43]. Immediately and independently, J.-F. Cardoso and P. Comon became enthusiastic about blind source separation, and the rest of the story is known.

We cannot give here the complete list of their contributions, but we would like to focus on the earliest ones. P. Comon adapted the concept of contrast function, inspired by Donoho's contrast function used in blind deconvolution, and formulated the ICA concept, first published at the HOS workshop in 1991 [22] and then in a famous paper published in Signal Processing [23]. J.-F. Cardoso introduced many concepts: a tensorial approach for conveniently representing and processing cumulants (together with P. Comon) [15], performance analysis of ICA [17,13], joint diagonalization [17] and the concept of equivariance [48,16] with the relative gradient (independently of Amari and Cichocki's natural gradient).

Finally, we believe that the success of the Signal Processing papers [44,26,70] has been due to a surprising performance with respect to the algorithm's simplicity. Half a day was sufficient to write and test the algorithm. However, clearly, the performance was dependent on the mixture's hardness. J.-F. Cardoso, with Laheld, were looking for algorithms enjoying invariant performances.

1.1.3.2 French and European supports

BSS and ICA benefited from the researcher interactivity inside the French signal processing community, first, through GRETSI conferences, which have brought together about 500 signal and image processing French researchers every other year since 1967. Moreover, in 1989, a French research group, funded by the French National Center for Research (CNRS) and the Ministry of Research, was created, for organizing scientific (informal) working groups on various topics. One of these working groups, first focusing on HOS and then on BSS and ICA, was supervised by J.-F. Cardoso from 1990 to 1997, who organized about three technical meetings per year, with, for each, on the average eight talks and 30 attendees. This working group was still active, supervised by E. Moreau and others, up to the middle of the 2000s. In parallel, the Working Group ATHOS (Advanced Topics in High Order Statistics) funded by the European Community and coordinated by P. Comon, contributed to promote BSS and ICA in the European signal processing community between 1991 and 1995.

It is surprising that American researchers addressed the problem relatively late. We believe, and we have had some hints through review reports of papers we submitted, that they considered the problem was simply a special case of blind multichannel equalization with trivial 0-order Moving Average filters. Of course, this is wrong, since in ICA (unlike blind equalization), the sources are not required to be temporally independently distributed in time (iid), but we often neglected to emphasize this point. In addition, MA models were generally assumed to be monic [72], which means that there is no spatial mixture at the zero delay.

1.1.3.3 From neural PCA to ICA

E. Oja, J. Karhunen et al. came to ICA and BSS by extending PCA neural networks which were popular at the end of the 1980s. In 2000, Karhunen explained to

Jutten: *"However, we knew that PCA can be realized more efficiently and accurately using standard numerical software, because the problem is linear. Therefore, we wanted to study nonlinear generalizations of PCA neural networks and learning rules. In those problems, there usually does not exist any such efficient conventional solution. We were also looking for nonlinear generalizations of PCA which could achieve something more than standard PCA. We developed several nonlinear neural extensions of linear PCA from different starting points. These developments are summarized in my two journal papers published in Neural Networks in 1994 and 1995... However, a problem with these extensions was that we had not at that time any convincing applications showing that nonlinear PCA is really useful and can provide something more than standard PCA. Independent component analysis is an extension of linear PCA, where uncorrelatedness assumptions are replaced by the stronger independence assumptions while relaxing the requirement of mutually orthogonal basis vectors. I was interested in that, especially after seeing your 1991 papers published in Signal Processing."*

1.1.3.4 From neural coding to ICA

The well known contribution of T. Bell and T. Sejnowsky [7] proposed some links between neural networks and entropy. However, the ideas which guided T. Bell were closer to theoretical biology. As Terry Sejnowsky said: *"Tony's motivation was based on a deep intuition that nature has used optimization principles to evolve nervous systems that need to self-organize channels in dendrites (his PhD thesis in Computer Science in Belgium) and in organizing the representations in the visual system. Although his 1995 paper in Neural Computation gets more citations, his 1997 paper in Vision Research is closer to his heart"* [8]. In the same spirit, J.-P. Nadal and N. Parga [57], from reflections on ICA, information theory and the concept of sparse neural coding introduced by Barlow at beginning of 1960s [5], very early made interesting, yet unrecognized, contributions [57].

1.1.3.5 From statistics to ICA

Clearly, independence is related to probability and statistics, and tools borrowed from these fields bring a lot to source separation. For instance, it appears that factorial analysis, intensively studied in statistics in the 1950s, is another way to formalize ICA, especially the separability problem. Although many results were available for many years, the Darmois results [27] have been brought to light by P. Comon in 1991 [22], and more recently, researchers used a few theorems published in the statistics book [46] published in 1973! We finish with an anecdote, which shows that it would have been possible to go faster.

In 1990, for administrative reasons, a PhD student had to register in statistics instead of signal processing post-graduate courses in Grenoble. However, he wanted absolutely to work on source separation (topics the student discovered thanks to the French working group), and D.T. Pham agreed to supervise him. J. Hérault and C. Jutten gave a short talk (2 hours) to D.T. Pham on source separation. Three days after, he sent them a 5 or 6-page note in which he sketched the Maximum Likelihood solution and emphasized the relevance of score functions in estimating equations. It was then realized that the nonlinear functions used in the first algorithm [44] correspond to a heuristic choice (fortunately robust), but are optimal for particular distributions. Since this date, Pham has brought valuable contributions [63,59,61] to the problem.

1.1.3.6 RIKEN laboratory contribution

Since 1990, Riken labs in Japan, especially Amari's and Cichocki's teams in Wako-shi near Tokyo, have been very active in source separation and ICA areas. In particular, many foreign researchers have been invited or hired for working in this domain. Finally, from 2002, many source separation and ICA softwares (but not all) have been merged in the ICAlab package which is available on line.

In 2000, A. Cichocki wrote to C. Jutten: "*I have started close and fruitful collaboration with Professor Amari and also other researchers [...] from April 1995 when I joined Frontier Research Program Riken, JAPAN, and I would like to mention that I have learned a lot from Professor Amari and his ideas. Before this fruitful collaboration I have started to study BSS/ICA since 1991 after reading several of your influential papers, including your Doctorate thesis and works of your Ph.D students. When I was in Germany in 1992–1994, at University Erlangen Nuremberg, we have published several brief papers (Electronics Letters 1992/94 IEEE Transaction on Circuits and Systems) and also in our book (in April 1993) we presented neural network approach to BSS: Neural Networks for Optimization and Signal Processing by A. Cichocki and R. Unbehauen (J. Wiley 1993 pp. 461–471).*"

Shun-Ichi Amari adds: "*I knew the Jutten-Hérault idea of source separation in late eighties, and had interest in, but did not do any work on that subject. It was in 1994 when Cichocki visited us and emphasized the importance of the subject that I had again interest. He showed me a number of papers in 1995, one of which is Bell-Sejnowski paper. I was impressed by that one, and thought that I could study more general mathematical aspects.*

One of the results is the idea of natural gradient, which we proposed in our 1995 NIPS paper (Amari-Cichocki-Yang, appeared in Proc. NIPS, 1996). The algorithm itself was proposed by Cichocki earlier, and also by Cardoso in 1995. But ours has a rigorous foundation based on the Lie group invariance and Riemannian metric derived therefrom.

From that on, I have carried out intensive research on this interesting subject, in particular, its mathematical foundations, in collaboration with Cichocki and many others.

One is its foundation from the point of semiparametric statistical models and information geometry. [...] I also have studied efficiency and super-efficiency of algorithms [3]. *There are a number of other ideas, but it is too much to state all of them.*"

1.2 PROBLEM FORMALIZATION

The BSS problem consists of retrieving unobserved sources, denoted in vector notation as $\mathbf{s}(t) = (s_1(t), \ldots, s_N(t))^T \in \mathbb{R}^N$, assuming zero mean and stationary from observed mixtures, $\mathbf{x}(t) = (x_1(t), \ldots, x_P(t))^T \in \mathbb{R}^P$, which can be written:

$$\mathbf{x}(t) = \mathscr{A}(\mathbf{s}(t)) \tag{1.3}$$

where $\mathscr{A}$ is an unknown mapping from $\mathbb{R}^N$ in $\mathbb{R}^P$, and where t denotes the sample index, which can stand for time for instance.

Various mixture models have been considered, initially linear instantaneous mixtures, then the linear *convolutive mixtures* at the beginning of the 1990s [21] (see Chapter 8), and more recently at the end of the 1990s the *nonlinear models* [73] (see Chapter 14). It is clear that, without extraneous assumptions, this problem is ill-posed. In order to overcome this problem, the usual assumption is the mutual statistical independence among the unknown sources. Although sometimes intricate to set up, this assumption is realistic and fully justified in many problems. However, other assumptions can be used successfully for ensuring mixture identifiability or source separability.

1.2.1 Invertible mixtures

If the mapping $\mathcal{A}$ is invertible (very often, the condition $N \leq P$ is then necessary, where N and P denote the number of sources and the number of sensors, respectively), identification of $\mathcal{A}$ or of its inverse $\mathcal{B}$, directly leads to source separation, i.e. provides estimated sources $y_i(t)$ such that:

$$y_i(t) = k_i(s_{\sigma(i)}(t)), \quad i = 1, 2, \ldots, N \tag{1.4}$$

where σ is a permutation in $\{1, 2, \ldots, N\}$, and k_i is a mapping corresponding to a residual distortion. This equation explicitly shows the typical indeterminacies of the BSS problem. According to the nature of the mixing, indeterminacies can be more or less severe: for simple linear instantaneous mixtures (i.e. when $\mathcal{A}$ is restricted to a simple matrix $\mathbf{A}$), sources are estimated up to a permutation and a scale.

1.2.2 Underdetermined mixtures

If the number of sources N is larger than the number of sensors P, the mixing is referred to as underdetermined (see Chapter 9), and is not invertible. In this case, the problems of identification of the mixing mapping $\mathcal{A}$ and of source restoration become two distinct problems. For linear memoryless mixtures, even if the mixing matrix is perfectly known, there exists an infinity of solutions. Priors are necessary (for instance sources can be discrete-valued, or sparse, i.e. with a small number of non-zero samples) in order to restore essential uniqueness of source inputs.

1.3 SOURCE SEPARATION METHODS

It is clear that, without additional assumptions, the BSS problem is ill-posed, even if a scale-permutation ambiguity is allowed. More precisely, for linear instantaneous mixtures, Darmois [27] showed that the problem has no solution for Gaussian and temporally iid sources. One can restore the well-posedness of the problem by imposing somehow a diversity between sources – among others, by the following two ideas [14]:

- One can assume sources are possibly temporally iid but non-Gaussian, which will lead to using higher (than 2)-order statistics, and correspond to ICA methods first developed.

- One can assume sources are possibly Gaussian but non-temporally iid, which only requires second-order statistics, and assumes sources have temporal structure.

1.3.1 Independent component analysis

The first idea, assuming that the sources are temporally iid and non-Gaussian, leads to methods referred to as independent component analysis (ICA). The fundamental assumption is that the unknown sources are statistically independent [22,23]. If the sources admit a probability density function (pdf), this assumption means that the joint pdf can be factorized as the product of the marginal pdf's:

$$p_s(s_1, s_2, \ldots, s_n) = p_1(s_1)p_2(s_2)\ldots p_n(s_n) \tag{1.5}$$

Basically, for linear instantaneous mixtures, ICA methods aim at estimating a demixing matrix $\mathbf{B}$ yielding estimated sources $\mathbf{y}(t) = \mathbf{B}\mathbf{x}(t)$, which are statistically independent. In fact, in that case, Comon showed [22,23] that output independence leads to a matrix $\mathbf{B}$ that satisfies $\mathbf{BA} = \mathbf{PD}$, where $\mathbf{P}$ is a permutation matrix and $\mathbf{D}$ is a diagonal matrix, i.e. estimated outputs are equal to the sources, up to a permutation and a scale. Actually the above independence criterion is not convenient, since it not only requires equality of two multivariate functions, but also requires their perfect knowledge to start with. Consequently, other independence measures based on the second characteristic function or the Kullback-Leibler divergence (see Chapter 2), lead to more convenient criteria and contrast functions (see Chapter 3), which always involve (explicitly or implicitly) higher-order statistics.

1.3.2 Non-temporally iid sources

In the initial methods [38,23], ICA does not use temporal structure of sources. It is based on two assumptions: (i) samples of each source $s_i(t)$ need to be identically distributed, according to a pdf $p_i(s_i)$, and (ii) any source $s_i(t)$ need to be independent of other sources, $s_j(t), j \neq i$. Strictly speaking, the iid assumption is not required, but the possible temporal structure of the sources is not considered: in ICA, sources may or may not be temporally iid.

However, another idea is based on non iid sources. At the end of the 1980s, a few algorithms were designed, based on the assumption that sources are temporally colored [75,9,54]: successive samples are then no longer independent, and the first "i" of the iid property is broken.

Later, algorithms exploiting source nonstationarity, i.e. breaking the "id" of the iid property, were designed [52,62].

These two approaches exploit properties of variance–covariance matrices, i.e. second-order statistics (see Chapter 7). These approaches have two main advantages with respect to the original ICA:

- They are simpler (second order statistics, only).
- They are able to separate Gaussian (or non-Gaussian) sources.

On the other hand, they need additional assumptions concerning linear independence between correlation profiles. Lastly, from an algorithmic point of view, these methods use joint diagonalization algorithms, which are now fast and efficient.

1.3.3 Other approaches

Exploiting other properties of sources, one can propose other methods. For instance, one can design algorithms based on geometrical properties of the joint distribution. This has been shown to be possible when sources have a bounded support [64,74, 60]. Moreover in digital communications, discrete-valued or cyclostationary sources are realistic assumptions which lead to other algorithms [34,41,18]. For the separation of speech signals, exploiting a coarse video of the speaker face allows extraction of a given source with improved performance [69,65].

From a general viewpoint, the principle of the BSS method is based on *diversity*, and initially spatial diversity (sensor array) was considered. However, one can enhance diversity by introducing new observations. In the linear case, one can obtain other observations, which preserve the linear mixing model, by applying a linear mapping on the observations: for instance, a Fourier transform (see Chapter 11) or a wavelet transform, and more generally any mapping $\mathcal{T}$ which preserves the linearity. In other words, denoting $\tilde{\mathbf{u}}(t) = \mathcal{T}[\mathbf{u}(t)]$, the mapping $\mathcal{T}$ satisfies:

$$\mathcal{T}[\mathbf{x}(t)] = \tilde{\mathbf{x}}(t) = \mathcal{T}[\mathbf{A}\mathbf{s}(t)] = \mathbf{A}\mathcal{T}[\mathbf{s}(t)] = \mathbf{A}\tilde{\mathbf{s}}(t). \tag{1.6}$$

One can also use a nonlinear mapping for generating virtual observations [24,19].

A current trend is to use all the available prior information (see Chapters 10, 12, 13 and 15) on sources as well as on mixing systems. As an example, for mixtures of images, or for spectroscopy applications, sources and mixture coefficients are essentially positive. Adding a positivity constraint (see Chapter 13) on sources and on mixing coefficients is a simple way to regularize the solutions: restricting the solution space may allow avoiding spurious solutions [55,56,50]. However, introducing constraints in BSS algorithms is not always easy. Currently, with this purpose, Bayesian approaches (see Chapter 12) are among the most satisfactory frameworks for modeling any constraints, provided that they can be formulated as probabilistic priors.

Finally, in a number of situations, sources can be considered as sparse (see Chapter 10), at least in a well suited space [35]. In such a space, looking for sparse solutions leads to very efficient methods, usually referred to as sparse component analysis (SCA), and especially able to restore the unknown sources in the underdetermined case, i.e. if there are more sources than sensors.

1.4 SPATIAL WHITENING, NOISE REDUCTION AND PCA

The use of PCA as a spatial whitening has often been seen as means to reduce the search for a separation matrix to the group of unitary matrices. However, it is much more than that, since it also allows to dramatically reduce the effect of additive noise.

In fact, consider that the observation model (1.3) is linear but noisy, and that the observations take the form:

$$\mathbf{x}(t) = \mathbf{A}\,\mathbf{s}(t) + \mathbf{v}(t)$$

where $\mathbf{v}(t)$ is a P-variate random vector independent of $\mathbf{s}(t)$. The covariance matrix of the observations, $\mathbf{R}_x$, can hence be written as

$$\mathbf{R}_x = \mathbf{A}\,\mathbf{R}_s\,\mathbf{A}^{\mathrm{H}} + \rho\,\mathbf{G} \tag{1.7}$$

where $\mathbf{R}_s$ denotes the (diagonal) covariance matrix of the sources, and $\rho\mathbf{G}$ the covariance of noise $\mathbf{v}(t)$. This format allows us to relate ρ to the signal to noise ratio, and $\mathbf{G}$ to what is called in antenna array processing the "noise spatial coherence".

If the noise spatial coherence $\mathbf{G}$ is known, or has been estimated beforehand [58,71], one can build an unbiased estimate of the signal whitening matrix, even if the signal to noise ratio is unknown. This is not new, and can be traced back to the 1980s [11,68]. Let's summarize how this can be implemented; other implementations, e.g. via a generalized singular value problem, are also described in [25], either in off-line or on-line forms.

First of all, since $\mathbf{A}$ and $\mathbf{R}_s$ are both unknown, one can legitimately assume that $\mathbf{R}_s = \mathbf{I}_N$, the $N \times N$ identity matrix; this is due to the scale indetermination inherent in the problem, already emphasized earlier in this chapter. In addition, assume the spatial noise coherence is full rank, which is generally not restrictive at all. Under these conditions, the observation covariance (1.7) may be written as $\mathbf{R}_x = \mathbf{A}\,\mathbf{A}^{\mathrm{H}} + \rho\,\mathbf{G}$. Since $\mathbf{G}$ is positive definite, it admits an inverse square root, $\mathbf{V}$, such that $\mathbf{V}\mathbf{G}\mathbf{V}^{\mathrm{H}} = \mathbf{I}_P$. As a consequence, (1.7) can be written in the simple form below:

$$\mathbf{V}\mathbf{R}_x\mathbf{V}^{\mathrm{H}} = \mathbf{V}\mathbf{A}\,\mathbf{A}^{\mathrm{H}}\mathbf{V}^{\mathrm{H}} + \rho\,\mathbf{I}_P \tag{1.8}$$

which shows that matrix $\mathbf{V}\mathbf{R}_x\mathbf{V}^{\mathrm{H}} - \rho\,\mathbf{I}_P$ has rank N, since it is equal to $\mathbf{V}\mathbf{A}\,\mathbf{A}^{\mathrm{H}}\mathbf{V}^{\mathrm{H}}$. If N and ρ are unknown, they can be estimated by computing all eigenvalues of matrix $\mathbf{V}\mathbf{R}_x\mathbf{V}^{\mathrm{H}}$, and by detecting the width and level of the plateau of eigenvalue profile.

Let λ_i denote the eigenvalues of matrix $\mathbf{V}\mathbf{R}_x\mathbf{V}^{\mathrm{H}} - \rho\,\mathbf{I}_P$, and $\mathbf{w}_i$, $1 \leqslant i \leqslant N$ the corresponding eigenvectors, spanning the source *signal subspace*. Now stack the latter eigenvectors in a $P \times N$ matrix $\mathbf{W}$.

Clearly, $\mathbf{W}^{\mathrm{H}}\mathbf{V}\mathbf{A}\,\mathbf{A}^{\mathrm{H}}\mathbf{V}^{\mathrm{H}}\mathbf{W} = \boldsymbol{\Lambda}$, the $N \times N$ diagonal matrix containing the λ_i's. We have thus found a *whitening* matrix, which not only reduces the separating matrix to be unitary, but also projects the observation onto the N-dimensional source signal subspace. This whitening matrix is not unique, and may be chosen to be $\boldsymbol{\Lambda}^{-1/2}\mathbf{W}^{\mathrm{H}}\mathbf{V}$.

We draw the attention of the reader to the fact that such techniques are quite old, and that they can certainly not be attributed to the literature of blind source separation. Finally, we note that the whitening matrix can be obtained equivalently by computing the PCA of $\mathbf{R}_x - \rho\mathbf{G}$.

1.5 APPLICATIONS

The BSS problem appears in many multi-sensor systems: antenna arrays in acoustics or electromagnetism, chemical sensor arrays, electrode arrays in electroencephalography, etc. This very wide set of possible applications is probably one reason for the success of ICA, and more generally of source separation methods. In particular, BSS methods have been intensively used in three domains:

- biomedical applications (see Chapter 18) like electrocardiography, electroencephalography, magnetoencephalography, magnetic resonance imaging;
- audio source separation (see Chapter 19), with especially applications for music and speech;
- communication applications (see Chapter 17).

In addition, BSS methods have also been used in (hyperspectral) image processing, watermarking, preprocessing for classification, monitoring of complex systems, etc. A good account of applications is available in [20].

Source separation methods essentially rely on parameter estimation, which usually requires a model of the separating system ($\mathscr{B}$), an objective criterion and an optimization algorithm (see Chapter 16). In order to achieve relevant results with an actual application (in which sources and mixtures are unknown), it is necessary that:

- sources satisfy the basic assumption: independence for ICA, positivity, sparsity, etc.;
- the separating system is suited to the mixing model, which assumes that the physical model that produces the observations is correct.

If these conditions are not satisfied, i.e. a wrong model or criterion is used, the optimization algorithm will provide estimated sources that are indeed optimal with respect to criterion and model, but whose relevance is not guaranteed.

1.6 CONTENT OF THE HANDBOOK

In Chapter 2, *Information*, D.-T. Pham considers mutual information as a criterion for blind source separation. In the first part, he explores its use in separating linear instantaneous mixtures, in which the criterion can be expressed in terms of entropies of the extracted sources. Entropy estimators are introduced and fast algorithms for their computation are developed. In a second part, he considers the use of the mutual information rate, which is needed when dealing with convolutive mixtures, but can also be applied in instantaneous mixtures to obtain better separation performance.

Blind Source Separation relies, explicitly or implicitly, on optimization criteria. Such criteria should enjoy basic properties in order to avoid the existence of non-separating solutions. In other words, the solution should be essentially unique, which means unique up to some trivial filter, such as scaling or permutation of sources. In Chapter 3, E. Moreau and P. Comon focus on the definition of such criteria, which are referred to as *contrasts*. Contrasts dedicated to static SISO and MISO mixtures are studied. Contrast criteria devoted to deflation procedures, extracting a single source at a time, are first analyzed. Other contrasts performing joint source extraction, handling convolutive mixtures or using reference signals are also eventually discussed.

Chapter 4 provides a *likelihood*-based theory of ICA. The three basic ingredients of an ICA model – linear mixture, independence, source models – are introduced in sequence. Each of these steps uncovers specific features in the ICA likelihood which are discussed as we proceed: equivariance, estimating equations, associated contrast functions (mutual information, entropy), robustness, adaptivity, performance bounds. By discussing source models (Gaussian and non-Gaussian, possibly noisy) only in the final stage, J.-F. Cardoso emphasizes the statistical features common to many ICA approaches, independently of the source models.

In Chapter 5, *Algebraic Methods after Prewhitening*, L. De Lathauwer discusses some popular algebraic methods for blind source separation. These methods start with a prewhitening of the data and rely on an approximate diagonalization of a fourth-order cumulant tensor by means of an orthogonal or unitary congruence transformation. The diagonalization procedure takes the form of a Jacobi iteration. In particular, the COM1, COM2, JADE and STOTD algorithms are discussed in detail. The chapter starts with a review of the basics of higher-order statistics, prewhitening and Jacobi iterations.

Chapter 6, by V. Zarzoso and A. Hyvärinen, surveys computational algorithms for solving the ICA problem. Most of these algorithms rely on gradient or Newton iterations for contrast function maximization, and can work either in batch or adaptive processing mode. After briefly summarizing the common tools employed in their design and analysis, the chapter reviews a variety of *iterative techniques* ranging from pioneering neural network approaches and relative (or natural) gradient methods to Newton-like fixed-point algorithms as well as methods based on some form of optimal step-size coefficient.

In Chapter 7, *Second-Order Methods Based on Color*, A. Yeredor provides an overview of separation methods which rely exclusively on Second-Order Statistics (SOS). While SOS alone are generally insufficient for BSS, they can be used whenever the sources exhibit sufficient temporal diversity, e.g., stationarity with different spectra. Identifiability issues, analytic performance measures and bounds are discussed first, and then two families of algorithms are presented: joint diagonalization based, as well as likelihood based methods are derived and analyzed.

In Chapter 8, *Convolutive Mixtures* by M. Castella, A. Chevreuil and J.-C. Pesquet, linear mixing models in which delayed sample values of the sources contribute to the observations are considered. The importance of temporal statistical properties is outlined, leading to specification of separating conditions. Approaches aiming at restoring all sources simultaneously are reviewed. Methods for extracting one source are then proposed and details concerning their use in an iterative deflation scheme are provided. Finally, the case of cyclostationary sources, which plays a key role in telecommunications, is addressed.

In Chapter 9, mixtures involving more sources than sensors are considered by P. Comon and L. De Lathauwer. Such mixtures are now referred to as *under-determined*.

The problem of blindly identifying such mixtures is addressed without resorting to additional assumptions, sparsity for instance. Statistical approaches exploit independence among sources, either in the strict sense, or at some order. On the other hand, deterministic approaches require the presence of some *diversity* in the data. Iterative and quasi-algebraic algorithms exist in both cases and are described in detail. Identifiability conditions appear to be specific for each algorithm, and are pointed out.

In Chapter 10, *Sparse Component Analysis*, R. Gribonval and M. Zibulevski describe sparsity-based source separation methods for possibly under-determined linear mixtures. The principle is to apply a sparsifying transform (such as a wavelet transform or a time-frequency transform), and to separate the sources in the transform domain before reconstructing them by an inverse transform. The sparsity of the sources is also exploited to estimate the mixing matrix. The chapter describes algorithms for sparse signal decomposition in overcomplete signal dictionaries, which is beyond the scope of source separation.

In Chapter 11, *Quadratic Time-frequency Domain Methods*, N. Thirion-Moreau and M. Amin address the problem of blind separation of deterministic signals and stochastic nonstationary processes incident on sensor arrays using spatial time-frequency distributions. Successful signal and feature separations, which are based on either prewhitening or original data, require identifying the signal power concentration points or regions in the time-frequency domain. The spatial quadratic time-frequency representations of the raw or pre-processed observations, at these time-frequency points are constructed and used in the estimation of the mixing matrix through the application of joint (zero-) diagonalization based algorithms.

In Chapter 12, *Bayesian Approaches*, A. Mohammad-Djafari and K. H. Knuth examine the process of deriving source separation algorithms using Bayesian probability theory. Given an explicitly defined probabilistic model for the observations and for the sources (*a priori* information), the authors propose to use the Bayesian inference framework to derive different source separation algorithms. Interestingly, many classical algorithms are obtained as special cases of the general Bayesian framework with different *a priori* models and hypotheses and different estimators. These techniques are illustrated through a variety of applications. By extending the signal models, the authors explore the relationship between source separation and source localization and characterization.

In Chapter 13, *Non-negative mixtures*, M. D. Plumbley, A. Cichocki and R. Bro present models and associated learning algorithms for non-negative matrix and tensor factorizations. They also explore a range of generalizations and extensions of these models, and alternative approaches and algorithms that also enforce non-negativity constraints, including special algorithms designed to handle large scale problems. They also mention some applications of non-negative methods, including chemometrics, text processing, image processing and audio analysis.

In Chapter 14, *Nonlinear Mixtures*, C. Jutten, M. Babaie-Zadeh and J. Karhunen address the source separation problem in nonlinear mixtures. Generally, in such mixtures, ICA fails in either identifying the mixtures or separating the sources. However, adding constraints on the sources (coloration, Markov model) or on the mixture structure (post-nonlinear model, etc.) allows separation to be achieved. A few algorithms based on mutual information minimization or on a Bayesian approach are proposed, as well as a few actual applications.

Semi-blind techniques described in Chapter 15 by V. Zarzoso and P. Comon arise as a judicious compromise, benefiting from the advantages of supervised and blind techniques. Algebraic (i.e., closed-form) solutions can provide perfect equalization in the absence of noise, and are shown to be connected to matrix and tensor algebra problems. Iterative semi-blind equalizers are useful in the presence of noise, and can be efficiently implemented by an *optimal step-size* gradient-based search. Any pattern of training symbols can readily be integrated in semi-blind direct equalization. The optimal combination of the training and blind criteria is also addressed therein.

In Chapter 16, *Overview of Source Separation Applications*, Y. Deville, Ch. Jutten and R. Vigario explain how to address a source separation problem in practice. In a general framework, these authors show the importance of a good understanding of the observed data, their possible physical/generative models and their statistical properties. This especially defines whether source separation should be based on independence or some other useful property. The authors also show that, in ICA, overfitting can lead to spurious "independent" components. Finally, the theoretical concepts are illustrated by a few examples using actual communications, astrophysical and biomedical datasets.

Spectrum monitoring of *radio communications* generally requires the estimation of many parameters belonging to spectrally overlapping sources which need to be blindly separated in a pre-processing step. Moreover, most radio communications sources are non-Gaussian and cyclostationary, and propagate through multipath channels which are often specular in time. The problem then consists of blindly separating specular convolutive mixtures of arbitrary cyclostationary non-Gaussian sources. Two approaches are considered and compared in Chapter 17 by P. Chevalier and A. Chevreuil. The first one is convolutive whereas the second, instantaneous, exploits the specularity property of the channels and aims at blindly separating the multiple paths of all the received sources before a potential post-processing.

The purpose of Chapter 18 consists first in showing the interest in using ICA in *biomedical applications* such as the analysis of human electromagnetic recordings. Next, the computational complexity of twelve of the most widespread ICA techniques is analyzed in detail by L. Albera et alterae, which allows to compare their performance when utilized in biomedical operational contexts. This chapter will hopefully be a useful reference for researchers from the biomedical community, especially for those who are not familiar with ICA techniques.

In Chapter 19, *Audio Applications*, E. Vincent and Y. Deville review past applications of convolutive ICA and SCA to the separation of audio signals. The applicability of independence and sparsity assumptions is discussed. Various BSS algorithms are then reviewed, with particular emphasis on the choice of a suitable contrast function and on the estimation of frequency-wise source permutations within frequency-domain convolutive ICA. The reported performance figures are compared over both recorded and synthesized audio mixtures.

References

[1] P. Duvaut (Ed.), Traitement du Signal, Special issue on nonlinear and non-Gaussian techniques, December 1990.
[2] L.B. Almeida, Separating a real-life nonlinear image mixture, J. Mach. Learn. Res. 6 (2005) 1199–1229.
[3] S. Amari, Superefficiency in blind source separation, IEEE Trans. Signal Process. 47 (1999) 936–944.
[4] B. Ans, J. Hérault, C. Jutten, Adaptive neural architectures: Detection of primitives, in: Proceedings of COGNITIVA'85, Paris, France, June 1985, pp. 593–597.
[5] H.B. Barlow, Possible principles underlying the transformation of sensory messages, in: W. Rosenblith (Ed.), Sensory Communication, MIT Press, 1961, p. 217.
[6] Y. Bar-Ness, J. Carlin, M. Steinberger, Bootstrapping adaptive interference cancelers: some practical limitations, in: Proc. The Globecom. Conference, 1982, pp. 1251–1255.
[7] T. Bell, T. Sejnowski, An information-maximization approach to blind separation and blind deconvolution, Neural Comput. 7 (1995) 1129–1159.
[8] T. Bell, T. Sejnowski, The "independent components" of natural scenes are edge filters, Vision Res. 37 (1997) 3327–3338.
[9] A. Belouchrani, K. Abed-Meraim, Séparation aveugle au second ordre de sources corrélées, in: GRETSI, Juan-Les-Pins, France, Sept. 1993, pp. 309–312.
[10] A. Benveniste, M. Goursat, G. Ruget, Robust identification of a non-minimum phase system, IEEE Trans. Auto. Contr. 25 (1980) 385–399.
[11] G. Bienvenu, L. Kopp, Optimality of high-resolution array processing using the eigensystem approach, IEEE Trans. ASSP 31 (1983) 1235–1248.
[12] G. Burel, Blind separation of sources: A nonlinear neural algorithm, Neural Netw. 5 (1992) 937–947.
[13] J.-F. Cardoso, On the performance of orthogonal source separation algorithm, in: Proceedings of EUSIPCO'94, 1994, pp. 776–779.
[14] J.-F. Cardoso, The three easy routes to independent component analysis: contrasts and geometry, in: Proceedings of ICA 2001, San Diego, USA, 2001.
[15] J.-F. Cardoso, P. Comon, Tensor based independent component analysis, in: Proceedings of EUSIPCO, Barcelona, Spain, September 1990, pp. 673–676.
[16] J.-F. Cardoso, B. Laheld, Equivariant adaptive source separation, IEEE Trans. SP 44 (1996) 3017–3030.
[17] J.-F. Cardoso, A. Souloumiac, An efficient technique for blind separation of complex sources, in: Proc. IEEE Signal Processing Workshop on Higher-Order Statistics, South Lac Tahoe, USA (CA), June 1993, pp. 275–279.
[18] M. Castella, Inversion of polynomial systems and separation of nonlinear mixtures of finite-alphabet sources, IEEE Trans. Signal Process. 56 (2008) 3905–3917.
[19] P. Chevalier, L. Albera, A. Ferreol, P. Comon, On the virtual array concept for higher order array processing, IEEE Trans. Signal Process. 53 (2005) 1254–1271.
[20] A. Cichocki, S.-I. Amari, Adaptive Blind Signal and Image Processing, Wiley, New York, 2002.
[21] P. Comon, Analyse en Composantes Indépendantes et identification aveugle, Traitement du Signal 7 (1990) 435–450. Special issue.
[22] P. Comon, Independent Component Analysis. Republished in Higher-Order Statistics, J.L. Lacoume (Ed.), Elsevier, 1992, pp. 29–38.
[23] P. Comon, Independent component analysis, a new concept? Signal Process. 36 (1994) 287–314.
[24] P. Comon, Blind identification and source separation in 2×3 under-determined mixtures, IEEE Trans. Signal Process. 52 (2004) 11–22.
[25] P. Comon, G.H. Golub, Tracking of a few extreme singular values and vectors in signal processing, Proc. IEEE 78 (1990) 1327–1343 (Published from Stanford report 78NA-89-01, February 1989).

[26] P. Comon, C. Jutten, J. Hérault, Blind separation of sources, Part II: Problem statement, Signal Process. 24 (1991) 11–20.
[27] G. Darmois, Analyse générale des liaisons stochastiques, Rev. Inst. Intern. Stat. 21 (1953) 2–8.
[28] D. Donoho, On minimum entropy deconvolution, in: Applied Time Series Analysis II, Tulsa, 1980, pp. 565–608.
[29] D. Donoho, Nature vs. math: Interpreting independent component analysis in light of computational harmonic analysis, in: Independent Component Analysis and Blind Source Separation, ICA2000, Helsinski, June 2000, pp. 459–470.
[30] D. Dugué, Analycité et convexité des fonctions caractéristiques, Ann. Inst. H. Poincaré XII (1951) 45–56.
[31] W. Feller, An Introduction to Probability Theory and its Applications, 3rd ed., Wiley, 1968.
[32] M. Gaeta, J.-L. Lacoume, Estimateurs du maximum de vraisemblance étendus à la séparation de sources non gaussiennes, Traitement du Signal 7 (1990) 419–434.
[33] D. Godard, Self recovering equalization and carrier tracking in two dimensional data communication systems, IEEE Trans. Com. 28 (1980) 1867–1875.
[34] O. Grellier, P. Comon, Blind separation of discrete sources, IEEE Signal Processing Letters 5 (1998) 212–214.
[35] R. Gribonval, S. Lesage, A survey of sparse components analysis for blind source separation: Principles, perspectives and new challenges, in: Proc. ESANN 2006, Bruges, Belgium, April 2006.
[36] J. Hérault, B. Ans, Circuits neuronaux à synapses modifiables: décodage de messages composites par apprentissage non supervisé, C.R. Acad. Sci. 299 (1984) 525–528.
[37] J. Hérault, C. Jutten, Space or time adaptive signal processing by neural networks models, in: Intern. Conf. on Neural Networks for Computing, Snowbird (Utah, USA), 1986, pp. 206–211.
[38] J. Hérault, C. Jutten, B. Ans, Détection de grandeurs primitives dans un message composite par une architecture de calcul neuromimétique en apprentissage non supervisé, in: Actes du Xème colloque GRETSI, Nice, France, May 1985, pp. 1017–1022.
[39] A. Hyvärinen, J. Karhunen, E. Oja, Independent Component Analysis, Wiley, 2001.
[40] A. Hyvärinen, P. Pajunen, Nonlinear independent component analysis: Existence and uniqueness results, Neural Netw. 12 (1999) 429–439.
[41] P. Jallon, A. Chevreuil, P. Loubaton, P. Chevalier, Separation of convolutive mixtures of cyclostationary sources: A contrast function based approach, in: C.G. Puntonet, A. Prieto (Eds.), Independent Component Analysis and Blind Signal Separation, ICA2004, Granada, Spain, September, Springer, Berlin, Heidelberg, New York, 2004, pp. 508–515.
[42] C. Jutten, Calcul neuromimétique et traitement du signal: analyse en composantes indépendantes, Thèse d'état ès sciences physiques, UJF-INP Grenoble, 1987.
[43] C. Jutten, J. Hérault, Analog implementation of a neuromimetic auto-adaptive algorithm, in: J. Hérault, F. Fogelman-Soulié (Eds.), Neuro-Computing: Algorithms, Architectures and Applications, Les Arcs, France, February 27–March 3, in: NATO ASI Series, Series F, vol. 68, Springer Verlag, 1989, pp. 145–152.
[44] C. Jutten, J. Hérault, Blind separation of sources, Part I: An adaptive algorithm based on a neuromimetic architecture, Signal Process. 24 (1991) 1–10.
[45] C. Jutten, A. Taleb, Source separation: From dusk till dawn, in: Proc. 2nd Int. Workshop on Independent Component Analysis and Blind Source Separation, ICA2000, Helsinki, Finland, 2000, pp. 15–26.
[46] A. Kagan, Y. Linnik, C. Rao, Characterization Problems in Mathematics Statistics, John Wiley & Sons, 1973.
[47] J.-L. Lacoume, P. Ruiz, Sources identification: A solution based on cumulants, in: IEEE ASSP Workshop, Mineapolis (USA), August 1988.
[48] B. Laheld, J.-F. Cardoso, Adaptative source separation without pre-whitening, in: Proc. European Signal Processing Conf. EUSIPCO 94, Edinburgh, Scotland, September 1994, pp. 183–186.
[49] Y. Le Cun, A learning scheme for assymetric threshold network, in: Proceedings of COGNITICA'85, Paris, France, 1985, pp. 599–604.

[50] J. Lee, F. Vrins, M. Verleysen, Non-orthogonal support width ICA, in: Proc. ESANN 2006, Bruges, Belgium, April 2006.
[51] S. Makino, T.-W. Lee, H. Sawada, Blind Speech Separation, in: Signals and Communication Technology, Springer, 2007.
[52] K. Matsuoka, M. Ohya, M. Kawamoto, A neural net for blind separation of nonstationary signals, Neural Netw. 8 (1995) 411–419.
[53] D. Mc Closkey, Kinesthetic sensibility, Physiol. Rev. 58 (1978) 763–820.
[54] L. Molgedey, H. Schuster, Separation of a mixture of independent signals using time delayed correlations, Phys. Rev. Lett. 72 (1994) 3634–3637.
[55] S. Moussaoui, D. Brie, C. Carteret, A. Mohammad-Djafari, Application of Bayesian non-negative source separation to mixture analysis in spectroscopy, in: Proceedings of the 24th International Workshop on Bayesian Inference and Maximum Entropy Methods in Science and Engineering, MaxEnt'2004, Garching, Germany, July 2004.
[56] S. Moussaoui, D. Brie, A. Mohammad-Djafari, C. Carteret, Separation of non-negative mixture of non-negative sources using a Bayesian approach and MCMC sampling, IEEE Trans. Signal Process. 54 (2006) 4133–4145.
[57] J.-P. Nadal, N. Parga, Nonlinear neurons in the low-noisy limit: a factorial code maximizes information transfer, Network 5 (1994) 565–581.
[58] A. Paulraj, T. Kailath, Eigenstructure methods for direction of arrival estimation in the presence of unknown noise field, IEEE Trans. Acoust. Speech Signal Process. 34 (1986) 13–20.
[59] D. Pham, Mutual information approach to blind separation of stationary sources, in: Proceedings of ICA'99, Aussois, France, January 1999, pp. 215–220.
[60] D. Pham, F. Vrins, Discriminacy of the minimum range approach to blind separation of bounded sources, in: Proc. ESANN 2006, Bruges, Belgium, April 2006.
[61] D.T. Pham, Blind separation of instantaneous mixture of sources based on order statistics, IEEE Trans. SP 48 (2000) 363–375.
[62] D.T. Pham, J.-F. Cardoso, Blind separation of instantaneous mixtures of nonstationary sources, IEEE Trans. Signal Process. 49 (2001) 1837–1848.
[63] D.T. Pham, P. Garat, Blind separation of mixture of independent sources through a quasimaximum likelihood approach, IEEE Trans. SP 45 (1997) 1712–1725.
[64] C. Puntonet, A. Mansour, C. Jutten, A geometrical algorithm for blind separation of sources, in: Actes du XV Colloque GRETSI 95, Juan-Les-Pins, France, September 1995.
[65] B. Rivet, L. Girin, C. Jutten, J.-L. Schwartz, Solving the indeterminations of blind source separation of convolutive speech mixtures, in: Proceedings of the International Conference of Acoustics, Speech and Signal Processing, ICASSP2005, Philadelphia (USA), March 2005.
[66] J.-P. Roll, Contribution à la proprioception musculaire, à la perception et au contrôle du mouvement chez l'homme, PhD thesis, Université d'Aix-Marseille 1, 1981.
[67] Y. Sato, A method of self recovering equalization for multilevel amplitude-modulation systems, IEEE Trans. Com. 23 (1975) 679–682.
[68] R.O. Schmidt, Multiple emitter location and signal parameter estimation, IEEE Trans. Antenna Propagation 34 (1986) 276–280.
[69] D. Sodoyer, L. Girin, C. Jutten, J.-L. Schwartz, Developing an audio-visual speech source separation algorithm, Speech Commun. 44 (2004) 113–125.
[70] E. Sorouchyari, Blind separation of sources, Part III: Stability analysis, Signal Process. 24 (1991) 21–29.
[71] P. Stoica, M. Cedervall, Detection tests for array processing in unknown correlated noise fields, IEEE Trans. Signal Process. 45 (1997) 2351–2362.
[72] A. Swami, G. Giannakis, S. Shamsunder, Multichannel ARMA processes, IEEE Trans. Signal Process. 42 (1994) 898–913.
[73] A. Taleb, C. Jutten, Source separation in post nonlinear mixtures, IEEE Trans. SP 47 (1999) 2807–2820.

[74] F. Theis, C. Puntonet, E. Lang, Nonlinear geometrical ICA, in: S.-I. Amari, A. Cichocki, S. Makino, N. Murata (Eds.), Proc. 4th Int. Symp. on Independent Component Analysis and Blind Signal Separation, ICA2003, Nara, Japan, April 2003, pp. 275–280.
[75] L. Tong, V. Soon, R. Liu, Y. Huang, Amuse: A new blind identification algorithm, in: Proc. ISCAS, New Orleans, USA, 1990.

CHAPTER

Information

2

D.T. Pham

2.1 INTRODUCTION

Blind source separation (BSS) deals typically with a mixing model of the form[1] $\mathbf{x}(\cdot) = \mathscr{A}\{\mathbf{s}(\cdot)\}$ where $\mathbf{s}(n)$ and $\mathbf{x}(n)$ represent the source and observed vectors at time n and $\mathscr{A}$ is a transformation, which can be instantaneous (operating on each $\mathbf{s}(n)$ to produce $\mathbf{x}(n)$), or global (operating on the whole sequence $\mathbf{s}(\cdot)$ of source vectors. The goal of BSS is to reconstruct the sources from the observations. Clearly, for this task to be possible $\mathscr{A}$ should not be completely unknown: it should belong to some class of transformations given $\mathbb{A}$ *a priori*. Most common classes are the class of linear (affine) instantaneous transformations and that of (linear) convolutions. They correspond to linear instantaneous mixtures and (linear) convolutive mixtures respectively. Nonlinear transformations have also been considered. For example, $\mathbb{A}$ may be constituted of linear instantaneous (or convolutive) transformation followed by nonlinear instantaneous transformation operating component-wise. The corresponding mixtures are called post-nonlinear (or convolutive post-nonlinear). This chapter deals primarily with linear mixtures; nonlinear mixtures are treated elsewhere (see Chapter 14).

In a blind context, the separation of sources can only rely on the basic knowledge which is their mutual independence. It is thus natural to try to achieve separation by minimizing an independence criterion between the components of $\mathscr{A}^{-1}\{\mathbf{x}(\cdot)\}$ among all $\mathscr{A} \in \mathbb{A}$, where $\mathscr{A}^{-1}$ denotes the inverse transformation of $\mathscr{A}$. We adopt here the mutual information and the independence criterion. This is a popular criterion and has many appeals. Firstly, it is invariant with respect to invertible transformation (see Lemma 2.1 below and what follows). In particular, it is scale invariant thus avoids a prewhitening step, which is needed in many other separation methods. Secondly, it is a very general and complete independence criterion: it is non-negative and can be zero if and only if there is independence. Some other criteria such as the cumulants are only partial: to ensure independence, one needs to check that all (cross) cumulants vanish, but in practice only a finite number of them can be considered. Finally, the mutual

[1] We do not consider the case of noisy mixtures. There are few methods which deal explicitly with noises. Often a preprocessing step is done to reduce noises, or a method designed for a noiseless model is found to be rather insensitive to noise and can thus be applied when some weak noises are present.

information can be interpreted in terms of entropy and Kullback-Leibler divergence [3], and is closely related to the expected log likelihood [1]. The downsize is that this criterion requires the knowledge of the joint density of the components of $\mathscr{A}^{-1}\{\mathbf{x}(\cdot)\}$, which is unknown in practice and hence must be replaced by some *nonparametric* estimate. The estimation procedure can be quite costly computationally. We shall however introduce some methods which are not much costlier than using simpler criteria (such as the cumulants).

The rest of this chapter contains two parts: the first one concerns the use of the mutual information between the observations at a given time, which is suitable for instantaneous (linear) mixtures, in which the temporal dependence of the source sequences are ignored (for simplicity or because it is weak). The second part concerns the use of the information rate between stationary processes, which is necessary to treat the case of convolutive (linear) mixtures, but can also be useful for the case of instantaneous mixtures when there is strong temporal dependence of the source sequences. Note that for the convolutive mixture, the sources can be recovered only up to filtering (as it will be seen later) and one may require temporal independence of the source to lift this ambiguity. The problem then may be viewed as the multi-channel blind deconvolution problem as it reduces to the well-known deconvolution problem when both source and observed sequences are scalar. We however call this problem blind separation-deconvolution as it aims to both recover the sources (separation) and make them temporally independence (deconvolution).

2.2 METHODS BASED ON MUTUAL INFORMATION

We first define mutual information and provide its main properties.

2.2.1 Mutual information between random vectors

Let $\mathbf{y}_1,\ldots,\mathbf{y}_P$, be P random vectors with joint *densities* $p_{\mathbf{y}_1,\ldots,\mathbf{y}_P}$ and marginal density $p_{\mathbf{y}_1},\ldots,p_{\mathbf{y}_P}$. The *mutual information* between these vectors is defined as the *Kullback-Leibler divergence* (or relative entropy) between the densities $\prod_{k=1}^P p_{\mathbf{y}_k}$ and $p_{\mathbf{y}_1,\ldots,\mathbf{y}_P}$:

$$I\{\mathbf{y}_1\ldots,\mathbf{y}_P\} = -\mathbb{E}\log\frac{p_{\mathbf{y}_1}(\mathbf{y}_1)\cdots p_{\mathbf{y}_P}(\mathbf{y}_P)}{p_{\mathbf{y}_1,\ldots,\mathbf{y}_P}(\mathbf{y}_1,\ldots,\mathbf{y}_P)}.$$

This measure is non-negative (but can be $+\infty$) and can vanish only if the random vectors are mutually independent [4]. It can also be written as:

$$I\{\mathbf{y}_1\ldots,\mathbf{y}_P\} = \sum_{i=1}^P H\{\mathbf{y}_k\} - H\{\mathbf{y}_1\ldots,\mathbf{y}_P\} \tag{2.1}$$

where $H(\mathbf{y}_1\ldots,\mathbf{y}_P)$ and $H(\mathbf{y}_1),\ldots,H(\mathbf{y}_P)$ are the joint *entropy* and the marginal entropies of $\mathbf{y}_1,\ldots,\mathbf{y}_P$:

$$H\{\mathbf{y}_1\ldots,\mathbf{y}_P\} = -\mathbb{E}\log p_{\mathbf{y}_1,\ldots,\mathbf{y}_P}(\mathbf{y}_1\ldots,\mathbf{y}_P) \tag{2.2}$$

and $H\{\mathbf{y}_k\}$ is defined in the same way with $p_{\mathbf{y}_k}$ in place of $p_{\mathbf{y}_1,\ldots,\mathbf{y}_P}$. The notation $H\{\mathbf{y}_1 \ldots, \mathbf{y}_P\}$ is the same as $H\{[\mathbf{y}_1^{\mathrm{T}} \cdots \mathbf{y}_P^{\mathrm{T}}]^{\mathrm{T}}\}$, ^{T} denoting the transpose.

The entropy possesses the following interesting property of equivariance with respect to invertible transformation.

LEMMA 2.1
Let $\mathbf{x}$ *be a random vector and* $\mathbf{y} = g(\mathbf{x})$ *where* g *is a differentiable invertible transformation with Jacobian (matrix of partial derivatives)* g'. *Then:*

$$H\{\mathbf{y}\} = H\{\mathbf{x}\} + \mathbb{E}\log|\det g'(\mathbf{x})|.$$

This result can be easily obtained from the definition (2.2) of entropy and the equality $p_{\mathbf{x}}(\mathbf{x}) = p_{\mathbf{y}}[g(\mathbf{x})]|\det g'(\mathbf{x})|$. It follows immediately from (2.1) that if g is a transformation operating component-wise (that is the i-th component of $g(\mathbf{x})$ depends only on the i-th component of $\mathbf{x}$) then the mutual information between the components of the transformed random vector $g(\mathbf{x})$ is the same as that between the components of the original random vector $\mathbf{x}$. Thus the mutual information between the random variables $y_1, \ldots, y_P$ remains unchanged when one applies to each of them an invertible transformation. Clearly, it is also unchanged if one permutes the random variables. This is consistent with the fact that independent random variables remain independent under the above operations. As a result, one can separate the sources from their linear mixtures only up to a scaling and a permutation (by exploiting their independence only).

The application of Lemma 2.1 is most interesting in the case of the linear (affine) mixture, since then one is dispensed with an expectation calculation: if $\mathbf{y} = \mathbf{B}\mathbf{x} + \mathbf{b}$ where $\mathbf{B}$ is an invertible matrix and $\mathbf{b}$ is a vector, then $H(\mathbf{y}) = H(\mathbf{x}) + \log|\det \mathbf{B}|$. In particular, the entropy is invariant with respect to translation and equivariant with respect to scaling: $H(ay + b) = H(y) + \log|a|$ for all real a and b.

2.2.2 The mixing model and separation criterion

It is clear that the use of mutual information between observations at a single time point is suitable only in the case of instantaneous mixtures, where the observed vector $\mathbf{x}(n)$ is an invertible transformation of the source vector $\mathbf{s}(n)$. We will focus on the case where this transformation is linear and thus is defined by a mixing matrix $\mathbf{A}$. Affine transformations may be considered but we already know that a translation has no effect on the mutual information, so there is no loss of generality by ignoring it. To separate the sources one tries to find a demixing matrix $\mathbf{B}$ such that components $y_1(n), \ldots, y_P(n)$ of the vector $\mathbf{y}(n) = \mathbf{B}\mathbf{x}(n)$ (the separated sources) are as independent as possible according to mutual information criterion. By (2.1) and the result in the last paragraph of the previous subsection, this criterion can be written up to a constant not depending on $\mathbf{B}$ as:

$$C(\mathbf{B}) = \sum_{k=1}^{P} H(y_k) - \log|\det \mathbf{B}|, \tag{2.3}$$

where we have suppressed the time index n in $H[y_k(n)]$ as this entropy does not depend on n by stationarity. It is worthwhile to note that the above criterion does not involve the joint entropy but only marginal entropies, so that the criterion no longer requires a multiple integration in $\mathbb{R}^P$.

2.2.3 Empirical criteria and entropy estimators

The criterion (2.3) is only theoretical since it involves the unknown entropies $H(y_k)$. In practice, these entropies have to be replaced by their estimates. Different estimators result in different empirical criteria.

A natural way to construct an entropy estimator is to replace the density in its definition by its estimate. In a blind context such estimate should be nonparametric as one does not have any *a priori* knowledge of the source distribution. A popular and well studied estimator of a density is the kernel estimate. The density p_y of a random variable y is estimated by

$$\hat{p}_y(v) = \frac{1}{Th\hat{\sigma}_y} \sum_{n=1}^{T} \mathcal{K}\left[\frac{v - y(n)}{h\hat{\sigma}_y}\right] \tag{2.4}$$

where $y(1), \ldots, y(T)$ is the observed sample, $\mathcal{K}$ is the *kernel* which is a density itself, h is the smoothing parameter and $\hat{\sigma}_y$ is the sample standard deviation of y: $\hat{\sigma}_y^2 = T^{-1}\sum_{n=1}^{T}[y(n) - \bar{y}]^2$, $\bar{y} = T^{-1}\sum_{n=1}^{T} y(n)$ being sample mean. In our definition, the smoothing is controlled by $h\hat{\sigma}_y$ (and not h) so that the estimator possesses the equivariance property with respect to scaling: $\hat{p}_{\alpha y}(v) = \hat{p}_y(v/\alpha)/\alpha$, which enjoys the true density. A possible entropy estimator of y is $-T^{-1}\sum_{n=1}^{T} \hat{p}_y[y(n)]$ but it is costly to be compute as the density estimate has to be evaluated on T random points. Instead we propose to estimate $H(y)$ by simply replacing the integral $\int \log[\hat{p}_y(v)]\hat{p}_y(v)dv$ by a Riemann sum. However, such an estimator will not in general possess the property of being invariant with respect to translation and equivariant with respect to scaling, as described in the last paragraph of subsection 2.2.1. Therefore we shall first standardize y to $y' = (y - \bar{y})/\hat{\sigma}_y$ then estimate $H(y)$ by relating it to $H(y')$, more precisely by

$$\hat{H}_\delta\{y\} = -\sum_{m=-\infty}^{\infty} \log[\hat{p}_{y'}(m\delta)]\hat{p}_{y'}(m\delta)\delta + \log\hat{\sigma}_y, \tag{2.5}$$

where δ is the discretization step. As $\hat{p}_{y'}(m\delta) = p_y(m\delta\hat{\sigma}_y + \bar{y})\hat{\sigma}_y$, one has

$$\begin{aligned}\hat{H}_\delta\{y\} = &-\sum_{m=-\infty}^{\infty} \log[\hat{p}_y(m\hat{\sigma}_y\delta + \bar{y})]\hat{p}_y(m\hat{\sigma}_y\delta + \bar{y})\hat{\sigma}_y\delta \\ &+ \log\sigma_y\left[1 - \sum_{m=-\infty}^{\infty} \hat{p}_y(m\hat{\sigma}_y\delta + \bar{y})\hat{\sigma}_y\delta\right].\end{aligned} \tag{2.6}$$

The expression in the last bracket [ ] is nearly zero as it is the difference between the integral $\int p_{h,y}(v)dv$ and its Riemann sum. It actually vanishes for the kernel which we use (see subsection 2.2.4.3 below). In this case $H_{h,\delta}\{y\}$ is the Riemann sum of $-\int \log[\hat{p}_{h,y}(v)]\hat{p}_{h,y}(v)dv$, based on the regular grid of origin $\bar{y}$ and spacing $\hat{\sigma}_y\delta$. Further, the chosen kernel has compact support so that this sum is a finite sum.

Another possibility is to circularize the density estimate. Supposing that we know that $\hat{p}_y$ almost vanishes outside some interval of length $M\hat{\sigma}_y$, then we may approximate $\hat{p}_y$ in this interval by the *circularized density estimate* $\hat{p}_y^M$ defined by $\hat{p}_y^M(v) = \sum_{m=-\infty}^{\infty} \hat{p}_y(v + mM\sigma_y)$. This yields the entropy estimator

$$\hat{H}^M\{y\} = -\int_0^{M\hat{\sigma}_y} [\log \hat{p}_y^M(v)]\hat{p}_y^M(v)dv = -\int_0^M [\log \hat{p}_{y'}^M(v)]\hat{p}_{y'}^M(v)dv + \log\hat{\sigma}_y, \quad (2.7)$$

since the integration in the first right hand side above remains the same over any interval of length $M\hat{\sigma}_y$. The last equality, which is obtained by a change of integration variable, shows that $\hat{H}^M$ is translation invariant and scale equivariant.

It can be seen that the Fourier coefficients of the periodic function $\hat{p}_y^M$ are precisely $\hat{\Phi}_y[2\pi m/(M\hat{\sigma}_y)], m \in \mathbb{Z}$, where $\hat{\Phi}_y(u) = \int e^{iuv}\hat{p}_y(v)dv$ is the *estimated characteristic function* of y, and therefore

$$\hat{p}_y^M(v) = \frac{1}{M\sigma_y}\sum_{m=-\infty}^{\infty} \hat{\Phi}_y\left(\frac{2\pi m}{M\sigma_y}\right)\exp\left(-\frac{2\pi m v}{M\sigma_u}\right). \quad (2.8)$$

It follows that one may compute $\hat{H}^M\{y\}$ by $-\sum_{m=-\infty}^{\infty} a_m^* \hat{\Phi}_y[2\pi m/(M\sigma_y)]$ where * denotes the complex conjugate and

$$a_m = \int_0^{M\sigma_Y} [\log \hat{p}_y^M(v)]\exp\left(\frac{i2\pi m v}{M\sigma_y}\right)\frac{dv}{M\sigma_y} \quad (2.9)$$

are the Fourier coefficients of $\log \hat{p}_y^M$. These coefficients may be computed *without integration* if $\mathcal{K}$ has been chosen such that its characteristic function $\Phi_{\mathcal{K}}$ has compact support hence the same holds for $\hat{\Phi}_y$ since

$$\hat{\Phi}_y(u) = \Phi_{\mathcal{K}}(uh\sigma_y)\sum_{n=1}^{T} e^{iuy(n)} \quad (2.10)$$

as can be easily seen. In this case the right hand side of (2.8) is a non-negative polynomial in $e^{\pm i2\pi v/(M\hat{\sigma}_y)}$; hence there exists a finite number of coefficients $b_0, \ldots, b_q$ such that it can be factorized as $|\sum_{m=0}^{q} e^{i2\pi m v/(M\sigma_y)} b_m|^2$ (see [20] for an algorithm to compute them). Further, the polynomial $\sum_{m=0}^{q} b_m z^m$ has no root inside the unit circle of the complex plane. Taking the derivative of both sides of the equality

$$\sum_{m=-\infty}^{\infty} \exp\left(\frac{i2\pi m v}{M\sigma_y}\right) a_m = \log\left|\sum_{m=0}^{r} \exp\left(\frac{i2\pi m v}{M\sigma_y}\right) b_m\right|^2$$

with repect to v, one gets

$$\sum_{m=1}^{\infty} \exp\left(\frac{i2\pi m v}{M\sigma_y}\right) m a_m = \frac{\sum_{m=1}^{r} e^{i2\pi m v/(M\sigma_y)} m b_m}{\sum_{m=0}^{r} e^{i2\pi m v/(M\sigma_y)} b_m},$$

noting that the last right hand side expands into a series with positive powers of $e^{i2\pi v/(M\sigma_y)}$ only. One then deduces the following recursion for computing the a_m:

$$m a_m b_0 + \sum_{l=1}^{m-1} (m-l) a_{m-l} b_l = m b_m, \quad m \geqslant 1,$$

with $a_0 = \log b_0^2$ and of course $a_{-m} = a_m^*$.

Finally, another entropy estimator can be constructed by expressing the entropy in terms of the *quantile function*. Recall that the quantile function Q_y of a random variable y is defined as

$$Q_y(u) = \inf\{v \in \mathbb{R} : P(y \leqslant v) \geqslant u\}, \quad 0 < u \leqslant 1,$$

and $Q_y(0) = \inf\{v \in \mathbb{R} : P(y \leqslant v) > 0\}$. The function Q_y is no other than the inverse function of the cumulative distribution function $F_y : F_y(v) = P(y \leqslant v)$ of y, in the case where this function is strictly increasing. A simple calculation shows that the derivative Q'_y of Q_y (if it exists) is related to the density p_y by

$$Q'_y(u) = 1/p_y[Q_y(u)] \quad \text{or} \quad p_y(v) = 1/Q'_y[F_y(v)].$$

It follows that the entropy of y can be also expressed as

$$H(y) = \int_{-\infty}^{\infty} \log Q'_y[F_y(v)]\ p_y(v) dv = \int_0^1 \log Q'_y(u)\, du.$$

This suggests approximating $H(y)$ by

$$H_{\{u\}}\{y\} = \sum_{l=2}^{L} \log\left[\frac{Q_y(u_l) - Q_y(u_{l-1})}{u_l - u_{l-1}}\right] \frac{u_l - u_{l-1}}{u_L - u_1} \tag{2.11}$$

where $\{u_1,\dots,u_L\}$ is a given strictly increasing sequence of numbers in $[0,1]$. Note that $u_1 = 0$ is not allowed in the case where the distribution of y has unbound lower support to avoid $Q_y(u_1) = -\infty$. Likewise $u_L = 1$ is not allowed if the distribution of y has unbound upper support. An entropy estimator is then obtained by replacing in the above expression Q_y by the *empirical quantile function* $\hat{Q}_y$. Let $y(1{:}T) \leqslant \cdots \leqslant y(T{:}T)$ be the *order statistics* from the sample $y(1),\dots,y(T)$, that is this sample arranged in non-decreasing order; the empirical quantile function is defined as the function which equals $y(i{:}T)$ on the interval $((i-1)/T, i/T]$. This definition is not always the one adopted in practice since it leads to a discontinuous function, but it is convenient for us, and since it is not important to choose $u_1,\dots,u_L$ precisely, one can always choose them such that the empirical quantile function at these point equals $y(n_1{:}T),\dots,y(n_L{:}T)$ for some sequence of integers $1 \leqslant n_1 < \cdots < n_L \leqslant T$. This yields the entropy estimator

$$\hat{H}_{\{u\}}\{y\} = \sum_{l=2}^{L} \log\left[\frac{y(n_l{:}T) - y(n_{l-1}{:}T)}{(n_l - n_{l-1})/T}\right] \frac{n_l - n_{l-1}}{n_L - n_1}. \tag{2.12}$$

The above formula shows clearly that $\hat{H}_{\{u\}}$ is invariant with respect to translation and the equivariant with respect to scaling, provided that change of sign is not allowed. If one would like $\hat{H}_{\{u\}}$ to be invariant with respect to sign change, that is $\hat{H}_{\{u\}}(-y) = \hat{H}_{\{u\}}\{y\}$, one must choose the u_l to be symmetric, that is $u_l = 1 - u_{L+1-l}, l = 1,\dots,L$.

In the sequel, we denote by $\hat{C}_\delta$, $\hat{C}^M$ and $\hat{C}_{\{u\}}$ the empirical criteria based on the entropy estimator $\hat{H}_\delta$, $\hat{H}^M$ and $\hat{H}_{\{u\}}$.

2.2.4 Computation of entropy estimators

The computation of the empirical criteria mostly reduces to that of the entropy estimators as that of $\log|\det \mathbf{B}|$ is straightforward. Since such calculations will be needed many times in a minimization algorithm, fast methods to compute them are desirable. We shall focus on the estimators $\hat{H}_{h,\delta}$ and $\hat{H}_h^M$ as $H_{\{u\}}$ is directly obtained from the order statistics.

The estimator $\hat{H}_{h,\delta}$ has been purposely defined such that its computation requires that of the kernel density estimate only *over a regular grid*. This allows the use of the *binning* technique which together with the use of a *cardinal spline* leads to reasonably fast computation (only several times that of fourth cumulants for example).

2.2.4.1 Cardinal spline

The cardinal spline of order r is defined as the indicator function of the interval $[0,1)$ convolved with itself r times, which we denote by $\mathbb{1}_{[0,1)}^{\star r}$. As this function is symmetric around $r/2$, we shift it by $r/2$, and thus we tale as kernel the function $\mathbb{1}_{[0,1)}^{\star r}(\cdot + r/2) = \mathbb{1}_{[-1/2,1/2)}^{\star r}$, the indicator function of the interval $[-1/2,1/2)$ convolved with itself r times. Such a kernel is not unusual. For $r = 1$, it is the rectangular kernel and for $r = 2$, the triangular kernel. For $r \to \infty$ one gets, after rescaling, the Gaussian kernel (since the sum of independent variables tends to be Gaussian). Even with r as low as 3, the function $\mathbb{1}^{\star 3}_{[1/2,1/2)}$ already has a very similar shape to the Gaussian density. We recommend taking $r=3$. The choice $r=1$ and $r=2$ introduces jumps, respectively in the density estimate and in its derivative. As a result, $\hat{H}_\delta\{\mathbf{bx}\}$ is not differentiable with respect to the row vector $\mathbf{b}$.

The cardinal spline $\mathbb{1}^{\star r}_{[0,1)}$ possesses some interesting features listed below.

1. It has compact support $[0,r)$ and has as Fourier transform the function $u \mapsto [(e^{iu}-1)/(iu)]^r$
2. It possesses the *partition of unity* property: $\sum_m \mathbb{1}^{\star r}_{[0,1)}(v+m)=0, \forall v$
3. The scaled kernel with an integer scale p, can be expressed as a convex combination of the same kernels, but not scaled and integer shifted:

$$\frac{1}{p}\mathbb{1}^{\star r}_{[0,1)}\left(\frac{u}{p}\right)=\sum_{l=0}^{pr-r} c_l^{p,r}\mathbb{1}^{\star r}_{[0,1)}(u-l)$$

where $c_l^{p,r}, l=0,\ldots,pr-r$ are the coefficients of the polynomial $(p^{-1}\sum_{l=0}^{p-1} z^l)^r$. This property can be obtained by noting that $\mathbb{1}_{[0,1)}(\cdot/p)=\sum_{l=0}^{p-1} B^l \mathbb{1}_{[0,1)}$ where B is the backward shift operator: $Bf=f(\cdot-1)$ and that this operator commutes with the convolution.

2.2.4.2 Binning technique

This technique [19,7] has been introduced to speed up the calculation of the kernel estimate over a regular grid. Initially, the kernel is approximated by a piecewise constant or linear function. But it has been shown [14] that no approximation is needed if the kernel can be expressed as a combination of easy to compute shifted kernels and the grid size is right. Although this technique can be very general, it is best applied to kernel $\mathcal{K}=\mathbb{1}^{\star r}_{[-1/2,1/2)}$, which from the property 3 above, can be expressed as

$$\mathcal{K}(u)=\sum_{l=0}^{pr-r} c_l^{p,r}\, p\mathbb{1}^{\star r}_{(0,1]}\left(pu+\frac{pr}{2}-l\right).$$

With the grid size $\delta=h/p$, a sub-multiple of h, one gets from formula (2.4)

$$\hat{p}_y\left[\bar{y}+\left(m-\frac{pr}{2}\right)\hat{\sigma}_y\delta\right]\hat{\sigma}_y\delta=\sum_{l=0}^{pr-r} c_l^{pr-r}\frac{1}{T}\sum_{n=1}^{T}\mathbb{1}^{\star r}_{[0,1)}\left[m-\frac{y(n)-\bar{y}}{\delta\hat{\sigma}_y}-l\right] \quad (2.13)$$

The computation of the above right hand side can be done in two steps. The first step, called *binning*, consists in computing the sequence $\tilde{\pi}_y(m-r/2)=T^{-1}\sum_{n=1}^{T}\mathbb{1}^{\star r}_{(0,1]}[m-\tilde{y}(n)], m\in\mathbb{Z}$, where $\tilde{y}(n)=[y(n)-\bar{y}/(\hat{\sigma}_y\delta)]$. The second step consists in smoothing this sequence via a moving average with coefficients $c_l^{pr-r}, l=0,\ldots,pr-r$. Usually p

is small so that this step is fast. Our experience reveals that one may even choose $p=1$, that is $\delta=h$, which eliminates this step altogether. Smaller value of δ does not seem to improve the estimator appreciably.

The binning is fast for small r. For $r=1$, it amounts to counting the fraction of time $\tilde{y}(n)$ falls into each interval $(m, m+1]$. This actually corresponds to the original binning technique. For general r, we note that $\mathbb{1}^{\star r}_{[0,1)}$ has support in $[0, r)$, hence the term $\mathbb{1}^{\star r}_{[0,1)}[m-\tilde{y}(n)]$ is non-zero only if $m \in \{\lfloor\tilde{y}(n)\rfloor+1, \ldots, \lfloor\tilde{y}(n)\rfloor+r\}$ where $\lfloor x \rfloor$ denotes the largest signed integer not exceeding x. Thus the non-zero $\tilde{\pi}_y(m-r/2)$ can be computed as follows.

1. Initialize $\tilde{\pi}_y(m-r/2)=0, m \in \{\min_{n\in T}\lfloor\tilde{y}(n)\rfloor+1, \ldots, \max_{n\in T}\lfloor\tilde{y}(n)\rfloor+r\}$.
2. For $n=1, \ldots, T$ add $\mathbb{1}^{\star r}_{[0,1)}[\lfloor\tilde{y}(n)\rfloor+j-\tilde{y}(n)]/T$ to $\tilde{\pi}_y(\lfloor\tilde{y}(n)\rfloor+j-r/2), j=1, \ldots, r$.

More details on binning can be found in [14,15].

2.2.4.3 Computation of $\hat{H}_\delta$

From the partition of unity property of $\mathbb{1}^{\star r}_{[0,1)}$, the $\tilde{\pi}_y(m-r/2)$ sum to 1 and can thus be interpreted as probabilities of the cells of length $\hat{\sigma}_y\delta$ centered at $(m-r/2)\hat{\sigma}_y\delta+\bar{y}$. Since the coefficients c_l^{pr-r} are positive and sum to 1, the right hand side of (2.13), denoted by $\hat{\pi}_y(m-pr/2)$, are also probabilities (of the same cells but centered at $(m-pr/2)\hat{\sigma}_y\delta+\bar{y})$. Hence by (2.6)

$$\hat{H}_\delta(y) = -\sum_m \hat{\pi}_y\left(m-\frac{pr}{2}\right)\log\hat{\pi}_y\left(m-\frac{pr}{2}\right)+\log(\hat{\sigma}_y\delta),$$

in the case where pr is even. In the case where it is odd one just changes slightly the definition (2.6) of $\hat{H}_\delta$ by summing with respect to half integers instead of integers.

2.2.4.4 Computation of $\hat{H}^M$

This estimator is best computed via the estimated characteristic function (2.10) using the formula $\hat{H}_h^M\{y\} = -\sum_{m=-\infty}^{\infty} a_m^* \Phi_y[2\pi m/(M\hat{\sigma}_y)]$, where a_m are the Fourier coefficients of $\log \hat{p}_y^M$ defined in (2.9). These coefficients can be computed by the method described in subsection 2.2.3, provided that the characteristic function $\Phi_{\mathcal{K}}$ of $\mathcal{K}$ has compact support. Note that the Fourier transform of $\mathbb{1}^{\star 2r}_{[-1/2,1/2)}$ is the function $v \mapsto [\sin(v/2)/(v/2)]^{2r}$ (see subsection 2.2.4.1), hence if we take $\mathcal{K}(v)=[\sin(v/2)/(v/2)]^{2r}/\mathbb{1}^{\star 2r}_{[-1/2,1/2)}(0)$, then $\Phi_K = \mathbb{1}^{\star 2r}_{[-1/2,1/2)}/\mathbb{1}^{\star 2r}_{[-1/2,1/2)}(0)$ which has compact support. We recommend choosing $r=2$: the choice $r=1$ is not allowed since then $\int v^2 K(v)dv=\infty$ and the choice $r>2$ is costly computationally.

2.2.5 Minimization of the empirical criteria

To minimize the criteria $\hat{H}_\delta$ and $\hat{H}^M$ a gradient descent, or better quasi Newton, algorithm may be used. The criterion $\hat{H}_{\{u\}}$ however, has jumps in its gradient and thus needs a special method for its minimization.

The quasi-Newton method requires the knowledge of the gradient and approximate Hessian. It is, however, more interesting to work with the *relative gradient* and *relative Hessian* [2], which are defined, at a point $\mathbf{B}$ and for a criterion $\hat{C}$, as the matrix of first and second derivatives of $\hat{C}(\mathbf{B}+\mathcal{E}\mathbf{B})$ with respect to the element of the matrix $\mathcal{E}$, evaluated at $\mathcal{E}=0$.

2.2.5.1 *Relative gradients of the empirical criteria*

Consider first the criterion $\hat{C}_\delta$. The matrix of derivatives of $\log|\det(\mathbf{B}+\mathcal{E}\mathbf{B})| = \log|\det\mathbf{B}| + \log|\det(\mathbf{I}+\mathcal{E})|$ with respect to $\mathcal{E}$ at $\mathcal{E}=0$ is simply the identity matrix. Therefore the relative gradient of $\hat{C}_\delta$ at $\mathbf{B}$ is

$$\hat{C}'_\delta(\mathbf{B}) = \begin{bmatrix} \partial\hat{H}_\delta\{y_1+\epsilon y_1\}/\partial\epsilon|_{\epsilon=0} & \cdots & \partial\hat{H}_\delta\{y_1+\epsilon y_P\}/\partial\epsilon|_{\epsilon=0} \\ \vdots & \ddots & \vdots \\ \partial\hat{H}_\delta\{y_P+\epsilon y_1\}/\partial\epsilon|_{\epsilon=0} & \cdots & \partial\hat{H}_\delta\{y_P+\epsilon y_P\}/\partial\epsilon|_{\epsilon=0} \end{bmatrix} - \mathbf{I} \quad (2.14)$$

where y_k are the components of $y=\mathbf{B}x$.

From the formula (2.5), one gets

$$\left.\frac{\partial\hat{H}_\delta\{y_j+\epsilon y_k\}}{\partial\epsilon}\right|_{\epsilon=0} = \sum_m \log\hat{p}_{y'_j}(m\delta)\left.\frac{\partial\hat{p}_{(y_j+\epsilon y_k)'}(m\delta)}{\partial\epsilon}\right|_{\epsilon=0}\delta + \frac{\partial\hat{\sigma}^2_{y_j+\epsilon y_k}/\partial\epsilon|_{\epsilon=0}}{2\hat{\sigma}^2_{y_j}},$$

where $(y_j+\epsilon y_k)' = [y_j-\bar{y}_j+\epsilon(y_k-\bar{y}_k)]/\hat{\sigma}_{y_j+\epsilon y_k}$, noting that $\sum_m \hat{p}_{(y_j+\epsilon y_k)'}(m\delta)=1$ and hence $\partial[\sum_m \hat{p}_{(y_j+\epsilon y_k)'}(m\delta)]/\partial\epsilon=0$. Since $\hat{\sigma}^2_{y_j+\epsilon y_k} = T^{-1}\sum_{n=1}^T[y_j(n)+\epsilon y_k(n)]^2 - \{T^{-1}\sum_{n=1}^T[y_j(n)+\epsilon y_k(n)]\}^2$,

$$\left.\frac{\partial\hat{\sigma}^2_{y_j+\epsilon y_k}}{\partial\epsilon}\right|_{\epsilon=0} = \frac{2}{T}\sum_{n=1}^{T}[y_j(n)-\bar{y}_j(n)]y_k(n)$$

On the other hand, from (2.4)

$$\hat{p}_{(y_j+\epsilon y_k)'}(m\delta) = \frac{1}{hT}\sum_{n=1}^{T}\mathcal{K}\left\{\frac{m\delta}{h} - \frac{y_j(n)-\bar{y}_j+\epsilon[y_k(n)-\bar{y}_k]}{h\hat{\sigma}_{y_j+\epsilon y_k}}\right\}.$$

Therefore its derivative with respect to ϵ at $\epsilon = 0$ equals

$$\frac{1}{h^2 T \hat{\sigma}_y} \sum_{n=1}^{T} \mathcal{K}' \left[\frac{m\delta - y_j'(n)}{h} \right] \left\{ y_k(n) - \bar{y}_k - [y_j - \bar{y}_j] \sum_{l=1}^{T} \frac{[y_j(l) - \bar{y}_j] y_k(l)}{T \hat{\sigma}^2_{y_j}} \right\}$$

where $\mathcal{K}'$ denotes the derivative of $\mathcal{K}$. From the above results, $\partial \hat{H}_\delta\{y_j + \epsilon y_k\}/\partial \epsilon|_{\epsilon=0} = T^{-1} \sum_{n=1}^{T} \hat{\varphi}_{\delta, y_j}[y_j(n)] y_k(n)$ where

$$\hat{\varphi}_{\delta, y_j}[y_j(n)] = \tilde{\varphi}_{\delta, y_j}[y_j(n)] - \frac{1}{T} \sum_{l=1}^{T} \tilde{\varphi}_{\delta, y_j}[y_j(l)] + (1 - \hat{\lambda}_{\delta, y_j}) \frac{y_j(n) - \bar{y}_j}{\sigma^2_{y_j}},$$

with

$$\tilde{\varphi}_{\delta, y_i}(u) = \sum_{m=-\infty}^{\infty} \log \hat{p}_{y_i'}(m\delta) \mathcal{K}' \left(\frac{m\hat{\sigma}_{y_j}\delta + \bar{y}_j - u}{h \hat{\sigma}_{y_j}} \right) \frac{\delta}{h^2 \hat{\sigma}_{y_j}}$$

$$\hat{\lambda}_{\delta, y_j} = \frac{1}{T} \sum_{n=1}^{T} \tilde{\varphi}_{\delta, y_j}[y_j(n)][y_j(n) - \bar{y}_j].$$

The function $\tilde{\varphi}_{\delta, y_i}$ can be rewritten as

$$\tilde{\varphi}_{\delta, y_i}(u) = -\frac{d}{du} \sum_{m=-\infty}^{\infty} \log \hat{p}_{y_i}(m\hat{\sigma}_{y_j}\delta + \bar{y}_j) \mathcal{K} \left(\frac{m\hat{\sigma}_{y_j}\delta + \bar{y}_j - u}{h \hat{\sigma}_{y_j}} \right) \frac{\delta}{h}.$$

The sum in the above right hand side can be viewed as a doubly smoothed estimator of the function $\log p_{y_i}$; hence $\tilde{\varphi}_{\delta, y_i}$ is an estimate of the logarithmic derivative φ_{y_j} of $-\log p_{y_i}$, called the *score function* of y_j. By integration by part the score function can be seen to satisfy: $\mathbb{E}\varphi_y(y) = 0$ and $\mathbb{E}[\varphi_y(y)y] = 1$. But its estimate $\tilde{\varphi}_{\delta,y}$ does not satisfy $T^{-1}\sum_{n=1}^{T} \tilde{\varphi}_{\delta,y}[y(n)] = 0$ and $T^{-1}\sum_{n=1}^{T} \tilde{\varphi}_{\delta,y}[y(n)]y(n) = 1$. The function $\hat{\varphi}_{\delta,y}$ is a corrected form of this estimator which by construction satisfies the above equalities. It follows that the diagonal of the gradient matrix (2.14) vanishes identically (which is actually a consequence of the equivariance property of $\hat{H}_\delta$ with respect to scaling).

Consider next the criterion $\hat{C}^M$. Its relative gradient is still given by the right hand side of (2.14) but with $\hat{H}^M$ in place of $\hat{H}_\delta$. By a similar argument as before using the formula (2.7), one gets

$$\left. \frac{\partial \hat{H}^M\{y_j + \epsilon y_k\}}{\partial \epsilon} \right|_{\epsilon=0} = \int_0^M \log \hat{p}^M_{y_j'}(v) \left. \frac{\hat{p}^M_{(y_j+\epsilon y_k)'}(v)}{\partial \epsilon} \right|_{\epsilon=0} + \frac{\partial \hat{\sigma}^2_{y_j+\epsilon y_k}/\partial \epsilon|_{\epsilon=0}}{2\hat{\sigma}^2_{y_j}}.$$

Since the function $\log \hat{p}^M_{y'_j}$ is periodic, the integration in the above formula can be extended to $(-\infty, \infty)$ by replacing $\hat{p}^M_{(y_j+\epsilon y_k)'}$ by $\hat{p}_{(y_j+\epsilon y_k)'}$. By a similar argument as before $\partial \hat{H}^M\{y_j + \epsilon y_k\}/\partial \epsilon|_{\epsilon=0} = T^{-1}\sum_{n=1}^T \hat{\varphi}^M_{y_j}[y_j(n)]y_k(n)$ where

$$\hat{\varphi}^M_{y_j}[y_j(n)] = \tilde{\varphi}^M_{y_j}[y_j(n)] - \frac{1}{T}\sum_{l=1}^{T}\tilde{\varphi}^M_{y_j}[y_j(l)] + (1-\hat{\lambda}^M_{y_j})\frac{y_j(n)-\bar{y}_j}{\hat{\sigma}^2_{y_j}},$$

with

$$\begin{aligned}\tilde{\varphi}^M_{y_i}(u) &= \int_{-\infty}^{\infty} \log \hat{p}^M_{y'_i}(v)\mathcal{K}'\left(\frac{v\hat{\sigma}_{y_j}+\bar{y}_j-u}{h\hat{\sigma}_{y_j}}\right)\frac{dv}{h^2\hat{\sigma}_{y_j}} \\ &= -\frac{d}{du}\int_{-\infty}^{\infty} \log \hat{p}^M_{y_j}(v)\mathcal{K}\left(\frac{v-u}{h\hat{\sigma}_{y_j}}\right)\frac{dv}{h\hat{\sigma}_{y_j}} \qquad (2.15) \\ \hat{\lambda}^M_{y_j} &= \frac{1}{T}\sum_{n=1}^{T}\tilde{\varphi}^M_{y_j}[y_j(n)][y_j(n)-\bar{y}_j].\end{aligned}$$

As before, $T^{-1}\sum_{n=1}^T \hat{\varphi}^M_y[y(n)] = 0$ and $T^{-1}\sum_{n=1}^T \hat{\varphi}^M_y[y(n)]y(n) = 1$; hence the gradient matrix of $\hat{C}^M$ has diagonal vanishing identically.

The computation of the function $\tilde{\varphi}^M_{y_j}$, even at the data points only, involves integration. This difficulty, however, can be avoided by working with its Fourier series. By (2.15), $\tilde{\varphi}^M_{y_j}$ is the derivative of the convolution of $-\log \hat{p}^M_{y_j}$ with the function $v \mapsto \mathcal{K}[-v/(h\sigma_{y_j})]/(h\sigma_{y_j})$, hence its Fourier coefficients equal $-ima_m\Phi_{\mathcal{K}}(-mh\hat{\sigma}_{y_j})$, where a_m are given in (2.9) and can be computed as described in subsection 2.2.3. The sum $T^{-1}\sum_{n=1}^T \hat{\varphi}^M_{y_j}[y_j(n)]y_k(n)$ is then obtained as $T^{-1}\sum_{n=1}^T \tilde{\varphi}^M_{y_j}[y_j(n)][y_k(n)-\bar{y}_k] + (1-\hat{\lambda}^M_{y_j})\widehat{\text{cov}}(y_j, y_k)$, where $\widehat{\text{cov}}(y_j, y_k)$ denotes the sample covariance between y_j and y_k and the first term is computed by representing $\tilde{\varphi}^M_{y_j}$ as a Fourier series:

$$\frac{1}{T}\sum_{n=1}^{T}\tilde{\varphi}^M_{y_j}[y_j(n)][y_k(n)-\bar{y}_k] = \sum_{m=-q}^{q} ima^*_m\Phi_{\mathcal{K}}(mh\hat{\sigma}_{y_j})\sum_{n=1}^{T}e^{imy_j(n)}[y_k(n)-\bar{y}_k].$$

The above formula can also be used for computing $\hat{\lambda}^M_{y_j}$ by putting $k=j$.

Note that since $\hat{H}_\delta\{y_j\}$ is a continuously differentiable function of the data $y_j(1), \ldots, y_j(T)$, one has $\partial\hat{H}_\delta\{y_j+\epsilon y_k\}/\partial\epsilon|_{\epsilon=0} = \sum_{n=1}^T[\partial\hat{H}_\delta\{y_j\}/\partial y_j(n)]y_k(n)$. Therefore by identification one gets $\hat{\varphi}_{\delta,y_j}[y_j(n)] = T\partial H_\delta\{y_j\}/\partial y_j(n)$. Similarly $\hat{\varphi}^M_{y_j}[y_j(n)] = T\partial \hat{H}^M\{y_j\}/\partial y_j(n)$.

We will not compute the gradient of $\hat{C}_{\{u\}}$ since it has jumps and is thus not useful.

2.2.5.2 Approximate relative Hessian of the empirical criteria

The relative Hessians of $\hat{C}_\delta$ and $\hat{C}^M$ are quite complex. Therefore we shall approximate them by that of the limiting criteria $C_{h,\delta}$ and C^M_δ obtained by letting $n \to \infty$ with h, δ and M fixed. To simplify further, we assume that the matrix $\mathbf{B}$ where the Hessians are computed (approximately) is truly demixing in the sense that $\mathbf{y} = \mathbf{B}\mathbf{x}$ has independent components.

We begin by computing the relative gradients of the limiting criteria $C_{h,\delta}$ and C^M_h. We may repeat the calculations in a previous subsection, replacing the sample average by the expectation. Thus $\hat{p}_y(v)$ is replaced by $p_{h,y}(v) = \mathbb{E}\{\mathcal{K}[(v-y)/(h\sigma_y)]/h\sigma_y\}$ where σ_y is the standard deviation of y. The relative gradients $C'_{h,\delta}$ and $C^{M\prime}_h$ of $C_{h,\delta}$ and C^M_h are again given by the right hand side of (2.14) but with $\hat{H}_\delta$ replaced respectively by $H_{h,\delta}(y) = -\sum_m \log[p_{h,y'}\{m\delta\}]p_{h,y'}(m\delta)\delta + \log\sigma_y$ and $H^M_h\{y\} = -\int_0^M \log[p^M_{h,y'}(v)]p^M_{h,y'}(v)dv + \log\sigma_y$. By a similar calculation as before, one gets $\partial H_{h,\delta}\{y_j + \epsilon y_k)/\partial\epsilon|_{\epsilon=0} = \mathbb{E}[\varphi_{h,\delta,y_j}(y_j)y_k]$ and $\partial H^M_h\{y_j + \epsilon y_k)/\partial\epsilon|_{\epsilon=0} = \mathbb{E}[\varphi^M_{y_j}(y_j)y_k]$ where

$$\varphi_{h,\delta,y_j}(u) = \tilde{\varphi}_{h,\delta,y_j}(u) - \mathbb{E}\tilde{\varphi}_{h,\delta,y_j}(y_j) + (1-\lambda_{h,\delta,y_j})(u-\mathbb{E}y_j)/\sigma^2_{y_j},$$
$$\varphi^M_{h,y_j}(u) = \tilde{\varphi}^M_{h,y_j}(u) - \mathbb{E}\tilde{\varphi}^M_{h,y_j}(y_j) + (1-\lambda^M_{h,y_j})(u-\mathbb{E}y_j)/\sigma^2_{y_j}$$

with

$$\tilde{\varphi}_{h,\delta,y_j}(u) = -\frac{d}{du}\sum_{m=-\infty}^{\infty}\log p_{h,\delta,y_i}(m\sigma_{y_j}\delta + \mathbb{E}y_j)\mathcal{K}\left(\frac{m\sigma_{y_j}\delta + \mathbb{E}y_j - u}{h\sigma_{y_j}}\right)\frac{\delta}{h},$$
$$\lambda_{h,\delta,y_j} = \mathbb{E}[\tilde{\varphi}_{h,\delta,y_j}(y_j)(y_j - \mathbb{E}y_j)],$$
$$\tilde{\varphi}^M_{h,y_i}(u) = -\frac{d}{du}\int_{-\infty}^{\infty}\log p^M_{h,y_j}(v)\mathcal{K}\left(\frac{v-u}{h\sigma_{y_j}}\right)\frac{dv}{h\sigma_{y_j}}$$
$$\lambda^M_{h,y_i}(u) = \mathbb{E}[\tilde{\varphi}^M_{h,y_i}(y_j)y_j].$$

Clearly, the relative gradient matrices of $C_{h,\delta}$ and C^M_h also have diagonal vanishing identically. Further they *vanish at any matrix* $\mathbf{B}$ *for which* $\mathbf{y} = \mathbf{B}\mathbf{x}$ *has independent components.*

From the definition of relative gradient, the matrix of (ordinary) derivatives of $C_{h,\delta}(\mathbf{B} + \mathcal{E}\mathbf{B})$ with respect to $\mathcal{E}$ is $C'_{h,\delta}(\mathbf{B} + \mathcal{E}\mathbf{B})(\mathbf{I} + \mathcal{E}^{\mathrm{T}})^{-1}$. Taking the derivative of the last expression at $\mathcal{E}=0$ and noting that $C'_{h,\delta}(\mathbf{B})=0$, one sees that the elements of the relative Hessian of $C_{h,\delta}$ at $\mathbf{B}$ are the derivatives of $C'_{h,\delta}(\mathbf{B}+\mathcal{E}\mathbf{B})$ at $\mathcal{E}=0$.

Since the diagonal of $C'_{h,\delta}(\mathbf{B}+\mathcal{E}\mathbf{B})$ vanish for all $\mathcal{E}$, the same holds for its derivatives with respect to $\mathcal{E}$. By the symmetry of the Hessian matrix, the derivative of $C'_{h,\delta}(\mathbf{B}+\mathcal{E}\mathbf{B})$ with respect to *any* diagonal element of $\mathcal{E}$ at $\mathcal{E}=0$ also vanishes. Thus, we need only compute the derivative of the i,j elements of $C'_{h,\delta}(\mathbf{B}+\mathcal{E}\mathbf{B})$, $i\neq j$, with respect to $\mathcal{E}_{kl}$, $k\neq l$, at $\mathcal{E}=0$. This element equals $\mathbb{E}[\varphi_{h,\delta,y_i+\mathcal{E}_{i\cdot}\mathbf{y}}(y_i+\mathcal{E}_{i\cdot}\mathbf{y})(y_j+\mathcal{E}_{j\cdot}\mathbf{y})]$ where $\mathcal{E}_{i\cdot}$ is the i-th row of $\mathcal{E}$. The above expression does not depend on $\mathcal{E}_{kl}$ for $k\notin\{i,j\}$; hence its derivative with respect to it vanishes. Using the fact that $\mathbf{y}$ has independent components, its derivative with respect to $\mathcal{E}_{jl}, l\neq j$ at $\mathcal{E}=0$ is

$$\mathbb{E}[\varphi_{h,\delta,y_i}(y_i)y_l]=\begin{cases}[\mathbb{E}\varphi_{h,\delta,y_i}(y_i)](\mathbb{E}y_k)=0 & \text{if } l\neq i\\ 1 & \text{if } l=i\end{cases}$$

and with respect to $\mathcal{E}_{il}, l\neq i$ at $\mathcal{E}=0$ is

$$[\mathbb{E}\varphi'_{h,\delta,y_i}(y_i)]\mathbb{E}(y_l y_j)+\mathbb{E}\left[\left.\frac{\partial\varphi_{h,\delta,y_i+\epsilon y_l}(y_i)}{\partial\epsilon}\right|_{\epsilon=0}\right](\mathbb{E}y_j),$$

where φ'_{h,δ,y_i} is the derivative of $\hat{\varphi}_{y_i}$. But $\mathbb{E}[\varphi_{h,\delta,y_i+\epsilon y_l}(y_i+\epsilon y_l)]=0$; hence by taking the derivative with respect to ϵ at $\epsilon=0$, $\mathbb{E}[\partial\varphi_{h,\delta,y_i+\epsilon y_l}(y_i)/\partial\epsilon|_{\epsilon=0}]=-\mathbb{E}[\varphi'_{h,\delta,y_i}(y_i)y_l]$. Therefore the above expression equals 0 for $l\neq j$ and $\mathbb{E}[\varphi'_{h,\delta,y_i}(y_i)\sigma^2_{y_j}]$ for $l=j$.

Finally the ij,kl element of the relative Hessian matrix of $C_{h,\delta}$ equals 0 if $(k,l)\notin\{(i,j),(j,i)\}$, $\mathbb{E}\varphi'_{h,\delta,y_i}(y_i)\sigma^2_{y_j}$ if $(k,l)=(i,j)$ and 1 if $(k,l)=(j,i)$.

The above calculations can also be applied to the limiting criterion C_h^M: one gets the same formula just by replacing φ_{h,δ,y_i} by φ^M_{h,y_i}.

In practice, the ij,ij element of the relative Hessian matrix of $C_{h,\delta}$ and C_δ^M would be estimated by $\hat{J}_\delta(y_i)\hat{\sigma}^2_{y_j}$ and $\hat{J}^M(y_i)\hat{\sigma}^2_{y_j}$, where

$$\hat{J}_\delta(y_i)=\frac{1}{T}\sum_{n=1}^{T}\hat{\varphi}'_{\delta,y_i}[y_i(n)],\qquad \hat{J}^M(y_i)=\frac{1}{T}\sum_{n=1}^{T}\hat{\varphi}^{M\prime}_{y_i}[y_i(n)], \tag{2.16}$$

φ'_{δ,y_i} and $\varphi^{M\prime}_{y_i}$ being the derivatives of $\hat{\varphi}_{\delta,y_i}$ and $\varphi^M_{y_i}$.

2.2.5.3 The quasi Newton algorithm

The Newton algorithm consists in replacing the function to be minimized by its Taylor expansion up to second order around the current point and then minimizing this expansion to get the new point. In the quasi Newton algorithm, the second order terms in the expansion are replaced by some approximations. In our case, as we work with relative gradient and Hessian, we consider the approximate expansion

$$\hat{C}_\delta(\mathbf{B}+\mathscr{E}\mathbf{B}) \approx \sum_{i\neq j}\left\{\mathscr{E}_{ij}\hat{\varphi}_{h,y_i}[y_i(n)]y_j(n)+\frac{1}{2}\mathscr{E}_{ij}^2\hat{J}_\delta(y_i)\hat{\sigma}_{y_j}^2+\mathscr{E}_{ij}\mathscr{E}_{ji}\right\},$$

where $\hat{J}_\delta(y_i)$ is given in (2.16). Minimizing the above expression yields

$$\begin{bmatrix}\mathscr{E}_{ij}\\ \mathscr{E}_{ji}\end{bmatrix} = \begin{bmatrix}\hat{J}_\delta(y_i)\hat{\sigma}_{y_j}^2 & 1\\ 1 & \hat{J}_\delta(y_j)\hat{\sigma}_{y_i}^2\end{bmatrix}^{-1}\frac{1}{T}\sum_{n=1}^{T}\begin{bmatrix}\hat{\varphi}_{\delta,y_i}[y_i(n)]y_j(n)\\ \hat{\varphi}_{\delta,y_j}[y_j(n)]y_i(n)\end{bmatrix}.$$

The matrix $\mathbf{B}$ is then updated by adding $\mathscr{E}\mathbf{B}$ to it; the diagonal element of $\mathscr{E}$ can be put to 0 as it mainly affects only the scale of the extracted sources. A similar algorithm is obtained for minimizing $\hat{C}^M$: one just replaces $\hat{J}_\delta(y_i)$ by $\hat{J}^M(y_i)$.

The 2×2 matrix in the above formula may not, however, be positive definite. It is important to have an approximate Hessian which is positive definite. In this case even if the criterion cannot be guaranteed to decrease at each step of the algorithm, it can be made to do so by reducing sufficiently the step size, therefore ensuring at least the convergence of the algorithm to a local minimum. A method to reduce the step size is described in [18], page 384. Note that $\hat{J}_\delta(y_i)$ is an estimate of $\mathbb{E}\varphi'_{h,\delta,y_i}(y_i)$, which for small h, δ is an approximation to $\mathbb{E}\varphi'_{y_i}(y_i)$. By integration by parts $\mathbb{E}\varphi'_{y_i}(y_i)$ is the same as $\mathbb{E}\varphi^2_{y_i}(y_i)$ and is known as the Fisher information $J(y_i)$ of y_i. The Fisher information is known to satisfy if y is not Gaussian; if y is Gaussian there is equality: $J(y) > 1/\sigma_y^2$. (This result can be obtained from the Schwartz inequality and the equality $\mathbb{E}[\varphi_y(y)(y-\mathbb{E}y)] = 1$.) Therefore $\mathbb{E}\varphi'_{h,\delta,y_i}(y_i) > 1/\sigma^2_{y_i}$ if h and δ are small enough. But we don't know how small is enough and besides because of statistical error, this inequality does not always translate into $\hat{J}_\delta(y_i) > 1/\hat{\sigma}^2_{y_i}$. To ensure that this equality holds, one may replace $\hat{J}_\delta(y)$ by $T^{-1}\sum_{n=1}^T \hat{\varphi}^2_{\delta,y}[y(n)]$, which is greater than $1/\hat{\sigma}_y^2$ by the Schwartz inequality and the equality $\sum_{n=1}^T \hat{\varphi}_{\delta,y}[y(n)][y(n)-\bar{y}] = 1$. Similarly, in the quasi Newton algorithm for minimizing $\hat{C}^M$, one may replace $\hat{J}^M(y_i)$ by $T^{-1}\sum_{n=1}^T \hat{\varphi}_y^{M\,2}[y(n)]$.

2.2.5.4 Minimization of the order statistic based criterion

The function $\mathbf{B}\mapsto \hat{C}_{\{u\}}(\mathbf{B})$ has jumps in its gradient, since, as it can be seen below, it is the case for the function $\mathbf{B}_{j\cdot}\to \hat{H}_{\{u\}}\{\mathbf{B}_{j\cdot}\mathbf{y}\}$ where $\hat{H}_{\{u\}}$ is defined in (2.12) and $\mathbf{B}_{j\cdot}$ is the j-th row of $\mathbf{B}$. Therefore the criterion $\hat{C}_{\{u\}}(\mathbf{B})$ cannot be minimized by the usual gradient based method. We present here a simple method based on the relaxation principle: one minimizes $C_{\{u\}}(\mathbf{B})$ successively with respect to a non-diagonal element of the matrix $\mathbf{B}$, keeping the other fixed. One can ignore the diagonal elements since the criterion is scale invariant, but to avoid a possible explosion of the non-diagonal elements on a row of $\mathbf{B}$ when the diagonal element of this row should be 0, it is recommended to renormalize the rows of $\mathbf{B}$, periodically or when they become too large. Thus, for a pair of indexes (i,j), one minimizes $\hat{C}_{\{u\}}(\hat{\mathbf{B}}+\beta\mathbf{E}_{ij})$ with respect to $\beta\in\mathbb{R}$, where $\hat{\mathbf{B}}$ is the current (estimated) separation matrix and $\mathbf{E}_{ij}$ is the matrix with 1 at the (i,j) place and 0 elsewhere. Letting β^* be the β realizing the minimum, one then replaces $\hat{\mathbf{B}}$ by $\hat{\mathbf{B}}+\beta^*\mathbf{E}_{ij}$ and continues with another pair of indexes. Another possibility is to minimize $\hat{C}_{\{u\}}(\hat{\mathbf{B}}+\beta\mathbf{E}_{ij}\hat{\mathbf{B}})$, and then replace $\hat{\mathbf{B}}$ by $\hat{\mathbf{B}}+\beta^*\mathbf{E}_{ij}\hat{\mathbf{B}}$, β^* being the β realizing the minimum.

Since one can always write $\mathbf{E}_{ij}$ in the form $(\mathbf{E}_{ij}\hat{\mathbf{B}}^{-1})\hat{\mathbf{B}}$, the above two minimization problems are particular cases of more general unidimensional problem, which is the minimization of $C_{\{u\}}(\hat{\mathbf{B}}+\beta\mathbf{E}_i\hat{\mathbf{B}})$ with respect to $\beta\in\mathbb{R}$ where $\hat{\mathbf{B}}$ and $\mathbf{E}_i$ are two given matrices with $\mathbf{E}_i$ having all rows other than the i-th row equal to zero. Putting $\hat{y}_1,\ldots,\hat{y}_P$ as the components of $\hat{\mathbf{B}}\mathbf{x}$ and $\hat{z}$ as the only non-zero (the i-th) component of $\mathbf{E}_i\hat{\mathbf{B}}\mathbf{x}$, one has:

$$\hat{C}_{\{u\}}(\hat{\mathbf{B}}+\beta\mathbf{E}_i\hat{\mathbf{B}})=\hat{C}_{\{u\}}(\hat{\mathbf{B}})$$
$$+\sum_{l=2}^{L}\log\left[\frac{(\hat{y}_i+\beta\hat{z})(n_l{:}T)-(\hat{y}_i+\beta\hat{z})(n_{l-1}{:}T)}{\hat{y}_i(n_l{:}T)-\hat{y}_i(n_{l-1}{:}T)}\right]\frac{n_l-n_{l-1}}{n_L-n_1}-\log|\det(\mathbf{I}+\beta\mathbf{E}_i)|$$

where $\hat{y}_i(k{:}T)$ and $(\hat{y}_i+\beta\hat{z})(k{:}T)$ are the k-th order statistics of $\hat{y}_i(1),\ldots,\hat{y}_i(T)$ and of $y_i(1)+\beta\hat{z}(1),\ldots,y_i(n)+\beta\hat{z}(n)$, respectively. Since only the i-th row of $\mathbf{E}_i$ is non-zero, $\det(\mathbf{I}+\beta\mathbf{E}_i)=1+c\beta$ where c is the i-th diagonal term of $\mathbf{E}_i$. Note that by taking $\mathbf{E}_i=\mathbf{E}_{ij}$ instead of $\mathbf{E}_i=\mathbf{E}_{ij}\hat{\mathbf{B}}^{-1}$ $(j\neq i)$ one has $\hat{z}=\hat{y}_j$ and $c=0$, which simplifies somewhat the computations.

The crucial point is that the order statistics $(\hat{y}_i+\beta\hat{z})(n_l{:}T)$, as functions of β, are piecewise linear continuous with a finite number of slope changes. Indeed, by definition, $(\hat{y}_i+\beta\hat{z})(j{:}T)$ is equal to $\hat{y}_i(k_j)+\beta\hat{z}_i(k_j)$, where $\{k_1,\ldots,k_T\}$ is a permutation of $\{1,\ldots,T\}$ such that:

$$\hat{y}_i(k_1)+\beta\hat{z}(k_1)\leqslant\cdots\leqslant\hat{y}_i(k_T)+\beta\hat{z}(k_T).$$

The indexes k_j depend of course on β, but if there is no tie among the $\hat{y}_i(1)+\beta\hat{z}(1),\ldots,\hat{y}_i(T)+\beta\hat{z}(T)$, the above inequality will be strict and it remains the same as β changes, as long as the change is small enough. Only when the change crosses a threshold for which a tie occurs among the $\hat{y}_i(1)+\beta\hat{z}(1),\ldots,\hat{y}_i(T)+\beta\hat{z}(T)$ can the k_j change. Let $\beta_1,\ldots,\beta_q$ the values of β, be arranged in increasing order, that produce such a tie, and put $\beta_0=-\infty$, $\beta_{q+1}=\infty$. Then for β in one of the open intervals $(\beta_0,\beta_1),\ldots,(\beta_q,\beta_{q+1})$ and for $j\in\{1,\ldots,T\}$, the expression $(\hat{y}_i+\beta\hat{z})(j{:}T)$ is a linear function of β of the form $\hat{y}_i(k_j)+\beta\hat{z}(k_j)$ for a certain k_j. One can show [9] that the function $\beta\mapsto\hat{C}_{\{u\}}(\hat{\mathbf{B}}+\beta\mathbf{E}_i\hat{\mathbf{B}})$, as given above, can attain its minimum (possibly $-\infty$) in such an interval only at its end points. It thus suffices to evaluate this function at these points, that is at $\beta_0,\ldots,\beta_{q+1}$, to find the its minimum point. (Note that a minimum point $\pm\infty$ is allowed; in this case the new $\hat{\mathbf{B}}$ is simply obtained by substituting its i-th row with $\pm$ the i-th row of $\mathbf{E}_i\mathbf{B}$.) In practice, one need not consider all the above β_k, but can restrict matters to those for which the non-decreasing sequence $(\hat{y}_i + \beta\hat{z})(1{:}T), \dots, (\hat{y}_i + \beta\hat{z})(T{:}T)$ has at least a tie at the n_1-th, ..., n_L-th places. A fast algorithm for this task has been developed in [9].

2.2.6 Statistical performance

We first examine the contrast property of the criteria, which ensures the separation of the sources at large sample size (at least if the criteria can be globally minimized). Then we look at the quality of the separation.

2.2.6.1 The contrast property of the criteria

The notion of *contrast* in blind source separation has been introduced by Comon [3] (see also Chapter 3, section 3.4 and [11]). In this context, a criterion is said to be the negative of a contrast if it can attain its global minimum if and only if the separation is exactly achieved *up to a scaling and a permutation*. This is a fundamental desirable property for a separation criterion. But it can hold for a theoretical criterion only and not for empirical criteria such as $\hat{C}_\delta$, $\hat{C}^M$ and $\hat{C}_{\{u\}}$ because of statistical fluctuation. To avoid this problem, we shall consider the limiting criteria obtained by letting the sample size T tend to infinity.

The empirical criteria $\hat{C}_\delta$, $\hat{C}^M$ and $\hat{C}_{\{u\}}$ would tend to C as $T \to \infty$ if we also let $h \to 0$ with an appropriately slow rate and $\delta \to 0$, $M \to \infty$ and the sequence $\{u\}$ becomes infinitely dense in $[0,1]$ in an appropriate way. The criterion C is the negative of a contrast if there is no more than one Gaussian source, since by the *Darmois Theorem* [3], no two sets of linear combinations of independent random variables can be independent unless at least two of these variables are Gaussian. We are, however, interested in the case where h, δ, M and the sequence $\{u\}$ are kept fixed (or converge to a finite limit) as $T \to \infty$. We shall show that the limiting criteria $C_{h,\delta}$, C_h^M and $C_{\{u\}}$ (the last being the limit of $\hat{C}_{\{u\}}$ as $n \to \infty$) are still the negative of a contrast, provided that h, δ and $1/M$ are small enough and $\{u\}$ is dense enough. This shows that the choice of h, δ, M and $\{u\}$ are not very important as long as h, δ is not too large, M is not too small and the sequence $\{u\}$ is not too coarse. Note that δ and $1/M$ can be arbitrarily small but not h, since the smaller the h, the larger the sample size needs to be to approach convergence.

The above contrast property results from the interesting property that the inverse mixing matrix $\mathbf{A}^{-1}$ is a stationary point of $C_{h,\delta}$, C_h^M and $C_{\{u\}}$ *regardless of the values of* h, δ, M *and the sequence* $\{u\}$. This can be seen from the formula for their relative gradient. Those for $C_{h,\delta}$ and C_h^M, have been derived in subsection 2.2.5.2; the one for $C_{\{u\}}$ will be derived later. We also need the invariance property of the above limiting criteria with respect to scaling and permutation and their convergence to C together with that of their relative Hessian as $(h,\delta) \to (0,0)$ and $(h,M) \to (0,\infty)$ and $\{u\}$ becomes infinitely dense. The convergence of the limiting criteria is not difficult to prove; that of their relative Hessian is expected and can also be seen from their formulae (in subsection 2.2.5.2 for $C_{h,\delta}$ and C_h^M and derived below for $C_{\{u\}}$).

We provide the proof for the above result for the criterion $C_{h,\delta}$ only in so far as those for other criteria are similar. Consider the Taylor expansion of $C_{h,\delta}(\mathbf{GA}^{-1})$ around $\mathbf{G}=\mathbf{I}$, up to second order:

$$C_{h,\delta}(\mathbf{GA}^{-1}) \approx C_{h,\delta}(\mathbf{A}^{-1}) + \frac{1}{2} \sum_{i\neq j,k\neq l} C''_{h,\delta;ij,kl} G_{ij} G_{kl}$$

where $C''_{h,\delta;ij,kl}$ constitute the elements of the relative Hessian of $C_{h,\delta}$ at $\mathbf{A}^{-1}$ and G_{ij} constitute the element of $\mathbf{G}$. The first order terms in $\mathbf{G}$ and those involving its diagonal are not present because the relative gradient of $C_{h,\delta}$ vanishes at $\mathbf{A}^{-1}$ and $C''_{h,\delta;ij,kl} = 0$ whenever $i = j$ or $k = l$, as shown in subsection 2.2.5.2. As $(h,\delta) \to (0,0)$, $C''_{h,\delta;ij,kl} \to C''_{ij,kl}$, the relative Hessian of C at $\mathbf{A}^{-1}$. Since $-C$ is a contrast, $\sum_{i\neq j,k\neq l} C''_{ij,kl}\mathcal{E}_{ij}\mathcal{E}_{kl}$ must be strictly positive unless the $\mathcal{E}_{ij}, i \neq j$ are all zero; hence so is $\sum_{i\neq j,k\neq l} C''_{h,\delta;ij,kl}G_{ij}G_{kl}$ for small enough h,δ. Thus for such h,δ, $C_{h,\delta}(\mathbf{GA}^{-1}) > C_{h,\delta}(\mathbf{A}^{-1})$ holds, for $\mathbf{G}$ close enough to but distinct from $\mathbf{I}$. Since the criterion is permutation invariant, this inequality also holds for $\mathbf{G}$ in an open neighborhood $\mathcal{P}$ of all permutation matrices but is not a permutation matrix. On the other hand, since $-C$ is a contrast $C(\mathbf{GA}^{-1}) > C(\mathbf{A}^{-1})$ for all $\mathbf{G} \in \mathcal{G}$, the set of matrices not in $\mathcal{P}$ with rows of unit norm. Since $\mathcal{G}$ is compact, the infimum of $C(\mathbf{GA}^{-1})$ over it is attained, implying that there is a $\epsilon > 0$ such that $C(\mathbf{GA}^{-1}) > C(\mathbf{A}^{-1}) + \epsilon$ for all $\mathbf{G} \in \mathcal{G}$. Finally since $C_{h,\delta}$ converges to C (uniformly on any compact set) as $(h,\delta) \to (0,0)$, for h,δ small enough $C_{h,\delta}(\mathbf{GA}^{-1}) > C_{h,\delta}(\mathbf{A}^{-1})$ for all $\mathbf{G} \in \mathcal{G}$. Together with the previous result and the invariance with respect to scaling of $C_{h,\delta}$, this shows that this criterion is the negative of a contrast.

We now compute the relative gradient of $C'_{\{u\}}$ of $C_{\{u\}}$. It is also given by the right hand side of (2.14) but with $\hat{H}_\delta(y)$ replaced by $H_{\{u\}}$, defined in (2.11). Thus we need to compute $\partial H_{\{u\}}\{y_j + \epsilon y_k)/\partial\epsilon|_{\epsilon=0}$ which leads us to consider the expansion of $Q_{y_j+\epsilon y_k}(u)$ around $\epsilon = 0$. We note that for any continuously differentiable function g,

$$\mathbb{E}g(y_j + \epsilon y_k) = \mathbb{E}g(y_j) + \epsilon\mathbb{E}[y_k g'(y_j)] + o(\epsilon),$$

where $o(\epsilon)$ denotes a term tending to 0 faster than ϵ as $\epsilon \to 0$, and that $\mathbb{E}[y_k g'(y_j)] = \int \mathbb{E}(y_k|y_j = v)g'(v)p_{y_j}(v)dv$ where $\mathbb{E}(\cdot|\cdot)$ denotes the conditional expectation operator. Therefore, letting g converge to the indicator function of an interval, one gets $F_{y_j+\epsilon y_k} = F_{y_j} - \epsilon\mathbb{E}(y_j|y_k)p_{y_j} + o(\epsilon)$. (This argument is only valid if the limit of the $o(\epsilon)$ term is still $o(\epsilon)$, but a more rigorous proof can be obtained.) In particular, for $0 < u < 1$ and $v = Q_{y_j+\epsilon y_k}(u)$:

$$u = F_{y_j}(v) - \epsilon\mathbb{E}(y_k|y_j = v)p_{y_j}(v) + o(\epsilon).$$

Note that $F_{y_j}(v) - u = F_{y_j}(v) - F_{y_j}[Q_{y_j}(u)] = p_{y_i}(\tilde{v})[v - Q_{y_i}(u)]$ where $\tilde{v}$ is a point between v and $Q_{y_j}(u)$. Since $v \to Q_{y_i}(u)$ as $\epsilon \to 0$, the same holds for $\tilde{v}$ and hence $p_{y_i}(\tilde{v}) \to p_{y_i}(v)$, and therefore

$$Q_{y_j+\epsilon y_k}(u) - Q_{y_j}(u) = \epsilon \mathbb{E}[y_k | y_j = Q_{y_j}(u)] + o(\epsilon),$$

provided that $p_{y_i}[Q_{y_i}(u)] \neq 0$. It follows that:

$$\left.\frac{\partial H_{\{u\}}\{y_j + \epsilon y_k\}}{\partial \epsilon}\right|_{\epsilon=0} = \sum_{l=2}^{L} \frac{\mathbb{E}[y_k | y_j = Q_{y_j}(u_l)] - \mathbb{E}[y_k | y_j = Q_{y_j}(u_{l-1})]}{Q_{y_j}(u_l) - Q_{y_j}(u_{l-1})} \frac{u_l - u_{l-1}}{u_L - u_1}.$$

It is clear that the above right hand side equals 1 if $k = j$; hence the gradient matrix $C'_{\{u\}}$ has diagonal vanishing identically.

We now compute the relative Hessian of $C_{\{u\}}$ (again at a truly demixing matrix **B**). By the same argument as in subsection 2.2.5.2, it is composed of the derivatives of $C'_{\{u\}}(\mathbf{B} + \mathcal{E}\mathbf{B})$ with respect to $\mathcal{E}\mathbf{B}$ at $\mathcal{E} = 0$, but that of a diagonal element $C'_{\{u\}}(\mathbf{B} + \mathcal{E}\mathbf{B})$ or with respect to a diagonal element of $\mathcal{E}$ is already known to be zero. The i, j element of $C'_{\{u\}}(\mathbf{B} + \mathcal{E}\mathbf{B})$ for $i \neq j$, is

$$\sum_{l=2}^{L} \frac{E(\mathcal{E}_{j\cdot}, \mathcal{E}_{i\cdot}; u_l) - E(\mathcal{E}_{j\cdot}, \mathcal{E}_{i\cdot}; u_{l-1})}{Q_{y_j}(u_l) - Q_{y_j}(u_{l-1})} \frac{u_l - u_{l-1}}{u_L - u_1},$$

where we have put $E(\mathcal{E}_{j\cdot}, \mathcal{E}_{i\cdot}; u) = \mathbb{E}[y_j + \mathcal{E}_{j\cdot}\mathbf{y} | y_j + \mathcal{E}_{i\cdot}\mathbf{y} = Q_{y_j + \mathcal{E}_{j\cdot}\mathbf{y}}(u)]$ for short. Hence its derivatives with respect to $\mathcal{E}_{kl}$ for $k \notin \{i, j\}$ vanish. The derivative of $\mathbb{E}(\mathcal{E}_{j\cdot}, \mathcal{E}_{i\cdot}; u)$ with respect to $\mathcal{E}_{jl}$ at $\mathcal{E} = 0$ is the same as that of $\mathbb{E}[y_j + \epsilon y_l | y_i = Q_{y_i}(u)]$ with respect to ϵ at $\epsilon = 0$, hence it is equal to $\mathbb{E}y_l$ if $l \neq j$, and to $Q_{y_i}(u_l)$ if $l = j$. The derivative of $E(\mathcal{E}_{j\cdot}, \mathcal{E}_{i\cdot}; u)$ with respect to $\mathcal{E}_{il}$ at $\mathcal{E} = 0$ is the same as that of $\mathbb{E}[y_j | y_i + \epsilon y_l = Q_{y_i + \epsilon y_l}(u)]$ with respect to ϵ at $\epsilon = 0$, which can be computed as $\lim_{\epsilon \to 0}\{\mathbb{E}[y_j | y_i + \epsilon y_l = Q_{y_i + \epsilon y_l}(u)] - \mathbb{E}y_j\}/\epsilon$. Therefore, we are led to consider the expansion of $\mathbb{E}(y_j | y_i + \epsilon y_j = v)$ around $\epsilon = 0$. Since y_j and y_i are independent, the joint density of $y_j, y_i + \epsilon y_j$ is $p_{y_j}(u) p_{y_i}(v - \epsilon u)$; hence

$$\mathbb{E}(y_j | y_i + \epsilon y_j = v) = \frac{\int u\, p_{y_j}(u) p_{y_i}(v - \epsilon u) du}{\int p_{y_j}(u) p_{y_i}(v - \epsilon u) du}.$$

Expanding $p_{y_i}(v - \epsilon u) = p_{y_i}(v) - \epsilon u\, p'_{y_i}(v) + o(\epsilon) = p_{y_i}(v)[1 + \epsilon u \varphi_{y_i}(v)] + o(\epsilon)$, one gets

$$\mathbb{E}(y_j | y_i + \epsilon y_j = v) = \frac{\mathbb{E}y_j + \epsilon \mathbb{E}(y_j^2)\varphi_{y_i}(v)}{1 + \epsilon(\mathbb{E}y_j)\varphi_{y_i}(v)} + o(\epsilon) = \mathbb{E}y_j + \sigma_{y_j}^2 \epsilon \varphi_{y_i}(v) + o(\epsilon).$$

Thus $\lim_{\epsilon\to 0}\{\mathbb{E}[y_j|y_i+\epsilon y_l = Q_{y_i+\epsilon y_l}(u)] - \mathbb{E}y_j\}/\epsilon = \sigma^2_{y_j}\varphi y_i[Q_{y_i}(u)]$.

Finally the ij,kl element of the relative Hessian matrix of $C_{\{u\}}$ equals 0 if $(k,l)\notin \{(i,j),(j,i)\}$, and 1 if $(k,l)=(j,i)$ and

$$\sigma^2_{y_j}\sum_{l=2}^{L}\frac{\varphi_{y_i}[Q_{y_i}(u_l)]-\varphi_{y_i}[Q_{y_i}(u_{l-1})]}{Q_{y_i}(u_l)-Q_{y_i}(u_{l-1})}\frac{u_l-u_{l-1}}{u_L-u_1}\quad \text{if } (k,l)=(i,j).$$

Note that the above sum is an approximation to the integral of $\varphi'_{y_i}p_{y_i}/(u_L-u_1)$ over $[Q_{y_i}(u_1), Q_{y_i}(u_L)]$. Therefore as the sequence $\{u\}$ becomes infinitely dense in $[0,1]$ it *generally* converges to $\mathbb{E}p'_{y_i}(y_i)=J(y_i)$.

The formulas for the relative Hessian of $C_{h,\delta}$, C_h^M and $C_{\{u\}}$ at $\mathbf{A}^{-1}$ show that they converge (as $(h,\delta)\to(0,0)$, $(h,M)\to(0,\infty)$ and $\{u\}$ becomes infinitely dense in $[0,1]$) to a positive definite matrix, as expected. This property has been used to obtain the contrast property of these criteria. There can be a problem with the criterion $C_{\{u\}}$ in the case where the sources have compact support though. The density is always considered as a function on $\mathbb{R}$ which vanishes outside the support of the random variable; therefore unless it admits a vanishing derivative at the boundary of the support the score function would be infinite there, which results in an infinite Fisher information. However, in the criterion $C_{\{u\}}$, if $u_1>0$ and $u_L<1$ then we will "miss" these infinite values of the score function and if $u_1=0$ or $u_L=1$ then our calculation is no longer valid (see below). Thus we can get a non-positive Hessian even if its presumed limit is infinite. As an example, consider the case where y_i has the uniform density over $[0,1]$. Its score function is zero inside $(0,1)$, and therefore from our formula, the ij,ij element of the relative Hessian of $C_{\{u\}}$ with $0<u_1<u_L<1$ would be 0; hence the relative Hessian is not positive definite. The criterion $H_{h,\delta}$ doesn't have this problem; the ij,ij element of the relative Hessian is $\sigma^2_{y_j}\mathbb{E}\varphi'_{h,\delta,y_i}$ and φ_{h,δ,y_i} is a well defined smooth function because of the smoothing introduced by the kernel. As $(h,\delta)\to(0,0)$, this function tends to $\pm\infty$ at 1 and 0, but this would only result in an infinite value of this element of the limiting relative Hessian.

In the case where the sources have compact support, it is preferable to take $u_1=0$ and $u_L=1$ in the criterion $\hat{C}_{\{u\}}$. The limiting criterion is not differentiable at $\mathbf{A}^{-1}$, but it can still be proved that it admits a *local* minimum there.

LEMMA 2.2 ([9])

In the case where the sources have means zero and compact support, then $H_{\{u\}}$ with $u_1=0$ and $u_L=1$ satisfies

$$H_{\{u\}}\{y_i+\mathcal{E}_{i\cdot}\mathbf{s}\}-H_{\{u\}}\{y_i\}=\sum_{j\neq i}\left[\frac{\max\{\mathcal{E}_{ij}Q_{y_j}(0),\mathcal{E}_{ij}Q_{y_j}(1)\}}{Q_{y_i}(1)-Q_{y_i}(u_{L-1})}(1-u_{L-1})\right.$$
$$\left.-\frac{\min\{\mathcal{E}_{ij}Q_{y_j}(0),\mathcal{E}_{ij}Q_{y_j}(1)\}}{Q_{y_i}(u_2)-Q_{y_i}(0)}u_2\right]+o(\|\mathcal{E}_{i\cdot}\|)$$

for all row vectors $\mathcal{E}_{i\cdot}$ with zero i-th component.[2]

The presence of the operators max and min in the above right hand side prevents it being approximated to first order by a linear form in $\mathcal{E}_{i\cdot}$. Hence $C_{\{u\}}$ is not differentiable at $\mathbf{A}^{-1}$. On the other hand, since $Q_{y_j}(0) < 0 < Q_{y_j}(1)$ because s_j has zero mean, the expression inside the square bracket in the above right hand side is non-negative and can vanish only if $\mathcal{E}_{i\cdot} = 0$. Therefore $\mathbf{A}^{-1}$ is a local minimum point of $C_{\{u\}}$.

The special case $L = 2$ is particularly interesting: $H_{\{0,1\}}\{y\}$ by (2.11), reduces to $\log R_y$ where $R_y = Q_y(1) - Q_y(0)$ is the *range* of y (supposed to have bounded support) and thus the criterion $C_{\{0,1\}}$ reduces to:

$$C_{\{0,1\}}(\mathbf{B}) = \sum_{k=1}^{P} \log R_{y_k} - \log|\det \mathbf{B}| = \log \frac{\prod_{k=1}^{P} R_{y_k}}{|\det \mathbf{B}|}. \tag{2.17}$$

Note that if the random variable y has a uniform distribution, then $\log R_y$ is exactly the entropy of y. Thus the criterion $C_{\{0,1\}}$ should work well when the sources are nearly uniform. It also possesses a simple interesting geometrical interpretation. Since $y_1, \ldots, y_P$ are the components of $\mathbf{Bx}$, the ratio in the above right hand side of (2.17) represents the volume of the set

$$\left\{\xi \in \mathbb{R}^P : \mathbf{B}\xi \in [Q_{y_1}(0), Q_{y_1}(1)] \times \cdots \times [Q_{y_P}(0), Q_{y_P}(1)]\right\}.$$

Moreover, one can easily see from the definition of the range, that this set is the smallest "hyper-parallelepiped" of sides parallel to the columns of $\mathbf{B}^{-1}$ which contains the support of the distribution of $\mathbf{x}$. Minimizing $C_{\{0,1\}}$ amounts to looking for the hyper-parallelepiped of smallest volume containing the support of the distribution of $\mathbf{x}$. Its sides then determine the columns of the inverse of the matrix $\mathbf{B}$ realizing the minimum of the criterion. As it is clear that these columns are only determined up to a permutation and a scaling, so are the rows $\mathbf{B}$, which is coherent with the fact that $C_{\{0,1\}}$ is invariant with respect to permutation and scaling. But the most interesting feature of this criterion, apart from its simplicity, is that it is *the negative of a contrast.*

PROPOSITION 2.3 ([9])
Let $\mathbf{x} = \mathbf{As}$ *where* $\mathbf{s}(n)$ *is a bounded random vector with independent components; then* $C_{\{0,1\}}$, *defined by* (2.17) *with* $y_1, \ldots, y_P$ *being the components of* $\mathbf{Bx}$, *attains its minimum if and only if* $\mathbf{BA}$ *is the product of a permutation and a diagonal matrix.*

[2] The constraint that the sources have zero means is for convenience only since otherwise one may subtract their means so that they are so. This would not change the above left hand side but would change the $Q_{s_i}(0)$ and $Q_{s_i}(1)$. Likewise, if $\mathcal{E}_{ii} \neq 0$ then one may write $H_{\{u\}}\{y_i + \mathcal{E}_{i\cdot}\mathbf{s}\} = \log|1 + \mathcal{E}_{ii}| + H_{\{u\}}\{y_i + \mathcal{E}'_{i\cdot}\mathbf{s}\}$ where $\mathcal{E}'_{ij} = \mathcal{E}_{ij}/(1 + \mathcal{E}_{ii})$ if $j \neq i, = 0$ otherwise.

2.2.6.2 Performance of the separation

It is of interest to note that setting the gradient matrix $\hat{C}_\delta$ or $\hat{C}^M$ to zero amounts to requiring that the nonlinear function $\hat{\varphi}_{\delta,y_j}$ or $\hat{\varphi}^M_{y_j}$ of y_j be empirically uncorrelated to y_k, for all $i \neq k$. Explicitly, the estimated demixing matrix $\mathbf{B}$ is the solution of the estimation equations

$$\frac{1}{T}\sum_{n=1}^{T}\psi_j[y_j(n)]y_k(n) = 0, \qquad 1 \leqslant j \neq k \leqslant P, \quad [y_1 \ \cdots \ y_P]^{\mathrm{T}} = \mathbf{Bx},$$

where ψ_j stands for $\hat{\varphi}_{\delta,y_j}$ or $\hat{\varphi}^M_{y_j}$. This is quite similar to the heuristic procedure in [8] and the quasi maximum likelihood method in [17]. The main difference is that in the above works, the functions ψ_j are given *a priori* (and deterministic) while here they are data driven (hence random). However, it can be expected (and in fact can be proved) that the results in [17] still apply by replacing ψ_j by the limits of $\hat{\varphi}_{\delta,y_j}$ or $\hat{\varphi}^M_{y_j}$ as $T \to \infty$, which in the case where h, δ, M are fixed, equal φ_{h,δ,y_j} or φ^M_{h,y_j}.

Letting $\mathbf{G} = \mathbf{BA}$, the separated sources can be expressed in terms of the original sources as: $y_j = \sum_{k=1}^{P} G_{ik}s_k$, G_{jk} denoting the elements of $\mathbf{G}$. We shall assume that the separated sources (or the original source) have been renumbered so that y_j estimate a multiple of the j-th source, and to eliminate scaling ambiguity, we assume that the y_j have been scaled in some way (for example by requiring having unit sample variance). Then G_{jk} would converge (as $T \to \infty$) to some limit $G^{\dagger}_{jj}$ if $k = j$ and 0 otherwise. If we change the scale of the sources by putting $s'_j = G^{\dagger}_{jj}s_j$, then $y_j = \sum_{k=1}^{P} G'_{ik}s'_k$ where $G'_{jk} = G_{jk}/G^{\dagger}_{kk}$ and $G_{jj} \to 1$. The coefficient G'_{jk}/G'_{jj} represents the contamination of the (rescaled) k-th source to the (rescaled) j-th source, but since these sources do not have the same variance (power) we shall consider instead $\delta_{jk} = (G'_{jk}/G'_{jj})\sigma'_k/\sigma'_j$ where σ'_j is the standard deviation of s'_j. In [17], it was shown, in the case of temporally independent sources, that the random vectors $[\delta_{jk} \ \ \delta_{kj}]^{\mathrm{T}}$, $1 \leqslant j < k \leqslant P$, are asymptotically independent with mean zero and covariance matrix

$$\frac{1}{T}\begin{bmatrix} \lambda_j^{-1} & 1 \\ 1 & \lambda_k^{-1} \end{bmatrix}^{-1} \begin{bmatrix} \rho_j^{-2} & 1 \\ 1 & \rho_k^{-2} \end{bmatrix} \begin{bmatrix} \lambda_j^{-1} & 1 \\ 1 & \lambda_k^{-1} \end{bmatrix}^{-1}$$

where

$$\rho_j = \frac{\mathbb{E}[\psi_j(s'_j)s'_j]}{\sqrt{\mathbb{E}\psi_j^2(s_j)}\sigma'_j} = \mathrm{corr}\{\psi_j(s'_j), s'_j\}, \qquad \lambda_j = \frac{\mathbb{E}[\psi_j(s'_j)s'_j]}{[\mathbb{E}\psi'_j(s'_j)]\sigma'_j}.$$

Note that δ_{jk} also equals $(G_{jk}/G_{jj})\sigma_k/\sigma_j$ where σ_j is the standard deviation of s_j, but the above formulas are expressed in terms of s'_j and not s_j. They do not depend on the scale of s'_j though, *if* $\psi_j = \varphi_{h,\delta,y_j}$ or $\psi_j = \varphi^M_{\delta,y_j}$.

It was also shown in [17] that the optimal choice of ψ_j is the score function φ_{y_j} of the j-th source. Thus our method is nearly optimal if $h, \delta, 1/M$ are small and remain fixed and is optimal if they converge to 0 with an appropriate slow rate (as $T \to \infty$). For $\psi_j = \varphi_{y_j}$, $\lambda_j = \rho_j^2$, hence the optimal covariance matrix of $[\delta_{jk} \;\; \delta_{kj}]^{\mathrm{T}}$ is the inverse of $\begin{bmatrix} \rho_j^{-2} & 1 \\ 1 & \rho_k^{-2} \end{bmatrix}$. The coefficient $1/\rho_j$ measures the nonlinearity of the score function: the larger it is, the more nonlinear the function. Nonlinearity of the score function is related to its non-Gausianity as Gaussian variable has linear score function. Thus the more non-Gaussian the sources are, the easier their separation.

2.3 METHODS BASED ON MUTUAL INFORMATION RATE

The use of the (ordinary) mutual information between the recovered sources at a same time point does not exploit the time dependence structure of the sources. The use of the mutual information rate can yield better performance of the algorithms by exploiting such dependence, and is also necessary to deal with convolutive mixtures.

2.3.1 Entropy and mutual information rate

We begin by defining these concepts and providing some of their properties.

2.3.1.1 Definitions and properties

Let $\mathrm{y}(\cdot) = \{\mathrm{y}(n), n \in \mathbb{Z}\}$ be a stationary process (possibly multivariate). Then [4]:

$$H\{\mathrm{y}(1),\ldots,\mathrm{y}(T)\} = H\{\mathrm{y}(1)\} + \sum_{n=2}^{T} H\{\mathrm{y}(n)|\mathrm{y}(n-1),\ldots,\mathrm{y}(1)\} \tag{2.18}$$

where $H\{\mathrm{y}(n)|\mathrm{y}(n-1),\ldots,\mathrm{y}(1)\}$ is the *conditional entropy* (which is the expectation of entropy of the conditional distribution) of $\mathrm{y}(n)$ given $\mathrm{y}(n-1),\ldots,\mathrm{y}(1)$. But since the conditioning diminishes the entropy [4]:

$$H\{\mathrm{y}(n+1)|\mathrm{y}(n),\ldots,\mathrm{y}(1)\} \leqslant H\{\mathrm{y}(n+1)|\mathrm{y}(n),\ldots,\mathrm{y}(2)\}$$

and, by stationarity, the above right hand side is the same as $H\{\mathrm{y}(n)|\mathrm{y}(n-1),\ldots,\mathrm{y}(1)\}$. One can then show that

$$H\{\mathrm{y}(1),\ldots,\mathrm{y}(T)\}/T \geqslant H\{\mathrm{y}(n)|\mathrm{y}(1),\ldots,\mathrm{y}(T-1)\} \tag{2.19}$$

and that the two sides of the above inequality are non-increasing with T and converge as $T \to \infty$ to a same limit (possibly infinite) which is called *entropy rate* of the stationary process [4] and is denoted by $H\{\mathrm{y}(\cdot)\}$. In the case where $\mathrm{y}(n)$ is a vector of components $y_1(n),\ldots,y_P(n)$, this limit will also be called *joint entropy rate* of the processes $y_1(\cdot),\ldots,y_P(\cdot)$ and denoted by $H\{y_1(\cdot),\ldots,y_P(\cdot)\}$.

One deduces easily from what precedes, the following important property.

LEMMA 2.4
For a stationary process $\{\mathbf{y}(n), n \in \mathbb{Z}\}$ *and an integer* $m > 1$

$$H\{\mathbf{y}(1)\} \leqslant H\{\mathbf{y}(m)|\mathbf{y}(m-1),\ldots,\mathbf{y}(1)\} \leqslant H\{y(1)\},$$

with all inequalities becoming equalities if and only if the process is temporally independent and with the first inequality becoming equality if the process is Markovian of order $m-1$.

The *mutual information rate* between the stationary processes $y_1(\cdot),\ldots,y_P(\cdot)$ can now be defined as

$$I\{y_1(\cdot),\ldots,y_P(\cdot)\} = \sum_{i=1}^{P} H\{y_i(\cdot)\} - H\{y_1(\cdot),\ldots,y_P(\cdot)\}. \tag{2.20}$$

The computation of the entropy (hence mutual information) rate is, however, difficult and costly. The only case where an analytical formula is available is the Gaussian case. For this reason, we introduce the concept of *Gaussian entropy rate*, defined as the entropy rate of a Gaussian process having the same autocovariance (hence spectral density) function. For a (vector) process $\mathbf{y}(\cdot)$, a direct calculation (see also [4]) shows that its Gaussian entropy rate equals:

$$H_g\{\mathbf{y}(\cdot)\} = \frac{1}{4\pi}\int_{-\pi}^{\pi} \log\det[4\pi^2 e\mathbf{f}_{\mathbf{y}}(\lambda)]d\lambda \tag{2.21}$$

where $\mathbf{f}_{\mathbf{y}}$ is the matrix of its spectral-interspectral densities and $e = \exp(1)$. Other useful results relating the entropy rate of a filtered process from entropy rate of the original process, are described in the next subsection.

2.3.1.2 Entropy rate and filtering

We will restrict ourselves to a class of "well behaved" filters. Specifically we consider (multi-input multi-output) filters with impulse response sequence belonging to a class $\mathscr{A}$ defined as follows: $\mathscr{A}$ consists of sequences $\{\mathbf{B}(l), l \in \mathbb{Z}\}$ satisfying $\sum_{l=-\infty}^{\infty} \|\mathbf{B}(l)\| < \infty$ and $\det[\sum_{l=-\infty}^{\infty} \mathbf{B}(l)e^{jl\lambda}] \neq 0$ for all λ.[3] It can be shown that the output of a such filter applied on a stationary process $\mathbf{x}(\cdot)$ admitting absolute moment[4] of order $\alpha \geqslant 1$, which is the convolution $(\mathbf{B} \star \mathbf{x})(\cdot)$, is well defined and is again a stationary processes admitting an absolute moment of order α. Moreover, the class $\mathscr{A}$ is closed with respect to convolution [13]: the convolution of two sequences in this class again belongs to this class. Even more interesting is that any sequence $\{\mathbf{B}(l), l \in \mathbb{Z}\}$ in this class admits an inverse[5] again in this class, which is precisely the sequence of Fourier coefficients

[3] The matrices $\mathbf{B}(l)$ are thus square.

[4] This is the moment of the absolute value.

[5] The inverse of $\{\mathbf{B}(l), l \in \mathbb{Z}\}$ is a sequence $\{\mathbf{B}^\dagger(l), l \in \mathbb{Z}\}$ such that $\sum_l \mathbf{B}^\dagger(l)\mathbf{B}(k-l) = \sum_l \mathbf{B}(l)\mathbf{B}^\dagger(k-l) = \mathbf{I}$ if $k = 1$, 0 otherwise.

of the function $\lambda \mapsto [\sum_{l=-\infty}^{\infty} \mathbf{B}(l)e^{il\lambda}]^{-1}$. This result can be deduced from a result of Wiener which says that if f is a 2π-periodic function, never vanishes and has absolutely summable Fourier coefficients, then the same is true for the function $1/f$ (see for example [21], p. 245).

Let $\{\mathbf{x}(n), n \in \mathbb{Z}\}$ be a vector stationary process and $\mathbf{y}(\cdot) = (\mathbf{B} \star \mathbf{x})(\cdot)$ its filtered version by a filter with impulse response in $\mathscr{A}$. It has been shown that [13] (under certain regularity conditions):

$$H\{\mathbf{y}(\cdot)\} = H\{\mathbf{x}(\cdot)\} + \int_{-\pi}^{\pi} \log \left| \det \sum_{l=-\infty}^{\infty} \mathbf{B}(l)e^{il\lambda} \right| \frac{d\lambda}{2\pi}. \tag{2.22}$$

A heuristic proof of this result can be put forward as follows. The transformation $\mathbf{x}(\cdot) \mapsto \mathbf{y}(\cdot)$ is *linear invertible*; hence as suggested by the Lemma 2.1 $H[\mathbf{y}(\cdot)]$ would differ from $H[\mathbf{x}(\cdot)]$ only by a "determinantal" term which does not depend on the distribution of the process $\{\mathbf{x}(n), n \in \mathbb{Z}\}$. By considering Gaussian processes and referring to the formula (2.21), one sees that this term is the same as the last term in (2.22).

The above result plays a fundamental role in the sequel as it describes how the entropy rate of a stationary process changes when it is filtered. In particular, it provides a means to compute the entropy rate of a *linear process* which is commonly used as a signal model in the literature. A random process is called linear if it is the output of a linear filter applied to a temporally independent process, and more precisely if it admits the representation

$$\mathbf{x}(n) = \sum_{l=-\infty}^{\infty} \mathbf{A}(l)\mathbf{e}(n-l) \tag{2.23}$$

where $\mathbf{e}(n)$ are independent identically distributed random vectors. Assuming further that the sequence $\{\mathbf{A}(l), l \in \mathbb{Z}\}$ is of class $\mathscr{A}$ the result (2.22) together with the equality $H\{\mathbf{e}(\cdot)\} = H\{\mathbf{e}(1)\}$ (by Lemma 2.4), yields immediately the entropy rate of the linear process (2.23). More generally the result (2.22) and Lemma 2.4 permit the computation of the entropy rate of a *filtered Markov process*, defined as (2.23) but with $\{\mathbf{e}(n), n \in \mathbb{Z}\}$ being a Markov process. Further, they entail the following important result.

The deconvolution principle: *let $\{\mathbf{y}(n), n \in \mathbb{Z}\}$ be a stationary vector process. Then for all sequences $\{\mathbf{B}(l), l \in \mathbb{Z}\}$ of class $\mathscr{A}$:*

$$H\{\mathbf{y}(\cdot)\} \leqslant H\{(\mathbf{B} \star \mathbf{y})(1)\} - \int_{-\pi}^{\pi} \log \left| \det \sum_{j=-\infty}^{\infty} \mathbf{B}(l)e^{il\lambda} \right| \frac{d\lambda}{2\pi} \tag{2.24}$$

with equality if and only if the process $\{(\mathbf{B} \star \mathbf{y})(n), n \in \mathbb{Z}\}$ is temporally independent. The same result holds with $H\{(\mathbf{B} \star \mathbf{y})(1)\}$ replaced by $H\{(\mathbf{B} \star \mathbf{y})(m) | (\mathbf{B} \star \mathbf{y})(m-1), \ldots, (\mathbf{B} \star \mathbf{y})(1)\}$ and then equality is attained if and only if the process $\{(\mathbf{B} \star \mathbf{y})(n), n \in \mathbb{Z}\}$ is Markovian of order $m-1$.

This result can be obtained by noting that the sum of $H\{\mathbf{y}(\cdot)\}$ and the integral in the above right hand side equals, by (2.22), precisely the entropy rate of the process $\{(\mathbf{B}\star\mathbf{y})(n), n\in\mathbb{Z}\}$, and then it suffices to apply Lemma 2.4.

It is clear that the above deconvolution principle remains valid if one replaces the class $\mathscr{A}$ by a smaller subclass. A subclass particularly interesting is the subclass $\mathscr{A}^+$ of all sequences $\{\mathbf{B}(l), l\in\mathbb{Z}\}$ in $\mathscr{A}$ which are *causal* and *minimum phase*; that is, such that $\mathbf{B}(l)=0$ for $l<0$ and $\det[\sum_{l=0}^{\infty}\mathbf{B}(l)z^l]\neq 0$ for all complex numbers z of modulus module not greater than 1. Then $[\sum_{l=0}^{\infty}\mathbf{B}(l)z^l]^{-1}$, as a function of the complex variable z, is holomorphic (analytic) in the unit disk $\{|z|\leqslant 1\}$ of the complex plane, and hence admits a development in series of non-negative powers of z. This shows that the inverse sequence of a sequence in $\mathscr{A}^+$ is also in $\mathscr{A}^+$. Since the constant term of this development is $\log\det\mathbf{B}(0)$, one has

$$\int_{-\pi}^{\pi}\log\left|\det\sum_{l=0}^{\infty}\mathbf{B}(l)e^{il\lambda}\right|\frac{d\lambda}{2\pi}=\log\det|\mathbf{B}(0)|, \tag{2.25}$$

as the above left hand side is also the integral of $\log\det[\sum_{l=0}^{\infty}\mathbf{B}(l)z^l]$ on the unit circle $\{|z|=1\}$. One can thus state the same deconvolution principle as before but with $\mathscr{A}$ replaced by $\mathscr{A}^+$ and the integral in (2.24) replaced by $\log|\det\mathbf{B}(0)|$.

2.3.2 Contrasts

The mutual information rate between random processes can be used, in the same way as the mutual information between the random variables, as the negative of a contrast for blind source separation. Such contrast imposes itself in a natural manner in the case of convolutive mixtures. But this doesn't rule out its use in the case of instantaneous mixtures, where it can yield better separation procedure since it allows exploitation of the temporal dependence of the source signals, especially when such dependence is strong.

The convolutive mixtures model is given by

$$\mathbf{x}(n)=\sum_{l=-\infty}^{\infty}\mathbf{A}(l)\mathbf{s}(n-l)=(\mathbf{A}\star\mathbf{s})(n) \tag{2.26}$$

where $\{\mathbf{A}(l), l\in\mathbb{Z}\}$ is an unknown matrix sequence. This model contains as a particular case the instantaneous mixture case where $\mathbf{A}(l)=0$ for all $l\neq 0$. The sources are then naturally recovered via another convolution:

$$\mathbf{y}(n)=\sum_{l=-\infty}^{\infty}\mathbf{B}(l)\mathbf{x}(n-l)=(\mathbf{B}\star\mathbf{x})(n). \tag{2.27}$$

In order that such recovery is theoretically possible, we suppose that the sequence $\{\mathbf{A}(l), l\in\mathbb{Z}\}$ is of class $\mathscr{A}$ so that the inverse sequence exists and is again of class $\mathscr{A}$. The blind separation method then consists in looking for a sequence $\{\mathbf{B}(l), l\in\mathbb{Z}\}$ in

$\mathscr{A}$ which minimizes an independence criterion between the components of the output process $\{\mathbf{y}(n), n \in \mathbb{Z}\}$. In the case of instantaneous mixtures, it is clear that one would look for such sequence only in the class of sequences with a single non-zero element at index 0.

From the formulas (2.20) and (2.22), the mutual information rate criterion can be written up to an additive constant as

$$\mathscr{C}_\infty[\mathbf{B}(\cdot)] = \sum_{k=1}^{P} H\{y_k(\cdot)\} - \int_{-\pi}^{\pi} \log \left| \det \sum_{l=-\infty}^{\infty} \mathbf{B}(l) e^{il\lambda} \right| \frac{d\lambda}{2\pi} \tag{2.28}$$

where $y_1(n), \ldots, y_P(n)$ are the components of $\mathbf{y}(\cdot)$ defined by (2.27). This criterion, by construction, is the negative of a contrast. Note that it is invariant with respect to permutation and filtering: it is unchanged when one pre-multiplies the sequence $\{\mathbf{B}(l), l \in \mathbb{Z}\}$ by a permutation matrix or pre-convolves this sequence with a sequence of diagonal matrices of class $\mathscr{A}$. This can be easily seen from (2.28) and (2.22). One thus can only recover the sources up to a permutation and a filtering, as one can expect, since these operations do not change their independence.

The advantage of the contrast $-\mathscr{C}_\infty$ is only theoretical since this contrast requires the knowledge of the distribution of the whole sequence $\{y_k(l), l \in \mathbb{Z}\}$ for every k, even if, thanks to the formula (2.22), one has already dispensed with their joint distribution. In the sequel, we shall investigate different ways to simplify this contrast to a practical useful form.

2.3.2.1 Contrasts based on the finite joint distribution

The natural way to construct a practical criterion out of $\mathscr{C}_\infty$ is to approximate $H[y_k(\cdot)]$ by $H[y_k(1), \ldots, y_k(m)]/m$. However, m can often be quite large for the approximation to be sufficiently accurate and then one is faced with the problem of estimation of a multivariate density in a high dimensional space. The difficulty is that the number of data needed to well estimate a multivariate density increases *exponentially* with the dimension (the curse of dimensionality). However, as it will be seen below, *in the case of an instantaneous mixture*, one can use $H\{y_k(1), \ldots, y_k(m)\}/m$ in place of $H\{y_k(\cdot)\}$ *even with small* m and still obtain the negative of a contrast (and a good criterion too). Indeed, the criterion

$$\mathscr{C}_m(\mathbf{B}) = \frac{1}{m} \sum_{k=1}^{P} H\{y_k(1), \ldots, y_k(m)\} - \log|\det \mathbf{B}|, \tag{2.29}$$

where $\mathbf{y}(n) = \mathbf{B}x(n)$ ($\mathbf{B}$ being the separation matrix), can be written as $1/m$ times the sum of the mutual information

$$I\{[y_1(1) \ \cdots \ y_1(m)]^{\mathrm{T}}, \ldots, [y_P(1) \ \cdots \ y_P(m)]^{\mathrm{T}}\} \tag{2.30}$$

and the constant term $H\{\mathbf{x}(1), \ldots, \mathbf{x}(m)\}$, since by Lemma 2.1:

$$H\{\mathbf{y}(1), \ldots, \mathbf{y}(m)\} = H\{\mathbf{x}(1), \ldots, \mathbf{x}(m)\} - m \log|\det \mathbf{B}|. \tag{2.31}$$

This result alas cannot be extended to the case of convolutive mixtures since one does not have a formula analogous to (2.31) for this case: the entropy rate of a finite segment of a filtered process cannot be computed from the finite joint distribution of the original process. In order to obtain a contrast, one should consider the negative of the mutual information (2.30) but then one is led back to the problem of estimating the density in a space of high dimension (when mP is large).

A more interesting approach would be to replace $H\{y_k(\cdot)\}$ in (2.28) by $H\{y_k(m)|y_k(m-1),\dots,y_k(1)\}$ with a small m, since the above conditional entropy, by (2.19), is a better approximation to $H[y_k(\cdot)]$ than $H[y_k(1),\dots,y_k(m)]/m$. In the case of instantaneous mixtures, this leads to the criterion:

$$\mathscr{C}_m^{\dagger}(\mathbf{B}) = \sum_{k=1}^{P} H\{y_k(m)|y_k(1),\dots,y_k(m-1)\} - \log|\det \mathbf{B}|. \tag{2.32}$$

One can show that the above criterion is, up to the additive constant $H\{\mathbf{x}(m)|\mathbf{x}(m-1),\dots,\mathbf{x}(1)\}$, the expectation of the Kullback-Leibler divergence between the conditional distribution of $\mathbf{y}(m)$ given $\mathbf{y}_k(m-1),\dots,\mathbf{y}(1)$ and the product of the conditional distribution of $y_k(m)$ given $y(m-1),\dots,y_k(1)$, $k=1,\dots,P$. Therefore it is minimized when the processes $\{y_k(n), n \in \mathbb{Z}\}$, $k=1,\dots,P$, are independent. But as before, this result cannot be extended to the case of convolutive mixtures. However, as it will be seen later, a convolutive version of the criterion (2.32) can be obtained, which is indeed the negative of a contrast *in the case where the sources are Markovian of order* $m-1$.

2.3.2.2 Gaussian mutual information rate

Another way to avoid the difficulty of estimating the mutual information rate is to replace it by the *Gaussian mutual information rate*, defined as the ordinary mutual information rate but with the processes involved replaced by a Gaussian process having the same covariance structure. By (2.20) and (2.21), the Gaussian mutual information rate between the processes $\{y_k(n), n \in \mathbb{Z}\}$, $k=1,\dots,P$, is:

$$I_g\{y_1(\cdot),\dots,y_P(\cdot)\} = \frac{1}{4\pi}\int_{-\pi}^{\pi} \{\log\det \mathrm{Diag}[\mathbf{f}_{\mathbf{y}}(\lambda)] - \log\det \mathbf{f}_{\mathbf{y}}(\lambda)\} d\lambda \tag{2.33}$$

where $\mathbf{f}_{\mathbf{y}}$ denotes the spectral-interspectral matrix density of the process $\{\mathbf{y}(n) = [y_1(n) \;\cdots\; y_P(n)]^{\mathrm{T}}\}$ and Diag denotes the operator which builds a diagonal matrix from the diagonal of its argument.

The criterion (2.33) is a *joint diagonalization* criterion: it is non-negative and can vanish if and only if $\mathbf{f}_{\mathbf{y}}$ is almost everywhere diagonal, by the Hadamard inequality (see for example [4], p. 502). Since this criterion only expresses the decorrelation between the reconstructed sources (at different times), *it is unable to separate convolutive mixtures of sources*. Indeed, it vanishes as soon as the sequence $\{\mathbf{B}(l), l \in \mathbb{Z}\}$ is chosen such that the matrix $[\sum_{l=-\infty}^{\infty} \mathbf{B}(l)e^{il\lambda}]\mathbf{f}_{\mathbf{x}}^{1/2}(\lambda)$ is unitary for almost all λ, $\mathbf{f}_{\mathbf{x}}^{1/2}$ being the Cholesky factor in the factorization $\mathbf{f}_{\mathbf{x}} = \mathbf{f}_{\mathbf{x}}^{1/2}(\mathbf{f}_{\mathbf{x}}^{1/2})^{\mathrm{T}}$. Nevertheless, in the case of instantaneous mixture, it is still the negative of a contrast, provided that the sources have non-proportional spectral densities [10].

However, the criterion (2.33) still involves an integration. Therefore it is of interest to degrade it by replacing the integral by a Riemann sum and the spectral density by a smooth version (for estimation purpose). In [10], the following criterion is introduced:

$$\frac{1}{2L}\sum_{l=1}^{L}\{\log\det \mathrm{Diag}[\mathbf{B}(k_M \star \mathbf{f}_{\mathbf{x}})(\lambda_l)\mathbf{B}^{\mathrm{T}}] - \log\det[\mathbf{B}(k_M \star \mathbf{f}_{\mathbf{x}})(\lambda_l)\mathbf{B}^{\mathrm{T}}]\}, \tag{2.34}$$

where k_M is a non-negative kernel function tending to the Dirac function as $M \to \infty$ and $\lambda_1, \dots, \lambda_L$ are L equispaced (angular) frequencies in $(-\pi, \pi]$.

Another possibility is to degrade the criterion (2.33) by introducing the Gaussian analogs of C_m and $C_m^{\dagger}$, obtained by replacing in their definitions (2.29) and (2.32), the random vectors involved by Gaussian vectors having the same covariance matrices. A direct calculation ignoring the constant terms leads to the criteria

$$\frac{1}{2m}\sum_{k=1}^{P}\log\det \mathrm{cov}\{[y_k(1) \ \cdots \ y_k(m)]^{\mathrm{T}}\} - \log|\det \mathbf{B}| \tag{2.35}$$

where $\mathrm{cov}(\cdot)$ denotes the covariance matrix and

$$\frac{1}{2}\sum_{k=1}^{P}\log \mathrm{var}[y_k(m) - y_k(m|1{:}m-1)] - \log|\det \mathbf{B}| \tag{2.36}$$

where $\mathrm{var}(\cdot)$ denotes the variance and $y_k(m|1{:}m-1)$ denotes the best linear predictor (in mean square sense) of $y_k(m)$ based on $y_k(1), \dots, y_k(m-1)$.

One can show that the negatives of the above criteria are contrasts for blind separation of instantaneous mixtures, under certain "non-proportionality" conditions [10].

2.3.2.3 Contrasts for linear or Markovian sources

The contrasts introduced in the previous subsection are not applicable for the separation of convolutive mixtures. The difficulty is that convolution is a global transformation operating on the whole source sequence and thus it is natural that a contrast must involve the distribution of the whole recovered source process and not the finite joint distribution only. However, this difficulty can be avoided in the case where the sources are linear or Markovian processes.

In the case where the sources are *linear processes*, one can separate them from their convolutive mixture by minimizing:

$$\mathscr{C}_1[\mathbf{B}(\cdot)] = \sum_{k=1}^{P} H\{y_k(1)\} - \int_{-\pi}^{\pi} \log\left|\det \sum_{l=-\infty}^{\infty} \mathbf{B}(l)e^{il\lambda}\right| \frac{d\lambda}{2\pi}. \tag{2.37}$$

To see this, note that by Lemma 2.4, $\mathscr{C}_1[\mathbf{B}(\cdot)] \geqslant \mathscr{C}_\infty[\mathbf{B}(\cdot)]$ with equality if and only if the processes $\{y_1(n), n \in \mathbb{Z}\}, \dots, \{y_P(n), n \in \mathbb{Z}\}$ are temporally independent. On the other hand, $\mathscr{C}_\infty[\mathbf{B}(\cdot)]$ is minimized if and only if these processes coincide with the source processes *up to a permutation and a filtering*, that is when the sequence $\{\mathbf{B}(l), l \in \mathbb{Z}\}$ is the inverse sequence of $\{\mathbf{A}(l), l \in \mathbb{Z}\}$ pre-multiplied by a permutation matrix and pre-convolved with some diagonal sequence $\{\mathbf{D}(l), l \in \mathbb{Z}\}$. Since the source processes are linear, one can find (at least) a sequence $\{\mathbf{B}(l), l \in \mathbb{Z}\}$ such that the processes $\{y_1(n), n \in \mathbb{Z}\}, \ldots, \{y_P(n), n \in \mathbb{Z}\}$ coincide with the source processes *up to a permutation and a filtering* and also are temporally independent, then the above results show that $\mathscr{C}_1[\mathbf{B}(\cdot)]$ is smallest when and only when such conditions are satisfied.

The above criterion $\mathscr{C}_1$ is often used to extract the sources which are temporally independent and have undergone a convolutive transformation before being convolutively mixed. It permits both separating the mixture into independent component processes and deconvolving them into temporally independent ones. It can be referred to as a separation-deconvolution criterion and can thus recover the sources up to a permutation, a scaling and a pure delay. (By *Darmois Theorem* [3], the only filter which preserves temporal independence is the scaling-and-delay filter.)

Supposing that the above convolutions, either in the source transformations or in the mixture model (2.26), are all of class $\mathscr{A}^+$ (hence of *minimum phase*), then one would restrict the separation-deconvolution filter to have impulse response $\{\mathbf{B}(l), l \in \mathbb{Z}\}$ also in $\mathscr{A}^+$. The criterion (2.37), by (2.25), then reduces to:

$$\mathscr{C}_1^+[\mathbf{B}(\cdot)] = \sum_{k=1}^{P} H\{y_k(1)\} - \log|\det \mathbf{B}(0)|. \tag{2.38}$$

However, the assumption of minimum phase is artificial and the minimum phase requirement of $\{\mathbf{B}(l), l \in \mathbb{Z}\}$ is not easy to be implemented.

Another assumption regarding the sources which can be considered is that each source is a *filtered Markov process*, defined as the output of a filter applied to a Markov process, of some order $m-1$. Such process can be viewed as a generalization of a linear process (which corresponds to $m = 1$). Note that Lemma 2.4 also yields

$$\mathscr{C}_m^\dagger[\mathbf{B}(\cdot)] = \sum_{k=1}^{P} H[y_k(m)|y_k(m-1), \ldots, y_k(1)] - \int_{-\pi}^{\pi} \log\left|\det \sum_{l=-\infty}^{\infty} \mathbf{B}(l)e^{il\lambda}\right| \frac{d\lambda}{2\pi} \tag{2.39}$$

is bounded below by $\mathscr{C}_\infty[\mathbf{B}(\cdot)]$, with the bound attained if and only if the processes $\{y_1(n), n \in \mathbb{Z}\}, \ldots, \{y_P(n), n \in \mathbb{Z}\}$ are Markovian of order $m-1$. Therefore by the same argument as before, using the fact that there exists at least a sequence $\{\mathbf{B}(l), l \in \mathbb{Z}\}$ for which the processes $\{y_1(n), n \in \mathbb{Z}\}, \ldots, \{y_P(n), n \in \mathbb{Z}\}$ are Markovian of order $m-1$ and coincide with the sources *up to a permutation and a filtering*, one obtains that the criterion $\mathscr{C}_m^+[\mathbf{B}(\cdot)]$ is smallest when and only when such a condition is satisfied.

Thus, minimizing $C_m^\dagger[\mathbf{B}(\cdot)]$ not only separates the sources, but extracts the underlying Markov processes which generate them. In practice, it is reasonable to assume that the sources themselves are Markovian of order $m-1$; in this case the above criterion permits recovering them up to a permutation, a scaling and a pure delay. The ambiguity relative to filtering has disappeared (more precisely reduced to scaling and delay), since one has *a priori* information that the sources are Markovian.

The above contrast (2.37) is designed for the case of convolutive mixtures of linear sources. But the assumption that the sources are linear processes can be also exploited to better separate them in the case of instantaneous mixtures. In this case, there is only a single demixing matrix $\mathbf{B}$, and $C_\infty(\mathbf{B})$, by the deconvolution principle (2.24), is bounded by:

$$\sum_{k=1}^{P} \inf_{b_k} \left\{ H\{(b_k \star y_k)(1)\} - \int_{-\pi}^{\pi} \log \left| \sum_{l=-\infty}^{\infty} b_k(l) e^{il\lambda} \right| \frac{d\lambda}{2\pi} \right\} - \log|\det \mathbf{B}| \tag{2.40}$$

where the infimum is searched among all scalar sequences $\{b_k(l), l \in \mathbb{Z}\}$ of class $\mathscr{A}$. Moreover the above bound can be attained if and only if the processes $\{y_k(n), n \in \mathbb{Z}\}$ are linear. One then deduces that the negative of (2.40) is a contrast since $-\mathscr{C}_\infty$ is a contrast and the sources are linear processes.

The deconvolution principle and (2.25) also yields that $\mathscr{C}_\infty(\mathbf{B})$ is bounded by

$$\sum_{k=1}^{P} \inf_{b_k} \{H\{(b_k \star y_k)(1)\} - \log|b_k(0)|\} - \log|\det \mathbf{B}| \tag{2.41}$$

where the infimum is searched among all scalar sequences $\{b_k(l), l \in \mathbb{Z}\}$ of class $\mathscr{A}^+$. Moreover the bound can be attained only if the processes $\{y_k(n), n \in \mathbb{Z}\}$ are linear causal of minimum phase, in the sense that it can be represented as the output of a filter with impulse response in $\mathscr{A}^+$ applied to a temporally independent process. Therefore the negative of (2.41) is a contrast *in the case where the sources are linear causal of minimum phase*.

Note that the coefficient $b_k(0)$ in (2.40) and (2.41) can be set to 1, since multiplying the $\{b_k(l), l \in \mathbb{Z}\}$ by a constant does not change the expressions inside the curly brackets $\{\}$ in (2.40) and (2.41).

2.3.3 Estimation and separation

The criteria introduced in subsection 2.3.2 are theoretical. In practice they have to be replaced by their estimates, obtained by replacing the unknown entropies in their definition by some estimate.

To estimate the criterion (2.29), one would need an estimator of the entropy of a random vector in $\mathbb{R}^m$. The estimator $\hat{H}_\delta$ introduced in subsection 2.2.3 can be extended for this case. One defines the multivariate kernel density estimate similarly to (2.4) using a product kernel. The entropy is then estimated by replacing the integral in its definition by a Riemann sum over a regular grid on $\mathbb{R}^m$. See [14] for more details. This work also explains how the binning method can be extended to this multivariate case. The computation cost is, however, high and *it increases exponentially with m*! The method is thus only reasonably affordable for very small m (less than 4). The estimator $\hat{H}^M$ can be generalized to the multivariate case but its computation is problematic since we do not have a factorization algorithm to compute the Fourier coefficients of the logarithm of the estimated (multivariate) circular density. The estimator $\hat{H}_{\{u\}}$ is not generalizable to the vector case as it is based on the order statistic.

For larger m, a possibility is to estimate the entropy $H\{y_k(\overline{1,m})\}$ of the vector $y_k(\overline{1,m}) = [y_k(1) \ \cdots \ y_k(m)]^{\mathrm{T}}$ by

$$\hat{H}\{y_k(\overline{1,m})\} = \frac{1}{T+1-m} \sum_{n=1}^{T+1-m} \log \hat{p}_{y_k(\overline{1,m})}[y_k(\overline{n,n+m-1})].$$

where $\hat{p}_{y_k(\overline{1,m})}$ is a multivariate kernel density estimator of $y_k(\overline{1,m})$. Since this density has to be evaluated at random points, the binning method cannot be used so the kernel need not be the product of cardinal spline kernels. The computation cost will be of the order $O(mT^2)$, which is considerable, but it increases only linearly with m.

The paper [14] also considers the estimation of the conditional entropy, which appears in the criterion $\mathscr{C}_m^\dagger$ defined in (2.32). The estimated criterion has been used for the separation of mixtures of Markovian sources in [6].

The computation simplifies considerably if one can be content oneself with the Gaussian mutual rate information criteria. The criteria (2.35) and (2.36) only involve the autocovariance function of the recovered sources and only at lags less than m. Therefore it suffices to estimate them by the corresponding sample autocovariance to obtain the estimated criteria. One can see, by a detailed calculation, that the resulting estimated criterion (2.36) is (asymptotically) equivalent to a negative multiple of the likelihood criterion when one models the sources as autoregressive process of order $m-1$. But the most interesting method is the one based on the criterion (2.34). Its estimate can be obtained easily by replacing the smoothed spectral density $k_M \star \mathbf{f_x}$ in its definition by a spectral density estimate (in particular, a smoothed periodogram). This criterion is in fact a *joint diagonalization* criterion, which can be minimized efficiently by a fast algorithm (see [10,12] for more details).

In the case where the sources are linear processes, the criteria (2.40) and (2.41) for separating instantaneous mixture, and (2.37) and (2.38) for blind separation-deconvolution, only involve the entropy of real variables. Such entropy can thus be estimated by the same methods described in subsection 2.2.3. Note however, such criteria require a deconvolution step which can be explicit as in the case of the criteria (2.40) and (2.41) where sequences $\{b_k(l), l \in \mathbb{Z}\}$ satisfying certain deconvolution criteria have to be found, or implicit as in the case of the criteria (2.37) and (2.38) in which the deconvolution is an integral part of the separation-deconvolution procedure.

2.3.3.1 Gradient of the criteria

All criteria, except the Gaussian criteria, involve entropy, and hence the computation of their gradient requires that of the entropy. As we have seen before, the derivative of the entropy leads to the score function. Here however, we are concerned with the entropy of a random *vector* and not variable, which will lead to.

For simplicity, we shall consider the theoretical entropy $H\{Y\}$ of a random vector Y of density function p_Y, and we are interested in the matrix of derivatives of $H\{Y+\mathcal{E}Z\}$ where Z is another random vector, with respect to the elements of the matrix $\mathcal{E}$ at $\mathcal{E}=0$. In [13], it was proved that

$$\left.\frac{\partial H\{Y+\mathcal{E}Z\}}{\partial \mathcal{E}_{ij}}\right|_{\mathcal{E}=0} = \mathbb{E}[\psi_{Y,i}(Y)z_j] \tag{2.42}$$

where $\mathcal{E}_{ij}$ denotes the ij element of $\mathcal{E}$, z_j the j-th component of Z and $\psi_{Y,i}$ the i-th component of ψ_Y, the gradient of $-\log p_Y$. A heuristic proof for this result can be obtained as follows. Instead of computing the above derivative, we shall consider the first-order Taylor expansion of $H(\mathbf{Y}+\mathcal{E}\mathbf{Z})$ with respect to the matrix $\mathcal{E}$ around 0. Write

$$H\{Y+\mathcal{E}Z\}-H\{Y\}=\mathbb{E}[\log p_Y(Y)-\log p_Y(Y+\mathcal{E}Z)]+\mathbb{E}\log\frac{p_Y(Y+\mathcal{E}Z)}{p_{Y+\mathcal{E}Z}(Y+\mathcal{E}Z)}$$

and note that the last term can be rewritten as

$$\int \log\frac{p_Y(u)}{p_{Y+\mathcal{E}Z}(u)}\,p_{Y+\mathcal{E}Z}(u)du=\int\left[\log\frac{p_Y(u)}{p_{Y+\mathcal{E}Z}(u)}-\frac{p_Y(u)}{p_{Y+\mathcal{E}Z}(u)}+1\right]p_{Y+\mathcal{E}Z}(u)du.$$

For small $\mathcal{E}$, the expression inside the above square bracket [ ] will be of the order $\|\mathcal{E}\|^2$ and one can expect that the same holds for its integral[6]. Since $\log p_Y(Y)-\log p_Y(Y+\mathcal{E}Z)=\psi_Y^{\mathrm{T}}(Y)\mathcal{E}Z+o(\|\mathcal{E}\|)$, one gets under the appropriate regularity condition:

$$H\{Y+\mathcal{E}Z\}=H\{Y\}+\mathbb{E}[\varphi_Y^{\mathrm{T}}(Y)\mathcal{E}Z]+o(\|\mathcal{E}\|),$$

which yields the desired result.

The function ψ_Y is called *joint score function*. Let $H\{Y\}$ be estimated by $\hat{H}\{Y\}$ based on a sample $Y(1),\ldots,Y(T)$, and assume that this estimator is continuously differentiable with respect to the data values; then $T\partial\hat{H}\{Y\}/\partial Y_i(n)$ where $Y_i(n)$ denotes the i-th component of $Y(n)$, can be used as an estimator of $\psi_{Y,i}[Y(n)]$, the i-th component of the joint score function at $Y(n)$. In [14], the entropy estimator $\hat{H}_\delta$ introduced in subsection 2.2.3 has been extended to the vector case and the associated joint score estimator has been provided explicitly.

[6]There is a technical difficulty here: this square bracket is of order $\|\mathcal{E}\|^2$ for a given u but not necessarily uniform in u, since p_Y and $p_{Y+\mathcal{E}Z}$ both converge to 0 at infinity. But this uniformity is not necessary either, because of the integration with respect to $p_{Y+\mathcal{E}Z}$.

2.3.3.2 The instantaneous mixtures case

Consider first the criterion $\mathscr{C}_m$ defined in (2.29). From the above result, its relative gradient matrix can be seen to have the general element:

$$\mathbb{E}\{y_j^{\mathrm{T}}(\overline{1,m})\psi^*_{y_k(\overline{1,m})}[y_k(\overline{1,m})]\} - \delta_{jk} \tag{2.43}$$

where $y_k(\overline{1,m})$ denotes the vector $[y_k(1) \ \cdots \ y_k(m)]^{\mathrm{T}}$ and δ_{jk} is the Kronecker term. For $j = k$, direct calculation using the definition of joint score function and integration by part, shows that the first term in (2.43) equals 1. Hence, the diagonal of the gradient matrix vanishes, which is expected, as the criterion is scale invariant.

Concerning the criterion $C_m^{\dagger}$ defined in (2.32), one notes that the conditional entropy can be written as the difference of two joint entropies

$$H\{y_k(m)|y_k(\overline{1,m-1})\} = H\{y_k(\overline{1,m})\} - H\{y_k(\overline{1,m-1})\}.$$

Therefore the relative gradient matrix of this criterion has general elements:

$$\mathbb{E}[y_j(\overline{1,m})\psi^*_{y_k(\overline{1,m})}[y_k(\overline{1,m})]\} - \delta_{ij} \tag{2.44}$$

where

$$\psi^*_{y_k(\overline{1,m})}[y_k(\overline{1,m})] = \psi_{y_k(\overline{1,m})}[y_k(\overline{1,m})] - \begin{bmatrix} \psi_{y_k(\overline{1,m-1})}[y_k(\overline{1,m-1})] \\ 0 \end{bmatrix}$$

is the conditional score function of $y_k(m)$ given $y_k(1{:}m-1)$. From the definition of the joint score function, one can see that the above conditional score function is the negative of the logarithmic derivative of the conditional density of $y_k(m)$ given $y_k(\overline{1,m-1})$. It can also be verified that for $j = k$ the first term in (2.44) equals 1.

We now compute the gradients of the "Gaussian" criteria (2.35) and (2.36). One notes that $\log|\det(\mathbf{M}+\boldsymbol{\delta})| - \log|\det(\mathbf{M})| = \log|\det(\mathbf{I}+\mathbf{M}^{-1}\boldsymbol{\delta})| = \mathrm{trace}(\mathbf{M}^{-1}\boldsymbol{\delta}) + o(\|\boldsymbol{\delta}\|)$ as $\boldsymbol{\delta} \to 0$. Then, by a direct calculation, the relative gradient matrix of the criterion (2.35) has the general element:

$$\frac{1}{m}\sum_{t=1}^{m} \mathrm{cov}\left\{y_j(t), \sum_{s=1}^{m} b_{k,m}(t,s) y_k(s)\right\} - \delta_{jk}, \tag{2.45}$$

$b_{k,m}(t,s)$ being the general element of the inverse of the covariance matrix of $y_k(\overline{1,m})^{\mathrm{T}}$. That of the criterion (2.36) has the general element:

$$\sum_{t=1}^{m} \mathrm{cov}\left\{y_j(t), a_{k,m-1}(m-t)\sum_{l=0}^{m-1} a_{k,m-1}(l) y_k(m-l)\right\} - \delta_{jk}, \tag{2.46}$$

$a_{k,m-1}(l)$ being the coefficients in the representation $\sum_{l=0}^{m-1} a_{k,m-1}(l) y_k(m-l)$ of $y_k(m) - y_k(m|1:m-1)$. They are given by $[a_{k,m-1}(0) \;\cdots\; a_{k,m-1}(m-1)] = [b_{k,m}(m,m) \;\cdots\; b_{k,m}(m,1)]/b_{k,m}(m,m)$. The gradient of the estimated criteria is given by the same formula, but with the covariance operator replaced by the sample covariance.

Finally, the relative gradient matrix of the criterion (2.34) has general element:

$$\frac{1}{L}\sum_{l=1}^{L}(k_M \star f_{y_k y_j})(\lambda_l)/(k_M \star f_{y_k y_k})(\lambda_l) - \delta_{jk} \tag{2.47}$$

$f_{y_k y_j}$ denoting the interspectrum between the processes $\{y_k(n), n \in \mathbb{Z}\}$ and $\{y_j(n), n \in \mathbb{Z}\}$.

Consider now the criteria (2.40) and (2.41) for linear sources. Putting

$$D_k(b_k) = H\{(b_k \star y_k)(1)\} - \int_{-\pi}^{\pi} \log\left|\sum_{l=-\infty}^{\infty} b_k(l) e^{il\lambda}\right| \frac{d\lambda}{2\pi}$$

then (2.40) can be written as $\sum_{k=1}^{P} \inf_{b_k} D_k(b_k) - \log\det \mathbf{B} = \sum_{k=1}^{P} D_k(b_k^*) - \log\det \mathbf{B}$, where $\{b_k^*(l), l \in \mathbb{Z}\}$ is the sequence which realizes the infimum of $D_k(b_k)$. In computing the relative gradient of $D_k(b_k^*)$, one should take into account that the $b_k^*(l)$ depends on $\mathbf{B}$ via the $y_k(n)$. However, the partial derivative of $D_k(b_k)$ with respect to b_k (for fixed $y_k(n)$) must vanish at b_k^*. Therefore in computing the relative gradient of $D_k(b_k^*)$ with respect to $\mathbf{B}$, one may treat the $b_k^*(l)$ as constant. Then by the same calculation as in subsection 2.2.5.1, but using the theoretical instead of estimate entropy and formula (2.42), one gets that the matrix of relative gradient of the criterion (2.40) has the general element:

$$\mathbb{E}\{(b_k^* \star y_j)(1)\varphi_{b_k^* \star y_k}[b_k^* \star y_k(1)]\} - \delta_{ik}. \tag{2.48}$$

One can seen that the computation of the above gradient requires a deconvolution stage in which the deconvolution filter needs to be found.

By a similar calculation, one obtains that the relative gradient of the criterion (2.41) is given by the same above formula, except that now the sequence $\{b_k^*(l), l \in \mathbb{Z}\}$ is the one that minimizes $H\{(b_k \star y_k)(1)\} - \log|b_k(0)|$ *under the constraint that it is of class* $\mathscr{A}^+$.

2.3.3.3 The convolutive mixtures case

This case needs some special considerations since the demixing sequence $\{\mathbf{B}(l), l \in \mathbb{Z}\}$ contains an infinite number of terms. One has either to truncate it, which amounts to assuming that it supports some given finite interval, $[-N_1, N_2]$ say, or to parameterize it by some vector parameter θ of fixed dimension. One may also adopt the nonparametric approach which considers the whole sequence but adds a penalty term in the criterion favoring sequences with smooth Fourier transforms. We shall focus on the approach which parameterizes the sequence $\{\mathbf{B}(l), l \in \mathbb{Z}\}$, noting that in the case where this sequence has support $[N_1, N_2]$, it can be parameterized by the elements of the matrices

$\mathbf{B}(-N_1),\ldots,\mathbf{B}(N_2)$. Another difficulty is that the source separation formula (2.27) requires the knowledge of the whole observation sequence but only a finite number of them is available. A simple and convenient way to overcome this difficulty is to extend the data periodically: one indexes them from 0 to $T-1$ and replaces the $\mathbf{x}(n)$ in (2.27) by $\mathbf{x}(n \bmod T)$. However, in the case where the sequence $\{\mathbf{B}(l), l \in \mathbb{Z}\}$ has given support, $[-N_1, N_2]$ say, which we refer to as the FIR (finite impulse response) case, one can compute $\mathbf{y}(n)$ as usual for $N_2 \leqslant n < T - N_2$ so that it is preferable to restrict oneself to using only those $\mathbf{y}(n)$.

The relative gradient is not a suitable tool here. Indeed, to define it one would consider an increment to $\mathbf{B}(\cdot)$ of the form $\mathcal{E} \star \mathbf{B}$, but in the FIR case the incremented sequence would have larger support thus violating the fixed support assumption. In the case where the $\mathbf{B}(l)$ are parameterized, the incremented sequence should belong to the same parametric family and such constraint would translate to very complex constraint on the $\mathcal{E}(l)$.

Since the criterion $\mathscr{C}_1[\mathbf{B}(\cdot)]$, defined in (2.37), involves only the entropy of a random variable, it can be estimated by any of the methods in subsection 2.2.3:

$$\hat{\mathscr{C}}_1[\mathbf{B}(\cdot)] = \sum_{k=1}^{P} H\{y_k\} - \int_{-\pi}^{\pi} \log \left| \det \sum_{l=-\infty}^{\infty} \mathbf{B}(l) e^{il\lambda} \right| \frac{d\lambda}{2\pi} \tag{2.49}$$

where $\hat{H}$ is either $\hat{H}_\delta$ or $\hat{H}^M$. (We don't use $\hat{H}_{\{u\}}$ because its derivative has jumps.) From the results of subsection 2.2.3

$$\frac{\partial}{\partial \theta_\mu} \sum_{k=1}^{P} \hat{H}\{y_k\} = \operatorname{trace}\left\{ \frac{1}{T} \sum_{n=0}^{T-1} \frac{\partial \mathbf{y}(n)}{\partial \theta_\mu} \hat{\varphi}_{\mathbf{y}}^{\mathrm{T}}[\mathbf{y}(n)] \right\} \tag{2.50}$$

where $\hat{\varphi}_{\mathbf{y}}^{\mathrm{T}}[\mathbf{y}(n)]$ denotes the vector with components $\hat{\varphi}_{y_k}[y_k(n)], k = 1, \ldots, P$, $\hat{\varphi}_{y_k}$ being either $\hat{\varphi}_{\delta, y_k}$ or $\hat{\varphi}_{y_k}^M$. In order to have the same formula in the FIR case where $\{\mathbf{B}(l), l \in \mathbb{Z}\}$ has support $[-N_1, N_2]$, we shall assume that the $\mathbf{x}(n)$ in this case are in fact available for $-N_2 \leqslant n < T + N_1$. Using

$$\frac{\partial \mathbf{y}(n)}{\partial \theta_\mu} = \sum_{l=-\infty}^{\infty} \frac{\partial \mathbf{B}(l)}{\partial \theta_\mu} \mathbf{x}(n - l \bmod T), \tag{2.51}$$

without the modulo in the FIR case, one gets

$$\frac{\partial \hat{\mathscr{C}}_1[\mathbf{B}(\cdot)]}{\partial \theta_\mu} = \operatorname{trace}\left\{ \sum_l \frac{\partial \mathbf{B}(l)}{\partial \theta_\mu} \frac{1}{T} \sum_{n=0}^{T-1} \mathbf{x}(n - l \bmod T) \hat{\varphi}_{\mathbf{y}}^{\mathrm{T}}[\mathbf{y}(n)] - \int_{-\pi}^{\pi} \frac{\check{\mathbf{B}}(\lambda)}{\partial \theta_\mu} \check{\mathbf{B}}^{-1}(\lambda) \frac{d\lambda}{2\pi} \right\}$$

where $\check{\mathbf{B}}(\lambda) = \sum_l \mathbf{B}(l)e^{-il\lambda}$ is the Fourier transform (series) of $\{\mathbf{B}(l), l \in \mathbb{Z}\}$. This formula is mostly useful in the FIR case, since then the sum with respect to l is a finite sum and the "mod T" is absent. For computational purposes the integral in (2.49) and hence in the above formula can be replaced by Riemann sums. But apart from this case, it is more interesting to work in the frequency domain. By the discrete Parseval Theorem, the sum in (2.50) can be written as $T^{-1}\sum_{k=0}^{T-1}\mathbf{d}_{\partial\mathbf{y}/\partial\theta_\mu}(2\pi k/T)\mathbf{d}^{\mathrm{H}}_{\hat{\varphi}_\mathbf{y}}(2\pi k/T)$ where ^{H} denotes the transpose conjugate and

$$\mathbf{d}_{\partial\mathbf{y}/\partial\theta_\mu}\left(\frac{2\pi k}{T}\right) = \sum_{n=0}^{T-1}\frac{\partial\mathbf{y}(n)}{\partial\theta_\mu}e^{-i2\pi kn/T}, \quad n = 0,\dots T-1$$

is the discrete Fourier transform of $\{\partial\mathbf{y}(n)/\partial\theta_\mu, n = 0,\dots,T-1\}$ and $\{\mathbf{d}_{\hat{\varphi}_\mathbf{y}}(2\pi n/T), n = 0,\dots,T-1\}$ is defined similarly. From (2.51), one gets

$$\begin{aligned}\mathbf{d}_{\partial\mathbf{y}/\partial\theta_\mu}\left(\frac{2\pi k}{T}\right) &= \sum_l e^{-2\pi kl/T}\frac{\partial\mathbf{B}(l)}{\partial\theta_\mu}\sum_{n=0}^{T-1}\mathbf{x}(n-l \bmod T)e^{-2\pi k(n-l)/T}\\ &= \frac{\partial\check{\mathbf{B}}(2\pi k/T)}{\partial\theta_\mu}\mathbf{d}_\mathbf{x}\left(\frac{2\pi k}{T}\right) = \frac{\partial\check{\mathbf{B}}(2\pi k/T)}{\partial\theta_\mu}\check{\mathbf{B}}^{-1}\left(\frac{2\pi k}{T}\right)\mathbf{d}_\mathbf{y}\left(\frac{2\pi k}{T}\right),\end{aligned}$$

the last equality following from $\mathbf{y}(n) = \sum_l \mathbf{B}(l)\mathbf{x}(n-l \bmod T)$. Finally, replacing the integral in (2.49) by a Riemann sum with spacing $2\pi/T$, one gets

$$\frac{\partial\hat{\mathscr{C}}_1[\mathbf{B}(\cdot)]}{\partial\theta_\mu} = \frac{1}{T}\sum_{n=0}^{T-1}\mathrm{trace}\left\{\frac{\partial\check{\mathbf{B}}(2\pi n/T)}{\partial\theta_\mu}\check{\mathbf{B}}^{-1}\left(\frac{2\pi n}{T}\right)\left[\hat{\mathbf{f}}_{\mathbf{y}\hat{\varphi}_\mathbf{y}}\left(\frac{2\pi n}{T}\right) - \mathbf{I}\right]\right\}, \quad (2.52)$$

where

$$\hat{\mathbf{f}}_{\mathbf{y}\hat{\varphi}_\mathbf{y}}\left(\frac{2\pi k}{T}\right) = \frac{1}{T}\mathbf{d}_\mathbf{y}\left(\frac{2\pi k}{T}\right)\mathbf{d}^{\mathrm{H}}_{\hat{\varphi}_\mathbf{y}}\left(\frac{2\pi k}{T}\right) \quad (2.53)$$

is the cross periodogram between $\mathbf{y}$ and $\hat{\varphi}_\mathbf{y}$. More comprehensive results can be found in [16]. In particular, a formula for an approximate Hessian of the criterion is provided. This formula, while too complex for general use, simplifies considerably in the univariate (pure deconvolution) case.

2.3.3.4 Estimating equations

By setting the gradient of the criteria, one obtains a system of equations that the recovered sources must satisfy. As we consider the theoretical criteria, such a system is only theoretical, the unknown functions need to estimated and the expectation operator is to be replaced by sample average.

In the case of instantaneous mixtures, by referring to the formulas (2.43)–(2.46), one sees that such equations are of the form:

$$\mathbb{E}\{[y_j(1) \ \cdots \ y_j(m)]^{\mathrm{T}} \psi_{k,m}[y_k(1),\ldots,y_k(m)]\} = 0, \ 1 \leqslant j \neq k \leqslant P \tag{2.54}$$

where $\psi_{k,m}$ is some function from $\mathbb{R}^m$ to $\mathbb{R}^m$. One can note that for the Gaussian criteria (2.35) and (2.36) for which the gradient are given in (2.45) and (2.46), these functions $\psi_{k,m}$ are linear and then the above equations take the form: $\mathbb{E}\{y_j(0)\sum_l b_k(l)y_k(-l)\} = 0$ where $\{b_k(l), l \in \mathbb{Z}\}$ is a sequence with only a finite number of non-zero terms. (Note that by stationarity, $\mathbb{E}\{y_j(l)y_k(n)\} = \mathbb{E}\{y_j(0)y_k(n-l)\}$.) The equations obtained by setting to zero the gradient of the criterion (2.47) can also be put in this form but the sequence $\{b_{k,l}, l \in \mathbb{Z}\}$ may contain an infinite number of non-zero terms. Similarly, the equations obtained by setting the gradient (2.48) to zero also involve an infinite sequence $\{b_k^*(l), l \in \mathbb{Z}\}$. if one truncates this sequence (which one needs to do anyway in practice), then the corresponding equations can again be put in the form (2.54), with $\psi_{k,m}$ taking the form:

$$\psi_{k,m}[y(1),\ldots,y(m)] = \begin{bmatrix} \beta_{k,1} \\ \vdots \\ \beta_{k,m} \end{bmatrix} \varphi_k \left[\sum_{l=1}^{m} b_k(l)y(l)\right] \tag{2.55}$$

for a certain real function of a real variable φ_k and for certain real numbers $b_k(l),\ldots, b_k(m)$ and $\beta_{k,1},\ldots,\beta_{k,m}$. In all cases, the functions $\psi_{k,m}$, which we will call *separating functions*, are specifically related to the densities of the recovered sources (and are linear in the Gaussian case).

Godambe [5] has introduced the concept of a system of *estimating equations*, the use of which is much more flexible than that of a contrast, since such a system need not be derived by *setting to zero the gradient of a contrast*. It is simply a system of equations which is satisfied as soon as the unknown parameter to be estimated equals the true parameter, or in the present context, as soon as the recovered sources are independent. Such a system typically involves the expectation operator, which one would replace by a sample average before solving it to obtain the estimate of the unknown parameters. One can easily see that any system of equations of the form (2.54) is a system of estimating equations, provided that the sources are centered or the random variables $\psi_{k,m}[y_k(1),\ldots,y_k(m)]$ have zero mean. Note however that if we insist on minimizing the introduced criteria, the functions $\psi_{k,m}[y_k(1),\ldots,y_k(m)]$ are related to the unknown distribution of the recovered sources and hence must be replaced by their estimators. And one cannot use any estimator; it must be compatible with the gradient of some estimate of the criterion. The derivative of an entropy estimator is an estimator of the score function but conversely an estimator of the score function may not be the derivative of any entropy estimator. The estimating equations approach is much more flexible; the functions $\psi_{k,m}[y_k(1),\ldots,y_k(m)]$ can simply be some *a priori* given guess of the score functions, or perhaps contain some parameters which can be adjusted from the data to obtain a more accurate guess.

The advantage of the estimating equations approach is its ease of implementation and its low computational cost. One needs only specify P separating functions $\psi_{1,m}, \ldots, \psi_{P,m}$ from $\mathbb{R}^m$ to $\mathbb{R}^m$, then solving the system of equations (2.54), after having replaced the expectation operator by sample means. The resolution of such systems can even be done on-line via an algorithm of adaptive type. Note that in the simplest case where $m = 1$ in (2.54) one obtains the estimating equations introduced in [17], which refines the heuristic procedure in the pioneer work [8]. On the other hand, by taking $\psi_{k,m}$ of the form (2.55) with φ_k being the identity function, one obtains the quasi maximum likelihood for the separation of correlated source in [17]. Moreover, many *ad hoc* methods for blind source separation are based on the cancellation of (higher order) cross cumulants (possibly with delay) between the recovered sources. This amounts essentially to solving a system of estimating equations.

However, the system of estimating equations method has some serious drawbacks. Firstly, these equations are only necessary but not sufficient for the separation, and thus can have multiple solutions and one may be led to recover spurious sources. By constructing these equations from a contrast, one can expect to have a better chance of avoiding such a problem. Further, one can compute the contrast and monitor it in the iteration algorithm to ensure that it increases at each step (by decreasing the step size for example). In this way, one can be sure to attain at least local maxima of the contrast. Secondly, the choice of the separating function can have great influence on the quality of the estimator: a bad choice can strongly degrade its performance. Due to the close relationship between the mutual information and the likelihood criterion [1] (see also Chapter 4, section 4.2.2, or Chapter 3, page 86), one may expect that minimizing a mutual information based criterion would produce (asymptotically) optimal separation performance (this was proved in the case of separation of instantaneous mixtures of temporally independent sources – see subsection 2.2.6).

In this regard, the results of the gradient calculation of the mutual information based contrasts would provide good candidates for the separating function, since these contrasts are closely linked to the maximum likelihood principle (see [1] and also Chapter 4, section 4.2.2, or Chapter 3, page 86). These functions involve the unknown distributions of the sources, but in practice, it is not necessary to know them exactly, and some crude estimation can suffice. Note that the general form (2.54) of the estimating equations require the specification of Pm real functions of m real numbers, which may be too much for our limited *a priori* knowledge or for the available data for even a crude estimation. If one has confidence in the assumption of linear process for the sources, one may adopt the separating functions of the form (2.55). One will then need to specify or to estimate only P real functions, but P linear filters are also needed to be specified or estimated.

2.4 CONCLUSION AND PERSPECTIVES

The important point which stands out is the generality of the mutual information approach. It can be applied to the case of instantaneous mixtures as well as convolutive mixtures. In the first case, it also permits taking into account and exploiting the temporal dependence of the source signals. We haven't considered here the case of nonlinear mixtures (for example post-nonlinear mixtures) where this approach is still applicable. Actually in the situation where nonlinear transformations are involved, it is more suitable than the approach based on higher order cumulants. The reason is that there is no easy way to relate the cumulants and cross cumulants of the components of a nonlinear transformation of a random vector from those of the original vector, while the entropies transform in a simple way via the formula of Lemma 2.1. Another point in favor of the mutual information approach is its asymptotic optimality, which we only mentioned briefly in the case of linear instantaneous mixtures. This optimality comes from the close relationship between mutual information based criteria and the maximum likelihood criterion (see [1]).

On the practical side, the implementation of the mutual information approach is not so costly as one may fear. Thanks to the use of discretization[7] introduced and the *binning* technique, the entropy can be computed with a reasonable cost. The computation and estimation of the joint entropy in high dimension, which approximates the entropy rate, remains, however, problematic. However, one should note that the case of convolutive mixtures is much more complex and difficult than that of instantaneous mixtures, especially if no *a priori* assumption is available. Under some weak assumptions on the sources such as that they are linear or Markovian, the calculation is manageable. Another possible approach is to develop a procedure for direct estimation of the entropy rate by exploiting its link with the optimal compression length in coding.

References

[1] J.-F. Cardoso, Blind signal separation: Statistical principles, in: Blind Estimation and Identification, Proc. IEEE 86 (1998) 2009–2025 (special issue).

[2] J.-F. Cardoso, B. Laheld, Equivariant adaptive source separation, IEEE Trans. Signal Process. 44 (1996) 3017–3030.

[3] P. Comon, Independent component analysis, a new concept? Signal Process. 36 (1994) 287–314.

[4] T. Cover, J. Thomas, Elements of Information Theory, Wiley, New York, 1991.

[5] V.P. Godambe, Conditional likelihood and unconditional optimum estimating equations, Biomtrika 61 (1976) 277–284.

[6] S. Hosseini, C. Jutten, D.T. Pham, Markovian source separation, IEEE Trans. Signal Process. 51 (2003) 3009–3019.

[7] M.C. Jones, Discretized and interpolated kernel density estimates, Am. Statist. Assoc. 84 (1989) 733–741.

[8] C. Jutten, J. Hérault, Blind separation of sources, part I: an adaptive algorithm based on neuromimetic structure, Signal Process. 24 (1991) 1–10.

[9] D. Pham, Blind separation of instantaneous mixtures of sources based on order statistics, IEEE Trans. Signal Process. 48 (2000) 363–375.

[10] D.T. Pham, Blind separation of instantaneous mixture of sources via the Gaussian mutual information criterion, Signal Process. 81 (2001) 850–870.

[7] In our experience, the discretization can be rather coarse.

[11] D.T. Pham, Contrast functions for blind separation and deconvolution of sources, in: Proceedings of ICA 2001 Conference, San Diego, USA, Dec. 2001, pp. 37–42.
[12] D.T. Pham, Joint approximate diagonalization of positive definite matrices, SIAM J. Matrix Anal. Appl. 22 (2001) 1136–1152.
[13] D.T. Pham, Mutual information approach to blind separation of stationary sourcs, IEEE Trans. Inform. Theory 48 (2002) 1935–1946.
[14] D.T. Pham, Fast algorithm for estimating mutual information, entropies and score functions, in: Proceedings of ICA 2003 Conference, Nara, Japan, Apr. 2003, pp. 17–22.
[15] D.T. Pham, Fast algorithms for mutual information based independent component analysis, IEEE Trans. Signal Process. 52 (2004) 2690–2700.
[16] D.-T. Pham, Generalized mutual information approach to multichannel blind deconvolution, Signal Process. 87 (2007) 2045–2060.
[17] D.T. Pham, P. Garat, Blind separation of mixtures of independent sources through a quasi maximum likelihood approach, IEEE Trans. Signal Process. 45 (1997) 1712–1725.
[18] W.H. Press, B.P. Flannery, S.A. Teukolsky, W.T. Vetterling, Numerical Recipes in C. The Art of Scientific Computing, Second Edition, Cambridge University Press, 1993.
[19] D.W. Scott, S.J. Sheather, Kernel density estimation with binned data, Commun. Statist. - Theory Meth. 14 (1985) 1353–1359.
[20] G.T. Wilson, Factorization of the covariance generating function of a pure moving average process, SIAM J. Numer. Anal. 6 (1969) 1–7.
[21] A. Zygmund, Trigonometric Series, vol. I, Cambridge Univ. Press, Cambridge, 1968.

CHAPTER

Contrasts

3

E. Moreau and P. Comon

3.1 INTRODUCTION

The Blind Source Separation (BSS) problem is defined by a mixture model, a set of source processes, and a set of assumptions. For instance, various types of mixture can be considered, as: (i) instantaneous (i.e. static) linear mixtures, (ii) convolutive mixtures, or (iii) nonlinear mixtures. The mixing operation is often assumed to be invertible, but it is not the case for *under-determined* mixtures described in Chapter 9.

Source signals can take their values in the real field or in the complex field, be of constant modulus or belong to a finite alphabet, *a priori* known or not. When source signals are considered to be random, several assumptions exist. Sources may be: (a) mutually statistically independent, (b) each identically and independently distributed in time (iid), or (c) cyclostationary.

Mixture models considered in this chapter are deterministic, and depend on some constant unknown parameters. Once this class of parametric models is fixed, the *Blind Identification* problem will consist of estimating these parameters only from the observation of the outputs of the system.

Contrast criteria. For this purpose, an optimization criterion needs to be defined. In Signal Processing, optimization criteria are often chosen to be a Mean Square Error (like in Wiener filtering). In the present BSS problem, such a criterion cannot be used since inputs are not observed. For this reason, one resorts to criteria called contrast functions, or simply *contrasts.*

In all cases, the question will always be the following. Given a mixture model and a set of assumptions:

1. Can the mixture be identified from the observations?
2. Can the source signals be separated, i.e. consistently estimated?
3. What algorithm can be used to perform these tasks?

Contrast functions answer the first two questions. In fact, a contrast function is subsequently defined as an optimization criterion such that all its global maxima correspond to a separation of all sources. We see that defining such a criterion is very important, because it will lead to a well-posed problem. Note that defining a contrast criterion is related to the property of source separability. Once the optimization criterion is defined, it will remain to devise a numerical algorithm to maximize it. The goal of this chapter is to propose several contrast criteria, well matched to various BSS problems; the numerical maximization is postponed to other chapters of this book, and will not be addressed in the present chapter.

3.1.1 Model and notation

Generally speaking, in many practical situations, one has at one's disposal several observation signals, that is, measurements $\mathbf{x}(n) \stackrel{\text{def}}{=} (x_1(n), \ldots, x_P(n))^T$, which depend on signals of interest $\mathbf{s}(n) \stackrel{\text{def}}{=} (s_1(n), \ldots, s_N(n))^T$. The observation model can hence be written:

$$\mathbf{x}(n) = \mathscr{A}\{\mathbf{s}(n)\} + \mathbf{b}(n) \tag{3.1}$$

where $\mathscr{A}\{\cdot\}$ is a multi-dimensional operator corresponding to the chosen mixture, or obtained from physical relationships (cf. Chapter 1). In Eq. (3.1) above, vector $\mathbf{b}(n)$ represents measurement noises and nuisance signals such as interferences. The definition of contrast criteria will be based on a noiseless observation model, since no statistical modeling of the noise $\mathbf{b}$ is supposed to be available. But of course, in the real world, blind source estimation and blind mixture identification will be performed in the presence of noise. On the other hand, if statistical properties of noise and sources were known, then it would be preferred to resort to the maximum likelihood criterion (see Chapter 4).

In the rest of this chapter, it will be assumed that the mixture is linear, and that the multivariate process $\mathbf{s}(n)$ is stationary and zero-mean, unless otherwise specified. More precisely, we distinguish between two cases of linear mixtures. The first one is instantaneous and can be written as:

$$x_i(n) = \sum_{j=1}^{N} A_{ij} s_j(n), \qquad \forall i \in \{1, \ldots, P\}. \tag{3.2}$$

The second one is convolutive and can be written:

$$x_i(n) = \sum_{j=1}^{N} \sum_{k} A_{i,j}(k) s_j(n-k), \qquad \forall i \in \{1, \ldots, P\}. \tag{3.3}$$

In the above observation models, N denotes the number of sources and P the number of sensors, as everywhere else in this book.

3.1.2 Principle of contrast functions

Our definition of *contrast* functions of course depends on the problem. It is important to distinguish at least two problems: (i) the *extraction* of a single source signal, which corresponds to a Multiple Input Single Output (*MISO*) filtering, and (ii) the joint estimation of the N source signals, which we call *separation* in the remainder, and which corresponds to a Multiple Input Multiple Output (*MIMO*) filtering.

In both cases the goal is the same: to build a criterion whose global maxima correspond to a solution (either extraction or separation). As emphasized earlier in this book, there is actually a whole equivalence class of solutions. Thus, it can be useful to impose which representative of this equivalence class we would like to find. In particular, a power normalization of the outputs (the estimated sources) fixes the scale indeterminacy.

In this chapter, principles controlling the definition of contrast criteria will be investigated in detail, and proofs of optimality will be given. As already pointed out, numerical algorithms will be studied in other chapters, and we shall here concentrate on the definition of contrasts.

3.1.3 Bibliographical remarks

The solution of the BSS problem by contrast maximization was first proposed in 1991 [13]. Five years later, we discovered the very nice paper of Donoho describing the solution to the Single Input Single Output (SISO) Blind Deconvolution by *Kurtosis Maximization (KM)* [26]. Donoho's work must be seen as being at the origin of the concept of contrast, even if he developed the idea in a different context, which was indeed specifically SISO. As will be seen later in this chapter, other criteria such as the *constant modulus (CM)* proposed by Godard [28], can be presented within the same framework. Note that several authors have devised numerical algorithms dedicated to blind SISO deconvolution without referring to contrasts, even if they were based on KM or CM [3,51,47].

The convolutive MIMO problem was first addressed in 1990 [11], via a cumulant matching approach. Both time and frequency domains were considered, and the continuity of the complex gain was proposed to get rid of the permutation ambiguity. But one had to wait until 1996 in order to see the first contrast functions usable for MIMO convolutive mixtures [18]. Then many other contributions followed [41,54,39,25,21,19].

In the MIMO problem, it can be envisaged to extract the sources one by one, or all together. In the former approach, one determines a series of MISO filters having P (or fewer) inputs, each aiming at extracting one source. This procedure is now referred to as *deflation*. It has specific advantages and drawbacks [23,51,54,49,44,48]. Contrast criteria adapted to deflation will be addressed in section 3.3.

3.2 CUMULANTS

Let $\mathbf{x}$ be a real random variable of dimension P. The mth order moments of $\mathbf{x}$ are (up to the factorial constants) the coefficients of degree m in the Taylor expansion of the first *characteristic function* of $\mathbf{x}$, $\Phi_{\mathbf{x}}(\mathbf{t}) = \mathbb{E}\{\exp(\jmath \mathbf{t}^{\mathrm{T}}\mathbf{x})\}$, in the neighborhood of the origin:

$$\mathbb{E}\{x_{i_1} x_{i_2} \ldots x_{i_m}\} = (-\jmath)^m \left. \frac{\partial^m \Phi_{\mathbf{x}}(\mathbf{t})}{\partial t_{i_1} \partial t_{i_2} \ldots \partial t_{i_m}} \right|_{\mathbf{t}=0}.$$

Similarly, mth order cumulants are defined with the help of the Taylor expansion of the second characteristic function $\Psi_{\mathbf{x}}(\mathbf{t}) = \log \Phi_{\mathbf{x}}(\mathbf{t})$ about the origin [33,4]:

$$\mathrm{cum}\{x_{i_1}, x_{i_2}, \ldots, x_{i_m}\} = (-\jmath)^m \left. \frac{\partial^m \Psi_{\mathbf{x}}(\mathbf{t})}{\partial t_{i_1} \partial t_{i_2} \ldots \partial t_{i_m}} \right|_{\mathbf{t}=0}.$$

Now if $\mathbf{x}$ is a complex random variable, it suffices to consider it as a real random variable of dimension twice larger (the real and imaginary parts) [33], so that the definition above applies [33]. Hence, in a more compact form, one can define first and second characteristic functions:

$$\Phi_{\mathbf{x}}(\mathbf{t}) = \mathbb{E}\{\exp(\jmath \Re[\mathbf{t}^{\mathrm{H}}\mathbf{x}])\}, \quad \Psi_{\mathbf{x}}(\mathbf{t}) = \log \Phi_{\mathbf{x}}(\mathbf{t}), \quad \mathbf{t} \in \mathbb{C}^P.$$

Cumulants of a scalar complex random variable are denoted as

$$\mathrm{cum}_{p,q}\{y\} = \mathrm{cum}\{\underbrace{y, \ldots, y}_{p\ termes}, \underbrace{y^*, \ldots, y^*}_{q\ termes}\} \tag{3.4}$$

whereas for multivariate complex random variables, the notation below is assumed:

$$\mathrm{cum}_{i\ldots k}^{\ell\ldots m}\{\mathbf{x}\} = \mathrm{cum}\{x_i, \ldots, x_k, x_\ell^*, \ldots, x_m^*\}. \tag{3.5}$$

The reader will remark that indices appearing as superscripts correspond to variables that are complex conjugated.

One can express cumulants as a function of moments by expanding the log function into Taylor expansion, and by reorganizing terms, see e.g. section 9.3.2. For instance, for zero-mean random variables, we have:

$$\begin{aligned} \mathrm{cum}_{ik}^{\ell m}\{\mathbf{x}\} &= \mathbb{E}\{x_i x_k x_\ell^* x_m^*\} - \mathbb{E}\{x_i x_k\}\mathbb{E}\{x_\ell^* x_m^*\} \\ &\quad - \mathbb{E}\{x_i x_\ell^*\}\mathbb{E}\{x_k x_m^*\} - \mathbb{E}\{x_i x_m^*\}\mathbb{E}\{x_k x_\ell^*\}. \end{aligned} \tag{3.6}$$

Components of the covariance matrix $\mathbb{E}\{\mathbf{x}\mathbf{x}^{\mathrm{H}}\} - \mathbb{E}\{\mathbf{x}\}\mathbb{E}\{\mathbf{x}\}^{\mathrm{H}}$ are second order cumulants. The higher order cumulants of $\mathbf{x}$ can be considered as a measure of the gap to Gaussianity since they are zero when variable $\mathbf{x}$ is normal.

The expression of high-order cumulants can be found in [32,38] in the real case, and in [2] in the complex case; these expressions are given in the appendix of Chapter 9, page 361.

However, the main interest in using cumulants rather than moments lies in the following fundamental property [4,38,33]:

PROPOSITION 3.1
If **x** *and* **y** *are mutually statistically independent random variables, then any cross-cumulant vanishes*

$$\text{cum}\{x_i, \ldots, x_j, y_k, \ldots, x_\ell\} = 0.$$

3.3 MISO CONTRASTS

3.3.1 MISO contrasts for static mixtures

Define the static MIMO mixture below:

$$\mathbf{x}(n) = \mathbf{A}\mathbf{s}(n) \tag{3.7}$$

and a *MISO extractor* filter defined by a vector **f** of size P, yielding a single output, $y(n) = \mathbf{f}^{\mathrm{H}}\mathbf{x}(n)$. This output can be written as a function of sources by introducing the *global filter* $\mathbf{g}^{\mathrm{H}} \stackrel{\text{def}}{=} \mathbf{f}^{\mathrm{H}}\mathbf{A}$:

$$y(n) = \mathbf{g}^{\mathrm{H}}\mathbf{s}(n).$$

The extractor does its job correctly if the global filter **g** yields an output that is proportional to one of the sources, that is, if: $\exists i : y(n) = \alpha s_i(n)$. The goal of this section is to explain how to find such a vector **g**, or the associated vector **f**.

Define the set $\mathcal{G}$ of filters that we utilize. We take it here to be the set of vectors of unit L^2 norm without restricting the generality. Define the subset $\mathcal{T}$ of $\mathcal{G}$ of *trivial filters* containing the N-dimensional vectors having exactly one non-zero entry. Next, denote $\mathcal{S}$ the linear space of random vectors of dimension N, with mutually statistically independent components, having finite cumulants of order $R = p + q$, among which at most one marginal cumulant is null. Last, denote $\mathcal{G} \cdot \mathcal{S}$ the set obtained by letting $\mathcal{G}$ act on $\mathcal{S}$.

DEFINITION 3.1
For every $y \in \mathcal{G} \cdot \mathcal{S}$ *and every integer triplet* (p, q, d), *let the optimization criterion:*

$$\Upsilon^d_{p,q}[y] \stackrel{\text{def}}{=} |\text{cum}_{p,q}\{y(n)\}|^d. \tag{3.8}$$

This criterion is an implicit function of **s** and **g**. In fact, according to the multi-linearity property of cumulants (see [33,4] or section 9.3.2.3 for further details), and because of the independence between sources, s_i, one gets:

$$\text{cum}_{p,q}\left\{\sum_i g_i^* s_i(n)\right\} = \sum_i g_i^{*p} g_i^q \,\text{cum}_{p,q}\{s_i(n)\}. \tag{3.9}$$

In practice, neither $\mathbf{s}(n)$ nor even generally $\text{cum}_{p,q}\{s_i(n)\}$ are known. Relation (3.9), even if it is very useful to derive the properties that we shall prove shortly, is totally unuseful to devise a numerical algorithm to extract sources. On the other hand, thanks to the multi-linearity of cumulants:

$$\text{cum}_{2,2}\left\{\sum_i f_i^* x_i(n)\right\} = \sum_{i,j,k\ell} f_i f_j f_k^* f_\ell^* \,\text{cum}\{x_i(n), x_j(n), x_k(n)^*, x_\ell(n)^*\}. \quad (3.10)$$

It is now clear that $\Upsilon^d_{2,2}[y]$ is a function of the extracting filter $\mathbf{f}$ and observation cumulants, which can be estimated. More generally, (3.8) may be seen as a polynomial of degree $(p+q)d$ in the P variables f_k and their complex conjugates. But before going further, it is necessary to check out whether this criterion is acceptable.

PROPOSITION 3.2

Let $\mathbf{s}$ *be a random vector with independent components, such that at least one of its marginal cumulants* $\text{cum}_{p,q}\{s_i\}$ *is non-zero. If* $p+q>2$ *and* $d \geq 1$, *then criterion* $\Upsilon^d_{p,q}$ *defined in* (3.8) *satisfies the three properties below:*

P1. *Invariance:*

$$\forall \mathbf{g} \in \mathcal{T}, \exists i : \Upsilon^d_{p,q}[\mathbf{g}^{\mathrm{H}}\mathbf{s}] = \Upsilon^d_{p,q}[s_i]. \quad (3.11)$$

P2. *Domination:*

$$\forall \mathbf{g} \in \mathcal{G}, \exists \mathbf{t} \in \mathcal{T}, \Upsilon^d_{p,q}[\mathbf{g}^{\mathrm{H}}\mathbf{s}] \leq \Upsilon^d_{p,q}[\mathbf{t}^{\mathrm{H}}\mathbf{s}]. \quad (3.12)$$

P3. *Discrimination:*

$$\text{If } \exists \mathbf{g} \in \mathcal{G} \textit{ such that}: \Upsilon^d_{p,q}[\mathbf{g}^{\mathrm{H}}\mathbf{s}] = \operatorname*{Max}_{\mathbf{t}\in\mathcal{T}} \Upsilon^d_{p,q}[\mathbf{t}^{\mathrm{H}}\mathbf{s}], \textit{ then } \mathbf{g} \in \mathcal{T}. \quad (3.13)$$

Property (3.11) tells that all solutions that deduce from each other by trivial filtering are equally acceptable; this property generates the whole equivalence class of solutions from a single one. Equation (3.12) ensures that mixing the sources decreases the criterion. Lastly, (3.13) guarantees that the global maximum is not reached at spurious points.

Actually, as further shown in the proof, the strict increase and the convexity of function $\phi(u) = u^d$ on $\mathbb{R}^+$, for $d \geq 1$, are fundamental in Proposition 3.2. In other words, the above properties remain true for criteria of the form:

$$\Upsilon^\phi_{p,q}[y] \overset{\text{def}}{=} \phi(|\text{cum}_{p,q}\{y(n)\}|) \quad (3.14)$$

where function ϕ is positive convex and strictly increasing.

Proof: If $\mathbf{g} \in \mathcal{T}$, then there exists a single non-zero component i in vector $\mathbf{g}$, so that $y(n) = \alpha s_i(n)$, for some α of unit modulus. Yet, thanks to the multi-linearity property (3.9), $\text{cum}_{p,q}\{y\} = |\alpha|^{(p+q)}|\text{cum}_{p,q}\{s_i\}|$. Property **P1** is readily obtained because α is of unit modulus.

Now take $\mathbf{g} \in \mathcal{G}$. Note that from (3.9), we have

$$|\text{cum}_{p,q}\{y\}| = \left|\sum_i g_i^{*p} g_i^q \, \text{cum}_{p,q}\{s_i\}\right| \leq \sum_i |g_i|^{p+q} |\text{cum}_{p,q}\{s_i\}|.$$

Yet for any $\mathbf{g} \in \mathcal{G}$, $||\mathbf{g}||^2 = 1$ (for the L^2 norm); hence $|g_i|^{p+q} \leq |g_i|^2, \forall i$, since $p+q > 2$. The sum is bounded by a convex linear combination, itself bounded by its largest term; thus $|\text{cum}_{p,q}\{y\}| \leq \text{Max}_i|\text{cum}_{p,q}\{s_i\}|$. Since ϕ is convex increasing, these inequalities can be transposed to $\phi(|\text{cum}_{p,q}\{y(n)\}|) \leq \phi(\text{Max}_i|\text{cum}_{p,q}\{s_i\}|)$, which proves **P2**.

Now suppose we have equality $\phi(|\text{cum}_{p,q}\{y(n)\}|) = \phi(\text{Max}_i|\text{cum}_{p,q}\{s_i\}|)$. Then all the previous inequalities are equalities. In particular, vector $\mathbf{g}$ is such that its norms L^2 and L^{p+q} are equal, whereas $p+q > 2$. This is possible only if a single entry of $\mathbf{g}$ is non-zero. Consequently $\mathbf{g} \in \mathcal{T}$, which proves **P3**. □

Thanks to the multi-linearity property of cumulants (3.10), criterion $\Upsilon^d_{p,q}[y]$ can be written as a polynomial function in components f_i and in their conjugates f_i^*; cumulants of order $p+q$ of variables x_k also enter linearly in this function. It is thus legitimate to denote this criterion:

$$\Upsilon^d_{p,q}[y] = \Upsilon^d_{p,q}(\mathbf{f}; p_{\mathbf{x}})$$

where $p_{\mathbf{x}}$ denotes the joint probability distribution function (pdf) of $\mathbf{x}$. Note that it is not necessary that this pdf be absolutely continuous; it is sufficient that the requested cumulants are finite (this can in particular apply to discrete variables with finite alphabets).

DEFINITION 3.2
An optimization criterion $\Upsilon(\mathbf{f}; p_{\mathbf{x}})$, depending only on $\mathbf{f}$ and on the pdf of the observation $\mathbf{x}(n)$, is called contrast if it satisfies the properties P1, P2 and P3 defined in Proposition 3.2.

Other examples of contrasts close to the one in (3.8) but considering so-called reference signals will be presented in section 3.3.3.3, page 77.

3.3.2 Deflation principle

When searching for the global maximum (or one of the global maxima if they are several) of $\Upsilon(\mathbf{f}; p_{\mathbf{x}})$, one obtains a filter allowing extraction of one of the sources s_i. Suppose, to simplify the notation and without restricting the generality, that $i = 1$. That is, assume the source extracted first is source number 1.

The *deflation* procedure consists of subtracting from observation $\mathbf{x}$ the contribution of the extracted source, $s^{[1]}$, by linear regression. A new "observation vector", $\mathbf{z}^{[1]}$, is formed, which is not correlated with $s^{[1]}$:

$$\mathbf{z}^{[1]} \stackrel{\text{def}}{=} \mathbf{x} - \mathbb{E}\{\mathbf{x}s^{[1]*}\}\mathbb{E}\{s^{[1]}s^{[1]*}\}^{-1}s^{[1]}.$$

The rank of the covariance matrix of $\mathbf{z}^{[1]}$ is thus smaller by 1 than that of the covariance of $\mathbf{x}$. One can then diminish the size of vector $\mathbf{z}^{[1]}$ by projecting it onto the dominant left singular space of dimension $P - 1$:

$$\mathbf{x}^{[2]} \stackrel{\text{def}}{=} \mathbf{P}^{[1]}\mathbf{z}^{[1]}.$$

Once one source is extracted, and a new observation $\mathbf{x}^{[2]}$ (of smaller dimension) is built, one can proceed similarly to the extraction of another source. In this manner, a series of extractors are built, $\mathbf{g}^{[1]}, \ldots, \mathbf{g}^{[N-1]}$, which extract the $N - 1$ first sources. The deflation procedure stops when the dimension of $\mathbf{x}^{[k]}$ reaches 1, and the estimate of $s^{[N]}$ is merely equal to $\mathbf{x}^{[N]}$.

In the absence of noise, there can be only $\min(N, P)$ iterations. In the presence of noise, the number of sources extracted can reach P, even if $P > N$; this means that some extracted sources may contain only noise. However, in practice, because the linear regression is not perfect, some sources may be extracted several times. Improvements to cope with this phenomenon can be found in [7].

In all cases, the number of deflation steps executed cannot exceed P, unless large numerical errors occur (e.g. for matrices of size 200 or larger), so that generally $N = P$. It is this type of procedure that was originally implemented in [23]. If the number of sources N is larger than P, the mixture is said to be *under-determined*, and the approach by contrast maximization described in this chapter does not apply (cf. Chapter 9). In fact, this approach is based on the search for a linear left inverse filter. But under-determined mixtures do not admit left inverses. Nevertheless, the contrast maximization procedure will allow extraction of P dominant sources; the estimates will be corrupted by the other sources, playing the role of noise.

3.3.3 MISO contrasts for convolutive mixtures

In this section, one now assumes the following dynamical model:

$$\mathbf{x}(n) = \sum_{k=0}^{L} \mathbf{A}(k)\mathbf{s}(n-k)$$

where $\mathbf{s}(n)$ is a process of dimension N, whose N components $s_i(n)$ are mutually statistically independent. matrices $\mathbf{A}(n)$ defining the impulse response of the mixture are each of size $P \times N$, with $P \geq N$.

In addition, one assumes that the Finite Impulse Response (FIR) filter $\{\mathbf{A}(n)\}$ admits an inverse, itself also FIR. In other words, after z-transform:

$$\exists \check{\mathbf{B}}(z) : \check{\mathbf{B}}(z)\check{\mathbf{A}}(z) = \mathbf{I}, \ \forall z \tag{3.15}$$

where $\check{\mathbf{A}}(z)$ and $\check{\mathbf{B}}(z)$ are both polynomial matrices. This tells us that for every z, the rank of $\check{\mathbf{A}}(z)$ remains full. In particular, it is necessary that $\check{\mathbf{A}}(z)$ contains only polynomials that are prime to each other. In fact for every column $\mathbf{a}(z)$ of $\check{\mathbf{A}}(z)$, the existence of a polynomial row $\mathbf{b}(z)$ such that $\sum_i b_i(z)a_i(z) = 1$ is then guaranteed by the Bézout identity.

Again, the deflation appraoch consists of extracting one source at a time, and hence of defining a filter $\check{\mathbf{b}}(z)$ whose output $y(n)$ maximizes some contrast criterion $\Upsilon(\mathbf{b}; p_{\mathbf{x}})$. As in the static case, one introduces the *global filter* $\check{\mathbf{g}}(z) \stackrel{\text{def}}{=} \check{\mathbf{b}}(z)\check{\mathbf{A}}(z)$. In other words, we have:

$$y(n) \stackrel{\text{def}}{=} \sum_k \mathbf{b}(n-k)\mathbf{x}(k) \stackrel{\text{def}}{=} \sum_\ell \mathbf{g}(n-\ell)\mathbf{s}(\ell). \tag{3.16}$$

However, another complication appears because of the dynamic character of the mixture. The approach is different depending on whether sources are white in the strong sense (iid), in which case they are *linear processes*, or not. These two cases will be consequently distinguished.

3.3.3.1 iid sources

Suppose first that sources are white in the strong sense, that is, each $s_i(n)$ is an iid sequence for every fixed i, $1 \le i \le N$. Trivial filters, preserving this whiteness and source independence, necessarily take the form:

$$\mathcal{T} : \mathbf{t}(n) = \alpha\delta(n - n_o)[0, \ldots, 0, 1, 0, \ldots, 0]^{\mathrm{T}}.$$

In other words, the outputs one should expect are the sources delayed by an integer multiple n_o of the sampling period, possibly permuted, and affected by a scalar factor α (non-zero complex scalar).

Several authors have proposed contrast functions in the context of dynamic mixtures of iid sources [18,53,41,54,52], some being dedicated to the separation (joint extraction) of sources (cf. section 3.5). A family of criteria are described here (section 3.3) that allow extraction of sources by deflation, and it will be shown that they are contrast functions.

LEMMA 3.3
The cumulant of the equalizer output $\check{\mathbf{b}}(z)$ *admits the following bound:*

$$|\mathrm{cum}_{p,q}\{y(n)\}| \le \max_i |\mathrm{cum}_{p,q}\{s_i(n)\}| \cdot \|\mathbf{g}\|_2^{p+q} \tag{3.17}$$

if $p + q \ge 2$ *and where* $\|\mathbf{g}\|_r \stackrel{\text{def}}{=} \left[\sum_{i,k} |g_i(k)|^r\right]^{1/r}$.

Proof: By multi-linearity of cumulants and thanks to the statistical independence of $s_i(n)$, we have

$$|\text{cum}_{p,q}\{y\}| \leq \sum_i |\text{cum}_{p,q}\{s\}| \sum_k |g_i(k)|^{p+q}.$$

Let $C_{s,max} = \max_i |\text{cum}_{p,q}\{s_i\}|$ to simplify the notation. Then $|\text{cum}_{p,q}\{y\}| \leq C_{s,max} \sum_{i,k} |g_i(k)|^{p+q}$. On the other hand, using inequalities between norms L^p, we have $||\mathbf{g}||_{p+q} \leq ||\mathbf{g}||_2$ because $p+q \geq 2$. Hence:

$$|\text{cum}_{p,q}\{y\}| \leq C_{s,max} \left[\sum_{i,k} |g_i(k)|^2 \right]^{(p+q)/2}$$

□

This is the expected result.

Remark. If $||\mathbf{g}||_2 = 1$, there exists a shorter proof, similar to the one given on page 71 in the static case. One writes that $|g_i(k)|^{p+q} \leq |g_i(k)|^2$, and that $\alpha_i \stackrel{\text{def}}{=} \sum_k |g_i(k)|^2$ form a convex linear combination. Then $|\text{cum}_{p,q}\{y\}| \leq \sum_i \alpha_i |\text{cum}_{p,q}\{s\}|$ is thus bounded by $C_{s,max}$.

Denote $\mathscr{S}$ the set of iid processes of dimension N, having mutually statistically independent components and finite non-zero cumulants of order $p+q$. Then, we have the following propositions.

PROPOSITION 3.4
If $p+q > 2$, criterion

$$\Upsilon^1_{p,q}[y] \stackrel{\text{def}}{=} |\text{cum}_{p,q}\{y(n)\}|$$

is a contrast on $\mathscr{H}.\mathscr{S}$, where $\mathscr{H}$ denotes filters with P inputs and one output, such that $\sum_{i,k} |g_i(k)|^2 = 1$.

Proof: The domination is a direct consequence of Lemma 3.3. If equality holds, then by inspecting the proof of the lemma we see that we necessarily have $||\mathbf{g}||_{p+q} = ||\mathbf{g}||_2$, which is possible (for $p+q > 2$) only if $\mathbf{g}$ contains a single non-zero entry. Hence $\mathbf{g}$ is a trivial filter, which proves the discrimination property **P3**. □

PROPOSITION 3.5
If $p+q>2$, criterion

$$\Upsilon^1_{p,q}[y] \stackrel{\text{def}}{=} \frac{|\text{cum}_{p,q}\{y(n)\}|}{\text{cum}_{1,1}\{y(n)\}^{(p+q)/2}}$$

is a contrast on $\mathcal{H} \times \mathcal{S}_1$, where $\mathcal{H}$ filters with P inputs and one output, finite norm, and $\mathcal{S}_1$ the subset of $\mathcal{S}$ of unit variance processes.

Proof: First remark that $\text{cum}_{1,1}\{y(n)\} = \sum_{i,k} |g_i(k)|^2$. Applying Lemma 3.3 gives us directly **P2**. In case of equality in the proof of the lemma, $||\mathbf{g}||_{p+q} = ||\mathbf{g}||_2$, which again proves that $\mathbf{g}$ is trivial. □

Criteria defined in Propositions 3.4 and 3.5 are then equivalent: fixing the source variances to 1, or the norm of the global filter to 1, amounts to the same thing, because it is part of the same indeterminacy, inherent in the problem.

PROPOSITION 3.6
Criterion $\Upsilon^{(d)}_{p,q} \stackrel{\text{def}}{=} |\text{cum}_{p,q}\{y(n)\}|^d$ is a contrast if $d \geq 1$.

Proof: This proposition is a direct consequence of the general result stated in Proposition 3.27 page 102, since function $\phi(u) = u^d$ is convex and strictly increasing if $d \geq 1$. □

Existence of the ideal filter. When the mixture has a Finite Impulse response (FIR), it is legitimate to ask oneself whether there always exists a FIR equalizer allowing exact extraction of every source in the absence of noise.

If $\mathbf{A}(z)$ is of full rank for every z, including for $z=\infty$, then there exist FIR equalizers. If additionally $\mathbf{A}(z)$ is of *reduced columns* (cf. for instance [24] or [31] for a definition), one can bound the degree of its inverse, which allows one to choose a sufficient equalizer length. For example, from [52, page 144], it suffices to choose an equalizer of length $NL-1$, if L denotes the length of the FIR mixture. The reader is invited to consult Chapter 8 for further information on this matter.

A simple case is encountered when the number of sources equals the number of sensors. In this case, matrix $\mathbf{A}(z)$ is square. If $\mathbf{A}(z)$ remains of full rank for all $z \neq 0$, it means that its determinant never vanishes, even at $z=\infty$. Thus matrix $\mathbf{A}(z)$, which is a polynomial in z^{-1}, is *unimodular*, i.e. its determinant is constant [31], and admits a polynomial inverse. Actually, this is rather obvious since the inverse is equal up to a constant factor to the transpose of the matrix of co-factors of $\mathbf{A}(z)$, and indeed a polynomial matrix.

3.3.3.2 Non iid sources

When sources are not iid, nothing changes if they are *linear processes*, that is, if they are obtained by linear filtering of stationary iid sequences. In fact, this scalar linear filtering is anyway part of the indeterminacy inherent in the problem, in the sense that it does not affect the statistical independence between sources. It suffices to extend the definition of trivial filters to:

DEFINITION 3.3

$$\mathcal{T} : \mathrm{t}(n) = \alpha(n)[0,\ldots,0,1,0,\ldots,0]^{\mathrm{T}}$$

where $\alpha(n)$ *is any SISO filter of finite* L^2 *norm.*

On the other hand, the problem is different if sources are nonlinear processes [49,48,8]. For instance, this case is encountered in transmissions via continuous phase modulation (CPM) [7], among others.

Some notations are useful. Denote $\rho_{s_i}(m)$ the auto-correlation of source s_i:

$$\rho_{s_i}(m) \stackrel{\text{def}}{=} \mathbb{E}\{s_i(n)s_i(n+m)^*\}.$$

Define the L^2 norm of a scalar filter $h(n)$ in the metric of ρ_{s_i} by:

$$||h||_i^2 \stackrel{\text{def}}{=} \sum_{k,\ell} h(k)h(\ell)^* \rho_{s_i}(\ell - k).$$

In particular, we have $\mathbb{E}\{|y(n)|^2\} = \sum_i ||g_i||_i^2$.

PROPOSITION 3.7

With these notations, for any pair of positive integers (p,q) *such that* $p+q>2$, *and for any convex real function* ϕ *null at the origin, the criterion*

$$\Upsilon_{p,q}^{\phi}(\mathbf{g};\mathbf{x}) \stackrel{\text{def}}{=} \phi(\mathrm{cum}_{p,q}\{y(n)\}) \tag{3.18}$$

is a contrast on the set $\mathcal{H}$ *of filters, of impulse response* $\mathbf{g}(n)$ *satisfying*

$$\sum_{i=1}^{P} ||g_i||_i^2 = 1 \tag{3.19}$$

and for sources having finite non-zero cumulants of order $p+q$.

Proof: Refer to Chapter 8, pages 301 and following for the proof of this proposition. □

3.3.3.3 Convolutive MISO contrasts with references

Even if iterative methods based on deflation procedures simplify the optimization problem in some respect, optimizing contrast function is not a simple problem. In all the above contrast examples, the most important factor is the order of cumulant. Indeed it implies that the parameters of the separating system within the criteria are found to be at a power at least equal to the corresponding cumulant order. In practice, this power is at least 4. One recent idea consists in reducing this high power dependence while keeping the advantages of cumulants. This has led to so-called referenced contrast functions which use auxiliary signals called reference signals through cross-cumulants. These reference signals have a double advantage. Not only do they allow the reduction of the parameters' power into the criterion but they also yield a supplementary degree of freedom. In particular one can obtain quadratic optimization algorithms whose solutions are rather classical.

A first interesting approach of this kind can be found in [10]. The reference signals, denoted $z_i(n), i \in \{1,\ldots,R-2\}$, are considered through the use of the following cross-cumulant

$$\kappa_{R,z}\{y(n)\} = \mathrm{Cum}\{y^{(*)_1}(n), y^{(*)_2}(n), z_1(n), \ldots, z_{R-2}(n)\} \tag{3.20}$$

where $(\cdot)^{(*)_i}$, $i = 1,2$ denote optional complex conjugations.

Ideally [1,10], these reference signals should correspond to a given source signal. Obviously, this ideal case cannot be met since the source signals are what we are looking for.

We introduce the following function

$$\Upsilon_{R,z}\{y(n)\} = |\kappa_{R,z}\{y(n)\}|, \quad R \geq 3. \tag{3.21}$$

The following proposition was shown in [10]:

PROPOSITION 3.8
In case of iid source signals, if there exists a couple of integers (j_0, k_0) such that

$$\max_{j=1}^{N} \sup_{k\in\mathbb{Z}} |\kappa_{R,z}\{s_j(n-k)\}| = |\kappa_{R,z}\{s_{j_0}(n-k_0)\}| < +\infty. \tag{3.22}$$

then the function $\Upsilon_{R,z}\{y(n)\}$ is a contrast if and only if the set

$$\mathcal{I} = \{(j,k) \in \{1,\ldots,N\} \times \mathbb{Z} \mid |\kappa_{R,z}\{s_j(n-k)\}| = |\kappa_{R,z}\{s_{j_0}(n-k_0)\}|\} \tag{3.23}$$

has a unique element.

It has to be noticed that in using complementary assumptions, the above function $\Upsilon_{R,z}\{y(n)\}$ is also a contrast for non iid but stationary source signals.

The validity of this contrast depends clearly on the reference signals. Mainly, these reference signals have to depend on source signals linearly together with additional constraints. In practice, the reference signals can simply be the observations. Moreover if the optimization procedure of the contrast is iterative, then one can use the current estimation of one extracted source as reference signal. Hence, there exist a lot of different possibilities.

From another point of view, it was recently shown in [9] that reference signals can be built from the source signals in using nonlinear transformations. In particular, this can be useful when *a priori* knowledge is available from source signals.

The algorithm developed for the optimization of the contrast in (3.21) is quadratic. This yields a very simple algorithm. However, it requires an auxiliary iterative procedure to constrain the considered reference signal to be well adapted. To overcome this last point, a second interesting approach can be found in [27]. It corresponds to the development of a cubic algorithm which requires considerably fewer contraints on the reference signal.

3.4 MIMO CONTRASTS FOR STATIC MIXTURES

3.4.1 Introduction

We have seen in section 3.3 how to build a contrast criterion for MISO blind equalizers, with the goal of extracting one of the inputs of the mixture. In this section, N sources are aimed to be simultaneously separated from P sensors. Since the blind equalizer has now several outputs, it is necessary to change the definitions introduced in section 3.3. In fact, an additional constraint must be imposed on MIMO blind equalizers: they should not yield outputs linked to the same source. Every source should ideally be associated injectively to a single output via a scalar filter. More precisely, we required the (scalar) output of the MISO blind equalizer to be of unit variance; we shall require now the (multivariate) output of the MIMO blind equalizer to have a covariance matrix equal to the identity.

It will be assumed in this section that $P = N$. Note that the case of *over-determined* systems, i.e. $P > N$, can generally be addressed under this assumption since it is the limiting case where some sources have a zero variance. This is why over-determined systems are often considered to satisfy $P \geqslant N$. Moreover, we shall restrict our attention to static systems (the dynamic case will be addressed in section 3.5).

DEFINITION 3.4

Let $\mathscr{H}$ be a set of static filters containing the identity $\mathbf{I}$, *$\mathscr{S}$ a set of source random variables, $\mathscr{H}\cdot\mathscr{S}$ the set of random variables obtained by the action of $\mathscr{H}$ on $\mathscr{S}$, and Υ a mapping from $\mathscr{H}\times\mathscr{H}\cdot\mathscr{S}$ to $\mathbb{R}$. Also denote by $\mathscr{T}$ the set of trivial filters of $\mathscr{H}$, which leave criterion Υ unchanged:*

P1. *Invariance:*

$$\forall \mathbf{x}\in\mathscr{H}\cdot\mathscr{S},\ \forall \mathbf{T}\in\mathscr{T},\ \Upsilon(\mathbf{T};\mathbf{x})=\Upsilon(\mathbf{I};\mathbf{x}). \tag{3.24}$$

Recall that trivial filters allow generation of the whole equivalence class of solutions from a single one.

DEFINITION 3.5
A mapping $\Upsilon(\mathbf{B};\mathbf{x})$ *from* $\mathscr{H} \times \mathscr{H} \cdot \mathscr{S}$ *onto* $\mathbb{R}$ *is a contrast if it depends solely on the pdf of* $\mathbf{x}$ *and if it satisfies, in addition to Property* **P1**, *the two properties below.*

P2. *Domination:*

$$\Upsilon(\mathbf{H};\mathbf{s}) \le \Upsilon(\mathbf{I};\mathbf{s}), \ \forall \mathbf{s} \in \mathscr{S}, \ \forall \mathbf{H} \in \mathscr{H}. \tag{3.25}$$

P3. *Discrimination:*

$$\forall s \in \mathscr{S}, \text{ if } \mathbf{H} \in \mathscr{H} \text{ satisfies } \Upsilon(\mathbf{H};\mathbf{s}) = \Upsilon(\mathbf{I};\mathbf{s}), \text{ then } \mathbf{H} \in \mathscr{T}. \tag{3.26}$$

Property (3.25) guarantees that mixing independent sources decreases the contrast. Consequently, the contrast is maximum when sources are separated. Property (3.26) avoids the existence of spurious solutions, that is, the existence of global maxima other than trivial filters. It may be seen as the reciprocal of the invariance condition (3.24). Note the differences with definitions given in the MISO case (page 70), especially concerning Property **P2**. The difference stems from the fact that one imposes the MIMO *global filter* to be bijective.

3.4.2 Contrasts for MIMO static mixtures

Consider again the output equations (3.7) and (3.15); the observation model is:

$$\mathbf{y}(n) = \mathbf{B}\mathbf{x}(n) = \underbrace{\mathbf{B}\mathbf{A}}_{\mathbf{G}}\,\mathbf{s}(n) \tag{3.27}$$

where matrices $\mathbf{B}$ and $\mathbf{A}$ are of size $N \times P$ and $P \times N$, respectively, with $P \ge N$, and where $\mathbf{G}$ denotes the *global filter* as before. The optimization criterion that we are looking for, $\Upsilon[\mathbf{y}]$, depends only on the pdf of observation $\mathbf{x}(n)$ and on the separator filter, $\mathbf{B}$, so that one can write it as $\Upsilon[\mathbf{y}] = \Upsilon(\mathbf{B}; p_{\mathbf{x}})$.

3.4.2.1 Trivial filters

In this section, in the static case in which we are interested, MIMO filters preserving the statistical independence of sources are necessarily of the form:

$$\mathbf{G} = \mathbf{\Lambda}\mathbf{P}, \tag{3.28}$$

where $\mathbf{\Lambda}$ is diagonal invertible and $\mathbf{P}$ is a permutation. The square $N \times N$ matrices have one, and only one, non-zero entry in each row and column.

3.4.2.2 Whitening at order two

Since we look for statistically independent outputs $y_i(n)$, it is natural to impose that they be uncorrelated at order 2 (even if it is known to be theoretically suboptimal). So we can apply first a transform $\mathbf{W}$ to observation vector $\mathbf{x}$ such that vector $\tilde{\mathbf{x}} = \mathbf{W}\mathbf{x}$ has a covariance matrix equal to identity.

Let $\mathbf{V_x}$ be the covariance of $\mathbf{x}$, which is a random variable of dimension P: $\mathbf{V_x} \stackrel{\text{def}}{=} \mathbb{E}\{\mathbf{x}\mathbf{x}^{\mathrm{H}}\}$. The rank of matrix $\mathbf{V_x}$ is equal to the number of sources, N. By a linear transform on $\mathbf{x}$, one can build a new variable $\tilde{\mathbf{x}}$ of dimension N with a covariance equal to identity. This operation is often called *spatial whitening* in sensor array processing, or *standardization* in statistics.

For this purpose, it suffices to consider the normalized Cholesky factorization (sometimes also referred to as the Crout factorization) of $\mathbf{V_x}$:

$$\mathbf{V_x} = \mathbf{L}\boldsymbol{\Delta}^2\mathbf{L}^{\mathrm{H}}$$

where $\mathbf{L}$ is lower triangular of dimensions $P \times N$ and having 1's on its diagonal, and $\boldsymbol{\Delta}$ is $N \times N$ diagonal with real strictly positive diagonal elements.

If $N \leq P$, there exists a full rank matrix $\mathbf{L}^{\dagger}$ of dimension $N \times P$ such that $\mathbf{L}^{\dagger}\mathbf{L} = \mathbf{I}$ ($\mathbf{L}^{\dagger}$ denoting a pseudo-inverse of $\mathbf{L}$). Then, random variable $\tilde{\mathbf{x}}$ may be built thanks to the relation:

$$\tilde{\mathbf{x}} = \boldsymbol{\Delta}^{-1}\mathbf{L}^{\dagger}\mathbf{x} \stackrel{\text{def}}{=} \mathbf{W}\,\mathbf{x}. \tag{3.29}$$

In practice, if $N < P$, the Cholesky factorization may turn out to be a poor means to estimate the rank of $\mathbf{V_x}$. The Eigen-Value Decomposition (EVD) of $\mathbf{V_x}$ should be preferred, and will yield a reliable estimate of the rank (see Chapter 1, page 14). However, given realizations $\mathbf{x}(n)$ of random variable $\mathbf{x}$, one can compute corresponding realizations $\tilde{\mathbf{x}}(n)$ of $\tilde{\mathbf{x}}$ without explicitly computing either $\mathbf{V_x}$ or its Cholesky factorization. It suffices to compute the Singular Value decomposition (SVD) of the data matrix $[\mathbf{x}(1), \ldots, \mathbf{x}(T)]$. Realizations $\tilde{\mathbf{x}}(n)$ are then given by the left singular vectors. This procedure has several advantages: it is more stable numerically, less computationally demanding, and more stable when $N < P$.

The whitening matrix $\mathbf{W}$ is not unique. We just saw that there exist several factorizations allowing us to compute it; but the matrices obtained by these factorizations will be *different*. Actually, there exist infinitely many ways to define the standardized vector $\tilde{\mathbf{x}}$. To see this, note that the covariance of $\mathbf{x}$ satisfies $\mathbf{W}\,\mathbf{V_x}\,\mathbf{W}^{\mathrm{H}} = \mathbf{I}$. This equality characterizes the spatial whitening, and still holds true if one multiplies the equation on the left by any unitary matrix $\mathbf{Q}$ and on the right by $\mathbf{Q}^{\mathrm{H}}$. In other words, for any unitary matrix $\mathbf{Q}$:

$$\mathbb{E}\{\mathbf{Q}\,\tilde{\mathbf{x}}\tilde{\mathbf{x}}^{\mathrm{H}}\,\mathbf{Q}^{\mathrm{H}}\} = \mathbf{Q}\,\mathbf{W}\,\mathbf{V_x}\,\mathbf{W}^{\mathrm{H}}\,\mathbf{Q}^{\mathrm{H}} = \mathbf{Q}\,\mathbf{I}\,\mathbf{Q}^{\mathrm{H}} = \mathbf{I}. \tag{3.30}$$

This result is by no means surprising, since second-order decorrelation does not yield statistical independence, except in the Gaussian case. If $\mathbf{s}$ is not Gaussian, vector $\mathbf{Q}^{\mathrm{H}}\mathbf{s}$ will have independent components only if $\mathbf{Q}^{\mathrm{H}}$ is trivial (that is, the product of a permutation and a regular diagonal matrix). The goal of the next section is precisely to determine optimization criteria allowing one to calculate matrix $\mathbf{Q}$ when $\mathbf{s}$ is not Gaussian.

Remark. In sensor array processing, it is very standard to assume, possibly after some pre-processing, that the covariance of the noise is proportional to the Identity, and that

$N \leqslant P - 1$. In the latter case, it becomes possible to perform the spatial whitening with the help of a denoised version of covariance $\mathbf{V}_\mathbf{x}$. The performances will be obviously much better; see Chapter 1, page 14.

3.4.2.3 Mutual information

Random variables $\{z_i\}$ are mutually statistically independent if and only if their joint probability distribution, $p_\mathbf{z}$, can be written as the product of the marginal ones, p_{z_i}. An initial idea is thus to measure a divergence between expressions $p_\mathbf{z}$ and $\prod_i p_{z_i}$ [12,13,15,46].

Assuming the Kullback divergence, one obtains the Mutual Information (MI) as a measure of statistical independence:

$$I(p_\mathbf{z}) \stackrel{\text{def}}{=} \int p_\mathbf{z}(\mathbf{u}) \frac{p_\mathbf{z}(\mathbf{u})}{\prod_i p_{z_i}(u_i)} d\mathbf{u}. \tag{3.31}$$

PROPOSITION 3.9

Criterion $\Upsilon^{IM}[\mathbf{y}] = \Upsilon^{IM}(\mathbf{B};\mathbf{x}) \stackrel{\text{def}}{=} -I(p_\mathbf{x})$ *is a contrast on* $\mathscr{H} \times \mathscr{H} \cdot \mathscr{S}$, *where* $\mathscr{H}$ *denotes the group of* $N \times N$ *invertible matrices, and* $\mathscr{S}$ *the set of finite-variance random variables, with independent components and having at most one Gaussian component.*

Proof: Condition **P1** is obviously satisfied because the definition of the MI itself does not depend on the way variables are sorted. In addition, MI is always invariant by scale change; this can be easily checked out by performing the variable change $(\mathbf{z},\mathbf{u}) \to (\Lambda\mathbf{z}, \Lambda\mathbf{u})$ in its definition.

The Kullback divergence between two probability distribution functions, $K(p_\mathbf{y}; p_\mathbf{z})$ is always positive, and vanishes if and only if $p_\mathbf{z}$ is equal to $p_\mathbf{y}$ almost everywhere (this can be easily proved by using the inequality $\log(u) \leq u - 1$). It can be deduced that $\Upsilon^{IM}(\mathbf{B};\mathbf{x}) \leq 0$ with equality if and only if $p_\mathbf{y}$ is equal to $\prod_i p_{y_i}$ almost everywhere, that is, if components of $\mathbf{y}$ are mutually statistically independent. Hence, vector $\mathbf{y}$ has independent components, but is also related to the sources by an invertible transform $\mathbf{y} = \mathbf{G}\mathbf{s}$. By invoking Theorem 11 of [15, p. 294], we know that matrix $\mathbf{G}$ is necessarily the product of a permutation and a diagonal matrix (this theorem is a direct consequence of the Darmois-Skitovic theorem stated in Chapter 9, page 330). We have proved **P2** and **P3**. □

Geometrical interpretation. Let $\mathbf{V}$ be the covariance matrix of $\mathbf{y} = \mathbf{B}\mathbf{x}$, and $g_\mathbf{y}$ the Gaussian pdf of covariance $\mathbf{V}$. Denote

$$I(g_\mathbf{y}) \stackrel{\text{def}}{=} \frac{1}{2} \log \frac{\prod_i V_{ii}}{\det \mathbf{V}}.$$

The latter represents the Mutual Information of zero-mean Gaussian random variables having the same covariance as $\mathbf{y}$. Define the *negentropy* of $\mathbf{y}$ as the gap between $p_{\mathbf{y}}$ and $g_{\mathbf{y}}$, in the sense of the Kullback divergence:

$$J(p_{\mathbf{y}}) \stackrel{\text{def}}{=} K(p_{\mathbf{y}}; g\mathbf{y}) = \int p_{\mathbf{y}}(\mathbf{u}) \log \frac{p_{\mathbf{x}}(\mathbf{u})}{g_{\mathbf{x}}(\mathbf{u})} d\mathbf{u}. \tag{3.32}$$

Then we have the following decomposition [13,15]:

PROPOSITION 3.10

$$I(p_{\mathbf{y}}) = I(g_{\mathbf{y}}) + J(p_{\mathbf{y}}) - \sum_i J(p_{y_i}). \tag{3.33}$$

Proof: First note that

$$\int (g_{\mathbf{y}}(\mathbf{u}) \log g_{\mathbf{y}}(\mathbf{u}) - p_{\mathbf{y}}(\mathbf{u}) \log g_{\mathbf{y}}(\mathbf{u}))\, d\mathbf{u} = 0 \tag{3.34}$$

since $\log g_{\mathbf{y}}$ only involves second order moments of $\mathbf{y}$, which are precisely the same for both distributions. Then, the same is true for marginal distributions, so that: $\int p_{y_i} \log g_{y_i} - \int g_{y_i} \log g_{y_i} = 0$. But in the last integral, only variable y_i appears. So we do not change anything by integrating with respect to all variables u_i, which leads to:

$$\sum_i \int p_{y_i}(u_i) \log g_{y_i}(u_i)\, du_i - \sum_i \int g_{\mathbf{y}}(\mathbf{u}) \log g_{y_i}(u_i) d\mathbf{u} = 0. \tag{3.35}$$

Now add these two (null) terms to the definition of the MI of $\mathbf{y}$. By changing the order of the terms, we get:

$$\begin{aligned} I(p_{\mathbf{y}}) = & \int \left(g_{\mathbf{y}} \log g_{\mathbf{y}} - \sum_i g_{\mathbf{y}} \log g_{y_i} \right) \\ & + \int (p_{\mathbf{y}} \log p_{\mathbf{y}} - p_{\mathbf{y}} \log g_{\mathbf{y}}) \\ & - \sum_i \int p_{\mathbf{y}} \log p_{y_i} + \sum_i \int p_{y_i} \log g_{y_i}. \end{aligned}$$

Lastly, by integrating on variables u_j, $j \neq i$, it is not difficult to see that the last term $\int p_{\mathbf{y}} \log p_{y_i}\, d\mathbf{u}$ can be rewritten $\int p_{y_i} \log p_{y_i}\, du_i$. And this is the expected result. □

Proposition 3.10 reveals the terms actually entering in the statistical dependence between components of $\mathbf{y}$. The first, $I(g_{\mathbf{y}})$, contains contributions of order 2, which can be canceled by standardization (spatial whitening) as already seen earlier in this chapter. The two other terms are sensitive to higher order cumulants. However, it can be shown that the joint negentropy, $J(p_{\mathbf{y}})$, is invariant by invertible transforms [22,15]. If the prior standardization is exact, then the only term that will be sensitive to unitary transforms will be the third one.

3.4.2.4 Sub-optimality of spatial whitening

The criterion to be maximized, $I(p_{\mathbf{y}}) = \Upsilon^{IM}(\mathbf{W}, \mathbf{Q})$, depends on two parameters: the whitening matrix $\mathbf{W}$ and the unitary separating matrix, $\mathbf{Q}$. By rewriting the three functions appearing in Proposition 3.10, and with obvious notations, one gets:

$$\Upsilon^{IM}(\mathbf{W},\mathbf{Q}) = I_g(\mathbf{W}) + J_p - \sum_i J_i(\mathbf{W},\mathbf{Q}).$$

The first approximation consists of using only I_g to estimate $\mathbf{W}$. The second, even more significant, consists of considering that $\mathbf{W}$ is fixed once and for all, and to take care only of the estimation of $\mathbf{Q}$ afterwards. Obviously, such a procedure is sub-optimal. Yet, it is widely used in practice, because of its simplicity. In fact, in many practical problems, there is little to gain by executing an exact (computationally heavy) optimization, jointly for the two variables $\mathbf{W}$ and $\mathbf{Q}$.

In practice, the maximization of Υ^{IM} is rather difficult to carry out [46], especially in rather large dimension. In fact, the estimation of probability densities is very costly, both in terms of computational time and number of realizations [17]. Fortunately, in the case of linear mixtures, much simpler solutions exist, and may be seen as approximations of the *mutual information* [12,13,34,15].

3.4.2.5 Contrasts based on standardized cumulants

In the Gaussian case, the minimization of the Mutual Information does not bring anything else but standardization. Higher order cumulants allow one to estimate the unitary part, $\mathbf{Q}$, explicitly or implicitly. By expanding the Mutual Information about the closest Gaussian distribution, the terms allowing identifiability of $\mathbf{Q}$ appear. More precisely, in the real standardized case, this expansion takes the form [13]:

$$I(p_{\mathbf{y}}) \approx J(p_{\mathbf{y}}) - \frac{1}{48}\sum_i 4\gamma_{iii}^2 + \gamma_{iiii}^2 + 7\gamma_{iii}^4 - 6\gamma_{iii}^2\gamma_{iii} \quad (3.36)$$

where γ_{iii} and γ_{iiii} denote marginal cumulants of variable y_i, of order 3 and 4 respectively; the approximation should be understood in the sense of the Central Limit Theorem [13,15].

In this expression, $J(p_y)$ is the negentropy, invariant under invertible transforms [15]. Minimizing the Mutual Information can thus be executed, at least in a first approximation, by maximizing marginal cumulants of order 3 or 4. This leads to the proposal of the optimization criterion below.

PROPOSITION 3.11
Given any pair of integers, (p,q) such that $p+q>2$, and for any integer $d \geq 1$, the criterion

$$\Upsilon^{(d)}_{p,q}(\mathbf{Q};\mathbf{x}) \stackrel{\text{def}}{=} \sum_i |\text{cum}_{p,q}\{y_i\}|^d \tag{3.37}$$

is a contrast from $\mathcal{H} \times \mathcal{H} \cdot \mathcal{S}$ to $\mathbb{R}$, where $\mathcal{H}$ denotes the group of unitary matrices, and $\mathcal{S}$ the set of random vectors with statistically independent components having at most one null marginal cumulant of order (p,q).

Instead of proving this proposition, we shall prove a more general result [41,45]:

PROPOSITION 3.12
Given any pair of integers, (p,q) such that $p+q>2$, and any convex strictly increasing function ϕ null at the origin, the criterion

$$\Upsilon^{\phi}_{p,q}(\mathbf{Q};\mathbf{x}) \stackrel{\text{def}}{=} \sum_i \phi(|\text{cum}_{p,q}\{y_i\}|) \tag{3.38}$$

is a contrast, under the same assumptions as Proposition 3.11.

It is useful to first prove the following lemma.

LEMMA 3.13
Let $\{\alpha_{ij}\}$ be a set of positive real numbers such that $\sum_i \alpha_{ij} \leq 1$ and $\sum_j \alpha_{ij} \leq 1$. Then if ϕ is convex and if $\phi(0)=0$, $\sum_i \phi\left(\sum_j \alpha_{ij} u_j\right) \leq \sum_j \phi(u_j)$.

Proof: If coefficients $\{a_j\}$ are such that $\sum_j a_j \leq 1$, it suffices to add a coefficient $a_0 = 1 - \sum_j a_j$ at zero in order to be able to apply Jensen's inequality. This gives us $\phi\left(\sum_j a_j u_j\right) \leq \sum_j a_j[\phi(u_j) - \phi(0)] + \phi(0)$ and, in particular, $\forall i$, and since $\phi(0)=0$, we have $\phi(\sum_j \alpha_{ij} u_j) \leq \sum_j \alpha_{ij}\phi(u_j)$. The result is then obtained by summing over i and by using the relation $\sum_i \alpha_{ij} \leq 1$. □

Let's now prove Proposition 3.12.

Proof: By hypothesis, $\mathbf{y} = \mathbf{G}\,\mathbf{s}$, where $\mathbf{G}$ is unitary. The multi-linearity of cumulants implies: $\text{cum}_{p,q}\{y_i\} = \sum_j G_{ij}^p G_{ij}^{*q} \text{cum}_{p,q}\{s_j\}$. Now use the triangle inequality of the modulus and the monotonicity of ϕ, and get: $\phi(|\text{cum}_{p,q}\{y_i\}|) \leq \phi(\sum_j |G_{ij}|^{p+q}|\text{cum}_{p,q}\{s_j\}|)$. But since $\mathbf{G}$ is unitary, $\sum_j |G_{ij}|^{p+q} \leq \sum_j |G_{ij}|^2 = 1$. We can thus apply Lemma 3.13 and obtain: $\sum_i \phi(|\text{cum}_{p,q}\{y_i\}|) \leq \sum_j \phi(|\text{cum}_{p,q}\{s_j\}|)$. We have just proved **P2**.

If the inequality is an equality, then by using the strict monotonicity of ϕ, one can deduce that $\sum_j |G_{ij}|^{p+q} u_j = \sum_j |G_{ij}|^2 u_j$. Hence $\sum_j (|G_{ij}|^2 - |G_{ij}|^{p+q}) u_j = 0$. Yet, every term in this sum is positive. Thus, for any j such that $u_j \neq 0$, $|G_{ij}|^{p+q} = |G_{ij}|^2$. Then for at least $N-1$ columns of $\mathbf{G}$, $\sum_i |G_{ij}|^{p+q} = \sum_i |G_{ij}|^2$. By inequality between L^r norms, and because $p+q > 2$, this is possible only if these $N-1$ columns of $\mathbf{G}$ each contain only a single non-zero element. But $\mathbf{G}$ is unitary. So its columns are orthogonal, and $\mathbf{G} = \mathbf{\Lambda}\,\mathbf{P}$, where $\mathbf{\Lambda}$ is diagonal with unit modulus entries, and $\mathbf{P}$ is a permutation. We have proved **P3**. □

The absolute value appearing in (3.37) prevents the contrast from being differentiable everywhere, which may raise problems if gradient algorithms are to be utilized. Fortunately, we have the following result [40]:

PROPOSITION 3.14
If $2p \geq 4$, if marginal source cumulants of order $2p$ are all of the same sign ϵ, and if at most one of them is null, then the criterion

$$\Upsilon^{\epsilon}_{p,p}(\mathbf{Q};\mathbf{x}) \stackrel{\text{def}}{=} \epsilon \sum_i \text{cum}_{p,p}\{y_i\}$$

is a contrast over the same sets as those of Proposition 3.11.

Proof: We shall show that $\Upsilon^{\epsilon}_{p,p} = \Upsilon^{(1)}_{p,p}$. By multi-linearity of cumulants, $\text{cum}_{p,p}\{y_i\} = \sum_k |G_{ik}|^p \text{cum}_{p,p}\{s_k\}$. Since source cumulants have the same sign ϵ, then $\epsilon\text{cum}_{p,p}\{y_i\} = \sum_k |G_{ik}|^p |\text{cum}_{p,p}\{s_k\}|$ is always positive. Then so is $\Upsilon^{\epsilon}_{p,p}$. □

Link with tensor diagonalization. Maximizing a contrast such as the one for Proposition 3.11 amounts to maximizing marginal cumulants of $\mathbf{y}$. But the Frobenius norm of the cumulant tensor of $\mathbf{y}$ is invariant under the action of unitary transform $\mathbf{Q}$. Consequently, maximizing $\Upsilon^{(2)}_{p,q}(\mathbf{Q};\mathbf{x})$ is equivalent to minimizing the sum of squared moduli of cross cumulants. In other words, we are attempting to put the cumulant tensor of $\mathbf{y}$ in a diagonal form, which is possible only if $\mathbf{y}$ is indeed a noiseless unitary mixture of independent sources.

From this point of view, the problem can be addressed in an approximate manner, by searching for a necessary (non-sufficient) condition of tensor diagonalization: the lower order slices of a diagonal tensor must themselves be diagonal. The goal of the next section is to exploit this property in order to define special-purpose contrast criteria.

3.4.2.6 Maximum likelihood

In the absence of noise, it is possible to relate *Likelihood* and *Mutual Information*. Denote $p_s = \prod_i p_{s_i}$ the source joint distribution, and p_x that of observations. The relation $\mathbf{x} = \mathbf{A}\,\mathbf{s}$ implies, by a simple change of coordinates, that the likelihood of a sample of T independent realizations, denoted $\mathbf{x}_T$, can be written as:

$$p_{\mathrm{x}|\mathbf{A}}(\mathbf{x}_T|\mathbf{A}) = \frac{1}{|\det \mathbf{A}|}\, p_s(\mathbf{A}^{-1}\mathbf{x}_T). \tag{3.39}$$

In the absence of noise, the separating filter is nothing else but $\mathbf{B} = \mathbf{A}^{-1}$. Denote $p_y(\cdot)$ the joint pdf of the outputs of the separating filter $\mathbf{B}$, i.e. $\mathbf{y} = \mathbf{Bx}$. Then we have $p_{\mathrm{x}|\mathbf{A}}(\mathbf{u}|\mathbf{A}) = p_{\mathrm{x}|\mathbf{B}}(\mathbf{u}|\mathbf{B}) = |\det \mathbf{B}|\, p_s(\mathbf{Bu})$. For an increasing number of independent observations, $\{\mathbf{x}_1, \ldots, \mathbf{x}_T\}$, the log-likelihood converges towards

$$\mathcal{L}_T(\mathbf{A}) \stackrel{\mathrm{def}}{=} \frac{1}{T} \log p(\mathbf{x}_1 \ldots \mathbf{x}_T|\mathbf{A}) \rightarrow \int p_x(\mathbf{u}) \log p_{\mathrm{x}|\mathbf{A}}(\mathbf{u}|\mathbf{A})\, d\mathbf{u}.$$

This is a familiar result. Now add and subtract $\log p_x(\mathbf{u})$ in order to be able to form *Kullback divergences*:

$$\mathcal{L}_\infty(\mathbf{A}) = -K\{p_x, p_{\mathrm{x}|\mathbf{A}}\} - S(p_x)$$

where $S(p_x) \stackrel{\mathrm{def}}{=} -\int p_x \log p_x$ denotes again the entropy of $\mathbf{x}$.

Then, because of the invariance of the Kullback divergence by invertible transforms, this mean likelihood can be written, up to the constant term $S(p_x)$ that does not depend on $\mathbf{B}$:

$$\Upsilon^{ML} \stackrel{\mathrm{def}}{=} -K\{p_y, p_s\}. \tag{3.40}$$

This equation is interesting. In fact, contrary to the Mutual Information, the maximization of the Likelihood minimizes the divergence between the distribution of the outputs $\mathbf{y}$ and that of sources $\mathbf{s}$. See Chapter 4 and references therein.

PROPOSITION 3.15

The mean log-likelihood can be decomposed into two parts:

$$K\{p_y, p_s\} = K\left\{p_z, \prod_i p_{z_i}\right\} + \sum_i K(z_i, s_i). \tag{3.41}$$

Proof: By definition, $K\{p_y, p_s\} = \int p_y \log p_y - \int p_y \log p_s$. Since $p_s = \prod_i p_{s_i}$, the second term rewrites $\sum_i \int p_{z_i} \log p_{s_i}$. Now add and subtract $\sum_i \int p_y \log p_{z_i}$, which yields $K\{p_y, p_s\} = \int p_y \log \frac{p_y}{\prod_i p_{z_i}} + \sum_i \int p_y \log \frac{p_{z_i}}{p_{s_i}}$. □

To conclude, the likelihood is the sum of the Mutual Information and the divergence between marginal distributions of sources and outputs. In the absence of knowledge of the source distributions, it is thus quite natural to resort to the Mutual Information. On the other hand, if the distributions of expected sources are accurately known, then the maximization of the likelihood should be preferred, since advantage will be taken of this additional knowledge.

3.4.3 Contrasts and joint diagonalization

In this section, it is assumed that observations have been spatially whitened at order 2; that is, it is assumed that **x** has been standardized according to (3.29).

A quite attractive class of contrasts, in the frame of static mixtures, concerns those that are equivalent to a criterion of joint approximate diagonalization of a set of matrices (or more generally tensors). These contrasts can lead to generally robust and efficient pairwise sweeping algorithms, reminiscent Jacobi-like algorithms developed for matrices.

Before talking about these contrasts, it is necessary to define the notion of *joint diagonalization*. The latter is usually considered via the criterion that we now introduce. Let $\mathbb{T}$ be a set of $T \geq 1$ cubic tensors $\mathbf{T}(m)$, $m = 1, \ldots, T$ of size $N \times \ldots \times N$ and order R. Consider the following function

$$\mathcal{D}(\mathbf{U}, \mathbb{T}) = \sum_{m=1}^{T} \left(\sum_{i=1}^{N} |T^{U}_{i,\ldots,i}(m)|^2 \right) \tag{3.42}$$

where

$$T^{U}_{i,\ldots,i}(m) = \sum_{n_1,\ldots,n_{R_1}=1}^{N} U_{i,n_1} \cdots U_{i,n_{R_1}} T_{n_1,\ldots,n_{R_1}}(m). \tag{3.43}$$

This function corresponds to the sum of squares of diagonal terms after linear transform.

DEFINITION 3.6
A joint diagonalizer of the tensor set $\mathbb{T}$ is defined by the unitary matrix

$$\hat{\mathbf{U}} = \arg\max_{\mathbf{U} \in \mathcal{U}} \mathcal{D}(\mathbf{U}, \mathbb{T}) \tag{3.44}$$

where $\mathcal{U}$ denotes the group of $N \times N$ unitary matrices.

The first contrast of this type had been proposed by Comon in [14,16] and can be written

$$\Upsilon_{com2}(\mathbf{y}) = \sum_{i=1}^{N} |\text{cum}\{\underbrace{y_i, \ldots, y_i}_{R\times}\}|^2. \tag{3.45}$$

The link between the previous contrast and a joint diagonalization criterion can be easily established. In fact, it corresponds to the case of a unique tensor in the set $\mathbb{T}$. In [12,13, 16], various pairwise sweeping algorithms have been proposed and aim at diagonalizing a cumulant tensor, mainly of order 3 or 4.

The case of a tensor set containing more than one tensor has been addressed by Cardoso in [5]. The criterion can be written as

$$\Upsilon_{jad}(\mathbf{y}) = \sum_{i,j,k=1}^{N} |\text{cum}\{y_i, y_i, y_j, y_k\}|^2. \tag{3.46}$$

The link between this contrast and joint diagonalization is now less obvious. In the simplest case, it suffices to define the set $\mathcal{M}_4$ of N^2 matrices as

$$\mathcal{M}_4 = \{(\mathbf{M}(i_3, i_4))_{i_1, i_2} = \text{cum}\{x_{i_1}, x_{i_2}, x_{i_3}, x_{i_4}\}\}. \tag{3.47}$$

One shows that, if $\mathbf{U}$ is a unitary matrix,

$$\Upsilon_{jad}(\mathbf{Ux}) = \mathcal{D}(\mathbf{U}, \mathcal{M}_4). \tag{3.48}$$

The maximization of criterion $\Upsilon_{jad}(\mathbf{y})$ via joint diagonalization of a set of matrices led to the JADE algorithm [5].

Later, a similar approach has been followed by DeLathauwer. In particular, he has shown in [35] that the following contrast function

$$\Upsilon_{stotd}(\mathbf{y}) = \sum_{i,j=1}^{N} |\text{cum}\{y_i, y_i, y_i, y_j\}|^2 \tag{3.49}$$

is equivalent to a joint diagonalization of a set of third order tensors. In fact, if we define the set $\mathbb{T}_4$ of N third order tensors:

$$\mathbb{T}_4 = \{(\mathbf{T}(i_2, i_3, i_4))_{i_1} = \text{cum}\{x_{i_1}, x_{i_2}, x_{i_3}, x_{i_4}\}\}, \tag{3.50}$$

one shows that if $\mathbf{U}$ is unitary, then

$$\Upsilon_{stotd}(\mathbf{Ux}) = \mathcal{D}(\mathbf{U}, \mathbb{T}_4). \tag{3.51}$$

More generally, one can show [39] that the previous approaches can be generalized for cumulants of any order. In order to do this, one first shows that

$$\Upsilon_{em1}(\mathbf{y}) = \sum_{i,i_1,\ldots,i_{R-O}=1}^{N} |\mathsf{C}_R^O\{y_i,\mathbf{y}\}|^2 \tag{3.52}$$

where

$$\mathsf{C}_R^O\{y_i,\mathbf{y}\} \stackrel{\text{def}}{=} \mathrm{cum}\{\underbrace{y_i,\ldots,y_i}_{O\times},y_{i_1},\ldots,y_{i_{R-O}}\} \tag{3.53}$$

is a contrast function if $R \geq 3$ and if $2 \leq O \leq R$. When $O = R$, only marginal cumulants appear in the previous sum. To see the link with joint diagonalization, if we define the set $\mathbb{T}_R^O$ of N^{R-O} tensors of order O as:

$$\mathbb{T}_R^O = \{(\mathbf{T}(i_{R-O+1},\ldots,i_R))_{i_1,\ldots,i_{R-O}} = \mathrm{cum}\{x_{i_1},\ldots,x_{i_R}\}\}, \tag{3.54}$$

one shows [39] that if $\mathbf{U}$ is unitary, then

$$\Upsilon_{em1}(\mathbf{U}\mathbf{x}) = \mathscr{D}(\mathbf{U},\mathbb{T}_R^O). \tag{3.55}$$

One can notice that contrast $\Upsilon_{com2}(\mathbf{y})$ corresponds to the case $O = R$, contrast $\Upsilon_{jad}(\mathbf{y})$ corresponds to $R = 4$ and $O = 2$, and contrast $\Upsilon_{stotd}(\mathbf{y})$ corresponds to the case $R = 4$ and $O = 3$. One of the advantages of the above generalization is to allow us to consider tensors of different orders simultaneously. Algebraic algorithms aiming at maximizing these contrast criteria are addressed in Chapter 5.

3.4.4 Non-symmetric contrasts

The above MIMO approaches correspond to symmetric contrast. That is, if the components of $\mathbf{y}$ are permuted then the contrast is unchanged. Even if it is a natural constraint for independent component analysis, for source separation it is not really a mandatory one. Indeed, all contrast functions as separating criteria must have their maxima which correspond to some separating solutions but not all. Notice that a single maximum with no local maxima is the ideal case for a criterion. Such a criterion has to be discovered. But it seems that it is not really a simple task.

Thus, in order to offer a greater flexibility, non-symmetric contrasts were introduced in [43] through a generalized definition. In particular, it has allowed weighted sums of cumulants. Finally, classical symmetric contrasts satisfy clearly the generalized definition. They thus correspond to a particular case.

As an example, it was shown in [43] that the function

$$\Upsilon_{\gamma}^{f}(\mathbf{y})=\sum_{i=1}^{N}\gamma_i f\left(\mathrm{C}_R\{y_i\}\right), \qquad R\in\mathbb{N}^*\setminus\{1,2\} \tag{3.56}$$

is a contrast if the real-valued coefficients γ_i satisfy for all i, $\gamma_i > 0$ and if $f(\cdot)$ is a convex and strictly increasing function. Hence, a weighted sum of cumulants of different order can be considered.

Under the supplementary assumption that all sources have a kurtosis with the same sign ε_4, it was also shown in [43] that the function

$$\Upsilon_{\gamma}'(\mathbf{y})=\varepsilon_4\sum_{i=1}^{N}\gamma_i\mathrm{C}_4\{y_i\} \tag{3.57}$$

is a contrast. It possesses a very interesting property. Indeed, an equivalence can be shown between this contrast and a *cumulant matching* criterion. This last kind of criterion corresponds to a quadratic measure between current values and desired values of some cumulants of given order. According to our problem and for fourth order cumulants, this is written as

$$\mathscr{K}_4(\mathbf{y})=\sum_{i_1,\ldots,i_4}\left(\mathrm{cum}\{y_{i_1},y_{i_2},y_{i_3},y_{i_4}\}-\mathrm{cum}\{s_{i_1},s_{i_2},s_{i_3},s_{i_4}\}\right)^2.$$

Now if we consider $\gamma_i = |\mathrm{C}_4\{s_i\}|$ for all i, into the non-symmetric contrast $\Upsilon_{\gamma}'(\mathbf{y})$, it can be shown that

$$\mathscr{K}_4(\mathbf{y})=g(\mathbf{s})-2\Upsilon_{\gamma}'(\mathbf{y})$$

where $g(\mathbf{s})$ depends only on statistics of source signals and it can then be considered to be constant. This last relation yields an interesting interpretation of the maximization of $\Upsilon_{\gamma}'(\mathbf{y})$. Hence the maximization of $\Upsilon_{\gamma}'(\mathbf{y})$ is equivalent to the minimization of $\mathscr{K}_4(\mathbf{y})$; that is the fourth order cumulants of the outputs of the separating system are fitted to those of the source signals.

Finally, in the case of two sources, an asymptotical statistical study can be drawn for contrast $\Upsilon_{\gamma}'(\mathbf{y})$. It was shown in [43] that the error variance of the searched parameter depends on source moments and on a free parameter characterizing the non-symmetry of the contrast. The minimization of this variance allows the calculation of an optimal value of this free parameter which leads to an optimal contrast in the class considered.

3.4.5 Contrasts with reference signals

The contrasts in section 3.4.3 using cross-cumulants have an interesting invariance property. Indeed, for all unitary matrices $\mathbf{V}$, if we consider $\mathbf{z}(n) = \mathbf{V}\mathbf{y}(n)$ then for all

$R \geq 3$ and all O such that $2 \leq O < R$, we have

$$\sum_{i,i_1,\ldots,i_{R-O}=1}^{N} |\mathsf{C}_R^O\{y_i,\mathbf{z}\}|^2 = \sum_{i,i_1,\ldots,i_{R-O}=1}^{N} |\mathsf{C}_R^O\{y_i,\mathbf{Vy}\}|^2$$
$$= \sum_{i,i_1,\ldots,i_{R-O}=1}^{N} |\mathsf{C}_R^O\{y_i,\mathbf{y}\}|^2 \tag{3.58}$$

where

$$\mathsf{C}_R^O\{y_i,\mathbf{z}\} = \mathrm{cum}\{\underbrace{y_i,\ldots,y_i}_{O\times},z_{i_1},\ldots,z_{i_{R-O}}\}. \tag{3.59}$$

Hence, defining the following function

$$\Upsilon_{em2}(\mathbf{y}) = \sum_{i,i_1,\ldots,i_{R-O}=1}^{N} |\mathsf{C}_R^O\{y_i,\mathbf{z}\}|^2 \tag{3.60}$$

the above equivalence shows clearly that it is again a contrast function when $\mathbf{z}(n)$ only depends on source signals through multiplication by any unitary matrix. In particular, one can consider $\mathbf{z}(n) = \mathbf{x}(n)$ or $\mathbf{z}(n) = \mathbf{s}(n)$. Notice that this last choice is not really useful in practice because if source signals are known, it is not really advantages to separate them. However, this allows us to justify the name given to $\mathbf{z}(n)$, i.e. reference signals. Indeed if we have a first estimation of source signals then this can easily be used as a reference signal. This procedure can then be iterated to obtain better performances.

Other such contrasts have been proposed. In the case of instantaneous mixture, it was shown in [1] that the following function is a contrast for any $R \geq 3$

$$\Upsilon_{R,\mathbf{z}}^{r2}(\mathbf{y}) = \sum_{i,j=1}^{N} |\mathsf{C}_R\{y_i,z_j\}|^2 \tag{3.61}$$

where

$$\mathsf{C}_R\{y_i,z_j\} = \mathrm{cum}\{y_i,y_i,\underbrace{z_j,z_j,\ldots}_{R-2\text{ termes}}\}. \tag{3.62}$$

However it requires that the signs of the R-th order cumulants of source signals are identical. That is the reason why the following contrast has been proposed

$$\Upsilon_{R,\mathbf{z}}^{r1}(\mathbf{y}) = \sum_{i=1}^{N} |\mathsf{C}_R\{y_i,z_i\}|. \tag{3.63}$$

This last one requires a condition that is less restrictive.

3.5 MIMO CONTRASTS FOR DYNAMIC MIXTURES

In this section, it is assumed that $\mathbf{x}(n)$ is the result of a linear and time-invariant filtering of a source process $\mathbf{s}(n)$. The output $\mathbf{y}(n)$ of the separating filter $\mathbf{B}(n)$ is written as:

$$\mathbf{y}(n)=\sum_{\ell}\mathbf{B}(n-\ell)\,\mathbf{x}(\ell)=\sum_{k}\underbrace{\sum_{\ell}\mathbf{B}(n-\ell)\,\mathbf{A}(\ell-k)}_{\mathbf{G}(n-k)}\,\mathbf{s}(k). \tag{3.64}$$

As before, $\mathbf{A}$ denotes the mixing filter, and $\mathbf{G}$ the global filter relating sources and outputs.

3.5.1 Space-time whitening

We have seen how to proceed to a static spatial whitening, by searching for a matrix $\mathbf{W}$ so that the covariance of $\mathbf{W}\,\mathbf{x}$ is the identity matrix. Here, it will be also attempted to whiten the output process in time. For this purpose, the filter $\mathbf{W}$ must be dynamic:

$$\tilde{\mathbf{x}}(n)=\sum_{\ell}\mathbf{W}(n-\ell)\,\mathbf{x}(\ell).$$

PROPOSITION 3.16
Any rational complex filter $\check{\mathbf{B}}(z)$ *can be decomposed into*

$$\check{\mathbf{B}}(z)=\check{\mathbf{U}}(z)\,\check{\mathbf{L}}(z)$$

where $\check{\mathbf{L}}(z)$ *is triangular with minimal phase (that is,* $\det\check{\mathbf{L}}(z)$ *has all its roots in the unit disk), and where* $\check{\mathbf{U}}(z)$ *is para-unitary , viz it verifies:*

$$\check{\mathbf{U}}(z)\,\check{\mathbf{U}}(1/z^*)^{\mathrm{H}}=\mathbf{I},\ \forall z \tag{3.65}$$

or, in the time domain

$$\sum_{k}\mathbf{U}(k)\mathbf{U}(k-n)^{\mathrm{H}}=\delta(n)\mathbf{I}. \tag{3.66}$$

Definitions, properties and usefulness of para-unitary filters can be found in [55,21,29].

Define the power spectral density matrix of $\mathbf{x}(n)$, assumed stationary, as the z-transform of its correlation matrix:

$$\check{\mathbf{\Gamma}}_x[z]\stackrel{\text{def}}{=}\sum_{n}\mathbf{\Gamma}_x[n]\,z^{-n},\quad \mathbf{\Gamma}_x[n]\stackrel{\text{def}}{=}\mathbb{E}\{\mathbf{x}(\ell)\mathbf{x}(\ell-n)^{\mathrm{H}}\}.$$

Then there exists a rational filter $\check{\mathbf{L}}(z)$ such that $\check{\mathbf{\Gamma}}_x[z]=\check{\mathbf{L}}(z)\check{\mathbf{L}}(1/z^*)^{\mathrm{H}}$. This filter is not unique, and is defined up to a *para-unitary* filter, $\check{\mathbf{U}}(z)$.

Suppose sources are statistically independent and temporally white at order 2; in other words, $\check{\mathbf{\Gamma}}_s[z]\stackrel{\text{def}}{=}\mathbf{I},\ \forall z$. Then $\check{\mathbf{\Gamma}}_x[z]=\check{\mathbf{A}}(z)\check{\mathbf{A}}(1/z^*)^{\mathrm{H}}$. Thus, there exists a para-unitary filter $\check{\mathbf{U}}(z)$ such that $\check{\mathbf{A}}(z) = \check{\mathbf{L}}(z)\check{\mathbf{U}}(z)$. The goal of the space-time whitening is to find a filter $\mathbf{W}(z)$ so that $\check{\mathbf{W}}(z)\check{\mathbf{L}}(z) = \check{\mathbf{U}}(z)$, where $\check{\mathbf{U}}(z)$ is para-unitary. The whitened process writes $\tilde{\mathbf{x}}(n) = \sum_\ell \mathbf{U}(n-\ell)\mathbf{s}(\ell)$, so that the source separation will be complete only when filter $\check{\mathbf{U}}(z)$ will have been inverted.

Existence of the inverse filter. If the mixture has a *Finite Impulse Response (FIR)*, $\check{\mathbf{A}}(z)$ is a polynomial matrix, and there exists a polynomial triangular matrix $\check{\mathbf{L}}(z)$. On the other hand, the existence of an inverse filter $\check{\mathbf{B}}(z) \stackrel{\text{def}}{=} \check{\mathbf{U}}(z)\check{\mathbf{W}}(z)$, which is itself polynomial, is not always guaranteed. nevertheless, we have the following proposition.

PROPOSITION 3.17
If the rank of $\check{\mathbf{A}}(z) = N,\ \forall z,\ N \leq P$, *then there exists a polynomial matrix* $\check{\mathbf{B}}(z)$ *such that* $\check{\mathbf{B}}(z)\check{\mathbf{A}}(z) = \mathbf{I}$.

Note by the way that if $N = P$, $\check{\mathbf{A}}(z)$ and $\check{\mathbf{B}}(z)$ are both unimodular (i.e. their determinant is constant for all z). If additionally $\check{\mathbf{A}}(z)$ is of reduced columns, as in the MISO case on page 75, one can bound the degree of $\check{\mathbf{B}}(z)$ [31,52,37,30,24], which theoretically allows us to choose the length of the equalizer. In Chapter 8, Theorem 8.7 page 294 establishes a precise result concerning the existence of the inverse.

It is also possible to look for inverse filters with Infinite Impulse Response (IIR), namely rational filters. Nevertheless, for the sake of convenience, only polynomial matrices will be used in this chapter (that is, MIMO FIR filters).

3.5.2 Contrasts for MIMO convolutive mixtures

As already pointed out in the introduction on page 78, the concepts in the MIMO convolutive (dynamic) case are similar to those of the MIMO static case. However, proofs are more complicated, which has motivated a separate presentation.

3.5.2.1 Trivial filters

First, suppose that each source is white in the strong sense, i.e. it is an iid stochastic process.

Then, the only filters $\mathbf{T}(k)$ that preserve both the whiteness and mutual (spatial) independence between components of $\sum_k \mathbf{T}(n-k)\mathbf{s}(k)$ are those that satisfy:

$$\forall i, \exists!(j,k) : T_{ij}(k) \neq 0 \qquad (3.67)$$

In other words, these filters induce pure delays, integer multiples of the sampling period, followed by a scaled permutation. Denote them as: $\check{\mathbf{T}}(z) = \mathbf{\Lambda}(z)\mathbf{P}$, where $\mathbf{\Lambda}$ is diagonal and contains only monomials in z (or z^{-1}) on its diagonal.

Now if sources are colored, it will be impossible to recover their color without additional information, based only on their mutual (spatial) statistical independence. The class of solutions is thus much larger than in the static case. If each source is a *linear process*, that is, if it may be obtained by filtering an iid process with a time-invariant linear filter, then a representative of the class of solutions is the filter yielding iid sources.

Trivial filters are, in that case, of the form $\check{\mathbf{T}}(z) = \mathbf{\Lambda}(z)\mathbf{P}$, where $\mathbf{\Lambda}(z)$ is this time not constrained to contain only monomials, but can contain more general SISO filters.

3.5.2.2 A more general mixture

Define the set $\mathscr{D}$ of *doubly normalized filters* $\check{\mathbf{G}}(z)$, such that [18]:

$$\forall i, \quad \sum_{jk} |G_{ij}(k)|^2 = 1, \tag{3.68}$$

$$\forall j, \quad \sum_{ik} |G_{ij}(k)|^2 = 1. \tag{3.69}$$

In particular, note that *para-unitary* filters belong to the latter set $\mathscr{D}$. Then our first contrast criterion is given by:

PROPOSITION 3.18
For any pair of positive integers (p,q) such that $p+q>2$, the criterion

$$\Upsilon^{(1)}_{p,q}(\mathbf{G};\mathbf{x}) \stackrel{\text{def}}{=} \sum_i |\text{cum}_{p,q}\{y_i\}| \tag{3.70}$$

is a contrast defined on $\mathscr{D} \times \mathscr{D} \cdot \mathscr{S}$, where $\mathscr{S}$ is the set of iid processes of dimension N, of mutually statistically independent components, having at most one null marginal cumulant of order (p,q).

Proof: The proof is similar to that given on page 85. By multi-linearity of cumulants, and by using the whiteness of sources, it comes about that $\text{cum}_{p,q}\{y_i\} = \sum_{jk} G_{ij}(k)^p G_{ij}(k)^{q^*} \text{cum}_{p,q}\{s_j\}$. Then by triangle inequality, $|\text{cum}_{p,q}\{y_i\}| = \sum_{jk} |G_{ij}(k)|^{p+q} |\text{cum}_{p,q}\{s_j\}|$. But since $\mathbf{G} \in \mathscr{D}$, we have in particular $|G_{ij}(k)| \leq 1$, and hence the majoration $|G_{ij}(k)|^{p+q} \leq |G_{ij}(k)|^2$. We eventually get:

$$|\text{cum}_{p,q}\{y_i\}| \leq \sum_{jk} |G_{ij}(k)|^2 \, |\text{cum}_{p,q}\{s_j\}|.$$

Now summing over index i in both sides, and using again the fact that $\mathbf{G} \in \mathscr{D}$, we get:

$$\Upsilon^{(1)}_{p,q}[\mathbf{y}] \leq \sum_j \sum_{ik} |G_{ij}(k)|^2 |\text{cum}_{p,q}\{s_j\}| = \sum_j |\text{cum}_{p,q}\{s_j\}| \stackrel{\text{def}}{=} \Upsilon^{(1)}_{p,q}[\mathbf{s}]$$

and this is nothing else but Property **P2** of contrasts criteria.

In order to prove **P3**, assume we have equality. Then, we must in particular:

$$\sum_{ijk}(|G_{ij}(k)|^2 - |G_{ij}(k)|^{p+q})|\mathrm{cum}_{p,q}\{s_j\}| = 0.$$

Yet, all the terms are positive real. They must then be each null, for all indices j such that $\mathrm{cum}_{p,q}\{s_j\} \neq 0$; and there are at least $N-1$ of them. We can deduce from this that $\check{\mathbf{G}}(z)$ is trivial for at least $N-1$ of its columns. But since $\mathbf{G} \in \mathcal{D}$, the last remaining column must also be trivial. □

This result can be easily generalized.

PROPOSITION 3.19
If ϕ is a strictly increasing convex function on $\mathbb{R}^+$, then under the same conditions as Proposition 3.18, the criterion

$$\Upsilon^{\phi}_{p.q} = (\mathbf{G};\mathbf{x}) \stackrel{\text{def}}{=} \sum_i \phi(|\mathrm{cum}_{p,q}\{y_i\}|)$$

is a contrast.

The proof goes along the same lines as for Proposition 3.11, and we leave it to the reader. Again, this proposition applies for instance with $\phi(u) = u^2$.

It turns out that these criteria are not contrasts anymore if sources are not linear processes (this case will be addressed in more detail in Chapter 8). But one can show that the following criterion [36] is a contrast:

$$\mathcal{J}^{(2)}_{2,2}(\mathbf{G};\mathbf{x}) \stackrel{\text{def}}{=} \sum_i \sum_{\tau_1,\tau_2,\tau_3} |\mathrm{cum}_{p,q}\{y_i(n), y_i(n+\tau_1), y_i(n+\tau_2), y_i(n+\tau_3)\}|.$$

Of course, this contrast is less discriminating than the previous one because:

$$\Upsilon^{(2)}_{2,2}[\mathbf{y}] \leq \mathcal{J}^{(2)}_{2,2}[\mathbf{y}] \leq \Upsilon^{(2)}_{2,2}[\mathbf{s}]$$

so that its use is not recommended when sources are white.

3.5.3 Contrasts and joint diagonalization

As in the case of static mixtures, contrast criteria can be defined, and allow one to separate convolutive mixtures via joint (possibly approximate) diagonalization of a set of matrices or tensors.

For the sake of convenience, our presentation will be less general than that of section 3.4.3; we shall indeed limit ourselves to joint diagonalization of matrix slices. However, our presentation extends to joint diagonalization of tensors as well. The results developed in the following have been originally presented in [20].

PROPOSITION 3.20
The functional below is a contrast

$$\mathcal{J}_{2,2}^{(2)}[\mathbf{y}] \stackrel{\text{def}}{=} \sum_{i} \sum_{j_1 j_2} \sum_{k_1 k_2} |\text{cum}\{y_i(n), y_i^*(n), y_{j_1}(n-k_1), y_{j_2}^*(n-k_2)\}|^2$$

defined over $\mathcal{H} \times \mathcal{H} \cdot \mathcal{S}$, where $\mathcal{H}$ denotes the group of para-unitary filters, and $\mathcal{S}$ is the set of iid processes with independent components, having at most one null marginal cumulant of order 4.

Proof: It suffices to consider the global filter, linking the MIMO equalizer outputs and sources; $\mathbf{y}(n) = \sum_m \mathbf{G}(m)\mathbf{s}(n-m)$. The squared modulus of the 4th order output cumulant can be written, thanks to multi-linearity, and using the fact that sources are iid:

$$\begin{aligned}
&|\text{cum}\{y_i(n), y_i^*(n), y_{j_1}(n-k_1), y_{j_2}^*(n-k_2)\}|^2 \\
&\quad = \sum_{qm} \sum_{q'm'} G_{iq}(m)^2 G_{iq'}(m')^{2^*} \; G_{j_1 q}(m-k_1) G_{j_1 q'}(m'-k_1)^*
\end{aligned}$$

$$G_{j_2 q}(m-k_2) G_{j_2 q'}(m'-k_2)^* \; \text{cum}\{s_q, s_q, s_q^*, s_q^*\} \, \text{cum}\{s_{q'}, s_{q'}, s_{q'}^*, s_{q'}^*\}.$$

Yet, $\mathbf{G}$ is para-unitary, which implies that $\sum_{jk} G_{jq}(m-k) G_{jq'}(m'-k) = \delta_{qq'} \delta_{mm'}$. When we sum up over indices (j_1, k_1) and (j_2, k_2), this propery can be utilized, which entails a simplified expression for squared cumulants:

$$\mathcal{J}_{2,2}^{(2)}[\mathbf{y}] = \sum_{i} \sum_{qm} |G_{iq}(m)|^8 \, |\text{cum}\{s_q, s_q, s_q^*, s_q^*\}|^2.$$

The para-unitarity of $\mathbf{G}$ gives $|G_{iq}(m)|^8 \le |G_{iq}(m)|^2$ for all (i, q, m), whose sum over $\{i, m\}$ is 1. We have consequently $\mathcal{J}_{2,2}^{(2)}[\mathbf{y}] \le \sum_q |\text{cum}\{s_q, s_q, s_q^*, s_q^*\}|^2$, which is nothing else but $\mathcal{J}_{2,2}^{(2)}[\mathbf{s}]$, and proves **P2**.

Now suppose that $\mathcal{J}_{2,2}^{(2)}[\mathbf{y}] = \mathcal{J}_{2,2}^{(2)}[\mathbf{s}]$. Then in particular

$$\sum_{i} \sum_{qm} (|G_{iq}(m)|^2 - |G_{iq}(m)|^8) \, |\text{cum}\{s_q, s_q, s_q^*, s_q^*\}|^2 = 0.$$

Since all terms are positive, they must individually vanish. This shows that for every index q such that the marginal cumulant of s_q is non-zero, $|G_{iq}(m)|^2 = |G_{iq}(m)|^8$, which is possible only if $|G_{iq}(m)|$ is equal to either 0 or 1. To conclude the proof, one uses for the last time the para-unitarity of **G** which, combined with the fact that at most one source cumulant is null, proves that **G** is trivial. □

Extension to higher orders. This result can be extended to cumulants of higher orders; the proof is very similar, but notations become really cumbersome. One can state that the criterion below is also a contrast [20,50]:

$$\mathcal{J}^{(2)}_{p_1,q_1;p_2,q_2} \stackrel{\text{def}}{=} \sum_i \sum_{\mathbf{j}} \sum_{\mathbf{k}} |\text{cum}\{\underbrace{y_i(n),\ldots y_i(n)^*}_{p_1+q_1 \text{ termes}}, \underbrace{y_{j_1}(n-k_1),\ldots y_{j_r}(n-k_r)^*}_{p_2+q_2 \text{ termes}}\}|^d \tag{3.71}$$

where in each of these cumulants, $y_i(n)$ appears p_1 times, its complex conjugate $y_i(n)^*$ appears q_1 times, and delayed versions $r \stackrel{\text{def}}{=} p_2 + q_2$ times, among which q_2 terms are complex conjugated. With these notations, it is necessary that $d \geq 1$, $p_1 + q_1 \geq 2$, $p_1+p_2+q_1+q_2 \geq 3$, and of course that the corresponding source cumulants are non-zero, for at least $N-1$ of them.

Extension to other convex functions. As in the static case, function $|\cdot|^d$ can be replaced by a convex increasing function, and a similar (more complicated) contrast may be obtained [41].

In order to obtain a criterion corresponding to the joint (approximate) diagonalization of several matrices, we must write the criterion in a smart manner. We shall describe this mechanism for the contrast criterion of Proposition 3.20, being understood that it can be extended to higher order cumulants [20], if the price of more complicated notation is agreed to be paid.

This time, we cannot use the global filter, since we want to involve equalizer $\mathbf{B}(n)$ and cumulants of observations $\mathbf{x}(n)$. To keep a rather light notation, let's denote in this section:

$$C_x(\mathbf{a},\boldsymbol{\alpha},\mathbf{b},\boldsymbol{\beta}) \stackrel{\text{def}}{=} \text{cum}\{x_{a_1}(n-\alpha_1),\, x^*_{a_2}(n-\alpha_2),\, x_{b_1}(n-\beta_1),\, x^*_{b_2}(n-\beta_2)\}$$

where bold indices represent pairs of indices; for instance, $\mathbf{a} \stackrel{\text{def}}{=} (a_1, a_2)$ and $\boldsymbol{\alpha} \stackrel{\text{def}}{=} (\alpha_1, \alpha_2)$.

PROPOSITION 3.21
Criterion $\mathcal{J}^{(2)}_{2,2}$ of Proposition 3.20 can be rewritten

$$\mathcal{J}^{(2)}_{2,2}(\mathbf{B};\mathbf{x}) = \sum_{\mathbf{b}} \sum_{\boldsymbol{\gamma}} ||\text{Diag}\{\mathbb{B}\,\mathbf{M}(\mathbf{b},\boldsymbol{\gamma})\mathbb{B}^{\text{H}}\}||^2$$

where $\mathbb{B} \stackrel{\text{def}}{=} [\mathbf{B}(0), \mathbf{B}(1), \ldots \mathbf{B}(L-1)]$, $\mathbf{b} = [b_1, b_2]$, $\boldsymbol{\gamma} = [\gamma_1, \gamma_2]$, *and where the family of matrices* $\{\mathbf{M}(\mathbf{b}, \boldsymbol{\gamma})\}$, *each of size* $NL \times NL$, *is defined in the following manner:*

$$M_{pq}(\mathbf{b}, \boldsymbol{\beta}) \stackrel{\text{def}}{=} C_x(\mathbf{a}, \boldsymbol{\alpha}, \mathbf{b}, \boldsymbol{\beta}), \quad p \stackrel{\text{def}}{=} \alpha_1 N + a_1, \quad q \stackrel{\text{def}}{=} \alpha_2 N + a_2. \tag{3.72}$$

The cumulant tensor $\mathbf{C}_x$ is actually of order 8, since it involves 8 indices, 4 of them being related to time delays. Its dimensions are $N \times N \times N \times N \times L \times L \times L_t \times L_t$, where one can choose L_t as being equal to the sum of channel and equalizer lengths. The family of matrices defined in (3.72) hence contains $N^2 L_t^2$ matrices of size $NL \times NL$.

Proof: The proof of Proposition 3.21, at order 4, can be found in [21]; its proof for higher orders is similar (but more complicated) and can be found in [20].

For fixed $(i, \mathbf{j}, \mathbf{k})$, the cumulant of $\mathbf{y}$ concerned by the proposition can be written as, according to (3.64):

$$\begin{aligned} &\mathrm{cum}\{y_i(n), y_i^*(n), y_{j_1}(n-k_1), y_{j_2}^*(n-k_2)\} \\ &\quad = \sum_{\mathbf{a},\mathbf{b}} \sum_{\boldsymbol{\alpha},\boldsymbol{\beta}} B_{ia_1}(\alpha_1) B_{ia_2}^*(\alpha_2) B_{j_1 b_1}(\beta_1) B_{j_2 b_2}^*(\beta_2) C_x(\mathbf{a}, \boldsymbol{\alpha}, \mathbf{b}, \boldsymbol{\beta} + \mathbf{k}). \end{aligned}$$

By denoting $\boldsymbol{\gamma} = \boldsymbol{\beta} + \mathbf{k}$ and by resorting to the para-unitarity of $\mathbf{B}$, criterion $\mathcal{J}_{2,2}^{(2)}(\mathbf{B}; \mathbf{x})$ resulting from the sum of these cumulants reduces to:

$$\mathcal{J}_{2,2}^{(2)}(\mathbf{B}; \mathbf{x}) = \sum_i \sum_{\mathbf{b}\boldsymbol{\gamma}} \left| \sum_{\mathbf{a}\boldsymbol{\alpha}} B_{ia_1}(\alpha_1) B_{ia_2}^*(\alpha_2) C_x(\mathbf{a}, \mathbf{b}, \boldsymbol{\alpha}, \boldsymbol{\gamma}) \right|^2.$$

Above, we have used the same simplification as in the proof of Proposition 3.20. Now let's utilize the definitions of matrix $\mathbb{B}$ and family (3.72). By changing indices as indicated in (3.72), we get:

$$\mathcal{J}_{2,2}^{(2)}(\mathbf{B}; \mathbf{x}) = \sum_i \sum_{\mathbf{b}\boldsymbol{\gamma}} \left| \sum_{pq} \mathbb{B}_{pi} \mathbb{B}_{qi}^* M_{pq}(\mathbf{b}, \boldsymbol{\gamma}) \right|^2.$$

And we indeed have diagonal terms of matrix $\mathbb{B}\mathbf{M}(\mathbf{b}, \boldsymbol{\gamma})\mathbb{B}^{\mathrm{H}}$. □

Remarks.

- Attempting to jointly maximize diagonal terms of these matrices amounts to "jointly diagonalizing" them, approximately of course because they rarely commute with each other. Furthermore, notice that index i only ranges from 1 to N, and that matrix $\mathbb{B}$ is rectangular. We are thus dealing with a joint *partial diagonalization*, which has been referred to as *PAJOD* in [21]. Since $\mathbf{B}$ is para-unitary, one can easily check that $\mathbb{B}$ is *semi-unitary*, that is:

$$\mathbb{B}\mathbb{B}^{\mathrm{H}} = \mathbf{I}.$$

- The contrast of Propositions 3.21 or 3.20 is much less distriminating than that of Propositions 3.18 and 3.19. In fact, it is easy to show that

$$\Upsilon_{2,2}^{(2)}[\mathbf{y}] \leq \mathcal{J}_{2,2}^{(2)}[\mathbf{y}] \leq \Upsilon_{2,2}^{(2)}[\mathbf{s}].$$

  So resorting to contrast $\mathcal{J}_{2,2}^{(2)}$ is recommended only if this allows the design of fast and efficient numerical algorithms.
- Some algorithms exist under the para-unitarity constraint [29,21], but to our knowledge, none under the constraint of "double orthonormalization" mentioned in Proposition 3.18. In fact, all research works carried out afterwards have been focussed mainly on para-unitary filters, which are only particular doubly normalized filters.

3.5.4 Non-symmetric contrasts

This subsection is a sort of generalization of section 3.4.4 concerning non-symmetrical contrasts.

We still assume throughout this section that the observations have been both spatially and temporally whitened at the second order, which means that $\mathbf{x}$ has been standardized and that the source signals are iid process (or linear filtering of such process). The demonstration of all the results of this paragraph can be found in [42].

For the following, we need another non-restrictive assumption. First we consider the set $\mathscr{C}_N$ of vectors $(u_1, \ldots, u_N)^T$ in $(\mathbb{R}_+)^N$ such that one of the two ensuing conditions is satisfied: $u_1 \geq u_2 \geq \cdots \geq u_N > 0$ or $u_1 \geq u_2 \geq \cdots \geq u_{N-1} > u_N = 0$ and the subset $\mathring{\mathscr{C}}_N$ of $\mathscr{C}_N$ defined by

$$\mathring{\mathscr{C}}_N = \{(u_1, \ldots, u_N) \in \mathbb{R}^N \mid u_1 > u_2 > \cdots > u_N \geq 0\}.$$

Denoting by $\mathrm{c}_R\{\mathbf{s}\}$ the vectors of the absolute values of the source signals order R cumulants:

$$\mathrm{c}_R\{\mathbf{s}\} = (|\mathrm{c}_R\{s_1\}|, \ldots, |\mathrm{c}_R\{s_N\}|)^T \tag{3.73}$$

we will assume in the following that $\mathrm{c}_R\{\mathbf{s}\} \in \mathscr{C}_N$.

We have the ensuing proposition

PROPOSITION 3.22
The function

$$\Upsilon_R^F(\mathbf{y}) = F(\mathrm{c}_R\{\mathbf{y}\}), \qquad R \in \mathbb{N}^* \setminus \{1,2\} \tag{3.74}$$

is a contrast function when the function $F(\cdot)$ *defined in* $(\mathbb{R}_+)^N$ *is a convex strictly increasing function such that whatever the permutation matrix* $\mathbf{P}$ *and whatever the vector* $\mathbf{u}$ *in* $\mathscr{C}_N$, *we have*

$$F(\mathbf{Pu}) \leq F(\mathbf{u}). \tag{3.75}$$

An interesting property is obtained when $\mathsf{C}_R\{\mathbf{s}\} \in \overset{\circ}{\mathscr{C}}_N$. In this case, one can show that:

PROPOSITION 3.23
The function $\Upsilon_R^F(\mathbf{y})$ *for* $R \geq 3$ *is a contrast solving the permutation problem when the function* $F(\cdot)$ *defined in* $(\mathbb{R}_+)^N$ *is a convex strictly increasing function such that whatever the permutation matrix* $\mathbf{P}$ *distinct from the identity matrix and whatever the vector* $\mathbf{u} \in \overset{\circ}{\mathscr{C}}_N$, *we have*

$$F(\mathbf{Pu}) < F(\mathbf{u}). \tag{3.76}$$

It implies that if the source signals have distinct order R cumulants (sorted in a decreasing order), the maximization of the aforementioned contrast enables recovery of the sources in the same order as for the initial one.

It is not a difficult task to find examples of contrasts satisfying the preceding hypothesis by choosing, for example, an additively separable function such as

$$F(\mathbf{u}) = \sum_{i=1}^{N} \gamma_i f(u_i) \tag{3.77}$$

where $(\gamma_1, \ldots, \gamma_N) \in (\mathbb{R}_+^*)^N$ and $f(\cdot)$ is a scalar function defined in $\mathbb{R}_+$. As a consequence, if $R \geq 3$, $\gamma_1 \geq \cdots \geq \gamma_N > 0$ and $f(\cdot)$ is a strictly increasing convex function, then the function

$$\Upsilon_R^f(\mathbf{y}) = \sum_{i=1}^{N} \gamma_i f(|\mathsf{C}_R\{y_i\}|) \tag{3.78}$$

is a contrast.

One can even show the following result, which is a direct consequence of Proposition 3.23:

PROPOSITION 3.24
If $R \geq 3$, $\gamma_1 > \cdots > \gamma_N > 0$ *and* $\mathsf{C}_R\{\mathbf{s}\} \in \overset{\circ}{\mathscr{C}}_N$, $\Upsilon_R^f(\cdot)$ *is a contrast solving the permutation problem.*

This latter contrast provides a parametric contrast family. It may lead to a more flexible use of this kind of contrast "at higher orders" by adapting the free parameters. However, it has to be noted that *a priori* knowledge about the sources statistics is required to obtain optimality.

We consider now a particular case of source signals conforming to

$$\forall i \in \{1,\ldots,N\}, \quad \text{sign}(\mathrm{C}_R\{s_i\}) = \epsilon_i \tag{3.79}$$

where $(\epsilon_1,\ldots,\epsilon_N) \in \{-1,1\}^N$ and where sign$(\cdot)$ is the sign function. The ensuing proposition emphasizes how the sign of the source signals cumulants can be taken into account:

PROPOSITION 3.25
If $R \geq 3$, $\gamma_1 \geq \cdots \geq \gamma_N > 0$ and $f : \mathbb{R} \longrightarrow \mathbb{R}$ is an odd strictly increasing function, non-negative and convex on $\mathbb{R}_+$, then the function

$$\Upsilon_R^{f'}(\mathbf{y}) = \sum_{i=1}^{N} \gamma_i \epsilon_i f(\mathrm{C}_R\{y_i\}) \tag{3.80}$$

is a contrast.

Thus, it implies that the function

$$\Upsilon_R^{(.)}(\mathbf{y}) = \sum_{i=1}^{N} \gamma_i \epsilon_i \, \mathrm{C}_R\{y_i\} \tag{3.81}$$

is a contrast. Finally, it is noticed that if the source signals R order cumulants have an identical sign $\epsilon = \epsilon_1 = \cdots = \epsilon_R$, the preceding contrast can be rewritten

$$\Upsilon_R^{(\cdot)}(\mathbf{y}) = \epsilon \sum_{i=1}^{N} \gamma_i \mathrm{C}_R\{y_i\}. \tag{3.82}$$

3.6 CONSTRUCTING OTHER CONTRAST CRITERIA

We have given numerous examples of contrasts, dedicated to the assumption of statistical independence between sources, and sometimes also their time whiteness. But it is also possible to build other contrasts by combining them together.

Linear combination. The simplest way to combine them is linearly:

PROPOSITION 3.26
Let $\Upsilon_k[y]$ be contrasts defined for $\mathbf{y} \in \mathcal{H} \cdot \mathcal{S}_k$, and $\{a_k\}$ a set of real strictly positive numbers. Then $\Upsilon[\mathbf{y}] = \sum_k \Upsilon_k[y]$ is also a contrast for $\mathbf{y} \in \mathcal{H} \cdot \cup_k \mathcal{S}_k$.

The proof is extremely simple, and may be found in [43] or [19].

Convex combination. Other ways of obtaining new contrast criteria consist of taking convex functions of contrasts:

Proposition 3.27
If $\Upsilon[\mathbf{y}]$ is a contrast, and if ϕ is a strictly increasing convex function on $\mathbb{R}^+$, then $\phi(\Upsilon[y])$ is also a contrast.

The reader is referred to [19] for a proof.

Discrete alphabets. At the low noise limit, it is theoretically possible to identify a mixture of N sources from a single sensor measurement, and to extract all of them under certain conditions avoiding (rare) ambiguities. It means that the discrete character induces a form of *diversity*. In the spirit of this chapter, we say that it is possible to define contrast functions that are associated with a given discrete source alphabet.

Contrary to the previously defined contrasts, this contrast criterion is deterministic [19], as shown below with Proposition 3.28.

Proposition 3.28
Let a source alphabet $\mathcal{A}$, not reduced to $\{0\}$, be defined in the complex plane by the roots of a polynomial $Q(z)=0$. Let $\mathcal{S}$ be the set of processes taking their values in alphabet $\mathcal{A}$, and $\mathcal{H}$ the set of $N \times N$ invertible FIR filters. Then the criterion below is a contrast

$$\Upsilon(\mathbf{A};\mathbf{y}) = -\sum_n \sum_i |Q(y_i(n))|^2 \tag{3.83}$$

The simplest case is $Q(z) = z^q - 1$, for which we have a PSK$-q$ constellation. In that case, trivial filters are given by [19]:

Proposition 3.29
Trivial filters of the PSK$-q$ alphabet are of the form $\mathbf{P}\,\check{\mathbf{D}}[z]$, where $\check{D}_{pp}[z]$ are rotations in the complex plane of an angle multiple of $2\pi/q$ combined with a pure delay, and $\mathbf{P}$ are permutations.

It is worth noting that the mutual statistical independence between sources is not required anymore in Proposition 3.28. It appears thus possible to separate *correlated sources* with the help of such a contrast.

Domination by a contrast. Lastly, one can mention the following beautiful property, which can be sometimes very useful:

Proposition 3.30
Let $\Upsilon_2[\mathbf{y}]$ be a contrast. If $\Upsilon_1[\mathbf{s}] = \Upsilon_2[\mathbf{s}]$ for all $\mathbf{s} \in \mathcal{S}$, and if $\Upsilon_1[\mathbf{y}] \leq \Upsilon_2[\mathbf{y}]$ for all $\mathbf{y} \in \mathcal{H}\cdot\mathcal{S}$, then $\Upsilon_1[\mathbf{y}]$ is a contrast for $\mathbf{y} \in \mathcal{H}\cdot\mathcal{S}$.

In other words, a function bounded above by a contrast is a contrast. We refer to [6, section 2.2.2] for a proof of this proposition.

3.7 CONCLUSION

Various important topics have had to be omitted for reasons of space. We mention some of them in this conclusion.

First, we have addressed in this chapter only the question of the choice of optimization criteria, and not of optimization algorithms. This will be the subject of several other chapters, including Chapters 5, 8, 9 and 15.

Blind identification or equalization in the presence of cyclostationary (or non-stationary) sources is another topic of interest. It is sufficiently complex to be the subject of a whole chapter, as shown in section 8.6 of Chapter 8, see also Chapter 17.

Last, the discrete character of sources, often encountered in practice, e.g. in digital communications, induces a form of diversity, as demonstrated by Proposition 3.28. However, the corresponding numerical algorithms can require a prior clustering, which is a quite time consuming operation. This would also deserve to be the subject of a whole chapter.

References

[1] A. Adib, E. Moreau, D. Aboutajdine, Source separation contrasts using a reference signal, IEEE Signal Process. Lett. 11 (2004) 312–315.

[2] L. Albera, P. Comon, Asymptotic performance of contrast-based blind source separation algorithms, in: Second IEEE Sensor Array and Multichannel Signal Processing Workshop, Rosslyn, VA, 4–6 August 2002.

[3] A. Benveniste, M. Goursat, G. Ruget, Robust identification of a non-minimum phase system, IEEE Trans. Auto. Contr. 25 (1980) 385–399.

[4] D.R. Brillinger, Time Series, Data Analysis and Theory, Holden-Day, 1981.

[5] J.F. Cardoso, A. Souloumiac, Blind beamforming for non-Gaussian signals, IEE Proceedings–Part F 140 (1993) 362–370. Special issue on Applications of High-Order Statistics.

[6] M. Castella, Séparation de sources non linéaires dans le cas des mélanges convolutifs, doctorat, Université de Marne la Vallée, Déc. 2004.

[7] M. Castella, P. Bianchi, A. Chevreuil, J.-C. Pesquet, A blind separation framework for detecting CPM sources mixed by a convolutive MIMO filter, in: Signal Processing, Elsevier, 2006, pp. 1950–1967.

[8] M. Castella, A. Chevreuil, J.-C. Pesquet, Separation aveugle d'un melange convolutif de sources non linéaires par une approche hierarchique, in: GRETSI, Paris, France, Sept. 2003.

[9] M. Castella, E. Moreau, Generalized identifiability conditions for blind convolutive mimo separation, IEEE Trans. Signal Process. 57 (7) (2009) 2846–2852.

[10] M. Castella, S. Rhioui, E. Moreau, J.-C. Pesquet, Quadratic higher-order criteria for iterative blind separation of a MIMO convolutive mixture of sources, IEEE Trans. Signal Process. (2007) 218–232.

[11] P. Comon, Analyse en Composantes Indépendantes et identification aveugle, Traitement du Signal 7 (1990) 435–450. Special Issue.

[12] P. Comon, Independent Component Analysis, in: Proc. Int. Sig. Proc. Workshop on Higher-Order Statistics, Chamrousse, France, July 1991, pp. 111–120. Keynote address.

[13] P. Comon, Independent component analysis, in: J.-L. Lacoume (Ed.), Higher Order Statistics, Elsevier, Amsterdam, London, 1992, pp. 29–38.

[14] P. Comon, Remarques sur la diagonalisation tensorielle par la méthode de Jacobi, in: XIVème Colloque Gretsi, Sept. 1993, pp. 125–128.

[15] P. Comon, Independent component analysis, a new concept? Signal Process. 36 (1994) 287–314. Special issue on Higher-Order Statistics.

[16] P. Comon, Tensor diagonalization, a useful tool in signal processing, in: M. Blanke, T. Soderstrom (Eds.), IFAC-SYSID, 10th IFAC Symposium on System Identification, vol. 1, Copenhagen, Denmark, July 1994, pp. 77–82. Invited session.

[17] P. Comon, Supervised classification, a probabilistic approach, in: Verleysen (Ed.), ESANN-European Symposium on Artificial Neural Networks, Brussels, Apr. 19–21, D facto Publ., 1995, pp. 111–128. Invited paper.
[18] P. Comon, Contrasts for multichannel blind deconvolution, IEEE Sig. Proc. Letters 3 (1996) 209–211.
[19] P. Comon, Contrasts, independent component analysis, and blind deconvolution, Int. J. Adapt. Control Signal Process. 18 (2004) 225–243.
[20] P. Comon, E. Moreau, Blind MIMO equalization and joint-diagonalization criteria, in: ICASSP'01, vol. 5, Salt Lake City, May 2001, pp. 2749–2752.
[21] P. Comon, L. Rota, Blind separation of independent sources from convolutive mixtures, IEICE Trans. on Fundamentals of Elec. Com. Comput. Sciences, E86-A (2003). Invited.
[22] T.M. Cover, J.A. Thomas, Elements of Information Theory, Wiley, 1991.
[23] N. Delfosse, P. Loubaton, Adaptive blind separation of independent sources: A deflation approach, Signal Process. 45 (1995) 59–83.
[24] F. Desbouvries, Systémes linéaires multivariables, in: M. Guglielmi (Ed.), Signaux Alatoires: Modelisation, Estimation, Detection, Hermes, Paris, 2004, (Ch. 5). Trait IC2, série Traitement du signal et de l'image.
[25] Z. Ding, Y. Li, Blind Equalization and Identification, Dekker, New York, 2001.
[26] D. Donoho, On minimum entropy deconvolution, in: Applied time-series analysis II, Academic Press, 1981, pp. 565–609.
[27] R. Dubroca, C.D. Luigi, E. Moreau, Cubic higher-order criterion and algorithm for blind extraction of a source signal, in: ICASSP'09, Taipei, Taiwan, Apr. 2009.
[28] D. Godard, Self recovering equalization and carrier tracking in two dimensional data communication systems, IEEE Trans. Com. 28 (1980) 1867–1875.
[29] S. Icart, P. Comon, L. Rota, Blind para-unitary equalization, Signal Process. 89 (2009) 283–290.
[30] Y. Inouye, R. Liu, A system-theoretic foundation for blind equalization of a FIR MIMO channel system, IEEE Trans. Circuits Syst. 49 (2002) 425–436.
[31] T. Kailath, Linear Systems, Prentice-Hall, 1980.
[32] M. Kendall, A. Stuart, The Advanced Theory of Statistics, Distribution Theory, vol. 1, 4th ed., C. Griffin, 1977.
[33] J.L. Lacoume, P.O. Amblard, P. Comon, Statistiques d'ordre supérieur pour le traitement du signal, Collection Sciences de l'Ingénieur, Masson, 1997.
[34] J.L. Lacoume, P. Ruiz, Separation of independent sources from correlated inputs, IEEE Trans. Signal Process. 40 (1992) 3074–3078.
[35] L.D. Lathauwer, B.D. Moor, J. Vandewalle, Independent component analysis and (simultaneous) third-order tensor diagonalization, IEEE Trans. Signal Process. 49 (2001) 2262–2271.
[36] R.-W. Liu, Y. Inouye, Blind equalization of MIMO-FIR channels driven by white but higher order colored source signals, IEEE Trans. Inform. Theory 48 (2002) 1206–1214.
[37] P. Loubaton, E. Moulines, P.A. Regalia, Subspace methods for blind identification and deconvolution, in: Giannakis, Hua, Stoica, Tong (Eds.), Signal Processing Advances in Wireless and Mobile Communications, Prentice-Hall, 2001 (Ch. 3).
[38] P. McCullagh, Tensor Methods in Statistics, in: Monographs on Statistics and Applied Probability, Chapman and Hall, 1987.
[39] E. Moreau, A generalization of joint-diagonalization criteria for source separation, IEEE Trans. Signal Process. 49 (2001) 530–541.
[40] E. Moreau, O. Macchi, High order contrasts for self-adaptive source separation, Int. J. Adapt. Control Signal Process. 10 (1996) 19–46.
[41] E. Moreau, J.-C. Pesquet, Generalized contrasts for multichannel blind deconvolution of linear systems, IEEE Signal Process. Lett. 4 (1997) 182–183.
[42] E. Moreau, J.-C. Pesquet, N. Thirion-Moreau, Blind signal separation based on asymmetrical contrast functions, IEEE Trans. Signal Process. 55 (1) (2007) 356–371.

[43] E. Moreau, N. Thirion-Moreau, Non symmetrical contrasts for sources separation, IEEE Trans. Signal Process. 47 (1999) 2241–2252.
[44] C.B. Papadias, Globally convergent blind source separation based on a multiuser kurtosis maximization criterion, IEEE Trans. Signal Process. 48 (2000) 3508–3519.
[45] J.-C. Pesquet, E. Moreau, Cumulant based independence measures for linear mixtures, IEEE Trans. Inform. Theory 47 (2001) 1947–1956.
[46] D.T. Pham, Blind separation of instantaneous mixture of sources via an independent component analysis, IEEE Trans. Signal Process. 44 (1996) 2768–2779.
[47] O. Shalvi, E. Weinstein, New criteria for blind deconvolution of nonminimum phase systems, IEEE Trans. Inform. Theory 36 (1990) 312–321.
[48] C. Simon, P. Loubaton, C. Jutten, Separation of a class of convolutive mixtures: a contrast function approach, Signal Process. 81 (2001) 883–887.
[49] C. Simon, P. Loubaton, C. Vignat, C. Jutten, G. D'urso, Blind source separation of convolutive mixtures by maximization of fourth-order cumulants: the non iid case, in: ICASSP, Seattle, May 1998.
[50] N. Thirion-Moreau, E. Moreau, Generalized criteria for blind multivariate signal equalization, IEEE Signal Process. Lett. 9 (2002) 72–74.
[51] J.R. Treichler, M.G. Larimore, New processing techniques based on the constant modulus adaptive algorithm, IEEE Trans. Acoust. Speech Signal Process. 33 (1985) 420–431.
[52] J. Tugnait, Adaptive blind separation of convolutive mixtures of independent linear signals, Signal Process. 73 (1999) 139–152.
[53] J.K. Tugnait, Blind spatio-temporal equalization and impulse response estimation for MIMO channels using a Godard cost function, IEEE Trans. Signal Process. 45 (1997) 268–271.
[54] J.K. Tugnait, Identification and deconvolution of multichannel non-Gaussian processes using higher order statistics and inverse filter criteria, IEEE Trans. Signal Process. 45 (1997) 658–672.
[55] P.P. Vaidyanathan, Multirate Systems and Filter Banks, Prentice-Hall, London, 1993.

CHAPTER

Likelihood

4

J.-F. Cardoso

This chapter is devoted to studying instantaneous mixtures of independent components. It is not intended to cover all existing methods described in the literature; rather, it focuses on elucidating the main statistical aspects of the problem by carefully analyzing the structure of the likelihood.

We examine successively three basic features of the ICA model: first, the idea of linear mixture which is associated with the equivariance property; second, the assumption of independence between components which, together with equivariance, determines the statistical structure of ICA; third, the specific models which can be designed for approximating the probability distribution of each component.

Each one of these three successive modeling steps induces specific features in the ICA likelihood, which are discussed as they are introduced: equivariance, related estimating equations, connection to contrast functions (mutual information, entropy), stability with respect to component modeling, adaptivity, and performance bounds.

The chapter emphasizes a unifying perspective on all these issues. It also includes a section on noisy models.

4.1 INTRODUCTION: MODELS AND LIKELIHOOD

Denote $\mathbf{X}$ a data matrix. We think of it as a $P \times T$ object collecting T samples of a $P \times 1$ vector and we set out to investigate the likelihood approach to analyze $\mathbf{X}$ as made of independent components. This brief introductory section sets the working context and outlines the "likelihood path" followed in this chapter.

Models. We shall restrict ourselves to *instantaneous models*. The instantaneous model of source separation postulates a number N (to be determined) of sources so that the data matrix $\mathbf{X}$ can be explained as

$$\mathbf{X} = \mathbf{AS} + \mathbf{N} \tag{4.1}$$

where $\mathbf{A}$ is the unknown $P \times N$ "mixing matrix", where $\mathbf{S}$ is an $N \times T$ matrix in which each row is called a "source signal" or a "component" and where matrix $\mathbf{N}$ represents an additive noise or some other form of measurement uncertainty. Quite often, the "horizontal dimension" is time, hence the term "instantaneous model" since (4.1) also reads $\mathbf{x}(t) = \mathbf{A}\mathbf{s}(t) + \mathbf{n}(t)$ at each time $t = 1, \ldots, T$. The term "instantaneous" stresses that mixing occurs across sensors, not across time but that terminology is too restrictive because index t could refer as well to a spatial dimension or even to a set without any particular structure or ordering.

No real data set can be perfectly captured by a mathematical model. In some contexts, it is necessary to include error terms in order to define a proper statistical model but this is not the case for blind source separation, as we shall see later. However, the decision whether to include a noise term or to ignore it has a strong impact on the mathematical structure of the likelihood. This is why this chapter addresses separately the cases of noisy models $\mathbf{X} = \mathbf{A}\mathbf{S} + \mathbf{N}$ and of noise-free models $\mathbf{X} = \mathbf{A}\mathbf{S}$.

Under-determined and over-determined mixtures. The mixture is said to be "under-determined" when it contains more sources than sensors ($N < P$); to be "over-determined" when it contains fewer sources than sensors; to be "determined" when matrix $\mathbf{A}$ is squared and invertible. In the analysis of *noisy* models, there is no mathematical difference in the likelihood between those three cases; indeed, they are considered in this way in section 4.7. The *noise-free* model offers a very different picture because its likelihood shows a very particular structure in the determined case $N = P$. As a matter of fact, the determined noise-free model belongs to a wider family of statistical "transformation models" which are associated with the concept of equivariance and with specific forms of the likelihood studied in section 4.2. We will focus on this class of models, referred to as *canonical models*:

$$\mathbf{X} = \mathbf{A}\mathbf{S} \quad \mathbf{A} \text{ is invertible } (P = N) \tag{4.2}$$

which are the baseline in source separation studies. The noise-free model will be considered only in the determined case. The noise-free *over-determined* case (more sensors than sources) is easily reduced to the determined case by projecting the observations onto the signal subspace. The dimensionality reduction can be achieved, for instance, by working on the N first principal components of $\mathbf{X}$. Regarding the *under-determined* noise-free models, their maximum likelihood estimation will not be addressed in this chapter.

The likelihood approach. The guideline throughout the chapter is the maximum likelihood principle. As soon as a probability distribution is specified for each source (and for the noise, in the case of noisy models), the distribution for the whole data set $\mathbf{X}$ is determined for any value of $\mathbf{A}$. This is denoted $p_{\mathbf{X}}(\mathbf{X}, \mathbf{A})$ and the maximum likelihood estimate of $\mathbf{A}$ is $\widehat{\mathbf{A}}_{\mathrm{ML}} = \arg\max_{\mathbf{A}} p_{\mathbf{X}}(\mathbf{X}, \mathbf{A})$ from which an estimate of the sources $\mathbf{S}$ can be obtained. To implement the maximum likelihood approach, one needs to choose source distribution models with the following properties: (1) The source model should make likelihood maximization algorithmically tractable. (2) It should be expressive enough to make $\mathbf{A}$ blindly identifiable. (3) It should be simple enough that analyzing the likelihood would guarantee blind identifiability even when the sources do not exactly follow the model (they rarely do!). (4) It should offer some adaptivity to the underlying source distributions in order to be able to separate a wide range of sources.

Points (3) and (4) above will become clearer in the next sections where we study a key property of the linear model: its robustness with respect to source models.

The two steps of estimating matrix $\mathbf{A}$ and separating sources $\mathbf{S}$ are not necessarily separated. Actually, they are intrinsically linked in the case of noise-free models for which $\mathbf{A}$ and $\mathbf{X}$ tend to vanish from the equations to the benefit of $\mathbf{A}^{-1}\mathbf{X}$. For noisy models, in contrast, it may be better to estimate the sources by a more efficient method than $\widehat{\mathbf{A}}^{-1}\mathbf{X}$, using for instance a Wiener filter (section 4.7).

Outline. We first examine the two ingredients entering the likelihood of the canonical model: the mixing operation (section 4.2) and the source independence (section 4.3). The asymptotic analysis (section 4.4) offers some results about blind identifiability, the Cramér-Rao bound, the Fisher efficiency, the robustness to source models, ... which are obtained in a very general manner, that is, without strong assumptions about the source models. The following two sections – 4.5 and 4.6 – examine in detail the two most important source models: non-Gaussian distributions with an iid property and Gaussian distributions with correlation. Noisy models are discussed in section 4.7. The chapter closes with a general comparison of these models.

Note. In this chapter, for the sake of exposition, all signals are assumed to have zero mean. We shall use the following notations: if $\mathbf{X}$ has size $N \times T$, then $\mathbf{x}_i$ denotes the ith row ($1 \times T$), and $\mathbf{x}(t)$ denotes the tth column ($N \times 1$) while $x_i(t)$ denotes entry (i, t) of $\mathbf{X}$.

4.2 TRANSFORMATION MODEL AND EQUIVARIANCE

The canonical model (4.2) of source separation is, in the first place, a *transformation model*: hypothetical source signals $\mathbf{S}$ are observed after they have undergone an invertible transform, namely left multiplication by an invertible matrix $\mathbf{A}$. Blind separation means that nothing is assumed about $\mathbf{A}$ except invertibility. Thus, the parameter to be identified can be any member of the set of invertible matrices of given $N \times N$ size. This is a multiplicative group, often called the *linear group* (of $\mathbb{R}^N$). Several important characteristics of our problem (such as *equivariance*) result from this group structure, independently of the assumption of source independence. This section is devoted to the exploration of this structure.

As we shall see, the key equations of the canonical source separation problem are more naturally expressed in terms of the product

$$\mathbf{Y} = \mathbf{A}^{-1}\mathbf{X} \tag{4.3}$$

rather than in terms of $\mathbf{A}$ and $\mathbf{X}$. The meaning of $\mathbf{Y}$ is clear: matrix $\mathbf{Y}$ represents the sources estimated from data $\mathbf{X}$ for a particular value of $\mathbf{A}$.

To derive likelihood equations, one postulates a probability density $p_{\mathbf{S}}(\mathbf{S})$ for matrix $\mathbf{S}$, supposed to be differentiable. Any source model $p_{\mathbf{S}}(\mathbf{S})$ is associated with the function

$\psi : \mathbb{R}^{N\times T} \mapsto \mathbb{R}^{N\times T}$:

$$[\psi(\mathbf{S})]_{it} = -\partial \log p_{\mathbf{S}}(\mathbf{S})/\partial s_i(t) \quad 1 \le i \le N, 1 \le t \le T \tag{4.4}$$

called the *score function* for model $p_{\mathbf{S}}$. In this section, we assume that this function exists and is known but we do not assume there is any structure to it.

4.2.1 Transformation likelihood

Transformation. If the probability density of a vector $\mathbf{s}$ is $p_s(\mathbf{s})$, then its transform $\mathbf{x} = \mathbf{A}\mathbf{s}$ by an invertible matrix $\mathbf{A}$ has density $p_x(\mathbf{x}) = p_s(\mathbf{A}^{-1}\mathbf{x})/|\det \mathbf{A}|$. Similarly, in the canonical model (4.2), the probability density $p_{\mathbf{X}}$ of data matrix $\mathbf{X}$ is related to the density $p_{\mathbf{S}}$ of $\mathbf{S}$ by[1]:

$$p_{\mathbf{X}}(\mathbf{X};\mathbf{A}) = \frac{1}{|\det \mathbf{A}|^T} p_{\mathbf{S}}(\mathbf{A}^{-1}\mathbf{X}). \tag{4.5}$$

Invariant form. Let us express the likelihood in invariant form, that is, as a sole function of $\mathbf{Y}$. The sample covariance matrices $\widehat{\mathbf{R}}_{\mathbf{X}}$ and $\widehat{\mathbf{R}}_{\mathbf{Y}}$ are defined by

$$\widehat{\mathbf{R}}_{\mathbf{X}} = \frac{1}{T}\sum_{t=1}^{N} \mathbf{x}(t)\mathbf{x}(t)^{\mathrm{T}} = T^{-1}\mathbf{X}\mathbf{X}^{\mathrm{T}}, \quad \widehat{\mathbf{R}}_{\mathbf{Y}} = \frac{1}{T}\sum_{t=1}^{N} \mathbf{y}(t)\mathbf{y}(t)^{\mathrm{T}} = T^{-1}\mathbf{Y}\mathbf{Y}^{\mathrm{T}}. \tag{4.6}$$

Since $\widehat{\mathbf{R}}_{\mathbf{X}} = \mathbf{A}\widehat{\mathbf{R}}_{\mathbf{Y}}\mathbf{A}^{\mathrm{T}}$, we have $\det \widehat{\mathbf{R}}_{\mathbf{X}} = |\det \mathbf{A}|^2 \, \det \widehat{\mathbf{R}}_{\mathbf{Y}}$. Combining that with the transformation law (4.5) yields

$$p_{\mathbf{X}}(\mathbf{X};\mathbf{A}) = e^{T\widehat{\Upsilon}(\mathbf{A}^{-1}\mathbf{X})} / \det(\widehat{\mathbf{R}}_{\mathbf{X}})^{T/2} \tag{4.7}$$

where the *empirical likelihood contrast* is the function $\widehat{\Upsilon} : \mathbb{R}^{N\times T} \mapsto \mathbb{R}$

$$\widehat{\Upsilon}(\mathbf{Y}) \stackrel{\text{def}}{=} \frac{1}{T}\log p_{\mathbf{S}}(\mathbf{Y}) + \frac{1}{2}\log\det \widehat{\mathbf{R}}_{\mathbf{Y}}. \tag{4.8}$$

This function of $\mathbf{Y}$ is only determined by the model $p_{\mathbf{S}}$ of source distribution. Hence, the likelihood depends on $\mathbf{A}$ only via $\mathbf{Y} = \mathbf{A}^{-1}\mathbf{X}$ since the denominator in Eq. (4.7) depends on the data only and not on the model. The maximum likelihood estimate for the canonical model therefore is

$$\widehat{\mathbf{A}}_{\mathrm{ML}} = \arg\max_{\mathbf{A}} \widehat{\Upsilon}(\mathbf{A}^{-1}\mathbf{X}).$$

[1] The exponent T in Eq. (4.5) is the number of samples, not to be confused here with the transposition operator. It results from applying $\mathbf{A}$ on the T columns of the data matrix.

4.2.2 Transformation contrast

What is Eq. (4.8) trying to tell us? Its meaning becomes clear when looking at its mean value. Assume that the data matrix $\mathbf{X}$ is a realization of a random variable so that, for any value of $\mathbf{A}$, one can define the mean value of the empirical contrast $\widehat{\Upsilon}(\mathbf{Y}) = \widehat{\Upsilon}(\mathbf{A}^{-1}\mathbf{X})$, denoted $\Upsilon(\mathbf{Y})$, defined as $\Upsilon(\mathbf{Y}) \stackrel{\text{def}}{=} \mathbb{E}\left\{\widehat{\Upsilon}(\mathbf{Y})\right\}$, and called the *likelihood contrast*. A simple calculation, generalizing [4] to the vector case, reveals that

$$T\mathbb{E}\left\{\widehat{\Upsilon}(\mathbf{Y})\right\} = -K\left[p_{\mathbf{Y}} \mid p_{\mathbf{S}}\right] + \text{cst} \tag{4.9}$$

where $K[p \mid q] = \int p \log p/q$ is the Kullback divergence and cst is a scalar constant depending only on the data distribution and *not* on the model built for it. In other words, it is a constant with respect to statistical inference. Hence, one should see $\widehat{\Upsilon}(\mathbf{Y})$ as an estimate of $\mathbb{E}\left\{\widehat{\Upsilon}(\mathbf{Y})\right\}$ so that maximizing the likelihood with respect to $\mathbf{A}$ is understood as minimizing an estimate of the divergence $K\left[p_{\mathbf{Y}} \mid p_{\mathbf{S}}\right]$ between the distribution $p_{\mathbf{Y}}$ of the transformed data $\mathbf{Y} = \mathbf{A}^{-1}\mathbf{X}$ and the distribution $p_{\mathbf{S}}$ of the source matrix $\mathbf{S}$. In other words, for a given source model $p_{\mathbf{S}}$, the likeliest mixing $\mathbf{A}$ is the one such that the empirical distribution of $\mathbf{A}^{-1}\mathbf{X}$ best fits the model distribution $p_{\mathbf{S}}$.

It must be stressed that $p_{\mathbf{Y}}$ and $p_{\mathbf{S}}$ are distributions of a very different nature. Distribution $p_{\mathbf{Y}}$ is the distribution of $\mathbf{A}^{-1}\mathbf{X}$ which thus depends on the mixing model via matrix $\mathbf{A}$ and on the distribution of $\mathbf{X}$ *whatever it is* (in particular, whether or not it follows a mixing model). In contrast, distribution $p_{\mathbf{S}}$ is the distribution used in defining a probabilistic model leading to a likelihood. Thus, $p_{\mathbf{S}}$ is freely chosen by the investigator and is not expected to be the "true" source distribution. It is an element of the model, not an element of the "reality". Still, the success of a likelihood procedure depends on our ability to design a source model $p_{\mathbf{S}}$ which is close enough to the data. The following sections show what that means in the source separation model.

4.2.3 Relative variations

To gain a first understanding of the shape of the likelihood, we look into the first and second derivatives of the empirical likelihood contrast $\widehat{\Upsilon}(\mathbf{Y}) = \widehat{\Upsilon}(\mathbf{A}^{-1}\mathbf{X})$ with respect to the parameter of interest: the mixing matrix $\mathbf{A}$. Since it belongs to a continuous group, it seems logical to use the appropriate notion of derivation: the Lie derivative. Fortunately, there is no need to resort to the abstract theory of Lie derivation. All the interesting results for the group of interest (the linear group of $\mathbb{R}^N$) can be obtained by defining "relative variations". We introduce below the notion of *relative gradient* and of *relative Hessian*.

Relative variation. A *relative variation* of a data matrix $\mathbf{Y}$ is an infinitesimal transformation of the form

$$\mathbf{Y} \longleftarrow (\mathbf{I} + \mathcal{E})\mathbf{Y} \tag{4.10}$$

where $\mathbf{I}$ is the $N \times N$ identity matrix and $\mathcal{E}$ is an infinitesimal matrix. If $\mathbf{Y} = \mathbf{A}^{-1}\mathbf{X}$, this transform corresponds to a change of $\mathbf{A}$ into $\mathbf{A}(\mathbf{I} + \mathcal{E})^{-1}$ which is equivalent to $\mathbf{A}(\mathbf{I} - \mathcal{E})$ at first order in $\mathcal{E}$.

Relative gradient. If $l(\mathbf{Y}):\mathbb{R}^{N\times T}\mapsto\mathbb{R}$ is a differentiable scalar function of a data block, the *relative gradient of* $l(\mathbf{Y})$ *at point* $\mathbf{Y}$ is the matrix of size $N\times N$ denoted $\nabla l(\mathbf{Y})$ and defined by

$$\nabla l(\mathbf{Y})=\frac{\partial l(\mathbf{Y}+\mathcal{E}\mathbf{Y})}{\partial\mathcal{E}}|_{\mathcal{E}=0}. \tag{4.11}$$

The first order relative variation of $l(\mathbf{Y})$ then is

$$l(\mathbf{Y}+\mathcal{E}\mathbf{Y})=l(\mathbf{Y})+\langle\nabla l(\mathbf{Y}),\mathcal{E}\rangle+o(||\mathcal{E}||) \tag{4.12}$$

where $\langle\cdot,\cdot\rangle$ is the scalar product between matrices: $\langle\mathbf{U},\mathbf{V}\rangle=\sum_{ij}\mathbf{U}_{ij}\mathbf{V}_{ij}$.

The connection between the relative gradient of the empirical contrast $\widehat{\Upsilon}(\mathbf{Y})$ and the usual regular gradient of the log-likelihood is found by noting that if $f(\mathbf{B})=l(\mathbf{BX})$, then $\nabla l(\mathbf{Y})=\frac{\partial f(\mathbf{B})}{\partial\mathbf{B}}\mathbf{B}^{-\mathrm{T}}$ at point $\mathbf{Y}=\mathbf{BX}$.

Relative gradient of the likelihood empirical contrast. The relative gradient of $\widehat{\Upsilon}(\mathbf{Y})$ is obtained by a first order expansion in $\mathcal{E}$ combining the gradient form (4.8), expansion (4.12), definition (4.4) of ψ and the standard properties $(\mathbf{I}+\mathcal{E})^{-1}=\mathbf{I}-\mathcal{E}+o(||\mathcal{E}||)$ and $\log\det(\mathbf{I}+\mathcal{E})=\mathrm{trace}(\mathcal{E})+o(||\mathcal{E}||)$. It yields

$$-\nabla\widehat{\Upsilon}(\mathbf{Y})=\frac{1}{T}\psi(\mathbf{Y})\mathbf{Y}^{\mathrm{T}}-\mathbf{I}. \tag{4.13}$$

Hence, stationary points of the likelihood are those matrices $\mathbf{A}$ such that the estimate $\mathbf{Y}=\mathbf{A}^{-1}\mathbf{X}$ verify the *estimating equation* $\nabla\widehat{\Upsilon}(\mathbf{Y})=0$, i.e.

$$T^{-1}\psi(\mathbf{Y})\mathbf{Y}^{\mathrm{T}}-\mathbf{I}=0. \tag{4.14}$$

Relative gradient algorithm. Efficient algorithms for solving estimating equations of the kind (4.14) in specific cases (i.e. depending on the form of ψ for specific source models) are studied later. However, there exists an elementary generic algorithm for solving $\nabla\widehat{\Upsilon}(\mathbf{Y})=0$ with some efficiency.

Cancelling the relative gradient $\nabla\widehat{\Upsilon}(\mathbf{Y})=0$ is a problem which suggests using a ... relative gradient algorithm. A relative variation $\mathcal{E}$ of $\mathbf{Y}$ into $(\mathbf{I}+\mathcal{E})\mathbf{Y}$ induces a variation of the log-likelihood equal to $\langle\nabla\widehat{\Upsilon}(\mathbf{Y}),\mathcal{E}\rangle$ at first order (Eq. (4.12)). Therefore, for this variation to be positive, is suffices to align the relative variation on the relative gradient, that is, to take $\mathcal{E}=\mu\nabla\widehat{\Upsilon}(\mathbf{Y})$ for some positive $\mu>0$ because that induces, at first order in μ, a variation of the log-likelihood equal to $\langle\nabla\widehat{\Upsilon}(\mathbf{Y}),\mathcal{E}\rangle=\langle\nabla\widehat{\Upsilon}(\mathbf{Y}),\mu\nabla\widehat{\Upsilon}(\mathbf{Y})\rangle=\mu||\nabla\widehat{\Upsilon}(\mathbf{Y})||^2$. That last quantity cancels only if $\nabla\widehat{\Upsilon}(\mathbf{Y})$ does.

Hence, the relative gradient algorithm iterates, for small enough μ, through

$$\mathbf{Y}\leftarrow(\mathbf{I}+\mu\nabla\widehat{\Upsilon}(\mathbf{Y}))\mathbf{Y}. \tag{4.15}$$

The algorithm stops progressing when the updating rule (4.15) does not produce any effect, that is, when $\nabla\widehat{\Upsilon}(\mathbf{Y})=0$.

4.2.4 The iid Gaussian model and decorrelation

The iid Gaussian model is half blind. If T samples are modeled as identically and independently distributed (iid) with a common Gaussian distribution $\mathcal{N}(0,\Sigma)$, the likelihood empirical is easily found to be given by

$$\widehat{\Upsilon}(\mathbf{Y}) = -k(\widehat{\mathbf{R}}_Y, \Sigma) + \text{cst} \tag{4.16}$$

where $k(\mathbf{R}_1, \mathbf{R}_2) \stackrel{\text{def}}{=} K[\mathcal{N}(\mathbf{R}_1) \,|\, \mathcal{N}(\mathbf{R}_2)]$ is the Kullback divergence between two Gaussian distributions with the same mean and covariance matrices $\mathbf{R}_1$ and $\mathbf{R}_2$. An explicit expression is

$$k(\mathbf{R}_1, \mathbf{R}_2) = \frac{1}{2}\left(\text{trace}(\mathbf{R}_1\mathbf{R}_2^{-1}) - \log\det \mathbf{R}_1\mathbf{R}_2^{-1} - P\right). \tag{4.17}$$

Hence, the likeliest Gaussian iid model is the one such that the empirical covariance matrix $\widehat{\mathbf{R}}_\mathbf{Y}$ of $\mathbf{Y}$ is as close as possible (as measured by the Kullback divergence (4.17)) to the common covariance matrix Σ postulated for the sources. Unfortunately, that does not provide a source separation principle because there are infinitely many values of $\mathbf{A}$ which solve this minimization problem exactly: the condition of equal covariances $\widehat{\mathbf{R}}_\mathbf{Y} = \Sigma$ can be fulfilled in infinitely many ways. Indeed, if it holds for a particular value of $\mathbf{A}$, it also holds for $\mathbf{A}\Sigma^{\frac{1}{2}}\mathbf{U}\Sigma^{-\frac{1}{2}}$ where $\mathbf{U}$ is any orthonormal matrix, i.e. such that $\mathbf{U}\mathbf{U}^\mathrm{T} = \mathbf{I}$.

It is interesting to see this lack of identifiability from a different angle. In the iid-Gaussian model, the score function is linear: $\psi(\mathbf{Y}) = \Sigma^{-1}\mathbf{Y}$ so that the estimating equation (4.14) becomes $0 = T^{-1}\psi(\mathbf{Y})\mathbf{Y}^\mathrm{T} - \mathbf{I} = T^{-1}\Sigma^{-1}\mathbf{Y}\mathbf{Y}^\mathrm{T} - \mathbf{I}$ which is equivalent to $\widehat{\mathbf{R}}_\mathbf{Y} = \Sigma$, as above. While N^2 constraints are needed to determine $\mathbf{A}$, the matrix equality $\widehat{\mathbf{R}}_\mathbf{Y} = \Sigma$ only provides $N(N+1)/2$ independent scalar constraints because covariance matrices are symmetric. Hence, the iid-Gaussian is half blind in the sense that it only provides (roughly) half the number of constraints needed to determine the mixture $\mathbf{A}$.

The above analysis shows that the source model must break the symmetry of $\psi(\mathbf{Y})\mathbf{Y}^\mathrm{T}$ observed in the iid-Gaussian case. In section 4.5, we investigate models which are iid but non-Gaussian and, in section 4.6, models which are Gaussian but not iid.

Orthogonal methods and decorrelation. The assumption of source independence implies that the covariance matrix Σ is diagonal. One can even choose it to be equal to the identity matrix, $\Sigma = \mathbf{I}$, since each source can be taken to have unit variance with its overall scale controlled by the corresponding column of $\mathbf{A}$. The likeliest iid-Gaussian model then is such that $\widehat{\mathbf{R}}_\mathbf{Y} = \mathbf{I}$. We just saw that this decorrelation condition for the rows of $\mathbf{Y}$ is not sufficient to determine $\mathbf{A}$ uniquely and that "stronger" models (in the sense that $\psi(\mathbf{Y})\mathbf{Y}^\mathrm{T}$ is not symmetric) are required. However, and maybe ironically, the stronger condition (4.14) does *not* entail empirical decorrelation. In other words, there is no reason in general why $T^{-1}\psi(\mathbf{Y})\mathbf{Y}^\mathrm{T} = \mathbf{I}$ would imply $T^{-1}\mathbf{Y}\mathbf{Y}^\mathrm{T} = \mathbf{I}$.

This fact may be seen as annoying: should one accept that a separation method which is supposed to go beyond decorrelation would not *in fine* produce sources which are "not even uncorrelated"? If that is considered unacceptable behavior, it is easily curbed by considering maximum likelihood *under the decorrelation constraint*, that is,

$$\max \widehat{\Upsilon}(\mathbf{Y}) \quad \text{under the constraint} \quad \mathbf{Y}=\mathbf{A}^{-1}\mathbf{X} \quad \text{and} \quad \widehat{\mathbf{R}}_Y=\mathbf{I}.$$

Those estimation methods which impose the empirical decorrelation of the recovered sources are called *orthogonal methods* and are said to operate "under the whiteness constraint".

We shall see in section 4.4.5 that the whiteness constraint increases the robustness with respect to source modeling errors. However, it may also limit estimation accuracy: the separation performance of any orthogonal method is limited by a "whitening bound" [3].

Maximum likelihood with the whiteness constraint. We examine how the whiteness constraint modifies the estimating equations and how it can be easily enforced using relative variations. The whiteness constraint $\widehat{\mathbf{R}}_Y=\mathbf{I}$ forbids some relative variations. If $\mathbf{Y}$ is white, i.e. $\widehat{\mathbf{R}}_Y=\mathbf{I}$, a relative variation $\mathcal{E}$ changes it into $(\mathbf{I}+\mathcal{E})\mathbf{Y}$ and changes its empirical covariance matrix into $(\mathbf{I}+\mathcal{E})\widehat{\mathbf{R}}_\mathbf{Y}(\mathbf{I}+\mathcal{E})^\mathrm{T}=(\mathbf{I}+\mathcal{E})(\mathbf{I}+\mathcal{E})^\mathrm{T}=\mathbf{I}+\mathcal{E}+\mathcal{E}^\mathrm{T}+\mathcal{E}\mathcal{E}^\mathrm{T}$ which, at first order, remains equal to the identity matrix if and only if $\mathcal{E}+\mathcal{E}^\mathrm{T}=0$. Hence, the whiteness constraint only allows relative variations which are *skew-symmetric*: $\mathcal{E}^\mathrm{T}=-\mathcal{E}$. Thus, the stationarity of the likelihood under the whiteness constraint requires that the relative variation – equal to $\langle\nabla\widehat{\Upsilon}(\mathbf{Y}),\mathcal{E}\rangle$ by Eq. (4.12) – vanishes for all skew-symmetric $\mathcal{E}$. This is equivalent to the cancellation of the skew-symmetric part of the relative gradient $\nabla\widehat{\Upsilon}(\mathbf{Y})$. Hence, under the whiteness constraint, the stationarity condition for the likelihood $\nabla\widehat{\Upsilon}(\mathbf{Y})=0$ becomes

$$\nabla\widehat{\Upsilon}(\mathbf{Y})-\nabla\widehat{\Upsilon}(\mathbf{Y})^\mathrm{T}=0 \quad \text{and} \quad \widehat{\mathbf{R}}_\mathbf{Y}=\mathbf{I}. \tag{4.18}$$

Using expression (4.14) of $\nabla\widehat{\Upsilon}(\mathbf{Y})$, the two conditions (4.18) can be combined into a single *estimating equation*

$$\frac{1}{T}(\psi(\mathbf{Y})\mathbf{Y}^\mathrm{T}-\mathbf{Y}\psi(\mathbf{Y})^\mathrm{T}+\mathbf{Y}\mathbf{Y}^\mathrm{T})-\mathbf{I}=0 \tag{4.19}$$

because the cancellation of the matrix on the l.h.s. of (4.19) implies the cancellation of both its symmetric part (which implies $\widehat{\mathbf{R}}_\mathbf{Y}=\mathbf{I}$) and of its skew-symmetric part (which implies $\nabla\widehat{\Upsilon}(\mathbf{Y})-\nabla\widehat{\Upsilon}(\mathbf{Y})^\mathrm{T}=0$). The constrained form (4.19) is to be compared to the unconstrained form (4.14).

Optimization under the whiteness constraint. The whiteness constraint can be satisfied explicitly by a two-step processing. In a first step, the data matrix $\mathbf{X}$ is left multiplied by any (matrix) square root of $\mathbf{R}_X^{-1}$, ensuring whiteness $\widehat{\mathbf{R}}_\mathbf{Y}=\mathbf{I}$. In a second step, the likelihood is maximized with respect to whiteness-preserving transformations, that is, with respect to rotation matrices. This can be achieved by the Jacobi method, as first proposed by Comon [8] for ICA) which proceeds pair-wise: all pairs of sources are selected in sequence and the criterion (here: the likelihood) is optimized iteratively by plane rotations (rotation of a pair of sources). Since a plane rotation depends on a single angle, each sub-problem reduces to a one-dimensional problem.

For gradient-based algorithms, one may proceed similarly: whitening followed by optimization over rotation matrices. However, a much simpler approach is possible in which the data are not explicitly whitened in a prior step. Rather, whitening is achieved by replacing the gradient $\nabla\widehat{\Upsilon}(\mathbf{Y}) = \frac{1}{T}(\psi(\mathbf{Y})\mathbf{Y}^{\mathrm{T}}) - \mathbf{I}$ with the form (4.19). In other words, the factor $\nabla\widehat{\Upsilon}(\mathbf{Y})$ in (4.15) is changed into $\frac{1}{T}(\psi(\mathbf{Y})\mathbf{Y}^{\mathrm{T}} - \mathbf{Y}\psi(\mathbf{Y})^{\mathrm{T}} + \mathbf{Y}\mathbf{Y}^{\mathrm{T}}) - \mathbf{I}$. Then, the algorithm will stop when that quantity vanishes, that is, when condition (4.19) is satisfied. Note that the whiteness condition is *not* necessarily satisfied as long as the algorithm has not converged.[2] This is not a weakness of the algorithm but rather a strong point: it is not necessary to enforce whiteness along the trajectory since it is automatically enforced at convergence. This results in a particularly simple form which is very appropriate for on-line algorithms [7].

4.2.5 Equivariant estimators and uniform performance

Estimating equations. We have established the estimating equations for the maximum likelihood with and without the whiteness constraint. In both cases, they take an invariant form, that is, they can be written as $\mathbf{H}(\mathbf{Y}) = 0$ for some *estimation function* $\mathbf{H} : \mathbb{R}^{N\times T} \to \mathbb{R}^{N\times N}$ with

$$\mathbf{H}(\mathbf{Y}) = \frac{1}{T}\psi(\mathbf{Y})\mathbf{Y}^{\mathrm{T}} - \mathbf{I} \qquad \text{(unconstrained)} \qquad (4.20)$$

$$\mathbf{H}(\mathbf{Y}) = \frac{1}{T}(\psi(\mathbf{Y})\mathbf{Y}^{\mathrm{T}} - \mathbf{Y}\psi(\mathbf{Y})^{\mathrm{T}} + \mathbf{Y}\mathbf{Y}^{\mathrm{T}}) - \mathbf{I} \qquad \text{(whiteness constraint)} \qquad (4.21)$$

One could consider more general estimating functions, which should at least satisfy $\mathbb{E}\mathbf{H}(\mathbf{S}) = 0$ when matrix $\mathbf{S}$ has independent rows. However, the above two forms are the only ones offering a clear interpretation in terms of likelihood.

Equivariant estimators. Any estimator of the mixing matrix can be seen as a function $\mathscr{A} : \mathbb{R}^{N\times T} \mapsto \mathbb{R}^{N\times N}$ returning an estimated matrix $\widehat{\mathbf{A}} = \mathscr{A}(\mathbf{X})$ when fed with data $\mathbf{X}$. An estimator is said to be *equivariant* if, for any invertible $N \times N$ matrix $\mathbf{M}$,

$$\mathscr{A}(\mathbf{M}\mathbf{X}) = \mathbf{M}\mathscr{A}(\mathbf{X}), \qquad (4.22)$$

that is, if *a linear transform of the data induces the same transform on the estimated matrix.*

Uniform performance. Consider a data matrix $\mathbf{X}$ from which a mixing matrix is estimated using an estimator $\mathscr{A}$. The estimated sources are $\widehat{\mathbf{S}} = \widehat{\mathbf{A}}^{-1}\mathbf{X} = \mathscr{A}(\mathbf{X})^{-1}\mathbf{X}$.

[2] Also note that the form (4.19) is *not*, in general, the relative gradient of a function of $\mathbf{Y}$.

Let us transform the data by an invertible matrix $\mathbf{M}$ to obtain a new data set $\tilde{\mathbf{X}} = \mathbf{M}\mathbf{X}$. The sources estimated from the transformed data set are $\mathscr{A}(\tilde{\mathbf{X}})^{-1}\tilde{\mathbf{X}} = \mathscr{A}(\mathbf{M}\mathbf{X})^{-1}\mathbf{M}\mathbf{X}$. If the estimator $\mathscr{A}$ is equivariant, those estimated sources also are $(\mathbf{M}\mathscr{A}(\mathbf{X}))^{-1}\mathbf{M}\mathbf{X} = \mathscr{A}(\mathbf{X})^{-1}\mathbf{X} = \widehat{\mathbf{S}}$, that is, the very same estimated sources are obtained from the original data $\mathbf{X}$ and from the transformed data $\mathbf{M}\mathbf{X}$.

Now, assume that the observations actually follow a mixing model: it exists as a "true" mixing matrix $\mathbf{A}_\star$ and "true" independent sources from which we observe a realization $\mathbf{S}$ of length T mixed by $\mathbf{A}_\star$, that is, $\mathbf{X} = \mathbf{A}_\star\mathbf{S}$. If $\mathbf{S}$ is estimated from $\mathbf{X}$ using an equivariant estimator, one obtains estimated sources:

$$\widehat{\mathbf{S}} = \mathscr{A}(\mathbf{X})^{-1}\mathbf{X} = \mathscr{A}(\mathbf{A}_\star\mathbf{S})^{-1}\mathbf{A}_\star\mathbf{S} = \mathscr{A}(\mathbf{S})^{-1}\mathbf{A}_\star^{-1}\mathbf{A}_\star\mathbf{S} = \mathscr{A}(\mathbf{S})^{-1}\mathbf{S}. \tag{4.23}$$

Therefore, the particular value $\mathbf{A}_\star$ of the mixing matrix has *no* effect on the estimated sources! The separation accuracy then is independent of the mixing matrix (and, in particular, of its conditioning). We shall see below that the (properly defined) Cramér-Rao bound for source separation does not depend on the mixing matrix. In that sense, "severely mixed sources" are no more difficult to separate than "lightly mixed sources". In particular, it is legitimate to test the statistical efficiency of an equivariant source separation algorithm using $\mathbf{A}_\star = \mathbf{I}$.

4.2.6 Summary

In the noise free case, the source separation model is, in the first place, a transformation model: the parameter of interest $\mathbf{A}$ is an invertible linear transform. Focusing on this sole parameter, and even before introducing the assumption of source independence, we have exhibited the equivariant structure. In particular, the likelihood contrast measures the deviation between the distribution of $\mathbf{Y} = \mathbf{A}^{-1}\mathbf{X} = \widehat{\mathbf{S}}$ and the model distribution for $\mathbf{S}$ (4.9). The (relative) gradient of the likelihood depends only on $\mathbf{Y} = \mathbf{A}^{-1}\mathbf{X}$; the stationarity condition for the likelihood, with or without the whiteness constraint, is a condition on $\mathbf{Y}$; the separation performance is independent of the actual mixture and therefore depends only on our ability to model the source distribution $p_{\mathbf{S}}$. The key feature of that distribution is the statistical independence of the rows of $\mathbf{S}$. It is that new ingredient which is studied in the next section.

4.3 INDEPENDENCE

The likeliest solution for a given source model is the maximum of an empirical contrast $\widehat{\Upsilon}(\mathbf{Y})$ (section 4.2.1) and that function is determined as soon as a source model $p_{\mathbf{S}}$ is specified, Eq. (4.8). Nothing specific has been said so far regarding the specific source distribution $p_{\mathbf{S}}$. In this section, we introduce the assumption of source independence, and only that, i.e. without specifying the distribution of each source. In some sense, we examine the pure effect of independence.

4.3.1 Score function and estimating equations

The source independence assumption reads

$$P_{\mathbf{S}}(\mathbf{S}) = \prod_{i=1}^{N} p_{\mathbf{s}_i}(\mathbf{s}_i) \tag{4.24}$$

where $\mathbf{s}_i$ is the ith row of $\mathbf{S}$, that is, the signal emitted by the ith source, modeled as having probability density $p_{\mathbf{s}_i}(\mathbf{s}_i)$. The empirical likelihood contrast (4.8) for the independent source model (4.24) then becomes

$$\widehat{\Upsilon}(\mathbf{Y}) = \frac{1}{T}\sum_{i=1}^{N} \log p_{\mathbf{s}_i}(\mathbf{y}_i) + \frac{1}{2}\log\det \widehat{\mathbf{R}}_{\mathbf{Y}}. \tag{4.25}$$

Independence expressed by the factorization (4.24) implies that each row ψ_i of matrix $\psi(\mathbf{S})$ depends only on corresponding row of $\mathbf{S}$:

$$\psi(\mathbf{S})_i = \varphi_i(\mathbf{s}_i) \quad \text{with} \quad [\varphi_i(\mathbf{s}_i)]_t = -\partial \log p_{\mathbf{s}_i}(\mathbf{s}_i)/\partial s_i(t) \quad (1 \le t \le T).$$

It translates into a stationarity condition of the likelihood showing *pair-wise decoupling* since the (i,j)th entry of the matrix equation $\nabla\widehat{\Upsilon}(\mathbf{Y}) = 0$ now reads

$$\frac{1}{T}\varphi_i(\mathbf{y}_i)\mathbf{y}_j^{\mathrm{T}} - \delta_{ij} = 0 \quad (1 \le i,j \le N) \tag{4.26}$$

which involves only $\mathbf{y}_i$ and $\mathbf{y}_j$. Even though nothing has been said so far of the specific source models to be considered later (that is, selecting $p_{\mathbf{s}_i}(\cdot)$), it is already possible to exhibit several characteristics of the ML estimator by analyzing the estimating equations (4.26). These are fundamental properties since they do not depend on source models.

For later reference, we note an important special case where one chooses for the row vector $\mathbf{s}_i$, a Gaussian distribution with zero mean and a $T \times T$ covariance matrix $\mathbf{R}_i$. Then, one readily finds the associated score function:

$$\varphi_i(\mathbf{s}_i) = \mathbf{s}_i \mathbf{R}_i^{-1} \quad \text{for Gaussian } \mathbf{s}_i \text{ with covariance } \mathbf{R}_i = \mathbb{E}\left\{\mathbf{s}_i^{\mathrm{T}}\mathbf{s}_i\right\}. \tag{4.27}$$

On the diagonal: scales. Consider first the estimating equations (4.26) for $i = j$, that is, the diagonal terms of $\nabla\widehat{\Upsilon}(\mathbf{Y}) = 0$. They also read

$$\frac{1}{T}\varphi_i(\mathbf{y}_i)\mathbf{y}_i^{\mathrm{T}} = 1 \quad (1 \le i \le N). \tag{4.28}$$

Each of these equations expresses the stationarity of the likelihood with respect to the relative variation induced by a term $\mathcal{E}_{ii}$. But the action of such a term on $\mathbf{Y}$ is to change the scale of $\mathbf{y}_i$ by a factor $(1 + \mathcal{E}_{ii})$. Hence, the role of equations (4.28) is to set the scale of each of the recovered sources to the level predicted by the model $p_{\mathbf{s}_i}$, and to fix the amplitude of the estimated sources.

The task of fixing scales seems of secondary importance for source separation. One could think of replacing each of the constraints (4.28) by a $T^{-1}\mathbf{y}_i\mathbf{y}_i^{\mathrm{T}} = \sigma_i^2$ (as in the Gaussian case discussed above), fixing the variance of the ith source to a given σ_i^2. However, it must be kept in mind that any scale-fixing choice may impact indirectly the off-diagonal estimating equations ($i \neq j$). This is because the condition $\boldsymbol{\varphi}_i(\mathbf{y}_i)\mathbf{y}_j^{\mathrm{T}} = 0$ is not *at all* equivalent to $\boldsymbol{\varphi}_i(\lambda\mathbf{y}_i)\mathbf{y}_j^{\mathrm{T}} = 0$ for a scale factor $\lambda \neq 1$ if the score $\boldsymbol{\varphi}_i$ is a non-linear function. Changing the scale factor thus *effectively* amounts to changing $\boldsymbol{\varphi}_i$ unless it has some specific scaling properties, i.e., if it exists as a function g such that $\boldsymbol{\varphi}_i(\lambda\mathbf{y}_i) = g(\lambda)\boldsymbol{\varphi}_i(\mathbf{y}_i)$.

Off diagonal: non-linear decorrelation. The $N(N-1)/2$ off-diagonal terms of $\nabla\widehat{\Upsilon}(\mathbf{Y}) = 0$ can be grouped pairwise. Using (4.26), this is:

$$\begin{bmatrix} \boldsymbol{\varphi}_i(\mathbf{y}_i)\mathbf{y}_j^{\mathrm{T}} \\ \boldsymbol{\varphi}_j(\mathbf{y}_j)\mathbf{y}_i^{\mathrm{T}} \end{bmatrix} = 0 \quad (1 \leq i < j \leq N). \tag{4.29}$$

For any two distinct sources $i \neq j$, the condition $\boldsymbol{\varphi}_i(\mathbf{y}_i)\mathbf{y}_j^{\mathrm{T}} = 0$ is the test of independence suggested by the maximum likelihood principle *resulting from choosing a model* $p_{\mathbf{s}_i}$. It corresponds to the property $\mathbb{E}\left\{\boldsymbol{\varphi}_i(\mathbf{s}_i)\mathbf{s}_j^{\mathrm{T}}\right\} = 0$ which is true (if $\mathbf{s}_j$ has zero mean) regardless of the distribution of $\mathbf{s}_i$ thanks to independence because then $\mathbb{E}\left\{\boldsymbol{\varphi}_i(\mathbf{s}_i)\mathbf{s}_j^{\mathrm{T}}\right\} = \mathbb{E}\{\boldsymbol{\varphi}_i(\mathbf{s}_i)\}\,\mathbb{E}\left\{\mathbf{s}_j^{\mathrm{T}}\right\}$. The stationarity condition of the likelihood $\boldsymbol{\varphi}_i(\mathbf{y}_i)\mathbf{y}_j^{\mathrm{T}} = 0$ can be understood as the empirical version of $\mathbb{E}\left\{\boldsymbol{\varphi}_i(\mathbf{s}_i)\mathbf{s}_j^{\mathrm{T}}\right\} = 0$.

Whiteness constraint. The form (4.18) of the estimating equations under the whiteness constraint similarly becomes

$$\begin{bmatrix} \mathbf{y}_i\mathbf{y}_j^{\mathrm{T}} \\ \boldsymbol{\varphi}_i(\mathbf{y}_i)\mathbf{y}_j^{\mathrm{T}} - \boldsymbol{\varphi}_j(\mathbf{y}_j)\mathbf{y}_i^{\mathrm{T}} \end{bmatrix} = 0 \quad (1 \leq i < j \leq N) \tag{4.30}$$

once the independence assumption is introduced and expressed by Eq. (4.28).

4.3.2 Mutual information

There is a direct connection between likelihood and mutual information. Recall from Eq. (4.9) that the mean empirical likelihood contrast is $T\mathbb{E}\left\{\widehat{\Upsilon}(\mathbf{Y})\right\} = -K\left[p_{\mathbf{Y}} \mid p_{\mathbf{S}}\right] + \text{cst}$. When source independence is introduced, the significant part, the Kullback divergence $K\left[p_{\mathbf{Y}} \mid p_{\mathbf{S}}\right]$ is found to decompose as:

$$K\left[p_{\mathbf{Y}} \mid p_{\mathbf{S}}\right] = K\left[p_{\mathbf{Y}} \mid \prod_i p_{\mathbf{s}_i}\right] = K\left[p_{\mathbf{Y}} \mid \prod_i p_{\mathbf{y}_i}\right] + K\left[\prod_i p_{\mathbf{y}_i} \mid \prod_i p_{\mathbf{s}_i}\right] \tag{4.31}$$

This is a known result (see e.g. [6]) in the case of $N \times 1$ vectors but it also obviously holds for $N \times T$ matrices. One of the above terms is the *mutual information* between rows of **Y**, defined as

$$I(\mathbf{Y}) \stackrel{\text{def}}{=} K\left[p_{\mathbf{Y}} \mid \prod_i p_{\mathbf{y}_i}\right]. \tag{4.32}$$

In addition, both distributions in the second term of (4.31) are distributions with independent rows, so that $K\left[\prod_i p_{\mathbf{y}_i} \mid \prod_i p_{\mathbf{s}_i}\right] = \sum_{i=1}^{N} K\left[p_{\mathbf{y}_i} \mid p_{\mathbf{s}_i}\right]$, yielding

$$K\left[p_{\mathbf{Y}} \mid p_{\mathbf{S}}\right] = I(\mathbf{Y}) + \sum_{i=1}^{N} K\left[p_{\mathbf{y}_i} \mid p_{\mathbf{s}_i}\right]. \tag{4.33}$$

Since maximizing the likelihood is minimizing an estimate of the likelihood contrast $K\left[p_{\mathbf{Y}} \mid p_{\mathbf{S}}\right]$ (section 4.2.2), decomposition (4.33) teaches us the following:

1. The likelihood approach amounts to maximizing the independence between components (by minimizing the mutual information $I(\mathbf{Y})$) while attempting to make the distributions of each component as close as possible to the postulated source distributions (by minimizing the sum of marginal divergences $K\left[p_{\mathbf{y}_i} \mid p_{\mathbf{s}_i}\right]$).
2. If the source distributions $p_{\mathbf{s}_i}$ are included among the parameters to be optimized over, the likelihood approach dictates minimizing each of the $K\left[p_{\mathbf{y}_i} \mid p_{\mathbf{s}_i}\right]$ terms, that is, estimating (at least in principle) the distribution $p_{\mathbf{s}_i}$ of each source based on the corresponding output $\mathbf{y}_i$.
3. If, for a given distribution of **Y**, the likeliest source model $p_{\mathbf{s}_i}$ is selected, that is, if (ideally) $p_{\mathbf{s}_i} = p_{\mathbf{y}_i}$, the $K\left[p_{\mathbf{y}_i} \mid p_{\mathbf{s}_i}\right] = 0$ so that the likelihood contrast reduces to mutual information:

$$\min_{p_{\mathbf{s}_1}, \ldots, p_{\mathbf{s}_N}} K\left[p_{\mathbf{Y}} \mid \prod_i p_{\mathbf{s}_i}\right] = K\left[p_{\mathbf{Y}} \mid \prod_i p_{\mathbf{y}_i}\right] = I(\mathbf{Y}). \tag{4.34}$$

Pierre Comon gave a formal definition of independent component analysis as a minimization of mutual information with respect to the mixing matrix. We see here that the likelihood approach leads to the same criterion if the likelihood is to be optimized with respect to both the mixing matrix *and* to the source models.

4.3.3 Mutual information, correlation and...

In this section (and the next), we keep on exploring the likelihood contrast and we shall see what the correlation lacks such as to be a measure of independence. Relying heavily on Kullback divergences, entropy, and other functionals of probability distributions, a shorter notation will come in handy: we write $K[\mathbf{Y} \mid \mathbf{S}]$ instead of $K[p_\mathbf{Y} \mid p_\mathbf{S}]$ and more generally $K[A \mid B] = K[p_A \mid p_B]$.

Let $\mathbf{Y}$ be a random $N \times T$ matrix with distribution $P_\mathbf{Y}$. Its *average correlation matrix* is the $N \times N$ covariance matrix defined as

$$\mathbf{R}_\mathbf{Y} = \frac{1}{T} \sum_{t=1}^{T} \mathbb{E}\left\{\mathrm{y}(t)\mathrm{y}(t)^\mathrm{T}\right\} = \mathbb{E}\left\{\widehat{\mathbf{R}}_\mathbf{Y}\right\} \tag{4.35}$$

where the average is over the columns: $T^{-1}\sum_t \cdots$. If $\mathbf{Y}$ is a zero-mean random matrix, we denote by $\mathbf{Y}^G$ a zero-mean Gaussian random matrix with independent columns, each with covariance matrix $\mathbf{R}_\mathbf{Y}$, as defined in (4.35). This matrix choice is the one for which $K\left[\mathbf{Y} \mid \mathbf{Y}^G\right]$ is minimal: in this sense, $\mathbf{Y}^G$ is, in distribution, the best iid-Gaussian approximation to $\mathbf{Y}$. The minimal value of the divergence then measures how far $\mathbf{Y}$ is from being iid-Gaussian. Set

$$G[\mathbf{Y}] \stackrel{\text{def}}{=} K\left[\mathbf{Y} \mid \mathbf{Y}^G\right] \tag{4.36}$$

and think of it as a measure of the "non-iid-Gaussianity" of $\mathbf{Y}$. In the vector case (matrices reduced to $N \times 1$ vectors), this quantity reduces to a plain non-Gaussianity measure. Definition (4.36) also applies to each row of $\mathbf{Y}$:

$$G[\mathrm{y}_i] = K\left[\mathrm{y}_i \mid \mathrm{y}_i^G\right]$$

where y_i^G then is an iid-Gaussian row vector of variance equal to the average variance $\frac{1}{T}\sum_t \mathbb{E}\left\{y_i(t)^2\right\}$.

Correlation. The distribution of $\mathbf{Y}^G$ depending only on $\mathbf{R}_\mathbf{Y}$, so does its mutual information. We call *correlation* of $\mathbf{Y}$, denoted $C[\mathbf{Y}]$, the mutual information of its approximation by $\mathbf{Y}^G$, that is,

$$C[\mathbf{Y}] \stackrel{\text{def}}{=} I[\mathbf{Y}^G]. \tag{4.37}$$

In other words, the correlation $C[\mathbf{Y}]$ measures the independence between the rows of $\mathbf{Y}$ when only the time-averaged second-order statistics are considered. It is not difficult to find its explicit expression as a function of $\mathbf{R}_\mathbf{Y}$:

$$C[\mathbf{Y}] = I[\mathbf{Y}^G] = T\, k(\mathbf{R}_\mathbf{Y}, \mathrm{diag}(\mathbf{R}_\mathbf{Y})) \tag{4.38}$$

where $\mathrm{diag}(\mathbf{R}_\mathbf{Y})$ denotes the diagonal matrix with the same diagonal as $\mathbf{R}_\mathbf{Y}$ and where $k(\mathbf{R}_1, \mathbf{R}_2)$ is defined in Eq. (4.17). If the rows of $\mathbf{Y}$ are mutually independent, then $\mathbf{R}_\mathbf{Y}$ is a diagonal matrix and $I(\mathbf{Y}^G) = 0$ but the converse is not true: $I(\mathbf{Y}^G) = 0$ only implies the diagonality of $\mathbf{R}_\mathbf{Y}$ which is far from implying the independence between the rows of $\mathbf{Y}$. One can see correlation $C[\mathbf{Y}]$ as the *iid-Gaussian* approximation of the mutual information $I[\mathbf{Y}]$.

Connection. Mutual information $I[\mathbf{Y}]$ and correlation $C[\mathbf{Y}]$, its much weaker counterpart, are connected via non-iid-Gaussianity (see proof in appendix) by

$$I[\mathbf{Y}] = C[\mathbf{Y}] - \sum_i G[\mathbf{y}_i] + G[\mathbf{Y}]. \tag{4.39}$$

However, the non-iid-Gaussianity $G[\mathbf{Y}]$ is left invariant, that under left multiplication by any invertible transform, that is, $G[\mathbf{TY}] = G[\mathbf{Y}]$ for any invertible $N \times N$ matrix $\mathbf{T}$ (proof in appendix). In particular, $G[\mathbf{Y}] = G[\mathbf{A}^{-1}\mathbf{X}] = G[\mathbf{X}]$ and therefore does not depend on $\mathbf{A}$ for fixed $\mathbf{X}$. Hence

$$I[\mathbf{Y}] = C[\mathbf{Y}] - \sum_i G[\mathbf{y}_i] + G[\mathbf{X}]. \tag{4.40}$$

Therefore, under linear transform, gauss $\mathbf{X}$ *being constant (independent of* $\mathbf{A}$*), minimizing mutual information* $I[\mathbf{Y}]$ *also is minimizing correlation* $C[\mathbf{Y}]$ *between sources while maximizing the deviation of each* $G[\mathbf{y}_i]$ *to iid-Gaussianity.*

Orthogonal contrasts and source extraction. We already stressed that the ML solution does not guarantee source decorrelation. If decorrelation is desired, it must be imposed explicitly. Decomposition (4.40) makes this more clear: mutual information gives equal weight to the decorrelation term and to marginal deviations to iid-Gaussianity. Imposing decorrelation amounts to putting an *infinite weight* to the decorrelation term. Constraining whiteness is imposing $C[\mathbf{Y}] = 0$; under this constraint, mutual information becomes

$$I[\mathbf{Y}] = -\sum_i G[\mathbf{y}_i] + \text{cst} \tag{4.41}$$

and thus reduces to a sum of the marginal criteria over all sources. Note however that these marginal criteria cannot be independently optimized since the rows $\mathbf{y}_i$ must remain uncorrelated.

4.3.4 Summary

Introducing the source independence hypothesis, as we did in this section, reveals new and simple structures in the maximum likelihood estimate and in the associated contrast function.

First, the estimating equations for the likeliest matrix $\widehat{\mathbf{A}}_{\text{ML}}$ appear as independence conditions between recovered sources. Further, these are *pairwise* independence conditions, with one pair of conditions for each pair of sources. These are the $N(N+1)/2$ pairs of estimating equations (4.29) or, when source decorrelation is requested, the alternative version (4.30). For a given pair $(i < j)$ of sources, the corresponding pair of conditions only involves the rows $\mathbf{y}_i$ and $\mathbf{y}_j$, and nothing else. We shall see in the next section, 4.4, that this decoupling persists in the Fisher information matrix, making possible a detailed statistical analysis of the ICA model. The decoupling is no longer enjoyed in the noisy models studied in section 4.7.

Regarding the likelihood contrast, the source independence assumption leads to its decomposition into mutual information plus a sum of marginal deviations. The latter could be ideally reduced to zero by perfect modeling/estimation of the source distributions, leaving only the ultimate ICA objective: mutual information. Hence, our analysis shows how mutual information is hiding behind the likelihood function. We also saw that, even though mutual information $I[\mathbf{Y}]$ is a "complicated" functional of $p_{\mathbf{Y}}$, it behaves in a simple way under linear transforms of $\mathbf{Y}$. Indeed, mutual information decomposes (4.39) in a correlation term which is simple (expressing only an average second-order property), plus deviations from iid-Gaussianity which are simple in that they are marginal (depending only on each of the sources and not on their joint distribution), plus a term which is constant under linear transforms.

4.4 IDENTIFIABILITY, STABILITY, PERFORMANCE

This section is a bit technical and can be omitted at first reading. By studying the shape of the likelihood in the vicinity of a separating point, we can answer questions about blind identifiability, the Fisher information contained in a sample, the expected performance of blind source separation, the stability of adaptive algorithms, and the design of Newton-like techniques for maximizing the likelihood.

Those questions make sense only if the data matrix $\mathbf{X}$ actually originates from a mixture $\mathbf{X} = \mathbf{A}_\star \mathbf{S}$ for some matrix $\mathbf{A}_\star$ and a source matrix $\mathbf{S}$ with independent rows. We assume that each row $\mathbf{s}_i$ of $\mathbf{S}$ is a realization of a random variable with "true" probability distribution $p^\star_{\mathbf{s}_i}(\mathbf{s}_i)$, associated with a "true" score function $\boldsymbol{\varphi}^\star_i(\mathbf{s}_i)$. However, we consider the case that the source models $p_{\mathbf{s}_i}$ used in the likelihood are *not necessarily equal* to the actual distribution $p^\star_{\mathbf{s}_i}(\mathbf{s}_i)$. So the score functions $\boldsymbol{\varphi}_i$ used in the ML estimator are not assumed to be equal to the "true" score functions $\boldsymbol{\varphi}^\star_i$. This is important since, in practice, it is often unrealistic to expect perfect source models.

A lemma. The "true score" $\boldsymbol{\varphi}^\star$ enjoys a remarkable property, known as Stein's lemma in the Gaussian case. Pham [17] uses a non-Gaussian extension which, in the vector case, goes as follows. If $\mathbf{s}$ is a $T \times 1$ variate whose distribution has an associated score function $\boldsymbol{\varphi}^\star(\mathbf{s})$ and if $g(\mathbf{s})$ is a real-valued function of $\mathbf{s}$ with gradient $\partial g(\mathbf{s})/\partial \mathbf{s}$, then under mild conditions, integration by parts reveals that

$$\mathbb{E}\{g(\mathbf{s})\boldsymbol{\varphi}^\star(\mathbf{s})\} = \mathbb{E}\left\{\frac{\partial g(\mathbf{s})}{\partial \mathbf{s}}\right\}, \tag{4.42}$$

that is, the covariance between the true score and a function of $\mathbf{s}$ is the mean gradient of that function. Two special cases are

$$\mathbb{E}\left\{\boldsymbol{\varphi}(\mathbf{s})^{\mathrm{T}}\boldsymbol{\varphi}^\star(\mathbf{s})\right\} = \mathbb{E}\left\{\frac{\partial \boldsymbol{\varphi}(\mathbf{s})}{\partial \mathbf{s}}\right\} \quad \text{and} \quad \mathbb{E}\left\{\mathbf{s}^{\mathrm{T}}\boldsymbol{\varphi}^\star(\mathbf{s})\right\} = \mathbf{I}. \tag{4.43}$$

Analysis will show that the behavior of the ML estimators is governed for each source i by two matrices of size $T \times T$: the covariance matrix $\mathbf{R}_i = \mathbb{E}\left\{\mathbf{s}_i^{\mathrm{T}}\mathbf{s}_i\right\}$ already introduced in (4.27) and the covariance matrix:

$$\mathbf{\Gamma}_i = \mathbb{E}\left\{\varphi_i^\star(\mathbf{s}_i)^{\mathrm{T}}\varphi_i^\star(\mathbf{s}_i)\right\}. \tag{4.44}$$

Matrix $\mathbf{\Gamma}_i$ is never smaller than $\mathbf{R}_i^{-1}$ since, using relation (4.43), one has:

$$\begin{aligned} 0 &\leq \mathrm{Cov}(\varphi_i^\star(\mathbf{s}_i) - \mathbf{s}_i\mathbf{R}_i^{-1}) && (4.45)\\ &= \mathbb{E}\left\{(\varphi_i^\star(\mathbf{s}_i) - \mathbf{s}_i\mathbf{R}_i^{-1})^{\mathrm{T}}(\varphi_i^\star(\mathbf{s}_i) - \mathbf{s}_i\mathbf{R}_i^{-1})\right\}\\ &= \mathbb{E}\left\{\varphi_i^{\star\mathrm{T}}\varphi_i^\star\right\} - \mathbb{E}\left\{\varphi_i^{\star\mathrm{T}}\mathbf{s}_i\right\}\mathbf{R}_i^{-1} - \mathbf{R}_i^{-1}\mathbb{E}\left\{\mathbf{s}_i^{\mathrm{T}}\varphi_i^\star\right\} + \mathbf{R}_i^{-1}\mathbb{E}\left\{\mathbf{s}_i^{\mathrm{T}}\mathbf{s}_i\right\}\mathbf{R}_i^{-1}\\ &= \mathbf{\Gamma}_i - \mathbf{I}\,\mathbf{R}_i^{-1} - \mathbf{R}_i^{-1}\mathbf{I} + \mathbf{R}_i^{-1}\mathbf{R}_i\mathbf{R}_i^{-1} = \mathbf{\Gamma}_i - \mathbf{R}_i^{-1}. && (4.46) \end{aligned}$$

Hence $\mathbf{\Gamma}_i \geqslant \mathbf{R}_i^{-1}$ with equality $\mathbf{\Gamma}_i = \mathbf{R}_i^{-1}$ if $\varphi_i^\star(\mathbf{s}_i) = \mathbf{s}_i\mathbf{R}_i^{-1}$, that is, if $\mathbf{s}_i$ is Gaussian. We shall see below that it is precisely the deviation from $\mathbf{\Gamma}_i$ to $\mathbf{R}_i^{-1}$ which quantifies the contribution of non-Gaussianity to blind separability.

In this section, we assume that $\mathbf{R}_i^{-1}$ exists; i.e. the covariance matrices of each source were assumed to be invertible. The following analysis does not apply otherwise. However, it suggests that the singularity of covariance matrices induces a performance singularity. This is indeed the case in particular for Gaussian models: in section 4.6, one sees the Fisher information tends towards infinity as $\mathbf{R}_i$ matrices get closer to singularity. This is a "super-efficient" behavior in which estimation error goes to zero at a rate faster than $T^{-1/2}$.

4.4.1 Elements of asymptotic analysis

We first briefly recall some basic results of asymptotic analysis in the case of a generic parametric model $p(x, \theta)$ where the probability distribution of a random variable x depends on a vector θ of parameters. Exposition is sketchy, mostly targeted to readers with some familiarity with the topic, with the main objective of introducing notations and of showing how standard arguments apply to ICA.

Consider estimating parameter θ as the maximizer $\hat{\theta}$ of an empirical contrast $g(x;\theta)$ which is *not necessarily* the log-likelihood $\log p(x;\theta)$. The estimate $\hat{\theta}$ based on a realization x is a solution of the estimating equation $\partial g(x;\hat{\theta})/\partial\theta = 0$. For the estimator to be consistent, it is required that $\mathbb{E}_\theta\{\partial g(x;\theta)/\partial\theta\} = 0$. This is the situation of interest for source separation where we shall use score functions which are not necessarily exact (i.e. derivatives of the "true" log-likelihood) but still lead to consistent estimators. We now sketch the computations leading to several types of result: **(a)** an equivalent for the estimation error: $\hat{\theta} - \theta$; **(b)** the computations behind Newton-like optimization methods; **(c)** stability conditions, that is, how different $g(x;\theta)$ can be from the true score while still having a maximum close to the true parameter value; **(d)** a bound on the variance of estimates; **(e)** a condition for identifiability.

a: First-order error equivalent. If the estimation error $\hat{\theta} - \theta$ is small enough that the estimating equation $\partial g(x;\hat{\theta})/\partial\theta = 0$ can be approximated by its first-order expansion $0 = \partial g(x;\hat{\theta})/\partial\theta \approx \partial g(x;\theta)/\partial\theta + \partial^2 g(x;\theta)/\partial\theta^2(\hat{\theta} - \theta)$, then $\hat{\theta} - \theta \approx -[\partial^2 g(x;\theta)/\partial\theta^2]^{-1}\partial g(x;\theta)/\partial\theta$. Hence, if the second derivative can be further approximated by its mean value, one gets

$$\hat{\theta} \approx \theta + \mathbf{F}_g(\theta)^{-1}\frac{\partial g(x;\theta)}{\partial\theta} \quad \text{where} \quad \mathbf{F}_g(\theta) \stackrel{\text{def}}{=} -\mathbb{E}_\theta\left\{\frac{\partial^2 g(x;\theta)}{\partial\theta^2}\right\} \tag{4.47}$$

and the covariance matrix for the error then is

$$\mathrm{Cov}(\hat{\theta}) \approx \mathbf{F}_g(\theta)^{-1}\bar{\mathbf{F}}_g(\theta)\mathbf{F}_g(\theta)^{-1} \quad \text{where} \quad \bar{\mathbf{F}}_g(\theta) \stackrel{\text{def}}{=} \mathrm{Cov}_\theta\left(\frac{\partial g(x;\theta)}{\partial\theta}\right). \tag{4.48}$$

b: Fisher scoring. The estimating equation $\partial g(x;\hat{\theta})/\partial\theta = 0$ can be solved by the Newton method which computes a sequence $\{\theta^{(n)}\}$ as

$$\theta^{(n+1)} = \theta^{(n)} - \frac{\partial^2 g(x;\theta^{(n)})}{\partial\theta^2}^{-1}\frac{\partial g(x;\theta^{(n)})}{\partial\theta}.$$

If the second derivative is replaced by its expectation, the update becomes

$$\theta^{(n+1)} = \theta^{(n)} + \mathbf{F}_g(\theta^{(n)})^{-1}\frac{\partial g(x;\theta^{(n)})}{\partial\theta} \tag{4.49}$$

using definition (4.47) and the algorithm is called "Fisher scoring". It boils down to a gradient algorithm if $\mathbf{F}_g(\theta)^{-1}$ is replaced by a small multiple of the identity.

c: Stability. The Newton method (or other gradient based method) can find a point where the gradient cancels $\mathbb{E}_\theta\left\{\partial g(x;\hat{\theta})/\partial\theta\right\} = 0$ only if this point is a local maximum (and not a saddle point). The theoretical approach to this problem is to determine "local stability" conditions by studying the positivity of matrix $\mathbf{F}_g(\theta)$.

d: Cramér-Rao bound. When the contrast function is the log-likelihood: $g(x;\theta) = \log p(x;\theta)$ and when the data x actually is distributed under $p(x;\theta)$, matrices $\mathbf{F}_g(\theta)$ and $\bar{\mathbf{F}}_g(\theta)$ are equal and their common value is the Fisher information matrix (FIM), denoted $\mathbf{F}(\theta)$:

$$\mathbf{F}(\theta) \stackrel{\text{def}}{=} \mathbb{E}_\theta\left\{\frac{\partial l(\theta)}{\partial\theta}\frac{\partial l(\theta)}{\partial\theta}^{\mathrm{T}}\right\} = -\mathbb{E}_\theta\left\{\frac{\partial^2 l(\theta)}{\partial\theta^2}\right\}. \tag{4.50}$$

In this case, the covariance of the estimation error is lower bounded through the inverse FIM. This is Cramér-Rao bound:

$$\mathrm{Cov}(\hat{\theta}) \geq \mathbf{CRB}(\theta) \stackrel{\text{def}}{=} \mathbf{F}(\theta)^{-1}. \tag{4.51}$$

e: Local identifiability. By local identifiability for parameter θ, we mean the strict positivity of the FIM $\mathbf{F}(\theta)$. The failure of this condition is a sign that the likelihood is flat in at least one direction in parameter space.

4.4.2 Fisher information matrix

All ingredients are available to evaluate the Fisher matrix for the data model $\mathbf{X} = \mathbf{A}_\star \mathbf{S}$. We use again local (or relative) parameters in the sense that any matrix $\mathbf{A}$ close to $\mathbf{A}_\star$ is (uniquely) represented by a matrix $\mathcal{E}$ defined by $\mathbf{A} = \mathbf{A}_\star(\mathbf{I}+\mathcal{E})^{-1}$. Then $\mathbf{Y} = (\mathbf{I}+\mathcal{E})\mathbf{S}$ according to Eq. (4.3) and

$$\log p_{\mathbf{X}}(\mathbf{X}; \mathbf{A}(\mathbf{I}+\mathcal{E})^{-1}) = T\,\widehat{\Upsilon}((\mathbf{I}+\mathcal{E})\mathbf{S}) + \text{cst}$$

according to the definition of the empirical log-likelihood Eq. (4.8). The derivative of the log-likelihood at the true value of the parameter, that is for $\mathcal{E} = 0$, thus is

$$\left.\frac{\partial \log p_{\mathbf{X}}(\mathbf{X}; \mathcal{E})}{\partial \mathcal{E}_{ij}}\right|_{\mathcal{E}=0} = [\nabla \widehat{\Upsilon}(\mathbf{S})]_{ij} = \frac{1}{T}\boldsymbol{\varphi}_i^\star(\mathbf{s}_i)\mathbf{s}_j^{\mathrm{T}} - \delta_{ij}.$$

Therefore, the Fisher matrix with respect to parameter $\mathrm{Vec}(\mathcal{E})$ is the covariance matrix of $\mathrm{Vec}(\nabla\widehat{\Upsilon}(\mathbf{S}))$. For each source i, define a scalar $\mathbf{F}^{(i)}$ and for each of the $N(N-1)/2$ pairs of sources $(1 \le i < j \le N)$, define a 2×2 matrix $\mathbf{F}^{(ij)}$

$$\mathbf{F}^{(i)} \stackrel{\text{def}}{=} \mathrm{Var}\left(\frac{1}{T}\boldsymbol{\varphi}_i^\star(\mathbf{s}_i)\mathbf{s}_i^{\mathrm{T}} - 1\right) \qquad \mathbf{F}^{(ij)} \stackrel{\text{def}}{=} \mathrm{Cov}\left(\frac{1}{T}\begin{bmatrix}\boldsymbol{\varphi}_i^\star(\mathbf{s}_i)\mathbf{s}_j^{\mathrm{T}} \\ \boldsymbol{\varphi}_j^\star(\mathbf{s}_j)\mathbf{s}_i^{\mathrm{T}}\end{bmatrix}\right). \tag{4.52}$$

Scalar $\mathbf{F}^{(i)}$ corresponds to the variable $\mathcal{E}_{ii}$ and matrix $\mathbf{F}^{(ij)}$ to the pair of variables $(\mathcal{E}_{ij}, \mathcal{E}_{ji})$. Thanks to source independence, it is easy to check that all other entries of the FIM (for instance those related to a pair $(\mathcal{E}_{ij}, \mathcal{E}_{kl})$ with $(i,j) \neq (k,l)$ and $(i,j) \neq (l,k)$) vanish. Hence, the Fisher information matrix is very sparse: it is block-diagonal with only 2×2 or 1×1 blocks. That shows that the problem of separating N sources is statistically decoupled into N scale estimation problems and into $N(N-1)/2$ pairwise source separation problems with fixed scale. Therefore, the statistical characterization essentially boils down to studying the 2×2 matrices $\mathbf{F}^{(ij)}$ (we shall not address the (marginally advantageous) issue of scale estimation, controlled by $\mathbf{F}^{(i)}$).

Focusing on the pairwise Fisher matrices $\mathbf{F}^{(ij)}$, we find (with $\boldsymbol{\varphi}_i^\star(\mathbf{s}_i) = \boldsymbol{\varphi}_i$)

$$\begin{aligned}
\mathbb{E}\left\{(\boldsymbol{\varphi}_i\mathbf{s}_j^{\mathrm{T}})^2\right\} &= \mathbb{E}\left\{\boldsymbol{\varphi}_i\mathbf{s}_j^{\mathrm{T}}\mathbf{s}_j\boldsymbol{\varphi}_i^{\mathrm{T}}\right\} = \mathbb{E}\left\{\mathrm{trace}(\mathbf{s}_j^{\mathrm{T}}\mathbf{s}_j\boldsymbol{\varphi}_i^{\mathrm{T}}\boldsymbol{\varphi}_i)\right\} \\
&= \mathrm{trace}\left(\mathbb{E}\left\{\mathbf{s}_j^{\mathrm{T}}\mathbf{s}_j\right\}\mathbb{E}\left\{\boldsymbol{\varphi}_i^{\mathrm{T}}\boldsymbol{\varphi}_i\right\}\right) = \mathrm{trace}(\mathbf{R}_j\boldsymbol{\Gamma}_i) \qquad \text{and} \\
\mathbb{E}\left\{(\boldsymbol{\varphi}_i\mathbf{s}_j^{\mathrm{T}})(\boldsymbol{\varphi}_j\mathbf{s}_i^{\mathrm{T}})\right\} &= \mathbb{E}\left\{\mathrm{trace}(\mathbf{s}_j^{\mathrm{T}}\boldsymbol{\varphi}_j\mathbf{s}_i^{\mathrm{T}}\boldsymbol{\varphi}_i)\right\} = \mathrm{trace}(\mathbb{E}\left\{\mathbf{s}_j^{\mathrm{T}}\boldsymbol{\varphi}_j\right\}\mathbb{E}\left\{\mathbf{s}_i^{\mathrm{T}}\boldsymbol{\varphi}_i\right\}) \\
&= \mathrm{trace}(\mathbf{I}\,\mathbf{I}) = \mathrm{trace}(\mathbf{I}) = T.
\end{aligned}$$

Therefore, with equation (4.44), the pairwise Fisher matrix reduces to

$$\mathbf{F}^{(ij)} = T \begin{bmatrix} f_{ij} & 1 \\ 1 & f_{ji} \end{bmatrix} \quad \text{where } f_{ij} \stackrel{\text{def}}{=} \frac{1}{T}\,\text{trace}(\boldsymbol{\Gamma}_i \mathbf{R}_j). \tag{4.53}$$

4.4.3 Blind identifiability

We are now able to determine the conditions allowing *blind* source separation. As already stated, the issue boils down to pairwise conditions. A given pair (i, j) of sources is blindly separable if and only if $\mathbf{F}^{(ij)} > 0$.

If $\mathbf{s}_i$ and $\mathbf{s}_j$ are Gaussian vectors with covariance matrices proportional to each other, then $\boldsymbol{\Gamma}_i = \mathbf{R}_i^{-1}$, $\boldsymbol{\Gamma}_j = \mathbf{R}_j^{-1}$, and $\mathbf{R}_j = \alpha \mathbf{R}_i$ for some scalar α. In that case, $f_{ij} = \alpha$ and $f_{ji} = 1/\alpha$ so that $\mathbf{F}^{(ij)}$ is *not* invertible. Therefore, a necessary condition for blind identifiability of two sources is that they are not Gaussian with proportional covariance matrices.

Let us see why this condition also is *sufficient* by assuming that $\mathbf{F}^{(ij)}$ is singular. That would imply a perfect correlation of $\boldsymbol{\varphi}_i^\star(\mathbf{s}_i)\mathbf{s}_j^{\mathrm{T}}$ with $\boldsymbol{\varphi}_j^\star(\mathbf{s}_j)\mathbf{s}_i^{\mathrm{T}}$. Therefore, it would exist as a fixed scalar α such that

$$\boldsymbol{\varphi}_i^\star(\mathbf{s}_i)\mathbf{s}_j^{\mathrm{T}} = \alpha \boldsymbol{\varphi}_j^\star(\mathbf{s}_j)\mathbf{s}_i^{\mathrm{T}}.$$

Since the right-hand side is linear in $\mathbf{s}_i$, the left-hand side must also be linear in $\mathbf{s}_i$ showing that the score $\boldsymbol{\varphi}_i^\star(\mathbf{s}_i)$ is a linear function. But the linearity of the score $\boldsymbol{\varphi}_i^\star$ implies that $\mathbf{s}_i$ is a Gaussian variable which entails $\boldsymbol{\varphi}_i^\star(\mathbf{s}_i) = \mathbf{s}_i \mathbf{R}_i^{-1}$. Hence, we get $\alpha \mathbf{s}_i \mathbf{R}_j^{-1} = \mathbf{s}_i \mathbf{R}_i^{-1}$. Left-multiplying this expression by $\mathbf{s}_i^{\mathrm{T}}$ and taking expectation over $\mathbf{s}_i$ yields $\alpha \mathbb{E}\left\{\mathbf{s}_i^{\mathrm{T}}\mathbf{s}_i\right\}\mathbf{R}_j^{-1} = \mathbb{E}\left\{\mathbf{s}_i^{\mathrm{T}}\mathbf{s}_i\right\}\mathbf{R}_i^{-1}$, so that $\alpha \mathbf{R}_i \mathbf{R}_j^{-1} = \mathbf{I}$. Hence, $\mathbf{s}_i$ is a Gaussian variable with a variance proportional to that of $\mathbf{s}_j$. Of course, applying the same reasoning symmetrically shows that $\mathbf{s}_j$ also is a Gaussian variable. Hence, two independent sources are blindly separable unless they are *both Gaussian with proportional covariance matrices.*

4.4.4 Asymptotic performance

Separation index. We start by defining a properly scaled performance index. If, given a data set $\mathbf{X} = \mathbf{A}_\star \mathbf{S}$, the mixing matrix is estimated at the value $\widehat{\mathbf{A}} = \mathbf{A}_\star(\mathbf{I} + \mathcal{E})^{-1}$, then the estimated sources are $\widehat{\mathbf{S}} = (\mathbf{I} + \mathcal{E})\mathbf{S}$. For a given value $\mathcal{E}$ of the relative error, the power of the i-th estimated source decomposes as

$$\mathbb{E}\left\{|\widehat{\mathbf{s}}_i|^2\right\} = \mathbb{E}\left\{\left(\sum_{j=1}^{N}(\mathbf{I}+\mathcal{E})_{ij}\mathbf{s}_j\right)^2\right\} = \sum_{j=1}^{N}(\mathbf{I}+\mathcal{E})_{ij}^2\,\mathbb{E}\left\{|\mathbf{s}_j|^2\right\}.$$

So the source of interest (source i) is found at level $(\mathbf{I}+\mathscr{E})^2_{ii}\mathbb{E}\{|\mathbf{s}_i|^2\} \approx \mathbb{E}\{|\mathbf{s}_i|^2\}$ while the level of a contaminating source (any source with $j \neq i$) is found at level $(\mathbf{I}+\mathscr{E})^2_{ij}\mathbb{E}\{|\mathbf{s}_j|^2\} \approx \mathscr{E}^2_{ij}\mathbb{E}\{|\mathbf{s}_j|^2\}$ where, in both cases, we kept only the leading order term in $\mathscr{E}$. Therefore, the performance index

$$\rho_{ij} \stackrel{\text{def}}{=} \mathbb{E}\left\{\mathscr{E}^2_{ij}\right\} \frac{\mathbb{E}\{|\mathbf{s}_j|^2\}}{\mathbb{E}\{|\mathbf{s}_i|^2\}} \tag{4.54}$$

measures an average interference-to-signal ratio. We call this quantity the "rejection rate" and use it to quantify the performance of source separation methods.

Cramér-Rao bound. The Cramér-Rao bound (CRB) for parameter $\mathscr{E}$ is given by the inverse FIM which can be computed in closed form thanks to its block-diagonal structure. In particular, the CRB for a pair $(\mathscr{E}_{ij}, \mathscr{E}_{ji})$ is

$$\operatorname{Cov}\left(\begin{bmatrix}\mathscr{E}_{ij}\\ \mathscr{E}_{ji}\end{bmatrix}\right) \geq \mathbf{F}^{(ij)-1} \tag{4.55}$$

so that the lowest achievable mean-square error on $\mathscr{E}_{ij}$ is the $(1,1)$ entry of this inverse. According to Eq. (4.53), one thus finds $\mathbb{E}\left\{\mathscr{E}^2_{ij}\right\} \geq \frac{1}{T} f_{ji}/(f_{ij}f_{ji}-1)$. Normalizing this result according to Eq. (4.54), we obtain the Cramér-Rao lower bound of the rejection rates:

$$\rho_{ij} \geq \rho^{\text{CRB}}_{ij} = \frac{1}{T}\frac{f_{ji}\,\sigma_j^2/\sigma_i^2}{f_{ij}f_{ji}-1}, \qquad \sigma_i^2 \stackrel{\text{def}}{=} \frac{1}{T}\mathbb{E}\{|\mathbf{s}_i|^2\}. \tag{4.56}$$

Note that, as expected by equivariance, this is strictly scale invariant since a scale factor $\lambda > 0$ applied to a source $\mathbf{s}_i$ will multiply σ_i by λ, $\mathbf{R}_i$ by λ^2 and $\mathbf{\Gamma}_i$ by λ^{-2}, leaving ρ^{CRB}_{ij} globally unchanged.

We note that some source distributions can produce arbitrarily good (low) rejection rates ρ_{ij} if f_{ij} diverges; some examples are given in the following sections.

4.4.5 When the source model is wrong

In many applications, one cannot expect the source model to capture the "true" source distribution and it is therefore necessary to understand how performance is affected when the score functions $\boldsymbol{\varphi}_i$ used in the estimating equations are *not* the true score functions $\boldsymbol{\varphi}_i^\star$. As sketched in section 4.4.1, the asymptotic behavior is then governed by matrices $\mathbf{F}_g(\theta)$ and $\bar{\mathbf{F}}_g(\theta)$ defined at Eqs (4.47) and (4.48) and our task now is to evaluate these quantities for the ICA likelihood. We shall soon find that the score functions appear in matrices

$$\mathbf{G}_i \stackrel{\text{def}}{=} \mathbb{E}\left\{\boldsymbol{\varphi}_i(\mathbf{s}_i)^{\mathrm{T}}\boldsymbol{\varphi}_i^\star(\mathbf{s}_i)\right\} \qquad \text{and} \qquad \bar{\mathbf{G}}_i \stackrel{\text{def}}{=} \mathbb{E}\left\{\boldsymbol{\varphi}_i(\mathbf{s}_i)^{\mathrm{T}}\boldsymbol{\varphi}_i(\mathbf{s}_i)\right\}, \tag{4.57}$$

similar to $\mathbf{\Gamma}_i$. These matrices appear in the moments

$$\phi_{ij} \stackrel{\text{def}}{=} \frac{1}{T}\text{trace}(\mathbf{G}_i\mathbf{R}_j) \qquad \bar{\phi}_{ij} \stackrel{\text{def}}{=} \frac{1}{T}\text{trace}(\bar{\mathbf{G}}_i\mathbf{R}_j) \tag{4.58}$$

similar to f_{ij}. We now evaluate $\mathbf{F}_g(\theta)$ and $\bar{\mathbf{F}}_g(\theta)$ in terms of these quantities.

The curvature matrix $\mathbf{F}_g(\theta) = -\mathbb{E}_\theta\left\{\partial^2 g(x;\theta)/\partial\theta^2\right\}$ defined at Eq. (4.47) is obtained by expanding $\mathbb{E}_\theta\{\partial g(x;\theta+\delta\theta)/\partial\theta\}$ to first order in $\delta\theta$, as follows. Consider the value of the element $\boldsymbol{\varphi}_i(\mathbf{y}_i)\mathbf{y}_j^{\mathrm{T}}$ of the likelihood gradient for $i \neq j$ evaluated at $\mathbf{Y} = (\mathbf{I}+\mathcal{E})\mathbf{S}$ and compute the expectation of its first order expansion in $\mathcal{E}$. The first order expansion of the score is $\boldsymbol{\varphi}_i(\mathbf{s}_i+\delta\mathbf{s}_i) = \boldsymbol{\varphi}_i(\mathbf{s}_i) + \delta\mathbf{s}_i\boldsymbol{\varphi}_i'(\mathbf{s}_i) + o(\|\delta\mathbf{s}_i\|)$ where $\boldsymbol{\varphi}_i'(\mathbf{s}_i)$ is the derivative of $\boldsymbol{\varphi}_i$, that is, the $T \times T$ matrix defined by

$$[\boldsymbol{\varphi}_i'(\mathbf{s}_i)]_{t't} = -\frac{\partial[\boldsymbol{\varphi}_i(\mathbf{s}_i)]_t}{\partial s_i(t')} \quad (1 \le t, t' \le T).$$

For $\mathbf{Y} = (\mathbf{I}+\mathcal{E})\mathbf{S}$ and $i \neq j$, we thus have

$$\begin{aligned}\boldsymbol{\varphi}_i(\mathbf{y}_i)\mathbf{y}_j^{\mathrm{T}} &= \boldsymbol{\varphi}_i\left(\mathbf{s}_i + \sum_a \mathcal{E}_{ia}\mathbf{s}_a\right)\left(\mathbf{s}_j + \sum_b \mathcal{E}_{jb}\mathbf{s}_b\right)^{\mathrm{T}} \\ &= \boldsymbol{\varphi}_i(\mathbf{s}_i)\mathbf{s}_j^{\mathrm{T}} + \sum_a \mathcal{E}_{ia}\mathbf{s}_a\boldsymbol{\varphi}_i'(\mathbf{s}_i)\mathbf{s}_j^{\mathrm{T}} + \sum_b \mathcal{E}_{jb}\boldsymbol{\varphi}_i(\mathbf{s}_i)\mathbf{s}_b^{\mathrm{T}} + o(\|\mathcal{E}\|^2).\end{aligned} \tag{4.59}$$

By source independence, $\mathbb{E}\left\{\mathbf{s}_a\boldsymbol{\varphi}_i'(\mathbf{s}_i)\mathbf{s}_j^{\mathrm{T}}\right\} = \delta_{ja}\mathbb{E}\left\{\mathbf{s}_j\boldsymbol{\varphi}_i'(\mathbf{s}_i)\mathbf{s}_j^{\mathrm{T}}\right\} = \delta_{ja}\text{trace}\mathbb{E}\left\{\boldsymbol{\varphi}_i'(\mathbf{s}_i)\mathbf{s}_j^{\mathrm{T}}\mathbf{s}_j\right\}$ $= \delta_{ja}\text{trace}(\mathbf{G}_i\mathbf{R}_j)$ where the last equality comes from definition (4.57) and from property (4.43) by which $\mathbb{E}\left\{\boldsymbol{\varphi}_i'(\mathbf{s}_i)\right\} = \mathbb{E}\left\{\boldsymbol{\varphi}_i(\mathbf{s}_i)^{\mathrm{T}}\boldsymbol{\varphi}_i^{\star}(\mathbf{s}_i)\right\}$. Similarly, $\mathbb{E}\left\{\boldsymbol{\varphi}_i(\mathbf{s}_i)\mathbf{s}_b^{\mathrm{T}}\right\} = \delta_{ib}$ $\mathbb{E}\left\{\boldsymbol{\varphi}_i(\mathbf{s}_i)\mathbf{s}_i^{\mathrm{T}}\right\} = T\delta_{ib}$ so that, for $i \neq j$, one finds $\mathbb{E}\left\{\boldsymbol{\varphi}_i(\mathbf{y}_i)\mathbf{y}_j^{\mathrm{T}}\right\} = \mathcal{E}_{ij}\text{trace}(\mathbf{G}_i\mathbf{R}_j) + T\mathcal{E}_{ji} + o(\|\mathcal{E}\|^2)$. Combining this relation for (i,j) with the symmetric relation for (j,i), one finally obtains

$$\mathrm{E}\left(\begin{bmatrix}\boldsymbol{\varphi}_i(\mathbf{y}_i)\mathbf{y}_j^{\mathrm{T}}\\ \boldsymbol{\varphi}_j(\mathbf{y}_j)\mathbf{y}_i^{\mathrm{T}}\end{bmatrix}\right) = \mathbf{F}_\phi^{(ij)} \times \begin{bmatrix}\mathcal{E}_{ij}\\ \mathcal{E}_{ji}\end{bmatrix} + o(\|\mathcal{E}\|^2) \quad \text{where } \mathbf{F}_\phi^{(ij)} \stackrel{\text{def}}{=} T\begin{bmatrix}\phi_{ij} & 1\\ 1 & \phi_{ji}\end{bmatrix} \tag{4.60}$$

Hence, even with the wrong scores, we find pairwise decoupling and the Hessian of the likelihood contrast remains block-diagonal with a 2×2 block $\mathbf{F}_\phi^{(ij)}$ for each pair $(1 \le i < j \le N)$ of sources. Since an infinitesimal transform $\mathcal{E}$ with null entries except $\mathcal{E}_{ij}$ and $\mathcal{E}_{ji}$ only mixes sources i and j, we have established that, under such a transform, the average likelihood contrast is, at leading order,

$$\mathbb{E}\left\{\widehat{\Upsilon}(\mathbf{S}+\mathscr{E}\mathbf{S})\right\} \approx \mathbb{E}\left\{\widehat{\Upsilon}(\mathbf{S})\right\} - \frac{1}{2}\begin{bmatrix}\mathscr{E}_{ij}\\ \mathscr{E}_{ji}\end{bmatrix}^{\mathrm{T}} \mathbf{F}_{\psi}^{(ij)} \begin{bmatrix}\mathscr{E}_{ij}\\ \mathscr{E}_{ji}\end{bmatrix}. \tag{4.61}$$

This expression is key to understanding the stability under wrong scores.

Stability. Stability for a pair (i,j) requires that the likelihood contrast has a local maximum at a separating point meaning, in view of Eq. (4.61), that matrix $\mathbf{F}_{\psi}^{(ij)}$ must be positive. Thus, local stability for a pair (i,j) requires $\phi_{ij} > 0$, $\phi_{ji} > 0$ and $\phi_{ij}\phi_{ji} > 1$. We shall see later the meaning of these conditions for specific source models. But, before proceeding, we consider stability *under the whiteness constraint*. Under the whiteness constraint, the variations of $\mathscr{E}$ are not arbitrary but necessarily skew-symmetric (section 4.2.4): if $\mathscr{E}_{ij} = \epsilon$, then $\mathscr{E}_{ij} = -\epsilon$ so that the quadratic form in (4.61) becomes $-\frac{1}{2}\epsilon^2 \begin{bmatrix}1\\ -1\end{bmatrix}^{\mathrm{T}} \mathbf{F}_{\psi}^{(ij)} \begin{bmatrix}1\\ -1\end{bmatrix} = -\frac{1}{2}\epsilon^2(\phi_{ij} + \phi_{ji} - 2)$. Therefore, the stability condition is $(\phi_{ij} + \phi_{ji} - 2) > 0$. That condition cannot be stronger than the unconstrained condition since the whiteness constraint only limits the possible variations of $\mathscr{E}$. Indeed, conditions $\phi_{ij} > 0$, $\phi_{ji} > 0$ and $\phi_{ij}\phi_{ji} > 1$ for unconstrained stability imply $(\phi_{ij} + \phi_{ji} - 2) > 0$ as seen from $(\phi_{ij} + \phi_{ji} - 2) = (\sqrt{\phi_{ij}} - \sqrt{\phi_{ji}})^2 + 2(\sqrt{\phi_{ij}\phi_{ji}} - 1)$.

Asymptotic rejection rates. Computations similar to those leading to Eq. (4.61), not reproduced here, yield the covariance matrix of the likelihood gradient (the equivalent of $\bar{\mathbf{F}}_{g}(\theta)$), with pairwise decoupling and lead to a block-diagonal matrix with, for each pair (i,j), a 2×2 block:

$$\operatorname{Cov}\left(\begin{bmatrix}\boldsymbol{\varphi}_i(\mathbf{s}_i)\mathbf{s}_j^{\mathrm{T}}\\ \boldsymbol{\varphi}_j(\mathbf{s}_j)\mathbf{s}_i^{\mathrm{T}}\end{bmatrix}\right) = T\begin{bmatrix}\bar{\phi}_{ij} & 1\\ 1 & \bar{\phi}_{ji}\end{bmatrix} \stackrel{\text{def}}{=} \bar{\mathbf{F}}_{\psi}^{(ij)} \tag{4.62}$$

with $\bar{\phi}_{ij}$ defined at Eq. (4.58). Combining with Eq. (4.60), the covariance matrix of the relative error for the (i,j) pair is given by $\mathbf{F}_{\psi}^{(ij)-1}\bar{\mathbf{F}}_{\psi}^{(ij)}\mathbf{F}_{\psi}^{(ij)-1}$, that is

$$\operatorname{Cov}\left(\begin{bmatrix}\widehat{\mathscr{E}}_{ji}\\ \widehat{\mathscr{E}}_{ij}\end{bmatrix}\right) = T^{-1}\begin{bmatrix}\phi_{ij} & 1\\ 1 & \phi_{ji}\end{bmatrix}^{-1}\begin{bmatrix}\bar{\phi}_{ij} & 1\\ 1 & \bar{\phi}_{ji}\end{bmatrix}\begin{bmatrix}\phi_{ij} & 1\\ 1 & \phi_{ji}\end{bmatrix}^{-1}. \tag{4.63}$$

Of course, with the true scores: $\boldsymbol{\varphi}_i = \boldsymbol{\varphi}_i^{\star}$, we fall back to $\mathbf{G}_i = \bar{\mathbf{G}}_i = \boldsymbol{\Gamma}_i$; $\phi_{ij} = \bar{\phi}_{ij} = f_{ij}$; $\mathbf{F}_{\psi}^{(ij)} = \bar{\mathbf{F}}_{\psi}^{(ij)} = \mathbf{F}^{(ij)}$ and the asymptotic covariance (4.63) equals the Cramér-Rao bound. With wrong scores, expression (4.63) quantifies the loss of accuracy induced by using approximate score functions.

4.4.6 Relative gradient and natural gradient

Equation (4.61) provides us with an equivalent for the curvature of likelihood contrast at a separating point via expressions (4.58) and (4.60). Hence, we have the necessary ingredients to maximize the likelihood by the Newton method or, more precisely, by its statistical variant, "Fisher" scoring (section 4.4.1).

Let us compare Fisher scoring with the simpler relative gradient. An equivariant iterative separation algorithm proceeds by updates of the form:

$$\mathbf{Y} \leftarrow (\mathbf{I} + \mathbf{D}(\mathbf{Y}))\mathbf{Y} \tag{4.64}$$

where $\mathbf{D}(\mathbf{Y})$ is an $N \times N$ matrix, depending on the current estimate $\mathbf{Y}$ of the sources. The relative gradient algorithm (4.15) consists in taking $\mathbf{D}(\mathbf{Y}) = \mu\widehat{\Upsilon}(\mathbf{Y})$ where μ is an adaptation step small enough to ensure stability. Fisher scoring amounts to "rectifying" the gradient by left multiplying it with the inverse (of an approximation) of Fisher matrix. But we just saw that, in the source context separation, when using the relative parametrization in terms of $\mathcal{E}$, the Fisher matrix decomposes in 2×2 blocks. The Fisher scoring then takes the form (4.64) with the off-diagonal terms of $\mathbf{D}$ given by

$$\begin{bmatrix} \mathbf{D}_{ij}(\mathbf{Y}) \\ \mathbf{D}_{ji}(\mathbf{Y}) \end{bmatrix} = \begin{bmatrix} \phi_{ij} & 1 \\ 1 & \phi_{ji} \end{bmatrix}^{-1} \begin{bmatrix} \widehat{\Upsilon}(\mathbf{Y})_{ij} \\ \widehat{\Upsilon}(\mathbf{Y})_{ji} \end{bmatrix}. \tag{4.65}$$

The difficulty here would not be to invert the Fisher matrix (since it is in reduced in small blocks) but to estimate it. It suffices to estimate the moments ϕ_{ij} but, according to expression (4.58), that seems to be quite a formidable task in the general case. However, we shall see that these moments have a very simple form in the models considered in the following sections (see for instance expression (4.75) in the iid-non Gaussian model).

Assume for a while that the distributions of the sources are identical and far from iid-Gaussianity, so that $\phi_{ij} = \phi_{ji} = \phi \gg 1$. Then, the matrix in Eq. (4.65) is close to $\phi^{-1}\mathbf{I}_2$ and $\mathbf{D}_{ij}(\mathbf{Y})$ is close to $\phi^{-1}\widehat{\Upsilon}(\mathbf{Y})_{ij}$ so that $\mathbf{D}(\mathbf{Y})$ is close to $\phi^{-1}\,\widehat{\Upsilon}(\mathbf{Y})$. In this limit, Fisher scoring becomes identical to the relative gradient algorithm with a step μ equal to ϕ^{-1}.

The method described above for implementing the Fisher scoring is described in [17] for the non-Gaussian iid model. Amari [1] has also proposed the "natural gradient" method for maximizing the likelihood. This method leads to rectifying the likelihood gradient by the inverse of the Fisher matrix and is thus equivalent to the scoring method described above. Note that many publications devoted to source separation refer to the natural gradient but in fact *ignore* the Fisher matrix, approximating it by a multiple of the identity matrix. These publications hence refer to the natural gradient but in fact implement the relative gradient [5].

4.4.7 Summary

We have sketched, with few claims for mathematical rigor, the main features of an asymptotic analysis for the noise free source separation problem in a very broad setting, that is without specifying the model of the sources, thus extending some results already available in more specific models (iid sources in [17] or Gaussian sources in [16]). It is not easy to go much further without considering more specific source models, as we shall do in the next two sections, 4.5 and 4.6.

4.5 NON-GAUSSIAN MODELS

We turn to studying iid models: each source signal is modeled as a sequence of T variables which are mutually independent and share the same (marginal) distribution. These models thus only require to specify for each source i, the distribution p_{s_i} of each $s_i(t)$ since then

$$p_{\mathbf{s}_i}(\mathbf{s}_i) = \prod_{t=1}^{T} p_{s_i}(s_i(t)). \tag{4.66}$$

The iid model ignores any time structure potentially present in the data. Hence, they are applicable in a data analysis context where the index $t = 1, \ldots, T$ is an arbitrary sample number. The separation results then are invariant under any permutation of the columns of $\mathbf{X}$.

In iid models, the distributions p_{s_i} cannot be chosen as Gaussian, as seen in section 4.4.3. Hence, the price to pay for the simplicity of iid models and ignoring any time structure is the use of non-Gaussian models for the marginals.

4.5.1 The likelihood contrast in iid models

In the iid framework, the contrast functions considered in section 4.3 take a simplified form: the probability of a block $\mathbf{Y}$ of size $N \times T$ is the product of the probability for each column, assumed to be identically distributed. If $\mathbf{Y}$ and $\mathbf{Y}'$ are two such blocks, then $K[\mathbf{Y} \mid \mathbf{Y}'] = TK[\mathbf{y} \mid \mathbf{y}']$ where $\mathbf{y}$ (resp. $\mathbf{y}'$) represents any column of $\mathbf{Y}$ (resp. $\mathbf{Y}'$). The mutual information for a block $\mathbf{Y}$ then reduces to

$$I[\mathbf{Y}] = K\left[\mathbf{Y} \mid \prod_i \mathbf{Y}_i\right] = TK\left[\mathbf{y} \mid \prod_i y_i\right] = TI[\mathbf{y}] \tag{4.67}$$

that is T times the mutual information of any given column.

Similarly, the measure $G[\mathbf{Y}]$ of non-iid-Gaussianity for a block reduces to T times the non-Gaussianity of a column:

$$G[\mathbf{Y}] = TG[\mathbf{y}] = TK\left[\mathbf{y} \mid \mathbf{y}^G\right] \tag{4.68}$$

where $\mathbf{y}^G$ is an $N \times 1$ Gaussian vector with the same mean and covariance matrix as $\mathbf{y}$. Hence, in the iid case, decomposition (4.40) becomes

$$I[\mathbf{y}] = C[\mathbf{y}] - \sum_i G[y_i] + \text{cst}.$$

The quantity $G[y]$ is sometimes called the "negentropy" (of the scalar y).

4.5.2 Score functions and estimating equations

The iid choice (4.66) yields simple score functions since

$$[\boldsymbol{\varphi}_i(\mathbf{s}_i)]_t = \varphi_i(s_i(t)) \quad \text{where} \quad \varphi_i(u) = -\frac{p'_{s_i}(u)}{p_{s_i}(u)}.$$

Function $\varphi_i : \mathbb{R} \to \mathbb{R}$ is called the *marginal score* for the ith source. It is a nonlinear function if p_{s_i} is on Gaussian. The estimating function $T^{-1}\boldsymbol{\psi}(\mathbf{Y})\mathbf{Y}^{\mathrm{T}} - \mathbf{I}$ corresponding to the maximum likelihood estimate then takes the form

$$\frac{1}{T}\sum_{t=1}^{T}\boldsymbol{\varphi}(\mathbf{y}(t))\mathbf{y}(t)^{\mathrm{T}} - \mathbf{I} \quad \text{where} \quad \boldsymbol{\varphi}(\mathbf{y}) = [\varphi_1(y_1), \ldots, \varphi_N(y_N)]^{\mathrm{T}}. \tag{4.69}$$

Hence the likeliest mixing matrix is such that, for any pair $(i \neq j)$ of sources,

$$\frac{1}{T}\sum_{t=1}^{T}\varphi_i(y_i(t))y_j(t) = 0, \tag{4.70}$$

that is, the maximum likelihood principle expresses source independence as the empirical decorrelation between each source and a non-linear version (via the marginal scores φ_i) of each other source.

Stochastic gradient ascent. With the gradient of the contrast taking the simple form (4.69), the relative gradient algorithm, summarized in Eq. (4.15), is seen to admit an on-line version for updating a separating matrix $\mathbf{B}_t$, as follows. Upon arrival of a sample $\mathbf{x}(t)$, the current estimate for the sources is $\mathbf{y}(t) = \mathbf{B}_t\mathbf{x}(t)$; this vector is used to form a stochastic (or instantaneous) version of the relative gradient: $\mathbf{G}_t = \boldsymbol{\varphi}(\mathbf{y}(t))\mathbf{y}(t)^{\mathrm{T}} - \mathbf{I}$; finally, this matrix is used to update the separating matrix by the relative update: $\mathbf{B}_{t+1} = (\mathbf{I} - \mu_t\mathbf{G}_t)\,\mathbf{B}_t$ for some small positive adaptation step μ_t. See [7].

4.5.3 Gaussianity index

In the iid model, the generic asymptotic analysis of section 4.4 takes the following form. Matrices $\mathbf{R}_i$ and $\boldsymbol{\Gamma}_i$ become multiples of the identity matrix, that is, $\mathbf{R}_i = \sigma_i^2\mathbf{I}$ and $\boldsymbol{\Gamma}_i = \sigma_i^{-2}\gamma_i\mathbf{I}$ with

$$\sigma_i^2 \stackrel{\text{def}}{=} \mathbb{E}\left\{s_i(t)^2\right\} \qquad \gamma_i \stackrel{\text{def}}{=} \sigma_i^2\mathbb{E}\left\{\varphi_i^\star(s_i(t))^2\right\} \tag{4.71}$$

where the definition of γ_i includes a normalizing term σ_i^2 which makes γ_i scale invariant. In section 4.4, we saw that $\boldsymbol{\Gamma}_i \geq \mathbf{R}_i^{-1}$ with equality when $\mathbf{s}_i$ is a Gaussian vector. In the iid case, that inequality combined with expressions (4.71), shows that $\gamma_i \geq 1$ with equality in the Gaussian limit[3]. Hence $\gamma_i - 1$ is the measure of non-Gaussianity which pops up from the asymptotic analysis. In particular, the lowest possible rejection rates depend only on γ_i, as shown by Eq. (4.74) below.

What is the connection between measuring non-Gaussianity by the excess of γ_i over 1 and the non-Gaussianity of definition (4.68) measured as the Kullback divergence from a distribution to its best Gaussian approximation? The link between the two concepts of non-Gaussianity is of differential nature: if s and t are two independent random variables, the non-Gaussianity of $s+t$ has a first order expansion in terms of the relative power $\rho = \mathrm{Var}(t)/\mathrm{Var}(s)$ as:

$$G[s+t] = G[s] - \frac{\rho}{2}(\gamma_s - 1) + o(\rho) \tag{4.72}$$

where γ_s is non-Gaussianity of s. This result can be obtained by combining the previous results but the reader is rather referred to [15] for rigorous proofs and the technical validity conditions. What expansion (4.72) shows is that a small independent contamination of a variable s always decreases its non-Gaussianity $G[s]$ proportionally to the relative power of the contamination and proportionally to the non-Gaussianity of s as measured by $\gamma_s - 1$.

4.5.4 Cramér-Rao bound

In the iid model, the elements f_{ij} of the Fisher matrix (4.53) become $f_{ij} = \gamma_i \sigma_j^2/\sigma_i^2$ and the FIM for the pair (i, j) becomes

$$\mathbf{F}^{(ij)} = T \begin{bmatrix} \sigma_j/\sigma_i & 0 \\ 0 & \sigma_i/\sigma_j \end{bmatrix} \begin{bmatrix} \gamma_i & 1 \\ 1 & \gamma_j \end{bmatrix} \begin{bmatrix} \sigma_j/\sigma_i & 0 \\ 0 & \sigma_i/\sigma_j \end{bmatrix}. \tag{4.73}$$

Therefore, the lowest possible rejection rates asymptotically in T, are

$$\rho_{ij}^{\mathrm{CRB}} = \frac{1}{T} \frac{\gamma_j}{\gamma_i \gamma_j - 1} \tag{4.74}$$

As already noted in the more general context of section 4.4.3, we see that it suffices that one of the two sources i or j is non-Gaussian for the denominator in (4.74) to be strictly positive.

4.5.5 Asymptotic performance

Consider now the asymptotic performance (as measured by rejection rates) for the maximum likelihood estimate based on score functions which are not necessarily correct, that is, when possibly $\varphi_i \neq \varphi_i^\star$. We can use the results of section 4.4.5 which become

[3] That $\gamma_i \geq 1$ is also seen by expanding $\mathbb{E}\left\{(\sigma_i \varphi_i(s_i) - \frac{s_i}{\sigma_i})^2\right\} \geq 0$ and using $\mathbb{E}\left\{\varphi_i^\star(s_i) s_i\right\} = 1$.

more tractable in the iid case. Indeed, the matrices $\mathbf{G}_i$ and $\bar{\mathbf{G}}_i$ defined at (4.57) become $\mathbf{G}_i = \sigma_i^{-2} g_i \mathbf{I}$ and $\bar{\mathbf{G}}_i = \sigma_i^{-2} \bar{g}_i \mathbf{I}$ with normalized moments

$$g_i \stackrel{\text{def}}{=} \sigma_i^2 \mathbb{E}\left\{\varphi_i'(s_i)\right\} = \sigma_i^2 \mathbb{E}\left\{\varphi_i(s_i)\varphi_i^\star(s_i)\right\} \qquad \bar{g}_i \stackrel{\text{def}}{=} \sigma_i^2 \mathbb{E}\left\{\varphi_i(s_i)^2\right\}. \tag{4.75}$$

We also have $\mathbf{R}_i = \sigma_i^2 \mathbf{I}$ in the iid case, so that the scalars defined at (4.58) take the form

$$\phi_{ij} = \frac{\sigma_j^2}{\sigma_i^2} g_i, \qquad \bar{\phi}_{ij} = \frac{\sigma_j^2}{\sigma_i^2} \bar{g}_i. \tag{4.76}$$

The good news here is further decoupling: the quantities defined at (4.58) now decouple in the sense that they are products of moments depending on i only or on j only. In particular, the stability conditions $\{\phi_{ij} > 0,\ \phi_{ji} > 0,\ \phi_{ij}\phi_{ji} > 1\}$ for a pair (i, j) become $\{g_i > 0,\ g_j > 0,\ g_i g_j > 1\}$ so that a sufficient stability condition is $g_i > 1$ for each source. Similarly, by expression (4.63), the rejection rates ρ_{ij} and ρ_{ji} appear as the diagonal terms of matrix

$$\frac{1}{T}\begin{bmatrix} g_i & 1 \\ 1 & g_j \end{bmatrix}^{-1} \begin{bmatrix} \bar{g}_i & 1 \\ 1 & \bar{g}_j \end{bmatrix} \begin{bmatrix} g_i & 1 \\ 1 & g_j \end{bmatrix}^{-1} = \begin{bmatrix} \rho_{ij} & \times \\ \times & \rho_{ji} \end{bmatrix}. \tag{4.77}$$

It is not very enlightening to work out this expression by explicit matrix inversion and multiplication but it does simplify when the scores are estimated by Pham's method, as seen next.

4.5.6 Adaptive scores

Since the best possible non-linear functions are the score functions associated with the true distribution of the sources, it is tempting to try to estimate adaptively the source densities and use the corresponding scores in the estimating equations. It is more straightforward, however, to estimate directly the score functions, as proposed by Pham [17]. We briefly review this nice approach.

The idea is to find the best score as a linear combination of some non-linear functions chosen in advance. Consider then approximating $\varphi_\star(s)$ by

$$\varphi(s;\alpha) = \sum_{j=1}^{J} \alpha_j \phi_j(s) = \phi(s)^{\mathrm{T}}\alpha \quad \text{where} \quad \phi(s) = [\phi_1(s), \ldots, \phi_J(s)]^{\mathrm{T}}. \tag{4.78}$$

The best linear combination, that is the best vector $\alpha = [\alpha_1, \ldots, \alpha_J]^{\mathrm{T}}$ of coefficients, is found by minimizing mean squared difference to the true score:

$$\mathbb{E}\left\{(\varphi_\star(s) - \varphi(s;\alpha))^2\right\} = \mathbb{E}\left\{\varphi_\star(s)^2\right\} - 2\mathbb{E}\left\{\varphi_\star(s)\phi^{\mathrm{T}}(s)\right\}\alpha + \alpha^{\mathrm{T}}\mathbb{E}\left\{\phi(s)\phi^{\mathrm{T}}(s)\right\}\alpha.$$

The minimum is reached at $\alpha_\star = \arg\min \mathbb{E}\left\{(\varphi_\star(s) - \varphi(s;\alpha))^2\right\}$ so that $\alpha_\star = \mathbb{E}\{\phi(s)\phi^{\mathrm{T}}(s)\}^{-1}\mathbb{E}\{\varphi_\star(s)\phi(s)\}$. This expression depends on the unknown score $\varphi_\star(s)$ but we can use the fact that $\mathbb{E}\{\varphi_\star(s)\phi(s)\} = \mathbb{E}\{\phi'(s)\}$ to obtain it as

$$\alpha_\star = \mathbb{E}\{\phi(s)\phi^{\mathrm{T}}(s)\}^{-1}\mathbb{E}\{\phi'(s)\}. \tag{4.79}$$

Therefore, a simple estimate of $\alpha_\star$ is obtained by substituting in (4.79) the expectations $\mathbb{E}\{\cdot\}$ by sample means.

This score estimation technique also offers the following two benefits: first, it guarantees that the stability conditions are met as soon as the identity function (which is nothing but the Gaussian score) is included in the function set; second, it yields a simple expression of the rejection rates. Let us see why.

The score approximation obtained by this method is, by construction, the orthogonal projection of the true score onto the function space spanned by the basis functions: the approximation error $\varphi_\star(s) - \varphi(s;\alpha_\star)$ for the best coefficients (4.79) is uncorrelated with (orthogonal to) the functions spanned by the basis. In particular, $\mathbb{E}\{(\varphi_\star(s) - \varphi(s;\alpha_\star))\,\varphi(s;\alpha_\star)\} = 0$ since $\varphi(s;\alpha_\star)$ is (by definition) spanned by the basis. Therefore $\mathbb{E}\{\varphi(s;\alpha_\star)^2\} = \mathbb{E}\{\varphi_\star(s)\varphi(s;\alpha_\star)\} = \mathbb{E}\{\varphi(s;\alpha_\star)'\}$ so that $g = \bar{g}$. Thus, expression (4.77) simplifies and one obtains the rejection rates in the form

$$\rho_{ij} = \frac{1}{T}\frac{g_j}{g_i g_j - 1}. \tag{4.80}$$

This is to be compared to the Cramér-Rao bound (4.74). For the bound to be reached for a pair (i,j), one should have $g_i = \gamma_i$ and $g_j = \gamma_j$. Since γ is the (normalized) variance of the true score while g is the variance of its projection on our function set, we see that the Cramér-Rao bound is met when the space spanned by the basis functions contains the true score. When this is not the case, we have $g < \gamma$ and the efficiency loss appears as a function of the angle between the true score and the function space spanned by the basis.

Regarding the stability conditions, we call upon the Cauchy-Schwarz inequality by which $\mathbb{E}\{s^2\}\mathbb{E}\{\varphi(s;\alpha_\star)^2\} \geq (\mathbb{E}\{\varphi(s;\alpha_\star)s\})^2$. On the left hand side, one finds the $\bar{g}$ moment while the right hand side equals 1 if the function space contains the identity function. This is because the identity function (or any member of the function space) is orthogonal to the approximation error: $\mathbb{E}\{(\varphi_\star(s) - \varphi(s;\alpha_\star))s\} = 0$, which implies $\mathbb{E}\{\varphi(s;\alpha_\star)s\} = \mathbb{E}\{\varphi_\star(s)s\} = 1$. Hence, including the identity function in the basis guarantees local stability.

4.5.7 Algorithms

The seminal paper [11] proposed solving the source separation problem by an iterative algorithm. The key idea already was to express independence by the decorrelation of non-linear versions of the sources. Pham [17] has clearly shown how to choose the decorrelation conditions from the maximum likelihood principle, how to guarantee the stability of separating points (section 4.5.6 above) and how to approximate the Hessian of the likelihood contrast to obtain a quasi-Newton algorithm. Cardoso [7] has shown how to build equivariant algorithms with or without the whiteness constraint, introducing the relative gradient.

Regarding the "orthogonal methods" based on pre-whitening followed by a rotation, Comon [8] has proposed a Jacobi algorithm which iterates through plane rotations towards the optimization of a cumulant-based approximation of mutual information. The JADE algorithm uses a similar contrast which also is a joint diagonality criterion of a set of cumulant matrices [6]. These methods are essentially equivalent to using a linear-cubic score function [6]. On those topics, see Chapters 3 and 5 in this volume.

Another algorithm based on the orthogonal approach is proposed by Hyvärinen in the fastICA algorithm, using fixed-point iterations (alternated with re-whitening steps) towards the minimization of marginal entropies. The resulting estimate can be shown to be statistically equivalent to likelihood maximization under the whiteness constraint. A good point of this approach is that the underlying score function can be freely chosen, in contrast to the cumulant-based methods which are implicitly restricted to linear-cubic scores. See Chapter 6 in this volume.

4.6 GAUSSIAN MODELS

In this section, we consider modeling each sequence $\{\mathbf{s}_i\}$ of source signals as a realization of a zero-mean Gaussian vector with a $T \times T$ covariance matrix $\mathbf{R}_i = \mathbb{E}\left\{\mathbf{s}_i^{\mathrm{T}}\mathbf{s}_i\right\}$. It is not possible to estimate all the $T(T+1)/2$ free parameters of the covariance matrices from T samples, so some restrictions must be imposed. We consider models where the covariance matrices are diagonal on some basis.

4.6.1 Diagonal Gaussian models

We call "diagonal Gaussian model" a model where all matrices $\mathbf{R}_i$ are diagonalized by a unique orthogonal transformation $\mathbf{T}$ of size $T \times T$, that is,

$$\mathbf{R}_i = \mathbb{E}\left\{\mathbf{s}_i^{\mathrm{T}}\mathbf{s}_i\right\} = \mathbf{T}\Sigma_i\mathbf{T}^{\mathrm{T}} \quad \text{with} \quad \Sigma_i = \operatorname{diag}\left(\bar{\sigma}_{i1}^2, \bar{\sigma}_{i2}^2, \cdots, \bar{\sigma}_{iT}^2\right) \quad \text{and} \quad \mathbf{T}\mathbf{T}^{\mathrm{T}} = \mathbf{I}.$$

Here and in the following, the quantities transformed by $\mathbf{T}$ are topped by a bar:

$$\bar{\mathbf{X}} = \mathbf{X}\mathbf{T}, \qquad \bar{\mathbf{s}}_i = \mathbf{s}_i\mathbf{T}, \qquad \bar{\sigma}_{i\tau}^2 = \mathbb{E}\left\{\bar{\mathbf{s}}_i^2\right\}, \quad \text{etc...}$$

The diagonality of Σ_i means that the entries of the row vector $\bar{\mathbf{s}}_i$ are uncorrelated. Denoting these entries $\bar{s}_{i\tau}$ for $\tau = 1, \ldots, T$, we may also write $\mathbb{E}\{\bar{s}_{i\tau}\bar{s}_{i\tau'}\} = \delta_{\tau\tau'}\bar{\sigma}_{i\tau}^2$. In a Gaussian model, decorrelation implies independence so that the joint model distribution for the sources reads

$$p_{\mathrm{S}}(\mathbf{S}) = \prod_{\tau=1}^{T} \phi(\bar{\mathbf{s}}(\tau); \Sigma(\tau)) \quad \text{with} \quad \Sigma(\tau) = \operatorname{diag}(\bar{\sigma}_{1\tau}^2, \ldots, \bar{\sigma}_{N\tau}^2) \tag{4.81}$$

where $\phi(\mathbf{y}; \mathbf{R})$ denotes the zero-mean Gaussian density with covariance $\mathbf{R}$. In the above, we have used $\det \mathbf{T} = 1$ for an orthonormal transform.

Such a diagonal Gaussian model is useful if the transform $\mathbf{T}$ is easily computed but it is not necessary that $\mathbf{T}^{\mathrm{T}}\mathbf{R}_i\mathbf{T}$ be exactly diagonal as analysis will show. The choice of $\mathbf{T}$ is discussed in section 4.6.5.

Estimating equations. Since the score of a row vector $\mathbf{s}_i$ modeled as Gaussian is $\boldsymbol{\varphi}_i(\mathbf{s}_i) = \mathbf{s}_i\mathbf{R}_i^{-1}$ (Eq. (4.27)), we have

$$\begin{aligned}\boldsymbol{\varphi}_i(\mathbf{s}_i)\mathbf{s}_j^{\mathrm{T}} &= \mathbf{s}_i\mathbf{R}_i^{-1}\mathbf{s}_j^{\mathrm{T}} = \mathbf{s}_i(\mathbf{T}\Sigma_i\mathbf{T}^{\mathrm{T}})^{-1}\mathbf{s}_j^{\mathrm{T}} = (\mathbf{s}_i\mathbf{T})\Sigma_i^{-1}(\mathbf{s}_j\mathbf{T})^{\mathrm{T}}\\ &= \bar{\mathbf{s}}_i\Sigma_i^{-1}\bar{\mathbf{s}}_j^{\mathrm{T}} = \sum_{\tau=1}^{T}\frac{\bar{s}_{i\tau}}{\bar{\sigma}_{i\tau}^2}\bar{s}_{j\tau} = \sum_{\tau=1}^{T}\left[\Sigma(\tau)^{-1}\bar{\mathbf{s}}(\tau)\bar{\mathbf{s}}(\tau)^{\mathrm{T}}\right]_{ij}\end{aligned}$$

so the relative gradient of the likelihood contrast (4.13) reads

$$-\nabla\widehat{\Upsilon}(\mathbf{Y}) = \frac{1}{T}\sum_{\tau=1}^{T}\Sigma(\tau)^{-1}\bar{\mathbf{y}}(\tau)\bar{\mathbf{y}}(\tau)^{\mathrm{T}} - \mathbf{I}. \tag{4.82}$$

We see that in diagonal Gaussian models, independence between two sources $i \neq j$ is expressed by

$$\frac{1}{T}\sum_{\tau=1}^{T}\frac{\bar{y}_{i\tau}}{\bar{\sigma}_{i\tau}^2}\bar{y}_{j\tau} = 0. \tag{4.83}$$

This condition is to be compared to condition (4.70) which expressed independence as the decorrelation between each source and a nonlinear version of each other source. Thus, moving from a non-Gaussian iid model to a Gaussian diagonal corresponds to moving from a nonlinear but fixed score function to a linear but variable (with sample number) score function or, so to speak, moving from "nonlinear decorrelation" to "non-stationary decorrelation".

4.6.2 Fisher information and diversity

For Gaussian variables, one has $\boldsymbol{\Gamma}_i = \mathbf{R}_i^{-1}$ (as shown at the beginning of this section), so that the elements f_{ij} of the FIM, defined at (4.53), become $f_{ij} = \mathrm{trace}((\mathbf{T}\Sigma_i\mathbf{T}^{\mathrm{T}})^{-1}(\mathbf{T}\Sigma_j\mathbf{T}^{\mathrm{T}}))/T = \mathrm{trace}(\Sigma_i^{-1}\Sigma_j)/T$, which also reads:

$$f_{ij} = \frac{1}{T}\sum_{\tau}\frac{\bar{\sigma}_{j\tau}^2}{\bar{\sigma}_{i\tau}^2}. \tag{4.84}$$

Inserting this last expression in (4.56) yields the Cramér-Rao bound for a Gaussian model with known variance profiles $\{\bar{\sigma}_{i\tau}^2\}$.

We saw in section 4.4.3 that the condition for blind separability of a pair $(i \neq j)$ in the Gaussian case is that matrices $\mathbf{R}_i$ and $\mathbf{R}_j$ should not be proportional. In the more specific case of *diagonal* Gaussian models where $\mathbf{R}_i = \mathbf{T}\Sigma_i\mathbf{T}^{\mathrm{T}}$, this condition reduces to the non-proportionality of the variance profiles $\{\bar{\sigma}^2_{i\tau}\}$ and $\{\bar{\sigma}^2_{j\tau}\}$. Hence, we find a *diversity* condition on the variance profiles.

The necessity of diversity appears clearly from Eq. (4.84): the lack of diversity between sources i and j means, by definition, that $\bar{\sigma}^2_{i\tau} = \alpha\bar{\sigma}^2_{j\tau}$ for some scalar α which entails $f_{ij}f_{ji} = \alpha^2\alpha^{-2} = 1$ and therefore the singularity of the Fisher matrix. Correspondingly, the condition (4.82) of decorrelation becomes symmetrical.

Incoherent profiles. Two variance profiles are said to be *incoherent* if

$$\frac{1}{T}\sum_\tau \frac{\bar{\sigma}^2_{j\tau}}{\bar{\sigma}^2_{i\tau}} = \left(\frac{1}{T}\sum_\tau \bar{\sigma}^2_{j\tau}\right)\left(\frac{1}{T}\sum_\tau \frac{1}{\bar{\sigma}^2_{i\tau}}\right).$$

In that case, the (ij) entry of the Fisher matrix becomes $f_{ij}\bar{\sigma}^2_j/\bar{\sigma}^2_i = v_i$ with

$$v_i \stackrel{\text{def}}{=} \left(\frac{1}{T}\sum_\tau \bar{\sigma}^2_{i\tau}\right)\left(\frac{1}{T}\sum_\tau \frac{1}{\bar{\sigma}^2_{i\tau}}\right)$$

which no longer depends on characteristics of the jth source. Hence, incoherent variance profiles yield the same kind of decoupling as seen in the non-Gaussian case in Eq. (4.76) and the Cramér-Rao bound then takes the same form (4.74) as in the non-Gaussian iid case with v_i in place of γ_i. Using again the Cauchy-Schwarz inequality, we find $v_i \geq 1$ with equality if and only if the variance profile is constant. In other words, it is a "non-stationarity" measure which, in a diagonal Gaussian model, plays the role of the non-Gaussianity measure found in iid models.

4.6.3 Gaussian contrasts

In a Gaussian model for an $N \times T$ data matrix $\mathbf{Y}$, where the columns of $\bar{\mathbf{Y}} = \mathbf{YT}$ are uncorrelated, where the τ-th column $\bar{\mathbf{y}}(\tau)$ has a covariance matrix $\mathrm{Cov}(\bar{\mathbf{y}}(\tau)) = \mathbf{R}_{\bar{\mathbf{y}}(\tau)}$, the log-density of $\mathbf{Y}$ is $\log p(\mathbf{Y}) = \sum_\tau \log\phi(\bar{\mathbf{y}}(\tau); \mathbf{R}_{\bar{\mathbf{y}}(\tau)})$. The contrasts discussed in section 4.3 take a suggestive form. Mutual information $I(\mathbf{Y})$ becomes

$$\begin{aligned} I[\mathbf{Y}] &= K\left[p_{\mathbf{Y}} \,|\, \prod_i p_{\mathbf{y}_i}\right] = K\left[p_{\bar{\mathbf{Y}}} \,|\, \prod_i p_{\bar{\mathbf{y}}_i}\right] = \sum_\tau K\left[p_{\bar{\mathbf{y}}(\tau)} \,|\, \prod_i p_{\bar{y}_i(\tau)}\right] \\ &= \sum_\tau k(\mathbf{R}_{\bar{\mathbf{y}}(\tau)}, \mathrm{diag}\mathbf{R}_{\bar{\mathbf{y}}(\tau)}). \end{aligned}$$

The first equality is the definition of mutual information between the rows of $\mathbf{Y}$; the second equality is created by invariance of the Kullback divergence under invertible transforms (here: under right multiplication by $\mathbf{T}$); the third equality results from the independence of the columns of $\bar{\mathbf{Y}}$; the last equality is created by the Gaussianity of each column $\bar{\mathbf{y}}(\tau)$. With expression (4.38) of correlation $C[\cdot]$, mutual information in diagonal Gaussian models hence appears as a sum of correlations:

$$I[\mathbf{Y}] = \sum_{\tau} C[\bar{\mathbf{y}}(\tau)]. \tag{4.85}$$

A striking comparison between correlation and dependence is obtained by defining a measure off($\mathbf{R}$) of (non-)diagonality for a positive matrix $\mathbf{R}$ by

$$\text{off}(\mathbf{R}) \stackrel{\text{def}}{=} k(\mathbf{R}, \operatorname{diag}\mathbf{R}) = \frac{1}{2}\log\frac{\det\operatorname{diag}\mathbf{R}}{\det\mathbf{R}} \tag{4.86}$$

and the sequence-averaging operator: $\text{ave}(f(\tau)) \stackrel{\text{def}}{=} \frac{1}{T}\sum_{\tau=1}^{T} f(\tau)$. With these notations, $C[\mathbf{Y}] = T\,k(\mathbf{R}_{\mathbf{Y}}, \operatorname{diag}\mathbf{R}_{\mathbf{Y}}) = T\,\text{off}(\mathbf{R}_{\mathbf{Y}}) = T\,\text{off}(\mathbf{R}_{\bar{\mathbf{Y}}}) = T\,\text{off}(\text{ave}(\mathbf{R}_{\bar{\mathbf{y}}(\tau)}))$ since $\mathbf{R}_{\mathbf{Y}} = \mathbf{R}_{\bar{\mathbf{Y}}}$ because $\mathbf{T}$ is orthogonal. We finally get

$$C[\mathbf{Y}] = T\,\text{off}(\text{ave}(\mathbf{R}_{\bar{\mathbf{y}}(\tau)})), \tag{4.87}$$

$$I[\mathbf{Y}] = T\,\text{ave}(\text{off}(\mathbf{R}_{\bar{\mathbf{y}}(\tau)})). \tag{4.88}$$

Hence, correlation $C[\mathbf{Y}]$ only measures the diagonality of the average covariance matrix while mutual information, in a diagonal Gaussian model, measures the average of the correlations at each τ.

The connection between correlation and mutual information was described in section 4.3 in the general case; let us see what becomes of it when applied to diagonal Gaussian models. If the distribution of $\bar{\mathbf{Y}}$ is diagonal Gaussian, its best iid-Gaussian approximation is by a diagonal Gaussian where all covariance matrices $\mathbf{R}_{\bar{\mathbf{y}}(\tau)}$ are identical to their mean $\mathbf{R}_{\mathbf{Y}} = \text{ave}(\mathbf{R}_{\bar{\mathbf{y}}(\tau)})$. The divergence to iid-Gaussianity for the block $\mathbf{Y}$ then simply is

$$G[\mathbf{Y}] = \sum_{\tau=1}^{T} k(\mathbf{R}_{\bar{\mathbf{y}}(\tau)}, \mathbf{R}_{\bar{\mathbf{Y}}}) = T\,\text{ave}(k(\mathbf{R}_{\bar{\mathbf{y}}(\tau)}, \text{ave}\mathbf{R}_{\bar{\mathbf{y}}(\tau)}))$$

which, after a simple manipulation, can be rewritten as

$$G[\mathbf{Y}] = \frac{T}{2}\frac{\log\det\text{ave}\mathbf{R}_{\bar{\mathbf{y}}(\tau)}}{\text{ave}\log\det\mathbf{R}_{\bar{\mathbf{y}}(\tau)}}.$$

The deviation to iid-Gaussianity for the i-th row of $\mathbf{Y}$ is similarly[4] given by

$$G[\mathbf{y}_i] = \sum_{\tau=1}^{T} k\left(\bar{\sigma}_{i\tau}^2, \frac{1}{T}\sum_{\tau}\bar{\sigma}_{i\tau}^2\right) = T\,\text{ave}\left(k(\bar{\sigma}_{i\tau}^2, \text{ave}\bar{\sigma}_{i\tau}^2)\right) = \frac{T}{2}\frac{\log\text{ave}\bar{\sigma}_{i\tau}^2}{\text{ave}\log\bar{\sigma}_{i\tau}^2}.$$

[4]The expression (4.17) of $k(\cdot,\cdot)$ is in particular valid when the arguments k are 1×1 matrices and then reduces to $k(\sigma_1^2, \sigma_2^2) = \sigma_1^2/\sigma_2^2 - \log\sigma_1^2/\sigma_2^2 - 1 \geq 0$.

Hence, in the diagonal Gaussian context, the general relation (4.40) becomes

$$I[\mathbf{Y}] = C[\mathbf{Y}] - T\sum_i \text{ave}(k(\bar{\sigma}^2_{i\tau}, \text{ave}\bar{\sigma}^2_{i\tau})) + \text{cst} \tag{4.89}$$

where the quantity $\text{ave}(k(\bar{\sigma}^2_{i\tau}, \text{ave}\bar{\sigma}^2_{i\tau}))$ appears as a measure, for the i-th source, of the deviation between a variance profile and its mean.

The decomposition (4.89) is thus understood as follows: in a diagonal Gaussian model, the linear transform which minimizes the mutual information also is the transform which meets an objective of joint decorrelation ($\min C[\mathbf{Y}]$) and of "de-uniformization" of each variance profile ($\max \text{ave}(k(\bar{\sigma}^2_{i\tau}, \text{ave}\bar{\sigma}^2_{i\tau}))$).

If $\mathbf{T}$ is the identity matrix, then $G[\mathbf{y}_i]$ is a measure of non-stationarity in the usual sense. If $\mathbf{T}$ is the Fourier transform (see below), the variance profile then corresponds to the spectrum so that $G[\mathbf{y}_i]$ is a measure of the "color" of the process (deviation of its spectrum to flatness or whiteness).

4.6.4 In practice: Localized Gaussian models

In a blind approach, the variance profiles $\{\bar{\sigma}^2_{i\tau}\}$ are usually not supposed to be known in advance: they must be estimated from the data. The issue of data-based score estimation has been addressed in the non-Gaussian case in section 4.5.6; for the Gaussian models considered here, it has a straightforward solution via local averaging.

Local averaging. In order to estimate the variance profiles $\bar{\sigma}^2_{i\tau}$, we shall assume some smoothness in their variation with respect to τ. The simplest approach (see [16] for other options) is to partition the interval $[1, T]$ in a number (Q) of sub-intervals: $[1, T] = \cup_{q=1}^{Q} \mathcal{I}_q$ and to assume the variance $\bar{\sigma}^2_{i\tau}$ to be constant over each of them, that is, with a slight abuse of notation: $\bar{\sigma}^2_{i\tau} = \bar{\sigma}^2_{iq}$ if $\tau \in \mathcal{I}_q$. One can then define local covariance matrices as

$$\widehat{\mathbf{R}}_{\bar{\mathbf{Y}}}(q) = \frac{1}{T_q}\sum_{\tau\in\mathcal{I}_q} \bar{\mathbf{y}}(\tau)\bar{\mathbf{y}}(\tau)^{\mathrm{T}} \tag{4.90}$$

where T_q is the number of data points in interval $\mathcal{I}_q$. In such a model, mutual information reduces to

$$I[\mathbf{Y}] = T\sum_q n_q \text{off}(\mathbf{R}_{\bar{\mathbf{Y}}}(q)) \tag{4.91}$$

where $n_q = \frac{T_q}{T}$ is the relative size of the q-th interval. Similarly, the empirical likelihood contrast (4.8) becomes

$$\widehat{\Upsilon}(\mathbf{Y}) = \sum_q n_q k(\widehat{\mathbf{R}}_{\bar{\mathbf{Y}}}(q), \Sigma(q)) + \text{cst} \tag{4.92}$$

with $\Sigma(q)=\Sigma(\tau)$ if $\tau \in \mathcal{I}_q$. For later reference, note that this expression is equivalent to

$$\widehat{\Upsilon}(\mathbf{Y})=\sum_q n_q k(\widehat{\mathbf{R}}_{\tilde{\mathbf{X}}}(q), \mathbf{R}_{\tilde{\mathbf{X}}}(q))+\text{cst} \tag{4.93}$$

with $\mathbf{R}_{\tilde{\mathbf{X}}}(q) = \mathbf{A}\Sigma(q)\mathbf{A}^{\mathrm{T}}$, which is a weighted measure of adjustment between the empirical matrices $\widehat{\mathbf{R}}_{\tilde{\mathbf{X}}}(q)$ estimated over each interval and the prediction of the model: $\mathbf{R}_{\tilde{\mathbf{X}}}(q)=\mathbf{A}\Sigma(q)\mathbf{A}^{\mathrm{T}}$.

Joint diagonalization. An important practical benefit of Gaussian modeling comes from formula (4.91), which shows that mutual information takes the form of a criterion of joint diagonality between the covariance matrices $\mathbf{R}_{\tilde{\mathbf{y}}(\tau)}$. Hence, it is straightforward to minimize it by using the appropriate joint diagonalization algorithm. For a thorough treatment, see Chapter 7 by A. Yeredor.

4.6.5 Choosing the diagonalizing transform

The most natural choices for transform $\mathbf{T}$ are as follows.

(a) Purely non-stationary model. A simple choice is to take $\mathbf{T}=\mathbf{I}$, that is, no transform at all! That means that each source sequence is seen as an iid Gaussian sequence modulated in amplitude by the scale factor $\bar{\sigma}_{i\tau}$. In other words, we are looking at a simplistic model of *non-stationary* process. An appropriate context for such a model would be for audio signals: the sequence $\bar{\sigma}_{i\tau}^2$ then represents the temporal variation of energy of the i-th source signal. Although this is a very crude model, it seems to be performing quite well with audio signals. See [13] for early publications in that direction and [16] for an exhaustive study.

(b) Purely stationary model. If $\mathbf{s}_i$ is the realization of a stationary process, its Fourier coefficients are asymptotically uncorrelated with a variance equal to the spectral density of the process. Therefore, for stationary sources, the natural choice is to take $\mathbf{T}$ to correspond to the matrix of the real discrete transform Fourier. In practice, of course, matrix $\mathbf{T}$ is not actually used: instead, its application to the data is computed via some fast Fourier transform. Spectral diversity entails the diversity of the correlation sequences: as a matter of fact, the separation methods for stationary signals originally resorted to using temporal correlation matrices rather than spectral covariance matrices. An initial approach [12] used only the algebraic structure of two correlation matrices. In [17], Pham introduces a likelihood approach in the time domain while [2] proposes a method of joint diagonalization of spectral covariance matrices but without the connection to the likelihood. It is finally Pham [14] who offers the most satisfying method, as outlined above, by moving to Fourier space: the log-likelihood in a Gaussian stationary model is well approximated by a criterion of joint diagonalization of spectral covariance matrices.

(c) Time-frequency and time-scale models. In between the original domain and the Fourier domain are wavelet domains. Wavelet coefficients are able to capture piecewise stationary statistics. They can be involved in a non-Gaussian iid framework (see Chapter 10 on sparsity by Gribonval and Zibulevski, Chapter 11 for a second-order approach in the time-frequency domain and Chapter 12 for the Bayesian approach.

4.7 NOISY MODELS

Accounting for additive noise in the source separation model dramatically changes the structure of the likelihood. In particular, the model is no longer equivariant and the likelihood contrast, after optimization with respect to the source parameters, can no longer be assimilated to mutual information. Hence this section has a very different flavor from those devoted to noise free models. We first address the issue of estimating the source signals assuming known parameters: known signal and noise distributions, known mixing matrix. Then, we look at the noisy likelihood in Gaussian models. Finally, we examine the operation of the Expectation-Maximization (EM) method for likelihood maximization in the noisy case.

4.7.1 Source estimation from noisy mixtures

Consider estimating a source vector $\mathbf{s}$ upon observation of a noisy mixture $\mathbf{x}$:

$$\mathbf{x} = \mathbf{A}\mathbf{s} + \mathbf{n} \tag{4.94}$$

assuming matrix $\mathbf{A}$ is known. The linear estimator of $\mathbf{s}$ based on $\mathbf{x}$ which minimizes the mean squared error is, by definition, given by $\mathbf{W}\mathbf{x}$ where matrix $\mathbf{W}$ of size $N \times P$ is the minimizer of $\mathbb{E}\{|\mathbf{W}\mathbf{x} - \mathbf{s}|^2\}$. It is easily (and classically) found to be $\mathbf{W} = \mathbb{E}\{\mathbf{s}\mathbf{x}^{\mathrm{T}}\}\mathbb{E}\{\mathbf{x}\mathbf{x}^{\mathrm{T}}\}^{-1}$. Denoting $\mathbf{R}_s = \mathbb{E}\{\mathbf{s}\mathbf{s}^{\mathrm{T}}\}$ and $\mathbf{R}_n = \mathbb{E}\{\mathbf{n}\mathbf{n}^{\mathrm{T}}\}$ and using $\mathbb{E}\{\mathbf{x}\mathbf{x}^{\mathrm{T}}\} = \mathbf{A}\mathbf{R}_s\mathbf{A}^{\mathrm{T}} + \mathbf{R}_n$ and $\mathbb{E}\{\mathbf{s}\mathbf{x}^{\mathrm{T}}\} = \mathbf{R}_s\mathbf{A}^{\mathrm{T}}$, one finds

$$\mathbf{W} = \mathbf{R}_s\mathbf{A}^{\mathrm{T}}(\mathbf{A}\mathbf{R}_s\mathbf{A}^{\mathrm{T}} + \mathbf{R}_n)^{-1}. \tag{4.95}$$

Some algebraic manipulations yield the alternative form:

$$\mathbf{W} = (\mathbf{A}^{\mathrm{T}}\mathbf{R}_n^{-1}\mathbf{A} + \mathbf{R}_s^{-1})^{-1}\mathbf{A}^{\mathrm{T}}\mathbf{R}_n^{-1} \tag{4.96}$$

which looks similar to the pseudo-inverse $\mathbf{A}^{\dagger} = (\mathbf{A}^{\mathrm{T}}\mathbf{A})^{-1}\mathbf{A}^{\mathrm{T}}$. In the high SNR regime, that is, if $\mathbf{A}^{\mathrm{T}}\mathbf{R}_n^{-1}\mathbf{A} \gg \mathbf{R}_s^{-1}$, the limiting form of (4.96) is

$$\mathbf{W}_{\star} = (\mathbf{A}^{\mathrm{T}}\mathbf{R}_n^{-1}\mathbf{A})^{-1}\mathbf{A}^{\mathrm{T}}\mathbf{R}_n^{-1} \tag{4.97}$$

which is equal to the pseudo-inverse $\mathbf{A}^{\dagger}$ for white noise: $\mathbf{R}_n = \sigma_n^2\mathbf{I}$. Matrix $\mathbf{W}_{\star}$ is a left-inverse matrix $\mathbf{A}$ since $\mathbf{W}_{\star}\mathbf{A} = \mathbf{I}$. One may check that $\mathbf{W}_{\star}$ also is the matrix which minimizes the mean quadratic error $\mathbb{E}\{|\mathbf{W}\mathbf{x} - \mathbf{s}|^2\}$ *under the constraint* that $\mathbf{W}\mathbf{A} = \mathbf{I}$.

Hence, linear separation of noisy mixtures faces a trade-off: either minimizing the noise level in the reconstructed sources while requiring a perfect separation $\mathbf{W}\mathbf{A} = \mathbf{I}$ or tolerating some contamination, that is, allowing $\mathbf{W}\mathbf{A} \neq \mathbf{I}$ to reach the lowest possible overall reconstruction error. The best choice depends on the application at hand but this is a real issue only in contexts with a significant noise level.

Let us recall that one may also consider *non-linear* source reconstruction. The function $g(\mathbf{x})$ which minimizes the mean squared error $\mathbb{E}\{|g(\mathbf{x}) - \mathbf{s}|^2\}$ is the conditional expectation of $\mathbf{s}$ given the data $\mathbf{x}$, that is, $g(\mathbf{x}) = \mathbb{E}\{\mathbf{s} \mid \mathbf{x}\}$. If $\mathbf{x}$ and $\mathbf{n}$ are jointly Gaussian, this is a linear function; one finds $g(\mathbf{x}) = \mathbb{E}\{\mathbf{s}|\mathbf{x}\} = \mathbf{W}\mathbf{x}$ with $\mathbf{W} = \mathbb{E}\left\{\mathbf{s}\mathbf{x}^{\mathrm{T}}\right\}\mathbb{E}\left\{\mathbf{x}\mathbf{x}^{\mathrm{T}}\right\}^{-1}$ as above. For *non-Gaussian* models, a better reconstruction is possible and the reconstruction error can be improved considerably for strongly non-Gaussian sources, in particular for discrete sources.

4.7.2 Noisy likelihood for localized Gaussian models

We now turn to the case of interest where matrix $\mathbf{A}$ is unknown and we start with the easiest case: the noisy version of the localized Gaussian models whose noise-free version was considered in section 4.6.4. Taking noise into account is a simple matter: the likelihood function is easily found to be

$$-\log p(\mathbf{X};\theta) = T\sum_{q=1}^{Q} n_q k\left(\widehat{\mathbf{R}}_{\tilde{\mathbf{X}}}(q),\, \mathbf{R}_{\tilde{\mathbf{X}}}(q;\theta)\right) + \mathrm{cst} \tag{4.98}$$

where θ is the set of all parameters which specify the localized covariance matrices, including typically the matrix $\mathbf{A}$, the variance profiles σ_{iq}^2 of the sources, and some (parametric) description of the noise covariance matrices. We note in the passing that if the noise is ignored and if $\mathbf{A}$ is invertible, then

$$k\left(\widehat{\mathbf{R}}_{\tilde{\mathbf{X}}}(q),\, \mathbf{R}_{\tilde{\mathbf{X}}}(q;\theta)\right) = k\left(\widehat{\mathbf{R}}_{\tilde{\mathbf{X}}}(q),\, \mathbf{A}\Sigma(q)\mathbf{A}^{\mathrm{T}}\right) = k\left(\widehat{\mathbf{R}}_{\tilde{\mathbf{Y}}}(q),\, \Sigma(q)\right),$$

which is the equivariant expression (4.92) of the noise free likelihood contrast.

Minimizing the form (4.98) of the likelihood with respect to unknown parameters can be considered for any localized Gaussian model, that is, for any parametrization $\theta \rightarrow \mathbf{R}_{\tilde{\mathbf{X}}}(q;\theta)$. In one of the simplest models, one may assume white uncorrelated noise so that $\mathbf{R}_{\tilde{\mathbf{X}}}(q;\theta) = \mathbf{A}\Sigma(q)\mathbf{A}^{\mathrm{T}} + \mathbf{R}_{\mathrm{n}}$. The parameter vector θ then is of size $(P+Q-1)\times N+P$, that is, P positive scalars for the noise variance on each sensor and, for each source, P mixing parameters (the column of $\mathbf{A}$) plus Q variance parameters (the variance of the source in each domain) minus one scale factor which can be exchanged between these two quantities.

The maximization of the noisy likelihood (4.98) can be conducted by a generic optimization algorithm based on the gradient of (4.98) and, possibly, on its Hessian (or an approximation of the latter by the Fisher information matrix). The derivative $\partial \log p(\mathbf{X};\theta)/\partial\theta$ can be computed by chaining the derivative $\partial\mathbf{R}_{\tilde{\mathbf{X}}}(q;\theta)/\partial\theta$ of the covariance matrices with respect to the unknown parameters with the derivative of the log-likelihood with respect to these matrices. One has to combine the specific expressions of $\partial\mathbf{R}_{\tilde{\mathbf{X}}}(q;\theta)/\partial\theta$ arising from a specific model (or parametrization) $\theta \rightarrow \mathbf{R}_{\tilde{\mathbf{X}}}(q;\theta)$ with expressions

$$\frac{\partial k(\widehat{\mathbf{R}}, \mathbf{R}(\theta))}{\partial\theta} = \frac{1}{2}\mathrm{trace}\left\{\mathbf{R}(\theta)^{-1}(\widehat{\mathbf{R}} - \mathbf{R}(\theta))\mathbf{R}(\theta)^{-1}\frac{\partial\mathbf{R}(\theta)}{\partial\theta}\right\}$$

$$\frac{\partial^2 k(\widehat{\mathbf{R}},\mathbf{R}(\theta))}{\partial\theta^2} = \frac{1}{2}\text{trace}\left\{\mathbf{R}(\theta)^{-1}\frac{\partial\mathbf{R}(\theta)}{\partial\theta}\mathbf{R}(\theta)^{-1}\frac{\partial\mathbf{R}(\theta)}{\partial\theta}\right\} + L(\widehat{\mathbf{R}}-\mathbf{R}(\theta))$$

where $L(\widehat{\mathbf{R}}-\mathbf{R}(\theta))$ is a linear term in $\widehat{\mathbf{R}}-\mathbf{R}(\theta)$ which thus may be neglected if $\widehat{\mathbf{R}} \approx \mathbf{R}(\theta)$. Neglecting that term amounts to approximating the Hessian of the log-likelihood by minus the Fisher information matrix. This is a standard approximation which has the benefit that it guarantees matrix positivity.

In is worth mentioning that the Fisher information matrix is sparse in simple models. For instance, if the variances of each source are adjusted independently over each domain, then the corresponding cross-terms in the FIM vanish.

Note that the assumption of source independence plays little role in this approach. Algorithmically, its most direct impact is that it simplifies the computation of the derivatives of the log-likelihood. In the next section, we consider instead the specific impact of the mixing model $\bar{\mathbf{X}} = \mathbf{A}\bar{\mathbf{S}} + \bar{\mathbf{N}}$.

4.7.3 Noisy likelihood and hidden variables

We now take quick look at the likelihood for mixtures observed in Gaussian noise. In that case, conditionally to the sources, the observations have a Gaussian distribution so that the EM algorithm can be used quite straightforwardly by considering the sources as the "hidden variables" of EM. After a short introduction, the non-Gaussian-iid model and the localized Gaussian model are briefly discussed.

Likelihood with hidden variables. Let $\mathbf{x}$ and $\mathbf{s}$ be two random variables whose joint distribution depends on a vector θ of parameters: $p(\mathbf{x},\mathbf{s};\theta)$. Using $p(\mathbf{x};\theta) = \int p(\mathbf{x},\mathbf{s};\theta)d\mathbf{s}$ and $p(\mathbf{x},\mathbf{s};\theta) = p(\mathbf{x}|\mathbf{s};\theta)p(\mathbf{s};\theta)$ and assuming that the order of derivation w.r.t. θ and integration w.r.t. $\mathbf{s}$ can be exchanged, one easily gets to

$$\frac{\partial \log p(\mathbf{x};\theta)}{\partial\theta} = \mathbb{E}\left\{\frac{\partial \log p(\mathbf{x},\mathbf{s};\theta)}{\partial\theta} \mid \mathbf{x};\theta\right\} \tag{4.99}$$

where the conditional expectation above is defined over $\mathbf{s}$, that is, for any function $g(\mathbf{x},\mathbf{s})$ of $\mathbf{x}$ and $\mathbf{s}$:

$$\mathbb{E}\{g(\mathbf{x},\mathbf{s}) \mid \mathbf{x};\theta\} = \int_{\mathbf{s}} g(\mathbf{x},\mathbf{s})\, p(\mathbf{s}|\mathbf{x};\theta)\, d\mathbf{s}$$

which is a function of $\mathbf{x}$ and θ.

If variable $\mathbf{s}$ is observed, the most likely value of θ is the solution of $\partial \log p(\mathbf{x}|\mathbf{s};\theta)/\partial\theta = 0$. If $\mathbf{s}$ is not observed (hidden variable), the most likely θ is the maximizer of $\log p(\mathbf{x};\theta)$ and thus a solution of $\partial \log p(\mathbf{x};\theta)/\partial\theta = 0$. Expression (4.99) shows that this is equivalent to averaging the estimating function $\partial \log p(\mathbf{x},\mathbf{s};\theta)/\partial\theta$ for the case of a visible $\mathbf{s}$ where the average is above all possible values of $\mathbf{s}$ with a weight given by the conditional probability $p(\mathbf{s}|\mathbf{x};\theta)$.

Unfortunately, since weight $p(\mathbf{s}|\mathbf{x};\theta)$ depends on θ, solving $\partial \log p(\mathbf{x};\theta)/\partial\theta = 0$ is not straightforward. Nonetheless, since formula (4.99) gives the gradient of the log-likelihood for any θ, one may implement a gradient-based algorithm for maximizing the likelihood provided the conditional expectation in (4.99) can be evaluated. See below for an example. Another possibility is to resort to the EM algorithm.

Expectation-Maximization for source separation. The structure of the noisy likelihood with visible data $\mathbf{x}$ and hidden data $\mathbf{s}$ calls for considering the Expectation-Maximization (EM) algorithm. The *EM functional* is defined by

$$\phi(\theta,\theta') = \mathbb{E}\left\{\log p(\mathbf{x},\mathbf{s};\theta) \mid \mathbf{x};\theta'\right\} \tag{4.100}$$

where the dependence on the observed variable $\mathbf{x}$ is not shown explicitly. In its simplest version, the EM algorithm builds a sequence $\{\theta^{(n)}, n = 0,1,2,\ldots\}$ of estimates of θ by the following iteration:

$$\theta^{(n+1)} = \arg\max_{\theta} \phi(\theta,\theta^{(n)}). \tag{4.101}$$

Hence, the estimate at step $n+1$ is related to the previous one by

$$0 = \mathbb{E}\left\{\frac{\partial \log p(\mathbf{x},\mathbf{s};\theta^{(n+1)})}{\partial\theta} \mid \mathbf{x};\theta^{(n)}\right\}, \tag{4.102}$$

stopping when $0 = \mathbb{E}\left\{\partial \log p(\mathbf{x},\mathbf{s};\hat{\theta})/\partial\theta \mid \mathbf{x};\hat{\theta}\right\}$. The identity (4.99) implies

$$\frac{\partial \log p(\mathbf{x};\theta)}{\partial\theta} = \left.\frac{\partial\phi(\theta,\theta')}{\partial\theta}\right|_{\theta'=\theta} \tag{4.103}$$

showing that a stationary point of EM also is a stationary point of the likelihood. EM iterations can be shown to increase the likelihood at each step, until convergence is reached.

Mixing parameters and source parameters. Let us now assume that the vector parameter θ is partitioned into two subvectors $\theta = (\theta_x, \theta_s)$ where θ_s governs the distribution of $\mathbf{s}$ while θ_x governs the distribution of $\mathbf{x}$ conditioned on $\mathbf{s}$. In other words, we assume the factorization $p(\mathbf{x},\mathbf{s};\theta) = p(\mathbf{x},\mathbf{s};\theta_x,\theta_s) = p(\mathbf{x}|\mathbf{s};\theta_x)\,p(\mathbf{s};\theta_s)$. Then, identity (4.99) yields

$$\begin{aligned}\frac{\partial \log p(\mathbf{x};\theta)}{\partial\theta_x} &= \mathbb{E}\left\{\frac{\partial \log p(\mathbf{x}|\mathbf{s};\theta_x)}{\partial\theta_x} \mid \mathbf{x};\theta\right\},\\ \frac{\partial \log p(\mathbf{x};\theta)}{\partial\theta_s} &= \mathbb{E}\left\{\frac{\partial \log p(\mathbf{s};\theta_s)}{\partial\theta_s} \mid \mathbf{x};\theta\right\}.\end{aligned} \tag{4.104}$$

Now, in the model $\mathbf{x} = \mathbf{As} + \mathbf{n}$ with a zero-mean Gaussian noise with covariance matrix $\mathbf{R_n}$, the distribution $p(\mathbf{x}|\mathbf{s})$ is a Gaussian with mean vector $\mathbf{As}$ and of covariance $\mathbf{R_n}$.

Hence,

$$\log p(\mathbf{x}|\mathbf{s};\theta) = -\frac{1}{2}(\mathbf{x}-\mathbf{A}\mathbf{s})^{\mathrm{T}}\mathbf{R}_{\mathrm{n}}^{-1}(\mathbf{x}-\mathbf{A}\mathbf{s}) - \frac{1}{2}\log\det\mathbf{R}_{\mathrm{n}} + \mathrm{cst}$$

and the derivatives of the log-likelihood are

$$\frac{\partial \log p(\mathbf{x};\theta)}{\partial \mathbf{A}} = \mathbf{R}_{\mathrm{n}}^{-1}\mathbb{E}\left\{(\mathbf{x}-\mathbf{A}\mathbf{s})\mathbf{s}^{\mathrm{T}} \mid \mathbf{x};\theta\right\} \tag{4.105}$$

$$\frac{\partial \log p(\mathbf{x};\theta)}{\partial \mathbf{R}_{\mathrm{n}}} = \frac{1}{2}\mathbf{R}_{\mathrm{n}}^{-1}\mathbb{E}\left\{(\mathbf{x}-\mathbf{A}\mathbf{s})(\mathbf{x}-\mathbf{A}\mathbf{s})^{\mathrm{T}} - \mathbf{R}_{\mathrm{n}} \mid \mathbf{x};\theta\right\}\mathbf{R}_{\mathrm{n}}^{-1}. \tag{4.106}$$

Hence, the derivatives of the noisy log-likelihood with respect to the mixing parameters are available as soon as the conditional expectations $\mathbb{E}\{\mathbf{s} \mid \mathbf{x};\theta\}$ and $\mathbb{E}\left\{\mathbf{s}\mathbf{s}^{\mathrm{T}} \mid \mathbf{x};\theta\right\}$ can be computed. The derivative with respect to the source parameters θ_s are model dependent.

Independent observations. Let us start with the case of independent observations $\mathbf{X} = [\mathbf{x}(1),\ldots,\mathbf{x}(T)]$ so that $\log p(\mathbf{X}) = \sum_{t=1}^{T}\log p(\mathbf{x}(t))$ and compute the derivatives of the log-likelihood and of the EM functional.

Combining derivatives (4.105) and (4.106) with the iid assumption yields

$$\frac{\partial \log p(\mathbf{X};\theta)}{\partial \mathbf{A}} = T\mathbf{R}_{\mathrm{n}}^{-1}\mathbf{G}_A(\theta), \qquad \frac{\partial \log p(\mathbf{X};\theta)}{\partial \mathbf{R}_{\mathrm{n}}} = \frac{T}{2}\mathbf{R}_{\mathrm{n}}^{-1}\mathbf{G}_N(\theta)\mathbf{R}_{\mathrm{n}}^{-1} \tag{4.107}$$

where matrices $\mathbf{G}_A(\theta)$ and $\mathbf{G}_N(\theta)$ are defined as

$$\mathbf{G}_A(\theta) = \widehat{\mathbf{R}}_{xs|x}(\theta) - \mathbf{A}\widehat{\mathbf{R}}_{ss|x}(\theta) \tag{4.108}$$

$$\mathbf{G}_N(\theta) = \widehat{\mathbf{R}}_{\mathbf{X}} - \mathbf{A}\widehat{\mathbf{R}}_{sx|x}(\theta) - \widehat{\mathbf{R}}_{xs|x}(\theta)\mathbf{A}^{\mathrm{T}} + \mathbf{A}\widehat{\mathbf{R}}_{ss|x}(\theta)\mathbf{A}^{\mathrm{T}} - \mathbf{R}_{\mathrm{n}} \tag{4.109}$$

with the following definitions of "θ-dependent conditional matrices":

$$\widehat{\mathbf{R}}_{sx|x}(\theta) = \frac{1}{T}\sum_{t=1}^{T}\mathbb{E}\{\mathbf{s} \mid \mathbf{x}(t);\theta\}\ \mathbf{x}(t)^{\mathrm{T}} \qquad \widehat{\mathbf{R}}_{ss|x}(\theta) = \frac{1}{T}\sum_{t=1}^{T}\mathbb{E}\left\{\mathbf{s}\mathbf{s}^{\mathrm{T}} \mid \mathbf{x}(t);\theta\right\} \tag{4.110}$$

The derivatives of the EM functional are identical provided matrices $\mathbf{G}_A(\theta)$ and $\mathbf{G}_N(\theta)$ are replaced by $\mathbf{G}_A(\theta')$ and $\mathbf{G}_N(\theta')$.

Those expressions lead to re-estimation formulas in various scenarios. The simplest case consists in re-estimating the mixing matrix $\mathbf{A}$ assuming a known noise covariance matrix $\mathbf{R}_{\mathrm{n}}$. Then, an EM step boils down to solving iteratively $\mathbf{G}_A(\theta) = 0$, that is, to iterate

$$\theta^{(n)} = (\mathbf{A}^{(n)}, \mathbf{R}_{\mathrm{n}}) \qquad \mathbf{A}^{(n+1)} = \widehat{\mathbf{R}}_{xs|x}(\theta^{(n)})\widehat{\mathbf{R}}_{ss|x}(\theta^{(n)})^{-1}. \tag{4.111}$$

Similarly, for a fixed mixing matrix $\mathbf{A}$, the EM re-estimation of the noise covariance matrix, *supposed to be diagonal*, consists in solving iteratively $\mathbf{G}_N(\theta) = 0$, that is, setting

$\theta^{(n)} = (\mathbf{A}, \mathbf{R}_\mathbf{n}^{(n)})$ and iterating through:

$$\mathbf{R}_\mathbf{n}^{(n+1)} = \mathrm{diag}(\widehat{\mathbf{R}}_\mathbf{X} - \mathbf{A}\widehat{\mathbf{R}}_{sx|x}(\theta^{(n)}) - \widehat{\mathbf{R}}_{xs|x}(\theta^{(n)})\mathbf{A}^\mathrm{T} + \mathbf{A}\widehat{\mathbf{R}}_{ss|x}(\theta^{(n)})\mathbf{A}^\mathrm{T}). \qquad (4.112)$$

Another option is the joint re-estimation of $\mathbf{A}$ and $\mathbf{R}_\mathbf{n}$; simple algebra yields the EM re-estimation rule:

$$\theta^{(n)} = (\mathbf{A}^{(n)}, \mathbf{R}_\mathbf{n}^{(n)}) \qquad (4.113)$$

$$\mathbf{A}^{(n+1)} = \widehat{\mathbf{R}}_{xs|x}(\theta^{(n)})\widehat{\mathbf{R}}_{ss|x}(\theta^{(n)})^{-1} \qquad (4.114)$$

$$\mathbf{R}_\mathbf{n}^{(n+1)} = \mathrm{diag}(\widehat{\mathbf{R}}_\mathbf{X} - \widehat{\mathbf{R}}_{xs|x}(\theta^{(n)})\widehat{\mathbf{R}}_{ss|x}(\theta^{(n)})^{-1}\widehat{\mathbf{R}}_{sx|x}(\theta^{(n)})). \qquad (4.115)$$

In all these cases, the EM algorithm is of straightforward implementation as soon as $\mathbf{x} \to \mathbb{E}\{\mathbf{s} \mid \mathbf{x}; \theta\}$ and $\mathbf{x} \to \mathbb{E}\left\{\mathbf{s}\mathbf{s}^\mathrm{T} \mid \mathbf{x}; \theta\right\}$ can be computed. We examine how that can be done in the two main source models: non-Gaussian-iid and localized Gaussian models.

Non-Gaussian models and EM. A popular method for running EM updates with non-Gaussian hidden data is to resort to *mixture of Gaussian distributions*. In the case of non-Gaussian-iid ICA models, it means that the marginal distribution $p_i(s)$ common to all the variables $s_i(t)$ is modeled as a superposition of L Gaussian densities:

$$p_i(s; \theta_i) = \sum_{l=1}^{L} \pi_{il} \phi(s; \mu_{il}, \sigma_{il}^2)$$

where $\phi(s; \mu, \sigma^2)$ denotes the Gaussian density with mean μ and variance σ^2 and where the weights π_{il} obey the constraint $\sum_l \pi_{il} = 1$. Hence, the distribution of s is that of a scalar variable obtained from first drawing a random index l in $[1, \ldots, L]$ with probability π_{il} and then drawing a real variable s under the Gaussian law $\phi(s; \mu_{il}, \sigma_{il}^2)$. The index l is a hidden variable and it turns out that the EM algorithm can also be used to estimate the parameters of the distribution of the i-th source, that is, $\theta_i = \{\pi_{il}, \mu_{il}, \sigma_{il}^2\}_{l=1}^{L}$. The conditional expectations $\mathbb{E}\{\mathbf{s} \mid \mathbf{x}(t); \theta\}$ and $\mathbb{E}\left\{\mathbf{s}\mathbf{s}^\mathrm{T} \mid \mathbf{x}(t); \theta\right\}$ are easily expressed in the mixture-of-Gaussian model but the resulting formulas may be computationally demanding. It may be a good idea to stick to the simplest models. For instance, in order to model a heavy-tailed distribution, it may be enough to take $L = 2$ and $\mu_{i1} = \mu_{i2} = 0$, leaving only two free parameters for a unit variance source (since one must have $\pi_1\sigma_{i1}^2 + \pi_2\sigma_{i2}^2 = 1$ in addition to the constraint $\pi_1 + \pi_2 = 1$). Similarly, in order to model symmetric sub-Gaussian distributions, one may use only two components with $\pi_{i1} = \pi_{i2} = 1/2$; $\mu_{i2} = -\mu_{i1}$; and $\sigma_{i1}^2 = \sigma_{i2}^2$.

Localized Gaussian models and EM. For the localized Gaussian models described in section 4.6.4, the EM re-estimation is straightforward. Indeed, if $\mathbf{x} = \mathbf{A}\mathbf{s} + \mathbf{n}$, we have already seen that $\mathbb{E}\{\mathbf{s}|\mathbf{x}; \theta\} = \mathbf{W}\mathbf{x}$ where $\mathbf{W}$ is the Wiener filter $\mathbf{W} = \mathbf{C}\mathbf{A}^\mathrm{T}\mathbf{R}_\mathbf{n}^{-1}$. Here, we defined matrix $\mathbf{C} = (\mathbf{A}^T\mathbf{R}_n^{-1}\mathbf{A} + \mathbf{R}_s^{-1})^{-1}$ which is nothing but the covariance of $\mathbf{s}$ conditioned on $\mathbf{x}$. In the Gaussian case, it does not depend on $\mathbf{x}$ and we get:

$$\mathbb{E}\left\{\mathbf{s}\mathbf{s}^T|\mathbf{x}\right\} = \mathbf{W}\mathbf{x}\mathbf{x}^T\mathbf{W}^T + \mathbf{C}.$$

Matrices $\mathbf{W}$ and $\mathbf{C}$ depend on the distribution of the sources; in our localized model, they are constant over each interval. Hence, for $\tau \in \mathscr{I}_q$, we have

$$\mathbb{E}\{\bar{\mathbf{s}} \mid \bar{\mathbf{x}}(\tau)\} = \mathbf{W}_q\bar{\mathbf{x}}(\tau) \quad \text{and} \quad \mathbb{E}\left\{\bar{\mathbf{s}}\bar{\mathbf{s}}^T \mid \bar{\mathbf{x}}(\tau)\right\} = \mathbf{W}_q\bar{\mathbf{x}}(\tau)\bar{\mathbf{x}}(\tau)^T\mathbf{W}_q^T + \mathbf{C}_q$$

where $\mathbf{C}_q$ and $\mathbf{W}_q$ are the local (interval-wise) versions of $\mathbf{C}$ and $\mathbf{W}$ defined above. Combining the previous results yields:

$$\widehat{\mathbf{R}}_{sx|x} = \sum_q n_q \mathbf{W}_q \mathbf{R}_{\bar{\mathbf{X}}}(q) \quad \text{and} \quad \widehat{\mathbf{R}}_{ss|x} = \sum_q n_q(\mathbf{W}_q \mathbf{R}_{\bar{\mathbf{X}}}(q)\mathbf{W}_q^T + \mathbf{C}_q).$$

Those expressions allow for a direct implementation of the EM method. See [9] for an application to the separation of multi-spectral images modeled as Gaussian stationary fields and the chapter by Mohammad-Djafari in this volume for several variations on this theme in a Bayesian framework [18].

4.8 CONCLUSION: A GENERAL VIEW

We conclude by first recapitulating the elements of the theory which are shared by the different source models, stressing the formal unity and then, by doing just the opposite, contrasting the differences induced by the choice of the source model and outlining the algorithmic consequences of this choice.

4.8.1 Unity

We saw how the source separation problem can be formulated in a likelihood approach. Writing down a likelihood function requires an explicit model for the distribution of each source. We attempted to develop the theory in the most general possible way by incorporating several ingredients in sequence: transformation, independence, source models.

The first ingredient is the mixing matrix $\mathbf{A}$: we focused on the ICA model as a "transformation model", leading to the notion of equivariance, of relative gradient, and of contrast function. It is the equivariance concept which explains why the difficulty of un-mixing a noise-free mixture does not depend *at all* on the particular mixing matrix. The methods which stem from a likelihood function are equivariant without even trying: the equivariant structure is somehow unavoidable. Any non-equivariant source separation method should be considered with suspicion.

Table 4.1 Summary of some compared features of source models. See text for explanations and comments.

Source model	iid-non-Gaussian	Gaussian non-stat.	Gaussian colored
Source parameter	marginal pdf	energy vs time	energy vs freq.
Diversity	not necessary	necessary	
Model size	$o(T)$	$O(T)$	$o(T)$
Exhaustive statistic	no (practically)	yes, but...	yes
Algorithms	gradient et al.	joint diagonalization	
Nuisance estimation	possible	automatic (built-in)	
Mutual information	not available	available	
Noise modeling	not very easy	rather easy	

The second ingredient is the assumption of source independence which leads to the decomposition of the likelihood contrast into a mutual information term supplemented by marginal terms expressing the (mis)adjustment of each source model to the recovered sources. Mutual information itself decomposes as a correlation term from which are subtracted marginal terms which measure the deviation of each source from iid-Gaussianity. Assuming that the distribution of each source is known, this makes it possible to obtain the Cramér-Rao bound for separation (in the equivariant form of a rejection rate) as well as conditions for blind identifiability and for the stability with respect to misspecification of the source distributions.

The third ingredient is the specification the source models. This is critical for obtaining tractable algorithms. We considered source models which are non-Gaussian but iid and source models which are Gaussian but correlated (with the strong condition that the covariance matrices of all sources can be diagonalized on a common basis: diagonal Gaussian models). In these two cases, specifying the distribution for a source requires specifying a scalar function: either a marginal density for the non-Gaussian-iid models or a variance profile for the Gaussian diagonal models. Even in these simple models it is not necessary or advisable to try to accurately estimate these scalar functions. It is often a better option to use simple approximations. In the non-Gaussian-iid case, one can for instance estimate the score function for each source as a linear combination of a small number of basis functions; in the Gaussian diagonal case, the variance profile can be estimated as a smooth function or as piece-wise constant.

4.8.2 Diversity

Beyond their common features, the various approaches to source modeling also exhibit significant differences which become very relevant in practice. The most salient features are now discussed with a tentative summary in Table 4.1.

The rows of the table can be read as follows.

Source model. This row lists the source models considered so far: iid non-Gaussian and Gaussian diagonal models described in section 4.6.5: the pure non-stationary model and the purely stationary model. The latter is labeled as "Gaussian colored", since each source is characterized by the color of its spectrum.

Source parameter. In all three models, the distribution of each source is described by a density. For the non-Gaussian models, it is a probability density. For the Gaussian models, it is a temporal or spectral energy density.

Diversity. In iid-non-Gaussian models, blind separability of any two sources does not require diversity: their probability distributions can be identical without entailing a loss of identifiability as long as they are not both Gaussian. In contrast, the Gaussian diagonal models cannot blindly separate two sources if their variance profiles (spectral or temporal) are proportional: diversity is required. In a given application, it will often be easy to determine if the requested diversity is plausible. For instance for the separation of speech signals, temporal diversity is to be expected.

Model size. The table row labeled "model size" refers to the number of free parameters in each source model. The row contains $o(T)$ or $O(T)$, to be understood as follows. In a plausible non-stationary model, the variance profile is expected to be smooth on some time scale. For instance, for speech separation, the variance would be estimated over time intervals of the typical duration of a phoneme. Thus, the number of parameters to describe a source would typically be proportional to the total signal length, hence the $O(T)$ typical behavior. Compare that to the Gaussian stationary model or to the model non-Gaussian-iid model in which a growing number of samples allows for refining the source model (the marginal probability density or the power spectral density). On the basis of T samples, one could try to estimate a growing number of parameters but, contrary to the previous case, there is no need to make this number proportional to the sample size T. It is probably wiser to use a fixed (or slowly increasing) number, which we denote $o(T)$. In this respect, see Chapter 2 by Pham, which considers the estimation of the marginal probability densities using kernels with T-dependent width.

Exhaustive statistic. For the Gaussian methods, the local covariance matrices defined at (4.90) form an exhaustive statistic: the likelihood (noisy or noise-free) depends on the data only through these matrices. This fact may lead to massive data compression, especially in the Gaussian stationary model for which the number of local matrices typically grows more slowly than the number of samples (*cf supra*). The non-Gaussian-iid model also offers an exhaustive statistic: the joint empirical distribution $\hat{p}(\mathbf{x}) = T^{-1}\sum_{t=1}^{T}\delta(\mathbf{x}-\mathbf{x}(t))$ but this object can only be represented in practice by the data set itself and therefore does not offer any practical compression. However, exhaustive statistics of relatively small size are available for the methods which approximate the mutual information using cumulants. For instance, the JADE algorithm uses a fixed size (independent of T) statistic made of all cumulants of orders 2 and 4.

Algorithms. For the diagonal Gaussian models, the algorithmic of the noise-free case is a no-brainer since the mutual information takes the form of a joint diagonality criterion of covariance matrices. Then, depending on whether or not one wants to enforce source decorrelation, one would use the Pham algorithm or the Cardoso-Souloumiac algorithm. For the non-Gaussian-iid case, the situation is less clear. If the mutual information is approximated by a function of the cumulants, then one may exploit their algebraic properties for devising a specific contrast optimization algorithm. However, the cumulant-based contrasts are only approximations of the likelihood contrast. In general, if the score functions are not polynomial, the minimization of $\widehat{\Upsilon}(\mathbf{Y})$ cannot be helped by algebra and one has to resort to generic optimization techniques. The relative gradient method may offer good performance but it requires at least the tuning of the learning step. Its acceleration by a Newton type method is not difficult if one can rely on the approximation of the Hessian explained in 4.4.6 but actual acceleration can only be guaranteed if the model holds. The fastICA method can be seen as belonging to that category [10].

Nuisance estimation. This row of the table is a reminder that the source parameters have the status of nuisance parameters. In the diagonal Gaussian models, these parameters are automatically estimated by the joint diagonalization procedure: the estimated power of each source in each interval appears directly after diagonalization on the diagonal of the corresponding covariance matrices. This is not true in non-Gaussian-iid models: if the estimation of the score functions φ_i is necessary or desired, it must be implemented explicitly.

Mutual information. In the non-Gaussian-iid models, mutual information $I[\mathbf{Y}]$ includes the joint non Gaussianity term $G[\mathbf{Y}] = TG[\mathbf{y}(t)]$ which is constant under linear transform and can thus be ignored in the minimization of $I[\mathbf{Y}]$. This is fortunate since estimating this term would require estimating the joint entropy $\mathbf{y}(t)$. However, ignoring this term prevents us from knowing the absolute value of $I[\mathbf{Y}]$ and we miss an indication that the likeliest value of $\mathbf{A}$ does correspond to the discovery of independent sources, that is, to a value of $I[\mathbf{Y}]$ compatible with 0. Gaussian contrasts offer this important benefit in that the mutual information is known explicitly as (4.87) which allows for a direct test of the model.

Noise. Introducing noise in the Gaussian models leaves the likelihood contrast in a relatively simple form (4.98) from which gradient and Hessian are easily calculated, so that standard optimization methods can be used to maximize the likelihood. Also, the EM algorithm offers simple and explicit re-estimation formulas. Adding noise in the non-Gaussian-iid model is more damaging. One can still use EM but, lacking exhaustive statistic, the computation cost can become prohibitive. The immunity to noise of cumulant based methods is real only for large data sets: the mean value of sample cumulants of order higher than 2 is not affected by additive Gaussian noise, but their estimation variance certainly is.

4.9 APPENDIX: PROOFS

Proof of Eq. (4.9): From the empirical contrast to its expectation.

$$
\begin{aligned}
T\mathbb{E}\left\{\widehat{\Upsilon}(\mathbf{Y})\right\} &= \mathbb{E}\{\log p_S(\mathbf{Y})\} + \frac{T}{2}\mathbb{E}\left\{\log\det\widehat{\mathbf{R}}_\mathbf{Y}\right\} \\
&= \mathbb{E}\left\{\log\frac{p_S(\mathbf{Y})}{p_\mathbf{Y}(\mathbf{Y})} + \log p_\mathbf{Y}(\mathbf{Y})\right\} + \frac{T}{2}\mathbb{E}\left\{\log\det\widehat{\mathbf{R}}_\mathbf{X} - 2\log|\det\mathbf{A}|\right\} \\
&= -K\left[p_\mathbf{Y} \mid p_S\right] - (H(\mathbf{Y}) + T\log|\det\mathbf{A}|) + \frac{T}{2}\mathbb{E}\left\{\log\det\widehat{\mathbf{R}}_\mathbf{X}\right\} \\
&= -K\left[p_\mathbf{Y} \mid p_S\right] + \mathrm{cst}
\end{aligned}
$$

where $\mathrm{cst} = -H(\mathbf{X}) + T/2\mathbb{E}\left\{\log\det\widehat{\mathbf{R}}_\mathbf{X}\right\}$ In this derivation, we used the definition of the Kullback divergence $K[\cdot \mid \cdot]$; that of Shannon entropy: $H(\mathbf{X}) = -\int p_\mathbf{X}(\mathbf{X})\log p_\mathbf{X}(\mathbf{X})$ and its transformation property: $H(\mathbf{X}) = H(\mathbf{AY}) = H(\mathbf{Y}) + T\log|\det\mathbf{A}|$ (if $\mathbf{A}$ is an invertible matrix $N \times N$ and $\mathbf{Y}$ an $N \times T$ matrix); and finally the fact that $\log\det\widehat{\mathbf{R}}_\mathbf{X} = \log\det\widehat{\mathbf{R}}_\mathbf{Y} + 2\log|\det\mathbf{A}|$ since $\widehat{\mathbf{R}}_\mathbf{X} = \mathbf{A}\widehat{\mathbf{R}}_\mathbf{Y}\mathbf{A}^\mathrm{T}$.

Proof of Eq. (4.39): Decomposition of mutual information. In order to establish (4.39), we introduce $\mathbf{Y}^{IG}$, a random matrix whose rows are independent, iid-Gaussian sequences with a variance equal to the mean variance of the corresponding rows of $\mathbf{Y}$, that is, $\mathbf{R}_{\mathbf{Y}^{IG}} = \mathrm{diag}(\mathbf{R}_\mathbf{Y})$. We shall decompose $K\left[\mathbf{Y} \mid \mathbf{Y}^{IG}\right]$ in two different ways. First, we decompose $K\left[\mathbf{Y} \mid \mathbf{Y}^{IG}\right]$ by "passing by $\mathbf{Y}^G$" and we shall see that

$$
K\left[\mathbf{Y} \mid \mathbf{Y}^{IG}\right] = K\left[\mathbf{Y} \mid \mathbf{Y}^G\right] + K\left[\mathbf{Y}^G \mid \mathbf{Y}^{IG}\right], \tag{4.116}
$$

that is, $K\left[\mathbf{Y} \mid \mathbf{Y}^{IG}\right] = G[\mathbf{Y}] + C(\mathbf{Y})$. Second, "passing $\mathbf{Y}^I$", we shall see that

$$
K\left[\mathbf{Y} \mid \mathbf{Y}^{IG}\right] = K\left[\mathbf{Y} \mid \mathbf{Y}^I\right] + K\left[\mathbf{Y}^I \mid \mathbf{Y}^{IG}\right], \tag{4.117}
$$

that is, $K\left[\mathbf{Y} \mid \mathbf{Y}^{IG}\right] = I[\mathbf{Y}] + \sum_i G[\mathbf{y}_i]$. The desired result (4.39) is the simple combination of relations (4.116) and (4.117) which we now establish. We have

$$
K\left[\mathbf{Y} \mid \mathbf{Y}^{IG}\right] = \int p_\mathbf{Y}\log\frac{p_\mathbf{Y}}{p_{\mathbf{Y}^{IG}}} = \int p_\mathbf{Y}\log\frac{p_\mathbf{Y}}{p_{\mathbf{Y}^G}} + \int p_\mathbf{Y}\log\frac{p_{\mathbf{Y}^G}}{p_{\mathbf{Y}^{IG}}}.
$$

The first term is the definition of $G[\mathbf{Y}]$ and the second is equal to $C(\mathbf{Y})$ since

$$
\int p_\mathbf{Y}\log\frac{p_{\mathbf{Y}^G}}{p_{\mathbf{Y}^{IG}}} = \int p_{\mathbf{Y}^G}\log\frac{p_{\mathbf{Y}^G}}{p_{\mathbf{Y}^{IG}}} = C(\mathbf{Y}),
$$

where the second equality is the definition of $C(\mathbf{Y})$ and the first equality stems from the fact $\log \frac{p_{\mathbf{Y}^G}(\mathbf{Y})}{p_{\mathbf{Y}^{IG}}(\mathbf{Y})}$ depends *linearly* on $\widehat{\mathbf{R}}_{\mathbf{Y}} = T^{-1}\sum_{t=1}^{T} \mathbf{y}(t)\mathbf{y}(t)^{\mathrm{T}}$. Now, by definition of $\mathbf{Y}^G$, the mean empirical covariance $\widehat{\mathbf{R}}_{\mathbf{Y}}$ has the same expectation under $P_{\mathbf{Y}}$ and under $P_{\mathbf{Y}^G}$, which concludes the proof of (4.116). Similarly,

$$K\left[\mathbf{Y} \mid \mathbf{Y}^{IG}\right] = \int p_{\mathbf{Y}} \log \frac{p_{\mathbf{Y}}}{p_{\mathbf{Y}^{IG}}} = \int p_{\mathbf{Y}} \log \frac{p_{\mathbf{Y}}}{p_{\mathbf{Y}^I}} + \int p_{\mathbf{Y}} \log \frac{p_{\mathbf{Y}^I}}{p_{\mathbf{Y}^{IG}}}.$$

The first term is $I[\mathbf{Y}]$ by definition and the second term is equal to $\sum_i G[\mathbf{y}_i]$ since

$$\int p_{\mathbf{Y}} \log \frac{p_{\mathbf{Y}^I}}{p_{\mathbf{Y}^{IG}}} = \int p_{\mathbf{Y}} \sum_i \log \frac{p_{\mathbf{y}_i}}{p_{\mathbf{y}_i^G}} = \sum_i \int p_{\mathbf{y}_i} \log \frac{p_{\mathbf{y}_i}}{p_{\mathbf{y}_i^G}} = \sum_i G[\mathbf{y}_i]$$

the first equality is because $p_{\mathbf{Y}^I}$ and $p_{\mathbf{Y}^{IG}}$ are two distributions of independent rows and the second equality comes from marginalizing $\int P_{\mathbf{Y}} f(\mathbf{y}_i) = \int p_{\mathbf{y}_i} f(\mathbf{y}_i)$; the last equality derives from definition of $G[\mathbf{y}_i]$. Hence, we have established Eq. (4.117).

Left invariance of $G[\mathbf{Y}]$. For an invertible matrix $\mathbf{T}$, the following sequence is valid:

$$G[\mathbf{TY}] = K\left[\mathbf{TY} \mid (\mathbf{TY})^G\right] = K\left[\mathbf{TY} \mid \mathbf{T}\mathbf{Y}^G\right] = K\left[\mathbf{Y} \mid \mathbf{Y}^G\right] = G[\mathbf{Y}].$$

The first and last equalities come from definition of G and the penultimate equality results from the invariance of the Kullback divergence under any invertible transform. It remains to validate the second equality by proving that $(\mathbf{TY})^G = \mathbf{T}\mathbf{Y}^G$. This last property stems from the fact the distribution of $\mathbf{Y}^G$ is determined only by $\mathbf{R}_{\mathbf{Y}}$ which transforms as $\mathbf{Y}$ in the sense that $\mathbf{R}_{\mathbf{TY}} = \mathbf{T}\mathbf{R}_{\mathbf{Y}}\mathbf{T}^{\mathrm{T}}$. This concludes the proof.

References

[1] S.-I. Amari, Natural gradient works efficiently in learning, Neural Comput. 10 (1998) 251–276.

[2] A. Belouchrani, K. Abed Meraim, J.-F. Cardoso, Éric Moulines, A blind source separation technique based on second order statistics, IEEE Trans. Signal Process. 45 (1997) 434–444.

[3] J.-F. Cardoso, On the performance of orthogonal source separation algorithms, in: Proc. EUSIPCO, Edinburgh, September 1994, pp. 776–779.

[4] J.-F. Cardoso, Infomax and maximum likelihood for source separation, IEEE Lett. Signal Process. 4 (1997) 112–114.

[5] J.-F. Cardoso, Learning in manifolds: The case of source separation, in: Proc. SSAP '98, 1998.

[6] J.-F. Cardoso, High-order contrasts for independent component analysis, Neural Comput. 11 (1999) 157–192.

[7] J.-F. Cardoso, B. Laheld, Equivariant adaptive source separation, IEEE Trans. Signal Process. 44 (1996) 3017–3030.

[8] P. Comon, Independent component analysis, a new concept? Signal Process. 36 (1994) 287–314. Elsevier, Special issue on Higher-Order Statistics.

[9] J. Delabrouille, J.-F. Cardoso, G. Patanchon, Multi–detector multi–component spectral matching and applications for CMB data analysis, Monthly Notices of the Royal Astronomical Society 346 (2003) 1089–1102. Also available as http://arXiv.org/abs/astro-ph/0211504.
[10] A. Hyvärinen, Fast and robust fixed-point algorithms for independent component analysis, IEEE Trans. Neural Netw. 10 (1999) 626–634.
[11] C. Jutten, J. Hérault, Une solution neuromimétique au problème de séparation de sources, Traitement du Signal 5 (1988) 389–403.
[12] L. Molgedey, H.G. Schuster, Separation of a mixture of independent signals using time delayed correlations, Phys. Rev. Lett. 72 (1994) 3634–3637.
[13] L. Parra, C. Spence, Convolutive blind source separation of non-stationary sources, IEEE Trans. Speech Audio Process. (2000) 320–327.
[14] D. Pham, Blind separation of instantaneous mixture of sources via the Gaussian mutual information criterion, Signal Process. 81 (2001) 855–870.
[15] D.-T. Pham, Entropy of a variable slightly contaminated with another, IEEE Signal Process. Lett. 12 (2005) 536–539.
[16] D.-T. Pham, J.-F. Cardoso, Blind separation of instantaneous mixtures of non stationary sources, IEEE Trans. Signal Process. 49 (2001) 1837–1848.
[17] D.-T. Pham, P. Garat, Blind separation of mixture of independent sources through a quasi-maximum likelihood approach, IEEE Trans. Signal Process. 45 (1997) 1712–1725.
[18] D. Rowe, A bayesian approach to blind source separation, J. Interdiscip. Math. 5 (2002) 49–76.

CHAPTER

Algebraic methods after prewhitening

5

L. De Lathauwer

5.1 INTRODUCTION

In this chapter we discuss four prewhitening-based algebraic algorithms for Independent Component Analysis (ICA). The prewhitening takes into account the structure of the covariance matrix of the observed data. This is typically done by means of an Eigenvalue Decomposition (EVD) or a Singular Value Decomposition (SVD). The prewhitening only allows one to find the mixing matrix up to an orthogonal or unitary factor. In this chapter, the unknown factor is estimated from the higher-order cumulant tensor of the observed data. This involves orthogonal tensor decompositions that can be interpreted as higher-order generalizations of the EVD.

In the remainder of this Introduction, we introduce some basic material that is needed in the further discussion. Section 5.1.1 deals with some basic concepts of multilinear algebra. Section 5.1.2 deals with Higher-Order Statistics (HOS). Section 5.1.3 recalls the principle of a Jacobi iteration. The rest of the chapter is structured as follows. Section 5.2 introduces the decompositions of covariance matrix and cumulant tensor on which the methods are based, and explains the common structure of the algorithms. Four specific algorithms are subsequently discussed in sections 5.3–5.6. The presentation is in terms of real-valued data. Wherever the generalization to the complex case is not straightforward, this will be explicitly stated.

Matlab code of the algorithms is available on the internet – see the corresponding entry in the References. Several variants of the algorithms can be thought of. In particular, it is possible to combine the decompositions of cumulant tensors of different order, as explained in [37].

5.1.1 Multilinear algebra

The outer product generalizes expressions of the type $\mathbf{a}\mathbf{b}^{\mathrm{T}}$, in which $\mathbf{a}$ and $\mathbf{b}$ are vectors.

DEFINITION 5.1
The outer product $\mathcal{A} \circ \mathcal{B} \in \mathbb{R}^{I_1 \times I_2 \times \cdots \times I_P \times J_1 \times J_2 \times \cdots \times J_Q}$ *of a tensor* $\mathcal{A} \in \mathbb{R}^{I_1 \times I_2 \times \cdots \times I_P}$ *and a tensor* $\mathcal{B} \in \mathbb{R}^{J_1 \times J_2 \times \cdots \times J_Q}$, *is defined by*

$$(\mathcal{A} \circ \mathcal{B})_{i_1 i_2 \ldots i_P j_1 j_2 \ldots j_Q} = a_{i_1 i_2 \ldots i_P} b_{j_1 j_2 \ldots j_Q},$$

for all values of the indices.

The entries of a Kth-order tensor $\mathcal{A}$, equal to the outer product of K vectors $\mathbf{u}^{(1)}$, $\mathbf{u}^{(2)},\ldots,\mathbf{u}^{(K)}$, are thus given by $a_{i_1 i_2 \ldots i_K} = u^{(1)}_{i_1} u^{(2)}_{i_2} \ldots u^{(K)}_{i_K}$.

The multiplication of a higher-order tensor with a matrix can be defined as follows.

DEFINITION 5.2

The mode-k product *of a tensor* $\mathcal{A} \in \mathbb{R}^{I_1 \times I_2 \times \cdots \times I_K}$ *by a matrix* $\mathbf{U} \in \mathbb{R}^{J_k \times I_k}$, *denoted by* $\mathcal{A} \bullet_k \mathbf{U}$, *is an* $(I_1 \times I_2 \times \cdots \times I_{k-1} \times J_k \times I_{k+1} \cdots \times I_K)$*-tensor defined by*

$$(\mathcal{A} \bullet_k \mathbf{U})_{i_1 i_2 \ldots j_k \ldots i_K} = \sum_{i_k} a_{i_1 i_2 \ldots i_k \ldots i_K} u_{j_k i_k},$$

for all index values.

The mode-k product allows us to express the effect of a basis transformation in $\mathbb{R}^{I_k}$ on the tensor $\mathcal{A}$. In the literature sometimes the symbol $\times_k$ is used instead of $\bullet_k$.

By way of illustration, let us have a look at the matrix product $\mathbf{A} = \mathbf{U}^{(1)} \cdot \mathbf{B} \cdot \mathbf{U}^{(2)^{\mathrm{T}}}$, involving matrices $\mathbf{B} \in \mathbb{R}^{I_1 \times I_2}$, $\mathbf{U}^{(1)} \in \mathbb{R}^{J_1 \times I_1}$, $\mathbf{U}^{(2)} \in \mathbb{R}^{J_2 \times I_2}$ and $\mathbf{A} \in \mathbb{R}^{J_1 \times J_2}$. Working with "generalized transposes" in the multilinear case (in which the fact that column vectors are transpose-free, would not have an inherent meaning), can be avoided by observing that the relationships of $\mathbf{U}^{(1)}$ and $\mathbf{U}^{(2)}$ (not $\mathbf{U}^{(2)^{\mathrm{T}}}$) with $\mathbf{B}$ are in fact the same. In the same way as $\mathbf{U}^{(1)}$ makes linear combinations of the rows of $\mathbf{B}$, $\mathbf{U}^{(2)}$ makes linear combinations of the columns. In the same way as the columns of $\mathbf{B}$ are multiplied by $\mathbf{U}^{(1)}$, its rows are multiplied by $\mathbf{U}^{(2)}$. In the same way as the columns of $\mathbf{U}^{(1)}$ are associated with the column space of $\mathbf{A}$, the columns of $\mathbf{U}^{(2)}$ are associated with the row space. This typical relationship is denoted by means of the $\bullet_k$-symbol: $\mathbf{A} = \mathbf{B} \bullet_1 \mathbf{U}^{(1)} \bullet_2 \mathbf{U}^{(2)}$. The symmetry is clear at the level of the individual entries: $a_{j_1 j_2} = \sum_{i_1=1}^{I_1} \sum_{i_2=1}^{I_2} u^{(1)}_{j_1 i_1} u^{(2)}_{j_2 i_2} b_{i_1 i_2}$, for all j_1, j_2. Note that $\mathbf{U}^{(1)} \cdot \mathbf{B} \cdot \mathbf{U}^{(2)^{\mathrm{H}}} = \mathbf{B} \bullet_1 \mathbf{U}^{(1)} \bullet_2 \mathbf{U}^{(2)^*}$.

5.1.2 Higher-order statistics

The basic HOS are higher-order moments and higher-order cumulants. Moment tensors of a real stochastic vector are defined as follows.

DEFINITION 5.3

The K*th-order* moment tensor $\mathcal{M}^{(K)}_{\mathbf{x}} \in \mathbb{R}^{I \times I \times \cdots \times I}$ *of a real stochastic vector* $\mathbf{x} \in \mathbb{R}^I$ *is defined by the element-wise equation*

$$(\mathcal{M}^{(K)}_{\mathbf{x}})_{i_1 i_2 \ldots i_K} = \mathrm{mom}\{x_{i_1}, x_{i_2}, \ldots, x_{i_K}\} = \mathbb{E}\{x_{i_1} x_{i_2} \ldots x_{i_K}\}. \tag{5.1}$$

Moments are obtained from the coefficients of the first characteristic function of the random vector [36]. The first-order moment is equal to the mean. The second-order moment is the correlation matrix (following the definition adopted in for example [29], in which the mean is not subtracted).

On the other hand, cumulants of a real stochastic vector are defined as follows.

DEFINITION 5.4
The Kth-order cumulant *tensor* $\mathcal{C}_{\mathbf{x}}^{(K)} \in \mathbb{R}^{I \times I \times \cdots \times I}$ *of a real stochastic vector* $\mathbf{x} \in \mathbb{R}^I$ *is defined by the element-wise equation*

$$(\mathcal{C}_{\mathbf{x}}^{(K)})_{i_1 i_2 \ldots i_K} = \mathrm{cum}\{x_{i_1}, x_{i_2}, \ldots, x_{i_K}\}$$
$$= \sum (-1)^{L-1}(L-1)! \mathbb{E}\left\{\prod_{i \in \mathtt{A}_1} x_i\right\} \mathbb{E}\left\{\prod_{i \in \mathtt{A}_2} x_i\right\} \ldots \mathbb{E}\left\{\prod_{i \in \mathtt{A}_L} x_i\right\}, \quad (5.2)$$

where the summation involves all possible partitions $\{\mathtt{A}_1, \mathtt{A}_2, \ldots, \mathtt{A}_L\}$ $(1 \leqslant L \leqslant K)$ *of the integers* $\{i_1, i_2, \ldots, i_K\}$*. For a real zero-mean stochastic vector* $\mathbf{x}$ *the cumulants up to order 4 are explicitly given by:*

$$(\mathbf{c}_{\mathbf{x}})_i = \mathrm{cum}\{x_i\} = \mathbb{E}\{x_i\}, \quad (5.3)$$
$$(\mathbf{C}_{\mathbf{x}})_{i_1 i_2} = \mathrm{cum}\{x_{i_1}, x_{i_2}\} = \mathbb{E}\{x_{i_1} x_{i_2}\}, \quad (5.4)$$
$$(\mathcal{C}_{\mathbf{x}}^{(3)})_{i_1 i_2 i_3} = \mathrm{cum}\{x_{i_1}, x_{i_2}, x_{i_3}\} = \mathbb{E}\{x_{i_1} x_{i_2} x_{i_3}\}, \quad (5.5)$$
$$(\mathcal{C}_{\mathbf{x}}^{(4)})_{i_1 i_2 i_3 i_4} = \mathrm{cum}\{x_{i_1}, x_{i_2}, x_{i_3}, x_{i_4}\} = \mathbb{E}\{x_{i_1} x_{i_2} x_{i_3} x_{i_4}\}$$
$$- \mathbb{E}\{x_{i_1} x_{i_2}\}\mathbb{E}\{x_{i_3} x_{i_4}\} - \mathbb{E}\{x_{i_1} x_{i_3}\}\mathbb{E}\{x_{i_2} x_{i_4}\} - \mathbb{E}\{x_{i_1} x_{i_4}\}\mathbb{E}\{x_{i_2} x_{i_3}\}. \quad (5.6)$$

For every component x_i *of* $\mathbf{x}$ *that has a non-zero mean,* x_i *has to be replaced in these formulas, except Eq.* (5.3) *and Eq.* (5.2) *when it applies to a first-order cumulant, by* $x_i - \mathbb{E}\{x_i\}$.

In the same way as moments are obtained from the coefficients of the first characteristic function of the random vector, cumulants are obtained from its second characteristic function [36]. It turns out that, again, the first-order cumulant is the mean of the stochastic vector. The second-order cumulant is the covariance matrix. The interpretation of a cumulant of order higher than 2 is not straightforward, but the powerful properties listed below will demonstrate its importance. For the moment it suffices to state that cumulants of a set of random variables give an indication of their mutual statistical dependence (completely independent variables resulting in a zero cumulant), and that higher-order cumulants of a single random variable are some measure of its non-Gaussianity (cumulants of a Gaussian variable, for $K > 2$, being equal to zero).

The definitions above are given for real-valued random vectors. In the case of complex random vectors, one or more of the arguments of $\mathrm{mom}\{\cdot\}$ and $\mathrm{cum}\{\cdot\}$ may be complex conjugated. In this chapter we will in particular use the so-called "circular" covariance matrix $\mathbf{C}_{\mathbf{x}} = \mathrm{cum}\{\mathbf{x}, \mathbf{x}^*\}$ and cumulant $\mathcal{C}_{\mathbf{x}}^{(4)} = \mathrm{cum}\{\mathbf{x}, \mathbf{x}^*, \mathbf{x}^*, \mathbf{x}\}$.

At first sight higher-order moments, because of their straightforward definition, might seem more interesting quantities than higher-order cumulants. However, cumulants have a number of important properties, which they do not share with higher-order moments, such that in practice cumulants are more frequently used. We enumerate some of the most interesting properties [38,39]:

1. *Symmetry:* real moments and cumulants are fully symmetric in their arguments, i.e.,

$$(\mathcal{M}_{\mathbf{x}}^{(K)})_{i_1 i_2 \ldots i_K} = (\mathcal{M}_{\mathbf{x}}^{(K)})_{\mathrm{P}(i_1 i_2 \ldots i_K)},$$
$$(\mathcal{C}_{\mathbf{x}}^{(K)})_{i_1 i_2 \ldots i_K} = (\mathcal{C}_{\mathbf{x}}^{(K)})_{\mathrm{P}(i_1 i_2 \ldots i_K)},$$

   in which P is an arbitrary permutation of the indices.
2. *Multilinearity:* if a real stochastic vector $\mathbf{x}$ is transformed into a stochastic vector $\tilde{\mathbf{x}}$ by a matrix multiplication $\tilde{\mathbf{x}} = \mathbf{A} \cdot \mathbf{x}$, then we have:

$$\mathcal{M}_{\tilde{\mathbf{x}}}^{(K)} = \mathcal{M}_{\mathbf{x}}^{(K)} \bullet_1 \mathbf{A} \bullet_2 \mathbf{A} \ldots \bullet_K \mathbf{A}, \tag{5.7}$$
$$\mathcal{C}_{\tilde{\mathbf{x}}}^{(K)} = \mathcal{C}_{\mathbf{x}}^{(K)} \bullet_1 \mathbf{A} \bullet_2 \mathbf{A} \ldots \bullet_K \mathbf{A}. \tag{5.8}$$

   (In the complex case, some of the matrices $\mathbf{A}$ may be complex conjugated, depending on the complex conjugation pattern adopted in the definition of the moment or cumulant.) The characteristic way in which moments and cumulants change under basis transformations, is actually why we can call them *tensors* [36]. Equations (5.7)–(5.8) are in fact more general: the matrix $\mathbf{A}$ does not have to represent a basis transformation; it does not even have to be square.
3. *Even distribution:* if a real random variable x has an even probability density function $p_x(x)$, i.e., $p_x(x)$ is symmetric about the origin, then the odd moments and cumulants of x vanish.
4. *Partitioning of independent variables:* if a subset of K stochastic variables x_1, x_2, $\ldots, x_K$ is independent of the other variables, then we have:

$$\mathrm{cum}\{x_1, x_2, \ldots, x_K\} = 0.$$

   This property does not hold in general for moments (e.g. for two mutually independent random variables x and y we find that $\mathrm{mom}\{x, x, y, y\} = \mathbb{E}\{x^2\} \cdot \mathbb{E}\{y^2\}$ does not vanish, unless one of the variables is identically equal to zero). A consequence of the property is that a higher-order cumulant of a stochastic vector that has mutually independent components, is a diagonal tensor, i.e. only the entries of which all the indices are equal can be different from zero. This very strong algebraic condition is the basis of all the ICA techniques that will be discussed in this chapter.
5. *Sum of independent variables:* if the stochastic variables $x_1, x_2, \ldots, x_K$ are mutually independent of the stochastic variables $y_1, y_2, \ldots, y_K$, then we have:

$$\mathrm{cum}\{x_1 + y_1, x_2 + y_2, \ldots, x_K + y_K\}$$
$$= \mathrm{cum}\{x_1, x_2, \ldots, x_K\} + \mathrm{cum}\{y_1, y_2, \ldots, y_K\}.$$

The cumulant tensor of a sum of independent random vectors is the sum of the individual cumulants. This property does not hold for moments either. As a matter of fact, it explains the term "cumulant". (One could expand mom$\{x_1 + y_1, x_2 + y_2, \ldots, x_K + y_K\}$ as a sum over all possible x-y combinations, but the cross-terms, containing x- as well as y-entries, do not necessarily vanish, as for cumulants – see the previous property.)

6. *Non-Gaussianity:* if y is a Gaussian variable with the same mean and variance as a given stochastic variable x, then we have for $K \geqslant 3$:

$$\mathcal{C}_x^{(K)} = \mathcal{M}_x^{(K)} - \mathcal{M}_y^{(K)}.$$

As a consequence, higher-order cumulants of a Gaussian variable are zero. In combination with the multilinearity property, we observe that higher-order cumulants have the interesting property of being blind for additive Gaussian noise. Namely, if a stochastic variable x is corrupted by additive Gaussian noise b, i.e.,

$$\hat{x} = x + b,$$

then we nevertheless find that

$$\mathcal{C}_{\hat{x}}^{(K)} = \mathcal{C}_x^{(K)} + \mathcal{C}_b^{(K)} = \mathcal{C}_x^{(K)}.$$

Generally speaking, it becomes harder to estimate HOS from sample data as the order increases, i.e. longer datasets are required to obtain the same accuracy [2,33]. Hence in practice the use of HOS is usually restricted to third- and fourth-order cumulants. For symmetric distributions fourth-order cumulants are used, since the third-order cumulants vanish, as mentioned in the third property.

5.1.3 Jacobi iteration

The ICA algorithms discussed in this chapter are of the Jacobi type. Therefore we repeat in this section the principle of a *Jacobi iteration*. We take the diagonalization of a Hermitean matrix as an example. For a more extensive discussion, we refer to [30].

Given a Hermitean matrix $\mathbf{A} \in \mathbb{C}^{I \times I}$, we want to find a unitary matrix $\mathbf{U} \in \mathbb{C}^{I \times I}$ such that

$$\mathbf{B} = \mathbf{U} \cdot \mathbf{A} \cdot \mathbf{U}^{\mathrm{H}} \tag{5.9}$$

is diagonal. In other words, we look for a unitary matrix $\mathbf{U}$ that minimizes the cost function

$$f(\mathbf{U}) = \mathrm{off}(\mathbf{B})^2 = \sum_{i \neq j} \left| b_{ij} \right|^2. \tag{5.10}$$

Because $\mathbf{U}$ is unitary, this is equivalent to maximizing the objective function

$$g(\mathbf{U}) = ||\text{diag}(\mathbf{B})||^2 = \sum_i |b_{ii}|^2. \tag{5.11}$$

Any unitary matrix can, up to multiplication by a diagonal matrix $\mathbf{D}$ of which the diagonal entries are unit-modulus, be written as a product of elementary Jacobi rotation matrices $\mathbf{J}(p,q,c,s)$, defined for $p < q$ by

$$\mathbf{J}(p,q,c,s) = \begin{array}{cc} \begin{array}{ccccccc} & & p & & q & & \end{array} & \\ \begin{pmatrix} 1 & \cdots & 0 & \cdots & 0 & \cdots & 0 \\ \vdots & \ddots & \vdots & & \vdots & & \vdots \\ 0 & \cdots & c & \cdots & -s & \cdots & 0 \\ \vdots & & \vdots & \ddots & \vdots & & \vdots \\ 0 & \cdots & \bar{s} & \cdots & c & \cdots & 0 \\ \vdots & & \vdots & & \vdots & \ddots & \vdots \\ 0 & \cdots & 0 & \cdots & 0 & \cdots & 1 \end{pmatrix} & \begin{array}{c} \\ \\ p \\ \\ q \\ \\ \\ \end{array} \end{array} \tag{5.12}$$

with $(c,s) \in \mathbb{R} \times \mathbb{C}$ such that $c^2 + |s|^2 = 1$. The matrix $\mathbf{D}$ does not change the cost function and can therefore be left aside. The following explicit parameterizations of $\mathbf{J}$ will be used in sections 5.3–5.6:

$$c = \cos\alpha \quad s = \sin\alpha \quad \alpha \in [0, 2\pi) \quad \text{(real case)} \tag{5.13}$$

$$c = \frac{1}{\sqrt{1+\theta^2}} \quad s = -\frac{\theta}{\sqrt{1+\theta^2}} \quad \theta \in \mathbb{R} \quad \text{(real case)} \tag{5.14}$$

$$c = \cos\alpha \quad s = \sin\alpha e^{i\phi} \quad \alpha, \phi \in [0, 2\pi) \quad \text{(complex case)} \tag{5.15}$$

$$c = \frac{1}{\sqrt{1+|\theta|^2}} \quad s = -\frac{\theta}{\sqrt{1+|\theta|^2}} \quad \theta \in \mathbb{C} \quad \text{(complex case).} \tag{5.16}$$

Due to the unitarity of $\mathbf{J}$, the only part of $\mathbf{J} \cdot \mathbf{A} \cdot \mathbf{J}^{\mathrm{H}}$ that affects the cost function is the (2×2) submatrix at the intersection of rows and columns p and q. An explicit expression for the elementary rotation that exactly diagonalizes the (2×2) submatrix is easy to derive. It can be verified that the cost function is periodic in the rotation angle with period $\pi/2$. Hence, one can always choose an inner rotation, i.e., a rotation over an angle in the interval $(-\pi/4, \pi/4]$. In this way, implicit re-ordering of columns and rows, which may hamper convergence, is avoided. A Jacobi iteration consists of successively applying Jacobi rotations to $\mathbf{A}$, for different (p,q), until the cost function has been minimized. In the process, the unitary matrix $\mathbf{U}$ is gradually built up as the product of all elementary rotations.

We now turn our attention to the choice of (p,q). From the standpoint of maximizing the reduction of the cost function $f(\mathbf{J})$, it makes sense to choose the pair (p,q) for which

Table 5.1 Cyclic Jacobi for the diagonalization of a Hermitean matrix

```
U = I
while off(A) > ε do
  for p = 1 : I − 1 do
    for q = p + 1 : I do
      Compute optimal Jacobi rotation J(p,q)
      A ← J · A · J^H
      U ← U · J^H
    end for
  end for
end while
```

$|a_{pq}|^2$ is maximal. This is what is done in the *classical Jacobi algorithm*. Since

$$\mathrm{off}(\mathbf{A})^2 \leqslant I(I-1)\left|a_{pq}\right|^2,$$

we have that

$$\mathrm{off}(\mathbf{J}\cdot\mathbf{A}\cdot\mathbf{J}^{\mathrm{H}})^2 \leqslant \left(1-\frac{2}{I(I-1)}\right)\mathrm{off}(\mathbf{A})^2, \tag{5.17}$$

such that the Jacobi algorithm for diagonalization of a Hermitean matrix converges at least linearly to a diagonal matrix. It can be proved that the asymptotic convergence rate is actually quadratic. However, finding the optimal (p,q) in each step is expensive. Therefore, one rather follows a fixed order when going through the different subproblems. This technique is called the *cyclic Jacobi algorithm*. Cyclic Jacobi also converges quadratically. The algorithm is outlined in Table 5.1.

5.2 INDEPENDENT COMPONENT ANALYSIS

5.2.1 Algebraic formulation

The basic linear ICA model is denoted by:

$$\mathbf{x} = \mathbf{A}\cdot\mathbf{s} + \mathbf{b} = \tilde{\mathbf{x}} + \mathbf{b}, \tag{5.18}$$

in which $\mathbf{x} \in \mathbb{R}^P(\mathbb{C}^P)$ is the *observation vector*, $\mathbf{s} \in \mathbb{R}^N(\mathbb{C}^N)$ is the *source vector* and $\mathbf{b} \in \mathbb{R}^P(\mathbb{C}^P)$ represents *additive noise*. $\tilde{\mathbf{x}} \in \mathbb{R}^P(\mathbb{C}^P)$ is the *signal part* of the observations. $\mathbf{A} \in \mathbb{R}^{P\times N}(\mathbb{C}^{P\times N})$ is the *mixing matrix*; its range is the *signal subspace*.

In this chapter we assume that the mixing vectors are linearly independent. This implies that the number of sensors P is not smaller than the number of sources N. It is actually possible to estimate the mixing matrix in cases where $P < N$. For algebraic approaches to this so-called *underdetermined* ICA problem we refer to [13,17,18,21–24] and references therein.

In this chapter we focus on the estimation of the mixing matrix. Since this matrix is nonsingular, with at least as many rows as columns, the sources may afterwards be estimated by premultiplying the observations with the pseudo-inverse of the estimate $\widehat{\mathbf{A}}$:

$$\widehat{\mathbf{s}} = \widehat{\mathbf{A}}^{\dagger} \cdot \mathbf{x}.$$

This equation implements in fact a *Linear Constrained Minimum Variance* (LCMV) beamformer, which minimizes the mutual interference of the sources. One may also follow other beamforming strategies [42].

Using the properties discussed in section 5.1.2, the model (5.18) leads in the real case to the following two key equations:

$$\mathbf{C_x} = \mathbf{A} \cdot \mathbf{C_s} \cdot \mathbf{A}^{\mathrm{T}} + \mathbf{C_b} = \mathbf{C_s} \bullet_1 \mathbf{A} \bullet_2 \mathbf{A} + \mathbf{C_b}, \tag{5.19}$$

$$\mathcal{C}_{\mathbf{x}}^{(K)} = \mathcal{C}_{\mathbf{s}}^{(K)} \bullet_1 \mathbf{A} \bullet_2 \mathbf{A} \ldots \bullet_K \mathbf{A} + \mathcal{C}_{\mathbf{b}}^{(K)}. \tag{5.20}$$

(In the complex case, some of the matrices $\mathbf{A}$ may be complex conjugated, depending on the complex conjugation pattern adopted for the covariance matrix and cumulant tensor.) The crucial point in (5.19) and (5.20) is that $\mathbf{C_s}$ and $\mathcal{C}_{\mathbf{s}}^{(K)}$ are diagonal, because of the statistical independence of the sources. Furthermore, $\mathcal{C}_{\mathbf{b}}^{(K)}$ vanishes if the noise is Gaussian.

Equations (5.19) and (5.20) can also be written as

$$\mathbf{C_x} = \sum_{n=1}^{N} \sigma_{s_n}^2 \mathbf{a}_n \mathbf{a}_n^{\mathrm{T}} + \mathbf{C_b} = \sum_{n=1}^{N} \sigma_{s_n}^2 \mathbf{a}_n \circ \mathbf{a}_n + \mathbf{C_b}, \tag{5.21}$$

$$\mathcal{C}_{\mathbf{x}}^{(K)} = \sum_{n=1}^{N} c_{s_n}^{(K)} \mathbf{a}_n \circ \mathbf{a}_n \circ \ldots \circ \mathbf{a}_n + \mathcal{C}_{\mathbf{b}}^{(K)}, \tag{5.22}$$

in which $\sigma_{s_n}^2$ and $c_{s_n}^{(K)}$ denote the variance and the Kth-order cumulant of the nth source, respectively, and in which $\mathbf{a}_n$ denotes the nth column of $\mathbf{A}$. Let us for the moment make abstraction of the noise term. Equation (5.21) is an expansion of the covariance of the observations in a sum of rank-1 terms, where each term consists of the contribution of one particular source. Equation (5.22) is an analogous decomposition of the higher-order cumulant of the observations. The latter is known as a *Canonical Decomposition (CanDecomp)* or *Parallel Factor Decomposition (ParaFac)* of the tensor $\mathcal{C}_{\mathbf{x}}^{(K)}$ [7,14,17,20,27, 31,35,40].

It is well-known that Eq. (5.19)/(5.21) does not contain enough information to allow for the full estimation of the mixing matrix. This will be explained in detail in section 5.2.2. However, the situation is very different at the tensor level. Under our conditions on the mixing matrix, and assuming that all source cumulants $c_{s_n}^{(K)}$ are non-zero, decomposition (5.20)/(5.22) is unique, up to trivial indeterminacies. This allows us to solve the ICA problem. The strategy behind the methods discussed in this chapter is as follows. First we extract as much information as possible from Eq. (5.19)/(5.21). How this can be done, is explained in section 5.2.2. The remaining free parameters are subsequently estimated using (5.20)/(5.22). This problem will in general terms be addressed in section 5.2.3. Specific algorithms are discussed in sections 5.3–5.6.

Other strategies are possible. As explained in the previous paragraph, the mixing matrix could in principle be estimated from Eq. (5.20)/(5.22) alone, without using second-order statistics. One could also combine the second-order and higher-order constraints by giving them each a finite weight. For such alternative approaches we refer to [13,19,22,44] and references therein.

In general, it makes sense to extract much information from Eq. (5.19)/(5.21). First, second-order statistics can be estimated more reliably than higher-order statistics, as already mentioned at the end of section 5.1.2 [2,33]. Second, after prewhitening the remaining problem consists of the determination of an orthogonal (unitary) matrix. That is, after prewhitening, the matrix that has to be determined, is optimally conditioned. Prewhitening may be less appropriate when the data are subject to Gaussian noise with unknown color. In that case, the effect of the noise on the second-order statistics cannot be mitigated in the way explained in section 5.2.2. An error that is introduced in a prewhitening step cannot be fully compensated for in the step that follows [4,28]. On the other hand, higher-order cumulants are theoretically insensitive to Gaussian noise, as explained in section 5.1.2.

5.2.2 Step 1: Prewhitening

Briefly, the goal of *prewhitening* is to transform the observation vector $\mathbf{x}$ into another stochastic vector, $\mathbf{z}$, which has unit covariance. This involves the multiplication of $\mathbf{x}$ with the inverse of a square root of its covariance matrix $\mathbf{C}_\mathbf{x}$. When $N < P$, a projection of $\mathbf{x}$ onto the signal subspace is carried out.

Let us discuss the problem in more detail. First, we observe that the covariance matrix $\mathbf{C}_\mathbf{x}$ takes the following form (for the moment, the noise term in Eq. (5.18) is neglected, for clarity):

$$\mathbf{C}_\mathbf{x} = \mathbf{A} \cdot \mathbf{C}_\mathbf{s} \cdot \mathbf{A}^T, \tag{5.23}$$

in which the covariance $\mathbf{C}_\mathbf{s}$ of $\mathbf{s}$ is diagonal, since the source signals are uncorrelated. (Throughout this section, the transpose is replaced by the Hermitean transpose in the complex case.) Assuming that the source signals have unit variance (without loss of generality, as we may appropriately rescale the mixing vectors as well), we have:

$$\mathbf{C}_\mathbf{x} = \mathbf{A} \cdot \mathbf{A}^T.$$

A first observation is that the number of sources can be deduced from the rank of $\mathbf{C}_\mathbf{x}$. Substitution of the SVD of the mixing matrix $\mathbf{A} = \mathbf{U} \cdot \boldsymbol{\Delta} \cdot \mathbf{Q}^T$ shows that the EVD of the observed covariance allows us to estimate the components $\mathbf{U}$ and $\boldsymbol{\Delta}$ while the factor $\mathbf{Q}$ remains unknown:

$$\mathbf{C}_\mathbf{x} = \mathbf{U} \cdot \boldsymbol{\Delta}^2 \cdot \mathbf{U}^T = (\mathbf{U}\boldsymbol{\Delta}) \cdot (\mathbf{U}\boldsymbol{\Delta})^T. \tag{5.24}$$

Hence the signal subspace can be estimated from the second-order statistics of the observations, but the actual mixing matrix remains unknown up to an orthogonal factor. Instead of working via the EVD of $\mathbf{C}_\mathbf{x}$, one may also obtain $\mathbf{U}$ and $\boldsymbol{\Delta}$ from the SVD of the data matrix. EVD/SVD-based prewhitening amounts to a *Principal Component Analysis* (PCA) [32].

The effect of the additive noise term $\mathbf{b}$ can be neutralized by replacing $\mathbf{C}_\mathbf{x}$ by the noise-free covariance $\mathbf{C}_\mathbf{x} - \mathbf{C}_\mathbf{b}$. In the case of spatially white noise (i.e. noise of which the components are mutually uncorrelated and have the same variance), $\mathbf{C}_\mathbf{b}$ takes the form of $\sigma_b^2 \mathbf{I}$, in which σ_b^2 is the variance of the noise on each data channel. In a more-sensors-than-sources setup, σ_b^2 can be estimated as the mean of the "noise eigenvalues", i.e. the smallest $P - N$ eigenvalues, of $\mathbf{C}_\mathbf{x}$. The number of sources is estimated as the number of significant eigenvalues of $\mathbf{C}_\mathbf{x}$; for a detailed procedure, we refer to [43].

After computation of $\mathbf{U}$ and $\boldsymbol{\Delta}$, a standardized random vector $\mathbf{z}$ can be defined as

$$\mathbf{z} = \boldsymbol{\Delta}^\dagger \cdot \mathbf{U}^T \cdot \mathbf{x}. \tag{5.25}$$

This vector has unit covariance. It is related to the source vector in the following way:

$$\mathbf{z} = \mathbf{Q}^T \cdot \mathbf{s} + \boldsymbol{\Delta}^\dagger \cdot \mathbf{U}^T \cdot \mathbf{b}. \tag{5.26}$$

5.2.3 Step 2: Fixing the rotational degrees of freedom using the higher-order cumulant

In this section we explain in general terms how the remaining degrees of freedom can be fixed, i.e. how the matrix $\mathbf{Q}$ that contains the right singular vectors of the mixing matrix can be estimated. As we have already exploited the information provided by the second-order statistics of the observations, we now resort to their HOS.

Assuming that the noise is Gaussian, we obtain from (5.26) that in the real case the higher-order cumulant of the standardized random vector $\mathbf{z}$ is given by

$$\mathcal{C}_\mathbf{z}^{(K)} = \mathcal{C}_\mathbf{s}^{(K)} \bullet_1 \mathbf{Q}^T \bullet_2 \mathbf{Q}^T \ldots \bullet_K \mathbf{Q}^T. \tag{5.27}$$

(In the complex case, some of the matrices $\mathbf{Q}$ may be complex conjugated, depending on the complex conjugation pattern adopted in the definition of the cumulant.) Let us write

$\mathbf{W} = \mathbf{Q}^{\mathrm{T}}$ for convenience. Then (5.27) can also be written as

$$\mathscr{C}_{\mathbf{z}}^{(K)} = \sum_{n=1}^{N} c_{s_n}^{(K)} \mathbf{w}_n \circ \mathbf{w}_n \circ \ldots \circ \mathbf{w}_n. \tag{5.28}$$

The standardized cumulant $\mathscr{C}_{\mathbf{z}}^{(K)}$ is related to the Kth-order cumulant of the observations by the multilinearity property:

$$\mathscr{C}_{\mathbf{z}}^{(K)} = \mathscr{C}_{\mathbf{x}}^{(K)} \bullet_1 (\mathbf{U} \cdot \boldsymbol{\Delta})^{\dagger} \bullet_2 (\mathbf{U} \cdot \boldsymbol{\Delta})^{\dagger} \ldots \bullet_K (\mathbf{U} \cdot \boldsymbol{\Delta})^{\dagger}. \tag{5.29}$$

We repeat that the source cumulant $\mathscr{C}_{\mathbf{s}}^{(K)}$ is theoretically a diagonal tensor, since the source signals are not only uncorrelated but also higher-order independent. Hence Eq. (5.27) is in fact a symmetric EVD-like tensor decomposition. It can be shown that this decomposition is up to trivial indeterminacies unique if at most one diagonal entry of $\mathscr{C}_{\mathbf{s}}^{(K)}$ is equal to zero [19, pp. 127–128]. Note again that this is very different from the situation for matrices. In the matrix case we have for instance $\mathbf{I} = \mathbf{Q}^{\mathrm{T}} \cdot \mathbf{I} \cdot \mathbf{Q} = \mathbf{V}^{\mathrm{T}} \cdot \mathbf{I} \cdot \mathbf{V}$ for arbitrary orthogonal $\mathbf{V}$. This is actually what makes us unable to find $\mathbf{Q}$ from the second-order statistics of the observations.

On the other hand, simply counting the degrees of freedom in Eq. (5.29) shows that in general a higher-order tensor cannot be diagonalized by means of orthogonal (unitary) transformations. In the real case for instance, the symmetric tensor $\mathscr{C}_{\mathbf{z}}^{(K)}$ contains $N(N+1)\ldots(N+K-1)/K!$ independent entries, while the decomposition allows only $N(N-1)/2$ (orthogonal factor $\mathbf{Q}$) + N (diagonal of $\mathscr{C}_{\mathbf{s}}^{(K)}$) degrees of freedom. This means that if $\mathscr{C}_{\mathbf{z}}^{(K)}$ is not perfectly known (due to finite datalength, non-Gaussian additive noise, etc.), its estimate can in general not be fully diagonalized. This leaves room for different algorithms, which deal with the estimation error in a different manner. Four approaches will be discussed in the following sections. The general structure of prewhitening-based algorithms that use the higher-order cumulant is summarized in Table 5.2.

5.3 DIAGONALIZATION IN LEAST SQUARES SENSE

The first reliable algebraic algorithm for ICA was derived by Comon [10]. This algorithm will be denoted as *COM2*. It works as follows. Since the higher-order cumulant cannot in general be diagonalized by means of orthogonal (unitary) transformations, we will make it as diagonal as possible in the least squares sense. Denote

$$\mathscr{C}' = \mathscr{C}_{\mathbf{z}}^{(K)} \bullet_1 \mathbf{U} \bullet_2 \mathbf{U} \bullet_3 \ldots \bullet_K \mathbf{U}. \tag{5.30}$$

Table 5.2 Prewhitening-based ICA using the higher-order cumulant

1. Prewhitening stage (PCA):
   - Estimate covariance $\mathbf{C_x}$.
   - EVD: $\mathbf{C_x} = \mathbf{U} \cdot \mathbf{\Delta}^2 \cdot \mathbf{U}^T$
   - Standardize data: $\mathbf{z} = \mathbf{\Delta}^\dagger \cdot \mathbf{U}^T \cdot \mathbf{x}$.

   (Section 5.2.2.)
2. Higher-order stage:
   - Estimate cumulant $\mathcal{C}_\mathbf{z}^{(K)}$.
   - (Approximate) diagonalization:
     $$\mathcal{C}_\mathbf{z}^{(K)} = \mathcal{C}_\mathbf{s}^{(K)} \bullet_1 \mathbf{Q}^T \bullet_2 \mathbf{Q}^T \dots \bullet_K \mathbf{Q}^T,$$
     in which $\mathbf{Q}$ is orthogonal. A class of algebraic algorithms:
     - COM2 (Section 5.3).
     - JADE (Section 5.4).
     - STOTD (Section 5.5).
     - COM1 (Section 5.6).

   (Section 5.2.3.)
3. Mixing matrix: $\mathbf{A} = \mathbf{U} \cdot \mathbf{\Delta} \cdot \mathbf{Q}^T$.
4. Source estimation by means of LCMV or other type of beamforming.

(Some of the matrices $\mathbf{U}$ may be complex conjugated in the complex case.) We look for an orthogonal (unitary) matrix $\mathbf{U}$ that minimizes the cost function

$$f_{K,2}(\mathbf{U}) = \text{off}(\mathcal{C}')^2 = \sum_{\neg(n_1=n_2=\dots=n_K)} \left| c'_{n_1 n_2 \dots n_K} \right|^2. \tag{5.31}$$

Because $\mathbf{U}$ is orthogonal (unitary), this is equivalent to maximizing the contrast function

$$g_{K,2}(\mathbf{U}) = \|\text{diag}(\mathcal{C}')\|^2 = \sum_n \left| c'_{nn\dots n} \right|^2. \tag{5.32}$$

The idea is visualized in Fig. 5.1.

The optimal $\mathbf{U}$ is obtained by means of a higher-order generalization of the Jacobi iteration described in section 5.1.3. Due to the orthogonality (unitarity) of $\mathbf{J}$ in (5.12), the only part of $\mathcal{C}_\mathbf{z}^{(K)}$ that affects the contrast function, is the $(2 \times 2 \times \dots \times 2)$-subtensor $\tilde{\mathcal{C}}$ formed by the entries of which all indices are equal to either p or q. Like $\mathcal{C}_\mathbf{z}^{(K)}$, $\tilde{\mathcal{C}}$ cannot in general be exactly diagonalized. Hence, contrary to the matrix case, we cannot say *a priori* which elementary subtensor will most increase the contrast function. Therefore the rotation pairs (p,q) are swept in a cyclic way. The maximal diagonalization of elementary subtensors gradually minimizes the cost function, so we can expect

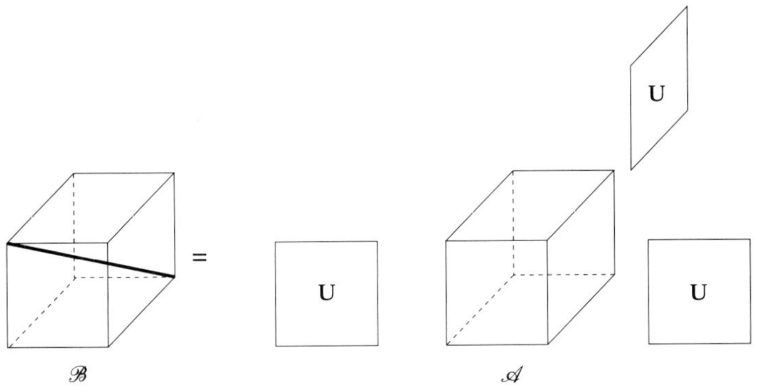

FIGURE 5.1

Visualization of the (approximate) diagonalization of a third-order tensor $\mathcal{A}$ by an orthogonal congruence transformation.

convergence to at least a local minimum. Contrary to the matrix case, the cost function can have spurious local minima. The lack of higher-order equivalent of (5.17) makes it hard to prove convergence to the global optimum, or to formally study the convergence speed. In practice, the algorithm seems to converge reasonably fast and local optima do not seem to pose a problem.

We now explain for a number of specific cases how the optimal Jacobi rotation can be determined.

5.3.1 Third-order real case

We repeat that the $(2 \times 2 \times 2)$ symmetric tensor to be diagonalized is denoted by $\tilde{\mathcal{C}}$. We define $\mathcal{C} = \tilde{\mathcal{C}} \bullet_1 \mathbf{J} \bullet_2 \mathbf{J} \bullet_3 \mathbf{J}$, in which $\mathbf{J}$ is an elementary (2×2) Jacobi rotation matrix. The original derivation started from the parameterization (5.14) [11]. Here we use the parameterization (5.13).

We construct a real symmetric (2×2)-matrix $\mathbf{B}$, as follows:

$$
\begin{aligned}
b_{11} &= a_1, \\
b_{12} &= 3a_4/2, \\
b_{22} &= 9a_2/4 + 3a_3/2 + a_1/4,
\end{aligned}
$$

in which the auxiliary variables a_1, a_2, a_3, a_4 are given by

$$
\begin{aligned}
a_1 &= \tilde{c}_{111}^2 + \tilde{c}_{222}^2, \\
a_2 &= \tilde{c}_{112}^2 + \tilde{c}_{122}^2, \\
a_3 &= \tilde{c}_{111}\tilde{c}_{122} + \tilde{c}_{112}\tilde{c}_{222}, \\
a_4 &= \tilde{c}_{122}\tilde{c}_{222} - \tilde{c}_{111}\tilde{c}_{112}.
\end{aligned}
$$

It can be proved that

$$\sum_{i=1}^{2} c_{iii}^2 = \mathbf{q}^T \cdot \mathbf{B} \cdot \mathbf{q},$$

in which $\mathbf{q} = (\cos(2\alpha)\ \sin(2\alpha))^T$ [26]. Hence the optimal rotation can be found by computing the dominant eigenvector of $\mathbf{B}$ and normalizing it to unit length. It is clear that the function $\sum_{i=1}^{2} c_{iii}^2$ is periodic in the rotation angle, with period $\pi/2$, as the sign of $\mathbf{q}$ is of no importance. The sign of the dominant eigenvector can be chosen such that $\mathbf{J}$ is an inner rotation. The actual elements of the optimal inner Jacobi rotation can then be obtained from the entries of $\mathbf{q}$ by using the basic goniometric relations $\cos\alpha = \sqrt{1+\cos(2\alpha)/2}$ and $\sin\alpha = \sin(2\alpha)/(2\cos\alpha)$.

5.3.2 Third-order complex case

In this section we assume that $\mathcal{C}_{\mathbf{z}}^{(3)}$ is defined by $(\mathcal{C}_{\mathbf{z}}^{(3)})_{ijk} = \text{cum}\{z_i, z_j^*, z_k^*\}$. The $(2 \times 2 \times 2)$ tensor that has to be diagonalized, is denoted by $\tilde{\mathcal{C}}$. We define $\mathcal{C} = \tilde{\mathcal{C}} \bullet_1 \mathbf{J} \bullet_2 \mathbf{J}^* \bullet_3 \mathbf{J}^*$, in which $\mathbf{J}$ is an elementary (2×2) complex Jacobi rotation matrix. We use the parameterization (5.15).

We construct a real symmetric (3×3)-matrix $\mathbf{B}$, as follows:

$$\begin{aligned}
b_{11} &= a_1, \\
b_{12} &= \text{Im}(v_1) + \text{Im}(v_2), \\
b_{13} &= \text{Re}(v_1) - \text{Re}(v_2), \\
b_{22} &= v_4 - \text{Re}(v_3), \\
b_{23} &= \text{Im}(v_3), \\
b_{33} &= v_4 + \text{Re}(v_3),
\end{aligned}$$

in which $\text{Re}(\cdot)$ and $\text{Im}(\cdot)$ denote, respectively, the real and complex part of a complex number, and in which the auxiliary variables are given by

$$\begin{aligned}
a_1 &= |\tilde{c}_{111}|^2 + |\tilde{c}_{222}|^2, \\
a_2 &= |\tilde{c}_{112}|^2 + |\tilde{c}_{212}|^2, \\
a_3 &= |\tilde{c}_{211}|^2 + |\tilde{c}_{122}|^2, \\
a_4 &= \tilde{c}_{111}\tilde{c}_{112}^*, \\
a_5 &= \tilde{c}_{111}\tilde{c}_{211}^*, \\
a_6 &= \tilde{c}_{222}\tilde{c}_{122}^*, \\
a_7 &= \tilde{c}_{222}\tilde{c}_{212}^*, \\
a_8 &= \tilde{c}_{111}\tilde{c}_{212}^* + \tilde{c}_{222}\tilde{c}_{112}^*, \\
a_9 &= \tilde{c}_{111}^*\tilde{c}_{122} + \tilde{c}_{222}\tilde{c}_{211}^*,
\end{aligned}$$

$$
\begin{aligned}
a_{10} &= \tilde{c}^*_{211}\tilde{c}_{112} + \tilde{c}_{122}\tilde{c}^*_{212},\\
v_1 &= a_7 - a_5/2,\\
v_2 &= a_4 - a_6/2,\\
v_3 &= a_9/2 + a_{10},\\
v_4 &= (a_1 + a_3)/4 + a_2 + \mathrm{Re}(a_8).
\end{aligned}
$$

It can be proved that

$$\sum_{i=1}^{2} |c_{iii}|^2 = \mathbf{q}^{\mathrm{T}} \cdot \mathbf{B} \cdot \mathbf{q},$$

in which $\mathbf{q} = (\cos(2\alpha)\ \sin(2\alpha)\sin\phi\ \sin(2\alpha)\cos\phi)^{\mathrm{T}}$ [26]. Like in the real case, the optimal rotation can be found by computing the dominant eigenvector of $\mathbf{B}$ and normalizing it to unit length. The actual elements of the optimal Jacobi rotation can be obtained from the entries of $\mathbf{q}$ by using the basic relations $\cos\alpha = \sqrt{1+\cos(2\alpha)/2}$ and $\sin\alpha e^{i\phi} = (\sin(2\alpha)\cos\phi + i\ \sin(2\alpha)\sin\phi)/(2\cos\alpha)$. The sign of the dominant eigenvector can be chosen such that $\mathbf{J}$ is an inner rotation.

5.3.3 Fourth-order real case

The $(2 \times 2 \times 2 \times 2)$ symmetric tensor to be diagonalized, is denoted by $\tilde{\mathcal{C}}$. We define $\mathcal{C} = \tilde{\mathcal{C}} \bullet_1 \mathbf{J} \bullet_2 \mathbf{J} \bullet_3 \mathbf{J} \bullet_4 \mathbf{J}$, in which $\mathbf{J}$ is an elementary (2×2) Jacobi rotation matrix. We use the parameterization (5.14) and work as in [10].

It turns out that in the real case the contrast function $g_{4,2}$ can be expressed as a function of the auxiliary variable $\xi = \theta - 1/\theta$. Setting the derivative w.r.t. ξ equal to zero, yields:

$$\sum_{k=0}^{4} d_k \xi^k = 0, \tag{5.33}$$

in which

$$
\begin{aligned}
d_4 &= \tilde{c}_{1111}\tilde{c}_{1112} - \tilde{c}_{1222}\tilde{c}_{2222},\\
d_3 &= u - 4(\tilde{c}^2_{1112} + \tilde{c}^2_{1222}) - 3\tilde{c}_{1122}(\tilde{c}_{1111} + \tilde{c}_{2222}),\\
d_2 &= 3v,\\
d_1 &= 3u - 2\tilde{c}_{1111}\tilde{c}_{2222} - 32\tilde{c}_{1112}\tilde{c}_{1222} - 36\tilde{c}^2_{1122},\\
d_0 &= -4(v + 4d_4),
\end{aligned}
$$

with

$$
\begin{aligned}
u &= \tilde{c}^2_{1111} + \tilde{c}^2_{2222},\\
v &= (\tilde{c}_{1111} + \tilde{c}_{2222} - 6\tilde{c}_{1122})(\tilde{c}_{1222} - \tilde{c}_{1112}).
\end{aligned}
$$

We first compute the real roots of Eq. (5.33). (Note that this can be done in a non-iterative way, since the polynomial is only of degree 4 [1].) After computation of these roots, the corresponding values of θ can be found from

$$\theta^2 - \xi\theta - 1 = 0. \tag{5.34}$$

This equation always shows precisely one solution in the interval $(-1, +1]$, corresponding to an inner rotation. If Eq. (5.33) has more than one real root, then the one corresponding to the maximum of the contrast function is selected.

5.3.4 Fourth-order complex case

In this section we assume that $\mathscr{C}_{\mathbf{z}}^{(4)}$ is defined by $(\mathscr{C}_{\mathbf{z}}^{(4)})_{ijkl} = \text{cum}\{z_i, z_j^*, z_k^*, z_l\}$.

Starting from parameterization (5.16), we can work as in section 5.3.3 [10]. It turns out that the contrast function $g_{4,2}$ can now be expressed as a function of the auxiliary variable $\xi = \theta - 1/\theta^*$. Differentiation of $g_{4,2}$ w.r.t. the modulus ρ and the argument ϕ of ξ yields a system of two polynomials of degree 3 and 4 in ρ, where the coefficients are polynomials in $e^{i\phi}$.

On the other hand, starting from (5.15), it was shown in [19] that the optimal Jacobi rotation can be found from the best rank-1 approximation of a real symmetric $(3 \times 3 \times 3 \times 3)$ tensor of which the entries are derived from $\tilde{\mathscr{C}}$. This best rank-1 approximation can be computed by means of the algorithms discussed in [25,34,35,45].

5.4 SIMULTANEOUS DIAGONALIZATION OF MATRIX SLICES

Call a $(J_1 \times J_2)$ matrix slice of a $(J_1 \times J_2 \times \cdots \times J_K)$ tensor $\mathscr{A}$ a matrix obtained by varying the first two indices of $\mathscr{A}$ and picking one particular value for the other indices. We will use Matlab notation $(\mathscr{A})_{:,:,j_3,\ldots,j_K}$ to denote such a matrix slice. From (5.28) we have that the slices $(\mathscr{C}_{\mathbf{z}}^{(K)})_{:,:,n_3,\ldots,n_K}$ have the following structure:

$$\begin{aligned}
(\mathscr{C}_{\mathbf{z}}^{(K)})_{:,:,n_3,\ldots,n_K} &= \sum_{n_1=1}^{N} c_{s_{n_1}}^{(K)} w_{n_1 n_3} \cdots w_{n_1 n_K} \mathbf{w}_{n_1} \circ \mathbf{w}_{n_1} \\
&= \sum_{n_1=1}^{N} c_{s_{n_1}}^{(K)} w_{n_1 n_3} \cdots w_{n_1 n_K} \mathbf{w}_{n_1} \mathbf{w}_{n_1}^{\mathrm{T}} \\
&= \mathbf{W} \cdot \mathbf{D}_{n_3,\ldots,n_K} \cdot \mathbf{W}^{\mathrm{T}},
\end{aligned} \tag{5.35}$$

in which the $(N \times N)$ matrices $\mathbf{D}_{n_3,\ldots,n_K}$ are diagonal. (The transpose may be replaced by the Hermitean transpose in the complex case.) Equation (5.35) is an EVD of $(\mathscr{C}_{\mathbf{z}}^{(K)})_{:,:,n_3,\ldots,n_K}$. We see that the different slices have a common eigenmatrix, namely $\mathbf{W} = \mathbf{Q}^{\mathrm{T}}$.

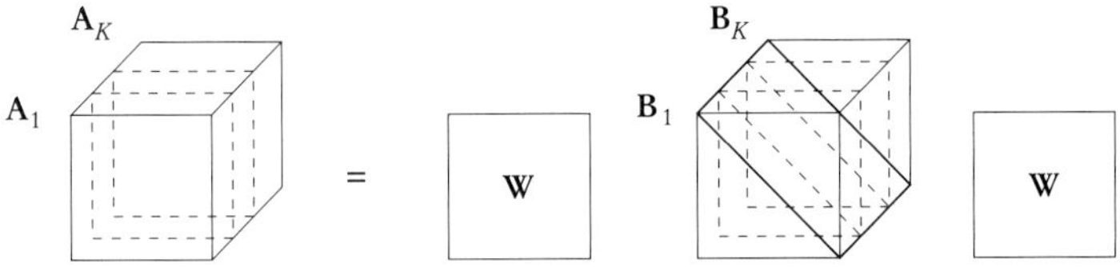

FIGURE 5.2

Visualization of the (approximate) simultaneous diagonalization of a set of matrices $\mathbf{A}_1,\ldots,\mathbf{A}_K$ by an orthogonal congruence transformation.

In the presence of noise, the different matrix slices may not have an eigenmatrix in common; i.e. a matrix that simultaneously diagonalizes all slices may not exist. Denote

$$\mathcal{C}'_{:,:,n_3,\ldots,n_K} = \mathbf{U}\cdot(\mathcal{C}^{(K)}_{\mathbf{z}})_{:,:,n_3,\ldots,n_K}\cdot\mathbf{U}^{\mathrm{T}}. \tag{5.36}$$

(Again, the transpose may be replaced by the Hermitean transpose in the complex case.) Cardoso and Souloumiac proposed to determine $\mathbf{Q}$ as the orthogonal (unitary) matrix that makes the matrices $\mathcal{C}'_{:,:,n_3,\ldots,n_K}$ jointly as diagonal as possible in least squares sense [5]. Formally, an orthogonal (unitary) matrix $\mathbf{U}$ is determined that minimizes the cost function

$$f_{K,SMD}(\mathbf{U}) = \sum_{n_3\ldots n_K} \mathrm{off}(\mathcal{C}'_{:,:,n_3,\ldots,n_K})^2 = \sum_{n_3\ldots n_K}\sum_{n_1\neq n_2}\left|c'_{n_1 n_2\ldots n_K}\right|^2. \tag{5.37}$$

Because $\mathbf{U}$ is orthogonal (unitary), this is equivalent to maximizing the contrast function

$$g_{K,SMD}(\mathbf{U}) = \sum_{n_3\ldots n_K} \|\mathrm{diag}(\mathcal{C}'_{:,:,n_3,\ldots,n_K})\|^2 = \sum_{n_1 n_3\ldots n_K}\left|c'_{n_1 n_1 n_3\ldots n_K}\right|^2. \tag{5.38}$$

The idea is visualized in Fig. 5.2.

The algorithm presented in [5] additionally exploits a second feature of the problem. Note that there are N^{K-2} matrix slices $(\mathcal{C}^{(K)}_{\mathbf{z}})_{:,:,n_3,\ldots,n_K}$. On the other hand, (5.35) shows that, in the absence of noise, these matrices are all linear combinations of the N rank-1 matrices $\mathbf{w}_{n_1}\mathbf{w}^{\mathrm{T}}_{n_1}$. It makes sense to work with a reduced set of N matrices that best represents the original set $\{(\mathcal{C}^{(K)}_{\mathbf{z}})_{:,:,n_3,\ldots,n_K}\}$. For the determination of the reduced set, let us consider the matrix slices $(\mathcal{C}^{(K)}_{\mathbf{z}})_{:,:,n_3,\ldots,n_K}$ as vectors in the $(N\times N)$ matrix space. These vectors are the columns of the $(N^2\times N^{K-2})$ matrix $\underline{\mathbf{C}}^{(K)}_{\mathbf{z}}$, defined by

$$\left(\underline{\mathbf{C}}^{(K)}_{\mathbf{z}}\right)_{(n_1-1)N+n_2,(n_3-1)N^{K-3}+\cdots+(n_{K-1}-1)N+n_K} = (\mathcal{C}^{(K)}_{\mathbf{z}})_{n_1,n_2,\ldots,n_K},$$

for all index entries. The dominant N-dimensional subspace of the column space of $\underline{\mathbf{C}}^{(K)}_{\mathbf{z}}$ is spanned by its N dominant left singular vectors. Denote these vectors by $\mathbf{e}_n$.

Re-matricizing yields the matrices $\mathbf{E}_n$ that will be simultaneously diagonalized:

$$(\mathbf{E}_n)_{n_1,n_2} = (\mathbf{e}_n)_{(n_1-1)N+n_2}.$$

In the case of fourth-order cumulants, $\underline{\mathbf{C}}_{\mathbf{z}}^{(K)}$ is a symmetric (Hermitean) matrix of which the singular vectors can be seen as "eigenmatrices" of $\mathcal{C}_{\mathbf{z}}^{(4)}$. The algorithm in [5] was consequently called *JADE*, standing for *Joint Approximate Diagonalization of Eigenmatrices*.

Denote

$$\mathbf{C}'_n = \mathbf{U} \cdot \mathbf{E}_n \cdot \mathbf{U}^{\mathrm{T}}. \tag{5.39}$$

(Again, the transpose may be replaced by the Hermitean transpose in the complex case.) JADE looks for an orthogonal (unitary) matrix $\mathbf{U}$ that minimizes the cost function

$$f_{JADE}(\mathbf{U}) = \sum_n \mathrm{off}(\mathbf{C}'_n)^2 = \sum_n \sum_{n_1 \neq n_2} \left|(\mathbf{C}'_n)_{n_1 n_2}\right|^2. \tag{5.40}$$

Because $\mathbf{U}$ is orthogonal (unitary), this is equivalent to maximizing the contrast function

$$g_{JADE}(\mathbf{U}) = \sum_n \|\mathrm{diag}(\mathbf{C}'_n)\|^2 = \sum_n \sum_{n_1} \left|(\mathbf{C}'_n)_{n_1 n_1}\right|^2. \tag{5.41}$$

The optimal $\mathbf{U}$ is obtained by means of a higher-order generalization of the Jacobi iteration described in section 5.1.3. Due to the orthogonality (unitarity) of $\mathbf{J}$ in (5.12), the only part of $\mathbf{E}_n$ that affects the contrast function, is the (2×2) submatrix $\tilde{\mathbf{C}}_n$ at the intersection of rows and columns p and q. Similar comments as in section 5.3 are in order. Rotation pairs (p, q) are swept in a cyclic way. Formal analysis of the convergence properties is difficult. Nevertheless, asymptotic quadratic convergence has been proven for the noise-free case [3]. In practice, the algorithm seems to converge reasonably fast and local optima do not seem to pose a problem. Note that the Jacobi iteration can be initialized with the eigenmatrix of one of the matrices that have to be diagonalized.

In [41] it has been proved that the asymptotical accuracy (for infinitesimal errors in the cumulant estimate) of COM2 and JADE are the same. In [8,9] several simulations were carried out to compare both methods in practical situations. The results seem to indicate that the methods have approximately the same accuracy provided that the number of sources N is not overestimated, while the performance of JADE degrades with respect to that of COM2 when N is overestimated.

We now explain how the optimal Jacobi rotation can be determined. It turns out that even in the complex case this rotation can be computed efficiently.

5.4.1 Real case

We repeat that the (2×2) symmetric matrices to be simultaneously diagonalized are denoted by $\tilde{\mathbf{C}}_n$. We define $\mathbf{C}_n = \mathbf{J} \cdot \tilde{\mathbf{C}}_n \cdot \mathbf{J}^T$, in which $\mathbf{J}$ is an elementary (2×2) Jacobi rotation matrix. We use parameterization (5.13) and work as in [5,6].

First define a $(2 \times N)$ matrix $\mathbf{G}$ of which the columns $\mathbf{g}_n$ are given by

$$\mathbf{g}_n = \begin{pmatrix} (\tilde{\mathbf{C}}_n)_{11} - (\tilde{\mathbf{C}}_n)_{22} \\ -(\tilde{\mathbf{C}}_n)_{12} - (\tilde{\mathbf{C}}_n)_{21} \end{pmatrix}.$$

It can be proved that the objective function is given by

$$\sum_n (\mathbf{C}_n)_{11}^2 + (\mathbf{C}_n)_{22}^2 = \mathbf{q}^T \cdot \mathbf{G}\mathbf{G}^T \cdot \mathbf{q},$$

in which $\mathbf{q} = (\cos(2\alpha) \ \sin(2\alpha))^T$. The vector $\mathbf{q}$ that maximizes this expression is the dominant eigenvector of $\mathbf{G}\mathbf{G}^T$, or, equivalently, the dominant left singular vector of $\mathbf{G}$. The actual elements of the optimal Jacobi rotation can be obtained from the basic relations $\cos\alpha = \sqrt{1 + \cos(2\alpha)/2}$ and $\sin\alpha = \sin(2\alpha)/(2\cos\alpha)$. The sign of the dominant eigenvector can be chosen such that $\mathbf{J}$ is an inner rotation.

5.4.2 Complex case

We repeat that the (2×2) Hermitean matrices to be simultaneously diagonalized are denoted by $\tilde{\mathbf{C}}_n$. We define $\mathbf{C}_n = \mathbf{J} \cdot \tilde{\mathbf{C}}_n \cdot \mathbf{J}^H$, in which $\mathbf{J}$ is an elementary (2×2) Jacobi rotation matrix. We use parameterization (5.15) and work as in [5,6].

First define a $(3 \times N)$ matrix $\mathbf{G}$ of which the columns $\mathbf{g}_n$ are given by

$$\mathbf{g}_n = \begin{pmatrix} (\tilde{\mathbf{C}}_n)_{11} - (\tilde{\mathbf{C}}_n)_{22} \\ -(\tilde{\mathbf{C}}_n)_{12} - (\tilde{\mathbf{C}}_n)_{21} \\ i(-(\tilde{\mathbf{C}}_n)_{12} + (\tilde{\mathbf{C}}_n)_{21}) \end{pmatrix}.$$

It can be proved that

$$\sum_n |(\mathbf{C}_n)_{11}|^2 + |(\mathbf{C}_n)_{22}|^2 = \mathbf{q}^T \cdot \mathrm{Re}(\mathbf{G}\mathbf{G}^H) \cdot \mathbf{q},$$

in which $\mathbf{q} = (\cos(2\alpha) \ \sin(2\alpha)\sin\phi \ \sin(2\alpha)\cos\phi)^T$. Hence the optimal rotation can be found via the dominant eigenvector of the real symmetric (3×3)-matrix $\mathrm{Re}(\mathbf{G}\mathbf{G}^H)$. The actual elements of the optimal Jacobi rotation can be obtained from the entries of $\mathbf{q}$ by using the basic relations $\cos\alpha = \sqrt{1 + \cos(2\alpha)/2}$ and $\sin\alpha e^{i\phi} = (\sin(2\alpha)\cos\phi + i\sin(2\alpha)\sin\phi)/(2\cos\alpha)$. The sign of the dominant eigenvector can be chosen such that $\mathbf{J}$ is an inner rotation.

5.5 SIMULTANEOUS DIAGONALIZATION OF THIRD-ORDER TENSOR SLICES

Instead of considering the tensor $\mathcal{C}_{\mathbf{z}}^{(K)}$ as a stack of N^{K-2} $(N \times N)$ matrices $(\mathcal{C}_{\mathbf{z}}^{(K)})_{:,:,n_3,\ldots,n_K}$, we could also consider it as a stack of N^{K-3} $(N \times N \times N)$ tensors $(\mathcal{C}_{\mathbf{z}}^{(K)})_{:,:,:,n_4,\ldots,n_K}$ and simultaneously diagonalize these. This can be done using the expressions in sections 5.3.1 and 5.3.2, taking into account that the cost function now consists of a sum of contributions associated with the different third-order tensor slices (i.e. the matrix $\mathbf{B}$ is now a sum of matrices associated with the different third-order tensor slices). We refer to [26] for more details. The technique is called *STOTD*, which stands for *Simultaneous Third-Order Tensor Diagonalization*.

Simulations in [26] indicate that the performance of JADE and STOTD are generally quite comparable, although in specific cases one method may perform better the other.

5.6 MAXIMIZATION OF THE TENSOR TRACE

It is well-known that the trace of a matrix is invariant under orthogonal (unitary) congruence transformations. Again, the situation is very different for higher-order tensors. In the orthogonal (unitary) congruence transformation (5.30) the trace of the tensor, i.e. the sum of its diagonal entries, does in general change with $\mathbf{U}$. Comon and Moreau realized that we can thus diagonalize the tensor by maximizing its trace, if we know that all source cumulants are positive [16]. If all cumulants are negative, then we can process $-\mathcal{C}_{\mathbf{z}}^{(K)}$ instead of $\mathcal{C}_{\mathbf{z}}^{(K)}$. The assumption that the cumulants of all sources have the same known sign is satisfied in many applications. For typical constellations used in telecommunication, the fourth-order cumulant is negative. Speech and audio signals typically have positive fourth-order cumulant. In this section we assume positive source cumulants.

Formally, we look for an orthogonal (unitary) matrix $\mathbf{U}$ that maximizes the contrast function

$$g_{K,1}(\mathbf{U}) = \sum_n c'_{nn\ldots n}. \tag{5.42}$$

The resulting algorithm will be denoted as *COM1*, where the "1" indicates that we maximize the 1-norm of the diagonal, instead of its 2-norm, like in COM2. The optimal $\mathbf{U}$ is again obtained by means of a Jacobi iteration. The fact that the diagonal entries are not squared, leads to simpler expressions than in COM2, such that the computation of an elementary rotation is easier.

Simulations indicate that the accuracy of COM1 is comparable to that of COM2 and JADE, provided (i) the number of sources is not over-estimated, (ii) the background noise is Gaussian, (iii) no source is Gaussian and (iv) the cumulants of the sources have the same sign [15]. COM1 seems to be less sensitive to non-Gaussian background noise than COM2 and JADE. When the number of sources is over-estimated, the convergence speed of the three methods decreases. COM2 is then the most reliable method whereas

JADE becomes the slowest. On the other hand, the performance of COM1 degrades with respect to that of COM2 and JADE when one of the sources is Gaussian or quasi-Gaussian. In addition, COM1 is unable to separate sources with different cumulant signs.

We now explain how an optimal Jacobi rotation can be determined. Due to space limitations we only consider the fourth-order complex case. We assume that $\mathcal{C}_{\mathrm{z}}^{(4)}$ is defined by $(\mathcal{C}_{\mathrm{z}}^{(4)})_{ijkl} = \mathrm{cum}\{z_i, z_j^*, z_k^*, z_l\}$. Like in section 5.3, the $(2\times2\times2\times2)$-subtensor that will be diagonalized, is denoted by $\tilde{\mathcal{C}}$. We define $\mathcal{C} = \tilde{\mathcal{C}} \bullet_1 \mathbf{J} \bullet_2 \mathbf{J}^* \bullet_3 \mathbf{J}^* \bullet_4 \mathbf{J}$, in which $\mathbf{J}$ is an elementary (2×2) Jacobi rotation matrix. An initial approach was based on the parameterization (5.16) [16]. Here we use parameterization (5.15) and we work as in [12].

We construct a real symmetric (3×3)-matrix $\mathbf{B}$ as follows:

$$\begin{aligned}
b_{11} &= \tilde{c}_{1111} + \tilde{c}_{2222},\\
b_{12} &= \mathrm{Im}(\tilde{c}_{1222} - \tilde{c}_{1211}),\\
b_{13} &= \mathrm{Re}(\tilde{c}_{1222} - \tilde{c}_{1211}),\\
b_{22} &= \frac{1}{2}(\tilde{c}_{1111} + \tilde{c}_{2222}) + 2\tilde{c}_{1212} - \mathrm{Re}(\tilde{c}_{1221}),\\
b_{23} &= \mathrm{Im}(\tilde{c}_{1221}),\\
b_{33} &= \frac{1}{2}(\tilde{c}_{1111} + \tilde{c}_{2222}) + 2\tilde{c}_{1212} + \mathrm{Re}(\tilde{c}_{1221}).
\end{aligned}$$

It can be shown that

$$\sum_{i=1}^{2} c_{iiii} = \mathbf{q}^{\mathrm{T}} \cdot \mathbf{B} \cdot \mathbf{q},$$

with $\mathbf{q} = (\cos(2\alpha)\ \sin(2\alpha)\sin\phi\ \sin(2\alpha)\cos\phi)^{\mathrm{T}}$. The optimal rotation can be found by computing the dominant eigenvector of $\mathbf{B}$ and scaling it to unit norm. The entries of the Jacobi rotation matrix are found from the entries of $\mathbf{q}$ by using the trigonometric relations $\cos\alpha = \sqrt{(1+\cos(2\alpha))/2}$ and $\sin\alpha e^{i\phi} = (\sin(2\alpha)\cos\phi + i\sin(2\alpha)\sin\phi)/(2\cos\alpha)$. The sign of the dominant eigenvector can be chosen such that $\mathbf{J}$ is a rotation over an angle in the interval $(-\pi/4, \pi/4]$.

References

[1] M. Abramowitz, I.E. Segun, Handbook of Mathematical Functions, Dover Publ, N.Y., 1968.

[2] C. Bourin, P. Bondon, Efficiency of high-order moment estimates, in: IEEE Signal Processing/ATHOS Workshop on Higher-Order Statistics, Girona, Spain, 1995, pp. 186–190.

[3] A. Bunse-Gerstner, R. Byers, V. Mehrmann, Numerical methods for simultaneous diagonalization, SIAM J. Matrix Anal. Appl. 14 (1993) 927–949.

[4] J.-F. Cardoso, On the performance of orthogonal source separation algorithms, in: Proc. VIIth European Signal Processing Conference, EUSIPCO-94, vol. 2, Edinburgh, Scotland, U.K., September 1994, pp. 776–779.
[5] J.-F. Cardoso, A. Souloumiac, Blind beamforming for non-Gaussian signals, IEE Proc.-F 140 (1994) 362–370. [Matlab code available at http://www.tsi.enst.fr/~cardoso/stuff.html].
[6] J.-F. Cardoso, A. Souloumiac, Jacobi angles for simultaneous diagonalization, SIAM J. Matrix Anal. Appl. 17 (1996) 161–164.
[7] J. Carroll, J. Chang, Analysis of individual differences in multidimensional scaling via an N-way generalization of "Eckart-Young" decomposition, Psychometrika 35 (1970) 283–319.
[8] P. Chevalier, Méthodes aveugles de filtrage d'antennes, Revue d'Electronique et d'Electricité 3 (1995) 48–58.
[9] P. Chevalier, On the performance of higher order separation methods, in: Proc. IEEE-ATHOS Workshop on Higher-Order Statistics, Begur, Spain, June 1995, pp. 30–34.
[10] P. Comon, Independent component analysis, a new concept?, Signal Process. 36 (1994) 287–314. Matlab code available at [http://www.i3s.unice.fr/~comon/].
[11] P. Comon, Tensor diagonalization, a useful tool in signal processing, in: 10th IFAC Symposium on System Identification, IFAC-SYSID'94, vol. 1, Copenhagen, Denmark, July 1994, pp. 77–82.
[12] P. Comon, From source separation to blind equalization, contrast-based approaches, in: Int. Conf. on Image and Signal Processing, ICISP'01, Agadir, Morocco, May 2001.
[13] P. Comon, Blind identification and source separation in 2×3 under-determined mixtures, IEEE Trans. Signal Process. 52 (2004) 11–22.
[14] P. Comon, Canonical tensor decompositions, tech. rep. RR-2004-17, Lab. I3S, Sophia-Antipolis, France, June 2004.
[15] P. Comon, P. Chevalier, V. Capdevielle, Performance of contrast-based blind source separation, in: SPAWC – IEEE Sig. Proc. Advances in Wireless Com., Paris, France, April 1997, pp. 345–348.
[16] P. Comon, E. Moreau, Improved contrast dedicated to blind separation in communications, in: Proc. ICASSP, Munich, Germany, April 1997, pp. 3453–3456. [Matlab code available at http://www.i3s.unice.fr/~comon/].
[17] P. Comon, B. Mourrain, Decomposition of quantics in sums of powers of linear forms, Signal Process. 53 (1996) 93–107.
[18] P. Comon, M. Rajih, Blind identification of under-determined mixtures based on the characteristic function, in: Proc. ICASSP'05, vol. 4, Philadelphia, March 2005, pp. 1005–1008.
[19] L. De Lathauwer, Signal Processing Based on Multilinear Algebra, PhD thesis, K.U. Leuven, E.E. Dept., Belgium, 1997.
[20] L. De Lathauwer, A link between the canonical decomposition in multilinear algebra and simultaneous matrix diagonalization, SIAM J. Matrix Anal. Appl. 28 (2006) 322–336.
[21] L. De Lathauwer, J. Castaing, Blind identification of underdetermined mixtures by simultaneous matrix diagonalization, IEEE Trans. Signal Process. 56 (2008) 1096–1105.
[22] L. De Lathauwer, J. Castaing, J.-F. Cardoso, Fourth-order cumulant based blind identification of underdetermined mixtures, IEEE Trans. Signal Process. 55 (2007) 2965–2973.
[23] L. De Lathauwer, P. Comon, B. De Moor, ICA algorithms for 3 sources and 2 sensors, in: Proc. Sixth Sig. Proc. Workshop on Higher Order Statistics, Caesarea, Israel, June 1999, pp. 116–120.
[24] L. De Lathauwer, B. De Moor, J. Vandewalle, An algebraic ICA algorithm for 3 sources and 2 sensors, in: Proc. Xth European Signal Processing Conference, EUSIPCO 2000, Tampere, Finland, September 2000.
[25] L. De Lathauwer, B. De Moor, J. Vandewalle, On the best rank-1 and rank-$(R_1, R_2, \ldots, R_n)$ approximation of higher-order tensors, SIAM J. Matrix Anal. Appl. 21 (2000) 1324–1342.
[26] L. De Lathauwer, B. De Moor, J. Vandewalle, Independent component analysis and (simultaneous) third-order tensor diagonalization, IEEE Trans. Signal Process. 49 (2001) 2262–2271. [Matlab code available at ftp://ftp.esat.kuleuven.ac.be/pub/SISTA/delathauwer/software/].

[27] L. De Lathauwer, B. De Moor, J. Vandewalle, Computation of the canonical decomposition by means of a simultaneous generalized Schur decomposition, SIAM J. Matrix Anal. Appl. 26 (2004) 295–327.

[28] L. De Lathauwer, B. De Moor, J. Vandewalle, A prewhitening-induced bound on the identification error in independent component analysis, IEEE Trans. Circuits Syst. I 52 (2005) 546–554.

[29] W. Gardner, Introduction to Random Processes, McGraw-Hill, 1990.

[30] G. Golub, C. Van Loan, Matrix Computations, 3rd ed., Johns Hopkins University Press, Baltimore, Maryland, 1996.

[31] R.A. Harshman, Foundations of the PARAFAC procedure: Model and conditions for an explanatory multi-mode factor analysis, UCLA Working Papers in Phonetics 16 (1970) 1–84.

[32] I. Jolliffe, Principal Component Analysis, 2nd ed., in: Springer Series in Statistics, Springer, N.Y., 2002.

[33] M. Kendall, A. Stuart, The Advanced Theory of Statistics, Griffin, London, 1977.

[34] E. Kofidis, P. Regalia, On the best rank-1 approximation of higher-order supersymmetric tensors, SIAM J. Matrix Anal. Appl. 23 (2002) 863–884.

[35] P. Kroonenberg, Applied Multiway Data Analysis, Wiley, 2008.

[36] P. McCullagh, Tensor Methods in Statistics, Chapman and Hall, 1987.

[37] E. Moreau, A generalization of joint-diagonalization criteria for source separation, IEEE Trans. Signal Process. 49 (2001) 530–541.

[38] C. Nikias, J. Mendel, Signal processing with higher-order spectra, IEEE Signal Process. Mag. 3 (1993) 10–37.

[39] C. Nikias, A. Petropulu, Higher-Order Spectra Analysis. A Nonlinear Signal Processing Framework, Prentice Hall, Englewood Cliffs, New Jersey, 1993.

[40] A. Smilde, R. Bro, P. Geladi, Multi-way Analysis. Applications in the Chemical Sciences, John Wiley and Sons, Chichester, UK, 2004.

[41] A. Souloumiac, J.-F. Cardoso, Performances en séparation de sources, in: Proc. GRETSI, Juan les Pins, France, 1993, pp. 321–324.

[42] B. Van Veen, K. Buckley, Beamforming: a versatile approach to spatial filtering, IEEE ASSP Mag. 5 (1988) 4–24.

[43] M. Wax, T. Kailath, Detection of signals by information theoretic criteria, IEEE Trans. Acoust., Speech Signal Process. 33 (1986) 387–392.

[44] A. Yeredor, Non-orthogonal joint diagonalization in the least-squares sense with application in blind source separation, IEEE Trans. Signal Process. 50 (2002) 1545–1553.

[45] T. Zhang, G. Golub, Rank-one approximation to high order tensors, SIAM J. Matrix Anal. Appl. 23 (2001) 534–550.

CHAPTER

Iterative algorithms

6

V. Zarzoso and A. Hyvärinen

6.1 INTRODUCTION

As seen in the preceding chapters of this book, the problems of blind source separation (BSS) and independent component analysis (ICA) can be solved by optimizing a so-called *contrast function* measuring at the separator output a known property of the sources or independent components. Specific approaches differ in how this optimization is carried out. The present chapter focuses on techniques performing the contrast optimization in an iterative fashion. An iterative technique is defined by an equation specifying how the separating filters are updated at each step of the algorithm. The update rule modifies the extracting filters so as to approach an optimum value of the contrast function and thus a valid separation solution; updates are repeated until convergence. Finding the right solution and the speed at which it is found generally depends on the contrast function itself, the type of update rule employed to optimize the contrast, the initialization of the separating filters and the selection of step size (learning rate or adaption coefficient) typically characterizing the iterative update.

Alternative BSS/ICA techniques based on closed-form solutions (feasible, e.g. in the two-signal case) or on matrix/tensor diagonalizations and decompositions presented in other chapters of this book and elsewhere in the literature. Tensor diagonalizations are associated with specific contrasts and usually involve Jacobi-type iterative procedures [38] (see also Chapter 5) to find the Givens rotations used to parameterize the separating matrix. Although no counterexample has been found to date, the theoretical convergence of the Jacobi technique in the tensor case remains to be proven. Unlike the tensorial approach, the iterative techniques studied in this chapter can be easily derived from practically any contrast function and, in general, do not require any special parameterization of the separating matrix. In addition, their local convergence properties can be studied by standard analysis tools; global convergence can even be shown theoretically for certain deflation algorithms such as Delfosse and Loubaton's [28] and FastICA with cubic nonlinearity [49]. Another attractive feature of iterative techniques is their ability to track (slow) variations of the mixing system, which renders them particularly useful in non-stationary environments.

From a historical perspective, most early BSS/ICA algorithms found in the literature are iterative in the sense of this chapter. These range from the pioneering neural network approaches by Jutten-Hérault [43,50,51], Comon [21], Moreau-Macchi [59,63],

Handbook of Blind Source Separation: Independent Component Analysis and Applications

Deville [29–31] and Cichocki [19,20] to the relative (or natural) gradient methods by Cardoso-Laheld [14,57] and Amari and colleagues [1,2,26]. The popular FastICA algorithm by Hyvärinen [47–49] as well as some of its variants [53,78] also fall into this class. More recent iterative techniques include the work of Regalia and co-workers [71, 72] and Zarzoso-Comon [84,82,85] based on some form of optimal value for the step size.

The chapter begins by recalling the model and goal of BSS as well as various contrast functions easily lending themselves to the iterative approach (sections 6.2–6.3). After presenting some common mathematical concepts about iterative optimization in section 6.4, section 6.5 considers iterative methods for the preliminary whitening of the data. The rest of the chapter presents a variety of iterative algorithms for BSS/ICA, summarizing the work of a good number of researchers in the field over two decades. Techniques for joint separation and sequential extraction are treated separately in sections 6.6–6.7 and sections 6.9–6.11, respectively, while the choice of nonlinear functions characterizing many of these techniques is addressed in section 6.8. We start the exposition by going through historically important methods in section 6.6. Then, emphasis is laid on iterative techniques that have provided satisfactory performance in many applications such as relative/natural gradient algorithms (section 6.7), deflation-based procedures (section 6.9) and the FastICA method (section 6.10). Finally, section 6.11 reports on techniques aiming at an optimal choice of the step size in their update rules. The presentation is intended to help the reader understand the derivation, working assumptions, connections, advantages, and limitations of some of the most relevant iterative methods for BSS/ICA found in the literature. The concluding remarks of section 6.12 bring the chapter to an end and outline some open problems deserving further investigation.

6.2 MODEL AND GOAL

Although some of the techniques surveyed in this chapter are readily extended to more involved scenarios such as the convolutive case, our focus is on instantaneous linear mixtures. In this scenario, the observed vector $\mathbf{x}(t) \in \mathbb{R}^P$ can be expressed as:

$$\mathbf{x}(t) = \mathbf{A}\mathbf{s}(t), \qquad t = 1, 2, \ldots \tag{6.1}$$

where $\mathbf{s}(t) \in \mathbb{R}^N$ is the unknown source vector, with $N \leqslant P$, and $\mathbf{A} \in \mathbb{R}^{P \times N}$ the unknown mixing matrix. The objective of BSS is to recover the source vector realizations $\{\mathbf{s}(t)\}$ from the corresponding realizations of the observed vector $\{\mathbf{x}(t)\}$. This process involves, explicitly or implicitly, the inversion of matrix $\mathbf{A}$, which is thus assumed to be full column rank. The separator output is computed as $\mathbf{y}(t) = \mathbf{B}\mathbf{x}(t)$, where $\mathbf{B} \in \mathbb{R}^{N \times P}$ is the separating matrix. A perfect separating matrix is such that the global filter $\mathbf{G} = \mathbf{B}\mathbf{A}$ can be written as the product of a non-singular diagonal matrix $\mathbf{D}$ and a permutation matrix $\mathbf{P} : \mathbf{G} = \mathbf{P}\mathbf{D}$. The diagonal elements of matrix $\mathbf{D}$ account for the unknown scales of the separated sources, whereas the permutation matrix $\mathbf{P}$ reflects the fact that the sources can appear in any order at the separator output. Such source scale and permutation indeterminacies are typically admissible in BSS/ICA.

To estimate the separating filters, many ICA methods compute an intermediate step called *whitening* or *sphering*. The idea is to find a so-called *whitening matrix* $\mathbf{W}$ such that the whitened signals

$$\mathbf{z}(t) = \mathbf{W}\mathbf{x}(t) \tag{6.2}$$

have an identity covariance matrix. As a result, we can write

$$\mathbf{z}(t) = \mathbf{Q}\mathbf{s}(t) \tag{6.3}$$

and the number of unknowns is reduced to a unitary mixing matrix $\mathbf{Q}$.

6.3 CONTRAST FUNCTIONS FOR ITERATIVE BSS/ICA

Most contrasts for BSS/ICA are based on concepts borrowed from information theory [25]. The derivation of these contrasts as well as their main relationships are briefly described next. This section follows closely [23,11,13] (see Chapters 2–4) but also reports more recent results.

6.3.1 Information-theoretic contrasts

Assuming T independent realizations, the normalized log-likelihood of observed vectors $\{\mathbf{x}(t)\}_{t=1}^{T}$ is given by

$$\frac{1}{T}\mathscr{L}(\mathbf{x}\,|\,\mathbf{A}) = \frac{1}{T}\sum_{t=1}^{T}\log p_s\left(\mathbf{A}^{-1}\mathbf{x}(t)\right) - \log|\det\mathbf{A}| \underset{T\to\infty}{\longrightarrow} -K(p_y; p_s) + \text{constant} \tag{6.4}$$

where p_s denotes the joint probability density function (pdf) of the sources and p_y represents the separator output pdf. Notation $K(p_y; p_s)$ stands for the Kullback-Leibler divergence between densities p_y and p_s:

$$K(p_y; p_s) = \int p_y(\mathbf{u})\log\frac{p_y(\mathbf{u})}{p_s(\mathbf{u})}d\mathbf{u}. \tag{6.5}$$

Consequently, finding the mixing matrix maximizing the probability of the observations – the *maximum likelihood* principle – is tantamount to estimating the separating matrix recovering the source pdf at the separator output.

A related concept that has received a lot of attention is the so-called *infomax* principle, applied to BSS by Bell and Sejnowski in [4]. Infomax consists in maximizing the functional

$$\Upsilon_{\text{IM}}(\mathbf{y}) = H\{\mathbf{r}(\mathbf{y})\} \tag{6.6}$$

where $H\{\cdot\}$ represents Shannon's (differential) entropy, defined as

$$H\{\mathbf{s}\} = -\int p_s(\mathbf{u})\log p_s(\mathbf{u})d\mathbf{u} \tag{6.7}$$

and $\mathbf{r}(\mathbf{y}) = [r_1(y_1), r_2(y_2), \ldots, r_N(y_N)]^{\mathrm{T}}$ is a vector of element-wise nonlinear functions of the separator outputs. If the derivatives $r_i'(\cdot)$ are constrained to be probability densities, straightforward algebraic manipulations involving Eqs (6.5) and (6.7) show that criterion (6.6) is equivalent to the minimization of the divergence $K(p_{\mathbf{y}}; \prod_{i=1}^N r_i')$. If the sources are assumed to be statistically independent, their joint pdf equals the product of their marginal pdfs: $p_{\mathbf{s}}(\mathbf{s}) = \prod_{i=1}^N p_{s_i}(s_i)$. Infomax thus amounts to maximum likelihood if the nonlinearities are matched to the source densities as $r_i'(y) = p_{s_i}(y)$; that is, $r_i(y)$ corresponds to the i-th source cumulative distribution function (cdf) [2,10].

Under the source independence assumption, the maximum likelihood criterion can also be shown to be equivalent to the minimization of the separator *mutual information*

$$I(p_{\mathbf{y}}) = K\left(p_{\mathbf{y}}; \prod_{i=1}^N p_{y_i}\right) = -H\{\mathbf{y}\} + \sum_{i=1}^N H\{y_i\} \tag{6.8}$$

up to a mismatch between the source and the separator output marginal densities. Yet the mutual information alone is a valid contrast for the blind separation of independent sources [23]. After prewhitening, the first term on the right-hand side of (6.8) becomes constant under orthogonal transformations. Hence, mutual information minimization and *marginal entropy* minimization are equivalent. Moreover, it can easily be shown that

$$H\{y_i\} = -K(p_{y_i}; p_{g_i}) + \text{constant}$$

where g_i is a zero-mean Gaussian random variable with the same variance as y_i. It follows that marginal entropy minimization is equivalent to the maximization of the separator output non-Gaussianity. This result is consistent with intuition and the Central Limit Theorem: as mixing increases Gaussianity, once should proceed in the opposite direction to achieve the separation. The divergence $K(p_{y_i}; p_{g_i})$ is also called (marginal) *negentropy*.

6.3.2 Cumulant-based approximations

The information-theoretic criteria summarized above involve pdfs, which are generally difficult to estimate and deal with in practice. To surmount this difficulty, densities can be approximated by their Edgeworth or Gram-Charlier expansions [77]. These involve the Hermite polynomials with the cumulants as coefficients. Using these expansions, the Kullback-Leibler divergence between two random vectors can be expressed as weighted square differences between their respective cumulants. For symmetric distributions $p_{\mathbf{u}}$ and $p_{\mathbf{v}}$, this cumulant-based Kullback-Leibler approximation reads:

$$K(p_{\mathbf{u}}; p_{\mathbf{v}}) \approx \frac{1}{4}\sum_{ij=1}^N \left(r_{ij}^{\mathbf{u}} - r_{ij}^{\mathbf{v}}\right)^2 + \frac{1}{48}\sum_{ijk\ell=1}^N \left(\kappa_{ijk\ell}^{\mathbf{u}} - \kappa_{ijk\ell}^{\mathbf{v}}\right)^2 \tag{6.9}$$

where $r_{ij}^{\mathbf{u}} = \mathrm{cum}\{u_i, u_j\}$ denotes the second-order cumulant (covariance) of (u_i, u_j) and $\kappa_{ijk\ell}^{\mathbf{u}} = \mathrm{cum}\{u_i, u_j, u_k, u_\ell\}$ the fourth-order cumulant of (u_i, u_j, u_k, u_ℓ). For mutually independent sources and after prewhitening, the maximum likelihood criterion (6.4) can thus be approximated as

$$\Upsilon^{o}_{\mathrm{ML}}(\mathbf{y}) \approx \frac{1}{24}\sum_{i=1}^{N} \kappa^{\mathbf{y}}_{iiii}\kappa_i \tag{6.10}$$

where κ_i denotes the fourth-order marginal cumulant of the i-th source. Symbol ^{o} stresses the fact that this function is an *orthogonal contrast*, as it must be maximized with respect to unitary transformations under the whitening constraint. Reference [86, 89] prove that cumulant approximation (6.10) is indeed a suitable contrast for the separation of independent sources. In a blind problem, however, the source statistics, and in particular their kurtoses, are unknown. Interestingly, the function is still a contrast even if the source kurtoses are replaced by coefficients $\{\alpha_i\}_{i=1}^N$:

$$\Upsilon^{o}_{\mathrm{KVP}}(\mathbf{y}) = \sum_{i=1}^{N} \alpha_i \kappa^{\mathbf{y}}_{iiii} \tag{6.11}$$

such that $\mathrm{sign}(\alpha_i) = \mathrm{sign}(\kappa_i)$; this is the *kurtosis value priors (KVP) contrast* [86]. In addition, if the source kurtoses are different and so are the weights, maximization of contrast (6.11) ensures that the estimated sources are arranged at the separator output according to their kurtosis values in the same order as the weights: the permutation ambiguity typical to ICA is thus eliminated. If all the sources have the same kurtosis sign, say ε, function (6.11) reduces to

$$\Upsilon^{o}_{\mathrm{M}}(\mathbf{y}) = \varepsilon\sum_{i=1}^{N} \kappa^{\mathbf{y}}_{iiii} = \varepsilon\sum_{i=1}^{N} \mathbb{E}\{y_i^4\} + \text{constant} \tag{6.12}$$

which had already proven to be a contrast in [14,63]. This contrast can be explicitly adapted to the separation of sub-Gaussian ($\varepsilon = -1$) and super-Gaussian ($\varepsilon = 1$) sources.

Recall that, under the whitening constraint, the mutual information (6.8) essentially reduces to the sum of marginal entropies. According to approximation (6.9), these two criteria can be expressed as the sum of squared fourth-order marginal cumulants:

$$\Upsilon^{o}_{\mathrm{MI}}(\mathbf{y}) = \Upsilon^{o}_{\mathrm{ME}}(\mathbf{y}) = \sum_{i=1}^{N} \left(\kappa^{\mathbf{y}}_{iiii}\right)^2. \tag{6.13}$$

Even if not directly derived from information-theoretical concepts, specific functions of the separator-output higher-order statistics are also proven to be valid contrasts for blind separation of independent sources. Such is the case of functional

$$\Upsilon^{o}_{1}(\mathbf{y}) = \sum_{i=1}^{N} |\kappa^{\mathbf{y}}_{iiii}| \tag{6.14}$$

if at most one of the sources has zero kurtosis [63]. In [66], functional (6.14) is called the *multiuser kurtosis (MUK)* maximization criterion and is derived as an extension of the blind equalization principles of [74] under the more restrictive assumption of sources with the same distribution. Reference [62] generalizes contrasts (6.12)–(6.14) to cumulants of any order $r > 2$ and to complex-valued (possibly non-circular) sources. See also section 3.4.2.5, page 83, for complementary details.

6.3.3 Contrasts for source extraction

The above contrasts aim at the simultaneous separation of all sources in the observed mixture, a process also referred to as *symmetric* or *joint separation*. An alternative is the *sequential extraction* or *deflationary separation* approach to BSS, whereby the sources are estimated one after another. The key advantage of the deflationary approach is that contrasts for single-source extraction can easily be proven to be free from spurious local extrema, so that all their local maxima are associated with valid extraction solutions. This attractive feature enables the design of iterative algorithms with (theoretical) *global* convergence to good separation settings. The main drawback, however, is that errors due to imperfect extractions (e.g. caused by finite sample size) tend to accumulate through successive deflation stages and, consequently, the estimation quality depends on the order in which a source is extracted. Specific iterative methods for sequential extraction based on the contrasts below will be reviewed in sections 6.9–6.11.

By virtue of Eq. (6.9), negentropy accepts the simple fourth-order approximation for symmetric densities [23]:

$$K(p_y; p_g) \approx \frac{1}{48}(\kappa_{\cdot}^{y} - \kappa_{\cdot}^{g})^2 = \frac{1}{48}(\kappa_{\cdot}^{y})^2 \tag{6.15}$$

since the statistics of orders higher than two of a Gaussian random variable are null. Accordingly, the absolute value of the fourth-order marginal cumulant is naturally linked to non-Gaussianity, a result that can easily be extended to any order greater than two. Hence, it is not surprising that a widely used contrast for the extraction of a single source is based on the *kurtosis*, or normalized fourth-order marginal cumulant:

$$\Upsilon_\kappa(y) \stackrel{\text{def}}{=} \varepsilon \frac{\kappa_{\cdot}^{y}}{(\sigma_y^2)^2} = \varepsilon \frac{\mathbb{E}\{|y|^4\} - 2\mathbb{E}^2\{|y|^2\} - |\mathbb{E}\{y^2\}|^2}{\mathbb{E}^2\{|y|^2\}} \tag{6.16}$$

where σ_y^2 denotes the variance of the extractor output $y = \mathbf{b}^{\mathrm{T}}\mathbf{x}$ and ε represents the sign of the fourth-order cumulant of the source of interest. The *kurtosis maximization* (KM) criterion based on contrast (6.16) started to receive attention with the pioneering work of Wiggins [80], Donoho [33] and Shalvi and Weinstein [74] on blind deconvolution of seismic signals and blind equalization of digital communication channels, and was later employed for source separation [28], even in the convolutive mixture scenario [79]. This criterion is quite general in that it does not require the observations to be prewhitened and can be applied to any kind of real- or complex-valued non-Gaussian sources without any modification. Moreover, Tugnait proved in [79] that contrast (6.16) is free from spurious extrema. For this result to hold, the data are assumed to follow exactly the noiseless ICA model (6.1) with infinite sample size.

In the real-valued case, Delfosse and Loubaton [28] showed that functional

$$\Upsilon^o_\kappa(y) = (\mathbb{E}\{y^4\} - 3)^2 \tag{6.17}$$

reminiscent of (6.13) and (6.15), is an orthogonal contrast for the extraction of a single source with non-zero fourth-order cumulant. The same was proven of function

$$\Upsilon^o_{\mathrm{m}}(y) = \varepsilon\mathbb{E}\{y^4\} \tag{6.18}$$

if the source of interest has a fourth-order cumulant with sign ε. Note that $\Upsilon^o_{\mathrm{m}}(y)$ can be seen as the extraction counterpart of the separation contrast $\Upsilon^o_{\mathrm{M}}(\mathbf{y})$ in Eq. (6.12). Again, orthogonal contrasts (6.17) and (6.18) do not present spurious local extrema for infinite sample size under the noiseless ICA model.

Similar contrasts for source extraction relying on marginal negentropy and related cumulant-based approximations are put forward and analyzed in [26]. Such contrasts are naturally extended to the simultaneous extraction of $M \leqslant N$ sources, but the absence of spurious local extrema is not proven.

6.3.4 Nonlinear function approximations

Cumulant-based approximations are implicitly based on polynomial functions, which usually grow fast with the signal amplitude. As a result, higher-order cumulants, including kurtosis, are rather sensitive to outliers, so that only a few possibly erroneous observations can have a drastic effect on the estimation quality. This limitation can be surmounted by using suitably chosen nonlinear functions that do not grow too fast; for example, the absolute value [45].

According to this idea, cumulant-based approximation (6.15) could be replaced by a number of general nonlinear functions, as proposed in [46]. Let us assume that we have fixed a number of nonlinear functions $f_k(\cdot), k = 1, \ldots, K$. The starting point in [46] was to compute a lower bound for negentropy, assuming we set the expectations $\mathbb{E}\{f_k(y)\}$ to some values. In practice, fixing the expectations means that we compute them from observed data; denote as $E_k = \sum_t f_k(y(t))$ these sample expectations. It was shown in [46] that, as a first-order approximation, a lower bound for negentropy can be obtained as

$$\sum_{k=1}^{K} c_k \left(E_k - \mathbb{E}\{f_k(\nu)\}\right)^2 \tag{6.19}$$

where ν is a standardized Gaussian variable, and y is standardized to zero mean and unit variance as well. The c_k are some constants which depend on the nonlinearities $f_k(\cdot)$ chosen. An intuitive interpretation of (6.19) is that the differences of the expectations of

$f_k(\cdot)$ for the data and a Gaussian variable give a reasonable measure of non-Gaussianity. Since negentropy is a measure of non-Gaussianity, it is not surprising that Eq. (6.19) is a valid lower bound. In [46] this expression was further shown to provide not just a lower bound but indeed a good approximation for suitably chosen functions $f_k(\cdot)$.

6.4 ITERATIVE SEARCH ALGORITHMS: GENERALITIES

To estimate the sources or independent components, most BSS/ICA methods rely on the optimization of functionals, such as contrasts (section 6.3), quantifying some property of the separator output. Due to the typical complexity of the functions involved, this maximization problem is often not tractable algebraically and iterative techniques are thus necessary. This section recalls the basic concepts of iterative optimization. The iterative BSS/ICA algorithms surveyed in this chapter rely, in one way or another, on these classical concepts.

6.4.1 Batch methods

Contrasts are functionals of the separator-output distributions but, with some abuse of notation, they can also be written in terms of the separating system. As a result, our objective can generally be expressed as the maximization of a real-valued function $\Upsilon(\mathbf{b})$ of a vector $\mathbf{b} = [b_1, b_2, \ldots, b_P]^{\mathrm{T}} \in \mathbb{R}^P$, which typically represents a *spatial filter* for the extraction of a single source, that is, a colum vector of matrix $\mathbf{B}^{\mathrm{T}}$. Results can easily be extended to the case of a matrix argument, such as the whole separating matrix $\mathbf{B}$, by a suitable redefinition of the scalar product (e.g. based on the Frobenius norm) as in section 6.7 [2,14]. For simplicity, we restrain the development to the real-valued case; extensions to the complex-case are often possible with minor modifications and suitable complex calculus definitions [7]. If $\Upsilon(\cdot)$ is a contrast, the extracting filter can be found as

$$\mathbf{b}_{\mathrm{o}} = \arg\max_{\mathbf{b}} \Upsilon(\mathbf{b}).$$

For the sake of conciseness and clarity, only function maximization is addressed, the case of minimization accepting totally analogous derivations. For orthogonal contrasts, the maximization is carried out with respect to unitary matrix $\mathbf{Q}$ or its column vectors $\{\mathbf{q}_i\}_{i=1}^N$ in (6.3).

As recalled in section 6.3, function $\Upsilon(\cdot)$ generally involves mathematical expectations. The so-called *batch*, *windowed* or *off-line methods* operate on a *block* of T observed vector samples, $\mathbf{X}_T = [\mathbf{x}(1), \mathbf{x}(2), \ldots, \mathbf{x}(T)]$, and estimate the ensemble averages by sample averages over the observed signal block. For instance, the gradient of orthogonal contrast (6.18) with respect to the extracting vector $\mathbf{q}$ is given, up to an irrelevant constant factor, by:

$$\nabla \Upsilon_{\mathrm{m}}^{o}(\mathbf{q}) = \varepsilon \mathbb{E}\{y^3 \mathbf{z}\} \tag{6.20}$$

with $y = \mathbf{q}^{\mathrm{T}}\mathbf{z}$, and can easily be approximated by its sample estimate over the observed signal block as

$$\nabla \Upsilon_{\mathrm{m}}^{o}(\mathbf{q}) \approx \varepsilon \frac{1}{T}\sum_{t=1}^{T} y(t)\mathbf{z}(t) = \frac{1}{T}\mathbf{Z}_T \mathbf{y}_T^{(3)} \tag{6.21}$$

where $\mathbf{Z}_T = [\mathbf{z}(1), \mathbf{z}(2), \ldots, \mathbf{z}(T)]$ and $\mathbf{y}_T^{(3)} = [y^3(1), y^3(2), \ldots, y^3(T)]^{\mathrm{T}}$. The generic iterative optimization techniques described in this section are particularly suited to the batch mode of operation.

Classical results on iterative optimization are based on the second-order Taylor series expansion of function $\Upsilon(\cdot)$ in the vicinity of point $\mathbf{b}$:

$$\Upsilon(\mathbf{b}^+) \approx \Upsilon(\mathbf{b}) + \nabla\Upsilon(\mathbf{b})^{\mathrm{T}}(\mathbf{b}^+ - \mathbf{b}) + \frac{1}{2}(\mathbf{b}^+ - \mathbf{b})^{\mathrm{T}}\mathbf{H}_{\Upsilon}(\mathbf{b})(\mathbf{b}^+ - \mathbf{b}). \tag{6.22}$$

In the above equation, the *gradient vector* is given by

$$\nabla\Upsilon(\mathbf{b}) = \left[\frac{\partial\Upsilon(\mathbf{b})}{\partial b_1}, \frac{\partial\Upsilon(\mathbf{b})}{\partial b_2}, \ldots, \frac{\partial\Upsilon(\mathbf{b})}{\partial b_P}\right]^{\mathrm{T}}$$

and $\mathbf{H}_{\Upsilon}(\mathbf{b}) \in \mathbb{R}^{P\times P}$ is the *Hessian matrix* of $\Upsilon(\cdot)$ computed at $\mathbf{b}$. The (i, j)-entry of this matrix contains the second-order derivatives of $\Upsilon(\cdot)$, defined as:

$$[\mathbf{H}_{\Upsilon}(\mathbf{b})]_{ij} = \frac{\partial^2\Upsilon(\mathbf{b})}{\partial b_i \partial b_j}.$$

The first-order necessary condition for $\mathbf{b}_0$ to be a *stationary* or *equilibrium point* of function $\Upsilon(\cdot)$ is that $\nabla\Upsilon(\mathbf{b}_0) = \mathbf{0}$. The second-order necessary condition states that, for a stationary point $\mathbf{b}_0$ to be a local maximum, matrix $\mathbf{H}_{\Upsilon}(\mathbf{b}_0)$ must be negative definite or, equivalently, the real parts of all its eigenvalues must be negative. This condition is easily derived from Taylor expansion (6.22).

6.4.1.1 Gradient search

Gradient-based search is probably the most basic technique for iterative optimization. It relies on the notion that the gradient vector is orientated in the direction of maximal local variation of the function at a given point $\mathbf{b}$. As a result, updating vector $\mathbf{b}$ in the direction of the gradient

$$\mathbf{b}^+ = \mathbf{b} + \mu\nabla\Upsilon(\mathbf{b}) \tag{6.23}$$

ensures that $\Upsilon(\mathbf{b}^+) > \Upsilon(\mathbf{b})$ if μ is sufficiently small. To see this, it suffices to express $\Upsilon(\mathbf{b}^+)$ in terms of its first-order Taylor series expansion (6.22) when $\mathbf{b}^+$ is given by (6.23):

$$\Upsilon(\mathbf{b}^+) \approx \Upsilon(\mathbf{b}) + \nabla\Upsilon(\mathbf{b})^{\mathrm{T}}(\mathbf{b}^+ - \mathbf{b}) = \Upsilon(\mathbf{b}) + \mu\nabla\|\Upsilon(\mathbf{b})\|^2.$$

The term μ is referred to as *step size*, *learning rate* or *adaption coefficient*. It determines a trade-off between convergence speed and solution accuracy. Convergence is usually faster for large μ, but the algorithm may not stabilize close enough to a valid solution, a phenomenon known as *misadjustment*; if μ is too large, the algorithm may even move away from the solution and not converge at all. Conversely, a small value of μ may guarantee convergence, but at the expense of slowing down the algorithm. Hence, an adequate choice of this parameter is crucial in tuning the performance of gradient-based methods. The topic of step-size selection will be addressed in section 6.11. The local convergence of gradient algorithms is linear: for two successive updates $\mathbf{b}$ and $\mathbf{b}^+$ in the vicinity of a stationary point $\mathbf{b}_0$, we have that $||\mathbf{b}^+ - \mathbf{b}_0|| < ||\mathbf{b} - \mathbf{b}_0||$.

6.4.1.2 Newton search

The main limitations of gradient-based optimization are the choice of step-size parameter μ and the often slow convergence. Newton algorithms can surmount these drawbacks at the expense of an increased cost due to the Hessian matrix computation. Again, the starting point of this iterative optimization technique is Taylor expansion (6.22). Update $\mathbf{b}^+$ is obtained as the maximizer of this approximation. The first-order necessary condition is obtained by cancelling the gradient with respect to $(\mathbf{b}^+ - \mathbf{b})$, given by $\nabla\Upsilon(\mathbf{b}) + \mathbf{H}_\Upsilon(\mathbf{b})(\mathbf{b}^+ - \mathbf{b})$. This leads to the Newton update:

$$\mathbf{b}^+ = \mathbf{b} - \mathbf{H}_\Upsilon^{-1}(\mathbf{b})\nabla\Upsilon(\mathbf{b}). \tag{6.24}$$

Newton algorithms show quadratic convergence: near a stationary point $\mathbf{b}_0$, two consecutive updates $\mathbf{b}$ and $\mathbf{b}^+$ verify $||\mathbf{b}^+ - \mathbf{b}_0|| < ||\mathbf{b} - \mathbf{b}_0||^2$. Remark that in gradient search the parameters must be adjusted in the "uphill" or "downhill" direction depending on whether the objective function is to be maximized or minimized, respectively. This ascent/descent choice is specified by the sign before the update term in Eq. (6.23). The Newton method, by contrast, searches for both local maxima and minima without distinction.

6.4.1.3 Constrained optimization

As recalled in section 6.3, some contrast functions – so-called orthogonal – need to be optimized after prewhitening subject to a constraint in the extracting vector such as $||\mathbf{q}|| = 1$. The Lagrange multipliers technique can then be used to perform the constrained optimization. To this end, a composite function, the so-called *Lagrangian*, is made up of the original contrast and an extra term expressing the constraint:

$$\tilde{\Upsilon}(\mathbf{q}) = \Upsilon(\mathbf{q}) + \frac{\beta}{2}(1 - ||\mathbf{q}||^2). \tag{6.25}$$

This function must be maximized with respect to $\mathbf{q}$ and *Lagrange multiplier* β. Cancelling the derivative of (6.25) with respect to β enforces the constraint, whereas its gradient with respect to $\mathbf{q}$ reads:

$$\nabla\tilde{\Upsilon}(\mathbf{q}) = \nabla\Upsilon(\mathbf{q}) - \beta\mathbf{q}. \tag{6.26}$$

Hence, the optimum pair $(\mathbf{q}_0, \beta_0)$ verifies the Lagrange condition:

$$\beta_0 \mathbf{q}_0 = \nabla \Upsilon(\mathbf{q}_0). \tag{6.27}$$

Replacing this equality into the constraint $\mathbf{q}_0^{\mathrm{T}} \mathbf{q}_0 = 1$ yields the optimum value of β:

$$\beta_0 = \nabla \Upsilon(\mathbf{q}_0)^{\mathrm{T}} \mathbf{q}_0. \tag{6.28}$$

On the other hand, the Hessian of Lagrangian (6.25) is given by:

$$\mathbf{H}_{\tilde{\Upsilon}}(\mathbf{q}) = \mathbf{H}_{\Upsilon}(\mathbf{q}) - \beta \mathbf{I} \tag{6.29}$$

where $\mathbf{I}$ is the identity matrix of appropriate dimensions. Accordingly, the eigenvalues of $\mathbf{H}_{\Upsilon}$ at a stationary point $\mathbf{q}_0$ are modified by β_0 given in Eq. (6.28). A sufficient condition for $\mathbf{q}_0$ to be a local maximum of (6.25) is that the real parts of the modified eigenvalues be negative. However, one should not forget that the optimization is carried out subject to the unit-norm constraint. To satisfy this constraint, the differential variations around $\mathbf{q}_0$ must be kept orthogonal to $\mathbf{q}_0$. Hence, the necessary condition of local maximality only involves the modified Hessian eigenvalues associated with the eigenvectors in the orthogonal complement of $\mathbf{q}_0$. Optimization under other constraints may lead to different conditions.

Using Eqs (6.26)–(6.29), constrained gradient- and Newton-based algorithms can be derived along the lines of the preceding sections; an example of constrained Newton algorithm is the FastICA method reviewed in section 6.10.

6.4.2 Stochastic optimization

A number of iterative methods for BSS/ICA update the separating system upon reception of a new observed vector sample. These kinds of iterative methods operating on a sample-by-sample basis are commonly referred to as *adaptive*, *recursive*, *stochastic*, *neural* or *on-line* implementations. They are based on an update rule with the general form:

$$\mathbf{b}(t+1) = \mathbf{b}(t) + \mu \boldsymbol{\phi}(\mathbf{b}(t), \mathbf{x}(t+1)). \tag{6.30}$$

Function $\boldsymbol{\phi}(\cdot,\cdot)$ quantifies the variation induced in the separating system by the arrival of a new observation sample at instant $(t+1)$. Note that block methods iterate over the same data block, whereas adaptive methods perform one iteration every time instant; hence the different notations for the current and updated vectors used in Eq. (6.30) compared to Eqs (6.23) and (6.24).

The convergence characteristic of stochastic algorithms can be analyzed with the aid of standard tools such as the *ordinary differential equation* (ODE) *method* [5]. The ODE method approximates the average trajectory of the discrete-time parameter $\mathbf{b}$ in (6.30) by the solution of the first-order differential equation

$$\frac{\partial \mathbf{b}}{\partial \tau} = \boldsymbol{\Phi}(\mathbf{b}) \tag{6.31}$$

evaluated at time instants $\tau_k = k\mu$, where τ represents the continuous-time variable. Function $\mathbf{\Phi}(\mathbf{b}) = [\Phi_1(\mathbf{b}), \Phi_2(\mathbf{b}), \ldots, \Phi_p(\mathbf{b})]^{\mathrm{T}}$ is called *mean field*, and is given by

$$\mathbf{\Phi}(\mathbf{b}) = \mathbb{E}\{\boldsymbol{\phi}(\mathbf{b}, \mathbf{x})\}$$

where the expectation is taken with respect to the distribution of $\mathbf{x}$. ODE approximation (6.31) is valid as long as the step size μ is sufficiently small and function $\mathbf{\Phi}(\cdot)$ verifies certain regularity conditions. The *equilibrium points* or *attractors* of the algorithm, denoted as $\mathbf{b}_0$, are the roots of the mean field:

$$\mathbf{\Phi}(\mathbf{b}_0) = 0.$$

The stability of an equilibrium point can be studied by means of the first-order Taylor expansion of $\Phi(\mathbf{b})$ around $\mathbf{b}_0$, given by

$$\mathbf{\Phi}(\mathbf{b}) \approx \mathbf{\Phi}(\mathbf{b}_0) + \mathbf{J}_{\mathbf{\Phi}}(\mathbf{b}_0)(\mathbf{b} - \mathbf{b}_0) = \mathbf{J}_{\mathbf{\Phi}}(\mathbf{b}_0)(\mathbf{b} - \mathbf{b}_0)$$

where $\mathbf{J}_{\mathbf{\Phi}}(\mathbf{b}_0)$ is the Jacobian matrix of $\mathbf{\Phi}(\cdot)$ computed at $\mathbf{b}_0$, with elements

$$[\mathbf{J}_{\mathbf{\Phi}}(\mathbf{b})]_{ij} = \frac{\partial \Phi_i(\mathbf{b})}{\partial b_j}.$$

Hence, for an equilibrium point $\mathbf{b}_0$ to be *locally asymptotically stable*, matrix $\mathbf{J}_{\mathbf{\Phi}}(\mathbf{b}_0)$ must be negative definite or, equivalently, the real parts of all its eigenvalues must be negative. Thus, the Jacobian $\mathbf{J}_{\mathbf{\Phi}}(\mathbf{b}_0)$ in stochastic optimization plays a similar role to that of the Hessian matrix in batch optimization (section 6.4.1).

The derivation of stochastic algorithms is particularly simple when contrast $\Upsilon(\cdot)$ can be written as the expectation of a function $\phi(y)$:

$$\Upsilon(y) = \mathbb{E}\{\phi(y)\}. \tag{6.32}$$

This is the case, for instance, for orthogonal contrast $\Upsilon^{o}_{\mathrm{m}}(y)$ in Eq. (6.18). The gradient of the contrast is also given by an expectation: $\nabla\Upsilon(y) = \mathbb{E}\{\boldsymbol{\phi}(y)\}$, with $\boldsymbol{\phi}(y) = \nabla\phi(y)$. Hence, the stochastic algorithm is obtained from the corresponding batch gradient algorithm

$$\mathbf{b}^{+} = \mathbf{b} + \mu\mathbb{E}\{\boldsymbol{\phi}(\mathbf{b})\}$$

by dropping the expectation operator and adjusting the parameters every time instant:

$$\mathbf{b}(t+1) = \mathbf{b}(t) + \mu\boldsymbol{\phi}(\mathbf{b}(t)).$$

Note that function $\boldsymbol{\phi}(\mathbf{b})$ also depends on $\mathbf{x}$ through the relationship $y = \mathbf{b}^{\mathrm{T}}\mathbf{x}$.

Finally, the same comments as in section 6.4.1.1 can be made concerning the importance of the step size coefficient μ in the trade-off between convergence speed and accuracy of stochastic algorithms.

6.4.3 Batch or adaptive estimates?

Most of the BSS/ICA algorithms surveyed in section 6.6 belong to the family of neural techniques. The main motivation behind adaptive methods is their ability to track variations of the mixing system in non-stationary environments, if the non-stationarity is mild enough. We have just seen that if the contrast function can be written as the expectation of some function of the separator output, this mode of operation is obtained by dropping the expectation operator and using a one-sample estimate instead. For instance, the stochastic approximation to gradient (6.20), would read

$$\nabla \Upsilon_{\mathrm{m}}^{o}(\mathbf{q}) \approx \varepsilon y^3(t)\mathbf{z}(t)$$

resulting in the stochastic update:

$$\mathbf{q}(t+1) = \mathbf{q}(t) + \mu\varepsilon y^3(t)\mathbf{z}(t) \tag{6.33}$$

with $y(t) = \mathbf{q}(t)^{\mathrm{T}}\mathbf{z}(t)$. These single-sample estimates often lead to rather slow convergence and compromise the tracking capabilities of adaptive techniques if the mixing system varies too rapidly.

On the other hand, sample approximation (6.21) would result in the gradient-based block algorithm

$$\mathbf{q}^{+} = \mathbf{q} + \mu\varepsilon \frac{1}{T}\sum_{i=1}^{T} y^3(t)\mathbf{z}(t) \tag{6.34}$$

where $y(t) = \mathbf{q}^{\mathrm{T}}\mathbf{z}(t)$, $t = 1, 2, \ldots, T$. Contrary to common belief, block implementations are not necessarily more costly than their adaptive counterparts, as they use more effectively the information contained in the observed signal block: the sample averaging used in block methods leads to more accurate gradient approximations than sample-by-sample estimates. On-line procedures use a given sample only once and, as a result, their asymptotic performance can never improve that of their batch counterparts, where all available data are reused as many times as required [1]. By processing a new data block upon convergence of the iterative algorithm on the current block, batch methods can still track variations in the mixing system as long as the channel remains stationary over the observed window length.

Indeed, these two modes of operation are not mutually exclusive but may actually be combined together in different fashions. For instance, a batch method may iterate over a block of signal samples while the observation window slides and refreshes the data block every few iterations. Even in the case where the whole data set is available before starting the iterations, many researchers have found it useful to seek a compromise between purely adaptive and purely block updates by using "mini-batches". According to this idea, the sample average in (6.34) would be computed over a small number of data points; for example, if the observation window consists of $T = 10^4$ samples, one could use 100 mini-batches of 100 samples each. In this case, random fluctuations in the gradient due to the small block size may be offset by the reduced complexity needed for the gradient computation at each iteration. On the other hand, it is often faster to compute one gradient update on a block of 100 samples than 100 sample-by-sample updates, especially in scientific computation environments, such as Matlab™, optimized for vector-wise operations.

Both types of iterative algorithms (batch and stochastic) are considered in the sequel, but one should keep in mind that most adaptive methods can also be implemented in batch mode, and vice versa, or even combined as explained above.

6.5 ITERATIVE WHITENING

Orthogonal contrasts require the data to be prewhitened. In the batch case, whitening is usually performed by conventional methods for computing the *eigenvalue decomposition* (EVD) of the data covariance matrix $\mathbf{R}_x = \mathbf{U}\boldsymbol{\Lambda}\mathbf{U}^T$, so that the whitening matrix in Eq. (6.2) can be identified as $\mathbf{W} = \boldsymbol{\Lambda}^{-\frac{1}{2}}\mathbf{U}^T$. This process is theoretically equivalent to computing the singular value decomposition of the (zero-mean) data matrix. If these highly efficient algorithms for EVD are not available, the following simple iterative algorithm can also be used instead [47]:

1. Normalize $\mathbf{W} \leftarrow \mathbf{W}/\sqrt{\|\mathbf{W}\mathbf{R}_x\mathbf{W}^T\|}$.
2. Repeat until convergence: $\mathbf{W} \leftarrow \frac{3}{2}\mathbf{W} - \frac{1}{2}\mathbf{W}\mathbf{R}_x\mathbf{W}^T\mathbf{W}$.

The norm in Step 1 is any proper matrix norm such as the spectral 2-norm, but not the Frobenius norm (which is not really matrix norm in the sense of linear algebra, but a vector norm applied on a matrix). However, this iterative method cannot compete with the speed of state-of-the-art EVD algorithms implemented in scientific computation software.

In the adaptive case, several possibilities exist. Obviously, one can estimate the covariance matrix on-line, and compute the EVD off-line using any of the methods given above, but a fully recursive EVD estimator can also be employed [42]. Adaptive implementations based on Gram-Schmidt orthogonalization, Cholesky decompositions of the data covariance matrix and linear prediction have also been proposed [22,28,63]. A widespread adaptive procedure is based on measuring the distance to "whiteness" with the objective function [14]:

$$\Psi(\mathbf{W}) = \frac{1}{2}(-\log\det(\mathbf{R}_z) + \mathrm{trace}(\mathbf{R}_z) - N)$$

where $\mathbf{R}_z = \mathbf{W}\mathbf{R}_x\mathbf{W}^T$; this is the Kullback-Leibler divergence between two zero-mean Gaussian distributions with covariance matrices $\mathbf{R}_z$ and $\mathbf{I}$, respectively. The above function can be minimized by the following stochastic iteration [14,57]:

$$\mathbf{W}(t+1) = \mathbf{W}(t) + \mu \left(\mathbf{I} - \mathbf{z}(t)\mathbf{z}(t)^{\mathrm{T}}\right) \mathbf{W}(t) \tag{6.35}$$

which is derived from the relative gradient (section 6.7). Other batch and adaptive whitening algorithms are derived and discussed in [18, Chapter 4].

6.6 CLASSICAL ADAPTIVE ALGORITHMS

Seminal neural-type methods for ICA were not explicitly based on contrasts but were mainly derived from heuristic ideas. In these methods, the update term is made up of nonlinear functions of the separator output. Due to the source independence, the average update terms cancel out at valid separation solutions. Non-separating solutions may also be attractors of the algorithm, so that additional conditions must be verified to render valid attractors stable and *spurious attractors* unstable. Such conditions depend on the choice of nonlinearities relative to the source distributions. Some of these classical methods are surveyed in sections 6.6.1 and 6.6.2. Other early adaptive algorithms, however, do rely on contrast maximization, either for joint separation (section 6.6.3) or for sequential extraction based on deflation (to be be discussed in section 6.9). Geometric ideas can also be exploited to perform adaptive source separation (section 6.6.4). All these algorithms use the conventional update rules recalled in section 6.4.

6.6.1 Hérault-Jutten algorithm

The pioneering BSS algorithm by Hérault and Jutten [43,50,51] is characterized by the input-output relationship:

$$\mathbf{y}(t) = (\mathbf{I} + \mathbf{C}(t))^{-1}\,\mathbf{x}(t). \tag{6.36}$$

This relationship stems from neuro-mimetic considerations leading to the *feedback architecture* $\mathbf{y}(t) = \mathbf{x}(t) - \mathbf{C}(t)\mathbf{y}(t)$. The elements $c_{ij}(t)$, $i \neq j$, of matrix $\mathbf{C}(t)$ are updated according to the adaptation rule:

$$c_{ij}(t+1) = c_{ij}(t) + \mu f\left(y_i(t)\right) g\left(y_j(t)\right) \tag{6.37}$$

where $f(\cdot)$ and $g(\cdot)$ are suitable nonlinear functions. Hence, according to section 6.4.2, the Hérault-Jutten (HJ) network seeks the matrix $\mathbf{C}$ cancelling the expectations $\mathbb{E}\{f(y_i(t))g(y_j(t))\}$, $i \neq j$. To impose enough constraints, functions $f(\cdot)$ and $g(\cdot)$ must be different. If a separating matrix leads to statistically independent sources, we find that

$$\mathbb{E}\{f(y_i)g(y_j)\} = \mathbb{E}\{f(y_i)\}\mathbb{E}\{g(y_j)\} \qquad i \neq j. \tag{6.38}$$

Taylor series expansions of functions $f(\cdot)$ and $g(\cdot)$ readily show that the right-hand side of Eq. (6.38) is equal to the product of higher-order cross-moments of the separator outputs. Nonlinearities have thus the ability to provide an implicit measure of higher-order independence. If functions $f(\cdot)$ and $g(\cdot)$ are odd, expression (6.38) involves moments of odd order only, and cancels if the sources are symmetrically distributed. Under the above conditions, it follows that any valid separating matrix is an equilibrium point of the HJ algorithm. The algorithm does not require prewhitening. Experimentally, its behavior is characterized by an initial phase of fast convergence to uncorrelated outputs followed by a second phase of slow convergence to independent components [43].

The convergence properties of the HJ network are studied at length in [24,29,76], at least for the case of $N = 2$ sources. Not surprisingly, convergence mainly depends on the choice of nonlinearities with respect to the source probability distributions. If $f(x) = x^3$ and $g(x) = x$, the equilibrium points associated with source separation solutions are stable if and only if

$$\mathbb{E}\{s_1^4\}\mathbb{E}\{s_2^4\} < 9\mathbb{E}\{s_1^2\}^2\mathbb{E}\{s_2^2\}^2. \tag{6.39}$$

Consequently, the HJ network with such nonlinearities is suitable for the separation of sub-Gaussian sources. If functions $f(\cdot)$ and $g(\cdot)$ are permuted, that is, if we take $f(x) = x$ and $g(x) = x^3$, the stability condition becomes

$$\mathbb{E}\{s_1^4\}\mathbb{E}\{s_2^4\} > 9\mathbb{E}\{s_1^2\}^2\mathbb{E}\{s_2^2\}^2 \tag{6.40}$$

thus allowing the separation of super-Gaussian sources. Stability conditions for general nonlinearities are given in [76].

Expression (6.36) describes a feedback or recurrent neural network architecture with linear transfer function $(\mathbf{I} + \mathbf{C}(t))^{-1}$. As a result, matrix $(\mathbf{I} + \mathbf{C}(t))$ needs to be inverted at each iteration. To avoid matrix inversion, a *feed-forward* or *direct structure*, characterized by $\mathbf{y}(t) = \mathbf{B}(t)\mathbf{x}(t)$, is proposed and analyzed in [59]. The entries of the separation matrix $\mathbf{B}$ are controlled by the HJ algorithm:

$$b_{ij}(t+1) = b_{ij}(t) + \mu f\left(y_i(t)\right) g\left(y_j(t)\right) \qquad i \neq j \tag{6.41}$$

whereas $b_{ii} = 1$, $1 \leqslant i \leqslant P$. A similar type of structure had been proposed a few years earlier in the context of cross-polarization suppression in satellite communications [3] and later extended to more general multiuser digital communications scenarios [32].

6.6.2 Self-normalized networks

The performance of the HJ algorithm is rather dependent on the mixing matrix conditioning and the initial guess of the separating matrix. A reason is that updates (6.37) and (6.41) lack control over the signal power of the separator output components. As a result, the nonlinear functions may actually be working in their linear regime if the output power is too small, thus possibly driving the algorithm to non-separating solutions. To circumvent this limitation, the so-called *self-normalized networks* for source separation are put forward by Cichocki *et al.* in [19]. The update rule of these algorithms can be expressed as:

$$\mathbf{B}(t+1) = \mathbf{B}(t) + \mu \left\{ \mathbf{I} - \mathbf{f}(\mathbf{y}(t)) \mathbf{g}^{\mathrm{T}}(\mathbf{y}(t)) \right\} \tag{6.42}$$

with $\mathbf{f}(\mathbf{y}) = [f(y_1), f(y_2), \ldots, f(y_N)]^{\mathrm{T}}$ and $\mathbf{g}(\mathbf{y}) = [g(y_1), g(y_2), \ldots, g(y_N)]^{\mathrm{T}}$. The main difference compared to Macchi's direct form of the HJ network (6.41) lies in the inclusion of updating terms

$$b_{ii}(t+1) = b_{ii}(t) + \mu \left\{ 1 - f(y_i(t))\, g(y_i(t)) \right\} \tag{6.43}$$

for the diagonal entries of $\mathbf{B}$. These terms guarantee that, at convergence, the scales of the separator outputs are normalized so as to verify $\mathbb{E}\{f(y)g(y)\} = 1$. The analysis carried out by Deville [30] reveals that the incorporation of the scaling terms does not alter the stability conditions of the original HJ networks (6.39) and (6.40). Extensions of update rule (6.42) are proposed in [19].

The output scale normalization enforced by Eq. (6.43) results in a constant output level, regardless of the actual short-term power of the sources. Although this feature enhances the convergence of the HJ network in ill-conditioned scenarios, it may present problems in non-stationary environments where the sources alternate between low- and high-power periods, such as in speech signals. A modified algorithm avoiding this problem is suggested in [31], which, for the direct structure reads:

$$b_{ij}(t+1) = b_{ij}(t) + \mu \frac{f(y_i(t))}{\sqrt{\mathbb{E}\{f^2(y_i)\}}} \frac{g(y_j(t))}{\sqrt{\mathbb{E}\{g^2(y_j)\}}} \qquad i \neq j.$$

The denominator terms, $N_{f,i} \stackrel{\text{def}}{=} \mathbb{E}\{f^2(y_i)\}$ and $N_{g,j} \stackrel{\text{def}}{=} \mathbb{E}\{g^2(y_j)\}$ are adaptively estimated via standard updates of the form

$$\begin{aligned} N_{f,i}(t+1) &= N_{f,i}(t) + \eta \left\{ f^2(y_i(t)) - N_{f,i}(t) \right\} \\ N_{g,j}(t+1) &= N_{g,j}(t) + \eta \left\{ g^2(y_j(t)) - N_{g,j}(t) \right\}. \end{aligned}$$

As in the original HJ network, the modified algorithms summarized in the preceding paragraphs do not require prewhitening.

6.6.3 Adaptive algorithms based on contrasts

In the above algorithms, the mean updating terms cancel if the separator-output signals are statistically independent. In other words, the set of equilibrium points contains the valid solutions separating independent sources. However, other non-separating transformations may also cancel the mean update terms; these are the spurious equilibria. The conditions for spurious equilibria to be locally unstable depend on the choice of nonlinearities and their relation to the source statistics, and determine the applicability scope of a given algorithm. For instance, we have seen that, in the two-signal case, the sources must verify condition (6.39) if $f(x) = x^3$ and $g(x) = x$, or condition (6.40) if $f(x) = x$ and $g(x) = x^3$, to guarantee that the only locally stable equilibria of algorithm (6.37) are valid separating solutions.

Contrast functions constitute an alternative methodology for deriving adaptive separation algorithms. By virtue of their three essential properties, namely, invariance, domination and discrimination, contrast functions are globally maximized by separating solutions only (Chapter 3). As a result, their maximization through one of the iterative procedures summarized in section 6.4 raises a very natural approach to source separation. This approach is also systematic in that iterative algorithms of this kind can easily be associated with each new contrast. However, one should keep in mind that, although the *global* maximization of a contrast guarantees source separation, an iterative algorithm employed for its maximization may still get trapped in local extrema.

6.6.3.1 Infomax

One of the earliest methods relying on some kind of contrast maximization is Bell and Sejnowski's neural algorithm based on the *informax* principle [4]. To derive its update equation, the entropy (6.6) can be expressed as:

$$\Upsilon_{\mathrm{IM}}(\mathbf{y}) = H\{\mathbf{r}(\mathbf{Bx})\} = H\{\mathbf{x}\} + \mathbb{E}\{\phi_{\mathrm{IM}}(\mathbf{y})\}$$

where $\phi_{\mathrm{IM}}(\mathbf{y}) \stackrel{\mathrm{def}}{=} \log|\det \mathbf{J}|$. Symbol $\mathbf{J}$ stands for the Jacobian matrix of the transformation $\mathbf{x} \mapsto \mathbf{r}(\mathbf{Bx})$, whose (i, j)-element is given by

$$[\mathbf{J}]_{ij} = \frac{\partial r_i(y_i)}{\partial x_j} = \frac{\partial r_i(y_i)}{\partial y_i}\frac{\partial y_i}{\partial x_j} = r_i'(y_i)b_{ij}.$$

Hence:

$$\phi_{\mathrm{IM}}(\mathbf{y}) = \sum_{i=1}^{N} \log r_i'(y_i) + \log|\det \mathbf{B}|.$$

Since the separator-input entropy $H\{\mathbf{x}\}$ is constant, maximizing $\Upsilon_{\mathrm{IM}}(\mathbf{y})$ is thus equivalent to maximizing the expectation $\mathbb{E}\{\phi_{\mathrm{IM}}(\mathbf{y})\}$. As explained at the end of section 6.4.2, the corresponding stochastic gradient update can easily be obtained by dropping the expectation operator in the gradient, leading to

$$\nabla\phi_{\mathrm{IM}}(\mathbf{y}) = \boldsymbol{\varphi}_{\mathrm{IM}}(\mathbf{y})\mathbf{x}^{\mathrm{T}} + (\mathbf{B}^{\mathrm{T}})^{-1}.$$

with $[\boldsymbol{\varphi}_{\mathrm{IM}}(\mathbf{y})]_i = \left(\log r_i'(y_i)\right)'$. As a result, we can finally write:

$$\mathbf{B}(t+1) = \mathbf{B}(t) + \mu\left\{\boldsymbol{\varphi}_{\mathrm{IM}}(\mathbf{y})\mathbf{x}^{\mathrm{T}} + \left(\mathbf{B}(t)^{\mathrm{T}}\right)^{-1}\right\} \tag{6.44}$$

Remark that this algorithm requires the inversion of $\mathbf{B}$ at each iteration. This matrix inversion can be avoided by using prewhitening or, as we will see in section 6.7.3, the relative gradient even without prewhitening [cf. Eqs (6.44) and (6.61)]. On the other hand, if nonlinearities $r_i(\cdot)$ are matched to the source cdfs, $1 \leqslant i \leqslant N$, vector $\boldsymbol{\varphi}_{\mathrm{IM}}(\mathbf{y})$ contains the score functions of the sources. This observation reflects the link between the infomax and the maximum likelihood criteria noticed in [2,10] and recalled in section 6.3.1.

6.6.3.2 Adaptive algorithms based on Givens parameterization

Other early algorithms for orthogonal contrast maximization rely on the parameterization of the unitary separating matrix in terms of $N(N-1)/2$ Givens planar rotations [42, 63]:

$$\mathbf{Q}^{\mathrm{T}} = \prod_{k<\ell} \mathbf{Q}_{ij}^{\mathrm{T}}(\theta_{k\ell}) \tag{6.45}$$

where $\mathbf{Q}_{k\ell}^{\mathrm{T}}(\theta_{k\ell})$ is an identity matrix except for entries (k,k), (k,ℓ), (ℓ,k) and (ℓ,ℓ), $1 \leqslant k,\ell \leqslant N$, which are given by

$$\begin{bmatrix} \cos\theta_{k\ell} & -\sin\theta_{k\ell} \\ \sin\theta_{k\ell} & \cos\theta_{k\ell} \end{bmatrix}. \tag{6.46}$$

This Givens parameterization spares the separating matrix orthogonalization required after each iteration of an algorithm such as (6.33). Moreau and Macchi [63] derive adaptive algorithms of this type from contrasts (6.12)–(6.14); that for the fourth-order moment sum contrast $\Upsilon_{\mathrm{M}}^{\mathrm{o}}(y)$ [Eq. (6.12)] can be summarized as follows. Up to an irrelevant constant factor, the derivative of $\mathbb{E}\{y_i^4\}$ with respect to $\theta_{k\ell}$ is $\mathbb{E}\{y_i^3 \partial y_i / \partial \theta_{k\ell}\}$. Dropping the expectation operator thus leads to the stochastic update:

$$\theta_{k\ell}(t+1) = \theta_{k\ell}(t) + \mu\varepsilon \sum_{i=1}^{N} y_i^3(t) \frac{\partial y_i(t)}{\partial \theta_{k\ell}}.$$

The algorithm is completed by noticing that

$$\frac{\partial y_i(t)}{\partial \theta_{k\ell}} = \prod_{pr} \mathbf{Q}_{pr}^{\mathrm{T}}(\theta_{pr}) \frac{\partial \mathbf{Q}_{k\ell}^{\mathrm{T}}}{\partial \theta_{k\ell}} \prod_{p'r'} \mathbf{Q}_{p'r'}^{\mathrm{T}}(\theta_{p'r'}) \mathbf{z}(t)$$

where the products extend over indices $p \leqslant k$, $r < \ell$ and $p' \geqslant k$, $r' > \ell$, respectively. Matrix $\partial \mathbf{Q}_{k\ell}^{\mathrm{T}}(\theta_{k\ell}) / \partial \theta_{k\ell}$ is composed of zeros everywhere except for entries (k,k), (k,ℓ), (ℓ,k) and (ℓ,ℓ), which are given by the derivatives of (6.46). In the two-signal case, the problem reduces to the estimation of a single angle $\theta = \theta_{12}$, whose update rule becomes:

$$\theta(t+1) = \theta(t) - \mu\varepsilon \left(y_1^3(t) y_2(t) - y_1(t) y_2^3(t) \right). \tag{6.47}$$

It is interesting to note that this update aims to cancel the cross-cumulant combination $(\kappa^{y}_{1112} - \kappa^{y}_{1222})$, which is indeed null when the separator outputs become independent. A very similar algorithm is proposed in [42] to adaptively maximize the mutual information based contrast $\Upsilon^{o}_{\mathrm{MI}}(y)$ in Eq. (6.13). Notice that Givens parameterization (6.45) is reminiscent of the Jacobi technique used the in the tensor diagonalization approach to ICA (Chapter 5).

6.6.4 Adaptive algorithms based on centroids

In the two-signal case, the whitened observations are related to the sources through a Givens transformation characterized by a single parameter θ [cf. Eqs (6.3) and (6.45)–(6.46)]. Consider the following complex-valued weighted sum of the whitened-signal fourth-order cumulants [87]:

$$\xi_{z} \stackrel{\text{def}}{=} (\kappa^{z}_{1111} - 6\kappa^{z}_{1122} + 4\kappa^{z}_{2222}) + j4(\kappa^{z}_{1112} - \kappa^{z}_{1222}) \tag{6.48}$$

with $j = \sqrt{-1}$. Due to the multilinearity property of cumulants [61], it is easy to prove that *centroid* (6.48) fulfils the relationship

$$\xi_{z} = \gamma e^{-j4\theta}$$

where $\gamma \stackrel{\text{def}}{=} (\kappa_{1} + \kappa_{2})$ represents the source kurtosis sum. It follows that parameter θ can be estimated as:

$$\hat{\theta} = -\frac{1}{4}\angle\{\mathrm{sign}(\gamma)\xi_{z}\} \tag{6.49}$$

where the symbol $\angle\{\cdot\}$ denotes the angle of its complex argument with respect to the positive real axis. Again thanks to the multilinearity of cumulants, the source kurtosis sum can be estimated from the whitened observations as $\gamma = (\kappa^{z}_{1111} + 2\kappa^{z}_{1122} + \kappa^{z}_{2222})$. Although mainly derived from geometrical ideas exploiting cumulant properties, this method is shown to generalize to the case $\gamma \neq 0$ the approximate maximum likelihood estimator of [41], initially derived under the assumption of symmetric sources with identical distribution and kurtosis in the range $[0, 4]$.

An adaptive estimator can easily be developed by noticing that

$$\xi_{z} = \mathbb{E}\{v^{4}\} \qquad \gamma = \mathbb{E}\{|v|^{4}\} - 8 \qquad \text{with } v \stackrel{\text{def}}{=} (z_{1} + jz_{2})^{4}.$$

Hence, it suffices to estimate ξ_{z} and γ through standard on-line rules:

$$\begin{aligned} \xi_{z}(t+1) &= (1-\mu)\xi_{z}(t) + \mu v(t+1) \\ \gamma(t+1) &= (1-\mu)\gamma(t) + \mu(|v(t+1)| - 8) \end{aligned} \tag{6.50}$$

and then plug back the updated values into Eq. (6.49). In [88], this adaptive estimator is proven to converge to a stable separation solution as long as $\gamma \neq 0$. Note that update (6.47) implicitly seeks the cancellation of the imaginary part of the separator-output centroid, which is null for independent signals. However, according to (6.49), the relevant parameter allowing the estimation of θ is the orientation of centroid ξ_z, rather than its exact position. This orientation is accurately estimated by (6.50) in just a few iterations if the centroid is initialized at the origin of the complex plane. Moreover, this high convergence speed is very robust to the particular choice of the adaptation coefficient [88]. This adaptive procedure can be extended to the general scenario of more than two signals through a pairwise approach inspired by Givens parameterization (6.45).

The neural method presented in [70] exploits other geometrical ideas limited to sub-Gaussian sources with bounded support.

6.7 RELATIVE (NATURAL) GRADIENT TECHNIQUES

A popular family of adaptive algorithms for BSS/ICA is based on the so-called *relative* or *natural gradient*. Both concepts, independently developed by Cardoso [8,14,57] and Amari [1,2], respectively, are closely related [12]. Although derived from different standpoints, they lead to identical adaptive algorithms for BSS/ICA, so that the terms "relative" and "natural" can be considered as synonyms in this context. An algorithm of this type had also been proposed, without derivations, in [20].

The relative gradient approach capitalizes on the particular structure of observation model (6.1), whereby the source vector is left-multiplied by the mixing matrix to generate the observed vector. Taking advantage of this multiplicative structure enables the design of iterative algorithms with *uniform performance*. By virtue of this property, the separation quality depends only on the source distribution and is essentially made independent of the mixing matrix. Moreover, relative gradient algorithms inherit some of the nice properties of Newton methods (e.g. local isotropic convergence) at the computational cost of classical gradient methods [12,14]. Increased robustness to saddle areas is reportedly another desirable feature of these techniques [1].

More details about the relative gradient and related concepts can be found in Chapter 4. The relative Newton method, which is an extension of the relative gradient algorithms summarized next, is presented in section 10.5 in the context of sparse component analysis.

6.7.1 Relative gradient and serial updating

As recalled in section 6.4, the classical gradient quantifies first-order variations of a function $\Upsilon(\cdot)$ for small perturbations around a given point. In the matrix case, this is mathematically expressed by the Taylor expansion:

$$\Upsilon(\mathbf{B}+\mathbf{E}) = \Upsilon(\mathbf{B}) + \mathrm{trace}\left(\nabla\Upsilon(\mathbf{B})^{\mathrm{T}}\mathbf{E}\right) + o(\mathbf{E}).$$

Hence, taking $\mathbf{E} = \mu\nabla\Upsilon(\mathbf{B})$ with μ sufficiently small guarantees that $\Upsilon(\mathbf{B}+\mathbf{E}) > \Upsilon(\mathbf{B})$, which naturally leads to the conventional gradient-based update

$$\mathbf{B}^+ = \mathbf{B} + \mu\nabla\Upsilon(\mathbf{B}).$$

The *relative gradient*, which we denote as $\tilde{\nabla}\Upsilon(\mathbf{B})$, quantifies first-order variations of function $\Upsilon(\cdot)$ in the vicinity of $\mathbf{B}$ due to perturbations of the form $(\mathbf{I}+\mathbf{E})\mathbf{B}$, that is:

$$\Upsilon(\mathbf{B}+\mathbf{E}\mathbf{B}) = \Upsilon(\mathbf{B}) + \text{trace}\left(\tilde{\nabla}\Upsilon(\mathbf{B})^{\mathrm{T}}\mathbf{E}\right) + o(\mathbf{E}). \tag{6.51}$$

Consequently, if $\mathbf{E} = \mu\tilde{\nabla}\Upsilon(\mathbf{B})$, for small enough μ, we have $\Upsilon(\mathbf{B}+\mathbf{E}\mathbf{B}) > \Upsilon(\mathbf{B})$. The resulting relative gradient update rule is then:

$$\mathbf{B}^+ = \mathbf{B} + \mu\tilde{\nabla}\Upsilon(\mathbf{B})\mathbf{B} = \left(\mathbf{I} + \mu\tilde{\nabla}\Upsilon(\mathbf{B})\right)\mathbf{B}. \tag{6.52}$$

Such a rule is referred to as *serial updating*, since it is obtained by pre-multiplying the previous value of $\mathbf{B}$ by the term $\mathbf{I} + \mu\tilde{\nabla}\Upsilon(\mathbf{B})$. This is consistent with the multiplicative structure of model (6.1). Now, if functional $\Upsilon(\mathbf{B})$ is a function of the separator-output probability distribution only, we can write, with some abuse of notation, $\Upsilon(\mathbf{B}) = \Upsilon(\mathbf{y})$. Post-multiplying (6.52) by the mixing matrix $\mathbf{A}$ and denoting as $\mathbf{G} = \mathbf{B}\mathbf{A}$ the global mixing-unmixing system such that $\mathbf{y} = \mathbf{G}\mathbf{s}$, we have:

$$\mathbf{G}^+ = \left(\mathbf{I} + \mu\tilde{\nabla}\Upsilon(\mathbf{G}\mathbf{s})\right)\mathbf{G}.$$

Therefore, the global matrix and thus the separator output depend exclusively on the source distribution and are independent of the actual value of the mixing matrix. Obviously, the mixing matrix determines the initial value of $\mathbf{G}$ and $\mathbf{y}$, but does not alter the asymptotic characteristics of the algorithm. This desirable feature is called *uniform performance* property [8,14,57]. It is shown in [8,9] that any source separation method (not necessarily iterative in the sense of this chapter) based only on the optimization of a function of the separator-output distribution is *equivariant* and thus offers uniform performance. Serial updating, through the relative gradient device, enables the derivation of iterative methods inheriting the uniform performance property. For specific forms of function $\Upsilon(\cdot)$, the resulting adaptive algorithms also possess this property, as will be seen in the following section. Note, however, that uniform performance requires the linear mixing model (6.1) to hold exactly.

The relationship between the two types of gradient can be established by expressing $\Upsilon(\mathbf{B}+\mathbf{E}\mathbf{B})$ in terms of the classical gradient:

$$\Upsilon(\mathbf{B}+\mathbf{E}\mathbf{B}) = \Upsilon(\mathbf{B}) + \text{trace}\left(\nabla\Upsilon(\mathbf{B})^{\mathrm{T}}\mathbf{E}\mathbf{B}\right) + o(\mathbf{E}). \tag{6.53}$$

Noting that, for matrices of suitable dimensions, $\mathrm{trace}(\nabla\Upsilon(\mathbf{B})^{\mathrm{T}}\mathbf{E}\mathbf{B}) = \mathrm{trace}(\mathbf{B}\nabla\Upsilon(\mathbf{B})^{\mathrm{T}}\mathbf{E})$ and comparing Eqs (6.51) and (6.53), one readily obtains:

$$\tilde{\nabla}\Upsilon(\mathbf{B}) = \nabla\Upsilon(\mathbf{B})\mathbf{B}^{\mathrm{T}}. \tag{6.54}$$

6.7.2 Adaptive algorithms based on the relative gradient

Let us assume that the contrast $\Upsilon(\cdot)$ can be expressed as the expectation of a function $\phi(\mathbf{y})$ of the distribution of $\mathbf{y}$, as in (6.32). Classical stochastic update rules for this type of contrast were derived at the end of section 6.4.2. According to relationship (6.54), the application of the relative gradient on function $\Upsilon(\mathbf{B})$ results in

$$\tilde{\nabla}\Upsilon(\mathbf{B}) = \mathbb{E}\{\tilde{\nabla}\phi(\mathbf{y})\} = \mathbb{E}\{\nabla\phi(\mathbf{y})\mathbf{B}^{\mathrm{T}}\} = \mathbb{E}\{\boldsymbol{\phi}(\mathbf{y})\mathbf{x}^{\mathrm{T}}\mathbf{B}^{\mathrm{T}}\} = \mathbb{E}\{\boldsymbol{\phi}(\mathbf{y})\mathbf{y}^{\mathrm{T}}\}$$

where $\boldsymbol{\phi}(\mathbf{y}) \stackrel{\mathrm{def}}{=} [\partial\phi(\mathbf{y})/\partial y_1, \partial\phi(\mathbf{y})/\partial y_2, \ldots, \partial\phi(\mathbf{y})/\partial y_N]^{\mathrm{T}}$. As expected from the previous section, the relative gradient of function $\Upsilon(\mathbf{B})$ depends exclusively on the separator-output distribution, and leads to an iterative batch algorithm with serial update and thus uniform performance. More importantly, due to the shape of $\Upsilon(\mathbf{B})$, its relative gradient is the expectation of a function of $\mathbf{y}$ only. By dropping the expectation operator, it is hence straightforward to derive the adaptive algorithm:

$$\mathbf{B}(t+1) = \left(\mathbf{I} + \mu\boldsymbol{\phi}(\mathbf{y}(t))\mathbf{y}(t)^{\mathrm{T}}\right)\mathbf{B}(t).$$

Thanks to serial updating, this algorithm inherits the same desirable properties of its batch equivalent. Note that the usual gradient would result in an adaptive algorithm with the update:

$$\mathbf{B}(t+1) = \mathbf{B}(t) + \mu\boldsymbol{\phi}(\mathbf{y}(t))\mathbf{x}(t)^{\mathrm{T}}$$

which does not enjoy uniform performance. If $\Upsilon(\cdot)$ is an orthogonal contrast, it must be optimized subject to a unitarity constraint on the separating matrix $\mathbf{B}$, which, after prewhitening, reduces to the orthogonal matrix $\mathbf{Q}^{\mathrm{T}}$ [cf. Eq. (6.3)]. To preserve orthogonality up to the first order, the update term must be projected on the space of skew-symmetric matrices, leading to the modified rule:

$$\mathbf{Q}^{\mathrm{T}}(t+1) = \left(\mathbf{I} + \mu\left[\boldsymbol{\phi}(\mathbf{y}(t))\mathbf{y}(t)^{\mathrm{T}} - \mathbf{y}(t)\boldsymbol{\phi}(\mathbf{y}(t))^{\mathrm{T}}\right]\right)\mathbf{Q}^{\mathrm{T}}(t).$$

This equation can be combined with the whitening algorithm (6.35), also based on the relative gradient, into a one-stage update for the whole separating system $\mathbf{B} = \mathbf{Q}^{\mathrm{T}}\mathbf{W}$:

$$\mathbf{B}(t+1) = \left(\mathbf{I} + \mu\mathbf{H}_{\boldsymbol{\phi}}(\mathbf{y}(t))\right)\mathbf{B}(t) \tag{6.55}$$

with

$$\mathbf{H}_{\boldsymbol{\phi}}(\mathbf{y}) = \mathbf{I} - \mathbf{y}\mathbf{y}^{\mathrm{T}} + \boldsymbol{\phi}(\mathbf{y})\mathbf{y}^{\mathrm{T}} - \mathbf{y}\boldsymbol{\phi}(\mathbf{y})^{\mathrm{T}}. \tag{6.56}$$

Expressions (6.55) and (6.56) define the family of *equivariant adaptive separation via independence (EASI)* algorithms developed at length in [14,57]. For orthogonal contrast $\Upsilon^o_{\rm M}(\mathbf{y})$ in (6.12), we have:

$$\phi_{\rm M}(\mathbf{y}) = \varepsilon \sum_{i=1}^{N} y_i^4 \qquad \boldsymbol{\phi}_{\rm M}(\mathbf{y}) = 4\varepsilon \left[y_1^3, y_2^3, \ldots, y_N^3\right]^{\rm T}. \tag{6.57}$$

In this case, $\boldsymbol{\phi}_{\rm M}(\mathbf{y}) = [g(y_1), g(y_2), \ldots, g(y_N)]^{\rm T}$ is a component-wise nonlinear function of $\mathbf{y}$, with $g(y) = 4\varepsilon y^3$. At a valid separation solution, the components of $\mathbf{y}$ are independent, and then $\mathbb{E}\{g(y_i)y_j\} = \mathbb{E}\{g(y_i)\}\mathbb{E}\{y_j\} = 0$, $\forall i \neq j$, so that $\mathbb{E}\{\boldsymbol{\phi}_{\rm M}(\mathbf{y})\mathbf{y}^{\rm T}\} = \mathbf{0}$ as well. According to section 6.4.2, it follows that any valid separation solution is a stationary point of the EASI algorithm based on (6.57).

This result can be generalized to arbitrary possibly different nonlinear functions $\{g_i(\cdot)\}_{i=1}^N$ by defining $\mathbf{g}(\mathbf{y}) = [g_1(y_1), g_2(y_2), \ldots, g_N(y_N)]^{\rm T}$ and using $\mathbf{H}_{\mathbf{g}}(\mathbf{y}) = \mathbf{I} - \mathbf{y}\mathbf{y}^{\rm T} + \mathbf{g}(\mathbf{y})\mathbf{y}^{\rm T} - \mathbf{y}\mathbf{g}(\mathbf{y})^{\rm T}$ instead of $\mathbf{H}_{\boldsymbol{\phi}}(\mathbf{y})$ in Eq. (6.55). To increase robustness against outliers while preserving uniform performance, a normalized form of the algorithm can be obtained with

$$\mathbf{H}_{\mathbf{g}}(\mathbf{y}) = \frac{\mathbf{I} - \mathbf{y}\mathbf{y}^{\rm T}}{1 + \mu \mathbf{y}^{\rm T}\mathbf{y}} + \frac{\mathbf{g}(\mathbf{y})\mathbf{y}^{\rm T} - \mathbf{y}\mathbf{g}(\mathbf{y})^{\rm T}}{1 + \mu |\mathbf{y}^{\rm T}\mathbf{g}(\mathbf{y})|}.$$

An asymptotic analysis along the lines of section 6.4.2 reveals that the stationary points of the generalized EASI algorithm are locally stable if, $\forall i \neq j$,

$$\lambda_i + \lambda_j < 0 \qquad \text{with } \lambda_i \stackrel{\rm def}{=} \mathbb{E}\{g_i'(s_i) - s_i g_i(s_i)\}.$$

For contrast (6.12), we have $\lambda_i = -4\varepsilon\kappa_i$, $1 \leqslant i \leqslant N$, where κ_i represents the i-th source kurtosis. Hence, the stability condition dictates that the kurtosis sum of any two sources be positive if $\varepsilon = 1$ or negative if $\varepsilon = -1$. Note that this asymptotic stability condition (all source pairs with the same kurtosis sum sign ε) is weaker than the validity condition of contrast (6.12) (all sources with the same kurtosis sign ε).

6.7.3 Likelihood maximization with the relative gradient

We conclude the section of relative/natural gradient methods with a unique case of fundamental importance: the maximum likelihood criterion (6.4). Under the assumption of statistically independent sources, this criterion is characterized by:

$$\phi_{\rm ML}(\mathbf{y}) = \sum_{i=1}^{N} \log p_{s_i}(y_i) + \log|\det \mathbf{B}| \tag{6.58}$$

$$\nabla\phi_{\rm ML}(\mathbf{y}) = \boldsymbol{\varphi}(\mathbf{y})\mathbf{x}^{\rm T} + (\mathbf{B}^{\rm T})^{-1} \tag{6.59}$$

$$\tilde{\nabla}\phi_{\rm ML}(\mathbf{y}) = \boldsymbol{\varphi}(\mathbf{y})\mathbf{y}^{\rm T} + \mathbf{I} \tag{6.60}$$

where $\boldsymbol{\varphi}(\mathbf{y}) \stackrel{\text{def}}{=} [\varphi_1(y_1), \varphi_2(y_2), \ldots, \varphi_N(y_N)]^{\mathrm{T}}$ and

$$\varphi_i(y) = \left(\log p_{s_i}(y)\right)' = \frac{p'_{s_i}(y)}{p_{s_i}(y)} \qquad 1 \leqslant i \leqslant N.$$

Equation (6.60) leads to the popular relative gradient update for maximizing the likelihood [1,2,9]:

$$\mathbf{B}(t+1) = \left(\mathbf{I} + \mu\left(\mathbf{I} + \boldsymbol{\varphi}(\mathbf{y})\mathbf{y}^{\mathrm{T}}\right)\right)\mathbf{B}(t). \tag{6.61}$$

Note that the matrix inversion is present in the classical gradient algorithm [cf. Eq. (6.59); see also the related infomax algorithm (6.44)] is avoided with the use of the relative gradient.

Nonlinear function $\varphi_i(\cdot)$ is the marginal score function of the i-th source. The scores are optimal in that, by virtue of the maximum likelihood principle, they guarantee asymptotic Fisher-efficiency, i.e. they lead to separating matrix estimates attaining the Cramér-Rao lower bound and are thus shown to minimize the interference-to-signal ratio at the separator output [14]. However, these functions depend on the source distributions, which in a truly blind problem are unknown. The estimation of the optimal nonlinearities from the available data is addressed next.

6.8 ADAPTING THE NONLINEARITIES

As we have just seen, relative gradient algorithms are essentially defined by a set of nonlinear functions $g_i(\cdot)$. In the general case, adaptation of the nonlinearities is necessary to make the algorithm converge to the right solution, i.e. to make the estimator consistent, and to guarantee local stability. In other algorithms such as FastICA (see the next section), adaptation of the nonlinearities will increase the statistical performance of the method.

As recalled at the end of the previous section, the optimal nonlinearities are given by the source score functions, $g_i^{\mathrm{opt}}(y) = \varphi_i(y)$. Hence, adapting the nonlinearities is fundamentally a problem of approximating the densities of the independent components. Thus, methods developed in the general framework of maximum likelihood estimation are applicable in this context. The theory is valid in the case of FastICA as well [45].

Pham and Garat [67] and Pham *et al.* [68] proposed approximating $g_i(\cdot)$ as a linear combination of fixed basis functions $\mathbf{f}(y) = [f_1(y), f_2(y), \ldots, f_K(y)]^{\mathrm{T}}$:

$$g_i(y) = \sum_{k=1}^{K} \alpha_{ik} f_k(y) = \boldsymbol{\alpha}_i^{\mathrm{T}} \mathbf{f}(y) \tag{6.62}$$

where the adaptation is determined by estimating the coefficients $\boldsymbol{\alpha}_i = [\alpha_{i1}, \alpha_{i2}, \ldots, \alpha_{iK}]^{\mathrm{T}}$, for $1 \leqslant i \leqslant N$. These authors proposed an ingenious method for estimating the coefficients $\boldsymbol{\alpha}_i$ based on linear regression. The optimal coefficients are found by minimizing the mean square error

$$\boldsymbol{\alpha}_i^{\text{opt}} = \arg\min_{\boldsymbol{\alpha}_i} \mathbb{E}\{(g_i(y) - \varphi_i(y))^2\} \tag{6.63}$$

with $g_i(y)$ given by (6.62). Cancelling the gradient of the expectation in (6.63) leads to the normal equations

$$\mathbb{E}\{\mathbf{f}(y)\mathbf{f}^{\mathrm{T}}(\mathbf{y})\}\boldsymbol{\alpha}_i^{\text{opt}} = \mathbb{E}\{\mathbf{f}(y)\varphi_i(y)\}. \tag{6.64}$$

The remarkable feature of this method is that it relies on functions $\{f_k(\cdot)\}_{k=1}^K$ only and does not require any prior knowledge on the source distributions. To prove this important result, the crucial point to realize is that

$$\mathbb{E}\{f_k(y)\varphi_i(y)\} = -\mathbb{E}\{f_k'(y)\} \tag{6.65}$$

for any differentiable function $f_k(\cdot)$ and density $p_{s_i}(\cdot)$ such that $f_k(y)p_{s_i}(y)$ cancels at infinity. Hence, $\mathbb{E}\{\mathbf{f}(y)\varphi_i(y)\}$ can be replaced by $-\mathbb{E}\{\mathbf{f}'(y)\}$ in Eq. (6.64), where $\mathbf{f}'(y) = [f_1'(y), f_2'(y), \ldots, f_K'(y)]^{\mathrm{T}}$. Moreover, if the basis functions include the identity, this optimal choice of $\boldsymbol{\alpha}_i$ is asymptotically efficient, in the sense of Fisher-efficiency; see section 4.5.6 for details. In practice, expectations are replaced by time averages over the separator output samples $\mathbf{y} = \mathbf{Q}^{\mathrm{T}}\mathbf{z}$. At every step of the algorithm (or every few iterations), the coefficients are re-estimated based on the current estimates of the independent components. This framework was theoretically further analyzed by Chen and Bickel [15], who called it *efficient ICA*.

Incidentally using the same name, a related variant of FastICA was developed in [53] based on a *generalized Gaussian distribution* (GGD) model:

$$p_{y_i}(y) \propto e^{-|y|^{\alpha_i}} \qquad 1 \leqslant i \leqslant N \tag{6.66}$$

which is essentially defined by parameters $\alpha_i > 0$. This parametric family can model sub-Gaussian and super-Gaussian distributions for $\alpha_i > 2$ and $\alpha_i < 2$, respectively, and reduces to the Gaussian density for $\alpha_i = 2$. Under this approximation, the optimal nonlinearities are given, up to an irrelevant scalar factor, by

$$g_i^{\text{opt}}(y) = \left(\log p_{y_i}(y)\right)' = -|y|^{\alpha_i - 1}\text{sign}(y) \qquad 1 \leqslant i \leqslant N$$

and are thus defined by the pdf model parameters. These can be adaptively estimated by the method of moments, that is, by fitting the separator-output statistics to their theoretical values under (6.66); e.g. the normalized fourth-order moment, given by $\Gamma(5/\alpha_i)\Gamma(1/\alpha_i)/\Gamma^2(3/\alpha_i)$, where $\Gamma(\cdot)$ denotes the gamma function. The same methodology was adopted a few years earlier to derive the adaptive *flexible ICA* method of [17]. These results were later extended to a more general exponential generative model in [90], where the parameters are adaptively found by optimizing the objective function with respect to the separating filters and the parameters themselves. Along the same lines, Karvanen and Koivunen [52] propose approximating the optimal $g(\cdot)$ by using *Pearson's system*, a classic family of densities with two parameters controlling its non-Gaussianity. In addition to sub-/super-Gaussianity, Pearson's family can also model, unlike the GGD, the skewness of the components.

6.9 ITERATIVE ALGORITHMS BASED ON DEFLATION

The algorithms reviewed in sections 6.6 and 6.7 aim at the joint or simultaneous separation of all sources in the observed mixture. The valid separation solutions are generally proven to be stationary points of contrasts for joint separation. Yet the absence of spurious (i.e. non-separating) local extrema is often not guaranteed due to the multimodality of contrasts based on higher-order statistics, even under ideal conditions such as infinite sample size and observed data perfectly satisfying the ICA model. Unfortunately, gradient methods can easily get trapped in spurious local extrema and saddle points. The relative gradient approach (section 6.7) may alleviate this shortcoming, but does not spare it completely. Similarly, the separator trajectories of Newton methods are not guaranteed to escape spurious equilibria. In addition, these methods require the computation and inversion of the Hessian matrix at each iteration; even if the extra computation can be afforded, an inaccurate Hessian estimation can lead to instability. These drawbacks have motivated significant research efforts towards more robust iterative algorithms for BSS/ICA.

As opposed to joint separation, the *sequential extraction* approach to BSS/ICA estimates one source after another. An important benefit of this approach is that contrasts for source extraction are easily proven, under ideal conditions, to be free of spurious local extrema. This feature allows the design of iterative algorithms with theoretical *global* convergence to valid separating filters. For instance, the globally convergent BSS method of [66] relies heavily on the fact that it can be uncoupled as separate algorithms for single source extraction working in parallel. To prevent the estimated source from being extracted more than once, its contribution to the observations must somehow be suppressed before searching for the next source; the observations are thus said to be "deflated". Several *deflation* techniques are summarized in the following paragraphs.

6.9.1 Adaptive deflation algorithm by Delfosse and Loubaton

In the real-valued case, Delfosse and Loubaton [28] prove that fourth-order extraction contrasts (6.17) and (6.18) are free from spurious extrema if maximized under the unit-norm constraint after prewhitening (section 6.3.3). As a result, the associated iterative algorithms are globally convergent to a valid extraction solution in ideal conditions. A similar result had previously been obtained by Shalvi and Weinstein [74] in the context of blind deconvolution of digital communication channels and was later extended by Papadias [66] to the blind separation of possibly complex-valued sources in instantaneous linear mixtures. Global convergence is a very desirable feature shared, for instance, by the kurtosis-based FastICA algorithm (section 6.10.6), but not enjoyed by most other algorithms surveyed in this chapter.

To derive the deflation algorithm of [28], the separating rotation is decomposed as

$$\mathbf{Q}^{\mathrm{T}}(\boldsymbol{\theta}) = \mathbf{Q}_1^{\mathrm{T}}(\theta_1)\mathbf{Q}_2^{\mathrm{T}}(\theta_2)\cdots\mathbf{Q}_{N-1}^{\mathrm{T}}(\theta_{N-1}) \tag{6.67}$$

where $\boldsymbol{\theta} = [\theta_1, \theta_2, \ldots, \theta_{N-1}]^{\mathrm{T}}$, with $\theta_i \in] - \pi/2, \pi/2[$, $1 \leqslant i \leqslant (N-1)$, and Givens rotation $\mathbf{Q}_i^{\mathrm{T}}(\theta_i)$ is defined as $\mathbf{Q}_{ij}^{\mathrm{T}}(\theta_{ij})$ in section 6.6.3 [Eqs (6.45)–(6.46)] with $j = N$. This matrix can be further divided in two parts:

$$\mathbf{Q}^{\mathrm{T}}(\boldsymbol{\theta}) = \begin{bmatrix} \tilde{\mathbf{Q}}^{\mathrm{T}}(\boldsymbol{\theta}) \\ \mathbf{q}^{\mathrm{T}}(\boldsymbol{\theta}) \end{bmatrix}$$

in which $\tilde{\mathbf{Q}}^{\mathrm{T}}(\boldsymbol{\theta}) \in \mathbb{R}^{(N-1)\times N}$ and

$$\mathbf{q}(\boldsymbol{\theta}) = [\sin\theta_1, \cos\theta_1 \sin\theta_2, \ldots, \\ \cos\theta_1 \cdots \cos\theta_{N-2} \sin\theta_{N-1}, \cos\theta_1 \cdots \cos\theta_{N-2} \cos\theta_{N-1}]^{\mathrm{T}} \in \mathbb{R}^N$$

represents the extracting vector for the source currently targeted as $y = \mathbf{q}^{\mathrm{T}}\mathbf{z}$. For orthogonal contrast $\Upsilon_{\mathrm{m}}^{o}(\cdot)$ [Eq. (6.18)], angular parameters $\boldsymbol{\theta}$ are estimated through the stochastic learning rule:

$$\boldsymbol{\theta}(t+1) - \boldsymbol{\theta}(t) + \mu\varepsilon y^3(t)\nabla y(t). \tag{6.68}$$

The gradient $\nabla y(t)$ is computed with respect to $\boldsymbol{\theta}$, and is given by

$$\nabla y = \nabla \mathbf{q}(\boldsymbol{\theta})^{\mathrm{T}}\mathbf{z} = \mathbf{D}(\boldsymbol{\theta})\tilde{\mathbf{Q}}^{\mathrm{T}}(\boldsymbol{\theta})\mathbf{z}$$

where $\mathbf{D}(\boldsymbol{\theta})$ denotes the $(N-1)\times(N-1)$ diagonal matrix with elements $d_{11} = 1$ and $d_{ii} = \prod_{k=1}^{i-1}\cos\theta_k$, $2 \leqslant i \leqslant (N-1)$. Calling $\tilde{\mathbf{z}} = \tilde{\mathbf{Q}}^{\mathrm{T}}(\boldsymbol{\theta})\mathbf{z} \in \mathbb{R}^{N-1}$, update (6.68) becomes:

$$\boldsymbol{\theta}(t+1) = \boldsymbol{\theta}(t) + \mu\varepsilon y^3(t)\mathbf{D}(\boldsymbol{\theta})\tilde{\mathbf{z}}(t).$$

To extract the next source, the algorithm is repeated using $\tilde{\mathbf{z}}$ instead of $\mathbf{z}$ and reducing the dimensions of $\mathbf{Q}$ and $\boldsymbol{\theta}$ accordingly. This dimensionality reduction, achieved by the special parameterization of matrix $\mathbf{Q}^{\mathrm{T}}$ in (6.67), alleviates the computational load after each deflation stage. Since, by construction, $\mathbf{q}$ and $\tilde{\mathbf{Q}}$ are orthogonal, vector $\tilde{\mathbf{z}}$ is uncorrelated with the source extracted by $\mathbf{q}$, so that the same source cannot be extracted more than once. A similar stochastic procedure is proposed to maximize contrast (6.17) and requires the adaptive estimation of the fourth-order marginal cumulant $(\mathbb{E}\{y^4\} - 3)$.

6.9.2 Regression-based deflation

As recalled in section 6.3.3, Tugnait [79] proved that contrast (6.16) does not present spurious extrema, even in the absence of prewhitening. Consequently, the contrast can be directly maximized through a conventional gradient-based update. Deflation is performed by means of *regression*, which can be summarized as follows. Once a source estimate, say $\widehat{s}$, has been obtained, its contribution to the observed signals is computed as the solution to the *minimum mean square error* (MMSE) problem

$$\widehat{\mathbf{a}} = \arg\min_{\mathbf{a}} \mathbb{E}\{\|\mathbf{x} - \mathbf{a}\widehat{s}\|^2\}$$

which is readily computed as

$$\widehat{\mathbf{a}} = \frac{\mathbb{E}\{\mathbf{x}\widehat{s}\}}{\mathbb{E}\{\widehat{s}^2\}}.$$

The observations are then "deflated" as

$$\mathbf{x} \leftarrow \mathbf{x} - \widehat{\mathbf{a}}\widehat{s}$$

before running the extraction algorithm again in search of the next source. This alternative approach to deflation can easily be extended to the convolutive mixture case [79]. Dimensionality reduction is also possible by projecting the data on the orthogonal complement of $\widehat{\mathbf{a}}$. An orthonormal basis of this subspace can be easily obtained, e.g. through the QR decomposition (Gram-Schmidt orthogonalization) of matrix $\tilde{\mathbf{A}} = [\widehat{\mathbf{a}}, \mathbf{I}_{P\times(P-1)}]$, given by $\tilde{\mathbf{A}} = \mathbf{QR}$, where $\mathbf{Q}$ is a unitary matrix and $\mathbf{R}$ an upper-diagonal matrix. Then, the deflated observations with reduced dimensionality can be computed as

$$\tilde{\mathbf{x}} \leftarrow \tilde{\mathbf{Q}}^{\mathrm{T}}\mathbf{x}$$

where $\tilde{\mathbf{Q}}$ is the $P \times (P-1)$ matrix composed of the last $(P-1)$ columns of $\mathbf{Q}$.

6.9.3 Deflationary orthogonalization

Although the FastICA algorithm is reviewed in detail in the next section, the deflation technique it employs can readily be used in conjunction with other iterative algorithms relying on prewhitening. For instance, this deflation scheme is combined with a gradient-based update based on contrast $\Upsilon_1^o(\mathbf{y})$ [Eq. (6.14)] to derive the globally convergent adaptive BSS algorithm of [66].

The idea behind *deflationary orthogonalization* is to guarantee that subsequent extracted components satisfy the second-order decorrelation imposed by the prewhitening assumption, that is, $\mathbb{E}\{\widehat{s}_i\widehat{s}_j\} = 0$, $\forall j < i$. This expectation is equivalent to $\mathbf{q}_i^{\mathrm{T}}\mathbf{R}_z\mathbf{q}_j = \mathbf{q}_i^{\mathrm{T}}\mathbf{q}_j$, so that decorrelation is fulfilled if and only if the extracting vectors are kept orthogonal. To this end, the classical *Gram-Schmidt orthogonalization* procedure can be used after each update of current extracting vector $\mathbf{q}_i$:

$$\mathbf{q}_i \leftarrow \mathbf{q}_i - \sum_{j<i} \mathbf{q}_j \mathbf{q}_j^{\mathrm{T}} \mathbf{q}_i.$$

This accepts an efficient implementation in matrix form as:

$$\mathbf{q}_i \leftarrow (\mathbf{I}_N - \mathbf{Q}_i \mathbf{Q}_i^{\mathrm{T}}) \mathbf{q}_i$$

where $\mathbf{Q}_i = [\mathbf{q}_1, \mathbf{q}_2, \ldots, \mathbf{q}_{i-1}]$.

6.10 THE FASTICA ALGORITHM

6.10.1 Introduction

On account of its high convergence speed and satisfactory performance in a wide variety of applications, *FastICA* is arguably one of the most popular iterative methods for ICA. The starting point is an *orthogonal contrast* of the form

$$\Upsilon_G(\mathbf{Q}) = \sum_{i=1}^{N} \mathbb{E}\{G_i(\mathbf{q}_i^{\mathrm{T}} \mathbf{z})\} \tag{6.69}$$

where the data $\mathbf{z}$ are prewhitened and matrix $\mathbf{Q}$ is constrained to be orthogonal. ICA estimation is performed by maximizing this contrast function. The functions $G_i(\cdot)$ can fundamentally be chosen in two ways:

- In an maximum likelihood framework, $G_i(y_i) = \log p_{s_i}(y_i)$ where p_{s_i} is the pdf of the i-th independent component [cf. Eqs (6.4) and (6.58)]. The functions $G_i(\cdot)$ can be fixed based on prior information of the independent components, or they can be estimated from the current estimates y_i using different methods discussed in Chapter 4 (see also section 6.8). In many applications, the independent components are known to be super-Gaussian (sparse), in which case all the functions $G_i(\cdot)$ are often set *a priori* as

$$G_i(y) = -\log\cosh y \qquad 1 \leqslant i \leqslant N. \tag{6.70}$$

  This corresponds to a symmetric super-Gaussian pdf and has the great benefit of being very smooth; a non-smooth $G(\cdot)$ can pose problems for any optimization method. As discussed in section 6.3, maximum likelihood estimation is equivalent, under some assumptions, to information-theoretic approaches such as minimum mutual information and minimum marginal entropy.
- A second framework consists in using cumulants. Indeed, if we choose

$$G_i(y) = y^4 \qquad 1 \leqslant i \leqslant N$$

the contrast in (6.69) is essentially the sum of kurtosis (6.12), since the separator outputs' variance is constant due to prewhitening and the orthogonality of $\mathbf{q}$. If we know *a priori* that the independent components are super-Gaussian, the maximization of this contrast function is a valid estimation method. If the sources are all sub-Gaussian, we switch the sign and choose $G_i(y) = -y^4$, which becomes again a valid contrast. This sign switching is automatically performed by ε in Eq. (6.12).

FastICA is a computationally efficient algorithm for optimizing the contrast (6.69). Function $G_i(\cdot)$ is implicitly expressed in the algorithm via its derivative, a nonlinearity $g_i(\cdot)$; that of (6.70) is the hyperbolic tangent function (tanh).

6.10.2 Implicit adaptation of the contrast in FastICA

FastICA has the interesting property that it chooses implicitly, for each component, whether $\mathbb{E}\{G_i(\mathbf{q}_i^T\mathbf{z})\}$ should be maximized or minimized. This is important since a contrast function of the form in (6.69) for fixed functions $G_i(\cdot)$ can only estimate independent components with certain distributions. As shown in [14,45], functional (6.69) can be a valid contrast only if

$$\mathbb{E}\{g_i(s_i)s_i - g_i'(s_i)\} > 0. \tag{6.71}$$

Indeed, a local analysis along the lines of section 6.4.1.3 reveals that this is the necessary condition for all extracting vectors to be local maxima of (6.69) under the unit-norm constraint. Actually, FastICA can be viewed as maximizing a contrast function of the form

$$\Upsilon_G(\mathbf{Q}) = \sum_i k_i \mathbb{E}\{G_i(\mathbf{q}_i^T\mathbf{z})\}$$

where $k_i = \pm 1$ is automatically chosen as

$$k_i = \text{sign}\left(\mathbb{E}\{g_i(s_i)s_i - g_i'(s_i)\}\right). \tag{6.72}$$

Thus, if the *nonlinear* (or *non-polynomial*) *cumulant* $\mathbb{E}\{g_i(s_i)s_i - g_i'(s_i)\}$ is negative for a component, its sign is implicitly changed in the contrast, so that effectively $\mathbb{E}\{G_i(\mathbf{q}_i^T\mathbf{z})\}$ is minimized. As will be explained in section 6.10.5, this automatic sign switching feature is achieved by the Newton iteration employed in FastICA, and can be considered as a generalization of the idea that we should, in general, maximize the squares or absolute values of the kurtoses. Maximizing the absolute values can be accomplished by estimating a sign as in (6.72), which then decides whether we maximize or minimize kurtosis. It is also possible to adapt functions $G_i(\cdot)$ to the data by modeling them by some family of functions; see section 6.10.7.1.

6.10.3 Derivation of FastICA as a Newton iteration

To begin with, we shall derive the FastICA algorithm for *one* component, using a "one-unit" contrast:

$$\Upsilon_G(\mathbf{q}) = \mathbb{E}\{G(\mathbf{q}^T\mathbf{z})\}.$$

The goal is to find the extrema of this contrast under the constraint $E\{(\mathbf{q}^T\mathbf{z})^2\} = ||\mathbf{q}||^2 = 1$. To this end, we form the Lagrangian as in section 6.4.1.3:

$$\tilde{\Upsilon}_G(\mathbf{q}) = \Upsilon_G(\mathbf{q}) + \frac{\beta}{2}(1 - ||\mathbf{q}||^2).$$

where β is a constant (the Lagrange multiplier). According to the Lagrange conditions [cf. Eq. (6.27)], the extrema are obtained at points where

$$\mathbb{E}\{\mathbf{z}g(\mathbf{q}^T\mathbf{z})\} - \beta\mathbf{q} = 0. \tag{6.73}$$

The optimum value of β can be easily evaluated as in Eq. (6.28) to give $\beta_0 = \mathbb{E}\{\mathbf{q}_0^T\mathbf{z}g(\mathbf{q}_0^T\mathbf{z})\}$, where $\mathbf{q}_0$ is the value of $\mathbf{q}$ at the optimum. Originally, FastICA was derived using a heuristic "fixed-point" idea [49]. This fixed-point approach is inspired by the Lagrange condition (6.27), which states that, when maximizing an objective function under the unit-norm constraint, the extracting vector is parallel to the gradient at the optimum solution. However, we shall not go into that derivation here. Instead, we use the approach in [47], which derives the iteration using an approximate Newton method.

Thus, let us try to solve (6.73) by the classical Newton method. The Hessian of $\tilde{\Upsilon}_G(\mathbf{q})$, denoted as $\mathbf{H}_{\tilde{\Upsilon}_G}(\mathbf{q})$, is obtained as in Eq. (6.29):

$$\mathbf{H}_{\tilde{\Upsilon}_G}(\mathbf{q}) = \mathbb{E}\{\mathbf{z}\mathbf{z}^T g'(\mathbf{q}^T\mathbf{z})\} - \beta\mathbf{I}. \tag{6.74}$$

To simplify the inversion of this matrix, we decide to approximate the first term in (6.74). An approximation that turns out to be very suitable here is

$$\mathbb{E}\{\mathbf{z}\mathbf{z}^T g'(\mathbf{q}^T\mathbf{z})\} \approx \mathbb{E}\{\mathbf{z}\mathbf{z}^T\}\mathbb{E}\{g'(\mathbf{q}^T\mathbf{z})\} = \mathbb{E}\{g'(\mathbf{q}^T\mathbf{z})\}\mathbf{I}. \tag{6.75}$$

Indeed, this approximation of the Hessian matrix can be inverted very easily:

$$\mathbf{H}^{-1}_{\tilde{\Upsilon}_G}(\mathbf{q}) \approx \left(E\{g'(\mathbf{q}^T\mathbf{z})\} - \beta\right)^{-1}\mathbf{I}.$$

We also approximate β by β_0 but using the current value of $\mathbf{q}$ instead of $\mathbf{q}_0$, that is, $\beta = E\{\mathbf{q}^T\mathbf{z}\,g(\mathbf{q}^T\mathbf{z})\}$. Combining this expression with Eqs (6.24), (6.26) and (6.29) leads to the following approximate Newton iteration:

$$\mathbf{q}^+ = \mathbf{q} - [\mathbb{E}\{\mathbf{z}g(\mathbf{q}^T\mathbf{z})\} - \beta\mathbf{q}]/[\mathbb{E}\{g'(\mathbf{q}^T\mathbf{z})\} - \beta]. \tag{6.76}$$

After every step, $\mathbf{q}^+$ is normalized by dividing it by its norm to project it back on the constraint set. This iteration can be further algebraically simplified to obtain the simpler form of the FastICA algorithm:

$$\mathbf{q}^+ = \mathbb{E}\{\mathbf{z}g(\mathbf{q}^T\mathbf{z})\} - \mathbb{E}\{g'(\mathbf{q}^T\mathbf{z})\}\mathbf{q}.$$

To reach this equation, it suffices to multiply both sides of (6.76) by $\mathbb{E}\{g'(\mathbf{q}^T\mathbf{z})\}-\beta$ and simplify the right-hand side; the multiplicative constant on the left-hand side is neutralized by the projection step.

6.10.4 Connection with gradient methods

There is a simple interesting connection between FastICA and gradient algorithms for ICA. This is not surprising since the Newton method is also closely related to gradient methods (section 6.4).

Let us consider the preliminary form of the algorithm in (6.76). Collecting the updates for all the rows of $\mathbf{Q}^T$ into a single equation, and using the orthogonality of $\mathbf{q}$, we obtain the following equivalent form of the iteration:

$$(\mathbf{Q}^T)^+ = \left(\mathbf{I}+\mathbf{D}\left[\mathrm{Diag}(-\beta_i)+\mathbb{E}\{g(\mathbf{y})\mathbf{y}^T\}\right]\right)\mathbf{Q}^T \tag{6.77}$$

where $\mathbf{y}=\mathbf{Q}^T\mathbf{z}$, with $\mathbf{Q}=[\mathbf{q}_1,\mathbf{q}_2,\ldots,\mathbf{q}_N]$ and

$$\begin{aligned}\beta_i &= \mathbb{E}\{y_i g(y_i)\}\\ \mathbf{D} &= \mathrm{Diag}(1/(\beta_i - E\{g_i'(y_i)\}))\\ g(\mathbf{y}) &= [g_1(y_1),\ldots,g_N(y_N)]^T.\end{aligned}$$

Note that the nonlinearity $g_i(\cdot)$ is now allowed to depend on i. This update can be compared to the natural gradient algorithm for maximization of the likelihood in section 6.7.3 [Eq. (6.61)]. We can see that the algorithms are very closely related. The main differences are that (a) in the relative gradient algorithm, the $-\beta_i$ are all set to one; (b) $\mathbf{D}$ is replaced by identity times the step size μ; (c) in maximum likelihood, the nonlinearities $g_i(\cdot)$ are adapted to the source score functions; and (d) maximum likelihood does not require prewhitening. Hence, $\mathbf{D}$ plays actually the role of a step size, although in the form of a diagonal matrix here, but it does not affect convergence to points where the average update term is null. Moreover, if the $g_i(\cdot)$ is indeed the score function of s_i, then the $-\beta_i$ are also (for infinite sample size) equal to one [48], as can easily be deduced from Eq. (6.65). In that case, then, both algorithms are theoretically equivalent under prewhitening.

However, this close connection with the natural gradient algorithm requires that the step size μ be positive. Here, the step size is given by the elements of $\mathbf{D}$. These terms are nothing else than estimates of the inverses of the nonlinear cumulants in (6.71). If the nonlinear cumulants are positive, FastICA is really taking steps in the same direction as the gradient method. The interesting point is that if such a nonlinear cumulant is negative, the algorithm is actually taking steps in the opposite direction. This is indeed equivalent to using an adaptive sign as explained in section 6.10.2.

Another interesting connection is found for the kurtosis-based FastICA algorithm, obtained with $g(y)=y^3$, whose update can be written as

$$\mathbf{q}^+ = \mathbf{q} - \frac{1}{3}\,\mathbb{E}\{\mathbf{z}(\mathbf{q}^T\mathbf{z})^3\}. \tag{6.78}$$

Due to the simplifying assumptions made in its derivation (section 6.10.3), the function optimized turns out to be actually the fourth-order moment $\Upsilon^o_{\rm m}(\mathbf{q}) = \varepsilon\mathbb{E}\{y^4\}$ [cf. Eq. (6.18)], with gradient $\nabla\Upsilon^o_{\rm m}(\mathbf{q}) = 4\varepsilon\mathbb{E}\{\mathbf{z}(\mathbf{q}^{\rm T}\mathbf{z})^3\}$. As a result, Eq. (6.78) is essentially a gradient-based update rule of the form

$$\mathbf{q}^+ = \mathbf{q} + \mu\nabla\Upsilon^o_{\rm m}(\mathbf{q})$$

with $\varepsilon = -1$ and a fixed value for the step size, $\mu = 1/12$. Hence, the kurtosis-based FastICA can be considered as a particular instance, using prewhitening and assuming sub-Gaussian sources, of the general family of gradient-based algorithms proposed in [79]. Though fixed to a constant value, FastICA's step-size choice is judicious in that it leads to cubic convergence of the algorithm for infinite sample size (section 6.10.6) and is able to maximize or minimize the objective, i.e. can extract sub-Gaussian and super-Gaussian sources. These properties are due to its connection with a Newton iteration, as discussed next.

6.10.5 Convergence of FastICA

It can be proven [47] that FastICA converges under the following assumptions:

1. The data exactly follow the noiseless ICA model (6.1), and the data size is infinite.
2. For the chosen $G_i(\cdot)$, the nonlinear cumulants on the left-hand side of (6.71) are non-zero.

What is remarkable is that local convergence is quadratic, whereas in some cases (as with the cubic non-linearity; see the next section) convergence is *global and cubic*. As recalled in section 6.4, quadratic convergence means that if at a given step the error (distance from true solution) is ϵ, in the next step it is $o(\epsilon^2)$. This means very fast convergence, since most optimization algorithms only provide linear convergence, i.e. in the next step the error is $\lambda\epsilon$ for some constant $\lambda < 1$. The Newton method does provide quadratic convergence, and thus we see that our approximation of the Hessian matrix in (6.75) was fortunate in the sense that it retains the quadratic convergence yet avoids the need for computing the inverse of the Hessian. For the kurtosis-based nonlinearity, the convergence is analyzed in detail in section 6.10.6; see [47] for a general analysis.

This super-fast convergence property is somewhat theoretical as it is only possible if the data follow the ICA model. For short sample sizes, however, convergence may slow down and even get trapped in saddle areas and local extrema, as has been noticed in [78] and further illustrated in [84]. See section 6.10.7.2 below for more details on this point.

Note that FastICA does *not* require the nonlinear cumulant to have a particular sign, thanks to the implicit adaptation of the contrast discussed in section 6.10.2. This adaptation can be understood from the following basic property of the Newton iteration briefly recalled in section 6.4.1.2: unlike gradient techniques, the Newton method simply searches for an extremum of the objective function, regardless of whether it is a maximum or a minimum. Thus, FastICA can find both kinds of extremum, which are valid solutions in ICA.

6.10.6 FastICA using cumulants

Shalvi and Weinstein [75] were probably the first to propose, in the context of blind deconvolution, an algorithm closely related to the cumulant-based FastICA. In fact, in the case of cumulants, convergence can be analyzed in much more detail, and it can be shown to be global, although it has to be assumed that the observed data follow exactly the ICA model and the sample size is infinite. The cumulant-based FastICA was actually proposed earlier [49] than the general method described above [47].

Consider the nonlinearity $g(y) = y^3$, in which case the algorithm is really using a simplified form of kurtosis as a contrast. The one-unit iteration (6.78) can also be expressed as

$$\mathbf{q}^+ = \mathbb{E}\left\{\mathbf{z}\left(\mathbf{q}^T\mathbf{z}\right)^3\right\} - 3\mathbf{q} \tag{6.79}$$

where $\mathbf{q}^+$ is normalized to unit norm after every iteration. Denote as $\mathbf{H} = \mathbf{Q}^T\mathbf{W}\mathbf{A}$ the total transformation matrix between the true sources and the estimated components (mixing-unmixing system). This is a simple transformation of $\mathbf{Q}$, so we can express the algorithm as a function of the columns of $\mathbf{H}^T$, $\{\mathbf{h}_i\}_{i=1}^N$. For each single unit, we have

$$\mathbf{h}^+ = \mathbb{E}\left\{\mathbf{s}\left(\mathbf{h}^T\mathbf{s}\right)^3\right\} - 3\mathbf{h}.$$

Denote as $\mathbf{h}(k)$ the value of $\mathbf{h}$ at the k-th iteration. Then

$$\mathbf{h}(k) = \mathbb{E}\left\{\mathbf{s}(\mathbf{h}(k-1)^T\mathbf{s})^3\right\} - 3\mathbf{h}(k-1).$$

Expanding the first term, we can calculate explicitly the expectation, and obtain for the i-th component of the vector $\mathbf{h}(k)$:

$$h_i(k) = \mathbb{E}\{s_i^4\}h_i(k-1)^3 + 3\sum_{j\neq i} h_j(k-1)^2 h_i(k-1) - 3h_i(k-1) \tag{6.80}$$

where we have used the fact that by the statistical independence of the s_i, we have $\mathbb{E}\{s_i^2 s_j^2\} = 1$, and $\mathbb{E}\{s_i^3 s_j\} = \mathbb{E}\{s_i^2 s_j s_\ell\} = \mathbb{E}\{s_i s_j s_\ell s_m\} = 0$ for four different indices (i, j, ℓ, m). Using $||\mathbf{h}(k)|| = ||\mathbf{q}(k)|| = 1$, Eq. (6.80) simplifies into

$$h_i(k) = \kappa_i h_i(k-1)^3 \tag{6.81}$$

where κ_i denotes the i-th source kurtosis. Note that the subtraction of $3\mathbf{h}(k-1)$ from the right side cancelled the term due to the cross-variances, enabling direct access to the fourth-order cumulants. In fact, the preceding derivations could have been simplified by noting that (6.79) is equal to the gradient of a simplified expression of kurtosis valid under the prewhitening assumption: $\mathbb{E}\{y^4\} - 3\mathbb{E}\{y^2\}^2 = \mathbb{E}\{y^4\} - 3||\mathbf{q}||^4$. Equation (6.81) is the fundamental "fixed-point" iteration in FastICA. Intuitively, it shows that small h_i go to zero very fast, due to the third power. For the total system, convergence is obtained when only one entry of $\mathbf{h}$ is equal to ± 1 and the others are zero.

Now, choosing j so that $\text{kurt}(s_j) \neq 0$ and $h_j(k-1) \neq 0$, we further obtain

$$\frac{|h_i(k)|}{|h_j(k)|} = \frac{|\kappa_i|}{|\kappa_j|}\left(\frac{|h_i(k-1)|}{|h_j(k-1)|}\right)^3 \qquad k > 0.$$

Next note that $|h_i(k)|/|h_j(k)|$ is not changed by the normalization step. It is therefore possible to solve explicitly $|h_i(k)|/|h_j(k)|$ from this recursive formula, which yields

$$\frac{|h_i(k)|}{|h_j(k)|} = \frac{\sqrt{|\kappa_j|}}{\sqrt{|\kappa_i|}}\left(\frac{\sqrt{|\kappa_i|}|h_i(0)|}{\sqrt{|\kappa_j|}|h_j(0)|}\right)^{3^k} \qquad k > 0. \tag{6.82}$$

For $j = \arg\max_p \sqrt{|\kappa_p|}|h_p(0)|$, we see that all the other components $h_i(k)$, $i \neq j$, quickly become small compared to $h_j(k)$. Taking the normalization $||\mathbf{h}(k)|| = ||\mathbf{q}(k)|| = 1$ into account, this means that $|h_j(k)| \to 1$ and $h_i(k) \to 0$ for all $i \neq j$. Hence, $\mathbf{q}(k)$ converges to the column of the prewhitened mixing matrix $\mathbf{WA}$ for which the kurtosis of the corresponding independent component s_j is not zero and which has not yet been found. This proves the *global* convergence of kurtosis-based FastICA. Moreover, convergence is cubic, which is even faster than quadratic, due to the 3^k exponent in (6.82). This fast convergence is achieved at the same computational cost per iteration as a simple gradient-based algorithm.

6.10.7 Variants of FastICA

6.10.7.1 Choosing and adapting the nonlinearity

As discussed in section 6.10.5, a fixed non-quadratic function $g(\cdot)$ provides convergence (i.e. a consistent estimator) for almost all non-Gaussian source distributions. However, it is useful, at least in theory, to adapt the nonlinearity $g(\cdot)$ in order to improve the statistical efficiency (reduce asymptotic variance) of the estimator. This adaptation can be obtained by using the theory of maximum likelihood estimation, which shows that the nonlinearities should be equal to the derivatives of the log-pdf's (scores) of the sources. Methods for approximating these functions have already been treated in section 6.8.

While adaptation of the nonlinearity improves, in principle, the statistical efficiency of the estimator, it may happen in practice that the statistically optimal nonlinearity is computationally difficult to obtain. Consider, for example, a very sparse (super-Gaussian) source whose pdf is very close to a GGD [Eq. (6.66)] with exponent $\alpha_i < 1$. Such a distribution leads to an optimal nonlinearity singular at the origin, and therefore computationally very problematic. This is why it may be preferable in practice to stick to a nonlinearity attaining a reasonable compromise between statistical and computational optimality. The tanh nonlinearity, the derivative of function (6.70), provides such a compromise, since it corresponds to a super-Gaussian pdf while being smooth; it has been found to to work well on a number of applications.

Another statistical consideration is also very important in practice: robustness (insensitivity) to outliers. As commented in section 6.3.4, kurtosis can be very sensitive to outliers, because the fourth power grows very fast as the argument moves away from zero [45,47]. The tanh nonlinearity has the further benefit over kurtosis of being more robust to outliers. Nevertheless, the use of kurtosis may be justified on the grounds of theoretical global convergence [28,49,66,79] (see also section 6.3.3) and, as exemplified by the RobustICA method of section 6.11.2, computational convenience.

6.10.7.2 Stabilized versions to prevent convergence problems

Convergence proofs for FastICA assume that the data follow the ICA model and that the sample size is infinite. Only under such theoretical conditions is quadratic or cubic (global) convergence guaranteed. Since in practice the sample size is finite, the convergence can only be close to such theoretical behavior. If the data do not have independent components, i.e. the model specifications are strongly violated, or if the sample size is very small, the convergence of FastICA can be slow. In the extreme case, the algorithm may not converge at all: it may oscillate between a number of different points, or get trapped in saddle areas [78,84].

To alleviate this problem, stabilized versions have been proposed [47]. They are based on the interpretation of FastICA as a gradient method as shown in Eq. (6.77). We can simply reduce the step size to make sure the algorithm converges, which leads to

$$(\mathbf{Q}^{\mathrm{T}})^{+} = \left(\mathbf{I} + \mu \mathbf{D}\left[\mathrm{Diag}(-\beta_i) + \mathbb{E}\{g(\mathbf{y})\mathbf{y}^{\mathrm{T}}\}\right]\right)\mathbf{Q}^{\mathrm{T}}$$

for some positive step size $\mu < 1$. Regalia and Kofidis [72] were further able to give exact conditions on the step size to guarantee that each step of the algorithm improves the contrast function (section 6.11).

6.10.7.3 FastICA for complex-valued data and dependent sources

The FastICA algorithm was originally developed for real-valued signals only. A first extension of FastICA to complex-valued data can be found in [6]. Since this extension is based on nonlinear functions of the form $\mathbb{E}\{G(|y|^2)\}$, the resulting algorithm is only valid for sources with circularly symmetric distributions, whose pseudo-covariance matrix is null: $\mathbb{E}\{\mathbf{s}\mathbf{s}^{\mathrm{T}}\} = \mathbf{0}$. Under this assumption, the convergence properties of the algorithm are quite similar to FastICA's for real-valued sources; in particular, the algorithm based on kurtosis was later shown to inherit the cubic global convergence property of its real counterpart [73]. In such a case, the update rule can be expressed as

$$\mathbf{q}^{+} = \mathbf{q} - \frac{1}{2}\mathbb{E}\{\mathbf{z}y^{*}|y|^{2}\} \tag{6.83}$$

where the extractor output is computed as $y = \mathbf{q}^{\mathrm{H}}\mathbf{z}$. In these equations, symbol $(\cdot)^{*}$ denotes complex conjugation and $(\cdot)^{\mathrm{H}}$ stands for the Hermitian (conjugate-transpose) operator. We can define the complex gradient operator as $\nabla_{\mathbf{q}} = \nabla_{\mathbf{q}_r} + j\nabla_{\mathbf{q}_i}$, where $\mathbf{q}_r$ and $\mathbf{q}_i$ represent the real and imaginary parts, respectively, of vector $\mathbf{q}$; this a scaled form of Brandwood's conjugate gradient [7]. Then, Eq. (6.83) is easily shown to be a gradient-descent algorithm on the fourth-order moment function $\mathbb{E}\{|y|^4\}$ with fixed step size $\mu = 1/8$. This result mimics the connection to gradient methods of the kurtosis-based FastICA algorithm in the real case made at the end of section 6.10.4.

More recent efforts have focused on extending the scope of the algorithm to non-circular sources [34,58,64,65]. Li and Adali [58] derive gradient, fixed-point and Newton-like algorithms based on the general definition of the fourth-order marginal cumulant, a non-normalized version of (6.16), valid for non-circular sources. The so-called *KM fixed-point (KM-F)* algorithm assigns the current gradient to the extracting vector

$$\mathbf{q}^+ = \mathbb{E}\{|y|^2 y^* \mathbf{z}\} - 2\mathbb{E}\{|y|^2\}\mathbb{E}\{y^* \mathbf{z}\} - \mathbb{E}^*\{y^2\}\mathbb{E}\{y\mathbf{z}\}$$

before the orthogonalization and normalization steps. A modification of [6] is proposed in [65] leading to the *non-circular FastICA (nc-FastICA)* algorithm:

$$\mathbf{q}^+ = -\mathbb{E}\{g(|y|^2)y^*\mathbf{z}\} + \mathbb{E}\{g'(|y|^2)|y|^2 + g(|y|^2)\}\mathbf{q} + \mathbb{E}\{\mathbf{z}\mathbf{z}^{\mathrm{T}}\}\mathbb{E}\{g'(|y|^2)y^{*2}\}\mathbf{q}^*$$

with $g(|y|^2) = G'(|y|^2)$. The incorporation of the whitened observation pseudo-covariance matrix $\mathbb{E}\{\mathbf{z}\mathbf{z}^{\mathrm{T}}\}$ guarantees local stability at the separating solutions even in the presence of non-circular sources. For the kurtosis-based nonlinearity, $g(|y|^2) = |y|^2$, the nc-FastICA update rule reads:

$$\mathbf{q}^+ = \mathbf{q} - \frac{1}{2}\mathbb{E}\{|y|^2 y^* \mathbf{z}\} + \frac{1}{2}\mathbb{E}\{\mathbf{z}\mathbf{z}^{\mathrm{T}}\}\mathbb{E}^*\{y^2\}\mathbf{q}^*. \tag{6.84}$$

The *complex fixed-point algorithm (CFPA)* of [34] turns out to rely on an update rule very similar to (6.84), but obtained through an ingenious alternative approach sparing differentiation. Similar algorithms are proposed in [64] through a negentropy-based family of cost functions with the general form $\mathbb{E}\{|G(y)|^2\}$ instead of $\mathbb{E}\{G(|y|^2)\}$. These nonlinear functions preserve phase information and are thus adapted to non-circular sources, but must be chosen in accordance with the source distributions to assure stability.

On the other hand, an extension of FastICA to independent subspace analysis (ISA), which is a model where the sources are no longer independent, was proposed in [54].

6.11 ITERATIVE ALGORITHMS WITH OPTIMAL STEP SIZE

6.11.1 Optimizing the step size

As explained throughout this chapter, the convergence properties of iterative techniques are to a large extent determined by the step size or learning rate employed in their update rules. Small step-size values may slow down convergence and thus reduce the tracking capabilities of the algorithm. Speed can be improved by increasing the step size, at the expense of larger oscillations around the optimal solution after convergence (misadjustment) or even the risk of not converging at all. This difficult trade-off between convergence speed and accuracy has spurred the development of iterative techniques based on some form of step-size optimization. Such research efforts are not exclusive to BSS/ICA, but include a variety of iterative techniques such as the ubiquitous LMS algorithm [35,44,55,56], which are not dealt with here; we rather restrain our attention to iterative algorithms for BSS/ICA.

In general, convergence proofs for stochastic algorithms [5] typically require that the learning parameter at the k-th iteration, denoted $\mu(k)$, verifies $\mu(k) \to 0$ and $\sum_k \mu(k) \to +\infty$ as $k \to +\infty$. A popular choice is $\mu(k) = \mu_0/k$, where μ_0 is the initial value of the step size. However, this choice approaches zero too quickly, and the stochastic algorithm may never reach the desired optimum in most cases. Ideally, $\mu(k)$ should be fitted to the degree of non-stationarity of the data, which may be difficult to determine beforehand. In some cases, these problems can be solved by the so-called *search-then-converge* strategy [27,40], where the learning parameter at iteration k is given by:

$$\mu(k) = \frac{\mu_0}{1+k/\tau} \qquad k = 1,2,\ldots$$

This technique guarantees that $\mu(k)$ stays close to a sufficiently large initial value μ_0 long enough to approach a solution. After this initial stage, the decreasing trend in $\mu(k)$ improves the final estimation accuracy. The search time constant τ is usually taken in the interval $100 \leqslant \tau \leqslant 500$, but its choice remains application dependent.

Regarding neural algorithms for ICA, Amari [1] and Amari and Cichocki [2] put forward adaptive rules for learning the step size in the context of neural algorithms. The idea is to make the step size depend on the gradient norm, in order to obtain a fast evolution at the beginning of the iterations and then a decreasing misadjustment as a stationary point is reached. These step-size learning rules, in turn, include other learning coefficients which must be set appropriately. Although the resulting algorithms are said to be robust to the choice of these coefficients, their optimal selection remains application dependent. Other guidelines for choosing the step size in natural gradient algorithms are given in [26], but are merely based on local stability conditions.

In the context of batch algorithms, Regalia [71] finds bounds for the step size guaranteeing monotonic convergence of the normalized fourth-order moment of the extractor output. These results are later extended in [72] to a more general class of functions valid for real-valued sources under prewhitening. Determining these step-size bounds for monotonic convergence involves the eigenspectrum of a Hessian matrix on a convex subset containing the unit sphere ($\|\mathbf{q}\| = 1$) and is thus a computationally expensive task.

While still ensuring monotonic convergence, an optimal step-size approach can easily be developed when the contrast can be expressed as a rational function of the learning rate at each iteration. When applied on the kurtosis contrast (6.16), this optimal step-size technique gives rise to the algorithm reviewed in the next section.

6.11.2 The RobustICA algorithm

Given a point $\mathbf{b}$, exact line search aims at finding the optimal step size leading to the *global maximum* of the objective function along the search direction $\mathbf{g}$:

$$\mu_{\text{opt}} = \arg\max_{\mu} \Upsilon(\mathbf{b} + \mu\mathbf{g}).$$

In this one-dimensional optimization problem, vectors $\mathbf{b}$ and $\mathbf{g}$ are fixed, so that $\Upsilon(\mathbf{b}+\mu\mathbf{g})$ becomes a function of the step size μ only and can thus be denoted (with some abuse of notation) as $\Upsilon(\mu)$. The exact line search technique is in general computationally intensive and presents other limitations [69], which explains why, despite being a well-known optimization method, it is very rarely used in practice. However, for the kurtosis contrast (6.16), $\Upsilon(\mu)$ is a rational function in μ [84,85]:

$$\Upsilon_\kappa(\mu) = \varepsilon\left(\frac{\mathbb{E}\{|y'|^4\} - |\mathbb{E}\{y'^2\}|^2}{\mathbb{E}^2\{|y'|^2\}} - 2\right) = \varepsilon\left(\frac{P(\mu)}{Q^2(\mu)} - 2\right) \tag{6.85}$$

where $y' = y + \mu g$, with $y = \mathbf{b}^{\mathrm{H}}\mathbf{x}$ and $g = \mathbf{g}^{\mathrm{H}}\mathbf{x}$. Polynomials $P(\cdot)$ and $Q(\cdot)$ are given by

$$P(\mu) = \sum_{k=0}^{4} h_k \mu^k \qquad Q(\mu) = \sum_{k=0}^{2} i_k \mu^k \tag{6.86}$$

with

$$\begin{aligned}
h_0 &= \mathbb{E}\{|a|^2\} - |\mathbb{E}\{a\}|^2 \\
h_1 &= 4\mathbb{E}\{|a|d\} - 4\mathbb{R}e(\mathbb{E}\{a\}\mathbb{E}^*\{c\}) \\
h_2 &= 4\mathbb{E}\{d^2\} + 2\mathbb{E}\{|a||b|\} - 4|\mathbb{E}\{c\}|^2 - 2\mathbb{R}e(\mathbb{E}\{a\}\mathbb{E}^*\{b\}) \\
h_3 &= 4\mathbb{E}\{|b|d\} - 4\mathbb{R}e(\mathbb{E}\{b\}\mathbb{E}^*\{c\}) \\
h_4 &= \mathbb{E}\{|b|^2\} - |\mathbb{E}\{b\}|^2 \\
i_0 &= \mathbb{E}\{|a|\}, \quad i_1 = 2\mathbb{E}\{d\}, \quad i_2 = \mathbb{E}\{|b|\}
\end{aligned} \tag{6.87}$$

and

$$a = y^2 \qquad b = g^2 \qquad c = yg \qquad d = \mathbb{R}e(yg^*). \tag{6.88}$$

As a result, the optimal step size can be obtained at low cost by finding the roots of a low-degree polynomial. This idea constitutes the basis for the *RobustICA* algorithm. The technique is also applicable to any other contrasts that can be expressed as polynomials or rational functions of μ, such as the constant modulus [37,83] or the constant power [39, 81] criteria used for blind equalization of digital communication channels, even in the semi-blind case (Chapter 15).

The RobustICA algorithm repeats the following steps until convergence:

(S1) Compute the search direction $\mathbf{g}$ at current point $\mathbf{b}$.
This direction is typically the gradient, $\mathbf{g} = \nabla\Upsilon_\kappa(\mathbf{b})$, which in the general case (including the possibility of complex-valued signals) is given by

$$\nabla\Upsilon_\kappa(\mathbf{b}) = \frac{4\varepsilon}{\mathbb{E}^2\{|y|^2\}}\left\{\mathbb{E}\{|y|^2y^*\mathbf{x}\} - \mathbb{E}\{y\mathbf{x}\}\mathbb{E}^*\{y^2\} - \frac{(\mathbb{E}\{|y|^4\} - |\mathbb{E}\{y^2\}|^2)\mathbb{E}\{y^*\mathbf{x}\}}{\mathbb{E}\{|y|^2\}}\right\}.$$

Newton directions can also be used at the expense of increased computational cost. To improve numerical conditioning, the search direction should be normalized to unit norm before passing to the next step.

(S2) Compute the optimal step-size polynomial coefficients.
The optimal step size is found among the roots of the polynomial in the numerator of $\partial \Upsilon_\kappa(\mu)/\partial \mu$. For the kurtosis contrast (6.85), this polynomial is given by:

$$p(\mu) = \sum_{k=0}^{4} a_k \mu^k. \tag{6.89}$$

Coefficients $\{a_k\}_{k=0}^4$ are easily derived in closed form after some algebraic manipulations [84,85]:

$$\begin{aligned}
a_0 &= -2h_0 i_1 + h_1 i_0 & a_1 &= -4h_0 i_2 - h_1 i_1 + 2h_2 i_0 \\
a_2 &= -3h_1 i_2 + 3h_3 i_0 & a_3 &= -2h_2 i_2 + h_3 i_1 + 4h_4 i_0 \\
a_4 &= -h_3 i_2 + 2h_4 i_1 & &
\end{aligned}$$

where $\{h_k\}_{k=0}^4$ and $\{i_k\}_{k=0}^2$ are given by Eqs (6.87) and (6.88).

(S3) Extract the optimal step-size polynomial roots $\{\mu_k\}_{k=1}^4$.
The step-size candidates are the real parts of the roots of polynomial (6.89). The roots of 4th-degree polynomials (quartics) can be found at practically no cost using standard algebraic procedures such as Ferrari's formula, known since the 16th century [69]. The computational complexity of this step is negligible compared with the calculation of the statistics required in the previous step.

(S4) Select the root leading to the *global maximum* of the contrast along the search direction:

$$\mu_{\text{opt}} = \arg\max_k \Upsilon_\kappa(\mathbf{b} + \mu_k \mathbf{g}). \tag{6.90}$$

To do so, we plug back the step-size candidates into Eqs (6.85) and (6.86). Again, this step is performed at a marginal cost from the coefficients computed in step S2. If the source kurtosis sign is not known, one just needs to replace (6.90) by

$$\mu_{\text{opt}} = \arg\max_k |\Upsilon_\kappa(\mathbf{b} + \mu_k \mathbf{g})| \tag{6.91}$$

as best root selection criterion. The algorithm can also be run by combining global line maximizations (6.90) and (6.91) for sources with known and unknown kurtosis sign, respectively, in any desired order.

(S5) Update $\mathbf{b}^+ = \mathbf{b} + \mu_{opt}\mathbf{g}$.

(S6) Normalize as $\mathbf{b}^+ \leftarrow \mathbf{b}^+/\|\mathbf{b}^+\|$.

As in [79], the extracting vector normalization in step S6 is performed to fix the ambiguity introduced by the scale invariance of contrast (6.16), and does not stem from prewhitening.

The advantages of RobustICA are many-fold:

- The generality of the kurtosis contrast (6.16) guarantees that real- and complex-valued signals can be treated by exactly the same algorithm without any modification. Both types of source signal can be present simultaneously in a given mixture, and complex sources need not be circular.
- Sequential extraction (deflation) can be performed via linear regression (section 6.9.2). As a result, prewhitening and the performance limitations it imposes [8] can be avoided. This feature may prove especially beneficial in ill-conditioned scenarios, the convolutive case, and underdetermined mixtures.
- The algorithm can target sub-Gaussian or super-Gaussian sources in the order defined by the user. This property enables the extraction of sources of interest when their Gaussianity character is known in advance, thus sparing a full separation of the observed mixture as well as the consequent unnecessary complexity and increased estimation error.
- The optimal step-size technique provides some robustness to the presence of saddle points and spurious local extrema in the contrast function.
- The method shows a very high convergence speed measured in terms of source extraction quality versus number of operations. In the real-valued two-signal case, the algorithm converges in a single iteration, even without prewhitening.

RobustICA's cost-efficiency and robustness are particularly remarkable for short sample length in the absence of prewhitening, where the method offers a superior quality-cost performance compared to other kurtosis-based algorithms such as FastICA (section 6.10.6) in the real case and the KM-F and nc-FastICA algorithms (section 6.10.7.3) in the complex non-circular case [82,84,85]. As all kurtosis-based methods, the main drawback of RobustICA is its sensitivity to outliers. This may be alleviated without sacrificing computational efficiency by means of more robust kurtosis estimators; e.g. simply discard samples over a certain amplitude threshold. Robust cumulant estimators are discussed in [60].

6.12 SUMMARY, CONCLUSIONS AND OUTLOOK

This chapter has intended to provide a general overview of some of the most significant iterative algorithms for BSS via ICA. Most methods for ICA being iterative, this survey has spanned over two decades of research into the topic. A wide number of batch and adaptive iterative algorithms have been derived from a rich variety of approaches, including heuristic ideas, geometrical notions and information-theoretical concepts. Although most of these methods rely on basic optimization techniques such as gradient and Newton iterations, exploiting the specific structural and statistical features of the BSS/ICA model yields techniques with improved performance such as the relative/natural gradient methods or deflationary algorithms like the popular FastICA and the more recent RobustICA. Due to space limitations, we could not treat other relevant iterative methods, discuss pertinent issues like complexity, or provide an experimental comparison of the different algorithms presented in the chapter. The computational complexity of the optimal step-size technique of section 6.11.2 and some of its variants is studied in section 15.7.2.5 of this book. Some iterative algorithms are compared in [16,36], among other works. A fair comparison should evaluate the separation quality versus computational cost trade-off of these techniques in different scenarios, as advocated in [82–85].

Several interesting avenues of research into iterative techniques for BSS/ICA remain open. The possibility of deriving globally convergent algorithms, along the lines of the deflation technique of [28] and the FastICA method [47], but with nonlinearities more adapted to the source distributions than kurtosis, should be ascertained. Even if a given method is shown to be globally convergent under some theoretical assumptions, spurious extrema can appear and hinder convergence to the right solution when real data is processed [78,84]. The existence of these spurious extrema should be analyzed for different contrasts and specific techniques put forward to improve the robustness to noise and finite sample effects. The optimal step-size methodology used in RobustICA signifies a first step in that direction, and could also help optimize the performance of stabilized FastICA (section 6.10.7.2); another is the exploitation of prior information as in semi-blind methods (Chapter 15). Some contrasts, such as the separator-output fourth-order moment sum, can also be maximized in batch mode by tensor-diagonalization based techniques. The comparative performance of the tensor diagonalization and iterative optimization approaches, in the cases where both are possible alternatives to contrast maximization, should receive closer attention.

References

[1] S. Amari, Natural gradient works efficiently in learning, Neural Computation 10 (1998) 251–276.

[2] S. Amari, A. Cichocki, Adaptive blind signal processing — Neural network approaches, Proceedings of the IEEE 86 (1998) 2026–2048.

[3] Y. Bar-Ness, J.W. Carlin, M.L. Steinberger, Bootstrapping adaptive cross pol cancelers for satellite communications, in: Proc. IEEE International Conference on Communications, vol. 2, Philadelphia, PA, June 13–17, 1982, pp. 4F.5.1–4F.5.5.

[4] A.J. Bell, T.J. Sejnowski, An information-maximization approach to blind separation and blind deconvolution, Neural Computation 7 (1995) 1129–1159.

[5] A. Benveniste, M. Métivier, P. Priouret, Algorithmes Adaptatifs et Approximations Stochastiques, Masson, Paris, 1987; English translation by S. S. Wilson, Adaptive Algorithms and Stochastic Approximations, Berlin: Springer-Verlag, 1990.

[6] E. Bingham, A. Hyvärinen, A fast fixed-point algorithm for independent component analysis of complex valued signals, International Journal of Neural Systems 10 (2000) 1–8.

[7] D.H. Brandwood, A complex gradient operator and its application in adaptive array theory, IEE Proceedings F: Communications Radar and Signal Processing 130 (1983) 11–16.
[8] J.-F. Cardoso, On the performance of orthogonal source separation algorithms, in: Proc. EUSIPCO-94, VII European Signal Processing Conference, Edinburgh, UK, September 13–16, 1994, pp. 776–779.
[9] J.-F. Cardoso, Estimating equations for source separation, in: Proc. ICASSP-97, 22nd IEEE International Conference on Acoustics, Speech and Signal Processing, vol. 5, Munich, Germany, April 21–24, 1997, pp. 3449–3452.
[10] J.-F. Cardoso, Infomax and maximum likelihood in blind source separation, IEEE Signal Processing Letters 4 (1997) 112–114.
[11] J.-F. Cardoso, Blind signal separation: Statistical principles, Proceedings of the IEEE 86 (1998) 2009–2025.
[12] J.-F. Cardoso, Learning in manifolds: the case of source separation, in: Proc. 9th IEEE Signal Processing Workshop on Statistical and Array Processing, Portland, OR, September 14–16, 1998.
[13] J.-F. Cardoso, Higher-order contrasts for independent component analysis, Neural Computation 11 (1999) 157–192.
[14] J.-F. Cardoso, B.H. Laheld, Equivariant adaptive source separation, IEEE Transactions on Signal Processing 44 (1996) 3017–3030.
[15] A. Chen, P.J. Bickel, Efficient independent component analysis, The Annals of Statistics 34 (2006) 2824–2855.
[16] P. Chevalier, L. Albera, P. Comon, A. Ferreol, Comparative performance analysis of eight blind source separation methods on radiocommunications signals, in: Proc. International Joint Conference on Neural Networks, Budapest, Hungary, July 25–29, 2004.
[17] S. Choi, A. Cichocki, S. Amari, Flexible independent component analysis, Journal of VLSI Signal Processing 26 (2000) 25–38.
[18] A. Cichocki, S.-I. Amari, Adaptive Blind Signal and Image Processing: Learning Algorithms and Applications, John Wiley & Sons, Inc., 2002.
[19] A. Cichocki, W. Kasprzak, S. Amari, Multi-layer neural networks with a local adaptive learning rule for blind separation of source signals, in: Proc. NOLTA-95, International Symposium on Nonlinear Theory and its Applications, Tokyo, Japan, 1995, pp. 61–65.
[20] A. Cichocki, R. Ubehauen, L. Moszczyński, E. Rummert, A new on-line adaptive learning algorithm for blind separation of source signals, in: Proc. International Symposium on Artificial Neural Networks, Taiwan, December 1994, pp. 406–411.
[21] P. Comon, Séparation de mélanges de signaux, in: Actes 12ème Colloque GRETSI, Juan les Pins, 12–16 juin 1989, pp. 137–140.
[22] P. Comon, Separation of stochastic processes, in: Proc. Workshop on Higher-Order Spectral Analysis, Vail, CO, June 28–30, 1989, pp. 174–179.
[23] P. Comon, Independent component analysis, a new concept?, Signal Processing 36 (1994) 287–314. Special Issue on Higher-Order Statistics.
[24] P. Comon, C. Jutten, J. Hérault, Blind separation of sources, part II: Problems statement, Signal Processing 24 (1991) 11–20.
[25] T.M. Cover, J.A. Thomas, Elements of Information Theory, John Wiley & Sons, New York, 1991.
[26] S. Cruces-Alvarez, A. Cichocki, S. Amari, From blind signal extraction to blind instantaneous signal separation: Criteria, algorithms, and stability, IEEE Transactions on Neural Networks 15 (2004) 859–873.
[27] C. Darken, J. Moody, Towards faster stochastic gradient search, in: J. Moody, S. Hanson, R. Lippmann (Eds.), Advances in Neural Information Processing Systems, vol. 4, Morgan Kaufmann, 1992, pp. 1009–1016.
[28] N. Delfosse, P. Loubaton, Adaptive blind separation of independent sources: A deflation approach, Signal Processing 45 (1995) 59–83.
[29] Y. Deville, A unified stability analysis of the Hérault-Jutten source separation neural network, Signal Processing 51 (1996) 229–233.

[30] Y. Deville, Analysis of the convergence properties of self-normalized source separation neural networks, IEEE Transactions on Signal Processing 47 (1999) 1272–1287.
[31] Y. Deville, O. Albu, N. Charkanin, New self-normalized blind source separation networks for instantaneous and convolutive mixtures, International Journal of Adaptive Control and Signal Processing 16 (2002) 753–761.
[32] A. Dinç, Y. Bar-Ness, Bootstrap: A fast blind adaptive signal separator, in: Proc. ICASSP-92, 17th IEEE International Conference on Acoustics, Speech and Signal Processing, vol. II, San Francisco, CA, March 23–26, 1992, pp. 325–328.
[33] D. Donoho, On minimum entropy deconvolution, in: Proc. 2nd Applied Time Series Analysis Symposium, Tulsa, OK, 1980, pp. 565–608.
[34] S.C. Douglas, Fixed-point algorithms for the blind separation of arbitrary complex-valued non-Gaussian signal mixtures, EURASIP Journal on Advances in Signal Processing (2007).
[35] N.D. Gaubitch, M. Kamrul Hasan, P.A. Naylor, Generalized optimal step-size for blind multichannel LMS system identification, IEEE Signal Processing Letters 13 (2006) 624–627.
[36] X. Giannakapoulos, J. Karhunnen, E. Oja, An experimental comparison of neural algorithms for independent component analysis and blind separation, International Journal of Neural Systems 9 (1999) 99–114.
[37] D.N. Godard, Self-recovering equalization and carrier tracking in two-dimensional data communication systems, IEEE Transactions on Communications 28 (1980) 1867–1875.
[38] G.H. Golub, C.F. Van Loan, Matrix Computations, 3rd ed., The Johns Hopkins University Press, Baltimore, MD, 1996.
[39] O. Grellier, P. Comon, Blind separation of discrete sources, IEEE Signal Processing Letters 5 (1998) 212–214.
[40] F. Ham, I. Kostanic, Principles of Neurocomputing for Science and Engineering, McGraw-Hill, 2001.
[41] F. Harroy, J.-L. Lacoume, Maximum likelihood estimators and Cramer-Rao bounds in source separation, Signal Processing 55 (1996) 167–177.
[42] F. Harroy, J.-L. Lacoume, M.A. Lagunas, A general adaptive algorithm for nonGaussian source separation without any constraint, in Proc. EUSIPCO-94, VII European Signal Processing Conference, Edinburgh, UK, September 13–16, 1994, pp. 1161–1164.
[43] J. Hérault, C. Jutten, B. Ans, Détection de grandeurs primitives dans un message composite par une architecture neuromimétique en apprentissage non supervisé, in: Actes 10ème Colloque GRETSI, Nice, France, May 20–24, 1985, pp. 1017–1022.
[44] Y. Huang, J. Benesty, J. Chen, Optimal step size of the adaptive multichannel LMS algorithm for blind SIMO identification, IEEE Signal Processing Letters 12 (3) (2005) 173–176.
[45] A. Hyvärinen, One-unit contrast functions for independent component analysis: A statistical analysis, in: Proc. IEEE Neural Networks for Signal Processing Workshop, Amelia Island, FL, 1997, pp. 388–397.
[46] A. Hyvärinen, New approximations of differential entropy for independent component analysis and projection pursuit, in: Advances in Neural Information Processing Systems, vol. 10, MIT Press, 1998, pp. 273–279.
[47] A. Hyvärinen, Fast and robust fixed-point algorithms for independent component analysis, IEEE Transactions on Neural Networks 10 (1999) 626–634.
[48] A. Hyvärinen, J. Karhunen, E. Oja, Independent Component Analysis, John Wiley & Sons, New York, 2001.
[49] A. Hyvärinen, E. Oja, A fast fixed-point algorithm for independent component analysis, Neural Computation 9 (1997) 1483–1492.
[50] C. Jutten, J. Hérault, Une solution neuromimétique au problème de séparation de sources, Traitement du signal 5 (1988) 389–403.
[51] C. Jutten, J. Hérault, Blind separation of sources, part I: An adaptive algorithm based on neuromimetic architecture, Signal Processing 24 (1991) 1–10.

[52] J. Karvanen, V. Koivunen, Blind separation methods based on Pearson system and its extensions, Signal Processing 82 (2002) 663–673.

[53] Z. Koldovský, P. Tichavský, E. Oja, Efficient variant of algorithm FastICA for independent component analysis attaining the cramér-rao lower bound, IEEE Transactions on Neural Networks 17 (2006) 1265–1277.

[54] U. Köster, A. Hyvärinen, FastISA: A fast fixed-point algorithm for independent subspace analysis, in Proc. ESANN-2006, European Symposium on Artificial Neural Networks, Bruges, Belgium, 2006.

[55] A.M. Kuzminskiy, Automatic choice of the adaption coefficient under nonstationary conditions, Radioelectronics and Communications Systems 25 (1982) 72–74.

[56] A.M. Kuzminskiy, A robust step size adaptation scheme for LMS adaptive filters, in: Proc. DSP'97, 13th International Conference on Digital Signal Processing, vol. 1, Santorini, Greece, July 2–4, 1997.

[57] B. Laheld, J.-F. Cardoso, Adaptive source separation with uniform performance, in: Proc. EUSIPCO-94, VII European Signal Processing Conference, Edinburgh, UK, September 13–16, 1994, pp. 183–186.

[58] H. Li, T. Adali, A class of complex ICA algorithms based on the kurtosis cost function, IEEE Transactions on Neural Networks 19 (2008) 408–420.

[59] O. Macchi, E. Moreau, Self-adaptive source separation, part I: Convergence analysis of a direct linear network controled by the Hérault-Jutten algorithm, IEEE Transactions on Signal Processing 45 (1997) 918–926.

[60] D. Mampel, A.K. Nandi, Robust cumulant estimation, in: A.K. Nandi (Ed.), Blind Estimation Using Higher-Order Statistics, Kluwer Academic Publishers, Boston, MA, 1999, chapter 5.

[61] P. McCullagh, Tensor Methods in Statistics, in: Monographs on Statistics and Applied Probability, Chapman and Hall, London, 1987.

[62] E. Moreau, Criteria for complex sources separation, in: Proc. EUSIPCO-96, VIII European Signal Processing Conference, Trieste, Italy, September 10–13, 1996, pp. 931–934.

[63] E. Moreau, O. Macchi, High-order contrasts for self-adaptive source separation, International Journal of Adaptive Control and Signal Processing 10 (1996) 19–46.

[64] M. Novey, T. Adali, Complex ICA by negentropy maximization, IEEE Transactions on Neural Networks 19 (2008) 596–609.

[65] M. Novey, T. Adali, On extending the complex FastICA algorithm to noncircular sources, IEEE Transactions on Signal Processing 56 (2008) 2148–2154.

[66] C.B. Papadias, Globally convergent blind source separation based on a multiuser kurtosis maximization criterion, IEEE Transactions on Signal Processing 48 (2000) 3508–3519.

[67] D.T. Pham, P. Garat, Blind separation of mixture of independent sources through a quasi-maximum likelihood approach, IEEE Transactions on Signal Processing 45 (1997) 1712–1725.

[68] D.T. Pham, P. Garat, C. Jutten, Separation of a mixture of independent sources through a maximum-likelihood approach, in: Proc. EUSIPCO-92, VI European Signal Processing Conference, Brussels, Belgium, 1992, pp. 771–774.

[69] W.H. Press, S.A. Teukolsky, W.T. Vetterling, B.P. Flannery, Numerical Recipes in C. The Art of Scientific Computing, 2nd ed., Cambridge University Press, Cambridge, UK, 1992.

[70] A. Prieto, C.G. Puntonet, B. Prieto, A neural learning algorithm for blind separation of sources based on geometric properties, Signal Processing 64 (1998) 315–331.

[71] P.A. Regalia, A finite-interval constant modulus algorithm, in: Proc. ICASSP-2002, 27th International Conference on Acoustics, Speech and Signal Processing, vol. III, Orlando, FL, May 13–17, 2002, pp. 2285–2288.

[72] P.A. Regalia, E. Kofidis, Monotonic convergence of fixed-point algorithms for ICA, IEEE Transactions on Neural Networks 14 (2003) 943–949.

[73] T. Ristaniemi, J. Joutsensalo, Advanced ICA-based receivers for block fading DS-CDMA channels, Signal Processing 82 (2002) 417–431.

[74] O. Shalvi, E. Weinstein, New criteria for blind deconvolution of nonminimum phase systems (channels), IEEE Transactions on Information Theory 36 (1990) 312–321.

[75] O. Shalvi, E. Weinstein, Super-exponential methods for blind deconvolution, IEEE Transactions on Information Theory 39 (1993) 504–519.

[76] E. Sorouchyari, Blind separation of sources, part III: Stability analysis, Signal Processing 24 (1991) 21–29.

[77] A. Stuart, K. Ord, Kendall's Advanced Theory of Statistics, 6th ed., vol. 1, Hodder Arnold, 1994.

[78] P. Tichavský, Z. Koldovský, E. Oja, Performance analysis of the FastICA algorithm and Cramér-Rao bounds for linear independent component analysis, IEEE Transactions on Signal Processing 54 (2006) 1189–1203.

[79] J.K. Tugnait, Identification and deconvolution of multichannel linear non-Gaussian processes using higher order statistics and inverse filter criteria, IEEE Transactions on Signal Processing 45 (1997) 658–672.

[80] R.A. Wiggins, Minimum entropy deconvolution, Geoexploration 16 (1978) 21–35.

[81] V. Zarzoso, P. Comon, Blind and semi-blind equalization based on the constant power criterion, IEEE Transactions on Signal Processing 53 (2005) 4363–4375.

[82] V. Zarzoso, P. Comon, Comparative speed analysis of FastICA, in: Proc. ICA-2007, 7th International Conference on Independent Component Analysis and Signal Separation, London, UK, September 9–12, 2007, pp. 293–300.

[83] V. Zarzoso, P. Comon, Optimal step-size constant modulus algorithm, IEEE Transactions on Communications 56 (2008) 10–13.

[84] V. Zarzoso, P. Comon, Robust independent component analysis by iterative maximization of the kurtosis contrast with algebraic optimal step size, IEEE Transactions on Neural Networks 21 (2010).

[85] V. Zarzoso, P. Comon, M. Kallel, How fast is FastICA?, in: Proc. EUSIPCO-2006, XIV European Signal Processing Conference, Florence, Italy, September 4–8, 2006.

[86] V. Zarzoso, P. Comon, R. Phlypo, A contrast function for independent component analysis without permutation ambiguity, Research Report I3S/RR-2009-04-FR, Laboratoire I3S, Sophia Antipolis, France, March 2009 (http://www.i3s.unice.fr/~mh/RR/2009/RR-09.04-V.ZARZOSO.pdf).

[87] V. Zarzoso, A.K. Nandi, Blind separation of independent sources for virtually any source probability density function, IEEE Transactions on Signal Processing 47 (1999) 2419–2432.

[88] V. Zarzoso, A.K. Nandi, Adaptive blind source separation for virtually any source probability density function, IEEE Transactions on Signal Processing 48 (2000) 477–488.

[89] V. Zarzoso, R. Phlypo, P. Comon, A contrast for independent component analysis with priors on the source kurtosis signs, IEEE Signal Processing Letters 15 (2008) 501–504.

[90] L. Zhang, A. Cichocki, S. Amari, Self-adaptive blind source separation based on activation functions adaptation, IEEE Transactions on Neural Networks 15 (2004) 233–244.

CHAPTER

Second-order methods based on color

7

A. Yeredor

7.1 INTRODUCTION

The previous chapters focused on the basic, classical ICA paradigm, in which one of the underlying assumptions is that each of the independent sources can be modeled as a sequence of *independent, identically distributed* (iid) random variables (possibly, but not necessarily, with a different distribution for each source). As such, the only key for separation in a truly blind scenario is the non-Gaussianity of the sources. Thus, as shown in previous chapters, second-order statistics (*SOS*) alone cannot yield consistent separation in such models, simply because they are unable to "capture" the non-Gaussianity. They may be used for attaining spatial *decorrelation* (a.k.a. spatial *whitening* or *sphering*), but then there remains an orthogonal (rotation) factor, which is indistinguishable by SOS.

However, in many practical BSS problems, the sources are not iid in time. They often exhibit some temporal structures, which may be characterized by statistical properties, such as temporal correlations. In such cases many of the classical ICA approaches may still be (and often are) used, simply ignoring these temporal structures. Fortunately, however, in such cases SOS may also be helpful (beyond spatial whitening) whenever the sources' second-order temporal statistics exhibit sufficient diversity. In fact, in such cases, it is also possible to base the separation on SOS alone, ignoring the higher-order features of the sources' distributions. A resulting fundamental difference from the classical iid model is that with sufficiently diverse temporal structures, it becomes possible to separate Gaussian sources, which are not separable using any of the classical approaches which ignore temporal structures. Moreover, the ability to employ the Gaussianity assumptions (one of the favorite assumptions in signal processing, which is nonetheless virtually despicable in the classical ICA context) enables one to conveniently construct estimators which are (at least asymptotically) optimal in this model – a desirable property, which is far less convenient to attain in classical ICA.

Random processes with second-order temporal statistical structures are classically divided into two broad classes: Wide-sense stationary (*WSS*) processes and nonstationary processes. WSS processes are characterized by the property that the statistical correlation between any two samples thereof depends only on the time-difference between the sampling instants. Such processes exhibit many appealing properties in the time and

Handbook of Blind Source Separation: Independent Component Analysis and Applications

frequency domains, which can be conveniently exploited for separation of their mixtures. Thus, the use of SOS for separation of WSS processes (often also called "colored" sources) is the main theme of this chapter.

SOS-based separation approaches for WSS sources can roughly be divided into two categories: approaches exploiting the special structure of the correlation matrices through (approximate) joint diagonalization; and approaches based on the principle of Maximum Likelihood (ML). While in general ML estimation is based on more than SOS (and is rather involved), under the assumption of Gaussian sources the ML estimate takes a relatively simple form (asymptotically) and is indeed based on SOS alone. Using the Gaussianity assumption, it is also possible to apply (asymptotically) optimal weighting to the joint-diagonalization based approach, thus obtaining estimates which are asymptotically optimal, and are thus asymptotically equivalent to the ML estimate.

The chapter is structured as follows. In the next section we provide some preliminary definitions and concepts associated with discrete-time WSS processes in general, and with parametric (Autoregressive (*AR*), Moving-Average (*MA*) and Autoregressive Moving-Average (*ARMA*)) processes in particular. In section 7.3 we outline the problem formulation and discuss issues of identifiability and of performance measures and bounds. In section 7.4 we present joint-diagonalization based methods, whereas in section 7.5 we present ML-based methods. In section 7.6 we discuss some supplementary issues, such as the effect of additive noise; some particular cases of nonstationary sources; and the framework of complex-valued signals.

7.2 WSS PROCESSES

Let $s(t)$, $t \in \mathbb{Z}$ denote a real-valued discrete-time random process, assumed for simplicity (and without loss of generality, for our purposes) to have zero mean. With vanishing first-order statistics, its second-order statistics are fully described by its correlation function

$$\breve{R}_s(t,\tau) \triangleq \mathbb{E}\{s(t+\tau)s(t)\} \quad \forall t,\tau \in \mathbb{Z}. \tag{7.1}$$

When $\breve{R}_s(t,\tau)$ depends only on τ, the process is called WSS, and we may simply denote $R_s(\tau) \triangleq \breve{R}_s(t,\tau)$. The Fourier transform of the correlation $R_s(\tau)$ is the *spectrum* (or spectral power density),

$$S_s(\nu) \triangleq \sum_{\tau=-\infty}^{\infty} R_s(\tau) e^{-j2\pi\nu\tau} \tag{7.2}$$

An appealing property of WSS processes is that since the covariance matrix of any set of T consecutive samples is a $T \times T$ Toeplitz matrix, it is asymptotically diagonalized by the Fourier matrix. Indeed, define the (zero-mean) normalized Fourier coefficients of the series $s(1),\ldots,s(T)$ at Fourier frequencies $\nu_k \triangleq \frac{k}{T}$,

$$\check{s}\left(\frac{k}{T}\right) \triangleq \frac{1}{\sqrt{T}} \sum_{t=1}^{T} s(t) e^{-j2\pi\frac{kt}{T}}, \quad k = 0, 1, \ldots, T-1. \tag{7.3}$$

Asymptotically, $\check{s}(\frac{k_1}{T})$ and $\check{s}(\frac{k_2}{T})$ are uncorrelated for $k_1 \neq k_2$,

$$\lim_{T\to\infty} \mathbb{E}\left\{\check{s}\left(\frac{k_1}{T}\right)\check{s}^*\left(\frac{k_2}{T}\right)\right\} = 0 \quad k_1 \neq k_2, \tag{7.4}$$

and the variance of $\check{s}(\frac{k}{T})$ is given by the spectrum at the respective frequency:

$$\lim_{T\to\infty} \mathbb{E}\{|\check{s}(k)|^2\} = S_s\left(\frac{k}{T}\right). \tag{7.5}$$

When $s(t)$ is a Gaussian process, these Fourier coefficients, being linear combinations of samples of $s(t)$, are jointly Gaussian as well (complex Gaussian, to be precise, except at $k = 0$ and, for even values of T, also at $k = \frac{T}{2}$). Therefore, since the (normalized) Fourier transform establishes an invertible, unit-Jacobian transformation between the time-samples and the normalized Fourier coefficients at $k = 0, \ldots, \frac{T}{2}$ (for a real-valued signal), the joint probability density function (*pdf*) of $s(1), \ldots, s(T)$ can be conveniently expressed in the Gaussian case in terms of the *pdf* of these coefficients, as follows (assuming T is even):

$$\begin{aligned} f(s(1),\ldots,s(T)) = & \frac{1}{\sqrt{2\pi S_s(0)}}\exp\left[-\frac{\check{s}^2(0)}{2S_s(0)}\right]\cdot\prod_{k=1}^{\frac{T}{2}-1}\frac{1}{\pi S_s(\frac{k}{T})}\exp\left[-\frac{|\check{s}(\frac{k}{T})|^2}{S_s(\frac{k}{T})}\right] \\ & \cdot\frac{1}{\sqrt{2\pi S_s(\frac{1}{2})}}\exp\left[-\frac{\check{s}^2(\frac{1}{2})}{2S_s(\frac{1}{2})}\right] \end{aligned} \tag{7.6}$$

Exploiting the periodicity, combined with the conjugate-symmetry of $\check{s}(\nu)$ and the symmetry of $S_s(\nu)$, we may also write this expression more compactly as

$$f(s(1),\ldots,s(T)) = \frac{1}{2}\prod_{k=0}^{T-1}\frac{1}{\sqrt{\pi S_s(\frac{k}{T})}}\exp\left[-\frac{|\check{s}(\frac{k}{T})|^2}{2S_s(\frac{k}{T})}\right] \tag{7.7}$$

Two processes $s_1(t)$ and $s_2(t)$ are said to be *jointly WSS* (*JWSS*) if they are each WSS, and if, in addition, their cross-correlation function $\breve{R}_{s1,s2}(t,\tau) \triangleq \mathbb{E}\{s_1(t+\tau)s_2(t)\}$ depends only on τ. A vector of signals $\mathbf{s}(\cdot) = [s_1(\cdot), \ldots, s_N(\cdot)]^T$ is said to be JWSS if every two signals in the vector are JWSS. In this case the correlation matrix $\mathbf{R}_s(\tau)$ and the spectrum $\mathbf{S}_s(\nu)$ of the vector signal are given by

$$\mathbf{R}_s(\tau) = \mathbb{E}\{\mathbf{s}(t+\tau)\mathbf{s}^T(t)\}, \quad \mathbf{S}_s(\nu) = \sum_{\tau=-\infty}^{\infty}\mathbf{R}_s(\tau)e^{-j2\pi\nu\tau}. \tag{7.8}$$

7.2.1 Parametric WSS processes

A parametric WSS process is a process whose spectrum depends on a finite number of parameters. A classical family of parametric processes are "linear processes", which can be modeled as the response of a rational, monic, linear, time-invariant (*LTI*) system to an iid input, which, as far as SOS are concerned, is a WSS process $w(t)$ with a flat spectrum $S_w(\nu) = \sigma_w^2$, namely an uncorrelated sequence with constant variance σ_w^2 (sometimes called the "*innovation sequence*"). When the LTI system is a finite impulse response (FIR) system, the resulting process is called a Moving Average (MA) process; when the LTI system is an all-poles (infinite impulse response (IIR)) system, the resulting process is called an Auto-Regressive (AR) process; and when the LTI system has both poles and zeros, the resulting process is called an Auto-Regressive Moving Average (ARMA) process. We provide a brief overview of the main features of these types of processes in the following subsections.

7.2.1.1 *MA processes*

Let the Z-transform of the LTI system's impulse response be given by

$$H(z)=B(z)=1+b_1z^{-1}+\cdots+b_Qz^{-Q}=\prod_{k=1}^{Q}(1-\varrho_k z^{-1}), \tag{7.9}$$

where the coefficients $b_0, b_1, \ldots, b_Q$ ($b_0 = 1$ takes care of the scaling convention, by which $H(z)$ is monic) are the impulse-response coefficients and $\varrho_1, \ldots, \varrho_Q$ are the Q zeros of $H(z)$. The output process $s(t)$ satisfies the difference equation

$$s(t)=\sum_{\ell=0}^{Q} b_\ell w(t-\ell). \tag{7.10}$$

Its spectrum is given by

$$S_s(\nu)=\sigma_w^2|B(e^{j2\pi\nu})|^2=\sigma_w^2\left|\sum_{\ell=0}^{Q} b_\ell e^{-j2\pi\nu\ell}\right|^2=\sigma_w^2\prod_{k=1}^{Q}|1-\varrho_k e^{-j2\pi\nu}|^2, \tag{7.11}$$

whereas its correlation is given by

$$R_s(\tau)=\sigma_w^2\sum_{\ell=0}^{Q} b_\ell b_{\ell+\tau} \quad \forall\tau \tag{7.12}$$

under the convention that $b_{\ell+\tau}$ is taken as zero for $\ell+\tau \notin \{0,\ldots,Q\}$. It is readily seen that $R_s(\tau)$ is a finite sequence, extending from $\tau=-Q$ to $\tau=Q$.

The resulting process is called an MA(Q) process.

7.2.1.2 AR processes

Now let the Z-transform of the LTI system's impulse response be given by

$$H(z) = \frac{1}{A(z)} = \frac{1}{1 + a_1 z^{-1} + \cdots a_P z^{-P}} = \frac{1}{\prod_{k=1}^{P}(1 - \xi_k z^{-1})}, \tag{7.13}$$

where the AR coefficients $a_0, a_1, \ldots, a_P$ ($a_0 = 1$ takes care of the scaling convention, by which $H(z)$ is monic) are the impulse response coefficients of the system's inverse, and $\xi_1, \ldots, \xi_P$ are the P poles of $H(z)$, assumed to be strictly within the unit-circle in the Z-plane (namely, $|\xi_k| < 1, k = 1, \ldots, P$). The output process $s(t)$ satisfies the difference equation

$$s(t) = -\sum_{\ell=1}^{P} a_\ell s(t - \ell) + w(t). \tag{7.14}$$

Its spectrum is given by

$$S_s(\nu) = \frac{\sigma_w^2}{|A(e^{j2\pi\nu})|^2} = \frac{\sigma_w^2}{\left|\sum_{\ell=0}^{P} a_\ell e^{-j2\pi\nu\ell}\right|^2} = \frac{\sigma_w^2}{\prod_{k=1}^{P} |1 - \xi_k e^{-j2\pi\nu}|^2}. \tag{7.15}$$

The correlation function satisfies, together with the AR coefficients and σ_w^2, the so-called *Yule-Walker equations*,

$$\begin{bmatrix} R_s(0) & R_s(-1) & \cdots & R_s(-P) \\ R_s(1) & R_s(0) & \ddots & R_s(-P+1) \\ \vdots & \ddots & \ddots & \vdots \\ R_s(P) & \cdots & R_s(1) & R_s(0) \end{bmatrix} \begin{bmatrix} a_0 \\ a_1 \\ \vdots \\ a_P \end{bmatrix} = \begin{bmatrix} \sigma_w^2 \\ 0 \\ \vdots \\ 0 \end{bmatrix}. \tag{7.16}$$

Since H(z) is an IIR system, $R_s(\tau)$ is an infinite sequence.

The resulting process is called an AR(p) process.

7.2.1.3 ARMA processes

Finally, let the Z-transform of the LTI system's impulse response be given by

$$H(z) = \frac{B(z)}{A(z)} = \frac{1 + b_1 z^{-1} + \cdots b_Q z^{-Q}}{1 + a_1 z^{-1} + \cdots a_P z^{-P}} = \frac{\prod_{k=1}^{Q}(1 - \varrho_k z^{-1})}{\prod_{k=1}^{P}(1 - \xi_k z^{-1})}, \tag{7.17}$$

where $a_0, a_1, \ldots, a_P$ and $b_0, b_1, \ldots, b_Q$ are, respectively, the AR and MA coefficient ($a_0 = 1$ and $b_0 = 1$ take care of the scaling convention, by which $H(z)$ is monic). The Q zeros are $\varrho_1,\ldots,\varrho_Q$, and the P poles are $\xi_1,\ldots,\xi_P$, which are assumed to be strictly within the unit-circle. The output process $s(t)$ satisfies the difference equation

$$s(t) = -\sum_{\ell=1}^{P} a_\ell s(t-\ell) + \sum_{\ell=0}^{Q} b_\ell w(t-\ell). \tag{7.18}$$

Its spectrum is given by

$$S_s(\nu) = \sigma_w^2 \frac{|B(e^{j2\pi\nu})|^2}{|A(e^{j2\pi\nu})|^2} = \sigma_w^2 \left| \frac{\sum_{\ell=0}^{Q} b_\ell e^{-j2\pi\nu\ell}}{\sum_{\ell=0}^{P} a_\ell e^{-j2\pi\nu\ell}} \right|^2 = \sigma_w^2 \frac{\prod_{k=1}^{Q} |1-\varrho_k e^{-j2\pi\nu}|^2}{\prod_{k=1}^{P} |1-\xi_k e^{-j2\pi\nu}|^2}. \tag{7.19}$$

The correlation function satisfies, together with the AR coefficients, σ_w^2 and b_Q, the so-called *modified Yule-Walker equations*,

$$\begin{bmatrix} R_s(Q) & R_s(Q-1) & \cdots & R_s(Q-P) \\ R_s(Q+1) & R_s(Q) & \ddots & R_s(Q-P+1) \\ \vdots & \ddots & \ddots & \vdots \\ R_s(Q+P) & \cdots & R_s(Q+1) & R_s(Q) \end{bmatrix} \begin{bmatrix} a_0 \\ a_1 \\ \vdots \\ a_P \end{bmatrix} = \begin{bmatrix} b_Q \sigma_w^2 \\ 0 \\ \vdots \\ 0 \end{bmatrix}. \tag{7.20}$$

Since H(z) is an IIR system, $R_s(\tau)$ is an infinite sequence.

The resulting process is called an ARMA(P,Q) process.

7.3 PROBLEM FORMULATION, IDENTIFIABILITY AND BOUNDS

The main theme of this chapter is concerned with linear instantaneous, square, invertible noiseless mixtures of real-valued sources. The extension of some (but not of all) of the presented approaches and algorithms to the case of complex-valued sources is rather straightforward, as briefly discussed in the last section (see also section 7.6). The possible presence of additive noise is a more subtle issue, also deferred to section 7.6. An "over-determined" model, in which the number of observed mixtures P is larger than the number of sources N, is useful only in the presence of noise (allowing the use of subspace methods, as we shall see in section 7.6). In the noiseless case, however, there is no advantage in using the additional (beyond N) mixtures, and any group of N out of the P mixtures can be used (as long as the associated $N \times N$ mixing matrix is invertible). As for the "under-determined" ($P < N$) case, as well as for cases of nonlinear or non-instantaneous (i.e. convolutive) mixtures, the reader is referred to Chapters 9 ("Under-determined Mixtures"), 14 ("Nonlinear Mixtures") and 8 ("Convolutive Mixtures"), respectively.

The observed discrete-time mixtures are given by

$$\mathbf{x}(t) = \mathbf{A}\mathbf{s}(t), \quad t = 1, \ldots, T \tag{7.21}$$

and the goal is to recover $\mathbf{s}(t)$ from $\mathbf{x}(t)$, most commonly via estimation of the $N \times N$ invertible mixing matrix A (or, equivalently, its inverse $\mathbf{B}$), using the observations' empirical SOS alone. The following model assumptions are employed regarding the sources:

The sources $s_n(\cdot), n = 1, \ldots, N$ are real-valued, zero-mean WSS processes, second-order ergodic and uncorrelated with each other.

The requirement of the sources being mutually uncorrelated is often called "spatial decorrelation". Note that this requirement is in general weaker than the classical ICA condition of mutual independence. Indeed, SOS-based separation algorithms do not require mutual independence, and rely exclusively on mutual decorrelation. Nevertheless, it has to be noted that for (asymptotic) optimality of some of the algorithms that will be presented in the sequel, the sources are assumed to be jointly Gaussian. Inevitably, sources which are jointly Gaussian and mutually uncorrelated are also mutually independent. Therefore, in the case of jointly Gaussian sources the decorrelation condition is equivalent to an independence condition.

The immediate consequence of the stationarity and spatial decorrelation is that the sources are JWSS, and their correlation matrices at different lags are all diagonal, given by

$$\mathbf{R}_s(\tau) = \mathbb{E}\{\mathbf{s}(t+\tau)\mathbf{s}^T(t)\} = \text{Diag}\{R_1(\tau), R_2(\tau), \ldots R_N(\tau)\} \tag{7.22}$$

(we shall use the notations $R_n(\tau)$ and $S_n(\nu)$ to denote the correlation sequence and the spectrum (resp.) of the n-th source). Consequently, the observations' correlation matrices are given by

$$\mathbf{R}_x(\tau) = \mathbb{E}\{\mathbf{x}(t+\tau)\mathbf{x}^T(t)\} = \mathbf{A}\mathbf{R}_s(\tau)\mathbf{A}^T. \tag{7.23}$$

The second-order ergodicity of the sources (combined with the instantaneous, invertible linear mixing) implies second-order ergodicity of the observations. In other words, $\mathbf{R}_x(\tau)$ can be estimated (for any fixed τ) to arbitrary precision from a sufficiently long observation interval (namely, a sufficiently large T).

7.3.1 Indeterminacies and identifiability

Before we turn to discuss the *identifiability* of $\mathbf{A}$ (or of $\mathbf{B}$) from SOS of the observations, we recall the inevitable fundamental limitations and indeterminacies of BSS. Basically, there are three *ambiguity factors*, each being a square ($N \times N$) matrix:

- $\mathbf{\Pi}$ is a *permutation factor*, with a single non-zero element (which equals 1) in each row and column.

- $\mathbf{\Delta}$ is a *scale factor*, with a diagonal matrix with strictly positive elements along its diagonal.
- $\boldsymbol{\varepsilon}$ is a *sign factor*, with a diagonal matrix with values of ± 1 along its diagonal.

The factors $\mathbf{\Delta}$ and $\boldsymbol{\varepsilon}$ can generally be joined into a single scale-sign factor $\mathbf{\Delta}\boldsymbol{\varepsilon}$, but we prefer to keep the distinction, due to one fundamental difference between these factors: the scaling ambiguity can be eliminated by employing some sources-scaling convention (see below) as a "working assumption". However, the sign ambiguity cannot be eliminated by any "working assumption" regarding the SOS of the sources, simply because SOS are "blind" to the signs of the sources.

Using these notations, we observe that

$$\mathbf{A}' = \mathbf{A}\mathbf{\Delta}\boldsymbol{\varepsilon}\mathbf{\Pi}, \quad \mathbf{s}'(\cdot) = \mathbf{\Pi}^T\boldsymbol{\varepsilon}\mathbf{\Delta}^{-1}\mathbf{s}(\cdot) \quad \Rightarrow \quad \mathbf{A}'\mathbf{s}'(\cdot) = \mathbf{A}\mathbf{s}(\cdot) \tag{7.24}$$

yielding the very same observations, whereas the modified sources $\mathbf{s}(\cdot)$ still satisfy the model assumptions. Therefore $\mathbf{A}'$ is not distinguishable from $\mathbf{A}$. We therefore say that $\mathbf{A}'$ is *equivalent* to $\mathbf{A}$ (and $\mathbf{B}' = \mathbf{\Pi}^T\boldsymbol{\varepsilon}\mathbf{\Delta}^{-1}\mathbf{B}$ is equivalent to $\mathbf{B}$).

In order to mitigate these ambiguities, we employ some "working assumptions", or conventions, which need not hold in practice, but would enable us to obtain a unique solution among the set of all equivalent solutions. The first assumption is a scaling assumption on the sources, aimed at fixing $\mathbf{\Delta}$. The most commonly used scaling assumption is to assume that all sources have unit-power, namely that $\mathbf{R}_s(0) = \mathbf{I}$. Such a scaling assumption is employed, for example, in AMUSE [52,51] and in SOBI [6] (see the following section for acronym interpretation and details). An alternative scaling assumption can be employed when the sources are assumed to be parametric (MA, AR, ARMA), by assuming unit variance of the innovation sequences, namely $\sigma_w^2 = 1$ for all sources. Such an assumption is employed, for example, in EML [15] and in WASOBI-AR [42]. A second assumption, aimed at mitigating the sign ambiguity, is that all elements of the first row of $\mathbf{A}$ are positive, assuming that they are all non-zeros. If any of them is (are) zero, then it can be assumed that the first non-zero element in the respective column is positive (in other words, the general assumption is that the leading non-zero element in each column of $\mathbf{A}$ is positive). Finally, to mitigate the permutation ambiguity, it can be further assumed that the columns of $\mathbf{A}$ are sorted in ascending order of the values of their leading (positive) elements.

We reiterate that these assumptions need not hold in practice. They simply provide us with means for picking up one particular (unique) solution from the set of all equivalent solutions. We shall show later that, in terms of any reasonable performance measure, such as the Interference to Source Ratio (*ISR*), all equivalent solutions are equally good (or bad), so selecting a "wrong" solution has no effect in terms of performance, as long as that solution belongs to the set of equivalent solutions.

Having established the basis for uniqueness of the solution, we now turn to discuss the identifiability. In the context of this chapter, we are interested in the identifiability of $\mathbf{A}$ (or of $\mathbf{B}$) *from the observations' empirical SOS only*. In "identifiability" of $\mathbf{A}$ we refer to the ability to estimate $\mathbf{A}$ (under the ambiguity-resolving working-assumptions above) to arbitrary precision from sufficiently many observations (namely, asymptotically in T).

Thanks to the above mentioned ergodicity assumption, this is equivalent to the ability to extract **A** (again, subject to the working assumptions) from the true SOS of the observations, namely from the observations' true $\mathbf{R}_x(\tau), \forall\tau$.

The general identifiability condition is described by the following theorem, which has appeared in various forms, e.g. in [51,50,1]

THEOREM 7.1
$\mathbf{A}$ and $\mathbf{B}$ are identifiable from SOS of the observations if and only if the correlation sequences of all sources are pairwise linearly independent.

To show this, assume the scaling assumption $\mathbf{R}_s = \mathbf{I}$, and let $\mathbf{A}_0$ denote any square root of $\mathbf{R}_x(0)$, such that

$$\mathbf{A}_0\mathbf{A}_0^T = \mathbf{R}_x(0) \quad \Rightarrow \quad \mathbf{B}_0\mathbf{R}_x(0)\mathbf{B}_0^T = \mathbf{R}_s(0) = \mathbf{I}, \tag{7.25}$$

where $\mathbf{B}_0 = \mathbf{A}_0^{-1}$. Since $\mathbf{R}_x(0) = \mathbf{A}\mathbf{A}^T$, this implies that $\mathbf{B}_0\mathbf{A}$ is an orthogonal matrix, which we shall denote $\mathbf{U}_0^T$. Now define

$$\tilde{\mathbf{R}}(1) \triangleq \mathbf{B}_0\mathbf{R}_x(1)\mathbf{B}_0^T, \tag{7.26}$$

and let $\mathbf{D}_1$ and $\mathbf{U}_1$ denote the eigenvalues and eigenvectors matrices (resp.) of $\tilde{\mathbf{R}}(1)$, such that $\tilde{\mathbf{R}}(1)\mathbf{U}_1 = \mathbf{U}_1\mathbf{D}_1$, implying that

$$\begin{aligned}\mathbf{D}_1 &= \mathbf{U}_1^T\tilde{\mathbf{R}}(1)\mathbf{U}_1 = \mathbf{U}_1^T\mathbf{B}_0\mathbf{R}_x(1)\mathbf{B}_0^T\mathbf{U}_1 \\ &= \mathbf{U}_1^T\mathbf{B}_0\mathbf{A}\mathbf{R}_s(1)\mathbf{A}^T\mathbf{B}_0^T\mathbf{U}_1 = \mathbf{U}_1^T\mathbf{U}_0^T\mathbf{R}_s(1)\mathbf{U}_0^T\mathbf{U}_1\end{aligned} \tag{7.27}$$

($\mathbf{U}_1$ is orthogonal because $\tilde{\mathbf{R}}(1)$ is symmetric). Since the eigenvalues are unique and $\mathbf{U}_0\mathbf{U}_1$ is orthogonal, $\mathbf{D}_1$ must contain the diagonal values of $\mathbf{R}_s(1)$. If all of these values are distinct, then $\mathbf{U}_1$ is unique, up to permutation and sign ambiguities. Therefore, in this case (unique values in $\mathbf{R}_s(1)$), it follows that $\mathbf{U}_1^T\mathbf{B}_0$ is equivalent to $\mathbf{B}$, and $\mathbf{A}_0\mathbf{U}_1$ is equivalent to $\mathbf{A}$. Resolving the permutation and sign ambiguities by the aforementioned conventions, $\mathbf{A}$ and $\mathbf{B}$ are uniquely identified.

However, if $\tilde{\mathbf{R}}(1)$ has any multiple eigenvalues, $\mathbf{U}_1$ is not unique: it can be multiplied (on the right) with any orthogonal matrix which applies any rotations to the eigenvalues with multiplicities, and still be an eigenvectors matrix of $\tilde{\mathbf{R}}(1)$ (with the same eigenvalues). In this case, we fix $\mathbf{U}_1$ so as to resolve the ambiguities in the columns of $\mathbf{A}_0\mathbf{U}_1$ corresponding to the distinct eigenvalues, and let J_1 denote the set of corresponding column indices, with $\bar{J}_1$ denoting the complementary set of indices, those of the multiple eigenvalues (such that $J_1 \bigcup \bar{J}_1 = \{1, \ldots, N\}$). Now, define

$$\tilde{\mathbf{R}}(2) \triangleq \mathbf{U}_1^T\mathbf{B}_0\mathbf{R}_x(2)\mathbf{B}_0^T\mathbf{U}_1, \tag{7.28}$$

and note that $\tilde{\mathbf{R}}(2)$ is already partly diagonal, at least in the indices (J_1,J_1). Its only non-diagonal blocks are (possibly) those associated with the indices $(\bar{J}_1,\bar{J}_1)$. Now let $\mathbf{D}_2$ and $\mathbf{U}_2$ denote the eigenvalues and eigenvectors matrices (resp.) of $\tilde{\mathbf{R}}(2)$, such that $\tilde{\mathbf{R}}(2)\mathbf{U}_2 = \mathbf{U}_2\mathbf{D}_2$, implying

$$\begin{aligned}\mathbf{D}_2 &= \mathbf{U}_2^T\tilde{\mathbf{R}}(2)\mathbf{U}_2 = \mathbf{U}_2^T\mathbf{U}_1^T\mathbf{B}_0\mathbf{R}_x(2)\mathbf{B}_0^T\mathbf{U}_1\mathbf{U}_2 \\ &= \mathbf{U}_2^T\mathbf{U}_1^T\mathbf{B}_0\mathbf{A}\mathbf{R}_s(2)\mathbf{A}^T\mathbf{B}_0^T\mathbf{U}_1\mathbf{U}_2 = \mathbf{U}_2^T\mathbf{U}_1^T\mathbf{U}_0^T\mathbf{R}_s(2)\mathbf{U}_0\mathbf{U}_1\mathbf{U}_2.\end{aligned} \tag{7.29}$$

At the indices (J_1,J_1), $\mathbf{U}_2$ equals the identity matrix (since that part of $\tilde{\mathbf{R}}(2)$ is diagonal). Its only nontrivial block(s) are in the indices $(\bar{J}_1,\bar{J}_1)$. Once again, the eigenvalues in $\mathbf{D}_2$ are the diagonal values of $\mathbf{R}_s(2)$, with exact correspondence at the diagonal indices J_1. Therefore, if (and only if) the diagonal values along the indices $\bar{J}_1$ are distinct, the nontrivial blocks of $\mathbf{U}_2$ (and hence this entire matrix) are unique, and therefore $\mathbf{A}_0\mathbf{U}_1\mathbf{U}_2$ is equivalent to $\mathbf{A}$ (and $\mathbf{U}_2^T\mathbf{U}_1^T\mathbf{B}_0$ is equivalent to $\mathbf{B}$).

If there are still eigenvalues with multiplicity along the indices $\bar{J}_1$ of the diagonal of $\mathbf{D}_2$, we repeat the process: first, we resolve the permutation and sign ambiguities along the indices of the distinct eigenvalues. The new indices set J_2 is the union of J_1 with the indices of the distinct values of $\mathbf{D}_2$ among $\bar{J}_1$; $\bar{J}_2$ is the complementary set; $\tilde{\mathbf{R}}(3)$ is constructed as

$$\tilde{\mathbf{R}}(3) \triangleq \mathbf{U}_2^T\mathbf{U}_1^T\mathbf{B}_0\mathbf{R}_x(3)\mathbf{B}_0^T\mathbf{U}_1\mathbf{U}_2, \tag{7.30}$$

and its eigenvectors and eigenvalues matrices $\mathbf{U}_3$ and $\mathbf{D}_3$ are extracted. The unique, nontrivial blocks of $\mathbf{U}_3$ along the indices $(\bar{J}_2,\bar{J}_2)$ are then extracted along the distinct eigenvalues in $\mathbf{D}_3$ within the indices $\bar{J}_2$.

The process is terminated at the K-th step, when all the eigenvalues of $\tilde{\mathbf{R}}(K)$ within the set $\bar{J}_{K-1}$ are distinct, and then, following resolution of the remaining ambiguities (via $\mathbf{U}_K$), we finally get the unique solution with $\mathbf{U}_K^T\mathbf{U}_{K-1}^T\cdots\mathbf{U}_1^T\mathbf{B}_0 = \mathbf{B}$ and $\mathbf{A}_0\mathbf{U}_1\mathbf{U}_2\cdots\mathbf{U}_K = \mathbf{A}$.

Evidently, this process can conclude if and only if none of the N scaled correlation sequences $R_n(\tau)$ (scaled such that all $R_n(0) = 1$) is identical for all τ. In fact, the procedure would terminate at the K-th step, if $K-1$ is the farthest lag to which two (or more) of the sources have identical scaled correlation sequences. Equivalently, in their unscaled versions, all correlation sequences have to be pairwise linearly independent.

Such a constructive proof proves the sufficiency of this condition, but does not prove its necessity (the fact that a particular procedure would not conclude if the condition is not satisfied, still does not mean that the condition is generally necessary). To show the necessity, assume that there are two sources with identical correlation sequences (after scaling). Without loss of generality, assume that these are $s_1(\cdot)$ and $s_2(\cdot)$. Then if any mixing matrix **A** is multiplied on the right with a matrix of the form

$$\mathbf{U} = \begin{bmatrix} \cos(\phi) & \sin(\phi) & \\ -\sin(\phi) & \cos(\phi) & \\ & & \mathbf{I}_{(N-2)\times(N-2)} \end{bmatrix} \tag{7.31}$$

(where ϕ is an arbitrary real-valued angle), then with the resulting mixing matrix $\mathbf{A}' = \mathbf{AU}$, the SOS of the observations $\mathbf{x}'(\cdot) = \mathbf{A}'\mathbf{s}(\cdot)$ will remain the same for all τ,

$$\mathbf{R}_{\mathbf{x}'}(\tau) = \mathbf{A}'\mathbf{R}_{\mathbf{s}}(\tau)\mathbf{A}'^T = \mathbf{AUR}_{\mathbf{s}}(\tau)\mathbf{U}^T\mathbf{A}^T = \mathbf{AR}_{\mathbf{s}}(\tau)\mathbf{A}^T = \mathbf{R}_{\mathbf{x}}(\tau). \tag{7.32}$$

This concludes the proof of the identifiability theorem. Note that due to the uniqueness of the Fourier transform, the condition in time-domain is also equivalent to a similar condition in the frequency-domain: **A** is identifiable by SOS if and only if the sources have distinct spectra, namely, if all spectra are pairwise linearly independent (i.e. no spectrum is a scaled version of another).

Note also that in [1] a similar condition is obtained for identification of **A** from estimated correlations over a pre-specified set of lags. The resulting condition is that the sources' correlations over that set of lags be pairwise linearly independent (which is, of course, more stringent than our general condition, since the estimator further constrains the permissible SOS).

We stress, in passing, that the identifiability condition above is certainly not a general condition for the general identifiability of **A** in our model. It is merely a condition for the identifiability of **A** with WSS sources from SOS alone. In general, other features of the sources (such as their higher-order statistics) may restore identifiability in cases where SOS alone are insufficient. However, when the sources are jointly Gaussian, their statistical distributions are fully characterized by their SOS, and therefore in that case our identifiability condition is strict: if **A** cannot be identified from SOS, then it cannot be identified at all.

7.3.2 Performance measures and bounds

The identifiability condition discussed in the previous subsection guarantees that **A** (or, equivalently, **B**) can be recovered exactly (subject to a scaling convention) from the true SOS (true correlation matrices) of the observed data, and, therefore, given the relevant ergodicity conditions, can be recovered to within arbitrary precision from sufficiently many observations T. Naturally, however, with any fixed observation length, estimation errors in **A** (or in **B**) are inevitable, and would subsequently affect the separation quality.

To quantify the estimation errors and the resulting separation performance, it is common practice to define the overall *global* mixing-unmixing matrix,

$$\mathbf{G} \triangleq \widehat{\mathbf{B}}\mathbf{A} = \widehat{\mathbf{A}}^{-1}\mathbf{A}. \tag{7.33}$$

Thus, when the estimated unmixing matrix is applied to the observations, the separator's output $\mathbf{y}(t)$ is a residual mixture of the sources with the global mixing $\mathbf{G}$. Perfect separation is obtained when $\mathbf{G}$ is a scaled permutation matrix, $\mathbf{G} = \boldsymbol{\Delta}\boldsymbol{\varepsilon}\boldsymbol{\Pi}$, but in practice $\mathbf{G}$ will (almost) always differ from this ideal form. For simplicity of the following discussion, we shall assume first that the permutation, sign and scale ambiguities are always resolved, such that a perfect $\mathbf{G}$ equals the identity matrix $\mathbf{I}$. We shall elaborate on the implications of unresolved ambiguities later on.

Each off-diagonal element of $\mathbf{G}$, say $G_{m,n}$, contains the residual contamination coefficient with which the n-th source is present in the reconstruction of the m-th source. If the scaling assumption $\mathbf{R}_s(0) = \mathbf{I}$ indeed holds true, then the squared value $G^2_{m,n}$ also reflects the power with which the n-th source is present in the reconstruction of the m-th source. Under a small-errors assumption we would have $G_{m,m} \approx 1$, and therefore $G^2_{m,n}$ also reflects the ratio obtained (in the reconstruction of the m-th source) between the powers of the n-th source (interference) and the m-th source. This quantity is often termed the n-to-m Interference to Source Ratio (*ISR*) *per realization*, also denoted $\text{rISR}_{m,n}$ (the "realization ISR").

Recall, however, that the scaling assumption $\mathbf{R}_s(0) = \mathbf{I}$ was merely used as a "working assumption", in order to identify a uniquely scaled solution. As mentioned earlier, this assumption need not hold in practice. Indeed, when it does not hold, the ideal form of $\mathbf{G}$ is no longer $\mathbf{I}$, but a general diagonal matrix, and $G^2_{m,n}$ no longer follows the interpretation of $\text{rISR}_{m,n}$. Therefore, to obtain the rISRs from $\mathbf{G}$, it is necessary to rescale the element of $\mathbf{G}$ and normalize by the respective power ratio,

$$\text{rISR}_{m,n} = \frac{G^2_{m,n}}{G^2_{m,m}} \cdot \frac{R_n(0)}{R_m(0)} \quad (m \neq n) \in \{1,\ldots,N\}. \tag{7.34}$$

With this "normalized" definition, any remaining scale (or sign) ambiguity in the columns of $\widehat{\mathbf{A}}$ (or in the rows of $\widehat{\mathbf{B}}$) becomes irrelevant for the rISR, since it merely implies rows scaling in $\mathbf{G}$, which cancels out in the rISR.

These are, however, only "realization ISRs", since $\widehat{\mathbf{A}}$ (or $\widehat{\mathbf{B}}$, resp.), and therefore also $\mathbf{G}$, depend on the particular observations $\mathbf{x}(t), t = 1,\ldots,T$. The rISRs obtained by a given algorithm with a given realization is of considerable interest; however, $\mathbf{G}$ is usually unknown in practice (since the true mixing $\mathbf{A}$ is unknown in practice), and does not characterize the general performance, or overall quality, of a given algorithm: the fact that with a given realization, algorithm A may yield a better (smaller) rISR than algorithm B, does not imply that algorithm A is better than B in any general sense.

To better characterize the performance, we are interested in statistical properties of the rISRs, e.g. in their statistical means, simply called the ISRs:

$$\text{ISR}_{m,n} = \mathbb{E}\{\text{rISR}_{m,n}\} = \mathbb{E}\left\{\frac{G^2_{m,n}}{G^2_{m,m}}\right\} \cdot \frac{R_n(0)}{R_m(0)} \quad (m \neq n) \in \{1,\ldots,N\}. \tag{7.35}$$

The ISR matrix is an $N \times N$ matrix, whose off-diagonal terms are composed of all $\mathrm{ISR}_{m,n}$, and whose diagonal terms are undefined, but can be taken as all 1s for compliance with (7.35).

It is important to note a general difference between the ISR matrix and a somewhat similar "mean contamination" matrix, defined as

$$\mathbf{C} \triangleq \mathbb{E}\{\mathbf{G} \odot \mathbf{G}\}. \tag{7.36}$$

where $\odot$ denotes Hadamard's (element-wise) product. This matrix is equivalent to the ISR matrix under the scaling convention $\mathbf{R}_s(0) = \mathbf{I}$, but if this scaling convention does not hold or is not employed, the two matrices may differ substantially.

An important difference in our context of stationary sources with SOS-based separation, is that for many algorithms the resulting ISR matrix depends only on the (normalized) spectral shapes of the sources. It does not depend on their overall powers or on the mixing matrix (see discussion of *equivariance* below). However, the contamination matrix usually depends also on the sources' powers.

Another possible performance measure is the mean square error in estimating the elements of the mixing matrix $\mathbf{A}$ or of the unmixing matrix $\mathbf{B}$. Under a small-errors assumption, the estimation errors in these matrices are linearly related to the elements of the global mixing matrix $\mathbf{G}$ as follows. Let us denote the estimation errors

$$\boldsymbol{\Delta}_{\mathrm{A}} \triangleq \widehat{\mathbf{A}} - \mathbf{A}, \quad \boldsymbol{\Delta}_{\mathrm{B}} \triangleq \widehat{\mathbf{B}} - \mathbf{B}, \quad \boldsymbol{\Delta}_{\mathrm{G}} \triangleq \mathbf{G} - \mathbf{I}. \tag{7.37}$$

Now, since

$$\mathbf{G} = \widehat{\mathbf{B}}\mathbf{A} = (\mathbf{B} + \boldsymbol{\Delta}_{\mathrm{B}})\mathbf{A} = \mathbf{I} + \boldsymbol{\Delta}_{\mathrm{B}}\mathbf{A} \tag{7.38}$$

and (using the approximation $(\mathbf{I} + \boldsymbol{\varepsilon})^{-1} \approx \mathbf{I} - \boldsymbol{\varepsilon}$ for $\boldsymbol{\varepsilon} \ll \mathbf{I}$)

$$\mathbf{G} = \widehat{\mathbf{A}}^{-1}\mathbf{A} = (\mathbf{A} + \boldsymbol{\Delta}_{\mathrm{A}})^{-1}\mathbf{A} = [\mathbf{A}(\mathbf{I} + \mathbf{B}\boldsymbol{\Delta}_{\mathrm{A}})]^{-1}\mathbf{A} = (\mathbf{I} + \mathbf{B}\boldsymbol{\Delta}_{\mathrm{A}})^{-1} \approx \mathbf{I} - \mathbf{B}\boldsymbol{\Delta}_{\mathrm{A}}, \tag{7.39}$$

we have

$$\boldsymbol{\Delta}_{\mathrm{G}} = \boldsymbol{\Delta}_{\mathrm{B}}\mathbf{A} \approx -\mathbf{B}\boldsymbol{\Delta}_{\mathrm{A}}. \tag{7.40}$$

Defining the vectorized error matrices

$$\boldsymbol{\delta}_{\mathrm{A}} \triangleq \mathrm{vec}(\boldsymbol{\Delta}_{\mathrm{A}}), \quad \boldsymbol{\delta}_{\mathrm{B}} \triangleq \mathrm{vec}(\boldsymbol{\Delta}_{\mathrm{B}}), \quad \boldsymbol{\delta}_{\mathrm{G}} \triangleq \mathrm{vec}(\boldsymbol{\Delta}_{\mathrm{G}}) \tag{7.41}$$

(where the $\mathrm{vec}(\cdot)$ operator concatenates the columns of its argument into a single vector) we have, using the relation $\mathrm{vec}(\mathbf{E}_1 \cdot \mathbf{E}_2 \cdot \mathbf{E}_3) = (\mathbf{E}_3^T \otimes \mathbf{E}_1) \cdot \mathrm{vec}(\mathbf{E}_2)$,

$$\boldsymbol{\delta}_{\mathrm{G}} = (\mathbf{A}^T \otimes \mathbf{I})\boldsymbol{\delta}_{\mathrm{B}} \approx (\mathbf{I} \otimes \mathbf{B})\boldsymbol{\delta}_{\mathrm{A}} \tag{7.42}$$

(where $\otimes$ denotes Kronecker's product).

To quantify and rank the performance of different algorithms it is sometimes necessary to reduce such matrix measures into a single scalar measure. A natural index in this context is the overall ISR index,

$$I_{\text{ISR}} = \sum_{\substack{m,n=1 \\ m \neq n}}^{N} \text{ISR}_{m,n}. \tag{7.43}$$

A similar definition of an overall contamination index is sometimes applied to **C** [6]. A related scalar performance index was proposed in [65], as

$$\begin{aligned} I_{ZAÏ} \triangleq \mathbb{E}\{\|\widehat{\mathbf{s}}(t) - \mathbf{s}(t)\|^2\} &= \mathbb{E}\{\|(\mathbf{G} - \mathbf{I})\mathbf{s}(t)\|^2\} \\ &= \text{trace}\{\mathbb{E}\{(\mathbf{G} - \mathbf{I})^T(\mathbf{G} - \mathbf{I})\mathbf{s}(t)\mathbf{s}^T(t)\}\} \\ &\approx \text{trace}\{\mathbb{E}\{(\mathbf{G} - \mathbf{I})^T(\mathbf{G} - \mathbf{I})\}\mathbf{R}_s(0)\} \end{aligned} \tag{7.44}$$

(the approximation in the last transition is due to a weakly justified and nearly harmless assumption of statistical independence between $\mathbf{s}(t)$ and **G**). With the scaling convention $\mathbf{R}_s(0) = \mathbf{I}$, this performance index reduces to

$$\begin{aligned} I_{ZAÏ} &= \mathbb{E}\left\{ \sum_{\substack{m,n=1 \\ m \neq n}}^{N} G_{m,n}^2 + \sum_{m=1}^{N} (G_{m,m} - 1)^2 \right\} \\ &= \sum_{m,n=1}^{N} \mathbb{E}\{\boldsymbol{\Delta}_{\mathbf{G}} \odot \boldsymbol{\Delta}_{\mathbf{G}}\} = \mathbb{E}\{\boldsymbol{\delta}_{\mathbf{G}}^T \boldsymbol{\delta}_{\mathbf{G}}\} \approx \sum_{m,n=1}^{N} C_{m,n} - N \end{aligned} \tag{7.45}$$

(the approximation in the last transition assumes $\mathbb{E}\{G_{m,m}\} = 1$).

Many BSS algorithms, in various contexts (all in a noiseless setting), share the appealing property of *equivariance* [8]. A separation algorithm is said to be when its separation performance does not depend on the mixing matrix **A** (as long as it is non-singular). Denote by $\mathbf{X}_T$ the collection of observed vectors $\mathbf{x}(t), t = 1, \ldots, T$ into an $N \times T$ matrix, and by $\widehat{\mathbf{B}}(\mathbf{X}_T)$ the (explicit or implicit) separation matrix estimated by a given algorithm, based on the observations $\mathbf{X}_T$. If for every invertible matrix **M**, the algorithm satisfies $\widehat{\mathbf{B}}(\mathbf{M}\mathbf{X}_T) = \widehat{\mathbf{B}}(\mathbf{X}_T)\mathbf{M}^{-1}$, then it essentially yields the same estimated source signals given a particular realization of the true source signals, regardless of the true mixing matrix. This holds per each realization of the source signals, and not only with respect to the mean performance. In other words, not only the ISR matrix, but also the rISRs are invariant to the mixing matrix.

Note that equivariance does not mean that the estimation errors $\boldsymbol{\Delta}_{\mathbf{B}}$ or $\boldsymbol{\Delta}_{\mathbf{A}}$ (in **B** or in **A**, resp.) are invariant in **A**. This property applies only to the error $\boldsymbol{\Delta}_{\mathbf{G}}$ in the global mixing **G**.

The equivariance property is appealing not only from a practical point of view, but also from a theoretical point of view. Once established, it enables one to analyze the performance of a given equivariant algorithm assuming a comfortably chosen mixing matrix (e.g. $\mathbf{A} = \mathbf{I}$), knowing that the results would be valid with any other mixing matrix as well.

The equivariant nature of the problem also induces equivariance on the attainable performance bounds. Before we turn to describe some of the leading algorithms (in the following sections), we shall review some of the performance bounds, which describe the best attainable performance (under some commonly assumed conditions on the estimator).

The *Cramér-Rao Lower Bound* (*CRLB*) is a lower bound on the mean square error (*MSE*) of any unbiased estimator. If $\mathbf{x}$ is a data vector whose probability distribution $f(\mathbf{x};\boldsymbol{\theta})$ depends on an unknown parameters vector $\boldsymbol{\theta}$, then the covariance matrix of any unbiased estimator $\widehat{\boldsymbol{\theta}}$ of $\boldsymbol{\theta}$ from $\mathbf{x}$ is bounded below as

$$\mathbb{E}\{(\widehat{\boldsymbol{\theta}}-\boldsymbol{\theta}_0)(\widehat{\boldsymbol{\theta}}-\boldsymbol{\theta}_0)^T\} \geqslant \mathbf{J}_{\boldsymbol{\theta}}^{-1}(\boldsymbol{\theta}_0) \tag{7.46}$$

where $\mathbf{J}_{\boldsymbol{\theta}}^{-1}(\boldsymbol{\theta}_0)$ denotes the Fisher Information Matrix (*FIM*), given by

$$\mathbf{J}_{\boldsymbol{\theta}}(\boldsymbol{\theta}_0) \triangleq \mathbb{E}\left[\frac{\partial^T \log f(\mathbf{x};\boldsymbol{\theta})}{\partial \boldsymbol{\theta}} \cdot \frac{\partial \log f(\mathbf{x};\boldsymbol{\theta})}{\partial \boldsymbol{\theta}};\boldsymbol{\theta}_0\right] = -\mathbb{E}\left[\frac{\partial^2 \log f(\mathbf{x};\boldsymbol{\theta})}{\partial \boldsymbol{\theta}\partial \boldsymbol{\theta}^T};\boldsymbol{\theta}_0\right]. \tag{7.47}$$

Here $\boldsymbol{\theta}_0$ denotes the true value of $\boldsymbol{\theta}$ and $\mathbb{E}[\cdot;\boldsymbol{\theta}_0]$ denotes the mean taken with $\boldsymbol{\theta} = \boldsymbol{\theta}_0$.

Under some mild conditions, the Maximum Likelihood Estimate (*MLE*) is asymptotically (in our model, "asymptotically" means $T \to \infty$) unbiased and attains the CRLB. In [38] Pham and Garat propose a Quasi Maximum Likelihood (*QML*) estimator for the separation of stationary sources with arbitrary spectra (the approach is reviewed in detail in section 7.5). When some of the processing parameters (filters) are chosen properly, the resulting estimator coincides with the true MLE in the case of Gaussian sources. Thus, its asymptotic performance (with optimal selection of the filters, which requires knowledge of the generally unknown sources' spectra) coincides with the CRLB. Therefore, the performance analysis results in [38] can also be interpreted as the CRLB for the problem of separating stationary Gaussian sources when their spectra are known. The bound on the resulting "mean contamination" matrix (which, under the scaling convention is also a bound on the ISR matrix) takes the form

$$\mathbb{E}\{G_{m,n}^2\} = C_{m,n} \geqslant \frac{1}{T} \cdot \frac{\phi_{m,n}}{\phi_{m,n}\phi_{n,m} - 1} \qquad m \neq n \tag{7.48}$$

where

$$\phi_{m,n} \triangleq \int_{-\frac{1}{2}}^{\frac{1}{2}} \frac{S_m(\nu)}{S_n(\nu)} d\nu. \tag{7.49}$$

In fact, as we shall point out shortly, it turns out that asymptotically, this is also the CRLB for the case of *unknown* sources' spectra.

In [15], Dégerine and Zaïdi derive the FIM for the case of Gaussian AR sources, where the unknown parameters $\boldsymbol{\theta}$ are the elements of the unmixing matrix $\mathbf{B}$ and the AR parameters of all sources, denoted here as $\{a_\ell^{(n)}\}$ for $n = 1, \ldots, N$ (the source index) and $\ell = 1, \ldots, P_n$ (the AR index), P_n denoting the AR order of the n-th source. The scaling convention in [15] is of unit variances of the innovation sequences $w_n(t)$ of all sources.

It is shown in [15] that asymptotically, the FIM is block-diagonal, with one block referring to the unmixing parameters and the other block referring to the AR parameters. This implies that (asymptotically) the bound for the mixing parameters' estimates $\widehat{\mathbf{B}}$ when the AR parameters are *unknown* equals the bound for $\widehat{\mathbf{B}}$ when the AR parameters are *known*. This property is well-expected, and is found to be in accordance with a similar block-diagonality structure of the FIM for BSS of sources with iid time-structure in [5] and also (in a slightly different context) in [41].

Therefore, one may obtain the asymptotic CRLB for estimation of the elements of $\mathbf{B}$ by inverting the respective block of the FIM. This is an $N^2 \times N^2$ matrix, whose elements associated with $B_{m,n}$ and $B_{s,t}$ are given by [15]:

$$\begin{aligned}[\mathbf{J}_{\mathbf{B}}(\mathbf{A})]_{(m,n),(s,t)} &= TA_{t,m}A_{n,s} \\ &\quad + \delta_{s,m} \sum_{i,j=0}^{P_s} (T-i-j) a_i^{(s)} a_j^{(s)} \sum_{k=1}^{N} A_{t,k} A_{n,k} R_k(i-j),\end{aligned} \tag{7.50}$$

where $\delta_{s,m}$ denotes Kronecker's delta function (zero for $s \neq m$, one for $s = m$). Note that we slightly abuse the notation convention of (7.47) in writing $\mathbf{J}_{\mathbf{B}}(\mathbf{A})$ instead of $\mathbf{J}_{\mathbf{B}}(\mathbf{B}, \{a_\ell^n\})$, so as to express the important direct dependence of $\mathbf{J}_{\mathbf{B}}$ on the mixing matrix $\mathbf{A}$. In [16], Doron *et al.* show that $\mathbf{J}_{\mathbf{B}}(\mathbf{A})$ defined in (7.50) satisfies the relation

$$\mathbf{J}_{\mathbf{B}}(\mathbf{A}) = (\mathbf{A} \otimes \mathbf{I}) \mathbf{J}_{\mathbf{B}}(\mathbf{I}) (\mathbf{A}^T \otimes \mathbf{I}). \tag{7.51}$$

Combined with the relation (7.42), this relation implies a simple bound on the covariance of the elements $\boldsymbol{\delta}_{\mathrm{G}}$ of the global matrix $\mathbf{G}$, through the bound on the covariance of the elements $\boldsymbol{\delta}_{\mathrm{B}}$ of $\mathbf{B}$:

$$\begin{aligned}\mathrm{Cov}\{\boldsymbol{\delta}_{\mathrm{G}}\} &= (\mathbf{A}^T \otimes \mathbf{I})\, \mathrm{Cov}\{\boldsymbol{\delta}_{\mathrm{B}}\} (\mathbf{A} \otimes \mathbf{I}) \\ &\geqslant (\mathbf{A}^T \otimes \mathbf{I})(\mathbf{A}^T \otimes \mathbf{I})^{-1} \mathbf{J}_{\mathbf{B}}^{-1}(\mathbf{I}) (\mathbf{A} \otimes \mathbf{I})^{-1} (\mathbf{A} \otimes \mathbf{I}) = \mathbf{J}_{\mathbf{B}}^{-1}(\mathbf{I}).\end{aligned} \tag{7.52}$$

Note that this result confirms the well-expected equivariance of the CRLB in this problem, as it implies that the mean contamination matrix $\mathbf{C}$ does not depend on $\mathbf{A}$.

By substituting $\mathbf{A} = \mathbf{I}$ in (7.50) and assuming asymptotic conditions ($T \rightarrow \infty$) we obtain

$$[\mathbf{J_B}(\mathbf{I})]_{(m,n),(s,t)} = T \cdot \begin{cases} 1+\varphi_{m,m} & m=t=n=s \\ 1 & m=t\neq n=s \\ \varphi_{n,m} & n=t\neq m=s \\ 0 & \text{otherwise} \end{cases} \tag{7.53}$$

where

$$\varphi_{m,n} \triangleq \sum_{i,j=0}^{P_s} a_i^{(n)} a_j^{(n)} R_m(i-j). \tag{7.54}$$

The expressions in (7.53) imply that $\mathbf{J_B}(\mathbf{I})$ can be rearranged in a block-diagonal structure comprised of 2×2 blocks, each involving the (m,n)-th and (n,m)-th elements (for all $m \neq m$, a total of $N(N-1)/2$ blocks), and a remaining diagonal part for all (n,n)-th elements. The block for the (m,n)-th and (n,m)-th elements takes the form

$$[\mathbf{J_B}(\mathbf{I})]_{\substack{(m,n)\\(n,m)}} = \begin{bmatrix} \varphi_{n,m} & 1 \\ 1 & \varphi_{m,n} \end{bmatrix}. \tag{7.55}$$

It then follows that the (asymptotic) bound on the MSE of an unbiased estimate of $B_{m,n}$ ($m \neq n$) when $\mathbf{A} = \mathbf{I}$, which is also a bound on $C_{m,n}$ with any $\mathbf{A}$, is given by

$$\mathbb{E}\{(\widehat{B}_{m,n} - B_{m,n})^2; \mathbf{A} = \mathbf{I}\} = C_{m,n} \geqslant \frac{1}{T} \cdot \frac{\varphi_{m,n}}{\varphi_{m,n}\varphi_{n,m} - 1} \quad m \neq n, \tag{7.56}$$

similar to Pham and Garat's expression (7.48). Moreover, by exploiting Fourier relations in the convolution and summation in (7.54), we obtain the expression

$$\varphi_{m,n} = \int_{-\frac{1}{2}}^{\frac{1}{2}} A_n(e^{j2\pi\nu}) A_n^*(e^{j2\pi\nu}) S_m(\nu) d\nu \tag{7.57}$$

(where $A_n(e^{j2\pi\nu}) = \sum_{k=0}^{P_n} a_k^{(n)} e^{-j2\pi\nu}$ denotes the Fourier transform of the inverse generating system of the n-th source; see subsection 7.2.1.2). Recalling the unit scaling convention on the variances of the sources' innovation sequences, we readily observe from (7.49) and (7.57) that $\varphi_{m,n} = \phi_{m,n}$, hence the two expressions (7.48) and (7.56) for the bound on $\mathbf{C}$ coincide. This is naturally expected, since AR processes are a particular case of general WSS processes. Recall, however, that Pham and Garat's expression was not presented as the CRLB in [38], but rather as the performance of the MLE when the source spectra are known. However, since, under very mild conditions any WSS process can be approximated as an AR process (e.g. see [32]), the bound in (7.56) confirms that (asymptotically) the bound in (7.48) is indeed the CRLB, also for a truly blind case of unknown spectra.

For completeness of the discussion, we observe that $\varphi_{n,n} = 1$, and therefore (from (7.53)) the diagonal part of $\mathbf{J_B}(\mathbf{I})$ is composed of $2N$ all along the diagonal. This implies that the bound on the variance in unbiased estimation of the diagonal elements of $\mathbf{B}$ when $\mathbf{A} = \mathbf{I}$ (or, equivalently, the variance in the resulting diagonal elements of $\mathbf{G}$, with any $\mathbf{A}$) is bounded below by $1/2N$, regardless of the sources' spectra.

In [16], Doron *et al.* generalize the bound on $\mathbf{C}$ to a bound on the ISR, relaxing the scaling convention:

$$\mathrm{ISR}_{m,n} \geqslant \frac{1}{T} \cdot \frac{\bar{\varphi}_{m,n}}{\bar{\varphi}_{m,n}\bar{\varphi}_{n,m} - 1} \cdot \frac{\sigma_m^2}{\sigma_n^2} \cdot \frac{R_n(0)}{R_m(0)} \qquad m \neq n, \tag{7.58}$$

with

$$\bar{\varphi}_{m,n} \triangleq \frac{\varphi_{m,n}}{\sigma_m^2} = \int_{-\frac{1}{2}}^{\frac{1}{2}} \frac{|A_n(e^{j2\pi\nu})|^2}{|A_m(e^{j2\pi\nu})|^2} d\nu, \tag{7.59}$$

where σ_m^2 and σ_n^2 denote the respective variances of the innovation sequences of the m-th and n-th sources. This bound is called the "*induced CRLB (iCRLB)*" (in [16]), since, strictly speaking, it does not bound the estimation error of a specific parameter, but rather asserts a bound on the resulting ISR, induced by the CRLB on the MSE in estimating the unmixing matrix.

Several key properties of the iCRLB are to be noted:

- **Equivariance**: As mentioned above, and as could be well-expected in noiseless BSS, this bound does not depend on the mixing matrix $\mathbf{A}$, but only on the sources' spectra.
- **Scale invariance**: The ISR bound is invariant to any scaling of the sources. Note that this property is not shared by the bound on $\mathbf{C}$: if, for example, source n is amplified by a certain factor, the bounds on all of the resulting $C_{m,n}$ $(m \neq n)$ would be reduced by the square of that factor; however, the bounds on all $\mathrm{ISR}_{m,n}$ would remain unchanged.
- **Invariance with respect to other sources**: The bound on $\mathrm{ISR}_{m,n}$ depends only on the spectral shapes of sources m and n, and is unaffected by the other sources.
- **Non-identifiability condition**: If sources m and n have similar spectral shapes (i.e. $S_m(\nu)$ is a scaled version of $S_n(\nu)$), then $\bar{\varphi}_{m,n} = 1/\bar{\varphi}_{n,m}$, implying (in (7.58)) an infinite bound on $\mathrm{ISR}_{m,n}$ and on $\mathrm{ISR}_{n,m}$ – which in turn implies non-identifiability of the relevant entries in $\mathbf{A}$ – as already discussed in subsection 7.3.1.
- **Resemblance to other bounds**: The general form of (7.48) is also shared by similar bounds developed for the case of sources with iid temporal structures (e.g. in [43,30]), with $\bar{\varphi}_{m,n}$ replaced by a quantity which depends only on the probability distribution function of the n-th source.

- **Limited relevance to non-Gaussian sources**: It is necessary to keep in mind that this bound was derived for Gaussian sources, and, strictly speaking, it does not bound the separation performance for non-Gaussian sources. Naturally, with non-Gaussian sources it should be possible to enhance the performance by using higher order statistics as well. But even if the separation methods are restricted to SOS-based methods, the relevant bound for non-Gaussian sources may be, in general, either higher or lower than this Gaussian bound. Nevertheless, if the sources' distributions are close to Gaussian, then it is reasonable to assume that this bound would still be indicative of the near-optimal performance that can be expected.

7.4 SEPARATION BASED ON JOINT DIAGONALIZATION

We now turn to provide an overview of some of the leading approaches and algorithms for SOS-based BSS for WSS processes. Naturally, we cannot cover all of the existing methods, so we shall try to focus on some of the most prevalent, or conceptually appealing. As mentioned in the Introduction, most existing approaches can be roughly divided into two classes: approaches based on joint diagonalization (JD) of correlation matrices, and approaches based on maximum likelihood principles. We shall review JD-based methods in this section, and likelihood-based methods in the next section.

The basis for all JD-based methods in our context is the relation (7.23), repeated here for convenience:

$$\mathbf{R}_{\mathrm{x}}(\tau) = \mathbb{E}\{\mathbf{x}(t+\tau)\mathbf{x}^T(t)\} = \mathbf{A}\mathbf{R}_{\mathrm{s}}(\tau)\mathbf{A}^T, \tag{7.60}$$

where $\mathbf{R}_{\mathrm{s}}(\tau)$ are diagonal for all τ, by virtue of the sources' spatial decorrelation. A straightforward estimate of the observations' correlation matrices from the available data is given by

$$\breve{\mathbf{R}}_{\mathrm{x}}(\tau) = \frac{1}{T}\sum_{t=1}^{T-\tau} \mathbf{x}(t+\tau)\mathbf{x}^T(t), \quad \tau = 0, 1, \ldots. \tag{7.61}$$

These estimates are known as "biased" correlation estimates, since the mean of $\breve{\mathbf{R}}_{\mathrm{x}}(\tau)$ is $(1-\frac{\tau}{T})\mathbf{R}_{\mathrm{x}}$. Nevertheless, they are more convenient to work with than their unbiased versions (in which division by T in (7.61) is substituted by division by $T-\tau$), and the bias is negligible for sufficiently long observation intervals (for $\tau \ll T$).

Although the true $\mathbf{R}_{\mathrm{x}}(\tau)$ are symmetric for all τ, their estimates in (7.61) are almost surely asymmetric (with the exception of $\breve{\mathbf{R}}_{\mathrm{x}}(0)$). It is therefore common practice to use "symmetrized" versions of these matrices,

$$\widehat{\mathbf{R}}_{\mathrm{x}}(\tau) = \frac{1}{2}\left[\breve{\mathbf{R}}_{\mathrm{x}}(\tau) + \breve{\mathbf{R}}_{\mathrm{x}}^T(\tau)\right]. \tag{7.62}$$

Before we proceed to describe the JD-based methods, let us elaborate on the concepts of exact JD (*EJD*) and approximate JD (*AJD*).

7.4.1 On exact and approximate joint diagonalization

Almost any two symmetric matrices can be exactly jointly diagonalized under some mild conditions. Consider two symmetric $N \times N$ matrices $\mathbf{\Omega}_0$ and $\mathbf{\Omega}_1$, at least one of which is nonsingular. Assuming, without loss of generality, that $\mathbf{\Omega}_0$ is (the) nonsingular, let the eigenvalue decomposition (EVD) of $\mathbf{\Omega}_1\mathbf{\Omega}_0^{-1}$ be denoted as

$$\mathbf{\Omega}_1\mathbf{\Omega}_0^{-1}\mathbf{V} = \mathbf{V}\mathbf{\Lambda} \tag{7.63}$$

where $\mathbf{V}$ denotes the eigenvectors matrix and $\mathbf{\Lambda} = \text{diag}\{\lambda_1, \lambda_2, \ldots, \lambda_N\}$ is the eigenvalues matrix. Note that since $\mathbf{\Omega}_1\mathbf{\Omega}_0^{-1}$ is *not* symmetric in general, some of its eigenvalues and eigenvectors may be complex-valued (and $\mathbf{V}$ is generally not an orthogonal matrix).

Straightforward algebraic manipulations on (7.63) lead to

$$\mathbf{V}^{-1}\mathbf{\Omega}_1\mathbf{V}^{-H} = \mathbf{\Lambda}\mathbf{V}^{-1}\mathbf{\Omega}_0\mathbf{V}^{-H} \tag{7.64}$$

(where $\mathbf{V}^{-H}$ denotes the inverse of $\mathbf{V}^H$). Denoting $\mathbf{D}_0 \triangleq \mathbf{V}^{-1}\mathbf{\Omega}_0\mathbf{V}^{-H}$ and $\mathbf{D}_1 \triangleq \mathbf{V}^{-1}\mathbf{\Omega}_1\mathbf{V}^{-H}$, we have $\mathbf{D}_1 = \mathbf{\Lambda}\mathbf{D}_0$. We shall now show that $\mathbf{D}_0$ and $\mathbf{D}_1$ are usually diagonal: Since, by construction, both are Hermitian, we get

$$\mathbf{\Lambda}\mathbf{D}_0 = \mathbf{D}_1 = \mathbf{D}_1^H = \mathbf{D}_0^H\mathbf{\Lambda}^* = \mathbf{D}_0\mathbf{\Lambda}^*, \tag{7.65}$$

or, equivalently,

$$\lambda_m D_0(m,n) = D_0(m,n)\lambda_n^* \quad (n,m) \in \{1,\ldots,N\}, \tag{7.66}$$

where $D_0(m,n)$ denotes the (m,n)-th element of $\mathbf{D}_0$. Therefore, whenever the eigenvalues of $\mathbf{\Omega}_1\mathbf{\Omega}_0^{-1}$ satisfy $\lambda_m \neq \lambda_n^*$ for all $m \neq n$, we get $D_0(m,n) = 0$ for all $m \neq n$, meaning that $\mathbf{D}_0$, and hence $\mathbf{D}_1$, are diagonal. Note that if all the eigenvalues $\lambda_1, \lambda_2, \ldots, \lambda_N$ are real-valued and distinct, then this condition is trivially satisfied.

Since $\mathbf{\Omega}_0 = \mathbf{V}\mathbf{D}_0\mathbf{V}^H$ and $\mathbf{\Omega}_1 = \mathbf{V}\mathbf{D}_1\mathbf{V}^H$, we conclude that the EJD of two symmetric matrices can be extracted from the EVD of the product of one with the inverse of the other, provided that all of the resulting eigenvalues are real-valued and distinct. $\mathbf{V}^{-1}$ is sometimes called the (exact) "joint diagonalizer" of $\mathbf{\Omega}_0$ and $\mathbf{\Omega}_1$.

Note that if $\mathbf{\Omega}_1\mathbf{\Omega}_0^{-1}$ has complex-valued eigenvalues (appearing in conjugate pairs), the condition $\lambda_m \neq \lambda_n^*$ would not be satisfied for some $m \neq n$. However, a sufficient (but not a necessary) condition for all eigenvalues to be real-valued is that at least one of the matrices (say $\mathbf{\Omega}_0$) be positive definite. To observe this, let $\mathbf{\Omega}_0\mathbf{U}_0 = \mathbf{U}_0\mathbf{\Lambda}_0$ denote the EVD of $\mathbf{\Omega}_0$, where $\mathbf{U}_0$ is orthogonal and $\mathbf{\Lambda}_0$ is diagonal with positive elements. From (7.63) we have

$$\mathbf{\Omega}_1\mathbf{\Omega}_0^{-1}\mathbf{V} = \mathbf{\Omega}_1\mathbf{U}_0\mathbf{\Lambda}_0^{-1}\mathbf{U}_0^T\mathbf{V} = \mathbf{V}\mathbf{\Lambda}, \tag{7.67}$$

Thus, defining $\bar{\mathbf{V}} \triangleq \mathbf{\Lambda}_0^{-1/2}\mathbf{U}_0\mathbf{V}$, we have

$$\mathbf{\Omega}_1\mathbf{U}_0\mathbf{\Lambda}_0^{-1/2}\bar{\mathbf{V}} = \mathbf{V}\mathbf{\Lambda} \quad \Rightarrow \quad \mathbf{\Lambda}_0^{-1/2}\mathbf{U}_0^T\mathbf{\Omega}_1\mathbf{U}_0\mathbf{\Lambda}_0^{-1/2}\bar{\mathbf{V}} = \bar{\mathbf{V}}\mathbf{\Lambda}. \tag{7.68}$$

Now, since $\Lambda_0^{-1/2}\mathbf{U}_0^T\Omega_1\mathbf{U}_0\Lambda_0^{-1/2}$ is symmetric (since $\Lambda_0^{-1/2}$ is real-valued), its eigenvalues matrix Λ (which is also the eigenvalues matrix of $\Omega_1\Omega_0^{-1}$) is real-valued.

In fact, this relation leads to an alternative, two-steps procedure, for finding the diagonalizing matrix $\mathbf{V}$: assuming that Ω_0 is positive-definite,

1. Find any square-root matrix $\mathbf{W}$ of Ω_0^{-1}, such that $\mathbf{W}^T\mathbf{W} = \Omega_0^{-1}$. For example, $\mathbf{W} = \Lambda_0^{-1/2}\mathbf{U}_0^T$ (but note that $\mathbf{W}$ is not unique, as it can be multiplied on the left by any orthogonal matrix and still be a square-root of Ω_0^{-1}).
2. Find the eigen-decomposition of the (symmetric) matrix $\mathbf{W}\Omega_1\mathbf{W}^T$, and denote its orthogonal eigenvectors matrix as $\mathbf{Q}^T$. Note that $\mathbf{Q}^T$ is unique (up to columns' signs and permutation) if and only if all eigenvalues of $\mathbf{W}\Omega_1\mathbf{W}^T$ are distinct.
3. The joint diagonalizer $\mathbf{V}^{-1}$ of Ω_0 and Ω_1 is given as $\mathbf{V}^{-1} = \mathbf{QW}$ (unique under the same condition). Note that the diagonalization operation can thus be viewed as applying "whitening" $\mathbf{W}$, followed by "rotation" $\mathbf{Q}$.

In fact, this procedure is the basis for the constructive identifiability proof, shown earlier in section 7.3.1.

Often in SOS-based BSS (and also in other EJD-based BSS methods; e.g. see [31] for an overview) Ω_0 is taken as the observations' empirical correlation matrix $\widehat{\mathbf{R}}_x[0]$. Thus, the matrix $\mathbf{W}$ above can be regarded as a "spatial whitening" matrix, since $\mathbf{W}\widehat{\mathbf{R}}_x[0]\mathbf{W}^T = \mathbf{I}$. Ω_1 is taken as the empirical estimate of some other matrix, whose underlying structure is $\mathbf{AD}_1\mathbf{A}^T$ (see [31] for common choices), such that $\mathbf{D}_1$ is diagonal by virtue of some statistical property of the sources (most commonly, statistical independence).

Assume, momentarily, that Ω_0 is taken as the true correlation matrix of the observations, $\Omega_0 = \mathbf{R}_x[0] = \mathbf{AR}_s(0)\mathbf{A}^T$, whereas Ω_1 is taken as some other observations-related matrix satisfying $\Omega_1 = \mathbf{AD}_1\mathbf{A}^T$, where $\mathbf{D}_1$ is diagonal. Then under the scaling convention $\mathbf{R}_s(0) = \mathbf{I}$, the condition (for a unique solution) on the eigenvalues of $\Omega_1\Omega_0^{-1} = \mathbf{AD}_1\mathbf{A}^{-1}$ reduces to a condition on the elements of $\mathbf{D}_1$ (which are exactly these eigenvalues). In practice, however, the true matrices are unavailable, and therefore Ω_0 and Ω_1 are taken as their empirical estimates, which almost surely satisfy the condition of distinct eigenvalues, even if the true matrices do not. Of course, this does not mean that the mixing matrix becomes identifiable just because we use estimated matrices instead of true matrices: if the true $\mathbf{D}_1$ does not have distinct eigenvalues, then the resulting (unique) estimate of $\mathbf{A} = \mathbf{V}$ (or of $\mathbf{B} = \mathbf{V}^{-1}$) would be erroneous (in the respective elements), with a variance which cannot be decreased to zero by extending the observation length.

When more than two matrices (say $\Omega_0, \Omega_1, \ldots, \Omega_M$, often called the "*target-matrices*") are involved, an exact joint diagonalizer of the entire set usually does not exist. It is, however, possible to look for "approximate joint diagonalization" (*AJD*), namely, to find a matrix $\mathbf{V}$ which, such that when $\mathbf{V}^{-1}$ is applied to the set of target matrices, the resulting matrices $\mathbf{V}^{-1}\Omega_0\mathbf{V}^{-T}, \ldots, \mathbf{V}^{-1}\Omega_M\mathbf{V}^{-T}$ are "as diagonal as possible", possibly subject to some necessary constraints.

In fact, AJD is a fundamental problem in data analysis and signal processing, finding applications in various contexts. An abundance of criteria for measuring the extent of (approximate) joint diagonalization, as well as some associated constraints and iterative algorithms for the resulting constrained optimization problem, has been proposed in the literature in the past decade (e.g. [9,35,53,57,24,25,68,63,60,4,54,28,13,46]). We shall now provide a succinct overview of three main approaches to AJD, to which we shall refer in the sequel. A more detailed overview of these and some other leading approaches can be found in Chapter 11 (on Time-Frequency Domain Methods), so the interested reader is referred to the relevant sections therein.

7.4.1.1 AJD based on whitening

The most natural extension of the EJD approach presented above is to first find a whitening matrix (usually taken as any square root of $\mathbf{\Omega}_0^{-1}$) and then transform all of the remaining "target-matrices" into their whitened version,

$$\widetilde{\mathbf{\Omega}}_m = \mathbf{W}\mathbf{\Omega}_m\mathbf{W}^T, \quad m = 1,\ldots,M. \tag{7.69}$$

Note that in a general BSS context, this stage is sometimes replaced by first whitening the observed data, and then re-estimating the (whitened) "target-matrices" from the whitened data. However, when these matrices are empirical correlation matrices, such an operation is equivalent to applying (7.69) to the original target-matrices, without involving computation of a whitened version of the observation.

Then the residual orthogonal (rotation) matrix $\mathbf{Q}$ has to be found. However, since it is now generally impossible to find a matrix which would jointly diagonalize *all* $\widetilde{\mathbf{\Omega}}_m$ simultaneously, $\mathbf{Q}$ is found by minimizing an off-diagonality criterion

$$\min_{\mathbf{Q}} \sum_{m=1}^{M} \text{off}(\mathbf{Q}\widetilde{\mathbf{\Omega}}_m\mathbf{Q}^T), \quad \text{s.t. } \mathbf{Q}\mathbf{Q}^T = \mathbf{I}, \tag{7.70}$$

where the off($\cdot$) operator is defined as the sum of squared off-diagonal elements of its argument matrix.

A computationally appealing algorithm for iterative computation of $\mathbf{Q}$ using successive Givens rotations (for the more general complex-valued case) is proposed by Cardoso and Souloumiac in [9]. The following is a brief overview of the version of this algorithm suited for the case of real-valued target matrices.

The idea is to iteratively update $\mathbf{Q}$ by multiplication on the left with an orthogonal rotation matrix, denoted $\mathbf{U}_{n,n'}(\theta)$ (for $(n \neq n') \in \{1,\ldots,N\}$), which takes the form of the $N \times N$ identity matrix $\mathbf{I}$, except in the following entries, taking the form

$$\begin{bmatrix} U_{n,n'}(n,n) & U_{n,n'}(n,n') \\ U_{n,n'}(n',n) & U_{n,n'}(n',n') \end{bmatrix} = \begin{bmatrix} \cos\theta & \sin\theta \\ -\sin\theta & \cos\theta \end{bmatrix} \tag{7.71}$$

where $U_{n,n'}(k,\ell)$ denotes the (k,ℓ)-th element of $\mathbf{U}_{n,n'}$, and θ is a (real-valued) rotation angle, selected so as to minimize the criterion

$$\min_{\theta} \sum_{m=1}^{M} \text{off}(\mathbf{U}_{n,n'}(\theta)\mathbf{M}_m\mathbf{U}_{n,n'}^T(\theta)), \tag{7.72}$$

in which $\mathbf{M}_m \triangleq \mathbf{Q}\widetilde{\boldsymbol{\Omega}}_m\mathbf{Q}^T$ $(m = 1, \ldots, M)$ are the transformed target matrices prior to the current update. Using straightforward trigonometric manipulations, this minimization problem can be reduced to the minimization of

$$a\cos^2 2\theta + b\sin^2 2\theta - 2c\cos 2\theta\sin 2\theta$$

with

$$a = \sum_{m=1}^{M}[M_m(n,n) - M_m(n',n')]^2, \quad b = 4\sum_{m=1}^{M} M_m^2(n,n'),$$
$$c = 2\sum_{m=1}^{M} M_m(n,n')[M_m(n,n) - M_m(n',n')] \tag{7.73}$$

(here $M_m(n,n')$ denotes the (n,n')-th element of $\mathbf{M}_m$). The minimizing θ is given by

$$\widehat{\theta} = \frac{1}{2}\operatorname{atan}\frac{a - b - \sqrt{(a-b)^2 + 4c^2}}{2c}. \tag{7.74}$$

The resulting $\mathbf{U}_{n,n'}(\widehat{\theta})$ is then used for updating $\mathbf{Q}$ (and, consequently, the matrices $\mathbf{M}_m$) as

$$\begin{aligned}\mathbf{Q} &\leftarrow \mathbf{U}_{n,n'}(\widehat{\theta})\mathbf{Q}\\ \mathbf{M}_m &\leftarrow \mathbf{U}_{n,n'}(\widehat{\theta})\mathbf{M}_m\mathbf{U}_{n,n'}^T(\widehat{\theta}), \quad m = 1, \ldots, M.\end{aligned} \tag{7.75}$$

The iterations proceed by applying repeated sweeps over all combinations of $(n \neq n') \in \{1, \ldots, N\}$ until convergence is attained (namely, until all $\hat{\theta}$ are sufficiently close to zero). Once $\mathbf{Q}$ is obtained, the combined approximate joint diagonalizer of the original (non-whitened) set of target matrices is obviously given by $\mathbf{V}^{-1} = \mathbf{QW}$.

7.4.1.2 AJD without whitening – least squares criteria

The whitening phase in the preceding approach attains exact joint diagonalization of $\boldsymbol{\Omega}_0$, possibly at the cost of poor diagonalization of the other matrices in the set. Thus, possible estimation errors in $\boldsymbol{\Omega}_0$ may have a severe effect on the resulting diagonalizer, since they can not be compensated for by better (or even exact) estimates of the other matrices. Therefore, such an approach, although computationally and conceptually appealing, is known to limit the resulting separation performance [7,57].

It was therefore proposed to apply AJD more "uniformly" to the entire set, by looking for a general (non-orthogonal) diagonalizing matrix, which would minimize an overall off-diagonality measure. One such natural measure (per matrix) is the Least-Squares (LS) based off(·) operator above. Evidently, however, simple extension of the criterion in (7.70) (to let the sum run from $m = 0$ and drop the constraint) cannot work: without an appropriate constraint on $\mathbf{Q}$ (substituting the orthogonality constraint), such a criterion would be trivially minimized by $\mathbf{Q} = 0$. Thus, various constraints and alternative criteria (mostly LS-based) have been proposed in recent years, each with an associated iterative optimization algorithm – e.g. [53,57,24,25,68,63,60,4,54,28].

Possible enhancement of the LS-based criteria can be attained by introducing proper weighting among elements of the target matrices (we shall elaborate on this possibility in the sequel). Related efficient algorithms for solving a weighted AJD problem can be found in [46,44].

7.4.1.3 Pham's AJD

An alternative, non-LS criterion for measuring off-diagonality was considered by Pham [35], mainly in the context of BSS based on maximum likelihood or minimization of mutual information (see more details in section 7.5 and 7.6). For each matrix $\mathbf{B}\boldsymbol{\Omega}_m\mathbf{B}^T$, the criterion measures the *Kullback-Leibler divergence* (*KLD*)[1] between two zero-mean multivariate Gaussian distributions: the first with covariance $\mathbf{B}\boldsymbol{\Omega}_m\mathbf{B}^T$, and the second with covariance $\mathrm{Ddiag}(\mathbf{B}\boldsymbol{\Omega}_m\mathbf{B}^T)$, where the $\mathrm{Ddiag}(\cdot)$ operator (shorthand for $\mathrm{Diag}(\mathrm{diag}(\cdot))$) nullifies all off-diagonal elements of its argument matrix. The overall off-diagonality criterion takes the form

$$J(\mathbf{B}) = \frac{1}{2}\sum_{m=0}^{M} \log\det \mathrm{Ddiag}(\mathbf{B}\boldsymbol{\Omega}_m\mathbf{B}^T) - \log\det \mathbf{B}\boldsymbol{\Omega}_m\mathbf{B}^T, \tag{7.76}$$

to be minimized with respect to the diagonalizing matrix $\mathbf{B}$ (free of constraints). A computationally efficient iterative algorithm for solving this minimization problem is provided in [35]. It consists of iterative updates of row-pairs in $\mathbf{B}$ as follows:

Let $\mathbf{b}_n^T$ and $\mathbf{b}_{n'}^T$ denote the n-th and n'-th rows of $\mathbf{B}$. They are updated as

$$\begin{bmatrix} \mathbf{b}_n^T \\ \mathbf{b}_{n'}^T \end{bmatrix} \leftarrow \begin{bmatrix} \mathbf{b}_n^T \\ \mathbf{b}_{n'}^T \end{bmatrix} - \frac{2}{1+\sqrt{1-4h_{n,n'}h_{n',n}}} \begin{bmatrix} 0 & h_{n,n'} \\ h_{n',n} & 0 \end{bmatrix} \begin{bmatrix} \mathbf{b}_n^T \\ \mathbf{b}_{n'}^T \end{bmatrix}, \tag{7.77}$$

where

$$\begin{bmatrix} h_{n,n'} \\ h_{n',n} \end{bmatrix} = \begin{bmatrix} \omega_{n,n'} & 1 \\ 1 & \omega_{n',n} \end{bmatrix}^{-1} \begin{bmatrix} g_{n,n'} \\ g_{n',n} \end{bmatrix} \tag{7.78}$$

with

$$g_{n,n'} = \frac{1}{M}\sum_{m=1}^{M} \frac{[\mathbf{B}\boldsymbol{\Omega}_m\mathbf{B}^T]_{n,n'}}{[\mathbf{B}\boldsymbol{\Omega}_m\mathbf{B}^T]_{n,n}}, \tag{7.79}$$

[1]The KLD between two *pdf*s $p_1(x)$ and $p_2(x)$ is given by $\int p_1(x)\log\frac{p_1(x)}{p_2(x)}dx$. The KLD is non-negative, and vanishes if and only if $p_1(x) = p_2(x)\ \forall x$. Note that the KLD is non-symmetric in $p_1(x)$ and $p_2(x)$.

$$\omega_{n,n'} = \frac{1}{M}\sum_{m=1}^{M}\frac{[\mathbf{B}\mathbf{\Omega}_m\mathbf{B}^T]_{n',n'}}{[\mathbf{B}\mathbf{\Omega}_m\mathbf{B}^T]_{n,n}}, \tag{7.80}$$

$[\mathbf{B}\mathbf{\Omega}_m\mathbf{B}^T]_{n,n'}$ denoting the (n,n')-th element of $\mathbf{B}\mathbf{\Omega}_m\mathbf{B}^T$. When the target-matrices are complex-valued Hermitian (which may happen even in the context of real-valued signals, if these matrices are taken in the frequency-domain, see subsection 7.5.3), the numerator in (7.79) should be replaced with its real part.

The iterations proceed by applying repeated sweeps over all combinations of $(n \neq n') \in \{1,\dots,N\}$ until convergence is attained (namely, until all $g_{n,n'}(\mathbf{B})$ are sufficiently close to zero). Like the other non-orthogonal AJD approaches discussed above, this procedure results in a matrix $\mathbf{B}$ which is (in general) not orthogonal. Note, however, that this criterion (and procedure) can only be used with positive-definite target matrices.

7.4.2 The first JD-based method

The first method relying on JD of estimated correlation matrices seems to have been proposed by Fêty and Van Uffelen in [21] (1988). It consists of estimating $\mathbf{A}$ by EJD of $\mathbf{\Omega}_0 = \widehat{\mathbf{R}}_x(0)$ and $\mathbf{\Omega}_1 = \widehat{\mathbf{R}}_x(1)$. As mentioned above, the identifiability condition (under the scaling convention) for such a scheme is that the diagonal values of the true $\mathbf{D}_1 = \mathbf{R}_s[1]$ be distinct. If any two (or more) of these values are identical (or are too close), then the resulting estimates would be rather poor. It was therefore proposed in [21] to resort to EJD of $\widehat{\mathbf{R}}_x(0)$ and $\widehat{\mathbf{R}}_x(\tau)$ in such cases, with $\tau > 1$.

7.4.3 AMUSE and its modified versions

The *Algorithm for Multiple Unknown Signals Extraction* (AMUSE) was proposed by Tong *et al.* [52,51] in 1990. Similar to the approach of Fêty and Van Uffelen [21], AMUSE relies on EJD of $\widehat{\mathbf{R}}_x(0)$ and $\widehat{\mathbf{R}}_x(\tau)$ for some $\tau \neq 0$. Note that the same idea was also proposed (four years later) by Molgedey and Schuster in [29]. The EJD in AMUSE is attained implicitly by applying a spatial whitening stage, which in effect modifies $\widehat{\mathbf{R}}_x(\tau)$ into $\widehat{\mathbf{R}}_z(\tau) \triangleq \widehat{\mathbf{W}}\widehat{\mathbf{R}}_x(\tau)\widehat{\mathbf{W}}^T$ ($\widehat{\mathbf{W}}$ being a square root of $\widehat{\mathbf{R}}_x^{-1}(0)$) and then finding the orthogonal eigenvalues matrix $\widehat{\mathbf{Q}}$ of $\widehat{\mathbf{R}}_z(\tau)$.

No specific strategy for selecting τ was provided. A possible natural approach (pursued, for example, by Zaïdi in [64]) is to look for the value of τ with which the eigenvalues of $\mathbf{R}_z(\tau)$ are the "most distinct", e.g. by using the distance between the closest eigenvalues as a measure of distinction. However, such an approach may be computationally demanding (requiring the eigen-decomposition of quite a few candidate matrices) and not very reliable, especially when the number of sources is large.

Several modifications of the basic EJD concept behind AMUSE have been proposed since. For example, one of the approaches proposed by Ziehe and Müller in the framework of the *Time-Delays based SEParation (TDSEP)* algorithm [67], is to construct $\mathbf{\Omega}_1$ as a sum of several time-lagged correlation matrices (whereas $\mathbf{\Omega}_0$ is taken as the zero-lag correlation). Another possible modification, which can be viewed as a generalization of the former, was proposed by the authors of AMUSE (although in a different context, [51]). The modification consists of applying the EJD to two matrices representing two different general linear combinations of lagged correlation matrices,

$$\mathbf{\Omega}_0 = \widehat{\mathbf{R}}_\alpha \triangleq \sum_{\tau=0}^{P} \alpha_\tau \widehat{\mathbf{R}}_\mathbf{x}(\tau) \quad \mathbf{\Omega}_1 = \widehat{\mathbf{R}}_\beta \triangleq \sum_{\tau=0}^{P} \beta_\tau \widehat{\mathbf{R}}_\mathbf{x}(\tau), \tag{7.81}$$

with different strategies for selection of the coefficients $\{\alpha_\tau\},\{\beta_\tau\}$. For example, when additive, temporally-white noise is present (see further discussion of the noisy case in section 7.6 below), the noise correlation induces bias on the estimate of $\mathbf{R}_\mathbf{x}(0)$, but not on the estimates of all other $\mathbf{R}_\mathbf{x}(\tau)$. It then makes sense to discard $\widehat{\mathbf{R}}_\mathbf{x}(0)$ (selecting $\alpha_0 = \beta_0 = 0$), and use a positive-definite linear combination of lagged correlation matrices for the whitening stage (with $\mathbf{\Omega}_0 = \widehat{\mathbf{R}}_\alpha$). A procedure for selecting the $\{\alpha_\tau\}$ coefficients such that $\widehat{\mathbf{R}}_\alpha$ be positive-definite is proposed by Tong *et al.* in [50] and also by Zaïdi in [66]. Once the $\{\alpha_\tau\}$ coefficients are selected, the $\{\beta_\tau\}$ coefficients are to be selected such that the eigenvalues of $\widehat{\mathbf{R}}_\beta \widehat{\mathbf{R}}_\alpha^{-1}$ (which are also the eigenvalues of $\widehat{\mathbf{W}}\widehat{\mathbf{R}}_\beta \widehat{\mathbf{W}}^T$, where $\widehat{\mathbf{W}}^T \widehat{\mathbf{W}} = \widehat{\mathbf{R}}_\alpha^{-1}$) are "as distinct as possible".

The use of two time-delayed correlation matrices as $\mathbf{\Omega}_0 = \widehat{\mathbf{R}}_\mathbf{x}(\tau_0)$ and $\mathbf{\Omega}_1 = \widehat{\mathbf{R}}_\mathbf{x}(\tau_1)$ with $\tau_1 \neq \tau_0 \neq 0$ was also proposed (in the noisy context) by Chang *et al.* [10,11]. A possible problem with such an approach is that $\mathbf{\Omega}_0$ is not guaranteed to be positive-definite, which might (or might not) affect the uniqueness of the EJD solution (see discussion in the previous subsection).

In [47,48], Tomé proposed using the empirical correlation matrix between linearly filtered versions of the observed signals as $\mathbf{\Omega}_1$ (with the zero-lagged correlation of the original observations as $\mathbf{\Omega}_0$). In fact, it is straightforward to show such a correlation matrix is merely a linear combination of lagged correlation matrices of the unfiltered observation, where the linear combination coefficients can be easily derived from the filter coefficients. Therefore, this modification is essentially equivalent to (7.81), with $\alpha_0 = 1$, $\alpha_\tau = 0, \forall \tau \neq 0$ and $\{\beta_\tau\}$ implicitly selected by selection of the filtering coefficients.

Another modification of AMUSE proposed by Tomé *et al.* (called "dAMUSE", [49]) consists of first augmenting the observations vectors, such that each entry $x_n(t)$ of the original vector is replaced with an $M \times 1$ vector (M is a selected parameter) $[x_n(t)\, x_n(t-1) \cdots x_n(t-M+1)]^T$ (for $n = 1,\ldots,N$). Then, the respectively augmented $\mathbf{\Omega}_0$ and $\mathbf{\Omega}_1$ (from dimensions $N \times N$ to dimensions $MN \times MN$) are empirical correlation matrices of the augmented observation vectors at two different lags. The result of applying EJD to these matrices is an augmented estimated unmixing matrix, which can unmix the N blocks from each other, but may still leave residual mixing within blocks (equivalent to generating filtered versions of the sources – which is acceptable in various applications, and can be resolved by applying additional constraints at the cost of increased computational complexity).

The original version of AMUSE, as well as most of the proposed modifications, does not specify explicitly an approach for choosing the lags (or, more generally, the linear combination coefficients $\{\alpha_\tau\}$ and $\{\beta_\tau\}$) so as to optimize the separation performance. Although some of the approaches propose to tune the coefficients so as to cause the generalized eigenvalues to be "as distinct as possible", there is no explicit relation to optimization of the performance. A more general SOS-based framework for the EJD, involving optimization of the performance, was recently considered by Yeredor in [61]. The matrices $\mathbf{\Omega}_0$ and $\mathbf{\Omega}_1$ can generally be viewed as constructed using

$$\mathbf{\Omega}_0 = \mathbf{X}_T \mathbf{P}_0 \mathbf{X}_T^T \quad \mathbf{\Omega}_1 = \mathbf{X}_T \mathbf{P}_1 \mathbf{X}_T^T, \tag{7.82}$$

where $\mathbf{X}_T$ is the $N \times T$ matrix of observation vectors and $\mathbf{P}_0$ and $\mathbf{P}_1$ are symmetric $T \times T$ matrices (called "association matrices"), which may be chosen so as to optimize the performance. For example, when choosing $\mathbf{P}_0 = \frac{1}{T}\mathbf{I}$, $\mathbf{\Omega}_0$ takes the form of the "standard" observations' empirical correlation matrix. When choosing $\mathbf{P}_1$ as a symmetric Toeplitz matrix with the value β_0/T along its main diagonal and $\beta_\tau/2T$ along its τ-th diagonals ($\tau = 1, \ldots, P$), $\mathbf{\Omega}_1$ takes the form of $\widehat{\mathbf{R}}_\beta$ defined in (7.81). By assuming Gaussian sources and applying small-error analysis, it is shown in [61] how the "association matrices" $\mathbf{P}_0$ and $\mathbf{P}_1$ can be chosen so as to minimize any (single) desired entry $\mathrm{ISR}_{m,n}$ of the resulting ISR matrix.

The approach in [61] applies to source signals of an arbitrary (but known[2]) temporal correlation structure, not necessarily a WSS structure. Thus, explicit derivation of the optimization procedure is rather involved and somewhat beyond the scope of this chapter, so we do not present these results in detail here.

We do note, however, that as shown in [61], an SOS-based two-matrices EJD approach is generally unable to optimize all entries of the ISR matrix simultaneously by any choice of association matrices. Moreover, even the optimized entry, $\mathrm{ISR}_{m,n}$, only takes the minimum value attainable by an EJD approach, which is generally higher than the respective iCRLB for that entry. In order to improve the performance for all ISRs and approach the iCRLB (at least in the Gaussian case), it is necessary to resort to AJD of several correlation matrices (or to other approaches), as detailed in the following subsection.

7.4.4 SOBI, TDSEP and modified versions

The *Second Order Blind Identification* (*SOBI*) algorithm was proposed by Belouchrani *et al.* in [6] in 1997. Unlike AMUSE, SOBI does not rely on the EJD of a pair of matrices, but rather on the AJD of a prescribed set of $K > 2$ correlation matrices.

[2]However, asymptotically, under sufficient ergodicity conditions, the prior knowledge in not necessary for attaining the optimal solution.

Since SOBI addresses the more general case of noisy mixtures with more observations than sources (to be discussed in section 7.6), it includes a preliminary phase aimed at identifying the noise-only subspace and projecting the observed signals onto the signal (+noise) subspace. However, for clarity of our exposition (for the noiseless case), we shall ignore this denoising phase, and assume, as before, that we observe N noiseless mixtures of the N sources. We also reduce the discussion to real-valued signals (although SOBI is originally aimed at complex-valued signals).

When presented in this context, the SOBI algorithm essentially applies whitening-based AJD to a set of matrices composed of the zero-lag correlation matrix (used for obtaining the whitening matrix $\widehat{\mathbf{W}}$ as a square-root of its inverse) and additional correlation matrices at different lags $\tau = 1,\ldots,M$. A set of spatially whitened observations $\mathbf{z}(t) = \widehat{\mathbf{W}}\mathbf{x}(t)$ is computed (for $t = 1,\ldots,T$), and the empirical symmetrized correlation matrices $\widehat{\mathbf{R}}_z(\tau)$ ($\tau = 1,\ldots,M$) of $\mathbf{z}(t)$ are obtained. Obviously, these matrices are also given by $\widehat{\mathbf{R}}_z(\tau) = \widehat{\mathbf{W}}\widehat{\mathbf{R}}_z(\tau)\widehat{\mathbf{W}}^T$ (so, as already mentioned above, in the noiseless case it is not really necessary to actually compute $\mathbf{z}(\cdot)$ in order to obtain these matrices). The process proceeds by looking for the orthogonal matrix $\widehat{\mathbf{Q}}$ which minimizes the off-diagonality criterion in (7.70), with $\mathbf{\Omega}_\tau = \widehat{\mathbf{R}}_z(\tau)$, $\tau = 1,\ldots,M$. The estimated separation matrix is then given by $\widehat{\mathbf{B}} = \widehat{\mathbf{Q}}\widehat{\mathbf{W}}$.

The TDSEP [67] approach (already mentioned above) operates along the same methodology as SOBI (but addresses the noiseless case only). TDSEP makes a distinction between an AMUSE-like approach (EJD of two matrices, one being the zero-lag correlation and the other being either one specific lagged correlation or a linear combination of several lagged correlations) and a SOBI-like approach (orthogonal diagonalization of lagged correlations of the spatially-whitened data). Other than slightly different formulations of the criteria and solutions, there are no essential differences between TDSEP on one hand and AMUSE/SOBI on the other hand.

7.4.4.1 Weighted AJD – WASOBI

Although capable of exploiting several lagged correlations, thereby possibly introducing a stronger averaging effect than AMUSE, there is no claim of optimality of the SOBI/TDSEP approach. As mentioned earlier, it is possible to turn the general (possibly constrained) LS optimization problem into a Weighted LS (WLS) problem. Optimal choice of the weights would allow optimal exploitation of the estimated correlation matrices in this framework. Indeed, the Weights-Adjusted SOBI (WASOBI) approach, proposed by Yeredor in 2000 [56], considers the case of Gaussian sources, for which the covariance matrix of the errors in estimating the lagged correlation matrices can be easily expressed in terms of the true correlation values.

For example, assume a slight modification of (7.61), in which the estimated correlation matrices (before symmetrization) are given by

$$\breve{\mathbf{R}}_x(\tau) = \frac{1}{T}\sum_{t=1}^{T}\mathbf{x}(t+\tau)\mathbf{x}^T(t), \quad \tau = 0,1,\ldots,M, \tag{7.83}$$

assuming that $T+M$ (rather than T) observations are available (the difference from (7.61) is in the upper limit of the summation). Then the correlation between the (m,n)-th element of $\check{\mathbf{R}}_{\mathbf{x}}(\tau_1)$ and the (p,q)-th element of $\check{\mathbf{R}}_{\mathbf{x}}(\tau_2)$ is given by

$$\begin{aligned}\mathbb{E}\{\check{R}_{m,n}(\tau_1)\check{R}_{p,q}(\tau_2)\} &= \frac{1}{T}\sum_{t=1}^{T}\sum_{s=1}^{T}\mathbb{E}\{x_m(t+\tau_1)x_n(t)x_p(s+\tau_2)x_q(s)\}\\ &= \frac{1}{T}\sum_{\tau=-T}^{T}\left(1-\frac{|\tau|}{T}\right)[R_{m,p}(\tau+\tau_1-\tau_2)R_{n,q}(\tau)+R_{m,q}(\tau+\tau_1)R_{n,p}(\tau-\tau_2)]\\ &\quad +R_{m,n}(\tau_1)R_{p,q}(\tau_2).\end{aligned} \tag{7.84}$$

Here $\check{R}_{m,n}(\tau)$ and $R_{m,n}(\tau)$ denote the (m,n)-th elements of $\check{\mathbf{R}}_{\mathbf{x}}(\tau)$ and $\mathbf{R}_{\mathbf{x}}(\tau)$ (resp.), and we have used the well-known property (e.g. [26]) that if w_1, w_2, w_3 and w_4 are any (real-valued) zero-mean jointly Gaussian random variables, then

$$\mathbb{E}\{w_1w_2w_3w_4\} = \mathbb{E}\{w_1w_2\}\mathbb{E}\{w_3w_4\} + \mathbb{E}\{w_1w_3\}\mathbb{E}\{w_2w_4\} + \mathbb{E}\{w_1w_4\}\mathbb{E}\{w_2w_3\}. \tag{7.85}$$

Thus, the covariance of $\check{R}_{m,n}(\tau_1)$ and $\check{R}_{p,q}(\tau_2)$ is obtained from (7.84) by eliminating the last term (which is the product of their means).

If we further assume that all sources are MA processes of a known order (say Q), the summation in (7.84) would extend from $-Q$ to Q, rather than from $-T$ to T (since, if all sources are MA processes of orders smaller or equal to Q, then so are their mixtures, hence their correlations at lags higher than Q must vanish). For sufficiently long observation intervals, the true (unknown) correlations in (7.84) may be replaced with their estimated counterparts. Thus, we may obtain good (consistent) approximations of the covariance matrix of all estimated correlations. Now, since the AJD (with an LS diagonality criterion) is basically a nonlinear LS fit to the estimated correlations, we may incorporate a weight matrix which is the inverse of the covariance matrix of these estimates for optimal estimation of **A** (asymptotically, in the sense of minimum MSE) – e.g. see [26].

An interesting generalization of the same idea (Yeredor and Doron, [62]) suggests that even if the source signals are (Gaussian) MA processes of maximal order Q, it would be helpful to incorporate estimated correlations of lags beyond Q into the WASOBI framework. Evidently, in the original SOBI (or TDSEP) framework, estimating correlations at such lags would be futile, since the true correlations there are zero and bear no information on the mixing matrix. However, since the estimation errors in these matrices are:

1. known, since the true correlation values (zeros) are known; and
2. correlated with the estimation errors in the smaller lags' correlations,

the estimates of these matrices can be used (implicitly in the optimally weighted scheme) to reduce the estimation errors in the smaller lags' correlations, thereby improving the overall performance in estimating **A**.

Although appealing in terms of the attainable performance improvement, this basic WASOBI approach may be prohibitively computationally demanding as the number of sources increases beyond five or six. Fortunately, however, using a modified version of WASOBI for (Gaussian) AR sources, it was shown by Tichavský *et al.* [42,44] that a computationally efficient scheme can be employed for applying the WASOBI approach in such cases. We shall not review this approach in detail in here, but we shall indicate that it is based on the following principles:

1. For Gaussian AR(P) sources, the empirical $P+1$ correlation matrices of the observations (at lags $\tau = 0,\ldots,P$) form (asymptotically) a *sufficient statistic* for the estimation of all of the mixing parameters and (nuisance) AR parameters of the sources. Therefore, estimation of these $P+1$ matrices is sufficient for attaining a solution which, when using optimal weighting, is equivalent to the ML solution, and is therefore asymptotically optimal.
2. When the sources are (nearly) separated, the resulting weight matrix for the AJD has a special block-diagonal structure, which enables us to significantly simplify the AJD solution, as well as the solution of the individual Yule-Walker equations, employed for estimation of the sources' AR parameters (which in turn serve for computation of the weights).
3. A fast and efficient specially-tailored Gauss-Newton-type solution for the nonlinear WLS problem of the weighted AJD can be employed.

Thus, the efficient AR-WASOBI algorithm proceeds iteratively as follows: first, the observations' empirical (symmetrized) correlation matrices $\widehat{\mathbf{R}}_{\mathbf{x}}(\tau)$ (up to lag $\tau = Q$) are computed. Then, an initial separation matrix $\widehat{\mathbf{B}}$ is obtained, e.g. by using plain (unweighted) SOBI. The estimated correlation matrices are updated as $\widehat{\mathbf{B}}\widehat{\mathbf{R}}_{\mathbf{x}}(\tau)\widehat{\mathbf{B}}^T$. Then, the separation is iteratively refined by:

1. estimating the AR parameters of each source from the updated sources' correlation sequences;
2. using the estimated AR parameters for computation of the block-diagonal weight matrix;
3. applying a single Gauss-Newton iteration to compute a "residual" separation matrix, denoted $\widehat{\mathbf{B}}_r$, via weighted AJD.

If $\widehat{\mathbf{B}}_r$ is still not close enough (to some prescribed tolerance) to the identity matrix $\mathbf{I}$, then the estimated correlation matrices are re-updated by further multiplication (with $\widehat{\mathbf{B}}_r$ on the left and $\widehat{\mathbf{B}}_r^T$ on the right), and the refinement process is repeated. Convergence is usually attained with very few iterations. The resulting overall separation matrix is naturally given by the implied reverse-product of all residual matrices with the initial separation matrix.

As mentioned earlier, this weighted AJD scheme attains (asymptotically) the same optimal performance as an ML-based approach, attaining the iCRLB. This is shown analytically (and demonstrated in simulation results) in [44].

7.4.4.2 Prediction-errors based methods

Methods based on the notion of prediction-errors were proposed by Dégerine and Malki in [14]. The first of these methods is very similar to SOBI, the only difference being that the empirical correlation matrices $\widehat{\mathbf{R}}_{\mathbf{x}}(\tau)$ are replaced with the empirical version of *partial autocovariance* matrices (denoted $\mathbf{\Delta}_{\mathbf{x}}(\tau)$). The second uses a particular form of the empirical *recursive symmetric partial autocorrelations* (denoted $\check{\mathbf{\Gamma}}_{\mathbf{x}}(\tau)$). In a sense, these two methods are equivalent to using different linear combinations of the empirical correlation matrices in SOBI, which in turn implies the introduction of particular weights into the AJD process. The third method exploits properties of the *canonical partial innovations* process. We shall review these methods briefly here, referring the interested reader to [14] for further details.

Partial autocovariance. Let the $n \times 1$ vector $\mathbf{e}_{\mathbf{x}}(t;k)$ denote the (progressive) linear prediction error of $\mathbf{x}(t)$ from its recent k samples,

$$\mathbf{e}_{\mathbf{x}}(t;k) \triangleq \sum_{\ell=0}^{k} \mathbf{A}_{\mathbf{x}}(\ell;k)\mathbf{x}(t-\ell), \quad \text{with} \quad \mathbf{A}_{\mathbf{x}}(0;k) = \mathbf{I}, \quad k \geqslant 0, \tag{7.86}$$

where the $N \times N$ coefficient matrices $\mathbf{A}_{\mathbf{x}}(\ell;k)$ (not to be confused with the mixing matrix $\mathbf{A}$) are selected (see detailed expressions below) so as to minimize the covariance matrices of these errors, denoted $\Sigma_{\mathbf{x}}^2(k) \triangleq \mathbb{E}\{\mathbf{e}_{\mathbf{x}}(t;k)\mathbf{e}_{\mathbf{x}}^T(t;k)\}$. Note that due to the stationarity, which implies *time-reversibility* of the sources (and therefore of the observations), the same coefficient matrices also minimize the covariance of the *backwards* prediction errors, denoted $\overleftarrow{\mathbf{e}}_{\mathbf{x}}(t;k)$, for which the prediction is based on the k immediate *future* samples of $\mathbf{x}(t)$. The resulting backwards-prediction covariance matrices are the same, $\Sigma_{\mathbf{x}}^2(k)$. The *partial autocovariance* matrices $\mathbf{\Delta}_{\mathbf{x}}(\cdot)$ are defined as:

$$\mathbf{\Delta}_{\mathbf{x}}(0) \triangleq \mathbb{E}\{\mathbf{x}(t)\mathbf{x}^T(t)\}, \quad \mathbf{\Delta}_{\mathbf{x}}(k) \triangleq \mathbb{E}\{\mathbf{e}_{\mathbf{x}}(t;k-1)\overleftarrow{\mathbf{e}}_{\mathbf{x}}^{T}(t-k;k-1)\}, \quad k \geqslant 1, \tag{7.87}$$

and offer an alternative representation of the SOS of $\mathbf{x}(\cdot)$. Naturally, these matrices can be uniquely determined from the "ordinary" correlation matrices $\mathbf{R}_{\mathbf{x}}(\cdot)$, e.g. using the following algorithm (a simplified version of Whittle's algorithm [55]):

1. Set $\mathbf{\Delta}_{\mathbf{x}}(0) = \mathbf{R}_{\mathbf{x}}(0) = \Sigma_{\mathbf{x}}^2(0)$ and $\mathbf{A}_{\mathbf{x}}(0,0) = \mathbf{I}$.
2. For $k = 1, 2, \ldots$ repeat:
   (a) Set $\mathbf{\Delta}_{\mathbf{x}}(k) = \mathbf{R}_{\mathbf{x}}(k) + \sum_{\ell=1}^{k-1} \mathbf{A}_{\mathbf{x}}(\ell;k-1)\mathbf{R}_{\mathbf{x}}(k-\ell)$;
   (b) Prepare the next coefficients matrices $\mathbf{A}_{\mathbf{x}}(\ell;k)$ (for $\ell = 1, \ldots, k$):

$$\mathbf{A}_{\mathbf{x}}(k;k) = -\mathbf{\Delta}_{\mathbf{x}}(k)[\Sigma_{\mathbf{x}}^2(k-1)]^{-1}$$
$$\mathbf{A}_{\mathbf{x}}(\ell;k) = \mathbf{A}_{\mathbf{x}}(\ell;k-1) + \mathbf{A}_{\mathbf{x}}(k;k)\mathbf{A}_{\mathbf{x}}(k-\ell;k-1);$$

   (c) Obtain the covariance matrix

$$\Sigma_{\mathbf{x}}^2(k) = \Sigma_{\mathbf{x}}^2(k-1) - \mathbf{\Delta}_{\mathbf{x}}(k)[\Sigma_{\mathbf{x}}^2(k-1)]^{-1}\mathbf{\Delta}_{\mathbf{x}}(k).$$

The resulting matrices $\mathbf{\Delta}_x(k)$ are all symmetric, and satisfy the condition that all resulting $\Sigma_x^2(k)$ are positive-definite.

Using similar definitions for the sources' prediction errors, it can be easily verified that

$$\mathbf{e}_x(t;k) = \mathbf{A}\mathbf{e}_s(t;k) \quad \overleftarrow{\mathbf{e}_x}(t;k) = \mathbf{A}\overleftarrow{\mathbf{e}_s}(t;k) \quad \forall k \geqslant 0, \tag{7.88}$$

(with $\mathbf{A}_s(\ell;k) = \mathbf{B}\mathbf{A}_x(\ell;k)\mathbf{A}$ for $\ell = 0,\ldots,k, \forall k \geqslant 0$) implying (via (7.87)) that the partial autocovariance matrices satisfy $\mathbf{\Delta}_x(k) = \mathbf{A}\mathbf{\Delta}_s(k)\mathbf{A}^T$, $k \geqslant 0$. Moreover, since the sources (and hence their prediction errors) are mutually uncorrelated, their partial autocorrelation matrices $\mathbf{\Delta}_s(k)$ are all diagonal. Thus, the matrices $\mathbf{\Delta}_x(k)$ (for several values of k) can be used as a substitutes to the "ordinary" correlation matrices in the SOBI framework. Obviously, however, these matrices first have to be estimated from the available observations $\mathbf{x}(\cdot)$. One possible estimation strategy is to use the Whittle algorithm above, with the true correlation matrices $\mathbf{R}_x(\cdot)$ replaced by their empirical (symmetrized) counterparts $\widehat{\mathbf{R}}_x(\cdot)$. An alternative, direct computation of the empirical partial autocovariance matrices from the data (based on recursive LS minimizations) is proposed in [14]. With this alternative, the data are first whitened, and then the partial autocorrelation matrices are estimated from the whitened data, and orthogonal AJD [9] is applied to the estimated matrix set.

Recursive symmetric partial autocorrelation. The second approach is based on the partial autocorrelation $\mathbf{\Gamma}(\cdot)$, defined as

$$\mathbf{\Gamma}_x(0) = \mathbf{\Delta}_x(0), \quad \mathbf{\Gamma}_x(k) \triangleq [\Sigma_x(k-1)]^{-1}\mathbf{\Delta}_x(k)[\Sigma_x(k-1)]^{-T}, \quad k \geqslant 1, \tag{7.89}$$

where $\Sigma_x(k)$ is (generally any) square-root of $\Sigma_x^2(k)$. Like $\mathbf{\Delta}_x(\cdot)$, the matrices $\mathbf{\Gamma}_x(\cdot)$ provide an alternative representation of the observations' SOS. However, unlike $\mathbf{\Delta}_x(\cdot)$, these matrices cannot (in general) serve as a substitute for the correlation matrices in SOBI: although the sources' partial autocorrelation matrices $\mathbf{\Gamma}_s(\cdot)$ are all diagonal, the observations' partial autocorrelation matrices generally do *not* satisfy the relation $\mathbf{\Gamma}_x(k) = \mathbf{A}\mathbf{\Gamma}_s(k)\mathbf{A}^T$. However, there exists a particular set of so-called *recursive symmetric* partial autocorrelation matrices, denoted $\mathbf{\Gamma}_x(\cdot)_s$, obtained (using (7.89)) with the respective symmetric square-roots of $\Sigma_x^2(\cdot)$, which are denoted $\Sigma_x(\cdot)_s$ and are obtained using the following recursion:

$$\begin{aligned}\Sigma_x(0)_s &= [\mathbf{\Delta}_x(0)]_s^{1/2}\\ \Sigma_x(k)_s &= \Sigma_x(k-1)_s[\mathbf{I} - \mathbf{\Gamma}_x^2(k)_s]_s^{1/2},\end{aligned} \tag{7.90}$$

where the operator $[\cdot]_s^{1/2}$ denotes the *symmetric* square-root of its matrix argument.

With this recursion (interlaced with (7.89)), the obtained observations' recursive symmetric partial autocorrelation matrices $\mathbf{\Gamma}_x(\cdot)_s$ satisfy the following relation to their sources-related counterparts:

$$\mathbf{\Gamma}_x(0)_s = \mathbf{A}\mathbf{A}^T, \quad \mathbf{\Gamma}_x(k)_s = \mathbf{Q}\mathbf{\Gamma}_s(\kappa)_s\mathbf{Q}^T, \quad k = 1,2\ldots \tag{7.91}$$

where $\mathbf{A} = \Sigma_\mathbf{x}(0)_s \mathbf{Q}^T$ and $\mathbf{Q}$ is an orthogonal matrix (namely, $\mathbf{Q} = \mathbf{A}^{-1}\Sigma_\mathbf{x}^{-1}(0)_s$ is the residual orthogonal separation matrix after whitening with $\mathbf{W} = \Sigma_\mathbf{x}^{-1}(0)_s$). Again, the unknown matrices $\mathbf{\Gamma}_\mathbf{x}(\cdot)_s$ can be estimated either by substituting the true correlation matrices with their empirical counterparts in the Whittle algorithm (and then applying (7.89) with (7.90)), or directly from the data, as proposed in [14].

Partial innovations canonical correlations. *Partial innovations* are normalized versions of the prediction errors, given by

$$\boldsymbol{\eta}_\mathbf{x}(t;k) = \Sigma_\mathbf{x}^{-1}(k)\mathbf{e}_\mathbf{x}(t;k), \quad \overleftarrow{\boldsymbol{\eta}}_\mathbf{x}(t;k) = \Sigma_\mathbf{x}^{-1}(k)\overleftarrow{\mathbf{e}}_\mathbf{x}(t;k), \quad k \geqslant 0. \tag{7.92}$$

Evidently, from (7.92), (7.89) and (7.87), the respective partial autocorrelation matrices can also be expressed as

$$\mathbf{\Gamma}_\mathbf{x}(0) = \mathbb{E}\{\mathbf{x}(t)\mathbf{x}^T(t)\}, \quad \mathbf{\Gamma}_\mathbf{x}(k) = \mathbb{E}\{\boldsymbol{\eta}_\mathbf{x}(t;k-1)\overleftarrow{\boldsymbol{\eta}}_\mathbf{x}(t-k;k-1)\}, \quad k \geqslant 1, \tag{7.93}$$

and they obviously depend (as before) on the choice of the square-root of $\Sigma_\mathbf{x}^2(\cdot)$ taken as $\Sigma_\mathbf{x}(\cdot)$ in (7.92), also affecting the relation between $\boldsymbol{\eta}_\mathbf{x}(\cdot)$ and $\boldsymbol{\eta}_\mathbf{s}(\cdot)$. When the *canonical recursive* version, denoted $\mathbf{\Gamma}_\mathbf{x}(\cdot)_c$ (to be defined immediately) is taken, it can be shown that $\boldsymbol{\eta}_\mathbf{x}(\cdot) = \boldsymbol{\eta}_\mathbf{s}(\cdot)$ for all $k \geqslant P$, where P is the smallest lag k such that the N sequences of diagonal values

$$\begin{gathered}[\Gamma_{\mathbf{s}(1,1)}(1), \Gamma_{\mathbf{s}(1,1)}(2), \ldots, \Gamma_{\mathbf{s}(1,1)}(k)] \\ [\Gamma_{\mathbf{s}(2,2)}(1), \Gamma_{\mathbf{s}(2,2)}(2), \ldots, \Gamma_{\mathbf{s}(2,2)}(k)] \\ \vdots \\ [\Gamma_{\mathbf{s}(N,N)}(1), \Gamma_{\mathbf{s}(N,N)}(2), \ldots, \Gamma_{\mathbf{s}(N,N)}(k)]\end{gathered}$$

(where $\Gamma_{\mathbf{s}(n,n)}(\cdot)$ denotes the (n,n)-th elements of $\mathbf{\Gamma}_\mathbf{s}(\cdot)$) are distinct. In fact, these sequences are the AR parameters obtained when trying to approximate each of the sources as an AR(k) process. Thus, for sources with distinct spectra such a value of P must exist.

The canonical recursive partial correlations are obtained by taking the EVD-based square-root as follows: let the decomposition

$$\mathbf{\Gamma}_\mathbf{x}(0)_c \triangleq \mathbf{\Delta}_\mathbf{x}(0) = \mathbf{L}(0)\mathbf{D}^2(0)\mathbf{L}(0)^T \tag{7.94}$$

denote the EVD of $\mathbf{\Delta}_\mathbf{x}(0)$, where $\mathbf{L}(0)$ is the orthogonal eigenvectors matrix and $\mathbf{D}(0)$ is a diagonal matrix of positive square-roots of the (positive) eigenvalues (subject to ordering conventions which ensure uniqueness). The implied canonical square root of $\mathbf{\Delta}_\mathbf{x}(0)$ is then $\Sigma_\mathbf{x}(0)_c \triangleq \mathbf{L}(0)\mathbf{D}(0)$. We then proceed for $k = 1, 2, \ldots$ with

$$\begin{aligned}\mathbf{\Gamma}_\mathbf{x}(k)_c &= \Sigma_\mathbf{x}^{-1}(k-1)_c\mathbf{\Delta}_\mathbf{x}(k)\Sigma_\mathbf{x}^{-T}(k-1)_c \triangleq \mathbf{L}(k)\mathbf{D}^2(k)\mathbf{L}(k)^T, \\ \Sigma_\mathbf{x}(k)_c &= \Sigma_\mathbf{x}(k-1)_c\mathbf{L}(k)[\mathbf{I} - \mathbf{D}^2(k)]_c^{1/2}\end{aligned} \tag{7.95}$$

(as before, $\mathbf{D}^2(k)$ and $\mathbf{L}(k)^T$ are eigenvalues and eigenvectors matrices obtained by EVD, and the canonical square-root of the diagonal matrix $[\mathbf{I}-\mathbf{D}^2(k)]$ is the diagonal matrix with positive square roots).

Now, defining $\mathbf{A}(k) \triangleq \mathbf{L}(0)\mathbf{D}(0)\mathbf{L}(1)\mathbf{L}(2)\cdots\mathbf{L}(k)$, it can be shown that for $k \geq P$ (P defined above) $\mathbf{A}(k)$ is equivalent (up to the standard ambiguities) to $\mathbf{A}$ (note that for $k > P$ all matrices $\mathbf{L}(k)$ reduce to permutation matrices). In practice, as before, the canonical recursive partial correlation matrices $\boldsymbol{\Gamma}_{\mathbf{x}}(\cdot)_c$ can either be estimated from the empirical correlation matrices using the Whittle algorithm and the recursive relations (7.95) above, or directly from the data, as shown in [14]. It is important to note that this approach requires prior knowledge of P. If P is not known in advance, but some upper-bound P_{max} is available,[3] it can be searched for by picking the value of $k \in \{1,\ldots,P_{max}\}$ for which the eigenvalues of $\boldsymbol{\Gamma}_{\mathbf{x}}(k)_c$ are "most dispersive" in some sense. While this approach works well in the case of two sources, it scales rather poorly, and may become practically ineffective as the number of sources grows.

7.5 SEPARATION BASED ON MAXIMUM LIKELIHOOD

A different approach to the problem of separating WSS sources can be taken using Maximum Likelihood estimation, in which the *pdf* (or its logarithm, often called the *log-likelihood*) of the observations is maximized with respect to the (estimated) mixing matrix $\mathbf{A}$ or unmixing matrix $\mathbf{B}$. In a truly blind scenario the sources' spectra are unknown, and therefore maximization of the log-likelihood must also involve maximization with respect to these spectra (which are considered *nuisance parameters* in this case). However, it is also possible to take a *Quasi ML* (*QML*) approach, in which the unknown spectra are substituted with some "educated guess" thereof, and the maximization proceeds with respect to the parameters of interest ($\mathbf{A}$ or $\mathbf{B}$) alone. QML is definitely not equivalent to ML, and, in particular, it generally does not enjoy the asymptotic optimality property of ML (unless the substituted nuisance parameters happen to take their true values).

Alternatively, it is also possible to take an *Exact ML* (*EML*) approach, and maximize the log-likelihood with respect to both the nuisance parameters (spectra) and the parameters of interest ($\mathbf{A}$ or $\mathbf{B}$). This can be a tractable optimization problem especially when the sources' spectra can be succinctly parameterized, such as in the case of parametric processes (in particular, AR sources).

In this section we shall concentrate on three methods which rely on ML principles. The first is the QML approach [38], proposed by Pham and Garat in 1997, which is based on expressing the observations' *pdf* in the frequency-domain, assuming arbitrary sources' spectra. The second is an EML approach [15], proposed by Dégerine and Zaïdi in 2004, which is based on expressing the observations' *pdf* in time-domain, assuming AR sources. The third is not exactly an ML-based approach, but it is closely related to ML,

[3]For example, if the sources are known to be AR processes, then P_{max} can be taken as the highest AR order.

being based on minimizing the *mutual information* (MI) between the estimated sources – as proposed by Pham in 2001 [34].

All three approaches rely on derivations which require the assumption of (jointly) Gaussian sources in order to be exact. Of course, the resulting estimation algorithms may also be applied when the sources are not Gaussian, but then the QML and EML approaches would no longer yield the "true" QML and EML estimates, and, similarly, the third approach would no longer minimize the MI, but rather the so-called *Gaussian MI* (*GMI*) between estimated sources (see details below, and also in Chapter 2, "Information").

7.5.1 The QML approach

Recall that the *pdf* of a segment $s(1), s(2), \ldots, s(T)$ from a sample-series of a (real-valued) stationary Gaussian process can be expressed in terms of its Fourier coefficients $\check{s}(\nu)$ taken at $\nu = \frac{k}{T}$ for $k = 0, \ldots, T-1$ as follows (see (7.7), repeated here for convenience):

$$f(s(1), \ldots, s(T)) = \frac{1}{2} \prod_{k=0}^{T-1} \frac{1}{\sqrt{\pi S_s(\frac{k}{T})}} \exp\left[-\frac{|\check{s}(\frac{k}{T})|^2}{2S_s(\frac{k}{T})}\right], \tag{7.96}$$

where $S_s(\nu)$ denotes the signal's spectrum. Likewise, the joint *pdf* of a segment $\mathbf{s}(1), \mathbf{s}(2), \ldots, \mathbf{s}(T)$ of N independent sources (all real-valued, stationary Gaussian) is given as the product of their individual *pdf*s, conveniently expressed as

$$f_{\mathbf{s}}(\mathbf{s}(1), \ldots, \mathbf{s}(T)) = \frac{1}{2^N} \prod_{k=0}^{T-1} \frac{1}{\det\left(\pi \mathbf{S}_s(\frac{k}{T})\right)^{\frac{1}{2}}} \exp\left[-\frac{1}{2}\check{\mathbf{s}}^H\left(\frac{k}{T}\right)\mathbf{S}_{\mathbf{s}}^{-1}\left(\frac{k}{T}\right)\check{\mathbf{s}}\left(\frac{k}{T}\right)\right], \tag{7.97}$$

where

$$\check{\mathbf{s}}\left(\frac{k}{T}\right) = \frac{1}{\sqrt{T}} \sum_{t=1}^{T} \mathbf{s}(t) e^{-j2\pi\frac{kt}{T}}, \quad k = 0, \ldots, T-1 \tag{7.98}$$

are the normalized vector Fourier coefficients, and $\mathbf{S}_{\mathbf{s}}(\nu) \triangleq \text{Diag}\{S_1(\nu), \ldots, S_N(\nu)\}$ is the sources' spectrum matrix. Now, since $\mathbf{x}(t) = \mathbf{A}\mathbf{s}(t)$, the joint *pdf* of the respective segment $\mathbf{x}(1), \ldots, \mathbf{x}(T)$ is given by the relation

$$f_{\mathbf{x}}(\mathbf{x}(1), \ldots, \mathbf{x}(T)) = \frac{1}{|\det \mathbf{A}|^T} f_{\mathbf{s}}(\mathbf{A}^{-1}\mathbf{x}(1), \ldots, \mathbf{A}^{-1}\mathbf{x}(T)), \tag{7.99}$$

which (combined with (7.97)) yields

$$f_{\mathbf{x}}(\mathbf{x}(1),\ldots,\mathbf{x}(T))=\frac{1}{2^N|\det\mathbf{A}|^T}$$
$$\cdot\prod_{k=0}^{T-1}\frac{1}{\det\left(\pi\mathbf{S}_s(\frac{k}{T})\right)^{\frac{1}{2}}}\exp\left[-\frac{1}{2}\left(\mathbf{A}^{-1}\check{\mathbf{x}}\left(\frac{k}{T}\right)\right)^H\mathbf{S}_s^{-1}\left(\frac{k}{T}\right)\left(\mathbf{A}^{-1}\check{\mathbf{x}}\left(\frac{k}{T}\right)\right)\right],$$

where

$$\check{\mathbf{x}}\left(\frac{k}{T}\right)=\frac{1}{\sqrt{T}}\sum_{t=1}^{T}\mathbf{x}(t)e^{-j2\pi\frac{kt}{T}},\quad k=0,\ldots,T-1 \tag{7.100}$$

are the normalized Fourier coefficients of the observations, satisfying $\mathbf{A}^{-1}\check{\mathbf{x}}(\nu)=\check{\mathbf{s}}(\nu)$. Exploiting the diagonality of $\mathbf{S}_s(\nu)$, the observations' log-likelihood can thus be expressed as

$$L_{\mathbf{x}}=-\frac{1}{2}\sum_{n=1}^{N}\sum_{k=0}^{T-1}\left[\frac{\left|\mathbf{e}_n^T\mathbf{A}^{-1}\check{\mathbf{x}}(\frac{k}{T})\right|^2}{S_n(\frac{k}{T})}+\log S_n\left(\frac{k}{T}\right)\right]-T\log|\det\mathbf{A}|+\text{Const.}, \tag{7.101}$$

where $\mathbf{e}_n$ denotes the n-th column of the $N\times N$ identity matrix $\mathbf{I}$. We reiterate the fact that this expression is exact only when the sources are jointly Gaussian (namely, each is Gaussian and they are statistically independent). If this is not the case, then this expression can at best serve as an approximation of the likelihood.

Since neither $\mathbf{A}$ nor the sources' spectra $S_n(\nu)$ are known (in a fully blind scenario), ML estimation requires maximization of $L_{\mathbf{x}}$ with respect to all. However, in the QML framework we assume that the sources' spectra are known, substituting their values with some presumed values, and maximizing $L_{\mathbf{x}}$ with respect to $\mathbf{A}$ alone. To this end, let $\hbar_n(\nu)$ denote the presumed spectrum of the n-th sources (substituting $S_n(\nu)$ in $L_{\mathbf{x}}$), and, using the relations $d(\mathbf{A}^{-1})=-\mathbf{A}^{-1}d\mathbf{A}\mathbf{A}^{-1}$ and $d\log\det\mathbf{A}=\text{trace}(\mathbf{A}^{-1}d\mathbf{A})$, observe that the perturbation $dL_{\mathbf{x}}$ in the likelihood resulting from a perturbation $d\mathbf{A}$ in $\mathbf{A}$ is given by

$$dL_{\mathbf{x}}=\sum_{n=1}^{N}\sum_{k=0}^{T-1}\frac{\mathbf{e}_n^T\mathbf{A}^{-1}d\mathbf{A}\mathbf{A}^{-1}\check{\mathbf{x}}(\frac{k}{T})\mathbf{e}_n^T\mathbf{A}^{-1}\check{\mathbf{x}}^*(\frac{k}{T})}{\hbar_n(\frac{k}{T})}-T\cdot\text{trace}(\mathbf{A}^{-1}d\mathbf{A}). \tag{7.102}$$

Now, denoting by $\partial_{m,n}$ the (m,n)-th element of the matrix $\mathbf{A}^{-1}d\mathbf{A}$, we may rewrite (7.102) as

$$dL_{\mathbf{x}}=\sum_{m=1}^{N}\sum_{n=1}^{N}\left[\sum_{k=0}^{T-1}\frac{\mathbf{e}_n^T\mathbf{A}^{-1}\check{\mathbf{x}}(\frac{k}{T})\mathbf{e}_m^T\mathbf{A}^{-1}\check{\mathbf{x}}^*(\frac{k}{T})}{\hbar_n(\frac{k}{T})}\right]\partial_{m,n}-T\sum_{n=1}^{N}\partial_{n,n}. \tag{7.103}$$

At the maximum of $L_{\mathbf{x}}$ with respect to $\mathbf{A}$, obtained at $\widehat{\mathbf{A}}$, this expression must vanish for all possible values of all $\partial_{m,n}$ $(m,n = 1,\ldots,N)$. This implies the following set of equations for $\widehat{\mathbf{A}}$:

$$\sum_{k=0}^{T-1} \frac{\mathbf{e}_n^T \widehat{\mathbf{A}}^{-1} \check{\mathbf{x}}(\frac{k}{T}) \mathbf{e}_m^T \widehat{\mathbf{A}}^{-1} \check{\mathbf{x}}^*(\frac{k}{T})}{h_n(\frac{k}{T})} = 0, \quad (m \neq n) \in \{1,\ldots,N\}, \tag{7.104}$$

and

$$\frac{1}{T} \sum_{k=0}^{T-1} \frac{\left| \mathbf{e}_n^T \widehat{\mathbf{A}}^{-1} \check{\mathbf{x}}(\frac{k}{T}) \right|^2}{h_n(\frac{k}{T})} = 1, \quad n = 1,\ldots,N, \tag{7.105}$$

which are also called *estimating equations.*

Let us now denote by $\widehat{s}_n(t) \triangleq \mathbf{e}_n \widehat{\mathbf{A}}^{-1} \mathbf{x}(t)$ the n-th estimated source, whose normalized Fourier coefficient is evidently given by $\widehat{\check{s}}_n(\frac{k}{T}) \triangleq \mathbf{e}_n^T \widehat{\mathbf{A}}^{-1} \check{\mathbf{x}}(\frac{k}{T})$. In addition, let us denote by $\phi_n(t)$ the (discrete-time) impulse response of the filter whose frequency response is $h_n^{-1}(\nu)$ (since all $h_n(\nu)$ are real-valued, all $\phi_n(t)$ are symmetric). Then applying Parseval's identity[4] to (7.104) and to (7.105) we get

$$\sum_{t=1}^{T} (\phi_n \star \widehat{s}_n)(t) s_m(t) = 0, \quad (m \neq n) \in \{1,\ldots,N\}, \tag{7.106}$$

and

$$\frac{1}{T} \sum_{t=1}^{T} (\phi_n \star \widehat{s}_n)(t) s_n(t) = 1, \quad n = 1,\ldots,N, \tag{7.107}$$

respectively, where $(\phi_n \star \widehat{s}_n)(t)$ denotes cyclic convolution (with cycle T) between the filter $\phi_n(t)$ and the n-th estimated source $\widehat{s}_n(t)$ (assuming that the filter's impulse response is shorter than T). Asymptotically (in T), this cyclic convolution can be regarded as a linear convolution, as end-effects become negligible.

The condition (7.106) implies that at the QML estimate of $\mathbf{A}$, the filtered version of $\widehat{s}_n(t)$ (filtered with $\phi_n(t)$) is empirically uncorrelated with all the other sources. This relation is satisfied for all $n = 1,\ldots,N$. The second condition (7.105) is merely a scaling condition, implying that the empirical correlation of the filtered version of $\widehat{s}_n(t)$ with $\widehat{s}_n(t)$ is 1. This condition is of marginal interest, since it only serves to resolve the scaling ambiguity when the sources spectra are indeed known, and imposes arbitrary (irrelevant) scaling in the more common case when the spectra $h_n(\nu)$ (and thus the filters $\phi_n(t)$) are chosen arbitrarily.

[4] Also known as Plancharel's identity, $\sum_{k=0}^{T-1} \check{f}(\frac{k}{T}) \check{g}^*(\frac{k}{T}) = \sum_{t=1}^{T} f(t) g^*(t)$.

Assume now that each $\phi_n(t)$ is a finite impulse response, vanishing for $t < -P$ and for $t > P$ (where $P < T/2$). In fact, this is equivalent to assuming that all sources are AR processes of orders less than or equal to P. Rewriting (7.104) more explicitly (and dividing by T), we obtain the condition

$$\begin{aligned}\frac{1}{T}\sum_{t=1}^{T}\sum_{\tau=-P}^{P}\phi_n(\tau)\widehat{s}_n(t-\tau)\widehat{s}_m(t) &= \sum_{\tau=-P}^{P}\phi_n(\tau)\left(\frac{1}{T}\sum_{t=1}^{T}\widehat{s}_n(t-\tau)\widehat{s}_m(t)\right)\\ &= \sum_{\tau=-P}^{P}\phi_n(\tau)\left(\frac{1}{T}\sum_{t=1}^{T}\mathbf{e}_n^T\widehat{\mathbf{A}}^{-1}\mathbf{x}(t-\tau)\mathbf{x}^T(t)\widehat{\mathbf{A}}^{-T}\mathbf{e}_m\right)\\ &= \mathbf{e}_n^T\widehat{\mathbf{A}}^{-1}\left[\sum_{\tau=-P}^{P}\phi_n(\tau)\left(\frac{1}{T}\sum_{t=1}^{T}\mathbf{x}(t-\tau)\mathbf{x}^T(t)\right)\right]\widehat{\mathbf{A}}^{-T}\mathbf{e}_m = 0, \quad m \neq n \end{aligned} \tag{7.108}$$

under the convention that when $t-\tau$ falls outside the range $[1, T]$, it is increased or decreased by T, so as to fit into that range – thereby realizing a cyclic convolution. However, asymptotically this convention can be replaced by simply zeroing out $\mathbf{x}(t-\tau)$ in such cases, with end-effects which become negligible as T increases. Thus, exploiting the symmetry of $\phi_n(\tau)$, the condition in (7.108) can also be expressed as

$$\mathbf{e}_n^T\widehat{\mathbf{A}}^{-1}\left[\sum_{\tau=0}^{P}\phi_n(\tau)\widehat{\mathbf{R}}_{\mathbf{x}}(\tau)\right]\widehat{\mathbf{A}}^{-T}\mathbf{e}_m = 0, \quad m \neq n, \tag{7.109}$$

where $\widehat{\mathbf{R}}_{\mathbf{x}}$ denotes the symmetrized empirical correlation matrix defined in (7.62). It is interesting to note that this condition is reminiscent of, yet fundamentally different from, a joint diagonalization condition.

Indeed, define the following set of "target matrices":

$$\mathbf{Q}_n \triangleq \sum_{\tau=0}^{P}\phi_n(\tau)\widehat{\mathbf{R}}_{\mathbf{x}}(\tau), \quad n = 1,\ldots,N. \tag{7.110}$$

Obviously, if the true (rather than empirical) correlations are used, then the matrices $\mathbf{A}^{-1}\mathbf{Q}_n\mathbf{A}^{-T}$ (where $\mathbf{A}$ is the true mixing matrix) are all diagonal, regardless of the choice of the $\phi_n(\tau)$ filters. However, the QML estimate $\widehat{\mathbf{A}}$ does not seek to jointly diagonalize the entire set of target matrices (which would generally be impossible when empirical correlations are used). Instead, $\widehat{\mathbf{A}}$ is chosen such that for each n, the respective transformed matrix $\widehat{\mathbf{A}}^{-1}\mathbf{Q}_n\widehat{\mathbf{A}}^{-T}$ has (at least) its n-th row (and n-th column, due to symmetry) all zeros, except for the diagonal ((n,n)-th) element (which would be 1 if the scaling equations (7.107) are incorporated). This condition can be seen as some kind of a hybrid exact-and-approximate joint diagonalization: each of the N matrices in the target set is to be exactly diagonalized in its respective (n-th) row and column, but may deviate from diagonality (arbitrarily) in all of its other rows and columns. Note, however, that in the case of $N = 2$ sources, the above condition amounts to exact joint diagonalization of $\mathbf{Q}_1$ and $\mathbf{Q}_2$.

The role of each of the filters $\phi_n(t)$ is merely to determine the linear combination coefficients in constructing the n-th target matrix $\mathbf{Q}_n$. Naturally, the choice of these coefficients affects the resulting performance, and optimal performance is obtained when the "true" filters (having frequency responses proportional to the inverse spectra of the respective sources) are chosen. In fact, for AR sources of orders smaller than or equal to P, there are at most $P+1$ coefficients in each filter. This implies that the set of empirical correlation matrices up to lag P are sufficient for ML estimation in this case, thus forming a sufficient statistic – as already mentioned in subsection 7.4.4.1 above.

An iterative procedure is proposed in [38] for solving the estimating equations (7.109). Denote by $\widehat{\mathbf{B}}$ a given estimate of $\mathbf{A}^{-1}$, assumed to be "near" a solution of the estimating equations. The idea is to apply multiplicative updates to $\widehat{\mathbf{B}}$, of the form

$$\widehat{\mathbf{B}} \leftarrow (\mathbf{I}-\mathbf{P}^T)\widehat{\mathbf{B}}, \tag{7.111}$$

where $\mathbf{P}$ is a "small" update matrix. Let us denote

$$\widetilde{\mathbf{Q}}_n \triangleq \widehat{\mathbf{B}}\mathbf{Q}_n\widehat{\mathbf{B}}^T, \quad n=1,\ldots,N \tag{7.112}$$

where $\widehat{\mathbf{B}}$ denotes the estimate of $\mathbf{B}$ from the previous iteration (and the $\mathbf{Q}_n$ matrices are as defined in (7.110) above). The estimating equations in terms of the update matrix $\mathbf{P}$ can now be expressed as

$$\mathbf{e}_n^T(\mathbf{I}-\mathbf{P}^T)\widetilde{\mathbf{Q}}_n(\mathbf{I}-\mathbf{P})\mathbf{e}_m = 0, \quad n \neq m, \tag{7.113}$$

which, when neglecting terms which are quadratic in the elements of $\mathbf{P}$, can also be written as

$$\mathbf{e}_n^T\widetilde{\mathbf{Q}}_n\mathbf{p}_m + \mathbf{e}_m^T\widetilde{\mathbf{Q}}_n\mathbf{p}_n = \widetilde{Q}_n(n,m), \quad n \neq m \tag{7.114}$$

where $\mathbf{p}_m$ and $\mathbf{p}_n$ denote (resp.) the m-th and n-th columns of $\mathbf{P}$, and where $\widetilde{Q}_n(n,m)$ denotes the (n,m)-th element of $\widetilde{\mathbf{Q}}_n$. Repeating this equation for all $N(N-1)$ combinations of $n \neq m$, we obtain $N(N-1)$ linear equations for the $N(N-1)$ off-diagonal elements of $\mathbf{P}$, arbitrarily setting their diagonal terms to zero.

Further simplification may be obtained if we assume that all $\widetilde{\mathbf{Q}}_n$ are already "nearly diagonal", so that the off-diagonal elements in each row are negligible with respect to the diagonal element in that row. Under this assumption, the linear equations (7.114) can be further approximated as

$$\widetilde{Q}_n(n,n)P(n,m) + \widetilde{Q}_n(m,m)P(m,n) = \widetilde{Q}_n(n,m), \quad n \neq m, \tag{7.115}$$

where $P(n,m)$ denotes the (n,m)-th element of $\mathbf{P}$. Rewriting this equation with the roles of m and n exchanged, we obtain two equations for the two unknowns $P(n,m)$

and $P(m,n)$, thus decoupling the original set of dimensions $N(N-1)\times N(N-1)$ into $N(N-1)/2$ sets of dimensions 2×2 each.

An alternative approach for the solution of a similar system of estimating functions (with respect to $\mathbf{B}$) was proposed by Dégerine and Zaïdi in the context of EML for Gaussian AR sources (in [15]), which is discussed in the following subsection.

7.5.2 The EML approach

Let $\mathbf{s}_n^T \triangleq [s_n(1), s_n(2), \ldots, s_n(T)]$ denote the $1\times T$ row-vector composed of the samples of the n-th source (for $n=1,\ldots,N$), and let the $T\times T$ matrix $\mathbf{R}_n \triangleq \mathbb{E}\{\mathbf{s}_n\mathbf{s}_n^T\}$ denote its correlation matrix. Evidently, if $s_n(\cdot)$ is a WSS process, then $\mathbf{R}_n$ is a Toeplitz matrix, whose (t_1,t_2)-th element is $R_n(t_1-t_2)$, defined earlier as the n-th source's autocorrelation sequence at lag t_1-t_2 (with $t_1,t_2\in[1,T]$). Let us also denote by $\mathbf{S}_T \triangleq [\mathbf{s}_1,\ldots,\mathbf{s}_N]^T = [\mathbf{s}(1),\ldots,\mathbf{s}(T)]$ the $N\times T$ matrix of concatenated source vectors (similar to the definition of $\mathbf{X}_T$ used above for the observation vectors).

If all sources are Gaussian, independent and WSS, their joint *pdf* (expressed using their time-domain representation, rather than their frequency-domain representation used in (7.97)) is given by

$$f_s(\mathbf{S}_T) = \prod_{n=1}^{N} \frac{1}{\det(2\pi\mathbf{R}_n)^{\frac{1}{2}}} \exp\left[-\frac{1}{2}\mathbf{s}_n^T\mathbf{R}_n^{-1}\mathbf{s}_n\right]$$
$$= (2\pi)^{-\frac{NT}{2}} \left(\prod_{n=1}^{N}(\det\mathbf{R}_n)^{-\frac{1}{2}}\right) \exp\left[-\frac{1}{2}\sum_{n=1}^{N}\mathbf{e}_n^T\mathbf{S}_T\mathbf{R}_n^{-1}\mathbf{S}_T^T\mathbf{e}_n\right]. \quad (7.116)$$

Now, since $\mathbf{S}_T = \mathbf{B}\mathbf{X}_T$, we may similarly express observations' *pdf* (in time domain) as

$$f_x(\mathbf{X}_T) = |\det\mathbf{B}|^T (2\pi)^{-\frac{NT}{2}} \left(\prod_{n=1}^{N}(\det\mathbf{R}_n)^{-\frac{1}{2}}\right)$$
$$\cdot \exp\left[-\frac{1}{2}\sum_{n=1}^{N}\mathbf{e}_n^T\mathbf{B}\mathbf{X}_T\mathbf{R}_n^{-1}\mathbf{X}_T^T\mathbf{B}^T\mathbf{e}_n\right]. \quad (7.117)$$

Assume now, that each source is a parametric WSS process, meaning that its correlation sequence depends on a small set of (unknown) parameters. For example, in [15] it is assumed that each (n-th) source is an AR(P_n) process, whose correlation depends on the variance σ_n^2 of its innovation sequence and on its AR parameters vector $\mathbf{a}_n \triangleq [a_1^{(n)},\ldots,a_{P_n}^{(n)}]^T$. The scaling assumption employed in [15] is that all $\sigma_n^2 = 1$, leaving $\mathbf{a}_n$ (for $n=1,\ldots,N$) as the only free parameters.

In fact, rather than use the AR parameters $\mathbf{a}_n$ directly, the authors of [15] preferred to use an alternative, equivalent set of parameters, called the *partial correlation* (or *parcor* for short) parameters, denoted (here) as $\boldsymbol{\gamma}_n$. There is a simple one-to-one relation between the AR parameters $\mathbf{a}_n$ and the parcor parameters $\boldsymbol{\gamma}_n$. In fact, this relation can be obtained by using a scalar version of the relations used in subsection 7.4.4.2 above: if $x(\cdot)$ denotes an arbitrary AR process of order P, then the relation between its AR parameters $a_1,\ldots,a_P$ and its parcor parameters $\gamma_1,\ldots,\gamma_P$ is the same as the relation between the matrix parameters $\mathbf{A}_{\mathbf{x}}(1;P),\ldots,\mathbf{A}_{\mathbf{x}}(P;P)$ and $\mathbf{\Gamma}_{\mathbf{x}}(1),\ldots,\mathbf{\Gamma}_{\mathbf{x}}(P)$ (respectively, using 1×1 matrices, namely scalars), as expressed in subsection 7.4.4.2. The motivation for using parcor parametrization is that the constraint on the standard AR parameters $a_1,\ldots,a_P$ (that all the roots of the polynomial $A(z)=1+a_1z^{-1}+\cdots+a_Pz^{-P}$ be inside the unit-circle) is much more difficult to manipulate (e.g. in constrained minimization) than the equivalent constraint on the parcor parameters $\gamma_1,\ldots,\gamma_P$ (which is $|\gamma_k|<1$, $k=1,\ldots,P$).

The correlation matrices $\mathbf{R}_n$ can therefore be written as $\mathbf{R}_n(\boldsymbol{\gamma}_n)$, so as to express their dependence on the respective parcor parameters. The complete set of unknown parameters therefore consists of the matrix $\mathbf{B}$ and all parcor parameter vectors $\boldsymbol{\gamma}_1,\ldots,\boldsymbol{\gamma}_N$. The MLE therefore consists of maximizing the likelihood

$$L_{\mathbf{x}}=T\log|\det\mathbf{B}|-\frac{1}{2}\sum_{n=1}^{N}\left(\mathbf{e}_n^T\mathbf{B}\mathbf{X}_T\mathbf{R}_n^{-1}(\boldsymbol{\gamma}_n)\mathbf{X}_T^T\mathbf{B}^T\mathbf{e}_n+\log\det\mathbf{R}_n(\boldsymbol{\gamma}_n)\right)+\text{Const.} \tag{7.118}$$

with respect to all these parameters.

The approach proposed in [15] for maximization of $L_{\mathbf{x}}$ is a *relaxation* approach, consisting of minimizing $L_{\mathbf{x}}$ separately in alternating directions, with respect to $\mathbf{B}$ and with respect to the parcor parameters $\{\boldsymbol{\gamma}_n\}$.

Maximization with respect to $\{\boldsymbol{\gamma}_n\}$ (assuming $\mathbf{B}$ is fixed at some $\widehat{\mathbf{B}}$) can be decomposed into N separate minimization problems, each minimizing (with respect to $\boldsymbol{\gamma}_n$)

$$\widehat{\mathbf{s}}_n^T\mathbf{R}_n^{-1}(\boldsymbol{\gamma}_n)\widehat{\mathbf{s}}_n+\log\det\mathbf{R}_n(\boldsymbol{\gamma}_n),$$

where $\widehat{\mathbf{s}}_n^T\triangleq\mathbf{e}_n^T\widehat{\mathbf{B}}\mathbf{X}_T$ denotes the estimated n-th source signal implied by the estimate $\widehat{\mathbf{B}}$ of $\mathbf{B}$. This minimization problem corresponds to ML estimation of the parcor parameters of an AR process from its sample realization $\widehat{\mathbf{s}}_n$. The solution can be obtained using a slightly modified version (due to differences in the scaling constraint) of a method proposed by Pham in [33]. In practice, the results are rather similar (asymptotically) to those obtained by using the more simple Yule-Walker solution for the parameters $\mathbf{a}_n$ (transforming to $\boldsymbol{\gamma}_n$ if desired).

With all AR parameters fixed at $\{\widehat{\boldsymbol{\gamma}}_n\}$, maximization of the log-likelihood with respect to $\mathbf{B}$ consists in maximizing

$$\log|\det\mathbf{B}|-\frac{1}{2}\sum_{n=1}^{N}\mathbf{e}_n^T\mathbf{B}\mathbf{Q}_n\mathbf{B}^T\mathbf{e}_n,$$

with

$$\mathbf{Q}_n \triangleq \frac{1}{T}\mathbf{X}_T\mathbf{R}_n(\hat{\boldsymbol{\gamma}}_n)\mathbf{X}_T^T. \tag{7.119}$$

Similar to the differentiation of $L_{\mathbf{x}}$ of (7.101) with respect to $\mathbf{A}$, it can be shown that the maximizing solution $\widehat{\mathbf{B}}$ must satisfy the set of estimating equations (similar to (7.109))

$$\mathbf{e}_n\widehat{\mathbf{B}}\mathbf{Q}_n\widehat{\mathbf{B}}^T\mathbf{e}_m = \delta_{n,m}, \quad n,m = 1,\ldots,N. \tag{7.120}$$

As already outlined in the previous subsection, one possible solution strategy (using multiplicative updates of $\widehat{\mathbf{B}}$) for this "hybrid exact-and-approximate" joint diagonalization problem was proposed by Pham and Garat in [38]. Dégerine and Zaïdi propose a different iterative approach in [15], which relies on relaxation (alternating directions minimization) with respect to the rows of $\widehat{\mathbf{B}}$, as follows. Let $\widehat{\mathbf{b}}_n^T$ denote the n-th row of $\widehat{\mathbf{B}}$. Consider a specific value of n, and define a scalar product $\langle \mathbf{u},\mathbf{v}\rangle_{\mathbf{Q}_n} \triangleq \mathbf{u}^T\mathbf{Q}_n\mathbf{v}$. Evidently, (7.120) implies that (with respect to this scalar product) $\widehat{\mathbf{b}}_n$ has unit-norm and is orthogonal to all other rows of $\widehat{\mathbf{B}}$. Thus, assuming all other rows of $\widehat{\mathbf{B}}$ are fixed, and given the latest value (from the previous iteration) of $\widehat{\mathbf{b}}_n$, we seek to update this vector by (obliquely) projecting it onto the subspace orthogonal to all other rows, and normalizing:

$$\begin{aligned}\widehat{\mathbf{b}}'_n &\leftarrow \widehat{\mathbf{b}}_n - \widehat{\mathbf{B}}_{(n)}^T\left(\widehat{\mathbf{B}}_{(n)}\mathbf{Q}_n\widehat{\mathbf{B}}_{(n)}^T\right)^{-1}\widehat{\mathbf{B}}_{(n)}\mathbf{Q}_n\widehat{\mathbf{b}}_n \\ \widehat{\mathbf{b}}_n &\leftarrow \widehat{\mathbf{b}}'_n/\sqrt{\widehat{\mathbf{b}}'^T_n\mathbf{Q}_n\widehat{\mathbf{b}}'_n},\end{aligned} \tag{7.121}$$

where $\widehat{\mathbf{B}}_{(n)}$ denotes the matrix $\widehat{\mathbf{B}}$ without its n-th row (an $(N-1)\times N$ matrix). When this procedure (applied sequentially for $n = 1,\ldots,N$) is repeated iteratively, convergence to a solution of (7.120) is attained within several sweeps.

Thus, the EML approach alternates between estimation of the AR parameters of the estimated sources (separated with the recent estimate of the unmixing matrix) and estimation of the separating matrix given the recent estimates of the sources' correlations. Asymptotically this approach is equivalent to QML if QML uses the true sources' spectra for $h_n(\nu)$ (in which case it can be easily shown that the matrices $\{\mathbf{Q}_n\}$ defined in (7.110) are the same matrices defined in (7.119)).

7.5.3 The GMI approach

A third approach, also closely related to the ML approach, was proposed by Pham in [34]. Let $\mathbf{Y}_T = [\mathbf{y}_1\ \mathbf{y}_2\ \cdots\ \mathbf{y}_N]^T$ denote an $N \times T$ matrix of some N signals of length T each.

The mutual information between these signals is defined as

$$I(\mathbf{Y}_T) = \sum_{n=1}^{N} H(\mathbf{y}_n) - H(\mathbf{Y}_T), \tag{7.122}$$

where

$$\begin{aligned} H(\mathbf{y}_n) &\triangleq -\mathbb{E}\{\log f_{\mathbf{y}_n}(\mathbf{y}_n)\}, \quad n = 1,\ldots,N \\ H(\mathbf{Y}_T) &\triangleq -\mathbb{E}\{\log f_{\mathbf{Y}_T}(\mathbf{Y}_T)\} \end{aligned} \tag{7.123}$$

are the entropies of the respective signal(s). The mutual information can be interpreted as the Kullback-Leibler divergence between the joint *pdf* of the N signals and the product of their individual (marginal) *pdf* s.

Generally, the *pdf* of a signal (or the joint *pdf* of several signals) is difficult to obtain and/or to express. However, if the signals are jointly Gaussian, their *pdf* can be easily expressed in terms of their mean and covariance. If they are also JWSS (and zero-mean), their *pdf* s, and hence their entropies, can be easily expressed (asymptotically) in terms of their spectra as follows:

$$\begin{aligned} H_g(y_n(\cdot)) &= \frac{1}{2}\int_0^1 \log(4\pi^2 S_{y_n}(\nu))d\nu + \frac{1}{2}, \quad n = 1,\ldots,N \\ H_g(\mathbf{y}(\cdot)) &= \frac{1}{2}\int_0^1 \log\det(4\pi^2 \mathbf{S}_\mathbf{y}(\nu))d\nu + \frac{N}{2}, \end{aligned} \tag{7.124}$$

where $S_{y_n}(\nu)$ denotes the spectrum of the n-th signal, and $\mathbf{S}_\mathbf{y}(\nu)$ denotes the joint spectral matrix of the signals. Due to the assumption of asymptotic conditions ($T \to \infty$), we substituted the finite-length notations $\{\mathbf{y}_n\}$ and $\mathbf{Y}_T$ with the general processes $\{y_n(\cdot)\}$ and $\mathbf{y}(\cdot) = [y_1(\cdot) \cdots y_N(\cdot)]^T$ (resp.). The subscript g denotes the presumed Gaussianity of the signals. When the signals are not Gaussian, these quantities can serve as the "Gaussian entropies", which are the respective entropies of Gaussian signals having the same SOS properties as the original signals. The implied (asymptotic) Gaussian mutual information, obtained by substituting the Gaussian entropies into (7.122), is

$$\begin{aligned} I_g(\mathbf{y}(\cdot)) &= \sum_{n=1}^{N} H_g(y_n(\cdot)) - H_g(\mathbf{y}(\cdot)) \\ &= \frac{1}{2}\int_0^1 [\log\det \mathrm{Ddiag}(\mathbf{S}_\mathbf{y}(\nu)) - \log\det \mathbf{S}_\mathbf{y}(\nu)]d\nu. \end{aligned} \tag{7.125}$$

The approach of GMI-based separation is to find a demixing matrix $\mathbf{B}$, such that the separated signals $\mathbf{y}(\cdot) \triangleq \mathbf{B}\mathbf{x}(\cdot)$ have the smallest possible GMI (namely, their joint *pdf* is as close as possible to the product of their individual *pdf* s – hence they are "as independent as possible"). Taking the frequency-domain GMI approach, we observe that

$\mathbf{S}_\mathbf{y}(\nu) = \widehat{\mathbf{B}}\mathbf{S}_\mathbf{x}(\nu)$; thus

$$I_g(\mathbf{y}(\cdot); \mathbf{B}) = \frac{1}{2}\int_0^1 [\log\det \mathrm{Ddiag}(\mathbf{B}\mathbf{S}_\mathbf{x}(\nu)\mathbf{B}^T) - \log\det \mathbf{B}\mathbf{S}_\mathbf{x}(\nu)\mathbf{B}^T]d\nu \tag{7.126}$$

is to be minimized with respect to $\mathbf{B}$. Obviously, the mixtures' spectral matrix $\mathbf{S}_\mathbf{x}(\nu)$ is unknown, but it can be estimated from the given observation, e.g. using a smoothed periodogram (or a windowed correlogram, a.k.a. Blackman-Tukey's spectral estimator [40]),

$$\widehat{\mathbf{S}}_\mathbf{x}(\nu) = \sum_{\tau=-L}^{L} w(\tau)\widehat{\mathbf{R}}_\mathbf{x}(\tau)e^{-j2\pi\nu\tau}, \tag{7.127}$$

where $w(\tau)$ is a selected symmetric window function, with support $-L \leqslant \tau \leqslant L$ (L is a selected integer, "much longer" than the longest correlation length of the sources), having a non-negative Fourier transform. For scaling purposes $w(0)$ equals 1. $\widehat{\mathbf{R}}_\mathbf{x}(\tau)$ are the symmetrized correlation estimators, defined in (7.62).

Further, we may substitute the integral (7.126) with a discrete sum with frequency spacing $\Delta\nu = 1/L$ (dictated by the resolution of the spectrum estimate), obtaining the following criterion for minimization with respect to $\mathbf{B}$:

$$\frac{1}{2L}\sum_{\ell=0}^{L-1}\left[\log\det \mathrm{Ddiag}\left(\mathbf{B}\widehat{\mathbf{S}}_\mathbf{x}\left(\frac{\ell}{L}\right)\mathbf{B}^T\right) - \log\det \mathbf{B}\widehat{\mathbf{S}}_\mathbf{x}\left(\frac{\ell}{L}\right)\mathbf{B}^T\right]d\nu.$$

This amounts to AJD of the set of spectrum matrices $\widehat{\mathbf{S}}_\mathbf{x}(\frac{\ell}{L})$ ($\ell = 0,\ldots,L-1$) using Pham's criterion (7.76), for which the computationally efficient algorithm of [35] was briefly outlined in section 7.4 above.

7.6 ADDITIONAL ISSUES

In this section we briefly discuss some issues that fall beyond the scope of this chapter, but are closely related to the topic of SOS-based separation.

7.6.1 The effect of additive noise

The noisy mixture model is given by

$$\mathbf{x}(t) = \mathbf{A}\mathbf{s}(t) + \mathbf{v}(t) \tag{7.128}$$

where $\mathbf{v}(t)$ denotes a vector of zero-mean noise signals, assumed to be statistically independent of the source signals. In addition, these signals are usually assumed to be WSS. The treatment of this problem depends on the additional assumptions made regarding the noise signals. Also, in the noisy mixture scenario it is sometimes assumed that the number of sensors (and noise signals) P is greater than the number of sources N (as opposed to the noiseless case, in which such redundancy has no practical implications).

Note first, that even if $\mathbf{A}$ is known, it is generally impossible to recover the sources from the given observations without distortion. If the inverse of $\mathbf{A}$ is applied to $\mathbf{x}(t)$, then the recovered sources are fully separated from each other, but are still noisy,

$$\hat{\mathbf{s}}(t) = \mathbf{B}\mathbf{x}(t) = \mathbf{s}(t) + \mathbf{B}\mathbf{v}(t). \tag{7.129}$$

While this may be a desired outcome in some applications, it might be of interest (in other applications) to recover the sources with minimum mean square error. If only an instantaneous (memoryless) estimator of the form $\hat{\mathbf{s}}(t) = \mathbf{H}\mathbf{x}(t)$ is considered, we may apply a (spatial) *Wiener filter* $\mathbf{H} = \mathbf{R}_{\mathbf{sx}}(0)\mathbf{R}_{\mathbf{xx}}^{-1}(0)$ with

$$\begin{aligned}\mathbf{R}_{\mathbf{sx}}(0) &= \mathbb{E}\{\mathbf{s}(t)\mathbf{x}^T(t)\} = \mathbf{R}_s(0)\mathbf{A}^T \\ \mathbf{R}_{\mathbf{xx}}(0) &= \mathbb{E}\{\mathbf{x}(t)\mathbf{x}^T(t)\} = \mathbf{A}\mathbf{R}_s(0)\mathbf{A}^T + \mathbf{R}_{\mathbf{v}}(0),\end{aligned} \tag{7.130}$$

where $\mathbf{R}_{\mathbf{v}}(0)$ denotes the noise vector's autocorrelation matrix. Therefore, if $\mathbf{R}_{\mathbf{v}}(0)$ and $\mathbf{R}_{\mathbf{s}}(0)$ (and $\mathbf{A}$) are known, the instantaneous minimum MSE estimate of $\mathbf{s}(t)$ is given by

$$\hat{\mathbf{s}}(t) = \mathbf{R}_s(0)\mathbf{A}^T\left[\mathbf{A}\mathbf{R}_s(0)\mathbf{A}^T + \mathbf{R}_{\mathbf{v}}(0)\right]^{-1}\mathbf{x}(t) \tag{7.131}$$

(this is the optimal *linear* instantaneous estimate, which is also the *optimal* (linear or nonlinear) instantaneous estimate when all sources and noise signals are jointly Gaussian). The resulting minimum (matrix) MSE (for $\mathbf{e}(t) \triangleq \hat{\mathbf{s}}(t) - \mathbf{s}(t)$) is given by

$$\mathbb{E}\{\mathbf{e}(t)\mathbf{e}^T(t)\} = \mathbf{R}_s(0) - \mathbf{R}_s(0)\mathbf{A}^T\left[\mathbf{A}\mathbf{R}_s(0)\mathbf{A}^T + \mathbf{R}_{\mathbf{v}}(0)\right]^{-1}\mathbf{A}\mathbf{R}_s(0). \tag{7.132}$$

Note, however, that this is still not the optimal (linear) minimum MSE estimator, since in general, the optimal estimator would exploit past and future samples of the observation for temporal filtering (in addition to spatial filtering) of the observations. Exploiting the joint stationarity of the sources and noise, the spatio-temporal Wiener filter can be conveniently expressed in the frequency-domain,

$$\check{\mathbf{H}}(\nu) = \mathbf{S}_{\mathbf{sx}}(\nu)\mathbf{S}_{\mathbf{xx}}^{-1}(\nu) = \mathbf{S}_s(\nu)\mathbf{A}^T\left[\mathbf{A}\mathbf{S}_s(\nu)\mathbf{A}^T + \mathbf{S}_{\mathbf{v}}(\nu)\right]^{-1}, \tag{7.133}$$

where $\mathbf{S}_{\mathbf{v}}(\nu)$ denotes the spectrum matrix of the noise signals. The time-domain version of $\check{\mathbf{H}}(\nu)$ is given by

$$\mathbf{H}(\tau) = \int_0^1 \check{\mathbf{H}}(\nu)e^{j2\pi\nu\tau}d\nu, \tag{7.134}$$

so that the linear minimum MSE estimate of the source signals is

$$\hat{\mathbf{s}}(t) = \sum_{\tau=-\infty}^{\infty} \mathbf{H}(\tau)\mathbf{x}(t-\tau), \tag{7.135}$$

and the spectrum of the JWSS estimation error process $\mathbf{e}(t)$ is given by

$$\mathbf{S}_{\mathbf{e}}(\nu) = \mathbf{S}_{\mathbf{s}}(\nu) - \mathbf{S}_{\mathbf{s}}(\nu)\mathbf{A}^T\left[\mathbf{A}\mathbf{S}_{\mathbf{s}}(\nu)\mathbf{A}^T + \mathbf{S}_{\mathbf{v}}(\nu)\right]^{-1}\mathbf{A}\mathbf{S}_{\mathbf{s}}(\nu). \tag{7.136}$$

Usually, however, $\mathbf{A}$ is unknown, and has to be estimated before any of the above can be applied. Generally, in the presence of significant noise, none of the approaches discussed in the previous section (under the assumption of a noiseless mixture) yields satisfactory results. Naturally, the bounds presented for the noiseless case are still valid as lower-bounds, but they become rather loose and usually unattainable. Even the celebrated equivariance property no longer holds in the presence of noise.

Nevertheless, depending on the assumptions and knowledge regarding the SOS of the noise signals, the following possible (partial) remedies can be taken:

- The easiest case: the SOS of the noise are fully known (e.g. if they are estimated during long periods of known silence of the sources, when only the noise is present). In that case, all of the AJD and ML approaches can be applied following subtraction of the known $\mathbf{R}_{\mathbf{v}}(\tau)$ from each $\widehat{\mathbf{R}}_{\mathbf{x}}(\tau)$, thus obtaining unbiased, consistent estimates of the correlation matrices of the unobserved noiseless mixtures. The resulting estimates of $\mathbf{A}$ (or of $\mathbf{B}$) are consistent, but obviously cannot match the performance of the noiseless case estimators: the subtraction of the known noise-correlations only eliminates the bias in the correlation estimates, but the presence of noise still increases the estimation variance.
- Temporally white noise: when the detailed noise statistics are unknown, it is common practice to at least assume that all noise signals are temporally white. In that case only $\widehat{\mathbf{R}}_{\mathbf{x}}(0)$ is susceptible to bias inflicted by the noise correlation at zero lag. Thus, algorithms which simply ignore $\widehat{\mathbf{R}}_{\mathbf{x}}(0)$, such as some of the variants of SOBI and AMUSE mentioned in section 7.4, can maintain consistency (although still suffering performance degradation).
- Temporally and spectrally white noise: another simplifying (and often justified) assumption is that the noise is not only temporally, but also spatially white, namely that its zero-lag correlation matrix is given by $\mathbf{R}_{\mathbf{v}}(0) = \sigma_v^2\mathbf{I}$ (with σ_v^2 unknown). In such cases it is often convenient to try to estimate the single noise-related parameter σ_v from the data, and then proceed as if the noise SOS are known, e.g. subtracting $\hat{\sigma}_v^2\mathbf{I}$ from $\widehat{\mathbf{R}}_{\mathbf{x}}(0)$. For example, when the number of observations P is larger than the number of sources N, a subspace approach can be taken, in which σ_v^2 is estimated as the average of the $P-N$ smallest eigenvalues of $\widehat{\mathbf{R}}_{\mathbf{x}}(0)$. Such an approach is taken, for example, in the "noisy" version of SOBI. If $P = N$, other approaches for estimating σ_v^2 may be applied, for example, by using the technique of [59] in the

GMI-based frequency-domain joint diagonalization. Of course, a similar approach of trying to estimate noise-related parameters from the data can also be applied with any parametric SOS of the noise, but this single-parameter case is probably the most prevalent.

7.6.2 Non-stationary sources, time-varying mixtures

Throughout this chapter we assumed that the sources are WSS and that the mixing matrix is constant in time. These two assumptions imply that the observations are also WSS, giving rise to the approaches reviewed in the previous sections.

However, nonstationary sources and/or time-varying mixtures may occur in practice. If the sources' SOS and/or the mixing-matrix are slowly-varying, then adaptive versions of some of the above-mentioned SOS-based approaches can be employed, as proposed, e.g. for the QML algorithm in [38], or simply by using exponentially-weighted estimates of the observations' correlation matrices,

$$\widehat{\mathbf{R}}_{\mathbf{x}}^{(t)}(\tau) = \lambda \widehat{\mathbf{R}}_{\mathbf{x}}^{(t)}(\tau) + \frac{1}{2}(1-\lambda)\left[\mathbf{x}(t)\mathbf{x}^T(t-\tau) + \mathbf{x}(t-\tau)\mathbf{x}^T(t)\right], \tag{7.137}$$

where $\widehat{\mathbf{R}}_{\mathbf{x}}^{(t)}(\tau)$ denotes the estimate of $\mathbf{R}_{\mathbf{x}}(\tau)$ at time-instant t, and $\lambda \in (0,1)$ is a "forgetting factor", selected according to the expected change-rate in the SOS (reflecting a trade-off between tracking ability and statistical stability). Then all of the SOS-based approaches may be employed, substituting the stationary correlation estimates with these nonstationary estimates, possibly with small multiplicative updates of the estimated mixing or unmixing matrix at each time-instant (rather than repeated batch-solutions).

When the sources are stationary but the mixing-matrix is time-varying, SOS-based methods which use a parameterized model for the temporal changes in the mixing matrix may be employed. For example, if the mixing matrix is assumed to be linearly time-varying (with an unknown change rate) over a fixed interval, the Time-Varying SOBI (TVSOBI) algorithm [58] may be employed. If the changes in the mixing matrix are modeled as time-periodic (with a known or unknown time-period), then the Second-Order Periodic Hypothesis Identification Algorithm (SOPHIA) approach can be useful (note that in this case the observations become *cyclo-stationary*, enabling the use of *cyclic frequency* estimation techniques for estimation of the time-period).

When the mixing matrix is fixed but the sources are not WSS, SOS-based separation is still an appealing option, provided that the sources' non-stationarity profiles exhibit sufficient diversity. For example, one possible statistical model considered by Pham and Cardoso in [37] is a block-stationary model, in which the time-segment $1,\ldots,T$ can be divided into several consecutive blocks, such that in each block each source has a constant (unknown) variance, and these variances vary between blocks. In that case, the observations' correlation matrices in each block can be estimated as

$$\widehat{\mathbf{R}}_{\mathbf{x}}^{[k]} = \frac{1}{T_k} \sum_{t=t_i^{[k]}}^{t_f^{[k]}} \mathbf{x}(t)\mathbf{x}^T(t), \quad k = 1,\ldots,K \tag{7.138}$$

where K is the number of blocks, the k-th block starting at $t_i^{[k]}$ and ending at $t_f^{[k]}$, being of length $T^{[k]} = t_f^{[k]} - t_i^{[k]} + 1$. Then, assuming Gaussian sources which are all temporally uncorrelated, Pham and Cardoso have shown in [37] that the ML estimate of B in this case (under the assumption of unknown source variances) is given by the joint diagonalizer of these estimated matrices via minimization of Pham's criterion,

$$\frac{1}{2T}\sum_{k=1}^{K} T_k \left[\log\det \mathrm{Ddiag}(\mathbf{B}\widehat{\mathbf{R}}_{\mathbf{x}}^{[k]}\mathbf{B}^T) - \log\det \mathbf{B}\widehat{\mathbf{R}}_{\mathbf{x}}^{[k]}\mathbf{B}^T\right].$$

As shown in [37], this estimate coincides (under similar model assumptions) with the estimate obtained by minimization of the GMI.

When the sources have a more general block-stationary time-structure (namely, when they are WSS but not necessarily white within each block), it is heuristically plausible to combine this approach with the stationarity-based approaches, e.g. by jointly diagonalizing correlation matrices at different lags, estimated over different blocks. However, such a solution would no longer be maximum likelihood (or minimum GMI). It was nevertheless shown by Tichavský *et al.* in [45], that if the sources are assumed to be (different) Gaussian AR processes within blocks, an asymptotically optimal weighted AJD scheme can be employed (similar to WASOBI), using a computationally efficient reweighting scheme (at a "near-separation" point) termed Block AutoRegressive Blind Identification (BARBI).

Another case of conveniently parameterized non-stationarity is the case of cyclostationary sources, considered (in the context of instantaneous mixtures), for example, in [2,3,20,22,23,36,12]. In a nutshell, a cyclostationary process is a nonstationary signal, whose time-dependent correlation $\check{R}(t,\tau)$ is periodic in t with fundamental frequency α. The Fourier series coefficients of this periodic pattern (at multiples of α) are called the "cyclic correlations" (all are functions of τ), for which consistent empirical estimates are readily obtained from an observed sample-sequence, given the cyclic frequency α (if α is unknown, it may be consistently estimated from the data as well). In the instantaneous, noiseless BSS framework, the observations' cyclic correlation matrices, denoted $\mathbf{R}_{\mathbf{x}}^{\alpha}(\tau)$, satisfy the same underlying relation

$$\mathbf{R}_{\mathbf{x}}^{\alpha}(\tau) = \mathbf{A}\mathbf{R}_{\mathbf{s}}^{\alpha}(\tau)\mathbf{A}^T, \tag{7.139}$$

where $\mathbf{R}_{\mathbf{s}}^{\alpha}(\tau)$ are the sources' cyclic correlation matrices, diagonal by virtue of the sources' spatial decorrelation. Therefore, AJD approaches can be applied to the observations' empirical cyclic (and ordinary) correlation matrices in order to attain consistent estimates of **A**.

More general non-stationarity patterns (separable by SOS-based approaches) are considered under the framework of time-frequency based methods, which is the subject of Chapter 11.

7.6.3 Complex-valued sources

Our discussions and analysis throughout this chapter concentrated on the case of real-valued sources and mixing matrix, giving rise to real-valued observations. In some applications (especially involving radio-communication signals) the source signals and/or the mixing matrix can be conveniently modeled as complex-valued. Naturally, the basic correlation-matrices relations remain the same as in the real-valued case, the only difference being that the *transpose* $(\cdot)^T$ operation is replaced with a *conjugate transpose* (Hermitian) $(\cdot)^H$:

$$\mathbf{R}_{\mathbf{x}}(\tau) = \mathbb{E}\{\mathbf{x}(t)\mathbf{x}^H(t-\tau)\} = \mathbf{A}\mathbf{R}_{\mathbf{s}}(\tau)\mathbf{A}^H, \tag{7.140}$$

where the sources' correlation matrices $\mathbf{R}_{\mathbf{s}}(\tau)$ are all diagonal (but, with the exception of $\mathbf{R}_{\mathbf{s}}(0)$, they may be complex-valued). Thus, all of the joint-diagonalization based approaches can be readily applied, possibly using complex-valued matrices for diagonalization. Both EJD and AJD approaches can be readily extended: a whitening matrix can found as a (complex-valued) square-root of $\widehat{\mathbf{R}}_{\mathbf{x}}(0)$, and then a remaining unitary factor can be found from the unitary eigenvectors matrix of a second ("whitened") correlation matrix (or a linear combination of whitened correlation matrices), or, in a unitary AJD framework, from the complex-valued unitary AJD [9] of several whitened correlation matrices. Alternatively, a complex-valued non-unitary AJD framework can be applied to the entire set.

The ML-based approaches can be similarly extended to the case of complex-valued sources and mixing. However, the analogy between the complex-valued and the real-valued cases is valid only when the complex-valued sources are "*circular*" (or "*proper*") (e.g. [39,19]), namely when the so-called "*covariation*" or "*pseudo-covariance*" matrices of the sources,

$$\mathbf{K}_{\mathbf{x}}(\tau) \triangleq \mathbb{E}\{\mathbf{s}(t)\mathbf{s}^T(t-\tau)\} \tag{7.141}$$

vanish for all τ (note the *transpose*, rather than the *conjugate transpose*). If these matrices do not vanish, then even if the sources are Gaussian, their *pdf*s or MI are not given by the respective expressions of section 7.5, which cannot account for their pseudo-covariances (and associated pseudo-spectra). Likewise, the optimal weighting scheme of WASOBI is no longer valid in this case.

However, "improper" complex-valued sources are by no means "more difficult" to separate (using SOS) than "proper" complex-valued sources. To the contrary: their "additional" SOS, in the form of their covariation matrices, can help find a consistent estimate of the mixing matrix even without exploiting lagged correlations. For example, Eriksson and Koivunen have shown [18,19] that under certain mild conditions $\mathbf{R}_\mathbf{x}(0)$ and $\mathbf{K}_\mathbf{x}(0)$ are sufficient for identifying $\mathbf{B}$ as the matrix which simultaneously whitens $\mathbf{R}_\mathbf{x}(0)$ via $\mathbf{B}\mathbf{R}_\mathbf{x}(0)\mathbf{B}^H$ and diagonalizes $\mathbf{K}_\mathbf{x}(0)$ via $\mathbf{B}\mathbf{K}_\mathbf{x}(0)\mathbf{B}^T$. Although this operation is reminiscent of EJD, it is essentially different (due to the use of a conjugate transpose in one transformation and a transpose in the other), and unfortunately no closed-form solution is currently known. The transformation is called a *strong uncorrelating transform* (*SUT*), and a possible iterative algorithm for SUT is proposed in [17].

Naturally, an alternative approach to separating complex-valued sources can be to represent the problem in terms of real-valued sources and a real-valued mixing-matrix, by augmenting the model dimensions from an $N \times N$ complex-valued model to a $2N \times 2N$ real-valued model. If the real and imaginary parts of each source are completely uncorrelated, then this is a degenerate case of "proper" sources, and, since all the new $2N$ sources are mutually uncorrelated, ordinary real-valued separation approaches may be employed (with possible problematic aspects in regrouping the separated sources into real and imaginary parts of the original sources). However, if for some or for all of the sources the real and imaginary parts are correlated (generally a case of "improper" sources), then the new sources can be viewed as groups of multi-component (two-dimensional) sources. An SOS-based separation approach for such a model (asymptotically optimal for Gaussian, temporally uncorrelated, block-stationary sources) can be found in [27].

References

[1] K. Abed-Meraim, Y. Xiang, Y. Hua, Generalized second order identifiability condition and relevant testing technique, in: Proc. ICASSP'00, 2000, pp. 2989–2992.

[2] K. Abed-Meraim, X. Yong, J.H. Manton, H. Yingbo, A new approach to blind separation of cyclostationary sources, in: Proc. SPAWC'99, 1999, pp. 114–117.

[3] K. Abed-Meraim, X. Yong, J.H. Manton, H. Yingbo, Blind source-separation using second-order cyclostationary statistics, IEEE Transactions on Signal Processing 49 (2001) 694–701.

[4] B. Afsari, Simple LU and QR based non-orthogonal matrix joint diagonalization, in: Lecture Notes in Computer Science (LNCS 3889): Proc. ICA'06, 2006, pp. 17–24.

[5] S.-I. Amari, J.-F. Cardoso, Blind source separation – semiparametric statistical approach, IEEE Transactions on Signal Processing 45 (1997) 2692–2700.

[6] A. Belouchrani, K. Abed-Meraim, J.-F. Cardoso, E. Moulines, A blind source separation technique using second-order statistics, IEEE Transactions on Signal Processing 45 (1997) 434–444.

[7] J.-F. Cardoso, On the performance of orthogonal source separation algorithms, in: Proc. EUSIPCO'94, 1994, pp. 776–779.

[8] J.-F. Cardoso, B. Laheld, Equivariant adaptive source separation, IEEE Transactions on Signal Processing 44 (1996) 3017–3030.

[9] J.-F. Cardoso, A. Souloumiac, Jacobi angles for simultaneous diagonalization, SIAM Journal on Matrix Analysis and Applications 17 (1996) 161–164.

[10] C. Chang, Z. Ding, F.Y. Sze, F.H. Chan, A matrix-pencil approach to blind source separation of non-white noise, in: Proc. ICASSP'98, 1998, pp. 2485–2488.

[11] C. Chang, Z. Ding, F.Y. Sze, F.H. Chan, A matrix-pencil approach to blind source separation of colored nonstationary signals, IEEE Transactions on Signal Processing 48 (2000) 900–907.

[12] N. CheViet, M. El Badaoui, A. Belouchrani, F. Guillet, Blind separation of cyclostationary sources using non-orthogonal approximate joint diagonalization, in: Proc. SAM'08, 2008, pp. 492–495.
[13] S. Dégerine, E. Kane, A comparative study of approximate joint diagonalization algorithms for blind source separation in presence of additive noise, IEEE Transactions on Signal Processing 55 (2007) 3022–3031.
[14] S. Dégerine, R. Malki, Second-order blind separation of sources based on canonical partial innovations, IEEE Transactions on Signal Processing 48 (2000) 629–641.
[15] S. Dégerine, A. Zaïdi, Separation of an instantaneous mixture of gaussian autoregressive sources by the exact maximum likelihood approach, IEEE Transactions on Signal Processing 52 (2004) 1492–1512.
[16] E. Doron, A. Yeredor, P. Tichavský, Cramér-rao lower bound for blind separation of stationary parametric gaussian sources, IEEE Signal Processing Letters 14 (2007) 417–420.
[17] S.C. Douglas, J. Eriksson, V. Koivunen, Adaptive estimation of the strong uncorrelating transform with applications to subspace tracking, in: Proc. ICASSP'06, 2006, pp. 941–944.
[18] J. Eriksson, V. Koivunen, Complex-valued ICA using second order statistics, in: Proc. MLSP'04, 2004, pp. 183–192.
[19] J. Eriksson, V. Koivunen, Complex random vectors and ICA models: identifiability, uniqueness, and separability information theory, IEEE Transactions on Information Theory 52 (2006) 1017–1029.
[20] A. Ferreol, P. Chevalier, L. Albera, Second-order blind separation of first- and second-order cyclostationary sources-application to AM, FSK, CPFSK, and deterministic sources, IEEE Transactions on Signal Processing 52 (2004) 845–861.
[21] L. Fêty, J.-P.V. Uffelen, New methods for signal separation, in: Proc. 14th Conf. on HF Radio Systems and Techniques, 1988, pp. 226–230.
[22] M.G. Jafari, S.R. Alty, J.A. Chambers, New natural gradient algorithm for cyclostationary sources, Vision, Image and Signal Processing, IEE Proceedings 151 (2004) 62–68.
[23] P. Jallon, A. Chevreuil, Separation of instantaneous mixtures of cyclostationary sources with application to digital communication signals, in: Proc. EUSIPCO'06, 2006.
[24] M. Joho, H. Mathis, Joint diagonalization of correlation matrices by using gradient methods with application to blind signal separation, in: Proc. SAM'02, 2002, pp. 273–277.
[25] M. Joho, K. Rahbar, Joint diagonalization of correlation matrices by using Newton methods with application to blind signal separation, in: Proc. SAM'02, 2002, pp. 403–407.
[26] S.M. Kay, Fundamentals of Statistical Signal Processing: Estimation Theory, Prentice-Hall, 1993.
[27] D. Lahat, J.-F. Cardoso, H. Messer, Optimal performance of second-order multidimensional ICA, in: Lecture Notes in Computer Science (LNCS 5441): Proc. ICA'09, 2009.
[28] X.-L. Li, X.D. Zhang, Nonorthogonal joint diagonalization free of degenerate solutions, IEEE Transactions on Signal Processing 55 (2007) 1803–1814.
[29] L. Molgedey, H.G. Schuster, Separation of a mixture of independent signals using time delayed correlations, Physical Review Letters 72 (1994) 3634–3637.
[30] E. Ollila, K. Hyon-Jung, V. Koivunen, Compact CramérRao bound expression for independent component analysis, IEEE Transactions on Signal Processing 56 (2008) 1421–1428.
[31] L. Parra, P. Sajda, Blind source separation via generalized eigenvalue decomposition, Journal of Machine Learning Research 4 (2003) 1261–1269.
[32] E. Parzen, Some recent advances in time series modeling, IEEE Transactions on Automatic Control 19 (1974) 723–730.
[33] D.-T. Pham, Maximum likelihood estimation of autoregressive model by relaxation on the reflection coefficients, IEEE Transactions on Acoustics, Speech, Signal Processing 38 (1988) 175–177.
[34] D.-T. Pham, Blind separation of instantaneous mixture of sources via the gaussian mutual information criterion, Signal Processing 81 (2001) 855–870.
[35] D.-T. Pham, Joint approximate diagonalization of positive definite matrices, SIAM Journal on Matrix Analysis and Applications 22 (2001) 1136–1152.

[36] D.-T. Pham, Blind separation of cyclostationary sources using joint block approximate diagonalization, in: Lecture Notes in Computer Science (LNCS 4666): Proc. ICA'07, 2007.
[37] D.-T. Pham, J.-F. Cardoso, Blind separation of instantaneous mixtures of non-stationary sources, IEEE Transactions on Signal Processing 49 (2001) 1837–1848.
[38] D.-T. Pham, P. Garat, Blind separation of mixture of independent sources through a quasi-maximum likelihood approach, IEEE Transactions on Signal Processing 45 (1997) 1712–1725.
[39] B. Picinbono, On circularity, IEEE Transactions on Signal Processing 42 (1994) 3473–3482.
[40] J.G. Proakis, D.G. Manolakis, Digital Signal Processing: Principles, Algorithms and Applications, Prentice-Hall, 1996.
[41] O. Shalvi, E. Weinstein, Maximum likelihood and lower bounds in system identification with non-gaussian inputs, IEEE Transactions on Information Theory 40 (1994) 328–339.
[42] P. Tichavský, E. Doron, A. Yeredor, J. Nielsen, A computationally affordable implementation of an asymptotically optimal BSS algorithm for AR sources, in: Proc. EUSIPCO'06, 2006.
[43] P. Tichavský, E. Koldovský, Z. Oja, Performance analysis of the FastICA algorithm and Cramér-Rao bounds for linear independent component analysis, IEEE Transactions on Signal Processing 54 (2006) 1189–1203.
[44] P. Tichavský, A. Yeredor, Fast approximate joint diagonalization incorporating weight matrices, IEEE Transactions on Signal Processing 57 (2009) 878–891.
[45] P. Tichavský, A. Yeredor, Z. Koldobský, A fast asymptotically efficient algorithm for blind separation of a linear mixture of block-wise stationary autoregressive processes, in: Proc. ICASSP'09, 2009.
[46] P. Tichavský, A. Yeredor, J. Nielsen, A fast approximate joint diagonalization algorithm using a criterion with a block diagonal weight matrix, in: Proc. ICASSP'08, 2008.
[47] A.M. Tomé, An iterative eigendecomposition approach to blind source separation problems, in: Proc. ICA'01, 2001.
[48] A.M. Tomé, The generalized eigendecomposition approach to the blind source separation problem, Digital Signal Processing 416 (2006) 288–302.
[49] A.M. Tomé, A.R. Teixeira, E.W. Lang, K. Stadlthanner, A.P. Rocha, R. Almeida, Blind source separation using time-delayed signals, in: Proc. IJCNN'04, 2004.
[50] L. Tong, Y. Inouye, R. Liu, A finite-step global convergence algorithm for the parameter estimation of multichannel MA processes, IEEE Transactions on Signal Processing 40 (1992) 2547–2558.
[51] L. Tong, R. Liu, Blind estimation of correlated source signals, in: Proc. 24th Asilomar Conf., 1990.
[52] L. Tong, V.C. Soon, Y.-F. Huang, R. Liu, AMUSE: A new blind identification algorithm, in: Proc. ISCAS'90, 1990, pp. 1784–1787.
[53] A.-J. van der Veen, Joint diagonalization via subspace fitting techniques, in: Proc. ICASSP'01, 2001.
[54] R. Vollgraf, K. Obermayer, Quadratic optimization for simultaneous matrix diagonalization, IEEE Transactions on Signal Processing 54 (2006) 3270–3278.
[55] P. Whittle, On the fitting of multivariate autoregressions and the approximate canonical factorization of a spectral density matrix, Biometrika 50 (1963) 129–134.
[56] A. Yeredor, Blind separation of gaussian sources via second-order statistics with asymptotically optimal weighting, IEEE Signal Processing Letters 7 (2000) 197–200.
[57] A. Yeredor, Non-orthogonal joint diagonalization in the least-squares sense with application in blind source separation, IEEE Transactions on Signal Processing 50 (2002) 1545–1553.
[58] A. Yeredor, TV-SOBI: An expansion of SOBI for linearly time-varying mixtures, in: Proc., ICA'03, 2003.
[59] A. Yeredor, On approximate joint "biagonalization" – a tool for noisy blind source separation, in: Proc., IEEE SSP Workshop '05, (2005).
[60] A. Yeredor, On using exact joint diagonalization for non-iterative approximate joint diagonalization, IEEE Signal Processing Letters 12 (2005) 645–648.
[61] A. Yeredor, On optimal selection of correlation matrices for matrix-pencil-based separation, in: Lecture Notes in Computer Science (LNCS 5441): Proc. ICA'09, 2009.

[62] A. Yeredor, E. Doron, Using farther correlations to further improve the optimally-weighted sobi algorithm, in: Proc. EUSIPCO'2002, 2002.
[63] A. Yeredor, A. Ziehe, K.-R. Müller, Approximate joint diagonalization using a natural-gradient approach, in: Lecture Notes in Computer Science (LNCS 3195): Proc. ICA'04, 2004, pp. 89–96.
[64] A. Zaïdi, Maximum de vraisemblance exact pour la séparation aveugle d'un mélange instantané de sources autorégressives Gaussiennes, PhD thesis, Univeristé Joseph Fourier, 2000.
[65] A. Zaïdi, Un nouvel indice de performance et sa borne inférieure, in: Proc. GRETSI'01, 2001.
[66] A. Zaïdi, Positive definite combination of symmetric matrices, IEEE Transactions on Signal Processing 53 (2005) 4412–4416.
[67] A. Ziehe, M. K.-R., TDSEP - an efficient algorithm for blind separation using time structure, in: Proc. ICANN'98, 1998, pp. 675–680.
[68] A. Ziehe, P. Laskov, G. Nolte, K.-R. Müller, A fast algorithm for joint diagonalization with non-orthogonal transformations and its application to blind source separation, Journal of Machine Learning Research 5 (2004) 777–800.

CHAPTER

Convolutive mixtures

8

M. Castella, A. Chevreuil, and J.-C. Pesquet

8.1 INTRODUCTION AND MIXTURE MODEL

The blind source separation problem is often considered in the case of instantaneous mixtures. This means that at a given moment in time, the observations depend on the values of several source signals at the same time. In the case of a linear mixture, which is most commonly encountered, any observation is a linear combination of the sources at the same time.

However, the sources may contribute to the mixture with several different delays: this may happen in several application domains such as telecommunications – where channel models often include multipath propagation; audio processing – reverberation may introduce delays; and biomedicine [42].

We first introduce the model of linear convolutive mixtures. The corresponding equations and notation are given. Figure 8.1 sums up the notations used in the whole chapter.

8.1.1 Model and notations

The vector of the source signals $\mathbf{s}(n) = (s_1(n),\ldots,s_N(n))^{\mathrm{T}}$ is assumed unknown and unobservable, and $\mathbf{x}(n) = (x_1(n),\ldots,x_P(n))^{\mathrm{T}}$ is the vector of the observations. Both $\mathbf{s}(n)$ and $\mathbf{x}(n)$ are possibly complex valued. The difference between the linear *convolutive* mixing model and the linear instantaneous one, is that delayed values of the source signals contribute to the output at a given time. More precisely, the mixing matrix is replaced by a multi-variate (or MIMO: multi-input/multi-output) linear time invariant (LTI) system with impulse response $(\mathbf{A}(n))_{n\in\mathbb{Z}}$. The observation signals are hence determined by the sources according to the following multichannel convolution model:

$$\forall n \in \mathbb{Z} \qquad \mathbf{x}(n) = \sum_{k\in\mathbb{Z}} \mathbf{A}(k)\mathbf{s}(n-k). \tag{8.1}$$

The blind separation problem which is treated here consists in recovering the source signals using only the observation signals. We assume that there is not any *a priori* information available, either on the sources or on the mixing system: we only assume the

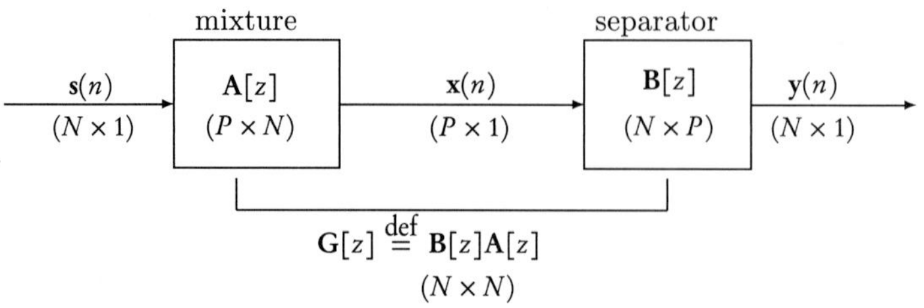

FIGURE 8.1

Considered system for source separation.

general structure given by (8.1). As explained in section 8.2, a structure such as (8.1) can generally be inverted by a MIMO-LTI system. Retrieving the sources is hence equivalent to finding a MIMO-LTI inverse system, called the separator. If its impulse response is denoted by $(\mathbf{B}(n))_{n\in\mathbb{Z}}$, the separated outputs are then given by:

$$\forall n \in \mathbb{Z} \qquad \mathbf{y}(n) = \sum_{k\in\mathbb{Z}} \mathbf{B}(k)\mathbf{x}(n-k). \tag{8.2}$$

Due to the convolutive context, we shall make use of the z-transforms of the LTI systems. For the mixing and separating systems, with respective impulse responses $(\mathbf{A}(n))_{n\in\mathbb{Z}}$ and $(\mathbf{B}(n))_{n\in\mathbb{Z}}$ we define:

$$\mathbf{A}[z] \triangleq \sum_{k\in\mathbb{Z}} \mathbf{A}(k)z^{-k} \qquad \text{and} \qquad \mathbf{B}[z] \triangleq \sum_{k\in\mathbb{Z}} \mathbf{B}(k)z^{-k}. \tag{8.3}$$

All z-transforms are defined similarly and will be denoted with square brackets, that is, for any filter with impulse response $(h(n))_{n\in\mathbb{Z}}$, the corresponding z-transform is $h[z]$.

It is often convenient to introduce the combined mixing-separating system. From equations (8.1) and (8.2), it is seen that the global output at the separator reads:

$$\forall n \in \mathbb{Z} \qquad \mathbf{y}(n) = \sum_{k\in\mathbb{Z}} \mathbf{G}(k)\mathbf{s}(n-k) \tag{8.4}$$

where the impulse response and z-transform of the global system $(\mathbf{G}(n))_{n\in\mathbb{Z}}$ are respectively given by the equations:

$$\forall n \in \mathbb{Z} \quad \mathbf{G}(n) = \sum_{k\in\mathbb{Z}} \mathbf{B}(n-k)\mathbf{A}(k) \quad \text{and:} \quad \mathbf{G}[z] = \mathbf{B}[z]\mathbf{A}[z].$$

8.1.2 Chapter organization

In the next section, we first bring together results concerning the invertibility of MIMO-LTI systems. The case of finite impulse response (FIR) systems is particularly given attention due to its importance in practice. Then, section 8.3 discusses the necessary assumptions, with a special emphasis on the specificities related to the convolutive model.

The methods which extract all sources simultaneously are presented in section 8.4 on joint methods. On the contrary, the methods which extract the sources one by one are explained in section 8.5 on sequential methods. Finally, the case of cyclostationary signals, which is common in digital communication, is addressed in section 8.6.

8.2 INVERTIBILITY OF CONVOLUTIVE MIMO MIXTURES

This section is concerned with the existence of a MIMO-LTI inverse system w.r.t the mixing system.[1] The available results come from the one-to-one correspondence given by (8.3) between the set of LTI systems and the set of z-transforms. In all this chapter, we shall assume that:

H8.1
The mixing filter is stable.

In stating the available results, some additionnal assumptions will be made. For example, it may often be assumed that:

H8.2
The mixing filter is causal.

In section 8.2.2, the following assumption will be made:

H8.3
The mixing filter has finite impulse response (FIR).

Let us briefly illustrate in the particular context of digital communications why the channel matrix $\mathbf{A}[z]$ may be seen as causal and FIR. We consider one of the entries of $\mathbf{A}[z]$ - say $A_{1,1}[z]$.

First, the propagation channel often follows a specular model (a typical example is that of high-frequency systems). It means that the contribution of the first source on the first sensor is merely a superimposition of delayed/scaled replica of $s_1(n)$ and the transfer function hence reads $\sum_{\ell=1}^{L} \lambda_\ell z^{-\tau_\ell}$ where L is the number of paths, and $(\lambda_\ell, \tau_\ell)$ is the attenuation/delay pair associated with the ℓ-th path. It can be seen that this model is FIR and that causality then follows from the fact that the delay τ_ℓ is positive.

Now, if the source $s_1(n)$ stems from the transmission of a linearly modulated sequence of symbols $(a_1(k))_{k\in\mathbb{Z}}$, it reads:

$$s_1(n) = \sum_k a_1(k)c_1(n-k)$$

where $(c_1(k))_{k\in\mathbb{Z}}$ is the impulse response of a square-root raised cosine. Rigorously, the filter $c_1[z]$ is neither FIR nor causal. But with reasonable approximation, there exists K

[1]This section can be omitted at the first reading.

such that $c_1[z] \approx \sum_{k=-K}^{K} c_1(k) z^{-k}$. This says that $c_1[z]$ can be approximated by a FIR filter, which can be considered causal by introducing a delay of K samples. Finally, the corresponding term of the channel matrix reads

$$A_{1,1}[z] = c_1[z] \sum_{\ell=1}^{L} \lambda_\ell z^{-\tau_\ell}$$

and satisfies assumptions H8.2 and H8.3 under the above approximation on $c_1[z]$.

The results concerning invertibility require one to use the z-transform of the mixing and separating system. It is known that the convergence domain of the z-transform $\mathbf{A}[z]$ depends on the assumptions made w.r.t the mixing system. More precisely:

- Under assumption H8.1, $\mathbf{A}[z]$ converges on an annulus $r_1 < |z| < r_2$ where r_1 and r_2 satisfy $r_1 < 1 < r_2$ (r_1 may be zero and r_2 infinite).
- Under assumptions H8.1 and H8.2, $\mathbf{A}[z]$ converges on a set $\{z \in \mathbb{C}, |z| > r\} \cup \{\infty\}$ where $0 \leq r < 1$.
- Under assumption H8.3 (which implies assumption H8.1), $\mathbf{A}[z]$ converges for all $z \in \mathbb{C}, z \neq 0$.

We will first give invertibility results in the case of an infinite impulse response (IIR). We will then restrict ourselves to the case of FIR systems for which additional results are known.

8.2.1 General results

PROPOSITION 8.1
Suppose $P = N$. The stable $N \times N$ filter $\mathbf{A}[z]$ is invertible by a stable filter if and only if $\det(\mathbf{A}[z]) \neq 0$ *for all* $z, |z| = 1$.

Proof: Let $\mathrm{Com}(\mathbf{A}[z])$ denote the matrix of cofactors of $\mathbf{A}[z]$ and let $\mathbf{B}[z]$ be the z transfer function given by:

$$\mathbf{B}[z] = \frac{1}{\det(\mathbf{A}[z])} \mathrm{Com}(\mathbf{A}[z])^{\mathrm{T}}. \tag{8.5}$$

Since $\mathbf{A}[z]$ is stable and $\det(\mathbf{A}[z]) \neq 0$ on the unit circle, $\mathbf{B}[z]$ defines an inverse of $\mathbf{A}[z]$ on an annulus containing the unit circle. It follows that the inverse $\mathbf{B}[z]$ is stable and its impulse response is given by:

$$\forall k \in \mathbb{Z} \qquad \mathbf{B}(k) = \frac{1}{\imath 2\pi} \oint z^{k-1} \mathbf{B}[z]\, dz \tag{8.6}$$

where the integral is calculated over the unit circle counterclockwise.

Conversely, if $\det(\mathbf{A}[z])$ vanishes on the unit circle, the exists no $\mathbf{B}[z]$ such that $\mathbf{B}[z]\mathbf{A}[z] = \mathbf{I}_N$ for all z, $|z| = 1$. □

PROPOSITION 8.2
Suppose that $P = N$ and that assumption H 8.2 holds. The stable and causal $N \times N$ filter $\mathbf{A}[z]$ is invertible by a stable and causal filter if and only if $\det(\mathbf{A}[z]) \neq 0$ in a domain $\{z \in \mathbb{C}, |z| > r\} \cup \{\infty\}$ where $0 \leq r < 1$.

The proof is identical to the preceding one, the only difference being that the convergence domain of $\mathbf{A}[z]$ and hence of $\mathbf{B}[z]$ is of the type $\{z \in \mathbb{C}, |z| > r\} \cup \{\infty\}$. It follows in this case that the inverse is both stable and causal.

PROPOSITION 8.3
The stable $P \times N$ filter $\mathbf{A}[z]$ is invertible by a stable filter if and only if $\text{rank}(\mathbf{A}[z]) = N$ for all $z, |z| = 1$.

Proof: Define $\mathbf{Q}[z] = \mathbf{A}^*[1/z]^T\mathbf{A}[z]$ (where $\mathbf{A}^*[.]$ is obtained from $\mathbf{A}[.]$ by conjugating all coefficients). The following z transfer function is defined on an annulus containing the unit circle:

$$\mathbf{B}[z] = \frac{1}{\det(\mathbf{Q}[z])}\text{Com}(\mathbf{Q}[z])^T\mathbf{A}^*[1/z]^T.$$

Since on the convergence domain $\mathbf{B}[z]\mathbf{A}[z] = \mathbf{I}_N$, it follows that a stable inverse of $\mathbf{A}[z]$ is given by the impulse response:

$$\forall k \in \mathbb{Z} \qquad \mathbf{B}(k) = \frac{1}{\imath 2\pi}\oint z^{k-1}\mathbf{B}[z]\,dz. \tag{8.7}$$

□

8.2.2 FIR systems and polynomial matrices

8.2.2.1 Coprimeness, irreducibility and invertibility

This section is concerned with systems satisfying assumption H8.3. Let us denote $\mathbb{C}[z^{\pm 1}]$ the ring of Laurent polynomials and $\mathbb{C}[z^{-1}]$ the ring of polynomials in the inderterminate z^{-1}. By definition, elements in $\mathbb{C}[z^{\pm 1}]$ are finite linear combinations of powers z^{-n} where n is any *relative* integer whereas elements in $\mathbb{C}[z^{-1}]$ are finite linear combinations of powers z^{-n} where n is a *non-negative* integer.

With these definitions, it can be seen from (8.3) that the z-transform establishes a one to one correspondence between MIMO FIR systems and polynomial matrices of the same size. The link is summarized in Fig. 8.2. Notice that the polynomial matrix corresponding to a causal system is in the subset $\mathbb{C}[z^{-1}]^{P\times N}$ of $\mathbb{C}[z^{\pm 1}]^{P\times N}$. Also, a causal system can be obtained from any FIR system by introducing a sufficient delay of $\ell \in \mathbb{N}$ samples, or equivalently by multiplying by $z^{-\ell}$.

DEFINITION 8.1
Let $\mathbf{A}[z]$ be a $P \times N$ polynomial matrix with elements in $\mathbb{C}[z^{-1}]$ (respectively $\mathbb{C}[z^{\pm 1}]$). $\mathbf{A}[z]$ is said to be coprime in $\mathbb{C}[z^{-1}]$ (respectively $\mathbb{C}[z^{\pm 1}]$) if its maximal minors are coprime polynomials in the considered ring.

$$
\begin{array}{ccc}
(\mathbf{A}(n))_{n\in\mathbb{Z}}:\ P\times N \text{ causal FIR} & \Longrightarrow & (\mathbf{A}(n))_{n\in\mathbb{Z}}:\ P\times N \text{ FIR} \\
\Big\downarrow z\text{-transform} & & \Big\downarrow z\text{-transform} \\
\mathbf{A}[z]\in\mathbb{C}[z^{-1}]^{P\times N} & \Longrightarrow & \mathbf{A}[z]\in\mathbb{C}[z^{\pm 1}]^{P\times N}
\end{array}
$$

$$
\begin{array}{ccc}
\exists\ell\in\mathbb{N}\quad (\mathbf{A}(n-\ell))_{n\in\mathbb{Z}}:\ P\times N \text{ causal FIR} & \Longleftrightarrow & (\mathbf{A}(n))_{n\in\mathbb{Z}}:\ P\times N \text{ FIR} \\
\Big\downarrow z\text{-transform} & & \Big\downarrow z\text{-transform} \\
\exists\ell\in\mathbb{N}\quad z^{-\ell}\mathbf{A}[z]\in\mathbb{C}[z^{-1}]^{P\times N} & \Longleftrightarrow & \mathbf{A}[z]\in\mathbb{C}[z^{\pm 1}]^{P\times N}
\end{array}
$$

FIGURE 8.2

Relations between polynomial matrices in rings $\mathbb{C}[z^{-1}], \mathbb{C}[z^{\pm 1}]$ and MIMO FIR systems of size $P \times N$.

It is a known fact of algebra that polynomials are coprime if and only if they have no common roots or equivalently if and only if there exists a Bezout relation between them. In the case of matrices, this fact reads as follows:

PROPOSITION 8.4
Assume $P \geq N$. Let $\mathbf{A}[z] \in \mathbb{C}[z^{-1}]^{P\times N}$ be a polynomial matrix. $\mathbf{A}[z]$ is coprime in $\mathbb{C}[z^{-1}]$ if and only if one of the two hereunder equivalent conditions is satisfied:

(i) $\mathbf{A}[z]$ *is irreducible, that is* $\text{rank}(\mathbf{A}[z]) = N$ *for all* $z \in \mathbb{C}\cup\{\infty\}, z \neq 0$.
(ii) $\mathbf{A}[z]$ *has a left inverse, that is there exists a matrix* $\mathbf{B}[z] \in \mathbb{C}[z^{-1}]^{N\times P}$ *such that:* $\mathbf{B}[z]\mathbf{A}[z] = \mathbf{I}_N$.

Now, if $\mathbf{A}[z] \in \mathbb{C}[z^{\pm 1}]^{P\times N}$, then $\mathbf{A}[z]$ is coprime in $\mathbb{C}[z^{\pm 1}]$ if and only if one of the two hereunder equivalent conditions is satisfied:

(i) $\mathbf{A}[z]$ *is irreducible, that is* $\text{rank}(\mathbf{A}[z]) = N$ *for all* $z \in \mathbb{C}, z \neq 0$.
(ii) $\mathbf{A}[z]$ *has a left inverse, that is there exists a matrix* $\mathbf{B}[z] \in \mathbb{C}[z^{\pm 1}]^{N\times P}$ *such that:* $\mathbf{B}[z]\mathbf{A}[z] = \mathbf{I}_N$.

The above proposition explains why some authors do not make any distinction between the notions of coprimeness and irreducibility: these two conditions are indeed both equivalent to the invertibility of the MIMO FIR system by another MIMO FIR system. The invertibility conditions are summed up in Fig. 8.3: note that all conditions can be expressed in a unified and simple manner by considering either the Laurent polynomial ring $\mathbb{C}[z^{\pm 1}]$ or the classical polynomial ring $\mathbb{C}[z^{-1}]$.

8.2.2.2 Practical results for FIR systems

The above paragraph gives a condition under which a MIMO FIR mixing system $(\mathbf{A}(n))_{n\in\mathbb{Z}}$ of size $P \times N$ admits a left inverse. Now, it is interesting to know to what extent a generic MIMO FIR system is likely to be invertible. The following proposition gives an answer [20]:

$$\begin{array}{ccc} (\mathbf{A}(n))_{n\in\mathbb{Z}} \text{ admits a left causal FIR inverse} & \Longrightarrow & (\mathbf{A}(n))_{n\in\mathbb{Z}} \text{ admits a left FIR inverse} \\ \Updownarrow & & \Updownarrow \\ \mathbf{A}[z] \text{ irreducible/invertible in } \mathbb{C}[z^{-1}] & \Longrightarrow & \mathbf{A}[z] \text{ irreducible/invertible in } \mathbb{C}[z^{\pm 1}] \end{array}$$

$$\begin{array}{ccc} (\mathbf{A}(n))_{n\in\mathbb{Z}} \text{ admits a left FIR inverse} & \Longleftrightarrow & \exists \ell\in\mathbb{N} \quad (\mathbf{A}(n-\ell))_{n\in\mathbb{Z}} \text{ admits a left causal FIR inverse} \\ \Updownarrow & & \Updownarrow \\ \mathbf{A}[z] \text{ irreducible/invertible in } \mathbb{C}[z^{\pm 1}] & \Longleftrightarrow & \exists \ell\in\mathbb{N} \quad z^{-\ell}\mathbf{A}[z] \text{ irreducible/invertible in } \mathbb{C}[z^{-1}] \end{array}$$

FIGURE 8.3

Invertibility conditions on a MIMO FIR systems of size $P \times N$.

PROPOSITION 8.5

Let $\mathbf{A}[z] \in \mathbb{C}[z^{-1}]^{P\times N}$ be a non-constant polynomial matrix whose coefficients have been drawn according to a continuous probability density. If $P > N$, then $\mathbf{A}[z]$ is almost surely left-invertible in $\mathbb{C}[z^{-1}]$. If $P \le N$, then $\mathbf{A}[z]$ is almost surely not left-invertible in $\mathbb{C}[z^{-1}]$.

In the case where a FIR left inverse exists, a legitimate question concerns the degree of the inverse.

PROPOSITION 8.6

If $\mathbf{A}[z] \in \mathbb{C}[z^{-1}]^{P\times N}$ is an irreducible polynomial matrix with $P \ge N$, then it admits a left-inverse $\mathbf{B}[z]$ whose degree (as a polynomial in z^{-1}) satisfies:

$$\deg(\mathbf{B}[z]) \le N \deg(\mathbf{A}[z]) - 1.$$

Under additional conditions, it is possible to get a tighter upper bound on the degree of the inverse. The case of a matrix in $\mathbb{C}[z^{\pm 1}]$ is also more difficult. The reader is referred to [20,16] for more details.

8.3 ASSUMPTIONS

In this section, we provide the assumptions required by convolutive BSS. The first category concerns the mixing matrix. The second one concerns the statistics of the sources. It turns out that, irrespective of the separation method, ambiguities which necessarily remain at the output of the separator can be *a priori* given. In general, the indeterminacies that affect the recovered sources are not as trivial as in the context of instantaneous mixtures. In particular, we will have to distinguish between linear sources and nonlinear ones.

8.3.1 Fundamental assumptions

First, it should be assumed that the mixing filter is left invertible. More precisely, we will assume that assumption H8.1 holds (stability of the mixing filter) and that the following condition holds:

H8.4
There exists a filter with z-transform $\mathbf{B}^0[z]$ such that: $\mathbf{B}^0[z]\mathbf{A}[z] = \mathbf{I}_N$.

The above condition corresponds merely to the invertibility by a filter, which is possibly IIR. This is the weakest assumption required. To derive and implement practical algorithms one generally also assumes that there exists a FIR separator $\mathbf{B}^0[z]$. According to the previous section, this is highly likely to happen for an FIR mixing system. In this case, the separator can even be made causal by introducing a time delay.

We will see in the following sections that it may be impossible to avoid a scalar filtering indeterminacy. In this case, the condition $\mathbf{B}^0[z]\mathbf{A}[z] = \mathbf{I}_N$ can be weakened to the condition that for one matrix $\mathbf{B}^0[z]$, $\mathbf{B}^0[z]\mathbf{A}[z]$ be a diagonal non-constant matrix: indeed, the blind context does not allow an observer to make any difference between a separator $\mathbf{B}[z]$ such that $\mathbf{B}[z]\mathbf{A}[z] = \mathbf{I}_N$ and another one such that $\mathbf{B}[z]\mathbf{A}[z]$ is only diagonal non-constant.

Since a blind context is considered, the sources are unobservable and it is assumed that no information is available on the mixing system $\mathbf{A}[z]$, except the general model given by (8.1). This lack of information needs to be compensated for by the following strong statistical assumption on the sources:

H8.5
The source signals $(s_1(n))_{n\in\mathbb{Z}}, \ldots, (s_N(n))_{n\in\mathbb{Z}}$ are mutually statistically independent, that is the vector $(\mathbf{s}(n))_{n\in\mathbb{Z}}$ has independent components.

We will not discuss further this assumption, which is a classical and fundamental one in blind source separation. We assume in addition that:

H8.6
The source signals are stationary, zero-mean and with unit variance.

This is the main context of the chapter. However, stationarity will be relaxed in section 8.4.3.3 (replaced by piecewise stationarity) and in section 8.6 (replaced by cyclo-stationarity).

8.3.2 Indeterminacies

Assumption H8.5 is not sufficient to obtain an exact copy of the sources $(\mathbf{s}(n))_{n\in\mathbb{Z}}$ from the observations $(\mathbf{x}(n))_n \in \mathbb{Z}$ only. At best, mutual independence allows one to recover the sources up to:

- the order of the sources: it can indeed be fixed arbitrarily and it is immaterial. With no additionnal information, the order of the sources hence cannot be recovered.
- a scalar filtering ambiguity: a scalar SISO filtering of the sources indeed does not modify assumption H8.5, from which follows this indeterminacy in the general case.

The first ambiguity is also found in the case of instantaneous mixtures. On the contrary, the second ambiguity is typical of the convolutive context: if the mixture is instantaneous (that is $\mathbf{A}[z]$ is a constant matrix), such a scalar filtering ambiguity reduces to a complex valued scaling ambiguity. The nature of the scalar filtering ambiguity is addressed next.

8.3.3 Linear and nonlinear sources

8.3.3.1 Definition

For a time series denoted by $(s(n))_{n\in\mathbb{Z}}$, we consider different assumptions, all in the context of assumption H8.6 (stationarity). The time series may first show no temporal dependence. We recall the following definition:

DEFINITION 8.2
The signal $(s(n))_{n\in\mathbb{Z}}$ is an independent and identically distributed (iid) sequence if the distributions of the samples are identical and mutually independent.

On the contrary, there may exist a temporal dependence between the successive samples. In this case, we distinguish the case of linear and nonlinear processes.

DEFINITION 8.3
A time series $(s(n))_{n\in\mathbb{Z}}$ is said to be:

- linear *if it can be written as the output of a filter driven by an iid sequence;*
- nonlinear *otherwise.*

From these definitions, it follows that an iid process is a particular case of linear process and that a nonlinear process is necessarily non-iid.

Instantaneous BSS methods apply equally to iid, linear or nonlinear sources. When dealing with a convolutive mixing model, attention must, however, be paid to these temporal characteristics since they induce consequences on the scalar filtering ambiguity. Many works consider the case of iid sources, but one should be aware that it is not the most general case and nonlinear sources are likely to appear in many practical situations. We give now a few examples of nonlinear but stationary processes to highlight the importance of nonlinear time series:

- A continuous phase modulation (CPM) signal corresponds to a linear modulation of a sequence of pseudo-symbols $(s(n))_{n\in\mathbb{Z}}$. The latter defines a nonlinear process and is a Markov chain defined by:

$$s(n) = e^{\imath\pi h a_n} s(n-1) \tag{8.8}$$

  where $0 < h < 1$ is the modulation index and $(a_n)_{n\in\mathbb{Z}}$ is an iid sequence of transmitted symbols with values in $\{-1,+1\}$.
- An autoregressive conditionally heteroscedastic (ARCH) process of order M is defined by

$$\forall n \in \mathbb{Z} \qquad s(n) = a(n)\xi(n) \tag{8.9}$$

  where $(\xi(n))_{n\in\mathbb{Z}}$ is an iid Gaussian zero-mean process with unit variance. $(a(n))_{n\in\mathbb{Z}}$ is positive and given by the following equation where $(\alpha_i)_{0\leq i\leq M}$ are real-valued coefficients:

$$\forall n \in \mathbb{Z} \qquad a(n)^2 = \alpha_0 + \sum_{i=1}^{M} \alpha_i s(n-i)^2. \tag{8.10}$$

Such processes are also nonlinear. They were introduced by Engle in 1982 and give rise to many applications and generalizations in finance and/or econometry.

- Another common situation where a nonlinear discrete-time source signal may be observed is when a linear continuous-time random process $S_a(t)$ is sampled. Indeed, assume that the trispectrum of $S_a(t)$ is defined and given by $H_a(-\omega_1 - \omega_2 - \omega_3)H_a^*(-\omega_1)H_a(\omega_2)H_a^*(-\omega_3)$ where H_a is the frequency response of some continuous-time linear filter. After sampling with period $D > 0$, we obtain, as a result of the Poisson formula, a discrete-time process with trispectrum

$$\frac{1}{D^3} \sum_{(l_1,l_2,l_3)\in\mathbb{Z}^3} H_a\left(-\omega_1 - \omega_2 - \omega_3 - 2\pi\frac{l_1+l_2+l_3}{D}\right) \times H_a^*\left(-\omega_1 - 2\pi\frac{l_1}{D}\right) H_a\left(\omega_2 + 2\pi\frac{l_2}{D}\right) H_a^*\left(-\omega_3 - 2\pi\frac{l_3}{D}\right). \tag{8.11}$$

So, if H_a is not band-limited (or if H_a is band-limited and $1/D$ is less than the Nyquist frequency), the resulting process is not linear as the above expression cannot be put into the form $H(-\omega_1 - \omega_2 - \omega_3)H^*(-\omega_1)H(\omega_2)H^*(-\omega_3)$ where H is a 2π-periodic function.

8.3.3.2 Scalar filtering ambiguities

As mentioned in section 8.3.2, each source can only be recovered up to a scalar filtering: this situation holds for *nonlinear* sources. This corresponds to the general case where no assumption is made on the temporal dependence of the source signals. We now specify the different situations where the general scalar filtering ambiguity has a simpler form.

- *iid sources*: it is known that blind equalization techniques exist for iid sources and indeed it is possible to reduce any scalar filtering ambiguity to a scaling and delay ambiguity by using temporal whiteness (see [35] for example or [8] for a MIMO convolutive context).
- *linear sources*: since the blind context does not permit us to make a difference between a source and any scalar filtering of it, it follows that linear sources can be considered in the frame of iid sources [31]. Indeed, write each linear source $(s_i(n))_{n\in\mathbb{Z}}$ as the output of a filter $f_i[z]$ driven by an iid process $(a_i(n))_{n\in\mathbb{Z}}$:

$$s_i(n) = f_i[z]a_i(n). \tag{8.12}$$

The model of Fig. 8.1 is equivalent to the model obtained by replacing the sources by $(a_i(n))_{n\in\mathbb{Z}}, i \in \{1,\dots,N\}$ and the mixing system by $\mathbf{A}[z]\mathrm{diag}(f_1[z],\dots,f_N[z])$.

This equivalent model corresponds to a mixture of independent iid source signals and the signals $(a_i(n))_{n\in\mathbb{Z}}, i \in \{1,\ldots,N\}$ can be recovered up to a scaling and a delay.

- *temporally white sources*: if the source signals are temporally *second-order* decorrelated (but not iid), the scalar filtering ambiguity can be reduced to an all-pass scalar filtering ambiguity.

If, in addition, the combined mixing-separating system is FIR, the scalar filtering ambiguity further reduces to be a scale factor and a delay. Indeed, the transfer function of any all-pass scalar FIR filter reads $\alpha z^{-\ell}$. The same reasoning actually holds as soon as the second-order statistics of the sources are known.

8.3.4 Separation condition

Generally speaking, we say that separation has been achieved when the sources have been recovered up to the previously described indeterminacies. The separation conditions need, however, to be precisely specified.

Nonlinear sources. According to the previous section the separation is sucessful when the global filter $(\mathbf{G}(n))_{n\in\mathbb{Z}}$ operates a scalar filtering on the sources followed by a permutation; and its z-transform hence reads:

$$\mathbf{G}[z] = \mathbf{P}\boldsymbol{\Lambda}[z] = \mathbf{P}\mathrm{Diag}(\lambda_1[z],\ldots,\lambda_N[z]) \tag{8.13}$$

where $\mathbf{P}$ is a permutation matrix and $\boldsymbol{\Lambda}[z] = \mathrm{Diag}(\lambda_1[z],\ldots,\lambda_N[z])$ is diagonal. This is the most general context and separation corresponds to separation up to unavoidable ambiguities. If the sources are temporally iid, these ambiguities can be partly solved by blind separating methods. For this reason, other separating conditions are introduced in this context.

iid sources. The scalar filtering indeterminacies reduce to a delay and scalar multiplication. This amounts to writing that for all $i \in \{1,\ldots,N\}$ we have $\lambda_i[z] = \alpha_i z^{-\ell_i}$ where $\alpha_i \in \mathbb{C}, \alpha_i \neq 0$ and $\ell_i \in \mathbb{Z}$ and hence we say that a filter is separating when it reads:

$$\mathbf{G}[z] = \mathbf{P}\boldsymbol{\Lambda}[z] = \mathbf{P}\mathrm{Diag}(\alpha_1 z^{-\ell_1},\ldots,\alpha_N z^{-\ell_N}). \tag{8.14}$$

The above definition of separation is the one considered in the next section for joint methods. In iterative methods, one is interested in the possibility of separating *one* single source, which constitutes a subtask of the global separation. This amounts to considering only one row of the separating filter or of the global filter $(\mathbf{G}(n))_{n\in\mathbb{Z}}$. Denote by $(\mathbf{g}(n))_{n\in\mathbb{Z}}$ the corresponding global row filter and $\mathbf{g}[z]$ its z-transform. In the nonlinear case, the filter is said to be effectively separating when only one component is non-zero, say the i_0th one:

$$\mathbf{g}[z] = \underbrace{(0,\ldots,0,\lambda_{i_0}[z],0,\ldots,0)}_{\text{only } i_0\text{th component non-zero}}. \tag{8.15}$$

In the case of iid sources, the corresponding non-zero component must correspond to a trivial filter:

$$\mathbf{g}[z] = \underbrace{(0,\ldots,0,\alpha_{i_0} z^{-\ell_{i_0}},0,\ldots,0)}_{\text{only } i_0\text{th component non-zero}}. \tag{8.16}$$

8.4 JOINT SEPARATING METHODS

Historically, it is mentioned in [7] how the context of convolutive mixtures can be faced: namely by considering a frequency approach (see section 8.4.3) or a second-order approach based on linear prediction ideas (see [16]). The first works specifically devoted to the convolutive case [45,46] set forth the strong ideas in the domain.

This section deals with separation methods which aim at recovering all the sources simultaneously. We refer to such approaches as joint separating methods. Iterative methods are addressed in section 8.5.

Joint approaches often rely on a prewhitening step, which is described in section 8.4.1. Time domain approaches are described in section 8.4.2. In section 8.4.3, frequency domain approaches are considered.

8.4.1 Whitening

In the instantaneous case, many BSS and ICA methods exploit the decorrelation of the sources in a whitening step which is performed prior to the proper separation. Indeed, imposing the decorrelation does not in general achieve the separation but it has been shown to yield simple constrast functions. Practically, after whitening, a unitary matrix remains to be identified, which can be done in general by considering some higher-order statistics.

One could wonder whether it is possible or not to extend the idea of whitening to the convolutive case. In this paragraph, we provide the theoretical background. However, we shall see that, contrary to the instantaneous case, the whitening is difficult to implement in a convolutive context: more information may be found in [1,28].

We consider second order temporally decorrelated sources and we denote by $\mathbf{W}[z]$ an $N \times P$ filter. Let us define the N-dimensional vector: $\tilde{\mathbf{x}}(n) \stackrel{\text{def}}{=} \mathbf{W}[z]\mathbf{x}(n)$. By definition, $\mathbf{W}[z]$ is a whitening filter if the components of $\tilde{\mathbf{x}}(n)$ are second-order decorrelated (or white) both in time and space. This means in the time domain:

$$\forall k \in \mathbb{Z} \qquad \mathbb{E}\{\tilde{\mathbf{x}}(n)\tilde{\mathbf{x}}(n-k)^{\mathrm{H}}\} = \delta_k \mathbf{I}_N. \tag{8.17}$$

Such a filter is not unique (consider the left multiplication of $\mathbf{W}[z]$ by any unitary matrix). Once a filter $\mathbf{W}[z]$ is computed, the remaining filter to be determined after whitening is denoted by $\mathbf{Q}[z]$. Figure 8.4 sums up the notations. It can be noticed that whitening amounts to splitting the separating filter in the following product:

$$\mathbf{B}[z] = \mathbf{Q}[z]\mathbf{W}[z]. \tag{8.18}$$

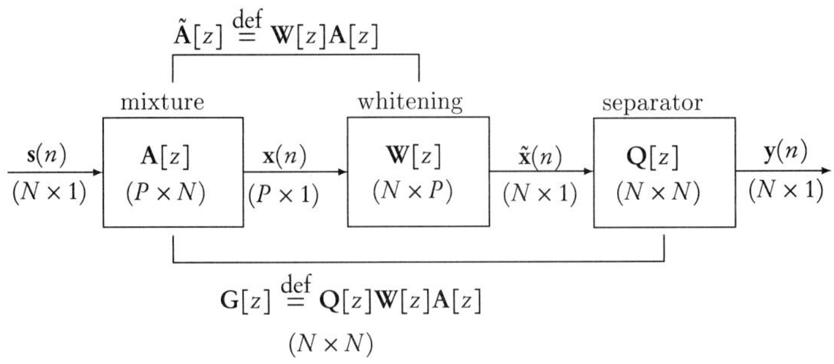

FIGURE 8.4

Whitening and para-unitary filter.

We define also the combined mixing-whitening filter $\tilde{\mathbf{A}}[z] \overset{\text{def}}{=} \mathbf{W}[z]\mathbf{A}[z]$.

The definition (8.17) is used in practice to determine $\mathbf{W}[z]$ and the corresponding whitened observations $\tilde{\mathbf{x}}(n)$. But it is not very insightful as far as the structure of $\mathbf{W}[z]$ is concerned. A simple calculus shows that for white sources, $\mathbf{W}[z]$ is a whitening filter if and only if $\tilde{\mathbf{A}}[z]$ is a para-unitary filter, the definition of which is given hereunder.

DEFINITION 8.4
The $N \times N$ filter $\tilde{\mathbf{A}}[z]$ is para-unitary if and only if one of the following equivalent conditions is satisfied:

$$\text{(i) } \tilde{\mathbf{A}}\left[\frac{1}{z^*}\right]^{\mathrm{H}} \tilde{\mathbf{A}}[z] = \mathbf{I}_N. \tag{8.19}$$

$$\text{(ii) } \sum_{l \in \mathbb{Z}} \tilde{\mathbf{A}}(l-k)^{\mathrm{H}} \tilde{\mathbf{A}}(l) = \delta_k \mathbf{I}_N, \qquad \forall k \in \mathbb{Z}. \tag{8.20}$$

$$\text{(iii) } \tilde{\mathbf{A}}[e^{\imath\omega}]^{\mathrm{H}} \tilde{\mathbf{A}}[e^{\imath\omega}] = \mathbf{I}_N, \qquad \forall \omega \in [0, 2\pi[. \tag{8.21}$$

One can obviously see that $\tilde{\mathbf{B}}[z] = \tilde{\mathbf{A}}[\frac{1}{z^*}]^{\mathrm{H}}$ is actually a function of z which defines a MIMO filter. Equation (8.19) can then be read as $\tilde{\mathbf{B}}[z]\tilde{\mathbf{A}}[z] = \mathbf{I}_N$. Therefore, $\tilde{\mathbf{A}}[z]$ admits a stable left inverse.[2] Besides, this left inverse is a para-unitary filter. Recalling the context of BSS, this means that if a whitening filter $\mathbf{W}[z]$ exists and is computed, the BSS may be processed w.r.t. the signal $\tilde{\mathbf{x}}(n)$ by looking for a separating filter $\mathbf{Q}[z]$ in the set of para-unitary filters.

The existence of a whitening filter is easily seen. Indeed, if $\mathbf{A}[z]$ admits a left stable inverse (which is the weakest required assumption), then the latter is a whitening filter. It remains to address practical considerations such as: if the mixing matrix $\mathbf{A}[z]$ is

[2]This inverse is not causal in general.

polynomial, may a whitening filter be also polynomial? The answer is provided in the following proposition, the proof of which may be found in [20].

THEOREM 8.7
Let $\mathbf{A}[z] \in \mathbb{C}[z^{-1}]^{P\times N}$ *with* $P \geq N$ *and let* $\Delta_{\mathbf{A}}[z]$ *be the greatest common divisor (GCD) of its maximal size* $N \times N$ *minors. Assume that*[3] $\Delta_{\mathbf{A}}[z] = z^{-\ell}$. *Then,* $\mathbf{A}[z]$ *can be factorized as:*

$$\mathbf{A}[z] = \mathbf{A}_I[z]\mathbf{A}_P[z] \tag{8.22}$$

where $\mathbf{A}_I[z] \in \mathbb{C}[z^{-1}]^{P\times N}$ *is irreducible with* $\deg \mathbf{A}_I[z] \leq \deg \mathbf{A}[z]$ *and* $\mathbf{A}_P[z] \in \mathbb{C}[z^{-1}]^{N\times N}$ *is para-unitary with* $\deg \mathbf{A}_P[z] = \ell$. *In addition, the above factorization is unique up to multiplication by a unitary matrix* $\mathbf{U} \in \mathbb{C}^{N\times N}$.

One can observe that the whitening filter $\mathbf{W}[z]$ corresponds to a left inverse of $\mathbf{A}_I[z]$ in Theorem 8.7. The whitening filter can thus be looked for in the set of FIR filters. Thanks to Proposition 8.6, the degree of the whitening filter should not exceed $N \deg \mathbf{A}_I[z]-1 \leq N \deg \mathbf{A}[z]-1$. This is of considerable importance: although there is an infinite number of whitening constraints in Eq. (8.17), $\mathbf{W}[z]$ can be searched in a *finite* dimensional set. Notice however that there is no reason why $\mathbf{A}_P[z]$ would be left-invertible by an $N \times N$ FIR filter. It means that the para-unitary filter $\mathbf{Q}[z]$ is not FIR in general, making its parametrization difficult. This makes the methods relying on the prewhitening difficult to implement.

It follows also from Theorem 8.7 that when the mixture is given by an irreducible matrix in $\mathbb{C}[z^{-1}]$, one can identify it up to a unitary matrix with second order statistics only [18]. Indeed, if $\mathbf{A}[z] \in \mathbb{C}[z^{-1}]$ is irreducible, it necessary follows from the uniqueness of decomposition (8.22) that $\mathbf{A}_I[z] = \mathbf{A}[z]\mathbf{U}$ and $\mathbf{A}_P[z] = \mathbf{U}^{\mathrm{H}}$ where $\mathbf{U}$ is a constant unitary matrix [20]. Therefore, only the unitary matrix $\mathbf{U}$ remains to be identified after whitening.

8.4.2 Time domain approaches

In [2], the separation of instantaneously mixed sources is addressed. It is achieved after maximizing certain contrast functions of the pre-whitened data. The approach of this section is a natural extension to the convolutive case. The results obtained are based on the fact that the mixture has been whitened prior to the separation; this crucial process is assumed to be perfectly done. This amouts to assuming that the mixing filter in this entire section 8.4.2 is represented by the para-unitary filter $\tilde{\mathbf{A}}[z]$ introduced in the former section. We introduce functions of the statistics of the data $(\tilde{\mathbf{x}}(n))_{n\in\mathbb{Z}}$ whose maximization achieves the separation, i.e. the arguments of the maxima are associated with global filters such that condition (8.13) holds.

[3]This condition is exactly the condition for left-invertibility of $\mathbf{A}[z]$ by a matrix in $\mathbb{C}[z^{\pm 1}]^{N\times P}$. This is generically true according to Proposition 8.5. In this case an inverse exists in $\mathbb{C}[z^{-1}]^{N\times P}$, an hence *a fortiori* in $\mathbb{C}[z^{\pm 1}]^{N\times P}$.

8.4.2.1 Higher-order contrast functions: The iid case

We denote by κ_{y_i} the fourth-order cumulant of y_i, the i-th output of the global filter $\mathbf{G}[z]$: $\kappa_{y_i} \triangleq \text{cum}\{y_i(n), y_i(n)^*, y_i(n), y_i(n)^*\}$. We introduce also functions such as

$$\Upsilon_f^N(\mathbf{y}) = \sum_{i=1}^{N} f(|\kappa_{y_i}|) \tag{8.23}$$

where the function f will be specified. Two particular choices for f are the square or identity functions, yielding:

$$\Upsilon_{(.)^2}^N(\mathbf{y}) = \sum_{i=1}^{N} |\kappa_{y_i}|^2 \quad \text{and:} \quad \Upsilon_{\text{id}}^N(\mathbf{y}) = \sum_{i=1}^{N} |\kappa_{y_i}|. \tag{8.24}$$

We assume that for $i \in \{1, \ldots, N\}$ at most one of the cumulants $\kappa_{s_i} \triangleq \text{cum}\{s_i(n), s_i(n)^*, s_i(n), s_i(n)^*\}$ is zero. It is shown in [8] that the above two expressions are contrasts under para-unitarity constraint of the global filter. It is quite insightful to understand why Υ_{id}^N is a contrast. In passing, one may notice why the para-unitary constraint is necessary for Υ_{id}^N to be a contrast. Indeed, if all the sources have the same statistics and if $\mathbf{G}[z]$ is an argument maximum of Υ_{id}^N, then the filter obtained after replacing all the rows of $\mathbf{G}[z]$ by the first one attains also the maximum and this filter is not separating, since it separates N times the same source. This situation is avoided under para-unitary constraint since Eqs (8.27) and (8.28) then hold necessarily.

Proof: Recalling (8.4), the i-th ($i \in \{1, \ldots, N\}$) output of the separator reads:

$$y_i(n) = \sum_{j=1}^{N} \sum_{k \in \mathbb{Z}} G_{ij}(k) s_j(n-k). \tag{8.25}$$

The multilinearity of cumulants, the fact that they vanish for independent variables and the independence of the sources yield:

$$\kappa_{y_i} = \sum_{j=1}^{N} \left(\sum_{k \in \mathbb{Z}} |G_{ij}(k)|^4 \right) \kappa_{s_j}. \tag{8.26}$$

Para-unitarity (see Definition 8.4; Eq. (8.20)) yields in addition:

$$\forall i \in \{1, \ldots, N\} \quad \sum_{j=1}^{N} \sum_{k \in \mathbb{Z}} |G_{ij}(k)|^2 = 1 \tag{8.27}$$

$$\forall j \in \{1,\ldots,N\} \quad \sum_{i=1}^{N}\sum_{k\in\mathbb{Z}} |G_{ij}(k)|^2 = 1 \tag{8.28}$$

and the terms $|G_{ij}(k)|$ are thus smaller than 1. Combining (8.26) and (8.28) yields the following upper-bound on the criterion:

$$\begin{aligned}\Upsilon_{\mathrm{id}}^N(\mathbf{y}) &= \sum_{i=1}^{N}\left|\sum_{j=1}^{N}\sum_{k\in\mathbb{Z}} |G_{ij}(k)|^4 \kappa_{s_j}\right| \le \sum_{j=1}^{N}\left(\sum_{i=1}^{N}\sum_{k\in\mathbb{Z}} |G_{ij}(k)|^4\right)|\kappa_{s_j}| \\ &\le \sum_{j=1}^{N}\left(\sum_{i=1}^{N}\sum_{k\in\mathbb{Z}} |G_{ij}(k)|^2\right)|\kappa_{s_j}| = \sum_{j=1}^{N}|\kappa_{s_j}| = \Upsilon_{\mathrm{id}}^N(\mathbf{s}).\end{aligned} \tag{8.29}$$

For any global filter of the form (8.14), the criterion Υ_{id}^N reaches its maximal value given by the above upper-bound. It remains to prove that only separating filters can reach this upper-bound. An equality in equation (8.29) implies $\sum_{i=1}^{N}\sum_{k\in\mathbb{Z}}|G_{ij}(k)|^4 = 1$. Combining this equality with (8.27) and (8.28) implies that the terms $\left(\sum_{k\in\mathbb{Z}}|G_{ij}(k)|^2\right)_{ij}$ are non-zero only for pairs of indices (i,j) corresponding to distinct rows and columns. In addition, for such an index pair (i,j), we have $|G_{ij}(k)| = 0$ except for one index k_0 for which $|G_{ij}(k_0)| = 1$. It follows that $(\mathbf{G}(k))_{k\in\mathbb{Z}}$ is made up of one single non-zero component in each column and in each row. This non-zero element corresponds to a delay and multiplication by a scalar of modulus one. $(\mathbf{G}(k))_{k\in\mathbb{Z}}$ is hence separating, which completes the proof. □

We have proved the validity of contrasts such as (8.24) in the case of 4-th order cumulants. It actually can be extended to functions of cumulants of order r greater than 2 ($r \ge 3$).[4] A proof of these results can be found in [8] and also in [19] which makes a connection between these contrasts and criteria used in iterative approaches. The general function $\Upsilon_f^N(\mathbf{y})$ defined in (8.23) has also been proved to be a contrast [29] when $f : \mathbb{R}^+ \to \mathbb{R}$ is a convex increasing function having a unique minimum (in zero). An extension can be done to functions like

$$\Upsilon_{f,(p,q)}^N(\mathbf{y}) = \sum_{i=1}^{N} f(|\kappa_{y_i}^{p,q}|)$$

where $\kappa_{y_i}^{p,q} \triangleq \mathrm{cum}\{\underbrace{y_i(t),\ldots,y_i(t)}_{p\text{ terms}},\underbrace{y_i(t)^*,\ldots,y_i(t)^*}_{q\text{ terms}}\}$ with $p+q \ge 3$. Other variants have

[4]Generally, the cumulant order is even ($r = 4, 6, \ldots$) because cumulants all identically vanish if r is odd and the distribution symmetric.

been introduced in [38] based on cross-cumulants and in [30] which introduces the notion of non-symmetrical contrast.

Finally, we have to mention that contrasts based on the notion of mutual information have been developed. They can theoretically be used with no prewhitening. We invite the reader to see [33] or the chapter on mutual information for more details.

8.4.2.2 Case of non-iid and nonlinear sources

Most of the works cited so far hold in the restrictive case of iid sources. As mentioned in section 8.3.3.2, the case of linear sources can be treated in the framework of iid sources. In addition, the contrasts derived from mutual information extend to linear and to Markovian sources. When using these contrasts, the original generating processes are recovered [33]. The behavior and local stability of similar information-theoretic contrasts when applied to nonlinear sources have been studied further in [31].

Under para-unitary constraint, contrast functions based on cumulants have been studied also in the case of non-iid sources. The non-linearity in particular implies that, for a given source, two samples at two different time lags are not independent. This is the reason why we introduce cross-cumulants with delayed signals. Specifically:

$$\kappa_{x_1,x_2^*,x_3,x_4^*}(t_1,t_2,t_3) \triangleq \text{cum}\{x_1(t), x_2(t+t_1)^*, x_3(t+t_2), x_4(t+t_3)^*\}. \tag{8.30}$$

By exploiting the para-unitary structure of the mixing system, it has been proved in [26] that the quantity:

$$\sum_{i_1,i_2,i_3,i_4=1}^{N} \sum_{(t_1,t_2,t_3)\in\mathbb{Z}^3} |\kappa_{x_{i_1},x_{i_2}^*,x_{i_3},x_{i_4}^*}(t_1,t_2,t_3)|^2 \tag{8.31}$$

is a constant independent of the mixing para-unitary filter. Based on this result, it has proved further in [26] that:

$$\Upsilon^N_{(.)^2,\mathbb{Z}^3}(\mathbf{y}) = \sum_{i=1}^{N} \sum_{(t_1,t_2,t_3)\in\mathbb{Z}^3} |\kappa_{y_i}(t_1,t_2,t_3)|^2 \tag{8.32}$$

is a contrast function. In the latter equation, we have let $\kappa_{y_i}(t_1,t_2,t_3)$ be $\kappa_{y_i,y_i^*,y_i,y_i^*}(t_1,t_2,t_3)$. The same result has been confirmed in [4].

The criterion (8.32) is the simplest known contrast in the general case of nonlinear sources: in particular, it has been shown that $\Upsilon^N_{(.)^2}$ in (8.24) is not always a contrast [4]. However, $\Upsilon^N_{(.)^2,\mathbb{Z}^3}$ is difficult to use in practice because of the infinite sum over $\mathbb{Z}^3$. It has been proposed in [4] to restrict the sum to a smaller finite set for a class of sources. In particular, the approach in [4] leads to the contrast $\Upsilon^N_{(.)^2}$ (8.24) in the particular case where the simplifying assumption of iid sources holds.

8.4.2.3 Practical difficulties

Theoretically, the maximization of the contrast functions previously described should lead to separation of the sources. We would like to list the difficulties associated in practice with such an approach:

- First and importantly, the prewhitening operation is a difficult task, as previously mentioned.
- Many of the contrast functions are only of theoretical interest because of the infinite sum. However, it is possible to use (8.24) and (8.32) if the sum can be restricted to a finite one [4].
- The contrast functions should be optimized on the set of para-unitary filters, which has a quite complicated structure. Some solutions have been proposed to overcome this problem: the para-unitary filters may be parameterized by Jacobi angles [41, 28]. In this case, it has been observed that spurious local maxima exist that are non-separating, which makes the separation difficult. It has also been proposed to take into account only a subset among the para-unitary constraints [9]. This simplification seems acceptable and, using a parameterization by Jacobi angles, it is possible to derive an optimization algorithm [10].

8.4.3 Frequency domain approaches

8.4.3.1 Basic ideas

In the convolutive case, the idea of a frequency approach to blind separation appeared quite early [7,45]. A convolutive mixture can be considered indeed as an instantaneous mixture at each frequency. As we will see, methods developed for the instantaneous model may thus be used in the frequency domain: the permutation indeterminacy at different frequency bins introduces an additional difficulty in this case. Alternatively, a frequency point of view opens up the possibility of designing criteria, which can be optimized with respect to the impulse response coefficients of the separator.

More precisely, let $\mathbf{S}_s[z]$ be the z-transform of the sources' autocorrelation sequence $(\mathbb{E}\{\mathbf{s}(n)\mathbf{s}(n-k)^H\})_{k\in\mathbb{Z}}$. We define similarly $\mathbf{S}_x[z]$ and $\mathbf{S}_y[z]$ respectively for the observations and the global outputs. From the input/output relation of the mixture, it follows that:

$$\mathbf{S}_x[z] = \mathbf{A}[z]\mathbf{S}_s[z]\mathbf{A}\left[\frac{1}{z^*}\right]^{\mathrm{H}} \tag{8.33}$$

and by independence of the sources, $\mathbf{S}_s[z]$ is diagonal. The basic idea in frequency methods consists in exploiting this fact at different frequencies. We will write ν the frequency corresponding to $z = e^{\imath 2\pi\nu}$ and we will write $\mathbf{S}_s(\nu) \triangleq \mathbf{S}_s[e^{\imath 2\pi\nu}]$. The matrices $\mathbf{S}_x(\nu)$ and $\mathbf{S}_y(\nu)$ are defined similarly. Equation (8.33) then also reads $\mathbf{S}_x(\nu) = \mathbf{A}(\nu)\mathbf{S}_s(\nu)\mathbf{A}(\nu)^{\mathrm{H}}$ and it is natural to search for separating matrices $\mathbf{B}(\nu)$ such that at the output and for any frequency, the matrix $\mathbf{S}_y(\nu) = \mathbf{B}(\nu)\mathbf{S}_x(\nu)\mathbf{B}(\nu)^{\mathrm{H}}$ should be as diagonal as possible: this corresponds indeed to the independence assumption on the sources and is the core idea in frequency approaches.

One has possibly noticed that only second-order statistics enter into Eq. (8.33). In this section, we describe indeed frequency domain methods which are based on second-order statistics only. Note that other second-order methods exist: for more details, we refer the reader to Chapter 7 on second-order methods.

The approaches introduced here seem interesting for several applications such as audio source separation. We should stress that second-order statistics at a given time instant are generally not sufficient to perform source separation. Only a prewhitening is possible, as described in section 8.4.1. To obtain the separation, additional information must be considered, which may be contained either in a spectral diversity of the sources or in their non-stationarity.[5] We describe below the main ideas.

8.4.3.2 Spectral diversity: Criteria based on a frequency approach

The identifiability of a MIMO convolutive mixture when the sources present *spectral diversity* has been proved in [18]. The two conditions to be satisfied are the following:

1. $\mathbf{A}[z]$ is a MIMO-FIR filter which is irreducible (in $\mathbb{C}[z^{\pm 1}]$).
2. $\mathbf{S}_s[z]$ is diagonal with distinct elements.

According to [24,25], the identification can be obtained by finding a global output such that $\mathbf{S}_y[z]$ be diagonal. This justifies the importance of "diagonality" criteria. The following diagonality criterion (which depends on the output $(\mathbf{y}(n))_{n\in\mathbb{Z}}$ of the separating system $\mathbf{B}[z]$) has been used by several authors:

$$\mathscr{C}_{\mathrm{y}}(\nu) = \frac{1}{2}\left[\sum_{i=1}^{N} \log S_{y_i}(\nu) - \log\det \mathbf{S}_{\mathrm{y}}(\nu)\right] \tag{8.34}$$

The matrix $\mathbf{S}_{\mathrm{y}}(\nu)$ being positive, this criterion follows from the Hadamard inequality. It can be shown that (8.34) is positive and vanishes only when $\mathbf{S}_{\mathrm{y}}(\nu)$ is diagonal. It follows that separation can be obtained by minimizing the following criterion, which stems from (8.34) by integrating over all frequencies:

$$\tilde{\Upsilon}^{\mathrm{freq}}(\mathbf{y}) = \int_0^1 \mathscr{C}_{\mathrm{y}}(\nu)\, d\nu. \tag{8.35}$$

Indeed, $\tilde{\Upsilon}^{\mathrm{freq}}(\mathbf{y})$ vanishes only when $\mathscr{C}_{\mathrm{y}}(\nu)$ is zero for almost all frequencies, which corresponds to a matrix $\mathbf{S}_{\mathrm{y}}(\nu)$ which is diagonal for almost all ν. A relative gradient optimization method can be derived from this criterion, which leads to the identification of the parameters of the impulse response of a separator.

Let us finally mention other works on frequency approaches. An identification method based on second-order statistics is proposed in [14]. The method is based on the

[5] Non-stationarity is assumed only in section 8.4.3.3.

diagonalization of covariance matrices of the observations after transformation in the frequency domain. Similar ideas are applied to higher order statistics in [6]: a method for identifying the mixing system is proposed based on a joint singular value decomposition of a set of matrices which are constructed based on polyspectra slices of the output. Another frequency approach is studied in [44]. Also, starting from a joint diagonalization criterion, frequency domain contrast have been shown in [4]. It is then possible to go back to known temporal contrasts and to generalize them to non-iid sources.

8.4.3.3 Non-stationarity and methods operating in the frequency domain

Using *non-stationarity*[6] seems particularly well-suited in specific situations such as audio or speech processing. In a context of non-stationarity, the different spectral matrices vary in time and it is possible to exploit their estimation on distinct temporal windows: in this case, piecewise stationarity is implicitly assumed. It has been proposed in [34] to minimize the following quantity for a given set of ν values:

$$\sum_t \widehat{\mathscr{C}}_{\mathrm{y}}(\nu, t) \tag{8.36}$$

where $\widehat{\mathscr{C}}_{\mathrm{y}}(\nu, t)$ is the estimated criterion (8.34) for a given temporal window indexed by t. The implementation of the method then faces the difficulty of solving the permutation ambiguity at each frequency.

Methods proposed in the instantaneous case have been extended [32]. Since the objective is to obtain at the output for all frequencies and temporal windows a spectral matrix $\mathbf{S}_{\mathrm{y}}(\nu, t)$ which is diagonal, the optimized criteria are based on the error between this matrix and a diagonal matrix:

$$\mathscr{E}_{\mathrm{y}}(\nu, t) = \boldsymbol{\Lambda}_{\mathrm{s}}(\nu, t) - \widehat{\mathbf{S}}_{\mathrm{y}}(\nu, t) \tag{8.37}$$

where $\widehat{\mathbf{S}}_{\mathrm{y}}(\nu, t)$ is the estimated interspectral matrix at the output and $\boldsymbol{\Lambda}_{\mathrm{s}}(\nu, t)$ represents a diagonal matrix which corresponds to the interspectral matrix of the sources. The different methods aim at minimizing (where $||.||_F$ is the Frobenius norm):

$$\sum_{t,\nu} ||\mathscr{E}_{\mathrm{y}}(\nu, t)||_F^2. \tag{8.38}$$

Similar weighted criteria can also be introduced.

Ideally, one could use for $\boldsymbol{\Lambda}_{\mathrm{s}}(\nu, t)$ the interspectral matrix of the sources. Since it is unknown, it has been proposed in [32] to minimize the criterion (8.38) with respect to $\boldsymbol{\Lambda}_{\mathrm{s}}(\nu)$ under a constraint on the separator to avoid the degenerate solution identically equal to zero. It is also possible [43] to replace $\boldsymbol{\Lambda}_{\mathrm{s}}(\nu, t)$ by the matrix composed only of the diagonal elements of $\widehat{\mathbf{S}}_{\mathrm{y}}(\nu, t)$, which leads to a joint diagonalization problem.

[6]Note that the term "non-stationarity" only applies to the sources and not to the mixing system which is assumed to be LTI.

8.5 ITERATIVE AND DEFLATION METHODS

In iterative deflation approaches, the sources are extracted one after the other [37,40]. This way of proceeding allows us to decompose the original signal separation problem in a sequence of problems which are more easily solved. We will first be interested in methods for extracting a single source, before describing how an iterative procedure can be used. In this section, the sources are assumed to be stationary.

8.5.1 Extraction of one source

8.5.1.1 Notation

In order to extract a source, we apply to the P observed scalar signals resulting from the convolutive mixture a filter $\mathbf{b}[z]$ with P inputs and 1 output. In this way, we obtain the scalar signal

$$y(n) = \sum_{k\in\mathbb{Z}} \mathbf{b}(k)\mathbf{x}(n-k), \quad n \in \mathbb{Z}. \tag{8.39}$$

We aim at rendering $y(n)$ equal to one of the sources $(s_i(n))_{n\in\mathbb{Z}}$, $i \in \{1,\dots,N\}$, up to a delay and a multiplicative factor (or up to a scalar filtering in the case of non-iid sources). Due to relation (8.1), the above equation can be rewritten as

$$y(n) = \sum_{k\in\mathbb{Z}} \mathbf{g}(k)\mathbf{s}(n-k), \quad n \in \mathbb{Z} \tag{8.40}$$

which involves the global filter with transfer function $\mathbf{g}[z] = \mathbf{b}[z]\mathbf{A}[z]$ having N inputs and one output. Instead of working on $\mathbf{b}[z]$, it is often more convenient to consider $\mathbf{g}[z]$. It can be noticed that the filters $\mathbf{b}[z]$ and $\mathbf{g}[z]$ may correspond to one of the lines of the filters $\mathbf{B}[z]$ and $\mathbf{G}[z]$ defined by (8.2) and (8.4), respectively.

8.5.1.2 Maximization of a cumulant-based contrast

Since it can only be expected to recover a source up to a multiplicative factor, without loss of generality, we can normalize the output power by imposing

$$\mathbb{E}\{|y(n)|^2\} = 1. \tag{8.41}$$

This condition introduces a constraint on the norm of the global filter. Indeed, let us introduce the scalar components of this filter by setting

$$\forall k \in \mathbb{Z} \qquad \mathbf{g}(k) = (g_1(k),\dots,g_N(k)) \tag{8.42}$$

and denote by $(\rho_{s_i}(n))_{n\in\mathbb{Z}}$, $i \in \{1,\dots,N\}$, the autocorrelation of the i-th source, which is here assumed to be definite positive. The weighted norm of the i-th component of the global filter is then given by

$$\|g_i\|_i = \left(\sum_{k\in\mathbb{Z}} g_i(k)g_i(\ell)^*\rho_{s_i}(k-\ell)\right)^{1/2} \tag{8.43}$$

which allows us to define the weighted norm of the global filter:

$$\|\mathbf{g}\| = \left(\sum_{i=1}^{N} \|g_i\|_i^2\right)^{1/2}. \tag{8.44}$$

With these definitions and the assumption of decorrelation between the sources, it is readily checked that the normalization in (8.41) is equivalent to

$$\|\mathbf{g}\| = 1. \tag{8.45}$$

By analogy with the instantaneous case, under this constraint we can maximize a function of the fourth-order cumulant of the output:

$$\Upsilon_f(\mathbf{g}) = f(\kappa_y) \tag{8.46}$$

where f is convex, defined on $\mathbb{R}$ such that $f(0) = 0$ and,

$$\kappa_y \triangleq \text{cum}\{y(n), y^*(n), y(n), y^*(n)\} \tag{8.47}$$

is assumed to be finite.[7] In addition, we make the assumption that one of the sources $(s_j(n))_{n\in\mathbb{Z}}$ where $j \in \{1,\ldots,N\}$ is such that

$$f(\kappa_{s_j}) > 0. \tag{8.48}$$

By using the multilinearity of the cumulants and the independence between the sources, we get the following relation

$$\begin{aligned}\Upsilon_f(\mathbf{g}) &= f\left(\sum_{i=1}^{N} \kappa_{g_i\star s_i}\right)\\ &= f\left(\sum_{i=1}^{N} \|g_i\|_i^4 \kappa_{\tilde{g}_i\star s_i}\right)\end{aligned} \tag{8.49}$$

[7]For this condition to be satisfied, it is sufficient that the fourth-order cumulant field $(\text{cum}\{s_i(n), s_i(n-k_1)^*, s_i(n-k_2), s_i(n-k_3)^*\})_{(k_1,k_2,k_3)\in\mathbb{Z}^3}$ of each source $(s_i(n))_{n\in\mathbb{Z}}$ is summable and the components $(g_i(k))_{k\in\mathbb{Z}}$ of the global filter are square summable.

where

$$\widetilde{g}_i = \begin{cases} \dfrac{g_i}{\|g_i\|_i} & \text{if } \|g_i\|_i \neq 0 \\ 0 & \text{otherwise.} \end{cases} \tag{8.50}$$

By using the normalization in (8.41) (or (8.45)), for all $i \in \{1,\ldots,N\}$, $\|g_i\|_i \leq 1$ and we have thus

$$\sum_{i=1}^{N} \|g_i\|_i^4 \leq \sum_{i=1}^{N} \|g_i\|_i^2 = \|\mathbf{g}\|^2 = 1. \tag{8.51}$$

The convexity of f allows us to deduce that

$$\begin{aligned} \Upsilon_f(\mathbf{g}) &= f\left(\sum_{i=1}^{N} \|g_i\|_i^4 \kappa_{\widetilde{g}_i \star s_i} + \left(1 - \sum_{i=1}^{N} \|g_i\|_i^4\right) 0\right) \\ &\leq \sum_{i=1}^{N} \|g_i\|_i^4 f(\kappa_{\widetilde{g}_i \star s_i}) + \left(1 - \sum_{i=1}^{N} \|g_i\|_i^4\right) f(0) \\ &\leq \max_{i \in \{1,\ldots,N\}} f(\kappa_{\widetilde{g}_i \star s_i}) \sum_{i=1}^{N} \|g_i\|_i^4. \end{aligned} \tag{8.52}$$

If, for all $i \in \{1,\ldots,N\}$, there exists $\widetilde{g}_i^\sharp$ maximizing $f(\kappa_{\widetilde{g}_i \star s_i})$ in the class of filters with impulse responses $\widetilde{g}_i$ such that $\|\widetilde{g}_i\|_i = 1$ and, if we denote by i_0 an index such that $\max_{i \in \{1,\ldots,N\}} f(\kappa_{\widetilde{g}_i^\sharp \star s_i}) = f(\kappa_{\widetilde{g}_{i_0}^\sharp \star s_{i_0}})$, we find that

$$\Upsilon_f(\mathbf{g}) \leq f(\kappa_{\widetilde{g}_{i_0}^\sharp \star s_{i_0}}) \sum_{i=1}^{N} \|g_i\|_i^4. \tag{8.53}$$

By invoking again inequality (8.51) and, according to (8.48), the fact that

$$f(\kappa_{\widetilde{g}_{i_0}^\sharp \star s_{i_0}}) \geq f(\kappa_{s_j}) > 0 \tag{8.54}$$

we conclude that

$$\Upsilon_f(\mathbf{g}) \leq f(\kappa_{\widetilde{g}_{i_0}^\sharp \star s_{i_0}}) \sum_{i=1}^{N} \|g_i\|_i^2 = f(\kappa_{\widetilde{g}_{i_0}^\sharp \star s_{i_0}}) = \Upsilon_f(\mathbf{g}^\sharp) \tag{8.55}$$

where $\mathbf{g}^\sharp$ is the filter having all its scalar components equal to zero except the component of index i_0, which is equal to $(\widetilde{g}_{i_0}^\sharp(k))_{k\in\mathbb{Z}}$. In other words, $\mathbf{g}^\sharp$ allows us to extract the source $(s_{i_0}(n))_{n\in\mathbb{Z}}$, up to the scalar filter of impulse response $(\widetilde{g}_{i_0}^\sharp(k))_{k\in\mathbb{Z}}$.

Conversely, if $\mathbf{g}$ is such that $\|\mathbf{g}\| = 1$ and $\Upsilon_f(\mathbf{g}) = \Upsilon_f(\mathbf{g}^\sharp)$, the chain of inequalities (8.52), (8.53) and (8.55) shows that we necessarily have

$$f(\kappa_{\tilde{g}_{i_0}^\sharp \star s_{i_0}}) \sum_{i=1}^{N} \|g_i\|_i^2 (1 - \|g_i\|_i^2) = 0. \tag{8.56}$$

Due to the strict inequality in (8.54), equality (8.56) is only possible when

$$\|g_i\|_i = \begin{cases} 1 & \text{if } i = i_0 \\ 0 & \text{otherwise.} \end{cases} \tag{8.57}$$

We thus obtain a filtered version of the source $(s_{i_0}(n))_n$. We have then shown that every global maximizer of Υ_f extracts one source.

Some comments are in order concerning this result:

- In a stronger sense, it was shown that, for IIR filters, if f is non-negative, every local maximum of Υ_f corresponds to a separating filter [37,3]. It must be emphasized the practical interest of this property which makes it possible to get rid of spurious local minima often encountered in global approaches.
- The required conditions for f, in particular (8.48), are satisfied if, for example, $f(\cdot) = |\cdot|^p$, $p \geq 1$, and at least one of the sources has a non-zero fourth-order cumulant. In the frequent case when one of the sources at least has a negative cumulant, we can choose $f = -\mathrm{Id}$, where Id is the identity function.
- The technical assumption that

$$\tilde{g}_i^\sharp = \arg \max_{\tilde{g}_i \,|\, \|\tilde{g}_i\|_i = 1} f(\kappa_{\tilde{g}_i \star s_i}) \tag{8.58}$$

  is not restrictive in practice. For example, it is satisfied as soon as we consider a restricted subset of global filters with impulse responses $(\mathbf{g}(k))_{k\in\mathbb{Z}}$ of finite length.
- In the case of iid sources [40], we have for every $i \in \{1, \ldots, N\}$,

$$\kappa_{\tilde{g}_i \star s_i} = \kappa_{s_i} \sum_{k\in\mathbb{Z}} |\tilde{g}_i(k)|^4. \tag{8.59}$$

  Since $\|\tilde{g}_i\|_i^2 = \sum_{k\in\mathbb{Z}} |\tilde{g}_i(k)|^2 = 1$, the resulting expression is maximal when

$$\tilde{g}_i(k) = \begin{cases} \Delta & \text{if } k = k_0 \\ 0 & \text{otherwise.} \end{cases} \tag{8.60}$$

  where $|\Delta| = 1$ and $k_0 \in \mathbb{Z}$. As mentioned in section 8.3.3.2, the indeterminacy in the recovery of a source simply reduces to a time shift and a multiplicative factor.
- In order to simplify our presentation, we have focused on the use of fourth-order auto-cumulants at zero delay. However, we can consider more general criteria of

the form

$$\sum_{(k_1,k_2,k_3)\in\mathcal{S}} f(\mathrm{cum}\{y(n), y(n-k_1)^*, y(n-k_2), y(n-k_3)\}^*)$$

where $f : \mathbb{C} \longrightarrow \mathbb{R}$ and $\mathcal{S}$ is a subset of $\mathbb{Z}^3$. We can also employ higher order cumulants, leading to contrasts of the form

$$f(\mathrm{cum}\{\underbrace{y(n),\ldots,y(n)}_{p \text{ times}}, \underbrace{y(n)^*,\ldots,y(n)^*}_{p \text{ times}}\})$$

where $p > 2$.

8.5.1.3 Use of the kurtosis

Instead of adopting a maximization approach subject to a unit power constraint, we can directly maximize a convex function of the kurtosis

$$\Upsilon_{\mathrm{kurt},f}(\mathbf{g}) = f\left(\frac{\kappa_y}{(\mathbb{E}\{|y(n)|^2\})^2}\right). \tag{8.61}$$

The equivalence with the previously considered criterion stems from the fact that

$$\Upsilon_{\mathrm{kurt},f}(\mathbf{g}) = \Upsilon_f(\tilde{\mathbf{g}}) \tag{8.62}$$

where

$$\forall k \in \mathbb{Z} \qquad \tilde{\mathbf{g}}(k) = \begin{cases} \dfrac{\mathbf{g}(k)}{\|\mathbf{g}\|} & \text{if } \|\mathbf{g}\| \neq 0 \\ 0 & \text{otherwise.} \end{cases} \tag{8.63}$$

We thus see that maximizing $\Upsilon_{\mathrm{kurt},f}$ is tantamount to setting the norm of the solution filter in an arbitrary manner and ensuring that the normalized filter maximizes Υ_f.

The equivalence between these two criteria is at the origin of the development of the usually employed extraction algorithm. It is important to keep in mind that the variable which, from a practical viewpoint, must be optimized is not $\mathbf{g}[z]$, but $\mathbf{b}[z]$. However, this does not create any major difficulty as we will see. Indeed, for real-valued or complex-valued sources, the kurtosis of $y(n)$ can be expressed as

$$\frac{\kappa_y}{(\mathbb{E}\{|y(n)|^2\})^2} = \frac{\mathbb{E}\{|y(n)|^4\}}{(\mathbb{E}\{|y(n)|^2\})^2} - \alpha \tag{8.64}$$

α being equal to 3 in the real case and to 2 in the complex circular one. The gradient of the kurtosis with respect to $\mathbf{b}(k)$ is thus given by

$$\frac{\partial}{\partial \mathbf{b}(k)}\left[\frac{\kappa_y}{(\mathbb{E}\{|y(n)|^2\})^2}\right] = \frac{4\mathbb{E}\{|y(n)|^2 y(n)\mathbf{x}(n-k)^{\mathrm{H}}\}}{(\mathbb{E}\{|y(n)|^2\})^2} - \frac{4\mathbb{E}\{|y(n)|^4\}\mathbb{E}\{y(n)\mathbf{x}(n-k)^{\mathrm{H}}\}}{(\mathbb{E}\{|y(n)|^2\})^3}. \tag{8.65}$$

To maximize $\Upsilon_{\mathrm{kurt},f}$, we can resort to a gradient algorithm when the function f is differentiable. Then, at iteration $p \in \mathbb{N}$, we compute

$$\forall n \quad y^{(p)}(n) = (\mathbf{b}^{(p)} \star \mathbf{x})(n) \tag{8.66}$$

$$\forall k \quad \widetilde{\mathbf{b}}^{(p)}(k) = \mathbf{b}^{(p)}(k) + \beta^{(p)} f'\left(\frac{\kappa_{y^{(p)}}}{(\mathbb{E}\{|y^{(p)}(n)|^2\})^2}\right)\frac{\partial}{\partial \mathbf{b}(k)}\left[\frac{\kappa_y}{(\mathbb{E}\{|y(n)|^2\})^2}\right] \tag{8.67}$$

$$\forall k \quad \mathbf{b}^{(p+1)}(k) = \widetilde{\mathbf{b}}^{(p)}(k) \tag{8.68}$$

where $\beta^{(p)} \in \mathbb{R}_+^*$. The step-size $\beta^{(p)}$ can be set to a small enough value to guarantee the algorithm convergence but it is more efficient to adjust it at each iteration so as to maximize $f(\kappa_{\widetilde{y}^{(p+1)}}/(\mathbb{E}\{|y^{(p+1)}(n)|^2\})^2)$. For example, when $f(\cdot) = |\cdot|^2$, this optimization leads to the search of the roots of a polynomial of degree 6, as the kurtosis of $\widetilde{y}^{(p+1)}(n)$ is a rational fraction of order 4 of $\beta^{(p)}$ [36].

From the recursive equations (8.67) and (8.68), a projected gradient algorithm can be easily deduced allowing us to maximize Υ_f subject to the unit power constraint for the extracted signal. The advantage over the previous algorithm is to be able to control the norm of the impulse response of the extraction filter. By noticing that the constraint in (8.41) can be rewritten as

$$\|\mathbf{b}\|_{\mathbf{R}_{\mathbf{x}}} = \left(\sum_{(k,\ell)\in\mathbb{Z}^2} \mathbf{b}(k)\mathbf{R}_{\mathbf{x}}(k-\ell)\mathbf{b}(\ell)^{\mathrm{H}}\right)^{1/2} = 1 \tag{8.69}$$

where, for all $k \in \mathbb{Z}$, $\mathbf{R}_{\mathbf{x}}(k) = \mathbb{E}\{\mathbf{x}(n)\mathbf{x}(n-k)^{\mathrm{H}}\}$, the desired normalization consists of replacing (8.68) by

$$\forall k \quad \mathbf{b}^{(p+1)}(k) = \frac{\widetilde{\mathbf{b}}^{(p)}(k)}{\|\widetilde{\mathbf{b}}^{(p)}\|_{\mathbf{R}_{\mathbf{x}}}}. \tag{8.70}$$

The update step (8.67) and, in particular, the computation of the gradient in (8.65) simplifies since $\mathbb{E}\{|y^{(p)}(n)|^2\} = 1$.

Note finally that the statistical moments which are used in the algorithm are usually empirically estimated from the available data (by averaging) and that the method is therefore all the more accurate as the number of signal samples is large.

8.5.1.4 CMA criterion

Another very popular method in digital communications is the CMA (Constant Modulus Algorithm) [15,39]. Then, the objective is to minimize

$$\Upsilon_{\text{CMA}}(\mathbf{g}) = \mathbb{E}\{(|y(n)|^2 - 1)^2\}. \tag{8.71}$$

Indeed, in the domain of telecommunications, the sources are often modulated signals and, when phase modulations are employed, the symbols take their values in an alphabet which constitutes a subset of the unit circle. Intuitively, we can expect to extract the symbols from the mixture if we recover this property of constant modulus.

Although this approach may appear different from the maximization of a contrast, we will see that it is closely related to the latter approach. To clarify this fact, let us rewrite the CMA criterion under the form:

$$\Upsilon_{\text{CMA}}(\mathbf{g}) = \mathbb{E}\{|(\tilde{\mathbf{g}} \star \mathbf{s})(n)|^4\}\|\mathbf{g}\|^4 - 2\|\mathbf{g}\|^2 + 1 \tag{8.72}$$

where $(\tilde{\mathbf{g}}(k))_{k\in\mathbb{Z}}$ is defined by (8.63). Let $(\mathbf{g}(k))_{k\in\mathbb{Z}}$ be such that $\mathbb{E}\{|(\tilde{\mathbf{g}} \star \mathbf{s})(n)|^4\} \neq 0$.[8] For $\Upsilon_{\text{CMA}}(\mathbf{g})$ to be minimum, we must have

$$\|\mathbf{g}\|^2 = \frac{1}{\mathbb{E}\{|(\tilde{\mathbf{g}} \star \mathbf{s})(n)|^4\}}. \tag{8.73}$$

The criterion then reduces to

$$\Upsilon_{\text{CMA}}(\mathbf{g}) = 1 - \frac{1}{\mathbb{E}\{|(\tilde{\mathbf{g}} \star \mathbf{s})(n)|^4\}}. \tag{8.74}$$

To minimize this expression, we must thus determine $(\tilde{\mathbf{g}}(k))_{k\in\mathbb{Z}}$ of unit norm minimizing $\mathbb{E}\{|(\tilde{\mathbf{g}} \star \mathbf{s})(n)|^4\}$. In addition, in the real or complex circular cases, relation (8.64) linking the fourth-order cumulants to the fourth-order moments can be reformulated as

$$\mathbb{E}\{|(\tilde{\mathbf{g}} \star \mathbf{s})(n)|^4\} = \alpha + \kappa_{\tilde{\mathbf{g}}\star\mathbf{s}} = \alpha - \Upsilon_{-\text{Id}}(\tilde{\mathbf{g}}), \tag{8.75}$$

where $\Upsilon_{-\text{Id}}$ is defined by (8.46), f being equal to minus identity. The minimization to be performed is therefore equivalent to maximizing $\Upsilon_{-\text{Id}}$ over the set of normalized filters with N inputs and 1 output. We have seen that the latter function is a contrast provided

[8] If such a sequence did not exist, the sources would be almost surely zero. Furthermore, when $\mathbb{E}\{|(\tilde{\mathbf{g}} \star \mathbf{s})(n)|^4\} = 0$, it can be easily checked that the value of Υ_{CMA} is equal to 1, which is greater than the value we will calculate.

that at least one of the sources has a negative cumulant, which is also a condition for the validity of the CMA.

In summary, under the considered conditions, the CMA criterion is minimized by

$$\mathbf{g}^{\sharp} = \frac{\tilde{\mathbf{g}}^{\sharp}}{(\mathbb{E}\{|(\tilde{\mathbf{g}}^{\sharp} \star \mathbf{s})(n)|^4\})^{1/2}}. \tag{8.76}$$

where $(\tilde{\mathbf{g}}^{\sharp}(k))_{k\in\mathbb{Z}}$ is the impulse response of a normalized filter maximizing $\Upsilon_{-\mathrm{Id}}$.

8.5.1.5 Some alternative approaches

Other iterative approaches have been developed for source extraction. In particular, one can mention the super-exponential approach [21], which is based on an appealing idea. In this method, a nonlinear recursive equation on the global filter is built, the convergence of which is guaranteed toward a filter extracting one of the sources. The separating filter with impulse response $(\mathbf{b}(k))_{k\in\mathbb{Z}}$ is deduced through a weighted least squares technique by using the source correlations as well as the cross-cumulants between the estimated source and the observations.

Recently, so-called *contrasts with reference* approaches have been introduced in [5]. The criterion to be maximized is the modulus of a cross-cumulant between the estimated source and filtered versions of the observations. Under some conditions, this contrast takes a quadratic form and it can therefore be easily maximized by eigenvalue decomposition techniques.

8.5.2 Deflation

Once one of the sources (or a filtered version of it) has been extracted, it is sufficient to subtract its contribution from the observations to obtain a mixture of $(N-1)$ sources. From this reduced size mixture (whose size has been *deflated*), we can extract another source by applying one of the methods described in section 8.5.1 and thus iteratively proceed so as to extract all the sources.

More precisely, the algorithm reads:

1. Initialization: $p=1$; extraction of the first source $(y_1(n))_{n\in\mathbb{Z}}$ from the observations $(\mathbf{x}^{(1)}(n))_n = (\mathbf{x}(n))_n$.
2. For $p \in \{2,\ldots,N\}$,
   (a) Subtraction of the extracted source at step $p-1$; we deduce a signal $(\mathbf{x}^{(p)}(n))_n$, having still P scalar components, corresponding to a reduced size mixture of $N-p+1$ sources.
   (b) Extraction of a p-th source $(y_p(n))_n$ from the modified observations $(\mathbf{x}^{(p)}(n))_n$.

In step 2a, in order to subtract from the mixture the contribution of the $(p-1)$th source, we compute

$$\mathbf{x}^{(p)}(n) = \mathbf{x}^{(p-1)}(n) - \sum_k \mathbf{t}^{(p)}(k)\, y_{p-1}(n-k), \quad n \in \mathbb{Z} \tag{8.77}$$

where $(\mathbf{t}^{(p)}(k))_{k\in\mathbb{Z}}$ is the impulse response of a filter with 1 input and P outputs. The sources being independent from the others, this impulse response can be derived as the solution of a classical mean square linear estimation problem by minimizing

$$\epsilon(\mathbf{t}^{(p)}) = \mathbb{E}\{|\mathbf{x}^{(p)}(n)|^2\}. \tag{8.78}$$

In practice, we often employ filters with finite impulse responses, which is theoretically founded when the mixture is also with finite length and it admits an FIR left inverse. The mean square linear estimation problem thus reduces to finding the solution of a linear system of equations having a finite number of unkowns.

Note that the deflation approach may lead to an accumulation of estimation errors which may become excessive after a certain number of source extractions. Improvements of the previous algorithm have been proposed to alleviate this problem (local post-optimization, decorrelation methods [3], ...).

8.6 NON-STATIONARY CONTEXT

8.6.1 Context

In this section, the restrictive assumption of stationarity of the source signals is partially relaxed. Indeed, we will assume that the components of the source vector $(\mathbf{s}(n))_{n\in\mathbb{Z}}$ are cyclostationary: this latter point is specified later on (see section 8.6.2). Besides the academic interest, the design of BSS algorithms able to cope with non-stationary environments is crucial for numerous fields such as – among others – digital communication and acoustic signals. Let us provide a few examples:

- "Cocktail party". The scenario is of two speakers talking to different people, hence saying different sentences, with different tempo, in the same room. The BSS problem consists in separating the contributions of the two speakers. The mutual independence between the speakers is, perhaps, a plausible assumption. As far as the stationarity of the two "transmitted" signals is concerned, this assumption is clearly not relevant. Indeed, at least two kinds of non-stationarities can be isolated: the middle-term non-stationarities (due to changes of syllables, switch from a sound to a silence) and the short-term non-stationarities (existence of a pitch for the vocal parts of a speech signal).
- Mixtures of digital communication signals. In a given frequency band, several sources can be assumed to interfere. This can be the case for High Frequency (HF) systems. Another example is provided by the context of mobile communications: in the neighborhood of a base station, the transmitted signal is dominant over the signals coming from other base stations. In a cooperative scenario, characterized by the fact that the main parameters are known to a terminal (transmission rates, learning sequences, error-correcting codes), the non-desired signals can be considered by the terminal as noise sources and their effect on the demodulation is not always crucial. Nevertheless, there are contexts where the crucial parameters

> evoked above are not accessible to the receiver, and hence have to be estimated for the demodulation of the different signals. This is the case for military applications (passive listening), but cooperative schemes can also be concerned (the transmission parameters are not all known by a given user). The complexity of this work is of course considerably reduced if, prior to this estimation, the BSS of the mixture is successfully processed. However, it is well known that digital communication signals are not stationary, but their statistics evolve with time even if the channel is stationary.

The above-mentioned signals are cyclostationary, e.g. the statistics of these signals show periodic or almost periodic variations with regard to the time index. This is specified in the next section.

As a preliminary remark, one should note that any of the BSS algorithms based on second-order statistics, originally designed for convolutive mixtures of wide-sense stationary data [24], may be considered; their capacity to separate the contributions is unaltered. Besides, the non-stationarities may even be profitably exploited, when these latter stem from a sudden change [34].

Nevertheless, we claim that no such direct conclusions can be drawn for the higher-order methods; we will shed light on this point. In particular, we focus our attention on how the methods may be adapted in a cyclostationary context to the sequential fourth-order method depicted (for stationary sources) later in section 8.5.1.

8.6.2 Some properties of cyclostationary time-series

The sources to be extracted are sampled versions of continuous-time processes. By notation, the continuous-time functions or processes will show the subscript a. A complex-valued process $(s_a(t))_{t\in\mathbb{R}}$ is cyclostationary with period T if, for all $L \in \mathbb{N}^*$ and $(\tau_1,\ldots,\tau_L) \in \mathbb{R}^L$, the joint cumulative distributions function of the vector[9] $s_a^{(*)}(t+\tau_1), s_a^{(*)}(t+\tau_2),\ldots,s_a^{(*)}(t+\tau_L)$, seen as functions of t, are periodic with period T.

EXAMPLE E8.1 (Digital communication signals)
Let $(a(n))_{n\in\mathbb{Z}}$ be transmitted symbols. They may be seen as outputs of an interleaver, and hence may be assumed to be iid. Moreover, its mean is considered to be zero. Its variance is σ^2. Call T the symbol-period, e.g. the delay between the transmission of two successive symbols. We call $1/T$ the symbol-rate. Assume that the modulation is linear; in this case, the model for the complex envelope of the transmitted signal is the convolution:

$$\forall t \in \mathbb{R} \qquad s_a(t) = \sum_{n\in\mathbb{Z}} a(n)c_a(t-nT) \tag{8.79}$$

where c_a is a (finite energy) shaping function. As, in particular, $(a_n)_{n\in\mathbb{Z}}$ is strictly stationary, $(s_a(t))_{t\in\mathbb{R}}$ given in (8.79) is cyclostationary with period T.

[9]The notation $s_a^{(*)}$ means that either s_a is considered or its conjugate.

EXAMPLE E8.2 (Simplified model for a vocal part of a speech signal)
Let m_a be a deterministic periodic function with period T and b_a, n_a two mutually independent processes, both strictly stationary. In this case, the signal

$$\forall t \in \mathbb{R} \qquad m_a(t)b_a(t) + n_a(t) \tag{8.80}$$

is cyclostationary with period T.

As a consequence, any moment of such cyclostationary processes is periodic (T is a period). For instance, we may have a look at the mean. For the communication signal (8.79), the mean is zero since $(a_n)_{n\in\mathbb{Z}}$ has zero mean. As far as the model (8.80) is concerned, the mean is $\mathbb{E}\{b_a(t)\}m_a(t)$ and is hence constant or T-periodic.

Let us focus on the second-order statistics: the auto-correlation function is defined as

$$\rho_{s_a}(t,\tau) = \mathbb{E}\{s_a(t+\tau)s_a(t)^*\}.$$

If s_a is cyclostationary with period T, the definition implies that $\rho_{s_a}(t,\tau)$, as a function of t, is periodic and T is a period. Conversely, a second-order process s_a such that $\rho_{s_a}(.,\tau)$ is, for all τ, periodic with period T is called a *periodically correlated* process.

In the following, we focus on digital communication signals such as (8.79). Simple calculations show that

$$\rho_{s_a}(t,\tau) = \sigma^2 \sum_{n\in\mathbb{Z}} c_a(t+\tau-nT)c_a(t-nT)^* \tag{8.81}$$

which illustrates that s_a is periodically correlated with period T. At least formally, $t \mapsto \rho_{s_a}(t,\tau)$ hence admits a Fourier expansion such as

$$\rho_{s_a}(t,\tau) = \sum_k R_{s_a}^{(k/T)}(\tau)e^{\imath 2\pi kt/T} \tag{8.82}$$

where

$$R_{s_a}^{(k/T)}(\tau) = \frac{1}{T}\int_0^T \rho_{s_a}(t,\tau)e^{-\imath 2\pi kt/T}\,dt.$$

The latter function is called the cyclic-correlation function at cyclic-frequency k/T. In view of (8.81) we show quite directly that

$$R_{s_a}^{(k/T)}(\tau) = \frac{\sigma^2}{T}\int_{\mathbb{R}} c_a(t+\tau)c_a(t)^* e^{-\imath 2\pi kt/T}\,dt. \tag{8.83}$$

It is insightful to consider the Fourier transform $S_{s_a}^{(k/T)}$ of $R_{s_a}^{(k/T)}$. We call it cyclo-spectrum of s_a at cyclo-frequency k/T. We easily show that

$$\forall \nu \in \mathbb{R} \qquad S_{s_a}^{(k/T)}(\nu) = \frac{\sigma^2}{T} \check{c}_a(\nu)\check{c}_a(\nu - k/T)^*,$$

where $\check{c}_a$ is the Fourier transform of the shaping function c_a.

Power spectral density function. We notice that the cyclo-correlation function at the null cyclo-frequency, $R_{s_a}^{(0)}$, is always a function of the positive type. By analogy with second-order stationary process, $S_{s_a}^{(0)}$ is called the power spectral density function. We obtain the famous expression for the power spectral density:

$$S_{s_a}^{(0)}(\nu) = \frac{\sigma^2}{T} |\check{c}_a(\nu)|^2.$$

Symmetry. As $\rho_{s_a}(t - \tau, \tau) = \rho_{s_a}(t, -\tau)^*$ for all $(t, \tau) \in \mathbb{R}^2$, one deduces that

$$R_{s_a}^{(-k/T)}(\tau) = R_{s_a}^{(k/T)}(-\tau)^* e^{-\imath 2\pi k \tau/T}.$$

In particular:

$$R_{s_a}^{(-k/T)}(0) = R_{s_a}^{(k/T)}(0)^*. \tag{8.84}$$

Effect of a filter. Let h_a be the impulse response of a filter (assumed to be square-summable). The process s_a being periodically correlated, the output $h_a \star s_a$ of the filter driven by s_a is also periodically correlated with the same period. We have the relation between the cyclo-spectra of s_a and those of $h_a \star s_a$:

$$S_{h_a \star s_a}^{(k/T)}(\nu) = \check{h}_a(\nu) S_{s_a}^{(k/T)}(\nu) \check{h}_a(\nu - k/T)^*. \tag{8.85}$$

At this point, we can deduce an important consequence: we show that, for the communication signal s_a given by (8.79), there are at most, three non-null cyclo-spectra, precisely those at the cyclic-frequencies $0, \pm\frac{1}{T}$. Indeed, due to reasons of spectral efficiency, the shaping filter c_a is such that the support of $\check{c}_a$ is included in an interval of the type:

$$\left[-\frac{1+\gamma}{2T}, \frac{1+\gamma}{2T} \right] \tag{8.86}$$

where γ, the excesss bandwidth factor, is such that $0 \le \gamma < 1$. Let h_a be such that $\check{h}_a$ is the indicator function of the interval (8.86). Obviously, we have: $s_a = h_a \star s_a$. The supports of $\nu \mapsto \check{h}_a(\nu)$ and $\nu \mapsto \check{h}_a(\nu - k/T)^*$ have an empty intersection for any $|k| > 1$. Thanks to (8.85), we deduce that $S_{s_a}^{(k/T)} = 0$ for any k such that $|k| > 1$. The bandwidth of the communication signal s_a implies that the Fourier expansion (8.82) is actually a finite summation with the three terms corresponding to the cyclic-frequencies $0, \frac{1}{T}$ and $-\frac{1}{T}$.

Cyclo-stationarity and sampling (communication signals). We denote by $T_e > 0$ some sampling period and let $s(n)$ be

$$s(n) = s_a(nT_e).$$

The auto-correlation function of $(s(n))_{n\in\mathbb{Z}}$ trivially verifies

$$\begin{aligned}\forall n, m \in \mathbb{Z} \quad \rho_s(n,m) &= \mathbb{E}\{s(n+m)s(n)^*\} \\ &= \rho_{s_a}(nT_e, mT_e)\end{aligned}$$

and hence has the Fourier expansion [see (8.82)]:

$$\rho_s(n,m) = \sum_{k=-1,0,1} R_s^{(k\alpha)}(m) e^{\imath 2\pi k\alpha n}, \tag{8.87}$$

where we have set $\alpha = \frac{T_e}{T}$ the unique cyclic-frequency and

$$R_s^{(k\alpha)}(m) = R_{s_a}^{(k/T)}(mT_e). \tag{8.88}$$

As a function of n, $\rho_s(n,m)$ is *not* in general periodic (except if α is a rational number: if the sampling period is randomly chosen, this is rarely the case), but is an almost-periodic series [11].

Remark R8.1

Suppose that $\alpha = p \in \mathbb{N}^*$. This particular case is associated with a constant $\rho_s(n,m)$ w.r.t. n which means that the time-series $(s(n))_{n\in\mathbb{Z}}$ is weakly stationary. This is not surprising, thanks to (8.79). Indeed, denote by $c_\ell(k) = c_a(pT+\ell)$, $C_\ell[z] = \sum_k c_\ell(k) z^{-k}$ and $a_\ell(k) = a(kp+\ell)$. Then $s(n) = \sum_{\ell=0}^{p-1} [C_\ell[z]] a_\ell(k)$ which shows that $(s(n))_{n\in\mathbb{Z}}$ is stationary (in the strict sense). For any β, we define

$$R_s^{(\beta)}(m) = \lim_{M\to\infty} \frac{1}{2M+1} \sum_{n=-(M-1)}^{M-1} \rho_s(n,m) e^{-\imath 2\pi n\beta}. \tag{8.89}$$

If $\alpha \notin \mathbb{N}$, this is consistent with the previous definition for $\beta = k\alpha$, $k = -0, \pm 1$ given in Eq. (8.88). Besides, we have for $\beta \notin \{0, \pm\alpha\}$: $R_s^{(\beta)}(m) = 0$.

We are ready to specify some points on the blind separation of cyclostationary sources.

8.6.3 Direct extension of the results of section 8.5 for the source extraction, and why it is difficult to implement

We consider a mixture of K sources of the type (8.79). The i-th source is denoted by $s_{a,i}$; it has a symbol-period T_i; the associated shaping filter (respectively excess bandwidth factor) is $c_{a,i}$ (respectively γ_i). We suppose that the receiver is equipped with an array of M sensors. As the transmission involves multi-path effects, we have to specify briefly the model of the observation. Denote $[-B,B]$ the common band of frequencies of the sources and set $\mathbf{s}_a(t)=(s_{a,1}(t),\ldots,s_{a,K}(t))^T$. The multi-variate observation, denoted by $(\mathbf{x}_a(t))_{t\in\mathbb{R}}$ follows the equation:

$$\mathbf{x}_a(t)=(\mathbf{A}_a\star\mathbf{s}_a)(t)$$

where $\mathbf{A}_a$ is an $M\times K$ matrix-valued impulse response: the component (i,j) represents the channel between source j and sensor i. The Fourier transform of any of its entries has a support included in $[-B,B]$. The observation is sampled and we denote by T_e the sampling period. Under the Shannon sampling condition

$$1/T_e>2B,$$

the model of the observation $\mathbf{y}(n)\triangleq\mathbf{y}_a(nT_e)$ is the same as this given in (8.1) where the i-th component of $\mathbf{s}(n)$ is $s_{a,i}(nT_e)$ and $A(k)=A_a(kT_e)$.

Following the same approach as in section 8.5, we aim at computing a scalar-valued signal $y(n)$ involving only one of the sources. In this respect, we consider a (variable) $1\times M$ MISO filter $\mathbf{b}[z]-\sum_k\mathbf{b}_k z^{-k}$ and we let $y(n)$ be

$$\forall n\in\mathbb{Z}\quad y(n)=\sum_k\mathbf{b}_k\mathbf{x}(n-k).$$

This can be written as

$$\forall n\in\mathbb{Z}\quad y(n)=\sum_k\mathbf{g}_k\mathbf{s}(n-k).$$

where the global filter $\mathbf{g}[z]$ is given by the relation $\mathbf{g}[z]=\mathbf{b}[z]\mathbf{A}[z]$. As $(y(n))_{n\in\mathbb{Z}}$ is a superimposition of filtered versions of the sources $(s_i(n))_{n\in\mathbb{Z}}$, $i=1,\ldots,K$, and these sources are cyclostationary and mutually decorrelated, we deduce that $(y(n))_{n\in\mathbb{Z}}$ is almost periodically correlated, and its set of cyclic-frequencies is

$$\mathcal{I}=\{0,\pm\alpha_1,\ldots,\pm\alpha_N\}$$

where $\alpha_i=\frac{T_e}{T_i}$.

The normalization of the mixture $(y(n))_{n\in\mathbb{Z}}$ as given in (8.41) does not make sense for cyclostationary sources; indeed, the energy $\mathbb{E}\{|y(n)|^2\}$ evolves almost peridiodically. We rather consider the normalization:

$$\frac{1}{\sqrt{\langle\mathbb{E}\{|y(n)|^2\}\rangle}}y(n),$$

where $\langle.\rangle$ is the time averaging operator defined, when it exists, as the Cesaro limit, i.e. for a series $(u_m)_{m\in\mathbb{Z}}$:

$$\langle u_n\rangle = \lim_{M\to\infty}\frac{1}{2M-1}\sum_{m=-(M-1)}^{M-1} u_m.$$

Considering that $\langle\mathbb{E}\{|y(n)|^2\}\rangle = 1$ is equivalent to imposing the constraint on the global filter $\mathbf{g}(z)$, $||\mathbf{g}|| = 1$ for example. Similarly to (8.43), we define:

$$||g_i||_i = \left(\sum_{k\in\mathbb{Z}} g_i(k)g_i(\ell)^* R_{s_i}^{(0)}(k-\ell)\right)^{1/2} \tag{8.90}$$

and

$$||\mathbf{g}|| = \left(\sum_{i=1}^{N}||g_i||_i^2\right)^{1/2}. \tag{8.91}$$

Following the results obtained in section 8.5.1 it is tempting to inspect the fourth-order cumulant $\kappa_y(n) \triangleq \text{cum}\{y(n), y(n)^*, y(n), y(n)^*\}$ which is in general a function of the time lag n. In order to get rid of the dependency in n, we consider the maximization of the function

$$f(\langle\kappa_y(n)\rangle), \tag{8.92}$$

where f is convex, $f(0) = 0$ and $f(\langle\kappa_{s_i(n)}\rangle) > 0$ for all[10] the indices i. As function f is convex, we may invoke the Jensen inequality and we obtain that:

$$f(\langle\kappa_y(n)\rangle) \leq \langle f(\kappa_y(n))\rangle.$$

[10]The condition "at least one i" is sufficient to ensure the extraction of one source. In general, however, it is wished to extract all the sources, which explains the more restrictive condition.

On the other hand, we can follow the inequalities given in section 8.5.1 after Eq. (8.46). We have, for all integers n:

$$f(\kappa_y(n)) = f\left(\sum_{i=1}^{N} ||g_i||_i^4 \kappa_{(\tilde{g}_i \star s_i)(n)}\right) \leq \sum_{i=1}^{N} ||g_i||_i^4 f(\kappa_{(\tilde{g}_i \star s_i)(n)}).$$

Taking the time average of both sides of the above inequality gives:

$$f(\langle \kappa_y(n) \rangle) \leq \langle f(\kappa_y(n)) \rangle \leq \sum_{i=1}^{N} ||g_i||_i^4 \langle f(\kappa_{(\tilde{g}_i \star s_i)(n)}) \rangle.$$

Eventually, the facts that $\sum_{i=1}^{N} ||g_i||_i^4 \leq (\sum_{i=1}^{N} ||g_i||_i^2)^2$ and $\sum_{i=1}^{N} ||g_i||_i^2 = 1$ allow us to write:

$$f(\langle \kappa_y(n) \rangle) \leq \max_{i=1,\ldots,N} \sup_{||\tilde{g}_i||=1} \langle f(\kappa_{(\tilde{g}_i \star s_i)(n)}) \rangle.$$

Conversely, the proposed upper bound is reached for filters $\mathbf{g}[z]$ having only one non-null component – say number i_0 – such that there exists a normalized filter $\tilde{g}_{i_0}^{\sharp}$ that attains

$$\max_{i=1,\ldots,N} \sup_{||\tilde{g}_i||=1} \langle f(\kappa_{(\tilde{g}_i \star s_i)(n)}) \rangle.$$

In conclusion, we have proved that $f(\langle \kappa_y(n) \rangle)$ is a constrast function.

For stationary sources, this key result has a straightforward application counterpart. Taking into account the expansion of the cumulant[11]

$$\kappa_y(n) = \mathbb{E}\{|y(n)|^4\} - 2(\mathbb{E}\{|y(n)|^2\})^2,$$

shows that this latter can be consistently estimated. Indeed, let M be the number of available data; if the data verify certain mixing conditions (always fufilled in digital communication frameworks) then, for $\ell \in \{2,4\}$ the following convergence in probability holds:

$$\frac{1}{M}\sum_{m=0}^{M-1} |y(m)|^{\ell} \longrightarrow \mathbb{E}\{|y(n)|^{\ell}\}.$$

[11] All the sources are supposed to be second-order complex circular, i.e. $\mathbb{E}\{s_i(n)s_i(m)\} = 0$ for all the indices n, m. This assumption is made from now on; note that it is verified for complex-valued sources such as QAM, M-PSK ($M > 2$).

This is used in the algorithm depicted in section 8.5.1.3 where, in all the equations, the expectation operator is replaced by the empirical estimate. However, this crucial remark on the consistency of the empirical estimator of the terms of the cumulant does not hold true, in general, for cyclostationary data. We show next that the consistent estimation of $\langle \kappa_y(n) \rangle$ requires much more *a priori* information on the sources. Recalling that the sources are assumed circularly, we have

$$\langle \kappa_y(n) \rangle = \langle \mathbb{E}\{|y(n)|^4\} \rangle - 2\langle (\mathbb{E}\{|y(n)|^2\})^2 \rangle. \tag{8.93}$$

The consistent estimation of $\langle \mathbb{E}|y(n)|^4 \rangle$ from $y(0), y(1), \ldots, y(M-1)$ still relies on the natural estimator

$$\frac{1}{M} \sum_{n=0}^{M-1} |y(n)|^4.$$

The difficulty is to estimate the term $\langle (\mathbb{E}\{|y(n)|^2\})^2 \rangle$. Denoting by $\mathcal{I} = \{0, \pm\alpha_1, \ldots, \pm\alpha_N\}$ the set of the second-order cyclic frequencies, we have:

$$\mathbb{E}\{|y(n)|^2\} = \sum_{\zeta \in \mathcal{I}} R_y^{(\zeta)}(0) e^{i2\pi\alpha n}.$$

The Parseval formula (for almost periodic series) is

$$\langle (\mathbb{E}\{|y(n)|^2\})^2 \rangle = \sum_{\zeta \in \mathcal{I}} |R_y^{(\zeta)}(0)|^2. \tag{8.94}$$

From the general symmetry formula (8.84), we deduce that $R_y^{(\zeta)}(0) = R_y^{(-\zeta)}(0)^*$. Hence

$$R_y^{(0)}(0)^2 + 2 \sum_{\zeta \in \mathcal{I}_+} |R_y^{(\zeta)}(0)|^2$$

where $\mathcal{I}_+$ is the set of all strictly positive cyclic frequencies of $y(n)$ (if the set is not empty, any element ζ of this set can be written $\zeta = \frac{T_e}{T_i}$ for a certain index i). Thanks to the normalized energy of $y(n)$, we finally obtain the equation:

$$\langle (\mathbb{E}\{|y(n)|^2\})^2 \rangle = 1 + 2 \sum_{\zeta \in \mathcal{I}_+} |R_y^{(\zeta)}(0)|^2. \tag{8.95}$$

On the other hand, the empirical estimate

$$\hat{R}_y^{(\zeta)}(0) = \frac{1}{M} \sum_{n=0}^{M-1} |y(n)|^2 e^{-i2\pi\zeta n}$$

of $R_y^{(\zeta)}(0)$ is consistent (see [13,12]).

Hence, the expansion (8.95) allows one to understand that the consistent estimation of the term $-2\langle(\mathbb{E}\{|y(n)|^2\})^2\rangle$ is achieved by computing

$$1+2\sum_{\zeta\in\mathcal{I}_+}\frac{1}{M}\sum_{m=0}^{M-1}|y(m)|^2 e^{-\imath 2\pi\zeta m}.$$

As can be seen in the previous expression, this requires the knowledge of all the strictly positive cyclic frequencies. As a matter of fact, the consistent estimation of the contrast function $f(\langle\kappa_y(n)\rangle)$ requires the knowledge (or consistent estimates) of the cyclic frequencies of the sources (or equivalently all the symbol periods in the case of digital communication signals). This fact discourages the use of this contrast. Indeed, it is not possible, in certain contexts such as the military one, to assume the symbol rates known; and the estimation of these parameters is a challenge (see [27] for the most simple case, e.g. when there is only one source).

In order to circumvent the difficulty raised above, it was proposed in [22] to consider a simpler function to optimize. By construction, this function does not involve any of the cyclic-frequencies of the mixture.

8.6.4 A function to minimize: A contrast?

Since the computation of a consistent estimate of $\langle\kappa_y(n)\rangle$ is not easy (or even unrealistic), we decide to drop the non-desired terms and substitute the function

$$\Upsilon_f(\mathbf{g})=f\left(\langle\mathbb{E}\{|y(n)|^4\}\rangle-2\langle\mathbb{E}\{|y(n)|^2\}\rangle^2\right)$$

for the function given in (8.92) which can be further simplified, due to the normalization constraint, to

$$\Upsilon_f(\mathbf{g})=f(\langle\mathbb{E}\{|y(n)|^4\}\rangle-2).$$

Remark. Though very restrictive (see the discussion above), the case when the mixture is stationary is interesting. Indeed, the function $\Upsilon_f(\mathbf{g})$ coincides, in this case, with $f(\kappa_y)$, which is the constrast function considered in section 8.5.1; this justifies why we have kept the same notation $\Upsilon_f(\mathbf{g})$.

The theoretical results obtained in the stationary case cannot be transposed directly in the cyclostationary case. This is why we investigate in the following if $\Upsilon_f(\mathbf{g})$ is a contrast function. In the case of a positive answer (which is in general the case), the algorithms used to achieve the separation are the same as those described in section 8.5.1.3.

On the one hand, we can express $\Upsilon_f(\mathbf{g})$ as a function of the cumulant $\kappa_y(n)$; indeed, we recall that, due to (8.93) and (8.95), we have:

$$\Upsilon_f(\mathbf{g})=f\left(\langle\kappa_y(n)\rangle+4\sum_{\zeta\in\mathcal{I}_+}|R_y^{(\zeta)}(0)|^2\right).$$

On the other hand, the mutual decorrelation of the sources implies that

$$R_y^{(\zeta)}(0) = \sum_{i=1}^{N} R_{g_i \star s_i}^{(\zeta)}(0) = \sum_{i=1}^{N} ||g_i||_i^2 R_{\tilde{g}_i \star s_i}^{(\zeta)}(0).$$

As a consequence:

$$\Upsilon_f(\mathbf{g}) = f\left(\sum_{i=1}^{N} ||g_i||_i^4 \beta_{\tilde{g}_i \star s_i} + 2\sum_{i<j} ||g_i||_i^2 ||g_j||_j^2 \lambda_{\tilde{g}_i \star s_i, \tilde{g}_j \star s_j}\right)$$

where we used the notation

$$\begin{aligned}\beta_{\tilde{g}_i \star s_i} &= \langle \kappa_{\tilde{g}_{\star s_i}}(n)\rangle + 4\sum_{\zeta \in \mathcal{I}_+} |R_{\tilde{g}_i \star s_i}^{(\zeta)}(0)|^2 \\ &= \langle \mathbb{E}\{|(\tilde{g}_i \star s_i(n))|^4\}\rangle - 2\end{aligned}$$

and

$$\lambda_{\tilde{g}_i \star s_i, \tilde{g}_j \star s_j} = 4\sum_{\zeta \in \mathcal{I}_+} \Re\left(R_{\tilde{g}_i \star s_i}^{(\zeta)}(0) R_{\tilde{g}_j \star s_j}^{(\zeta)}(0)^*\right).$$

By construction,

$$\left(\sum_{i=1}^{N} ||g_i||_i^2\right)^2 = 1 = \sum_{i} ||g_i||_i^4 + 2\sum_{i<j} ||g_i||_i^2 ||g_j||_j^2.$$

It can be deduced from this equation and the convexity of f that

$$\Upsilon_f(\mathbf{g}) \le \sum_{i=1}^{N} ||g_i||_i^4 f(\beta_{\tilde{g}_i \star s_i}) + 2\sum_{i<j} ||g_i||_i^2 ||g_j||_j^2 f(\lambda_{\tilde{g}_i \star s_i, \tilde{g}_j \star s_j}). \tag{8.96}$$

We may introduce the quantity

$$\beta_{\max} = \max_{i \in \{1,\ldots,N\}} \sup_{||\tilde{g}||_i = 1} f(\beta_{\tilde{g} \star s_i}).$$

If

$$\forall i,j \quad \text{and} \quad \forall ||\tilde{g}_i||_i = 1, ||\tilde{g}_j||_j = 1 \quad f(\lambda_{\tilde{g}_i \star s_i, \tilde{g}_j \star s_j}) \le \beta_{\max} \tag{8.97}$$

then it can be deduced from (8.96) that

$$\Upsilon_f(\mathbf{g}) \le \beta_{\max} ||g_i||_i^4.$$

Besides, if

$$\beta_{\max} > 0, \tag{8.98}$$

it can be concluded that

$$\Upsilon_f(\mathbf{g}) \leq \beta_{\max}. \tag{8.99}$$

Denote by i_0 an index such that $\beta_{\max}$ is reached (see the discussion of Eq. (8.58)), and $\tilde{g}^\sharp_{i_0}$ a filter in which the upper bound is attained. The inequality (8.99) becomes an equality if and only if the components of $\mathbf{g}$ are all null except that in position i_0, and $\tilde{g}_{i_0} = \tilde{g}^\sharp_{i_0}$. As a consequence, if conditions (8.97) and (8.98) hold, then $\Upsilon_f(\mathbf{g})$ is a contrast function.

We focus on the validity of conditions (8.97) and (8.98) when the sources are linearly modulated communication signals.

If the sources have different cyclic-frequencies. As seen previously, each source (say number k) has at most one strictly positive cyclic-frequency which is $\alpha_k = T_k/T_e$ where T_e is the sampling period. The assumption reads: the rates of the sources are all different. The assumption implies that, for $i \neq j$, any ζ cannot be simultaneously the cyclic-frequency of the source i and the source j. The consequence is that for all the filters $\tilde{g}_i$, $\tilde{g}_j$ we have

$$\sum_{\zeta \in \mathcal{I}_+} R^{(\zeta)}_{\tilde{g}_i \star s_i}(0) R^{(\zeta)}_{\tilde{g}_j \star s_j}(0)^* = 0$$

hence

$$\lambda_{\tilde{g}_i \star s_i, \tilde{g}_j \star s_j} = 0.$$

For sake of clarity, we shall assume that the convex function f is of the type $f(.) = |.|^\theta$ for a certain $\theta \in [1, +\infty[$. The condition (8.97) reads here as:

$$\beta_{\max} \geq 0,$$

which is weaker than (8.98). Of course, this latter is verified when there exists i for which

$$|\beta_{s_i}| > 0.$$

In order to check whether this condition is verified or not, we suggest giving an analytical expression for β_{s_i}. As the analysis is focused on a single source, we shall drop the index i in order to clarify the equations. We recall that $\beta_s = \langle \mathbb{E}\{|s(n)|^4\}\rangle - 2$, which can also be written

$$\beta_s = \langle \kappa_{s(n)} \rangle + 2|R^{(\alpha)}_s(0)|^2.$$

It is straightforward to show that

$$R_s^{(\alpha)}(0) = \sigma^2 \sum_{k\in\mathbb{Z}} \langle |c_a(nT_e - kT)|^2 e^{-\imath 2\pi\alpha n} \rangle_n.$$

The above expression is rather untractable and we work it out. In this respect, we recall that for any $t \in \mathbb{R}$, we have:

$$\mathbb{E}\{|s_a(t)|^2\} = \sum_{k=-1,0,1} R_{s_a}^{(k/T)}(0) e^{\imath 2\pi kt/T}.$$

After substituting nT_e to the time index t and applying the operator $\langle . e^{-\imath 2\pi\alpha n}\rangle$ on both sides of the equation, it yields

$$R_s^{(\alpha)}(0) = R_{s_a}^{(1/T)}(0)$$

at least if $\alpha = T_e/T$ is not a multiple 1/2. Due to (8.83), it can be deduced that the condition $T_e/T \notin \frac{1}{2}\mathbb{N}$ implies that

$$R_s^{(\alpha)}(0) = \sigma^2 \int_{\mathbb{R}} |c_a(tT)|^2 e^{-\imath 2\pi t} dt.$$

In particular, this shows that $R_s^{(\alpha)}(0)$ does not depend either on the sampling period T_e, or on T: it is a function of σ^2 and γ, the excess band-width factor. The same kind of calculus shows that

$$\langle \kappa_s(n) \rangle = \kappa \int_{\mathbb{R}} |c_a(tT)|^4 dt$$

when T_e does not coincide with $T, \frac{T}{2}, \frac{T}{3}, \frac{2T}{3}$ (or a multiple: we mention the reference [17] for further details), where κ is the kurtosis of the symbol sequence in question.

Thanks to these expressions, it is possible to analyse β_s. In particular, it is proved in [23] that

$$\beta_s < 0$$

when $\kappa < 0$ (this condition is always fulfilled for standard modulations). This result is rather technical and is out of the scope of this chapter. The consequence is that $\Upsilon_f(\mathbf{g})$ is a contrast function.

General case. We have for all indices $i \neq j$:

- either the sources i, j have distinct symbol periods; hence

$$\lambda_{\tilde{g}_i \star s_i, \tilde{g}_j \star s_j} = 0$$

- or they have the same symbol-period; hence

$$\lambda_{\tilde{g}_i\star s_i,\tilde{g}_j\star s_j} = 4\Re\left(R^{(\alpha)}_{s_i}(0)R^{(\alpha)}_{s_j}(0)^*\right)$$

  if α denotes the common positive cyclic-frequency.

Let us consider two sources i, j having the same symbol-period and we set $\alpha = T_e/T$. Due to (8.97), we inquire whether

$$4|R^{(\alpha)}_{\tilde{g}_i\star s_i}(0)R^{(\alpha)}_{\tilde{g}_j\star s_j}(0)| < \beta_{\max}.$$

is valid or not. Unfortunately, it is possible to exhibit filters $\tilde{g}_i, \tilde{g}_j$ such that the above inequality is not true. Again, this point is rather technical. It, however, proves that the majorizations (8.96) and (8.99) are probably too loose. And we cannot conclude directly that $\Upsilon_f(\mathbf{g})$ is a contrast function. We should mention the contribution [23]: at least for the choice $f = -\mathrm{Id}$, the function $\Upsilon_f(\mathbf{g})$ is a contrast for various scenarios.

REMARK R8.2
The Matlab codes of the algorithms of convolutive BSS evoked in this chapter can be found at the following address:
http://www-public.it-sudparis.eu/~castella/toolbox/

References

[1] A. Belouchrani, A. Cichocki, Robust whitening procedure in blind separation context, Electronic Letters 36 (2000) 2050–2051.

[2] J.-F. Cardoso, Blind signal separation: statistical principles, Proceedings of the IEEE 9 (1998) 2009–2026. Special issue on blind identification and estimation.

[3] M. Castella, P. Bianchi, A. Chevreuil, J.-C. Pesquet, A blind source separation framework for detecting CPM sources mixed by a convolutive MIMO filter, Signal Processing 86 (2006) 1950–1967.

[4] M. Castella, J.-C. Pesquet, A.P. Petropulu, A family of frequency- and time-domain contrasts for blind separation of convolutive mixtures of temporally dependent signals, IEEE Transactions on Signal Processing 53 (2005) 107–120.

[5] M. Castella, S. Rhioui, E. Moreau, J.-C. Pesquet, Source separation by quadratic contrast functions: a blind approach based on any higher-order statistics, in: Proc. IEEE Int. Conf. on Acoustics, Speech and Signal Processing, ICASSP, Philadelphia, USA, March 2005.

[6] B. Chen, A.P. Petropulu, Frequency domain blind MIMO system identification based on second- and higher order statistics, IEEE Transactions on Signal Processing 49 (2001) 1677–1688.

[7] P. Comon, Analyse en composantes indépendantes et identification aveugle, Traitement du Signal 7 (1990) 435–450.

[8] P. Comon, Contrasts for multichannel blind deconvolution, IEEE Signal Processing Letters 3 (1996) 209–211.

[9] P. Comon, E. Moreau, Blind MIMO equalization and joint-diagonalization criteria, in: Proc. IEEE Int. Conf. on Acoustics, Speech and Signal Processing, ICASSP, Salt Lake City, USA, May 2001.

[10] P. Comon, L. Rota, Blind separation of independent sources from convolutive mixtures, IEICE Trans. on Fundamentals of Electronics, Communications, and Computer Sciences, E86-A (2003), pp. 542–549.
[11] C. Corduneanu, Almost periodic functions, Chelsea, New York, 1989.
[12] A. Dandawatté, G. Giannakis, Statistical tests for presence of cyclostationarity, IEEE Transactions on Signal Processing 42 (1994).
[13] D. Dehay, H. Hurd, Representation and estimation for periodically and almost periodically correlated random processes, in: W.A. Gardner (Ed.), Cyclostationarity in Communications and Signal Processing, IEEE Press, 1993.
[14] K.I. Diamantaras, A.P. Petropulu, B. Chen, Blind two-input-two-output FIR channel identification based on frequency domain second-order statistics, IEEE Transactions on Signal Processing 48 (2000) 534–542.
[15] D. N. Godard, Self-recovering equalization and carrier tracking in two-dimensional data communication systems, IEEE Transactions on Communications 28 (1980) 1867–1875.
[16] A. Gorokhov, P. Loubaton, Subspace based techniques for blind separation of convolutive mixtures with temporally correlated sources, IEEE Transactions on Circuits and Systems I 44 (1997) 813–820.
[17] S. Houcke, A. Chevreuil, P. Loubaton, Blind equalization: case of an unknown symbol period, IEEE Transactions on Signal Processing 51 (2003) 781–793.
[18] Y. Hua, J.K. Tugnait, Blind identifiability of FIR-MIMO systems with colored input using second order statistics, IEEE Signal Processing Letters 7 (2000) 348–350.
[19] Y. Inouye, Criteria for blind deconvolution of multichannel linear time-invariant systems, IEEE Transactions on Signal Processing 46 (1998) 3432–3436.
[20] Y. Inouye, R.-W. Liu, A system-theoretic foundation for blind equalization of an FIR MIMO channel system, IEEE Transactions on Circuits and Systems – I 49 (2002) 425–436.
[21] Y. Inouye, K. Taneae, Super-exponential algorithms for multichannel blind deconvolution, IEEE Transactions on Signal Processing 48 (2000) 881–888.
[22] P. Jallon, A. Chevreuil, P. Loubaton, P. Chevalier, Separation of convolutive mixtures of cyclostationary sources: a contrast function based approach, in: Proc. of ICA'04, September 2004.
[23] P. Jallon, A. Chevreuil, P. Loubaton, P. Chevalier, Separation of convolutive mixtures of linear modulated signals using constant modulus algorithm, in: Proc. of ICASSP, Philadelphia, USA, 2005.
[24] M. Kawamoto, Y. Inouye, Blind deconvolution of MIMO-FIR systems with colored inputs using second-order statistics, IEICE Transactions on Fundamentals of Electronics, Communications, and Computer Sciences E86-A (2003) 597–604.
[25] M. Kawamoto, Y. Inouye, Blind separation of multiple convolved colored signals using second-order statistics, in: Proc. of ICA'03, Nara, Japan, April 2003.
[26] R. Liu, Y. Inouye, Blind equalization of MIMO-FIR chanels driven by white but higher order colored source signals, IEEE Transactions on Information Theory 48 (2002) 1206–1214.
[27] L. Mazet, P. Loubaton, Cyclic correlation based symbol rate estimation, in: Proc. of ASILOMAR, 1999.
[28] J.G. McWhirter, P.D. Baxter, T. Cooper, S. Redif, J. Foster, An EVD algorithem for para-hermitian polynomial matrices, IEEE Transactions on Signal Processing 55 (2007) 2158–2169.
[29] E. Moreau, J.-C. Pesquet, Generalized contrasts for multichanel blind deconvolution of linear systems, IEEE Signal Processing Letters 4 (1997) 182–183.
[30] E. Moreau, J.-C. Pesquet, N. Thirion-Moreau, An equivalence between non symmetrical contrasts and cumulant matching for blind signal separation, in: First Int. Workshop on Independent Component Analysis and Signal Separation, Aussois, France, January 1999.
[31] M. Ohata, K. Matsuoka, Stability analysis of information-theoretic blind separation algorithms in the case where the sources are nonlinear processes, IEEE Transactions on Signal Processing 50 (2002) 69–77.
[32] L. Parra, C. Spence, Convolutive blind separation of non-stationary sources, IEEE Transactions on Speech and Audio Processing 8 (2000) 320–327.

[33] D.T. Pham, Mutual information approach to blind separation of stationary sources, IEEE Transactions on Information Theory 48 (2002) 1935–1946.
[34] D.-T. Pham, C. Servière, H. Boumaraf, Blind separation of convolutive audio mixtures using nonstationarity, in: Proc. of ICA'03, Nara, Japan, 2003, pp. 975–980.
[35] O. Shalvi, E. Weinstein, New criteria for blind deconvolution of nonminimum phase systems (channels), IEEE Transactions on Information Theory 36 (1990) 312–321.
[36] C. Simon, Séparation aveugle des sources en mélange convolutif, PhD thesis, Université de Marne-la-Vallée, November 1999.
[37] C. Simon, P. Loubaton, C. Jutten, Separation of a class of convolutive mixtures: a contrast function approach, Signal Processing, (2001), pp. 883–887.
[38] N. Thirion-Moreau, E. Moreau, Generalized criteria for blind multivariate signal equalization, IEEE Signal Processing Letters 9 (2002) 72–74.
[39] J.K. Tugnait, Blind spatio-temporal equalization and impulse response estimation for MIMO channels using a Godard cost function, IEEE Transactions on Signal Processing 45 (1997) 268–271.
[40] J. K. Tugnait, Identification and deconvolution of multichannel linear non-gaussian processes using higher order statistics and inverse filter criteria, IEEE Transactions on Signal Processing 45 (1997) 658–672.
[41] P.P. Vaidyanathan, Multirate Systems and Filter Banks, Prentice Hall, Englewood Cliffs, 1993.
[42] R. Vinjamuri, D.J. Crammond, D. Kondziolka, H.-N. Lee, Z.-H. Mao, Extraction of sources of tremor in hand movements of patients with movement disorders, IEEE Transactions on Information Technology in Biomedicine 13 (2009) 49–56.
[43] W. Wang, S. Sanei, J.A. Chambers, Penalty function-based joint diagonalization approach for convolutive blind separation of nonstationary sources, IEEE Transactions on Signal Processing 53 (2005) 1654–1669.
[44] H.-C. Wu, J.C. Principe, Simultaneous diagonalization in the frequency domain (SDIF) for source separation, in: Proc. of ICA'99, Aussois, France, January 1999, pp. 245–250.
[45] D. Yellin, E. Weinstein, Criteria for multichannel signal separation, IEEE Transactions on Signal Processing 42 (1994) 2158–2168.
[46] D. Yellin, E. Weinstein, Multichannel signal separation: Methods and analysis, IEEE Transactions on Signal Processing 44 (1996) 106–118.

CHAPTER

Algebraic identification of under-determined mixtures

9

P. Comon and L. De Lathauwer

As in the previous chapters, we consider the linear statistical model below

$$\mathbf{x} = \mathbf{A}\,\mathbf{s} + \mathbf{b} \tag{9.1}$$

where $\mathbf{x}$ denotes the P-dimensional vector of observations, $\mathbf{s}$ the N-dimensional source vector, $\mathbf{A}$ the $P \times N$ mixing matrix and $\mathbf{b}$ an additive noise, which stands for background noise as well as modeling errors. Matrix $\mathbf{A}$ is unknown deterministic, whereas $\mathbf{s}$ and $\mathbf{b}$ are random and also unobserved. All quantities involved are assumed to take their values in the real or complex field. It is assumed that components s_n of vector $\mathbf{s}$ are statistically mutually independent, and that random vectors $\mathbf{b}$ and $\mathbf{s}$ are statistically independent.

The particularity of this chapter is that the number of sources, N, is assumed to be strictly larger than the number of sensors, P. Even if the mixing matrix were known, it would in general be quite difficult to recover the sources. In fact the mixing matrix does not admit a left inverse, because the linear system is *under-determined*, which means that it has more unknowns than equations. The goal is to identify the mixing matrix $\mathbf{A}$ from the sole observation of realizations of vector $\mathbf{x}$. The recovery of sources themselves is not addressed in the present chapter.

Note that other approaches exist that do not assume statistical independence among sources s_n. One can mention non-negativity of sources and mixture (see [51] and Chapter 13), finite alphabet (see [19] and Chapters 6 and 12) with possibly a sparsity assumption on source values (see [41,35] and Chapter 10).

This chapter is organized as follows. General assumptions are stated in section 9.1. Necessary conditions under which the identification problem is well posed are pointed out in section 9.2. Various ways of posing the problem in mathematical terms are described in section 9.3. Then tensor tools are introduced in section 9.4 in order to describe numerical algorithms in section 9.5.

9.1 OBSERVATION MODEL

In (9.1), mainly three cases can be envisaged concerning the noise:

H9.1
First, noise $\mathbf{b}$ may be assumed to be Gaussian. If one admits that noise is made of a large number of independent contributions, invoking the central limit theorem justifies this assumption.

H9.2
Second, it may be assumed to have independent components b_p. Or, more generally, it may be assumed to derive linearly from such a noise, so that it can be written as $\mathbf{b} = \mathbf{A}_2\,\mathbf{v}$, for some unknown matrix $\mathbf{A}_2$ and some random vector $\mathbf{v}$ with independent components.

H9.3
Third, the noise may not satisfy the assumptions above, in which case it is assumed to be of small variance.

Under hypotheses H9.1 or H9.2, (9.1) can be rewritten as a *noiseless* model. In fact, we have the following:

$$\mathbf{x} = [\mathbf{A},\, \mathbf{A}_2] \begin{pmatrix} \mathbf{s} \\ \mathbf{v} \end{pmatrix} \tag{9.2}$$

where the random vector in the right hand side may be viewed as another source vector with statistically independent components. The price to pay is an increase in the number of sources.

On the other hand, models (9.1) or (9.2) will be approximations under hypothesis H9.3. We shall subsequently see that this leads to two different problems: the noiseless case corresponds to an exact fit of statistics such as cumulants, whereas the latter leads to an approximate fit.

In the algorithms developed in this chapter, we shall be mainly concerned by hypothesis H9.3, which is more realistic. However, identifiability results are known under hypothesis H9.2.

9.2 INTRINSIC IDENTIFIABILITY

Before we look at the identification problem, it is useful to examine the identifiability conditions that are inherent in the problem. It may happen that the actual necessary conditions that need to be satisfied in order to identify the mixing matrix are algorithm dependent, and eventually significantly stronger.

Linear mixtures of independent random variables have been studied for years in Statistics [38,44], and the oldest result is probably due to Dugué (1951), Darmois (1953) and Skitovich (1954). However, the latter results concern mainly identifiability and uniqueness, and were not constructive, in the sense that no numerical algorithm could be built from their proofs.

9.2.1 Equivalent representations

Before addressing the general case, it is convenient to have a look at the case where mixing matrix $\mathbf{A}$ has two (or more) collinear columns. Without restricting the generality, assume the Q first columns of $\mathbf{A}$, denoted $\mathbf{a}^{(q)}$, are collinear to the first one, that is:

$$\mathbf{a}^{(q)} = \alpha_q \mathbf{a}^{(1)}, \quad 1 \leqslant q \leqslant Q, \alpha_1 \stackrel{\text{def}}{=} 1.$$

Then equation (9.1) can obviously be rewritten as

$$\mathbf{x} = \left[\mathbf{a}^{(1)}, \mathbf{a}^{(Q+1)}, \ldots, \mathbf{a}^{(N)}\right] \cdot \begin{pmatrix} \sum_q \alpha_q s_q \\ s_{Q+1} \\ \vdots \\ s_N \end{pmatrix} + \mathbf{b}.$$

We end up with a linear statistical model similar to (9.1), but of smaller size, $N - Q + 1$, which satisfies the same independence assumption. It is clear that identifying the α_q is not possible without resorting to additional assumptions. Even if techniques do exist to solve this problem, they are out of the scope of this chapter. With the hypotheses we have assumed, only the direction of vector $\mathbf{a}^{(1)}$ can be estimated. Hence from now on, we shall assume that

H9.4
No columns of matrix $\mathbf{A}$ are collinear.

Now assume $\mathbf{y}$ admits two noiseless representations

$$\mathbf{y} = \mathbf{A}\,\mathbf{s} \quad \text{and} \quad \mathbf{y} = \mathbf{B}\,\mathbf{z}$$

where components of $\mathbf{s}$ (resp. $\mathbf{z}$) have statistically independent components, and $\mathbf{A}$ (resp. $\mathbf{B}$) have pairwise noncollinear columns. Then we introduce the definition below [44]:

DEFINITION 9.1
Two representations $(\mathbf{A}, \mathbf{s})$ *and* $(\mathbf{B}, \mathbf{z})$ *are* equivalent *if every column of* $\mathbf{A}$ *is proportional to some column of* $\mathbf{B}$*, and vice versa.*

If all representations of $\mathbf{y}$ *are equivalent, they are said to be* essentially unique*, that is, they are equal up to permutation and scaling.*

9.2.2 Main theorem

Then, we have the following identifiability theorem [44]:

THEOREM 9.1 (Identifiability)
Let $\mathbf{y}$ *be a random vector of the form* $\mathbf{y} = \mathbf{A}\,\mathbf{s}$*, where* s_p *are independent, and* $\mathbf{A}$ *does not have any collinear columns. Then* $\mathbf{y}$ *can be represented as* $\mathbf{y} = \mathbf{A}_1\,\mathbf{s}_1 + \mathbf{A}_2\,\mathbf{s}_2$*, where* $\mathbf{s}_1$ *is non-Gaussian,* $\mathbf{s}_2$ *is Gaussian independent of* $\mathbf{s}_1$*, and* $\mathbf{A}_1$ *is essentially unique.*

This theorem is quite difficult to prove, and we refer the readers to [44] for further readings. However, we shall give a proof in the case of dimension $P = 2$ in section 9.2.4, page 330.

REMARK R9.1
If $\mathbf{s}_2$ is 1-dimensional, then $\mathbf{A}_2$ is also essentially unique, because it has a single column.

REMARK R9.2
Note that Theorem 9.1 does not tell anything about the uniqueness of the source vector itself. It turns out that if, in addition, the columns of $\mathbf{A}_1$ are linearly independent, then the distribution of $\mathbf{s}_1$ is unique up to scale and location indeterminacies [44]. But in our framework, the number of sources exceeds the number of sensors so that this condition cannot be fulfilled. We just give two examples below in order to make this issue clearer.

EXAMPLE E9.1 (Uniqueness)
Let s_i be independent with no Gaussian component, and b_i be independent Gaussian. Then the linear model below is identifiable, but $\mathbf{A}_2$ is not essentially unique whereas $\mathbf{A}_1$ is:

$$\begin{pmatrix} s_1 + s_2 + 2\,b_1 \\ s_1 + 2\,b_2 \end{pmatrix} = \mathbf{A}_1\,\mathbf{s} + \mathbf{A}_2 \begin{pmatrix} b_1 \\ b_2 \end{pmatrix} = \mathbf{A}_1\,\mathbf{s} + \mathbf{A}_3 \begin{pmatrix} b_1 + b_2 \\ b_1 - b_2 \end{pmatrix}$$

with

$$\mathbf{A}_1 = \begin{pmatrix} 1 & 1 \\ 1 & 0 \end{pmatrix}, \quad \mathbf{A}_2 = \begin{pmatrix} 2 & 0 \\ 0 & 2 \end{pmatrix} \quad \text{and} \quad \mathbf{A}_3 = \begin{pmatrix} 1 & 1 \\ 1 & -1 \end{pmatrix}.$$

Hence the distribution of $\mathbf{s}$ is essentially unique. But $(\mathbf{A}_1, \mathbf{A}_2)$ is not equivalent to $(\mathbf{A}_1, \mathbf{A}_3)$.

EXAMPLE E9.2 (Non uniqueness)
Let s_i be independent with no Gaussian component, and b_i be independent Gaussian. Then the linear model below is identifiable, but the distribution of $\mathbf{s}$ is not unique [36]:

$$\begin{pmatrix} s_1 + s_3 + s_4 + 2\,b_1 \\ s_2 + s_3 - s_4 + 2\,b_2 \end{pmatrix} = \mathbf{A} \begin{pmatrix} s_1 \\ s_2 \\ s_3 + b_1 + b_2 \\ s_4 + b_1 - b_2 \end{pmatrix} = \mathbf{A} \begin{pmatrix} s_1 + 2\,b_1 \\ s_2 + 2\,b_2 \\ s_3 \\ s_4 \end{pmatrix}$$

with

$$\mathbf{A} = \begin{pmatrix} 1 & 0 & 1 & 1 \\ 0 & 1 & 1 & -1 \end{pmatrix}.$$

Further details may be found in [36,8] and [44, ch.10]. In particular, it is pointed out in [36] that the source distributions can be obtained in a unique fashion, even when sources cannot be extracted, provided various sufficient conditions are satisfied.

9.2.3 Core equation

The first characteristic function of a real random variable is defined as the conjugated Fourier transform of its probability distribution, that is, for a real random variable x with distribution dF_x, it takes the form: $\Phi_x(t) = \int_u e^{\jmath tu}\, dF_x(u)$, which is nothing else but $\mathbb{E}e^{\jmath tx}$. If the random variable is multi-dimensional, the Fourier transform is taken on all variables, leading to:

$$\Phi_{\mathbf{x}}(\mathbf{t}) \stackrel{\text{def}}{=} \mathbb{E}e^{\jmath \mathbf{t}^{\mathrm{T}}\mathbf{x}} = \int_{\mathbf{u}} e^{\jmath \mathbf{t}^{\mathrm{T}}\mathbf{u}}\, dF_{\mathbf{x}}(\mathbf{u}). \tag{9.3}$$

The second characteristic function is defined as the logarithm of the first one: $\Psi_x(\mathbf{t}) \stackrel{\text{def}}{=} \log \Phi_x(\mathbf{t})$. It always exists in the neighborhood of the origin, and is hence uniquely defined as long as $\Phi_x(\mathbf{t}) \neq 0$. An important property of the second characteristic function is given by the *Marcinkiewicz* theorem that we recall below without proof:

THEOREM 9.2 (Marcinkiewicz, 1938)
If a second characteristic function $\Psi_x(t)$ is a polynomial, then its degree is at most 2 and x is Gaussian.

This theorem will be used in the proof of Theorem 9.5, together with the following:

THEOREM 9.3 (Cramér, 1939)
If a finite sum of independent random variables is Gaussian, then each of them is Gaussian.

Another basic property will be useful for our further developments. If x and y are two statistically independent random variables, then the joint second c.f. splits into the sum of the marginals:

$$\Psi_{x,y}(u,v) = \Psi_x(u) + \Psi_y(v). \tag{9.4}$$

This property can be easily proved by direct use of the definition $\Psi_{x,y}(u,v) = \log[\mathbb{E}\exp \jmath (ux+vy)]$, which yields $\Psi_{x,y}(u,v) = \log[\mathbb{E}\exp(\jmath\, ux)\mathbb{E}\exp(\jmath\, vy)]$ as soon as x and y are independent.

Denote by $\Psi_x(\mathbf{u})$ the joint second characteristic function of the observed random vector $\mathbf{x}$, and $\psi_p(v_p)$ the marginal second characteristic function of source s_p. Then, the following core equation is established:

PROPOSITION 9.4
If s_p are mutually statistically independent random variables, and if $\mathbf{x} = \mathbf{A}\,\mathbf{s}$, then we have the core equation:

$$\Psi_x(\mathbf{u}) = \sum_p \Psi_{s_p}\left(\sum_q u_q A_{qp}\right) \tag{9.5}$$

for any $\mathbf{u}$ in a neighborhood of the origin.

Proof: First notice that, by definition (9.3) of the characteristic function, we have $\Psi_x(\mathbf{u}) = \Psi_{As}(\mathbf{u}) = \Psi_s(\mathbf{A}^T\mathbf{u})$. Then the remainder of the proof is an immediate consequence of property (9.4). To see this, just notice that because s_p are independent, $\Psi_s(\mathbf{v}) = \sum_p \Psi_{s_p}(v_p)$. Replacing v_p by its value, i.e. the p-th row of $\mathbf{A}^T\mathbf{u}$, yields (9.5). □

In (9.5), we see that identifying the mixing matrix $\mathbf{A}$ amounts to decomposing the multivariate joint characteristic function of $\mathbf{x}$ into a sum of univariate functions, or in other words, to finding the linear combinations entering latter univariate functions. Various approaches to this problem will be surveyed in section 9.3.

9.2.4 Identifiability in the 2-dimensional case

We are now in a position to state the proof of the identifiability theorem 9.1 in the case where $P = 2$.

THEOREM 9.5 (Darmois-Skitovic)
Let s_i be statistically independent *random variables, and two linear statistics:*

$$y_1 = \sum_i a_i\, s_i \quad and \quad y_2 = \sum_i b_i\, s_i.$$

If y_1 and y_2 are statistically independent, then random variables s_k for which $a_k\, b_k \neq 0$ are Gaussian.

Proof: It does not restrict the generality to assume that column vectors $[a_k, b_k]^T$ are not collinear, as pointed out earlier in this section. This is equivalent to saying that (1) one can group variables that are mixed with collinear columns, and (2) one makes a change of variable by summing the corresponding sources together. Note that if two variables s_p and s_q are grouped together, into a new Gaussian variable $s_p + s_q$, then from Cramér's theorem (page 329), both s_p and s_q are Gaussian.

To simplify the proof and make it more readable, we also assume that ψ_p are differentiable, which is not a necessary assumption. Assume the notations below for the characteristic functions involved:

$$\begin{aligned}
\Psi_{1,2}(u, v) &= \log \mathbb{E} \exp(\jmath\, y_1\, u + \jmath\, y_2\, v) \\
\Psi_k(w) &= \log \mathbb{E} \exp(\jmath\, y_k\, w), \quad k \in \{1, 1\} \\
\psi_p(w) &= \log \mathbb{E} \exp(\jmath\, s_p\, w), \quad p \in \{1, \ldots, N\}.
\end{aligned}$$

From (9.4), the statistical independence between s_p's implies:

$$\begin{aligned}\Psi_{1,2}(u,v) &= \sum_{k=1}^{N} \psi_k(u\,a_k + v\,b_k)\\ \Psi_1(u) &= \sum_{k=1}^{N} \psi_k(u\,a_k)\\ \Psi_2(v) &= \sum_{k=1}^{N} \psi_k(v\,b_k)\end{aligned}$$

which are in fact core equations similar to (9.5), whereas statistical independence between y_1 and y_2 implies

$$\Psi_{1,2}(u,v) = \Psi_1(u) + \Psi_2(v).$$

Hence we have $\sum_{k=1}^{N} \psi_p(u\,a_k + v\,b_k) = \sum_{k=1}^{N} \psi_k(u\,a_k) + \psi_k(v\,b_k)$. The equality is trivial for terms for which $a_k\,b_k = 0$. So from now on, one can restrict the sum to terms corresponding to $a_k\,b_k \neq 0$. Since the equations hold true for any (u,v), write this equation at $u + \alpha/a_N$ and $v - \alpha/b_N$ for an arbitrary α:

$$\sum_{k=1}^{N} \psi_k\left(u\,a_k + v\,b_k + \alpha\left(\frac{a_k}{a_N} - \frac{b_k}{b_N}\right)\right) = f(u) + g(v).$$

Now perform subtraction so as to cancel the N-th term, divide the result by α, and let $\alpha \to 0$; we obtain:

$$\sum_{k=1}^{N-1}\left(\frac{a_k}{a_N} - \frac{b_k}{b_N}\right)\psi_k^{(1)}(u\,a_k + v\,b_k) = f^{(1)}(u) + g^{(1)}(v)$$

for some *univariate functions* $f^{(1)}(u)$ and $g^{(1)}(u)$.

Hence, we now have a similar expression, but with one term less than before in the sum. The idea is then to repeat the procedure $(N-1)$ times in order to eventually get:

$$\prod_{j=2}^{N}\left(\frac{a_1}{a_j} - \frac{b_1}{b_j}\right)\psi_1^{(N-1)}(u\,a_1 + v\,b_1) = f^{(N-1)}(u) + g^{(N-1)}(v).$$

As a consequence, $\psi_1^{(N-1)}(u\,a_1 + v\,b_1)$ is linear, as a sum of two univariate functions (in fact $\psi_1^{(N)}$ is a constant because $a_1 b_1 \neq 0$). By succesive integrations, we eventually see that ψ_1 is a polynomial. Lastly invoke Theorem 9.2 to conclude that s_1 is Gaussian.

Now, the reasoning we have made for s_1 holds valid for any s_p. By repeating the proof for any ψ_p such that $a_p b_p \neq 0$, we would also prove that s_p is Gaussian. This concludes the proof. □

REMARK R9.3
The proof found in the literature is given when ψ_p are not all differentiable, but is more complicated [44,38]. The proof can also be extended to infinitely many variables.

REMARK R9.4
The proof was derived above for real variables, but it also holds true in the complex field. Some other interesting remarks concerning specificities of the complex field may be found in [37]. Some issues related to complex variables will be stated in the following sections.

9.3 PROBLEM FORMULATION

In Eq. (9.5), $\Psi_x(\mathbf{u})$ is written as a sum of contributions of the individual sources, i.e. the sources have been separated. This remains true when we apply a linear transformation to both sides of the equation. One could for instance compute derivatives at the origin. This leads to cumulant-based methods; see section 9.3.2. Specific algorithms are discussed in sections 9.5.2.2, 9.5.4 and 9.5.5. One could also compute derivatives in different points than the origin; see section 9.3.1. Specific algorithms are discussed in sections 9.5.2.3 and 9.5.6.

9.3.1 Approach based on derivatives of the joint characteristic function

The goal is hence to produce simple equations from the core equation (9.5), whose coefficients can be estimated from realizations of observation $\mathbf{x}$.

The idea discussed in this section has its roots in the proof of the theorem of Darmois-Skitovic [44,38]. It was first proposed in [72] and further developed in [25]. A variant has been proposed in [77]; see also section 9.5.3.

Consider the core equation (9.5) for values $\mathbf{u}$ belonging to some finite set $\mathcal{G}$ of cardinality L. Assume source characteristic functions ψ_p all admit finite derivatives up to order r in a neighborhood of the origin containing $\mathcal{G}$. Then, taking $r = 3$ as a working example [25]:

$$\frac{\partial^3 \Psi_x}{\partial u_i \partial u_j \partial u_k}(\mathbf{u}) = \sum_{p=1}^{N} A_{ip} A_{jp} A_{kp}\, \psi_p^{(3)} \left(\sum_{q=1}^{P} u_q A_{qp} \right). \tag{9.6}$$

Now denote $\mathbf{u}(\ell)$ the L points of the grid $\mathcal{G}$, $1 \leqslant \ell \leqslant L$, and define $B_{\ell p} \stackrel{\text{def}}{=} \psi_p^{(3)}(\sum_{q=1}^{P} u_q(\ell) A_{qp})$. The array of 3rd order derivatives may be estimated from the observations of $\mathbf{x}$. Hence the problem has now been translated into the following: given an array $T_{ijk\ell}$, find two matrices $\mathbf{A}$ and $\mathbf{B}$ such that

$$T_{ijk\ell} \approx \sum_{p} A_{ip} A_{jp} A_{kp} B_{\ell p}. \tag{9.7}$$

In sections 9.5.2.3 and 9.5.6, algorithms will be proposed to perform this matching, ignoring the dependence of $\mathbf{B}$ on $\mathbf{A}$.

Up to now, random variables have been assumed to take their values in the real field. However, in a number of applications, e.g. telecommunications, identification problems are posed in the complex field. For complex random variables, one can work with proper generalizations of the characteristic functions.

9.3.2 Approach based on cumulants

In this section, we shall show how the blind identification problem can be seen, in a first approximation, as a cumulant matching problem, which will allow us to solve it in different ways, in sections 9.5.2.2, 9.5.4 and 9.5.5.

9.3.2.1 Definitions

If Φ can be expanded in Taylor series about the origin, then its coefficients are related to *moments*:

$$\mu'^{x}_{(r)} \stackrel{\text{def}}{=} \mathbb{E}X^r = (-\jmath)^r \left. \frac{\partial^r \Phi(t)}{\partial t^r} \right|_{t=0}. \tag{9.8}$$

It is usual to introduce *central moments* $\mu^{x}_{(r)}$ as the moments of the centered variable $x - \mu'^{x}_{(1)}$.

Similarly, if Ψ may be expanded in Taylor series about the origin, then its coefficients are the *cumulants*:

$$\kappa^{x}_{(r)} \stackrel{\text{def}}{=} \text{cum}\{\underbrace{X, X, \ldots, X}_{r\ times}\} = (-\jmath)^r \left. \frac{\partial^r \Psi(t)}{\partial t^r} \right|_{t=0}. \tag{9.9}$$

The relation between moments and cumulants can be obtained by expanding the logarithm and grouping terms of the same order together.

EXAMPLE E9.3 (Cumulants of order 2, 3 and 4)
The cumulant of 2nd order, $\kappa_{(2)}$, is nothing else but the variance: $\mu'_{(2)} - {\mu'_{(1)}}^2 = \kappa_{(2)}$. And for zero-mean random variables, cumulants of order 3 and 4 are related to moments by: $\kappa_{(3)} = \mu_{(3)}$ and $\kappa_{(4)} = \mu_{(4)} - 3\,{\mu_{(2)}}^2$.

EXAMPLE E9.4 (Skewness and Kurtosis)
The *skewness* is a 3rd order normalized cumulant: $\mathcal{K}_{(3)} \stackrel{\text{def}}{=} \kappa_{(3)}/{\kappa_{(2)}}^{3/2}$. The *kurtosis* is a normalized 4th order cumulant $\mathcal{K}_{(4)} \stackrel{\text{def}}{=} \kappa_{(4)}/{\kappa_{(2)}}^2$.

Skewness and kurtosis are null for any Gaussian random variable. These quantities can serve as measures of deviation from Gaussianity. In fact, random variables having a negative (resp. positive) kurtosis can be called *platykurtic* (resp. *leptokurtic*) [45]. Conversely, random variables having zero kurtosis (referred to as *mesokurtic*) are not necessarily Gaussian.

For multivariate random variables, denote the cumulants

$$\kappa_{ij..\ell} = \text{cum}\{X_i, X_j, \ldots X_\ell\}.$$

As explained above, expressions of moments as a function of cumulants can be obtained by expanding the logarithm in the definition of the second characteristic function and grouping terms of same order together. This yields for instance:

$$\begin{aligned} \mu'_i &= \kappa_i, \\ \mu'_{ij} &= \kappa_{ij} + \kappa_i \kappa_j, \\ \mu'_{ijk} &= \kappa_{ijk} + [3]\,\kappa_i \kappa_{jk} + \kappa_i \kappa_j \kappa_k. \end{aligned} \tag{9.10}$$

In the relation above, we have used McCullagh's *bracket notation* [55] defined below.

Bracket notation. A sum of k terms that can be deduced from each other by permutation of indices is denoted by the number k between brackets followed by a single monomial describing the generic term. This is McCullagh's *bracket notation* [55].

EXAMPLE E9.5
Simple examples will serve better than a long explanation.

$$[3]\delta_{ij}\,\delta_{kl} = \delta_{ij}\,\delta_{kl} + \delta_{ik}\,\delta_{jl} + \delta_{il}\,\delta_{jk}, \tag{9.11}$$

$$[3]a_{ij}\,b_k\,c_{ijk} = a_{ij}\,b_k\,c_{ijk} + a_{ik}\,b_j\,c_{ijk} + a_{jk}\,b_i\,c_{ijk}. \tag{9.12}$$

The presence of the bracket yields an implicit summation; all terms with r indices are completely symmetric order-r tensors. The number of distinct monomials that may be obtained by permutation is equal to the integer appearing between brackets. As additional examples, the following expressions are consistent:

$$[3]\,a_i\,\delta_{jk},\quad [6]\,a_i\,a_j\,\delta_{kl},\quad [10]\,b_i\,b_j\,b_k\,\delta_{lm},\quad [35]\,A_{ijk}\,B_{abcd}\,C_{ijkabcd}.$$

9.3.2.2 Relations between moments and cumulants

Relations (9.10) can be inverted in order to obtain cumulants as a function of moments [45,55]. In the case of non-central random variables, *multivariate cumulants* of order 3 and 4 can be given in a compact form as a function of *multivariate moments* as:

$$\kappa_{ij} = \mu'_{ij} - \mu'_i\,\mu'_j, \tag{9.13}$$

$$\kappa_{ijk} = \mu'_{ijk} - [3]\,\mu'_i\,\mu'_{jk} + 2\mu'_i\,\mu'_j\,\mu'_k, \tag{9.14}$$

$$\kappa_{ijkl} = \mu'_{ijkl} - [4]\,\mu'_i\,\mu'_{jkl} - [3]\mu'_{ij}\,\mu'_{kl} + 2\,[6]\,\mu'_i\,\mu'_j\,\mu'_{kl} - 6\mu'_i\,\mu'_j\,\mu'_k\,\mu'_l. \tag{9.15}$$

On the other hand, if variables are all zero-mean, then the expressions simplify to:

$$\kappa_{ij} = \mu_{ij}, \tag{9.16}$$

$$\kappa_{ijk} = \mu_{ijk}, \tag{9.17}$$

$$\kappa_{ijkl} = \mu_{ijkl} - [3]\,\mu_{ij}\,\mu_{kl}. \tag{9.18}$$

At order 5 and 6 we have:

$$\kappa_{ijklm} = \mu_{ijklm} - [10]\,\mu_{ij}\,\mu_{klm},$$

$$\kappa_{ijklmn} = \mu_{ijklmn} - [15]\,\mu_{ij}\,\mu_{klmn} - [10]\,\mu_{ijk}\,\mu_{lmn} + 2\,[15]\,\mu_{ij}\,\mu_{kl}\,\mu_{mn}.$$

9.3.2.3 Properties of cumulants

Cumulants enjoy several useful properties [56]. Some of them are shared by moments, but others are not. See also Chapter 5.

First of all, moments and cumulants enjoy the so-called *multi-linearity property*:

PROPERTY 9.6
If two random variables are related by a linear transform, $\mathbf{y} = \mathbf{A}\,\mathbf{x}$, *then their cumulants are related multi-linearly:*

$$\mathrm{cum}\{y_i, y_j, \ldots, y_k\} = \sum_{p,q,\ldots,r} A_{ip}\,A_{jq}\ldots A_{kr}\,\mathrm{cum}\{x_p, x_q, \ldots, x_r\}. \tag{9.19}$$

A similar property holds for moments.

This property actually enables us to call moments and cumulants *tensors* [55]. As a corollary, we have the following:

$$\mathrm{cum}\{\alpha x_1, x_2, .., x_n\} = \alpha\,\mathrm{cum}\{x_1, x_2, .., x_n\}, \tag{9.20}$$

$$\mathrm{cum}\{x_1 + y_1, x_2, .., x_n\} = \mathrm{cum}\{x_1, x_2, .., x_n\} + \mathrm{cum}\{y_2, x_2, .., x_n\}.$$

Another obvious property directly results from the definition, namely that of invariance by permutation of indices:

$$\text{cum}\{X_1, X_2, ..X_r\} = \text{cum}\{X_{\sigma(1)}, X_{\sigma(2)}, ..X_{\sigma(r)}\}.$$

In other words, r-th order cumulants (and moments) are fully *symmetric* r-th order tensors.

Let's now turn to properties that are specific to cumulants. First, they are invariant with respect to translation; this means that $\forall r > 1$ and $\forall h$ is a constant:

$$\text{cum}\{X_1 + h, X_2, .., X_r\} = \text{cum}\{X_1, X_2, .., X_r\}.$$

This property is sometimes referred to as the *shift invariance of cumulants*. Next, cumulants of a set of random variables are null as soon as this set can be split into two statistically independent subsets:

$$\{X_1, \ldots, X_p\} \text{ independent of } \{Y_1, \ldots, Y_q\} \Rightarrow \text{cum}\{X_1, \ldots, X_p, Y_1, \ldots, Y_q\} = 0.$$

A consequence of this property is the *additivity of cumulants.*

PROPERTY 9.7
Let **X** *and* **Y** *be (possibly multivariate) statistically independent random variables. Then*

$$\text{cum}\{X_1 + Y_1, X_2 + Y_2, .., X_r + Y_r\} = \text{cum}\{X_1, X_2, .., X_r\} + \text{cum}\{Y_1, Y_2, .., Y_r\}. \quad (9.21)$$

9.3.2.4 Cumulants of complex random variables

Up to now, random variables have been assumed to take their values in the real field. Moments and cumulants of complex random variables can be defined starting from proper generalizations of the characteristic functions. Contrary to real variables, there is not a unique way of defining a cumulant (or a moment) of order r of a complex random variable; in fact, it depends on the number of conjugated terms. It is thus necessary to be able to distinguish between complex random variables that are conjugated and those that are not. For this purpose, one introduces a specific notation, with superscripts:

$$\text{cum}\{X_i, ..X_j, X_k^*, ..X_\ell^*\} \stackrel{\text{def}}{=} \kappa_{i..j}^{x\,k..\ell}. \quad (9.22)$$

EXAMPLE E9.6
The covariance matrix of a complex random variable is

$$\mathbb{E}z_i\, z_j^* - \mathbb{E}z_i\, \mathbb{E}z_j{}^* = \kappa_i^{z\,j}.$$

Among all the possible definitions, only one is called *circular cumulant*, namely the one having exactly half of its arguments conjugated. All other cumulants may be called *non-circular* cumulants. Note that there exist circular cumulants only at even orders. For instance, the cumulant below is circular

$$\kappa_{ij}^{zk\ell} = \text{cum}\{z_i, z_j, z_k^*, z_\ell^*\}.$$

Exact expressions of cumulants of multivariate complex random variables are given in the Appendix.

9.3.2.5 Blind identification via cumulant matching

Combining model (9.1) with the multi-linearity (9.19) and additivity (9.21) properties of cumulants leads to a system of equations, which we shall solve in section 9.5. In order to fix the ideas, consider first cumulants of order 4 of the observed random vector **x**:

$$\kappa_{i\ell}^{\mathbf{x}jk} = \sum_{n=1}^{N} A_{in} A_{jn}^* A_{kn}^* A_{\ell n}\, \kappa_{nn}^{\mathbf{s}nn}, \tag{9.23}$$

$$\kappa_{ijk}^{\mathbf{x}\ell} = \sum_{n=1}^{N} A_{in} A_{jn} A_{kn} A_{\ell n}^*\, \kappa_{nnn}^{\mathbf{s}n}, \tag{9.24}$$

$$\kappa_{ijk\ell}^{\mathbf{x}} = \sum_{n=1}^{N} A_{in} A_{jn} A_{kn} A_{\ell n}\, \kappa_{nnnn}^{\mathbf{s}}. \tag{9.25}$$

In the equations above, the notation introduced in (9.22) has been used.

In general, the use of the circular cumulant $\kappa_{ij}^{\mathbf{x}k\ell}$ is preferred. But cumulant $\kappa_{ijk\ell}^{\mathbf{x}}$ has also been successfully used in the identification of communication channels [18,30,31, 40]. On the other hand, cumulant $\kappa_{ijl}^{\mathbf{x}\ell}$ is rarely used because it is generally close to zero. The choice of the cumulants to use depends on the prior information we have on the source cumulants.

In section 9.5.5, the circular cumulant of order 6,

$$\kappa_{ijn}^{\mathbf{x}k\ell m} \stackrel{\text{def}}{=} \text{cum}\{x_i, x_j, x_k^*, x_\ell^*, x_m^*, x_n\},$$

will be also used. Because again of properties (9.19) and (9.21), the equation to solve for **A** is then the following, for all 6-uplet of indices:

$$\kappa_{ijp}^{\mathbf{x}k\ell m} = \sum_{n=1}^{N} A_{in} A_{jn} A_{kn}^* A_{\ell n}^* A_{mn}^* A_{pn}\, \kappa_{nnn}^{\mathbf{s}nnn}. \tag{9.26}$$

9.4 HIGHER-ORDER TENSORS

For our purposes, a tensor may be assimilated to its array of coordinates. Justifications may be found in [22, sec. 2.1]; see also [26, sec. 2.1]. Hence we shall not make the distinction in the remainder. A tensor of order r defined on the tensor product of r

vector spaces of dimension N_i, $1 \leqslant i \leqslant r$, will be represented by an array of numbers, of dimensions $N_1 \times N_2 \times \cdots \times N_r$.

9.4.1 Canonical tensor decomposition

9.4.1.1 Definition

DEFINITION 9.2
If an r-th order tensor $[\![T_{ij\ldots\ell}]\!]$ can be written as an outer product $T_{ij\ldots\ell} = u_i\, v_j \ldots w_\ell$, then it is called a rank-1 tensor.

In a compact form, one shall write rank-1 tensors as $\mathbf{T} = \mathbf{u} \otimes \mathbf{v} \otimes \ldots \otimes \mathbf{w}$, where $\otimes$ denotes the outer (tensor) product.

DEFINITION 9.3 (Tensor rank)
A tensor $[\![T_{ij\ldots\ell}]\!]$ can always be decomposed into a sum of rank-1 tensors as

$$\mathbf{T} = \sum_{n=1}^{r} \mathbf{u}_n \otimes \mathbf{v}_n \otimes \ldots \otimes \mathbf{w}_n. \tag{9.27}$$

The minimal value of r for which the equality holds is called the rank of tensor $\mathbf{T}$.

For order-2 tensors, which are actually matrices, this definition coincides with the usual definition of matrix rank.

When r equals rank$\{\mathbf{T}\}$, the decomposition given in (9.27) is often referred to as the *Canonical Decomposition (CanDecomp)* (CAND) of tensor $\mathbf{T}$ [13]. Other names appear in the literature, such as PARAFAC in psychometrics and chemometrics [42,47,50,67], or POLYADIC FORM [43] in mathematical physics. In linear algebra, the acronym CP is now often used, and stands for CANDECOMP-PARAFAC; see [26] and references therein.

An alternative representation of (9.27) is obtained by imposing each vector to be of unit modulus, and by inserting a scale factor λ_p in the decomposition, which yields:

$$\mathbf{T} = \sum_{n=1}^{r} \lambda_n\, \mathbf{u}_n \otimes \mathbf{v}_n \otimes \ldots \otimes \mathbf{w}_n. \tag{9.28}$$

9.4.1.2 Symmetry

A tensor is *symmetric* if its entries do not change when permuting the indices. The terminology of "supersymmetry" should be avoided [21]. When decomposing a symmetric tensor, it may be relevant to impose all vectors in each outer product of (9.28) to be the same. This leads to a CAND of the form

$$\mathbf{T} = \sum_{n=1}^{r} \lambda_n\, \mathbf{u}_n \otimes \mathbf{u}_n \otimes \ldots \otimes \mathbf{u}_n. \tag{9.29}$$

Up to now, it has not yet been proved that the rank defined above under the symmetry constraint is the same as that defined in (9.28). Hence, the rank defined above is called the *symmetric rank* of **T** [21]. In this chapter, by the rank of a structured tensor it will be always meant the structured rank of that tensor. This definition applies to both real or complex *symmetric tensors*.

In the complex field, Hermitian symmetry is also quite important, and often more useful than plain symmetry. Similar definitions can be introduced for various symmetry properties. For instance, for the 4th order cumulant tensor defined in (9.23), the rank is the minimal integer r such that the following CAND holds:

$$\mathbf{T} = \sum_{n=1}^{r} \lambda_n \, \mathbf{u}_n \otimes \mathbf{u}_n^* \otimes \mathbf{u}_n^* \otimes \mathbf{u}_n \tag{9.30}$$

and similarly for the 6th order cumulant tensor (9.26),

$$\mathbf{T} = \sum_{n=1}^{r} \lambda_n \, \mathbf{u}_n \otimes \mathbf{u}_n \otimes \mathbf{u}_n^* \otimes \mathbf{u}_n^* \otimes \mathbf{u}_n^* \otimes \mathbf{u}_n. \tag{9.31}$$

In the CAND above, vectors $\mathbf{u}_n$ are wished to be equal to some normalized column of matrix **A** up to a scaling factor.

9.4.1.3 Link with homogeneous polynomials

The linear space of symmetric tensors of order d and dimension P can be bijectively mapped to the space of homogeneous polynomials of degree d in P variables. This property will be useful in section 9.5.2.1. As pointed out in [17,55], there exist two different notations in the literature; we recall them below and relate them.

Let **x** be an array of unknowns of size P, and **j** a multi-index of the same size. One can assume the notation $\mathbf{x}^{\mathbf{j}} \stackrel{\text{def}}{=} \prod_{k=1}^{P} x_k^{j_k}$ and $|\mathbf{j}| \stackrel{\text{def}}{=} \sum_k j_k$. Then for homogeneous monomials of degree d, $\mathbf{x}^{\mathbf{j}}$, we have $|\mathbf{j}| = d$.

EXAMPLE E9.7
Take the example of cubics in 4 variables to fix the ideas. One can associate every entry T_{ijk} of a 3rd order symmetric tensor with a monomial $T_{ijk}x_i x_j x_k$. For instance, T_{114} is associated with $T_{114}x_1^2 x_4$, and thus to $T_{114}\mathbf{x}^{[2,0,0,1]}$; this means that we have a map $f([1,1,4]) = [2,0,0,1]$.

More generally, the d-dimensional vector index $\mathbf{i} \in \{1,\ldots,P\}^d$ can be associated with a P-dimensional vector index $\mathbf{f}(\mathbf{i})$ containing the number of times each variable x_k appears in the associated monomial [17]. Whereas the d entries of **i** take their values in $\{1,\ldots,P\}$, the P entries of $\mathbf{f}(\mathbf{i})$ take their values in $\{1,\ldots,d\}$ with the constraint that they sum up to d: $\sum_k f_k(\mathbf{i}) = d, \forall \mathbf{i}$.

As a consequence, in order to define the bijection, it suffices to associate every polynomial $p(\mathbf{x})$ with the symmetric tensor $\mathbf{T}$ as:

$$p(\mathbf{x}) = \sum_{|\mathrm{f}(\mathrm{i})|=d} T_{\mathrm{i}} \mathbf{x}^{\mathrm{f}(\mathrm{i})} \tag{9.32}$$

where T_{i} are the entries of tensor $\mathbf{T}$. The dimension of these spaces is $\binom{P+d-1}{d}$, and one can choose as a basis the set of monomials: $\mathscr{B}(P;d) = \{\mathbf{x}^{\mathrm{j}}, |\mathrm{j}| = d\}$.

EXAMPLE E9.8

Let p and q be two homogeneous polynomials in P variables, associated with tensors $\mathbf{P}$ and $\mathbf{Q}$, possibly of different orders. Then, polynomial pq is associated with $\mathbf{P}\otimes\mathbf{Q}$. In fact we have [17]:

$$p(\mathbf{x})q(\mathbf{x}) = \sum_{\mathrm{i}}\sum_{\mathrm{j}} P_{\mathrm{i}} Q_{\mathrm{j}} \mathbf{x}^{\mathrm{f}(\mathrm{i})+\mathrm{f}(\mathrm{j})} = \sum_{[\mathrm{ij}]} [\mathbf{P}\otimes\mathbf{Q}]_{[\mathrm{ij}]}\, \mathbf{x}^{\mathrm{f}([\mathrm{ij}])}.$$

9.4.2 Essential uniqueness

Let us, for the sake of convenience, consider the third-order version of (9.27):

$$\mathbf{T} = \sum_{n=1}^{r} \mathbf{u}_n \otimes \mathbf{v}_n \otimes \mathbf{w}_n. \tag{9.33}$$

Obviously, the order in which terms enter the sum is not relevant. The consequence is that the triplet of matrices $(\mathbf{U},\mathbf{V},\mathbf{W})$ is defined up to a common permutation of columns. Next, if $(\mathbf{U},\mathbf{V},\mathbf{W})$ corresponds to a CanD, then so does $(\mathbf{U}\boldsymbol{\Delta}_U, \mathbf{V}\boldsymbol{\Delta}_V, \mathbf{W}\boldsymbol{\Delta}_U^{-1}\boldsymbol{\Delta}_V^{-1})$, for any pair of diagonal invertible matrices $(\boldsymbol{\Delta}_U, \boldsymbol{\Delta}_V)$.

We see that, in the absence of additional assumptions, the best we can do is to calculate one representative of this equivalence class of CanD solutions. In the literature, uniqueness up to scale and permutation is sometimes referred to as *essential uniqueness*. So our first goal will be to identify the conditions under which essential uniqueness is met.

9.4.2.1 Necessary conditions for uniqueness

If tensor $\mathbf{T}$ is of size $I \times J \times K$, and has rank r, then the number of degrees of freedom on the left hand side of (9.33) is IJK, whereas it is equal to $(I+J+K-2)r$ on the right hand side. So the CanD will not be essentially unique if the number of unknowns is larger than the number of equations, that is, if $IJK \geqslant (I+J+K-2)r$. More generally, we have the necessary condition:

PROPOSITION 9.8
A d-th order tensor **T** *of dimensions* $P_1 \times P_2 \times \cdots P_d$ *may have an essentially unique CanD only if:*

$$\text{rank}\{\mathbf{T}\} \leqslant \frac{\prod_{k=1}^{d} P_k}{1-d+\sum_{k=1}^{d} P_k} \stackrel{\text{def}}{=} \rho. \tag{9.34}$$

In other words, the rank of **T** should not exceed an upper bound ρ. The closest integer larger than ρ, $\lceil \rho \rceil$, is called the *expected rank*. Similar reasoning can be carried out for symmetric tensors, using the normalized CanD expression (9.29), and leads to:

PROPOSITION 9.9
A d-th order symmetric tensor **T** *of dimension* P *may have an essentially unique CanD only if:*

$$\text{rank}\{\mathbf{T}\} \leqslant \frac{\binom{P+d-1}{d}}{P} \stackrel{\text{def}}{=} \rho_s. \tag{9.35}$$

Two remarks are in order here. The first is that even when rank$\{\mathbf{T}\} \leqslant \rho$ (equality may occur if ρ is integer), there may be infinitely many solutions. In section 9.4.2.2 we will present conditions such that there is just one solution, up to permutation and scaling. The second remark is that, for certain dimensions, even in the generic case, more than $\lceil \rho \rceil$ terms might be necessary to form the tensor; in other words, the generic rank might be strictly larger than the expected rank. See [22,20] for an easily accessible summary of these odd properties.

9.4.2.2 Sufficient conditions for uniqueness

When studying the arithmetic complexity of the product of two matrices, Kruskal obtained a sufficient condition for the essential uniqueness of the CanD of a 3rd order tensor. This condition involved the notion of *k-rank* of a set of column vectors [49]:

DEFINITION 9.4 (Kruskal's rank)
A matrix **A** *has k-rank* k_A *if and only if every subset of* k_A *columns of* **A** *is full column rank, and this does not hold true for* $k_A + 1$. *The k-rank of a matrix* **A** *will be denoted by* k-rank$\{\mathbf{A}\}$.

REMARK R9.5
It is important to distinguish between the k-rank and the rank of a matrix. Remember that in a matrix of rank r, there is *at least* one subset of r linearly independent columns. In a matrix of k-rank k_A, *every* subset of k_A columns is of rank k_A. Note that the k-rank is also related to what is sometimes called the kernel of the set of column vectors, or the *spark* of the matrix.

The sufficient condition developed in [49] has been later extended to tensors of arbitrary order [64,70], and can be stated as follows:

THEOREM 9.10
A d-th order tensor $\mathbf{T}$ *of dimensions* $P_1 \times P_2 \times \cdots P_d$ *admits an essentially unique CanD, as* $\mathbf{T} = \sum_{n=1}^{\text{rank}\{\mathbf{T}\}} \lambda_n \, \mathbf{a}_n^{(1)} \otimes \mathbf{a}_n^{(2)} \otimes \ldots \mathbf{a}_n^{(d)}$, *if*

$$2\,\text{rank}\{\mathbf{T}\} + d - 1 \leqslant \sum_{k=1}^{d} \text{k-rank}\{\mathbf{A}^{(k)}\}. \tag{9.36}$$

REMARK R9.6
For generic $P \times r$ matrices, the k-rank equals $\min(P, r)$. Hence, if matrices $\mathbf{A}^{(k)}$ all have more columns than rows, the sufficient condition (9.36) can be generally simplified to

$$2\,\text{rank}\{\mathbf{T}\} + d - 1 \leqslant \sum_{k=1}^{d} P_k, \tag{9.37}$$

which gives an explicit upper bound on the rank of $\mathbf{T}$.

When we can assume that at least one of the tensor dimensions is "large" (meaning that it is larger than $\text{rank}\{\mathbf{T}\}$), a more relaxed sufficient uniqueness condition can be derived. Let us first consider a third-order tensor $\mathbf{T}$ of dimensions $P_1 \times P_2 \times P_3$, with $P_3 \geqslant \text{rank}\{\mathbf{T}\}$. To be able to formulate the condition, we need to introduce the following matrix rank-1 detection tool, which is a variant of the tool introduced in [9]. The proof of the Theorem is given in [27].

THEOREM 9.11
Consider the mapping Γ: $(\mathbf{X}, \mathbf{Y}) \in \mathbb{C}^{P_1 \times P_2} \times \mathbb{C}^{P_1 \times P_2} \longmapsto \Gamma(\mathbf{X}, \mathbf{Y}) = \mathscr{P} \in \mathbb{C}^{P_1 \times P_2 \times P_1 \times P_2}$ *defined by:*

$$p_{ijkl} = x_{ij} y_{kl} + y_{ij} x_{kl} - x_{il} y_{kj} - y_{il} x_{kj}$$

for all index values. Given $\mathbf{X} \in \mathbb{C}^{P_1 \times P_2}$, $\Gamma(\mathbf{X}, \mathbf{X}) = 0$ *if and only if the rank of* $\mathbf{X}$ *is at most one.*

We now have the following sufficient condition for CanD uniqueness [27].

THEOREM 9.12
A third-order tensor $\mathbf{T}$ *of dimensions* $P_1 \times P_2 \times P_3$ *admits an essentially unique CanD, as* $\mathbf{T} = \sum_{n=1}^{\text{rank}\{\mathbf{T}\}} \lambda_n \, \mathbf{a}_n^{(1)} \otimes \mathbf{a}_n^{(2)} \otimes \mathbf{a}_n^{(3)}$, *if the following two conditions are satisfied:*

1. $\mathbf{A}^{(3)}$ *is full column rank.*
2. *The tensors* $\Gamma(\mathbf{a}_u^{(1)} \mathbf{a}_u^{(2)\mathrm{T}}, \mathbf{a}_v^{(1)} \mathbf{a}_v^{(2)\mathrm{T}})$, $1 \leqslant u < v \leqslant \text{rank}\{\mathbf{T}\}$, *are linearly independent.*

REMARK R9.7

The generic version of Theorem 9.12 depends on the symmetry of $\mathbf{T}$. If $\mathbf{T}$ is unsymmetric, then it generically admits an essentially unique CanD if (i) $P_3 \geqslant \text{rank}\{\mathbf{T}\}$, and (ii) $\text{rank}\{\mathbf{T}\}(\text{rank}\{\mathbf{T}\}-1) \leqslant P_1P_2(P_1-1)(P_2-1)/2$ [27]. The second condition implies that $\text{rank}\{\mathbf{T}\}$ is roughly bounded by the product of P_1 and P_2. On the other hand, if $P_3 \geqslant \text{rank}\{\mathbf{T}\}$, then the generic version (9.37) of Kruskal's condition reduces in the third-order case to $\text{rank}\{\mathbf{T}\}+2 \leqslant P_1+P_2$, which implies that $\text{rank}\{\mathbf{T}\}$ is roughly bounded by the sum of P_1 and P_2. We conclude that, if $\mathbf{A}^{(3)}$ is full column rank, Theorem 9.12 is an order of magnitude more relaxed than Kruskal's condition in the third-order case. Moreover, the proof of Kruskal's condition is not constructive, while the proof of Theorem 9.12 yields an algorithm. This algorithm will be further discussed in section 9.5.3.

REMARK R9.8

As mentioned above, the generic version of Theorem 9.12 depends on the symmetry of $\mathbf{T}$. If $\mathbf{A}^{(1)}=\mathbf{A}^{(2)}$, then we have generic essential uniqueness if $P_3 \geqslant \text{rank}\{\mathbf{T}\}$, and

$$\frac{\text{rank}\{\mathbf{T}\}(\text{rank}\{\mathbf{T}\}-1)}{2} \leqslant \frac{P_1(P_1-1)}{4}\left(\frac{P_1(P_1-1)}{2}+1\right) - \frac{P_1!}{(P_1-4)!4!}1_{\{P_1\geqslant 4\}}, \quad (9.38)$$

where

$$1_{\{P_1\geqslant 4\}} = \begin{cases} 0 & \text{if } P_1<4 \\ 1 & \text{if } P_1 \geqslant 4, \end{cases}$$

as conjectured in [71]. The latter paper also presents an algorithm with which, for any given value of P_1, it can be checked whether expression (9.38) is correct.

Now let us turn to fourth-order tensors. We first introduce a third-order tensor rank-1 detection tool [27].

THEOREM 9.13

Consider the mappings $\Omega_1 : (\mathbf{X},\mathbf{Y}) \in \mathbb{C}^{P_1\times P_2\times P_3} \times \mathbb{C}^{P_1\times P_2\times P_3} \to \Omega_1(\mathbf{X},\mathbf{Y}) \in \mathbb{C}^{P_1\times P_1\times P_2\times P_2\times P_3\times P_3}$, $\Omega_2 : (\mathbf{X},\mathbf{Y}) \in \mathbb{C}^{P_1\times P_2\times P_3} \times \mathbb{C}^{P_1\times P_2\times P_3} \to \Omega_2(\mathbf{X},\mathbf{Y}) \in \mathbb{C}^{P_1\times P_1\times P_2\times P_2\times P_3\times P_3}$ *and* $\Omega : (\mathbf{X},\mathbf{Y}) \in \mathbb{C}^{P_1\times P_2\times P_3} \times \mathbb{C}^{P_1\times P_2\times P_3} \to \Omega(\mathbf{X},\mathbf{Y}) \in \mathbb{C}^{P_1\times P_1\times P_2\times P_2\times P_3\times P_3\times 2}$, *defined by*

$$\begin{aligned}(\Omega(\mathbf{X},\mathbf{Y}))_{ijklmn1} &= (\Omega_1(\mathbf{X},\mathbf{Y}))_{ijklmn} \\ &= x_{ikm}y_{jln}+y_{ikm}x_{jln}-x_{jkm}y_{iln}-y_{jkm}x_{iln} \qquad (9.39)\\ (\Omega(\mathbf{X},\mathbf{Y}))_{ijklmn2} &= (\Omega_2(\mathbf{X},\mathbf{Y}))_{ijklmn} \\ &= x_{ikm}y_{jln}+y_{ikm}x_{jln}-x_{ilm}y_{jkn}-y_{ilm}x_{jkn}. \qquad (9.40)\end{aligned}$$

Then we have $\Omega(\mathbf{X},\mathbf{X})=0$ *if and only if* $\mathbf{X}$ *is at most rank-1.*

We now have the following sufficient condition for CanD uniqueness [27].

THEOREM 9.14

A fourth-order tensor $\mathbf{T}$ *of dimensions* $P_1\times P_2\times P_3\times P_4$ *admits an essentially unique CanD, as* $\mathbf{T}\sum_{n=1}^{\text{rank}\{\mathbf{T}\}} \lambda_n\, \mathbf{a}_n^{(1)}\otimes\mathbf{a}_n^{(2)}\otimes\mathbf{a}_n^{(3)}\otimes\mathbf{a}_n^{(4)}$, *if the following two conditions are satisfied:*

1. $\mathbf{A}^{(4)}$ *is full column rank.*
2. *The tensors* $\{\Omega(\mathbf{a}_t^{(1)} \otimes \mathbf{a}_t^{(2)} \otimes \mathbf{a}_t^{(3)}, \mathbf{a}_u^{(1)} \otimes \mathbf{a}_u^{(2)} \otimes \mathbf{a}_u^{(3)})\}_{1 \leqslant t < u \leqslant \mathrm{rank}\{\mathbf{T}\}}$ *are linearly independent.*

REMARK R9.9
The generic version of Theorem 9.14 is that $\mathbf{T}$ admits an essentially unique CanD if (i) $P_4 \geqslant \mathrm{rank}\{\mathbf{T}\}$, and (ii) $\mathrm{rank}\{\mathbf{T}\}(\mathrm{rank}\{\mathbf{T}\} - 1) \leqslant P_1 P_2 P_3 (3P_1 P_2 P_3 - P_1 P_2 - P_2 P_3 - P_3 P_1 - P_1 - P_2 - P_3 + 3)/4$ [27]. The second condition implies that $\mathrm{rank}\{\mathbf{T}\}$ is roughly bounded by the product of P_1, P_2 and P_3. On the other hand, if $P_4 \geqslant \mathrm{rank}\{\mathbf{T}\}$, then the generic version (9.37) of Kruskal's condition roughly bounds $\mathrm{rank}\{\mathbf{T}\}$ by the sum of P_1, P_2 and P_3. We conclude that, if $\mathbf{A}^{(4)}$ is full column rank, Theorem 9.14 is two orders of magnitude more relaxed than Kruskal's condition in the fourth-order case. Moreover, the proof of Theorem 9.14 is constructive.

In Theorems 9.12 and 9.14 we assumed that at least one of the tensor dimensions is larger than $\mathrm{rank}\{\mathbf{T}\}$. One can work in a similar way if the rank of one matrix representation of $\mathbf{T}$ is larger than $\mathrm{rank}\{\mathbf{T}\}$ (see section 9.5.1). For an example, we refer to the discussion of the FOOBI algorithm in section 9.5.4.

REMARK R9.10
The uniqueness properties discussed in this section apply to exact data. In practice, the number of mixing vectors that can be handled is limited by the number of available samples, the noise level and the condition of the mixture. Moreover, in certain antenna array applications, the characteristics of the antennas and the geometry of the array may induce a structure in the data that by itself bounds the number of sources that can effectively be dealt with [15,14,34].

9.4.3 Computation

As explained above, underdetermined mixtures are estimated in this chapter by formulating the problem in terms of a CanD of a (partially) symmetric higher-order tensor. This decomposition may in principle be computed by means of any general-purpose optimization method that minimizes the norm of the difference between the given tensor and its approximation by a sum of rank-1 terms. We mention the popular Alternating Least Squares (ALS) algorithm [42,47,67,78], conjugate gradient [60], Levenberg-Marquardt [74,57,22] and minimization of the least absolute error [76]. It is interesting to note that, due to the multilinear structure of the problem, the size of the optimal step in a given search direction may be computed explicitly [58,63]. The symmetric CanD of symmetric $(2 \times 2 \times \ldots \times 2)$ tensors can be found by means of an algorithm developed by Sylvester; see section 9.5.2.1. Overdetermined CanD may be computed by means of the algebraic algorithms proposed in [32,52,75]. These algorithms can also be used in the core iteration of the algebraic algorithm for underdetermined CanD presented in [27]. One of the advantages of algebraic algorithms is that they will find the global optimum when the data are exact. Strictly speaking, there are some iterative procedures inside (e.g. the Singular Value Decomposition (SVD)), but these are well mastered.

Variants of these algorithms, designed for underdetermined ICA, are discussed in section 9.5.

Concerning fitting a sum of rank-1 terms to a higher-order tensor, a comment is in order. Because the set of tensors of rank N is not closed, the error generally does not admit a minimum but only an infimum [53,21]. In other words, in general there is no best solution for the canonical factors unless the rank of $\mathbf{T}$ is exactly N. In cases where there is no minimum but only an infimum, some entries in the factor matrices will tend to infinity as the algorithm approaches the infimum. This has been called *CP-degeneracy* by some authors [50,61,73,22]. This phenomenon has been the object of study over the last few years [21,33,46,68,69]. The algebraic algorithms in [32,27,52,75] do not suffer from degeneracy.

9.5 TENSOR-BASED ALGORITHMS

9.5.1 Vector and matrix representations

Matrices of size $M \times N$ can be stored in 1-dimensional arrays of size MN. We adopt the following conventions relating $\mathbf{X} \in \mathbb{C}^{M\times N}$ and $\mathbf{x} \in \mathbb{C}^{MN}$:

$$\mathbf{x} = \mathrm{vec}(\mathbf{X}) \quad \text{and} \quad \mathbf{X} = \mathrm{unvec}(\mathbf{x}) \quad \Leftrightarrow \quad (\mathbf{x})_{(m-1)N+n} = (\mathbf{X})_{mn}, \\ 1 \leqslant m \leqslant M, \quad 1 \leqslant n \leqslant N.$$

Similarly, tensors can be stored in lower order tensors, e.g. in matrices or vectors. For later use, we now define two such arrangements. First, 4th order cumulant tensors (9.23)

$$\mathbf{C}_4^{\mathbf{x}} \overset{\text{def}}{=} \mathrm{mat}([\![\kappa_{i\ell}^{\mathbf{x}\,jk}]\!]) \tag{9.41}$$

are stored in a $P^2 \times P^2$ Hermitian matrix, sometimes called the *quadricovariance* of $\mathbf{x}$. More precisely, if $\kappa_{i\ell}^{\mathbf{x}\,jk}$ is defined as in (9.23), matrix $\mathbf{C}_4^{\mathbf{x}}$ is defined as follows:

$$(\mathbf{C}_4^{\mathbf{x}})_{(i-1)P+j,(k-1)P+\ell} = \kappa_{i\ell}^{\mathbf{x}\,jk}.$$

Second, 6th order cumulant tensors (9.26) can also be stored in a $P^3 \times P^3$ Hermitian matrix, sometimes called *hexacovariance* of $\mathbf{x}$:

$$\mathbf{C}_6^{\mathbf{x}} \overset{\text{def}}{=} \mathrm{mat}([\![\kappa_{ijk}^{\mathbf{x}\,\ell mn}]\!]). \tag{9.42}$$

In that case, matrix $\mathbf{C}_6^{\mathbf{x}}$ is defined as

$$(\mathbf{C}_6^{\mathbf{x}})_{P(P(i-1)+j-1)+n,P(P(\ell-1)+m-1)+k} = \kappa_{ijk}^{\ell mn}.$$

Consider two matrices $\mathbf{A}$ and $\mathbf{B}$, of dimensions $I \times J$ and $K \times L$, respectively. The *Kronecker product*, denoted $\mathbf{A} \otimes \mathbf{B}$, is the $IK \times JL$ matrix defined by:

$$\mathbf{A} \otimes \mathbf{B} \stackrel{\text{def}}{=} \begin{pmatrix} A_{11}\,\mathbf{B} & A_{12}\,\mathbf{B} & \dots \\ A_{21}\,\mathbf{B} & A_{22}\,\mathbf{B} & \dots \\ \vdots & \vdots & \\ & & \end{pmatrix}.$$

It is clear from the above that matrix $\mathbf{A} \otimes \mathbf{B}$ contains the coordinates of tensor $\mathbf{A} \otimes \mathbf{B}$; but they should not be confused, since the former is a matrix and the latter a 4th order tensor.

Another useful product that helps in manipulating tensor coordinates in matrix form is the column-wise Kronecker product, also often called the *Khatri-Rao product*. Now let matrices $\mathbf{A}$ and $\mathbf{B}$ have the same number of columns J, and denote $\mathbf{a}_j$ and $\mathbf{b}_j$ their j-th column, respectively. Then the Khatri-Rao product $\mathbf{A} \odot \mathbf{B}$ is defined as:

$$\mathbf{A} \odot \mathbf{B} \stackrel{\text{def}}{=} \left(\mathbf{a}_1 \otimes \mathbf{b}_1,\, \mathbf{a}_2 \otimes \mathbf{b}_2,\, \dots\, \mathbf{a}_J \otimes \mathbf{b}_J\right).$$

REMARK R9.11
As demonstrated in [14] the way cumulants are stored in a matrix array is important, and has an impact on the number of sources that can be localized.

9.5.2 The 2-dimensional case

9.5.2.1 Sylvester's theorem

In this section, we concentrate on symmetric tensors of order d and dimension 2. Such tensors are bijectively associated with binary quantics, that is, homogeneous polynomials in two variables.

THEOREM 9.15 (Sylvester, 1896)
A binary quantic $p(x_1, x_2) = \sum_{i=0}^{d} \binom{d}{i} c_i\, x_1^i\, x_2^{d-i}$ can be written as a sum of d-th powers of r distinct linear forms in $\mathbb{C}$ as:

$$p(x_1, x_2) = \sum_{j=1}^{r} \lambda_j (\alpha_j\, x_1 + \beta_j\, x_2)^d, \tag{9.43}$$

if and only if **(i)** *there exists a vector $\mathbf{q}$ of dimension $r+1$, with components q_ℓ, such that*

$$\begin{bmatrix} c_0 & c_1 & \cdots & c_r \\ \vdots & & & \vdots \\ c_{d-r} & \cdots & c_{d-1} & c_d \end{bmatrix} \mathbf{q} = 0 \tag{9.44}$$

and **(ii)** *the polynomial $q(x_1, x_2) = \sum_{i=0}^{r} q_i\, x_1^i\, x_2^{r-i}$ admits r distinct roots, i.e. it can be written as $q(x_1, x_2) = \prod_{j=1}^{r} (\beta_j^*\, x_1 - \alpha_j^*\, x_2)$.*

The proof of this theorem is fortunately constructive [23,21] and yields Algorithm 9.1, as described in [7] for instance. Given a binary polynomial $p(x_1, x_2)$ of degree d with coefficients $a_i = \binom{d}{i} c_i$, $0 \leqslant i \leqslant d$, define the Hankel matrix $H[r]$ of dimensions $d - r + 1 \times r + 1$ with entries $H[r]_{ij} = c_{i+j-2}$. Then the following algorithm outputs coefficients λ_j and coefficients of the linear forms $\boldsymbol{\ell}_j^{\mathrm{T}}\mathbf{x}$, for $1 \leqslant j \leqslant \operatorname{rank}\{p\}$.

ALGORITHM 9.1 (SYLVESTER)

1. Initialize $r = 0$.
2. Increment $r \leftarrow r + 1$.
3. If the column rank of $H[r]$ is full, then go to step 2.
4. Else compute a basis $\{\boldsymbol{\ell}_1, \ldots, \boldsymbol{\ell}_l\}$ of the right kernel of $H[r]$.
5. Specialization:
   - Take a generic vector $\mathbf{q}$ in the kernel, e.g. $\mathbf{q} = \sum_i \mu_i \boldsymbol{\ell}_i$ by drawing randomly coefficients μ_i.
   - Compute the roots of the associated polynomial $q(x_1, x_2) = \sum_{i=0}^{r} q_i\, x_1^i\, x_2^{d-i}$. Denote them $(\beta_j, -\alpha_j)$, where $|\alpha_j|^2 + |\beta_j|^2 = 1$.
   - If the roots are not distinct in the projective space, try another specialization. If distinct roots cannot be obtained, go to step 2.
   - Else if $q(x_1, x_2)$ admits r distinct roots then compute coefficients λ_j, $1 \leqslant j \leqslant r$, by solving the linear system below, where a_i denotes $\binom{d}{i} c_i$

$$\begin{bmatrix} \alpha_1^d & \cdots & \alpha_r^d \\ \alpha_1^{d-1}\beta_1 & \cdots & \alpha_r^{d-1}\beta_r \\ \alpha_1^{d-2}\beta_1^2 & \cdots & \alpha_r^{d-1}\beta_r^2 \\ \vdots & \vdots & \vdots \\ \beta_1^d & \cdots & \beta_r^d \end{bmatrix} \boldsymbol{\lambda} = \begin{bmatrix} a_0 \\ a_1 \\ a_2 \\ \vdots \\ a_d \end{bmatrix}.$$

6. The decomposition is $p(x_1, x_2) = \sum_{j=1}^{r} \lambda_j \boldsymbol{\ell}_j(\mathbf{x})^d$, where $\boldsymbol{\ell}_j(\mathbf{x}) = (\alpha_j\, x_1 + \beta_j\, x_2)$.

Note that step 5 is a specialization only if the dimension of the right kernel is strictly larger than 1.

9.5.2.2 Sylvester's algorithm applied to cumulants

From (9.32), we know that any symmetric tensor can be associated with a homogeneous polynomial. First, consider in this section the case of 4th order cumulants. Decomposing the 4th order cumulant tensor of 2 random variables is equivalent to decomposing a homogeneous polynomial of degree 4 in 2 variables. As a consequence, we can use Sylvester's theorem described in the previous section.

But for $P = 2$ sensors, the necessary condition (9.35) is never satisfied for under-determined mixtures [23,17]. That's why it has been proposed in [16, sec. 3.2][18, sec. III] to fix the indeterminacy remaining in the decomposition of a tensor by using jointly two

4th order tensors of dimension 2, namely (9.23) and (9.25). This is explained in detail in the algorithm below. Variants of the algorithm are presented in [30,31].

ALGORITHM 9.2 (ALGECUM)

1. Compute an estimate of the two 4th order cumulant arrays $\kappa_{ij}^{\mathbf{x}k\ell}$ and $\kappa_{ijk\ell}^{\mathbf{x}}$, defined in section 9.3.2.4.
2. Compute the 2×4 Hankel matrix defined in (9.44), built on $\kappa_{ijk\ell}^{\mathbf{x}}$.
3. Compute two 4-dimensional vectors $\mathbf{v}_1$ and $\mathbf{v}_2$ forming a basis of its null space.
4. Associate the 4-way array $\kappa_{ij}^{\mathbf{x}k\ell}$ with a 4th degree polynomial in 4 real variables; this polynomial lives in a linear space of dimension 35, and can be expressed in a (arbitrarily chosen) basis of the latter.
5. For a finite subset of values $\theta \in [0, \pi)$ and $\varphi \in [0, 2\pi)$, compute vector $\mathbf{g}(\theta, \varphi) \stackrel{\text{def}}{=} \mathbf{v}_1 \cos\theta + \mathbf{v}_2 \sin\theta\, e^{\jmath\varphi}$.
6. For each value (θ, φ), compute the three linear forms $\ell_j(x|\theta, \varphi)$, $1 \leq j \leq 3$ associated with $\mathbf{g}(\theta, \varphi)$.
7. Express $|\ell_j(x|\theta, \varphi)|^4$ in the chosen basis of the 35-dimensional linear space, and denote $\mathbf{u}_j(\theta, \varphi)$ their coordinate vector.
8. Detect the value (θ_o, φ_o) of (θ, φ) for which $\kappa_{ij}^{\mathbf{x}k\ell}$ is the closest to the linear space spanned by $\{\mathbf{u}_1, \mathbf{u}_2, \mathbf{u}_3\}$.
9. Set $L_j = \ell_j(\theta_o, \varphi_o)$, and $\mathbf{A} = [L_1, L_2, L_3]$, where L_j are expressed by their two coordinates.

9.5.2.3 Sylvester's algorithm applied to characteristic functions

In the previous section, we have used Sylvester's algorithm in order to decompose the 4th order cumulant tensor of a 2-dimensional random variable. We shall see in this section how to apply the latter theorem to the characteristic function itself.

The starting point is to take any two derivatives of the core equation (9.5), and to combine them so as to cancel one term in the sum in (9.5). More precisely, for any triplet of indices, (n, i, j), $n \leq N$, $i, j \leq P$, define the differential operator below:

$$D_{n,i,j} \stackrel{\text{def}}{=} A_{in} \frac{\partial}{\partial u_j} - A_{jn} \frac{\partial}{\partial u_i}.$$

Then, it is clear that for any triplet (n, i, j), $D_{n,i,j} \Psi_x(\mathbf{u})$ does not depend on n, because that term involving ψ_n vanishes (recall that we denoted $\psi_n \stackrel{\text{def}}{=} \Psi_{s_n}$ for the sake of simplicity).

Thus, by applying such an operator N times for all the successive values of n, one eventually gets zero. Of course, the problem is not yet solved, because we don't know the entries of $\mathbf{A}$, and hence the exact form of operator $D_{n,i,j}$. However, in dimension 2, the pair (i,j) is necessarily kept fixed to $(1,2)$. As a consequence, after N successive actions of D, we have:

$$\left\{\prod_{n=1}^{N} D_{n,i,j}\right\}\Psi_x(\mathbf{u}) = \sum_{k=0}^{N} q_k \frac{\partial^N \Psi_x(\mathbf{u})}{\partial u_j^{N-k}\partial u_i^k} = 0, \quad \forall \mathbf{u}\in\Omega \tag{9.45}$$

where q_k is a known function of the unknown matrix $\mathbf{A}$.

By inserting the core equation (9.5) into (9.45), we obtain:

$$\sum_{n=1}^{N}\left[\sum_{k=0}^{N} q_k A_{jn}^{N-k} A_{in}^{k}\right]\psi_n^{(N)}\left(\sum_p A_{pn} u_p\right) = 0 \tag{9.46}$$

where $\psi_n^{(N)}$ denotes the N-th derivative of ψ_n. Since it is true for any $\mathbf{u}$ in the neighborhood of the origin, we have eventually:

$$\sum_{k=0}^{N} q_k A_{jn}^{N-k} A_{in}^{k} = 0, \quad \forall n. \tag{9.47}$$

The latter equation may be seen as a polynomial in A_{in}/A_{jn}, and can be rooted, which would yield the N solutions for this ratio if q_k were known. So let's now concentrate on how to estimate q_k.

As suggested in section 9.3.1, consider equation (9.45) on a grid $\mathcal{G}$ of L values $\{\mathbf{u}[1],\ldots,\mathbf{u}[L]\}$. One can then build the over-determined linear system $\mathbf{H}[N]\,\mathbf{q} = 0$, where $\mathbf{H}[N]$ is the $K \times N+1$ matrix of N-th order derivatives given below:

$$H[N] \stackrel{\text{def}}{=} \begin{pmatrix} \frac{\partial^N \Psi_x(\mathbf{u}[1])}{\partial u_j^N} & \frac{\partial^N \Psi_x(\mathbf{u}[1])}{\partial u_j^{N-1}\partial u_i} & \cdots & \frac{\partial^N \Psi_x(\mathbf{u}[1])}{\partial u_i^N} \\ \frac{\partial^N \Psi_x(\mathbf{u}[2])}{\partial u_j^N} & \frac{\partial^N \Psi_x(\mathbf{u}[2])}{\partial u_j^{N-1}\partial u_i} & \cdots & \frac{\partial^N \Psi_x(\mathbf{u}[2])}{\partial u_i^N} \\ \vdots & \vdots & \vdots & \vdots \\ \frac{\partial^N \Psi_x(\mathbf{u}[K])}{\partial u_j^N} & \frac{\partial^N \Psi_x(\mathbf{u}[K])}{\partial u_j^{N-1}\partial u_i} & \cdots & \frac{\partial^N \Psi_x(\mathbf{u}[K])}{\partial u_i^N} \end{pmatrix}. \tag{9.48}$$

The latter system allows one to estimate components q_k of vector $\mathbf{q}$. These results consequently lead us to the following algorithm [25,24], able to estimate two rows of matrix $\mathbf{A}$ up to a scale factor:

ALGORITHM 9.3 (ALGECAF1)

1. Fix the number N of sources sought (the algorithm can increment on N, starting with $N = P$).

2. Select two sensor indices $[i,j]$, $1 \le [i,j] \le P$.
3. Define a grid $\mathcal{G}$ of L values $\mathbf{u}[\ell]$ in the neighborhood of the origin in $\mathbb{R}^P$, $1 \le \ell \le L$.
4. Estimate the N-th order derivatives of the joint second characteristic function of observation $[x_i, x_j]$, $\Psi_x(\mathbf{u})$ on this grid, and store them in a matrix $\mathbf{H}[N]$ as defined in (9.48).
5. Compute the right singular vector $\mathbf{q}$ of $\mathbf{H}[N]$ associated with the smallest singular value.
6. Root the N-th degree polynomial whose coefficients are q_k, $0 \le k \le N$ in the projective space (that is, include infinity if necessary).
7. Associate each root with the ratio A_{in}/A_{jn}.

This algorithm can be made more robust by observing that similar equations can be obtained if derivatives of order higher than N are computed. In fact, (9.45) is still null if further derivatives are taken. For one additional derivative, we have:

$$\frac{\partial}{\partial u_\ell}\sum_{k=0}^{N} q_k \frac{\partial^N \Psi_x(\mathbf{u})}{\partial u_j^{N-k}\,\partial u_i^k} = 0, \tag{9.49}$$

where $\ell \in \{i,j\}$. Denote $\mathbf{H}[N+1,i]$ and $\mathbf{H}[N+1,j]$ the two $K \times N+1$ matrices built from (9.49), and $\ell \in \{i,j\}$. Then $\mathbf{q}$ satisfies the following linear system:

$$\begin{bmatrix} \mathbf{H}[N] \\ \mathbf{H}[N+1,i] \\ \mathbf{H}[N+1,j] \end{bmatrix} \cdot \mathbf{q} = 0 \tag{9.50}$$

where matrices $\mathbf{H}[N,\ell]$, $\ell \in \{i,j\}$, are defined by (9.49). We then obtain the following:

ALGORITHM 9.4 (ALGECAF2)

1. Run steps 1 to 4 of algorithm ALGECAF1.
2. Build matrices $\mathbf{H}[N+1,i]$ and $\mathbf{H}[N+1,j]$.
3. Compute the right singular vector $\mathbf{q}$ of (9.50).
4. Run steps 6 and 7 of algorithm ALGECAF1.

REMARK R9.12
Note that if the grid of values contains only the origin (which implies $L = 1$), only cumulants of order N are required in Algorithm 9.3, and additional cumulants of order $N+1$ in Algorithm 9.4. There is no need to estimate derivatives of the characteristic function. However, taking a few values in the neighborhood of the origin has been shown to improve the results. Computational details are omitted and can be found in [25].

9.5.3 SOBIUM family

In this section we assume that the sources are mutually uncorrelated but individually correlated in time. The presentation is in terms of complex data, but the method works also for real data, under somewhat more restrictive conditions on the number of sources; e.g. see Remark R9.8. Up to a noise term, the spatial covariance matrices of the observations satisfy [5] (see also Chapter 7):

$$
\begin{aligned}
\mathbf{C}_1 &\stackrel{\text{def}}{=} \mathbb{E}\mathbf{x}_t\mathbf{x}_{t+\tau_1}^{\mathrm{H}} = \mathbf{A}\cdot\mathbf{D}_1\cdot\mathbf{A}^{\mathrm{H}} \\
&\vdots \\
\mathbf{C}_K &\stackrel{\text{def}}{=} \mathbb{E}\mathbf{x}_t\mathbf{x}_{t+\tau_K}^{\mathrm{H}} = \mathbf{A}\cdot\mathbf{D}_K\cdot\mathbf{A}^{\mathrm{H}}
\end{aligned}
\tag{9.51}
$$

in which $\mathbf{D}_k \stackrel{\text{def}}{=} \mathbb{E}\mathbf{s}_t\mathbf{s}_{t+\tau_k}^{\mathrm{H}}$ is diagonal, $k = 1,\ldots,K$. One of the delays τ_k can be equal to zero. The approach in this section applies to any ICA technique that is based on a simultaneous matrix diagonalization like (9.51). Besides spatial covariance matrices for different time lags, matrices $\{\mathbf{C}_k\}$ may correspond to spatial covariance matrices measured at different time instances, in the case of non-stationary sources subject to a constant mixing [62]. They may also correspond to spatial time-frequency distributions [6]. They could for instance also be Hessian matrices of the second characteric function of the observations, sampled at different working points [77].

Stack the matrices $\mathbf{C}_1,\ldots,\mathbf{C}_K$ in a tensor $\mathcal{T}\in\mathbb{C}^{P\times P\times K}$ as follows: $(\mathcal{T})_{p_1p_2k} \stackrel{\text{def}}{=} (\mathbf{C}_k)_{p_1p_2}$, $p_1=1,\ldots,P$, $p_2=1,\ldots,P$, $k=1,\ldots,K$. Then (9.51) can be rewritten as follows:

$$
\mathcal{T}=\sum_{n=1}^{N}\mathbf{a}_n\otimes\mathbf{a}_n^{*}\otimes\mathbf{d}_n, \tag{9.52}
$$

in which

$$
(\mathbf{d}_n)_k \stackrel{\text{def}}{=} (\mathbf{D}_k)_{nn}, \qquad 1\leqslant n\leqslant N, \quad 1\leqslant k\leqslant K.
$$

For later use, we define $\mathbf{D}\in\mathbb{C}^{K\times N}$ as follows:

$$
(\mathbf{D})_{kn} \stackrel{\text{def}}{=} (\mathbf{D}_k)_{nn}, \qquad 1\leqslant n\leqslant N, \quad 1\leqslant k\leqslant K.
$$

The key observation is that decomposition (9.52) is a CanD of tensor $\mathcal{T}$. The CanD uniqueness properties (see section 9.4.2) allow one to determine mixtures even in the underdetermined case. One can in principle use any CanD algorithm. This approach is called *SOBIUM*, which stands for *Second-Order Blind Identification of Underdetermined Mixtures* [28].

A powerful technique can be derived if we can assume that $\mathbf{A}\odot\mathbf{A}^*$ and $\mathbf{D}$ are tall (i.e. have at least as many rows as columns) and that they are full column rank. Let us define a matrix representation $\bar{\mathbf{T}}\in\mathbb{C}^{P^2\times K}$ of $\mathcal{T}$ as follows:

$$
(\bar{\mathbf{T}})_{(p_1-1)P+p_2,k} = (\mathcal{T})_{p_1p_2k} \qquad p_1=1,\ldots,P, \quad p_2=1,\ldots,P, \quad k=1,\ldots,K.
$$

Then Eq. (9.52) can be written in a matrix format as:

$$\bar{\mathbf{T}} = (\mathbf{A} \odot \mathbf{A}^*) \cdot \mathbf{D}^{\mathrm{T}}. \tag{9.53}$$

The full rank property of the two factors in this product implies that N is equal to the rank of $\bar{\mathbf{T}}$. This is an easy way to estimate the number of sources, even in the underdetermined case.

Let the "economy size" SVD of $\bar{\mathbf{T}}$ be given by:

$$\bar{\mathbf{T}} = \mathbf{U} \cdot \boldsymbol{\Sigma} \cdot \mathbf{V}^{\mathrm{H}}, \tag{9.54}$$

in which $\mathbf{U} \in \mathbb{C}^{P^2 \times N}$ and $\mathbf{V} \in \mathbb{C}^{K \times N}$ are column-wise orthonormal matrices and in which $\boldsymbol{\Sigma} \in \mathbb{R}^{N \times N}$ is positive diagonal. Combination of (9.53) and (9.54) yields that there exists an *a priori* unknown non-singular matrix $\mathbf{F} \in \mathbb{C}^{N \times N}$ that satisfies:

$$\mathbf{A} \odot \mathbf{A}^* = \mathbf{U} \cdot \boldsymbol{\Sigma} \cdot \mathbf{F}. \tag{9.55}$$

Matrix $\mathbf{F}$ can be found by imposing the Khatri-Rao structure in the left-hand side of this equation. This Khatri-Rao structure is a vectorized form of a matrix rank-1 structure. We work as follows.

Define $\mathbf{H} = \mathbf{U}\boldsymbol{\Sigma} \in \mathbb{C}^{P^2 \times N}$ and $\mathbf{H}_n = \mathrm{unvec}(\mathbf{h}_n) \in \mathbb{C}^{P \times P}$, $n = 1, \ldots, N$. Equation (9.55) can now be written as:

$$\mathbf{H}_n = \sum_{t=1}^{N} \left(\mathbf{a}_t \mathbf{a}_t^{\mathrm{H}}\right) (\mathbf{F}^{-1})_{tn}. \tag{9.56}$$

The rank-1 structure of $\mathbf{a}_t \mathbf{a}_t^{\mathrm{H}}$ is exploited by using the mapping Γ, defined in Theorem 9.11. From the set of matrices $\{\mathbf{H}_n\}$, we construct the set of N^2 tensors $\{\mathbf{P}_{rs} \stackrel{\mathrm{def}}{=} \Gamma(\mathbf{H}_r, \mathbf{H}_s)\}_{1 \leqslant r,s \leqslant N}$. Due to the bilinearity of Γ, we have from (9.56):

$$\mathbf{P}_{rs} = \sum_{t,u=1}^{N} (\mathbf{F}^{-1})_{tr} (\mathbf{F}^{-1})_{us} \Gamma\left(\mathbf{a}_t \mathbf{a}_t^{\mathrm{H}}, \mathbf{a}_u \mathbf{a}_u^{\mathrm{H}}\right). \tag{9.57}$$

We now check whether there exists a symmetric matrix $\mathbf{M} \in \mathbb{C}^{N \times N}$ that is a solution to the following set of homogeneous linear equations (it will soon become clear that such a solution indeed exists):

$$\sum_{r,s=1}^{N} M_{rs} \mathbf{P}_{rs} = \mathbf{O}. \tag{9.58}$$

Substitution of (9.57) and (9.58) yields:

$$\sum_{r,s=1}^{N}\sum_{t,u=1}^{N}(\mathbf{F}^{-1})_{tr}(\mathbf{F}^{-1})_{us}M_{rs}\Gamma\left(\mathbf{a}_t\mathbf{a}_t^{\mathrm{H}},\ \mathbf{a}_u\mathbf{a}_u^{\mathrm{H}}\right)=\mathbf{O}. \tag{9.59}$$

Using the symmetry of $\mathbf{M}$, the fact that Γ is symmetric in its arguments and Theorem 9.11, (9.59) can be reduced to:

$$\sum_{r,s=1}^{N}\sum_{\substack{t,u=1\\ t<u}}^{N}(\mathbf{F}^{-1})_{tr}(\mathbf{F}^{-1})_{us}M_{rs}\Gamma\left(\mathbf{a}_t\mathbf{a}_t^{\mathrm{H}},\ \mathbf{a}_u\mathbf{a}_u^{\mathrm{H}}\right)=\mathbf{O}. \tag{9.60}$$

We now assume, like in the second condition of Theorem 9.12, that the tensors $\Gamma\left(\mathbf{a}_t\mathbf{a}_t^{\mathrm{H}},\ \mathbf{a}_u\mathbf{a}_u^{\mathrm{H}}\right)$, $1 \leqslant t < u \leqslant N$, are linearly independent. Then (9.60) can only hold if the coefficients $\sum_{r,s=1}^{N}(\mathbf{F}^{-1})_{tr}(\mathbf{F}^{-1})_{us}M_{rs}$ vanish when $t \neq u$. (If $t = u$, then $\Gamma\left(\mathbf{a}_t\mathbf{a}_t^{\mathrm{H}},\ \mathbf{a}_u\mathbf{a}_u^{\mathrm{H}}\right) = \mathbf{O}$ because of Theorem 9.11.) This can be expressed in matrix terms as follows:

$$\mathbf{M}=\mathbf{F}\cdot\mathbf{\Lambda}\cdot\mathbf{F}^{\mathrm{T}}, \tag{9.61}$$

in which $\mathbf{\Lambda}$ is diagonal. It is easy to verify that *any* diagonal matrix $\mathbf{\Lambda}$ generates a matrix $\mathbf{M}$ that satisfies Eq. (9.58). Hence, solving (9.58) yields N linearly independent matrices $\{\mathbf{M}_n\}$, which can be decomposed as

$$\begin{aligned}\mathbf{M}_1 &= \mathbf{F}\cdot\mathbf{\Lambda}_1\cdot\mathbf{F}^{\mathrm{T}}\\ &\ \ \vdots\\ \mathbf{M}_N &= \mathbf{F}\cdot\mathbf{\Lambda}_N\cdot\mathbf{F}^{\mathrm{T}},\end{aligned} \tag{9.62}$$

in which $\mathbf{\Lambda}_1,\ldots,\mathbf{\Lambda}_N$ are diagonal. Note that, if the covariance matrices $\{\mathbf{C}_k\}$ are such that tensor $\mathbf{T}$ satisfies the two conditions in Theorem 9.12, then the exact solution of the underdetermined problem may be found by means of an Eigenvalue Decomposition (EVD). Indeed, $\mathbf{F}$ can be found from the EVD

$$\mathbf{M}_1\cdot\mathbf{M}_2^{-1}=\mathbf{F}\cdot(\mathbf{\Lambda}_1\cdot\mathbf{\Lambda}_2^{-1})\cdot\mathbf{F}^{-1}.$$

If $\mathbf{M}_2$ is singular, or if $\mathbf{M}_1\cdot\mathbf{M}_2^{-1}$ has coinciding eigenvalues, then one may work with linear combinations of $\{\mathbf{M}_n\}$. In the case of inexact data, it is preferable to take all matrices in (9.62) into account. Equation (9.62) may be solved by means of any method for joint approximate non-orthogonal matrix diagonalization, such as the algorithms presented in [32,60,75,78] and references therein.

Once matrix $\mathbf{F}$ has been found from (9.62), the mixing matrix $\mathbf{A}$ can be found from (9.55). Define $\tilde{\mathbf{A}} = \mathbf{A}\odot\mathbf{A}^* \in \mathbb{C}^{P^2\times N}$ and $\tilde{\mathbf{A}}_n = \mathrm{unvec}(\tilde{\mathbf{a}}_n) \in \mathbb{C}^{P\times P}$, $n = 1,\ldots,N$.

Then the rank of $\tilde{\mathbf{A}}_n$ is theoretically equal to 1: $\tilde{\mathbf{A}}_n = \mathbf{a}_n \mathbf{a}_n^{\mathrm{H}}$. Consequently, $\mathbf{a}_n$ can, up to an irrelevant scaling factor, be determined as the left singular vector associated with the largest singular value of $\tilde{\mathbf{A}}_n$, $n = 1, \ldots, N$.

ALGORITHM 9.5 (SOBIUM, CASE $N \leqslant K$)

1. Estimate the covariance matrices $\mathbf{C}_1, \ldots, \mathbf{C}_K$. Define $\bar{\mathbf{T}} = [\mathrm{vec}(\mathbf{C}_1) \cdots \mathrm{vec}(\mathbf{C}_K)]$.
2. Compute the SVD $\bar{\mathbf{T}} = \mathbf{U} \cdot \boldsymbol{\Sigma} \cdot \mathbf{V}^{\mathrm{H}}$. $\mathbf{H} = \mathbf{U} \cdot \boldsymbol{\Sigma}$. The number of sources N equals $\mathrm{rank}(\bar{\mathbf{T}})$.
3. Compute $\mathbf{P}_{st} = \Gamma(\mathbf{H}_s, \mathbf{H}_t)$, $1 \leqslant s \leqslant t \leqslant N$.
4. Compute N linearly independent symmetric matrices $\mathbf{M}_n$ that (approximately) satisfy $\sum_{r,s=1}^{N} M_{rs} \mathbf{P}_{rs} = \mathbf{O}$.
5. Compute non-singular $\mathbf{F}$ that best simultaneously diagonalizes the matrices $\mathbf{M}_n$.
6. Compute $\tilde{\mathbf{A}} = \mathbf{U} \cdot \boldsymbol{\Sigma} \cdot \mathbf{F}$.
7. Estimate mixing vector $\mathbf{a}_n$ as the dominant left singular vector of $\tilde{\mathbf{A}}_n = \mathrm{unvec}(\tilde{\mathbf{a}}_n)$, $n = 1, \ldots, N$.

9.5.4 FOOBI family

In this section we work with the quadricovariance, defined in sections 9.3.2.5 and 9.5.1. The presentation is in terms of complex data, but the method works also for real data, under somewhat more restrictive conditions on the number of sources. Eq. (9.23) is a CanD of the fourth-order cumulant tensor. The CanD uniqueness properties (see section 9.4.2) allow one to determine mixtures even in the underdetermined case. This approach is called *FOOBI*, which stands for *Fourth-Order-Only Blind Identification* [9,29].

Equation (9.23) can be written in a matrix format as follows:

$$\mathbf{C}_4^{\mathbf{x}} = (\mathbf{A} \odot \mathbf{A}^*) \cdot \boldsymbol{\Delta}_4 \cdot (\mathbf{A} \odot \mathbf{A}^*)^{\mathrm{H}}, \tag{9.63}$$

where $\boldsymbol{\Delta}_4$ is the $N \times N$ diagonal matrix containing the 4th order source marginal cumulants. We have the following theorem [29].

THEOREM 9.16

Consider a tensor $\mathbf{T} \in \mathbb{C}^{P \times P \times P \times P}$, satisfying the symmetries $t_{klij} = t^*_{ijkl}$ and $t_{jilk} = t^*_{ijkl}$, and its matrix representation $\bar{\mathbf{T}} = \mathrm{mat}(\mathbf{T}) \in \mathbb{C}^{P^2 \times P^2}$. The matrix $\bar{T}$ can be eigen-decomposed as

$$\bar{\mathbf{T}} = \mathbf{E} \cdot \boldsymbol{\Lambda} \cdot \mathbf{E}^{\mathrm{H}}, \tag{9.64}$$

in which $\mathbf{E} \in \mathbb{C}^{P^2 \times N}$ is column-wise orthonormal, with $e_{(p_1-1)P+p_2,n} = e^*_{(p_2-1)P+p_1,n}$, $p_1, p_2 = 1, \ldots, P$, $n = 1, \ldots, N$, and in which $\boldsymbol{\Lambda} \in \mathbb{C}^{N \times N}$ is a diagonal matrix of which the diagonal elements are real and non-zero. N is the rank of $\bar{\mathbf{T}}$.

From this theorem we have that $\mathbf{C}_4^{\mathrm{x}}$ can be decomposed as

$$\mathbf{C}_4^{\mathrm{x}} = \mathbf{E} \cdot \mathbf{\Lambda} \cdot \mathbf{E}^{\mathrm{H}}. \tag{9.65}$$

This matrix EVD may be easily computed. Note that N is equal to the rank of $\mathbf{C}_4^{\mathrm{x}}$. This is an easy way to estimate the number of sources, even in the underdetermined case.

We now assume that all sources have strictly positive kurtosis. The more general situation will be addressed in Remark R9.13. From Eq. (9.63) it follows that $\mathbf{C}_4^{\mathrm{x}}$ is positive (semi)definite. We have the following theorem [29].

THEOREM 9.17
Let $\mathbf{C}_4^{\mathrm{x}}$ *be positive (semi)definite and assume that it can be decomposed as in* (9.63) and (9.65). *Then we have:*

$$(\mathbf{A} \odot \mathbf{A}^*) \cdot (\mathbf{\Delta}_4)^{1/2} = \mathbf{E} \cdot \mathbf{\Lambda}^{1/2} \cdot \mathbf{V}, \tag{9.66}$$

in which $\mathbf{V}$ *is real* $(N \times N)$ *orthogonal.*

Equation (9.66) is analogous to the SOBIUM equation (9.55). The matrix $\mathbf{V}$ can be determined by exploiting the Khatri-Rao structure of $\mathbf{A} \odot \mathbf{A}^*$. The following rank-1 detection tool was used in the original FOOBI algorithm [9,29]. $\mathbb{H}^{P \times P}$ represents the space of $(P \times P)$ Hermitean matrices.

THEOREM 9.18
Consider the mapping $\tilde{\Gamma} : (\mathbf{X}, \mathbf{Y}) \in \mathbb{H}^{P \times P} \times \mathbb{H}^{P \times P} \to \tilde{\Gamma}(\mathbf{X}, \mathbf{Y}) \in \mathbb{C}^{P \times P \times P \times P}$ *defined by*

$$(\tilde{\Gamma}(\mathbf{X}, \mathbf{Y}))_{ijkl} = x_{ij} y_{kl}^* + y_{ij} x_{kl}^* - x_{ik} y_{jl}^* - y_{ik} x_{jl}^*. \tag{9.67}$$

Then we have: $\tilde{\Gamma}(\mathbf{X}, \mathbf{X}) = 0$ *if and only if* $\mathbf{X}$ *is at most rank-1.*

Analogous to SOBIUM, we define a matrix $\mathbf{H} \stackrel{\text{def}}{=} \mathbf{E} \cdot \mathbf{\Lambda}^{1/2}$, Hermitean matrices $\mathbf{H}_s \stackrel{\text{def}}{=} \text{unvec}(\mathbf{h}_s)$ and we construct a set of N^2 fourth-order tensors $\{\mathbf{P}_{rs} \stackrel{\text{def}}{=} \tilde{\Gamma}(\mathbf{H}_r, \mathbf{H}_s)\}_{1 \leqslant r,s \leqslant N}$. Then we look for a real symmetric matrix $\mathbf{M} \in \mathbb{R}^{N \times N}$ that is a solution to the following set of homogeneous linear equations:

$$\sum_{r,s=1}^{P} M_{rs} \mathbf{P}_{rs} = \mathbf{O}. \tag{9.68}$$

We assume that the tensors $\tilde{\Gamma}\left(\mathbf{a}_t \mathbf{a}_t^{\mathrm{H}}, \mathbf{a}_u \mathbf{a}_u^{\mathrm{H}}\right)$, $1 \leqslant t < u \leqslant N$, are linearly independent. This corresponds to the second condition of Theorem 9.12. It turns out that, under this condition, (9.68) has N linearly independent solutions, which can be decomposed as:

$$\begin{aligned} \mathbf{M}_1 &= \mathbf{V} \cdot \mathbf{D}_1 \cdot \mathbf{V}^{\mathrm{T}} \\ &\vdots \\ \mathbf{M}_N &= \mathbf{V} \cdot \mathbf{D}_N \cdot \mathbf{V}^{\mathrm{T}}, \end{aligned} \tag{9.69}$$

in which $\mathbf{D}_1, \ldots, \mathbf{D}_N \in \mathbb{R}^{N\times N}$ are diagonal. The difference with (9.62) is that all matrices are real, even when $\mathbf{C}_4^{\mathbf{x}}$ is complex, and that $\mathbf{V}$ is orthogonal. Note that, in the case of exact data, the solution of the underdetermined problem may be found by means of an EVD. Indeed, every equation in (9.69) is an EVD of a real symmetric matrix. In the case of inexact data, it is preferable to take all matrices in (9.69) into account. The joint approximate orthogonal matrix diagonalization problem may be solved by means of the Jacobi iteration presented in [11,12]. Once matrix $\mathbf{V}$ has been found from (9.69), the mixing matrix $\mathbf{A}$ can be found from (9.66) in a similar way as in SOBIUM.

ALGORITHM 9.6 (FOOBI)

1. Estimate the quadricovariance $\mathbf{C}_4^{\mathbf{x}}$ of the data.
2. Compute the EVD $\mathbf{C}_4^{\mathbf{x}} = \mathbf{E}\cdot\mathbf{\Lambda}\cdot\mathbf{E}^{\mathrm{H}}$. $\mathbf{H} = \mathbf{E}\cdot\mathbf{\Lambda}^{1/2}$. The number of sources N equals rank($\mathbf{C}_4^{\mathbf{x}}$). Normalize the eigenvectors such that $\mathbf{H}_n = \text{unvec}(\mathbf{h}_n)$, $1 \leqslant n \leqslant N$, is Hermitean.
3. Compute $\mathbf{P}_{st} = \tilde{\Gamma}(\mathbf{H}_s, \mathbf{H}_t)$, $1 \leqslant s \leqslant t \leqslant N$.
4. Compute N linearly independent real symmetric matrices $\mathbf{M}_n$ that (approximately) satisfy $\sum_{r,s=1}^{N} M_{rs}\mathbf{P}_{rs} = \mathbf{O}$.
5. Compute orthogonal $\mathbf{V}$ that best simultaneously diagonalizes the matrices $\mathbf{M}_n$.
6. Compute $\mathbf{F} = \mathbf{E}\cdot\mathbf{\Lambda}^{1/2}\cdot\mathbf{V}$.
7. Estimate mixing vector $\mathbf{a}_n$ as the dominant left singular vector of unvec($\mathbf{f}_n$), $n = 1, \ldots, N$.

REMARK R9.13
In the derivation above, we have assumed that all sources are strictly leptokurtic. If all the sources have strictly negative kurtosis, then we simply process $-\mathscr{C}_4^{\mathbf{x}}$. When not all kurtosis values have the same sign, (9.66) can be replaced by

$$\mathbf{A} \odot \mathbf{A}^* = \mathbf{E}\cdot\mathbf{V}, \tag{9.70}$$

in which now $\mathbf{V}$ is real non-singular instead of orthogonal. Equation (9.69) then becomes a non-orthogonal joint diagonalization problem, like in SOBIUM.

In the remainder of this section, we assume that all the sources are strictly leptokurtic. (If all the sources are strictly platykurtic, then we process $-\mathbf{C}_4^{\mathbf{x}}$ instead of $\mathbf{C}_4^{\mathbf{x}}$.)

Generically, as long as $N \leqslant P^2$ (complex case) or $N \leqslant P(P+1)/2$ (real case), the number of sources corresponds to the rank of $\mathbf{C}_4^{\mathbf{x}}$ and Theorem 9.17 still applies. We now introduce a new rank-1 detecting tool [29].

THEOREM 9.19
Consider the mapping $\Theta : (\mathbf{X}, \mathbf{Y}) \in \mathbb{H}^{P\times P} \times \mathbb{H}^{P\times P} \to \Theta(\mathbf{X}, \mathbf{Y}) \in \mathbb{H}^{P\times P}$ *defined by*

$$\Theta(\mathbf{X}, \mathbf{Y}) = \mathbf{XY} - \text{trace}(\mathbf{X})\mathbf{Y} + \mathbf{YX} - \text{trace}(\mathbf{Y})\mathbf{X}. \tag{9.71}$$

Then we have that $\Theta(\mathbf{X}, \mathbf{X}) = 0$ *if and only if* $\mathbf{X}$ *is at most rank-1.*

Define $\mathbf{H} \in \mathbb{C}^{P^2 \times N}$ and $\mathbf{H}_s = \text{mat}(\mathbf{h}_s) \in \mathbb{C}^{P \times P}$, $s = 1, \ldots, N$, as above. Also define symmetric matrices $\mathbf{B}_{p_1 p_2} \in \mathbb{C}^{N \times N}$ by

$$(\mathbf{B}_{p_1 p_2})_{st} = (\Theta(\mathbf{H}_s, \mathbf{H}_t))_{p_1 p_2}, \quad 1 \leqslant p_1, p_2 \leqslant P, \quad 1 \leqslant s, t \leqslant N.$$

The following theorem leads to a new algorithm for the computation of $\mathbf{V}$ [29].

THEOREM 9.20
The matrix $\mathbf{V}$ *in Eq.* (9.66) *satisfies*

$$\begin{aligned} &\text{diag}(\mathbf{V}^{\mathrm{T}} \cdot \text{Real}(\mathbf{B}_{p_1 p_2}) \cdot \mathbf{V}) = 0, \quad 1 \leqslant p_1 \leqslant p_2 \leqslant P, \\ &\text{diag}(\mathbf{V}^{\mathrm{T}} \cdot \text{Imag}(\mathbf{B}_{p_1 p_2}) \cdot \mathbf{V}) = 0, \quad 1 \leqslant p_1 < p_2 \leqslant P. \end{aligned} \tag{9.72}$$

This theorem shows that the matrix $\mathbf{V}$ can be computed by means of simultaneous off-diagonalization of a number of real symmetric matrices. One can use a simple variant of the Jacobi algorithm derived in [11,12], the difference being that one should chose in each step the Jacobi rotation that minimizes (instead of maximizes) the sum of the squared diagonal entries. Simultaneous orthogonal off-diagonalization was also used in [4].

ALGORITHM 9.7 (FOOBI-2)

1. Estimate the quadricovariance $\mathbf{C}_4^{\mathrm{x}}$ of the data.
2. Compute the EVD $\mathbf{C}_4^{\mathrm{x}} = \mathbf{E} \cdot \mathbf{\Lambda} \cdot \mathbf{E}^{\mathrm{H}}$. $\mathbf{H} = \mathbf{E} \cdot \mathbf{\Lambda}^{1/2}$. The number of sources N equals rank($\mathbf{C}_4^{\mathrm{x}}$). Normalize the eigenvectors such that $\mathbf{H}_n = \text{unvec}(\mathbf{h}_n)$, $1 \leqslant n \leqslant N$, is Hermitean.
3. Compute $\Theta(\mathbf{H}_s, \mathbf{H}_t)$, $1 \leqslant s \leqslant t \leqslant N$. Stack the results in $\mathbf{B}_{p_1 p_2}$, $1 \leqslant p_1 \leqslant p_2 \leqslant P$.
4. Compute orthogonal $\mathbf{V}$ that best simultaneously off-diagonalizes the matrices $\{\text{Real}(\mathbf{B}_{p_1 p_2})\}$ and $\{\text{Imag}(\mathbf{B}_{p_1 p_2})\}$.
5. Compute $\mathbf{F} = \mathbf{E} \cdot \mathbf{\Lambda}^{1/2} \cdot \mathbf{V}$.
6. Estimate mixing vector $\mathbf{a}_n$ as the dominant left singular vector of unvec($\mathbf{f}_n$), $n = 1, \ldots, N$.

9.5.5 BIOME family

The algorithm we describe in this section, BIRTH, is using 6th order cumulants as in [2], but other orders can be considered as shown in [3], where a general family of algorithms called BIOME$-2q$ is described. We shall restrict our attention to the former, which actually corresponds to BIOME-6.

The basic idea is that the hexacovariance $\mathbf{C}_6^{\mathrm{y}}$ defined in (9.42) enjoys the property below

$$\mathbf{C}_6^{\mathrm{y}} = \mathbf{A}^{\odot 3} \, \mathbf{\Delta}_6 \, \mathbf{A}^{\odot 3 \mathrm{H}}$$

where $\mathbf{\Delta}_6$ is the $N \times N$ diagonal matrix containing the 6th order source marginal cumulants. In fact, this is another way of writing the multi-linearity property 9.6, with the help of the Khatri-Rao product. If we define a full rank square root matrix of $\mathbf{C}_6^{\mathbf{y}}$, denoted $\mathbf{C}_6^{\mathbf{y}\,1/2}$, then it is related to $\mathbf{A}^{\odot 3}$ up to an (unknown) unitary matrix $\mathbf{V}$:

$$\mathbf{C}_6^{\mathbf{y}\,1/2} = \mathbf{A}^{\odot 3}\boldsymbol{\Delta}_6^{1/2}\mathbf{V}.$$

The problem is that none of the three matrix factors in the right-hand side is known. But there is a redundancy in the left-hand side, which can be exploited. This is done in the algorithm below, by rewriting the above equation for each $P^2 \times N$ submatrix $\boldsymbol{\Gamma}[n]$ of $\mathbf{C}_6^{\mathbf{y}\,1/2}$, as:

$$\boldsymbol{\Gamma}[n] = (\mathbf{A} \odot \mathbf{A}^{\mathrm{H}})\,\mathbf{D}[n]\boldsymbol{\Delta}_6^{1/2}\mathbf{V}$$

where $\mathbf{D}[n]$ is the diagonal matrix containing the n-th row of $\mathbf{A}$, $1 \leqslant n \leqslant P$. Hence matrices $\boldsymbol{\Gamma}[n]$ share the same common right singular subspace defined by $\mathbf{V}$. Yet, if we notice that we can get rid of the unknown factor $(\mathbf{A} \odot \mathbf{A}^{\mathrm{H}})$ by computing the product between a pseudo-inverse $\boldsymbol{\Gamma}^{\dagger}[m]$ and $\boldsymbol{\Gamma}[n]$, then we see that $\mathbf{V}$ can be eventually obtained via the EVD of the Hermitian matrix

$$\boldsymbol{\Theta}_{m,n} \stackrel{\mathrm{def}}{=} \boldsymbol{\Gamma}_m^{\dagger}\boldsymbol{\Gamma}_n = \mathbf{V}^{\mathrm{H}}\left(\mathbf{D}[m]^{-1}\mathbf{D}[n]\right)\mathbf{V}.$$

This can be done for any pair of indices (m,n), $m \neq n$. But it is more stable to do it for all of them simultaneously. Once matrix $\mathbf{V}$ has been obtained this way, matrix $\mathbf{A}^{\odot 3}$ is obtained up to a scale factor with the product $\mathbf{C}_6^{\mathbf{y}\,1/2}\mathbf{V}^{\mathrm{H}}$. As a conclusion, the whole BIRTH algorithm can be described as follows

ALGORITHM 9.8 (BIRTH)

1. Estimate 6th order cumulants of observation $\mathbf{y}$ and store them in the hexacovariance matrix $\mathbf{C}_6^{\mathbf{y}}$, of size $P^3 \times P^3$ as defined in (9.42).
2. Compute a square root $\mathbf{C}_6^{\mathbf{y}\,1/2}$ of $\mathbf{C}_6^{\mathbf{y}}$, of size $P^3 \times N$, e.g. via an EVD.
3. Cut $\mathbf{C}_6^{\mathbf{y}\,1/2}$ into P blocks $\boldsymbol{\Gamma}_n$ each of size P^2N.
4. Compute the $P(P-1)/2$ products $\boldsymbol{\Theta}_{m,n} \stackrel{\mathrm{def}}{=} \boldsymbol{\Gamma}_m^{\dagger}\boldsymbol{\Gamma}_n$ and jointly diagonalize them (approximately), so that $\boldsymbol{\Theta}_{m,n} \approx \mathbf{V}\boldsymbol{\Lambda}[m,n]\mathbf{V}^{\mathrm{H}}$, where $\mathbf{V}$ is $N \times N$ unitary.
5. Compute $\widehat{\mathbf{A}^{\odot 3}} = (\mathbf{C}_6^{\mathbf{y}})^{1/2}\mathbf{V}^{\mathrm{H}}$.
6. Store each of the N columns of $\widehat{\mathbf{A}^{\odot 3}}$ in a vector $\mathbf{b}_n$ of size P^3.
7. Transform each vector $\mathbf{b}_n$ in a family of P matrices of size $P \times P$. Compute the common dominant eigenvector, $\hat{\mathbf{a}}_n$, of these Hermitian matrices.
8. Stack column vectors $\hat{\mathbf{a}}_n$ to form matrix $\hat{\mathbf{A}}$.

Several variants are described in [2], each allowing one to obtain vectors $\hat{\mathbf{a}}_n$ with better accuracy levels, at the price of increased computational complexity.

9.5.6 ALESCAF and LEMACAF

In this section, we elaborate on the calculation of the solution of (9.7). The goal is to minimize the fitting error [25,24]:

$$\Upsilon = \left\| \mathbf{T} - \sum_{n=1}^{\text{rank}\{\mathbf{T}\}} \lambda_n \, \mathbf{a}_n \otimes \mathbf{a}_n \otimes \mathbf{a}_n \otimes \mathbf{b}_n \right\|^2 \tag{9.73}$$

where $\boldsymbol{\Lambda}$ is an $N \times N \times N \times N$ diagonal tensor containing ones in its diagonal.

Beside the fact that this tensor has a partial symmetry (in fact in the first three modes), this minimization problem is similar to that for general tensors:

$$\Upsilon = \left\| \mathbf{T} - \sum_{n=1}^{\text{rank}\{\mathbf{T}\}} \lambda_n \, \mathbf{a}_n \otimes \mathbf{b}_n \otimes \mathbf{c}_n \otimes \mathbf{d}_n \right\|^2 . \tag{9.74}$$

So let's look at this slightly more general problem first and solve it by means of an ALS scheme.

ALGORITHM 9.9 (ALESCAF [25])

1. Define a grid $\mathcal{G}$ containing L points of $\mathbb{C}^P$.
2. For every point $\mathbf{u}[\ell]$ of $\mathcal{G}$, compute the derivative tensor $\mathbf{T}(\mathbf{u}[\ell])$.
3. Initialize, possibly randomly, three sets of vectors $\mathbf{a}_n$. Denote $\mathbf{A}[1]$, $\mathbf{A}[2]$ and $\mathbf{A}[3]$ matrices containing initial values.
4. Execute the ALS algorithm from $t = 1$ and until the stopping criterion is reached:

$$\begin{aligned}
\mathbf{B}^{\mathrm{T}}[3t] &= (\mathbf{A}[3t] \odot \mathbf{A}[3t-1] \odot \mathbf{A}[3t-2])^{-1}\, \mathbf{T}_{N^3 \times L} \\
\mathbf{A}^{\mathrm{T}}[3t+1] &= (\mathbf{A}[3t] \odot \mathbf{A}[3t-1] \odot \mathbf{B}[3t])^{-1}\, \mathbf{T}_{N^2 L \times N} \\
\mathbf{A}^{\mathrm{T}}[3t+2] &= (\mathbf{A}[3t+1] \odot \mathbf{A}[3t] \odot \mathbf{B}[3t])^{-1}\, \mathbf{T}_{N^2 L \times N} \\
\mathbf{A}^{\mathrm{T}}[3t+3] &= (\mathbf{A}[3t+2] \odot \mathbf{A}[3t+1] \odot \mathbf{B}[3t])^{-1}\, \mathbf{T}_{N^2 L \times N}.
\end{aligned}$$

One could also regularize the optimization criterion by adding a penalty term involving the norm of the columns, as in [59], which yields the new objective:

$$\Upsilon_R = \left\| \mathbf{T} - \sum_{p=1}^{\text{rank}\{\mathbf{T}\}} \lambda_p \, \mathbf{a}_p \otimes \mathbf{b}_p \otimes \mathbf{c}_p \otimes \mathbf{d}_p \right\|^2 + \eta \sum_{p=1}^{N} ||\mathbf{a}_p||^2 + ||\mathbf{b}_p||^2 + ||\mathbf{c}_p||^2 + ||\mathbf{d}_p||^2 \tag{9.75}$$

if we decide to assign the same weight η to every penalty term.

Another way is to leave the diagonal tensor $\boldsymbol{\Lambda}$ to take free values, but to constraint loading matrices $\{\mathbf{A}, \mathbf{B}, \mathbf{C}, \mathbf{D}\}$ to have a unit norm. It is easy to see that the two latter approaches are the same, as pointed out in [53].

Now let's go back to our original problem (9.73) and give the example of the Levenberg-Marquardt implementation presented in [22].

ALGORITHM 9.10 (LEMACAF)

1. Define a grid $\mathcal{G}$ containing L points of $\mathbb{C}^P$.
2. For every point $\mathbf{u}$ of $\mathcal{G}$, compute the derivative tensor $\mathbf{T}(\mathbf{u})$.
3. Initialize, possibly randomly, vectors $\mathbf{b}_n$ and $\mathbf{a}_n$. Denote $\mathbf{B}[0]$ and $\mathbf{A}[0]$ the matrices containing these initial values.
4. Arrange all vectors $\mathbf{b}_n$ and $\mathbf{a}_n$ in a single vector $\mathbf{p}$ of size $(P+L)N$. The penalty term in (9.75) is now just $\eta\,||\mathbf{p}||^2$. For $t=1$ and until stop, do the following:
   - Compute the gradient $\mathbf{g}[t]$ of Υ_R at $\mathbf{p}[t]$.
   - Compute the Jacobian $\mathbf{J}[t]$ of Υ_R at $\mathbf{p}[t]$.
   - Compute $\mathbf{p}[t+1]=\mathbf{p}[t]-[\mathbf{J}[t]^{\mathrm{H}}\mathbf{J}[t]+\lambda[t]\mathbf{I}]^{-1}\mathbf{g}[t]$.

The delicate issue in such algorithms is the choice of parameter $\lambda[t]$. We refer to optimization textbooks for this important question [54,39].

9.5.7 Other algorithms

We have reviewed algorithms that exploit the mutual statistical independence between sources. All these algorithms are based on the fact that the problem can be formulated in terms of a CanD of a tensor that enjoys symmetries. In some cases there exist closed form solutions.

Other algorithms have not been mentioned. In particular, the decomposition of symmetric tensors in the complex field has been addressed in [7] for dimensions larger than 2. Bayesian approaches are addressed in Chapter 12. Algorithms exploiting sparseness properties or nonnegativity constraints are studied in Chapters 10 and 13, respectively.

An important class of approaches, which we consider as quite promising, consists of working directly on the data arranged in tensor format. This avoids resorting to statistics of the data, and hence does not need the assumption of source independence. On the other hand, the availability of some diversity in the measurements is required. To build a data tensor from measurements is often the actual challenge, but this has been successfully addressed in a number of applications [65–67].

In this context, several *deterministic* algorithms devoted to blind identification have been developed. They range from descent or Newton algorithms, Alternating Least Squares (ALS) [42,67,22], Levenberg-Marquardt [57,22,74] or conjugate gradient [60], to global line search techniques, e.g. the so-called *Enhanced Line Search (ELS)* algorithm [63,58]. Note that uniqueness conditions, already tackled in section 9.4.2, and reported in [22], can sometimes be improved [27,26].

9.6 APPENDIX: EXPRESSIONS OF COMPLEX CUMULANTS

In this appendix, we only give expressions of cumulants for zero-mean complex variables that are distributed symmetrically with respect to the origin. However, they do not need to be circularly distributed. Below, cumulants are denoted with κ and moments with μ. As before, superscripts correspond to variables that are complex conjugated, and McCullagh's bracket notation [55] is used, with an obvious extension to the complex case [1]; note that in [10] the cumulant expressions are incomplete, and we assume that random variables are circularly distributed. The notation $[\bar{2}]$ means twice the real part of the argument. We have for orders 4 and 6:

$$\begin{aligned}
\kappa_{ijk\ell} &= \mu_{ijk\ell} - [3]\mu_{ij}\mu_{k\ell},\\
\kappa_{ijk}^{\ell} &= \mu_{ijk}^{\ell} - [3]\mu_{ij}\mu_{k}^{\ell},\\
\kappa_{ij}^{k\ell} &= \mu_{ij}^{k\ell} - [2]\mu_{i}^{k}\mu_{j}^{\ell} - \mu_{ij}\mu^{k\ell},\\
\kappa_{ijk\ell mn} &= \mu_{ijk\ell mn} - [15]\mu_{ijk\ell}\mu_{mn} + 2[15]\mu_{ij}\mu_{k\ell}\mu_{mn},\\
\kappa_{ijk\ell m}^{n} &= \mu_{ijk\ell m}^{n} - [5]\mu_{ijk\ell}\mu_{m}^{n} - [10]\mu_{ijk}^{n}\mu_{\ell m} + 2[15]\mu_{ij}\mu_{k\ell}\mu_{m}^{n},\\
\kappa_{ijk\ell}^{mn} &= \mu_{ijk\ell}^{mn} - \mu_{ijk\ell}\mu^{mn} - [8]\mu_{ijk}^{m}\mu_{\ell}^{n} - [6]\mu_{ij}^{mn}\mu_{k\ell}\\
&\quad + [6]\mu_{ij}\mu_{k\ell}\mu^{mn} + 2[12]\mu_{ij}\mu_{k}^{m}\mu_{\ell}^{n},\\
\kappa_{ijk}^{\ell mn} &= \mu_{ijk}^{\ell mn} - [3]\mu_{ijk}^{\ell}\mu^{mn} - [9]\mu_{ij}^{\ell m}\mu_{k}^{n} - [3]\mu_{ij}\mu_{k}^{\ell mn}\\
&\quad + 2[9]\mu_{ij}\mu_{k}^{\ell}\mu^{mn} + 2[6]\mu_{i}^{\ell}\mu_{j}^{m}\mu_{k}^{n},
\end{aligned}$$

and eventually for order 8:

$$\begin{aligned}
\kappa_{ijk\ell mnpq} &= \mu_{ijk\ell mnpq} - [28]\mu_{ijk\ell mn}\mu_{pq} - [35]\mu_{ijk\ell}\mu_{mnpq}\\
&\quad + 2[210]\mu_{ijk\ell}\mu_{mn}\mu_{pq} - 6[105]\mu_{ij}\mu_{k\ell}\mu_{mn}\mu_{pq},\\
\kappa_{ijk\ell mnp}^{q} &= \mu_{ijk\ell mnp}^{q} - [7]\mu_{ijk\ell mn}\mu_{p}^{q} - [21]\mu_{ijk\ell m}^{q}\mu_{np}\\
&\quad - [35]\mu_{ijk\ell}\mu_{mnp}^{q} + 2[105]\mu_{ijk}^{q}\mu_{\ell m}\mu_{np}\\
&\quad + 2[105]\mu_{ijk\ell}\mu_{mn}\mu_{p}^{q} - 6[105]\mu_{ij}\mu_{k\ell}\mu_{mn}\mu_{p}^{q},\\
\kappa_{ijk\ell mn}^{pq} &= \mu_{ijk\ell mn}^{pq} - \mu_{ijk\ell mn}\mu^{pq} - [12]\mu_{ijk\ell m}^{p}\mu_{n}^{q}\\
&\quad - [15]\mu_{ijk\ell}^{pq}\mu_{mn} - [15]\mu_{ijk\ell}\mu_{mn}^{pq} - [20]\mu_{ijk}^{p}\mu_{\ell mn}^{q}\\
&\quad + 2[15]\mu_{ijk\ell}\mu_{mn}\mu^{pq} + 2[30]\mu_{ijk\ell}\mu_{m}^{p}\mu_{n}^{q}\\
&\quad + 2[120]\mu_{ijk}^{p}\mu_{\ell m}\mu_{n}^{q} + 2[45]\mu_{ij}^{pq}\mu_{k\ell}\mu_{mn}\\
&\quad - 6[15]\mu_{ij}\mu_{k\ell}\mu_{mn}\mu^{pq} - 6[90]\mu_{ij}\mu_{k\ell}\mu_{m}^{p}\mu_{n}^{q},\\
\kappa_{ijk\ell m}^{npq} &= \mu_{ijk\ell m}^{npq} - [3]\mu_{ijk\ell m}^{n}\mu^{pq} - [15]\mu_{ijk\ell}^{np}\mu_{m}^{q}\\
&\quad - [10]\mu_{ijk}^{npq}\mu_{\ell m} - [5]\mu_{ijk\ell}\mu_{m}^{npq} - [30]\mu_{ijk}^{n}\mu_{\ell m}^{pq}\\
&\quad + 2[15]\mu_{ijk\ell}\mu_{m}^{n}\mu^{pq} + 2[30]\mu_{ijk}^{n}\mu_{\ell m}\mu^{pq}\\
&\quad + 2[60]\mu_{ijk}^{n}\mu_{\ell}^{p}\mu_{m}^{q} + 2[90]\mu_{ij}^{np}\mu_{k\ell}\mu_{m}^{q}
\end{aligned}$$

$$
\begin{aligned}
&+2[15]\mu_{i}^{npq}\mu_{jk}\mu_{\ell m}\\
&-6[45]\mu_{ij}\mu_{k\ell}\mu_{m}^{n}\mu^{pq}-6[60]\mu_{ij}\mu_{k}^{n}\mu_{\ell}^{p}\mu_{m}^{q},\\
\kappa_{ijk\ell}^{mnpq} &= \mu_{ijk\ell}^{mnpq}-[6]\mu_{ijk\ell}^{mn}\mu^{pq}-[6]\mu_{ij}^{mnpq}\mu_{k\ell}-[16]\mu_{ijk}^{mnp}\mu_{\ell}^{q}\\
&-[16]\mu_{ijk}^{m}\mu_{\ell}^{npq}-[18]\mu_{ij}^{mn}\mu_{k\ell}^{pq}-\mu_{ijk\ell}\mu^{mnpq}\\
&+2[\bar{2}]([3]\mu_{ijk\ell}\mu^{mn}\mu^{pq}+[48]\mu_{ijk}^{m}\mu_{\ell}^{n}\mu^{pq})\\
&+2[36]\mu_{ij}^{mn}\mu_{k\ell}\mu^{pq}+2[72]\mu_{ij}^{mn}\mu_{k}^{p}\mu_{\ell}^{q}\\
&-6[72]\mu_{ij}\mu_{k}^{m}\mu_{\ell}^{n}\mu^{pq}-6[24]\mu_{i}^{m}\mu_{j}^{n}\mu_{k}^{p}\mu_{\ell}^{q}\\
&-6[9]\mu_{ij}\mu_{k\ell}\mu^{mn}\mu^{pq}.
\end{aligned}
$$

References

[1] L. Albera, P. Comon, Asymptotic performance of contrast-based blind source separation algorithms, in: Second IEEE Sensor Array and Multichannel Signal Processing Workshop, Rosslyn, VA, 4–6 August 2002.

[2] L. Albera, P. Comon, P. Chevalier, A. Ferreol, Blind identification of underdetermined mixtures based on the hexacovariance, in: ICASSP'04, vol. II, Montreal, May 17–21, 2004, pp. 29–32.

[3] L. Albera, A. Ferreol, P. Comon, P. Chevalier, Blind identification of overcomplete mixtures of sources (BIOME), Lin. Alg. Appl. 391 (2004) 1–30.

[4] A. Belouchrani, K. Abed-Meraim, M.G. Amin, A.M. Zoubir, Joint antidiagonalization for blind source separation, in: Proc. IEEE Int. Conf. on Acoustics, Speech, and Signal Processing, ICASSP-01, vol. 5, Salt Lake City, May 2001, pp. 2789–2792.

[5] A. Belouchrani, K. Abed-Meraim, J.-F. Cardoso, E. Moulines, A blind source separation technique using second order statistics, IEEE Trans. Signal Process. 45 (1997) 434–444.

[6] A. Belouchrani, M.G. Amin, Blind source separation based on time-frequency signal representations, IEEE Trans. Signal Process. 46 (1998) 2888–2897.

[7] J. Brachat, P. Comon, B. Mourrain, E. Tsigaridas, Symmetric tensor decomposition, Lin. Alg. Appl., (2009) (submitted) hal:inria-00355713, arXiv:0901.3706.

[8] X.R. Cao, R.W. Liu, General approach to blind source separation, IEEE Trans. Signal Process. 44 (1996) 562–570.

[9] J.-F. Cardoso, Super-symmetric decomposition of the fourth-order cumulant tensor. Blind identification of more sources than sensors, in: Proc. IEEE Int. Conf. on Acoustics, Speech, and Signal Processing, ICASSP-91, Toronto, 1991, pp. 3109–3112.

[10] J.-F. Cardoso, E. Moulines, Asymptotic performance analysis of direction-finding algorithms based on fourth-order cumulants, IEEE Trans. Signal Process. 43 (1995) 214–224.

[11] J.-F. Cardoso, A. Souloumiac, Blind beamforming for non-Gaussian signals, IEE Proc.-F 140 (1994) 362–370 [Matlab code available at http://www.tsi.enst.fr/~cardoso/stuff.html].

[12] J.-F. Cardoso, A. Souloumiac, Jacobi angles for simultaneous diagonalization, SIAM J. Matrix Anal. Appl. 17 (1996) 161–164.

[13] J. Carroll, J. Chang, Analysis of individual differences in multidimensional scaling via an N-way generalization of Eckart-Young decomposition, Psychometrika 35 (1970) 283–319.

[14] P. Chevalier, L. Albera, A. Ferreol, P. Comon, On the virtual array concept for higher order array processing, IEEE Trans. Signal Process. 53 (2005) 1254–1271.

[15] P. Chevalier, A. Ferréol, On the virtual array concept for the fourth-order direction finding problem, IEEE Trans. Signal Process. 47 (1999) 2592–2595.

[16] P. Comon, Blind channel identification and extraction of more sources than sensors, in: SPIE Conference, San Diego, July 19–24, 1998, pp. 2–13. Keynote address.

[17] P. Comon, Tensor decompositions, state of the art and applications, in: J.G. McWhirter, I.K. Proudler (Eds.), Mathematics in Signal Processing V, Clarendon Press, Oxford, UK, 2002, pp. 1–24. arXiv:0905.0454v1.

[18] P. Comon, Blind identification and source separation in 2 × 3 under-determined mixtures, IEEE Trans. Signal Process. 52 (2004) 11–22.

[19] P. Comon, Contrasts, independent component analysis, and blind deconvolution, Int. J. Adapt. Control Signal Process. 18 (2004) 225–243 (Special issue on Signal Separation: http://www3.interscience.wiley.com/cgi-bin/jhome/4508. Preprint: I3S Research Report RR-2003-06).

[20] P. Comon, Tensors, usefulness and unexpected properties, in: IEEE Workshop on Statistical Signal Processing, SSP'09, Cardiff, UK, August 31–September 3, 2009. Keynote.

[21] P. Comon, G. Golub, L.-H. Lim, B. Mourrain, Symmetric tensors and symmetric tensor rank, SIAM J. Matrix Anal. Appl. 30 (2008) 1254–1279.

[22] P. Comon, X. Luciani, A.L.F. De Almeida, Tensor decompositions, alternating least squares and other tales, J. Chemomet. 23 (2009) 1254–1279.

[23] P. Comon, B. Mourrain, Decomposition of quantics in sums of powers of linear forms, Signal Process. 53 (1996) 93–107. Special issue on High-Order Statistics.

[24] P. Comon, M. Rajih, Blind identification of under-determined complex mixtures of 2 independent sources, in: 3rd IEEE SAM Workshop, Barcelona, Spain, July 18–21, 2004.

[25] P. Comon, M. Rajih, Blind identification of under-determined mixtures based on the characteristic function, Signal Process. 86 (2006) 2271–2281.

[26] P. Comon, J.M.F. ten Berge, L. De Lathauwer, J. Castaing, Generic and typical ranks of multi-way arrays, Lin. Alg. Appl. 430 (2009) 2997–3007.

[27] L. De Lathauwer, A link between the canonical decomposition in multilinear algebra and simultaneous matrix diagonalization, SIAM J. Matrix Anal. Appl. 28 (2006) 642–666.

[28] L. De Lathauwer, J. Castaing, Blind identification of underdetermined mixtures by simultaneous matrix diagonalization, IEEE Trans. Signal Process. 56 (2008) 1096–1105.

[29] L. De Lathauwer, J. Castaing, J.-F. Cardoso, Fourth-order cumulant-based identification of underdetermined mixtures, IEEE Trans. Signal Process. 55 (2007) 2965–2973.

[30] L. De Lathauwer, P. Comon, B. De Moor, ICA algorithms for 3 sources and 2 sensors, in: Proc. IEEE Signal Processing Workshop on Higher-Order Statistics, HOS'99, Caesarea, Israel, June 14–16, 1999, pp. 116–120.

[31] L. De Lathauwer, B. De Moor, J. Vandewalle, Algebraic ICA algorithm for 3 sources and 2 sensors, in: Proc. Xth European Signal Processing Conference, EUSIPCO 2000, Tampere, Finland, September 5–8, 2000.

[32] L. De Lathauwer, B. De Moor, J. Vandewalle, Computation of the canonical decomposition by means of a simultaneous generalized Schur decomposition, SIAM J. Matrix Anal. Appl. 26 (2004) 295–327.

[33] V. De Silva, L.-H. Lim, Tensor rank and the ill-posedness of the best low-rank approximation problem, SIAM J. Matrix Anal. Appl. 30 (2008) 1084–1127.

[34] M.C. Dogan, J.M. Mendel, Applications of cumulants to array processing – Part I: Aperture extension and array calibration, IEEE Trans. Signal Process. 43 (1995) 1200–1216.

[35] D. Donoho, M. Elad, N. Temlyakov, Stable recovery of sparse overcomplete representations in the presence of noise, IEEE Trans. Inform. Theory 52 (2006) 6–18.

[36] J. Eriksson, V. Koivunen, Identifiability, separability and uniquness of linear ICA models, IEEE Signal Process. Lett. (2004) 601–604.

[37] J. Eriksson, V. Koivunen, Complex random vectors and ICA models: Identifiability, uniqueness and separability, IEEE Trans. Inform. Theory 52 (2006) 1017–1029.

[38] W. Feller, An Introduction to Probability Theory and its Applications, vol. II, Wiley, 1966; 1971, 2nd ed.
[39] P.E. Gill, W. Murray, M.H. Wright, Practical Optimization, Academic Press, 1981.
[40] O. Grellier, P. Comon, B. Mourrain, P. Trebuchet, Analytical blind channel identification, IEEE Trans. Signal Process. 50 (2002) 2196–2207.
[41] R. Gribonval, E. Bacry, Harmonic decomposition of audio signals with matching pursuit, IEEE Trans. Signal Process. 51 (2003) 101–111.
[42] R.A. Harshman, Foundations of the Parafac procedure: Model and conditions for an "explanatory" multi-mode factor analysis, UCLA Working Papers in Phonetics 16 (1970) 1–84.
[43] F.L. Hitchcock, The expression of a tensor or a polyadic as a sum of products, J. Math. Phys. 6 (1927) 165–189.
[44] A.M. Kagan, Y.V. Linnik, C.R. Rao, Characterization problems in mathematical statistics, in: Probability and Mathematical Statistics, Wiley, New York, 1973.
[45] M. Kendall, A. Stuart, The Advanced Theory of Statistics, Distribution Theory, vol. 1–3, C. Griffin, 1977.
[46] W.P. Krijnen, T.K. Dijkstra, A. Stegeman, On the non-existence of optimal solutions and the occurrence of degeneracy in the Candecomp/Parafac model, Psychometrika 73 (2008) 431–439.
[47] P.M. Kroonenberg, Applied Multiway Data Analysis, Wiley, 2008.
[48] P.M. Kroonenberg, J.D. Leeuw, Principal component analysis of three-mode data, Psychometrika 45 (1980) 69–97.
[49] J.B. Kruskal, Three-way arrays: Rank and uniqueness of trilinear decompositions, Lin. Alg. Appl. 18 (1977) 95–138.
[50] J.B. Kruskal, R.A. Harshman, M.E. Lundy, How 3-MFA data can cause degenerate Parafac solutions, among other relationships, in: R. Coppi, S. Bolasco (Eds.), Multiway Data Analysis, Elsevier Science, North-Holland, 1989, pp. 115–121.
[51] D.D. Lee, H.S. Seung, Learning the parts of objects by non-negative matrix factorization, Nature 401 (1999) 788–791.
[52] S. Leurgans, R.T. Ross, R.B. Abel, A decomposition for three-way arrays, SIAM J. Matrix Anal. Appl. 14 (1993) 1064–1083.
[53] L.-H. Lim, P. Comon, Nonnegative approximations of nonnegative tensors, J. Chemomet. 23 (2009) 432–441.
[54] K. Madsen, H.B. Nielsen, O. Tingleff, Methods for non-linear least squares problems, Informatics and Mathematical Modelling, Technical University of Denmark, DTU, Richard Petersens Plads, Building 321, DK-2800 Kgs. Lyngby, 2004.
[55] P. McCullagh, Tensor Methods in Statistics, in: Monographs on Statistics and Applied Probability, Chapman and Hall, 1987.
[56] C.L. Nikias, A.P. Petropulu, Higher-Order Spectra Analysis. A Nonlinear Signal Processing Framework, Prentice Hall, Englewood Cliffs, New Jersey, 1993.
[57] D. Nion, L. De Lathauwer, Block component model based blind DS-CDMA receivers, IEEE Trans. Signal Process. 56 (2008) 5567–5579.
[58] D. Nion, L. De Lathauwer, An enhanced line search scheme for complex-valued tensor decompositions. Application in DS-CDMA, Signal Process. 88 (2008) 749–755.
[59] P. Paatero, A weighted non-negative least squares algorithm for three-way Parafac factor analysis, Chemometrics Intell. Lab. Syst. 38 (1997) 223–242.
[60] P. Paatero, The multilinear engine: A table-driven, least squares program for solving multilinear problems, including the n-way parallel factor analysis model, J. Comput. Graph. Statist. 8 (1999) 854–888.
[61] P. Paatero, Construction and analysis of degenerate Parafac models, J. Chemomet. 14 (2000) 285–299.
[62] D.-T. Pham, J.-F. Cardoso, Blind separation of instantaneous mixtures of non-stationary sources, IEEE Trans. Signal Process. 49 (2001) 1837–1848.

[63] M. Rajih, P. Comon, R.A. Harshman, Enhanced line search: A novel method to accelerate Parafac, SIAM J. Matrix Anal. Appl. 30 (2008) 1148–1171.

[64] N.D. Sidiropoulos, R. Bro, On the uniqueness of multilinear decomposition of N-way arrays, J. Chemomet. 14 (2000) 229–239.

[65] N.D. Sidiropoulos, R. Bro, G.B. Giannakis, Parallel factor analysis in sensor array processing, IEEE Trans. Signal Process. 48 (2000) 2377–2388.

[66] N.D. Sidiropoulos, G.B. Giannakis, R. Bro, Blind Parafac receivers for DS-CDMA systems, IEEE Trans. Signal Process. 48 (2000) 810–823.

[67] A. Smilde, R. Bro, P. Geladi, Multi-Way Analysis, Wiley, 2004.

[68] A. Stegeman, Degeneracy in Candecomp/Parafac and Indscal explained for several three-sliced arrays with a two-valued typical rank, Psychometrika 72 (2007) 601–619.

[69] A. Stegeman, L. De Lathauwer, A method to avoid diverging components in the Candecomp/Parafac model for generic $I \times J \times 2$ arrays, SIAM J. Matrix Anal. Appl. 30 (2009) 1614–1638.

[70] A. Stegeman, N.D. Sidiropoulos, On Kruskal's uniqueness condition for the Candecomp/Parafac decomposition, Lin. Alg. Appl. 420 (2007) 540–552.

[71] A. Stegeman, J.M.F. ten Berge, L. De Lathauwer, Sufficient conditions for uniqueness in Candecomp/Parafac and Indscal with random component matrices, Psychometrika 71 (2006) 219–229.

[72] A. Taleb, C. Jutten, On underdetermined source separation, in: ICASSP99, Phoenix, Arizona, March 15–19, 1999.

[73] J.M.F. ten Berge, Kruskal's polynomial for $2 \times 2 \times 2$ arrays and a generalization to $2 \times n \times n$ arrays, Psychometrika 56 (1991) 631–636.

[74] G. Tomasi, R. Bro, A comparison of algorithms for fitting the Parafac model, Comp. Stat. & Data Anal. 50 (2006) 1700–1734.

[75] A.-J. van der Veen, A. Paulraj, An analytical constant modulus algorithm, IEEE Trans. Signal Process. 44 (1996) 1136–1155.

[76] S.A. Vorobyov, Y. Rong, N.D. Sidiropoulos, A.B. Gershman, Robust iterative fitting of multilinear models, IEEE Trans. Signal Process. (2005) 2678–2689.

[77] A. Yeredor, Blind source separation via the second characteristic function, Signal Process. 80 (2000) 897–902.

[78] A. Yeredor, Non-orthogonal joint diagonalization in the least-squares sense with application in blind source separation, IEEE Trans. Signal Process. 50 (2002) 1545–1553.

CHAPTER

Sparse component analysis 10

R. Gribonval and M. Zibulevsky

10.1 INTRODUCTION

Sparse decomposition techniques for signals and images underwent considerable development during the flourishing of wavelet-based compression and denoising methods [75] in the early 1990s. Some ten years elapsed before these techniques began to be fully exploited for blind source separation [72,106,66,70,116,61,15,111]. Their main impact is that they provide a relatively simple framework for separating a number of sources exceeding the number of observed mixtures. Also they greatly improve quality of separation in the case of square mixing matrix.

The underlying assumption is essentially geometrical [98], differing considerably from the traditional assumption of independent sources [19]. This can be illustrated using the simplified model of speech discussed by Van Hulle [106]. One wishes to identify the mixing matrix and extract the sources based on the observed mixture:

$$\mathbf{x}(t) = \mathbf{A}\mathbf{s}(t) + \mathbf{b}(t), \qquad 1 \le t \le T. \tag{10.1}$$

For this purpose, one assumes that at each point t (variable designating a time instant for temporal signals, a pixel coordinate for images, etc.), a single source is significantly more active than the others. If $\Lambda_n \subset \{1,\dots,T\}$ denotes the set of points where the source with index n is most active (we will refer to the temporal *support* of source n), then for all $t \in \Lambda_n$ one has by definition $|s_n(t)| \gg |s_m(t)|$ for $m \neq n$, and therefore:

$$\mathbf{x}(t) \approx s_n(t)\mathbf{A}_n, \qquad t \in \Lambda_n \tag{10.2}$$

where $\mathbf{A}_n = (A_{pn})_{1 \le p \le P}$ is the (unknown) n-th column of the mixing matrix. It follows that the set of points $\{\mathbf{x}(t) \in \mathbb{C}^P, t \in \Lambda_n\}$ is more or less aligned along the straight line passing through the origin and directed by vector $\mathbf{A}_n$. As shown in Fig. 10.1(a) for a stereophonic mixture ($P = 2$) of three audio sources ($N = 3$), this alignment can be observed in practice on the *scatter plot* $\{\mathbf{x}(t) \in \mathbb{C}^P, 1 \le t \le T\}$. In this figure in dimension $P = 2$, the scatter plot is the collection of points representing the pairs of values $(x_1(t), x_2(t))$ observed for all T samples of the mixture signal.

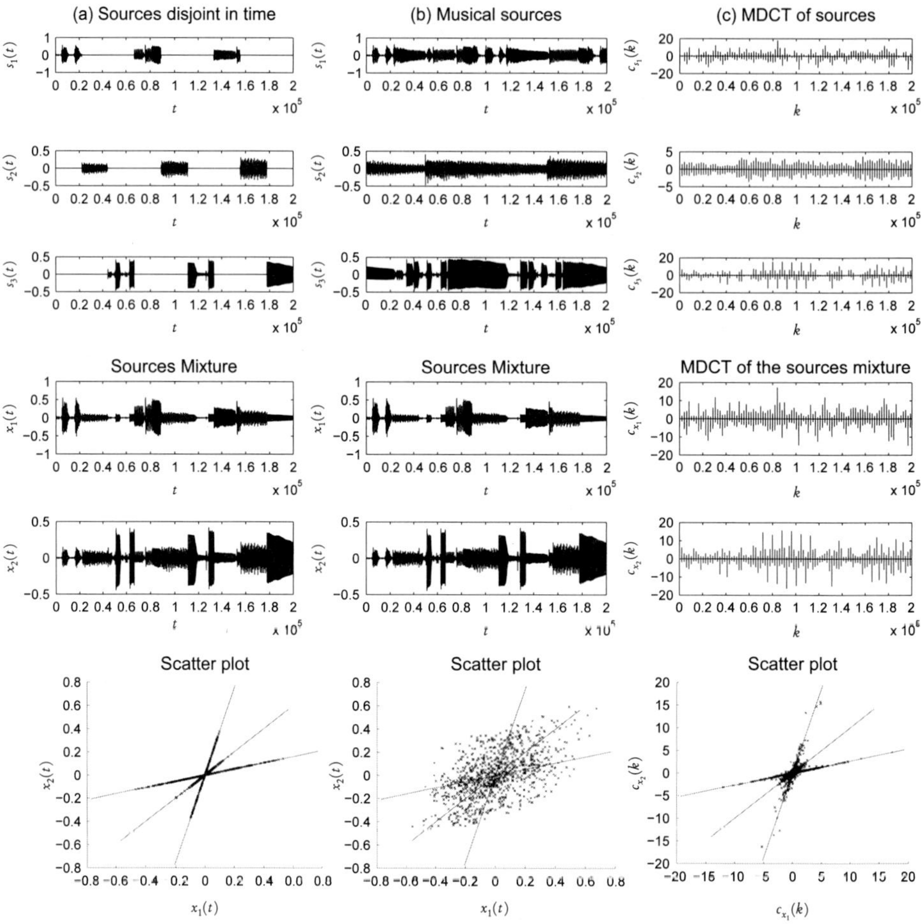

FIGURE 10.1

"Geometric" principle of source separation by sparse decomposition. Sources (top), the corresponding mixtures (middle) and the resulting scatter plots (bottom). (a) – For sources with disjoint temporal supports, the temporal scatter plot shows alignments along the columns of the mixing matrix. (b) – For more complex signals such as musical signals (center), the temporal scatter plot is not legible. (c) – An appropriate orthogonal time-frequency transform (MDCT) applied to the same musical signals (right) results in a scatter plot in which the column directions of the mixing matrix are clearly visible.

It is thus conceivable to estimate the mixing matrix $\widehat{\mathbf{A}}$ using a *clustering* algorithm [110] – the scatter plot is separated into N clusters of points $\{\mathbf{x}(t), t \in \widehat{\Lambda}_n\}$ – and to estimate the direction $\widehat{\mathbf{A}}_n$ of each cluster. If the mixture is (over)-determined ($N \leq P$), the sources can be estimated with the pseudo-inverse [47] of the estimated mixing matrix $\widehat{\mathbf{s}}(t) := \widehat{\mathbf{A}}^{\dagger}\mathbf{x}(t)$, but even in the case of an *under-determined* mixture ($N > P$), by assuming the sources have disjoint supports, they can be estimated using least squares[1] from the obtained clusters:

$$\hat{s}_n(t) := \begin{cases} \langle \mathbf{x}(t), \hat{\mathbf{A}}_n \rangle / \|\hat{\mathbf{A}}_n\|_2^2 & \text{if } t \in \hat{\Lambda}_n, \\ 0 & \text{if not.} \end{cases}$$

Although simple and attractive, Van Hulle's technique generally cannot be applied in its original form, as illustrated in Fig. 10.1(b), which shows the temporal scatter plot of a stereophonic mixture ($P = 2$) of $N = 3$ musical sources. It is clear that the sources cannot be considered to have disjoint temporal supports, and the temporal scatter plot does not allow visual distinguishing of the column directions of the mixing matrix.

This is where sparse signal representations come into play. Figure 10.1(c) shows the coefficients $c_{s_n}(k)$ of an orthogonal time-frequency transform applied to the sources (MDCT [87]), the coefficients $c_{x_p}(k)$ of the mixture, and finally the time-frequency scatter plot of points $(c_{x_1}(k), c_{x_2}(k))$, $1 \le k \le T$. On the time-frequency scatter plot, one can once again observe the directions of columns of the mixing matrix and apply the strategy described previously for estimating $\mathbf{A}$. The MDCT coefficients for the original sources on the same figure explain this "miracle": the coefficients of the MDCTs for the different sources have (almost) disjoint supports, given that for each source only a small number of MDCT coefficients are of significant amplitude. The above-mentioned approach can once again be used to estimate $\hat{\mathbf{A}}$ and the coefficients of the time-frequency transform applied to the sources. The sources themselves are estimated by applying an inverse transform.

Figure 10.2 shows the structure of sparsity-based source separation methods, for which we have just introduced the main ingredients:

1. A step for changing the representation of the observed mixture, to make the supports of the source coefficients in the new representation as disjoint as possible. For this purpose, *joint sparse representation* techniques are used.
2. The column directions of the mixing matrix are estimated using the mixture representation. In addition to classical ICA algorithms, *clustering algorithms* can be applied to the *scatter plot*.
3. Finally, *estimating the (sparse) source representations* makes it possible to *reconstruct* the sources.

Chapter outline

The rest of this chapter provides a detailed discussion of the main approaches for each step of sparse separation. Sparse signal representation and dictionaries are introduced in section 10.2 with relevant examples. The formalism and main algorithms for calculating joint sparse representations of mixtures $\mathbf{x}(t)$ are described in section 10.3, which

[1]The notation $\langle \cdot, \cdot \rangle$ is used for the scalar product between two vectors, e.g. for column vectors $\langle \mathbf{x}(t), \hat{\mathbf{A}}_n \rangle := \mathbf{x}(t)\hat{\mathbf{A}}_n^{\mathrm{H}} = \sum_{p=1}^{P} x_p(t)\hat{A}_{pn}^*$ and for row vectors $\langle s_n, s_{n'} \rangle = \sum_{t=1}^{T} s_n(t) s_{n'}^*(t)$.

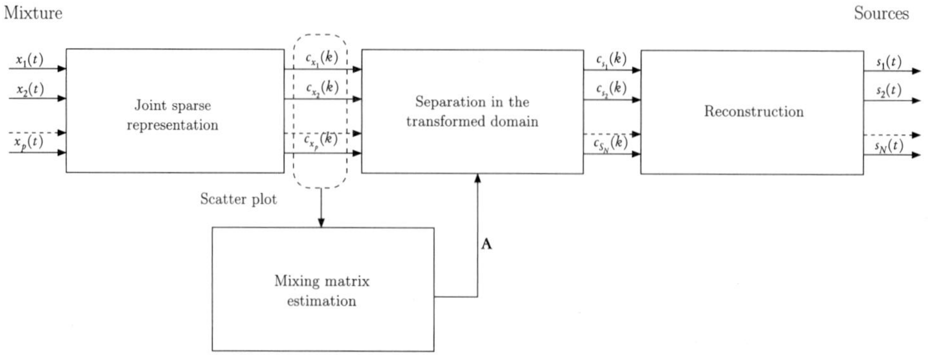

FIGURE 10.2

Block diagram of sparsity-based source separation methods.

concludes by reviewing the main characteristics, advantages and disadvantages of these algorithms, the scope of which extends well beyond source separation. Section 10.4 examines actually using the sparsity of joint mixture representations for the second step of source separation, in which the mixing matrix is estimated from scatter plots, while section 10.5 discusses the Relative Newton framework for the special case of a square mixing matrix. Finally, section 10.6 focuses on the algorithms for the actual separation step in the "transform" domain, to estimate the sources once the mixing matrix has been estimated. The possible choices for the different steps of sparse source separation are summarized in section 10.7, which also discusses the factors that may determine these choices. To conclude, in section 10.8 we assess the new directions and challenges opened up by sparse source separation, such as using new diversity factors between sources, blind estimation of suitable dictionaries and separation of under-determined convolutive mixtures.

10.2 SPARSE SIGNAL REPRESENTATIONS

The basic assumption underlying all sparse source separation methods is that each source can be (approximately) represented by a linear combination of a few elementary signals φ_k, known as *atoms*[2] that are often assumed to be of unit energy. Specifically it is assumed that each (unknown) source signal can be written:

$$s(t) = \sum_{k=1}^{K} c(k)\, \varphi_k(t) \tag{10.3}$$

[2]Although the notations φ (or ψ) and $\boldsymbol{\Phi}$ (or $\boldsymbol{\Psi}$) are used elsewhere in this book to denote the marginal score functions (or the joint score function) of the sources and the first (or second) characteristic function, they will be used in this chapter to denote atoms with regard to signal synthesis and analysis and the corresponding dictionaries/matrices.

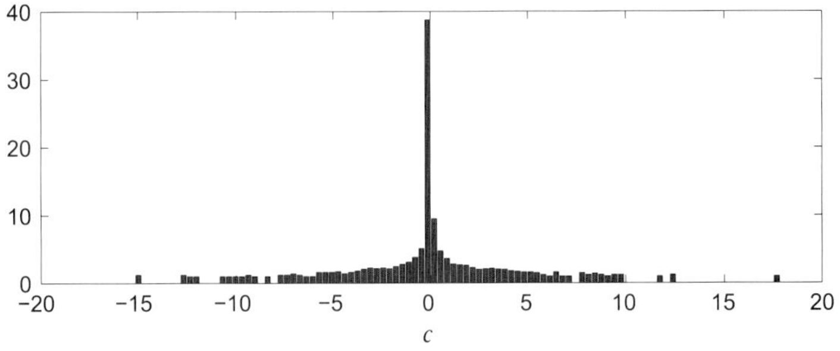

FIGURE 10.3

Histogram of MDCT coefficients of the first source in the mixture of musical signals in Fig. 10.1(c).

where only few entries $c(k)$ of the vector $\mathbf{c}$ are significant, most of them being negligible. In the case of Van Hulle's simplified speech model [106], the considered atoms are simply Dirac functions, $\varphi_k(t) = \delta(t-k)$, i.e. time samples. Many audio signal processing applications involve the short-term Fourier transform (STFT, also called Gabor transform or spectrogram) [75] where the index $k = (n, f)$ denotes a time-frequency point, $c(k)$ is the value of the STFT applied to the signal s at this point, and (10.3) corresponds to the reconstruction formula for s based on its STFT using the *overlap-add* (OLA) method [89]. It is possible to use other families of atoms known as *dictionaries*, which are better suited to representing a given type of signal. In our context, the sparseness of a signal's representation in a dictionary determines the dictionary's relevance.

10.2.1 Basic principles of sparsity

Informally, a set of coefficients $c(k)$ is considered sparse if most of the coefficients are zero or very small and only a few of them are significant. In statistical (or probabilistic) terms, coefficients $c(k)$ are said to have a sparse or super-Gaussian distribution if their histogram (or probability density) has a strong peak at the origin with heavy tails, as shown in Fig. 10.3.

The most common model of sparse distribution is the family of generalized Gaussian distributions:

$$p_c(c) \propto \exp(-\alpha |c|^{\tau}) \tag{10.4}$$

for $0 < \tau \leq 1$, where the $\tau = 1$ case corresponds to the Laplace distribution. Certain Bayesian source separation techniques that assume sparsity use Student-t distributions to exploit their analytical expression in terms of the Gaussian Mixture Model [40]. For the same reasons, the bi-Gaussian model, where one Gaussian distribution has low variance and the other high variance, is often used as well [30].

For the generalized Gaussian model, the set of coefficients $\mathbf{c} = \{c(k)\}_{k=1}^{K}$, assumed statistically independent, has a distribution of $p_c(\mathbf{c}) \propto \exp(-\alpha \|\mathbf{c}\|_{\tau}^{\tau})$ where $\|\mathbf{c}\|_{\tau}$ denotes the ℓ^τ "norm"[3]:

$$||\mathbf{c}||_\tau := \left(\sum_{k=1}^{K} |c(k)|^\tau\right)^{1/\tau}. \tag{10.5}$$

The likelihood of a set of coefficients thus increases as its ℓ^τ norm decreases. The norm $||\mathbf{c}||_\tau^\tau$ can then be used to quantify the sparsity of a representation $\mathbf{c}$, which is the approach taken in the rest of this chapter.

10.2.2 Dictionaries

In general, one considers complete dictionaries, that is to say dictionaries that allow the reconstruction of any signal $s \in \mathbb{C}^T$; in other words, the space generated by the atoms of the dictionary is $\mathbb{C}^T$ in its entirety. Limiting the approach presented in this chapter to the case of real signals poses no particular problems. With $\boldsymbol{\Phi}$ denoting the $K \times T$ matrix whose K rows consist of atoms $\varphi_k(t), 1 \leq t \leq T, k = 1,\ldots K$, the sparse model (10.3) of sources can be written in matrix form[4]:

$$s = c_s \cdot \boldsymbol{\Phi} \tag{10.6}$$

where $c_s = (c_s(k))_{k=1}^K$ is a row vector of K coefficients.

If the selected dictionary is a basis, $K = T$ and a one-to-one correspondence exists between the source signal s and its coefficients $c_s = s \cdot \boldsymbol{\Phi}^{-1}$ ($c_s = s \cdot \boldsymbol{\Phi}^{\mathrm{H}}$ for an orthonormal basis). This is the case for a dictionary corresponding to an orthogonal basis of discrete wavelets, or the orthogonal discrete Fourier basis [75], or a MDCT (*Modified Discrete Cosine Transform* [87]) type basis.

By contrast, if the dictionary is redundant ($K > T$), there is an infinite number of possible coefficient sets for reconstructing each source. This is the case for a Gabor dictionary, comprising atoms expressed as

$$\varphi_{n,f}(t) = w(t - nT)\exp(2j\pi f t) \tag{10.7}$$

– where w is called an analysis window – and used to compute the STFT of a signal as $STFT_s(n,f) = \langle s, \varphi_{n,f}\rangle$. The construction and study of redundant dictionaries for sparse signal representation is based on the idea that, among the infinitely many possibilities, one can choose a representation particularly well suited to a given task.

[3]Strictly speaking, this is a norm only for $1 \leq \tau \leq \infty$, with the usual modification $||\mathbf{c}||_\infty = \sup_k |c(k)|$. We will also use the term "norm", albeit improperly, for $0 < \tau < 1$. For $\tau = 0$, with the convention $c^0 = 1$ if $c > 0$ and $0^0 = 0$, the ℓ^0 norm is the number of non-zero coefficients of $\mathbf{c}$.

[4]With the usual source separation notation, the mixing matrix $\mathbf{A}$ acts to the left on the matrix of source signals $\mathbf{s}$, which is why the dictionary matrix $\boldsymbol{\Phi}$ must act *to the right* on the coefficients c_s for signal synthesis. This notation will be used throughout this chapter.

Since the early 1990s, building dictionaries has been the focus of several harmonic analysis studies, still being pursued today. This has led to the construction of several wavelet families as well as libraries of orthogonal wavelet packet bases or local trigonometric functions [75] and other more sophisticated constructions (curvelets, etc. [95]) too numerous to cite here. Sparse coding [41,70,69,1] is closer to the source separation problem and also aims at the learning of a dictionary (redundant or non-redundant) maximizing the sparse representation of a training data set.

10.2.3 Linear transforms

In a given dictionary $\mathbf{\Phi}$, the traditional approach for calculating a representation of s consists of calculating a linear *transform*:

$$c_s^{\Psi} := s \cdot \mathbf{\Psi}, \tag{10.8}$$

and the calculated coefficients are thus the values of the correlations

$$c_s^{\Psi}(k) = \langle s, \psi_k \rangle \tag{10.9}$$

of s with a set of dual atoms $\psi_k(t)$, which form the columns of the matrix $\mathbf{\Psi}$. One can thus consider the atoms $\varphi_k(t)$ as related to signal synthesis, as opposed to the atoms related to signal analysis $\psi_k(t)$.

For the transformation of a signal s (analysis, (10.8)), followed by its reconstruction (synthesis, (10.6)), the transform must satisfy $s \cdot \mathbf{\Psi\Phi} = s$ for every signal s to result in perfect reconstruction, i.e. $\mathbf{\Psi\Phi} = \mathbf{I}$ ($\mathbf{I}$ is the identity matrix, here $T \times T$ in size). For example, in the case of a Gabor dictionary (10.7), a set of dual atoms compatible with the reconstruction condition (10.8) takes the form:

$$\psi_{n,f}(t) = w'(t - nT)\exp(2j\pi f t), \tag{10.10}$$

where the analysis window $w'(t)$ and the reconstruction window $w(t)$ must satisfy the condition $\sum_n w'(t - nT)w(t - nT) = 1$ for every t. Therefore, analyzing a signal with the analysis atoms $\psi_{n,f}$ is akin to calculating its STFT with the window w', whereas reconstruction with the synthesis atoms $\varphi_{n,f}$ corresponds to *overlap-add* reconstruction based on the STFT [89].

When the dictionary $\mathbf{\Phi}$ is a basis for the signal space, as is the discrete orthonormal Fourier basis, or a bi-orthogonal wavelet basis [75], the only transform $\mathbf{\Psi}$ allowing perfect reconstruction is $\mathbf{\Psi} = \mathbf{\Phi}^{-1}$ (a fast Fourier transform or a fast wavelet transform correspondingly).

For these types of non-redundant dictionaries that correspond to "critically sampled" transforms, the relation $\mathbf{\Phi\Psi} = \mathbf{I}$ also holds; that is, synthesis of a signal based on a representation c_s, followed by analysis of the signal, makes it possible to find the exact coefficients used. This does not hold true in the case of redundant linear transforms, such as the STFT or the continuous wavelet transform.

In the case of any redundant dictionary, there is an infinite number of "dual" matrices $\boldsymbol{\Psi}$ that satisfy the reconstruction condition $\boldsymbol{\Psi\Phi} = \mathbf{I}$, but none satisfies the identity $\boldsymbol{\Phi\Psi} = \mathbf{I}$. The pseudo-inverse [47] $\boldsymbol{\Psi} = \boldsymbol{\Phi}^{\dagger} = (\boldsymbol{\Phi}^{\mathrm{H}}\boldsymbol{\Phi})^{-1}\boldsymbol{\Phi}^{\mathrm{H}}$ of $\boldsymbol{\Phi}$ offers a specific choice of linear transform, providing the least squares solution to the representation problem $s = c_s\boldsymbol{\Phi}$:

$$c_s^{\boldsymbol{\Phi}^{\dagger}} = \arg \min_{c_s | c_s\boldsymbol{\Phi}=s} \sum_{k=1}^{K} |c_s(k)|^2. \tag{10.11}$$

10.2.4 Adaptive representations

The reason for using a redundant dictionary is to find a particularly sparse representation from among the infinite possibilities for a given signal. The dictionary's redundancy is aimed at offering a broad range of atoms likely to represent the typical signal structures in a suitable way, so that the signal can be approximated by a linear combination of a small, carefully selected atom set from the dictionary.

Paradoxically, the more redundant the dictionary, the higher the number of non-zero coefficients in a representation obtained by a linear transform. Thus, for the STFT associated with a Gabor dictionary, redundancy increases with overlap between adjacent windows (which increases the number of analysis frames) and with *zero-padding* (which increases the number of discrete analysis frequencies), but since a given atom's "neighboring" atoms in time-frequency are highly correlated to it, the number of non-negligible coefficients observed during analysis of a Gabor atom (10.7) also increases with the transform's redundancy.

To fully leverage the potential of redundant dictionaries and obtain truly sparse signal representation, linear transforms must be replaced by adaptive techniques. The principle of these techniques, which emerged in the 1990s, is to select a limited subset Λ of atoms from the dictionary based on the analyzed signal. Consequently, only the selected atoms are assigned a non-zero coefficient, resulting in a sparse representation or a sparse approximation of the signal. The main techniques to obtain adaptive signal representation are described in the following section, in the slightly more general context of joint sparse representation of mixtures, which is the first step in sparse source separation.

10.3 JOINT SPARSE REPRESENTATION OF MIXTURES

The first step in a system of sparse source separation (see Fig. 10.2) consists in calculating a joint sparse representation of the observed mixture $\mathbf{x}$, to obtain a representation that facilitates both the estimation of the mixing matrix and that of the sources by clustering algorithms. While this type of representation can be obtained with linear transforms (STFT, wavelet transforms, etc.), greater sparsity may be achieved by using algorithms for joint adaptive decomposition of the mixture.

10.3.1 Principle

An approximate sparse representation of a mixture $\mathbf{x}$ takes the form $\mathbf{x} \approx C_{\mathbf{x}}\mathbf{\Phi}$, where:

$$C_{\mathbf{x}} = \begin{bmatrix} c_{x_1} \\ \cdots \\ c_{x_P} \end{bmatrix} = \left[C_{\mathbf{x}}(1) \; \dots \; C_{\mathbf{x}}(K)\right] \tag{10.12}$$

is a matrix of dimensions $P \times K$ in which each row c_{x_p} is a row vector of K coefficients $c_{x_p}(k)$ representing one of the P channels of the mixture. Calculating $C_{\mathbf{x}}$ for a mixture $\mathbf{x} \approx \mathbf{As}$ is aimed at reducing the problem to a separation problem $C_{\mathbf{x}} \approx \mathbf{A}C_{\mathbf{s}}$, which is expected to be easier to solve if $C_{\mathbf{s}}$ is sparse.

By combining the sparse model of sources (see (10.3) or (10.6)) with the noisy instantaneous linear mixture model (10.1), one obtains a global sparse model of the mixture:

$$\mathbf{x} = \mathbf{A}C_{\mathbf{s}}\mathbf{\Phi} + \mathbf{b}, \qquad \text{where} \qquad C_{\mathbf{s}} = \begin{bmatrix} c_{s_1} \\ \cdots \\ c_{s_N} \end{bmatrix} = \left[C_{\mathbf{s}}(1) \; \dots \; C_{\mathbf{s}}(K)\right] \tag{10.13}$$

is an $N \times K$ matrix in which each row c_{s_n} is a sparse vector consisting of K coefficients $c_{s_n}(k)$ representing one of the N sources, and each column $C_{\mathbf{s}}(k)$, $1 \le k \le K$, indicates the degree of "activity" of the different sources for a given atom. If a representation $C_{\mathbf{s}}$ of the sources exists where they have disjoint supports (each atom only activated for one source at most), then $C_{\mathbf{x}} := \mathbf{A}C_{\mathbf{s}}$ is an admissible (approximate) representation of the (noisy) mixture $\mathbf{x}$ and should enable its separation via the scatter plot.

When $\mathbf{\Phi}$ is an orthonormal basis such as the MDCT, a unique representation exists for the mixture (and the sources) and can thus be calculated simply by inverse linear transform, necessarily satisfying $C_{\mathbf{x}} \approx \mathbf{A}C_{\mathbf{s}}$. We saw an example of this with the musical sources, where the scatter plot of the columns of the representation $C_{\mathbf{x}}$ obtained by linear transform with an orthonormal MDCT-type basis $\mathbf{\Phi}$ clearly highlighted the column directions of the mixing matrix (see Fig. 10.1(c)), making it possible to pursue the separation. By contrast, when $\mathbf{\Phi}$ is a redundant dictionary, the calculated representation $C_{\mathbf{x}}$ depends on the chosen algorithm and does not necessarily satisfy the identity $C_{\mathbf{x}} \approx \mathbf{A}C_{\mathbf{s}}$, which is crucial for the subsequent steps of sparse separation.

Selecting a relevant joint sparse approximation algorithm for the mixture thus depends on the algorithm's ability to provide a representation that satisfies the approximation $C_{\mathbf{x}} \approx \mathbf{A}C_{\mathbf{s}}$ with sufficient accuracy, provided the sources allow a representation $C_{\mathbf{s}}$ with sufficiently disjoint supports. This important point will be discussed for each of the joint sparse mixture approximation algorithms described in this section together with their main properties.

10.3.2 Linear transforms

The simplest joint representation of a mixture involves applying the same linear transform (e.g. STFT) to all channels. Formally, by juxtaposing the transforms $c_{x_p}^{\mathbf{\Psi}} := x_p\mathbf{\Psi}$ of each of the mixture components x_p, $1 \leq p \leq P$, one obtains a matrix:

$$C_{\mathbf{x}}^{\Psi} := \mathbf{x}\Psi \tag{10.14}$$

called the transform of mixture **x**. Its numerical calculation is particularly simple but, as discussed above, this type of transform does not necessarily take full advantage of the dictionary's redundancy to obtain representations with high sparseness.

Algorithms

The linear transforms Ψ applied to a mixture are often associated with fast algorithms: Fourier transform, STFT, orthogonal wavelet transform, etc. Another possibility [29] is that the dictionary Φ consists in the union of a certain number of orthonormal bases associated with fast transforms (wavelet, local cosine, etc.), $\Phi = [\Phi_1 \ldots \Phi_J]$, in which case the pseudo-inverse $\Phi^{\dagger} = J^{-1}\Phi^H$ provides a dual transform $\Psi := \Phi^{\dagger}$ that is simple and quick to apply by calculating each of the component fast transforms and concatenating the coefficients obtained. It is also possible to simultaneously calculate two STFTs for the same mixture, one with a small analysis window and the other with a large window, to obtain a sort of multi-window transform [82].

Properties and limitations

By the linearity of the transform, the initial source separation problem (10.1) has an exact analog in the transform domain (see (10.8) and (10.14)):

$$C_{\mathbf{x}}^{\Psi} - \mathbf{A}C_{\mathbf{s}}^{\Psi} + C_{\mathbf{b}}^{\Psi}. \tag{10.15}$$

Separation in the transform domain, to estimate the source transforms $\widehat{C}_{\mathbf{s}}$ based on $C_{\mathbf{x}}$ and ultimately reconstruct $\widehat{\mathbf{s}} := \widehat{C}_{\mathbf{s}}\Phi$, is thus relevant if the supports of the sources in the transform domain $C_{\mathbf{s}}^{\Psi}$ are reasonably disjoint.

Despite the advantages of a simple, efficient numerical implementation, transforms are far from ideal in terms of sparse representation when the dictionary is redundant. In particular, the least squares representation $C_{\mathbf{x}}^{\Phi^{\dagger}} := \mathbf{x}\Phi^{\dagger}$ is generally not as sparse as representations minimizing other criteria more appropriate to sparsity, such as ℓ^{τ} criteria, with $0 \leq \tau \leq 1$. This is why adaptive approaches based on minimizing the ℓ^{τ} norm where $0 \leq \tau \leq 1$ have elicited intense interest since the early 1990s [48,23], in a context extending well beyond that of source separation [83,43,90,29,96]. Most early efforts have focused on adaptive methods for sparse representation of a given signal or image; interest in equivalent methods for joint representation of several signals has developed subsequently [49,68,50,27,104,102,57]. We will discuss joint representation below, first in the general case of ℓ^{τ} criteria where $0 \leq \tau \leq \infty$. We will then look more specifically at the corresponding algorithms for values of $\tau \leq 2$ which can be used to obtain sparse representations.

10.3.3 Principle of ℓ^τ minimization

Rather than using linear transforms, *joint sparse representations* can be defined that minimize an ℓ^τ criterion, for $0 \le \tau \le \infty$. Before defining them in detail, we will review the definition of the corresponding "single-channel" sparse representations: among all the admissible representations of a signal x_1 (i.e. satisfying $x_1 = c_{x_1}\mathbf{\Phi}$), the representation $c^\tau_{x_1}$ with the smallest ℓ^τ norm is selected, which amounts to selecting the most likely representation in terms of an assumed generalized Gaussian distribution of the coefficients (10.4).

In the multi-channel case, the optimized criterion takes the form:

$$C^\tau_{\mathbf{x}} := \arg \min_{C_{\mathbf{x}} | C_{\mathbf{x}}\mathbf{\Phi}=\mathbf{x}} \sum_{k=1}^{K} \left(\sum_{p=1}^{P} \left| c_{x_p}(k) \right|^2 \right)^{\tau/2} \tag{10.16}$$

for $0 < \tau < \infty$, with an obvious modification for $\tau = \infty$. For $\tau = 0$, with the convention $c^0 = 1$ if $c > 0$ and $0^0 = 0$, the ℓ^0 norm affecting the criteria (10.16) is the number of non-zero columns $C_{\mathbf{x}}(k)$, i.e. the number of atoms used in the representation. Minimizing the ℓ^0 norm is simply a means of representing the mixture with as few atoms as possible, which intuitively corresponds to looking for a sparse representation.

This particular form of the optimized criterion defines a representation of x that actually takes into account the *joint* properties of the different channels x_p, in contrast to a criterion such as $\sum_k \sum_p |c_{x_p}|^\tau$, which may be more intuitive but decouples optimization over the different channels. In section 10.3.4, we will examine a Bayesian interpretation explaining the form taken by the criterion (10.16).

Due to the intrinsic noise $\mathbf{b}$ of the mixing model (10.1) and the possible inaccuracy of the sparse source model (10.6), it is often more useful to look for an approximate sparse representation – also known as a *joint sparse approximation* – via the optimization of the criterion:

$$C^{\tau,\lambda}_{\mathbf{x}} := \arg\min_{C_{\mathbf{x}}} \left\{ ||\mathbf{x} - C_{\mathbf{x}}\mathbf{\Phi}||^2_F + \lambda \sum_{k=1}^{K} \left(\sum_{p=1}^{P} \left| c_{x_p}(k) \right|^2 \right)^{\tau/2} \right\} \tag{10.17}$$

where $||\mathbf{y}||^2_F := \sum_{pt} |y_p(t)|^2$ is the Frobenius norm of the $\mathbf{y}$ matrix, i.e. the sum of the energies of its rows. The first term of the criterion is a data fidelity term that uses least squares to measure the quality of the approximation obtained. The second term measures the "joint" sparsity of the multi-channel representation $C_{\mathbf{x}}$, relative to the criterion (10.16), and the λ parameter determines the respective importance attributed to the quality of the approximation and to sparsity. When λ tends to zero, the obtained solution, $C^{\tau,\lambda}_{\mathbf{x}}$, approaches $C^\tau_{\mathbf{x}}$; when λ is sufficiently large, $C^{\tau,\lambda}_{\mathbf{x}}$ tends to zero.

10.3.4 Bayesian interpretation of ℓ^τ criteria

The specific form of the ℓ^τ criterion used to select a multi-channel joint representation (10.16)–(10.17), which couples the selection of coefficients for the different channels, is not chosen arbitrarily. The following paragraphs aim to provide an overview of its derivation in a Bayesian context. Readers more interested in the algorithmic and practical aspects can skip to section 10.3.6.

Whether an adaptive or transform approach is used, a Bayesian interpretation is possible for optimizing the criteria in (10.11), (10.16) or (10.17): (10.11) corresponds to calculating the most likely coefficients for reconstructing $\mathbf{x} = C_{\mathbf{x}}\mathbf{\Phi}$ under the assumption of an iid Gaussian distribution $p_{C_{\mathbf{x}}}(\cdot)$ with variance $\sigma_{\mathbf{x}}^2$. Under this assumption, the distribution of $C_{\mathbf{x}}$ is expressed (see Eq. (10.4) for $\tau = 2$):

$$p_{C_{\mathbf{x}}}(C_{\mathbf{x}}) \propto \exp\left(-\frac{\|C_{\mathbf{x}}\|_F^2}{2\sigma_{\mathbf{x}}^2}\right) = \exp\left(-\frac{\sum_{k=1}^{K}\|C_{\mathbf{x}}(k)\|_2^2}{2\sigma_{\mathbf{x}}^2}\right). \tag{10.18}$$

In the specific context of source separation, the distribution $p_{C_{\mathbf{x}}}(C_{\mathbf{x}})$ of $C_{\mathbf{x}} = \mathbf{A}C_{\mathbf{s}}$ depends on the distribution $p_{C_{\mathbf{s}}}(C_{\mathbf{s}})$ of the source coefficients $C_{\mathbf{s}}$ and on the distribution $p_{\mathbf{A}}(\mathbf{A})$ of the mixing matrix. In the absence of prior knowledge of the mixing matrix, for blind separation its distribution is naturally assumed invariant by arbitrary spatial rotations, i.e. $p_{\mathbf{A}}(\mathbf{A}) = p_{\mathbf{A}}(\mathbf{UA})$ for any unitary matrix $\mathbf{U}$. This is the case if the columns $\mathbf{A}_n$ are normalized $\|\mathbf{A}_n\|_2 = 1$, independent and uniformly distributed over the unit sphere of $\mathbb{C}^P$. This assumption alone implies that the prior distribution of the coefficients $C_{\mathbf{x}}$ of the noiseless mixture for any atom k is radial; that is, for any unitary matrix $\mathbf{U}$:

$$p(C_{\mathbf{x}}(k)) = p(\mathbf{U}C_{\mathbf{x}}(k)) = r_k(\|C_{\mathbf{x}}(k)\|_2) \tag{10.19}$$

where $r_k(\cdot) \geq 0$ is a function of the distributions $p_{C_{\mathbf{s}}(k)}(C_{\mathbf{s}}(k))$ and $p_{\mathbf{A}}(\mathbf{A})$.

If the coefficients $C_{\mathbf{s}}(k)$ are iid for different k, then the form of the prior distribution of $C_{\mathbf{x}}$ can be deduced:

$$p(C_{\mathbf{x}}) = \prod_{k=1}^{K} r(\|C_{\mathbf{x}}(k)\|_2). \tag{10.20}$$

A *maximum-likelihood* (ML) approach makes it possible to define a decomposition of the noiseless mixture as:

$$C_{\mathbf{x}}^r := \arg\max_{C_{\mathbf{x}}|C_{\mathbf{x}}\mathbf{\Phi}=\mathbf{x}} p(C_{\mathbf{x}}) = \arg\min_{C_{\mathbf{x}}|C_{\mathbf{x}}\mathbf{\Phi}=\mathbf{x}} \sum_{k=1}^{K} -\log r(\|C_{\mathbf{x}}(k)\|_2). \tag{10.21}$$

In the noisy case, assuming the noise $\mathbf{b}$ is iid Gaussian of variance $\sigma_{\mathbf{b}}^2$, a *maximum a posteriori (MAP)* approach maximizes the posterior probability, i.e. it minimizes:

$$-\log p(C_{\mathbf{x}} \mid \mathbf{x}) = \frac{1}{2\sigma_{\mathbf{b}}^2}\|\mathbf{x} - C_{\mathbf{x}}\mathbf{\Phi}\|_F^2 + \sum_{k=1}^{K} -\log r(\|C_{\mathbf{x}}(k)\|_2). \tag{10.22}$$

The function r, which in theory can be calculated from the distributions $p_{\mathbf{A}}(\mathbf{A})$ of the mixing matrix and $p_c(c)$ of the sparse source coefficients, is sometimes difficult to use in solving the corresponding optimization problem. Since the distributions on which this function depends are not necessarily well understood themselves, it is reasonable to solve the global optimization problem (10.21) or (10.22) with an assumed specific form of the r function, such as $r(z) := \exp(-\alpha |z|^\tau)$ for an exponent $0 \leq \tau \leq 2$, similarly to the adaptive approaches (10.16)–(10.17).

10.3.5 Effect of the chosen ℓ^τ criterion

The choice of the ℓ^τ norm is important *a priori*, since it can considerably influence the sparsity of the resulting solution. Figure 10.4 illustrates this point schematically. Figure 10.4(a)-(b)-(c) show, in dimension 2, the contour lines of ℓ^τ norms for $\tau = 2$, $\tau = 1$ and $\tau = 0.5$. In Fig. 10.4(d), a continuous line represents the set of admissible representations ($C_\mathbf{x}\mathbf{\Phi} = \mathbf{x}$) of a mixture $\mathbf{x}$, which forms an affine sub-space of the space $\mathbb{R}^K$ of all possible representations. The least ℓ^τ norm representation of $\mathbf{x}$ is the intersection of this line with the "smaller" contour line of the ℓ^τ norm that intersects it. These intersections have been indicated for $\tau = 2$, $\tau = 1$ and $\tau = 0$ (for $\tau = 0$ the intersections are with the axes). One can see that none of the coordinates of the least squares solution ($\tau = 2$) is non-zero, whereas the least ℓ^0 or ℓ^1 norm solutions are situated along the axes. Here, the least ℓ^1 norm solution, which is unique, also coincides with the least ℓ^τ norm solutions for $0 < \tau \leq 1$. Geometrically, this is explained by the fact that the contour lines of the ℓ^τ norms, for $0 < \tau \leq 1$ have "corners" on the axes, in contrast to the contour lines of the ℓ^2 norm (and of the ℓ^τ norms for $1 < \tau \leq 2$), which are convex and rounded.

As a result, in the single-channel context, any representation $C_\mathbf{x}^\tau \in \mathbb{R}^K$ (or $C_\mathbf{x}^{\tau,\lambda}$) that solves the optimization problem (10.16) (or (10.17)) for $0 \leq \tau \leq 1$, is *truly sparse* insofar as it has no more than T non-zero coefficients, where T is the dimension of the analyzed signal $\mathbf{x} \in \mathbb{R}^T$.[5] This property is due to the "concavity" of the ℓ^τ norm [64] for $0 \leq \tau \leq 1$, as shown in Fig. 10.4(c). We will refer to this property repeatedly throughout this chapter, for the multi-channel case as well. It can be defined as follows:

DEFINITION 10.1
The representation $C_\mathbf{x}^\tau$ (or $C_\mathbf{x}^{\tau,\lambda}$) is a truly sparse representation if it implies no more than T atoms of the dictionary (where T is the dimension of the signals), i.e. if there are no more than T indices k whereby the column $C_\mathbf{x}(k)$ is non-zero.

This definition is given here in the multi-channel case, but it generalizes an equivalent idea initially proposed for the single-channel case by Kreutz-Delgado *et al.* [64].

For $\tau > 1$, since ℓ^τ is a real norm in the mathematical sense of the term, it is convex, and its minimization does not lead to truly sparse representations, as shown in Fig. 10.4(a). However, the convexity of the corresponding criteria greatly facilitates their optimization and can favor their practical use.

[5]Here, the fact that we are considering real-valued signals and coefficients seems to be important, since examples with complex coefficients show that the ℓ^τ minimizer can be non "truly sparse" [108].

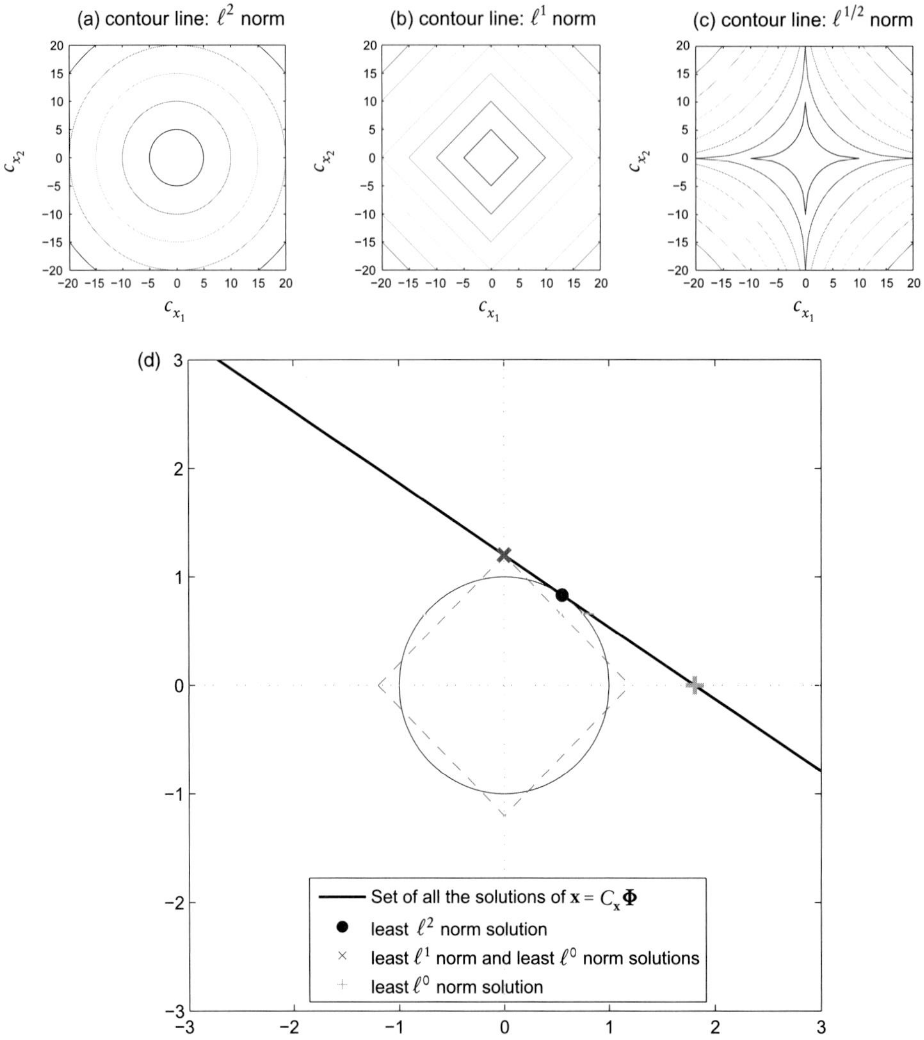

FIGURE 10.4

(a)-(b)-(c): contour lines of ℓ^τ norms for $\tau = 2, 1, 1/2$. (d) Schematic comparison of the sparsity of representations obtained using least squares (non-adaptive linear transform) and low-dimensional ℓ^1 or ℓ^0 norm minimization (adaptive representations). The continuous line symbolizes all the possible representations $C_{\mathbf{x}} = (c_{x_1}, c_{x_2})$ of a given mixture $\mathbf{x}$. The intersection of this line with the circle centered on the origin which is tangentine to it provides the least squares solution. Its intersections with the x-axis ("+" symbol) and the y-axis ("×" symbol) correspond to two solutions with a unique non-zero coefficient, and thus a minimal ℓ^0 norm. The y-axis solution is also the unique intersection between all the possible representations and the smallest "oblique" square centered on the origin (dotted line) that intersects it, making it the least ℓ^1 norm solution as well. One can see that the least squares solution has more non-zero coefficients than the least ℓ^0 or ℓ^1 norm solutions.

Intensive theoretical work, initiated by the results of Donoho and Huo [33], has shown in the single-channel case that the specific choice of the ℓ^τ norm optimized with $0 \leq \tau \leq 1$ has little (or no) influence on the representation obtained [55,54,52] when the analyzed signal $\mathbf{x}$ is adequately approximated with sufficiently sparse coefficients $C_\mathbf{x}$. Similar results for the multi-channel case [102] show that, when the same assumptions are made, the ℓ^0 and ℓ^1 criteria essentially lead to the same representations. Extending these results to the ℓ^τ criteria, $0 \leq \tau \leq 1$, will undoubtedly point to the same conclusion.

In light of these theoretical results, the choice of a specific ℓ^τ criterion to optimize for $0 \leq \tau \leq 1$ depends primarily on the characteristics of the algorithms available for optimization and their numerical properties. The description of optimization algorithms for the slightly broader range of $0 \leq \tau \leq 2$ is the subject of the following section.

10.3.6 Optimization algorithms for ℓ^τ criteria

Adaptive representation algorithms that optimize a criterion (10.17) or (10.16) for $0 \leq \tau \leq 2$ have received a great deal of attention, with regard to theoretical analysis as well as algorithms and applications.

For $\tau = 2$, both (10.16) and (10.17) can be solved with simple linear algebra, but for $\tau = 0$, the problem is a combinatorial one [31,78] and *a priori* untractable. A body of theoretical work initiated by Donoho and Huo [33] suggests that its solution is often equal or similar to that of the corresponding problems with $0 < \tau \leq 1$ [55]. The algorithms able to optimize these criteria are of particular interest, despite the specific difficulties of the non-convex optimization of $\tau < 1$, and in particular the risk of obtaining a local optimum. Finally, the algorithms for optimizing strictly convex ℓ^τ criteria with $1 < \tau \leq 2$ offer the advantage of simplicity, even though the theoretical elements linking the representations they provide to those obtained with ℓ^τ criteria for $0 \leq \tau \leq 1$ are currently lacking. Their main advantage, owing to the strict convexity of the optimized criterion, is the guaranteed convergence to the unique global optimum of the criterion.

For all the algorithms presented below, selecting an order of magnitude for the regularization parameter λ is also a complex issue in blind mode. An excessively high value for this parameter will over-emphasize the sparsity of the obtained representation $C_\mathbf{x}^{\tau,\lambda}$, compromising the quality of the reconstruction $\|\mathbf{x} - C_\mathbf{x}^{\tau,\lambda}\mathbf{\Phi}\|_2$. An excessively low value will amount to assuming that the sparse source model is accurate and the mixture is noiseless, which could result in artefacts. The selection of parameters τ and λ, which define the optimized criterion, should thus represent a compromise between the sparsity of the representation and the numerical complexity of the minimization. In typical applications, λ would be chosen to obtain a reconstruction error of the order of the noise, if the magnitude of the latter is known.

Algorithm for $\tau = 2$: regularized least squares

For $\tau = 2$, the exact representation $C_\mathbf{x}^2$ solves the least squares problem $\min \|C_\mathbf{x}\|_F^2$ under the constraint $C_\mathbf{x}\mathbf{\Phi} = \mathbf{x}$ and is obtained analytically by linear transform, using the pseudo-inverse of the dictionary, i.e. $C_{\mathbf{x}}^{2} = C_{\mathbf{x}}^{\mathbf{\Phi}^{\dagger}} = \mathbf{x}\mathbf{\Phi}^{\dagger}$. As shown in Fig. 10.4, $C_{\mathbf{x}}^{2}$ is generally not a "truly sparse" representation. The approximate representations $C_{\mathbf{x}}^{2,\lambda}$, which solve the regularized least squares problem $\min ||\mathbf{x} - C_{\mathbf{x}}\mathbf{\Phi}||_F^2 + \lambda ||C_{\mathbf{x}}||_F^2$ for $\lambda > 0$, are also obtained linearly as $C_{\mathbf{x}}^{2,\lambda} = \mathbf{x}\mathbf{\Psi}_{\lambda}$, where $\mathbf{\Psi}_{\lambda}$ has two equivalent expressions:

$$\mathbf{\Psi}_{\lambda} := (\mathbf{\Phi}^{\mathrm{H}}\mathbf{\Phi} + \lambda \mathbf{I}_T)^{-1}\mathbf{\Phi}^{\mathrm{H}} = \mathbf{\Phi}^{\mathrm{H}}(\mathbf{\Phi}\mathbf{\Phi}^{\mathrm{H}} + \lambda \mathbf{I}_K)^{-1} \tag{10.23}$$

with $\mathbf{I}_M$ as the $M \times M$ identity matrix. When λ tends to zero, $\mathbf{\Psi}_{\lambda}$ tends to the pseudo-inverse $\mathbf{\Phi}^{\dagger}$ of the dictionary, and for any λ value the linearity of the representation guarantees that $C_{\mathbf{x}}^{2,\lambda} = \mathbf{A}C_{\mathbf{s}}^{2,\lambda} + C_{\mathbf{b}}^{2,\lambda}$.

Algorithm for $0 < \tau \leq 2$: M-FOCUSS

For $0 < \tau \leq 2$ the *M-FOCUSS* algorithm [27] (an *iterative reweighted least squares* or IRLS algorithm) and its regularized version are iterative techniques that converge extremely quickly towards a local minimum of the criterion to be optimized. This algorithm is derived from the FOCUSS algorithm initially defined for the single-channel case [48]. Given the non-convexity of the optimized criterion, it is difficult to predict whether the global optimum is reached, which in any case depends on the chosen initialization. It has not yet been demonstrated (except in the $P = 1$ single-channel case) that the algorithm always converges on a "truly sparse" solution, although this is indeed the case experimentally [27]. Starting with an initialization $C_{\mathbf{x}}^{(0)}$, the value of which can have a decisive impact on the algorithm's point of convergence, a weighted mean square criterion is minimized iteratively:

$$C_{\mathbf{x}}^{(m)} := \arg\min_{C_{\mathbf{x}}} \left\{ ||\mathbf{x} - C_{\mathbf{x}}\mathbf{\Phi}||_F^2 + \frac{\lambda|\tau|}{2} ||C_{\mathbf{x}}(W^{(m)})^{-1}||_F^2 \right\} \tag{10.24}$$

where the diagonal weighting matrix:

$$W^{(m)} := \mathrm{diag}\left(||C_{\mathbf{x}}^{(m-1)}(1)||_2^{1-\tau/2}, \ldots, ||C_{\mathbf{x}}^{(m-1)}(K)||_2^{1-\tau/2} \right) \tag{10.25}$$

tends to drive to zero those columns whose energy $||C_{\mathbf{x}}^{(m-1)}(k)||_2$ is low. In practice, as we saw in section 10.3.6, each iteration thus involves updating the diagonal weighting matrix $W^{(m)}$ and the weighted dictionary $\mathbf{\Phi}^{(m)} := W^{(m)}\mathbf{\Phi}$ to obtain:

$$C_{\mathbf{x}}^{(m)} := \mathbf{x} \left((\mathbf{\Phi}^{(m)})^{\mathrm{H}} \, \mathbf{\Phi}^{(m)} + \frac{\lambda|\tau|}{2} \mathbf{I}_T \right)^{-1} (\mathbf{\Phi}^{(m)})^{\mathrm{H}} \, W^{(m)}. \tag{10.26}$$

While the M-FOCUSS algorithm converges very rapidly towards a local minimum of the optimized criterion [27] and often produces very sparse representations, it is difficult to determine when convergence to the global minimum is reached, depending on the chosen initialization. The initialization most often used, and which in practice gives satisfactory results, is the least squares representation, obtained using the pseudo-inverse $\mathbf{\Phi}^{\dagger}$ of the dictionary. Even if the number of iterations necessary before convergence is low, each iteration can be quite costly numerically because of the inversion of a square matrix of size at least T which is required, cf. (10.26).

Because the algorithm is globally nonlinear, it is difficult *a priori* to determine whether its representation $C_{\mathbf{x}}$ of the mixture satisfies the identity $C_{\mathbf{x}} \approx \mathbf{A}C_{\mathbf{s}}$ with reasonable accuracy. Theoretical results [52,102] on optimization criteria similar to (10.17) suggest this to be true when the representations of the sources are sufficiently sparse and disjoint, but actual proof is still lacking for the optimum $C_{\mathbf{x}}^{\tau,\lambda}$ of the criterion (10.17), and for the approximate numerical solution $\lim_{m\to\infty} C_{\mathbf{x}}^{(m)}$ calculated with M-FOCUSS.

Algorithms for $\tau = 1$ (single-channel case)

Optimizing (10.16) in the single-channel case ($P = 1$) for $\tau = 1$ is known as *Basis Pursuit* and optimizing (10.17) is known as *Basis Pursuit Denoising* [23]. Proposed by Donoho and his colleagues, Basis Pursuit and Basis Pursuit Denoising are closer to principles than actual algorithms, since the optimization method is not necessarily specified. Several experimental studies and the intense theoretical work initiated by Donoho and Huo [33] have shown the relevance of sparse representations obtained using the ℓ^1 criterion; for example, they guarantee the identity $C_{\mathbf{x}} \approx \mathbf{A}C_{\mathbf{s}}$ when $C_{\mathbf{s}}$ is sufficiently sparse [102].

However, the generic numerical optimization methods of *linear programming*, *quadratic programming*, *conic programming*, etc. [12] for calculating them (in both the real and complex cases and the single-channel and multi-channel cases) are often computationally costly. In order to reduce computations, Truncated Newton strategy was used by [23] in the framework of Primal-Dual Interior Point method. Algorithms such as LARS (*Least Angle Regression*) [35] that take more specific account of the optimization problem [105, 35,74,85,45] are able to further simplify and accelerate the calculations. The next challenge is to adapt these efficient algorithms to the multi-channel case.

Algorithms for $1 \le \tau \le 2$: iterative thresholding (single-channel case)

For the values $1 \le \tau \le 2$, the criterion to optimize (10.17) is convex. Efficient iterative numerical methods were proposed [28,36,42] for calculating the optimum $c_x^{\lambda,\tau}$, for the $P = 1$ single-channel case. Starting with an arbitrary initialization $c_x^{(0)}$ (typically $c_x^{(0)} = 0$), the principle is to iterate the following thresholding step:

$$c_x^{(m)}(k) = S_{\lambda,\tau}\left(c_x^{(m-1)}(k) + \langle x - c_x^{(m-1)}\mathbf{\Phi}, \varphi_k\rangle\right), \qquad 1 \le k \le K \tag{10.27}$$

where $S_{\lambda,\tau}$ is a thresholding function, plotted in Fig. 10.5 for $\lambda = 1$ and a few values of $1 \le \tau \le 2$.

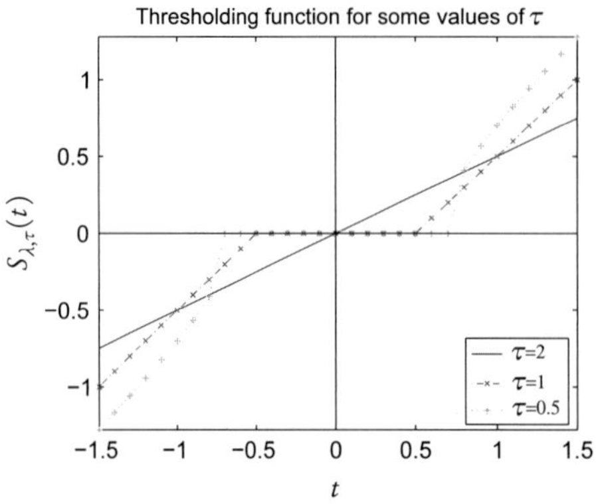

FIGURE 10.5

Curves of the thresholding function $S_{\lambda,\tau}(t)$ over the interval $t \in [-1.5,\ 1.5]$ for $\lambda = 1$ and a few values of τ.

For $\tau = 1$, the soft thresholding function is used:

$$S_{\lambda,1}(t) := \begin{cases} t + \frac{\lambda}{2}, & \text{if } t \leq -\frac{\lambda}{2} \\ 0, & \text{if } |t| < \frac{\lambda}{2} \\ t - \frac{\lambda}{2}, & \text{if } t \geq \frac{\lambda}{2} \end{cases} \tag{10.28}$$

whereas for $\tau > 1$:

$$S_{\lambda,\tau}(t) := u, \text{ where } u \text{ satisfies } u + \frac{\lambda \cdot \tau}{2} \cdot \text{sign}(u) \cdot |u|^{\tau-1} = t. \tag{10.29}$$

This thresholding function comes from the optimization of criterion (10.17) when $\mathbf{\Phi}$ is an orthonormal basis (recall that iterative thresholding is defined for the $P = 1$ single-channel case), because each coordinate must then be optimized independently:

$$\min_{c(k)} |\langle x, \varphi_k \rangle - c(k)|^2 + \lambda |c(k)|^\tau .$$

When the value of the optimum $c_x^{\lambda,\tau}(k)$ is non-zero, the derivative of the optimized criterion exists and cancels outs, which implies:

$$2(c_x^{\lambda,\tau}(k) - \langle x, \varphi_k \rangle) + \lambda \cdot \text{sign}(c_x^{\lambda,\tau}(k)) \cdot \left| c_x^{\lambda,\tau}(k) \right|^{\tau-1} = 0$$

that is, according to the expression of the thresholding function (10.29):

$$c_x^{\lambda,\tau}(k) = S_{\lambda,\tau}(\langle x, \varphi_k \rangle).$$

For $1 \leq \tau \leq 2$, Daubechies, Defrise and De Mol proved [28, theorem 3.1] that if the dictionary satisfies $||\mathbf{\Phi} c||_2^2 \leq ||c||_2^2$ for all c, then $c^{(m)}$ converges strongly to a minimizer $c_x^{\lambda,\tau}$ of (10.17), regardless of the initialization. If $\mathbf{\Phi}$ is also a basis (not necessarily orthonormal), this minimizer is unique, and the same is true by strict convexity if $\tau > 1$. However, this minimizer is generally a "truly sparse" solution only for $\tau = 1$. In addition, when the dictionary consists of atoms of unit energy, then at best $||\mathbf{\Phi} c||_2^2 \leq \mu ||c||_2^2$ for a certain number $\mu < \infty$. The algorithm must then be slightly modified to ensure convergence, by iterating the thresholding step below:

$$c_x^{(m)}(k) = S_{\frac{\lambda}{\mu},\tau}\left(c_x^{(m-1)}(k) + \frac{\langle x - c_x^{(m-1)}\mathbf{\Phi}, \varphi_k \rangle}{\mu} \right). \tag{10.30}$$

While there is currently no known complexity analysis of this algorithm, it often takes few iterations to obtain an excellent approximation of the solution. Additional significant acceleration can be achieved via sequential optimization over subspaces spanned by the current thresholding direction and directions of a few previous steps (SESOP) [37]. Many algorithmic variants exist and the approach is intimately related to the notion of proximal algorithm. Each iteration is relatively low-cost given that matrix inversion is not necessary, unlike for the algorithm M-FOCUSS, which has been examined previously. These thresholding algorithms can be extended to the multi-channel case [44]. To use them for sparse source separation, it remains to be verified whether they guarantee approximate identification $\mathbf{C_x} \approx \mathbf{A}\mathbf{C_s}$.

10.3.7 Matching pursuit

Matching Pursuit proposes a solution to the problem of joint sparse approximation $\mathbf{x} \approx \mathbf{C_x}\mathbf{\Phi}$ not through the global optimization of a criterion, but through an iterative *greedy algorithm*. Each step of the algorithm involves finding the atom in the dictionary that, when added to the set of already selected atoms, will reduce the approximation error to the largest extent. We assume here that the atoms in the dictionary are of unit energy. To approximate the multi-channel signal $\mathbf{x}$, a residual $\mathbf{r}^{(0)} := \mathbf{x}$ is initialized, and then the following steps are repeated, starting at $m = 1$:

1. Selection of most correlated atom:

$$k_m := \arg\max_k \sum_{p=1}^{P} |\langle \mathbf{r}_p^{(m-1)}, \varphi_k \rangle|^2. \tag{10.31}$$

2. Updating of residual:

$$\mathbf{r}_p^{(m)} := \mathbf{r}_p^{(m-1)} - \langle \mathbf{r}_p^{(m-1)}, \varphi_{k_m} \rangle \varphi_{k_m}, \qquad 1 \le p \le P. \tag{10.32}$$

After M iterations, one obtains the decomposition:

$$\mathbf{x} = \sum_{m=1}^{M} \begin{pmatrix} c_1(m)\varphi_{k_m} \\ \cdots \\ c_P(m)\varphi_{k_m} \end{pmatrix} + \mathbf{r}^{(M)} \tag{10.33}$$

with $c_p(m) := \langle \mathbf{r}_p^{(m-1)}, \varphi_{k_m} \rangle$.

This algorithm was initially introduced for single-channel processing by Mallat and Zhang [76]. The generalization to the multi-channel case [49,50,27,68] that we present here converges in the same manner as the original single-channel version [49,50] in that the residual norm tends to zero as the number of iterations tends to infinity. In contrast, the convergence rate, whether single-channel or multi-channel, is not fully understood [68]. In one of the variants of this algorithm, the residuals can be updated at each step by orthogonal projection of the analyzed signal on the linear manifold spanned by all the selected atoms (referred to as orthogonal *matching pursuit*). It is also possible to select the best atom at each step according to other criteria [68,27,104], or to select several atoms at a time [79,14].

The *matching pursuit* presented here, which involves no matrix inversion, can be implemented quite rapidly by using the structure of the dictionary and fast algorithms for calculating and updating $\mathbf{r}^{(m)}\mathbf{\Phi}^{\mathrm{H}}$ [53,65]. The main parameter is the choice of the stopping criterion; the two main approaches set either the number of iterations or the targeted approximation error $||\mathbf{r}^{(m)}||_F$ beforehand. A typical choice is to stop when the residual error is of the order of magnitude of the noise level, if the latter is known.

Like the approaches based on ℓ^1 norm minimization [102], *matching pursuit* has been the subject of several experimental and theoretical studies, which have shown [104] that it can guarantee the identity $C_{\mathbf{x}} \approx \mathbf{A}C_{\mathbf{s}}$ when $C_{\mathbf{s}}$ is sufficiently sparse [56].

10.3.8 Summary

Table 10.1 summarizes the main numerical characteristics of the algorithms for joint sparse signal representation presented in this section. Their field of application extends far beyond source separation, since the need to jointly represent several signals exists in a number of areas. For example, color images consist of three channels (red-green-blue) whose intensities are correlated, and the use of joint sparse representation techniques (based on *matching pursuit* [43]) has proven effective for low-bit rate image compression. Readers interested in further details on these algorithms will find their original definitions in the bibliographic references and, in some cases therein, an analysis of certain theoretical properties that have been proven.

Table 10.1 Comparative summary of algorithms for joint sparse mixture representation discussed in section 10.3

Method	**Transform**	**M-FOCUSS**	**Basis Pursuit**	**Iterative thresholding**	**Matching Pursuit**
Criterion	(10.17)	(10.17)	(10.17)	(10.17)	(10.31)–(10.32)
Parameters	$\tau = 2$ λ	$0 \leq \tau \leq 2$ λ	$\tau = 1$ λ	$1 \leq \tau \leq 2$ λ	M iterations
Advantages	–very fast	–high sparsity ($\tau \leq 1$)	–high sparsity	–fast –good sparsity	–fast –good sparsity
Issues and difficulties	–choice of λ –limited sparsity	–choice of λ –initialization –memory cost –computing time (inversion of large matrices)	–choice of λ –memory cost –computing time (linear, quadratic, conic progr.)	–choice of λ	–choice of M
References					
$P = 1$		[48] [55]	[23] [33]	[28]	[76] [101]
$P > 1$		[52] [27]	[103] [102]	[44]	[58] [49] [104]

In practice, the joint sparse representation of a mixture implies two choices:

1. selection of a dictionary Φ in which the mixture channels are likely to allow a sufficiently sparse joint representation;
2. selection of an algorithm and its parameters for calculating a representation.

For example, for audio signals, the Gabor dictionary (10.7) is often used to take account of the frequency components that may appear at different times in the mixture, and the representation most often used is the STFT, a linear transform. While we have attempted in this section to present useful information for selecting an algorithm, selecting the dictionary currently mostly depends on expert knowledge of the analyzed signals in most cases. One can also learn a dictionary automatically from a set of signal realizations [70,71,80,4,92].

10.4 ESTIMATING THE MIXING MATRIX BY CLUSTERING

The algorithms discussed above can be used to analyze a mixture $\mathbf{x}$ in order to calculate a joint sparse representation (or approximation) $C_{\mathbf{x}}$, which is the first step in a source separation system based on sparsity, as described in Fig. 10.2. The second step of this system, for separating instantaneous linear mixtures, consists in estimating the mixing matrix $\mathbf{A}$ by means of the scatter plot of the coefficients $\{C_{\mathbf{x}}(k)\}_{1\leq k\leq K}$.

For determined mixtures, it is of course possible to use any ICA method to estimate $\mathbf{A}$, which means that the other techniques described in this book can be employed as well. Here we concentrate on approaches specifically based on the assumed sparsity of the sources.

The two main classes of approaches used make assumptions about the sparsity of the sources but not about the type of mixture, i.e. determined or undetermined. The first approach, based on Bayesian (or variational) interpretation, aims to maximize the likelihood of the estimated matrix and involves (approximate) optimization of the global criteria; the second is more geometric and relies on clustering algorithms.

Variational approaches

Variational approaches [70,115] generally involve joint iterative optimization of the matrix and the source representations. This task is very challenging computationally, and the quality of solution may suffer from the presence of spurious local minima. However, in the case of square mixing matrix, the problem is reduced to the optimization with respect to unmixing matrix only, which can be treated efficiently using the Relative Newton method [112,113] as described in section 10.5.

Geometric approach: clustering from scatter plots

The geometric approach is very intuitive and capable of estimating the number of sources [8]. In addition, certain clustering algorithm variants do not make the strong sparsity assumption in order to estimate the mixing matrix using simple algorithms, even when the source representations are almost non-disjoint [2,8].

Most of the approaches presented below are applicable in theory, regardless of the number of sensors $P \geq 2$ [107,98,2,8]. However, we will only explain these approaches where $P = 2$, corresponding for example to stereophonic audio mixtures, because the geometric intuitiveness is simpler and more natural in such cases. Since generalizing to the $P \geq 3$ case poses no particular conceptual problems, we leave this to the reader. However, the computing requirements become excessive for certain methods, particularly those based on calculating histograms.

10.4.1 Global clustering algorithms

Figure 10.6(a), an enlargement of Fig. 10.1(c), shows a typical scatter plot of the coefficients $\{C_{\mathbf{x}}(k)\} \subset \mathbb{R}^2$ for a stereophonic mixture ($P = 2$). Alignments of points are observed along the straight lines generated by the columns $\mathbf{A}_n$ of the mixing matrix. In this example, each point $C_{\mathbf{x}}(k) = \rho(k) \cdot [\cos\theta(k), \sin\theta(k)]^T \in \mathbb{R}^2$ can also be shown with polar coordinates:

$$\begin{cases} \rho(k) & := & (-1)^{\varepsilon}\sqrt{|c_{x_1}(k)|^2 + |c_{x_2}(k)|^2}, \\ \theta(k) & := & \arctan(c_{x_2}(k)/c_{x_1}(k)) + \varepsilon\pi, \end{cases} \tag{10.34}$$

where $\varepsilon \in \{0, 1\}$ is selected such that $0 \leq \theta(k) < \pi$, as shown in Fig. 10.6(b). The points then accumulate around the angles θ_n corresponding to the directions of the columns $\mathbf{A}_n$. A natural idea [100,15] is to exploit the specific geometric structure of the scatter plot in order to find the directions $\widehat{\theta}_n$, making it possible to estimate $\widehat{\mathbf{A}}_n := [\cos\widehat{\theta}_n \sin\widehat{\theta}_n]^T$.

Raw histogram

A naive approach consists of calculating a histogram of $\theta(k)$ angles, assuming the accumulation of points corresponding to the directions of interest will result in visible peaks. The resulting raw histogram, shown in Fig. 10.7(a) and based on points from Fig. 10.6(b), does make it possible to see (and detect by thresholding) one of the directions of interest, but trying to find the other directions with it would be futile.

The scatter plot often contains a number of points of low amplitude $\rho(k)$, whose direction $\theta(k)$ is not representative of the column directions of the mixing matrix. These points, which correspond to the numerous atoms (time-frequency points in this example) where all the sources are approximately inactive, have a "flattening" effect on the histogram of directions, making it unsuitable for detecting the direction of the sources.

Weighted histograms

A weighted histogram can be calculated in a more efficient manner, and may be smoothed with "potential functions", also called Parzen windows. The weighting aims to obtain a histogram that depends on the small number of points with large amplitude rather than the majority of points with negligible amplitude. The following function is

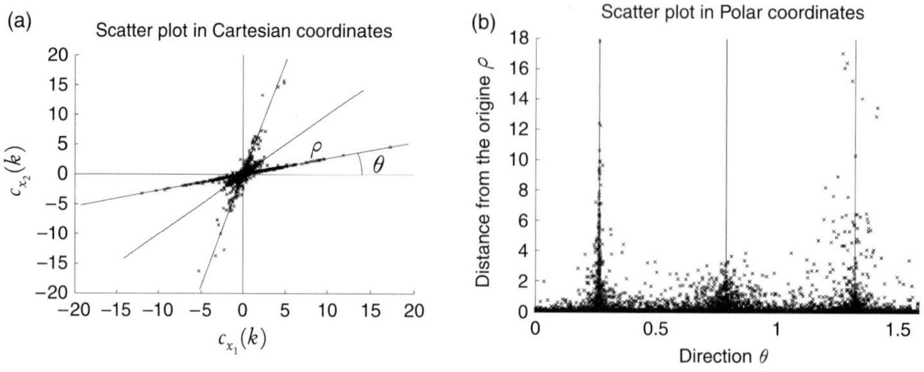

FIGURE 10.6

Scatter plot of MDCT coefficients for the mixture on the right of Fig. 10.1, with Cartesian coordinates (a) and polar coordinates (b). Accumulations of points, shown as dotted lines, are observed along the directions of the columns of the mixing matrix.

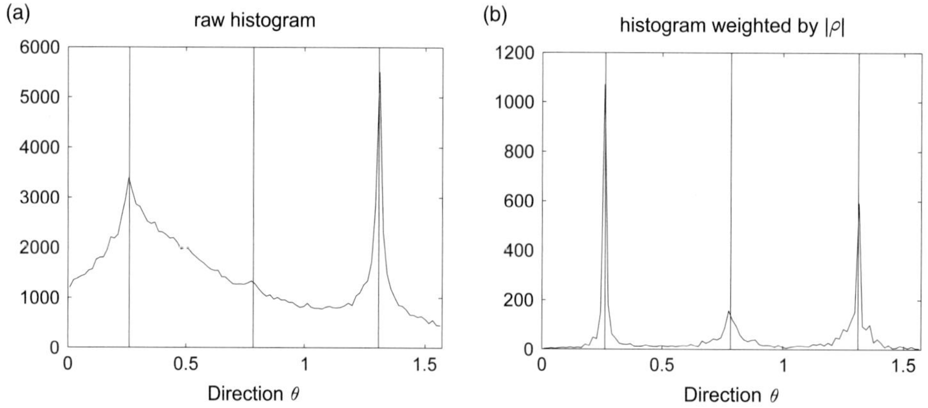

FIGURE 10.7

Histograms based on the scatter plot with polar coordinates in Fig. 10.6(b): (a) raw histogram measuring the frequencies of occurrence at a given θ angle; (b) histogram weighted by the distance from the origin $|\rho|$. Weighting based on distance from the origin results in strong peaks close to the directions of the actual sources (indicated by solid lines) whereas only one peak is visible on the raw histogram.

calculated (for a set of discrete angles on a grid $\theta_\ell = \ell\pi/L, 0 \leq \ell \leq L$):

$$H(\theta_\ell) := \sum_{k=1}^{K} f(|\rho(k)|) \cdot w(\theta_\ell - \theta(k)) \tag{10.35}$$

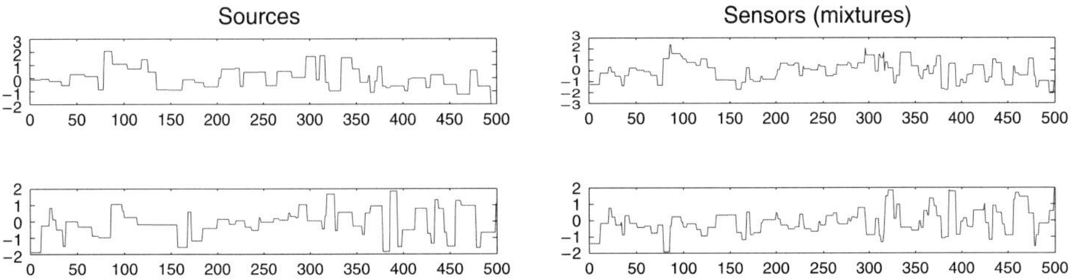

FIGURE 10.8

Time plots of block signals.

where $f(\rho)$ is a weighting function that gives a different role to the scatter plot points depending on their distance from the origin $|\rho(k)|$, and $w(\cdot)$ is a smoothing window, also known as a potential function [15] or Parzen window.

Figure 10.7(b) illustrates the histogram thus obtained with a rectangular smoothing window of size π/L (corresponding to a *non-smoothed* histogram) and respectively $f(|\rho|) = |\rho|$. Taking account of ρ to weight the histogram proves to be critical in this example for the success of histogram-based techniques, since it highlights peaks in the histogram close to angles corresponding to the actual directions of the sources in the mixture. Weighting according to the distance from the origin $|\rho(k)|$ is more satisfactory here than using a raw histogram, because it more effectively highlights these peaks. Other forms of weighting are possible [111], but no theoretical or experimental comparative studies were found examining the effect of the chosen smoothing window and weighting function on the quality of the estimation of **A**.

10.4.2 Scatter plot selection in multiscale representations

In many cases, especially in wavelet-related decompositions, there are distinct groups of coefficients, in which sources have different sparsity properties. The idea is to select those groups of features (coefficients) which are best suited for separation, with respect to the following criteria: (1) sparsity of coefficients; (2) separability of sources' features [117, 114,62]. After the best groups are selected, one uses only these in the separation process, which can be accomplished by any ICA method, including the Relative Newton method (in the case of square mixing matrix), or by clustering.

10.4.2.1 Motivating example: random blocks in the Haar basis

Typical block functions are shown in Fig. 10.8. They are piecewise constant, with random amplitude and duration of each constant piece. Let us take a close look at the Haar wavelet coefficients at different resolutions. Wavelet basis functions at the finest resolution are obtained by translation of the Haar mother wavelet:

$$\varphi_j(t) = \begin{cases} -1 & \text{if } t = 0 \\ 1 & \text{if } t = 1 \\ 0 & \text{otherwise.} \end{cases} \tag{10.36}$$

	Raw signals	All wavelet coefficients	Hi resolution wavelet coefficients
BS-Infomax	13.9%	4.2%	0.69%
C-means clustering	13.3%	2.4%	0.41%

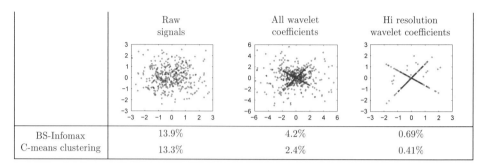

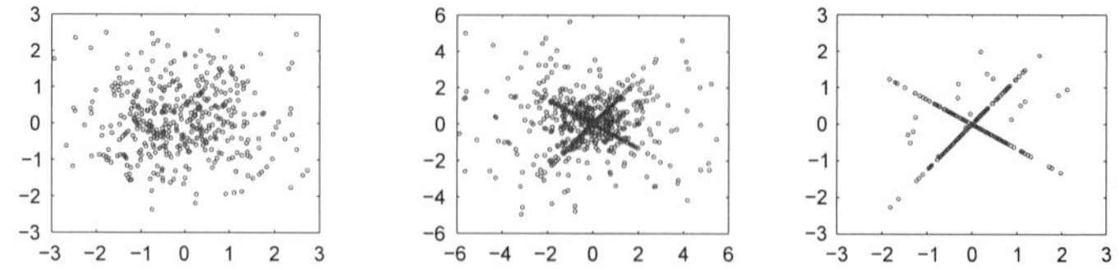

FIGURE 10.9

Separation of block signals: scatter plots of sensor signals and mean-squared separation errors (%).

Taking a scalar product of a function $s(t)$ with the wavelet $\varphi_j(t-\tau)$, one produces a finite differentiation of the function $s(t)$ at the point $t=\tau$. This means that the number of non-zero coefficients at the finest resolution for a block function will correspond roughly to the number of jumps it has. Proceeding to the next, coarser resolution level

$$\varphi_{j-1}(t) = \begin{cases} -1 & \text{if } t=-1,-2 \\ 1 & \text{if } t=0,1 \\ 0 & \text{otherwise} \end{cases} \tag{10.37}$$

the number of non-zero coefficients still corresponds to the number of jumps, but the total number of coefficients at this level is halved, and so is the sparsity. If we proceed further in this direction, we will achieve levels of resolution where the typical width of a wavelet $\varphi_j(t)$ is comparable to the typical distance between jumps in the function $s(t)$. In this case, most of the coefficients are expected to be non-zero, and, therefore, sparsity will fade out.

To demonstrate how this influences the accuracy of blind source separation, two block-signal sources were randomly generated (Fig. 10.8, left), and mixed by the matrix

$$\mathbf{A} = \begin{pmatrix} 0.8321 & 0.6247 \\ -0.5547 & 0.7809 \end{pmatrix}. \tag{10.38}$$

The resulting mixtures, $x_1(t)$ and $x_2(t)$ are shown in Fig. 10.8, right. Figure 10.9, first column, shows the scatter plot of $x_1(t)$ versus $x_2(t)$, where there are no visible distinct features. In contrast, the scatter plot of the wavelet coefficients at the highest resolution (Fig. 10.9, third column) shows two distinct orientations, which correspond to the columns of the mixing matrix.

Results of separation of the block sources are presented in Fig. 10.9. The largest error (13%) was obtained on the raw data, and the smallest (below 0.7%) — on the wavelet coefficients at the highest resolution, which have the best sparsity. Use of all wavelet coefficients leads to intermediate sparsity and performance.

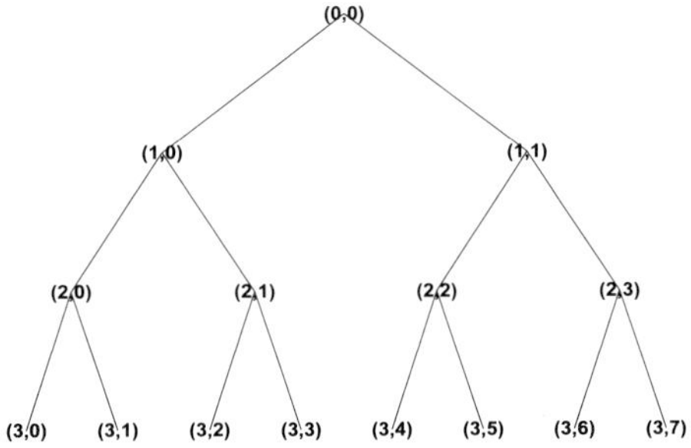

FIGURE 10.10

Wavelet packets tree.

10.4.2.2 Adaptive wavelet packet selection

Multiresolution analysis. Our choice of a particular wavelet basis and of the sparsest subset of coefficients was obvious in the above example: it was based on knowledge of the structure of piecewise constant signals. For sources having oscillatory components (like sounds or images with textures), other systems of basis functions, for example, wavelet packets [26], or multiwavelets [109], might be more appropriate. The wavelet packets library consists of the triple-indexed family of functions:

$$\varphi_{jnk}(t)=2^{j/2}\varphi_n(2^j t-k), \quad j,k\in\mathbb{Z},\, n\in\mathbb{N}. \tag{10.39}$$

As in the case of the wavelet transform, j,k are the scale and shift parameters, respectively, and n is the frequency parameter, related to the number of oscillations of a particular generating function $\varphi_n(t)$. The set of functions $\varphi_{jn}(t)$ forms a (j,n) wavelet packet. This set of functions can be split into two parts at a coarser scale: $\varphi_{j-1,2n}(t)$ and $\varphi_{j-1,2n+1}(t)$. It follows that these two form an orthonormal basis of the subspace which spans $\{\varphi_{jn}(t)\}$. Thus, one arrives at a family of wavelet packet functions on a binary tree (Fig. 10.10). The nodes of this tree are numbered by two indices: the depth of the level $j=0,1,\ldots,J$, and the number of nodes $n=0,1,2,3,\ldots,2^j-1$ at the specified level j. Using wavelet packets allows one to analyze given signals not only with a scale-oriented decomposition but also on frequency sub-bands. Naturally, the library contains the wavelet basis.

The decomposition coefficients $c_{jnk}=\langle s,\varphi_{jnk}\rangle$ also split into (j,n) sets corresponding to the nodes of the tree, and there is a fast way to compute them using banks of *conjugate mirror filters*, as is implemented in the fast wavelet transform.

Choice of the best nodes in the tree. A typical example of scatter plots of the wavelet packet coefficients at different nodes of the wavelet packet tree is shown in Fig. 10.11 (frequency-modulated signals were used as sources.) The upper left scatter plot, labeled "C", corresponds to the set of coefficients at all nodes. The remainder are the scatter

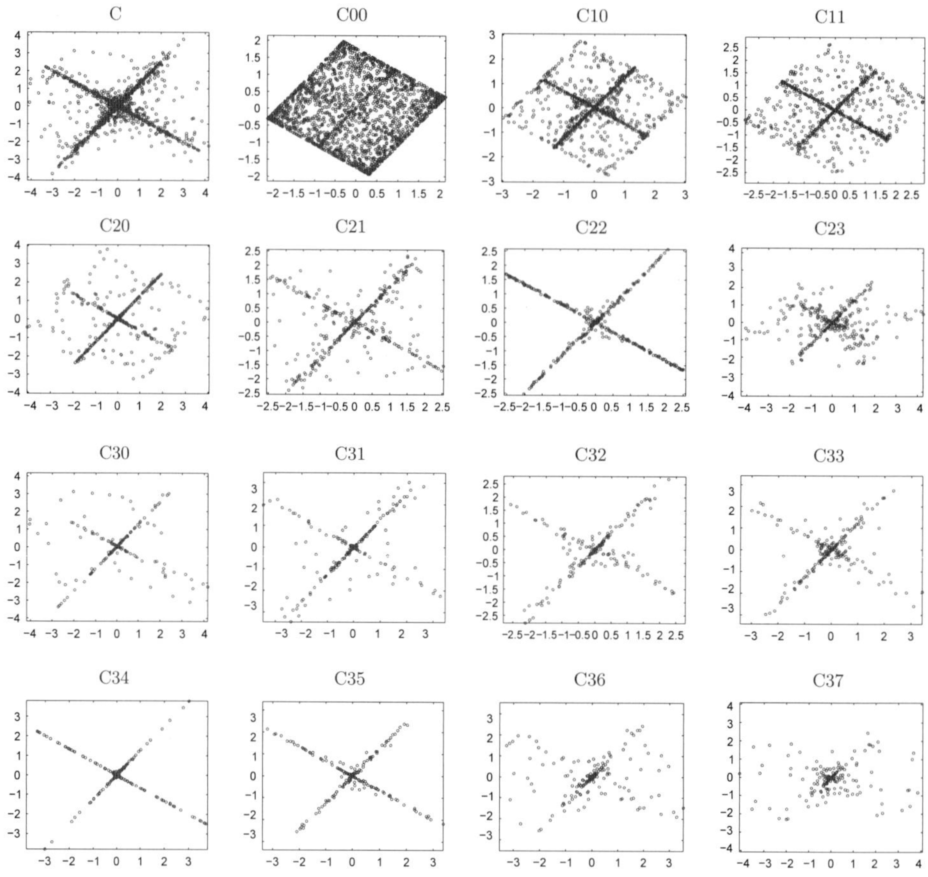

FIGURE 10.11

Scatter plots of the WP coefficients of the FM mixtures.

plots of sets of coefficients indexed in a wavelet packets tree. Generally speaking, the more distinct the directions appearing on these plots, the more precise the estimation of the mixing matrix, and, therefore, the better the separation.

It is difficult to decide in advance which nodes contain the sparsest sets of coefficients. That is why one can use the following simple adaptive approach.

First, apply a clustering algorithm for every node of the tree and compute a measure of clusters' distortion (for example, the mean squared distance of data points to the centers of their own clusters. Here again, the weights of the data points can be incorporated. Second, choose a few best nodes with the minimal distortion, combine their coefficients into one data set, and apply a separation algorithm to these data.

More sophisticated techniques dealing with adaptive choice of best nodes, as well as their number, can be found in [63,62].

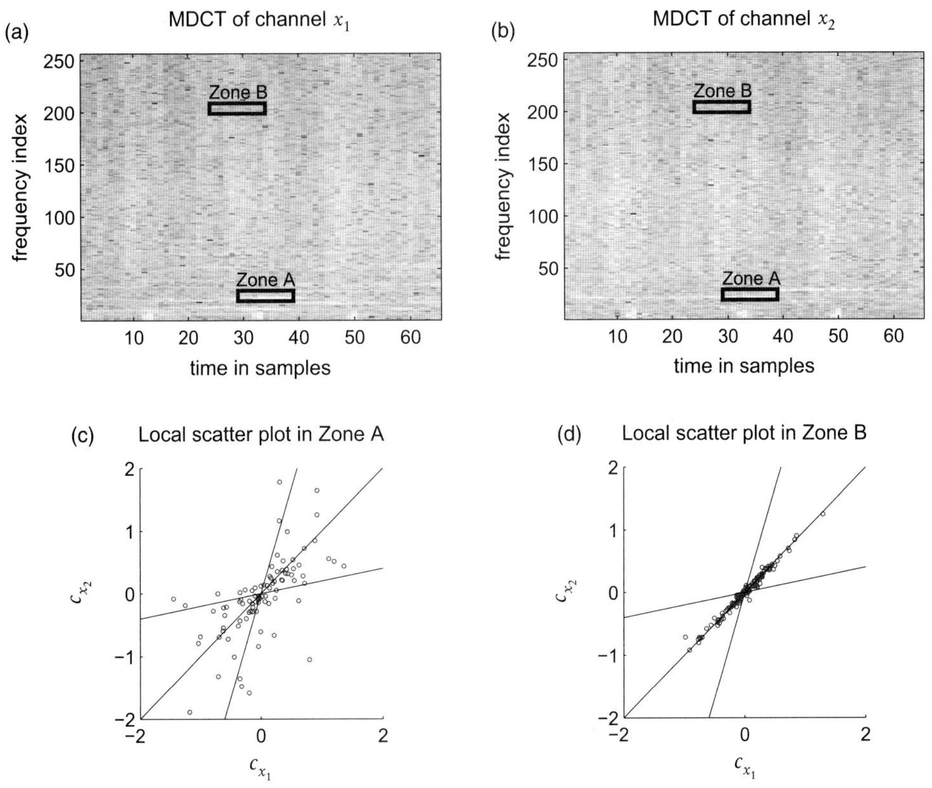

FIGURE 10.12

Top: MDCTs c_{x_1} and c_{x_2} of the two channels x_1 and x_2 of the noiseless stereophonic mixture with three sources represented in Fig. 10.1(c), with grey-level used to show intensity (from black for time-frequency zones with low energy to white for zones with higher energy). Two time-frequency zones (Zone A and Zone B) are indicated. Bottom: corresponding local scatter plots: (a) Zone A, where several sources are simultaneously active; (b) Zone B, where one source is dominant. On average, Zone A points have higher energies than Zone B points, therefore contributing more to the weighted histograms. However, Zone B points provide a more reliable estimation of direction.

10.4.3 Use of local scatter plots in the time-frequency plane

Despite the visible presence of alignments along the mixing matrix columns in the global scatter plot, the most traditional clustering algorithms such as K-means [110] can prove difficult to use for estimating the matrix. While a K-means algorithm is particularly simple to implement, the results depend on proper initialization and correctly determining beforehand the number of sources present. Using the global scatter plot is even more difficult when the mixed sources contribute with very diverse intensities to the mixture, or when their representations are not sufficiently disjoint, as when several musical instruments play in time and in harmony, thereby producing simultaneous activity at certain times and frequencies.

Figure 10.12 illustrates this phenomenon, showing the time-frequency representations (MDCT) of each channel of a stereophonic mixture, as well as the *local scatter plots*

$\{C_{\mathbf{x}}(k)\}_{k\in\Lambda}$ where Λ corresponds to a subset of time-frequency atoms close to a central time and frequency. In the first local scatter plot, one cannot see any distinct alignment along a column of the underlying mixing matrix. In the second plot, on the contrary *all* the points are distinctly aligned around *a single* mixture direction, that of the source whose activity is dominant throughout the corresponding time-frequency zone. The direction of this source can thus be identified using the local scatter plot.

If this type of zone can be detected, $\mathbf{A}$ can be estimated, which is the principle of the methods proposed by Deville *et al.* [3,38,2,88]: the variance of the ratio $c_{x_1}(k)/c_{x_2}(k)$ is calculated for each region considered, then regions with lower variance are selected to estimate the corresponding column directions of $\mathbf{A}$. A similar approach, but one where the different channels play a more symmetrical role, consists in performing Principle Component Analysis (PCA) based on the local scatter plot and selecting the zones whose main direction is the most dominant relative to the others [8]. Techniques based on selecting "simple autoterms" of bilinear time-frequency transforms [39] are also very close to this principle.

10.5 SQUARE MIXING MATRIX: RELATIVE NEWTON METHOD

When the mixing matrix is square non-degenerative, one can perform BSS by any of the methods presented in other chapters of this book. However, use of sparsity prior can greatly improve separation quality in this case as well [115,117]. We consider the model

$$C_{\mathbf{x}} = \mathbf{A}C_{\mathbf{s}} \tag{10.40}$$

where the mixture coefficients $C_{\mathbf{x}}$ can be obtained via analysis transform of the mixture signals (10.14) or, even better, by joint sparse approximation (10.17). The equation above can be treated as a standard BSS problem with $C_{\mathbf{s}}$, $C_{\mathbf{x}}$ and the coefficient index k substituted by S, X and time index t:

$$X = \mathbf{A}S. \tag{10.41}$$

Now we just say that the sources S are sparse themselves. We will use log-likelihood BSS model [81,19] parameterized by the separating matrix $\mathbf{B} = \mathbf{A}^{-1}$. If we assume the sources to be iid, stationary and white, the normalized negative-log-likelihood of the mixtures is

$$L(\mathbf{B};X) = -\log|\det\mathbf{B}| + \frac{1}{T}\sum_{i,k} h(b_i x(t)), \tag{10.42}$$

where b_i is i-th row of $\mathbf{B}$, $h(\cdot) = -\log f(\cdot)$, and $f(\cdot)$ is the probability density function (pdf) of the sources. A consistent estimator can be obtained by the minimization of (10.42), also when $h(\cdot)$ is not exactly equal to $-\log f(\cdot)$ [84]. Such *quasi-ML estimation* is practical when the source pdf is unknown, or is not well-suited for optimization. For example, when the sources are sparse, the absolute value function or its smooth approximation is a good choice for $h(\cdot)$ [23,80,71,115,117,114]. Among other convex smooth approximations to the absolute value, one can use

$$h_1(c) = |c| - \log(1+|c|) \tag{10.43}$$
$$h_\lambda(c) = \lambda h_1(c/\lambda) \tag{10.44}$$

with λ a proximity parameter: $h_\lambda(c) \to |c|$ as $\lambda \to 0^+$. The widely accepted natural gradient method does not work well when the approximation of the absolute value becomes too sharp. Below we describe the Relative Newton method [112,113], which overcomes this obstacle. The algorithm is similar in part to the Newton method of Pham and Garat [84], while enriched by positive definite Hessian modification and line search, providing global convergence.

10.5.1 Relative optimization framework

Suppose one has some current estimate of the separating matrix $\mathbf{B}_k$. One can think about the current source estimate

$$S_k = \mathbf{B}_k X \tag{10.45}$$

as a mixture in some new BSS problem

$$S_k = \mathbf{A}_k S$$

where S are the actual sources to be found. This new problem can be approximately solved by one or several steps of some separation method, leading to the next BSS problem, so on. One uses such an idea for the minimization of quasi-ML function (10.42):

- Start with an initial estimate $\mathbf{B}_0$ of the separation matrix;
- **For** $k = 0, 1, 2, \ldots$, Until convergence
  1. Compute current source estimate $S_k = \mathbf{B}_k X$;
  2. For the new BSS problem $S_k = \mathbf{A}_k S$, decrease sufficiently its quasi-ML function $L(V; S_k)$. Namely, get "local" separation matrix V_{k+1} by one or a few steps of a conventional optimization method (starting with $V = I$).
  3. Update the overall separation matrix

$$\mathbf{B}_{k+1} = V_{k+1}\mathbf{B}_k; \tag{10.46}$$

The Relative (or Natural) Gradient method [25,7,22], described in this book in Chapter 4, *Likelihood* and Chapter 6, *Iterative Algorithms*, is a particular instance of this approach, when the standard gradient descent step is used in item 2. The following remarkable *invariance* property of the Relative Gradient method is also preserved in general:

Given current source estimate S_k, the trajectory of the method and the final solution do not depend on the original mixing matrix.

This means that even ill-conditioned mixing matrix does not influence the convergence of the method any more than does a starting point, and does not influence the final solution at all. Therefore one can analyze local convergence of the method assuming the mixing matrix to be close to the identity. In this case, when $h(c)$ is smooth enough, the Hessian of (10.42) is well-conditioned, and high linear convergence rate can be achieved even with the Relative Gradient method. However, if we are interested in $h(c)$ being less smooth (for example, using small λ in modulus approximation (10.43)), use of a Newton step in item 2 of Relative Optimization become rewarding.

A natural question arises: how can one be sure about the global convergence in terms of "global" quasi-log-likelihood $L(\mathbf{B};X)$ given by (10.42) just reducing the "local" function $L(V;S_k)$? The answer is:

One-step reduction of the "local" quasi-log-likelihood leads to the equal reduction of the "global" one:

$$L(\mathbf{B}_k;X)-L(\mathbf{B}_{k+1};X)=L(I;S_k)-L(V_{k+1};S_k).$$

This can be shown by subtracting the following equalities obtained from (10.42), (10.45) and (10.46)

$$L(\mathbf{B}_k;X)=-\log|\det \mathbf{B}_k|+L(I;S_k) \tag{10.47}$$

$$L(\mathbf{B}_{k+1};X)=-\log|\det \mathbf{B}_k|+L(V_k;S_k). \tag{10.48}$$

10.5.2 Newton method

The Newton method is an efficient tool of unconstrained optimization. It converges much faster than the gradient descent when the function has a narrow valley, and provides a quadratic rate of convergence. However, its iteration may be costly, because of the necessity to compute the Hessian matrix of the mixed second derivatives and to solve the corresponding system of linear equations. In the next subsection we will see how this difficulty can be overcome using the Relative Newton method, but first we describe the Modified Newton method in a general setting.

Modified Newton method with a line search

Suppose that we minimize an arbitrary twice-differentiable function $f(x)$ and x_k is the current iterate with the gradient $g_k=\nabla f(x_k)$ and the Hessian matrix $H_k=\nabla^2 f(x_k)$. The Newton step attempts to find a minimum of the second-order Taylor expansion of f around x_k

$$y_k=\arg\min_y\left\{q(x_k+y)=f(x_k)+g_k^T y+\frac{1}{2}y^T H_k y\right\}.$$

The minimum is provided by solution of the Newton system

$$H_k y_k=-g_k.$$

When $f(x)$ is not convex, the Hessian matrix is not necessarily positive definite; therefore the direction y_k is not necessarily a direction of descent. In this case we use the Modified Cholesky factorization[6] [46], which automatically finds a diagonal matrix R such that the matrix $H_k + R$ is positive definite, providing a solution to the modified system

$$(H_k + R)y_k = -g_k. \tag{10.49}$$

After the direction y_k is found, the new iterate x_{k+1} is given by

$$x_{k+1} = x_k + \alpha_k y_k \tag{10.50}$$

where the step size α_k is determined by exact line search

$$\alpha_k = \arg\min_{\alpha} f(x_k + \alpha y_k) \tag{10.51}$$

or by a backtracking line search (see for example [46]):

$$\alpha := 1; \;\; \text{WHILE } f(x_k + \alpha y_k) > f(x_k) + \beta\alpha g_k^T y_k, \quad \alpha := \gamma\alpha$$

where $0 < \beta < 1$ and $0 < \gamma < 1$. The use of line search guarantees monotone decrease of the objective function at every Newton iteration. Our typical choice of the line search constants is $\beta = \gamma = 0.3$. It may also be reasonable to give β a small value, like 0.01.

10.5.3 Gradient and Hessian evaluation

The likelihood $L(\mathbf{B};X)$ given by (10.42) is a function of a matrix argument $\mathbf{B}$. The corresponding gradient with respect to $\mathbf{B}$ is also a matrix

$$G(\mathbf{B}) = \nabla L(\mathbf{B};X) = -\mathbf{B}^{-T} + \frac{1}{T} h'(\mathbf{B}X)X^T, \tag{10.52}$$

where $h'(\cdot)$ is used as an element-wise matrix function. The Hessian of $L(\mathbf{B};X)$ is a linear mapping (4D tensor) $\mathcal{H}$ defined via the differential of the gradient $dG = \mathcal{H} d\mathbf{B}$. We can also express the Hessian in standard matrix form converting $\mathbf{B}$ into a long vector $b = \text{vec}(\mathbf{B})$ using row stacking. We will denote the reverse conversion $\mathbf{B} = \text{mat}(b)$. Denote

$$\hat{L}(b, X) \equiv L(\text{mat}(b), X), \tag{10.53}$$

with the gradient

$$g(b) = \nabla \hat{L}(b;X) = \text{vec}(G(\mathbf{B})). \tag{10.54}$$

[6] MATLAB code of modified Cholesky factorization by Brian Borchers, available at http://www.nmt.edu/ ~ borchers/ldlt.html.

The Hessian of $-\log|\det \operatorname{mat}(b)|$ is determined using i-th column A^i and j-th row A_j of $\mathbf{A} = \mathbf{B}^{-1}$. Namely, the r-th column of H, $r = (i-1)N + j$, contains the matrix $(A^i A_j)^T$ stacked row-wise (see [112,113]):

$$H^r = \operatorname{vec}(A^{j^T} A_i^T). \tag{10.55}$$

The Hessian of the second term in $\hat{L}(b, X)$ is a block-diagonal matrix with the following $N \times N$ blocks constructed using the rows B_m of $\mathbf{B}$:

$$\frac{1}{T}\sum_t h''(B_m x(t))x(t)x^T(t), \quad m = 1, \ldots, N. \tag{10.56}$$

Hessian simplifications in Relative Newton method

Relative Newton iteration k uses the Hessian of $L(I, S^k)$. Hessian of $-\log|\det \mathbf{B}|$ given by (10.55), becomes very simple and sparse, when $\mathbf{B} = \mathbf{A} = I$. Each column of H contains only one non-zero element, which is equal to 1:

$$H^r = \operatorname{vec}(e_j e_i^T), \tag{10.57}$$

where e_j is an N-element standard basis vector, containing 1 at the j-th position and zeros at others. The second block-diagonal term of the Hessian (10.56) also simplifies greatly when $X = S^k \to S$. When we approach the solution, and the sources are independent and zero mean, the off-diagonal elements in (10.56) converge to zero as sample size grows. As a result, we have only two non-zero elements in every row of the Hessian: one from $\log|\det(\cdot)|$ and another from the diagonal of the second term. After reordering the variables

$$v = [V_{11}, V_{12}, V_{21}, V_{13}, V_{31}, \ldots, V_{22}, V_{23}, V_{32}, V_{24}, V_{42}, \ldots V_{NN}]^T$$

we get the Hessian with diagonal 2×2 and 1×1 blocks, which can be inverted very fast.

In order to guarantee a descent direction and avoid saddle points, we substitute the Hessian with a positive definite matrix: change the sign of the negative eigenvalues (see for example [46]), and force the small eigenvalues to be above some threshold (say, 10^{-8} of the maximal one).

10.5.4 Sequential optimization

When the sources are sparse, the quality of separation greatly improves with reduction of smoothing parameter λ in the modulus approximation (10.44). On the other hand, the optimization of the likelihood function becomes more difficult for small λ. Therefore, we first optimize the likelihood with some moderate λ, then reduce λ by a constant factor (say, 10 or 100), and perform optimization again, and so on. This *Sequential Optimization* approach reduces the overall computations considerably.

Smoothing Method of Multipliers (SMOM)

Even gradual reduction of the smoothing parameter may require a significant number of Newton steps after each update of λ. A more efficient way to achieve an accurate solution of a problem involving a sum of absolute value functions is to use the SMOM method, which is an extension of the Augmented Lagrangian technique [11,86,10] used in constrained optimization. It allows us to obtain an accurate solution without forcing the smoothing parameter λ to go to zero. SMOM can be efficiently combined with the Relative Optimization. We refer the reader to [113] for the exposition of this approach.

10.5.5 Numerical illustrations

Two data sets were used. The first group of sources was random sparse data with Gaussian distribution of non-zero samples, generated by the MATLAB function SPRANDN. The second group of sources consisted of four natural images from [24]. The mixing matrix was generated randomly with uniform iid entries.

Relative Newton method

Figure 10.13 shows the typical progress of different methods applied to the artificial data with 5 mixtures of 10k samples. The Fast Relative Newton method (with 2×2 block-diagonal Hessian approximation) converges in about the same number of iterations as the Relative Newton with exact Hessian, but significantly outperforms it in time. Natural gradient in batch mode requires many more iterations, and has difficulty in converging when the smoothing parameter λ in (10.44) becomes too small.

In the second experiment, we demonstrate the advantage of the batch-mode quasi-ML separation, when dealing with sparse sources. We compared the Fast Relative Newton method with stochastic natural gradient [25,7,22], Fast ICA [60] and JADE [20]. All three codes are available at public web sites [73,59,21]. Stochastic natural gradient and Fast ICA used tanh nonlinearity. Figure 10.14 shows separation of artificial stochastic sparse data: 5 sources of 500 samples, 30 simulation trials. As we see, Fast Relative Newton significantly outperforms other methods, providing practically ideal separation with the smoothing parameter $\lambda = 10^{-6}$ (sequential update of the smoothing parameter was used here). Timing is of about the same order for all the methods, except of JADE, which is known to be much faster with relatively small matrices.

Sequential Optimization vs Smoothing Method of Multipliers (SMOM)

In the third experiment we have used the first stochastic sparse data set: 5 mixtures, 10k samples. Figure 10.15 demonstrates the advantage of the SMOM combined with the frozen Hessian strategy [113]: the Hessian is computed once and then used in several Newton steps, while they are effective enough. As we see, the last six outer iterations do not require new Hessian evaluations. At the same time the Sequential Optimization method without Lagrange multipliers, requires 3 to 8 Hessian evaluations per outer iteration towards the end. As a consequence the method of multipliers converges much faster.

In the fourth experiment, we separated four natural images [24], presented in Fig. 10.16. Sparseness of images can be achieved via various wavelet-type transforms [115,117,114],

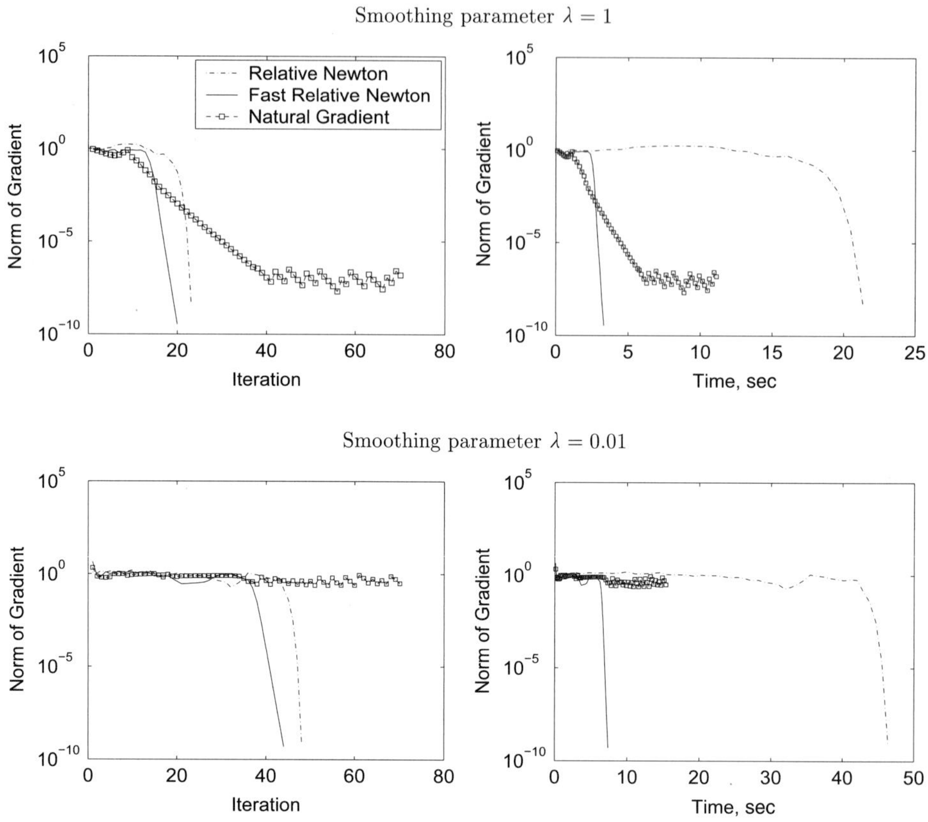

FIGURE 10.13

Separation of artificial sparse data with 5 mixtures by 10k samples: Relative Newton with exact Hessian (dashed line), Fast Relative Newton (continuous line), Natural Gradient in batch mode (squares).

but even simple differentiation can be used for this purpose, since natural images often have sparse edges. Here we used the stack of horizontal and vertical derivatives of the mixture images as an input to the separation algorithms. Figure 10.16 shows the separation quality achieved by stochastic natural gradient, Fast ICA, JADE, the Fast Relative Newton method with $\lambda = 10^{-2}$ and the SMOM. Like in the previous experiments, SMOM provides practically ideal separation with ISR of about 10^{-12}. It outperforms the other methods by several orders of magnitude.

10.5.6 Extension of Relative Newton: blind deconvolution

We only mention two important extensions of the Relative Newton method. First, the block-coordinate version of the method [16] has an improved convergence rate for large problems. Second, the method can be modified for blind deconvolution of signals and images [17,18].

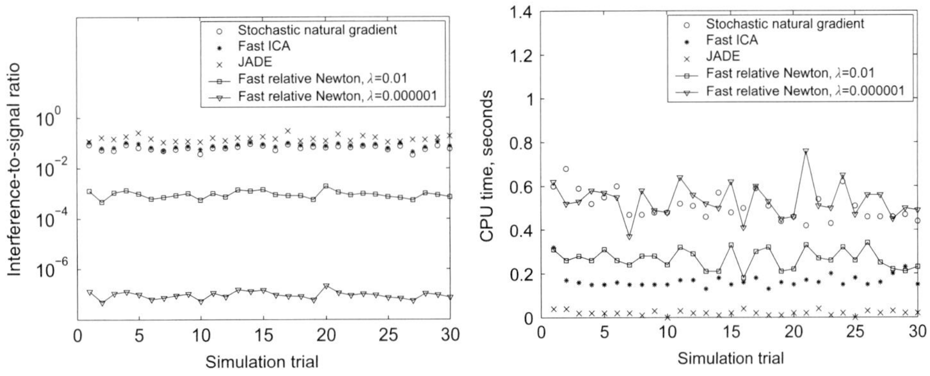

FIGURE 10.14

Separation of stochastic sparse data: 5 sources of 500 samples, 30 simulation trials. Left – interference-to-signal ratio, right – CPU time.

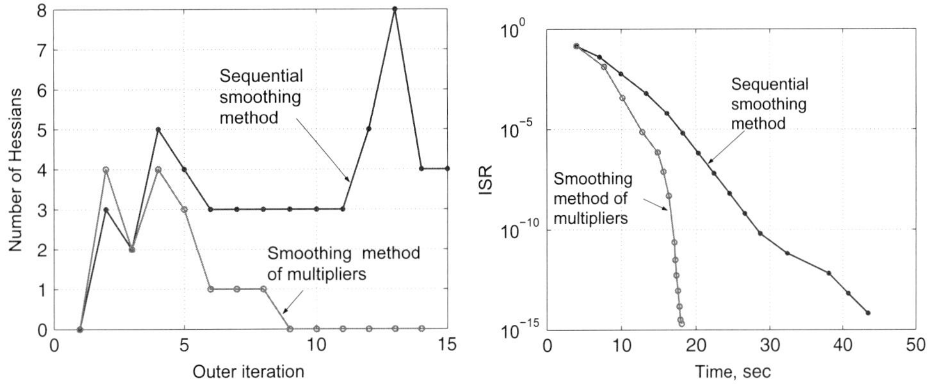

FIGURE 10.15

Relative Newton method with frozen Hessian. Left – number of Hessian evaluations per outer iteration (the first data set); Right – interference-to-signal ratio. Dots correspond to outer iterations.

10.6 SEPARATION WITH A KNOWN MIXING MATRIX

The first two steps of a sparse source separation system (Fig. 10.2) involve calculating:

- a sparse representation/approximation $C_{\mathbf{x}}$ such that $\mathbf{x} \approx C_{\mathbf{x}}\mathbf{\Phi}$;
- an estimation of the mixing matrix $\widehat{\mathbf{A}}$.

The sources are finally separated in the last step. According to the type of mixture (determined or not), several approaches are possible, depending more or less on the sparsity and disjointedness of the source representations.

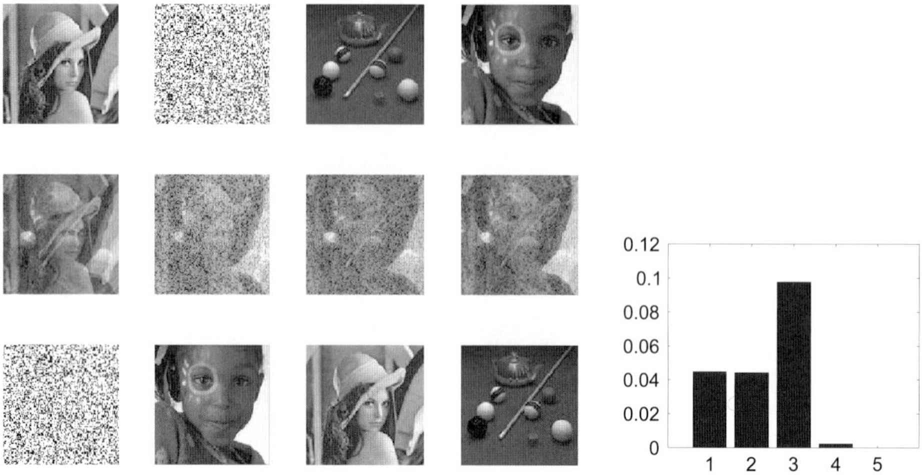

FIGURE 10.16

Separation of images with preprocessing by differentiation. Left: top – sources, middle – mixtures, bottom – separated. Right: Interference-to-signal ratio (ISR) of image separation. 1 – stochastic natural gradient; 2 – Fast ICA; 3 – JADE; 4 – Relative Newton with $\lambda = 10^{-2}$; 5 – SMOM (bar 5 is not visible because of very small ISR, of order 10^{-12}.)

10.6.1 Linear separation of (over-)determined mixtures

When $\widehat{\mathbf{A}}$ is an invertible square matrix, its inverse $\mathbf{B} := \widehat{\mathbf{A}}^{-1}$ can be applied to the mixture $\mathbf{x} = \mathbf{A}\mathbf{s}$, assumed noisefree, to obtain $\widehat{s} := \widehat{\mathbf{A}}^{-1}\mathbf{x} = \widehat{\mathbf{A}}^{-1}\mathbf{A}\mathbf{s}$. An alternative consists in first estimating a representation of the sources $C_{\widehat{s}} := \widehat{\mathbf{A}}^{-1} C_{\mathbf{x}}$, then estimating the sources by reconstruction, $\widehat{\mathbf{s}} := C_{\widehat{s}}\boldsymbol{\Phi}$. In the over-determined case where the number of sensors exceeds the number of sources, the pseudo-inverse $\mathbf{B} := \widehat{\mathbf{A}}^{\dagger}$ can be used. These two alternatives (separation of the initial mixture or its representation) are strictly identical provided that $C_{\mathbf{x}}$ is an *exact representation* of the mixture $\mathbf{x}$. This is not true when $C_{\mathbf{x}}$ is only a *sparse approximation* of mixture $\mathbf{x} \approx C_{\mathbf{x}}\boldsymbol{\Phi}$; whereas the first approach guarantees that $\mathbf{x} = \widehat{\mathbf{A}}\widehat{\mathbf{s}}$, the second approach only provides the approximation $\mathbf{x} \approx C_{\mathbf{x}}\boldsymbol{\Phi} = \widehat{\mathbf{A}}C_{\widehat{s}}\boldsymbol{\Phi} = \widehat{\mathbf{A}}\widehat{\mathbf{s}}$. For (over-)determined separation problems, separation in the sparse domain is potentially interesting only when the mixture to separate is noisy, in which case obtaining an approximation $\mathbf{x} \approx \widehat{A}\widehat{\mathbf{s}}$ can effectively denoise the mixture.

Sparse source separation can be particularly useful for separating under-determined mixtures where the number of sources N exceeds the number of sensors P. In this case, it is often much more efficient than any of the linear separation methods, which estimate the sources using a separation matrix $\mathbf{B}$ taking the form $\widehat{\mathbf{s}} = \mathbf{B}\mathbf{x}$. Simple linear algebra arguments show that it is not possible to perfectly separate the sources linearly in the under-determined case, and more specifically [51] that the estimation of at least one of the sources necessarily undergoes a level of residual interference from the other sources of around $10\log_{10}(N/P-1)$ decibels [51, Lemma 1]. In the following discussion, we focus on sparse methods that can overcome the limits of linear separation. They all start by estimating a representation $C_{\hat{\mathbf{s}}}$ of the sources before reconstructing the estimated sources as $\hat{\mathbf{s}} = C_{\hat{\mathbf{s}}}\mathbf{\Phi}$. Depending on the case, this representation is obtained directly from the initial mixture $\mathbf{x}$ and the estimated matrix $\widehat{\mathbf{A}}$ (sections 10.6.5 to 10.6.8), or from $\widehat{\mathbf{A}}$ and the joint sparse representation $C_{\mathbf{x}}$ of the mixture (sections 10.6.2 to 10.6.4), which assumes that the identity $C_{\mathbf{x}} \approx \widehat{\mathbf{A}}C_{\mathbf{s}}$ is reasonably verified.

10.6.2 Binary masking assuming a single active source

The source model with disjoint support representations described in the introduction corresponds to estimating the representation of each source by means of masking, based on the calculated representation of the mixture $C_{\mathbf{x}}$:

$$c_{\hat{s}_n}(k) := \chi_n(k) \cdot \frac{\widehat{\mathbf{A}}_n^H C_{\mathbf{x}}(k)}{\|\widehat{\mathbf{A}}_n\|_2^2} \tag{10.58}$$

where the mask $\chi_n(k)$ is defined as:

$$\chi_n(k) := \begin{cases} 1 & \text{if } n = n(k); \\ 0 & \text{if not.} \end{cases} \qquad \text{with } n(k) := \arg\max_n \frac{|\widehat{\mathbf{A}}_n^H C_{\mathbf{x}}(k)|}{\|\widehat{\mathbf{A}}_n\|_2^2}; \tag{10.59}$$

that is, only the most active source is attributed a non-zero coefficient, whereas the coefficients of the other sources are set to zero, or "masked". The degree of a source's activity, which determines the mask, is measured as the correlation between the corresponding column $\widehat{\mathbf{A}}_n$ of the estimated mixing matrix and the component $C_{\mathbf{x}}(k)$ of the mixture.

Successful masking depends on the sources mainly having disjoint representations, effectively resulting in the approximation $C_{\mathbf{x}}(k) \approx \widehat{\mathbf{A}}_{n(k)} c_{s_n}(k)$. Although a strong assumption, this can be checked for many audiophonic mixtures [9] in the time-frequency domain. It is the main reason for the good results obtained with the DUET algorithm [111], which performs a variant of masking in the time-frequency domain where the mask $\chi_n(k)$ is applied to one of the channels c_{x_p} of the mixture, rather than to the linear combination of all of the channels $\widehat{\mathbf{A}}_n^H C_{\mathbf{x}}$. Inversely, masking $\widehat{\mathbf{A}}_n^H C_{\mathbf{x}}$ has the advantage of guaranteeing that the source to be estimated is in fact present in the masked "virtual channel", whereas a source not present in a given channel cannot be estimated by masking this channel.

10.6.3 Binary masking assuming $M < P$ active sources

When the number of mixed sources N is high, the mean number of sources simultaneously active can exceed 1, despite the sparse representation of each source. Assuming the number of active sources does not exceed M, where $M < P$, an approach [50,6] generalizing the binary masking described above involves:

1. determining the set $I_M(k) \subset \{1,\ldots,N\}$ of indices of M active sources in the $C_{\mathbf{x}}(k)$ component;
2. using least squares to calculate the components of the active sources.

Denoting as $\widehat{\mathbf{A}}_I$ the matrix composed of $\widehat{\mathbf{A}}$ columns with index $k \in I$, if the set of active sources is I, the estimated components of the sources are:

$$c_{\widehat{s}_n}(k) := 0, \qquad k \notin I, \tag{10.60}$$

$$\left(c_{\widehat{s}_n}(k)\right)_{k\in I} := \widehat{\mathbf{A}}_I^{\dagger} C_{\mathbf{x}}. \tag{10.61}$$

The set $I_M(k)$ of active sources is thus chosen to minimize the mean square error of reconstructing $C_{\mathbf{x}}$:

$$I_M(k) := \arg \min_{I|\mathrm{card}(I)\leq M} ||C_{\mathbf{x}} - \widehat{\mathbf{A}}_I \widehat{\mathbf{A}}_I^{\dagger} C_{\mathbf{x}}||_2^2 \tag{10.62}$$

$$= \arg \max_{I|\mathrm{card}(I)\leq M} ||\widehat{\mathbf{A}}_I \widehat{\mathbf{A}}_I^{\dagger} C_{\mathbf{x}}||_2^2. \tag{10.63}$$

When $M = 1$, the principle of binary masking applies exactly as described above.

10.6.4 Local separation by ℓ^τ minimization

When the number of sources assumed active can reach $M = P$, it is no longer possible to define the set $I(k)$ of active sources by mean square error minimization since generically, with any subset I of P active sources, a perfect reconstruction is obtained, $C_{\mathbf{x}} = \widehat{\mathbf{A}}_I \widehat{\mathbf{A}}_I^{-1} C_{\mathbf{x}}$.

An alternative is to exploit the sparsity of the sources by a maximum likelihood approach that assumes a generalized Gaussian distribution of the source coefficients:

$$C_{\widehat{\mathbf{s}}}(k) := \arg \min_{C_{\mathbf{s}}(k)|\widehat{\mathbf{A}}C_{\mathbf{s}}(k)=C_{\mathbf{x}}(k)} \sum_{n=1}^{N} |c_{s_n}(k)|^\tau. \tag{10.64}$$

which for $0 \leq \tau \leq 1$ results in estimating a source representation where, for each atom φ_k, no more than P sources are "active" [64]. In practice, rather than use the heavy artillery of iterative algorithms for ℓ^τ norm optimization (FOCUSS, etc.) to solve (10.64), a combinatorial approach can be employed in low dimension, which involves selecting the best set of P active sources as:

$$I(k) := \arg \min_{I|\mathrm{card}(I)=P} ||\widehat{\mathbf{A}}_I^{-1} C_{\mathbf{x}}||_\tau. \tag{10.65}$$

For greater numerical efficiency, the $\binom{N}{P}$ possible inverse matrices $\widehat{\mathbf{A}}_I^{-1}$ can be precalculated as soon as the mixing matrix is estimated.

This approach is widely used, particularly for separating audio sources based on their STFTs, and provides good results overall. For stereophonic audio mixtures, Bofill and Zibulevsky [15] proposed this sort of local optimization for $\tau = 1$, and Saab *et al.* [93] experimentally studied how the choice of τ affects the quality of the results, for sound mixtures. For the examples they considered, their results are satisfactory in terms of separation quality, and do not vary substantially with the chosen exponent τ.

For noisy mixtures, the separation criterion (10.68) can be replaced to advantage by the following, where the λ parameter adjusts the compromise between faithful reconstruction of the mixture coefficients and sparsity of the estimated sources:

$$C_{\hat{s}}(k) := \arg\min_{C_s(k)} \left\{ \frac{1}{2\sigma_b^2} \|C_x(k) - \widehat{A} C_s(k)\|_F^2 + \lambda \sum_{n=1}^{N} |c_{s_n}(k)|^\tau \right\}. \tag{10.66}$$

Depending on the value of λ, the number of active estimated sources varies from zero to P. As the reader may have noticed, for an exponent $\tau = 0$, the system of masking presented in section 10.6.3 applies, but the number of active sources can vary for each component.

10.6.5 Principle of global separation by ℓ^τ minimization

When the $\mathbf{\Phi}$ dictionary is an orthonormal basis and a simple orthogonal transform is used to calculate the joint mixture representation, the independent estimation of source components according to the criterion (10.66) actually corresponds to optimizing a more global criterion:

$$C_{\hat{s}} := \arg\min_{C_s} \left\{ \|\mathbf{x} - \widehat{A} C_s \mathbf{\Phi}\|_F^2 + \lambda \sum_{nk} \left| c_{s_n}(k) \right|^\tau \right\}. \tag{10.67}$$

This results in the equation:

$$\begin{aligned} \|\mathbf{x} - \widehat{A} C_s \mathbf{\Phi}\|_F^2 &= \|\mathbf{x}\mathbf{\Phi}^H - \widehat{A} C_s\|_F^2 = \|C_x^{\mathbf{\Phi}^H} - \widehat{A} C_s\|_F^2 \\ &= \sum_{k=1}^{K} \|C_x^{\mathbf{\Phi}^H}(k) - \widehat{A} C_s(k)\|_2^2. \end{aligned}$$

For an arbitrary dictionary, calculating sources via the criterion (10.67) amounts to a *maximum a posteriori (MAP)* estimation with the assumption of a generalized Gaussian distribution of source coefficients (10.4). In the same vein, Zibulevsky and Pearlmutter [115] popularized the separation of under-determined mixtures by using a sparse source model and MAP estimation:

$$C_{\hat{s}} := \arg\max_{C_s} p(C_s|\mathbf{x}, \widehat{A}) = \arg\min_{C_s} \left\{ \frac{1}{2\sigma_b^2} \|\mathbf{x} - \widehat{A} C_s \mathbf{\Phi}\|_F^2 + \sum_{nk} h(c_{s_n}(k)) \right\} \tag{10.68}$$

with $h(c) = -\log p_c(c)$ where $p_c(\cdot)$ is the common distribution of source coefficients, assumed independent. Note that the estimation of $C_{\hat{s}}$ does not involve the representation

C_x of the mixture, calculated in an initial step (see Fig. 10.2). Here this representation only serves to estimate the mixing matrix. Since optimization of a criterion such as (10.68) or (10.67) is only a principle, we will now provide specific details on algorithms that apply it.

10.6.6 Formal links with single-channel traditional sparse approximation

It may come as a surprise that optimization problems such as (10.67) are formally more conventional than the problems of joint sparse mixture representation (10.16) and (10.17) mentioned in section 10.3. This formal link, while not essential for understanding the definition of iterative thresholding or *Matching Pursuit* algorithms for separation, is more critical for describing the FOCUSS algorithm in this context. Readers more interested in the description of algorithms that are conceptually and numerically simpler can skip to section 10.6.7.

Given that:

$$\widehat{\mathbf{A}}C_s\mathbf{\Phi} = \sum_{nk} c_{s_n}(k)\widehat{\mathbf{A}}_n\varphi_k,$$

the optimization (10.67) consists in approximating the mixture $\mathbf{x}$ (considered as a $P \times T$ matrix) with a sparse linear combination of $P \times T$ matrices taking the form $\{\widehat{\mathbf{A}}_n\varphi_k\}_{nk}$.

In the vector space of the $P \times T$ matrices, the problem is one of sparse approximation of the "vector" $\mathbf{x}$ based on the dictionary $\mathit{\Phi}$[7] comprising $N \times K$ "multi-channel atoms" $\{\widehat{\mathbf{A}}_n\varphi_k\}_{nk}$. Thus, the optimization algorithms for sparse representation of a *one-dimensional* signal do not require substantial modification to their formal expressions: iterative thresholding techniques [28,36] for $1 \leq \tau \leq 2$, linear, quadratic or conic programming [23,12] for $\tau = 1$, and iterative reweighted least squares (such as FOCUSS) for $0 < \tau \leq 2$. The detailed description of these algorithms makes use of the elements described in section 10.3, considering the specific case of $P = 1$, i.e. where the representations are no longer *joint*.

10.6.7 Global separation algorithms using ℓ^τ minimization

While things are simple on paper, the same is not necessarily true numerically: the thresholding algorithms are feasible because they involve no matrix inversion, but minimizing the ℓ^1 norm (*basis pursuit*) or the ℓ^τ norm (FOCUSS) quickly becomes impracticable because of the repeated inversions of $(P \times T) \times (P \times T)$ square matrices. By contrast, we will see that *matching pursuit* is easy to apply, with an algorithm that is efficient in practice.

Algorithm for $0 < \tau \leq 2$: FOCUSS and basis pursuit

The FOCUSS algorithm for minimizing the criterion (10.67) is iterative. Starting from an initialization $C_s^{(0)}$, the value of which can decisively influence the algorithm's point

[7]The italicized notation $\mathit{\Phi}$ is used here to distinguish between this dictionary of matrices and the dictionary of atoms $\mathbf{\Phi}$ from which it is built. Whereas $\mathbf{\Phi}$ is a $K \times T$ matrix, $\mathit{\Phi}$ should hereafter be considered as a $(N \times K) \times (P \times T)$ matrix.

of convergence, the weighted mean square criterion is iteratively minimized:

$$C_s^{(m)} := \arg\min_{C_s} \left\{ \|\mathbf{x} - \widehat{\mathbf{A}} C_s \boldsymbol{\Phi}\|_F^2 + \frac{\lambda|\tau|}{2} \sum_{nk} \left| (w_{nk}^{(m)})^{-1} c_{s_n}(k) \right|^2 \right\} \tag{10.69}$$

where $w_{nk}^{(m)} := |c_{s_n}^{(m-1)}(k)|^{1-\tau/2}$. Denoting as $\mathcal{W}^{(m)}$ the $(N \times K) \times (N \times K)$ square weighting matrix whose diagonal elements are the $N \times K$ coefficients $w_{nk}^{(m)}$, each iteration consists in updating $\mathcal{W}^{(m)}$ as well as the weighted dictionary $\Phi^{(m)}$ in which the $N \times K$ atoms are the multi-channel signals $w_{nk}^{(m)} \widehat{\mathbf{A}}_n \varphi_k$, each of size $P \times T$. Using the description of the M-FOCUSS algorithm in the single-channel case, section 10.3.6, we then calculate

$$\mathcal{C}_s^{(m)} := \mathcal{X} \left((\Phi^{(m)})^{\mathrm{H}} \, \Phi^{(m)} + \frac{\lambda|\tau|}{2} \mathbf{I}_{P \times T} \right)^{-1} (\Phi^{(m)})^{\mathrm{H}} \, \mathcal{W}^{(m)} \tag{10.70}$$

where $\mathcal{X}$ is none other than the $P \times T$ column vector obtained by reorganizing the coefficients of $\mathbf{x}$, and $\mathcal{C}$ is an $N \times K$ column vector whose reorganization as a matrix is precisely $C_s^{(m)}$. This algorithm, which requires prior inversion of a $(P \times T) \times (P \times T)$ square matrix at each iteration can only be used in practice if the dictionary is sufficiently structured to permit blockwise inversion, which is the case when $\boldsymbol{\Phi}$ is an orthonormal basis. The quadratic programming algorithms used for *basis pursuit* raise the same difficulties.

Algorithms for $1 \leq \tau \leq 2$: iterative thresholding

Given the estimated mixing matrix $\widehat{\mathbf{A}}$ and the dictionary $\boldsymbol{\Phi}$ considered, the first step is to determine a real $\mu < \infty$ such that, for any $N \times K$ matrix C_s, $\|\widehat{\mathbf{A}} C_s \boldsymbol{\Phi}\|_F^2 \leq \mu \|C_s\|_F^2$. Based on some initialization $C_s^{(0)}$ (often chosen as zero), this is followed by iteratively defining:

$$c_{s_n}(k)^{(m)} = S_{\frac{\lambda}{\mu},\tau} \left(c_{s_n}^{(m-1)}(k) + \frac{\langle \widehat{\mathbf{A}}_n^{\mathrm{H}} (\mathbf{x} - \widehat{\mathbf{A}} C_s^{(m-1)} \boldsymbol{\Phi}), \varphi_k \rangle}{\mu} \right) \tag{10.71}$$

and the resulting sequence necessarily converges to a minimizer of (10.67) as soon as $1 \leq \tau \leq 2$ [28]. If $\tau > 1$ or if $\boldsymbol{\Phi}$ *and* $\widehat{\mathbf{A}}$ are *both* invertible square matrices, this minimizer is also unique [28].

10.6.8 Iterative global separation: demixing pursuit

Algorithms for solving *global* ℓ^τ optimization problems such as (10.67) for $0 < \tau \leq 1$ are extremely time and memory hungry, especially if the data to be separated are high-dimensional and the computing time or power requirements are considerable. A much faster alternative approach is *demixing pursuit* [56,67], a type of *matching pursuit* applied directly to the "vector" $\mathbf{x}$ with the atom dictionary $\{\widehat{\mathbf{A}}_n\varphi_k\}_{nk}$. The definition of this algorithm assumes that the atoms are of unit energy, which amounts to assuming the single-channel atoms φ_k are of unit energy and the columns $\widehat{\mathbf{A}}_n$ of the mixing matrix are also of unit norm. This last assumption is not restrictive since it is compatible with the natural indeterminations of source separation [19]. Theoretical research [56] indicates that if (10.67) has a sufficiently sparse solution, then *demixing pursuit* as well as ℓ^1 norm minimization can be used to find it.

More explicitly, to decompose and separate a multi-channel mixture using *demixing pursuit*, the first step is to initialize a residual $\mathbf{r}^{(0)} = \mathbf{x}$, followed by iteration of the following steps starting with $m = 1$:

1. Select the most correlated mixing matrix column and atom:

$$(n_m, k_m) := \arg\max_{nk} |\langle \widehat{\mathbf{A}}_n^{\mathrm{H}} \mathbf{r}_p^{(m-1)}, \varphi_k \rangle|. \tag{10.72}$$

2. Update the residual:

$$\mathbf{r}^{(m)} := \mathbf{r}^{(m-1)} - \langle \widehat{\mathbf{A}}_{n_m} \mathbf{r}_p^{(m-1)}, \varphi_{k_m} \rangle \widehat{\mathbf{A}}_{n_m} \varphi_{k_m}. \tag{10.73}$$

The decomposition below is obtained after M iterations:

$$\mathbf{x} = \sum_{m=1}^{M} c(m) \widehat{\mathbf{A}}_{n_m} \varphi_{k_m} + \mathbf{r}^{(M)} \tag{10.74}$$

with $c(m) := \langle \widehat{\mathbf{A}}_{n_m} \mathbf{r}^{(m-1)}, \varphi_{k_m} \rangle$.

Variants similar to orthogonal *matching pursuit* are possible, and in terms of numerical calculation, *demixing pursuit* can be implemented very efficiently by exploiting the dictionary's structure and the rapid updates for finding the largest scalar product from one iteration to the next. An open, rapid implementation is available [53,65].

10.7 CONCLUSION

In this chapter we have described the main steps of source separation methods based on sparsity. One of the primary advantages these methods offer is the possibility to separate under-determined mixtures, which involves two independent steps in most methods [99]: (a) estimating the mixing matrix; (b) separation using the known mixing matrix. In practice, as shown in Fig. 10.2, these methods usually involve four steps:

1. joint sparse representation $C_{\mathbf{x}}$ of the channels of the mixture $\mathbf{x}$;
2. estimation of $\mathbf{A}$ based on the scatter plot $\{C_{\mathbf{x}}(k)\}$;
3. sparse separation in the transform domain;
4. reconstruction of the sources.

Implementing a method of sparse source separation requires making a certain number of choices.

Choosing the dictionary Φ. This choice is essentially based on expert knowledge of the class of signals considered. A typical choice for audio mixtures or seismic signals is a Gabor time-frequency dictionary, or a wavelet time-scale dictionary; this is also true for biological signals such as cardiac or electro-encephalogram signals. For natural images, dictionaries of wavelets, curvelets or anisotropic Gabor atoms are generally used. Implementing methods to automatically choose a dictionary remains an open problem, even if techniques for learning dictionaries based on reference data sets are now available. The very concept of matching a dictionary to a class of signals is in need of further clarification.

Choosing the algorithm for the joint sparse representation C_x of the mixture. This choice depends *a priori* on several factors. Available computing power can rule out certain algorithms such as M-FOCUSS or *basis pursuit* when the mixture is high-dimensional (many sensors P or samples/pixels/voxels/ etc. T). In addition, depending on the level of noise $\mathbf{b}$ added to the mixture, the distribution of source coefficients in the selected dictionary, etc., the parameters for optimal algorithm performance (exponent τ, penalty factor λ, stop criterion for *matching pursuit*, etc.) can vary. How easy they are to adjust, via a development data set, may then be taken into consideration.

Choosing the separation method. Separation methods do not all have the same robustness in the case of an inaccurately estimated mixing matrix. A comparison of several methods [67] based on time-frequency dictionaries and stereophonic audio mixtures ($P = 2$) showed that, while less effective than when P active sources are assumed (section 10.6.4) and the mixing matrix $\mathbf{A}$ is perfectly known, binary masking separation assuming a single active source (section 10.6.2) is more robust, and thus more effective, than when only an imperfect estimation of $\mathbf{A}$ is available.

Choosing the algorithm for estimating $\mathbf{A}$. This is undoubtedly the most difficult choice; while any "reasonable" choice for the dictionary, for the joint sparse mixture representation algorithm or for the separation method makes it possible to separate the sources *grosso modo* when the mixing matrix, even if under-determined, is known, an excessive error in estimating the mixing matrix has a catastrophic effect on the results, which become highly erratic. In particular, if the number of sources is not known in advance, an algorithm that provides a very robust estimation is an absolute necessity. This tends to eliminate approaches based on detecting peaks in a histogram in favor of those, like the algorithm DEMIX [8], that exploit redundancies (e.g. temporal and frequential persistence in an STFT) in the mixture representation to make estimating the column directions of $\mathbf{A}$ more robust (with measurement of reliability). If the source supports are not disjoint enough, it becomes necessary to use the approaches in section 10.6.4, which are not very robust if $\mathbf{A}$ is not accurately estimated, making a robust estimation of the mixing matrix all the more critical.

10.8 OUTLOOK

To date, most blind source separation techniques based on sparsity have relied on a sparse model of all the sources in a common dictionary, making it possible to exploit their *spatial diversity* in order to separate them. The principle of these techniques consists in decomposing the mixture into the basic components, grouping these components by the similarity of their spatial characteristics, and finally recombining them to reconstruct the sources. This principle may be applied to more general decompositions [97], provided that an appropriate similarity criterion enables regrouping components from the same source. Below we discuss some of the possibilities that are starting to be explored.

Spatial diversity and morphological diversity: looking for diversity in sources. While sources have sparse decompositions in *different* dictionaries, sparsity can also be exploited to separate them. This is a valid approach, including for the separation of sources from a single channel, when no spatial diversity is available. This type of approach, called Morphological Component Analysis (MCA) [97] because it uses differences in the *waveforms* of the sources to separate them, has been successfully applied to separating tonal and transient "layers" in musical sounds [29] for compression, and for the separation of texture from smoother content in images [96]. It is also possible to jointly exploit the spatial diversity and the *morphological diversity* of sources [13], which amounts to looking for a mixture decomposition with the form:

$$\mathbf{x} = \sum_{n=1}^{N} \mathbf{A}_n \cdot c_{s_n} \cdot \mathbf{\Phi}^{(n)} + \mathbf{b} = \sum_{n=1}^{N}\sum_{k=1}^{K} c_{s_n}(k)\mathbf{A}_n \varphi_k^{(n)} + \mathbf{b} \tag{10.75}$$

where $c_{s_n} \cdot \mathbf{\Phi}^{(n)}$ is a sparse approximation of s_n in a specific dictionary $\mathbf{\Phi}^{(n)}$. The estimated sources are then reconstructed as $\widehat{s_n} := \sum_k c_{s_n}(k)\varphi_k^{(n)}$. The validity of such an approach outside a purely academic context is based on the existence of morphological diversity between sources. Beyond its implementation, the real challenge of MCA lies in determining source-specific dictionaries (by expert knowledge or learning).

Separation of under-determined convolutive mixtures. So far, we have mainly discussed the separation of instantaneous mixtures, but sparsity also makes it possible to process simple forms of convolutive mixtures as well as anechoic mixtures (i.e. with a delay) [111,88]. The analysis of real mixtures raises problems of convolutive separation and possibly under-determination, $\mathbf{x} = \mathbf{A} \star \mathbf{s} + \mathbf{b}$. To adapt and extend sparsity-based techniques to this context, the conditions of mixture $\mathbf{A}$ must be identified, and separation must be performed when $\mathbf{A}$ is known or estimated. The first task remains challenging, despite some interesting approaches [77], but the second is easier and should be addressed by an approach similar to the deconvolution methods based on ℓ^1 norm minimization [94,34]. Formally, once $\mathbf{A}$ is estimated, a decomposition of the mixture is sought with the form:

$$\mathbf{x} = \sum_{n=1}^{N} \mathbf{A}_n \star (c_n \cdot \mathbf{\Phi}^{(n)}) + \mathbf{b} = \sum_{n=1}^{N}\sum_{k=1}^{K} c_n(k) \cdot \mathbf{A}_n \star \varphi_k^{(n)} + \mathbf{b} \tag{10.76}$$

and the sources $\widehat{s_n} := \sum_k c_{s_n}(k)\varphi_k^{(n)}$ are reconstructed. The success of such an approach, still to be demonstrated in practice, depends on the global diversity (combined morphological and spatial diversity) of the multi-channel waveforms $\mathbf{A}_n \star \varphi_k^{(n)}$.

Exploring the broad application potential of sparse source separation methods remains substantially limited by the algorithmic complexity of the optimizations involved. It is thus critical to develop and implement new algorithms which ideally combine three crucial properties:

- propensity to guarantee or ensure high probability of good sparse approximations;
- high numerical efficiency, i.e. economic use of computing resources (memory, arithmetic operations, etc.);
- robustness in the case of approximate estimation of the mixing matrix.

Algorithms with "certifiable" performance. A number of sparse decomposition algorithms have been proposed, and we have described the main examples in this chapter. They include the iterative algorithms of the *matching pursuit* category and the techniques based on minimizing ℓ^1 criteria, studied extensively at the initiative of Donoho and Huo [33] (see for example [103,56,32,52] and the references cited therein). These algorithms perform well – in a precise mathematical sense that makes it possible to "certify" their performance – provided the sources are close enough to a sparse model. Nonconvex techniques for ℓ^τ optimization, $0 < \tau < 1$, are currently being analyzed [52], but somewhat surprisingly, there is less interest in convex ℓ^τ optimization techniques for $1 < \tau \leq 2$. "Certification" using theorems on the quality of results from joint approximation algorithms (see section 10.3) or separation algorithms (see section 10.6) is thus an area in which numerous questions are still open.

In the single-channel case, Gribonval and Nielsen [54,55] proved that while a highly sparse representation exists (implying a sufficiently small number of atoms), it is the only solution common to all the optimization problems (10.16) for $0 \leq \tau \leq 1$. Similar results [52] lead to the conclusion that all the single-channel sparse approximation problems (10.17) have very similar solutions for $0 \leq \tau \leq 1$, provided that the analyzed signal can be approximated effectively (in terms of mean square error) by combining a sufficiently small number of atoms from the dictionary. Tropp *et al.* [104,102] have shown that for $\tau = 1$, these results extend to the noisy multi-channel case.

Numerically efficient algorithms. From a numerical point of view, we have discussed the difficulties of solving the optimization problems (10.17) or (10.67) for $0 \leq \tau \leq 1$, which can involve inverting large-size matrices. Other approaches, such as the *matching pursuit* family, offer an interesting alternative in terms of algorithmic complexity, both for joint approximation and separation with a known mixing matrix. A fast implementation of complexity $\mathcal{O}(T \log T)$ called MPTK (*the matching pursuit toolkit*) was proposed and is freely available [53,65]. With redundant dictionaries such as multiscale Gabor dictionaries, it is then possible to calculate good sparse approximations in a timeframe close to the signal's duration, even when very long. Other proposals such as iterative thresholding algorithms [28,42,36,44] are also extremely promising in terms of numerical efficiency.

Estimating the mixing matrix... and the dictionary! When sparsity is assumed, the problem of identifiability (and identification) of the mixing matrix is not a trivial one, given its direct link to the robustness of separation algorithms as a function of the accuracy of mixing matrix estimation. While we are starting to understand how to analyze and "certify" the performance of sparse approximation algorithms, initial results from similar assessments of mixing matrix identification algorithms [5], although encouraging, are only valid under restrictive sparse models that cannot tolerate noise. The choice of dictionary(/ies) $\mathbf{\Phi}/\mathbf{\Phi}^{(n)}$ is to some extent a dual problem, whose impact on separation performance is yet to be understood. While current practice involves choosing the analysis dictionary(/ies) from a library of traditional dictionaries, based on prior available knowledge of the sources, *learning* the dictionary based on the data to separate, or *sparse coding* is formally equivalent to the problem of estimating $\mathbf{A}$ (aside from the transposition given that $\mathbf{x}^{\mathrm{H}} = \mathbf{\Phi}^{\mathrm{H}} \mathrm{C}_{\mathrm{s}}^{\mathrm{H}} \mathbf{A}^{\mathrm{H}} + \mathbf{b}^{\mathrm{H}}$). The approaches may differ considerably in practice because of the substantial difference in dimension of the matrices to estimate ($\mathbf{A}$ is $P \times N$, $\mathbf{\Phi}$ is $K \times T$, and generally $K \geq T \gg N \approx P$), which fundamentally changes the problem's geometry and one's intuitive grasp of it.

The idea of representing a mixture in a domain where it is sparse in order to facilitate separation has made it possible to successfully tackle the problem of separating under-determined instantaneous linear mixtures. The methods most frequently used today [91] rely on traditional dictionaries (Gabor, wavelets) identical for all sources, and representations based on linear transforms (STFT, etc.). To go further and determine whether sparsity – promising *a priori* for handling more complex and more realistic source separation problems – can move beyond this stage in practice, it is clear that certain technical obstacles identified above must be overcome. But above all, a serious effort to comparatively evaluate and analyze sparsity-based approaches is now more critical than ever.

Acknowledgements

R. Gribonval would like to warmly thank Simon Arberet for his precious help in preparing many of the figures in this chapter, as well as Gilles Gonon, Sacha Krsulović, Sylvain Lesage and Prasad Sudhakar for carefully reading an initial draft and offering many useful suggestions, comments and criticisms, which enabled substantial improvement of the organization of this chapter.

References

[1] S. Abdallah, M. Plumbley, If edges are the independent components of natural images, what are the independent components of natural sounds? in: Proc. Int. Conf. Indep. Component Anal. and Blind Signal Separation, ICA2001, San Diego, California, December 2001, pp. 534–539.

[2] F. Abrard, Y. Deville, A time-frequency blind signal separation method applicable to underdetermined mixtures of dependent sources, Signal Processing 85 (2005) 1389–1403.

[3] F. Abrard, Y. Deville, P. White, From blind source separation to blind source cancellation in the underdetermined case: a new approach based on time-frequency analysis, in: Proc. Int. Workshop on Independent Component Analysis and Blind Signal Separation, ICA 2001, San Diego, California, December 2001.
[4] M. Aharon, M. Elad, A. Bruckstein, The K-SVD: An algorithm for designing of overcomplete dictionaries for sparse representation, IEEE Transactions on Signal Processing 54 (2006) 4311–4322.
[5] M. Aharon, M. Elad, A. Bruckstein, On the uniqueness of overcomplete dictionaries, and a practical way to retrieve them, Journal of Linear Algebra and Applications 416 (2006) 48–67.
[6] A. Aïssa-El-Bey, K. Abed-Meraim, Y. Grenier, Underdetermined blind source separation of audio sources in time-frequency domain, in: Proc. First Workshop on Signal Processing with Sparse/Structured Representations, SPARS'05, Rennes, France, November 2005.
[7] S. Amari, A. Cichocki, H.H. Yang, A new learning algorithm for blind signal separation, in: Advances in Neural Information Processing Systems, vol. 8, MIT Press, 1996.
[8] S. Arberet, R. Gribonval, F. Bimbot, A robust method to count and locate audio sources in a stereophonic linear instantaneous mixture, in: J.P. Rosca, D. Erdogmus, S. Haykin (Eds.), Proc. of the Int'l. Workshop on Independent Component Analysis and Blind Signal Separation, ICA 2006, Charleston, South Carolina, USA, in: LNCS Series, vol. 3889, Springer, March 2006, pp. 536–543.
[9] R. Balan, J. Rosca, Statistical properties of STFT ratios for two channel systems and applications to blind source separation, in: Proc. Int. Workshop on Independent Component Analysis and Blind Signal Separation, ICA 2000, Helsinki, Finland, June 2000, pp. 429–434.
[10] A. Ben-Tal, M. Zibulevsky, Penalty/barrier multiplier methods for convex programming problems, SIAM Journal on Optimization 7 (1997) 347–366.
[11] D. Bertsekas, Constrained Optimization and Lagrange Multiplier Methods, Academic Press, New York, 1982.
[12] D. Bertsekas, Non-Linear Programming, 2nd ed., Athena Scientific, Belmont, MA, 1995.
[13] J. Bobin, Y. Moudden, J.-L. Starck, M. Elad, Multichannel morphological component analysis, in: Proc. First Workshop on Signal Processing with Sparse/Structured Representations, SPARS'05, Rennes, France, November 2005.
[14] J. Bobin, J.-L. Starc, J. Fadili, Y. Moudden, D.L. Donoho, Morphological component analysis: An adaptative thresholding strategy, IEEE Transactions on Image Processing 16 (2007) 2675–2681.
[15] P. Bofill, M. Zibulevsky, Underdetermined blind source separation using sparse representations, Signal Processing 81 (2001) 2353–2362.
[16] A.M. Bronstein, M.M. Bronstein, M. Zibulevsky, Blind source separation using block-coordinate relative Newton method, Signal Processing 84 (2004) 1447–1459.
[17] A.M. Bronstein, M.M. Bronstein, M. Zibulevsky, Relative optimization for blind deconvolution, IEEE Transactions on Signal Processing 53 (2005) 2018–2026.
[18] A.M. Bronstein, M.M. Bronstein, M. Zibulevsky, Y.Y. Zeevi, Blind deconvolution of images using optimal sparse representations, IEEE Transactions on Image Processing 14 (2005) 726–736.
[19] J.-F. Cardoso, Blind signal separation: Statistical principles, Proceedings of the IEEE 9 (1998) 2009–2025. Special Issue on Blind Identification and Estimation.
[20] J.-F. Cardoso, High-order contrasts for independent component analysis, Neural Computation 11 (1999) 157–192.
[21] J.-F. Cardoso, JADE for real-valued data, tech. report, ENST, 1999. http://sig.enst.fr:80/~cardoso/guidesepsou.html.
[22] J.-F. Cardoso, B. Laheld, Equivariant adaptive source separation, IEEE Transactions on Signal Processing 44 (1996) 3017–3030.
[23] S.S. Chen, D.L. Donoho, M.A. Saunders, Atomic decomposition by basis pursuit, SIAM Journal on Scientific Computing 20 (1998) 33–61.
[24] A. Cichocki, S. Amari, K. Siwek, ICALAB toolbox for image processing – benchmarks, tech. report, The Laboratory for Advanced Brain Signal Processing, RIKEN Brain Science Institute, 2002. http://www.bsp.brain.riken.go.jp/ICALAB/ICALABImageProc/benchmarks/.

[25] A. Cichocki, R. Unbehauen, E. Rummert, Robust learning algorithm for blind separation of signals, Electronics Letters 30 (1994) 1386–1387.

[26] R.R. Coifman, Y. Meyer, M.V. Wickerhauser, Wavelet analysis and signal processing, in: B. Ruskai, et al. (Eds.), Wavelets and their Applications, Jones and Barlett, Boston, 1992.

[27] S. Cotter, B. Rao, K. Engan, K. Kreutz-Delgado, Sparse solutions to linear inverse problems with multiple measurement vectors, IEEE Transactions on Signal Processing 53 (2005) 2477–2488.

[28] I. Daubechies, M. Defrise, C. De Mol, An iterative thresholding algorithm for linear inverse problems with a sparsity constraint, Communications on Pure and Applied Mathematics 57 (2004) 1413–1457.

[29] L. Daudet, B. Torrésani, Hybrid representations for audiophonic signal encoding, Signal Processing 82 (2002) 1595–1617. Special Issue on Coding Beyond Standards.

[30] M. Davies, N. Mitianoudis, A sparse mixture model for overcomplete ICA, IEE Proceedings - Vision Image and Signal Processing 151 (2004) 35–43. Special issue on Nonlinear and Non-Gaussian Signal Processing.

[31] G. Davis, S. Mallat, M. Avellaneda, Adaptive greedy approximations, Construction Approximation 13 (1997) 57–98.

[32] D. Donoho, M. Elad, V. Temlyakov, Stable recovery of sparse overcomplete representations in the presence of noise, IEEE Transactions on Information Theory 52 (2006) 6–18.

[33] D. Donoho, X. Huo, Uncertainty principles and ideal atomic decompositions, IEEE Transactions on Information Theory 47 (2001) 2845–2862.

[34] C. Dossal, Estimation de fonctions géométriques et déconvolution, PhD thesis, École Polytechnique, Palaiseau, France, 2005.

[35] B. Efron, T. Hastie, I. Johnstone, R. Tibshirani, Least angle regression, Annals of Statistics 32 (2004) 407–499.

[36] M. Elad, Why simple shrinkage is still relevant for redundant representations? IEEE Transactions on Information Theory 52 (2006) 5559–5569.

[37] M. Elad, B. Matalon, M. Zibulevsky, Coordinate and subspace optimization methods for linear least squares with non-quadratic regularization, Applied and Computational Harmonic Analysis 23 (2007) 346–367.

[38] Y.F. Abrard, Blind separation of dependent sources using the "time-frequency ratio of mixtures" approach, in: ISSPA, IEEE, Paris, France, 2003.

[39] C. Févotte, C. Doncarli, Two contributions to blind source separation using time-frequency distributions, IEEE Signal Processing Letters 11 (2004) 386–389.

[40] C. Févotte, S.J. Godsill, A Bayesian approach for blind separation of sparse sources, IEEE Transactions on Audio, Speech and Language Processing 14 (6) (2006) 2174–2188.

[41] D. Field, B. Olshausen, Emergence of simple-cell receptive field properties by learning a sparse code for natural images, Nature 381 (1996) 607–609.

[42] M. Figueiredo, R. Nowak, An EM algorithm for wavelet-based image restoration, IEEE Transactions on Image Processing 12 (2003) 906–916.

[43] R.M. Figueras i Ventura, P. Vandergheynst, P. Frossard, Low rate and flexible image coding with redundant representations, IEEE Transactions on Image Processing 15 (2006) 726–739.

[44] M. Fornasier, H. Rauhut, Recovery algorithms for vector valued data with joint sparsity constraints, Tech. Report 27, Johns Radon Institute for Computational and Applied Mathematics, Austrian Academy of Sciences, 2006.

[45] J.-J. Fuchs, Some further results on the recovery algorithms, in: Proc. First Workshop on Signal Processing with Sparse/Structured Representations, SPARS'05, Rennes, France, November 2005, pp. 67–70.

[46] P.E. Gill, W. Murray, M.H. Wright, Practical Optimization, Academic Press, New York, 1981.

[47] G.H. Golub, C. Van Loan, Matrix Computations, 2nd ed., The Johns Hopkins University Press, Baltimore and London, 1989.

[48] I.F. Gorodnitsky, B.D. Rao, Sparse signal reconstruction from limited data using focuss: A reweighted norm minimization algorithm, IEEE Transactions on Signal Processing 45 (1997) 600–616.

[49] R. Gribonval, Sparse decomposition of stereo signals with matching pursuit and application to blind separation of more than two sources from a stereo mixture, ICASSP'02, in: Proc. Int. Conf. Acoust. Speech Signal Process., vol. 3, IEEE, Orlando, Florida, 2002, pp. III/3057–III/3060.

[50] R. Gribonval, Piecewise linear source separation, in: M. Unser, A. Aldroubi, A. Laine, (Eds.), Proc. SPIE '03, in: Wavelets: Applications in Signal and Image Processing X, vol. 5207, San Diego, CA, August 2003, pp. 297–310.

[51] R. Gribonval, L. Benaroya, E. Vincent, C. Févotte, Proposals for performance measurement in source separation, in: Proc. 4th Int. Symp. on Independent Component Anal. and Blind Signal Separation, ICA2003, Nara, Japan, April 2003, pp. 763–768.

[52] R. Gribonval, R.M. Figueras i Ventura, P. Vandergheynst, A simple test to check the optimality of sparse signal approximations, EURASIP Signal Processing 86 (2006) 496–510. Special issue on Sparse Approximations in Signal and Image Processing.

[53] R. Gribonval, S. Krstulović, MPTK, The Matching Pursuit Toolkit, 2005.

[54] R. Gribonval, M. Nielsen, On the strong uniqueness of highly sparse expansions from redundant dictionaries, in: Proc. Int. Conf. Independent Component Analysis, ICA'04, in: LNCS, Springer-Verlag, Granada, Spain, 2004.

[55] R. Gribonval, M. Nielsen, Highly sparse representations from dictionaries are unique and independent of the sparseness measure, Appl. Comput. Harm. Anal. 22 (2007) 335–355.

[56] R. Gribonval, M. Nielsen, Beyond sparsity: Recovering structured representations by ℓ^1-minimization and greedy algorithms, Advances in Computational Mathematics 28 (2008) 23–41.

[57] R. Gribonval, H. Rauhut, K. Schnass, P. Vandergheynst, Atoms of all channels, unite! Average case analysis of multi-channel sparse recovery using greedy algorithms, Journal of Fourier Analysis and Applications 14 (2008) 655–687.

[58] R. Gribonval, P. Vandergheynst, On the exponential convergence of Matching Pursuits in quasi-incoherent dictionaries, IEEE Transactions on Information Theory 52 (2006) 255–261.

[59] A. Hyvärinen, The Fast-ICA MATLAB package, tech. report, HUT, 1998. http://www.cis.hut.fi/~aapo/.

[60] A. Hyvärinen, Fast and robust fixed-point algorithms for independent component analysis, IEEE Transactions on Neural Networks 10 (1999) 626–634.

[61] A. Jourjine, S. Rickard, O. Yilmaz, Blind separation of disjoint orthogonal signals: Demixing n sources from 2 mixtures, in: Proc. IEEE Conf. Acoustics Speech and Signal Proc., ICASSP'00, vol. 5, Istanbul, Turkey, June 2000, pp. 2985–2988.

[62] P. Kisilev, M. Zibulevsky, Y. Zeevi, A multiscale framework for blind separation of linearly mixed signals, The Journal of Machine Learning Research 4 (2003) 1339–1363.

[63] P. Kisilev, M. Zibulevsky, Y.Y. Zeevi, B.A. Pearlmutter, Multiresolution framework for sparse blind source separation, tech. report, Department of Electrical Engineering, Technion. Haifa, Israel, 2000. http://ie.technion.ac.il/~mcib/.

[64] K. Kreutz-Delgado, B. Rao, K. Engan, T.-W. Lee, T. Sejnowski, Convex/schur-convex (csc) log-priors and sparse coding, in: 6th Joint Symposium on Neural Computation, Institute for Neural Computation, 1999, pp. 65–71.

[65] S. Krstulovic, R. Gribonval, MPTK: Matching Pursuit made tractable, in: Proc. Int. Conf. Acoust. Speech Signal Process., ICASSP'06, vol. 3, Toulouse, France, May 2006, pp. III–496 - III–499.

[66] T.-W. Lee, M.S. Lewicki, M. Girolami, T.J. Sejnowski, Blind source separation of more sources than mixtures using overcomplete representations, IEEE Signal Processing Letters 6 (1999) 87–90.

[67] S. Lesage, S. Krstulovic, R. Gribonval, Under-determined source separation: comparison of two approaches based on sparse decompositions, in: J.P. Rosca, D. Erdogmus, S. Haykin (Eds.), Proc. of the Int'l. Workshop on Independent Component Analysis and Blind Signal Separation, ICA 2006, Charleston, South Carolina, USA, in: LNCS Series, vol. 3889, Springer, 2006, pp. 633–640.

[68] D. Leviatan, V. Temlyakov, Simultaneous approximation by greedy algorithms, Tech. Report 0302, IMI, Dept. of Mathematics, University of South Carolina, Columbia, SC 29208, 2003.
[69] M. Lewicki, Efficient coding of natural sounds, Nature Neuroscience 5 (2002) 356–363.
[70] M. Lewicki, T. Sejnowski, Learning overcomplete representations, Neural Computation 12 (2000) 337–365.
[71] M.S. Lewicki, B.A. Olshausen, A probabilistic framework for the adaptation and comparison of image codes, Journal of the Optical Society of America 16 (1999) 1587–1601.
[72] J. Lin, D. Grier, J. Cowan, Faithful representation of separable distributions, Neural Computation 9 (1997) 1305–1320.
[73] S. Makeig, ICA toolbox for psychophysiological research, Computational Neurobiology Laboratory, the Salk Institute for Biological Studies, 1998. http://www.cnl.salk.edu/~ica.html.
[74] D. Malioutov, M. Cetin, A. Willsky, Homotopy continuation for sparse signal representation, in: Proceedings of the IEEE International Conference on Acoustics, Speech, and Signal Processing, ICASSP'05, vol. V, March 2005, pp. 733–736.
[75] S. Mallat, A Wavelet Tour of Signal Processing, Academic Press, San Diego, CA, 1998.
[76] S. Mallat, Z. Zhang, Matching pursuit with time-frequency dictionaries, IEEE Transactions on Signal Processing 41 (1993) 3397–3415.
[77] T. Melia, S. Rickard, Extending the DUET blind source separation technique, in: Proc. First Workshop on Signal Processing with Sparse/Structured Representations, SPARS'05, Rennes, France, November 2005, pp. 67–70.
[78] B. Natarajan, Sparse approximate solutions to linear systems, SIAM Journal on Computing 25 (1995) 227–234.
[79] D. Needell, J.A. Tropp, Cosamp: Iterative signal recovery from incomplete and inaccurate samples, tech. report, Caltech, 2008.
[80] B.A. Olshausen, D.J. Field, Sparse coding with an overcomplete basis set: A strategy employed by v1? Vision Research 37 (1997) 3311–3325.
[81] B.A. Pearlmutter, L.C. Parra, Maximum likelihood blind source separation: A context-sensitive generalization of ICA, in: Advances in Neural Information Processing Systems, vol. 9, MIT Press, 1997.
[82] E. Pearson, The Multiresolution Fourier Transform and its application to Polyphonic Audio Analysis, PhD thesis, University of Warwick, September 1991.
[83] L. Peotta, L. Granai, P. Vandergheynst, Image compression using an edge adapted redundant dictionary and wavelets, Signal Processing 86 (2006) 444–456.
[84] D. Pham, P. Garat, Blind separation of a mixture of independent sources through a quasi-maximum likelihood approach, IEEE Transactions on Signal Processing 45 (1997) 1712–1725.
[85] M. Plumbley, Geometry and homotopy for ℓ^1 sparse signal representations, in Proc. First Workshop on Signal Processing with Sparse/Structured Representations, SPARS'05, Rennes, France, November 2005, pp. 67–70.
[86] R. Polyak, Modified barrier functions: Theory and methods, Mathematical Programming 54 (1992) 177–222.
[87] J. Princen, A. Bradley, Analysis/synthesis filter bank design bases on time domain aliasing cancellation, IEEE Transactions on Acoustics, Speech and Signal Proc. ASSP-34 (1986) 1153–1161.
[88] M. Puigt, Y. Deville, Time-frequency ratio-based blind separation methods for attenuated and time-delayed sources, Mechanical Systems and Signal Processing 19 (2005) 1348–1379.
[89] L. Rabiner, R. Schafer, Digital Processing of Speech Signals, Prentice Hall, 1978.
[90] V.P. Rahmoune A, F. P, Flexible motion-adaptive video coding with redundant expansions, IEEE Transactions on Circuits and Systems for Video Technology 16 (2006) 178–190.
[91] S. Rickard, R. Balan, J. Rosca, Real-time time-frequency based blind source separation, in: 3rd International Conference on Independent Component Analysis and Blind Source Separation, ICA2001, San Diego, CA, December 2001.
[92] R. Rubinstein, M. Zibulevsky, M. Elad, Double sparsity: Learning sparse dictionaries for sparse signal approximation, IEEE Transactions on Signal Processing (2009) (in press).

[93] R. Saab, Ö Yilmaz, M. McKeown, R. Abugharbieh, Underdetermined sparse blind source separation with delays, in Proc. First Workshop on Signal Processing with Sparse/Structured Representations, SPARS'05, Rennes, France, November 2005, pp. 67–70.
[94] F. Santosa, W. Symes, Linear inversion of band-limited reflection sismograms, SIAM J. Sci. Statistic. Comput. 7 (1986) 1307–1330.
[95] J.L. Starck, E.J. Candès, D.L. Donoho, The curvelet transform for image denoising, IEEE Transactions on Image Processing 11 (2000) 670–684.
[96] J.-L. Starck, M. Elad, D. Donoho, Image decomposition : Separation of textures from piecewise smooth content, in: M. Unser, A. Aldroubi, A. Laine (Eds.), Wavelet: Applications in Signal and Image Processing X, in: Proc. SPIE '03, vol. 5207, SPIE (The International Society for Optical Engineering), San Diego, CA, August 2003, pp. 571–582.
[97] J.-L. Starck, Y. Moudden, J. Bobin, M. Elad, D. Donoho, Morphological component analysis, in: Proceedings of the SPIE conference wavelets, vol. 5914, July 2005.
[98] F. Theis, A. Jung, C. Puntonet, E. Lang, Linear geometric ICA: Fundamentals and algorithms, Neural Computation 15 (2003) 419–439.
[99] F.J. Theis, E.W. Lang, Formalization of the two-step approach to overcomplete BSS, in: Proc. SIP 2002, Kauai, Hawaii, USA, 2002, pp. 207–212.
[100] F.J. Theis, C. Puntonet, E.W. Lang, A histogram-based overcomplete ICA algorithm, in: Proc. 4th Int. Symp. on Independent Component Anal. and Blind Signal Separation, ICA2003, Nara, Japan, 2003, pp. 1071–1076.
[101] J. Tropp, Greed is good: Algorithmic results for sparse approximation, IEEE Transactions on Inform. Theory 50 (2004) 2231–2242.
[102] J. Tropp, Algorithms for simultaneous sparse approximation. Part II: Convex relaxation, Signal Processing 86 (2006) 589–602. Special issue on Sparse Approximations in Signal and Image Processing.
[103] J. Tropp, Just relax: Convex programming methods for identifying sparse signals in noise, IEEE Transactions on Information Theory 52 (2006) 1030–1051.
[104] J. Tropp, A. Gilbert, M. Strauss, Algorithms for simultaneous sparse approximation. Part I: Greedy pursuit, Signal Processing 86 (2006) 572–588. Special issue on Sparse Approximations in Signal and Image Processing.
[105] B. Turlach, On algorithms for solving least squares problems under an ℓ^1 pernalty or an ℓ^1 constraint, in: 2004 Proc. of the American Statistical Association, vol. Statistical Computing Section [CDROM] Alexandria, VA, American Statistical Association, 2005, pp. 2572–2577.
[106] M. Van Hulle, Clustering approach to square and non-square blind source separation. in: IEEE Workshop on Neural Networks for Signal Processing, NNSP99, 1999, pp. 315–323.
[107] L. Vielva, D. Erdogmus, J. Principe, Underdetermined blind source separation using a probabilistic source sparsity model, in: Proc. Int. Conf. on ICA and BSS, ICA2001, San Diego, California, 2001, pp. 675–679.
[108] E. Vincent, Complex nonconvex l_p norm minimization for underdetermined source separation, in: Proc. Int. Conf. Indep. Component Anal. and Blind Signal Separation, ICA2001, Springer, 2007, pp. 430–437.
[109] D. Weitzer, D. Stanhill, Y.Y. Zeevi, Nonseparable two-dimensional multiwavelet transform for image coding and compression, Proc. SPIE 3309 (1997) 944–954.
[110] R. Xu, D. Wunsch II, Survey of clustering algorithms, IEEE Transactions on Neural Networks 16 (2005) 645–678.
[111] O. Yilmaz, S. Rickard, Blind separation of speech mixtures via time-frequency masking, IEEE Transactions on Signal Processing 52 (2004) 1830–1847.
[112] M. Zibulevsky, Blind source separation with Relative Newton method, Proceedings ICA-2003, (2003), pp. 897–902.
[113] M. Zibulevsky, Relative Newton and smoothing multiplier optimization methods for blind source separation, in: S. Makino, T. Lee, H. Sawada (Eds.), Blind Speech Separation, in: Springer Series: Signals and Communication Technology XV, Springer, 2007.

[114] M. Zibulevsky, P. Kisilev, Y.Y. Zeevi, B.A. Pearlmutter, Blind source separation via multinode sparse representation, in: Advances in Neural Information Processing Systems, vol. 12, MIT Press, 2002.

[115] M. Zibulevsky, B.A. Pearlmutter, Blind source separation by sparse decomposition in a signal dictionary, Neural Computation 13 (2001) 863–882.

[116] M. Zibulevsky, B.A. Pearlmutter, Blind source separation by sparse decomposition, Tech. Report CS99-1, Univ. of New Mexico, July 1999.

[117] M. Zibulevsky, B.A. Pearlmutter, P. Bofill, P. Kisilev, Blind source separation by sparse decomposition, in: S.J. Roberts, R.M. Everson (Eds.), Independent Components Analysis: Principles and Practice, Cambridge University Press, 2001.

CHAPTER

Quadratic time-frequency domain methods

11

N. Thirion-Moreau and M. Amin

11.1 INTRODUCTION

In this chapter, we present the principles of blind separation and recovery of nonstationary signals incident on sensor arrays, specifically those characterized by their instantaneous frequencies. Both classes of mono- and multi-components signals are considered. The chapter provides the fundamental approach to nonstationary source separation based on spatial quadratic time-frequency distributions[1] applied to deterministic signals as well as nonstationary stochastic processes. The main advantage of this class of methods is that they utilize the time-frequency signatures underlying nonstationary signals. Further, these methods do not necessarily rely on the classical statistical independence assumption like the methods presented in the previous chapters. As a consequence, the problem of the separation of deterministic signals as well as correlated stochastic processes can be tackled.

Successful signal separations involve three main steps. First, spatial quadratic time-frequency representations (SQTFR or spectra (SQTFS)) of the observations across the array must be constructed. Second, time-frequency regions of signal power concentration and localization should be properly determined and time-frequency points of high values must be selected. Third, the mixing matrix should be properly estimated from these time-frequency regions so as to undo the mixing of signals at the multi-sensor receiver.

The chapter covers various key techniques for blind source separations geared towards the nonstationary class of signals. These techniques can be primarily put into two different categories: those that require a pre-whitening step and others which proceed without pre-processing. We also discuss the signal separation techniques which only target single time-frequency points and those which incorporate multiple terms, whether they correspond to source auto-terms or source cross-terms, or a combined arrangement of the two. Unitary and non-unitary based joint diagonalization (JD) and joint zero diagonalization (JZD) methods for mixing matrix estimation are delineated. Alternative

[1]It is not intended to cover all existing time-frequency based methods: separation methods based on linear distributions will not be presented here (but the reader can find a description of some of them in Chapters 10 and 19 of this book).

Handbook of Blind Source Separation: Independent Component Analysis and Applications

methods achieving the same task by combining an algebraic method together with a classification algorithm are also presented.

Finally, we provide several examples demonstrating the performance of the various steps underlying nonstationary blind source separation.

11.2 PROBLEM STATEMENT

11.2.1 Model and assumptions

We consider the BSS problem where $N \in \mathbb{N}\backslash\{0,1\}$ signals called the "sources" are received on an antenna of $P \in \mathbb{N} \setminus \{0,1\}$ sensors. We assume that $P \geq N$, i.e. the over-determined case is considered (for the under-determined case which is outside the scope of this chapter, one may refer the reader to [10,20,22,30,42,43,54] for example or Chapters 9 and 10 of this book). Using matrix notations, the relation linking the input and the output of the mixing system, for the noiseless case is written:

$$\mathbf{x}(t) = \mathbf{A}\mathbf{s}(t), \tag{11.1}$$

where the $P \times N$ matrix $\mathbf{A}$ is referred to as the *mixing matrix*, the $P \times 1$ "observations" vector is denoted by $\mathbf{x}(t) = (x_1(t), \ldots, x_P(t))^{\mathrm{T}}$, the $N \times 1$ "source" vector is $\mathbf{s}(t) = (s_1(t), \ldots, s_N(t))^{\mathrm{T}}$, and $(\cdot)^{\mathrm{T}}$ stands for the transposition operator.

We will use the following assumption about the sources:

H1. The sources $s_i(t)$, $i = 1, \ldots, N$ are either unknown and unobservable deterministic signals or unknown and unobservable (stationary, cyclostationary or nonstationary) random processes. If the sources are random processes, they are neither assumed independent nor uncorrelated.

We further assume that:

- the mixing matrix $\mathbf{A}$ is unknown but full column rank;
- the number of sources N is known.

11.2.2 Indeterminacies and sources estimation

Considering that the source signals are not observable and that the mixing matrix is unknown, the goal is to estimate an "inverse" of the mixing system to recover the contributions of the different sources. It is well-known that such a problem possesses some indeterminacies: without additional assumptions about the sources, they will be recovered only up to a scale and a permutation. These indeterminacies are inherent to the underlying problem.

The BSS problem can be reformulated into finding a matrix $\mathbf{B}$, called *separation matrix*, such that the restored signals $\mathbf{y}(t)$ can be written as:

$$\mathbf{y}(t) = \mathbf{B}\mathbf{x}(t) = \mathbf{B}\mathbf{A}\mathbf{s}(t) = \mathbf{P}\mathbf{D}\mathbf{s}(t) = \widehat{\mathbf{s}}(t), \tag{11.2}$$

with $\mathbf{D}$ a diagonal invertible matrix and $\mathbf{P}$ a permutation matrix.

With the estimate of the mixing matrix $\widehat{\mathbf{A}}$, the source signals can be provided by $\widehat{\mathbf{A}}^{\dagger}\mathbf{x}(t)$, where $\mathbf{A}^{\dagger}$ stands for the pseudo-inverse of $\mathbf{A}$. Finally, due to the above assumptions and the indeterminacies linked to the BSS problem, one may, without loss of generality, assume that all sources are of unit power.

11.2.3 Spatial whitening

It is rather typical to begin with a *normalization* stage, namely the *spatial whitening of the observations*. As established in [11,23], such a preliminary step establishes a bound with regard to the best reachable performances in the context of BSS. As discussed in section 11.4.2, the uncorrelated sources assumption can be relaxed, as such, and the whitening stage can be eliminated. The spatial whitening of the observations consists of a decorrelation of the signals paired with a unit power constraint. A vector whose components are uncorrelated and whose length is equal to a unit value qualifies as "white". This preliminary processing stage (on which several array processing methods are based [45,40]), enables a reduction of the problem dimension due to a projection onto the "signal subspace".

The singular value decomposition [39] of the regular mixing matrix $\mathbf{A}$ is given by: $\mathbf{A} = \mathbf{V}\mathbf{D}^{\frac{1}{2}}\mathbf{U}$, with $\mathbf{V}$ and $\mathbf{U}$ representing unitary matrices of dimension $P \times P$ and $N \times N$, respectively (i.e. $\mathbf{V}\mathbf{V}^{\mathrm{H}} = \mathbf{V}^{\mathrm{H}}\mathbf{V} = \mathbf{I}_P$ with $\mathbf{I}_P$ the $P \times P$ identity matrix, $\mathbf{U}\mathbf{U}^{\mathrm{H}} = \mathbf{U}^{\mathrm{H}}\mathbf{U} = \mathbf{I}_N$ and $(.)^{\mathrm{H}}$ is the transpose conjugate operator), and $\mathbf{D}^{\frac{1}{2}}$ is a $P \times N$ diagonal matrix[2] ($\mathbf{D} = \mathrm{Diag}\{\delta_1, \delta_2, \ldots, \delta_N\}$, with $(\mathbf{D})_{ii} = \delta_i$ for all $i = 1, \ldots, N$ and $\delta_1 \geq \delta_2 \geq \ldots \geq \delta_N \geq 0$). As a consequence:

$$\mathbf{A}\mathbf{A}^{\mathrm{H}} = \mathbf{V}\mathbf{\Delta}\mathbf{V}^{\mathrm{H}}, \tag{11.3}$$

where $\mathbf{\Delta} = \mathbf{D}^{\frac{1}{2}}(\mathbf{D}^{\frac{1}{2}})^{\mathrm{H}}$ is a square $P \times P$ diagonal matrix ($\mathbf{\Delta} = \mathrm{Diag}\{\delta_1, \ldots, \delta_N, 0, \ldots, 0\}$).

Define a time smoothing operator:

$$\langle z(t) \rangle = \lim_{T \to +\infty} \frac{1}{T} \int_{-\frac{T}{2}}^{\frac{T}{2}} z(t)dt. \tag{11.4}$$

When the time smoothing operator argument is a matrix, the result is a matrix defined component-wise as:

$$(\langle \mathbf{R}(t) \rangle)_{ij} = \langle R_{ij}(t) \rangle \qquad \forall i, j. \tag{11.5}$$

[2]The top $N \times N$ square submatrix is diagonal, and the remaining $P - N$ rows are zeros.

Nonstationary uncorrelated random sources case

Considering two random complex signal vectors $\mathbf{z}_1(t)$ and $\mathbf{z}_2(t)$, their *cross-correlation matrix* is defined as:

$$\mathbf{R}_{\mathbf{z}_1\mathbf{z}_2}(t,\tau) = \mathbb{E}\left\{\mathbf{z}_1\left(t+\frac{\tau}{2}\right)\mathbf{z}_2^{\mathrm{H}}\left(t-\frac{\tau}{2}\right)\right\}. \tag{11.6}$$

For the special case of $\mathbf{z}_1 = \mathbf{z}_2 = \mathbf{z}$, the *correlation matrix* for a single random complex signal vector $\mathbf{z}(t)$ becomes,

$$\mathbf{R}_{\mathbf{z}}(t,\tau) = \mathbb{E}\left\{\mathbf{z}\left(t+\frac{\tau}{2}\right)\mathbf{z}^{\mathrm{H}}\left(t-\frac{\tau}{2}\right)\right\}. \tag{11.7}$$

where $\mathbb{E}\{\cdot\}$ is the mathematical expectation. When nonstationary sources are considered, it is necessary to introduce what is called the *mean correlation matrix* $\overline{\mathbf{R}}_{\mathbf{z}}(\tau)$, defined as:

$$\overline{\mathbf{R}}_{\mathbf{z}}(\tau) = \langle \mathbf{R}_{\mathbf{z}}(t,\tau) \rangle. \tag{11.8}$$

In the case of stationary sources, $\mathbf{R}_{\mathbf{z}}(t,\tau)$ is independent of time and is only a function of τ. It becomes sufficient, in what follows, to directly consider $\mathbf{R}_{\mathbf{z}}(\tau)$ (which also happens to be equal to $\overline{\mathbf{R}}_{\mathbf{z}}(\tau)$).

Using Eqs (11.1) and (11.3) and invoking the unit power and uncorrelated source assumptions, the observations mean correlation matrix $\overline{\mathbf{R}}_{\mathbf{x}}(\tau)$ (whose size is $P \times P$) is given, for $\tau = 0$, by:

$$\overline{\mathbf{R}}_{\mathbf{x}}(0) = \mathbf{A}\overline{\mathbf{R}}_{\mathbf{s}}(0)\mathbf{A}^{\mathrm{H}} = \mathbf{A}\mathbf{A}^{\mathrm{H}} = \mathbf{V}\boldsymbol{\Delta}\mathbf{V}^{\mathrm{H}}, \tag{11.9}$$

which is recognized as the eigenvalue decomposition (ED) of the matrix $\overline{\mathbf{R}}_{\mathbf{x}}(0)$. The matrix $\boldsymbol{\Delta}$ can be estimated from the data and used to provide the diagonal elements of matrix $\mathbf{D}$.

The $N \times P$ *spatial whitening matrix* $\mathbf{W}$ is defined as $\mathbf{W} = (\mathbf{D}^{\frac{1}{2}})^{\dagger}\mathbf{V}^{\mathrm{H}}$ with $(\mathbf{D}^{\frac{1}{2}})^{\dagger}$ denoting the pseudo-inverse of $\mathbf{D}^{\frac{1}{2}}$ whose dimension is $N \times P$. The $N \times 1$ *whitened observation* vector is given by:

$$\mathbf{z}(t) = \mathbf{W}\mathbf{x}(t) = \mathbf{W}\mathbf{A}\mathbf{s}(t) = \mathbf{U}\mathbf{s}(t). \tag{11.10}$$

It is clear that $\overline{\mathbf{R}}_{\mathbf{z}}(0) = \mathbf{U}\overline{\mathbf{R}}_{\mathbf{s}}(0)\mathbf{U}^{\mathrm{H}} = \mathbf{I}_N$. The spatial whitening stage corresponds to a projection onto the signal subspace and a normalization. This step makes it possible to consider a unitary mixture of sources. In this regard, the unitary matrix $\mathbf{U}$ must be estimated to effectively separate the sources.

Due to the indeterminacies inherent to the blind source separation problem, if one considers the vector $\mathbf{y}(t) = \mathbf{Q}\mathbf{z}(t)$, a perfect estimate of the sources is obtained when the matrix $\mathbf{Q}$ assumes the following form $\mathbf{PDU}^{\mathrm{H}}$. The estimated mixing matrix $\widehat{\mathbf{A}}$ is then given by $\widehat{\mathbf{A}} = \mathbf{W}^{\dagger}\widehat{\mathbf{U}}$.

Deterministic source case

When deterministic finite mean power sources are considered, the time smoothing operator, defined in Eq. (11.4), may be applied. For time limited signals with finite energy, or periodic signals of finite power with period T, the smoothing operator is defined as,

$$\langle z(t)\rangle = \frac{1}{T}\int_{-\frac{T}{2}}^{\frac{T}{2}} z(t)dt. \tag{11.11}$$

The source correlation matrix is then defined as $\mathbf{C_s}(\tau) = (C_{s_is_j}(\tau))$, whose elements are equal to:

$$C_{s_is_j}(\tau) = \left\langle s_i\left(t+\frac{\tau}{2}\right)s_j^*\left(t-\frac{\tau}{2}\right)\right\rangle \quad \forall i,j. \tag{11.12}$$

The corresponding $P \times P$ observation correlation matrix $\mathbf{C_x}(\tau)$ is:

$$\mathbf{C_x}(\tau) = \mathbf{AC_s}(\tau)\mathbf{A}^{\mathrm{H}}. \tag{11.13}$$

For $\tau = 0$, and due to the signal strength assumptions, we have:

$$\mathbf{C_x}(0) = \mathbf{AA}^{\mathrm{H}}, \tag{11.14}$$

which resembles Eq. (11.9), but for the class of deterministic signals. Similarly, as in the random process case, the $N \times P$ whitening matrix $\mathbf{W}$ can be defined for deterministic sources using the eigenvalue decomposition of matrix $\mathbf{C_x}(0)$. Since the mathematical developments are the same as in the previous case, they are not pursued any further.

11.2.4 A generalization to the noisy case

In a noisy context, the mixing mixture model becomes:

$$\mathbf{x}(t) = \mathbf{As}(t) + \mathbf{b}(t). \tag{11.15}$$

New assumptions have to be introduced:

H2. The noise signals $b_i(t)$, $i = 1,\ldots,P$ are random processes, which are white, stationary, zero-mean, independent from the sources, and mutually independent. The noise correlation matrix $\mathbf{R_b}(\tau)$ is:

$$\mathbf{R_b}(t,\tau) = \mathbf{R_b}(\tau) = \mathbb{E}\{\mathbf{b}(t+\tau/2)\mathbf{b}^{\mathrm{H}}(t-\tau/2)\} \tag{11.16}$$

with:

$$\begin{cases} \mathbf{R_b}(\tau) = 0, & \tau \neq 0, \\ \mathbf{R_b}(\tau) = \sigma_b^2\mathbf{I}_P, & \tau = 0. \end{cases} \tag{11.17}$$

The problem amounts to searching for a matrix $\mathbf{B}$ such that the restored signals $\mathbf{y}(t)$ have the following form:

$$\mathbf{y}(t) = \mathbf{Bx}(t) = \mathbf{BAs}(t) + \mathbf{Bb}(t) = \mathbf{Gs}(t) + \mathbf{Bb}(t) = \mathbf{PDs}(t) + \mathbf{Bb}(t). \quad (11.18)$$

For noisy environments, Eq. (11.18) replaces Eq. (11.2). The sources signals can no more be perfectly recovered since they are perturbed by an additional noise. However, in each signal $y_i(t)$, one finds the contribution of one single source.

Spatial whitening in the noisy case

The mean correlation matrix of the observed signals $\overline{\mathbf{R}}_{\mathbf{x}}(\tau)$[3] is given in $\tau = 0$, by:

$$\overline{\mathbf{R}}_{\mathbf{x}}(0) = \mathbf{A}\overline{\mathbf{R}}_{\mathbf{s}}(0)\mathbf{A}^{\mathrm{H}} + \mathbf{R}_{\mathbf{b}}(0) = \mathbf{A}\overline{\mathbf{R}}_{\mathbf{s}}(0)\mathbf{A}^{\mathrm{H}} + \sigma_b^2\mathbf{I}_P \quad (11.19)$$

which simplifies into $\overline{\mathbf{R}}_{\mathbf{x}}(0) = \mathbf{AA}^{\mathrm{H}} + \sigma_b^2\mathbf{I}_P$. From Eq. (11.3),

$$\overline{\mathbf{R}}_{\mathbf{x}}(0) = \mathbf{V}\boldsymbol{\Delta}\mathbf{V}^{\mathrm{H}} + \sigma_b^2\mathbf{I}_P = \mathbf{V}(\boldsymbol{\Delta} + \sigma_b^2\mathbf{I}_P)\mathbf{V}^{\mathrm{H}} \quad (11.20)$$

which represents the eigenvalue decomposition of matrix $\overline{\mathbf{R}}_{\mathbf{x}}(0)$.

Two cases have to be distinguished:

- Case $P > N$: the noise power σ_b^2 can be estimated from the $P - N$ smallest eigenvalues. The matrix $\boldsymbol{\Delta}$ can then be estimated as well as the matrix $\mathbf{D}$.
- Case $P = N$: the noise power σ_b^2 cannot be estimated from the eigenvalues of $\overline{\mathbf{R}}_{\mathbf{x}}(0)$. In this case, the estimation of $\boldsymbol{\Delta}$, and subsequently $\mathbf{D}$, requires the *a priori* knowledge of σ_b^2.

The $N \times P$ spatial whitening matrix $\mathbf{W}$ is still defined as $\mathbf{W} = (\mathbf{D}^{\frac{1}{2}})^{\dagger}\mathbf{V}^{\mathrm{H}}$. The $N \times 1$ vector of the whitened signals is then given by:

$$\mathbf{z}(t) = \mathbf{Wx}(t) = \mathbf{WAs}(t) + \mathbf{Wb}(t) = \mathbf{Us}(t) + \mathbf{Wb}(t). \quad (11.21)$$

Recent efforts have been put into the development of solutions to the BSS problem which do not rely upon the use of covariance matrices or cumulants tensors, but rather on the use of spatial quadratic time-frequency distributions (SQTFD) or spatial quadratic time-frequency spectra (SQTFS) [1,8,5,9,16,17,28,26,27,29,31,32,37,38,35,42,43,50]. Both cases of non-stationary random and deterministic signal mixture have been considered.

[3]For stationary signals, one simply has to consider $\mathbf{R}_{\mathbf{x}}(\tau)$.

11.3 SPATIAL QUADRATIC t-f SPECTRA AND REPRESENTATIONS

11.3.1 Bilinear and quadratic transforms

Consider two distinct, scalar, real or complex, continuous-time signals, denoted by $x(t)$ and $y(t)$. The associated *bilinear transform* (BT) and *quadratic transform* (QT) are defined as follows [19,33]:

$$(x(t), y(t)) \overset{\text{BT}}{\longleftrightarrow} D_{xy}(t, \nu; R) = \int_{\mathbb{R}^2} x(\theta) y^*(\theta') R(\theta, \theta'; t, \nu) d\theta d\theta', \tag{11.22}$$

where $(\cdot)^*$ is the conjugate of a complex number and $R(\theta, \theta'; t, \nu)$ is the kernel of the bilinear transform. The kernel depends on four parameters and is assumed data-independent.

The quadratic transform is a case of the BT with $y(t) = x(t)$,

$$\begin{aligned}(x(t), x(t)) \overset{\text{QT}}{\longleftrightarrow} D_{xx}(t, \nu; R) &= D_x(t, \nu; R) \\ &= \int_{\mathbb{R}^2} x(\theta) x^*(\theta') R(\theta, \theta'; t, \nu) d\theta d\theta'. \end{aligned} \tag{11.23}$$

11.3.2 Spatial bilinear and quadratic transforms

Consider two distinct continuous-time signals, which may be represented by real or complex vectors of the same length $N \times 1$ and denoted by $\mathbf{x}(t) = (x_1(t), \ldots, x_N(t))^{\mathrm{T}}$ and $\mathbf{y}(t) = (y_1(t), \ldots, y_N(t))^{\mathrm{T}}$. The *spatial bilinear transform* (SBT), parameterized by a kernel $\mathbf{R}$, is defined as,

$$(\mathbf{x}(t), \mathbf{y}(t)) \overset{\text{TSB}}{\longleftrightarrow} \mathbf{D}_{\mathbf{xy}}(t, \nu; \mathbf{R}) = \int_{\mathbb{R}^2} \mathbf{x}(\theta) \mathbf{y}^{\mathrm{H}}(\theta') \odot \mathbf{R}(\theta, \theta'; t, \nu) d\theta d\theta', \tag{11.24}$$

where $\odot$ stands for the Hadamard product [41] (or term-by-term product of the elements of two matrices of same size). The integral operates on each term of the matrix, resulting in an $N \times N$ matrix. This equation can be developed and written in the following matrix form:

$$\mathbf{D}_{\mathbf{xy}}(t, \nu; \mathbf{R}) = \begin{pmatrix} D_{x_1 y_1}(t, \nu; R_{11}) & \ldots & D_{x_1 y_N}(t, \nu; R_{1N}) \\ \vdots & \ddots & \vdots \\ D_{x_N y_1}(t, \nu; R_{N1}) & \ldots & D_{x_N y_N}(t, \nu; R_{NN}) \end{pmatrix},$$

with $D_{x_i y_j}(t, \nu; R_{ij})$ representing the bilinear transform associated with the scalar signals $(x_i(t), y_j(t))$, $\forall\, i, j \in \{1, \ldots, N\}$. It is parameterized by the kernel $R_{ij}(\theta, \theta'; t, \nu)$.

In the following, we opt to using the same kernel for all pairs $(x_i(t), y_j(t))$, $\forall\, i,j \in \{1,\ldots,N\}$. Thus, we simply have,

$$\mathbf{D}_{\mathbf{xy}}(t,\nu;R) = \int_{\mathbb{R}^2} \mathbf{x}(\theta)\mathbf{y}^{\mathrm{H}}(\theta')R(\theta,\theta';t,\nu)d\theta d\theta'. \tag{11.25}$$

We will also omit the notational dependency on the kernel, replacing $\mathbf{D}_{\mathbf{xy}}(t,\nu;R)$ by $\mathbf{D}_{\mathbf{xy}}(t,\nu)$. The *spatial quadratic transform* (SQT) is defined, like in the scalar case, as the SBT applied to a single signal, i.e. $\mathbf{y}(t) = \mathbf{x}(t)$:

$$(\mathbf{x}(t),\mathbf{x}(t)) \overset{\text{SQT}}{\longleftrightarrow} \mathbf{D}_{\mathbf{xx}}(t,\nu;R) = \mathbf{D}_{\mathbf{x}}(t,\nu;R) = \mathbf{D}_{\mathbf{x}}(t,\nu)$$
$$= \int_{\mathbb{R}^2} \mathbf{x}(\theta)\mathbf{x}^{\mathrm{H}}(\theta')R(\theta,\theta';t,\nu)d\theta d\theta'. \tag{11.26}$$

The diagonal terms of this matrix are termed *auto-terms*, as they correspond to the quadratic terms associated with each component of the vector $\mathbf{x}(t)$. The off-diagonal terms are termed *cross-terms*, since they correspond to the bilinear transforms associated with two different components of this vector.

11.3.3 (Spatial) quadratic time-frequency representations

A QT, $D_x(t,\nu;R)$ (paired with $x(t) \overset{\text{FT}}{\longleftrightarrow} \check{x}(\nu)$, $\check{x}(\nu)$ stands for the Fourier transform (FT)) is called "energetic" if, $\forall x$, it satisfies: $\int_{\mathbb{R}^2} D_x(t,\nu;R)dtd\nu = E_x = \int_{\mathbb{R}} |x(t)|^2 dt$ $(= \int_{\mathbb{R}} |\check{x}(\nu)|^2 d\nu)$ with E_x representing the signal energy. One can dispatch different *quadratic energetic transforms* into different classes, depending on the properties they possess. For example, the *Cohen class* is the class of quadratic time-frequency distributions (TFD) covariant by time and frequency translations whereas the *affine class* is the class of quadratic TFD covariant by time translation and scale changes (these two classes are not disjoint). Their members, associated with $x(t)$ and respectively denoted by ψ_x and ϖ_x, conform to the definition condition of the considered class and can be written in a general form that characterizes the class (*cf.* Table 11.1[4]). They also have to satisfy the *energetic* condition implying that their kernels, respectively $K(t,\tau)$ and $K'(t,\tau)$, exhibit specific properties (*cf.* Table 11.1). By changing the kernel, it is then possible to obtain the different members of a given class (*cf.* Table 11.2). Finally, one can provide a few examples of distributions belonging either to the Cohen class: the Spectrogram (Sp), the Pseudo Wigner (PW), Pseudo Wigner-Ville (PWV), smoothed Pseudo Wigner (sPW), smoothed Pseudo Wigner-Ville (sPWV), Born-Jordan (BJ) distributions etc..., or to the Affine class: the Scalogram (Sc) or to both classes: the Wigner (W), Wigner-Ville (WV) and Choï-Williams (CW) distributions (*cf.* [19,33] for more details and information).

[4]Where $\imath^2 = -1$.

Table 11.1 A summary of the definition and the principal properties of the Cohen and Affine classes

Cohen class definition $x(t) \leftrightarrow \psi_x(t,\nu) \quad \Rightarrow y(t) = x(t-\theta)e^{2\iota\pi\eta t} \quad \leftrightarrow \quad \psi_y(t,\nu) = \psi_x(t-\theta,\nu-\eta)$ Affine class definition $x(t) \leftrightarrow \varpi_x(t,\nu) \quad \Rightarrow y(t) = \lvert a\rvert^{\frac{1}{2}} x(a(t-\tau)) \quad \leftrightarrow \quad \varpi_y(t,\nu) = \varpi_x(a(t-\tau), \frac{\nu}{a})$
General form of TFD belonging to the Cohen class $\psi_{xy}(t,\nu) = \int_{\mathbb{R}^2} x(\theta+\frac{\tau}{2})y^*(\theta-\frac{\tau}{2})K(t-\theta,\tau)e^{-2\iota\pi\nu\tau}d\tau d\theta$ $\psi_x(t,\nu) = \psi_{xx}(t,\nu)$ General form of TFD belonging to the Affine class $\varpi_{xy}(t,\nu) = \int_{\mathbb{R}^2} x(\theta+\frac{\tau}{2})y^*(\theta-\frac{\tau}{2})\lvert\nu\rvert K'(\nu(t-\theta),-\nu\tau)d\tau d\theta$ $\varpi_x(t,\nu) = \varpi_{xx}(t,\nu)$
Energetic condition satisfied by members of: The Cohen class: $\int K(t,0)dt = 1$. The Affine class: $\int_{\mathbb{R}^3} \frac{1}{\lvert\nu\rvert}K'(t,\tau)e^{-2\iota\pi\nu\tau}dt d\tau d\nu = 1$.

Table 11.2 Kernels of some QTFD

Transform	Kernel $K(t,\tau)$ (Cohen class)	Kernel $K'(t,\tau)$ (Affine class)
WV	$\delta(t)$ where $\delta(\cdot)$ is the Dirac distribution	$\delta(t)e^{2\iota\pi\tau}$
PWV	$\delta(t)H(\tau)$, $H(\tau) = p^*(\frac{\tau}{2})p(-\frac{\tau}{2})$	/
sPWV	$G(t)H(\tau)$	/
Sp	$H(-\frac{\tau}{2}-t)H^*(\frac{\tau}{2}-t)$	/
Sc	/	$\lvert\frac{1}{\nu_0}\rvert H_0^*(\frac{-t-\frac{\tau}{2}}{\nu_0})H_0(\frac{-t+\frac{\tau}{2}}{\nu_0})$
CW	$\sqrt{\frac{\sigma}{4\pi}}\frac{1}{\lvert\tau\rvert}e^{-\frac{\sigma t^2}{4\tau^2}}$	$\sqrt{\frac{\sigma}{4\pi}}\frac{1}{\lvert\tau\rvert}e^{-\frac{\sigma t^2}{4\tau^2}}e^{2\pi\iota\tau}$
BJ	$\begin{cases} \frac{1}{\tau}, & \lvert t/\tau\rvert < 1/2 \\ 0, & \lvert t/\tau\rvert > 1/2 \end{cases}$	/

The spatial Wigner distribution of the signal vector $\mathbf{z}(t)$ (SW, denoted by $\mathbf{D}_{\mathrm{SW,z}}(t,\nu)$) is

$$\mathbf{D}_{\mathrm{SW,z}}(t,\nu) = \int_{\mathbb{R}} \mathbf{z}\left(t+\frac{\tau}{2}\right)\mathbf{z}^{\mathrm{H}}\left(t-\frac{\tau}{2}\right)\mathrm{e}^{-2\iota\pi\nu\tau}d\tau, \tag{11.27}$$

which is a member of the spatial quadratic time-frequency distributions (SQTFD) class. The spatial Wigner-Ville distribution (SWV, denoted by $\mathbf{D}_{\mathrm{SWV,z}}(t,\nu)$), is in essence the spatial Wigner distribution applied to the analytic (complex) signal $\mathbf{z}_A(t)$ (where $\mathbf{z}_A(t) = \mathbf{z}(t) + \imath \mathbf{z}_Q(t)$, $\mathbf{z}_Q(t) = \mathcal{H}\{\mathbf{z}(t)\}$, $\mathcal{H}\{\cdot\}$ is the Hilbert transform).

Most of the simulations in section 11.6 are performed with the spatial Pseudo Wigner-Ville distribution (SPWV, denoted by $\mathbf{D}_{\mathrm{SPWV},\mathbf{z}}(t,\nu)$) given by,

$$\mathbf{D}_{\mathrm{SPWV},\mathbf{z}}(t,\nu) = \mathbf{D}_{\mathrm{SPW},\mathbf{z}_A}(t,\nu) = \int_{\mathbb{R}} \mathbf{z}_A\left(t+\frac{\tau}{2}\right)\mathbf{z}_A^{\mathrm{H}}\left(t-\frac{\tau}{2}\right) h(\tau) e^{-2\imath\pi\nu\tau} d\tau \quad (11.28)$$

where $\mathbf{D}_{\mathrm{SPW},\mathbf{z}}(t,\nu)$ is the spatial Pseudo Wigner distribution (SPW) and $h(\tau)$ is the smoothing (short-time) window (by considering $h(\tau) = 1$, for all τ, the spatial Wigner-Ville distribution is recovered).

11.3.4 (Spatial) bilinear and quadratic time-frequency spectra

The cross-correlation matrix $\mathbf{R}_{\mathbf{z}_1\mathbf{z}_2}(t,\tau)$ of two complex random signal vectors $\mathbf{z}_1(t)$ and $\mathbf{z}_2(t)$ is defined in Eq. (11.6). One can define the corresponding *spatial bilinear time-frequency spectrum* (SBTFS) as:

$$\mathbf{D}_{\mathbf{z}_1\mathbf{z}_2}(t,\nu) = \int_{\mathbb{R}^2} \mathbf{R}_{\mathbf{z}_1\mathbf{z}_2}(\theta,\theta')K(\theta,\theta';t,\nu)d\theta d\theta'. \quad (11.29)$$

The *spatial quadratic time-frequency spectrum* (SQTFS) for the complex random signal vector $\mathbf{z}(t)$ is given in terms of the correlation matrix $\mathbf{R}_{\mathbf{z}}(t,\tau)$, of Eq. (11.7), as:

$$\mathbf{D}_{\mathbf{z}}(t,\nu) = \int_{\mathbb{R}^2} \mathbf{R}_{\mathbf{z}}(\theta,\theta')K(\theta,\theta';t,\nu)d\theta d\theta'. \quad (11.30)$$

To illustrate, consider the spatial Wigner spectrum (SWS, denoted by $\mathbf{D}_{\mathrm{SWS},\mathbf{z}}(t,\nu)$) [33]:

$$\mathbf{D}_{\mathrm{SWS},\mathbf{z}}(t,\nu) = \int_{\mathbb{R}} \mathbf{R}_{\mathbf{z}}(t,\tau)e^{-2\imath\pi\nu\tau}d\tau. \quad (11.31)$$

It is noted that the spatial Wigner distribution, given in Eq. (11.27), and obtained by considering a deterministic signal, can be recovered by discarding the mathematical expectation in Eq. (11.31).

11.3.5 Descriptions of key properties and model structure; additional assumptions about the sources

The choice of the time-frequency kernel allows us to consider SQTFD or SQTFS (SQTFD (or S)), for instance, which exhibits a *Hermitian symmetry*:

$$\mathbf{D}_{\mathbf{x}}(t,\nu) = \mathbf{D}_{\mathbf{x}}^{\mathrm{H}}(t,\nu). \quad (11.32)$$

In such case, the auto-terms of the SQTFD under consideration assume real values and the cross-terms possess a hermitian symmetry. This property is satisfied by many SQTFD, among which are the SW, SWV, SCW, SPW, SPWV, SSPWL, SSPWVL (with an additional assumption about one of the two smoothing windows being real-valued), SBJ, etc.

Using the Eqs (11.1), (11.7) and (11.30) or (11.1) and (11.26), whether the source vector is random or deterministic, the observations SQTFS or the observations SQTFD $\mathbf{D}_x(t,\nu)$ admit the following decomposition [63,3,4]:

$$\mathbf{D}_x(t,\nu) = \mathbf{A}\mathbf{D}_s(t,\nu)\mathbf{A}^H, \tag{11.33}$$

where $\mathbf{D}_s(t,\nu)$, respectively, represents the source SQTFS or the source SQTFD for a random or a deterministic signal vector. With additive noise, the above equation becomes:

$$\begin{aligned}\mathbf{D}_x(t,\nu) &= \mathbf{A}\mathbf{D}_s(t,\nu)\mathbf{A}^H + \mathbf{D}_b(t,\nu) + \mathbf{A}\mathbf{D}_{sb}(t,\nu) + \mathbf{D}_{bs}(t,\nu)\mathbf{A}^H \\ &= \mathbf{A}\mathbf{D}_s(t,\nu)\mathbf{A}^H + \mathbf{D}_b(t,\nu),\end{aligned} \tag{11.34}$$

since the noise is zero-mean and independent of the source signals.

Using the Eqs (11.10), (11.7) and (11.30) or (11.10) and (11.26), the whitened observations SQTFD (or S) $\mathbf{D}_z(t,\nu)$ admit the following decomposition [8]:

$$\mathbf{D}_z(t,\nu) = \mathbf{U}\mathbf{D}_s(t,\nu)\mathbf{U}^H. \tag{11.35}$$

In the noisy case,

$$\begin{aligned}\mathbf{D}_z(t,\nu) &= \mathbf{U}\mathbf{D}_s(t,\nu)\mathbf{U}^H + \mathbf{W}\mathbf{D}_b(t,\nu)\mathbf{W}^H + \mathbf{U}\mathbf{D}_{sb}(t,\nu)\mathbf{W}^H + \mathbf{W}\mathbf{D}_{bs}(t,\nu)\mathbf{U}^H \\ &= \mathbf{U}\mathbf{D}_s(t,\nu)\mathbf{U}^H + \mathbf{W}\mathbf{D}_b(t,\nu)\mathbf{W}^H.\end{aligned} \tag{11.36}$$

In general, matrix $\mathbf{D}_s(t,\nu)$ does not exhibit a particular algebraic structure. However, if random signals are considered, and if the sources satisfy the independence assumption (or even simply the decorrelation assumption) like in [8,32], then $\mathbf{D}_s(t,\nu)$ is a diagonal matrix for all t and ν. Accordingly, to estimate matrix $\mathbf{A}$, after a preliminary whitening stage of the observations, one should find a unitary matrix that joint diagonalizes the matrices $\mathbf{D}_z(t,\nu)$ for a given set of t-f points. A similar approach can be pursued with less restrictive assumptions on the signals or sources. To perform BSS, two additional assumptions on the source signals are introduced. They are stated as follows:

HD. There exist points in the t-f plane where each point corresponds to an auto-term of a single source signal. In other words, if $D_{s_i s_j}(t,\nu) = (\mathbf{D}_s(t,\nu))_{ij}$, then there exist (t_k,ν_k) such that,

$$D_{s_i s_j}(t_k,\nu_k) = \delta_{i,j} D_{i,j,k}. \tag{11.37}$$

That is, for $j = i$ there is at least a k-th t-f point, such that $D_{i,i,k} \neq 0$.

HZ. There exist points in the t-f plane where each point corresponds to a cross-term. That is, at (t_k, ν_k),

$$D_{s_i s_j}(t_k, \nu_k) = (1 - \delta_{i,j}) D_{i,j,k}. \tag{11.38}$$

Notice that each of the above assumptions for deterministic signals plays the role of the classical statistical independence assumption for random signals. It is clear that a "known" discriminating property for source signals is always required for a blind separation. Here, we consider signals whose quadratic time-frequency distribution (or spectra) do not highly overlap. In other words, the signatures of the sources in the t-f plane are "sufficiently" different to be able to find t-f points satisfying the aforementioned assumptions.

An analytical example in 11.3.6 and computer simulations (all along section 11.3) are provided to illustrate the validity of the above assertions.

11.3.6 Example

Consider a simple, yet interesting, analytic example. We consider two multi-components random sources given by:

$$\begin{aligned} s_1(t) &= \exp(\imath(2\pi f_0 t + \phi_0)) + \exp(\imath(2\pi f_1 t + \phi_1)), \\ s_2(t) &= \exp(\imath(2\pi f_2 t + \phi_2)) + \exp(\imath(2\pi f_3 t + \phi_3)), \end{aligned} \tag{11.39}$$

where $\phi_k, \forall k = 0, \ldots, 3$ are random, uniformly distributed over $[0, 2\pi[$ variables and $f_0, f_1, f_2\ f_3 \neq 0$ ($\phi_0 \equiv \phi_1 \equiv \phi_2 \equiv \phi_3 \equiv 0$, leads to the case of multi-components deterministic sources; finally if $f_0 = f_1$ and $f_2 = f_3$ the case of mono-component sources is treated). The calculation of the different terms of the correlation matrix, leads to:

$$\begin{aligned} R_{s_1 s_2}(t,\tau) = {} & \mathbb{E}\{\exp(\imath(\phi_0 - \phi_2))\} \exp(2\imath\pi t(f_0 - f_2)) \times \exp\left(2\imath\pi\tau\left(\frac{f_0 + f_2}{2}\right)\right) \\ & + \mathbb{E}\{\exp(\imath(\phi_0 - \phi_3))\} \exp(2\imath\pi t(f_0 - f_3)) \times \exp\left(2\imath\pi\tau\left(\frac{f_0 + f_3}{2}\right)\right) \\ & + \mathbb{E}\{\exp(\imath(\phi_1 - \phi_2))\} \exp(2\imath\pi t(f_1 - f_2)) \times \exp\left(2\imath\pi\tau\left(\frac{f_1 + f_2}{2}\right)\right) \\ & + \mathbb{E}\{\exp(\imath(\phi_1 - \phi_3))\} \exp(2\imath\pi t(f_1 - f_3)) \times \exp\left(2\imath\pi\tau\left(\frac{f_1 + f_3}{2}\right)\right) \end{aligned} \tag{11.40}$$

$$\begin{aligned} R_{s_1}(t,\tau) = {} & \exp(2\imath\pi f_0 \tau) + \exp(2\imath\pi f_1 \tau) + 2\exp\left(2\imath\pi \frac{(f_0 + f_1)}{2}\tau\right) \\ & \times \mathbb{E}\{\cos(2\pi(f_1 - f_0)t + \phi_1 - \phi_0)\} \end{aligned} \tag{11.41}$$

$$\begin{aligned} R_{s_2}(t,\tau) = {} & \exp(2\imath\pi f_2 \tau) + \exp(2\imath\pi f_3 \tau) + 2\exp\left(2\imath\pi \frac{(f_2 + f_3)}{2}\tau\right) \\ & \times \mathbb{E}\{\cos(2\pi(f_3 - f_2)t + \phi_3 - \phi_2)\} \end{aligned} \tag{11.42}$$

where $R_{s_i s_j}(t,\tau) = (\mathbf{R}_\mathbf{s}(t,\tau))_{ij}$, $R_{s_i}(t,\tau) = R_{s_i s_i}(t,\tau)$ and $R_{s_j s_i}(t,-\tau) = R^*_{s_i s_j}(t,\tau)$. We assume the two sources different, i.e. $f_0 \neq f_1$ and $f_2 \neq f_3$ and $f_0 \neq f_2$. If $\phi_0 \equiv \phi_1$, the components of the signal $s_1(t)$ are correlated. The same is true for $s_2(t)$ if we consider $\phi_2 \equiv \phi_3$. The two signals s_1 and s_2 are zero mean. In the case of $\phi_0 \equiv \phi_2$ or $\phi_0 \equiv \phi_3$ or $\phi_1 \equiv \phi_2$ or $\phi_1 \equiv \phi_3$, the two signals become correlated, whereas if ϕ_0, ϕ_1, ϕ_2 and ϕ_3 are statistically independent, the two signals are uncorrelated on average. Considering the Wigner spectrum in Eq. (11.31), we obtain

$$D_{\mathrm{SWS},s_1}(t,\nu) = [\delta(\nu - f_0) + \delta(\nu - f_1)] \otimes 1_t + 2\delta\left(\nu - \left(\frac{f_0+f_1}{2}\right)\right) \times \mathbb{E}\{\cos(2\pi(f_1 - f_0)t + \phi_1 - \phi_0)\} \tag{11.43}$$

$$D_{\mathrm{SWS},s_2}(t,\nu) = [\delta(\nu - f_2) + \delta(\nu - f_3)] \otimes 1_t + 2\delta\left(\nu - \left(\frac{f_2+f_3}{2}\right)\right) \times \mathbb{E}\{\cos(2\pi(f_3 - f_2)t + \phi_3 - \phi_2)\} \tag{11.44}$$

$$\begin{aligned} D_{\mathrm{SWS},s_1 s_2}(t,\nu) = {} & \delta\left(\nu - \left(\frac{f_0+f_2}{2}\right)\right) \exp(2\imath\pi t(f_0 - f_2))\mathbb{E}\{\exp(\imath(\phi_0 - \phi_2))\} \\ & + \delta\left(\nu - \left(\frac{f_1+f_2}{2}\right)\right) \exp(2\imath\pi t(f_1 - f_2))\mathbb{E}\{\exp(\imath(\phi_1 - \phi_2))\} \\ & + \delta\left(\nu - \left(\frac{f_0+f_3}{2}\right)\right) \exp(2\imath\pi t(f_0 - f_3))\mathbb{E}\{\exp(\imath(\phi_0 - \phi_3))\} \\ & + \delta\left(\nu - \left(\frac{f_1+f_3}{2}\right)\right) \exp(2\imath\pi t(f_1 - f_3))\mathbb{E}\{\exp(\imath(\phi_1 - \phi_3))\} \end{aligned} \tag{11.45}$$

where 1_t stands for the distribution which equals 1 for all t, $\otimes$ is the product of two distributions, $D_{\mathrm{SWS},s_i s_j}(t,\nu) = (\mathbf{D}_{\mathrm{SWS},\mathbf{s}}(t,\nu))_{ij}$ and $D_{\mathrm{SWS},s_j s_i}(t,\nu) = D^*_{\mathrm{SWS},s_i s_j}(t,\nu)$ (the Hermitian symmetry property given by Eq. (11.32) is thus satisfied).

We note that when the signals $s_1(t)$ and $s_2(t)$ are uncorrelated, they imply that the matrix $\mathbf{D}_{\mathrm{SWS},\mathbf{s}}(t,\nu)$ is diagonal for all t and ν. Otherwise, the term $D_{\mathrm{SWS},s_1 s_2}(t,\nu)$ is, generally, not zero for the frequency values $\nu = (f_1 + f_2)/2$ and $\forall t$, if $\phi_2 \equiv \phi_1$, $\nu = (f_0 + f_2)/2$, if $\phi_0 \equiv \phi_2$, $\nu = (f_0 + f_3)/2$ if $\phi_0 \equiv \phi_3$, and for $\nu = (f_1 + f_3)/2$, if $\phi_1 \equiv \phi_3$. As a consequence, the matrix $\mathbf{D}_{\mathrm{SWS},\mathbf{s}}(t,\nu)$ is not diagonal for all t and ν, implying that a joint diagonalization (JD) procedure for the estimation of $\mathbf{A}$ is no longer possible $\forall t,\nu$. However, there exist t-f points for which the matrix $\mathbf{D}_{\mathrm{SWS},\mathbf{s}}(t,\nu)$ assumes a very specific algebraic structure, despite the fact that the considered signals are correlated.

It would be much too difficult, giving space limitations, to distinguish and address all possible cases. Herein, we will focus on two cases. In the first case, we assume that $f_0 = 0.1, f_1 = 0.2, f_2 = 0.3$ and $f_3 = 0.4$, and $\phi_0 \equiv \phi_1 \equiv \phi_2 \equiv \phi_3 \equiv 0$ (deterministic sources with disjoint signatures in the time-frequency plane). In the second case, we assume that $f_0 = f_2 = 0.1, f_1 = 0.2$ and $f_3 = 0.4$, and $\phi_0 \equiv \phi_1 \equiv \phi_2 \equiv \phi_3 \equiv 0$ (deterministic sources with signatures that partially overlap in the t-f plane). Since the sources are assumed deterministic, the spatial Wigner spectrum ($\mathbf{D}_{\mathrm{SWS}}(t,\nu)$) is replaced

Table 11.3 The SQTFD algebraic structure versus the considered t-f point (t,ν) in the first case

Considered t-f points: $\forall t$ and $\nu =$	0.1	0.15	0.2	0.25	0.3	0.35	0.4
Sources auto-terms present	yes	yes	yes	no	yes	yes	yes
Sources cross-terms present	no	no	yes	yes	yes	no	no

Table 11.4 The SQTFD algebraic structure versus the considered t-f point (t,ν) in the second case

Considered t-f points: $\forall t$ and $\nu =$	0.1	0.15	0.2	0.25	0.3	0.4
Sources auto-terms present	yes	yes	yes	yes	no	yes
Sources cross-terms present	yes	yes	no	yes	yes	no

by the spatial Wigner distribution ($\mathbf{D}_{\mathsf{SW}}(t,\nu)$). In the first case, for $\nu = f_0 = 0.1$ and $\forall t$, $D_{\mathsf{SW},s_1}(t,\nu) \neq 0$ and, $D_{\mathsf{SW},s_2}(t,\nu) = D_{\mathsf{SW},s_1s_2}(t,\nu) = 0$. In the same manner, for $\nu = f_3$ and $\forall t$, $D_{\mathsf{SW},s_2}(t,\nu) \neq 0$ and $D_{\mathsf{SW},s_1}(t,\nu) = D_{\mathsf{SW},s_1s_2}(t,\nu) = 0$. For $\nu = \frac{f_0+f_1}{2} = 0.15$ and $\forall t$, $D_{\mathsf{SW},s_1}(t,\nu) \neq 0$ and $D_{\mathsf{SW},s_2}(t,\nu) = D_{\mathsf{SW},s_1s_2}(t,\nu) = 0$. For $\nu = \frac{f_2+f_3}{2} = 0.35$ and $\forall t$, $D_{\mathsf{SW},s_2}(t,\nu) \neq 0$ and $D_{\mathsf{SW},s_1}(t,\nu) = D_{\mathsf{SW},s_1s_2}(t,\nu) = 0$. For these t-f points, the source SQTFD $\mathbf{D}_{\mathsf{SW,s}}(t,\nu)$ is real (due to the Hermitian symmetry as stated by Eq. (11.32)), effectively diagonal and rank one, which suggests the use of a JD procedure.

In the first case, for $\nu = \frac{f_0+f_3}{2} = \frac{f_1+f_2}{2} = 0.25$ and $\forall t$, $D_{\mathsf{SW},s_1}(t,\nu) = D_{\mathsf{SW},s_2}(t,\nu) = 0$ and $D_{\mathsf{SW},s_1s_2}(t,\nu) \neq 0$. For such t-f points, $\mathbf{D}_{\mathsf{SW,s}}(t,\nu)$ is generally complex and zero-diagonal,[5] which suggests the use of a joint zero-diagonalization (JZD) procedure.

In the second case (i.e. $f_0 = f_2 = 0.1, f_1 = 0.2, f_3 = 0.4$), the same kind of reasoning is used to derive the values of $D_{\mathsf{SW},s_1}, D_{\mathsf{SW},s_2}$ and D_{SW,s_1,s_2} for the different values of (t,v).

We have summarized, in Tables 11.3 and 11.4, what occurs in each of these two configurations.

These tables suggest the existence of four different types of t-f points (t,ν):

- Type I: the t-f points that correspond to source auto-terms only (Assumption HD). Namely, $\forall t$ and $\nu = 0.1, 0.15, 0.35, 0.4$ in the first case and $\forall t$ and $\nu = 0.2, 0.4$ in the second case; for those points, the source SQTFD (or S) is a diagonal matrix. This property is generally lost because of the linear mixture; yet, in such t-f points after mixing, one still finds source auto-terms only.
- Type II: the t-f points that correspond to source cross-terms only (Assumption HZ). Namely, $\forall t$ and $\nu = 0.25$ in the first case and $\forall t$ and $\nu = 0.3$ in the second case. For

[5] A zero-diagonal matrix is a square matrix whose diagonal terms are all null, i.e. $\mathbf{Z} = (Z_{i,j})$ is a zero-diagonal matrix if and only if $Z_{i,i} = 0$ for all i.

those t-f points, the source SQTFD (or S) is a zero-diagonal matrix, this property is then generally lost because of the linear mixture; yet, in such points, after mixing, one still finds source cross-terms only.

- Type III: the t-f points that correspond to both source cross- and auto-terms. Namely $\forall t$ and $\nu = 0.2, 0.3$ in the first case and $\forall t$ and $\nu = 0.1, 0.15, 0.25$ in the second case. For those t-f points, the source SQTFD (or S) does not exhibit an algebraic structure that could be directly exploited.
- Type IV: the t-f points where there are neither source cross-terms nor source auto-terms (all the other t-f points!).

Only the first two types of t-f points (I & II) are of interest with regard to the BSS.

Remark 1. It is important to notice that the *rank-1 diagonal matrix* property is in general true when we consider t-f points that correspond to source auto-terms only. In fact, due to the "middle point rule" [33], which defines the interference geometry, if two sources could simultaneously co-exist in the same t-f points then these two "auto-terms" would interfere in their middle (here in the same t-f point) to provide a third term which happens to be a source cross-term. This means that the source SQTFD (or S), estimated at this t-f point, will no longer be a diagonal matrix (unless (albeit improbable) this source cross-term is null). It finally implies that the diagonal matrix with one single non-zero term on the diagonal (rank one diagonal matrix) is nearly the only possibility of a diagonal matrix.

Remark 2. We note that when the sources are not uncorrelated, a classical whitening (diagonalization of the correlation matrix at the null time delay) is no more possible or warrantable. Further, since the SQTFD (or S) for the t-f points that correspond to source auto-terms only are diagonal but rank one, they cannot be used for whitening. That is also why we are motivated to discuss, in the following, an alternative solution to estimate the mixing matrix $\mathbf{A}$ in the context of non-unitary joint matrix decomposition algorithms (*cf.* 11.4.2).

11.4 TIME-FREQUENCY POINTS SELECTION

Nonstationary signal separation techniques based on SQTFD (or S) require a preliminary stage of selecting (in an automatic or manual way) particular t-f points: those corresponding to source auto-terms only (type I) and/or to source cross-terms only (type II). This step is important to construct the set of matrices for joint diagonalization (this set will be denoted by $\mathcal{M}_{\mathrm{JD}}$) and/or joint zero-diagonalization (this set will be denoted by $\mathcal{M}_{\mathrm{JZD}}$). The existence of such t-f points is ensured from assumptions HD and HZ. This problem remains a complicated one, since a property about the unknown matrix $\mathbf{D}_{\mathrm{s}}(t,\nu)$ (namely the source SQTFD (or S)) has to be determined only from the observations SQTFD (or S)) matrix $\mathbf{D}_{\mathrm{x}}(t,\nu)$.

11.4.1 Automatic time-frequency points selection in a whitened context

In context of whitening, several automatic t-f points selection procedures have been suggested [5,29,32,36,37]. Most of them take advantage of the fact that the trace, trace$\{\cdot\}$, is invariant under unitary transform. That is, $\text{trace}\{\mathbf{D}_z(t,\nu)\} = \text{trace}\{\mathbf{D}_s(t,\nu)\}$, making it possible to decide whether source auto-terms are present or not.

Under the additional assumption that the SQTFD (or S) exhibits a Hermitian symmetry, one possible way to select the matrices (stemming from the whitened observations SQTFD (or S)) for joint zero-diagonalization, is to look for those of small absolute value ($|\cdot|$) of the trace (smaller than a given threshold ε_2; there should not be any source auto-term), while the Euclidian (Frobenius) norm ($\|\cdot\|$) of their imaginary part ($\Im\{\cdot\}$) keeps on being high enough (higher than a given value of threshold ε_1; there should be source cross-terms). A way to select the (whitened observations) SQTFD (or S) matrices for joint diagonalization is to look for those whose absolute value of the trace is high enough (higher than a given threshold ε_4; there should be source auto-terms) while the Euclidian norm of their imaginary part must be small (smaller than a given threshold ε_3; there should not be source cross-terms). To sum up:

For joint zero-diagonalization ($\mathcal{M}_{\mathsf{JZD}}$), choose SQTFD (or S) matrices corresponding to t-f points (t,ν) such that:

$$\begin{cases} \|\Im\{\mathbf{D}_z(t,\nu)\}\| > \varepsilon_1 \\ |\text{trace}\{\mathbf{D}_z(t,\nu)\}| < \varepsilon_2. \end{cases} \tag{11.46}$$

For joint-diagonalization ($\mathcal{M}_{\mathsf{DC}}$) choose SQTFD (or S) corresponding to t-f points (t,ν) such that:

$$\begin{cases} \|\Im\{\mathbf{D}_z(t,\nu)\}\| < \varepsilon_3 \\ |\text{trace}\{\mathbf{D}_z(t,\nu)\}| > \varepsilon_4. \end{cases} \tag{11.47}$$

This detection procedure ((11.46) and/or (11.47)) is denoted by C_{G} and has been suggested in [36].

Remark 3. The above decision rule is challenged by the fact that t-f points may exist where the source cross-terms are present but have a zero imaginary part and a non-zero real part. In such t-f points, if there are source cross-terms only, the matrices should be joint zero-diagonalized (whereas it is decided not to diagonalize), and if there are both source auto- and cross-terms, the decision should be not to diagonalize, whereas it is decided to joint diagonalize. With regard to joint zero-diagonalization, mis-labeling of the t-f points is not considered a problem, since it only means that the formulated set is smaller than the set available. With regard to joint diagonalization, it becomes problematic, since wrong matrices are now members of the used set. That is why it has been subsequently suggested in [29,37], to introduce another constraint (in the form of an additional threshold ε_3') to partially clear this ambiguity. The constraint is concerned with the imaginary part of the SQTFD (or S) matrices. The goal is to make a distinction between areas where the source cross-terms are zero within the neighborhood of the underlying t-f point (t,ν) and areas where the source cross-terms are coincidentally cancelling in (t,ν). Let $G(t,\nu)$ denote a 2D lowpass filter. The decision rule for joint diagonalization becomes:

For $\mathcal{M}_{\mathsf{DC}}$, choose SQTFD (or S) matrices corresponding to t-f points (t,ν) such that:

$$\begin{cases} \|\Im\{\mathbf{D}_{\mathrm{z}}(t,\nu)\}\| < \varepsilon_3 \\ |\mathrm{trace}\{\mathbf{D}_{\mathrm{z}}(t,\nu)\}| > \varepsilon_4 \\ \|\{G * \Im\{\mathbf{D}_{\mathrm{z}}\}\}(t,\nu)\| < \varepsilon_3' \end{cases} \tag{11.48}$$

where $\{\cdot * \cdot\}(t,\nu)$ stands for the convolution product. This detection procedure ((11.46) and/or (11.48)) is denoted by C_{GM}. In [37], a noisy model has also been explicitly considered: the values of the thresholds involved in the detector are no more constant but depend on the signal to noise ratio (SNR). Such an issue has also been studied in [15,16,18].

Other detectors can be found in the literature [5,15,17,32] in the whitened context. Whereas [32,17] propose detectors for t-f points corresponding to source auto-terms only, [5,15] provide two kinds of detectors (source auto-terms and source cross-terms). Notice that [15,17,18] deal more specifically with the modifications to take the noise into account (decision rules linked to the noise statistics and/or modification of the decision thresholds).

In [5], the following procedure is suggested:

For $\mathcal{M}_{\mathsf{JZD}}$, choose SQTFD (or S) matrices corresponding to the t-f points (t,ν) such that:

$$\frac{\mathrm{trace}\{\mathbf{D}_{\mathrm{z}}(t,\nu)\}}{\|\mathbf{D}_{\mathrm{z}}(t,\nu)\|} < \varepsilon_6. \tag{11.49}$$

For $\mathcal{M}_{\mathsf{JD}}$, choose SQTFD (or S) matrices corresponding to t-f points (t,ν) such that:

$$\frac{\mathrm{trace}\{\mathbf{D}_{\mathrm{z}}(t,\nu)\}}{\|\mathbf{D}_{\mathrm{z}}(t,\nu)\|} > \varepsilon_6 \tag{11.50}$$

where ε_6 is a small positive constant. In [43,44], a pre-selection of t-f is realized through the following:

Keep t-f points (t,ν) such that,

$$\|\mathbf{D}_{\mathrm{x}}(t,\nu)\| > \varepsilon_5 \tag{11.51}$$

with $\varepsilon_5 \approx 0.05 \times \|\mathbf{D}_{\mathrm{x}}(t_h,\nu_h)\|$ and (t_h,ν_h) the t-f of highest energy. It enables reduction in computational time by discarding the t-f points with negligible energy (it is also a rather more different, but easy, way to deal with the problem of the noise than the one suggested in [15,17,37]). This detection procedure ((11.49) and/or (11.50) + (11.51)) is denoted by C_{B}.

One can notice a potential problem with the above detector concerning the t-f points selected for joint-diagonalization. If a non negligible value of the trace qualifies the presence of source auto-terms, it does not necessarily mean the absence of source cross-terms. Besides, if both source auto-terms and cross-terms are simultaneously present at a given t-f point, the source SQTFD (or S) $\mathbf{D}_s(t,\nu)$ will not have any specific algebraic structure. Further, even if the source signatures are disjoint in the t-f plane, there is no guarantee of the total absence of t-f points where source auto-terms and source cross-terms are simultaneously present (*cf.* analytical example given in 11.3.6).

The detector suggested in [32] exploits the fact that the sources SQTFD (or S) is a diagonal rank one matrix at a t-f point where there are only source auto-terms (for instance, there can be only one single source auto-term). As a consequence, the whitened observations SQTFD (or S) is a rank one matrix at such t-f point. The following procedure is thus suggested:

Keep t-f points (t,ν) such that:

$$\text{trace}\{\mathbf{D}_x(t,\nu)\} \geq \varepsilon_7, \quad \varepsilon_7 = \text{mean}\{\text{trace}\{\mathbf{D}_x(t,\nu)\}\} \quad \forall t, \forall \nu. \tag{11.52}$$

where mean$\{\cdot\}$ stands for the mean value.

For $\mathcal{M}_{JD}$, we keep SQTFD (or S) matrices corresponding to t-f points (t,ν) such that:

$$\mathcal{F}(t,\nu) = \frac{\max_k |\lambda_k(t,\nu)|}{\sum_{k=1}^{N} |\lambda_k(t,\nu)|} > 1 - \varepsilon_8 \tag{11.53}$$

where $\{\lambda_k(t,\nu); k = 1,\ldots,N\}$ are the eigenvalues of $\mathbf{D}_z(t,\nu)$ and ε_8 is a small positive constant. This procedure ((11.52) + (11.53)) is denoted by C_F.

11.4.2 Automatic time-frequency points selection in a non-whitened context

In a nonwhitening context, alternative t-f points detectors have been suggested [29, 38,35,43,44]. The one presented in [35], takes advantage of the fact that the source SQTFD (or S) is a diagonal rank one matrix at a t-f point where only one single source auto-term exists. As a consequence the observations SQTFD (or S) is a rank one matrix as well. A singular value decomposition can be used (SVD) leading to $\mathbf{D}_x(t,\nu) = \mathbf{V}(t,\nu)\mathbf{\Lambda}(t,\nu)\mathbf{U}^H(t,\nu)$, with $\mathbf{V}(t,\nu)$ and $\mathbf{U}(t,\nu)$ representing unitary $P \times P$ matrices, and $\mathbf{\Lambda}(t,\nu) = \text{Diag}\{\boldsymbol{\lambda}(t,\nu)\}$ is a diagonal matrix whose diagonal elements are positive. Denoting by $\boldsymbol{\lambda}(t,\nu) = (\lambda_1(t,\nu),\ldots,\lambda_P(t,\nu))^T$ and assuming that the singular values are sorted in a decreasing order: $\lambda_1(t,\nu) \geq \lambda_2(t,\nu) \geq \ldots \geq \lambda_P(t,\nu) \geq 0$, then one way to check whether a matrix is rank one or not is given by the following:

For $\mathcal{M}_{JD}$ choose SQTFD (or S) corresponding to t-f points (t,ν) such that:

$$\begin{cases} \lambda_1(t,\nu) > \varepsilon_9 \\ \sum_{k=2}^{P} \lambda_k(t,\nu) < \varepsilon_{10} \end{cases} \tag{11.54}$$

where ε_9 is a (sufficiently) high constant and ε_{10} is a (sufficiently) small positive constant.

Another way is given by the following:

For $\mathcal{M}_{\text{JD}}$ choose SQTFD (or S) corresponding to t-f points (t,ν) such that:

$$\begin{cases} \mathscr{C}(t,\nu) = \dfrac{\lambda_1(t,\nu)}{\sum\limits_{k=1}^{P} \lambda_k(t,\nu)} > 1 - \varepsilon_{11} \\ \sum\limits_{k=1}^{P} \lambda_k(t,\nu) > \varepsilon_{12} \end{cases} \tag{11.55}$$

where ε_{11} and ε_{12} are small positive constant values. This constitutes a generalization of the detector given in [32] to the non-whitened case (the test on the trace is replaced here by a test on the sum of the eigenvalues). This detection procedure (11.55) is denoted by C_{GG}.

In [38], the aim is to consider a few matrices for joint diagonalization in the selection of the t-f points. Under the additional assumption that the mixing matrix and the sources signals are not simultaneously complex and that the SQTFD (or S) kernel exhibits Hermitian symmetry, the following detection procedure was suggested:

For $\mathcal{M}_{\text{JD}}$, choose the SQTFD (or S) corresponding to t-f points (t,ν) such that:

$$\begin{cases} \|\Im\{\mathbf{D}_{\mathbf{x}}(t,\nu)\}\| < \varepsilon_{14} \\ \|\Re\{\mathbf{D}_{\mathbf{x}}(t,\nu)\}\| > \varepsilon_{15} \\ \|\{G * \Im\{\mathbf{D}_{\mathbf{x}}\}\}(t,\nu)\| < \varepsilon'_{14} \end{cases} \tag{11.56}$$

where $G(t,\nu)$ denotes a 2D lowpass filter, ε_{14} and ε'_{14} are sufficiently small positive constants and ε_{15} is a sufficiently high positive constant. This detection procedure (11.56) is denoted by C_{S}.

For the same reasons as before, this decision rule is not restrictive enough, which leads to the need for introducing an additional constraint (in the form of an additional threshold operating after images filtering).

Another detector has been suggested in [43], for *quasi-disjoint* sources (the signatures of the sources in the t-f plane can overlap at very few t-f points). It takes advantage of the rank one property. The selection procedure reads:

Keep t-f points (t,ν) such that

$$\|\mathbf{D}_{\mathbf{x}}(t,\nu)\| > \varepsilon_5, \quad \varepsilon_5 \approx 0.05 \times \|\mathbf{D}_{\mathbf{x}}(t_h,\nu_h)\| \tag{11.57}$$

and (t_h,ν_h) is the t-f point of higher energy.

For $\mathcal{M}_{\text{JZD}}$, choose SQTFD (or S) corresponding to t-f points (t,ν) such that:

$$\left| \frac{\lambda_1(t,\nu)}{\|\mathbf{D}_{\mathbf{x}}(t,\nu)\|} - 1 \right| > \varepsilon_{13}. \tag{11.58}$$

For $\mathcal{M}_{\mathrm{JD}}$, choose SQTFD corresponding to time-frequency points (t,ν) such that:

$$\left|\frac{\lambda_1(t,\nu)}{\|\mathbf{D}_{\mathbf{x}}(t,\nu)\|}-1\right|<\varepsilon_{13} \tag{11.59}$$

where ε_{13} is a small positive constant (typically $\varepsilon_{13}=0.8$) and where $\lambda_1(t,\nu)$ stands for the highest singular value of $\mathbf{D}_{\mathbf{x}}(t,\nu)$. This detection procedure ((11.57) + (11.58) and/or (11.59)) is denoted by $\mathrm{C_L}$.

In [29], a slight modification was suggested: a different threshold (ε_{13}) is considered for JD and JZD (else, it means that one considers the existence of three classes of t-f points instead of four). ε_{13} is replaced by ε'_{13} in (11.59), leading to the following:

For $\mathcal{M}_{\mathrm{JD}}$, choose t-f points (t,ν) such that

$$\left|\frac{\lambda_1(t,\nu)}{\|\mathbf{D}_{\mathbf{x}}(t,\nu)\|}-1\right|<\varepsilon'_{13}. \tag{11.60}$$

This detection procedure ((11.57) + (11.58) and/or (11.60)) is denoted by $\mathrm{C_{LM}}$.

The aforementioned t-f point detectors enable the definition of one or sometimes two sets ($\mathcal{M}_{\mathrm{JD}}$ and $\mathcal{M}_{\mathrm{JZD}}$). The following section deals with the different separation algorithms, a preliminary whitening stage being applied or not.

11.5 SEPARATION ALGORITHMS

From the two matrices sets $\mathcal{M}_{\mathrm{JD}}$ and/or $\mathcal{M}_{\mathrm{JZD}}$, the problem consists of:

- directly estimating the mixing matrix $\mathbf{A}$ or the separation matrix $\mathbf{B}$ (Eq. (11.2)),
- or – if a preliminary whitening stage is used – estimating a *unitary* matrix $\mathbf{U}$ (resp. *orthogonal* if $\mathbf{A}$ possesses real values) required for Eq. (11.10).

All the matrices belonging to $\mathcal{M}_{\mathrm{JD}}$ admit the decomposition given by Eq. (11.33) (resp. Eq. (11.35) in a whitened context) where the source SQTFD (or S) matrices $\mathbf{D}_{\mathbf{s}}(t,\nu)$ assume a very specific algebraic structure, being diagonal matrices. As a consequence, a first natural manner to tackle the separation problem consists of opting for a JD algorithm. Such approach is rather classical in BSS and many techniques are based on it. One can cite AMUSE [55], SOBI [7], JADE [12], STOTD [23], CHESS [59], ACMA [57] for example. The main difference appears in the construction of the matrix set $\mathcal{M}_{\mathrm{JD}}$, which is comprised of:

- fourth order cumulant matrices in JADE, third order tensors, third or fourth order cumulants in STOTD or their generalization (matrices of cumulants of any order, third (resp. m) order tensors of cumulants of any order (resp. n with $n>m$)) in [46];
- correlation matrices of different time-lag or different filtered signals in SOBI or OFI;

- SQTFD matrices at selected t-f points in [8];
- algebraically derived matrices subject to the constant modulus sources hypothesis in ACMA;
- Hessian of the joint characteristic function in CHESS;
- cross-spectral matrices at different frequencies in [51,47] or higher order cumulant matrices for distinct times in [21] when the case of convolutive mixtures is considered.

Since all the matrices of $\mathcal{M}_{\text{JZD}}$ admit the same decomposition given by Eq. (11.33) (resp. Eq. (11.35) in a whitened context) where the sources SQTFD (or S) matrices $\mathbf{D}_{\text{s}}(t,\nu)$ possess a particular algebraic structure (they happen to be zero-diagonal in that case), the JZD constitutes an alternative to the first approach (note that joint zero-diagonalization finds applications in other domains, among which are cryptography [58] and telecommunications). We will see further that other solutions, combining JD with JZD or purely algebraic methods followed by a classification algorithm, can also be proposed.

11.5.1 Joint diagonalization and/or joint zero-diagonalization criteria

Consider two matrix sets linked to the problem of the joint diagonalization and joint zero diagonalization. The first set of K_{JD}, $K_{\text{JD}} \in \mathbb{N}^*$ matrices $\mathbf{M}_{\text{D},k} \in \mathbb{C}^{P\times P}$, $k \in \{1,\ldots,K_{\text{JD}}\}$, is denoted by $\mathcal{M}_{\text{JD}}$. These matrices admit the same decomposition: there exist a $P \times N$, $(P \geq N)$, full column rank matrix $\mathbf{A}_{\text{D}}$ and a set $\mathcal{D}$ of K_{JD} $N \times N$ diagonal matrices $\mathbf{D}_k \in \mathbb{C}^{N\times N}$, $k \in \{1,\ldots,K_{\text{JD}}\}$, such that

$$\mathbf{M}_{\text{D},k} = \mathbf{A}_{\text{D}}\mathbf{D}_k\mathbf{A}_{\text{D}}^{\text{H}}, \qquad \forall k \in \{1,\ldots,K_{\text{JD}}\}. \tag{11.61}$$

The second set of K_{JZD}, $K_{\text{JZD}} \in \mathbb{N}^*$, matrices $\mathbf{M}_{\text{Z},k} \in \mathbb{C}^{P\times P}$, $k \in \{1,\ldots,K_{\text{JZD}}\}$, is denoted by $\mathcal{M}_{\text{JZD}}$. Those matrices all admit the following decomposition: there exist a $P \times N$, $P \geq N$, full column rank matrix $\mathbf{A}_{\text{Z}}$ and a set $\mathcal{Z}$ of K_{JZD} $N \times N$ zero-diagonal matrices $\mathbf{Z}_k \in \mathbb{C}^{N\times N}$, $k \in \{1,\ldots,K_{\text{JZD}}\}$, such that:

$$\mathbf{M}_{\text{Z},k} = \mathbf{A}_{\text{Z}}\mathbf{Z}_k\mathbf{A}_{\text{Z}}^{\text{H}}, \qquad \forall k \in \{1,\ldots,K_{\text{JZD}}\}. \tag{11.62}$$

The *joint diagonalization* (JD) problem can be stated as follows: estimating $\mathbf{A}_{\text{D}}$ and $\mathbf{D}_i$, $i \in \{1,\ldots,K_{\text{JD}}\}$ using the matrix set $\mathcal{M}_{\text{JD}}$ only. Equivalently, the *joint zero-diagonalization* (JZD) problem can be stated as follows: estimating $\mathbf{A}_{\text{Z}}$ and $\mathbf{Z}_i$, $i \in \{1,\ldots,K_{\text{JZD}}\}$ using the matrix set $\mathcal{M}_{\text{JZD}}$ only. If $\mathbf{A}_{\text{D}} = \mathbf{A}_{\text{Z}} = \mathbf{A}_{\text{DZ}}$, one can define the joint diagonalization/zero-diagonalization (JD/JZD) problem which consists of estimating the matrices $\mathbf{A}_{\text{DZ}}$, $\mathbf{D}_i$, $i \in \{1,\ldots,K_{\text{JD}}\}$ and $\mathbf{Z}_i$, $i \in \{1,\ldots,K_{\text{JZD}}\}$ using the two matrix sets $\mathcal{M}_{\text{JD}}$ and $\mathcal{M}_{\text{JZD}}$. Since the factorization of the matrices belonging to those two sets is known, a rather classical way to solve the aforementioned JD and JZD problems is to minimize the two following quadratic cost functions:

$$\mathcal{I}_{\text{D}}(\mathbf{A},\{\mathbf{D}_i\}) = \sum_{i=1}^{K_{\text{JD}}} \|\mathbf{M}_{\text{D},i} - \mathbf{A}\mathbf{D}_i\mathbf{A}^{\text{H}}\|^2 \tag{11.63}$$

and

$$\mathscr{I}_{\mathsf{Z}}(\mathbf{A},\{\mathbf{Z}_i\})=\sum_{i=1}^{K_{\mathsf{JZD}}}||\mathbf{M}_{\mathsf{Z},i}-\mathbf{A}\mathbf{Z}_i\mathbf{A}^{\mathrm{H}}||^2 \tag{11.64}$$

where $||\cdot||$ stands for the Frobenius norm of the matrix given in the argument and $\{\mathbf{D}_i\}$ and $\{\mathbf{Z}_i\}$ are the sets of diagonal and zero-diagonal matrices considered. One can find a slightly modified version of the first cost function $\mathscr{I}_{\mathsf{D}}$ in Eq. (11.63) [60]:

$$\mathscr{I}_{\mathsf{D,YER}}(\mathbf{A},\{\mathbf{D}_i\})=\sum_{i=1}^{K_{\mathsf{JD}}}\omega_i||\mathbf{M}_{\mathsf{D},i}-\mathbf{A}\mathbf{D}_i\mathbf{A}^{\mathrm{H}}||^2 \tag{11.65}$$

where $\omega_1,\omega_2,\ldots,\omega_{K_{\mathsf{JD}}}\in\mathbb{R}^{+*}$ are positive ponderation coefficients.

However, one can proceed rather differently, by premultiplying $\mathbf{M}_{\mathsf{D},i}$ in Eq. (11.61) (resp. $\mathbf{M}_{\mathsf{Z},i}$ in Eq. (11.62)) by the pseudo-inverse (Moore-Penrose generalized inverse) $\mathbf{A}_{\mathsf{D}}^{\dagger}$ of $\mathbf{A}_{\mathsf{D}}$ (resp. $\mathbf{A}_{\mathsf{Z}}^{\dagger}$ of $\mathbf{A}_{\mathsf{Z}}$) and by postmultiplying by $(\mathbf{A}_{\mathsf{D}}^{\dagger})^{\mathrm{H}}$ (resp. $(\mathbf{A}_{\mathsf{Z}}^{\dagger})^{\mathrm{H}}$):

$$\mathbf{A}_{\mathsf{D}}^{\dagger}\mathbf{M}_{\mathsf{D},i}(\mathbf{A}_{\mathsf{D}}^{\dagger})^{\mathrm{H}}=\mathbf{D}_i, \qquad \forall i\in\{1,\ldots,K_{\mathsf{JD}}\} \tag{11.66}$$

and

$$\mathbf{A}_{\mathsf{Z}}^{\dagger}\mathbf{M}_{\mathsf{Z},i}(\mathbf{A}_{\mathsf{Z}}^{\dagger})^{\mathrm{H}}=\mathbf{Z}_i, \qquad \forall i\in\{1,\ldots,K_{\mathsf{JZD}}\}. \tag{11.67}$$

In order to directly estimate the pseudo-inverse of the matrices $\mathbf{A}_{\mathsf{D}}$ or $\mathbf{A}_{\mathsf{Z}}$ (in the BSS context, the separation matrix instead of the mixing matrix), one can consider two other quadratic cost functions:

$$\mathscr{C}_{\mathsf{D}}'(\mathbf{B},\{\mathbf{D}_i\})=\sum_{i=1}^{K_{\mathsf{JD}}}||\mathbf{B}\mathbf{M}_{\mathsf{D},i}\mathbf{B}^{\mathrm{H}}-\mathbf{D}_i||^2 \tag{11.68}$$

and

$$\mathscr{C}_{\mathsf{Z}}'(\mathbf{B},\{\mathbf{Z}_i\})=\sum_{i=1}^{K_{\mathsf{JZD}}}||\mathbf{B}\mathbf{M}_{\mathsf{Z},i}\mathbf{B}^{\mathrm{H}}-\mathbf{Z}_i||^2. \tag{11.69}$$

By defining

$$\widehat{\mathbf{D}}_i=\arg\min_{\mathbf{D}_i}\mathscr{C}_{\mathrm{D}}'(\mathbf{B},\{\mathbf{D}_i\}), \tag{11.70}$$

one easily finds that, if $\mathbf{B}$ is fixed:

$$\begin{aligned}\mathscr{C}'_{\mathsf{D}}(\mathbf{B},\{\mathbf{D}_i\}) &= \sum_{i=1}^{K_{\mathsf{JD}}} \text{trace}\{(\mathbf{B}\mathbf{M}_{\mathsf{D},i}\mathbf{B}^{\mathrm{H}} - \mathbf{D}_i)^H(\mathbf{B}\mathbf{M}_{\mathsf{D},i}\mathbf{B}^{\mathrm{H}} - \mathbf{D}_i)\} \\ &= \sum_{i=1}^{N_m} \text{trace}\{\mathbf{B}\mathbf{M}_{\mathsf{D},i}^{\mathrm{H}}\mathbf{B}^{\mathrm{H}}\mathbf{B}\mathbf{M}_{\mathsf{D},i}\mathbf{B}^{\mathrm{H}} - \mathbf{D}_i^{\mathrm{H}}\mathbf{B}\mathbf{M}_{\mathsf{D},i}\mathbf{B}^{\mathrm{H}} \\ &\quad - \mathbf{B}\mathbf{M}_{\mathsf{D},i}^{\mathrm{H}}\mathbf{B}^{\mathrm{H}}\mathbf{D}_i + \mathbf{D}_i^{\mathrm{H}}\mathbf{D}_i\}. \end{aligned} \tag{11.71}$$

$$\begin{aligned}\frac{\partial}{\partial \mathbf{D}_i}\mathscr{C}'_{\mathsf{D}}(\mathbf{B},\{\mathbf{D}_i\}) &= -2(\mathbf{B}\mathbf{M}_{\mathsf{D},i}\mathbf{B}^{\mathrm{H}} - \mathbf{D}_i) = 0 \\ &\Rightarrow \widehat{\mathbf{D}}_i = \mathsf{Diag}\{\mathbf{B}\mathbf{M}_{\mathsf{D},i}\mathbf{B}^{\mathrm{H}}\} \quad \forall i = 1,\dots,K_{\mathsf{JD}}, \end{aligned} \tag{11.72}$$

since the matrices $\mathbf{D}_i$ for all $i = 1,\dots,K_{\mathsf{JD}}$ are diagonal. $\mathsf{Diag}\{\mathbf{M}\}$ is a matrix whose elements $(\mathsf{Diag}\{\mathbf{M}\})_{ij} = \delta_{ij}M_{ij}$ where $\delta_{ij} = 1$ if $i = j$ and 0 otherwise, and $\mathsf{Offdiag}\{\mathbf{M}\} = \mathbf{M} - \mathsf{Diag}\{\mathbf{M}\}$ is a matrix with $(\mathsf{Offdiag}\{\mathbf{M}\})_{ij} = (1-\delta_{ij})M_{ij}$. It finally leads to a simple solution:

$$\mathscr{C}'_{\mathsf{D}}(\mathbf{B},\{\widehat{\mathbf{D}_i}\}) = \sum_{i=1}^{K_{\mathsf{JD}}} ||\mathsf{Offdiag}\{\mathbf{B}\mathbf{M}_{\mathsf{D},i}\mathbf{B}^{\mathrm{H}}\}||^2 \stackrel{\text{def}}{=} \mathscr{C}_{\mathsf{D}}(\mathbf{B}). \tag{11.73}$$

In the same way, one can show that

$$\widehat{\mathbf{Z}}_i = \mathsf{Offdiag}\{\mathbf{B}\mathbf{M}_{\mathsf{D},i}\mathbf{B}^{\mathrm{H}}\} \quad \forall i = 1,\dots,K_{\mathsf{JZD}}, \tag{11.74}$$

since matrices $\mathbf{Z}_i$ for all $i = 1,\dots,K_{\mathsf{JZD}}$ are zero-diagonal. It finally leads to a simple solution:

$$\mathscr{C}'_{\mathsf{Z}}(\mathbf{B},\{\widehat{\mathbf{Z}_i}\}) = \sum_{i=1}^{K_{\mathsf{JZD}}} ||\mathsf{Diag}\{\mathbf{B}\mathbf{M}_{\mathsf{Z},i}\mathbf{B}^{\mathrm{H}}\}||^2 \stackrel{\text{def}}{=} \mathscr{C}_{\mathsf{Z}}(\mathbf{B}). \tag{11.75}$$

If $\mathbf{A}_{\mathsf{D}}$ (resp. $\mathbf{A}_{\mathsf{Z}}$) is a unitary matrix $\mathscr{I}_{\mathsf{D}} = \mathscr{C}'_{\mathsf{D}}$ (resp. $\mathscr{I}_{\mathsf{Z}} = \mathscr{C}'_{\mathsf{Z}}$) since the Euclidian (Frobenius) norm is invariant by multiplication by a unitary matrix of the matrix given in argument. The studied cost functions are generally different in the other cases.

11.5.2 Whitened-based separation algorithms

Joint diagonalization under unitary constraint. Consider matrix $\mathbf{A}_{\mathsf{D}}$ in Eq. (11.61) to be a unitary $N \times N$ matrix (orthogonal if it has real values). In [12], the cost function defined in Eq. (11.63) is rewritten as:

$$\begin{aligned}\mathscr{I}_{\mathsf{D}}(\mathbf{A},\{\mathbf{D}_i\}) &= \sum_{i=1}^{K_{\mathsf{JD}}} ||\mathbf{M}_{\mathsf{D},i}||^2 - \sum_{i=1}^{K_{\mathsf{JD}}} ||\mathsf{Diag}\{\mathbf{A}^{\mathrm{H}}\mathbf{M}_{\mathsf{D},i}\mathbf{A}\}||^2 \\ &= \sum_{i=1}^{K_{\mathsf{JD}}} ||\mathsf{Offdiag}\{\mathbf{A}^{\mathrm{H}}\mathbf{M}_{\mathsf{D},i}\mathbf{A}\}||^2 \stackrel{\text{def}}{=} \mathscr{I}'_{\mathsf{D}}(\mathbf{A}) \end{aligned} \tag{11.76}$$

under the constraint $\mathbf{A}\mathbf{A}^{\mathrm{H}} = \mathbf{I}_N$. The minimization of the cost function $\mathcal{I}'_{\mathrm{D}}(\mathbf{A})$ amounts to the maximization of its opposite $-\mathcal{I}'_{\mathrm{D}}(\mathbf{A})$ and the unitary constraint on $\mathbf{A}$ makes it possible to avoid the trivial solution $\mathbf{A} = \mathbf{0}$.

The matrix $\widehat{\mathbf{A}}$ that joint diagonalizes the matrix set $\mathcal{M}_{\mathrm{JD}}$ is defined as the argument of a maximum of the criteriom $-\mathcal{I}'_{\mathrm{D}}(\mathbf{A})$ over the set of the unitary matrices $\mathcal{U}$:

$$\widehat{\mathbf{A}} = \arg\max_{\mathbf{A}\in\mathcal{U}}(-\mathcal{I}'_{\mathrm{D}}(\mathbf{A})). \tag{11.77}$$

The JD under the unitary constraint consists of determining a unitary matrix that maximizes the sum of the squared diagonal elements of all the matrices of the set $\mathcal{M}_{\mathrm{JD}}$ after linear transform (it amounts to minimizing the sum of the squared off-diagonal terms, which constitutes a natural measure of the deviation with regard to a diagonal matrix). Such a matrix is called a *joint diagonalizer*. The method suggested in [13] to optimize this criteriom is used in the JADE algorithm [12]. It can be viewed as a generalization of the Jacobi method used for the diagonalization of one single matrix. This solution is an iterative algorithm based on a parameterization of the matrix $\mathbf{A}$ in the form of a product of plane or Givens rotations. In 11.6, the method based on this algorithm is denoted by $\mathrm{JD_B}$.

Joint zero-diagonalization under unitary constraint. Consider matrix $\mathbf{A}_{\mathrm{Z}}$ in Eq. (11.62) to be a unitary $N \times N$ matrix (orthogonal if it has real values). In [5], the cost function given in Eq. (11.64) is rewritten as:

$$\mathcal{I}_{\mathrm{Z}}(\mathbf{A}, \{\mathbf{Z}_i\}) = \sum_{i=1}^{K_{\mathrm{JZD}}} \|\mathrm{Diag}\{\mathbf{A}^{\mathrm{H}}\mathbf{M}_{\mathrm{Z},i}\mathbf{A}\}\|^2 \stackrel{\mathrm{def}}{=} \mathcal{I}'_{\mathrm{Z}}(\mathbf{A}). \tag{11.78}$$

under the same constraint $\mathbf{A}\mathbf{A}^{\mathrm{H}} = \mathbf{I}_N$. It is noted that the minization of the cost function $\mathcal{I}'_{\mathrm{Z}}(\mathbf{A})$ amounts to the maximization of its opposite $-\mathcal{I}'_{\mathrm{Z}}(\mathbf{A})$. A matrix $\widehat{\mathbf{A}}$ that joint zero-diagonalizes the set $\mathcal{M}_{\mathrm{JZD}}$ is defined as an argument of a maximum of the criteriom $-\mathcal{I}'_{\mathrm{Z}}(\mathbf{A})$ over the set of unitary matrices $\mathcal{U}$:

$$\widehat{\mathbf{A}} = \arg\max_{\mathbf{A}\in\mathcal{U}}(-\mathcal{I}'_{\mathrm{Z}}(\mathbf{A})). \tag{11.79}$$

The unitary matrix $\widehat{\mathbf{A}}$ is referred to as a *joint zero-diagonalizer*. The algorithm used to optimize this criteriom is the same as the one used in the case of the joint diagonalization under unitary constraint. In 11.6, the method based on this algorithm is denoted by $\mathrm{JZD_B}$.

Joint diagonalization and zero-diagonalization under unitary constraint. When $\mathbf{A}_{\mathrm{D}} = \mathbf{A}_{\mathrm{Z}} = \mathbf{A}_{\mathrm{DZ}}$, the two matrices sets $\mathcal{M}_{\mathrm{JD}}$ and $\mathcal{M}_{\mathrm{JZD}}$ can be simultaneously considered and the two cost functions can be combined. The aim is then to minimize the following criteriom $\mathscr{I}'_{\mathsf{DZ}}(\mathbf{A})$ (or to maximize its opposite):

$$\mathscr{I}_{\mathsf{DZ}}(\mathbf{A}, \{\mathbf{D}_i\}, \{\mathbf{Z}_i\}) = -\sum_{i=1}^{K_{\mathsf{JD}}} ||\mathsf{Diag}\{\mathbf{A}^{\mathrm{H}}\mathbf{M}_{\mathsf{D},i}\mathbf{A}\}||^2 + \sum_{i=1}^{K_{\mathsf{JZD}}} ||\mathsf{Diag}\{\mathbf{A}^{\mathrm{H}}\mathbf{M}_{\mathsf{Z},i}\mathbf{A}\}||^2 \stackrel{\mathrm{def}}{=} \mathscr{I}'_{\mathsf{DZ}}(\mathbf{A}) \tag{11.80}$$

under the same constraint $\mathbf{A}\mathbf{A}^{\mathrm{H}} = \mathbf{I}_N$. A matrix $\widehat{\mathbf{A}}$ that both joint diagonalizes the matrices set $\mathscr{M}_{\mathsf{JD}}$ and joint zero-diagonalizes the matrices set $\mathscr{M}_{\mathsf{JZD}}$ is defined as an argument of a maximum of the criteriom $-\mathscr{I}'_{\mathsf{DZ}}(\mathbf{A})$ over the set of all unitary matrices $\mathscr{U}$:

$$\widehat{\mathbf{A}} = \arg\max_{\mathbf{A}\in\mathscr{U}}(-\mathscr{I}'_{\mathsf{DZ}}(\mathbf{A})). \tag{11.81}$$

In this case, the unitary matrix $\widehat{\mathbf{A}}$ is called a *joint diagonalizer/zero-diagonalizer*. The algorithm used to optimize this criteriom remains the same as the one used for the joint diagonalization under unitary constraint. In 11.6, the method based on this algorithm is denoted by $\mathsf{JD}/\mathsf{JZD}_{\mathsf{B}}$ [6].

The main advantage of joint (zero-) diagonalization is that the matrices of the considered set neither have to be simultaneously exactly (zero-) diagonalizable, nor individually (zero-) diagonalizable. The optimization ensures the minimization of the sum of the elements that are expected to be of zero values. As such, the minimization makes it possible to obtain “approximatively (zero-) diagonal” matrices and is robust to estimation errors.

A summary of the different methods based on a preliminary whitening stage

1. Estimation of the whitening matrix and of the whitened observations vector.
2. Estimation of the whitened observations SQTFD (or S) for a given set of t-f points.
3. Determination of the matrix sets corresponding to useful t-f points. Those t-f points can be determined according to one of the automatic selection procedure described in 11.3.
4. Estimation of the unknown unitary matrix by unitary joint diagonalization and/or unitary joint zero-diagonalization.
5. Estimation of the sources vector from the estimated mixing matrix.

11.5.3 Non-whitened based separation algorithms

The spatial whitening makes it possible to reformulate the BSS problem into a problem of estimating of a unitary matrix (orthogonal in the real case). It is based on the assertion that the sources are uncorrelated. When such a hypothesis is not checked, a preliminary whitening stage is no longer applicable. It is because of this reason, matrix decomposition algorithms that are not based on the unitary constraint are of interest.

Joint diagonalization without unitary constraint (orthogonality for real values matrices). In [29], it was determined to use the cost function $\mathscr{C}_{\mathsf{D}}(\mathbf{B})$ given in Eq. (11.73) which depends on $\mathbf{B} \in \mathbb{C}^{N\times P}$, instead of $\mathbf{A}$. In this case, the joint diagonalizer $\widehat{\mathbf{B}}$ does not provide an estimate of the mixing matrix, but rather an estimate of the separation matrix (up to a scale factor and up to a permutation onto the columns). Two optimization methods have been suggested: the first one is based on an algebraic optimization scheme [29] (a good initialization is essential with regard to the convergence of the algorithm). The cost function $\mathscr{C}_{\mathsf{D}}(\mathbf{B})$ is rewritten as:

$$\mathscr{C}_{\mathsf{D}}(\mathbf{B}) = \sum_{\ell=1}^{N} \mathbf{b}_\ell \mathbf{R}_{\mathsf{D}}(\mathbf{B}_{\bar{\ell}})\mathbf{b}_k^{\mathrm{H}} \tag{11.82}$$

where $\mathbf{R}_{\mathsf{D}}(\mathbf{B}_{\bar{\ell}})$ is defined as

$$\mathbf{R}_{\mathsf{D}}(\mathbf{B}_{\bar{\ell}}) = \sum_{i=1}^{K_{\mathsf{JD}}} \mathbf{M}_{\mathsf{D},i} \left(\sum_{\substack{k=1 \\ k\neq \ell}}^{N} \mathbf{b}_k^{\mathrm{H}} \mathbf{b}_k \right) \mathbf{M}_{\mathsf{D},i}^{\mathrm{H}}. \tag{11.83}$$

The minimization of $\mathscr{C}_{\mathsf{D}}(\mathbf{B})$ is achieved line by line. For a given line, i.e. ℓ is fixed, one simple way to find a minimum of the quadratic form given by the Eq. (11.73) is to search for a unit-norm vector $\mathbf{R}_{\mathsf{D}}(\mathbf{B}_{\bar{\ell}})$ associated with the smallest (non-null) eigenvalue. However, since $\mathbf{R}_{\mathsf{D}}(\mathbf{B}_{\bar{\ell}})$ for a given ℓ depends on $\mathbf{b}_\ell$ and since the optimization is performed line by line, it finally leads to the following iterative algorithm:

> Given $\mathbf{B}^{(0)}$ a (good) initial matrix whose lines are normalized, for all $i \in \mathbb{N}_*$, for all $\ell \in \{1,\ldots,N\}$, do (1) and (2)
>
> **(1)** Calculate $\mathbf{R}_\ell(\mathbf{B}^{(i-1)})$.
>
> **(2)** Find the Nth smallest eigenvalue $\lambda_\ell^{(i)}$ and the associated unit-norm eigenvector $\mathbf{b}_\ell^{(i)}$ of the matrix $\mathbf{R}_\ell(\mathbf{B}^{(i-1)})$.
>
> Stop when $|\lambda_\ell^{(i)} - \lambda_\ell^{(i-1)}| \leq \varepsilon$ where ε is a small enough threshold value.

The existence of a solution that minimizes $\mathscr{C}_{\mathsf{D}}(\mathbf{B})$ is guaranteed because the criterion is a continuous function of $\mathbb{C}^{N\times M} \mapsto \mathbb{R}$ and it is bounded by zero. The non-increasing nature of the criterion is also locally ensured at each iteration (i.e. $\mathscr{C}_{\mathsf{D}}(\mathbf{B}^{(n)}) \leq \mathscr{C}_{\mathsf{D}}(\mathbf{B}^{(n-1)})$. However, this is not sufficient (as well-known) to ensure its global convergence to any global minima. That is why the choice of the initial point is key to acceptable performance. One possible initialization is to consider the solution given by the orthogonal joint diagonalization [5] and to start in the neighborhood of the solution. Another way is to consider the solution given by the generalized eigendecomposition of two matrices coming from the set $\mathscr{D}$ (or built from matrices of this set). In 11.6, the method based on that algorithm is denoted by $\mathsf{JD}_{\mathsf{NO,ALG}}$. It is worth mentioning that the JD_A algorithm is close to the one proposed in [24]. However, the matrix set under consideration, in this case, is not "whitened".

Another approach to optimize the considered cost function is proposed in [25]. It is based on the use of a Newton algorithm (Levenberg-Marquardt in the rectangular case, because of eventual problems of inversion of the Hessian). In this algorithm the columns of the separation matrix are parameterized like in [56]. It is rather computationally demanding since both the Gradient and the Hessian of the cost function have to be calculated. In 11.6, the method based on this algorithm will be denoted by $\text{JD}_{\text{NO,PAR}}$.

Many other non-unitary joint diagonalization algorithms have been suggested [56,24, 60,64,62,61]. Except for the algorithm suggested in [48] which is based on a criterion using the Kullback-Leibler divergence, they are all based on one of the two least squares criteria given in Eq. (11.63) or in Eq. (11.68), weighted or not (for weighted criterion, Eq. (11.65) then replaces Eq. (11.63)). The main issue is then to optimize the used criterion.

In [64,62], two other solutions are proposed to optimize $\mathscr{C}_\text{D}(\mathbf{B})$ given in Eq. (11.73). In their case, the optimization is subject to the fact that $\mathbf{B}$ can be written as:

$$\mathbf{B} = \left[\prod_{k=1}^{M}(\mathbf{I} + \mathbf{W}^{(k)})\right]\mathbf{B}^{(0)}, \tag{11.84}$$

where $\mathbf{B}^{(0)}$ is a non-singular initial matrix, M is the iteration number and the $\mathbf{W}^{(k)}$ matrices are update matrices, with small norm, required to possess null diagonal elements. Such a constraint ensures the invertibility of the joint diagonalizer $\mathbf{B}$ and avoids the trivial solution $\mathbf{B} = 0$. Authors of [64] use a Quasi-Newton algorithm and the resulting algorithm is called FFDIAG. The main inconvenience associated with this algorithm is that nothing guarantees the fact that $\mathbf{B}$ is effectively a minimum (even local) of $\mathscr{C}_\text{D}(\mathbf{B})$. In [62], the authors use the same cost function subject to the same constraint, but they now consider an optimal step size natural gradient algorithm [2], which enables them to avoid the previous problem. The resulting algorithm is called DOMUNG (Diagonalization of Matrices Using Natural Gradient).

In [56,60], two alternative algorithms can be found to optimize the cost function $\mathscr{I}_\text{D}(\mathbf{A}, \{\mathbf{D}_i\})$ given in Eq. (11.63) or $\mathscr{I}_{\text{D,YER}}(\mathbf{A}, \{\mathbf{D}_i\})$ given in Eq. (11.65) when the weighted version is considered. The algorithm described in [60] enables the estimation of a joint diagonalizer $\widehat{\mathbf{A}} \in \mathbb{C}^{P\times N}$ by minimizing the cost function given in Eq. (11.65), with respect to $\mathbf{A}$ and versus the set of diagonal matrices $\mathscr{D}$ i.e. $\{\mathbf{D}_i\}$:

$$\widehat{\mathbf{A}} = \arg\min_{\mathbf{A},\{\mathbf{D}_i\}} (\mathscr{I}_{\text{D,YER}}(\mathbf{A}, \{\mathbf{D}_i\})). \tag{11.85}$$

This criteriom is optimized using an iterative algorithm that alternates two minimization stages. The first stage consists of minimizing the criteriom with respect to one of the columns of the matrix $\mathbf{A}$ (the other columns and the diagonal matrices are fixed); the second stage consists of minimizing with respect to the diagonal matrices, $\mathbf{A}$ being fixed. This algorithm is called AC−DC (Alternating Columns-Diagonal Centers) in [60]. In [61], the author suggests a non-iterative solution based on the exact diagonalization of two matrices which can eventually be used as initial value in the AC − DC algorithm. The methods developed in [37,35] are based upon this algorithm which is denoted by $\text{JD}_{\text{NO,YER}}$, in 11.6.

When $w_1 = w_2 = \cdots = w_{K_{\text{JD}}} = 1$, one obtains the cost function $\mathcal{J}_{\text{D}}(\mathbf{A},\{\mathbf{D}_i\})$ given in Eq. (11.63) and which is used in [56]. The joint diagonalization problem is reformulated in terms of a sub-spaces adjustment problem. By imposing a supplementary normalization constraint on the columns of the matrix $\mathbf{A}$ ($||\mathbf{a}_k|| = 1$), the author shows that the cost function can be rewritten as:

$$\mathcal{J}_{\text{D,VAN}}(\mathbf{A},\{\mathbf{D}_i\}) = ||\mathbf{P}_{\underline{\mathbf{A}}}^{\perp}\mathbf{M}||^2 \tag{11.86}$$

where $\mathbf{P}_{\underline{\mathbf{A}}}^{\perp} = \mathbf{I} - \underline{\mathbf{A}}\underline{\mathbf{A}}^{\dagger}$, with $\underline{\mathbf{A}} = \mathbf{A}^* \circ \mathbf{A}$, $\circ$ stands for the Khatri-Rao product and $\mathbf{M} = (\mathbf{m}_1,\ldots,\mathbf{m}_{K_{\text{JD}}})$ with $\mathbf{m}_i = \text{vec}(\mathbf{M}_i)$ and $\text{vec}(\cdot)$ is the operator which concatenates the columns of a matrix into a vector. This cost function is then optimized based on a Gauss-Newton algorithm (which may lead to problems, among which the fact that the convergence of the algorithm depends on a good initial value). To that aim, the author uses the same parameterization of the columns of the matrix $\mathbf{A}$ as the one suggested in [52] (a unit-norm vector can always be written as a product of Givens rotations).

Finally in [48], a different cost function is used:

$$\begin{aligned}
\mathcal{C}'_{\text{PHA,D}}(\mathbf{B},\{\mathbf{D}_i\}) &= \sum_{i=1}^{K_{\text{JD}}} ||\text{Offdiag}_2\{\mathbf{B}\mathbf{M}_{\text{D},i}\mathbf{B}^{\text{H}}\}||^2 \\
&= \sum_{i=1}^{K_{\text{JD}}} ||\log\{\det\{\text{Diag}\{\mathbf{B}\mathbf{M}_{\text{D},i}\mathbf{B}^{\text{H}}\}\}\} - \log\{\det\{\mathbf{B}\mathbf{M}_{\text{D},i}\mathbf{B}^{\text{H}}\}\}||^2 \\
&\stackrel{\text{def}}{=} \mathcal{C}_{\text{PHA,D}}(\mathbf{B})
\end{aligned} \tag{11.87}$$

where $\text{Offdiag}_2\{\mathbf{M}\}$ measures the Kullback-Leibler divergence between the square matrix $\mathbf{M}$ and the diagonal matrix with the same diagonal terms as the matrix $\mathbf{M}$. The method presented in [9] relies upon the algorithm developed in [48,50], which implies that the matrices of the set $\mathcal{M}_{\text{JD}}$ are definite positive matrices. In [49], such a constraint is partially avoided by means of conditions about the matrix $\mathbf{B}$.

Joint zero-diagonalization without unitary constraint (orthogonality in the real case). The algorithm suggested in [29], enables the estimation of a joint zero-diagonalizer by minimizing the following cost function $\mathcal{C}_{\text{Z}}(\mathbf{B})$ given in Eq. (11.75) written versus the matrix $\mathbf{B}$. The joint zero-diagonalizer provides an estimate of the separation matrix. Three methods have been suggested to optimize this cost function. The first one is based on an algebraic optimization scheme [29], where the cost function is rewritten as:

$$\mathcal{C}_{\text{Z}}(\mathbf{B}) = \sum_{\ell=1}^{N} \mathbf{b}_\ell \mathbf{R}_{\text{Z}}(\mathbf{b}_\ell)\mathbf{b}_k^{\text{H}} \tag{11.88}$$

where $\mathbf{R}_Z(\mathbf{b}_\ell)$ is defined as:

$$\mathbf{R}_Z(\mathbf{b}_\ell) = \sum_{i=1}^{N_Z} \mathbf{M}_{Z,i}(\mathbf{b}_\ell^H \mathbf{b}_\ell)\mathbf{M}_{Z,i}^H. \tag{11.89}$$

The optimization is performed line by line (like in the joint-diagonalization algorithm). In section 11.6, the method based on this algorithm is denoted by $\text{JZD}_{\text{NO,ALG}}$. The second one [25,26] uses a Newton algorithm (Levenberg-Marquardt in the rectangular case) with a parameterization of the columns of the matrix $\mathbf{B}$. In section 11.6, the method based on this algorithm is denoted by $\text{JZD}_{\text{NO,PAR}}$. Finally a third non-iterative solution has been suggested in [14] (eventually, the solution can be used as initial value in the two previous iterative algorithms too). In section 11.6, the method based on this algorithm is denoted by $\text{JZD}_{\text{NO,CHA}}$.

Joint diagonalization and zero-diagonalization without unitary constraint. To write a more general cost function, a combined cost function can be used [27,29]:

$$\mathscr{C}_{DZ}(\mathbf{B}, \{\widehat{\mathbf{D}_i}\}, \{\widehat{\mathbf{Z}_i}\}) = \alpha\mathscr{C}_D(\mathbf{B}) + (1-\alpha)\mathscr{C}_Z(\mathbf{B}) \stackrel{\text{def}}{=} \mathscr{C}_{DZ}(\mathbf{B}), \qquad \alpha \in [0,1]. \tag{11.90}$$

It takes into account both joint diagonalization and joint zero-diagonalization aspects. A matrix $\widehat{\mathbf{B}}$ that joint diagonalizes the set $\mathscr{M}_{JD}$ and joint zero-diagonalizes the set $\mathscr{M}_{JZD}$ is defined as the argument of a minimum of the criteriom $\mathscr{C}_{DZ}(\mathbf{B}, \{\mathbf{D}_i\}, \{\mathbf{Z}_i\})$ with respect to the matrices set $\mathbb{C}^{N\times P}$:

$$\widehat{\mathbf{B}} = \arg \min_{\mathbf{B}\in\mathbb{C}^{N\times P}} \mathscr{C}_{DZ}(\mathbf{B}). \tag{11.91}$$

The matrix $\widehat{\mathbf{B}}$ is then called a *joint diagonalizer/zero-diagonalizer*. Two methods have been suggested to optimize this cost function: the first one is based on an algebraic optimization scheme [29]. In section 11.6, the method based on this algorithm is denoted by $\text{JD/JZD}_{\text{NO,ALG}}$. The second one [25] uses a Newton algorithm (Levenberg-Marquardt in the rectangular case) with a parameterization of the columns of the separation matrix. In section 11.6, the method based on this algorithm is denoted by $\text{JD/JZD}_{\text{NO,PAR}}$.

11.5.4 Algebraic methods and classification

In the previous section, it has been noticed that a "good" sources auto-terms detector is a detector of t-f points corresponding to one single source.

We assume that $N_p = K_1 + K_2 + \cdots + K_N$, $(K_1, \ldots, K_N) \in (\mathbb{N}^*)^N$ t-f points denoted by $(t_{\ell,m_\ell}, \nu_{\ell,m_\ell})$ have been detected such that for all $(i,j) \in (1,\ldots,N)^2$, one has:

$$\forall \ell \in (1,\ldots,N), \qquad \forall m_\ell \in (1,\ldots,K_\ell), \\ D_{s_i s_j}(t_{\ell,m_\ell}, \nu_{\ell,m_\ell}) = \delta_{i,j,\ell} D_{s_i}(m_i) \tag{11.92}$$

with $D_{s_i}(m_i) \neq 0$ for all $i \in (1,\ldots,N)$ and $m_i \in (1,\ldots,K_i)$ and $\delta_{i,j,\ell} = 1$ if $i = j = \ell$ and 0 otherwise.

The index l refers to one of the N sources, K_l corresponds to the number of t-f points associated to the l-th source and the m_l index enables us to focus on one of these K_l t-f points. Two different methods can be considered to directly estimate the columns of the mixing matrix **A**. We will assume that the mixing matrix and the sources are both real and that the SQTFD (or S) have a Hermitian symmetry.

The first method is termed the **direct method**. It is based on the fact that all the columns of the matrix $\mathbf{D}_{\mathbf{x}}(t_{i,m_i}, \nu_{i,m_i})$ for a given $i \in (1,\ldots,N)$ are collinear. Thus, by denoting by $(\mathbf{d}_{\mathbf{x}}(t_{i,m_i}, \nu_{i,m_i}))_k$ the k-th column of the matrix $\mathbf{D}_{\mathbf{x}}(t_{i,m_i}, \nu_{i,m_i})$, one can easily show that for all $i \in (1,\ldots,N)$

$$(\mathbf{d}_{\mathbf{x}}(t_{i,m_i}, \nu_{i,m_i}))_k = D_{s_i}(m_i) A_{ki} \mathbf{a}_i, \qquad \forall m_i \in (1,\ldots,K_i). \tag{11.93}$$

As a consequence, a non-null column of the matrix $\mathbf{D}_{\mathbf{x}}(t_{i,m_i}, \nu_{i,m_i})$ can be used as an estimate (up to a multiplicative coefficient which is not a problem because of the inherent indetermination of the BSS problem) of the i-th column of **A**.

Rather than applying that operation on each column of $\mathbf{D}_{\mathbf{x}}(t_{i,m_i}, \nu_{i,m_i})$, $i \in (1,\ldots,N)$, a single one can be used: the one with the higher Euclidian norm. The advantage is to avoid problems linked to eventual null coefficients of the mixing matrix. In 11.6, this method will be denoted by A_D.

The second approach is qualified as the SVD **based method**. Since the SQTFD (or S) have a Hermitian symmetry, the matrices $\mathbf{D}_{\mathbf{s}}(t_{i,m_i}, \nu_{i,m_i})$ are real, $\forall i \in (1,\ldots,N)$ and $\forall m_i \in (1,\ldots,K_i)$. Since the matrix **A** is real too, all the matrices $\mathbf{D}_{\mathbf{x}}(t_{i,m_i}, \nu_{i,m_i})$ are real and symmetric. Since all the matrices $\mathbf{D}_{\mathbf{x}}(t_{i,m_i}, \nu_{i,m_i})$, $i \in (1,\ldots,N)$, $m_i \in (1,\ldots,K_i)$, are rank one and symmetric, they can be decomposed into:

$$\begin{aligned}\mathbf{D}_{\mathbf{x}}(t_{i,m_i}, \nu_{i,m_i}) &= \mathbf{V}(i,m_i)\mathbf{\Lambda}(i,m_i)\mathbf{V}^{\mathrm{T}}(i,m_i) \\ &= \lambda(i,m_i)\mathbf{v}_{\lambda}(i,m_i)\mathbf{v}_{\lambda}^{\mathrm{T}}(i,m_i)\end{aligned} \tag{11.94}$$

where $\mathbf{V}(i,m_i)$ and $\mathbf{\Lambda}(i,m_i)$ are respectively orthogonal and diagonal matrices; $\lambda(i,m_i)$ is the only real non-null singular value and $\mathbf{v}_{\lambda}(i,m_i)$ is the singular vector related to that singular value.

For a given value of $i \in (1,\ldots,N)$, the vectors $\mathbf{v}_{\lambda}(i,m_i)$, $m_i \in (1,\ldots,K_i)$ are collinear to the i-th column of **A**.

One of the advantages of this second method stems from the fact that the matrices $\mathbf{D}_{\mathbf{x}}(t_{i,m_i}, \nu_{i,m_i})$, $i \in (1,\ldots,N)$, $m_i \in (1,\ldots,K_i)$ being estimated, are likely not to be rank one. The use of the SVD to estimate the singular vector related to the highest singular value, corresponds to the best estimate in terms of least mean squares of the one obtained in the case of a matrix which is a rank-one matrix. In section 11.6, this method will be denoted by A_{SVD}.

From a practical point of view, two problems subsist. The first one comes from the fact that if we know for sure that a given t-f point is assigned to a source signal, we do not know which one it is. More importantly we do not know if we were able to collect t-f points corresponding to all N source signals. The second one comes from the fact that, since matrices $\mathbf{D}_s(t_{i,m_i}, \nu_{i,m_i})$, $i \in (1, \ldots, N)$, $m_i \in (1, \ldots, K_i)$ are "only" estimated ones, the different vectors estimating one column of $\mathbf{A}$ are very likely noncollinear. Thus, there is a certain dispersion around their theoretical value.

To solve the two aforementioned points, the following solutions were suggested. A way to improve the estimation stage of the columns of the mixing matrix is to compute the inertia centers of each clusters of points by merging to the nearest points in order to finally obtain one single point (*cf.* Fig. 11.8). This automatic hierarchical ascendent classification method is known under the name of unweighed pair-group method of aggregation using arithmetic averages, UPGMA [53]. Basically, weights are assigned to each points representing the number of already amalgamated points (weights being initialized to 1). The Euclidian distance between two points is then measured. If this distance is lower than a given threshold, say δ, the weighted average of these two points replaces them. The new points' weight corresponds to the sum of the weights of the old ones. When a point does not have neighbors any more it is considered as an inertia center. These two methods have been presented in [34]. We note that in [42–44], one can find a very similar idea as the one developed in the SVD based method. However, different assumptions are made about the sources (they have to be disjoint) and a rather simple classification is used since it is based on a calculus of the angles between the vectors.

Considering the first point, among all found inertia centers, only the N ones corresponding to the strongest weights are selected (see the right of the Fig. 11.8). These N associated vectors enable the estimation of the N columns of the mixing matrix, up to their order. This does not have any importance with regard to our estimation problem for which a column order indetermination remains inherent. We can finally remark that such an approach could also make it possible to estimate the number of sources if this one would not have been supposed known.

Summary of the different methods, without pre-whitening

1. Estimation of the observations SQTFD (or S) for a given set of t-f points.
2. Selection of useful t-f points using an automatic time-frequency points detection procedure described in section 11.3.
3. Determination of the corresponding matrix set(s).
4. For algebraic methods: using t-f points corresponding to one single auto-term, to calculate the estimated column vectors of the mixing matrix (direct or SVD based method).
5. For non-unitary joint decomposition based methods: estimation of the mixing matrix or of its pseudo-inverse (the separation matrix) using a non-unitary joint (zero-) diagonalization algorithm.
6. Recovery of the sources by means of the estimated mixing matrix.

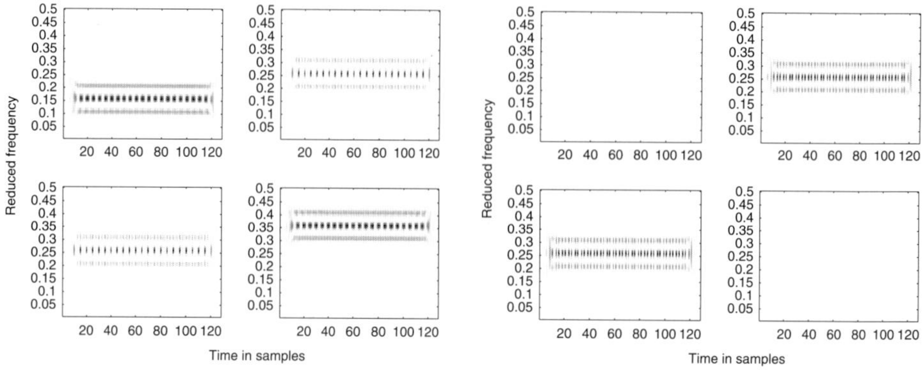

FIGURE 11.1

Sources spatial PWV distribution (set J_1 of signals), left: real part, right: imaginary part.

11.6 PRACTICAL AND COMPUTER SIMULATIONS

A comparaison between the different algorithms is performed using computer simulations. In this respect, we use the following performance index [46]:

$$I(\widehat{\mathbf{A}}^{\dagger}\mathbf{A}) = \frac{1}{N(N-1)}\left[\sum_{i=1}^{N}\left(\sum_{j=1}^{N}\frac{|(\widehat{\mathbf{A}}^{\dagger}\mathbf{A})_{ij}|^2}{\max_l |(\widehat{\mathbf{A}}^{\dagger}\mathbf{A})_{il}|^2} - 1\right) + \sum_{j=1}^{N}\left(\sum_{i=1}^{N}\frac{|(\widehat{\mathbf{A}}^{\dagger}\mathbf{A})_{ij}|^2}{\max_l |(\widehat{\mathbf{A}}^{\dagger}\mathbf{A})_{lj}|^2} - 1\right)\right] \tag{11.95}$$

where N is the dimension of the considered matrix $\widehat{\mathbf{A}}^{\dagger}\mathbf{A}$. This index is given in dB in the different tables: $I(\cdot)$ dB $= 10\log(I(\cdot))$. The best results are obtained when the index performance $I(\cdot)$ is found to be close to 0 in linear scale ($-\infty$ in logarithmic scale).

11.6.1 Synthetic source signals

Set J_1 of sources

In this first example, the two multi-components sources of the analytical example detailed in 11.3.6 are considered (each source is a sum of two cisoids (reduced frequencies 0.1 and 0.2 for the first one and 0.3 and 0.4 for the second one)). The length of the two sources is 128 time samples. The spatial Pseudo Wigner-Ville distribution (SPWV) is used. The real part (resp. the imaginary part) of the sources SPWV distribution is displayed on the left (resp. on the right) of Fig. 11.1. It is computed over 64 frequency bins and with a Hamming window of size 33. One can check that the diagonal terms of the SPWV distribution have a real value since they correspond to the QTFD of each of the 2 sources. With regard to the off-diagonal terms, they are generally complex: they correspond to the BTFD of the two distinct sources.

Set J_2 of sources

We consider $N = 3$ sources signals with a time duration of 256 samples. The first source is the same as the first source of set J_1, the second one is linear modulation of frequency and the third one is a sinusoidal modulation of frequency. The spatial pseudo Wigner-Ville SPWV distribution is used. Its real part (resp. its imaginary part) is displayed on the top (resp. on the bottom) of Fig. 11.2. It has been calculated over 128 frequency channels and with a Hamming window of length 65.

11.6.2 Mixture

We consider the three following mixing matrices:

$$\mathbf{A}_1 = \begin{pmatrix} 1 & 0.75 \\ -0.5 & 1 \end{pmatrix}, \quad \mathbf{A}_2 = \begin{pmatrix} 1 & 1 & 1 \\ 0 & -2 & 2 \\ -2 & 0 & -1 \end{pmatrix}, \quad \mathbf{A}_3 = \begin{pmatrix} 1 & 1 & 1 \\ 0 & -2 & 2 \\ -2 & 0 & -1 \\ 1 & 0.5 & 1 \end{pmatrix}.$$

The real part (resp. the imaginary part) of the SPWV distribution of the signals observed after the mixing of the sources belonging to set J_1 by the mixing matrix $\mathbf{A}_1$ is given on the left (resp. on the right) of Fig. 11.3.

The real part (resp. the imaginary part) of the SPWV distribution of the signals observed after the mixing of the sources belonging to set J_2 by the mixing matrix $\mathbf{A}_2$ is given on the top (resp. on the bottom) of Fig. 11.4, and that by the mixing matrix $\mathbf{A}_3$ is given on the top (resp. on the bottom) of Fig. 11.5.

11.6.3 Time-frequency points selection

The detected t-f points used by all the compared algorithms are represented by a "+" superimposed on the trace of the sources SQTFD for the joint diagonalization and with the sum of the off-diagonal terms of the imaginary part of the sources SQTFD for the joint zero-diagonalization (*cf.* the two colored charts on the top left of Fig. 11.6). Yet, sometimes, they are represented by a black point (*cf.* the two black and white charts on the top right of Fig. 11.6). For each set of curves, one can find the t-f points selected for the JD on the left and for the JZD on the right. To be able to compare the different t-f points selection procedures, we consider different scenarios.

The threshold values and the resulting number of selected matrices for the JD and/or the JZD are depicted in Table 11.5 for the set J_1 of sources and in Table 11.6 for the set J_2 of sources. The selected t-f points are displayed in Fig. 11.6 for the source set J_1 and in Fig. 11.7 for the source set J_2.

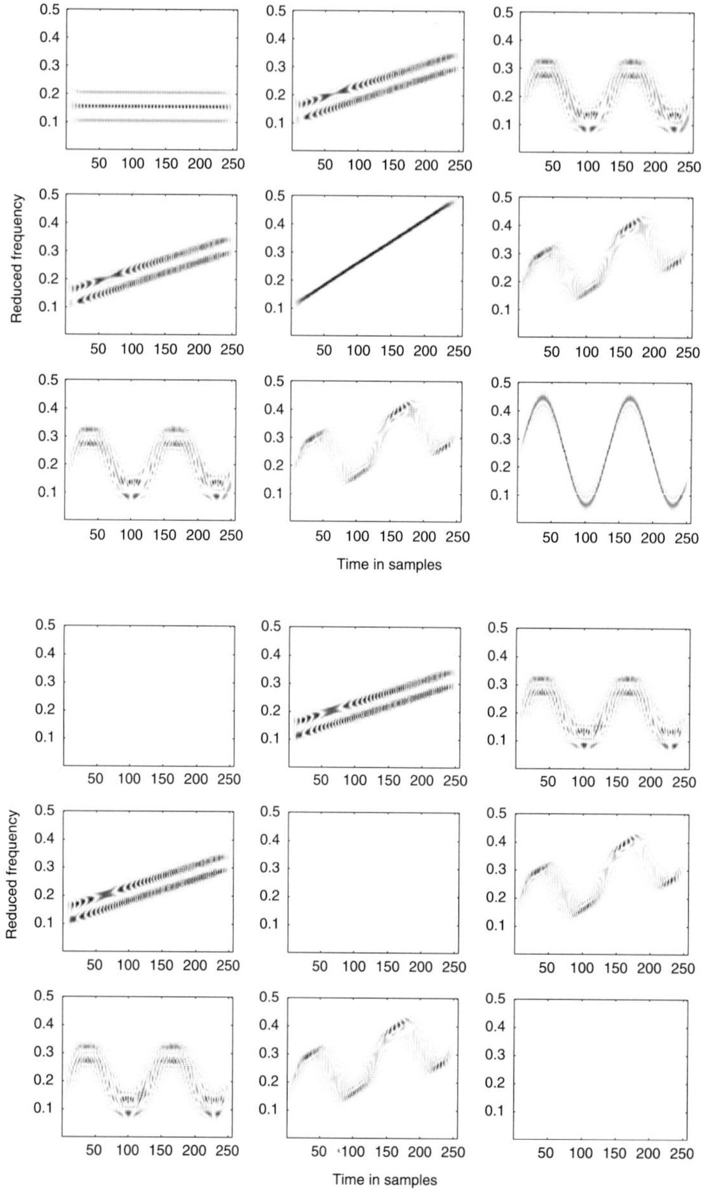

FIGURE 11.2

Sources spatial PWV distribution (set J_2 of signals), top: real part, bottom: imaginary part.

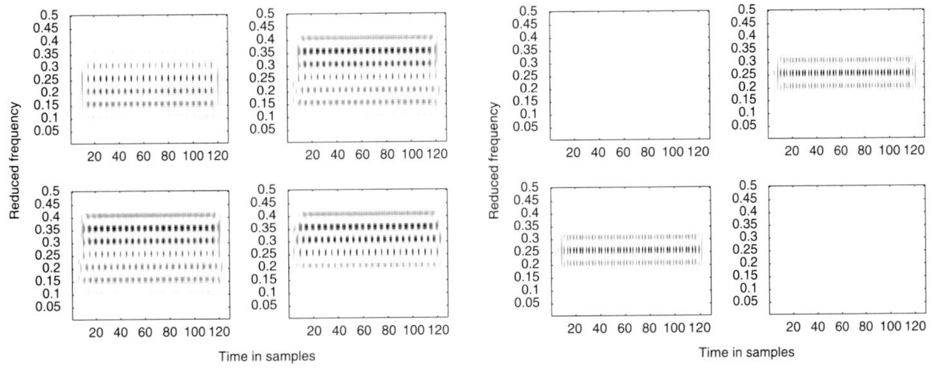

FIGURE 11.3

Spatial PWV distribution of the non-whitened observations (set J_1 of signals after mixing by matrix $\mathbf{A}_1$), left: real part, right: imaginary part.

In the first scenario, the t-f points are selected by hand. In the second scenario, they are obtained using detector C_G. In the third scenario, detector C_{GM} is used whereas in the 4th scenario, detector C_B is applied. Detector C_F is used in the 5th scenario and detector C_{GG} is used in the 6th. In the 7th scenario, detector C_S is used. In the 8th scenario, detector C_L is used. Finally in the 9th scenario, detector C_{LM} is applied.

11.6.4 Results

Using the aforementioned performance index, the results obtained with the different methods, for the source set J_1, are given in Table 11.7. The results for the source set J_2 are given in Table 11.8. In each table, we compare unitary joint-diagonalization (JD_B) and unitary joint zero-diagonalization (JZD_B) of the matrices selected after a pre-whitening stage. We also compare non-unitary joint diagonalization algorithms ($JD_{NO,YER}$, $JD_{NO,PAR}$ and $JD_{NO,ALG}$), and non-unitary joint zero-diagonalization ($JZD_{NO,ALG}$, $JZD_{NO,PAR}$ and $JZD_{NO,CHA}$) and purely algebraic methods combined with a classification method (A_D and A_{SVD}). All these algorithms operate without pre-whitening. With regard to this latter class of methods (*cf.* remarks in 11.5.4), the inertia centers of the clouds of points are evaluated using a threshold $\delta = 0.1$.

In the first example (set J_1 of sources), the best results are obtained with the two algebraic methods combined with a classification algorithm. In a general point of view, they are often better when no pre-whitening stage is applied except when wrong matrices have been selected in $\mathcal{M}_{JD}$ and/or $\mathcal{M}_{JZD}$ (*cf.* C_G, C_B for the JD). The techniques based on a pre-whitening are more robust to possible errors in the selection of t-f points used in the constructing of the matrices sets $\mathcal{M}_{JD}$ and/or $\mathcal{M}_{JZD}$. They give quite similar results, irrespective of the considered detector. They are bounded in terms of performance because of the spatial whitening (this theoretical bound is given in the last line of Tables 11.7 and 11.8). The same conclusions as for the sources set J_2 apply.

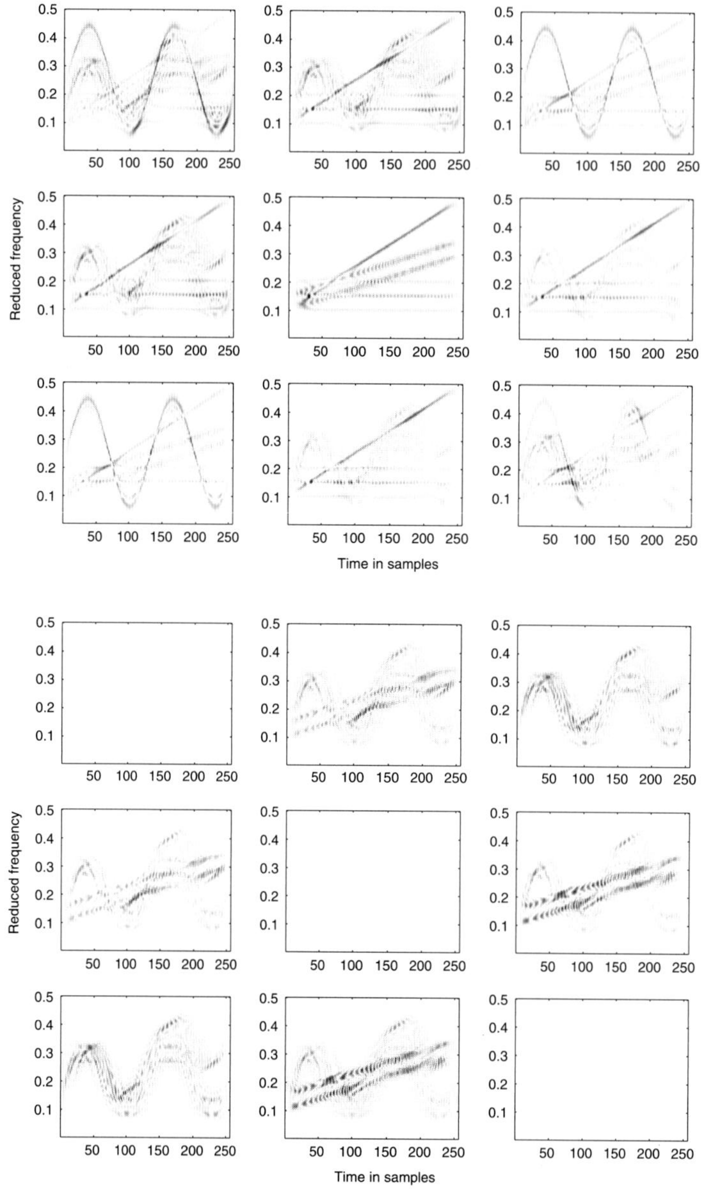

FIGURE 11.4

Spatial PWV distribution of the non-whitened observations (set J_2 of signals after the mixing by the mixing matrix $\mathbf{A}_2$), top: real part, bottom: imaginary part.

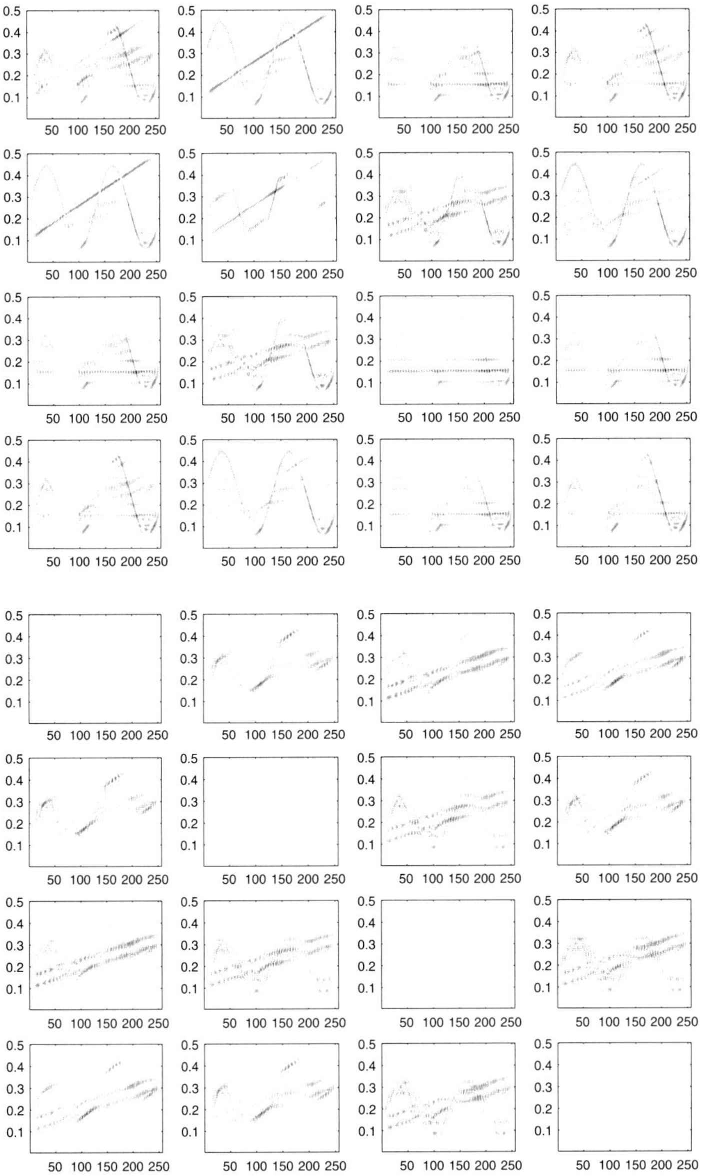

FIGURE 11.5

Spatial PWV distribution of the non-whitened observations (set J_2 of signals after mixing by mixing matrix $\mathbf{A}_3$), top: real part, bottom: imaginary part.

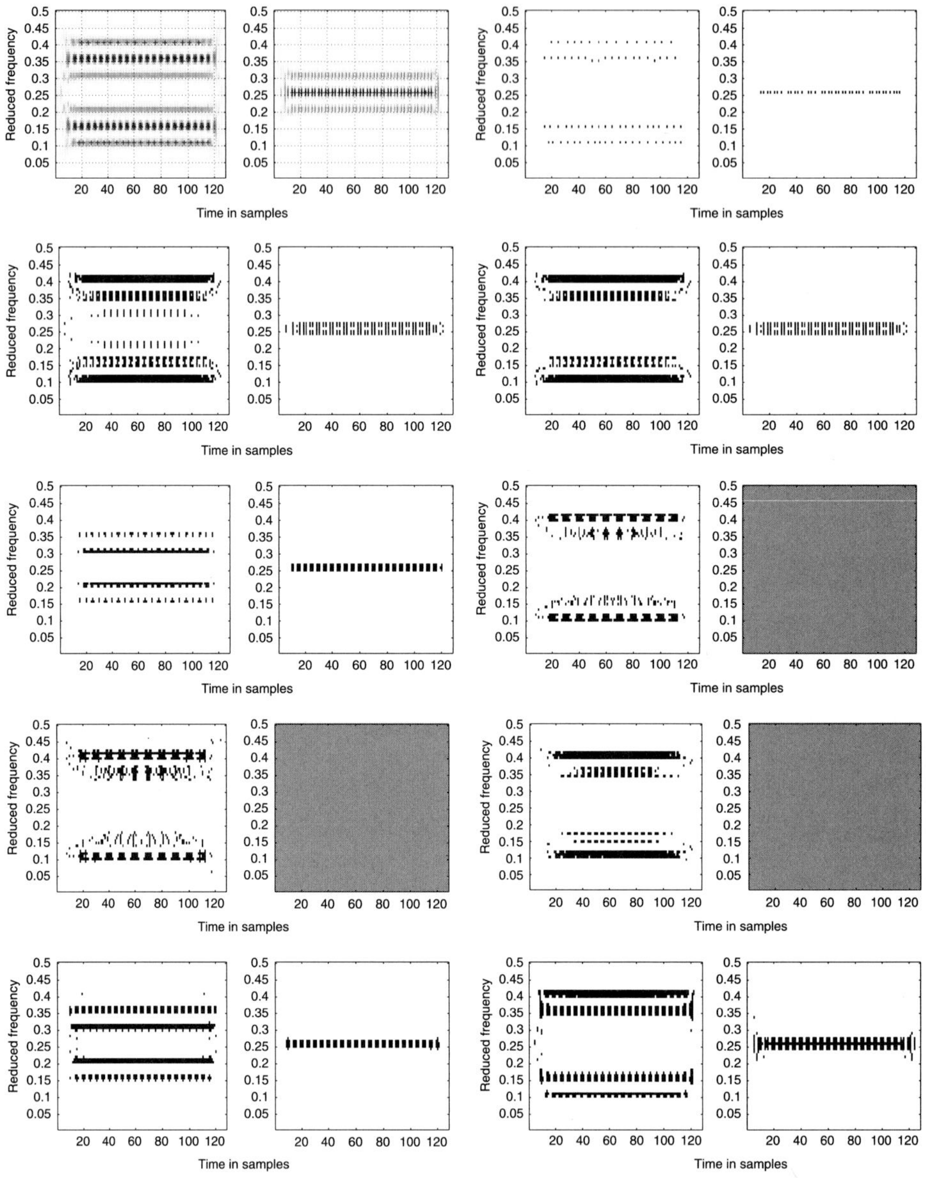

FIGURE 11.6

Set J_1 of sources, t-f points selected by hand (first line); with c_G (left of second line); with c_{GM} (right of second line); with c_B (left of third line); with c_F (right of third line); with c_{GG} (left of fourth line); with c_S (right of fourth line); with c_L (left of fifth line); with c_{LM} (right of fifth line).

Table 11.5 Threshold values and number of matrices selected for joint diagonalization and/or joint zero diagonalization versus the time-frequency points detector, set J_1 of sources

	Manual select	C_G	C_{GM}	C_B	C_F
Threshold value	\	$\varepsilon_1 = 10$	$\varepsilon_1 = 10$	$\varepsilon_5 = 30$	\
Threshold value	\	$\varepsilon_2 = 1$	$\varepsilon_2 = 1$	$\varepsilon_6 = 0.5$	$\varepsilon_8 = 0.004$
Threshold value	\	$\varepsilon_3 = 0.5$	$\varepsilon_3 = 0.5$	\	\
Threshold value	\	\	$\varepsilon'_3 = 1$	\	\
Threshold value	\	$\varepsilon_4 = 12$	$\varepsilon_4 = 12$	\	\
Matrices for JD	77	1131	1011	394	569
Matrices for JZD	38	197	197	199	\

	C_{GG}	C_S	C_L	C_{LM}
Threshold value	$\varepsilon_{11} = 0.004$	$\varepsilon_{14} = 0.5$	$\varepsilon_5 = 32$	$\varepsilon_5 = 32$
Threshold value	$\varepsilon_{12} = 5$	$\varepsilon_{15} = 12$	$\varepsilon_{13} = 0.005$	$\varepsilon'_{13} = 0.005$
Threshold value	\	$\varepsilon'_{14} = 0.5$	\	$\varepsilon_{13} = 0.15$
Matrices for JD	716	793	345	345
Matrices for JZD	\	\	680	211

Table 11.6 Threshold values and number of matrices selected for joint diagonalization and/or joint zero diagonalization versus the time-frequency points detector, set J_2 of sources

	C_G	C_{GM}	C_F	C_{GG}	C_S
Threshold value	$\varepsilon_1 = 13$	$\varepsilon_1 = 13$	\	$\varepsilon_{11} = 0.003$	$\varepsilon_{14} = 1$
Threshold value	$\varepsilon_2 = 1$	$\varepsilon_2 = 1$	$\varepsilon_8 = 0.01$	$\varepsilon_{12} = 5$	$\varepsilon_{15} = 20$
Threshold value	$\varepsilon_3 = 0.5$	$\varepsilon_3 = 0.5$	\	\	$\varepsilon'_{14} = 1$
Threshold value	\	$\varepsilon'_3 = 1$	\	\	\
Threshold value	$\varepsilon_4 = 20$	$\varepsilon_4 = 20$	\	\	\
Matrices for JD, mixing A_2	789	531	472	169	343
Matrices for JZD, mixing A_2	953	953	\	\	\
Matrices for JD, mixing A_3	780	536	472	178	343
Matrices for JZD, mixing A_3	952	952	\	\	\

Table 11.7 A comparison of the performances obtained with 13 different methods on different time-frequency points sets built by means of the different detectors (set J_1 of sources)

Algo. \ Perf.	Manual select.	C_G	C_{GM}	C_B	C_F
JD_B (dB)	−35.65	−36.02	−36.06	−20.5	−36.05
JZD_B (dB)	−36.07	−36.07	−36.07	−36.07	\
JD/JZD_B (dB)	−35.77	−36.03	−36.06	−34.63	\
$JD_{NO,YER}$ (dB)	−39.25	−16.80	−53.72	−3.12	−48.00
$JD_{NO,ALG}$ (dB)	−38.28	−20.11	−53.94	+2.01	−47.68
$JD_{NO,PAR}$ (dB)	−38.29	−20.11	−53.94	+2.0	−47.68
A_D (dB)	−43.67	−57.15	−62.72	−9.63	−66.03
A_{SVD} (dB)	−44.49	−58.76	−63.89	−8.35	−68.46
$JZD_{NO,ALG}$ (dB)	−40.61	−48.31	−48.31	−49.64	\
$JZD_{NO,PAR}$ (dB)	−40.61	−48.31	−48.31	−49.64	\
$JZD_{NO,CHA}$ (dB)	−40.4	−46.04	−46.04	−49.81	\
$JD/JZD_{NO,ALG}$ (dB)	−38.51	−20.81	−56.35	−14.4	\
$JD/JZD_{NO,PAR}$ (dB)	−38.51	−20.81	−56.35	−14.4	\
JADE (dB)			2.92		
SOBI (dB)			−36.07		
Whitening limit (dB)			−36.07		

Algo. \ Perf.	C_{GG}	C_S	C_L	C_{LM}
JD_B (dB)	−36.05	−35.92	−35.93	−35.93
JZD_B (dB)	\	\	−35.93	−36.07
JD/JZD_B (dB)	\	\	−35.63	−36.04
$JD_{NO,YER}$ (dB)	−47.57	−59.17	−45.21	−45.21
$JD_{NO,ALG}$ (dB)	−47.25	−59.36	−45.21	−45.21
$JD_{NO,PAR}$ (dB)	−47.25	−59.36	−45.21	−45.21
A_D (dB)	−72.14	−73.75	−49.67	−49.67
A_{SVD} (dB)	−73.82	−68.62	−48.11	−48.11
$JZD_{NO,ALG}$ (dB)	\	\	−21.06	−50.28
$JZD_{NO,PAR}$ (dB)	\	\	−21.06	−50.28
$JZD_{NO,CHA}$ (dB)	\	\	−21.69	−50.73
$JD/JZD_{NO,ALG}$ (dB)	\	\	−24.66	−54.06
$JD/JZD_{NO,PAR}$ (dB)	\	\	−24.66	−54.06
JADE (dB)		2.92		
SOBI (dB)		−36.07		
Whitening limit (dB)		−36.07		

Table 11.8 A comparison of the performances obtained with 13 different methods on different time-frequency points sets built by means of five different detectors (set J_2 of sources, square mixing matrix $\mathbf{A}_2$ and rectangular mixing matrix $\mathbf{A}_3$)

Performances	C_G		C_{GM}		C_F	
Method \ Mixing	$\mathbf{A}_2$	$\mathbf{A}_3$	$\mathbf{A}_2$	$\mathbf{A}_3$	$\mathbf{A}_2$	$\mathbf{A}_3$
JD (dB)	−29.41	−29.77	−29.62	−29.73	−29.95	−29.94
JZD (dB)	−29.94	−29.94	−29.94	−29.94	\	\
JD/JZD (dB)	−29.73	−29.90	−29.94	−29.97	\	\
$JD_{NO,YER}$ (dB)	−30.94	−33.35	−39.74	−41.59	−41.16	−42.36
$JD_{NO,ALG}$ (dB)	−30.53	−30.40	−38.76	−41.47	−35.18	−34.79
$JD_{NO,PAR}$ (dB)	−30.53	−30.40	−38.76	−41.47	−35.18	−34.79
A_D (dB)	−45.71	−46.4	−48.60	−48.25	−53.31	−50.03
A_{SVD} (dB)	−48.22	−47.24	−48.60	−49.58	−51.65	−52.3
$JZD_{NO,ALG}$ (dB)	−36.08	−35.75	−35.95	−35.75	\	\
$JZD_{NO,PAR}$ (dB)	−36.08	−35.75	−35.95	−35.75	\	\
$JZD_{NO,CHA}$ (dB)	−29.83	−30.12	−29.83	−30.12	\	\
$JD/JZD_{NO,ALG}$ (dB)	−32.3	−32.38	−38.55	−38.80	\	\
$JD/JZD_{NO,PAR}$ (dB)	−32.3	−32.38	−38.55	−38.80	\	\
JADE (dB)	−23.77	−23.76	−23.77	−23.76	−23.77	−23.76
SOBI (dB)	−26.92	−19.51	−26.92	−19.51	−26.92	−19.51
Whitening limit (dB)			−30.15			

Performances	C_S		C_{GG}	
Method \ Mixing	$\mathbf{A}_2$	$\mathbf{A}_3$	$\mathbf{A}_2$	$\mathbf{A}_3$
JD (dB)	−29.46	−29.46	−29.5	−29.95
JZD (dB)	\	\	\	\
JD/JZD (dB)	\	\	\	\
$JD_{NO,YER}$ (dB)	−55.28	−38.79	−39.51	−39.32
$JD_{NO,ALG}$ (dB)	−52.69	−53.39	−33.54	−29.16
$JD_{NO,PAR}$ (dB)	−52.69	−53.39	−33.54	−29.16
A_D (dB)	−40.91	−42.98	−47.55	−44.36
A_{SVD} (dB)	−41.51	−41.23	−47.03	−45.63
$JZD_{NO,ALG}$ (dB)	\	\	\	\
$JZD_{NO,PAR}$ (dB)	\	\	\	\
$JZD_{NO,CHA}$ (dB)	\	\	\	\
$JD/JZD_{NO,ALG}$ (dB)	\	\	\	\
$JD/JZD_{NO,PAR}$ (dB)	\	\	\	\
JADE (dB)	−23.77	−23.76	−23.77	−23.76
SOBI (dB)	−26.92	−19.51	−26.92	−19.51
Whitening limit (dB)		−30.15		

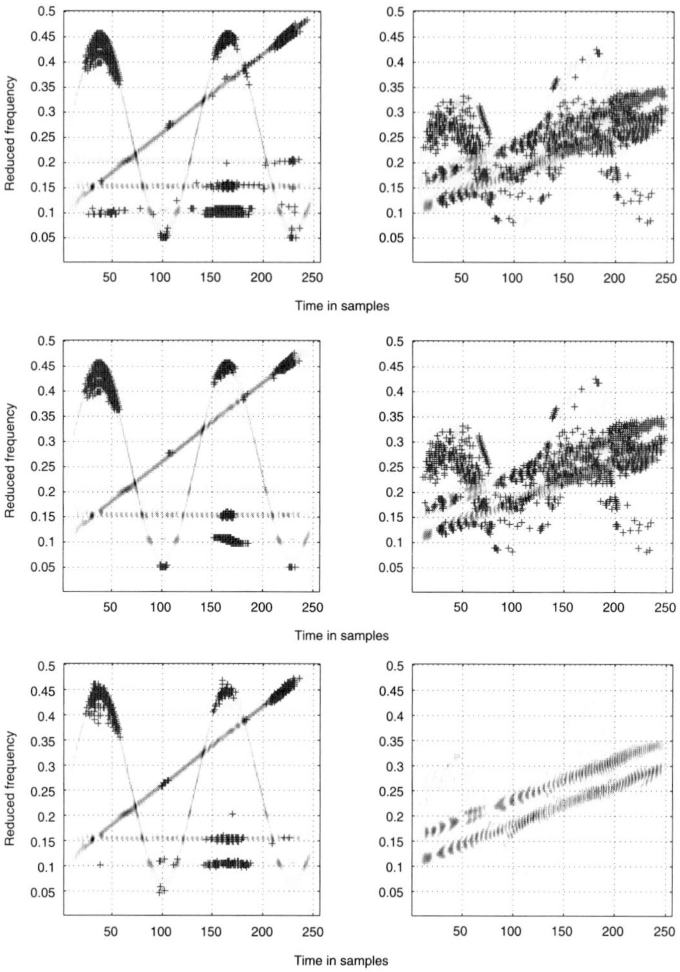

FIGURE 11.7

Set J_2 of sources, t-f points selected with C_G (first line); with C_{GM} (second line); with C_F (third line); with C_{GG} (fourth line); with C_S (fifth line).

11.7 SUMMARY AND CONCLUSION

This chapter is dedicated to provide the fundamentals of source separations based on spatial quadratic time-frequency distributions applied to nonstationary deterministic signals as well as nonstationary stochastic processes. The techniques employed for this purpose do not require the classical statistical independence assumption. For each technique, three main steps are typically performed. These steps, which are detailed and described in the chapter, are the construction of the spatial quadratic time-frequency representations or

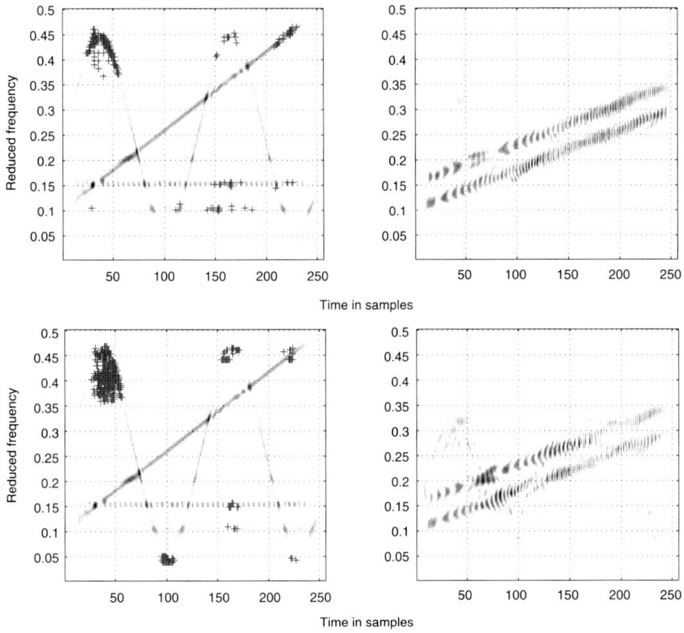

FIGURE 11.7

(*continued*)

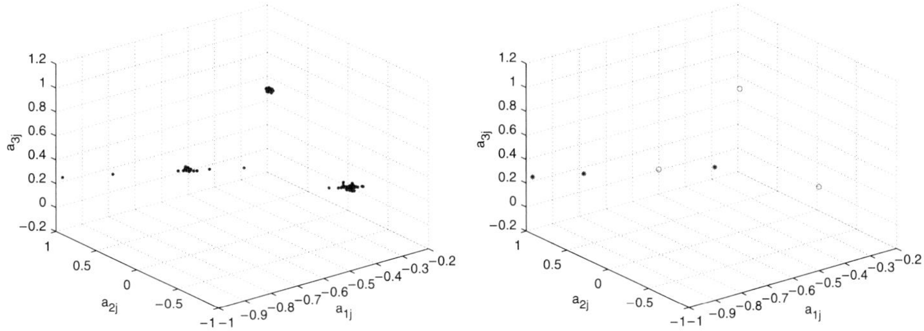

FIGURE 11.8

Columns of the mixing matrix $\mathbf{A}$ estimated using A_{SVD} on the matrices set built thanks to C_{GM} (set J_2 of sources and mixing matrix $\mathbf{A}_2$). Left: before classification, right: after classification (the 3 kept inertia centers are represented by o, the others by $*$).

spectra of the observations across the array, the specific time-frequency points selection, the separation algorithms used to estimate the mixing matrix.

Practical applications of these techniques, the generalization to the convolutive case and other approaches exploiting signal nonstationarity are important and relevant topics, but have not been included in this chapter due to space limitations.

References

[1] A. Abed-Meraïm, K. Belouchrani, R. Leyman, Time-frequency signal analysis and processing: a comprehensive reference, in: B. Boashash (Ed.), Blind Source Separation Using Time-Frequency Distributions, Prentice-Hall, Oxford, UK, 2003.

[2] S.I. Amari, S. Douglas, Why natural gradient, in: International Conference on Acoustic Speech and Signal Processing (ICASSP), vol. 2, Seattle, Washington, USA, May 1998, pp. 1213–1216.

[3] M. Amin, Y. Zhang, Spatial time-frequency distributions and their applications, in: Proceedings of the International symposium on Signal Processing and its Applications (ISSPA), Kuala-Lumpur, Malaysia, August 2001, pp. 254–255.

[4] M. Amin, Y. Zhang, Time-frequency signal analysis and processing: a comprehensive reference, in: B. Boashash (Ed.), Spatial Time-Frequency Distributions and their Applications, Prentice-Hall, Oxford, UK, 2003.

[5] A. Belouchrani, K. Abed-Meraïm, M.G. Amin, A. Zoubir, Joint anti-diagonalization for blind source separation, in: International Conference on Acoustic Speech and Signal Processing (ICASSP), Utah, USA, May 2001, pp. 2196–2199.

[6] A. Belouchrani, K. Abed-Meraïm, M.G. Amin, A.M. Zoubir, Blind separation of nonstationary sources, IEEE Signal Processing Letters 11 (2004) 605–608.

[7] A. Belouchrani, K. Abed-Meraïm, J.-F. Cardoso, E. Moulines, A blind source separation technique using second order statistics, IEEE Transactions on Signal Processing 45 (1997) 434–444.

[8] A. Belouchrani, M.G. Amin, Blind source separation based on time-frequency signal representations, IEEE Transactions on Signal Processing 46 (1998) 2888–2897.

[9] A. Bousbia-Salah, A. Belouchrani, H. Bousbia-Salah, A one step time-frequency blind identification, in: International symposium on Signal Processing and its Applications (ISSPA), vol. 1, Paris, France, July 2003, pp. 581–584.

[10] J.-F. Cardoso, Super-symmetric decomposition of the fourth order cumulant tensor. Blind identification of more sources than sensors, in: International Conference on Acoustic Speech and Signal Processing, Toronto, Canada, April 1991, pp. 3109–3112.

[11] J.-F. Cardoso, On the performance of orthogonal sources separation algorithms, in: EUSIPCO, Edinburgh, Scotland, UK, September 1994, pp. 776–779.

[12] J.-F. Cardoso, A. Souloumiac, Blind beamforming for non-gaussian signals, IEE Proceedings-F 40 (1993) 362–370.

[13] J.-F. Cardoso, A. Souloumiac, Jacobi angles for simultaneous diagonalization, SIAM Journal on Matrix Analysis and Applications 17 (1996) 161–164.

[14] G. Chabriel, J. Barrère, N. Thirion-Moreau, E. Moreau, Algebraic joint zero-diagonalization and blind source separation, IEEE Transactions on Signal Processing 56 (2008) 980–989.

[15] A. Cirillo, L. Zoubir, N. Ma, M. Amin, Automatic classification of auto- and cross-terms of time-frequency distributions in antenna arrays, in: International Conference on Acoustic Speech and Signal Processing (ICASSP), vol. II, Orlando, USA, May 2002, pp. 1445–1448.

[16] L. Cirillo, Narrowband array signal processing using time-frequency distributions, Ph.D. thesis, Technischen Universität Darmstadt, May 2007.

[17] L. Cirillo, M. Amin, Auto-term detection using time-frequency array processing, in: International Conference on Acoustic Speech and Signal Processing (ICASSP), vol. VI, Hong-Kong, April 2003, pp. 465–468.

[18] L. Cirillo, A. Zoubir, M. Amin, Blind source separation in the time-frequency domain based on multiple hypothesis testing, IEEE Transactions on Signal Processing 56 (2008) 2267–2279.

[19] L. Cohen, Time-frequency analysis, in: Alan V. Oppenheimem (Ed.), Prentice Hall Signal Processing Series, 1995.

[20] P. Comon, Blind identification and source separation in 2×3 under-determined mixtures, IEEE Transactions on Signal Processing 52 (2004) 11–22.

[21] P. Comon, E. Moreau, Blind mimo equalization and joint diagonalization criteria, in: International Conference on Acoustic Speech and Signal Processing (ICASSP), vol. 5, Salt Lake City, USA, May 2001, pp. 2749–2752.

[22] P. Comon, M. Rajih, Blind identification of complex under-determined mixtures, in: International Conference on Independent Component Analysis (ICA), Granada, Spain, September 2004, pp. 105–112.
[23] L. De Lathauwer, Signal processing based on multilinear algebra, Ph.D. thesis, Université catholique de Leuven, Belgique, September 1997.
[24] S. Dégerine, E. Kane, A comparative study of approximate joint diagonalization algorithms for blind source separation in presence of additive noise, IEEE Transactions on Signal Processing 55 (2007) 3022–3031.
[25] E.-M. Fadaili, Décompositions matricielles conjointes et séparation aveugle de sources, Ph.D. thesis, Université du Sud Toulon-Var, Juillet 2006.
[26] E.-M. Fadaili, N. Thirion-Moreau, E. Moreau, Algorithme de zéro-diagonalisation conjointe pour la séparation de sources déterministes, in: 20ème Colloque GRETSI, Louvain-La-Neuve, Belgique, Septembre 2005.
[27] E.-M. Fadaili, N. Thirion-Moreau, E. Moreau, Combined non-orthogonal joint zero-diagonalization and joint diagonalization for source separation, in: IEEE Workshop on Statistical Signal Processing (SSP'2005), Bordeaux, France, Juillet 2005.
[28] E.-M. Fadaili, N. Thirion-Moreau, E. Moreau, Non orthogonal zero-diagonalization for sources separation based on time-frequency representation, in: International Conference on Acoustic Speech and Signal Processing (ICASSP), vol. V, Philadelphia, USA, March 2005, pp. 297–300.
[29] E.-M. Fadaili, N. Thirion-Moreau, E. Moreau, Non-orthogonal joint diagonalization/zero-diagonalization for source separation based on time-frequency distributions, IEEE Transactions on Signal Processing 55 (2007).
[30] A. Ferréol, L. Albera, P. Chevalier, Fourth order blind identification of under-determined mixtures of sources (FOBIUM), in: International Conference on Acoustic Speech and Signal Processing, Hong Kong, China, April 2003, pp. 41–44.
[31] C. Févotte, C. Doncarli, A unified presentation of blind separation methods for convolutive mixtures using block-diagonalization, in: Independent Component Analysis (ICA), Nara, Japan, April 2003, pp. 349–354.
[32] C. Févotte, C. Doncarli, Two contributions to blind source separation using time-frequency distributions, IEEE Signal Processing Letters 11 (2004) 386–389.
[33] P. Flandrin, Temps-fréquence, Edition Hermès, collection traitement du signal, Paris, 2ème édition, 1998.
[34] L. Giulieri, Séparation aveugle de sources basée sur l'utilisation des transformées spatiales quadratiques, Ph.D. thesis, Université de Toulon et du Var, Décembre 2003.
[35] L. Giulieri, H. Ghennioui, N. Thirion-Moreau, E. Moreau, Non orthogonal joint diagonalization of spatial quadratic time-frequency matrices for sources separation, IEEE Signal Processing Letters 12 (2005) 415–418.
[36] L. Giulieri, N. Thirion-Moreau, P.-Y. Arquès, Blind sources separation using bilinear and quadratic time-frequency representations, in: International Conference on Independent Component Analysis (ICA), San Diego, USA, December 2001, pp. 486–491.
[37] L. Giulieri, N. Thirion-Moreau, P.-Y. Arquès, Blind sources separation based on bilinear time-frequency representations: a performance analysis, in: International Conference on Acoustic Speech and Signal Processing (ICASSP), Orlando, USA, May 2002, pp. 1649–1652.
[38] L. Giulieri, N. Thirion-Moreau, P.-Y. Arquès, Blind sources separation based on quadratic time-frequency representations: a method without pre-whitening, in: International Conference on Acoustic Speech and Signal Processing (ICASSP), vol. V, Hong-Kong, China, April 2003, pp. 289–292.
[39] G.H. Golub, C.F. Van Loan, Matrix Computations, Johns Hopkins University Press, Baltimore, 1989.
[40] S. Haykin, in: S. Haykin (Ed.), Advances in Spectrum Analysis and Array Processing, Vols II and III, Prentice Hall, Englewood Cliffs, 1991, and 1992.
[41] R. Horn, C. Johnson, Matrix Analysis, Cambridge University Press, 1990.
[42] N. Linh-Trung, A. Belouchrani, K. Abed-Meraïm, B. Boashash, Separating more sources than sensors using time-frequency distributions, in: Proceedings of the International Symposium on Signal Processing and its Applications (ISSPA), Kuala-Lumpur, Malaysia, August 2001, pp. 583–586.

[43] N. Linh-Trung, A. Belouchrani, K. Abed-Meraïm, B. Boashash, Time-frequency signal analysis and processing: a comprehensive reference, in: B. Boashash (Ed.), Undetermined Blind Sources Separation for FM-Like Signals, Prentice-Hall, Oxford, UK, 2003, pp. 3357–3366 (Chapter 8.5).
[44] N. Linh-Trung, A. Belouchrani, K. Abed-Meraïm, B. Boashash, Separating more sources than sensors using time-frequency distributions, Eurasip Journal on Applied Signal Processing (2005), 2828–2847.
[45] S. Marcos, Les méthodes à haute résolution, traitement d'antenne et analyse spectrale (Collection Traitement du Signal), Hermès, Marcos S. (Ed.), 1998.
[46] E. Moreau, A generalization of joint-diagonalization criteria for source separation, IEEE Transactions on Signal Processing 49 (2001) 530–541.
[47] J.-C. Pesquet, B. Chen, A. Petropulu, Frequency domain blind mimo contrast functions for separation of convolutive mixtures, in: International Conference on Acoustic Speech and Signal Processing (ICASSP), vol. 5, Salt Lake City, USA, May 2001, pp. 2765–2768.
[48] D.-T. Pham, Joint approximate diagonalization of positive definite matrices, SIAM Journal on Matrix Analysis and Applications 22 (2001) 1136–1152.
[49] D.T. Pham, Exploiting source non stationary and coloration in blind source separation, in: 14th International Conference on Digital Signal Processing (DSP), vol. 1, Santorini, Grce, 2002, pp. 151–154.
[50] D.-T. Pham, J.-F. Cardoso, Blind separation of instantaneous mixtures of non-stationary sources, IEEE Transactions on Signal Processing 49 (2001) 1837–1848.
[51] K. Rahbar, J.P. Reilly, Blind source separation of convolved sources by joint approximate diagonalization of cross-spectral denisty matrices, in: International Conference on Acoustic Speech and Signal Processing (ICASSP), vol. 5, Salt Lake City, USA, May 2001, pp. 2745–2748.
[52] P. Régalia, An adaptive unit norm filter with applications to signal analysis and Karhunen Loeve transformations, IEEE Transactions on Circuits and Systems 37 (1990) 646–649.
[53] P. Sneath, R. Sokal, Numerical Taxonomy: The Principles and Practice of Numerical Classification, Freeman, San Francisco, 1973.
[54] A. Taleb, C. Jutten, On under-determined source separation, in: International Conference on Acoustic Speech and Signal Processing, Phoenix, Arizona, USA, March 1999.
[55] L. Tong, V.C. Soon, Y.F. Huang, R. Liu, Amuse: a new blind identification algorithm, in: IEEE ISCAS, New Orleans, USA, May 1990, pp. 1784–1787.
[56] A.J. Van Der Veen, Joint diagonalization via subspace fitting techniques, in: International Conference on Acoustic Speech and Signal Processing (ICASSP), Salt Lake City, USA, May 2001, pp. 2773–2776.
[57] A.J. Van Der Veen, A. Paulraj, An analytical constant modulus algorithm, IEEE Transactions on Signal Processing 44 (1996) 1–19.
[58] J. Walgate, A.J. Short, L. Hardy, V. Vedral, Local distinguishability of multipartite orthogonal quantum states, Physical Review Letter 85 (2000) 4972–4975.
[59] A. Yeredor, Blind source separation via the second characteristic function, Signal Processing 80 (2000) 897–902.
[60] A. Yeredor, Non-orthogonal joint-diagonalization in the least squares sense with application in blind sources separation, IEEE Transactions on Signal Processing 50 (2002) 1545–1553.
[61] A. Yeredor, On using exact joint diagonalization for non iterative approximate joint diagonalization, IEEE Signal Processing Letters 12 (2005) 645–648.
[62] A. Yeredor, A. Ziehe, K.-R. Müller, Approximate joint diagonalization using natural gradient approach, in: Lecture Notes in Computer Science (LNCS 3195): Independent Component Analysis and Blind Sources Separation, in: Proceedings ICA, Granada, Espagne, September 2004, pp. 89–96.
[63] Y. Zhang, M. Amin, Subspace analysis of spatial time-frequency distribution matrices, IEEE Transactions on Signal Processing 49 (2001) 747–759.
[64] A. Ziehe, P. Laskov, G. Nolte, K.-R. Müller, A fast algorithm for joint diagonalization with non-orthogonal transformations and its application to blind source separation, Journal of Machine Learning Research 5 (2004) 801–818.

CHAPTER

Bayesian approaches

12

A. Mohammad-Djafari and K.H. Knuth

12.1 INTRODUCTION

Source separation is an inherently *ill-posed inverse problem*, and is unsolvable without *prior information*. The *Bayesian approach* is unique in that it treats the problem as an *inference problem*, and incorporates prior information in both the signal model and the prior probabilities of the model parameters. The advantage is that the source separation algorithm is *derived* from Bayes' theorem by computing the *posterior probability* of the model parameters based on both the prior information and the recorded mixtures. Unsatisfactory performance of the algorithm can be traced back to either the signal model, the assigned *prior probabilities*, or any simplifying assumptions made during the derivation. This approach enables one to construct generic source separation algorithms based on a minimum amount of prior information, as well as informed source separation algorithms that are fine-tuned to specific applications. Modification of a derived algorithm is typically straightforward as all the signal model, the assigned prior probabilities and simplifying assumptions are made explicit.

Bayes' theorem can be used as a mathematical tool to update our state of knowledge about the problem prior to making observations to a posterior state of knowledge after making observations. Being Bayesian is just a logical way to handle a problem where we have unknown quantities (sources $\mathbf{s}$ and mixing matrix $\mathbf{A}$), observations (measured mixed signals or images $\mathbf{x}$), a *forward model* $\mathcal{M}$ linking them ($\mathcal{M}: \mathbf{x} = \mathbf{A}\mathbf{s}$ or $\mathbf{x} = \mathbf{A}\mathbf{s} + \boldsymbol{\epsilon}$), and a prior state of knowledge about the unknowns $p(\mathbf{s}, \mathbf{A}|\mathcal{M})$. Bayes' theorem then gives the expression of the posterior probability $p(\mathbf{s}, \mathbf{A}|\mathbf{x}, \mathcal{M})$ which represents our state of knowledge about those unknown quantities by combining the information contained in the model, any *prior information* we have about the model parameter values and the information obtained by making observations.

One common misconception about the Bayesian approach is focused on what it means to assign a probability to the model parameter values. The fact that we assign a probability distribution to describe the possible values of the model parameter does not mean that that parameter is a result of any random process, also referred to as a random variable. Instead, the probability distribution merely describes all we know about the possible values of this parameter. In a given problem the parameter has a particular value; however, we just do not yet know that value.

Handbook of Blind Source Separation: Independent Component Analysis and Applications

Before delving into a detailed exposition of Bayesian source separation, we first consider the development and evolution of these ideas and their relation to the field of source separation in general. There are always many useful ways to look at a problem, and it is the fact that different viewpoints have both advantages and disadvantages that allow us to utilize these viewpoints to expand and extend our ideas in new directions. As this field has developed, many ways of visualizing source separation have been brought forward. From the earliest work, source separation was viewed as a neural network [24,32, 20,19,18,5,52] where the units varied their weights to unmix the signals so that the outputs minimized the mutual information that they carried, or equivalently maximized the information they provide. By generalizing the operation of a neural network, the idea of minimizing a contrast function was introduced [14]. Meanwhile, the analogies to principal component analysis [33] and decorrelation led to the viewpoint where cumulant tensors are diagonalized [11,19,21], and the more general problem of factor analysis [1]. The contrast function was found to be extremely useful in terms of quantifying the mutual information of the separated signals, and this continued to spin off information-theoretic techniques. By working with contrast functions, one is choosing a function to optimize. The likelihood is a useful function to maximize, and this naturally gave rise to the maximum likelihood perspective [41,56,12,42,54]. By incorporating prior information, or including additional model parameters, one is led to the Bayesian viewpoint, which can be implemented using Bayes' theorem directly [35,34,63,36,45,65,66,68,71,23] or by applying variational Bayes by using relative entropies, [16,78,79] and ensemble learning [43,44].

This is only a sample of the techniques used to design source separation algorithms. As with all evolutionary explosions, there was in source separation a rapid diversification of techniques that may yet further diversify, but will eventually settle down into a set of useful methods that will be selected based on their successes and failures when applied to real-world problems.

Source separation methods based on the Bayesian approach and the incorporation of prior information have demonstrated successful results in real-world problems in situations where classical blind source separation methods have failed. The content of this chapter first focuses on the basics of the Bayesian approach to source separation and then demonstrates the application of the Bayesian methodology to a variety of real-world problems.

12.2 SOURCE SEPARATION FORWARD MODEL AND NOTATIONS

In this chapter we will discuss the application of Bayesian methods to the two data types commonplace in source separation, namely time-series and digital images. We will discuss how the Bayesian methodology readily enables one to extend these methods to source localization and characterization.

We begin by introducing some standard notations for the 1D case, as well as the more general cases:

$x_i(t)$ or $x_i(\mathbf{r})$	observation recorded from sensor number i at time t or at a position $\mathbf{r}$, $i = 1, \ldots, m$
$s_j(t)$ or $s_j(\mathbf{r})$	unknown source number j at time t or at position $\mathbf{r}$, $j = 1, \ldots, n$
$\mathbf{x}(t)$ or $\mathbf{x}(\mathbf{r})$	the vector of observations recorded at time t or at position $\mathbf{r}$
$\mathbf{s}(t)$ or $\mathbf{s}(\mathbf{r})$	the vector of the source signals emitted at time t or from position $\mathbf{r}$
$\mathcal{T} = \{1, \ldots, T\}$	represents a time interval
$\mathcal{R} = \{\mathbf{r}_i = (x_i, y_i)\}$	represents the set of pixel positions of an image
$\mathbf{X} = \{\mathbf{x}(t), t \in \mathcal{T}\}$ or $\{\mathbf{x}(\mathbf{r}), \mathbf{r} \in \mathcal{R}\}$	a matrix representing all the observations
$\mathbf{S} = \{\mathbf{s}(t), t \in \mathcal{T}\}$ or $\{\mathbf{s}(\mathbf{r}), \mathbf{r} \in \mathcal{R}\}$	a matrix representing all the sources
$\mathbf{A}$	a matrix representing the mixing matrix

Note that the positions $\mathbf{r}$ in the case of image analysis refers to image coordinates (x, y), but can be easily generalized to higher-dimensional data analysis or to refer to additional signal characteristics.

Typically, we will consider relatively simple signal models. In the 1D case of time series analysis, we assume instantaneous linear mixing with additive noise

$$x_i(t) = \sum_{j=1}^{n} A_{i,j} s_j(t) + \epsilon_i(t), \quad t \in \mathcal{T} \longrightarrow \mathbf{x}(t) = \mathbf{A}\,\mathbf{s}(t) + \boldsymbol{\epsilon}(t) \tag{12.1}$$

and similarly in the 2D case we have

$$x_i(\mathbf{r}) = \sum_{j=1}^{n} A_{i,j} s_j(\mathbf{r}) + \epsilon_i(\mathbf{r}), \quad \mathbf{r} \in \mathcal{R} \longrightarrow \mathbf{x}(\mathbf{r}) = \mathbf{A}\,\mathbf{s}(\mathbf{r}) + \boldsymbol{\epsilon}(\mathbf{r}). \tag{12.2}$$

In general we can write

$$\mathbf{X} = \mathbf{A}\mathbf{S} + \mathbf{E} \tag{12.3}$$

where $\mathbf{E} = \{\boldsymbol{\epsilon}(t), t \in \mathcal{T}\}$ or $\{\boldsymbol{\epsilon}(\mathbf{r}), \mathbf{r} \in \mathcal{R}\}$ represent the errors, which can be attributed to inadequacies of the model or measurement process, and are often called *noise*.

In the following section, we introduce the Bayesian methodology by focusing on the 1D case of time-series analysis. When necessary, we will highlight details specific to 2D image analysis.

12.3 GENERAL BAYESIAN SCHEME

The general scheme of the Bayesian approach in source separation can be summarized in the following steps:

- Select a *signal model* (or *forward model*) $\mathcal{M}$ that describes how the emitted source signals $\mathbf{S}$ generate the recorded mixtures $\mathbf{X}$. One must then assign a probability distribution that quantifies how probable it would have been for a hypothesized source signal to have resulted in the recorded mixtures. For example, in a forward model like $\mathbf{X} = \mathbf{A}\mathbf{S} + \mathbf{E}$, where the noise is assumed to be additive, we would assign $p(\mathbf{X}|\mathbf{A}, \mathbf{S}) = p_E(\mathbf{X} - \mathbf{A}\mathbf{S})$ where p_E is the probability law assigned to the additive errors.
- Assign the prior probabilities to the model parameters according to some available information. For example, we would need to assign $p(\mathbf{S})$ for the sources and $p(\mathbf{A})$ for the elements of the mixing matrix $\mathbf{A}$.
- Use *Bayes' theorem* to obtain the posterior probability of all the unknown model parameters $p(\mathbf{A}, \mathbf{S}|\mathbf{X})$:

$$p(\mathbf{A}, \mathbf{S}|\mathbf{X}) = \frac{p(\mathbf{X}|\mathbf{A}, \mathbf{S})\, p(\mathbf{A})\, p(\mathbf{S})}{p(\mathbf{X})} \propto p(\mathbf{X}|\mathbf{A}, \mathbf{S})\, p(\mathbf{A})\, p(\mathbf{S}) \tag{12.4}$$

where $\propto$ means *proportional to* and:

$$p(\mathbf{X}) = \iint p(\mathbf{X}|\mathbf{A}, \mathbf{S})\, p(\mathbf{A})\, p(\mathbf{S})\, \mathrm{d}\mathbf{S}\, \mathrm{d}\mathbf{A}. \tag{12.5}$$

This posterior probability represents our state of knowledge about the values of the unknown model parameters after taking account the recorded mixtures.
- Use this posterior probability to make *inferences* about the values of $\mathbf{A}$ or $\mathbf{S}$, or all parameters jointly.

We may note here that Bayes' theorem gives the posterior probability of all the unknown model parameters. In the case where we have only one or two scalar unknowns, we could just plot this probability distribution and look at it. However, when there are more than a few unknowns, evaluating this posterior probability requires visualizing it in multiple dimensions, which is not practical. In these cases, we typically summarize the posterior probability with summary quantities such as the mode, mean, or median or by defining a region of relatively high probability [30,59,60]. When summarizing the posterior probability by its mode, it is often easier to find the mode of the logarithm of the posterior probability. The logarithm often simplifies the mathematical expressions to be evaluated while simultaneously smoothing out local optima, which greatly facilitates search algorithms. Another more general approach to summarize a multivariate posterior law is to approximate it with a separable one. This is the essence of the Variational Bayes Approximation (VBA) [55,3,2,4,69,58]. Within this approximation, each parameter is associated with its own posterior density, which can then be described crudely by its mean and variance. There the variance provides a measure of the reliability of the estimate of the parameter in question.

In even the most simple source separation problem, we have two sets of unknowns, **A** and **S**, which constitute a high-dimensional space. In these situations, there are typically three approaches we may take:

1. Joint Estimation
   In this case we work to estimate the values of all of the model parameters $(\widehat{\mathbf{A}}, \widehat{\mathbf{S}})$ using $p(\mathbf{A}, \mathbf{S}|\mathbf{X})$. If we decide to estimate the most probable values, we are left with

$$(\widehat{\mathbf{A}}, \widehat{\mathbf{S}}) = \arg\max_{(\mathbf{A},\mathbf{S})} \{J(\mathbf{A}, \mathbf{S}) = \ln p(\mathbf{A}, \mathbf{S}|\mathbf{X})\}. \tag{12.6}$$

2. Estimation of **A**
   We may be interested only in identifying the mixing matrix, and not interested in the source signals. To obtain a probability distribution for the mixing matrix alone we integrate $p(\mathbf{A}, \mathbf{S}|\mathbf{X})$ over all possible values of the source signals **S** to obtain the marginal distribution $p(\mathbf{A}|\mathbf{X})$. The marginal distribution can be used to make inferences about **A**, such as estimating the most probable values of the mixing matrix

$$\widehat{\mathbf{A}} = \arg\max_{\mathbf{A}} \{J(\mathbf{A}) = \ln p(\mathbf{A}|\mathbf{X})\}. \tag{12.7}$$

3. Estimation of **S**
   We may be interested only in the source signals and not at all care about the mixing matrix **A**. In this case, we can integrate $p(\mathbf{A}, \mathbf{S}|\mathbf{X})$ over all possible values of the mixing matrix **A** to obtain $p(\mathbf{S}|\mathbf{X})$ and use it to make inferences about **S**. For example, the most probable source signals can be found by

$$\widehat{\mathbf{S}} = \arg\max_{\mathbf{S}} \{J(\mathbf{S}) = \ln p(\mathbf{S}|\mathbf{X})\}. \tag{12.8}$$

12.4 RELATION TO PCA AND ICA

To illustrate the basic steps of the Bayesian approach we examine its relationship with classical methods used in data reduction and source separation, such as *Principal Component Analysis* (PCA) and *Independent Component Analysis* (ICA). We start with a simple case where the *forward model* is given by $\mathbf{x}(t) = \mathbf{A}\mathbf{s}(t) + \boldsymbol{\epsilon}(t)$ and assume that the errors $\epsilon_i(t)$ are independent and Gaussian distributed with zero mean and known variances $\sigma_{\epsilon i} = \sigma_\epsilon, \forall i$. The likelihood term is then given by

$$p(\mathbf{X}|\mathbf{A}, \mathbf{S}) = \prod_t \mathcal{N}(\mathbf{x}(t) - \mathbf{A}\mathbf{s}(t), \Sigma_\epsilon) \tag{12.9}$$

where $\Sigma_\epsilon = \sigma_\epsilon \mathbf{I}$.

The next step is to assign the prior distributions $p(\mathbf{A})$ and $p(\mathbf{S})$. The prior probability $p(\mathbf{A})$ quantifies our prior knowledge about the possible values of the mixing matrix, whereas our prior knowledge about the amplitude distribution of the source signals $\mathbf{S} = \mathbf{s}_{1..T}$ is quantified by $p(\mathbf{S})$. It is common to assume that the sources are mutually independent. This enables us to factor the joint prior probability into prior probabilities for each of the individual sources $p(\mathbf{S}) = \prod_j p(s_{j_{1..T}})$. Furthermore, it is also common to assume that the source statistics do not vary in time, which means that we can further factor the source prior into a product of prior probabilities for the source amplitude at any point in time $p(\mathbf{S}) = \prod_t p(\mathbf{s}(t))$. By combining these two hypotheses we can write

$$p(\mathbf{S}) = p(\mathbf{s}_{1..T}) = \prod_t p(\mathbf{s}(t)) = \prod_t \prod_j p_j(s_j(t)) \tag{12.10}$$

where p_j is the prior probability of amplitude of the signal emitted by the j-th source $s_j(t)$.

The advantage of the Bayesian approach is that these assumptions are made explicit, and can easily be modified for situations where they are inappropriate [35,34,45]. We now consider some of the assumptions that are implicitly made by PCA and ICA.

Concerning the mixing matrix, we initially assume that we know very little about the values of its elements. We can quantify our ignorance by assigning $p(\mathbf{A}) = \prod_i \prod_j p(a_{i,j})$ a uniform prior probability $p(a_{i,j}) = \text{constant}$ or a Gaussian prior probability $p(a_{i,j}) = \mathcal{N}(0, \sigma_a^2)$ with a very large value for the variance σ_a^2 [36,61,62,45].

Later in the paper, we will revisit these assumptions, but for now we have all the necessary elements to apply Bayes' rule

$$\begin{aligned} p(\mathbf{A}, \mathbf{S}|\mathbf{X}) &\propto p(\mathbf{X}|\mathbf{A}, \mathbf{S})\, p(\mathbf{S})\, p(\mathbf{A}) \\ &\propto \prod_t \left[\mathcal{N}(\mathbf{x}(t) - \mathbf{A}\mathbf{s}(t), \Sigma_\epsilon)\, p(\mathbf{s}(t))\right]\, p(\mathbf{A}). \end{aligned} \tag{12.11}$$

With the posterior probability in hand, we can now examine the possible solutions.

1. Known A, estimation of S

The posterior probability of $\mathbf{S}$ based on known $\mathbf{A}$ is given by

$$p(\mathbf{S}|\mathbf{A}, \mathbf{X}) \propto p(\mathbf{X}|\mathbf{A}, \mathbf{S})\, p(\mathbf{S}) \propto \prod_t \left[\mathcal{N}(\mathbf{x}(t) - \mathbf{A}\mathbf{s}(t), \Sigma_\epsilon)\, p(\mathbf{s}(t))\right]. \tag{12.12}$$

By taking the log of the posterior probability, we have

$$-\ln p(\mathbf{S}|\mathbf{A}, \mathbf{X}) = \sum_t \frac{1}{\sigma_\epsilon^2} ||\mathbf{x}(t) - \mathbf{A}\mathbf{s}(t)||^2 - \ln p(\mathbf{s}(t)) + c \tag{12.13}$$

where c is a constant independent of $\mathbf{S}$ and $||\cdot||^2$ is the Frobenius norm. To complete the calculation, we need to assign $p(\mathbf{s}(t))$. A simple, but potentially reasonable, assignment is $p(\mathbf{s}(t)) = \mathcal{N}(0, \sigma_s^2 \mathbf{I})$. Then, the posterior $p(\mathbf{S}|\mathbf{A}, \mathbf{X})$ is also Gaussian

$$p(\mathbf{S}|\mathbf{A}, \mathbf{X}) = \prod_t p(\mathbf{s}(t)|\mathbf{A}, \mathbf{x}(t)) = \prod_t \mathcal{N}(\widehat{\mathbf{s}}(t), \widehat{\Sigma}). \tag{12.14}$$

The most probable, or the Maximum A Posteriori (MAP), and Posterior Mean (PM) estimates are the same and can be computed easily by searching the space of all possible $\mathbf{S}$ to find the result $\widehat{\mathbf{S}}$ which maximizes $\ln p(\mathbf{S}|\mathbf{A}, \mathbf{X})$

$$\widehat{\mathbf{s}}(t) = (\mathbf{A}'\mathbf{A} + \lambda_s \mathbf{I})^{-1} \mathbf{A}'\mathbf{x}(t) \quad \text{or} \quad \widehat{\mathbf{S}} = (\mathbf{A}'\mathbf{A} + \lambda_s \mathbf{I})^{-1} \mathbf{A}'\mathbf{X} \tag{12.15}$$

where $\lambda_s = {\sigma_\epsilon}^2/\sigma_s^2$. Note that $\widehat{\Sigma} = {\sigma_\epsilon}^2 (\mathbf{A}'\mathbf{A} + \lambda_s \mathbf{I})^{-1}$ and if $\sigma_s^2 >> {\sigma_\epsilon}^2$, then $\lambda_s \simeq 0$ and $\widehat{\mathbf{s}}(t) = \mathbf{A}^+ \mathbf{x}(t)$ where $\mathbf{A}^+$ is the generalized inverse of $\mathbf{A}$.

2. Known sources $\mathbf{S}$, estimation of $\mathbf{A}$

In this case we look at the posterior probability of $\mathbf{A}$ based on known $\mathbf{S}$

$$p(\mathbf{A}|\mathbf{S}, \mathbf{X}) \propto p(\mathbf{X}|\mathbf{A}, \mathbf{S})\, p(\mathbf{A}) \propto \prod_t \mathcal{N}(\mathbf{x}(t) - \mathbf{A}\mathbf{s}(t), \Sigma_\epsilon)\, p(\mathbf{A}). \tag{12.16}$$

Hypothesizing that $p(a_{i,j}) = \mathcal{N}(0, \sigma_a^2), \forall i, j$, it is easy to show that $p(\mathbf{A}|\mathbf{S}, \mathbf{X})$ is also Gaussian such that

$$-\ln p(\mathbf{A}|\mathbf{S}, \mathbf{X}) = \sum_t \frac{1}{{\sigma_\epsilon}^2} \|\mathbf{x}(t) - \mathbf{A}\mathbf{s}(t)\|^2 - \ln p(\mathbf{A}) + c \tag{12.17}$$

where c is a constant independent of $\mathbf{A}$. Again, the MAP and PM estimates of $\mathbf{A}$ are the same and can be computed easily by finding $\widehat{\mathbf{A}}$ which maximizes $\ln p(\mathbf{A}|\mathbf{S}, \mathbf{X})$ with respect to $\mathbf{A}$

$$\widehat{\mathbf{A}} = \left(\sum_t \mathbf{s}(t)\mathbf{s}'(t) + \lambda_a \mathbf{I} \right)^{-1} \sum_t \mathbf{x}(t)\mathbf{s}'(t) = (\mathbf{S}\mathbf{S}' + \lambda_a \mathbf{I})^{-1} \mathbf{X}\mathbf{s}' \tag{12.18}$$

where $\lambda_a = {\sigma_\epsilon}^2/\sigma_a^2$. Note that when $\sigma_a^2 >> {\sigma_\epsilon}^2$, we have $\lambda_a \simeq 0$ and $\widehat{\mathbf{A}} = (\mathbf{S}\mathbf{S}')^{-1} \mathbf{X}\mathbf{s}'$.

3. Joint estimation of $\mathbf{A}$ and $\mathbf{S}$

Here we begin with the joint density $p(\mathbf{A}, \mathbf{S}|\mathbf{X})$, which when taking the logarithm, gives

$$-\ln p(\mathbf{A}, \mathbf{S}|\mathbf{X}) = \sum_t \left[\frac{1}{{\sigma_\epsilon}^2} \|\mathbf{x}(t) - \mathbf{A}\mathbf{s}(t)\|^2 + \ln p(\mathbf{s}(t)) \right] - \ln p(\mathbf{A}) + c \tag{12.19}$$

where c is a constant independent of $\mathbf{A}$ and $\mathbf{S}$. If we try to obtain the Joint MAP estimate, we have to look at the criterion $J(\mathbf{A}, \mathbf{S}) = -\ln p(\mathbf{A}, \mathbf{S}|\mathbf{X})$. A naive, but simple, algorithm is to search in an alternate way:

$$\begin{cases} \widehat{\mathbf{S}} &= \arg\max_{\mathbf{S}} \left\{ p(\mathbf{S}|\widehat{\mathbf{A}},\mathbf{X}) \right\} = \arg\min_{\mathbf{S}} \left\{ \|\mathbf{X}-\widehat{\mathbf{A}}\mathbf{S}\|^2 - \sigma_\epsilon{}^2 \ln p(\mathbf{S}) \right\} \\ \widehat{\mathbf{A}} &= \arg\max_{\mathbf{A}} \left\{ p(\mathbf{A}|\widehat{\mathbf{S}},\mathbf{X}) \right\} = \arg\min_{\mathbf{A}} \left\{ \|\mathbf{X}-\mathbf{A}\widehat{\mathbf{S}}\|^2 - \sigma_\epsilon{}^2 \ln p(\mathbf{A}) \right\} \end{cases} \tag{12.20}$$

However, in general, there is no guarantee that this algorithm converges to the joint MAP solution.

This can be simplified if we consider the particular case of a Gaussian prior

$$-\ln p(\mathbf{A},\mathbf{S}|\mathbf{X}) = \sum_t \left[\frac{1}{\sigma_\epsilon{}^2} \|\mathbf{x}(t) - \mathbf{A}\mathbf{s}(t)\|^2 - \frac{1}{\sigma_s^2} \|\mathbf{s}(t)\|^2 \right] + \frac{1}{\sigma_a^2} \|\mathbf{A}\|^2 + c.$$

One can see that this criterion is quadratic in **S** when **A** is fixed and quadratic in **A** when **S** is fixed, and we have analytical solutions for each step of the search:

$$\begin{cases} \widehat{\mathbf{S}} = (\widehat{\mathbf{A}}'\widehat{\mathbf{A}} + \lambda_1 \mathbf{I})^{-1} \widehat{\mathbf{A}}' \mathbf{X}, \\ \widehat{\mathbf{A}} = \left(\widehat{\mathbf{S}}\widehat{\mathbf{S}}' + \lambda_2 \mathbf{I} \right)^{-1} \mathbf{X}\widehat{\mathbf{S}}' \end{cases} \quad \text{with} \quad \begin{cases} \lambda_1 = \dfrac{\sigma_\epsilon{}^2}{\sigma_s^2} \\ \lambda_1 = \dfrac{\sigma_\epsilon{}^2}{\sigma_a^2} \end{cases} \tag{12.21}$$

or alternatively

$$\begin{cases} \widehat{\mathbf{s}}(t) = (\widehat{\mathbf{A}}'\widehat{\mathbf{A}} + \lambda_s \mathbf{I})^{-1} \widehat{\mathbf{A}}' \mathbf{x}(t), \\ \widehat{\mathbf{A}} = \left(\sum_t \widehat{\mathbf{s}}(t) \widehat{\mathbf{s}}'(t) + \lambda_a \mathbf{I} \right)^{-1} \sum_t \mathbf{x}(t) \widehat{\mathbf{s}}'(t). \end{cases} \tag{12.22}$$

However, one should note that even if the criterion is quadratic in **A** given **S** and quadratic in **S** given **A**, it is bilinear in the couple $(\mathbf{A},\mathbf{S})$.

When $\lambda_a = \lambda_s = 0$, this algorithm is equivalent to an alternate optimization of the Least Squares criterion $J(\mathbf{A},\mathbf{S}) = \|\mathbf{X} - \mathbf{A}\mathbf{S}\|^2$ with respect to **S** and to **A** which is used in spectrometry [9,10,1,51,49,50,46]. In practice, this algorithm may converge to any local minimum of the criterion function and a satisfactory solution can only be obtained by imposing other constraints such as positivity of the elements of the mixing matrix or the sources

$$\begin{cases} \widehat{\mathbf{s}}(t) = (\widehat{\mathbf{A}}'\widehat{\mathbf{A}})^{-1} \widehat{\mathbf{A}}' \mathbf{x}(t), & \text{then apply the constraints on } \widehat{\mathbf{s}}(t) \\ \widehat{\mathbf{A}} = \left(\sum_t \widehat{\mathbf{s}}(t) \widehat{\mathbf{s}}'(t) \right)^{-1} \sum_t \mathbf{x}(t) \widehat{\mathbf{s}}'(t), & \text{then apply the constraints on } \widehat{\mathbf{A}}. \end{cases}$$

Studying the convergence properties of such algorithms is not easy. However, it is much easier to assign specific priors $p(\mathbf{s}(t))$ and $p(\mathbf{A})$. For example, choosing truncated

Gaussians is a convenient way to impose the positivity. To search for the joint MAP solution, we may use the following algorithm

$$\begin{cases} \widehat{\mathbf{s}}(t) = (\widehat{\mathbf{A}}'\widehat{\mathbf{A}} + \lambda_s \mathbf{I})^{-1}\widehat{\mathbf{A}}'\mathbf{x}(t), & \text{then apply the constraints on } \widehat{\mathbf{s}}(t) \\ \widehat{\mathbf{A}} = \left(\sum_t \widehat{\mathbf{s}}(t)\widehat{\mathbf{s}}'(t) + \lambda_a \mathbf{I}\right)^{-1} \sum_t \mathbf{x}(t)\widehat{\mathbf{s}}'(t), & \text{then apply the constraints on } \widehat{\mathbf{A}} \end{cases}$$

where λ_s and λ_a are, respectively, proportional to the variances σ_s^2 and σ_a^2:

$$p(a_{i,j}) = 0 \text{ if } a_{i,j} < 0 \quad \text{and} \quad p(a_{i,j}) = 2\mathcal{N}(0, \sigma_a^2) \text{ if } a_{i,j} \geqslant 0 \tag{12.23}$$

and the sources are

$$p(s_j(t)) = 0 \text{ if } s_j(t) < 0 \quad \text{and} \quad p(s_j(t)) = 2\mathcal{N}(0, \sigma_s^2) \text{ if } s_j(t) \geqslant 0. \tag{12.24}$$

4. Estimation of **A** with marginalization of **S**

The main idea here is to integrate $p(\mathbf{A}, \mathbf{S}|\mathbf{X})$ with respect to $\mathbf{S}$ to obtain the marginal $p(\mathbf{A}|\mathbf{X})$ and use it to infer $\mathbf{A}$. This can also be done by using the relation $p(\mathbf{A}|\mathbf{X}) \propto p(\mathbf{X}|\mathbf{A})p(\mathbf{A})$, which means that we need to find $p(\mathbf{X}|\mathbf{A})$. To obtain this, we may note that when $\mathbf{s}(t)$ and $\boldsymbol{\epsilon}(t)$ are assumed to be mutually independent and stationary, then $\mathbf{x}(t)$ is also stationary and we have: $p(\mathbf{X}|\mathbf{A}) = \prod_t p(\mathbf{x}(t)|\mathbf{A})$ where:

$$p(\mathbf{x}|\mathbf{A}) = \int p(\mathbf{x}, \mathbf{s}|\mathbf{A})\, d\mathbf{s} = \int p(\mathbf{x}|\mathbf{A}, \mathbf{s})\, p(\mathbf{s})\, d\mathbf{s}.$$

Note also that when $p(\mathbf{A})$ = constant, we have $p(\mathbf{A}|\mathbf{X}) \propto p(\mathbf{X}|\mathbf{A})$ which is the likelihood of data $\mathbf{X}$ given $\mathbf{A}$.

It is easy to show that many classical methods of PCA and ICA are based on different approximations of the expression of this likelihood. In particular, in two cases, we can have simple expressions for this likelihood: (i) the Gaussian likelihood case and (ii) the case of a mixing model without noise.

The Gaussian likelihood case and the connection to PCA. If we assume $\mathbf{s}(t)$ and $\boldsymbol{\epsilon}(t)$ to be Gaussian distributed, we have

$$\begin{array}{ll} p(\mathbf{x}(t)|\mathbf{A}, \mathbf{s}(t), \Sigma_\epsilon) & = \mathcal{N}(\mathbf{A}\mathbf{s}(t), \Sigma_\epsilon) \\ p(\mathbf{s}(t)|\Sigma_s) & = \mathcal{N}(0, \Sigma_s) \end{array} \longrightarrow p(\mathbf{x}(t)|\mathbf{A}, \Sigma_\epsilon, \Sigma_s) = \mathcal{N}(0, \mathbf{A}\Sigma_s\mathbf{A}' + \Sigma_\epsilon). \tag{12.25}$$

The mixing matrix $\mathbf{A}$, which maximizes the likelihood

$$p(\mathbf{X}|\mathbf{A}, \Sigma_\epsilon, \Sigma_s) = \prod_t p(\mathbf{x}(t)|\mathbf{A}, \Sigma_\epsilon, \Sigma_s),$$

FIGURE 12.1

Source separation model where the sources are considered as the hidden variables.

is then equivalent to the PCA solution which is found from the covariance matrix of the observations $\mathbf{x}_{1..T}$ by $\Sigma_x = \frac{1}{T}\sum_t \mathbf{x}(t)\mathbf{x}'(t)$ and its identification with $\mathbf{A}\Sigma_s\mathbf{A}'$ in the case of a model without noise ($\Sigma_\epsilon = 0$) or with $\mathbf{A}\Sigma_s\mathbf{A}' + \Sigma_\epsilon$ for the general case.

Non-Gaussian case and the relation to ICA. In general it is not possible to obtain an analytical solution for the maximum likelihood estimator in the case where the likelihood is non-Gaussian. The reason is that the multiple integrals are not analytically solvable. An exception is the delta function likelihood, which represents the case of a mixture model without noise ($\boldsymbol{\epsilon} = 0$). In this case the delta functions solve the integrals trivially and yield an analytic solution in the case where the mixing matrix is invertible. A great number of ICA methods have been developed which focus on this analytically-tractable case [32,20,19,57,14,31,26,7,6,15,12,13]. See also the books [17,25,62].

In fact, in the case $\mathbf{x}(t) = \mathbf{A}\mathbf{s}(t)$ where $\mathbf{A}$ is invertible $\mathbf{A} = \mathbf{B}^{-1}$, it is easy to show that

$$\ln p(\mathbf{X}|\mathbf{B}) = \frac{T}{2}\ln|\mathbf{B}| + \sum_t\sum_j \ln p_j([\mathbf{B}\mathbf{x}(t)]_j) + c \qquad (12.26)$$

where c is a constant independent of $\mathbf{B}$ and where p_j is the *a priori* distribution of the source amplitudes s_j. Noting that $\mathbf{y}(t) = \mathbf{B}\mathbf{x}(t)$, an iterative gradient-like algorithm which maximizes $\ln p(\mathbf{x}_{1..T}|\mathbf{B})$ can be formulated as

$$\mathbf{B}^{(k+1)} = \mathbf{B}^{(k)} - \gamma\mathbf{H}(\mathbf{y}) \quad \text{with} \quad \mathbf{H}(\mathbf{y}) = \boldsymbol{\phi}(\mathbf{y})\mathbf{y}^t - \mathbf{I}, \qquad (12.27)$$

where $\boldsymbol{\phi}(\mathbf{y}) = [\phi_1(y_1), \ldots, \phi_n(y_n)]^t$ with $\phi_j(z) = -\frac{p_j'(z)}{p_j(z)}$.

Note that the expression of $\phi_j(y_j)$ depends on both the prior distribution of the source amplitudes and their derivatives. Several particular cases are shown in Table 12.1.

The non-Gaussian case and the EM algorithm. In general, we do not have an analytical expression for the likelihood. A class of methods designed to obtain the maximum likelihood solution is based on the EM algorithm [22], where the main idea is to consider the sources as the hidden variables, observations $\mathbf{X}$ as incomplete data, and the ensemble $(\mathbf{X}, \mathbf{S})$ as the complete data (Fig. 12.1).

The EM algorithm can then be summarized as follows:

$$\begin{cases} \text{E:} \quad Q(\mathbf{A}, \widehat{\mathbf{A}}) = \int J(\mathbf{x}, \mathbf{s})\, p(\mathbf{x}, \mathbf{s}|\widehat{\mathbf{A}})\, d\mathbf{s} \\ \text{M:} \quad \widehat{\mathbf{A}} = \arg\max_{\mathbf{A}} \left\{ Q(\mathbf{A}, \widehat{\mathbf{A}}) \right\} \end{cases} \qquad (12.28)$$

Table 12.1 This table lists several source amplitude distributions $p(z)$ and the function $\phi(z) = -\frac{p'(z)}{p(z)}$ which are used in a great number of ICA algorithms

Gauss	$p(z) \propto \exp\left[-\alpha z^2\right]$	$\phi(z) = 2\alpha z$
Laplace	$p(z) \propto \exp\left[-\alpha\|z\|\right]$	$\phi(z) = \alpha \mathrm{sign}(z)$
Cauchy	$p(z) \propto \dfrac{1}{1+(z/\alpha)^2}$	$\phi(z) = \frac{2z/\alpha^2}{1+(z/\alpha)^2}$
Gamma	$p(z) \propto z^{\alpha} \exp\left[-\beta z\right]$	$\phi(z) = -\alpha/z + \beta$
sub-Gaussian	$p(z) \propto \exp\left[-\frac{1}{2}z^2\right] \mathrm{sech}^2(z)$	$\phi(z) = z + \tanh(z)$
Mixture of Gaussians	$p(z) \propto \exp\left[-\frac{1}{2}(z-\alpha)^2\right] + \exp\left[-\frac{1}{2}(z+\alpha)^2\right]$	$\phi(z) = \alpha z - \alpha \tanh(\alpha z)$

where $J(\mathbf{x}, \mathbf{s}) = \ln p(\mathbf{x}, \mathbf{s}|\mathbf{A}) = \ln p(\mathbf{x}|\mathbf{A}, \mathbf{s}) + \ln p(\mathbf{s})$ is the log-likelihood of the complete data. However, in general, the E-step computation cannot be done analytically. It is interesting to detail these two steps for the Gaussian case where $J(\mathbf{x}, \mathbf{s}) = \|\mathbf{x} - \mathbf{A}\mathbf{s}\|^2 + \lambda_s \|\mathbf{s}\|^2$. Applying this to the expression of $Q(\mathbf{A}, \widehat{\mathbf{A}})$ and keeping only the parts depending on $\mathbf{A}$, we obtain the following algorithm:

$$\begin{cases} \text{E:} \ \widehat{\mathbf{S}} = \mathbb{E}\left\{\mathbf{S}|\widehat{\mathbf{A}}, \mathbf{X}\right\} \\ \text{M:} \ \widehat{\mathbf{A}} = \arg\max_{\mathbf{A}} \left\{p(\mathbf{A}|\widehat{\mathbf{S}}, \mathbf{X})\right\} \end{cases} \tag{12.29}$$

which is called *restoration-maximization* and can be compared to the joint MAP algorithm where the first maximization with respect to $\mathbf{S}$ is replaced with the computation of the posterior mean. The two algorithms are then equivalent in the Gaussian case.

Integrating out A. In the same way that we integrated out $\mathbf{S}$ from $p(\mathbf{A}, \mathbf{S}|\mathbf{X})$ to obtain $p(\mathbf{A}|\mathbf{X})$, we can also integrate out $\mathbf{A}$ from $p(\mathbf{A}, \mathbf{S}|\mathbf{X})$ to obtain $p(\mathbf{S}|\mathbf{X})$. Again, except in the Gaussian case, we will not have analytical expression for $p(\mathbf{S}|\mathbf{X})$ and, again, using the EM algorithm in the Gaussian case, we will obtain the same *restoration-maximization* algorithm of the previous section.

12.5 PRIOR AND LIKELIHOOD ASSIGNMENTS

An important step in any source separation method is the appropriate modeling of the sources $p(\mathbf{S})$. Whichever model is selected, it will depend on a set of parameters $\boldsymbol{\theta}_s$, which are typically called hyperparameters as they are extra parameters needed to describe our model parameters. These hyperparameters may need to be estimated as well. This gives us a joint source prior that can be written as

$$p(\mathbf{S},\boldsymbol{\theta}_s) = p(\mathbf{S}|\boldsymbol{\theta}_s)p(\boldsymbol{\theta}_s). \tag{12.30}$$

The situation is the same for the mixing matrix $p(\mathbf{A}|\boldsymbol{\theta}_A)$ as well as the errors and noise $p(\mathbf{E}|\boldsymbol{\theta}_e)$, which consequently affects the likelihood $p(\mathbf{X}|\mathbf{A},\mathbf{S},\boldsymbol{\theta}_e)$. Typically, these hyperparameters are grouped together and denoted with a single vector symbol $\boldsymbol{\theta} = (\boldsymbol{\theta}_\epsilon,\boldsymbol{\theta}_s,\boldsymbol{\theta}_A)$. In a supervised learning problem, we may assume that these hyperparameters are known, but in general practice, the hyperparameters are considered to be unknowns to be estimated. As indicated in the joint source prior above, the Bayesian approach requires that we assign them a prior law $p(\boldsymbol{\theta})$, which can be factored into a series of independent terms

$$p(\mathbf{A},\mathbf{S},\boldsymbol{\theta}|\mathbf{X}) \propto p(\mathbf{X}|\mathbf{A},\mathbf{S},\Sigma_\epsilon)\, p(\mathbf{A}|\boldsymbol{\theta}_A)\, p(\mathbf{S}|\boldsymbol{\theta}_s)\, p(\boldsymbol{\theta}). \tag{12.31}$$

12.5.1 General assignments

Often prior and likelihood assignments can be made in relatively generic cases. For example, if the observation errors $\boldsymbol{\epsilon}_i$ are assumed to be independent, centered and Gaussian, we have

$$\begin{aligned} p(\mathbf{X}|\mathbf{A},\mathbf{S},\Sigma_\epsilon) &= \prod_t \mathcal{N}(\mathbf{A}\mathbf{s}(t),\Sigma_\epsilon) \\ &\propto \exp\left[\frac{-1}{2}\sum_t \left[(\mathbf{x}(t)-\mathbf{A}\mathbf{s}(t))'\Sigma_\epsilon^{-1}(\mathbf{x}(t)-\mathbf{A}\mathbf{s}(t))\right]\right] \\ &\propto \exp\left[\frac{-1}{2}\sum_t \text{trace}\left\{(\mathbf{x}(t)-\mathbf{A}\mathbf{s}(t))\,\Sigma_\epsilon^{-1}(\mathbf{x}(t)-\mathbf{A}\mathbf{s}(t))'\right\}\right]. \end{aligned}$$

The elements of the mixing matrix $\mathbf{A}$ reflect the coupling between the sources and detectors. In a clearly physical situation, these matrix elements depend on the physical transmission from the sources to the detectors. We will discuss prior assignments in these cases later. For now we consider the situation where the mixing process is modeled generically, or is not well-understood. The following choices are typical:

- Gaussians:

$$p(\mathbf{A}_{ij}) = \mathcal{N}(M_{ij},\sigma^2_{a,ij}) \longrightarrow p(\mathbf{A}|\boldsymbol{\theta}_A) = \mathcal{N}(\mathbf{M}_A,\Sigma_A).$$

A first choice is $M_{i,j} = 0, \sigma^2_{a,ij} = 1/\gamma, \forall i,j$, which gives

$$p(\mathbf{A}) \propto \exp\left[-\gamma||\mathbf{A}||^2\right].$$

Another choice is $M_{i,j} = 1$, for $i = j$ and $M_{i,j} = 0$, for $i \neq j$ with $\sigma^2_{a,ij} = \gamma, \forall i, j$, which gives

$$p(\mathbf{A}) \propto \exp\left[-\gamma||\mathbf{I} - \mathbf{A}||^2\right].$$

- Generalized Gaussian:

$$p(A_{i,j}) \propto \exp\left[-\gamma|A_{i,j}|\right] \longrightarrow p(\mathbf{A}) \propto \exp\left[-\gamma \sum_i \sum_j |A_{i,j}|^\alpha\right]$$

  with $1 \leqslant \alpha \leqslant 2$ and which has the capability of modeling sparsity of the mixing matrix elements.
- Gamma:

$$p(A_{i,j}) = \mathcal{G}(\alpha_j, \beta_j) \propto A_{i,j}^{\alpha_j - 1} \exp\left[-\beta_j A_{i,j}\right] I_{[0,+\infty]}$$

  which ensures non-negativity of the matrix elements in situations where inversion cannot occur, such as in spectrometry.
- Uniform: Uniform priors can be assigned in the general uninformative case.

For generic hyperparameters where a physical connection is not obvious, for convenience one often chooses a conjugate prior, such as the Inverse Gamma for variances, the Inverse Wishart for covariance matrices and Gaussians for means [74]. Next we discuss how to incorporate physical information into the separation algorithm via prior assignment.

12.5.2 Physical priors

In many problems, the hyperparameters have a *physical significance* which is also of importance for the researcher. In these cases, the physical laws governing signal propagation can be used to derive *relevant priors*. In doing so, this places physical constraints on the solutions, sometimes even solving the indeterminacy problems related to source scaling. In many situations, these hyperparameters are related to the source characteristics, such as position or orientation. We will later see that this enables us to derive algorithms that simultaneously perform source separation and localization, or source characterization.

Here we will work out a straightforward example where the mixing matrix elements depend on the signal propagation from source to detector. Given a propagation law, we can derive appropriate prior probabilities. Consider a sound propagation problem where the sound energy follows an inverse-square propagation law. If we know the distance r_{ij} from source j to detector i, the mixing matrix element $A_{i,j}$ must follow this inverse-square propagation law

$$A_{i,j} = \frac{1}{4\pi r_{ij}^2}. \tag{12.32}$$

Let us assume that we are completely ignorant of the position of the source within the three-dimensional sphere of radius R surrounding the detector. This ignorance can be quantified by assigning a uniform probability for its position in spherical coordinates (r, θ, ϕ) within any volume element of that space

$$p(r, \theta, \phi | I) = \frac{1}{V} = \frac{3}{4\pi R^3}, \tag{12.33}$$

where V is the spherical volume of radius R surrounding the detector. This is the prior probability that the source is at any position with respect to the detector.

However, we only need the probability that the source is some distance r from the detector. This can be obtained by marginalizing over all possible values of the angular coordinates

$$p(r|I) = \int_0^{2\pi} d\phi \int_0^{\pi} \sin\theta d\theta r^2 p(r, \theta, \phi | I) \tag{12.34}$$

$$= \int_0^{2\pi} d\phi \int_0^{\pi} \sin\theta d\theta r^2 \frac{3}{4\pi R^3} \tag{12.35}$$

$$= \frac{3r^2}{R^3}.$$

The prior on the source-detector distance is very reassuring since it is naturally invariant with respect to coordinate rescaling (change of units). Specifically if we introduce a new coordinate system so that $\rho = ar$ and $P = aR$ with $a > 0$, we find equating the probabilities around $\rho + d\rho$ and $r + dr$ that

$$|p(\rho|I) d\rho| = |p(r|I) dr| \tag{12.36}$$

$$p(\rho|I)|d\rho| = p(r|I)|dr|$$

$$p(\rho|I) = p(r|I) \left|\frac{d\rho}{dr}\right|^{-1}$$

$$p(\rho|I) = \frac{3r^2}{R^3} \left|\frac{d\rho}{dr}\right|^{-1}$$

$$p(\rho|I) = \frac{3a\rho^2}{P^3} |a|^{-1}$$

$$p(\rho|I) = \frac{3\rho^2}{P^3}.$$

Since this prior is invariant with respect to coordinate rescaling, we can measure distances using any units we wish.

We can now use this to derive a prior for the mixing matrix element. First write the joint probability using the product rule

$$p(A_{i,j}, r|I) = p(r|I)p(A_{i,j}|r, I). \tag{12.37}$$

The first term on the right is the source-detector distance prior, and the second term is a delta function described by the hard constraint of the inverse-square law. Some readers may wonder why we go through the difficulty of using delta functions rather than computing the Jacobians and just performing a change of variables with the probability densities as we did before when demonstrating invariance with respect to rescaling. The reason is that in more complex problems where the parameter of interest depends on multiple other parameters, the change of variables technique becomes extremely difficult. Care must be taken when using delta functions however, since the argument needs to be written so that it is solved for the parameter of interest, in this case $A_{i,j}$ rather than another parameter such as r.

These assignments give

$$p(A_{i,j}, r|I) = \frac{3r^2}{R^3}\delta(A_{i,j} - (4\pi r^2)^{-1}). \tag{12.38}$$

We now marginalize over all possible values of r

$$p(A_{i,j}|I) = \int_0^R dr \frac{3r^2}{R^3}\delta(A_{i,j} - (4\pi r^2)^{-1}). \tag{12.39}$$

To do this we will need to make a change of variables again by defining $u = (A_{i,j} - (4\pi r^2)^{-1})$, so that

$$r^2 = [(4\pi)(A_{i,j} - u)]^{-1} \tag{12.40}$$

$$r^3 = [(4\pi)(A_{i,j} - u)]^{-3/2} \tag{12.41}$$

and $du = (2\pi r^3)^{-1} dr$, which can be rewritten as

$$dr = 2^{-3/2}(2\pi)^{-1/2}(A_{i,j} - u)^{-3/2} du \tag{12.42}$$

giving us

$$p(A_{i,j}|I) = 2^{-4}\pi^{-3/2}\frac{3}{R^3}\int_{-\infty}^{u_{max}} du (A_{i,j} - u)^{-5/2}\delta(u), \tag{12.43}$$

where $u_{max} = A_{i,j} - (4\pi R^2)^{-1}$. The delta function will select $u = 0$ as long as it is true that $u_{max} > 0$ or equivalently that $r < R$. If this is not true, the integral will be zero.

If this hard constraint of the sources being within a distance R of the detectors causes problems, your choice for R was too small. The result is

$$p(A_{i,j}|I) = \frac{3}{16\pi^{3/2}R^3} A_{i,j}^{-5/2}. \tag{12.44}$$

so that the prior for the mixing matrix elements is proportional to $A_{i,j}^{-5/2}$. Readers more familiar with statistics may note (and perhaps worry) that this prior is improper since it blows up as $A_{i,j}$ goes to zero. This is not a practical concern since the values of $A_{i,j}$ are bounded on the left by the maximal value of $r < R$ and on the right by the fact that we are working in the far-field regime so that the source cannot be too close to the detector.

12.6 SOURCE MODELING

We now consider the rich topic of *modeling the sources*. To be able to separate sources, it is important that each source differs from the other in some respect. It is the respect in which the sources differ that will lead to drastically different source separation algorithms. It has been established that for Gaussian-distributed sources, separation is not possible if they do not differ in either their temporal or spectral behavior. This is also true for non-Gaussian sources, which are inseparable unless they differ in their prior probabilities or likelihood. Many non-Gaussian models have been used in a wide array of applications. Among the most common source models is the *Generalized Gaussian* (GG) [70, 75,28], *gamma distribution* [49] and *mixture of Gaussians* (MoG) [75,64,28,27,29].

In the following, we distinguish two cases of *stationary* and *non-stationary* sources. First we consider the GG, Gamma and mixture of Gaussian (MoG) models, giving more emphasis on the latter, which naturally introduces the notion of a second level of *hidden variables*. Second, we propose models with hidden variables, which gives the possibility of accounting for the *non-stationarity*. These models include a mixture of Gaussian models with *hidden Markovian models*.

12.6.1 Modeling stationary white sources

1. Generalized Gaussian

In this model, the probability law of the source $s_j(t)$ is:

$$p(s_j|\gamma_j, \beta_j) \propto \exp\left[-\beta_j |s_j|^{\gamma_j}\right], \quad 1 \leqslant \gamma_j \leqslant 2 \tag{12.45}$$

So, the joint probability law of the sources is parameterized by the vector of parameters $\boldsymbol{\beta} = \{\beta_1, \ldots, \beta_n\}$ and $\boldsymbol{\gamma} = \{\gamma_1, \ldots, \gamma_n\}$. Figure 12.2 shows an example of such sources compared to the Gaussian case.

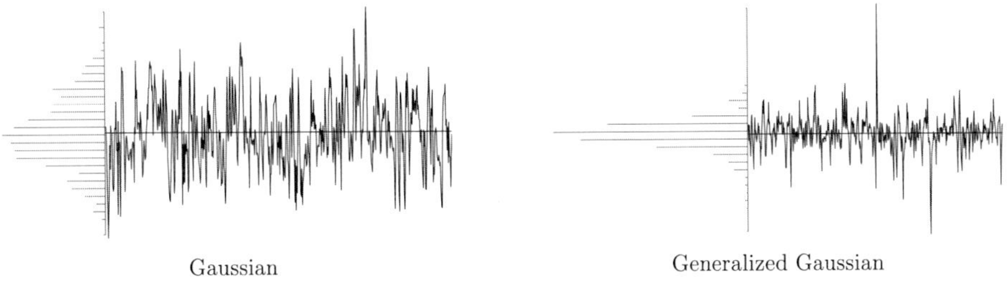

$$p(s_j(t)) = \mathcal{N}(0, \sigma_j^2) \propto \exp\left[-\frac{1}{\sigma_j^2}|s_j(t)|^2\right] \qquad p(s_j(t)) \propto \exp\left[-\beta_j |s_j(t)|^{\gamma_j}\right]$$

$$p(s_j(t), t = 1..T) \propto \exp\left[-\frac{1}{\sigma_j^2}\sum_t |s_j(t)|^2\right] \qquad p(s_j(t), t = 1..T) \propto \exp\left[-\beta_j \sum_t |s_j(t)|^{\gamma_j}\right]$$

FIGURE 12.2

Examples of Gaussian and GG sources.

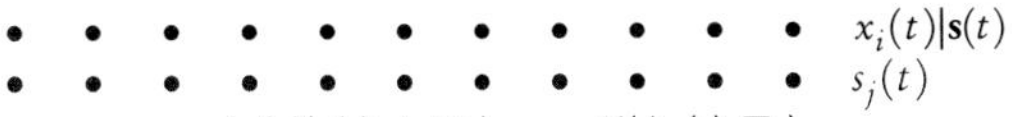

$$\begin{aligned} p(\mathbf{x}(t)|\mathbf{s}(t), \mathbf{A}, \Sigma_\epsilon) &= \mathcal{N}(\mathbf{A}\mathbf{s}(t), \Sigma_\epsilon) \\ p(s_j(t)|\gamma_j, \beta_j) &\propto \exp\left[-\beta_j |s_j|^{\gamma_j}\right] \end{aligned}$$

FIGURE 12.3

A model of source separation where the source hyperparameters are considered to be hidden variables.

By assigning a Gaussian likelihood, we have a model where the source hyperparameters are considered to be hidden variables as shown in Fig. 12.3. It is easy to show that:

$$-\ln p(\mathbf{S}, \mathbf{A}|\mathbf{X}, \boldsymbol{\gamma}, \boldsymbol{\beta}) \propto -\sum_t \left[(\mathbf{x}(t) - \mathbf{A}\mathbf{s}(t))' \boldsymbol{\Sigma}_\epsilon^{-1} (\mathbf{x}(t) - \mathbf{A}\mathbf{s}(t)) + \sum_j \beta_j |s_j(t)|^{\gamma_j}\right]$$

and if we can fix *a priori* the hyperparameters, then we can use the following algorithm:

$$\begin{cases} \widehat{s}_j^{(k+1)}(t) = \arg\min_{s_j(t)} \left\{J_1(s_j(t)) = -\ln p\left(s_j(t)|\mathbf{x}(t), \widehat{\mathbf{A}}^{(k)}, \widehat{\mathbf{s}}_{k\neq j}^{(k)}(t)\right)\right\} \\ \widehat{\mathbf{A}}^{(k+1)} = \arg\min_{\mathbf{A}} \left\{J_2(\mathbf{A}) = -\ln p(\mathbf{A}|\widehat{\mathbf{S}}^{(k)}, \mathbf{X})\right\} \end{cases}$$

where:

$$\begin{cases} J_1(s_j(t)) = \sum_i \left| x_i(t) - A_{i,j} s_j(t) - \sum_{j' \neq j} A_{ij'} \widehat{s}_{j'}^{(k)}(t) \right|^2 + \beta_j \sum_j |s_j(t)|^{\gamma_j} \\ J_2(\mathbf{A}) = \sum_t \|\mathbf{x}(t) - \mathbf{A}\widehat{\mathbf{s}}(t)\|^2 - \lambda_A \ln p(\mathbf{A}). \end{cases}$$

One can use different optimization algorithms such as fixed point or gradient based to realize these two successive optimization steps.

When we choose a uniform or Gaussian prior law for the elements of the mixing matrix $\mathbf{A}$, the second criterion $J_2(\mathbf{A})$ becomes a quadratic function of $\mathbf{A}$ and its optimum has an explicit expression. With other choices, in general, there is not an analytical solution, but the solution can be computed numerically using an optimization algorithm.

Conjugate priors are a common choice for assigning prior probabilities to the mixing matrix and the other hyperparameters $\boldsymbol{\theta} = (\mathbf{A}, \Sigma_\epsilon, \boldsymbol{\gamma}, \boldsymbol{\beta})$. This is done mainly because it ensures that the posterior probability will have the same functional form as the likelihood. This results in the posterior probability

$$p(\mathbf{A}, \Sigma_\epsilon, \boldsymbol{\gamma}, \boldsymbol{\beta} | \mathbf{x}_{1..T}) \propto p(\mathbf{x}_{1..T} | \mathbf{A}, \Sigma_\epsilon, \boldsymbol{\gamma}, \boldsymbol{\beta})\, p(\mathbf{A})\, p(\Sigma_\epsilon)\, p(\boldsymbol{\gamma})\, p(\boldsymbol{\beta}). \tag{12.46}$$

This allows us first to estimate $(\mathbf{A}, \Sigma_\epsilon, \boldsymbol{\gamma}, \boldsymbol{\beta})$ and then use the results for the estimation of the sources. However, obtaining an expression for $p(\mathbf{A}, \Sigma_\epsilon, \boldsymbol{\gamma}, \boldsymbol{\beta} | \mathbf{x}_{1..T})$ is difficult, and again we rely on the EM algorithm. The main difficulty again is the computation of the E-step, which can also be done using approximations or with Markov chain Monte Carlo (MCMC) sampling (Rao-Blackwell or SEM [70]).

Another possibility is to use variational Bayes methods which attempt to approximate the posterior probability with a separable distribution

$$\left[p(\mathbf{A}, \Sigma_\epsilon, \boldsymbol{\gamma}, \boldsymbol{\beta} | \mathbf{x}_{1..T}) \simeq \prod_t p(\mathbf{A}, \Sigma_\epsilon | \mathbf{x}(t), \mathbf{s}(t))\; p(\boldsymbol{\gamma} | \mathbf{s}(t))\; p(\boldsymbol{\beta} | \mathbf{s}(t)). \right]$$

Such an approach is demonstrated in [29].

2. Mixture of Gaussians

The mixture of Gaussians (MoG) model describes the source amplitude distribution $s_j(t)$ as a linear combination of Gaussians

$$p(s_j | \alpha_{jk}, m_{jk}, \sigma_{jk}) = \sum_{k=1}^{K_j} \alpha_{jk} \mathcal{N}(m_{jk}, \sigma_{jk}^2), \quad \text{with} \sum_{k=1}^{K_j} \alpha_{jk} = 1 \tag{12.47}$$

which is parameterized by $3 \times K_j$, parameters $\mathbf{m}_{j.} = \{m_{jk}, k = 1, \ldots, K_j\}$ and $\boldsymbol{\sigma}_{j.} = \{\sigma_{jk}^2, k = 1, \ldots, K_j\}$. The sources are then described by $\mathbf{K} = \{K_1, \ldots, K_n\}$, $\boldsymbol{\alpha} = \{\alpha_{jk}\}$, $\mathbf{m} = \{\mathbf{m}_{j.}, j = 1, \ldots, n\}$ and $\boldsymbol{\sigma} = \{\boldsymbol{\sigma}_{j.}, j = 1, \ldots, n\}$.

An important feature of MoG modeling is interpretation of $\alpha_{jk} = P(z_j(t) = k)$ as the probabilities of hidden discrete-valued variables $z_j(t)$, such that

$$p(s_j(t) | z_j(t) = k, m_{jk}, \sigma_{jk}) = \mathcal{N}(m_{jk}, \sigma_{jk}^2) \quad \text{and} \quad P(z_j(t) = k) = \alpha_{jk}. \tag{12.48}$$

$$
\begin{aligned}
p(\mathbf{x}(t)|\mathbf{s}(t),\mathbf{A},\Sigma_\epsilon) &= \mathcal{N}(\mathbf{A}\mathbf{s}(t),\Sigma_\epsilon) \\
p(s_j(t)|z_j(t)=k,m_{jk},\sigma_{jk}) &= \mathcal{N}(m_{jk},\sigma_{jk}) \\
P(z_j(t)=k) &= \alpha_{jk}
\end{aligned}
$$

FIGURE 12.4

Mixture of Gaussians (MoG) and related hidden variables.

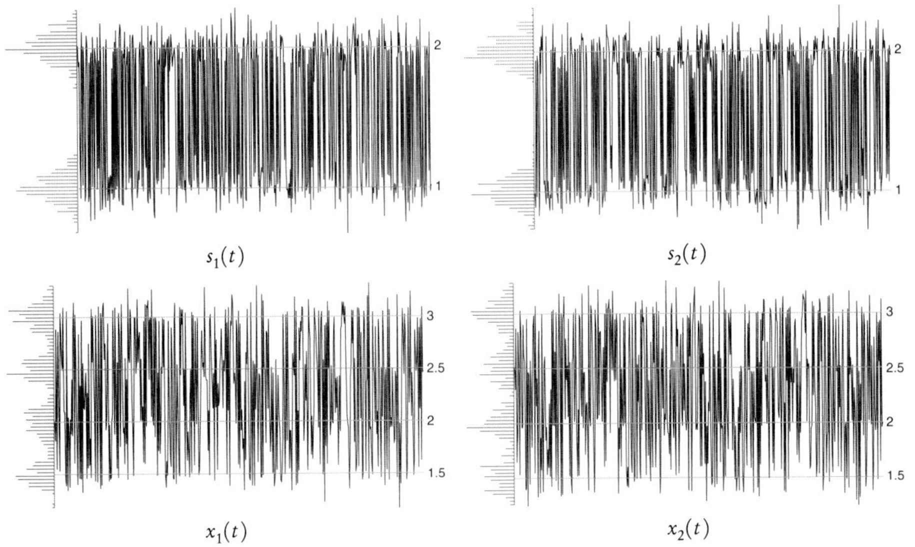

FIGURE 12.5

Conservation of the MoG model. The sources s_1 and s_2 are modeled as mixtures of two Gaussians. As a result, the observations x_1 and x_2 are also MoG distributed with four components instead of two.

The structure of this model is illustrated in Fig. 12.4. So, the problem then reduces to an estimation of $\mathbf{s}(t) = \{s_1(t),\ldots,s_n(t)\}$, $\mathbf{z}(t) = \{z_1(t),\ldots,z_n(t)\}$ and the hyperparameters $(\mathbf{A},\Sigma_\epsilon)$ and $\boldsymbol{\theta}_s = (\mathbf{m},\boldsymbol{\sigma})$.

We may note that when the sources are modeled by MoG, the observations $x_i(t)$ are also described by a MoG

$$p\left(x_i(t)|\mathbf{A},\mathbf{z}(t),\mathbf{m},\boldsymbol{\sigma}\right) = \mathcal{N}\left(\sum_j A_{i,j} m_{jz_j(t)}, \sum_j A_{i,j}\sigma_{jz_j(t)}A_{ji} + \sigma^2_{\epsilon_i}\right) \tag{12.49}$$

with $K_i = \prod_j K_j$ components. This effect is illustrated in Fig. 12.5.

We have used the MoG model in BSS in a variety of applications. Here, we mention an algorithm using this model designed to perform a joint estimate of the sources $\mathbf{S}$, the hidden variables $\mathbf{Z}$, the mixing matrix $\mathbf{A}$ and the hyperparameters m_{jk} and σ_{jk} by maximizing $p(\mathbf{S},\mathbf{Z},\mathbf{A},\boldsymbol{\theta}|\mathbf{X})$, with respect to its arguments

$$\begin{cases} \widehat{\mathbf{s}}(t) &= \arg\min_{\mathbf{s}(t)} \left\{ J_1(\mathbf{s}(t)) = -\ln p\left(\mathbf{s}(t)|\mathbf{x}(t),\widehat{\mathbf{z}}(t),\widehat{\mathbf{A}},\widehat{\mathbf{s}}(t)\right)\right\} \\ \widehat{\mathbf{A}} &= \arg\min_{\mathbf{A}} \left\{ J_2(\mathbf{A}) = -\ln p\left(\mathbf{A}|\widehat{\mathbf{S}},\mathbf{X}\right)\right\} \\ \widehat{\mathbf{z}}(t) &= \arg\min_{\mathbf{z}(t)} \left\{ J_3(\mathbf{z}(t)) = -\ln p\left(\mathbf{z}(t)|\widehat{\mathbf{s}}(t),\widehat{\boldsymbol{\theta}}\right)\right\} \\ \widehat{\boldsymbol{\theta}} &= \arg\min_{\boldsymbol{\theta}} \left\{ J_4(\boldsymbol{\theta}) = -\ln p\left(\boldsymbol{\theta}|\widehat{\mathbf{s}}(t),\widehat{\mathbf{z}}(t)\right)\right\}. \end{cases} \tag{12.50}$$

We have also used the same joint posterior probability for computing other estimators such the posterior means using a MCMC and Gibbs sampling scheme using the appropriate marginals $p(\mathbf{S}|\mathbf{A},\mathbf{Z},\boldsymbol{\theta},\mathbf{X})$, $p(\mathbf{Z}|\mathbf{S},\boldsymbol{\theta})$, $p(\mathbf{A}|\mathbf{S},\boldsymbol{\theta},\mathbf{X})$ and $p(\boldsymbol{\theta}|\mathbf{A},\mathbf{S},\mathbf{X})$

$$\begin{cases} \widehat{\mathbf{s}}(t) &\sim p(\mathbf{s}(t)|\widehat{\mathbf{z}}(t),\widehat{\mathbf{A}},\widehat{\boldsymbol{\theta}},\mathbf{x}(t)) \\ \widehat{\mathbf{A}} &\sim p(\mathbf{A}|\widehat{\mathbf{S}},\mathbf{X}) \\ \widehat{\mathbf{z}}(t) &\sim p(\mathbf{z}(t)|\widehat{\mathbf{s}}(t),\widehat{\boldsymbol{\theta}}) \\ \widehat{\boldsymbol{\theta}} &\sim p(\boldsymbol{\theta}|\widehat{\mathbf{s}}(t),\widehat{\mathbf{z}}(t)). \end{cases} \tag{12.51}$$

For more details, see [72,73,70,75].

12.6.2 Accounting for temporal correlations of the sources

It is not always realistic to assume that the sources $\mathbf{s}(t)$ are temporally white. The fact that sources are often not white is the main reason that whitening is performed before applying the source separation algorithm. However, performing this whitening step and then applying BSS is not optimal. In the Bayesian approach, it is possible to account for this directly through the prior modeling using Markovian models both in one and two dimensions. The case of Gauss-Markov models is interesting, because they are well understood through the correlation functions and power spectral density notions.

A Gauss-Markov model of order 1 (Fig. 12.6), in the one-dimensional case is

$$\begin{aligned} p(s_j(t)|s_j(t-1)) &= \mathcal{N}(\alpha_j s_j(t-1),\sigma_j^2) \\ &\propto \exp\left[\frac{-1}{2\sigma_j^2}\left(s_j(t)-\alpha_j s_j(t-1)\right)^2\right], \end{aligned} \tag{12.52}$$

which can also be written as an Auto-Regressive (AR) process of order 1

$$s_j(t)=\alpha_j s_j(t-1)+\eta_j(t), \quad p(\eta_j(t))=\mathcal{N}(0,\sigma_j^2). \tag{12.53}$$

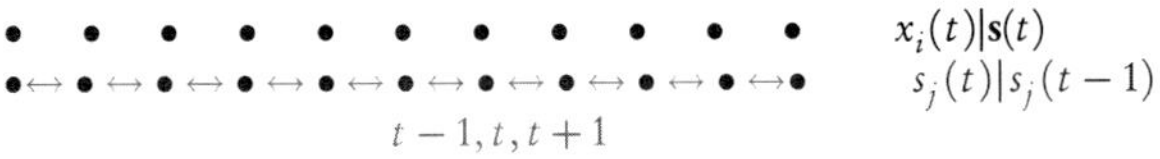

FIGURE 12.6

A Gauss-Markov model for the sources.

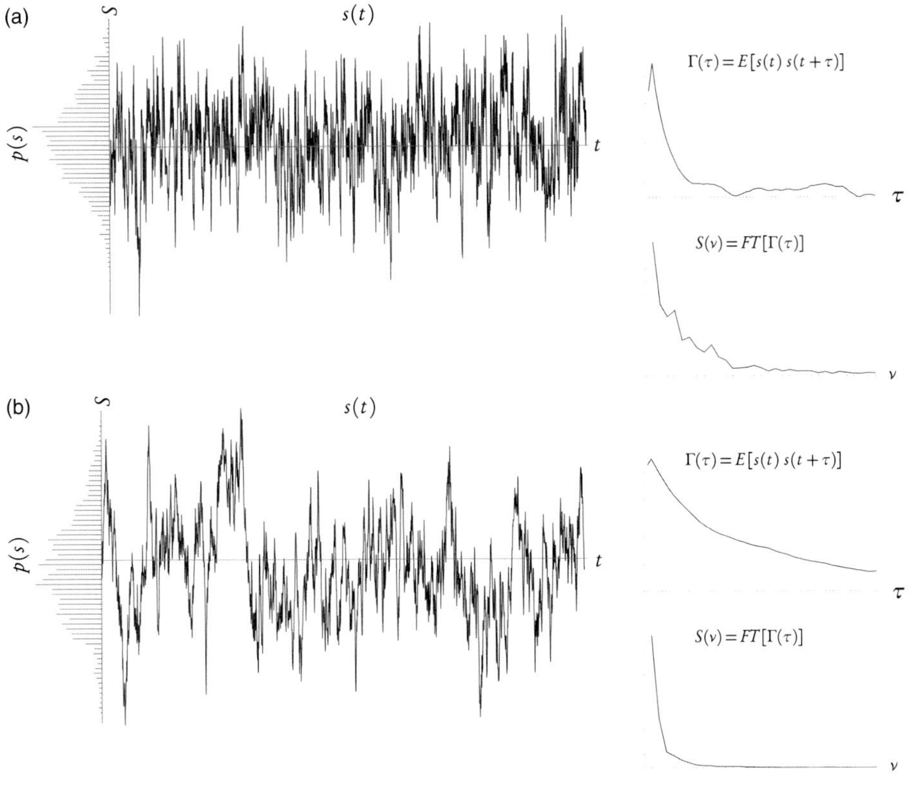

FIGURE 12.7

Two AR signals with parameters .7 and .95. On the right, we have their autocorrelation functions $\Gamma(\tau)$ and their psd $S(\nu)=TF[\Gamma(\tau)]$.

An example of two AR-modeled signals is illustrated in Fig. 12.7. These models are characterized by two parameters α and σ^2, but very often a model with only one parameter is used

$$p(s_j(t)|s_j(t-1)) = \mathcal{N}\left(s_j(t-1), 1/(2\beta_j)\right) \\ \propto \exp\left[-\beta_j\left(s_j(t)-s_j(t-1)\right)^2\right], \tag{12.54}$$

which can also be written as

$$p(s_j(t), t \in \mathcal{T}) \propto \exp\left[-\beta_j \sum_j \left(s_j(t) - s_j(t-1)\right)^2\right]. \tag{12.55}$$

The Gauss-Markov model can be extended to images by considering

$$p(s_j(\mathbf{r})|s_j(\mathbf{r}'), \mathbf{r}' \in \mathcal{V}(\mathbf{r})) \propto \exp\left[-\beta_j \left(s_j(\mathbf{r}) - \sum_{\mathbf{r}' \in \mathcal{V}(\mathbf{r})} s_j(\mathbf{r}')\right)^2\right] \tag{12.56}$$

$$p(s_j(\mathbf{r}), \mathbf{r} \in \mathcal{R}) \propto \exp\left[-\beta_j \sum_{\mathbf{r} \in \mathcal{R}} \left(s_j(\mathbf{r}) - \sum_{\mathbf{r}' \in \mathcal{V}(\mathbf{r})} s_j(\mathbf{r}')\right)^2\right]. \tag{12.57}$$

12.6.3 Modeling non-stationary sources

Often the assumption of stationarity of the sources and observations is inappropriate. This is particularly true when the sources are images with contours and regions. In these cases, it is more appropriate to use Gauss-Markov models with hidden variables describing contours or region labels.

12.6.3.1 Markov chains with jumps

Discontinuous changes in the source behavior can be modeled using a binary-valued hidden variable $q_j(t)$. Here we describe two useful models

$$p(s_j(t)|s_j(t-1), q_j(t)) = \begin{cases} \mathcal{N}(\alpha_j s_j(t-1), \sigma_j^2) & \text{if } q_j(t) = 0, \\ \mathcal{N}(0, \sigma_j^2) & \text{if } q_j(t) = 1, \end{cases} \tag{12.58}$$

with two parameters (α_j, σ_j^2), and

$$\begin{aligned} p(s_j(t)|s_j(t-1), q_j(t)) &= \begin{cases} \mathcal{N}(s_j(t-1), 1/(2\beta_j)) & \text{if } q_j(t) = 0, \\ \mathcal{N}(0, 1/(2\beta_j)) & \text{if } q_j(t) = 1, \end{cases} \\ &\propto \begin{cases} \exp\left[-\beta_j (s_j(t) - s_j(t-1))^2\right] & \text{if } q_j(t) = 0, \\ \exp\left[-\beta_j (s_j(t))^2\right] & \text{if } q_j(t) = 1, \end{cases} \end{aligned}$$

with one parameter $\beta_j = 2/\sigma_j^2$. The structure of this model is illustrated in Fig. 12.8, and an example is illustrated in Fig. 12.9.

To model these hidden variables, we can consider two situations. First, we assume that the $q_j(t)$ are independent. This results in the Bernoulli-Gaussian model

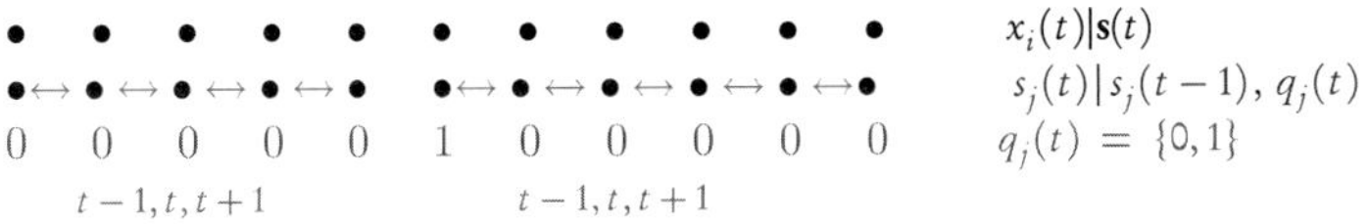

$$p(s_j(t)|s_j(t-1), q_j(t)) \propto \exp\left[-\beta_j(s_j(t) - (1-q_j(t))s_j(t-1))^2\right]$$
$$p(s_j(t), t \in \mathcal{T}|\mathbf{q}) \propto \exp\left[-\beta_j \textstyle\sum_t (s_j(t) - (1-q_j(t))s_j(t-1))^2\right]$$

FIGURE 12.8

Markov chain with hidden binary variables modeling jumps and contours.

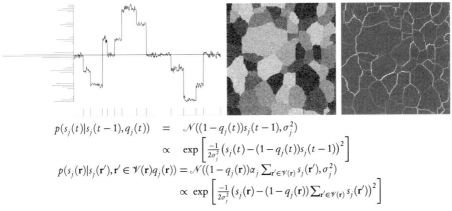

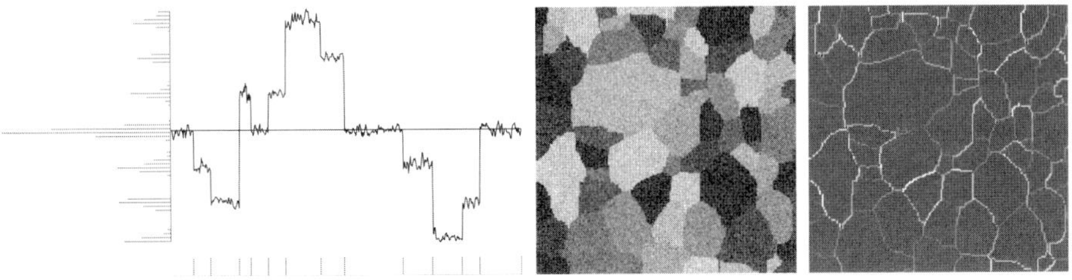

$$\begin{aligned} p(s_j(t)|s_j(t-1), q_j(t)) &= \mathcal{N}((1-q_j(t))s_j(t-1), \sigma_j^2) \\ &\propto \exp\left[\tfrac{-1}{2\sigma_j^2}(s_j(t) - (1-q_j(t))s_j(t-1))^2\right] \end{aligned}$$
$$\begin{aligned} p(s_j(\mathbf{r})|s_j(\mathbf{r}'), \mathbf{r}' \in \mathcal{V}(\mathbf{r}) q_j(\mathbf{r})) &= \mathcal{N}((1-q_j(\mathbf{r}))\alpha_j \textstyle\sum_{\mathbf{r}' \in \mathcal{V}(\mathbf{r})} s_j(\mathbf{r}'), \sigma_j^2) \\ &\propto \exp\left[\tfrac{-1}{2\sigma_j^2}(s_j(\mathbf{r}) - (1-q_j(\mathbf{r}))\textstyle\sum_{\mathbf{r}' \in \mathcal{V}(\mathbf{r})} s_j(\mathbf{r}'))^2\right] \end{aligned}$$

FIGURE 12.9

Markov models with hidden variables. The one-dimensional case is shown on the left and the two-dimensional case is shown on the right.

$$\begin{cases} p(q_j(t) = 1) &= \lambda_j, \\ p(q_j(t) = 0) &= 1 - \lambda_j, \end{cases} \longrightarrow p(q_j(t), t \in \mathcal{T}) = \lambda_j^{\sum_t q_j(t)} (1-\lambda_j)^{\sum_t (1-q_j(t))}.$$

If we assume that the $q_j(t)$ are Markovian we then have

$$p(q_j(t) = k | q_j(t-1) = l) = p_{j_{kl}}, \quad k, l = \{0, 1\}, \quad p_{j_{kl}} = \begin{bmatrix} \alpha_j & 1-\beta_j \\ 1-\alpha_j & \beta_j \end{bmatrix}$$

for the one-dimensional case, and

$$p(q_j(\mathbf{r}), \mathbf{r} \in \mathcal{R}) \propto \exp\left[-\alpha_j \sum_{\mathbf{r}} \sum_{\mathbf{r}' \in \mathcal{V}(\mathbf{r})} \delta(q_j(\mathbf{r}) - q_j(\mathbf{r}'))\right] \tag{12.59}$$

for two-dimensional images.

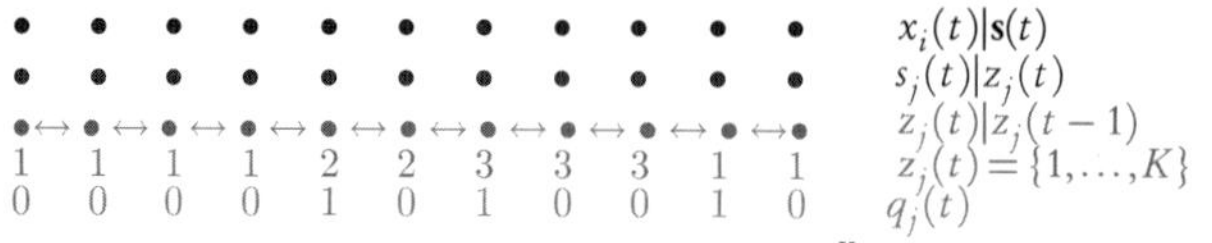

$$p(s_j(t)|z_j(t)=k)=\mathcal{N}(m_{jk},\sigma_{jk}) \longrightarrow p(s_j(t)) = \sum_{k=1}^{K} p(z_j(t)=k)\,\mathcal{N}(m_{jk},\sigma_{jk}),$$

$$p(z_j(t) = k) = p_{jk}, \qquad p(z_j(t) = k|z_j(t-1) = l) = p_{jkl}$$

FIGURE 12.10

A MoG model with hidden Markov labels (segments).

12.6.3.2 Hierarchical Markovian models with hidden classification variables

An important class of signals and images are those which are piecewise homogeneous, and described by a finite number K of classes. To model such sources we introduce a hidden discrete-valued variable $z_j(t)$, which represents the classification labels taking the values $\{1,2,\ldots,K\}$. The most simple model for all the pixels in a given class k is

$$p(s_j(t)|z_j(t)=k)=\mathcal{N}(m_{jk},\sigma_{jk}) \tag{12.60}$$

which gives

$$p(s_j(t))=\sum_{k=1}^{K} p(z_j(t)=k)\,\mathcal{N}(m_{jk},\sigma_{jk}), \tag{12.61}$$

which is a MoG model.

Again, we can consider two cases where $z_j(t)$ evaluated at different times t is independent or described as a Markov chain

$$p(z_j(t)=k)=p_{jk}, \qquad p(z_j(t)=k|z_j(t-1)=l)=p_{jkl}. \tag{12.62}$$

The structure of this MoG model with hidden Markov labels is illustrated in Fig. 12.10. Similarly, for images, we have

$$p(s_j(\mathbf{r})|z_j(\mathbf{r})=k)=\mathcal{N}(m_{jk},\sigma_{jk}) \tag{12.63}$$

where $z_j(\mathbf{r})$ is modeled by a Markov field

$$p(z_j(\mathbf{r}),\mathbf{r}\in\mathcal{R})\propto \exp\left[-\alpha_j \sum_{\mathbf{r}} \sum_{\mathbf{r}'\in\mathcal{V}(\mathbf{r})} \delta(z_j(\mathbf{r})-z_j(\mathbf{r}'))\right]. \tag{12.64}$$

This model, illustrated in Fig. 12.11 is called a Potts model, which reduces to the Ising model when the number of classes $K=2$.

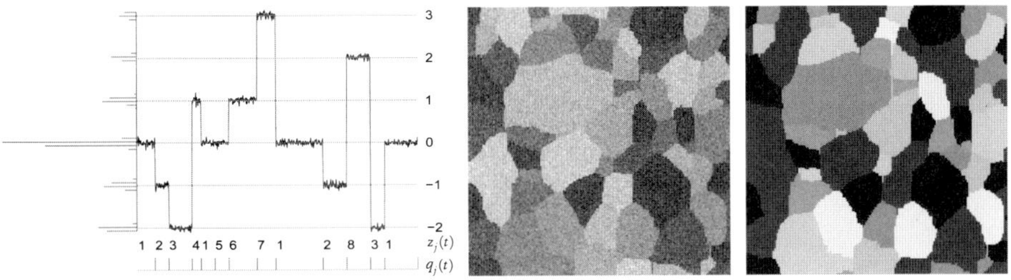

$$p(s_j(\mathbf{r})|z_j(\mathbf{r}) = k) = \mathcal{N}(m_{jk}, \sigma_{jk})$$

$$p(z_j(\mathbf{r}), \mathbf{r} \in \mathcal{R}) \propto \exp\left[-\alpha_j \sum_{\mathbf{r}\in\mathcal{R}} \sum_{\mathbf{r}'\in\mathcal{V}(\mathbf{r})} \delta(z_j(\mathbf{r}) - z_j(\mathbf{r}'))\right]$$

FIGURE 12.11

A MoG field with hidden Markov labels (regions).

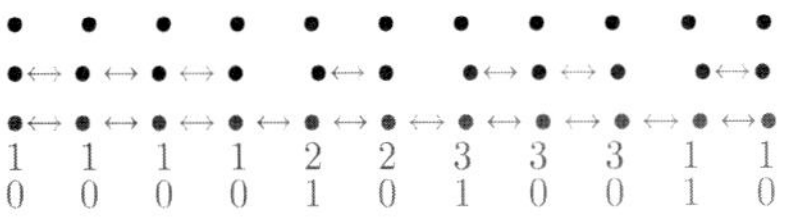

$x_i(t)|\mathbf{s}(t)$
$s_j(t)|s_j(t-1), z_j(t), z_j(t-1)$
$z_j(t)|z_j(t-1)$
$z_j(t) = \{1, \ldots, K\}$
$q_j(t) = 1 - \delta(z_j(t) - z_j(t-1))$

FIGURE 12.12

Markov chain with a hidden Potts classification model.

We may note that $z_j(\mathbf{r})$ represents a segmentation of the image, and in this case we have a deterministic link between $z_j(t)$ and $q_j(t)$

$$q_j(t) = \delta(z_j(t) - z_j(t-1)),$$

where $\delta(t) = 1$ if $t = 0$ and $\delta(t) = 0$ if $t \neq 0$. The structure of a Markov chain with a hidden Potts classification model is illustrated in Fig. 12.12.

12.6.3.3 Gauss-Markov-Potts model

An extension of the previous models is a doubly Markovian model, which can be considered as a Mixture of Gauss-Markov fields. Here we discuss six different variations of this basic model, which are illustrated in Fig. 12.13.

Model 1 corresponds to the models used in many classical source separation methods, such as PCA and ICA. Different choices for the source amplitude prior $p(s_j)$ results in different algorithms. Generalized Gaussian (GG) models have been used successfully either directly or when the source separation problem is expressed in the Fourier or wavelet domains where sparsity is a relevant characteristic.

Model 2 is the MoG model and is also used in classic solutions. The MoG model is ideal for modeling non-Gaussian sources. We may note that sources s_j are iid only conditionally on z_j.

Model 1: A simple iid model for $s_j(t)$

$x_i(t)|\mathbf{s}(t)$
$s_j(t)$

Model 2: An iid model for $s_j(t)|z_j(t)$ with iid hidden variables $z_j(t)$. This is equivalent to Model 1 with an MoG model for $s_j(t)$

$x_i(t)|\mathbf{s}(t)$
$s_j(t)|z_j(t)$
$z_j(t) \in \{1,\ldots,K\}$

Model 3: A simple Markovian model for the sources $s_j(t)$ which accounts for the temporal dynamics of the observations.

$x_i(t)|\mathbf{s}(t)$
$s_j(t)|s_j(t-1)$

$t-1, t, t+1$

Model 4: A compound Markovian model with hidden binary contour variables $q_j(t)$.

$x_i(t)|\mathbf{s}(t)$
$s_j(t)|s_j(t-1), q_j(t)$
$q_j(t) = \{0, 1\}$

Model 5: A compound Markovian model where $s_j(t)|z_j(t)$ are iid, and the hidden labels $z_j(t)$ are modeled by the Potts model. This model translates the homogeneity inside each region. The contours $q_j(t)$ can be obtained in a deterministic way from $z_j(t)$, and are closed by construction.

$x_i(t)|\mathbf{s}(t)$
$s_j(t)|z_j(t)$
$z_j(t)|z_j(t-1)$
$z_j(t) = \{1,\ldots,K\}$
$q_j(t)$

Model 6: Gauss-Markov-Potts model.

$x_i(t)|\mathbf{s}(t)$
$s_j(t)|s_j(t-1), z_j(t), z_j(t-1)$
$z_j(t)|z_j(t-1)$
$z_j(t) = \{1,\ldots,K\}$
$q_j(t) = 1 - \delta(z_j(t) - z_j(t-1))$

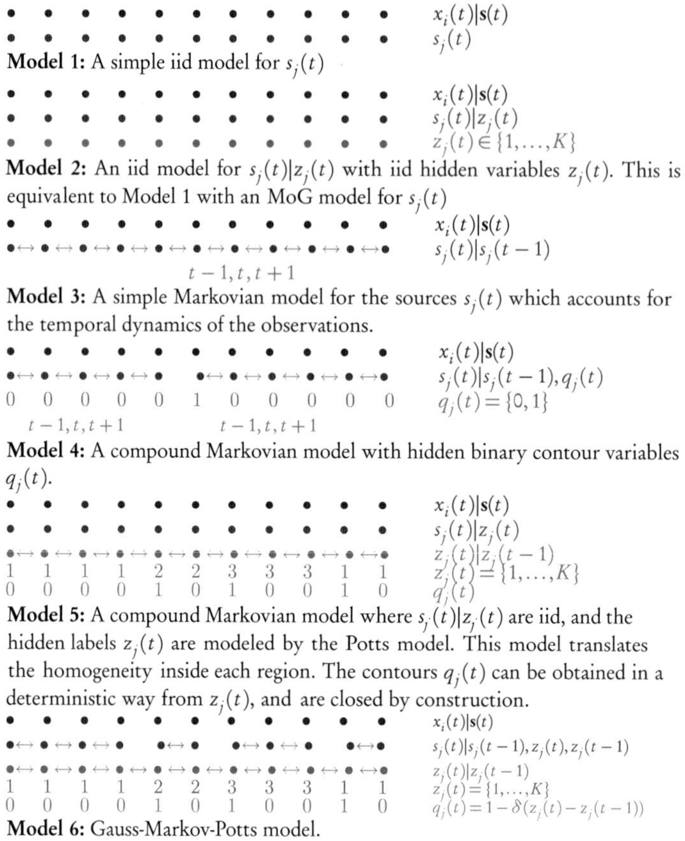

FIGURE 12.13

A variety of source models.

Model 3 accounts for temporal dynamic of the sources. However, these sources are stationary (at least to order two). Many methods use this feature to develop source separation methods that rely on the spectral diversity of the sources.

Model 4 takes into account the discontinuities in the sources. In this case the hidden binary-valued variable $q_j(t)$ represents contours in images. If we compare this model with Model 3 with some particular non-Gaussian $p(s_j(t)|s_j(t-1))$, we can find, in an implicit way, a link between this model and a hidden variable model. For example, a Markovian model with a non-quadratic or non-convex potential function implicitly introduces a hidden variable like $q_j(t)$ which is either binary (with a truncated quadratic potential function) or real-valued. However, in all cases this variable is related to contours.

Model 5 accounts more explicitly for the presence of classes (homogeneous regions) in the images. In this model, the hidden variables $z_j(t)$ represent the classification variables. The set of all the pixels in a given class are assumed to be iid and share the same mean and variance. They are grouped into compact regions according to the Potts-Markov model for those labels. Using such a model in source separation problems makes it possible to jointly estimate sources $s_j(\mathbf{r})$ and their classification or segmentation labels $z_j(\mathbf{r})$. We have used this model for classification, segmentation and data reduction in hyperspectral images.

Model 6 is a more sophisticated version of Model 5 where the pixels are not assumed to be iid. Here it is assumed that they have a Gauss-Markov structure and they share the same mean and the same covariance structure.

12.7 ESTIMATION SCHEMES

After the derivation of the posterior probability, one must select an estimation scheme. This choice will dictate the final structure of the source separation algorithm, and can significantly impact both tractability and performance with respect to a given application. Here we outline a comprehensive set of possible estimation schemes.

Given the posterior probability $p(\mathrm{S}, \mathrm{A}, \boldsymbol{\theta}|\mathrm{X})$, the most probable values for the source signal S, the mixing matrix elements $\widehat{\mathrm{A}}$ and the hyperparameters $\widehat{\boldsymbol{\theta}}$ can be found in various ways depending on whether one chooses to integrate out the effects of some of the parameters.

$$
\begin{array}{lllll}
\textbf{Alg. 1:} & \widehat{\mathrm{S}} \sim p(\mathrm{S}|\widehat{\mathrm{A}},\widehat{\boldsymbol{\theta}},\mathrm{X}) & \longrightarrow \widehat{\mathrm{A}} \sim p(\mathrm{A}|\widehat{\mathrm{S}},\widehat{\boldsymbol{\theta}},\mathrm{X}) & \longrightarrow \widehat{\boldsymbol{\theta}} \sim p(\boldsymbol{\theta}|\widehat{\mathrm{A}},\widehat{\mathrm{S}},\mathrm{X}) \\
\textbf{Alg. 2:} & \widehat{\mathrm{S}} \sim p(\mathrm{S}|\widehat{\mathrm{A}},\widehat{\boldsymbol{\theta}},\mathrm{X}) & \longrightarrow \widehat{\mathrm{A}} \sim p(\mathrm{A}|\widehat{\boldsymbol{\theta}},\mathrm{X}) & \longrightarrow \widehat{\boldsymbol{\theta}} \sim p(\boldsymbol{\theta}|\widehat{\mathrm{A}},\mathrm{X}) \\
\textbf{Alg. 3:} & \widehat{\mathrm{S}} \sim p(\mathrm{S}|\widehat{\mathrm{A}},\widehat{\boldsymbol{\theta}},\mathrm{X}) & \longrightarrow \widehat{\mathrm{A}} \sim p(\mathrm{A}|\widehat{\boldsymbol{\theta}},\mathrm{X}) & \longrightarrow \widehat{\boldsymbol{\theta}} \sim p(\boldsymbol{\theta}|\mathrm{X}).
\end{array}
$$

In each of these schemes, $\sim$ represents either *equal to the argument which maximizes* or *compute the mean* or *sample using*. The first example corresponds to performing the joint MAP estimate. Due to the high-dimensionality, this requires an effective optimization algorithm. The second example corresponds to the Posterior Mean (PM), which requires an effective integration computation to integrate over the source signals. The third example requires multiple integrations, which can only effectively be performed using MCMC sampling techniques.

The first example is probably the simplest as one can estimate each of the parameters by

$$
\begin{aligned}
p(\mathrm{S}|\mathrm{A},\boldsymbol{\theta},\mathrm{X}) &\propto p(\mathrm{X}|\mathrm{S},\mathrm{A},\boldsymbol{\theta}_\epsilon)\, p(\mathrm{S}|\boldsymbol{\theta}_s), \\
p(\mathrm{A}|\mathrm{S},\boldsymbol{\theta},\mathrm{X}) &\propto p(\mathrm{X}|\mathrm{S},\mathrm{A},\boldsymbol{\theta}_\epsilon)\, p(\mathrm{A}|\boldsymbol{\theta}_A), \\
p(\boldsymbol{\theta}|\mathrm{A},\mathrm{S},\mathrm{X}) &\propto p(\mathrm{X}|\mathrm{S},\mathrm{A},\boldsymbol{\theta}_\epsilon)\, p(\boldsymbol{\theta}).
\end{aligned}
\tag{12.65}
$$

The second example is computationally more difficult since it includes an integration

$$\begin{aligned} p(\mathbf{A}|\boldsymbol{\theta},\mathbf{X}) &\propto p(\mathbf{X}|\mathbf{A},\boldsymbol{\theta}_\epsilon,\boldsymbol{\theta}_s)\,p(\mathbf{A}|\boldsymbol{\theta}_A),\\ p(\boldsymbol{\theta}|\mathbf{A},\mathbf{X}) &\propto p(\mathbf{X}|\mathbf{A},\boldsymbol{\theta}_\epsilon,\boldsymbol{\theta}_s)\,p(\boldsymbol{\theta})\\ \text{with}\quad p(\mathbf{X}|\mathbf{A},\boldsymbol{\theta}_\epsilon,\boldsymbol{\theta}_s) &= \int p(\mathbf{X}|\mathbf{S},\mathbf{A},\boldsymbol{\theta}_\epsilon)\,p(\mathbf{S}|\boldsymbol{\theta}_s)\,\mathrm{d}\mathbf{S} \end{aligned} \tag{12.66}$$

and obtaining an analytic expression for $p(\mathbf{X}|\mathbf{A},\boldsymbol{\theta}_\epsilon,\boldsymbol{\theta}_s)$ is not always possible.

The third example needs a marginalization with respect to **S** to obtain $p(\mathbf{A}|\boldsymbol{\theta},\mathbf{X})$ and another step of marginalization with respect to **A** to obtain $p(\boldsymbol{\theta}|\mathbf{X})$:

$$\begin{aligned} p(\boldsymbol{\theta}|\mathbf{X}) &\propto p(\mathbf{X}|\boldsymbol{\theta}_\epsilon,\boldsymbol{\theta}_s)\,p(\boldsymbol{\theta}),\\ \text{with}\quad p(\mathbf{X}|\boldsymbol{\theta}_\epsilon,\boldsymbol{\theta}_s,\boldsymbol{\theta}_A) &= \int p(\mathbf{X}|\mathbf{A},\boldsymbol{\theta}_\epsilon,\boldsymbol{\theta}_s)\,p(\mathbf{A}|\boldsymbol{\theta}_A)\,\mathrm{d}\mathbf{A}. \end{aligned} \tag{12.67}$$

This is best handled using MCMC sampling techniques.

12.8 SOURCE SEPARATION APPLICATIONS

We have used a variety of models in a wide array of source separation applications. Here we will summarize some of our interesting results.

12.8.1 Spectrometry

This application is focused on a synthetic example related to near-infrared spectroscopy. Figure 12.14 shows a simulated set of ten mixtures of three chemical substances denoted A, B and C on Fig. 12.14. It is assumed that the chemicals do not interact so that the linear mixing model is valid. In this application, we have additional prior information that both the source concentrations and the mixing matrix elements are positive quantities.

In this problem, we used Model 1 with a Gamma prior probability for both the source concentrations and the mixing matrix elements. The implementation relies on an optimization algorithm that maximizes the joint posterior probability of the sources, the mixing matrix and the hyperparameters [47,48].

12.8.2 Source separation in astrophysics

One of the applications of source separation in astrophysics is in radio astronomy where one aims to separate spatially distributed sources at different wavelengths. Of particular interest is the isolation of the Cosmic Microwave Background (CMB). Here, the source separation model Model 3 is applied in the Fourier domain. Figure 12.15 illustrates the simulated results of three sources with six observations.

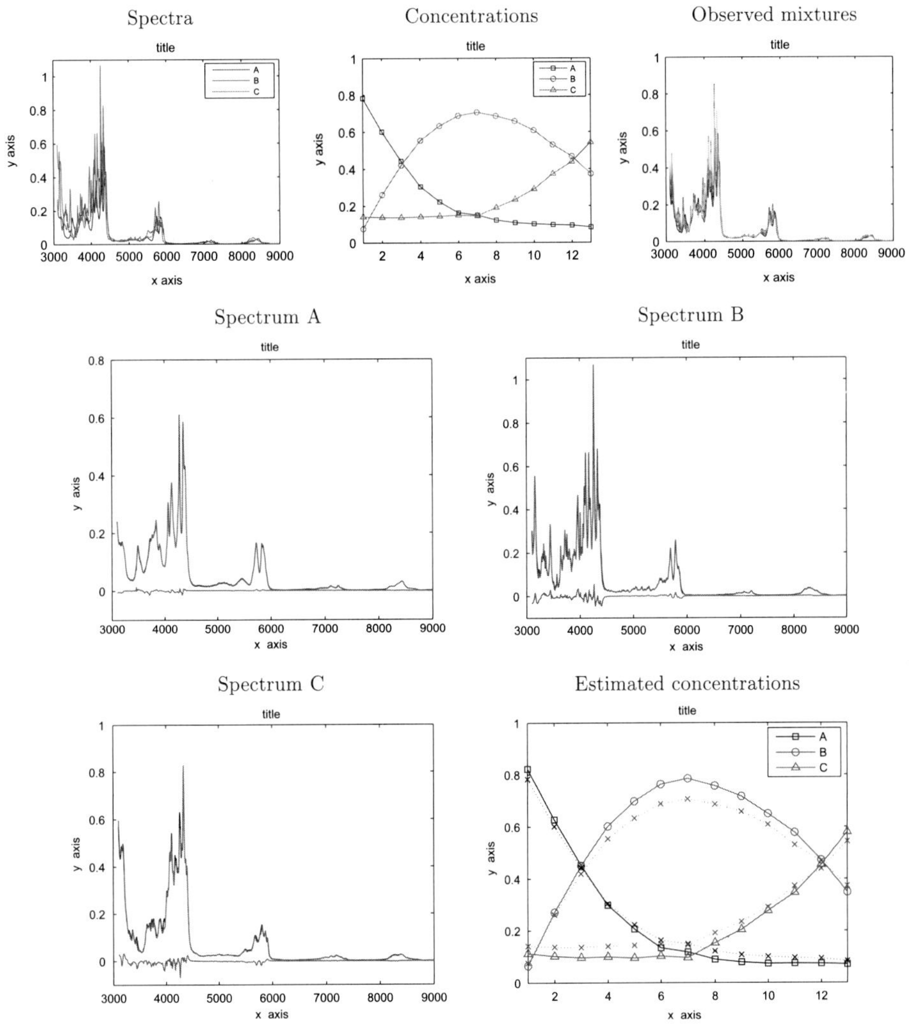

FIGURE 12.14

Source separation in spectrometry. The upper three plots show the three spectral sources, mixing matrix elements, and ten observations. The lower four plots show the estimated source spectra as well as the mixing matrix elements.

12.8.3 Source separation in satellite imaging

Here we consider two examples related to astronomical and satellite imaging. The objective here is illustrative in the sense that we separate the synthetic mixture of two images using a source model based on a mixture of two Gaussians with a two-level Potts

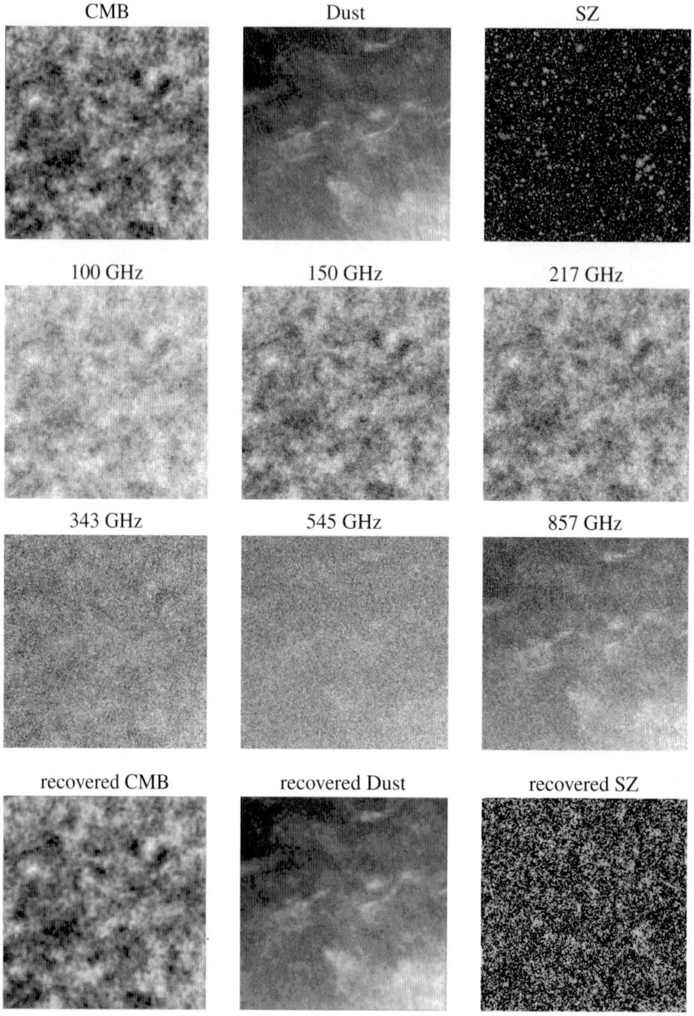

FIGURE 12.15

Source separation in astrophysics: The top three plots illustrate the three sources $\mathbf{S}$, which represent the Cosmic Microwave Background, Dust, and noise. The center six plots show the observations. The three lower plots reveal the estimated radio sources $\widehat{\mathbf{S}}$.

model [70]. Figure 12.16 shows a satellite imaging example where the ultimate goal is segmentation.

Figure 12.17 illustrates an example which involves source separation in the wavelet domain. The main advantage to working in the wavelet domain is that an MoG model

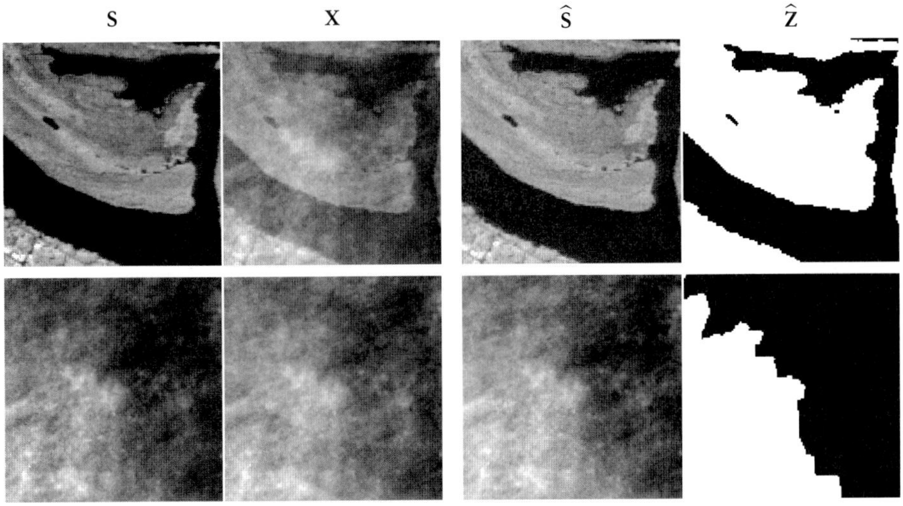

FIGURE 12.16

Source separation in satellite imaging. The column of images on the left illustrates the source images $\mathbf{S}$. These are followed by two observations $\mathbf{X}$ and the separated sources $\widehat{\mathbf{S}}$. On the right is an illustration of the segmentation of each source $\widehat{\mathbf{Z}}$.

with only two Gaussians is sufficient to describe the wavelet coefficients. One can also use Model 1 or Model 2 with (GG) or (MoG).

12.8.4 Data reduction, classification and separation in hyperspectral imaging

Hyperspectral data consists of a multidimensional data set, often a 3D data set or a data cube, which can be presented either as a set of spectra $x_{\mathrm{r}}(\omega)$ [67,53,3,2] or as a set of images $x_{\omega}(\mathbf{r})$. The spectral data in these images are highly redundant so that one of the main goals in this application is data reduction.

If we consider hyperspectral data as the set of spectra, then the data reduction problem can be written as

$$x_{\mathrm{r}}(\omega)=\sum_{k=1}^{K} A_{\mathrm{r},k}\, s_k(\omega)+\epsilon_{\mathrm{r}}(\omega) \tag{12.68}$$

where $s_k(\omega)$ are the K most independent spectral sources. In this model the columns of the mixing matrix $\mathbf{A}$ are the images $A_k(\mathbf{r})$. In the ideal case, each of these images has non-zero values only in the regions corresponding to the class of material associated with the spectra $s_k(\omega)$.

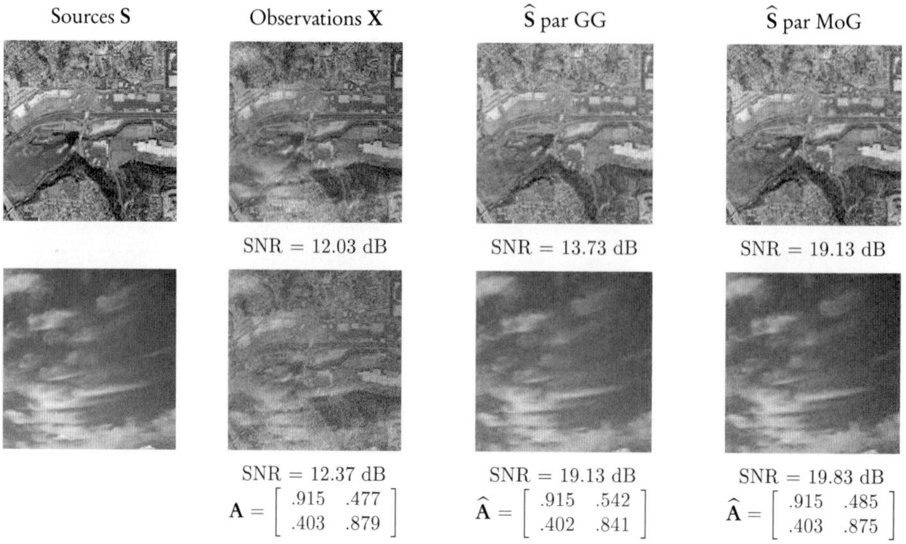

FIGURE 12.17

Source separation using Model 1 and Model 2. The left column illustrates the sources $\mathbf{S}$. This is followed by the observations $\mathbf{X} = \mathbf{AS} + \mathbf{E}$. The last two columns from left to right show the separation results using the GG model and the MoG model.

If we consider hyperspectral data as a set of images $x_\omega(\mathbf{r})$, then the data reduction problem can be written as

$$x_\omega(\mathbf{r}) = \sum_{j=1}^{N} A_{\omega,j} s_j(\mathbf{r}) + \epsilon_\omega(\mathbf{r}) \tag{12.69}$$

where the sources $s_\omega(\mathbf{r})$ are the N independent source images and the columns of the mixing matrix $\mathbf{A}$ are the spectra $A_j(\omega)$. Again, in an ideal case, each of source images $s_j(\mathbf{r})$ has non-zero values in those regions corresponding to the associated spectrum $A_j(\omega)$.

This becomes more clear, if in both models, we choose $K = N$, which means that the columns of $A_{\mathbf{r},k}$ in the first model and the source images $s_j(\mathbf{r})$ in the second model represent the same quantities. The same is true for the columns of $A_{\omega,j}$ in the first model and the sources $s_k(\omega)$ in the second model.

A priori one can handle these problems independently. However, we can also account for this particular feature. Indeed, in the second case, the columns of the mixing matrix to be estimated are the spectra. So, it is at least known that their values must be positive.

In the results shown here we consider the second model where we assume that the sources $s_j(\mathbf{r})$ are mutually independent. However, a particular feature of these images is that they all are composed of a finite number of homogeneous regions. To account for this, we see that Model 5 or Model 6 of the previous section are appropriate.

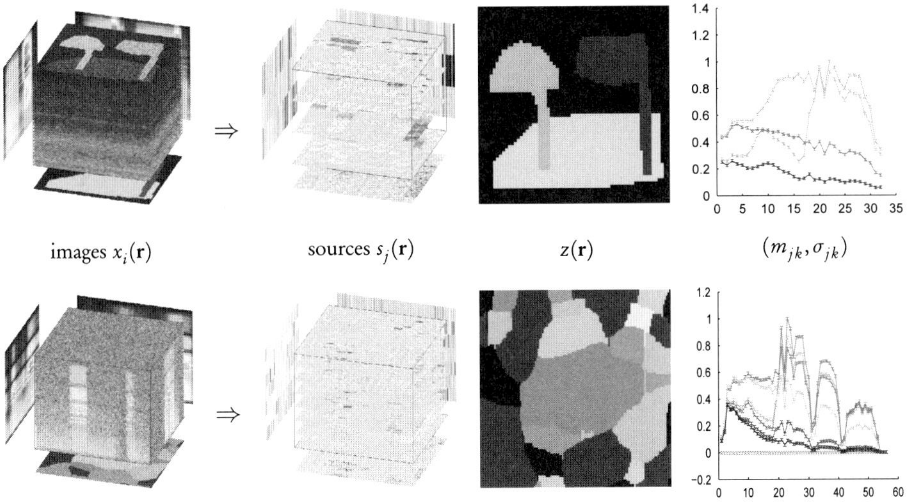

FIGURE 12.18

Data reduction and classification of hyperspectral data. From left to right: 32 images $x_i(\mathbf{r})$ of a hyperspectral data set, 8 estimated sources $s_j(\mathbf{r})$, estimated common segmentation $z(\mathbf{r})$, original means m_{jk} and variances σ_{jk} and along with their estimated values. The means m_{jk} represent the columns of the mixing matrix $\mathbf{A}$ and estimated σ_{jk} represent their diversity.

We used these models with the extra property that all these images share the same common $z(\mathbf{r})$, which represents the common segmentation of those images [2]. Figure 12.18 illustrates the results obtained by applying this approach.

12.9 SOURCE CHARACTERIZATION

The Bayesian approach enables one to extend the problem of source separation to the more general problem of source characterization where the source signals are estimated in conjunction with their associated hyperparameters. In this case, the hyperparameters describe physical characteristics of the sources and provide much more information than source separation alone.

12.9.1 Source separation and localization

In many problems the relevant hyperparameters describe the relative positions and orientations of the sources with respect to the detectors. Solving this problem will not only separate the sources, but will also localize them. This requires that one be able to model the source signal propagation, which will involve a function that describes how the signal amplitude changes as a function of both the distance from the source and the orientation with respect to the source. Since in this model the recorded signal strength is a function of the source-detector distance, the overall amplitudes of the sources can be recovered. It is only in this situation that the scaling indeterminacy of the source signal amplitude is resolved.

As hyperparameters, we include the source positions $\mathbf{p}$ as well as the orientations $\mathbf{q}$, assuming that the source emission may be directional. Signal propagation velocity c may be finite and relatively slow, as in the case of sound propagation, or the speed of light, as in the case of electromagnetic signals. In cases of slow propagation speeds or great distances, time delays τ_{ij}, which depend on the source-detector distance r_{ij},

$$\tau_{ij} = \frac{r_{ij}}{c} \tag{12.70}$$

must be accounted for. This means that the observed signals $x(t)$ depend on the source-detector distance

$$x_i(t) = A_{i,j} s_j(t - \tau_{ij}) \tag{12.71}$$

as do the mixing matrix elements $A_{i,j}$ due to the fact that the signal amplitude may vary as a function of source-detector distance. We saw this previously in the case of far-field sound propagation where

$$A_{i,j} = \frac{1}{4\pi r_{ij}^2}. \tag{12.72}$$

The necessary introduction of time delays in this class of problems illustrates the intimate relationship between source separation and source deconvolution, since the source signals here can be modeled as an instantaneous signal convolved with a delta function at a delayed time. Extension to simultaneous source separation and deconvolution is feasible in cases where there are reflections in the environment. However, this is only practical when the reflection models are sufficiently simple. Modeling reflectance and absorption of sound signals in a real room environment is difficult in practice and learning the model parameters is even more difficult due to both the high-dimensionality of the parameter space and the large number of computations that need to be performed to solve the forward problem.

In the case of biological electromagnetic signals, the source-detector distances are typically sufficiently small that the signal propagation can be considered instantaneous. In this case, one need not model the time delays, although the mixing matrix dependence on the source position and orientation remains critical. Relevant applications include cardiac depolarization signals in an electrocardiogram (ECG) where the mixing matrix is known as the lead field matrix, and electric and magnetic brain signals in electroencephalography (EEG) and magnetoencephalography (MEG).

It is true that many of these problems can be solved in a "blind" fashion where one does not take into account these physical hyperparameters. However, by doing so, one passes up opportunities to learn more about the signals. In some cases it is practical first to perform a blind estimation of the source signals, and then to estimate the hyperparameters. While computationally convenient, one runs the risk of obtaining a sub-optimal solution to the source signals, which then propagates errors into the hyperparameter estimation. One must remember that the acceptable level of error in signal processing applications is not always the same as the acceptable level of error in scientific applications. As a result, what may seem like a good strategy in terms of conventional signal processing may be disastrous in a scientific application.

12.9.2 Neural source estimation

As an example of the relationship between source separation and localization, we will consider the problem of separating and localizing electromagnetic neural activity in the human brain. We will show how different approximations and solution strategies can lead to two distinct standard algorithms: Infomax ICA [5,35] and the standard electromagnetic source estimation solution (ESE) [38].

The neural sources in the brain generate dipolar current fields, which can be detected by using electrodes to record electric potential differences on the surface of the scalp as in EEG or by using superconducting quantum interference devices (SQUIDs) to detect magnetic fields above the surface of the head as in MEG. These dipole sources are described by their positions $\mathbf{p}$ and orientations $\mathbf{q}$ in the brain, as well as the time course of their activity $\mathbf{s}(t)$. The mixing matrix $\mathbf{A}$ describes how each source signal contributes to the signal $\mathbf{x}(t)$ recorded by the detectors.

To solve this problem, we can write the joint posterior probability as being proportional to the likelihood times the prior

$$p(\mathbf{A},\mathbf{p},\mathbf{q},\mathbf{s}(t)|\mathbf{x}(t)) \propto p(\mathbf{x}(t)|\mathbf{A},\mathbf{p},\mathbf{q},\mathbf{s}(t))\, p(\mathbf{A},\mathbf{p},\mathbf{q},\mathbf{s}(t)), \tag{12.73}$$

where the likelihood describes what we know about the forward problem, which is the signal propagation from the sources to the detectors, and the joint prior describes what we know about the characteristics of the sources. We can simplify the joint prior by factoring it into several terms that reflect the interdependencies of the hyperparameters

$$p(\mathbf{A},\mathbf{p},\mathbf{q},\mathbf{s}(t)) = p(\mathbf{A}|\mathbf{p},\mathbf{q})\, p(\mathbf{q}|\mathbf{p})\, p(\mathbf{p})\, p(\mathbf{s}(t)). \tag{12.74}$$

This factorization reflects the fact that the mixing matrix depends on the source positions and orientations and that the source orientations can depend on their positions. This gives us a general posterior for this problem

$$p(\mathbf{A},\mathbf{p},\mathbf{q},\mathbf{s}(t)|\mathbf{x}(t)) \propto p(\mathbf{x}(t)|\mathbf{A},\mathbf{p},\mathbf{q},\mathbf{s}(t))\, p(\mathbf{A}|\mathbf{p},\mathbf{q})\, p(\mathbf{q}|\mathbf{p})\, p(\mathbf{p})\, p(\mathbf{s}(t)). \tag{12.75}$$

For our first algorithm, we will assume that we know nothing about the source positions and orientations, nor do we know anything about the nature of the mixing. We can describe this ignorance about the problem by assigning uniform distributions for the prior probabilities $p(\mathbf{A}|\mathbf{p},\mathbf{q})$, $p(\mathbf{q}|\mathbf{p})$, and $p(\mathbf{p})$ defined over a sufficiently large range of parameter values. The posterior then can be written as

$$p(\mathbf{A},\mathbf{s}(t)|\mathbf{x}(t)) \propto p(\mathbf{x}(t)|\mathbf{A},\mathbf{s}(t))\; p(\mathbf{s}(t)). \tag{12.76}$$

If we believe the external noise level to be sufficiently small and the number of sources equals the number of detectors, then the inverse of the mixing matrix $\mathbf{A}$ can be used to obtain the source signals. The source signals $\mathbf{s}(t)$ can be treated as a set of nuisance parameters and removed by marginalization as described in section 12.3

$$p(\mathbf{A}|\mathbf{x}(t)) \propto \int d\mathbf{s} \quad p(\mathbf{x}(t)|\mathbf{A},\mathbf{s}(t))\; p(\mathbf{s}(t)). \tag{12.77}$$

We assign a delta function likelihood to express the noise-free condition, and assume the sources to be independent by factoring the joint source prior into a product of source amplitude priors for each independent source. Performing the integral results in the standard Infomax ICA algorithm to determine the mixing matrix $\mathbf{A}$ [5,35–37].

For our second algorithm, we focus on the fact that the location and orientation of the sources determine the mixing matrix. We will use the fact that we know a great deal about the forward problem, which is described by the signal propagation from the sources to the detectors. This enables us to predict the values of the mixing matrix from the detector positions and the source hyperparameters

$$A_{i,j} = F(d_i, p_j, q_j). \tag{12.78}$$

Note that the function F may not be analytically known and may need to be computed numerically. We encode the prior information (12.78) by assigning a delta function prior for $\mathbf{A}$

$$p(\mathbf{A}|\mathbf{p},\mathbf{q}) = \prod_{i,j} \delta(A_{i,j} - F(d_i, p_j, q_j)). \tag{12.79}$$

Note here that we assume the detector positions $\mathbf{d}$ to be known with certainty. Again, these also could be known only with a given precision in which case they too would be hyperparameters. However, if they were completely unknown the problem would have infinitely many solutions and would remain unsolvable.

We treat the mixing matrix A as a nuisance parameter and marginalize over it to obtain

$$\begin{aligned} p(\mathbf{p},\mathbf{q},\mathbf{s}(t)|\mathbf{x}(t)) \propto\; & p(\mathbf{q}|\mathbf{p})\; p(\mathbf{p})\; p(\mathbf{s}(t)) \\ & \times \int d\mathbf{A} \quad p(\mathbf{x}(t)|\mathbf{A},\mathbf{p},\mathbf{q},\mathbf{s}(t)) \prod_{i,j} \delta(A_{i,j} - F(d_i, p_j, q_j)). \end{aligned} \tag{12.80}$$

This time we are able to accommodate some uncertainty in the forward model by assigning a Gaussian likelihood

$$p(\mathbf{x}(t)|\mathbf{A}, \mathbf{p}, \mathbf{q}, \mathbf{s}(t)) \propto \prod_{k,t} \exp\left(-\frac{(x_{kt} - \sum_m A_{km} s_{mt})^2}{2\sigma_k^2}\right) \tag{12.81}$$

where we explicitly express the vector of time series data as x_{kt} where k denotes the k-th data channel and t denotes the t-th time point. The sum in the exponential implements the matrix multiplication written in component form. The integral

$$\int d\mathbf{A} \prod_{k,t} \exp\left(-\frac{(x_{kt} - \sum_m A_{km} s_{mt})^2}{2\sigma_k^2}\right) \prod_{i,j} \delta(A_{i,j} - F(d_i, p_j, q_j)) \tag{12.82}$$

is easily performed with the delta functions and results in

$$\prod_{k,t} \exp\left(-\frac{(x_{kt} - \sum_m F_{km} s_{mt})^2}{2\sigma_k^2}\right) \tag{12.83}$$

where $F_{km} = F(d_k, p_m, q_m)$. We further simplify the notation by introducing $\hat{x}_{kt} = \sum_m F_{km} s_{mt}$ as the signal predicted by the forward model. This gives us the marginal posterior

$$p(\mathbf{p}, \mathbf{q}, \mathbf{s}(t)|\mathbf{x}(t)) \propto p(\mathbf{q}|\mathbf{p})\, p(\mathbf{p})\, p(\mathbf{s}(t)) \prod_{k,t} \exp\left(-\frac{(x_{kt} - \hat{x}_{kt})^2}{2\sigma_k^2}\right). \tag{12.84}$$

Here again detailed prior information about the source characteristics can be encoded. However, for illustrative purposes we will assign uniform priors that reflect an ignorance about the source locations and orientations. If we aim to find the MAP estimate, we can take the logarithm of the marginal posterior to get

$$\log p(\mathbf{p}, \mathbf{q}, \mathbf{s}(t)|\mathbf{x}(t)) = -\sum_k \sum_t \frac{(x_{kt} - \hat{x}_{kt})^2}{2\sigma_k^2} + C. \tag{12.85}$$

where the additive constant C is the logarithm of the implicit multiplicative constant in the proportionality. Note that this is equivalent to minimizing the chi-squared cost function

$$\chi^2 = -\sum_k \sum_t \frac{(x_{kt} - \hat{x}_{kt})^2}{2\sigma_k^2}, \tag{12.86}$$

which is what is often done in the most common source localization techniques.

Here again we see the versatility of the Bayesian approach. From a complete model of the source separation problem, we can obtain a wide variety of algorithms based on what we know about the problem, what we care to know about the problem, and the approximations we are willing to make. In one case we chose to focus on the source behavior and the mixing matrix and with some additional assumptions we derive the Infomax ICA algorithm. In the other case we chose to focus on the fact that the mixing matrix is determined by the relative positions and orientations of the sources to the detector, and with some additional assumptions we obtained the usual chi-squared fitting.

12.9.3 Source characterization in biophysics

The source hyperparameters are not always related to source position and orientation. In some cases, hyperparameters may be related to other relevant characteristics of the source signals. These problems are referred to as source characterization. In the next section, we consider an example where the hyperparameters describe how the sources vary their behavior from one experimental trial to the next. The result is a source separation algorithm called differentially variable component analysis (dVCA) [40,77,39].

The problem of neural source estimation is focused on both separating and characterizing neural sources in the brain. Neuroscience experiments typically rely on multiple experimental trials during which neural activity is evoked by an external stimulus. These evoked responses are typically generated by an unknown number of sources in the brain, each of which is at an unknown location. Moreover, it is not guaranteed that the evoked response from a given neural source will be the same from trial to trial. In fact, our research has shown that evoked responses vary both in amplitude and latency (time delay) [76].

The typical data analysis method relies on averaging the data recorded by each detector across the set of trials to decrease the effects of the noise, which consists of the uninteresting brain signals. However, averaging implicitly assumes a signal model where there is only a single neural source whose evoked response is constant

$$x_r(t) = s(t) + \eta(t) \tag{12.87}$$

where $x_r(t)$ is the signal recorded by the detector during the r-th trial, $s(t)$ is the evoked response or source signal, and $\eta(t)$ is additive noise. To estimate the evoked response, we write the posterior probability

$$p(s|\mathbf{x}) \propto p(s(t))\, p(\mathbf{x}(t)|s(t)), \tag{12.88}$$

where $\mathbf{x}$ is a vector of recordings over R experimental trials. The most straightforward probability assignments involve a uniform prior for the source amplitude and a Gaussian likelihood. These assignments result in

$$p(s|\mathbf{x}) \propto \exp\left(-\frac{\left(\sum_{r=1}^{R}\sum_{t=1}^{T} s(t) - x_r(t)\right)^2}{2\sigma^2}\right). \tag{12.89}$$

The MAP estimate of the evoked response is then found by taking the derivative of the logarithm of the posterior with respect to the response at a given time q and setting it equal to zero

$$\frac{\partial}{\partial s(q)} p(s|\mathbf{x}) = -\frac{1}{\sigma^2}\sum_{r=1}^{R}(s(q) - x_r(q)) = 0, \tag{12.90}$$

and solving for $s(q)$

$$\hat{s}(q) = -\frac{1}{R}\sum_{r=1}^{R} x_r(q), \tag{12.91}$$

which is an average of the recorded signals over all experimental trials. While numerically convenient, this result only applies if there is a single neural source that responds in the same way to each experimental trial.

However, we know that there are multiple sources and that their responses vary. By choosing a more appropriate signal model,

$$x_{mr}(t) = \sum_{n=1}^{N} C_{mn}\alpha_{nr}s_n(t - \tau_{nr}) + \eta_{mr}(t) \tag{12.92}$$

we now have a source separation problem to solve. In the equation above, C_{mn} is the mixing matrix which describes how each source is electromagnetically coupled to the detector, α_{nr} is a hyperparameter that describes the relative amplitude of the n-th source during the r-th trial, and τ_{nr} is a hyperparameter that describes the relative latency (time) shift of the n-th source's signal during the r-th trial. There is some indeterminacy in the model, which is handled by imposing the following constraints: $\langle\alpha\rangle = 1$ and $\langle\tau\rangle = 0$. We now have a model rich with hyperparameters, which will be more difficult to solve in general. However, these hyperparameters will provide additional information about the problem due to both constraints imposed by their interrelationships and any prior assignments we choose.

The probability assignments are kept simple. We assign uniform priors and a Student-t likelihood, since the variance is unknown. The logarithm of the posterior is of the form

$$p(\mathbf{C}, \mathbf{s}(t), \boldsymbol{\alpha}, \boldsymbol{\tau}|\mathbf{x}(t)) \propto Q^{-MRT/2} \tag{12.93}$$

where

$$Q = \sum_{m=1}^{M}\sum_{r=1}^{R}\sum_{t=1}^{T}\left(x_{mr}(t) - \sum_{n=1}^{N} C_{mn}\alpha_{nr}s_n(t-\tau_{nr})\right), \tag{12.94}$$

M is the number of detectors, R is the number of trials, T is the number of time samples per trial, and N is the number of sources. The dVCA solution is obtained by finding the joint MAP estimate for all of the parameters. By taking the derivative of (12.94) above with respect to each of the parameters in turn, we obtain a set of equations that can be iterated in a fixed point iteration scheme.

The optimal source waveshapes are given by

$$\hat{s}_j(q) = \frac{\sum\limits_{m=1}^{M}\sum\limits_{r=1}^{R} W_{mrj}C_{mj}\alpha_{jr}}{\sum\limits_{m=1}^{M}\sum\limits_{r=1}^{R}(C_{mj}\alpha_{jr})^2} \tag{12.95}$$

where

$$W_{mrj} = x_{mr}(q+\tau_{jr}) - \sum_{n=1, n\neq j}^{N} C_{mn}\alpha_{nr}s_n(q-\tau_{nr}+\tau_{jr}), \tag{12.96}$$

which takes the data after they have been shifted by the latency of the component being estimated and subtracts from them all of the other components after they have been appropriately scaled and shifted. The MAP estimate for the source waveshape then takes the weighted average of the W's which are trial estimates of the source.

The optimal single trial amplitude scale is given by

$$\hat{\alpha}_{jp} = \frac{\sum\limits_{m=1}^{M}\sum\limits_{t=1}^{T} U_{mtjp}V_{mtjp}}{\sum\limits_{m=1}^{M}\sum\limits_{r=1}^{R} V_{mtjp}^2} \tag{12.97}$$

where

$$U_{mtjp} = x_{mp}(t) - \sum_{n=1, n\neq j}^{N} C_{mn}\alpha_{np}s_n(t-\tau_{np}), \tag{12.98}$$

and

$$V_{mtjp} = C_{mj}s_j(t-\tau_{jp}). \tag{12.99}$$

This result is found by projecting the detector-scaled time-shifted component onto the data after subtracting off the other components as in a matching-filter [80].

The optimal mixing matrix is found similarly by

$$\hat{C}_{ij} = \frac{\sum_{r=1}^{R}\sum_{t=1}^{T} X_{rtij} Y_{rtij}}{\sum_{m=1}^{M}\sum_{r=1}^{R} Y_{rtij}^{2}} \tag{12.100}$$

where

$$X_{rtij} = x_{ir}(t) - \sum_{n=1, n\neq j}^{N} C_{in}\alpha_{nr}s_n(t-\tau_{nr}), \tag{12.101}$$

and

$$Y_{rtij} = \alpha_{jr}s_j(t-\tau_{jr}). \tag{12.102}$$

Last, the estimation of the latency τ leads to a complex solution that is most easily handled by finding the value of τ_{jp} that minimizes

$$Z = \sum_{m=1}^{M}\sum_{t=1}^{T} C_{mj}\alpha_{jp}s_j(t-\tau_{jp})B, \tag{12.103}$$

where

$$B = x_{mp}(t) - \sum_{n=1, n\neq j}^{N} C_{mn}\alpha_{np}s_n(t-\tau_{np}). \tag{12.104}$$

This is the cross-correlation between the estimated source and the data after the contributions from the other sources have been subtracted off. This is then averaged over all M detectors. In practice, since a discrete model is being used for the source waveshapes, we utilize a discrete set of latencies with resolution equal to the sampling rate. Iterating these equations over all sources and trials completes the algorithm. It should be noted that iterative fixed point algorithms are not guaranteed to converge to a single solution much less the correct solution. For this reason, the algorithm becomes unstable for large numbers of sources.

To demonstrate this algorithm, we examine intra-cortically recorded local field potentials from macaque striate cortex during a visuomotor GO-NOGO pattern recognition task [8]. The data consist of an ensemble of $R = 222$ trials recorded from a single electrode during the GO response. Figure 12.19(left) shows three examples of recorded single trials (noisy waveforms). The average event-related potential (AERP), found by taking the ensemble average of the waveforms in the 222 trials, has been overlaid on top of each

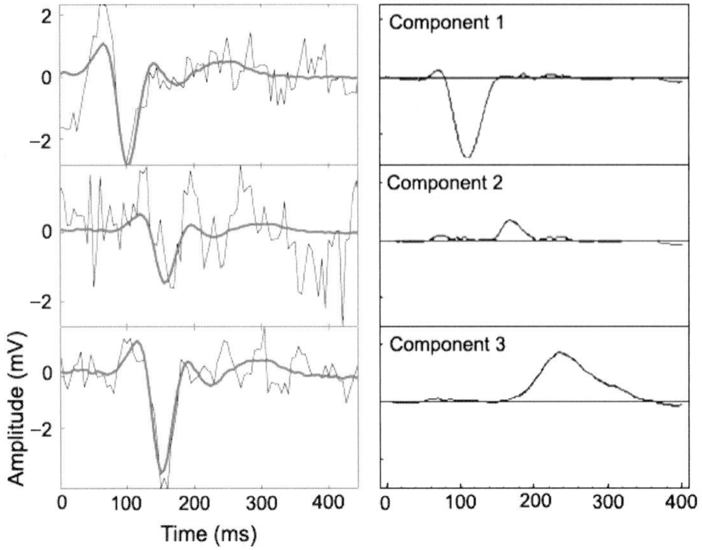

FIGURE 12.19

(Left) Examples of three single-trial recordings with the average evoked response overlaid to demonstrate the degree of trial-to-trial variability. (Right) Three extracted component waveforms each of which displays unique variability in both amplitude and latency.

single-trial waveform after appropriate amplitude scaling and latency shifting according to (12.97) and (12.103) where the AERP is used as the sole source waveform. Examination of the shape of the AERP suggests the contribution of multiple neural sources.

In this example we set the number of components to be identified to three, and utilized the local shapes of the AERP around the three extrema as initial guesses for the component waveforms in the MAP estimation. The dVCA algorithm relies on the fact that each component exhibits a unique pattern of variability. Figure 12.19(right) shows the resulting evoked component waveform estimates. As expected the three components exhibit different variances in their single-trial amplitudes, $\sigma_\alpha^2 = \{0.05, 1.0, 0.14\}$ and latencies $\sigma_\tau^2 = \{24, 123, 132.6 \text{ ms}^2\}$ for the first, second and third components respectively. An examination of the residual variance, as described in [76], shows that these more detailed models better account for the event-related variability. More detailed examples of the application of dVCA are published elsewhere [39].

12.10 CONCLUSION

The Bayesian approach explicitly breaks the source separation problem into three pieces: the signal model, the cost function, and the search algorithm. One begins by choosing an appropriate signal model for the physical problem. We have seen that this signal model can contain hyperparameters that enable the researcher to learn much more about the signals. These hyperparameters place additional constraints on the separation solution, and in doing so they effectively bring additional information to the problem and aid in separation.

Once a signal model has been chosen and the relevant model parameters identified, Bayesian probability theory is used to derive the form of the posterior probability for the model parameters. Additional prior information enters the problem through the specific probability assignments, which again constrain the separation solution. At this stage a series of approximations and strategies may be employed. Uninteresting model parameters can be marginalized out to simplify the form of the posterior or reduce the dimensionality of the parameter space. The result is a marginal posterior probability that can act as a cost function.

Last, a search algorithm must be employed to explore the behavior of this cost function in the parameter space. One may choose to maximize the posterior probability and produce a MAP estimate, or compute the expected value of the parameter by integrating over the marginal posterior. This can be performed analytically or numerically. Each of these strategies can produce a radically different algorithm with a variety of speed and convergence properties.

We have presented a variety of examples including both time series and image analysis applied to scientific disciplines ranging from astronomy to neuroscience. These examples illustrate the scope of problems that can be handled simply by choosing the appropriate signal model. By including hyperparameters that describe other source characteristics, one can easily shift focus from a source separation problem, to source deconvolution, source localization, and source characterization. The Bayesian methodology highlights that the difference between these various problems depends mainly on the selection of the relevant model parameters.

The structured approach offered by the Bayesian methodology takes a lot of guess-work out of algorithm design, and allows the researcher to focus instead on choosing the signal model, incorporating relevant prior information, making useful approximations, and selecting an appropriate search algorithm.

References

[1] H. Attias, Independent factor analysis, Neural Computation 11 (1999) 803–851.
[2] N. Bali, A. Mohammad-Djafari, Approximation en champ moyen pour la séparation de sources appliqué aux images hyperspectrales, in: GRETSI, Louvains, Belgique, Septembre 2005.
[3] N. Bali, A. Mohammad-Djafari, Mean Field Approximation for BSS of images with compound hierarchical Gauss-Markov-Potts model, in: MaxEnt05, San José CA, US, American Institute of Physics (AIP), August 2005.
[4] M. Beal, Z. Ghahramani, Variational Bayesian learning of directed graphical models with hidden variables, Bayesian Statistics 1 (2006) 793–832.
[5] A. Bell, T. Sejnowski, An information-maximization approach to blind separation and blind deconvolution, Neural Computation 7 (1995) 1129–1159.
[6] A. Belouchrani, Séparation autodidacte de sources: algorithmes, performances et application à des signaux expérimentaux, thèse, Ecole Nationale Supérieure des Télécommunications, 1995.

[7] A. Belouchrani, J.-F. Cardoso, Maximum likelihood source separation for discrete sources, in: EUSIPCO'94, 1994.
[8] S. Bressler, R. Coppola, R. Nakamura, Episodic multiregional cortical coherence at multiple frequencies during visual task performance, Nature 266 (1993) 153–156.
[9] R. Bro, S. De Jong, A fast non-negativity constrained least squares algorithm, Journal of Chemometrics 11 (1997) 393–401.
[10] S.-I. Campbell, G.-D. Poole, Computing non-negative rank factorizations, Linear Algebra and its Applications 35 (1981) 175–182.
[11] J.-F. Cardoso, Fourth-order cumulant structure forcing: Application to blind array processing, in: IEEE Sixth SP Workshop on Statistical Signal and Array Processing, IEEE, 1992, pp. 136–139.
[12] J.-F. Cardoso, Infomax and maximum likelihood for source separation, IEEE Letters on Signal Processing 4 (1997) 112–114.
[13] J.-F. Cardoso, Blind signal separation: statistical principles, Proceedings of IEEE 9 (1998) 2009–2025.
[14] J.-F. Cardoso, High-order contrasts for independent component analysis, Neural Computation 11 (1999) 157–192.
[15] J.-F. Cardoso, B. Labeld, Equivariant adaptative source separation, IEEE Transactions on Signal Processing 44 (1996) 3017–3030.
[16] R.A. Choudrey, W.D. Penny, S.J. Roberts, An ensemble learning approach to independent component analysis, in: Proceedings of Neural Networks for Signal Processing, Sydney, Australia, 2000.
[17] A. Cichocki, S. Amari, Adaptive blind signal and image processing: Learning Algorithms and Applications, Wiley, 2002.
[18] A. Cichocki, R. Unbehauen, L. Moszczynski, E. Rummert, A new on-line adaptive learning algorithm for blind separation of source signals, in: Proc. ISANN-94, 1994, pp. 406–411.
[19] P. Comon, Independent component analysis – A new concept? Signal Processing 36 (1994) 287–314.
[20] P. Comon, C. Jutten, J. Hérault, Blind separation of sources, Part II: Problems statement, Signal Processsing 24 (1991) 11–20.
[21] L. De Lathauwer, B. De Moor, J. Vandewalle, Independent component analysis based on higher-order statistics only, in: Signal Processing Workshop on Statistical Signal and Array Processing, IEEE, 1996, p. 356.
[22] B. Delyon, M. Lavielle, E. Moulines, Convergence of a stochastic approximation version of the EM algorithm, The Annals of Statistics 27 (1999) 94–128.
[23] C. Fevotte, S.J. Godsill, A Bayesian approach for blind separation of sparse sources, IEEE Transactions on Audio, Speech and Language Processing 14 (2006) 2174–2188.
[24] J. Hérault, C. Jutten, Space or time adaptive signal processing by neural network models, in: AIP Conference Proceedings 151 on Neural Networks for Computing, Woodbury, NY, USA, American Institute of Physics Inc., 1987, pp. 206–211.
[25] A. Hyvärinen, J. Karhunen, E. Oja, Independent Component Analysis, John Wiley and Sons, 2001.
[26] A. Hyvarinen, E. Oja, Independent component analysis: Algorithms and applications, Neural Networks 13 (2000) 411–430.
[27] M. Ichir, A. Mohammad-Djafari, Bayesian wavelet based statistical signal and image separation, in: Bayesian Inference and Maximum Entropy Methods, Jackson Hole (WY), USA, Aug, MaxEnt Workshops, American Institute of Physics, 2003, pp. 417–428. Maxent03.
[28] M. Ichir, A. Mohammad-Djafari, Séparation de sources modélisées par des ondelettes, in: GRETSI, Paris, France, September 2003.
[29] M. Ichir, A. Mohammad-Djafari, Wavelet domain blind image separation, in: M.A. Unser, A. Aldroubi, A.F. Laine (Eds.), Wavelets: Applications in Signal and Image Processing X (International Conference on Electronic Imaging, vol. 5207, SPIE, Bellingham, WA, August 2003), 2003, pp. 361–370.
[30] E. Jaynes, Prior probabilities, IEEE Transactions on SSC, SSC-4 (1968) 227–241.

[31] C. Jutten, Source separation: From dusk till dawn, in: Proc. of 2nd Int. Workshop on Independent Component Analysis and Blind Source Separation, ICA'2000, Helsinki, Finland, July 2000, pp. 15–26.

[32] C. Jutten, J. Hérault, Blind separation of sources, Part I: An adaptive algorithm based on neuromimetic architecture, Signal Processing 24 (1991) 1–10.

[33] J. Karhunen, P. Pajunen, E. Oja, The nonlinear PCA criterion in blind source separation: Relations with other approaches, Neurocomputing 22 (1–3) (20 November, 1998) 5–20.

[34] K.H. Knuth, Bayesian source separation and localization, in: A. Mohammad-Djafari (Ed.), SPIE'98 Proceedings: Bayesian Inference for Inverse Problems, San Diego, USA, SPIE, 1998, pp. 147–158.

[35] K.H. Knuth, Difficulties applying recent blind source separation techniques to EEG and MEG, in: G. Erickson, J. Rychert, C. Smith (Eds.), Maximum Entropy and Bayesian Methods, MaxEnt'97, Kluwer, 1998, pp. 209–222.

[36] K.H. Knuth, A Bayesian approach to source separation., in: J.-F. Cardoso, C. Jutten, P. Loubaton (Eds.), Proceedings of International Workshop on Independent Component Analysis and Signal Separation, ICA'99, Aussois, France, 1999, pp. 283–288.

[37] K.H. Knuth, Informed source separation: A Bayesian tutorial., in: E. Çetin, M. Tekalp, E. Kuruoǧlu, B. Sankur, (Eds.), Proceedings of the 13th European Signal Processing Conference, EUSIPCO 2005, Antalya, Turkey, 2005.

[38] K.H. Knuth, H.G. Vaughan Jr., Convergent Bayesian formulations of blind source separation and electromagnetic source estimation, in: W. von der Linden, V. Dose, R. Fischer, R. Preuss (Eds.), Maximum Entropy and Bayesian Methods, MaxEnt'98, Kluwer, 1999, pp. 217–226.

[39] K.H. Knuth, A. Shah, W. Truccolo, S. Bressler, M. Ding, C. Schroeder, Differentially variable component analysis (dVCA): Identifying multiple evoked components using trial-to-trial variability, Journal of Neurophysiology 95 (2006) 3257–3276.

[40] K.H. Knuth, W. Truccolo, S. Bressler, M. Ding, Separation of multiple evoked responses using differential amplitude and latency variability, in: T.-W. Lee, T.-P. Jung, S. Makeig, T.J. Sejnowski (Eds.), Proceedings of International Workshop on Independent Component Analysis and Signal Separation, ICA'01, San Diego, USA, 2001, pp. 463–468.

[41] M.M. Gaeta, J.L. Lacoume, Source separation without *a priori* knowledge: The maximum likelihood solution, in: Proc. EUSIPCO, 1990, pp. 621–624.

[42] D.J.C. MacKay, Information Theory, Inference and Learning Algorithms, Cambridge University Press, 2003.

[43] J. Miskin, D. MacKay, Ensemble learning for blind image separation and deconvolution, in: M. Girolami (Ed.), Advances in Independent Component Analysis, Springer, 2000, pp. 93–121.

[44] J. Miskin, D. MacKay, Ensemble learning for blind source separation, in: S. Roberts, R. Everson (Eds.), Independent Component Analysis: Principles and Practice, Cambridge University Press, 2001, pp. 209–233.

[45] A. Mohammad-Djafari, A Bayesian approach to source separation, in: Proceedings of International Workshop on Bayesian Inference and Maximum Entropy Methods in Science and Engineering, MaxEnt'99, vol. 567, American Institute of Physics (AIP) proceedings, 1999, pp. 221–244.

[46] S. Moussaoui, D. Brie, C. Carteret, A. Mohammad-Djafari, Application of Bayesian non-negative source separation to mixture analysis in spectroscopy, in: International Workshop on Bayesian Inference and Maximum Entropy Methods in Science and Engineering, MaxEnt'2004, R. Fischer, R. Preuss, and U. von Toussaint, (Eds.), vol. 735, Munich, Allemagne, July 25–30, 2004, American Institute of Physics (AIP), pp. 237–244.

[47] S. Moussaoui, D. Brie, A. Mohammad-Djafari, C. Carteret, Separation of non-negative mixture of non-negative sources using a Bayesian approach and MCMC sampling, IEEE Transactions on Signal Processing 54 (2006) 4133–4145.

[48] S. Moussaoui, C. Carteret, D. Brie, A. Mohammad-Djafari, Bayesian analysis of spectral mixture data using Markov Chain Monte Carlo methods sampling, Chemometrics and Intelligent Laboratory Systems 81 (2006) 137–148.

[49] S. Moussaoui, C. Carteret, A. Mohammad-Djafari, O. Caspary, D. Brie, B. Humbert, Approche bayésienne pour l'analyse de mélanges en spectroscopie, in: Chimiométrie'2003, Paris, France, December 2003.
[50] S. Moussaoui, A. Mohammad-Djafari, D. Brie, O. Caspary, A Bayesian method for positive source separation, in: IEEE International Conference on Acoustics, Speech, and Signal Processing, ICASSP'2004, Montreal, Canada, May 17–21, 2004, pp. 485–488.
[51] S. Moussaoui, A. Mohammad-Djafari, D. Brie, O. Caspary, C. Carteret, B. Humbert, A Bayesian method for positive source separation: Application to mixture analysis in spectroscopy, in: IAR'2003, Duisburg, Germany, November 2003.
[52] D. Obradovic, G. Deco, Information maximization and independent component analysis: Is there a difference? Neural Computation 10 (1998) 2085–2101.
[53] L. Parra, C. Spence, A. Ziehe, K.-R. Mueller, P. Sajda, Unmixing hyperspectral data, in: Advances in Neural Information Processing Systems 13, NIPS'2000, MIT Press, 2000, pp. 848–854.
[54] B.A. Pearlmutter, L.C. Parra, Maximum likelihood blind source separation: A context-sensitive generalization of ICA, in: Advances in Neural Information Processing Systems 9, MIT Press, 1997, pp. 613–619.
[55] W. Penny, S. Kiebel, K. Friston, Variational Bayesian inference for FMRI time series, NeuroImage 19 (2003) 727–741.
[56] D.T. Pham, Separation of a mixture of independent sources through a maximum likelihood approach, in: Proc. EUSIPCO, 1992, pp. 771–774.
[57] D.-T. Pham, Blind separation of instantaneous mixture sources via independent component analysis, IEEE Transactions on Signal Processing 44 (1996) 2768–2779.
[58] T. Raiko, H. Valpola, M. Harva, J. Karhunen, Building blocks for variational Bayesian learning of latent variable models, Journal of Machine Learning Research 8 (2007) 155–201.
[59] C. Robert, L'analyse statistique bayésienne, Economica, Paris, France, 1992.
[60] C. Robert, The Bayesian Choice. A Decision-Theoretic Motivation, in: Springer Texts in Statistics, Springer Verlag, New York, NY, USA, 1997.
[61] C. Robert, Monte Carlo Statistical Methods, Springer-Verlag, Berlin, 1999.
[62] C. Robert, The Bayesian Choice, 2nd ed., Springer-Verlag, 2001.
[63] S.J. Roberts, Independent component analysis: Source assessment separation, a Bayesian approach, in: IEE Proceedings, Vision, Image and Signal Processing, 1998, pp. 149–154.
[64] S.-J. Roberts, R. Choudrey, Data decomposition using independent component analysis with prior constraints, Pattern Recognition 36 (2003) 1813–1825.
[65] D.B. Rowe, The general blind source separation model and a Bayesian approach with correlation, in: IEE Proceedings – Vision, Image, and Signal Processing, 1999.
[66] D.B. Rowe, A Bayesian approach to blind source separation, Journal of Interdisciplinary Mathematics 5 (1) (2002) 49–76.
[67] K. Sasaki, S. Kawata, S. Minami, Component analysis of spatial and spectral patterns in multispectral images. I. Basics, Journal of the Optical Society of America. A 4 (1987) 2101–2106.
[68] S. Senecal, P.-O. Amblard, Bayesian separation of discrete sources via gibbs sampling, in: Proc. Int. Conf. ICA 2000, 2000, pp. 566–572.
[69] V. Smídl, A. Quinn, The variational Bayes method in signal processing, Springer, Berlin, New York, 2006.
[70] H. Snoussi, Approche bayésienne en séparation de sources. Applications en imagerie, thèse, Université de Paris-Sud, Orsay, France, September 2003.
[71] H. Snoussi, J. Idier, Bayesian blind separation of generalized hyperbolic processes in noisy and underdeterminate mixtures, IEEE Transactions on Signal Processing 54 (2006) 3257–3269.
[72] H. Snoussi, A. Mohammad-Djafari, Bayesian source separation with mixture of Gaussians prior for sources and Gaussian prior for mixture coefficients, in: A. Mohammad-Djafari (Ed.), Bayesian Inference and Maximum Entropy Methods, Gif-sur-Yvette, France, July, Proc. of MaxEnt, American Institute of Physics, 2000, pp. 388–406.

[73] H. Snoussi, A. Mohammad-Djafari, Unsupervised learning for source separation with mixture of Gaussians prior for sources and Gaussian prior for mixture coefficients, in: D.J. Miller (Ed.), Neural Networks for Signal Processing XI, IEEE workshop, IEEE, September2001, pp. 293–302.
[74] H. Snoussi, A. Mohammad-Djafari, Information geometry and prior selection, in: C. Williams (Ed.), Bayesian Inference and Maximum Entropy Methods, MaxEnt Workshops, American Institute of Physics, August 2002, pp. 307–327.
[75] H. Snoussi, A. Mohammad-Djafari, Fast joint separation and segmentation of mixed images, Journal of Electronic Imaging 13 (2004) 349–361.
[76] W. Truccolo, M. Ding, K.H. Knuth, R. Nakamura, S. Bressler, Trial-to-trial variability of cortical evoked responses: Implications for the analysis of functional connectivity, Clinical Neurophysiology 113 (2002) 206–226.
[77] W. Truccolo, K.H. Knuth, A. Shah, S. Bressler, C. Schroeder, M. Ding, Estimation of single-trial multi-component ERPS: Differentially variable component analysis, Biological Cybernetics 89 (2003) 426–438.
[78] H. Valpola, M. Harva, J. Karhunen, An unsupervised ensemble learning method for nonlinear dynamic state-space models, Neural Computation 14 (2002) 2647–2692.
[79] H. Valpola, M. Harva, J. Karhunen, Hierarchical models of variance sources, Signal Processing 84 (2004) 267–282.
[80] C. Woody, Characterization of an adaptive filter for the analysis of variable latency neuroelectric signals, Medical Biological Engineering 5 (1967) 539–553.

CHAPTER

Non-negative mixtures

13

M.D. Plumbley, A. Cichocki, and R. Bro

13.1 INTRODUCTION

Many real-world unmixing problems involve inherent non-negativity constraints. Most physical quantities are non-negative: lengths, weights, amounts of radiation, and so on. For example, in the field of air quality, the amount of a particulate from a given source in a particular sample must be non-negative; and in musical audio signal processing, each musical note contributes a non-negative amount to the signal power spectrum. This type of non-negativity constraint also arises in, for example, hyperspectral image analysis for remote sensing, positron emission tomography (PET) image sequences in medical applications, or semantic analysis of text documents.

Often we lose this non-negativity constraint when, for example, we subtract the mean from the data, such as when we perform the usual prewhitening process for independent component analysis (ICA). However, we need to be aware that doing this may lose us important information that could help find the solution to our unmixing problem. Even where the non-negativity constraint is not inherently part of the problem, analogies with biological information processing systems suggest that this is an interesting direction to investigate, since information in neural systems is typically communicated using spikes, and the spike rate is a non-negative quantity.

In this chapter we discuss some algorithms for the use of non-negativity constraints in unmixing problems, including *positive matrix factorization (PMF)* [71], *nonnegative matrix factorization (NMF)*, and their combination with other unmixing methods such as *non-negative ICA* and sparse non-negative matrix factorization. The 2-D models can be naturally extended to multiway array (tensor) decompositions, especially Non-negative Tensor Factorization (NTF) and Non-negative Tucker Decomposition (NTD).

13.2 NON-NEGATIVE MATRIX FACTORIZATION

Suppose that our sequence of observation vectors $\mathbf{x}_t$, $1 \leqslant t \leqslant T$ is approximated by a linear mixing model

$$\mathbf{x}_t \approx \mathbf{A}\mathbf{s}_t = \sum_n \mathbf{a}_n s_{nt}$$

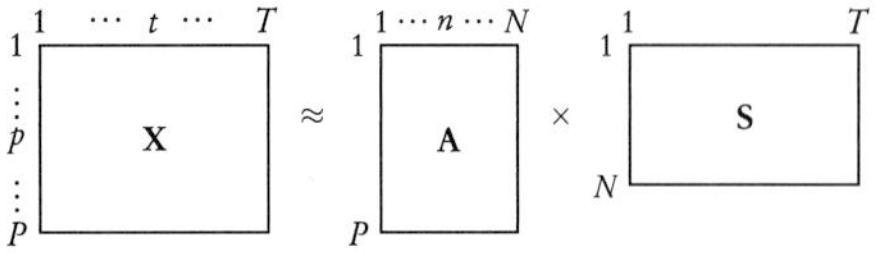

FIGURE 13.1

Basic NMF model $\mathbf{X} \approx \mathbf{AS}$.

or in matrix notation

$$\mathbf{X} \approx \mathbf{AS} = \mathbf{A}\,\mathbf{V}^T = \sum_n \mathbf{a}_n \mathbf{v}_n^T \tag{13.1}$$

where $\mathbf{X} = [x_{pt}]$ is a data matrix of observations x_{pt} for the p-th source at the t-th sample, $\mathbf{A} = [a_{pn}] = [\mathbf{a}_1, \mathbf{a}_2, \ldots, \mathbf{a}_N] \in \mathbb{R}^{P \times N}$ is a mixing matrix giving the contribution of the n-th source to the p observation, and $\mathbf{S} = [s_{nt}]$ is a source matrix giving the value for the n-th mixture at the t-th sample (Fig. 13.1) and for convenience we use $\mathbf{V} = \mathbf{S}^T = [\mathbf{v}_1, \mathbf{v}_2, \ldots, \mathbf{v}_N] \in \mathbb{R}^{T \times N}$.

In this chapter, we are interested in the conditions where the sources $\mathbf{S}$ and/or the mixing contributions $\mathbf{A}$ are *non-negative*. The problem of finding $\mathbf{A}$ and $\mathbf{S}$ given only the observed mixtures $\mathbf{X}$ when both $\mathbf{A}$ and $\mathbf{S}$ are non-negative was first analyzed by Leggett [59] under the name *curve-resolution* and later by Paatero and Tapper [71] as the *positive matrix factorization* (PMF). Although the method was commonly used in certain fields, it was later re-invented and popularized by Lee and Seung as the *non-negative matrix factorization* (NMF) [56]. In the ten years since the Lee and Seung paper appeared in *Nature*, there have been hundreds of papers describing algorithms and applications of NMF.[1]

In "plain" NMF we only assume non-negativity of $\mathbf{A}$ and $\mathbf{S}$. Unlike blind source separation methods based on independent component analysis (ICA) we do not assume that the sources s_n are independent, although we will introduce other assumptions or constraints on $\mathbf{A}$ or $\mathbf{S}$ later. We notice that this symmetry of assumptions leads to a symmetry in the factorization: for (13.1) we could just as easily write

$$\mathbf{X}^{\mathrm{T}} \approx \mathbf{S}^{\mathrm{T}}\mathbf{A}^{\mathrm{T}} \tag{13.2}$$

so the meaning of "source" and "mixture" are somewhat arbitrary.

The standard NMF model has been extended in various ways, including Semi-NMF, Multi-layer NMF, Tri-NMF, Orthogonal NMF, Non-smooth NMF and Convolutive NMF. We shall explore some of these extensions later (section 13.3).

[1] While the terms *curve-resolution* and *PMF* pre-date NMF, we will prefer *NMF* in this chapter due to its widespread popular use in the source separation literature.

13.2.1 Simple gradient descent

Let us first develop a simple alternating gradient descent method to solve the standard NMF problem (13.1) for $\mathbf{A}$ and $\mathbf{S}$ given the observations $\mathbf{X}$. Consider the familiar Euclidean distance cost function

$$J_{\mathrm{E}} = D_{\mathrm{E}}(\mathbf{X}; \mathbf{AS}) = \frac{1}{2}\|\mathbf{X} - \mathbf{AS}\|_{\mathrm{F}}^2 = \frac{1}{2}\sum_{pt}(x_{pt} - [\mathbf{AS}]_{pt})^2 \tag{13.3}$$

where $[\mathbf{M}]_{pt}$ is the (p, t)-th element of the matrix $\mathbf{M}$. For a simple gradient descent step for $\mathbf{S}$, we wish to update $\mathbf{S}$ according to

$$\mathbf{S} \leftarrow \mathbf{S} - \eta \frac{\partial J_{\mathrm{E}}}{\partial \mathbf{S}} \tag{13.4}$$

where η is a small update factor and $[\partial J_{\mathrm{E}}/\partial \mathbf{S}]_{nt} = \partial J_{\mathrm{E}}/\partial s_{nt}$, or as individual terms

$$s_{nt} \leftarrow s_{nt} - \eta_{nt} \frac{\partial J_{\mathrm{E}}}{\partial s_{nt}} \tag{13.5}$$

where we now allow η_{nt} to take on different values for each combination of (n, t).

In order to calculate the partial derivative, consider that our cost function

$$J_{\mathrm{E}} = \frac{1}{2}\|\mathbf{X} - \mathbf{AS}\|_{\mathrm{F}}^2 = \frac{1}{2}\mathrm{trace}((\mathbf{X} - \mathbf{AS})^{\mathrm{T}}(\mathbf{X} - \mathbf{AS})) \tag{13.6}$$

obtains an infinitesimal change $J_{\mathrm{E}} \leftarrow J_{\mathrm{E}} + \partial J_{\mathrm{E}}$ due to an infinitesimal change to $\mathbf{S}$,

$$\mathbf{S} \leftarrow \mathbf{S} + \partial \mathbf{S}. \tag{13.7}$$

Differentiating (13.6) w.r.t. this infinitesimal change $\partial \mathbf{S} = [\partial s_{nt}]$ we get

$$\partial J_{\mathrm{E}} = -\mathrm{trace}((\mathbf{X} - \mathbf{AS})^{\mathrm{T}}\mathbf{A}\partial \mathbf{S}) \tag{13.8}$$

$$= -\mathrm{trace}((\mathbf{A}^{\mathrm{T}}\mathbf{X} - \mathbf{A}^{\mathrm{T}}\mathbf{AS})^{\mathrm{T}}\partial \mathbf{S}) \tag{13.9}$$

$$= -\sum_{nt}[\mathbf{A}^{\mathrm{T}}\mathbf{X} - \mathbf{A}^{\mathrm{T}}\mathbf{AS}]_{nt}\partial s_{nt} \tag{13.10}$$

and hence

$$\frac{\partial J_{\mathrm{E}}}{\partial s_{nt}} = -[\mathbf{A}^{\mathrm{T}}\mathbf{X} - \mathbf{A}^{\mathrm{T}}\mathbf{AS}]_{nt} = -([\mathbf{A}^{\mathrm{T}}\mathbf{X}]_{nt} - [\mathbf{A}^{\mathrm{T}}\mathbf{AS}]_{nt}). \tag{13.11}$$

Substituting (13.11) into (13.5) we get

$$s_{nt} \leftarrow s_{nt} + \eta_{nt}([\mathbf{A}^{\mathrm{T}}\mathbf{X}]_{nt} - [\mathbf{A}^{\mathrm{T}}\mathbf{AS}]_{nt}) \tag{13.12}$$

or gradient update step for $s_{nt} = [\mathbf{S}]_{nt}$. Due to the symmetry between $\mathbf{S}$ and $\mathbf{A}$, a similar procedure will derive

$$a_{pn} \leftarrow a_{pn} + \eta_{pn}([\mathbf{X}\mathbf{S}^{\mathrm{T}}]_{pn} - [\mathbf{A}\mathbf{S}\mathbf{S}^{\mathrm{T}}]_{pn}) \tag{13.13}$$

as the gradient update step for $a_{pn} = [\mathbf{A}]_{pn}$. A simple gradient update algorithm would therefore be to alternate between applications of (13.12) and (13.13) until convergence, while maintaining the non-negativity of the elements a_{pn} and s_{nt}, i.e. we would actually apply

$$s_{nt} \leftarrow \left[s_{nt} + \eta_{nt}([\mathbf{A}^{\mathrm{T}}\mathbf{X}]_{nt} - [\mathbf{A}^{\mathrm{T}}\mathbf{A}\mathbf{S}]_{nt})\right]_{+} \tag{13.14}$$

where $[s]_{+} = \max(0, s)$ is the rectification function, and similarly for a_{pn}.

13.2.2 Multiplicative updates

While gradient descent is a simple procedure, convergence can be slow, and the convergence can be sensitive to the step size. In an attempt to overcome this, Lee and Seung [57] applied *multiplicative update rules*, which have proved particularly popular in NMF applications since then.

To construct a multiplicative update rule for s_{nt}, we can choose η_{nt} such that the first and third terms on the RHS of (13.12) cancel, i.e. $s_{nt} = \eta_{nt}[\mathbf{A}^{\mathrm{T}}\mathbf{A}\mathbf{S}]_{nt}$ or $\eta_{nt} = s_{nt}/[\mathbf{A}^{\mathrm{T}}\mathbf{A}\mathbf{S}]_{nt}$. Substituting this back into (13.12) we get

$$s_{nt} \leftarrow s_{nt} \frac{[\mathbf{A}^{\mathrm{T}}\mathbf{X}]_{nt}}{[\mathbf{A}^{\mathrm{T}}\mathbf{A}\mathbf{S}]_{nt}} \tag{13.15}$$

which is now in the form of a multiplicative update to s_{nt}. Repeating the process for a_{pn} we get the update rule pair

$$a_{pn} \leftarrow a_{pn} \frac{[\mathbf{X}\mathbf{S}^{\mathrm{T}}]_{pn}}{[\mathbf{A}\mathbf{S}\mathbf{S}^{\mathrm{T}}]_{pn}} \qquad s_{nt} \leftarrow s_{nt} \frac{[\mathbf{A}^{\mathrm{T}}\mathbf{X}]_{nt}}{[\mathbf{A}^{\mathrm{T}}\mathbf{A}\mathbf{S}]_{nt}}. \tag{13.16}$$

An alternative pair of update rules can be derived by starting from the (generalized) Kullback-Leibler divergence,

$$J_{\mathrm{KL}} = D_{\mathrm{KL}}(\mathbf{X}; \mathbf{A}\mathbf{S}) = \sum_{pt} \left(x_{pt} \log \frac{x_{pt}}{[\mathbf{A}\mathbf{S}]_{pt}} - x_{pt} + [\mathbf{A}\mathbf{S}]_{pt} \right) \tag{13.17}$$

which reduces to the usual KL divergence between probability distributions when $\sum_{pt} x_{pt} = \sum_{pt} [\mathbf{A}\mathbf{S}]_{pt} = 1$. Repeating the derivations above for this (13.17) we obtain the gradient descent update rules

$$a_{pn} \leftarrow \left[a_{pn} + \eta_{pn} \left(\sum_t s_{nt} x_{pt} / [\mathbf{AS}]_{pt} - \sum_t s_{nt} \right) \right]_+ \tag{13.18}$$

$$s_{nt} \leftarrow \left[s_{nt} + \eta_{nt} \left(\sum_p a_{pn} x_{pt} / [\mathbf{AS}]_{pt} - \sum_p a_{pn} \right) \right]_+ \tag{13.19}$$

and the corresponding multiplicative update rules

$$a_{pn} \leftarrow a_{pn} \frac{\sum_t s_{nt} x_{pt} / [\mathbf{AS}]_{pt}}{\sum_t s_{nt}} \qquad s_{nt} \leftarrow s_{nt} \frac{\sum_p a_{pn} x_{pt} / [\mathbf{AS}]_{pt}}{\sum_p a_{pn}}. \tag{13.20}$$

(In practice, a small positive ϵ is added to the denominator of each of these updates in order to avoid divide-by-zero problems.)

In fact we can obtain even simpler update equations if we introduce a sum-to-1 constraint on the columns of $\mathbf{A}$

$$\lambda_n \triangleq \sum_p a_{pn} = 1. \tag{13.21}$$

We can always obtain this from any factorization $\mathbf{AS}$ by mapping $\mathbf{A}' \leftarrow \mathbf{A}\Lambda$, $\mathbf{S}' \leftarrow \mathbf{S}\Lambda^{-1}$ where $\Lambda = \text{Diag}(\lambda_1, \ldots, \lambda_N)$ is the $N \times N$ diagonal matrix with the sums of the columns of $\mathbf{A}$ as its diagonal entries.

We can impose this constraint after (13.20) with a further update step

$$a_{pn} \leftarrow \frac{a_{pn}}{\sum_p a_{pn}} \tag{13.22}$$

which in turn makes the division by $\sum_t s_{nt}$ in (13.20) redundant, since it will appear inside both the numerator and denominator of (13.22). So, using this together with the constraint $\sum_p a_{pn} = 1$ in the right hand equation in (13.20), we get the simpler update equations

$$\begin{aligned} a_{pn} &\leftarrow a_{pn} \sum_t s_{nt} (x_{pt} / [\mathbf{AS}]_{pt}) \\ a_{pn} &\leftarrow \frac{a_{pn}}{\sum_p a_{pn}} \\ s_{nt} &\leftarrow s_{nt} \sum_p a_{pn} (x_{pt} / [\mathbf{AS}]_{pt}) \end{aligned} \tag{13.23}$$

which is the algorithm presented in [56].

These multiplicative update rules have proved to be attractive since they are simple, do not need the selection of an update parameter η, and their multiplicative nature and non-negative terms on the RHS ensure that the elements cannot become negative. They do also have some numerical issues, including that it is possible for the denominators to become zero, so practical algorithms often add a small offset term to prevent divide-by-zero errors [2]. There are also now a number of alternative algorithms available which are more efficient, and we shall consider some of these later.

13.2.3 Alternating least squares (ALS)

Rather than using a gradient descent direction to reduce the Euclidean cost function J_E in (13.3), we can use a Newton-like method to find alternately the **S** and **A** that directly minimizes J_E.

Let us first consider the update to **S** for a fixed **A**. Writing the derivative in (13.11) in matrix form we get

$$\frac{\partial J_E}{\partial \mathbf{S}} = -(\mathbf{A}^T\mathbf{X} - \mathbf{A}^T\mathbf{A}\mathbf{S}) \tag{13.24}$$

which must be zero at the minimum, i.e. the equation

$$(\mathbf{A}^T\mathbf{A})\mathbf{S} = \mathbf{A}^T\mathbf{X} \tag{13.25}$$

must hold at the **S** that minimizes J_E. We can therefore solve (13.25) for **S**, either using $\mathbf{S} = (\mathbf{A}^T\mathbf{A})^{-1}\mathbf{A}^T\mathbf{X}$, or through more efficient linear equation solver methods such as the Matlab function `linsolve`. Similarly for **A** we minimize J_E by solving $(\mathbf{S}\mathbf{S}^T)\mathbf{A}^T = \mathbf{S}\mathbf{X}^T$ for **A**.

Now these least squared solutions do not themselves enforce the non-negativity of **S** and **A**. The simplest way to do this is to project the resulting optimal values into the positive orthant, producing the resulting sequence of steps:

$$\mathbf{S} \leftarrow [(\mathbf{A}^T\mathbf{A})^{-1}\mathbf{A}^T\mathbf{X}]_+ \tag{13.26}$$

$$\mathbf{A} \leftarrow [\mathbf{X}\mathbf{S}^T(\mathbf{S}\mathbf{S}^T)^{-1}]_+ \tag{13.27}$$

where $[\mathbf{M}]_+$ sets all negative values of the matrix to zero. While the removal of the negative values by projection onto the positive orthant means that there are no theoretical guarantees on its performance [49], this procedure has been reported to perform well in practice [91,2].

Rather than using ad hoc truncation of least squares solutions it is also possible to use the NNLS (non-negativity constrained least squares) algorithm of Hanson and Lawson [32]. This is an active-set algorithm which in a finite number of steps will give the least squares solution subject to the non-negativity constraints. In the context of the ALS algorithm, the original algorithm can be speeded up substantially by using the current active set as a starting point. In practice, the active set does not change substantially during iterations, so the cost of using the NNLS algorithm in this way is typically less than unconstrained least squares fitting. Further speed-up is possible by exploiting the structure of the ALS updates [7].

Recently algorithms have been introduced to reduce the computational complexity of these ALS algorithms by performing block-wise or separate row/column updates instead of updating the whole matrices of the whole factor matrices **A** and **S** each step [15,16,21]. We will return to these large-scale NMF algorithms in section 13.4.3.

13.3 EXTENSIONS AND MODIFICATIONS OF NMF

The basic NMF method that we have introduced in the previous section has been modified in many different ways, either through the introduction of costs and/or penalties on the factors, inclusion of additional structure, or extension to multi-factor and tensor factorization.

13.3.1 Constraints and penalties

It is often useful to be able to modify the standard NMF method by imposing certain constraints or penalties to favor particular types of solutions. For example, in (13.21) we have already seen that Lee and Sung [56] included sum-to-1 constraint on the columns $\mathbf{a}_n$ of **A**

$$\sum_p a_{pn} = 1$$

as an option as part of their method, to remove the scaling redundancy between columns $\mathbf{a}_n$ of **A** and the rows of **S**. Since all the elements a_{pn} are non-negative, $a_{pn} \geqslant 0$, we notice also that $\sum_p a_{pn} = \sum_p |a_{pn}| \equiv ||\mathbf{a}_n||_1$, so this also imposes a unit ℓ_1 norm on each of the columns of **A**.

13.3.1.1 Sparseness

Hoyer [42] introduced a modification to the NMF method to include a *sparseness* penalty on the elements of **S**, which he called *non-negative sparse coding*. He modified the Euclidean cost function (13.3) to include an additional penalty term:

$$D_{\mathrm{ESS}}(\mathbf{X};\mathbf{AS}) = \frac{1}{2}||\mathbf{X}-\mathbf{AS}||_{\mathrm{F}}^2 + \lambda \sum_{nt} s_{nt} \tag{13.28}$$

for some weight $\lambda \geqslant 0$. Hoyer also required a unit ℓ_1 norm on the columns of **A**, $||\mathbf{a}_n||_1 = 1$.

From a probabilistic perspective, Hoyer and Hyvärinen [44] pointed out that (13.28) is equivalent to a maximum log-likelihood approach where we assume that the noise $\mathbf{E} =$

$\mathbf{X} - \mathbf{AS}$ has a normal distribution, while the sources have an exponential distribution, $p(s_{nt}) \propto \exp(-s_{nt})$.

Hoyer showed that this new cost function was non-increasing under the **S** update rule

$$s_{nt} \leftarrow s_{nt} \frac{[\mathbf{A}^{\mathrm{T}}\mathbf{X}]_{nt}}{[\mathbf{A}^{\mathrm{T}}\mathbf{AS}]_{nt} + \lambda} \tag{13.29}$$

which is a very simple modification of the original Lee-Sung multiplicative update rule (13.15). A similar rule was not available for the update to **A**, so he instead suggested a projected gradient method

$$\begin{aligned} a_{pn} &\leftarrow \left[a_{pn} - \eta([\mathbf{ASS}^{\mathrm{T}}]_{pn} - [\mathbf{XS}^{\mathrm{T}}]_{pn})\right]_{+} \\ a_{pn} &\leftarrow a_{pn}/\|\mathbf{a}_n\|_2 \end{aligned} \tag{13.30}$$

so that the complete algorithm is to repeat (13.30) and (13.29) until convergence.

Hoyer and Hyvärinen [44] demonstrated that NMF with this sparsity penalty can lead to learning of higher-level contour coding from complex cell outputs [44]. Sparsity constraints are also useful for text mining applications [74].

As an alternative way to include sparseness constraints in the NMF method, Hoyer [43] also introduced the idea of maintaining a fixed level of sparseness for the columns of **A** and rows of **S**, where this is defined as

$$\text{sparseness}(\mathbf{u}) = \frac{\sqrt{N} - \|\mathbf{u}\|_1/\|\mathbf{u}\|_2}{\sqrt{N} - 1} \tag{13.31}$$

where N is the number of elements of the vector $\mathbf{u}$. This measure of sparseness (13.31) is defined so that a vector $\mathbf{u}_{\mathrm{S}}$ with a single non-zero element has sparseness$(\mathbf{u}_{\mathrm{S}}) = 1$, and a vector $\mathbf{u}_{\mathrm{NS}}$ with all N components equal (disregarding sign changes) has sparseness $(\mathbf{u}_{\mathrm{NS}}) = 0$.

The idea of the method is to iteratively update **A** and **S** while maintaining fixed levels of sparseness, specifically sparseness$(\mathbf{a}) = S_{\mathrm{A}}$ for the columns of **A**, and sparseness$(\mathbf{s}) = S_{\mathrm{S}}$ for the rows of **S**. (An additional unity ℓ_1 norm constraint on the rows of **S**, $\|s_n\|_2 = 1$, is used to avoid scaling ambiguities.)

Updating with these sparseness constraints is achieved with a sequence of projected gradient updates

$$\mathbf{a}_n \leftarrow P_{\mathrm{A}}\left[\mathbf{a}_n - \eta_{\mathrm{A}}([\mathbf{ASS}^{\mathrm{T}}]_{\bullet n} - [\mathbf{XS}^{\mathrm{T}}]_{\bullet n})\right] \tag{13.32}$$

$$\mathbf{s}_n \leftarrow P_{\mathrm{S}}\left[\mathbf{s}_n - \eta_{\mathrm{S}}([\mathbf{A}^{\mathrm{T}}\mathbf{AS}]_{n\bullet} - [\mathbf{A}^{\mathrm{T}}\mathbf{X}]_{n\bullet})\right] \tag{13.33}$$

where $[\mathbf{M}]_{\bullet n}$ is the n-th column vector of **M**, $[\mathbf{M}]_{n\bullet}$ is the n-th row vector of **M**, and $P_{\mathrm{A}}[\cdot]$ and $P_{\mathrm{S}}[\cdot]$ are special projection operators for columns of **A** and rows of **S** respectively which impose the required level of sparseness. The projection operator $P_{\mathrm{A}}[\mathbf{a}]$ projects the column vector **a** so that it is (a) non-negative, (b) has the same ℓ_1

norm $||\mathbf{a}||_2$, and (c) has the required sparseness level, sparseness($\mathbf{a}$) $= S_A$. Similarly, the projection operator $P_S[\mathbf{s}]$ projects the row vector $\mathbf{s}$ so that it is (a) non-negative, (b) has unit ℓ_1 norm $||\mathbf{s}||_2 = 1$, and (c) has the required sparseness level, sparseness($\mathbf{s}$) $= S_S$. These projection operators are implemented by an iterative algorithm which solves this joint constraint problem: for details see [43].

Hoyer demonstrated that this method was able to give parts-based representations of image data, even when the images were not so well aligned, and where the original NMF algorithm would give a global representation [43].

13.3.1.2 "Smoothness"

Another common penalty term is a so-called "smoothness" constraint, obtained by penalizing the (squared) Frobenius norm of e.g. $\mathbf{A}$ [76]:

$$||\mathbf{A}||_F^2 = \sum_{pn} a_{pn}^2. \tag{13.34}$$

The term "smoothness" is perhaps a little misleading: it does not refer to any "blurring" or "smoothing" between, for example, neighboring pixels in an image; it merely refers to the penalization of large values a_{pn}, so the resulting matrix is less "spiky" and hence more "smooth".

If we add this non-smoothness penalty (13.34) into the Euclidean cost function (13.3) we obtain a new cost function

$$J = D(\mathbf{X}; \mathbf{AS}) = \frac{1}{2}||\mathbf{X} - \mathbf{AS}||_F^2 + \frac{1}{2}\alpha \sum_{pn} a_{pn}^2 \tag{13.35}$$

which will act to reduce the tendency to produce large elements in $\mathbf{A}$. From a probabilistic perspective we can regard this as imposing a Gaussian prior on the elements a_{pn} of $\mathbf{A}$. This modifies the derivative of J w.r.t. $\mathbf{A}$, giving

$$\frac{\partial J}{\partial a_{pn}} = -([\mathbf{XS}^T]_{pn} - [\mathbf{ASS}^T]_{pn}) + \alpha a_{pn} \tag{13.36}$$

giving a new gradient update step of

$$a_{pn} \leftarrow a_{pn} + \eta_{pn}([\mathbf{XS}^T]_{pn} - [\mathbf{ASS}^T]_{pn} - \alpha a_{pn}) \tag{13.37}$$

and again using $\eta_{pn} = a_{pn}/[\mathbf{ASS}^T]_{pn}$ we obtain the multiplicative update

$$a_{pn} \leftarrow a_{pn} \frac{[\mathbf{XS}^T]_{pn} - \alpha a_{pn}}{[\mathbf{ASS}^T]_{pn}} \tag{13.38}$$

for which J in (13.35) is non-increasing [76]. To ensure a_{pn} remains non-negative in this multiplicative update, we can set negative values to a small positive ϵ. (If we were simply to set negative elements to zero, the multiplicative update would never be able make that element non-zero again if required.)

Similarly, we can separately or alternatively apply such a non-smoothness penalty to **S**, obtaining a similar adjustment to the update steps for s_{nt}.

13.3.1.3 Continuity

In the context of audio source separation, Virtanen [94] proposed a *temporal continuity* objective along the rows (t-direction) of **S** (or alternatively, along the columns of **A**, as in Virtanen's original paper [94]). This temporal continuity is achieved by minimizing a total variation (TV) cost to penalize changes in the values of s_{nt} in the t ("time") direction

$$C_{\mathrm{TV}t}(\mathbf{S}) = \frac{1}{2}\sum_{nt} |s_{n,(t-1)} - s_{n,t}| \tag{13.39}$$

where t is summed from 2 to T. Total variation has also been applied for image reconstruction in the Compressed Sensing literature, where it is used in a 2-dimensional form [62], and an earlier approach for smoothness (in this sense) was developed and showcased in spectroscopy [5].

The derivative of $C_{\mathrm{TV}t}$ is straightforward:

$$\frac{\partial s_{nt} C_{\mathrm{TV}t}(\mathbf{S})}{\partial s_{nt}} = \begin{cases} -1 & \text{if } s_{n,t} < s_{n,(t-1)} \text{ and } s_{n,t} < s_{n,(t+1)}, \\ +1 & \text{if } s_{n,t} > s_{n,(t-1)} \text{ and } s_{n,t} > s_{n,(t+1)}, \\ 0 & \text{otherwise.} \end{cases} \tag{13.40}$$

(apart from the boundary cases $t = 1$ and $t = T$) so this can be incorporated into a steepest-descent update method for **S**.

Chen and Cichocki [11] introduced a different smoothness measure based on the difference between s_{nt} and a "temporally smoothed" (low-pass-filtered) version

$$\bar{s}_n(t) = \alpha \bar{s}_n(t-1) + \beta s_n(t) \tag{13.41}$$

where $\beta = 1 - \alpha$, and we write $s_n(t) \equiv s_{nt}$ to clarify the time dimension. We can write this in matrix notation for the rows $\mathbf{s}_n$ of **S** as

$$\bar{\mathbf{s}}_n = \mathbf{T}\mathbf{s}_n, \qquad \mathbf{T} = \begin{bmatrix} \beta & 0 & \cdots & 0 \\ \alpha\beta & \beta & \cdots & 0 \\ \vdots & \ddots & \ddots & \vdots \\ \alpha^{T-1}\beta & \ldots & \alpha\beta & \beta \end{bmatrix} \tag{13.42}$$

where **T** is a $T \times T$ Toeplitz matrix that we can simplify to retain only, for example, the diagonal and first 4 subdiagonals by neglecting terms in $\alpha^k\beta$ for $k > 4$.

By incorporating a cost

$$R = \frac{1}{T}||\mathbf{s}_n - \bar{\mathbf{s}}_n||_2^2 = ||(\mathbf{I} - \mathbf{T})\mathbf{s}_n||_2^2 \tag{13.43}$$

and a unit-variance (fixed ℓ_1-norm) constraint on the rows $\mathbf{s}_n$, they obtain a modification to the Euclidean cost (13.3)

$$J = \frac{1}{2}||\mathbf{X} - \mathbf{AS}||_F^2 + \frac{\lambda}{2T}\sum_n ||(\mathbf{I} - \mathbf{T})\mathbf{s}_n||_2^2 \tag{13.44}$$

where λ is a regularization coefficient, and hence a new multiplicative update step for $\mathbf{S}$ as

$$s_{nt} \leftarrow s_{nt} \frac{[\mathbf{A}^T\mathbf{X}]_{nt}}{[\mathbf{A}^T\mathbf{AS}]_{nt} + \lambda[\mathbf{SQ}]_{nt}} \tag{13.45}$$

where $\mathbf{Q} = \frac{1}{T}(\mathbf{I} - \mathbf{T})^T(\mathbf{I} - \mathbf{T})$.

13.3.2 Relaxing the non-negativity constraints

We can consider relaxing or replacing the non-negativity constraints on the factors. For example, if we remove all non-negativity constraints from (13.1) and instead impose an orthogonality and unit norm constraint on the columns of $\mathbf{A}$, minimizing the mean squared error (13.3) will find the *principal subspace*, i.e. the subspace spanned by the principal components of $\mathbf{S}$.

13.3.2.1 Semi-NMF

In *Semi-NMF* [23] we assume that only one factor matrix $\mathbf{A}$ or $\mathbf{S}$ is non-negative, giving for example $\mathbf{X} \approx \mathbf{AS}$ where $\mathbf{S}$ is non-negative, but $\mathbf{A}$ can be of mixed sign.

To achieve uniqueness of factorization we need to impose additional constraints such as mutual independence, sparsity or semi-orthogonality. This leads, for example, to non-negative ICA, non-negative sparse coding, or non-negative PCA.

13.3.2.2 Non-negative ICA

Suppose that we relax the non-negativity on $\mathbf{A}$, and instead suppose that the rows $\mathbf{s}_n$ of $\mathbf{S}$ are sampled from N independent non-negative sources $s_1, \cdots, s_N$. In other words, we suppose we have an independent component analysis (ICA) model, with an additional constraint of non-negativity on the sources s_n: we refer to this as *non-negative independent component analysis* (NNICA).

If we wish, we can always solve NNICA using classical ICA approaches, then change the sign of any negative sources [14]. However, we can also consider the NNICA model directly. Suppose we whiten the observation vectors $\mathbf{x}$ to give

$$\mathbf{z} = \mathbf{W}\mathbf{x} \tag{13.46}$$

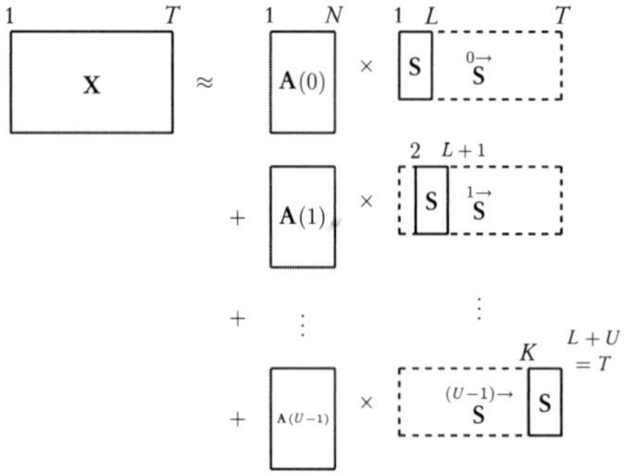

FIGURE 13.2

Convolutive NMF model for Non-negative Matrix Factor Deconvolution (NMFD).

so that $\mathbf{z}$ has identity covariance $\mathbb{E}\{\mathbf{z}\mathbf{z}^T\} = \mathbf{I}$, but do this whitening without subtracting the mean $\bar{\mathbf{z}}$ of $\mathbf{z}$. Then to find the independent components (factors) $\mathbf{s}$ it is sufficient to look for an orthonormal rotation matrix $\mathbf{Q}$ such that $\mathbf{Q}\mathbf{Q}^T = \mathbf{I}$ such that the resulting output $\mathbf{y} = \mathbf{Q}\mathbf{z} = \mathbf{Q}\mathbf{W}\mathbf{x}$ is non-negative [78]. This leads to simple algorithms such as a *non-negative PCA* method [79,81] related to the nonlinear PCA rule for standard ICA [67], as well as constrained optimization approaches based on the Lie Group geometry of the set of orthonormal matrices [80].

13.3.3 Structural factor constraints

In certain applications, the factors $\mathbf{A}$ and $\mathbf{S}$ may have a natural structure that should be reflected in the parametrizations of the factors. For example, Smaragdis [87,88] and Virtanen [95] introduced a *Convolutive NMF* model, whereby our model becomes

$$x_{pt} \approx \sum_{n,u} a_{pn}(u) s_{n,t-u} \tag{13.47}$$

which we can write in a matrix convolution form as (Fig. 13.2)

$$\mathbf{X} = \sum_{u=0}^{U-1} \mathbf{A}(u) \overset{u\rightarrow}{\mathbf{S}} \tag{13.48}$$

where the $\overset{u\rightarrow}{\cdot}$ matrix notation indicates that the contents of the matrix are shifted u places to the right

$$[\overset{u\rightarrow}{\mathbf{S}}]_{nt} = [\mathbf{S}]_{n,t-u}. \tag{13.49}$$

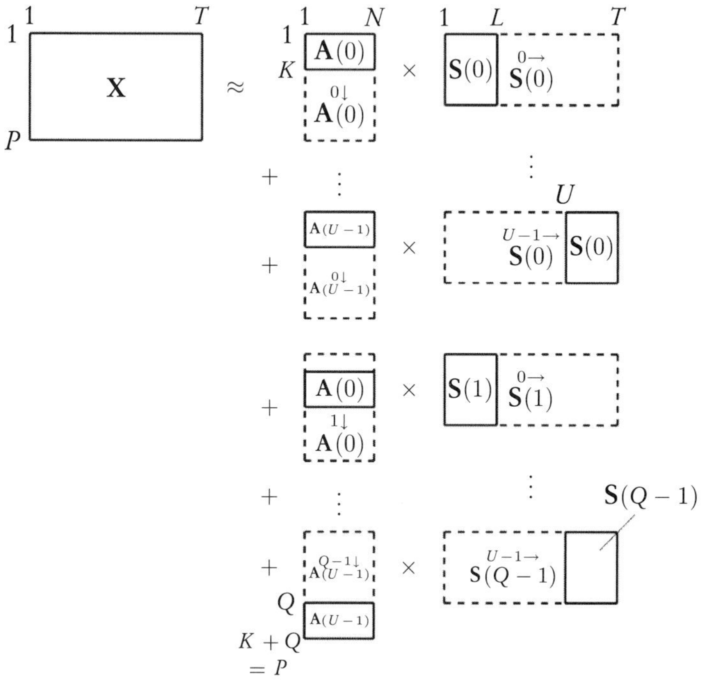

FIGURE 13.3

Two-dimensional convolutive NMF model (NMF2D).

Finding non-negative $\mathbf{A}(u)$ and $\mathbf{S}$ from (13.47) is also known as *non-negative matrix factor deconvolution* (NMFD).

Schmidt and Mørup [86] extended the convolutive model to a 2-dimensional convolution

$$x_{pt} \approx \sum_{n,q,u} a_{p-q,n}(u) s_{n,t-u}(q) \tag{13.50}$$

which we can write in a matrix convolution form as (Fig. 13.3)

$$\mathbf{X} = \sum_{q=0}^{Q-1} \sum_{u=0}^{U-1} \overset{q\downarrow}{\mathbf{A}(u)} \, \overset{u\rightarrow}{\mathbf{S}(q)} \tag{13.51}$$

where the $\overset{q\downarrow}{\cdot}$ matrix notation indicates that the contents of the matrix are shifted q places down

$$[\overset{q\downarrow}{\mathbf{A}}]_{pn} = [\mathbf{A}]_{p-q,n}. \tag{13.52}$$

Alternatively, if we change notation a little to write

$$a_n(p-q,u) \equiv a_{p-q,n}(u) \qquad s_n(q,t-u) \equiv s_{n,t-u}(q) \tag{13.53}$$

we could write (13.51) as

$$\mathbf{X} = \sum_{n=1}^{N} \mathbf{X}_n \tag{13.54}$$

where

$$[\mathbf{X}_n]_{pt} = \sum_{q=0}^{Q-1} \sum_{u=0}^{U-1} a_n(p-q,t-u) s_n(q,u) \equiv a_n(p,t) * s_n(p,t) \tag{13.55}$$

with $*$ as a 2-D convolution operator. So this can be viewed as a sum of N elementary 2D "objects" $s_n(p,t)$ convolved with "filters" $a_n(p,t)$, or vice-versa.

This type of 1-D and 2-D convolutive model has been applied to the analysis of audio spectrograms. For example, Smaragdis [87] used the 1-D model to analyze drum sounds, on the basis that drum sounds produce a characteristic time-frequency pattern that repeats whenever the drum is "hit". On the other hand, Schmidt and Mørup [86] applied the 2-D model to analysis of spectrograms of pitched sounds on a log-frequency scale. Here a time shift (u) corresponds to onset time of the note, while the frequency shift (q) corresponds to adding a constant log-frequency offset, or multiplying all pitches in the "object" by a constant factor.

In a more general case, we can consider transform-invariant factorization [97]

$$\mathbf{X} = \sum_{u} \mathbf{A}^{(u)} \mathbf{T}^{(u)}(\mathbf{S}) \tag{13.56}$$

where $\{\mathbf{T}^{(u)}, u = 1, \ldots, U\}$ is a set of matrix transformation functions. This can include 1-D and 2-D convolutions (if u ranges over a 2-D set) but could represent more general transforms.

As a further generalization, Schmidt and Laurberg [85] introduce the idea that the matrices $\mathbf{A}$ and $\mathbf{S}$ can be determined by underlying parameters. Their model is given by

$$\mathbf{X} \approx \mathbf{A}(\mathbf{a})\mathbf{S}(\mathbf{s}) \tag{13.57}$$

where $\mathbf{a}$ and $\mathbf{s}$ are parameters which determine the generation of the matrix-valued functions $\mathbf{A}(\mathbf{a})$ and $\mathbf{S}(\mathbf{s})$. In their paper they model $\mathbf{a}$ and $\mathbf{s}$ as Gaussian processes.

13.3.4 Multi-factor and tensor models

The standard NMF model (13.1) is sometimes known as the *Two-Way Factor Model*, being a product of two matrices. There are many different ways to extend this to models with three or more factors, or to models which include tensors as factors, i.e. where each element has more than two indices. For example, we could have order 3 tensors, which have elements x_{ijk} with 3 indices, instead of the usual matrices which have elements x_{ij} with 2 indices (i.e. our usual matrices are order 2 tensors) [36].

13.3.4.1 Multi-layer NMF

In multi-layer NMF the basic matrix **A** is replaced by a set of cascaded (factor) matrices. Thus, the model can be described as [17,13]

$$\mathbf{X} \approx \mathbf{A}_1 \mathbf{A}_2 \cdots \mathbf{A}_K \mathbf{S}. \tag{13.58}$$

Since the model is linear, all the matrices $\mathbf{A}_k (k = 1, 2, \ldots, K)$ can be merged into a single matrix **A** if no any additional constraints are imposed upon the individual matrices $\mathbf{A}_k$. However, if we impose sparsity constraints for each individual matrix $\mathbf{A}_k$, then multi-layer NMF can be used to considerably improve the performance of standard NMF algorithms due to distributed structure and alleviating the problem of local minima. To improve the performance of the NMF algorithms (in particular, for ill-conditioned and badly-scaled data) and to reduce the risk of getting stuck in local minima of a cost function due to non-convex alternating minimization, we use a multi-stage procedure combined with a multi-start initialization, in which we perform a sequential decomposition of non-negative matrices as follows. In the first step, we perform the basic approximate decomposition $\mathbf{X} \approx \mathbf{A}_1 \mathbf{S}_1$ using any available NMF algorithm with sparsity constraint imposed on matrix $\mathbf{A}_1$. In the second stage, the results obtained from the first stage are used to build up a new input data matrix $\mathbf{X} \leftarrow \mathbf{S}_1$, that is, in the next step, we perform a similar decomposition $\mathbf{S}_1 \approx \mathbf{A}_2 \mathbf{S}_2$, using the same or different update rules. We continue our decomposition taking into account only the last obtained components. The process can be repeated for an arbitrary number of times until some stopping criteria are satisfied. Physically, this means that we build up a distributed system that has many layers or cascade connections of K mixing subsystems. The key point in this approach is that the update process to find parameters of matrices $\mathbf{S}_k$ and $\mathbf{A}_k (k = 1, 2, \ldots, K)$ is performed sequentially, i.e. layer-by-layer, where each layer is randomly initialized with different initial conditions.

Tri-NMF, also called three factor NMF, can be considered as a special case of the multi-layer NMF and can take the following general form [23]:

$$\mathbf{X} \approx \mathbf{AMS} \tag{13.59}$$

where non-negativity constraints are imposed on all, or on selected, factor matrices. Note that if we do not impose any additional constraints on the factors (besides non-negativity), the three-factor NMF can be reduced to the standard (two-factor) NMF by imposing the following mapping $\mathbf{A} \leftarrow \mathbf{AM}$ or $\mathbf{S} \leftarrow \mathbf{MS}$.

However, the three-factor NMF is not equivalent to the standard NMF if we apply additional constraints or conditions. For example, in orthogonal Tri-NMF we impose additional orthogonality constraints upon the matrices **A** and **S**, $\mathbf{A}^T\mathbf{A} = \mathbf{I}$ and $\mathbf{S}\mathbf{S}^T = \mathbf{I}$,

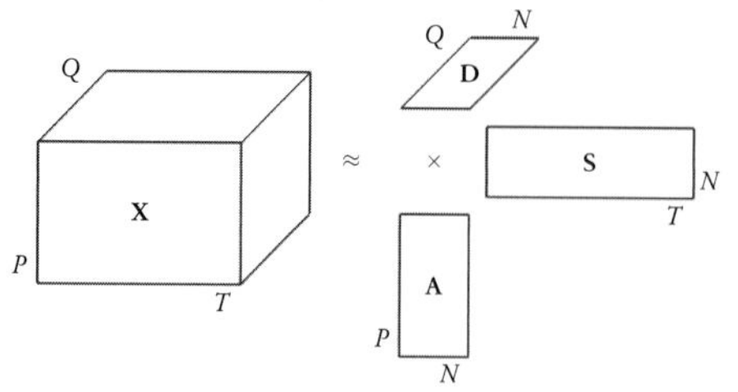

FIGURE 13.4

Three-Way PARAFAC factor model.

while the matrix **M** can be an arbitrary unconstrained matrix (i.e. it has both positive and negative entries). For uni-orthogonal Tri-NMF only one matrix **A** or **S** is orthogonal. Non-smooth NMF (nsNMF) was proposed by Pascual-Montano *et al.* [73] and is a special case of the three-factor NMF model in which the matrix **M** is fixed and known, and is used for controlling the sparsity or smoothness of the factor matrix **S** and/or **A**.

13.3.4.2 Non-negative tensor factorization

In early work on matrix factorization without non-negativity constraints, Kruskal [55] considered "three way arrays" (order 3 tensors) of the form (Fig. 13.4)

$$x_{ptq} = \sum_n a_{pn} s_{nt} d_{qn} = \sum_n a_{pn} v_{tn} d_{qn} = \tag{13.60}$$

which can be written in matrix notation using the frontal slices of data tensor as

$$\mathbf{X}_q \approx \mathbf{A}\mathbf{D}_q\mathbf{S} = \mathbf{A}\mathbf{D}_q\mathbf{V}^T \tag{13.61}$$

where $[\mathbf{X}_q]_{pt} = x_{ptq}$ represent frontal slices of **mX** and $\mathbf{D}_q$ is the $N \times N$ diagonal matrix with elements $[\mathbf{D}_q]_{nn} = d_{nq}$. This model is known as the *PARAFAC* or *CANDECOMP* (CANonical DECOMPosition) model [36]. A non-negative version of PARAFAC was first introduced by Carroll *et al.* [9] and Krijnen & ten Berge [54]. Later, more efficient approaches were developed by Bro (1997) [4] based on the modified NNLS mentioned earlier and Paatero [70], who generalized his earlier 2-way positive matrix factorization (PMF) method to the 3-way PARAFAC model, referring to the result as *PMF3* (3-way positive matrix factorization). The non-negatively constrained PARAFAC is also sometimes called *non-negative tensor factorization* (NTF). In some cases NTF methods may increase the number of factors and add complexity. However, in many contexts they do not lead to an increase in the number of factors, (they maintain them) and quite often they lower the complexity – because NNLS is cheaper than LS in iterative algorithms. In addition, this approach can result in a reduced number of active parameters yielding a clearer "parts-based" representation [63]. Non-negatively constrained PARAFAC has been used in numerous applications in environmental analysis, food studies, and pharmaceutical analysis and in chemistry in general [6].

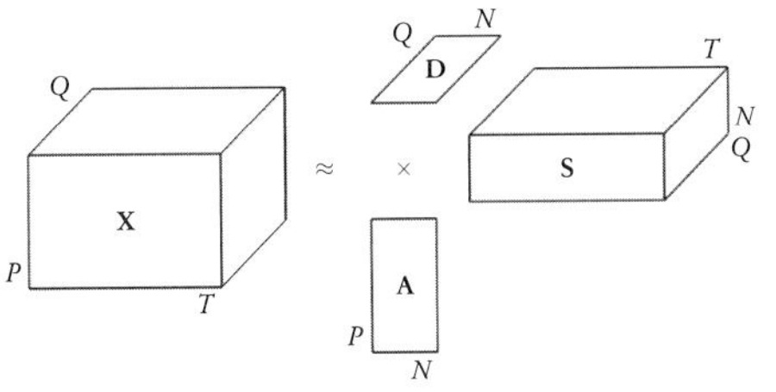

FIGURE 13.5

PARAFAC2/NTF2 factor model.

Later Welling and Weber [96] also discussed a factorization of an order R tensor $x_{i_1,\ldots,i_R}$ into a product of r order 2 tensors

$$x_{p_1,\ldots,p_R} \approx \sum_{n=1}^{N} a^{(1)}_{p_1,n} a^{(2)}_{p_2,n} \cdots a^{(R)}_{p_R,n} \tag{13.62}$$

subject to the constraint that the parameters are non-negative. They called the result *positive tensor factorization* (PTF) or *non-negative tensor factorization* (NTF). NTF can be presented in vector-matrix form as follows

$$\mathbf{X} \approx \sum_{n=1}^{N} \mathbf{a}^{(1)}_n \circ \mathbf{a}^{(2)}_n \circ \cdots \circ \mathbf{a}^{(R)}_n = \mathbf{I} \times_1 \mathbf{A}^{(1)} \times_2 \mathbf{A}^{(2)} \cdots \times_R \mathbf{A}^{(R)} \tag{13.63}$$

where $\circ$ denotes outer product and $\times_r$ denotes r-mode multiplication of tensor via matrix and $\mathbf{I}$ is R-order identity tensor (with one on the superdiagonal). Welling and Weber developed update rules for NTF which are analogous to the Lee and Seung [57] multiplicative update rules.

Ding *et al.* [24] also considered adding orthogonality constraints to the 3-way factor model (13.60). They showed that this additional constraint leads to a clustering model, and demonstrated its application to document clustering.

A further extension of these tensor models is to allow one or more of the factors to also be a higher-order tensor. For example, the *PARAFAC2* model [35,48] includes an order 3 tensor in the factorization (Fig. 13.5):

$$x_{ptq} \approx \sum_{n} a_{pn} s_{ntq} d_{nq}. \tag{13.64}$$

In matrix notation we can write (13.64) as

$$\mathbf{X}_q \approx \mathbf{A}\mathbf{D}_q\mathbf{S}_q \tag{13.65}$$

with $\mathbf{X}_q$ and $\mathbf{D}_q$ as for the PARAFAC/PMF3 model above, and $[\mathbf{S}_q]_{nt} = s_{ntq}$. In addition to Eq. (13.64), the PARAFAC2 model includes extra constraints on the $\mathbf{S}_q$ matrices

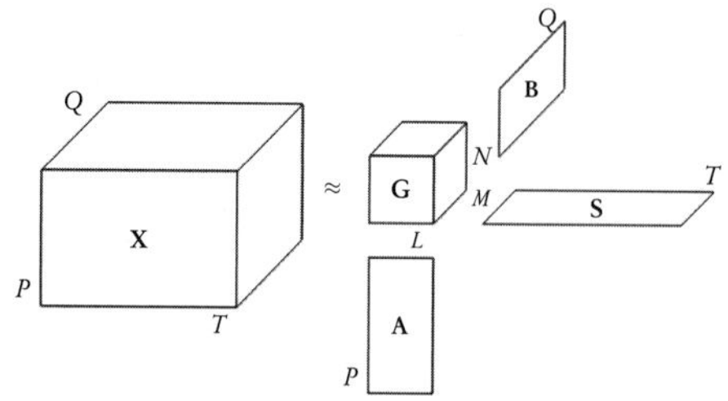

FIGURE 13.6

Three-way Tucker model.

to obtain a unique solution. The first non-negative algorithm for PARAFAC2 was introduced in [5]. Cichocki *et al.* [20,19] call the model in Eq. (13.64) *NTF2* to distinguish it from the PARAFAC-based non-negative tensor factorization (NTF) model (13.60).

FitzGerald *et al.* [26] combined convolutive NMF models (NMFD/NMF2D) with tensor factorization, leading to shift-invariant non-negative tensor factorization. They applied this to musical audio source separation, where the tensor **X** is of order 3, representing spectrograms with frequency p, time t and channel q.

Another multi-way model is the *Tucker model* (Fig. 13.6)

$$x_{pqt} \approx \sum_{lmn} g_{lmn} a_{pl} s_{tm} b_{qn} \tag{13.66}$$

which in its general form is

$$x_{p_1,p_2,\ldots,p_R} \approx \sum_{n_1,\ldots,n_R} g_{n_1,\ldots,n_R} a_{p_1,n_1} \times \cdots \times a_{p_R,n_R} \tag{13.67}$$

where the *Tucker core* $g_{n_1,\ldots,n_r}$ controls the interaction between the other factors. The Tucker model has also been implemented in non-negative versions, where it is sometimes called Non-negative Tucker Decomposition (NTD). The first implementations of non-negative Tucker as well as a number of other constraints were given in [47] and in [5]. Several researchers have recently applied non-negative Tucker models to EEG analysis, classifications and feature extractions, and have demonstrated encouraging results [63,50,51,75].

13.3.5 ALS Algorithms for non-negative tensor factorization

Almost all existing NMF algorithms can be relatively easily extended for R-order non-negative tensor factorization by using the concept of matricizing or unfolding. Generally speaking, the unfolding of an R-th order tensor can be understood as the process of construction of a matrix containing all the r-mode vectors of the tensor. The order of the columns is not unique and in this book it is chosen in accordance with Kolda and

Bader [53]. The mode-r unfolding of tensor $\mathbf{X} \in \mathbb{R}^{I_1 \times I_2 \times \cdots \times I_R}$ is denoted by $\mathbf{X}_{(r)}$ and arranges the mode-r fibers into columns of a matrix.

Using the concept of unfolding an R-order NTF can represented as the following set of non-negative matrix factorizations

$$\mathbf{X}_{(r)} \approx \mathbf{A}^{(r)} \mathbf{Z}_{(-r)}, \qquad (r = 1, 2, \ldots, R) \tag{13.68}$$

where $\mathbf{X}_{(r)} \in \mathbb{R}_+^{I_r \times I_1 \cdots I_{r-1} I_{r+1} \cdots I_R}$ is the r-mode unfolded matrix of the R-order tensor $\mathbf{X} \in \mathbb{R}_+^{I_1 \times I_2 \times \cdots \times I_R}$ and

$$\mathbf{Z}_{(-r)} = \left[\mathbf{A}^{(R)} \odot \cdots \odot \mathbf{A}^{(r+1)} \odot \mathbf{A}^{(r-1)} \odot \cdots \odot \mathbf{A}^{(1)}\right]^T \in \mathbb{R}_+^{N \times I_1 \cdots I_{r-1} I_{r+1} \cdots I_R} \tag{13.69}$$

where $\odot$ denotes the Khatri-Rao product [53].

Using this model we can derive standard (global) ALS update rules:

$$\mathbf{A}^{(r)} \leftarrow \left[\mathbf{X}_{(r)} \mathbf{Z}_{(-r)}^T \left(\mathbf{Z}_{(-r)}^T \mathbf{Z}_{(-r)}\right)^{-1}\right]_+, \quad (r = 1, 2, \ldots, R). \tag{13.70}$$

By defining the residual tensor as

$$\begin{aligned}
\mathbf{X}^{(n)} &= \mathbf{X} - \sum_{j \neq n} \mathbf{a}_j^{(1)} \circ \mathbf{a}_j^{(2)} \circ \cdots \circ \mathbf{a}_j^{(R)} \\
&= \mathbf{X} - \sum_{j=1}^{N} \left(\mathbf{a}_j^{(1)} \circ \mathbf{a}_j^{(2)} \circ \cdots \circ \mathbf{a}_j^{(R)}\right) + \left(\mathbf{a}_n^{(1)} \circ \mathbf{a}_n^{(2)} \circ \cdots \circ \mathbf{a}_n^{(R)}\right), \\
&= \mathbf{X} - \widehat{\mathbf{X}} + \left(\mathbf{a}_n^{(1)} \circ \mathbf{a}_n^{(2)} \circ \cdots \circ \mathbf{a}_n^{(R)}\right), \qquad (n = 1, 2, \ldots, N)
\end{aligned} \tag{13.71}$$

we can derive local ALS updates rules [75]:

$$\mathbf{a}_n^{(r)} \leftarrow \left[\mathbf{X}_{(r)}^{(n)} \left(\mathbf{a}_n^{(R)} \odot \cdots \odot \mathbf{a}_n^{(r+1)} \odot \mathbf{a}_n^{(r-1)} \odot \cdots \odot \mathbf{a}_n^{(1)}\right)\right]_+, \tag{13.72}$$

for $r = 1, 2, \ldots, R$ and $n = 1, 2, \ldots, N$ and with normalization (scaling) $\mathbf{a}_n^{(r)} \leftarrow ||\mathbf{a}_n^{(r)} / \mathbf{a}_n^{(r)}||_2$ for $r = 1, 2, \ldots, R-1$. The local ALS update can be expressed in equivalent tensor notation:

$$\mathbf{a}_n^{(r)} \leftarrow \left[\mathbf{X}^{(n)} \times_1 \mathbf{a}_n^{(1)} \cdots \times_{r-1} \mathbf{a}_n^{(r-1)} \times_{r+1} \mathbf{a}_n^{(r+1)} \cdots \times_R \mathbf{a}_n^{(R)}\right]_+, \tag{13.73}$$

$$(r = 1, 2, \ldots, R) \quad (n = 1, 2, \ldots, N). \tag{13.74}$$

In similar way we can derive global and local ALS updates rules for Non-negative Tucker Decomposition [21,75].

13.4 FURTHER NON-NEGATIVE ALGORITHMS

In section 13.2 we briefly developed three simple and popular algorithms for NMF. It is arguably the very simplicity of these algorithms, and in particular the Lee-Seung multiplicative algorithms (13.16) and (13.23), which have led to the popularity of the NMF approach.

Nevertheless, in recent years researchers have gained an improved understanding of the properties and characteristics of these NMF algorithms. For example, while Lee and Seung [57] claimed that their multiplicative algorithm (13.16) converges to a stationary point, this is now disputed [31], and in any case Lin [61] also points out that a stationary point is not necessarily a minimum. For more on these alternative approaches, see for example [12,90,2,61,19].

In addition, there has previously been interest in the effect of non-negative constraints in neural network learning (e.g. [29,34,89]). Another approach is the use of geometric constraints, based on looking for the edges or bounds of the scattering matrix [3,38,1]. Recent work has also investigated alternative algorithms specifically designed for large-scale NMF problems [15,21]. In this section we will investigate some of these alternative approaches.

13.4.1 Neural network approaches

Given an input $\mathbf{X} = [x_{pt}]$, representing a sequence of input vectors $\mathbf{x}_1, \ldots, \mathbf{x}_T$, we can construct a simple linear "neural network" model

$$\mathbf{Y} = \mathbf{B}\mathbf{X} \tag{13.75}$$

where $\mathbf{B}$ is a $Q \times P$ linear weight matrix and $\mathbf{Y} = [y_{qt}]$ is the output from neuron q for sample t. We can write (13.75) in its pattern-by-pattern form as

$$\mathbf{y}(t) = \mathbf{B}\mathbf{x}(t) \quad t = 1, 2, \ldots \tag{13.76}$$

Without any non-negativity constraints, the network (13.75) has been widely studied for the task of principal component analysis (PCA) or PCA subspace analysis (PSA): e.g. see [66,41]. For example, Williams [98] described his *Symmetric Error Correction* (SEC) network, based on the idea of reducing the mean squared error reconstruction. A similar method was suggested independently by Oja and Karhunen [68] to find the principal subspace of a matrix. For the learning algorithm in the SEC network, the weight matrix $\mathbf{B}$ is updated on a pattern-by-pattern basis according to

$$\mathbf{B}(t+1) = \mathbf{B}(t) + \eta(t)[\mathbf{x}(t) - \hat{\mathbf{x}}(t)]\mathbf{y}^{\mathrm{T}}(t) \tag{13.77}$$

where $\hat{\mathbf{x}}(t) = \mathbf{B}^{\mathrm{T}}\mathbf{y}(t)$ is considered to be an approximate reconstruction of the input $\mathbf{x}$ using the weights $\mathbf{B}$. Alternatively, the following batch update rule can be used:

$$\mathbf{B}(t+1) = \mathbf{B}(t) + \eta(t)[\mathbf{X} - \hat{\mathbf{X}}]\mathbf{Y}^{\mathrm{T}} \tag{13.78}$$

where $\hat{\mathbf{X}} = \mathbf{B}^T\mathbf{Y}$ is the approximate reconstruction. With $m \leqslant n$ outputs, and without any non-negativity constraints, update rule (13.77) finds the minimum of the mean squared reconstruction error

$$J_E = D_E(\mathbf{X};\hat{\mathbf{X}}) = \frac{1}{2}||\mathbf{X} - \hat{\mathbf{X}}||_F^2 \tag{13.79}$$

and hence finds the principal subspace of the input, i.e. the space spanned by the principal eigenvectors of $\mathbf{XX}^T$ [99].

Harpur and Prager [34] suggested modifying this network to include a non-negativity constraint on the output vector $\mathbf{y}$, so that its activity is determined by

$$y_q(t) = [\mathbf{b}_q^T\mathbf{x}(t)]_+ \tag{13.80}$$

where $\mathbf{b}_q = (b_{q1}, \ldots, b_{qP})^T$, and use this non-negative $\mathbf{Y}$ to form the reconstruction $\hat{\mathbf{X}}$ in (13.78). They showed that this *recurrent error correction* (REC) network, with the non-negativity constraint on the output, could successfully separate out individual horizontal and vertical bars from images in the "bars" problem introduced by Földiák [28], while the network without the non-negativity constraint would not.

Harpur noted that this *recurrent error correction* (REC) network might be under-constrained when fed with a mixture of non-negative sources, illustrating this for $n = m = 2$ [33, p. 68]. He suggests that this uncertainty could be overcome by starting learning with weight vectors inside the "wedge" formed by the data, but points out that this would be susceptible to any noise on the input. Plumbley [77] attempted to overcome this uncertainty by incorporating anti-Hebbian lateral inhibitory connections between the output units, a modification of Földiák's Hebbian/anti-Hebbian network [27].

Charles and Fyfe [10], following on from earlier work of Fyfe [29], investigated a range of non-negative constraints on the weights and/or outputs of a PCA network. Their goal was to find a *sparse coding* of data, with most values being zero or near zero [69]. With non-negative constraints on the outputs, they noted that update equation (13.78) is a special case of the nonlinear PCA algorithm [46], and so their learning algorithm minimizes the residual error at the input neurons. They also tested their network on the "bars" problem, using various nonlinearities (threshold linear, sigmoid and exponential) as well as pre-processing to equalize the input variances $E(x_i^2)$. They found that performance was most reliable with non-negative constraint on weights b_{qp} as well as the outputs $y_q(t)$.

13.4.2 Geometrical methods

13.4.2.1 Edge vectors

Several non-negative methods have been inspired by a geometric approach to the problem. Much of the earliest work in NMF in the Seventies and Eighties was based on such approaches (e.g. see [3] and references therein). Consider the 2-dimensional case $P = N = 2$. If the sources s_{nt} are non-negative, we can often see this clearly on a scatter plot of

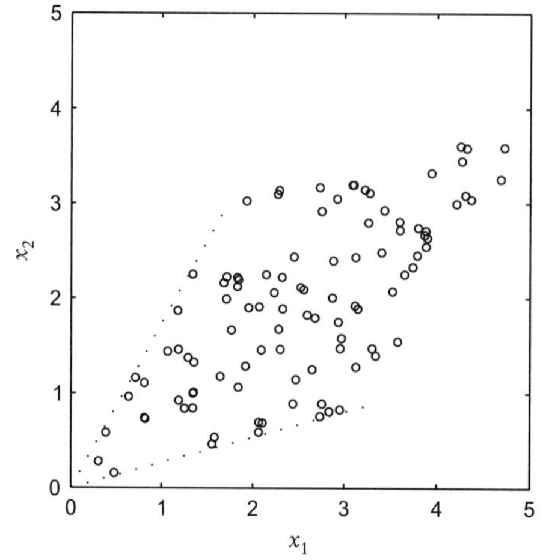

FIGURE 13.7

Scatter plot for observations of weighted non-negative sources.

x_{1t} against x_{2t} (Fig. 13.7). This scatter plot suggests that we could identify the underlying generating factors by looking for the *edges* in the distribution [38]. For example, suppose sample t' of source $p' = 2$ were zero, i.e. $s_{2,t'} = 0$. Then we immediately have

$$x_{pt'} = a_{p,1} s_{1,t'} \tag{13.81}$$

meaning that we can solve for the *basis vector* $\mathbf{a}_1 = (a_{1,1}, a_{2,1})$ apart from a scaling ambiguity [37]. This condition occurs along the edge of the scatter plot, so if we find observed vectors $\mathbf{x}_t$ on both of these edges, so-called *edge vectors*, then we can estimate the original mixing matrix **A**, and hence the source matrix **S**.

This approach has been generalized to more than two dimensions using the concept of an *extremal polyhedral cone*, finding a few spanning vectors that fix the edges of the data [93] (see also the review by Henry [39]) and Henry [40] introduces a related *extreme vector algorithm* (EVA) that searches for N-dimensional edges in the data. The geometrical approach can also give insights into issues of uniqueness of NMF, which has been investigated by Donoho and Stodden [25] and Klingenberg *et al.* [52].

13.4.2.2 Bounded pdf approaches

Some geometrical algorithms have also been introduced for cases where the sources have an additional constraint of being bounded from above as well as bounded from below (as in the non-negative case).

Puntonet *et al.* [83] and Puntonet and Prieto [84] developed separation algorithms for sources with such a bounded pdf. Their algorithm operates as each data vector arrives, updating the weights to minimize an *angular proximity*, and they also consider adjustments to their algorithm to cope with noise, which might give rise to observed data vectors which lie outside the basis vectors [84,82]. In contrast to normal ICA-based measures, which require independent sources, they found that their approach can be used to separate non-independent bounded sources. For good separation, Puntonet *et al.* [84] note that it is important to obtain *critical vectors* that map to the edges of the hyper-parallelepiped, analogous to the *edge vectors* in the geometrical NMF/PMF methods.

Yamaguchi, Hirokawa and Itoh [101,45] independently propose a similar approach for bounded data. They proposed an algebraic method for ICA of images pairs, based on the extremum points on a scatter diagram. This uses the upper- and lower-boundedness of source image values, and does not use independence. They also note that the algorithm relies on critical vectors at the apexes of the scatter diagram, so signals with low pdf at their extrema will be more difficult to separate.

Finally, Basak and Amari [1] considered the special case of bounded source signals with uniform pdf. After prewhitening, the data fills a hypercube. The hypercube is rigidly rotated using a matrix exponential $\mathbf{B} = \exp(\eta\mathbf{Z})$ to generate special (determinant 1) orthogonal matrices $\mathbf{B} \in SO(N)$ with a local learning rule used to bring data points into the unit hypercube by minimizing a 1-norm distance outside of this unit hypercube. This leads to a type of nonlinear PCA-type learning rule [46] with nonlinearity $g(y) = \text{sgn}(y)$ if y is outside the hypercube.

13.4.3 Algorithms for large-scale NMF problems

For large scale NMF problems, where the data matrix $\mathbf{X}$ is very large, the computation complexity and memory required for standard NMF algorithms can become very large. Recently new algorithms have been introduced which reduce these, e.g. through block-wise or row/column-wise updates.

13.4.3.1 ALS for large-scale NMF

If the data matrix $\mathbf{X}$ is of large dimension ($P \gg 1$ and $T \gg 1$), but where the number of non-negative components N is relatively small, ($N \ll P$ and $N \ll T$), we can reduce the computational complexity and memory allocation by taking a block-wise approach, where we select only very few rows and columns of the data matrix $\mathbf{X}$. In this approach, instead of performing a single large-scale factorization $\mathbf{X} \approx \mathbf{AS}$, we sequentially perform two (much smaller dimensional) non-negative matrix factorizations:

$$\mathbf{X}_R \approx \mathbf{A}_R\,\mathbf{S} \tag{13.82}$$

$$\mathbf{X}_C \approx \mathbf{A}\,\mathbf{S}_C \tag{13.83}$$

where $\mathbf{X}_R \in \mathbb{R}^{R\times T}$ and $\mathbf{X}_C \in \mathbb{R}^{P\times C}$ are data matrices constructed from the preselected rows and columns of the data matrix $\mathbf{X} \in \mathbb{R}^{P\times T}$, respectively. Analogously, we can construct the reduced matrices: $\mathbf{A}_R \in \mathbb{R}^{R\times N}$ and $\mathbf{S}_C \in \mathbb{R}^{N\times C}$ by using the same indices for the columns and rows as those used for the construction of the data sub-matrices $\mathbf{X}_R$ and $\mathbf{X}_C$, respectively.

There are several strategies to choose the columns and rows of the input data matrix. The simplest scenario is to randomly select rows and columns from a uniform distribution. Another heuristic option is to choose those rows and columns that provide the largest l_p-norm, especially the Chebyshev-norm, $p = \infty$.

This approach can be applied to any NMF algorithm. In the special case, for squared Euclidean distance (Frobenius norm), instead of alternately minimizing the cost function $J_E = ||\mathbf{X} - \mathbf{A}\,\mathbf{S}||_F^2$, we can minimize sequentially a set of two cost functions:

$$J_{ES} = ||\mathbf{X}_R - \mathbf{A}_R\,\mathbf{S}||_F^2 \quad \text{for fixed} \quad \mathbf{A}_R \tag{13.84}$$
$$J_{EA} = ||\mathbf{X}_C - \mathbf{A}\,\mathbf{S}_C||_F^2 \quad \text{for fixed} \quad \mathbf{S}_C \tag{13.85}$$

This leads to the following ALS updates rules for large-scale NMF [15,21]

$$\mathbf{S} \leftarrow [(\mathbf{A}_R^T\mathbf{A}_R)^{-1}\mathbf{A}_R^T\mathbf{X}_R]_+ \tag{13.86}$$
$$\mathbf{A} \leftarrow [\mathbf{X}_C\mathbf{S}_C^T(\mathbf{S}_C\mathbf{S}_C^T)^{-1}]_+. \tag{13.87}$$

13.4.3.2 Hierarchical ALS

An alternative fast local ALS algorithm, called *Hierarchical ALS* (HALS), sequentially estimates the individual columns $\mathbf{a}_n$ of $\mathbf{A}$ and rows $\mathbf{s}_n$ of $\mathbf{S}$ instead of directly computing the whole factor matrices $\mathbf{A}$ and $\mathbf{S}$ in each step.[2] The HALS algorithm is often used for multi-layer models (see section 13.3.4.1) in order to improve performance.

The basic idea is to define the residual matrix [5,18,30]:

$$\mathbf{X}^{(n)} = \mathbf{X} - \sum_{j \neq n} \mathbf{a}_j\,\mathbf{s}_j^T = \mathbf{X} - \mathbf{A}\,\mathbf{S} + \mathbf{a}_n\mathbf{s}_n^T, \quad (n = 1,2,\ldots,N) \tag{13.88}$$

and to minimize the set of squared Euclidean cost functions:

$$J_{EA}^{(n)} = ||\mathbf{X}^{(n)} - \mathbf{a}_n\,\mathbf{s}_n^T||_F^2 \quad \text{for fixed} \quad \mathbf{s}_n \tag{13.89}$$
$$J_{EB}^{(n)} = ||\mathbf{X}^{(n)} - \mathbf{a}_n\,\mathbf{s}_n^T||_F^2 \quad \text{for fixed} \quad \mathbf{a}_n \tag{13.90}$$

subject to constraints $\mathbf{a}_n \geq 0$ and $\mathbf{s}_n \geq 0$ for $n = 1,2,\ldots,N$. In order to estimate the stationary points, we simply compute the gradients of the above local cost functions with respect to the unknown vectors $\mathbf{a}_n$ and $\mathbf{s}_n$ (assuming that other vectors are fixed) and equalize them to zero:

$$\frac{\partial J_{EA}^{(n)}}{\partial \mathbf{a}_n} = \mathbf{a}_n\,\mathbf{s}_n^T\,\mathbf{s}_n - \mathbf{X}^{(n)}\,\mathbf{s}_n = 0 \tag{13.91}$$
$$\frac{\partial J_{EB}^{(n)}}{\partial \mathbf{s}_n} = \mathbf{a}_n\,\mathbf{a}_n^T\,\mathbf{s}_n - \mathbf{X}^{(n)\,T}\,\mathbf{a}_n = 0. \tag{13.92}$$

[2]The HALS algorithm is "Hierarchical" since we sequentially minimize a set of simple cost functions which are hierarchically linked to each order via residual matrices which approximate rank-one bilinear decomposition.

Hence, we obtain the local ALS algorithm:

$$\mathbf{a}_n \leftarrow \frac{1}{\mathbf{s}_n^T \mathbf{s}_n} \left[\mathbf{X}^{(n)} \mathbf{s}_n\right]_+ \tag{13.93}$$

$$\mathbf{s}_n \leftarrow \frac{1}{\mathbf{a}_n^T \mathbf{a}_n} \left[\mathbf{X}^{(n)\,T} \mathbf{a}_n\right]_+ . \tag{13.94}$$

In practice, we usually normalize the column vectors $\mathbf{a}_n$ and $\mathbf{s}_n$ to unit length vectors (in l_2-norm sense) at each iteration step. In this case, the above local ALS update rules can be further simplified by ignoring the denominators and imposing a vector normalization after each iterative step, to give a simplified scalar form of the HALS updated rules:

$$a_{pn} \leftarrow \left[\sum_t v_{tn} x_{pt}^{(n)}\right]_+, \qquad a_{pn} \leftarrow a_{pn}/\|\mathbf{a}_n\|_2^2 \tag{13.95}$$

$$v_{tn} \leftarrow \left[\sum_p a_{pn} x_{pt}^{(n)}\right]_+ \tag{13.96}$$

where $x_{pt}^{(n)} = x_{pt} - \sum_{j \neq n} a_{pj}\, b_{tj}$. The above updates rules are extremely simple and quite efficient and can be further optimized for large scale NMF [15,16,21].

13.5 APPLICATIONS

NMF has been applied to a very wide range of tasks such as air quality analysis, text document analysis, and image processing. While it would be impossible to fully survey every such application here, we will select a few here to illustrate the possibilities, and as pointers for further information.

13.5.1 Air quality and chemometrics

As discussed by Henry [39] in the field of air quality, s_{jk} represents the amount of a particulate from source j in sample k, and so must be non-negative. Similarly, a_{ij} is the mass fraction of chemical constituent (or *species*) i in source j, which again must be positive. This leads to an interpretation of (13.1) as a chemical mass balance equation, where x_{ik} are the total amount of species i observed in sample k. This is known as a *multivariate receptor model* [37] where a_{ij} are called the *source compositions*, and s_{jk} are called the *source contributions*.

In geochemistry, this model could also represent the composition of geological samples modeled as a mixture of N pure components. In chemometrics, the spectrum of a mixture is represented as a linear combination of the spectra of pure components. Again, the nature of the physical process leading to the observations requires that all of these quantities are non-negative [39].

13.5.2 Text analysis

Text mining usually involves the classification of text documents into groups or clusters according to their similarity in semantic characteristics. For example, a web search engine often returns thousands of pages in response to a broad query, making it difficult for users to browse or to identify relevant information. Clustering methods can be used to automatically group the retrieved documents into a list of meaningful topics. The NMF approach is attractive for document clustering, and usually exhibits better discrimination for clustering of partially overlapping data than other methods such as Latent Semantic Indexing (LSI).

Preprocessing strategies for document clustering with NMF are very similar to those for LSI. First, the documents of interest are subjected to stop-word removal and word streaming operations. Then, for each document a weighted term-frequency vector is constructed that assigns to each entry the occurrence frequency of the corresponding term. Assuming P dictionary terms and T documents, the sparse term-document matrix $\mathbf{X} \in \mathbb{R}^{P \times T}$ is constructed from weighted term-frequency vectors, that is

$$x_{pt} = f_{pt} \log\left(\frac{T}{T_p}\right) \tag{13.97}$$

where f_{pt} is the frequency of occurrence the p-th term in the t-th document, and T_p is the number of documents containing the p-th term. The entries of $\mathbf{X}$ are always non-negative and equal to zero when the p-th term either does not appear in the t-th document or appears in all the documents.

The aim is to factorize the matrix $\mathbf{X}$ into the non-negative basis matrix $\mathbf{A}$ and the non-negative topic-document matrix $\mathbf{X} \in \mathbb{R}_+^{N \times T}$ where N denotes the number of topics. The position of the maximum value in each column-vector in $\mathbf{S}$ informs us to which topic a given document can be classified. The columns of $\mathbf{A}$ refer to the cluster centers, and the columns in $\mathbf{S}$ are associated with the cluster indicators. A more general scheme for simultaneous clustering both with respect to terms and documents can be modeled by Tri-NMF.

The application of NMF to document clustering has also been discussed by many researchers. For example B. Xu *et al.* [100] propose to use orthogonality constraints in their Constrained NMF algorithm, where the orthogonality of lateral components is enforced by the additional penalty terms added to the KL I-divergence and controlled by the penalty parameters.

In language modeling, Novak and Mammone [65] used non-negative matrix factorization as an alternative to Latent Semantic Analysis for language modeling in an application directed at automatic speech transcription of biology lectures. Tsuge *et al.* [92] also applied NMF to dimensionality reduction of document vectors applied to document retrieval of MEDLINE data. They minimize either Euclidean distance or Kullback-Leibler divergence of the reconstruction, showing that NMF gave better performance than the conventional vector space model.

13.5.3 Image processing

Image analysis often includes non-negativity, corresponding to, for example, the non-negative amount of light falling on a surface and a non-negative reflectance of an illuminated surface. In their now-classic paper, Lee and Seung [56] showed that NMF could discover a "parts-based" representation of face images. The found parts like the eyes and mouth would be represented by different NMF basis images, unlike other analysis approaches such as PCA which would tend to produce global basis images which covered the whole face image. However, this parts-based representation may be strongly dependent on the background and content color, and may not always be obtained [43].

The non-negativity constraint also arises in, for example, hyperspectral image analysis for remote sensing [72,60,64] where **A** is considered to model the amount of substances at each pixel, with **S** the spectral signatures of those substances.

Buchsbaum and Bloch [8] also applied NMF to Munsell color spectra, which are widely used in color naming studies. The basis functions that emerged corresponded to spectra representing familiar color names, such as "Red", "Blue", and so on.

NMF has also been applied to sequences of images. Lee *et al.* [58] applied NMF to dynamic myocardial PET (positron emission tomography) image sequences. They were able to extract basis images that corresponded to major cardiac components, together with time-activity curves with shapes that were similar to those observed in other studies.

13.5.4 Audio analysis

While audio signals take both positive and negative samples when represented as a raw time series of samples, non-negativity constraints arise when represented as a power or magnitude spectrogram. Due to the time-shift-invariant nature of audio signals, convolutive NMF models (section 13.3.3) are suitable for these. They have been used to discover, for example, drum sounds in an audio stream [87], and for separation of speech [88] and music [95]. To allow for pitch-invariant basis functions, Schmidt and Mørup [86] extended the convolutive model to a 2-dimensional convolution using a spectrogram with a log-frequency scale, so that changes in fundamental frequency become shifts on the log-frequency axis.

13.5.5 Gene expression analysis

NMF has also been increasingly used recently in analyzing DNA microarrays. Here the rows of **X** represent the expression levels of genes, while the columns represent the different samples. NMF is then used to search for "metagenes", helping for example to identify functionally related genes. For a recent review of this area, see, for example, [22].

13.6 CONCLUSIONS

In this chapter we have briefly presented basic models and associated learning algorithms for non-negative matrix and tensor factorizations. Currently the most efficient and promising algorithms seem to be those based on the alternating least squares (ALS) approach: these implicitly exploit the gradient and Hessian of the cost functions and provide high convergence speed if they are suitably designed and implemented. Multiplicative algorithms are also useful where the data matrix and factor matrices are very sparse. We have also explored a range of generalizations and extensions of these models, and alternative approaches and algorithms that also enforce non-negativity constraints, including special algorithms designed to handle large scale problems. Finally we touched on a few applications of non-negative methods, including chemometrics, text processing, image processing and audio analysis.

With non-negativity constraints found naturally in many real-world signals, and with the improved theoretical understanding and practical algorithms produced by recent researchers, we consider that the non-negative methods we have discussed in this chapter are a very promising direction for future research and applications.

Acknowledgements

MP is supported by EPSRC Leadership Fellowship EP/G007144/1 and EU FET-Open project FP7-ICT-225913 "Sparse Models, Algorithms, and Learning for Large-scale data (SMALL)".

References

[1] J. Basak, S.-I. Amari, Blind separation of a mixture of uniformly distributed source signals: A novel approach, Neural Computation 11 (1999) 1011–1034.
[2] M.W. Berry, M. Browne, A.N. Langville, V.P. Pauca, R.J. Plemmons, Algorithms and applications for approximate nonnegative matrix factorization, Computational Statistics & Data Analysis 52 (2007) 155–173.
[3] O.S. Borgen, B.R. Kowalski, An extension of the multivariate component-resolution method to three components, Analytica Chimica Acta 174 (1985) 1–26.
[4] R. Bro, PARAFAC: Tutorial and applications, Chemometrics and Intelligent Laboratory Systems 38 (1997) 149–171.
[5] R. Bro, Multi-way analysis in the food industry. models, algorithms, and applications, Ph.D. thesis, University of Amsterdam, the Netherlands, 1998.
[6] R. Bro, Review on multiway analysis in chemistry – 2000–2005, Critical Reviews In Analytical Chemistry 36 (2006) 279–293.
[7] R. Bro, S. de Jong, A fast non-negativity-constrained least squares algorithm, Journal of Chemometrics 11 (1997) 393–401.
[8] G. Buchsbaum, O. Bloch, Color categories revealed by non-negative matrix factorization of Munsell color spectra, Vision Research 42 (2002) 559–563.

[9] J.D. Carroll, G. de Soete, S. Pruzansky, Fitting of the latent class model via iteratively reweighted least squares CANDECOMP with nonnegativity constraints, in: R. Coppi, S. Bolasco (Eds.), Multiway Data Analysis, Elsevier, Amsterdam, 1989, pp. 463–472.

[10] D. Charles, C. Fyfe, Modelling multiple-cause structure using rectification constraints, Network: Computation in Neural Systems 9 (1998) 167–182.

[11] Z. Chen, A. Cichocki, Nonnegative matrix factorization with temporal smoothness and/or spatial decorrelation constraints, Paper preprint, 2005.

[12] M. Chu, F. Diele, R. Plemmons, S. Ragni, Optimality, computation, and interpretation of nonnegative matrix factorizations, 18 October 2004. Preprint.

[13] A. Cichocki, S. Amari, R. Zdunek, R. Kompass, G. Hori, Z. He, Extended SMART algorithms for non-negative matrix factorization, in: LNAI-4029, vol. 4029, Springer, 2006, pp. 548–562.

[14] A. Cichocki, P. Georgiev, Blind source separation algorithms with matrix constraints, IEICE Transactions on Fundamentals of Electronics, Communications and Computer Sciences E86-A (2003) 522–531.

[15] A. Cichocki, A. Phan, Fast local algorithms for large scale nonnegative matrix and tensor factorizations, IEICE (invited paper), 2009.

[16] A. Cichocki, A. Phan, C. Caiafa, Flexible HALS algorithms for sparse non-negative matrix/tensor factorization, in: Proc. of 18th IEEE workshops on Machine Learning for Signal Processing, Cancun, Mexico, 16–19, October 2008.

[17] A. Cichocki, R. Zdunek, Multilayer nonnegative matrix factorization, Electronics Letters 42 (2006) 947–948.

[18] A. Cichocki, R. Zdunek, S. Amari, Hierarchical ALS algorithms for nonnegative matrix and 3D tensor factorization, in: Lecture Notes in Computer Science, vol. LNCS-4666, 2007, pp. 169–176.

[19] A. Cichocki, R. Zdunek, S.-I. Amari, Nonnegative matrix and tensor factorization, IEEE Signal Processing Magazine 25 (2008) 142–145.

[20] A. Cichocki, R. Zdunek, S. Choi, R. Plemmons, S. Ichi Amari, Novel multi-layer non-negative tensor factorization with sparsity constraints, Adaptive and Natural Computing Algorithms (2007) 271–280.

[21] A. Cichocki, R. Zdunek, A. Phan, S. Amari, Nonnegative Matrix and Tensor Factorizations, Wiley, Chichester, 2009.

[22] K. Devarajan, Nonnegative matrix factorization: An analytical and interpretive tool in computational biology, PLoS Computational Biology 4 (2008) e1000029.

[23] C. Ding, T. Li, M.I. Jordan, Convex and semi-nonnegative matrix factorizations, IEEE Transactions on Pattern Analysis and Machine Intelligence, 2010 (in press).

[24] C. Ding, T. Li, W. Peng, H. Park, Orthogonal nonnegative matrix tri-factorizations for clustering, in: KDD '06: Proceedings of the 12th ACM SIGKDD International Conference on Knowledge Discovery and Data Mining, ACM, New York, NY, USA, 2006, pp. 126–135.

[25] D. Donoho, V. Stodden, When does non-negative matrix factorization give a correct decomposition into parts? in: S. Thrun, L. Saul, B. Schölkopf (Eds.), Advances in Neural Information Processing Systems, vol. 16, MIT Press, Cambridge, MA, 2004.

[26] D. FitzGerald, M. Cranitch, E. Coyle, Extended nonnegative tensor factorisation models for musical sound source separation, Computational Intelligence and Neuroscience 2008 (2008) Article ID 872425.

[27] P. Földiák, Adaptive network for optimal linear feature extraction, in: Proceedings of the IEEE/INNS International Joint Conference on Neural Networks, IJCNN-89, vol. 1, Washington DC, 18–22 June 1989, pp. 401–405.

[28] P. Földiák, Forming sparse representations by local anti-Hebbian learning, Biological Cybernetics 64 (1990) 165–170.

[29] C. Fyfe, Positive weights in interneurons, in: G. Orchard (Ed.), Neural Computing: Research and Applications II. Proceedings of the-Third Irish Neural Networks Conference, Belfast, Northern Ireland, 1–2 Sept 1993, Irish Neural Networks Association, Belfast, NI, 1994, pp. 47–58.

[30] N. Gillis, F. Glineur, Nonnegative matrix factorization and underapproximation, in: 9th International Symposium on Iterative Methods in Scientific Computing, Lille, France, 2008.
[31] E.F. Gonzalez, Y. Zhang, Accelerating the Lee-Seung algorithm for nonnegative matrix factorization, Tech. Report TR05-02, Dept. of Computational and Applied Mathematics, Rice University, 3 March 2005.
[32] R.J. Hanson, C.L. Lawson, Solving Least Squares Problems, Prentice-Hall, Inc., Englewood Cliffs, 1974.
[33] G.F. Harpur, Low Entropy Coding with Unsupervised Neural Networks, Ph.D. thesis, Department of Engineering, University of Cambridge, February, 1997.
[34] G.F. Harpur, R.W. Prager, Development of low entropy coding in a recurrent network, Network: Computation in Neural Systems 7 (1996) 277–284.
[35] R.A. Harshman, PARAFAC2: Mathematical and technical notes, UCLA Working Papers in Phonetics 22 (1972) 30–47.
[36] R.A. Harshman, M.E. Lundy, The PARAFAC model for three-way factor analysis and multidimensional scaling, in: H.G. Law, J.C.W. Snyder, J. Hattie, R.P. McDonald (Eds.), Research Methods for Multimode Data Analysis, Praeger, New York, 1984, pp. 122–215.
[37] R.C. Henry, History and fundamentals of multivariate air quality receptor models, Chemometrics and Intelligent Laboratory Systems 37 (1997) 37–42.
[38] R.C. Henry, Receptor model applied to patterns in space (RMAPS) part I: Model description, Journal of the Air & Waste Management Association 47 (1997) 216–219.
[39] R.C. Henry, Multivariate receptor models - current practice and future trends, Chemometrics and Intelligent Laboratory Systems 60 (2002) 43–48.
[40] R.C. Henry, Multivariate receptor modeling by n-dimensional edge detection, Chemometrics and Intelligent Laboratory Systems 65 (2003) 179–189.
[41] K. Hornik, C.-M. Kuan, Convergence analysis of local feature extraction algorithms, Neural Networks 5 (1992) 229–240.
[42] P.O. Hoyer, Non-negative sparse coding, in: Neural Networks for Signal Processing XII (Proc. IEEE Workshop on Neural Networks for Signal Processing), Martigny, Switzerland, 2002, pp. 557–565.
[43] P.O. Hoyer, Non-negative matrix factorization with sparseness constraints, Journal of Machine Learning Research 5 (2004) 1457–1469.
[44] P.O. Hoyer, A. Hyvärinen, A multi-layer sparse coding network learns contour coding from natural images, Vision Research 42 (2002) 1593–1605.
[45] K. Itoh, Blind signal separation by algebraic independent component analysis, in: Proceedings of the 13th Annual Meeting of the IEEE Lasers and Electro-Optics Society, LEOS 2000, vol. 2, Rio Grande, Puerto Rico, 13–16 November 2000, pp. 746–747.
[46] J. Karhunen, J. Joutsensalo, Representation and separation of signals using nonlinear PCA type learning, Neural Networks 7 (1994) 113–127.
[47] H.A.L. Kiers, A.K. Smilde, Constrained three-mode factor analysis as a tool for parameter estimation with second-order instrumental data, Journal of Chemometrics 12 (1998) 125–147.
[48] H.A.L. Kiers, J.M.F. Ten Berge, R. Bro, PARAFAC2 — Part I. A direct fitting algorithm for the PARAFAC2 model, Journal of Chemometrics 13 (1999) 275–294.
[49] D. Kim, S. Sra, I. S. Dhillon, Fast Newton-type methods for the least squares nonnegative matrix approximation problem, in: Proceedings of the SIAM Conference on Data Mining, 2007.
[50] Y.-D. Kim, S. Choi, Nonnegative Tucker decomposition, in: Proc. of Conf. Computer Vision and Pattern Recognition, CVPR-2007, Minneapolis, Minnesota, June 2007.
[51] Y.-D. Kim, A. Cichocki, S. Choi, Nonnegative Tucker decomposition with alpha divergence, in: Proceedings of the IEEE International Conference on Acoustics, Speech, and Signal Processing, ICASSP2008, Nevada, USA, 2008.
[52] B. Klingenberg, J. Curry, A. Dougherty, Non-negative matrix factorization: Ill-posedness and a geometric algorithm, Pattern Recognition 42 (2009) 918–928.

[53] T.G. Kolda, B.W. Bader, Tensor decompositions and applications, SIAM Review 51 (2009) 455–500.

[54] W. Krijnen, J. ten Berge, Contrastvrije oplossingen van het CANDECOMP/PARAFAC-model, Kwantitatieve Methoden 12 (1991) 87–96.

[55] J.B. Kruskal, Three-way arrays: Rank and uniqueness of trilinear decompositions, with application to arithmetic complexity and statistics, Linear Algebra and its Applications 18 (1977) 95–138.

[56] D.D. Lee, H.S. Seung, Learning the parts of objects by non-negative matrix factorization, Nature 401 (1999) 788–791.

[57] D.D. Lee, H.S. Seung, Algorithms for non-negative matrix factorization, in: T.K. Leen, T.G. Dietterich, V. Tresp (Eds.), Advances in Neural Information Processing Systems, vol. 13, MIT Press, 2001, pp. 556–562.

[58] J.S. Lee, D.D. Lee, S. Choi, D.S. Lee, Application of non-negative matrix factorization to dynamic positron emission tomography, in: T.-W. Lee, T.-P. Jung, S. Makeig, T.J. Sejnowski (Eds.), Proceedings of the International Conference on Independent Component Analysis and Signal Separation, ICA2001, San Diego, California, December 9–13, 2001, pp. 629–632.

[59] D.J. Leggett, Numerical analysis of multicomponent spectra, Analytical Chemistry 49 (1977) 276–281.

[60] M. Lennon, G. Mercier, M.C. Mouchot, L. Hubert-Moy, Independent component analysis as a tool for the dimensionality reduction and the representation of hyperspectral images, in: Proceedings of the IEEE 2001 International Geoscience and Remote Sensing Symposium, IGARSS'01, vol. 6, IEEE, 9–13 July, 2001, pp. 2893–2895.

[61] C.-J. Lin, Projected gradient methods for nonnegative matrix factorization, Neural Computation 19 (2007) 2756–2779.

[62] M. Lustig, D. Donoho, J.M. Pauly, Sparse MRI: The application of compressed sensing for rapid MR imaging, Magnetic Resonance in Medicine 58 (2007) 1182–1195.

[63] M. Mørup, L.K. Hansen, S.M. Arnfred, Algorithms for sparse nonnegative Tucker decompositions, Neural Computation 20 (2008) 2112–2131.

[64] H.H. Muhammed, P. Ammenberg, E. Bengtsson, Using feature-vector based analysis, based on principal component analysis and independent component analysis, for analysing hyperspectral images, in: Proceedings of the 11th International Conference on Image Analysis and Processing, Palermo, Italy, 26–28 September 2001, pp. 309–315.

[65] M. Novak, R. Mammone, Use of non-negative matrix factorization for language model adaptation in a lecture transcription task, in: Proceedings of the 2001 IEEE International Conference on Acoustics, Speech, and Signal Processing, vol. 1, Salt Lake City, UT, USA, 7–11 May, 2001, pp. 541–544.

[66] E. Oja, Neural networks, principal components, and subspaces, International Journal of Neural Systems 1 (1989) 61–68.

[67] E. Oja, The nonlinear PCA learning rule in independent component analysis, Neurocomputing 17 (1997) 25–45.

[68] E. Oja, J. Karhunen, On stochastic approximation of the eigenvectors and eigenvalues of the expectation of a random matrix, Journal of Mathematical Analysis and Applications 106 (1985) 69–84.

[69] B.A. Olshausen, D.J. Field, Emergence of simple-cell receptive-field properties by learning a sparse code for natural images, Nature 381 (1996) 607–609.

[70] P. Paatero, Least squares formulation of robust non-negative factor analysis, Chemometrics and Intelligent Laboratory Systems 37 (1997) 23–35.

[71] P. Paatero, U. Tapper, Positive matrix factorization: A non-negative factor model with optimal utilization of error estimates of data values, Environmetrics 5 (1994) 111–126.

[72] L. Parra, C. Spence, P. Sajda, A. Ziehe, K.-R. Müller, Unmixing hyperspectral data, in: Advances in Neural Information Processing Systems 12 (Proc. NIPS'99), MIT Press, 2000, pp. 942–948.

[73] A. Pascual-Montano, J.M. Carazo, K. Kochi, D. Lehmean, R. Pacual-Marqui, Nonsmooth nonnegative matrix factorization (nsNMF), IEEE Transaction Pattern Analysis and Machine Intelligence 28 (2006) 403–415.

[74] V.P. Pauca, F. Shahnaz, M.W. Berry, R.J. Plemmons, Text mining using non-negative matrix factorizations, in: Proc. of the Fourth SIAM International Conference on Data Mining, Lake Buena Vista, FL, 2004, pp. 452–456.

[75] A. Phan, A. Cichocki, Fast and efficient algorithms for nonnegative Tucker decomposition, in: Proc. of The Fifth International Symposium on Neural Networks, in: LNCS, vol. 5264, Springer, Beijing, China, 2008, pp. 772–782.

[76] J. Piper, V.P. Pauca, R.J. Plemmons, M. Giffin, Object characterization from spectral data using nonnegative factorization and information theory, in: In Proc. AMOS Technical Conf., Maui, HI, September 2004.

[77] M.D. Plumbley, Adaptive lateral inhibition for non-negative ICA, in: T.-W. Lee, T.-P. Jung, S. Makeig, T.J. Sejnowski (Eds.), Proceedings of the International Conference on Independent Component Analysis and Signal Separation, ICA2001, San Diego, California, December 9–13, 2001, pp. 516–521.

[78] M.D. Plumbley, Conditions for nonnegative independent component analysis, IEEE Signal Processing Letters 9 (2002) 177–180.

[79] M.D. Plumbley, Algorithms for nonnegative independent component analysis, IEEE Transactions on Neural Networks 14 (2003) 534–543.

[80] M.D. Plumbley, Geometrical methods for non-negative ICA: Manifolds, Lie groups and toral subalgebras, Neurocomputing 67 (2005) 161–197.

[81] M.D. Plumbley, E. Oja, A "nonnegative PCA" algorithm for independent component analysis, IEEE Transactions on Neural Networks 15 (2004) 66–76.

[82] A. Prieto, C.G. Puntonet, B. Prieto, A neural learning algorithm for blind separation of sources based on geometric properties, Signal Processing 64 (1998) 315–331.

[83] C.G. Puntonet, A. Mansour, C. Jutten, Geometrical algorithm for blind separation of sources, in: Actes du XVème Colloque GRETSI, Juan-Les-Pins, France, 18–21 September, 1995, pp. 273–276.

[84] C.G. Puntonet, A. Prieto, Neural net approach for blind separation of sources based on geometric properties, Neurocomputing 18 (1998) 141–164.

[85] M.N. Schmidt, H. Laurberg, Nonnegative matrix factorization with Gaussian process priors, Computational Intelligence and Neuroscience 2008 (2008) Article ID 361705.

[86] M.N. Schmidt, M. Mørup, Nonnegative matrix factor 2-D deconvolution for blind single channel source separation, in: Independent Component Analysis and Signal Separation, International Conference on, in: Lecture Notes in Computer Science (LNCS), vol. 3889, Springer, 2006, pp. 700–707.

[87] P. Smaragdis, Non-negative matrix factor deconvolution: Extraction of multiple sound sources from monophonic inputs, in: Independent Component Analysis and Blind Signal Separation: Proceedings of the Fifth International Conference, ICA 2004, Granada, Spain, September 22–24, 2004, pp. 494–499.

[88] P. Smaragdis, Convolutive speech bases and their application to supervised speech separation, IEEE Transactions on Audio, Speech, and Language Processing 15 (2007) 1–12.

[89] M.W. Spratling, Pre-synaptic lateral inhibition provides a better architecture for self-organizing neural networks, Network: Computation in Neural Systems 10 (1999) 285–301.

[90] S. Sra, I.S. Dhillon, Nonnegative matrix approximation: Algorithms and applications, Tech. Report TR-06-27, Dept. of Computer Sciences, University of Texas at Austin, Austin, TX 78712, USA, 21 June 2006.

[91] R. Tauler, E. Casassas, A. Izquierdo-Ridorsa, Self-modelling curve resolution in studies of spectrometric titrations of multi-equilibria systems by factor analysis, Analytica Chimica Acta 248 (1991) 447–458.

[92] S. Tsuge, M. Shishibori, S. Kuroiwa, K. Kita, Dimensionality reduction using non-negative matrix factorization for information retrieval, in: IEEE International Conference on Systems, Man, and Cybernetics, Tucson, AZ, USA, 7–10 October, 2001, pp. 960–965, vol. 2.

[93] J.M. van den Hof, J.H. van Schuppen, Positive matrix factorization via extremal polyhedral cones, Linear Algebra and its Applications 293 (1999) 171–186.

[94] T. Virtanen, Sound source separation using sparse coding with temporal continuity objective, in: H.C. Kong, B.T.G. Tan (Eds.), Proceedings of the International Computer Music Conference, ICMC 2003, Singapore, 29 September–4 October, 2003, pp. 231–234.

[95] T. Virtanen, Separation of sound sources by convolutive sparse coding, in: Proceedings of the ISCA Tutorial and Research Workshop on Statistical and Perceptual Audio Processing, SAPA 2004, Jeju, Korea, 3 October, 2004.

[96] M. Welling, M. Weber, Positive tensor factorization, Pattern Recognition Letters 22 (2001) 1255–1261.

[97] H. Wersing, J. Eggert, E. Körner, Sparse coding with invariance constraints, in: Proceedings of the International Conference on Artificial Neural Networks, ICANN 2003, Istanbul, 2003, pp. 385–392.

[98] R.J. Williams, Feature discovery through error-correction learning, ICS Report 8501, Institute for Cognitive Science, University of California, San Diego, May 1985.

[99] L. Xu, Least mean square error reconstruction principle for self-organizing neural-nets, Neural Networks 6 (1993) 627–648.

[100] W. Xu, X. Liu, Y. Gong, Document clustering based on non-negative matrix factorization, in: Proceedings of the 26th Annual International ACM SIGIR Conference on Research and Development in Informaion Retrieval, SIGIR'03, 2003, pp. 267–273.

[101] T. Yamaguchi, K. Hirokawa, K. Itoh, Independent component analysis by transforming a scatter diagram of mixtures of signals, Optics Communications 173 (2000) 107–114.

CHAPTER

Nonlinear mixtures

14

C. Jutten, M. Babaie-Zadeh, and J. Karhunen

14.1 INTRODUCTION

Blind source separation problems in linear mixtures have been intensively investigated and are now well understood; see for example [26,53,29]. Since 1985, many methods have been proposed [60,28,30,21,27,10,54], especially those based on independent component analysis (ICA) [31] exploiting the assumption of statistical independence of the source signals. These methods have been applied in various domains, and many examples can be found in application; see Chapters 16–19 of this book.

The linear mixing model, either without memory (instantaneous mixtures) or with memory (convolutive mixtures), is an approximated model, which is valid provided that the nonlinearities in the mixing system are weak, or the amplitudes of the signals are limited. In various situations, this approximation does not hold, for instance when one uses sensors with hard nonlinearities, or when signal levels lead to saturation of conditioning electronic circuits. Thus, it is relevant to consider the blind source separation problem in the more general framework of nonlinear mixtures.

This problem has been sketched in the early works by Jutten [59] where the best linear separating solution was estimated, by Burel [25] for known nonlinearities with unknown parameters, and by Parra *et al.* [84,83] for nonlinear transforms with Jacobian equal to 1. The essential problem of the existence of solutions was considered at the end of the 1990s by Hyvärinen and Pajunen [55] in a general framework, and by Taleb and Jutten [99] for particular nonlinear mixtures.[1] In addition to the theoretical interest, nonlinear mixtures are relevant for a few realistic applications, for example in image processing [6,77,44] and in instrumentation [92,20,22,37,36,38].

Assume now that there are available T samples of a random vector $\mathbf{x}$ with P components, which is modeled by

$$\mathbf{x} = \mathscr{A}(\mathbf{s}) + \mathbf{b}, \tag{14.1}$$

where $\mathbf{s}$ is a random source vector in $\mathbb{R}^N$, whose components are the source signals $s_1, s_2, \ldots, s_N$, assumed to be statistically independent, $\mathscr{A}$ is a nonlinear transform from $\mathbb{R}^N$ to $\mathbb{R}^P$, and $\mathbf{b}$ is an additive noise vector which is independent of the source signals.

[1] In the framework of blind source separation. One can find results in statistics [34,62] on the related problem of factorial decomposition from the beginning of 1950s.

In this chapter, we assume, unless other assumptions are explicitly made, that (i) there are as many observations as sources: $P = N$, (ii) the nonlinear transform $\mathscr{A}$ is invertible, and (iii) the noise vector is neglectible ($\mathbf{b} = 0$).

The nonlinear blind source separation problem is then the following: Is it possible to estimate the sources $\mathbf{s}$ from the nonlinear observations $\mathbf{x}$ only? Without extra prior information, this problem is ill-posed and has no solution. For achieving a solution, we can add assumptions on the statistics of the source signals, or prior information on their coloredness or nonstationarity.

This chapter is organized as follows. The next section 14.2 is devoted to the existence and uniqueness of solutions provided by ICA in the general nonlinear framework. In section 14.3, we consider the influence of structural constraints on identifiability and separability. In section 14.4, we address regularization effects due to priors on sources. In section 14.5, we focus on some properties of mutual information as independence criteria and of a quadratic criterion. A Bayesian approach for general nonlinear mixtures is considered in section 14.6. In section 14.7, we briefly discuss other methods introduced for nonlinear mixtures. The interested reader can find more information on them in the references given. We finish this chapter with a short presentation of a few applications (section 14.8) of nonlinear BSS before conclusions (section 14.9).

14.2 NONLINEAR ICA IN THE GENERAL CASE

In this section, we present theoretical considerations which justify that statistical independence alone is not sufficient for solving the blind source separation problem for general nonlinear mixtures. These results clearly show that nonlinear ICA is a highly non-unique concept.

14.2.1 Nonlinear independent component analysis (ICA)

A natural extension of the linear ICA method to the nonlinear case consists in estimating, by using only observations $\mathbf{x}$, a nonlinear transform $\mathscr{B}$ from $\mathbb{R}^N \to \mathbb{R}^N$ such that the random vector

$$\mathbf{y} = \mathscr{B}(\mathbf{x}) \tag{14.2}$$

has mutually independent components.

Does the mapping $\mathscr{B} \circ \mathscr{A}$, which preserves independence of the components of the source vector, allow one to separate the sources? If it is possible, what are the necessary or sufficient conditions?

14.2.2 Definitions and preliminary results

We do not recall the definition of mutual independence of random vectors, which is based on factorization of probability density functions. We just provide the definition of the σ-diagonal transform.

DEFINITION 14.1
A bijective function $\mathcal{H}$ from $\mathbb{R}^n$ to $\mathbb{R}^n$ is called σ-diagonal if it preserves independence of any random vector.

This definition implies that *any* random vector[2] $\mathbf{x} \in \mathbb{R}^n$ with mutually independent components is transformed to a random vector $\mathbf{y} = \mathcal{H}(\mathbf{x})$ with mutually independent components, too. The set of σ-diagonal transforms will be denoted $\mathfrak{T}$. One can prove the following theorem [96].

THEOREM 14.1
A bijective function $\mathcal{H}$ from $\mathbb{R}^n$ to $\mathbb{R}^n$ is σ-diagonal if and only if its components h_i, $i = 1, \ldots, n$, satisfy:

$$\mathcal{H}_i(u_1, \ldots, u_n) = h_i(u_{\sigma(i)}), \quad i = 1, \ldots, n, \tag{14.3}$$

where the functions h_i are from $\mathbb{R}$ to $\mathbb{R}$, and σ denotes a permutation in the set $\{1, \ldots, n\}$.

This theorem has the following corollary.

COROLLARY 14.2
A bijective function $\mathcal{H}$ from $\mathbb{R}^n$ to $\mathbb{R}^n$ is σ-diagonal if and only if its Jacobian matrix is diagonal, up to any permutation.

In the following, transforms whose Jacobian matrices are diagonal up to any permutation σ, will be called σ-diagonal. *A priori*, ICA provides a transform which preserves mutual independence only for particular sources, the sources involved in the mixing. It means that we cannot claim that the transform preserves independence for other distributions, and especially for *any* distributions, so that the transform is σ-diagonal. Consequently, separation cannot be guaranteed. Moreover, even when the sources are separated, this happens only up to any permutation (which is not a problem) and up to a unknown nonlinear function (which is much more annoying). In fact, if u and v are two independent random variables, then for any invertible transforms f and g, the random variables $f(u)$ and $g(v)$ are independent, too. Therefore, due to the distortions, the estimated sources can strongly differ from original sources. Of course, sources are separated, but the remaining indeterminacy is very undesirable for restoring the sources.

14.2.3 Existence and uniqueness of transforms preserving independence

14.2.3.1 The problem

Following Darmois [33,34], consider the factorial representation of a random vector $\mathbf{x}$ in a random vector $\boldsymbol{\zeta}$ with mutually independent components ζ_i:

$$\mathbf{x} = \mathcal{H}_1(\boldsymbol{\zeta}). \tag{14.4}$$

[2]That is, whatever its distribution is.

For studying the uniqueness of the representation, one can look for another factorial representation of the random vector $\mathbf{x}$ in a random vector $\boldsymbol{\omega}$ with mutually independent components ω_i such that:

$$\mathbf{x} = \mathcal{H}_1(\boldsymbol{\zeta}) = \mathcal{H}_2(\boldsymbol{\omega}). \tag{14.5}$$

If there exist two factorial representations of $\mathbf{x}$, with two different random vectors $\boldsymbol{\zeta}$ and $\boldsymbol{\omega}$, then there is no uniqueness.

14.2.3.2 Existence

Generally, for any random vector $\mathbf{x} = \mathcal{A}(\mathbf{s})$ with components without particular properties (especially not mutually independent), one can design a transform $\mathcal{H}$ such that $\mathcal{H} \circ \mathcal{A}$ preserves independence, but is not σ-diagonal, so that $\mathcal{H}$ is not a separating transform as defined by (14.3). This result, which is based on a constructive method similar to a Gram-Schmidt procedure to be discussed soon, has been proved in the 1950s by Darmois [33]. This result has also been used in [55] for designing parametric families of solutions for nonlinear ICA.

A simple example. We now present a simple example of a transform which preserves independence while being still mixing [99]. Let s_1 be a Rayleigh distributed random variable (with values in $\mathbb{R}^+$) with a probability density function (pdf) $p_{s_1}(s_1) = s_1 \exp(-s_1^2/2)$, and s_2 a random variable independent of s_1, with a uniform distribution $s_2 \in [0, 2\pi)$. Let us then consider the nonlinear transform

$$\begin{aligned}[y_1, y_2] &= \mathcal{H}(s_1, s_2) \\ &= [s_1 \cos(s_2), s_1 \sin(s_2)]\end{aligned} \tag{14.6}$$

whose Jacobian matrix is non-diagonal:

$$\mathbf{J} = \begin{pmatrix} \cos(s_2) & -s_1 \sin(s_2) \\ \sin(s_2) & s_1 \cos(s_2) \end{pmatrix}. \tag{14.7}$$

The joint pdf of y_1 and y_2 is

$$\begin{aligned} p_{y_1,y_2}(y_1, y_2) &= \frac{p_{s_1,s_2}(s_1, s_2)}{|\det(\mathbf{J})|} \\ &= \frac{1}{2\pi} \exp\left(\frac{-y_1^2 - y_2^2}{2}\right) \\ &= \left(\frac{1}{\sqrt{2\pi}} \exp\frac{-y_1^2}{2}\right)\left(\frac{1}{\sqrt{2\pi}} \exp\frac{-y_2^2}{2}\right). \end{aligned}$$

The joint pdf can be factorized, and thus one can conclude that the random variables y_1 and y_2 are independent. However, it is clear that these variables are still mixtures of random variables s_1 and s_2. The transform $\mathcal{H}$ preserves independence for the random variables s_1 and s_2 (Rayleigh and uniform), but not for *any* random variables. Other examples can be found in the literature (for instance in [73]), or can be easily invented.

Method for designing non-separating transforms preserving independence. Let $\mathbf{x}$ be a random vector, resulting for instance from a mixture $\mathbf{x} = \mathcal{A}(\mathbf{s})$ of mutually independent random variables $\mathbf{s}$, where $\mathcal{A}$ is a transform with a non-diagonal Jacobian.[3] We propose to design an invertible transform $\mathcal{B}$ (independent of $\mathcal{A}$) which preserves independence while being still mixing, i.e. such that the random vector $\mathbf{y} = \mathcal{B}(\mathbf{x})$ provided by the invertible transform $\mathcal{B}$ has mutually independent components and that $\mathcal{B} \circ \mathcal{A}$ has a non-diagonal Jacobian.

Since the transform $\mathcal{B}$ is invertible, one can write:

$$p_{\mathbf{y}}(\mathbf{y}) = p_{\mathbf{x}}(\mathbf{x})/|\det J_{\mathcal{B}}(\mathbf{x})|, \tag{14.8}$$

where $J_{\mathcal{B}}(\mathbf{x})$ is the Jacobian matrix related to the transform $\mathcal{B}$. Without loss of generality, one can assume the random variables y_i, $i = 1, \ldots, n$, are uniformly distributed in $[0, 1]$. Moreover, because the variables y_i are assumed to be independent, they satisfy $p_{\mathbf{y}}(\mathbf{y}) = \prod_i p_{y_i}(y_i) = 1$, and Eq. (14.8) simplifies to

$$p_{\mathbf{x}}(\mathbf{x}) = |J_{\mathcal{B}}(\mathbf{x})|. \tag{14.9}$$

Looking for solutions with the following form:

$$\begin{cases} \mathcal{B}_1(\mathbf{x}) = h_1(x_1) \\ \mathcal{B}_2(\mathbf{x}) = h_2(x_1, x_2) \\ \vdots \\ \mathcal{B}_n(\mathbf{x}) = h_n(x_1, x_2, \ldots, x_n), \end{cases} \tag{14.10}$$

Equation (14.9) becomes:

$$p_{\mathbf{x}}(\mathbf{x}) = \prod_{i=1}^{n} \frac{\partial \mathcal{B}_i(\mathbf{x})}{\partial x_i} \tag{14.11}$$

or, using Bayes' theorem:

$$p_{x_1}(x_1) p_{x_2|x_1}(x_1, x_2) \ldots p_{x_n|x_1,\ldots,x_{n-1}}(x_1, \ldots, x_n) = \prod_{i=1}^{n} \frac{\partial \mathcal{B}_i(\mathbf{x})}{\partial x_i}. \tag{14.12}$$

[3]In fact, in that case, the random vector $\mathbf{x}$ would have independent components.

By integrating (14.12), one gets the following solution:

$$\begin{cases} \mathcal{B}_1(x_1) & = F_{x_1}(x_1) \\ \mathcal{B}_2(x_1, x_2) & = F_{x_2|x_1}(x_1, x_2) \\ \vdots & \\ \mathcal{B}_n(x_1, x_2, \ldots, x_n) & = F_{x_n|x_1,\ldots,x_{n-1}}(x_1, x_2, \ldots, x_n) \end{cases} \tag{14.13}$$

where F_{x_1} is the cumulative probability density function of the random variable x_1, and $F_{x_{k+1}|x_1,\ldots,x_k}$ is the conditional cumulative probability density function of the random variable x_{k+1}, conditionally to $x_1, \ldots, x_k$. Generally, the transform $\mathcal{B} \circ \mathcal{A}$ is a nonlinear transform which is non σ-diagonal since its Jacobian matrix is not diagonal, but it transforms any random vector $\mathbf{x}$ to a random vector $\mathbf{y}$ with independent components.

This result by Darmois is negative since it shows that for any mixtures $\mathbf{x} = \mathcal{A}(\mathbf{s})$, there exists at least one non σ-diagonal transform $\mathcal{H}$[4] which is a mixture of the variables $\mathbf{x}$ although it preserves statistical independence. Consequently, using statistical independence without other constraints or priors, one gets a transform which preserves statistical independence but will not be a separating transform (that is, σ-diagonal).

14.2.3.3 Conclusion

Independent components estimated from (14.1) can then be very different from the actual sources. Generally, using ICA for solving the source separation problem in nonlinear mixtures requires additional information on sources or constraints, in order to look for a transform in a restricted manifold $\mathcal{G} = \mathcal{B} \circ \mathcal{A}$ which regularizes the solutions.

14.3 ICA FOR CONSTRAINED NONLINEAR MIXTURES

In this section we first present, using the previous results, the theoretical ICA framework when the transforms $\mathcal{A}$ and $\mathcal{B}$ have structural constraints. We then study a few constraints on mixtures for which ICA is able to identify the separating σ-diagonal transforms.

14.3.1 Structural constraints

Let us assume that we restrict the transforms in a set denoted $\mathfrak{Q}$. For characterizing the indeterminacies for transforms $\mathcal{G} \in \mathfrak{Q}$, one must solve the tricky equation for independence preservation that can be written

$$\forall E \in \mathfrak{M}_N, \quad \int_E dF_{s_1} dF_{s_2} \cdots dF_{s_N} = \int_{\mathcal{G}(E)} dF_{y_1} dF_{y_2} \cdots dF_{y_N}. \tag{14.14}$$

[4] $\mathcal{H}$ depends on $\mathbf{x}$, and, generally is not a transform preserving independence for other random vectors $\mathbf{u} \neq \mathbf{x}$.

There $\mathfrak{M}_N$ is the set of all the measurable compacts in $\mathbb{R}^N$, and F_u is the cumulative density probability function of the random variable u.

Let us denote by $\mathfrak{T}$ the set of σ-diagonal mappings. Then, one can define the set[5] $\mathfrak{P}$:

$$\mathfrak{P} = \{(F_{s_1}, F_{s_2}, \ldots, F_{s_N}) / \exists \mathcal{G} \in \mathfrak{Q} \setminus \mathfrak{T} : \mathcal{G}(\mathbf{s}) \text{ has independent components}\}. \tag{14.15}$$

This is the set of all the distributions for which there exist non σ-diagonal transforms $\mathcal{G} \in \mathfrak{Q}$ (not belonging to the set $\mathfrak{T}$), and which preserves mutual independence of the components of the source vector **s**.

Ideally, $\mathfrak{P}$ should be an empty set, and $\mathfrak{T} \cap \mathfrak{Q}$ should contain an identity function as a unique element. However, this is generally not satisfied:

1. Source separation is possible if the distributions of the sources belong to the set $\bar{\mathfrak{P}}$, which is the complementary set of $\mathfrak{P}$.
2. The sources are then restored up to a σ-diagonal transform, that is, a transform belonging to the set $\mathfrak{T} \cap \mathfrak{Q}$.

Solving (14.14), or defining the set of distributions $\mathfrak{P}$, is generally a tricky problem, except for particular models $\mathfrak{Q}$ such as linear invertible transforms.

14.3.2 Smooth transforms

Recently, multi-layer perceptron (MLP) neural networks [47] have been used in [114,8,7] for estimating the nonlinear separating transform $\mathcal{B}$. For justifying this choice, in addition to the universal approximation property of MLP, Almeida claims that restricting the target transforms to the set of smooth transforms[6] generated by an MLP provides regularized solutions which ensures that nonlinear ICA leads to source separation. However, the following example [12] proves that the smoothness property is not a sufficient condition for this purpose.

Without loss of generality, let us consider two independent random variables $\mathbf{s} = (s_1, s_2)^T$ uniformly distributed in the interval $[-1, 1]$, and the smooth nonlinear transform represented by the matrix

$$\mathbf{M} = \begin{pmatrix} \cos(\theta(r)) & -\sin(\theta(r)) \\ \sin(\theta(r)) & \cos(\theta(r)) \end{pmatrix} \tag{14.16}$$

where $r \triangleq \sqrt{s_1^2 + s_2^2}$. This transform is a rotation whose rotating angle $\theta(r)$ depends on the radius r:

$$\theta(r) = \begin{cases} \theta_0 (1-r)^q, & 0 \le r \le 1 \\ 0, & r > 1 \end{cases} \tag{14.17}$$

[5] In Eq. (14.15), the difference of two sets $\setminus$ is defined as: $\mathfrak{Q} \setminus \mathfrak{T} = \{x / x \in \mathfrak{Q} \text{ and } x \notin \mathfrak{T}\}$.

[6] f is a smooth transform if its derivatives of any order exist and are continuous.

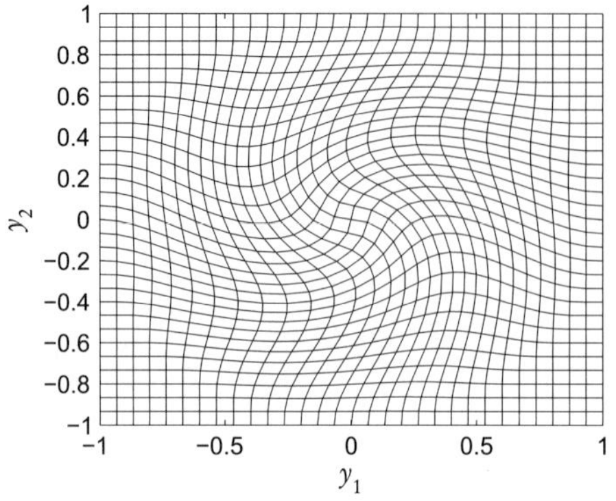

FIGURE 14.1

The smooth transform $\mathbf{M}$ is a mixing transform which preserves independence of a uniformly distributed vector. In the figure, the curves y_1 and y_2 that are images of lines s_1 and s_2 after the transform $\mathbf{M}$, respectively, show clearly these properties.

where $q \geq 2$. Figure 14.1 shows the image of the region $\{-1 \leqslant s_1 \leqslant 1, -1 \leqslant s_2 \leqslant 1\}$ by the transformation for $q = 2$ and $\theta_0 = \pi/2$. One can easily compute the Jacobian matrix of this transform [12]:

$$\mathbf{J}_M(r) = \begin{pmatrix} \cos(\theta(r)) & -\sin(\theta(r)) \\ \sin(\theta(r)) & \cos(\theta(r)) \end{pmatrix} \begin{pmatrix} 1 - s_2 \dfrac{\partial \theta}{\partial s_1} & -s_2 \dfrac{\partial \theta}{\partial s_2} \\ s_1 \dfrac{\partial \theta}{\partial s_1} & 1 + s_1 \dfrac{\partial \theta}{\partial s_2} \end{pmatrix}. \tag{14.18}$$

Computing the determinant, one gets

$$\det \mathbf{J}_\mathbf{M}(r) = 1 + s_1 \frac{\partial \theta}{\partial s_2} - s_2 \frac{\partial \theta}{\partial s_1} \tag{14.19}$$

and since

$$s_1 \frac{\partial \theta}{\partial s_2} = s_2 \frac{\partial \theta}{\partial s_1} = \frac{s_1 s_2}{r} \theta'(r) \tag{14.20}$$

one finally gets $\det \mathbf{J}_\mathbf{M}(r) = 1$, and:

$$p_{y_1 y_2}(y_1, y_2) = p_{s_1 s_2}(s_1, s_2). \tag{14.21}$$

From Eq. (14.18), one can deduce that the Jacobian matrix of this smooth transform is non-diagonal, and consequently that the transform is mixing. However, from

Eq. (14.21), this transform preserves independence of the two random variables uniformly distributed on $[-1,1]$. This counterexample proves that restricting nonlinear transforms to the set of smooth transforms is not a sufficient condition for separation.

14.3.3 Example of linear mixtures

For linear invertible mixtures, the transform $\mathscr{A}$ is linear and can be represented by a square regular mixing matrix **A**. In this case, it is sufficient to constrain the separating model $\mathscr{B}$ to belong to the set of square invertible matrices. Source separation is obtained by estimating a matrix **B** such that $\mathbf{y} = \mathbf{Bx} = \mathbf{Gs}$ has mutually independent components. The global mapping $\mathscr{G} = \mathbf{G}$ is then an element of the set of square invertible matrices $\mathfrak{Q}$.

The set of σ-diagonal *linear* transforms, $\mathfrak{T} \cap \mathfrak{Q}$, is the set of matrices equal to the product of a permutation matrix and a diagonal matrix. From the Darmois-Skitovich theorem [33], it is clear that the set $\mathfrak{P}$ contains distributions with at least two Gaussian components. This result is similar to Comon's identifiability theorem [31] which proves that for a linear invertible mixture, independence allows separation of sources up to a permutation and a diagonal matrices, provided that there is at most one Gaussian source.

14.3.4 Conformal mappings

Hyvärinen and Pajunen [55] show that if one restricts the nonlinear mappings to the set of conformal mappings ($\mathfrak{Q}$), ICA allows one to separate the sources.

DEFINITION 14.2
A conformal mapping is a mapping which preserves orientated angles.

Conformal mappings are often considered in the framework of functions of complex-valued variables, which are restricted to plane (two-dimensional) mappings.[7] We then have the following theorem:

THEOREM 14.3
Let $Z = f(z)$ be a holomorphic function defined in a domain D. If $\forall z \in D$, $f'(z) \neq 0$, the mapping $Z = f(z)$ is a conformal mapping.

This result shows that the set of conformal mappings is contained in the set of smooth mappings, due to the property of algebraic angle preservation. Hyvärinen and Pajunen prove that ICA is able to estimate a separating mapping, up to a rotation, provided that the following conditions hold:

- The mixing mapping $\mathscr{A}$ is a conformal mapping, such that $\mathscr{A}(0) = 0$.
- Each source has a known bounded support.

[7]This kind of mapping is frequently used for solving problems with intricate geometry, in a transformed domain where the geometry becomes simple. For instance, Joukovski mapping is a classical example for studying profiles of plane wings in aeronautics.

It seems that the extension of this result to conformal mappings in N dimensions has not been considered. Of course, the angle preservation condition seems very restrictive. In particular, it is not very realistic in the framework of nonlinear mappings associated to a nonlinear sensor array.

14.3.5 Post-nonlinear (PNL) mixtures

Initially, post-nonlinear mixtures are inspired by devices for which the sensors and their amplifiers used for signal conditioning are assumed to be nonlinear, for example due to saturation. In addition to its relevance for most sensor arrays, as we shall see, this model has the nice property that it is separable using ICA with weak indeterminacies.

14.3.5.1 PNL model

In post-nonlinear (PNL) mixtures, observations are mixtures with the following form (Fig. 14.2):

$$x_i(t) = f_i\left(\sum_{j=1}^{N} a_{ij}s_j(t)\right), \quad i = 1,\ldots,N. \tag{14.22}$$

The PNL mixture consists of a linear mixture $\mathbf{As}(t)$, followed on each channel by a nonlinear mapping f_i. In addition, we assume that the linear mixing matrix $\mathbf{A}$ is regular with $P = N$, and that the nonlinear mappings f_i are invertible. In the following, a PNL mixture $\mathbf{x}(t) = \mathbf{f}(\mathbf{As}(t))$ will be simply denoted by $(\mathbf{A}, \mathbf{f})$. In addition to its theoretical interest, the PNL model belongs to the class of L-ZMNL[8] models, which suits perfectly to many realistic applications. For instance, one can meet such models in sensor arrays [82,20], in satellite communication systems [89], and in biological systems [65].

As we explained above, the main issue concerns the identifiability of the mixture model (leading to the separability, if $\mathbf{A}$ is regular and $\mathbf{f}$ invertible) from the statistical independence assumption. For this purpose, it is first necessary to constrain the separation structure $\mathcal{B}$ so that:

1. $\mathcal{B}$ is able to invert the mixture in the sense of Eq. (14.3);
2. $\mathcal{B}$ is as simple as possible for reducing residual distortions g_i, using only source independence.

Under these two constraints, we choose as a separation structure $\mathcal{B}$ the *mirror* structure of the mixing structure $\mathcal{A} = (\mathbf{A}, \mathbf{f})$ (see Fig. 14.2). We denote the post-nonlinear separating structure $\mathbf{y}(t) = \mathbf{B}\,\mathbf{h}(\mathbf{x}(t))$ by $(\mathbf{h}, \mathbf{B})$. The global mapping $\mathcal{G}$ is then an element of the set $\mathcal{Q}$ of mappings which are the cascade of a linear invertible mixture (regular matrix $\mathbf{A}$) followed by stepwise invertible nonlinear mappings, and then by another invertible linear mapping (regular matrix $\mathbf{B}$).

[8] L for Linear and ZMNL for Zero-Memory NonLinearity: it is then a separable system, with a linear stage followed by a nonlinear mapping.

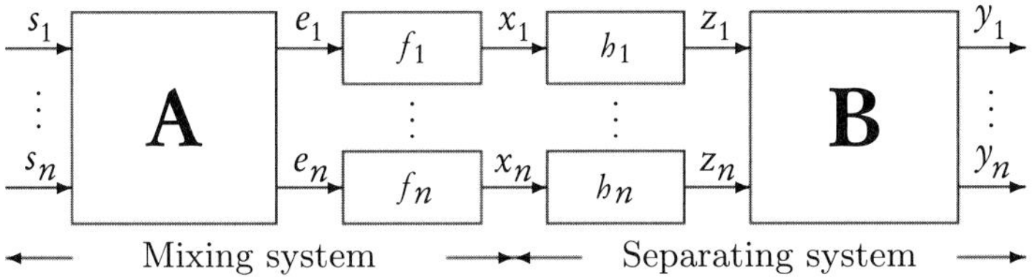

FIGURE 14.2

Mixing and separating models for post-nonlinear (PNL) model.

14.3.5.2 Separability using ICA

In [99], it is shown that independence of the components of the output vector $\mathbf{y}$ is sufficient to identify PNL mixtures with the same indeterminacies as for linear mixtures.

LEMMA 14.4
Consider a PNL model $(\mathbf{A}, \mathbf{f})$ *and a separating structure* $(\mathbf{h}, \mathbf{B})$ *such that (H1)* $\mathbf{A}$ *is a regular matrix having at least two non-zero entries in each row or each column, (H2) the functions* f_i *are invertible, (H3)* $\mathbf{B}$ *is a regular matrix, (H4)* $g_i = h_i \circ f_i$ *satisfies* $g_i'(u) \neq 0, \forall i, \forall u \in \mathbb{R}$, *(H5) of which at most one source* s_i *is Gaussian, and each source has a pdf which is equal to zero on a interval. Then the vector* $\mathbf{y} = \mathscr{B} \circ \mathscr{A}(\mathbf{s})$ *has mutually independent components if and only if the mappings* h_i *are affine mappings and if* $\mathbf{B}$ *satisfies* $\mathbf{BA} = \mathbf{\Pi\Delta}$, *where* $\mathbf{\Pi}$ *is a permutation matrix and* $\mathbf{\Delta}$ *is a diagonal matrix.*

The condition on the mixing matrix $\mathbf{A}$ shows that the estimation of nonlinear mappings is possible only if the mixing is "sufficiently" mixing. Although this seems surprising at first glance, this result is easy to understand. Let us assume that the mixing matrix $\mathbf{A}$ is diagonal, and the observations $f_i(a_{ii}s_i)$ are mutually independent random variables. Sources are then already separated and consequently it is impossible (without extra *prior*) to estimate sources with a weaker indeterminacy than an unknown nonlinear mapping.

The condition that the probability density function (pdf) must be equal to zero on an interval is a technical condition used in the proof, but it does not seems to be theoretically necessary. In fact, Achard and Jutten [1] extended this result by relaxing this assumption.

LEMMA 14.5
Let $(\mathbf{A}, \mathbf{f})$ *be a PNL mixture and* $(\mathbf{h}, \mathbf{B})$ *a PNL separating structure such that (H1)* $\mathbf{A}$ *is invertible, and* $\forall i, j$ such that $a_{ij} \neq 0, \exists k \neq j / a_{ik} \neq 0$, or $\exists l \neq i$ such that $a_{lj} \neq 0$, *(H2) mappings* $h_i(.)$ *are differentiable and invertible, (H3) the random vector* $\mathbf{s}$ *has independent components* s_i, *of which at most one is Gaussian, (H4) the pdf of each source is differentiable and its derivative is continuous on its support. Then the random vector* $\mathbf{y} = \mathscr{B} \circ \mathscr{A}(\mathbf{s})$ *has independent components if and only if* $h_i \circ f_i$, $\forall i = 1, \ldots, N$ *are linear functions and* $\mathbf{BA} = \mathbf{\Pi\Delta}$, *where* $\mathbf{\Pi}$ *is a permutation matrix and* $\mathbf{\Delta}$ *is a diagonal matrix.*

Assumption H1 is a necessary condition, which ensures that the mixing matrix is sufficiently mixing. If this condition is satisfied, there is no non-zero isolated entry a_{ij}, i.e. without other non-zero entry in the row i or in the column j. If a non-zero and isolated entry a_{ij} exists, the mixing $x_i = f_i(a_{ij}s_j)$ would be independent of all the other mixings: the source s_j would then be already separated in $x_i = f_i(a_{ij}s_j)$, and it would not be possible to estimate the inverse of the function f_i and consequently to retrieve the source s_j up to a scaling factor.

As a conclusion, PNL mixtures are identifiable using ICA for sources of which at most one is Gaussian (the set $\mathfrak{P}$ contains the multivariate distributions which have at most two Gaussian components), with the same indeterminacies than linear mixtures (the set of *linear* σ-diagonal mappings $\mathfrak{T} \cap \mathfrak{Q}$ is the set of square matrices which are the product of a permutation matrix and a diagonal matrix) if the mixing matrix $\mathbf{A}$ is "sufficiently" mixing.

14.3.5.3 Extension to nonlinear mixtures with memory

Identifiability of PNL mixtures can be generalized to convolutive PNL mixtures (CPNL), in which the scalar mixing matrix $\mathbf{A}$ is replaced by a matrix $\mathbf{A}(z)$, whose entries are linear filters, and each source is independent and identically distributed (iid) [14]. In fact, denoting $\mathbf{A}(z) = \sum_k \mathbf{A}_k z^{-k}$, and using the following notations:

$$\mathbf{s} \triangleq \left(\ldots, \mathbf{s}^T(k-1), \mathbf{s}^T(k), \mathbf{s}^T(k+1), \ldots\right)^T \tag{14.23}$$

$$\mathbf{x} \triangleq \left(\ldots, \mathbf{x}^T(k-1), \mathbf{x}^T(k), \mathbf{x}^T(k+1), \ldots\right)^T, \tag{14.24}$$

one gets:

$$\mathbf{x} = \mathbf{f}\left(\bar{\mathbf{A}}\mathbf{s}\right) \tag{14.25}$$

where the function $\mathbf{f}$ acts componentwise, and the matrix:

$$\bar{\mathbf{A}} = \begin{bmatrix} \cdots & \cdots & \cdots & \cdots & \cdots \\ \cdots & \mathbf{A}_{k+1} & \mathbf{A}_k & \mathbf{A}_{k-1} & \cdots \\ \cdots & \mathbf{A}_{k+2} & \mathbf{A}_{k+1} & \mathbf{A}_k & \cdots \\ \cdots & \cdots & \cdots & \cdots & \cdots \end{bmatrix}. \tag{14.26}$$

The iid nature of each source, i.e. the temporal independence of samples $s_i(k), i = 1, \ldots, T$, ensures the spatial independence of $\mathbf{s}$. Thus, CPNL mixtures can be considered as particular PNL mixtures (the mixing matrix $\bar{\mathbf{A}}$ is a block-Toeplitz matrix). For mixing matrices $\mathbf{A}(z)$ whose entries are finite impulse response (FIR) filters, Eq. (14.25) is associated to a PNL mixture of finite dimension and separation results (for instantaneous PNL mixtures) hold. For mixing matrices whose entries are infinite impulse response (IIR) filters, Eq. (14.25) is equivalent to a PNL mixture of infinite dimension, for which we have no separability proof, and we just conjecture the separability.

Moreover, by using a suitable parameterization, Wiener systems (Fig. 14.3) can be viewed as particular PNL mixtures. Consequently, identifiability of PNL mixtures leads to invertibility of Wiener systems [100].

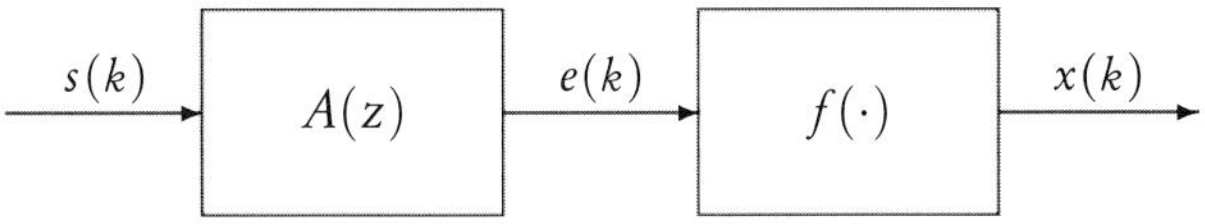

FIGURE 14.3

A Wiener system consists of a filter $A(z)$ followed by a nonlinear mapping $f(.)$.

14.3.6 Bilinear mixtures

S. Hosseini and Y. Deville have addressed [50,51] the separation of "bilinear" (or "linear-quadratic") mixing models, that is, mixing systems of the form:

$$\begin{cases} x_1 = s_1 - l_1 s_2 - q_1 s_1 s_2 \\ x_2 = s_2 - l_2 s_1 - q_2 s_1 s_2. \end{cases} \tag{14.27}$$

The Jacobian of the above mixing model is:

$$J = 1 - l_1 l_2 - (q_2 + l_2 q_1) s_1 - (q_1 + l_1 q_2) s_2. \tag{14.28}$$

In their works, Hosseini and Deville have shown that the nonlinear mapping (14.27) is invertible if the values of the parameters (l_1, l_2, q_1, q_2) and the range of variations of s_1, s_2 are such that either $J > 0$ for all values of (s_1, s_2), or $J < 0$ for all values of (s_1, s_2). On the other hand, if $J > 0$ for some values of (s_1, s_2) and $J < 0$ for some other values of (s_1, s_2), then the above bilinear transformation is not bijective, and we cannot determine (s_1, s_2) from (x_1, x_2) even if the values of the parameters l_1, l_2, q_1, q_2 were known. Figure 14.4 shows the transformation of the region $s_i \in [-0.5, 0.5]$ for two different sets of values of parameters, one of which corresponds to an invertible mapping and the other to a non-invertible mapping. Note that one may consider the non-invertible case as a "highly nonlinear" mixture, which is not separable.

Assuming that for a problem at hand the bilinear model is invertible, Hosseini and Deville have proposed that the recurrent structure of Fig. 14.5 (which is inspired by the early work of Hérault *et al.* [48]) is able to retrieve the source signals. In fact, this recurrent structure (where k is the iteration number) can be written as:

$$\begin{cases} y_1^{(k+1)}(\cdot) = x_1(\cdot) + l_1 y_2^{(k)}(\cdot) + q_1 y_1^{(k)}(\cdot) y_2^{(k)}(\cdot) \\ y_2^{(k+1)}(\cdot) = x_2(\cdot) + l_2 y_1^{(k)}(\cdot) + q_2 y_1^{(k)}(\cdot) y_2^{(k)}(\cdot) \end{cases} \tag{14.29}$$

Comparing the above equation with (14.27), one sees that if $y_i^{(k)}(\cdot) = s_i(\cdot), i \in \{1,2\}$, then $y_i^{(k+1)}(\cdot) = y_i^{(k)}(\cdot), i \in \{1,2\}$, that is, the above iteration has converged. They have also studied the stability of the above recurrent structure and shown that it is stable at the point $(y_1, y_2) = (s_1, s_2)$ if and only if the absolute values of the two eigenvalues of the Jacobian matrix of the mapping (14.27) are smaller than one.

In the previous paragraphs it has been assumed that all the parameters l_1, l_2, q_1, q_2 are known. In blind source separation, however, these parameters have to be estimated from

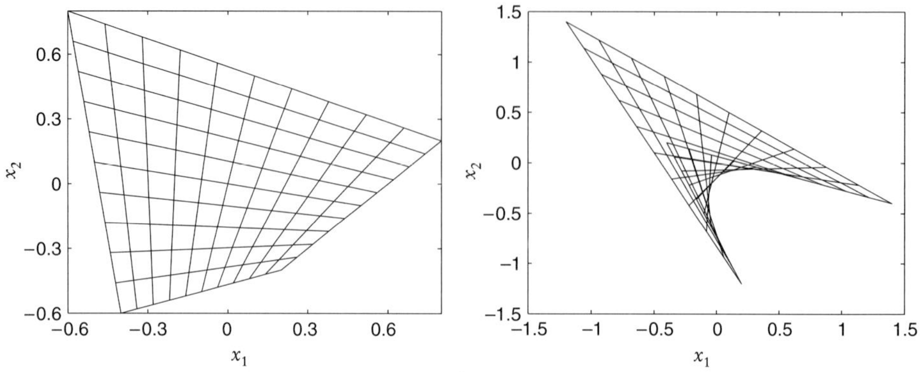

FIGURE 14.4

The transformation of the region $s_i \in [-0.5, 0.5]$ of (s_1, s_2) plane to the (x_1, x_2) plane, by the bilinear transformation (14.27) for $l_2 = -l_1 = 0.5$, and (left) $q_2 = -q_1 = 0.8$, (Right) $q_2 = -q_1 = 3.2$. The left transform is bijective ($J > 0$ everywhere) and hence invertible, while the right transform is not bijective and hence non-invertible.

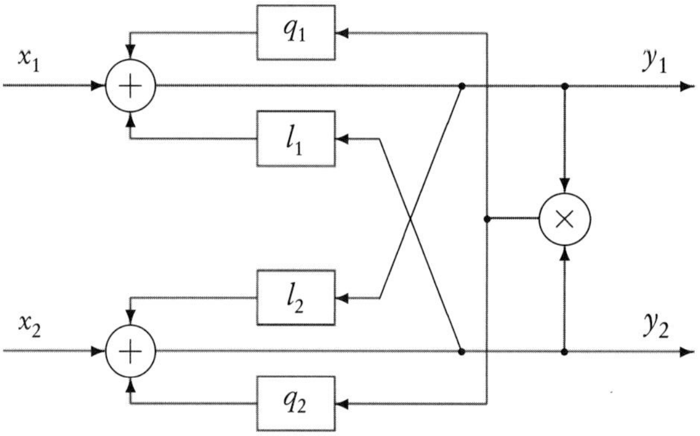

FIGURE 14.5

Recurrent structure used by Hosseini and Deville [50,51] for inverting bilinear mixtures.

the data. Hosseini and Deville then assume that these parameters can be estimated by the statistical independence of the outputs.[9] Then, they propose an iterative algorithm of the form:

[9]Note that here Hosseini and Deville have implicitly assumed that the invertibility of the bilinear mapping (14.27) insures its separability. In fact, invertibility of (14.27) means that if l_1, l_2, q_1, q_2 are known one can obtain (s_1, s_2) from (x_1, x_2), while (blind) separability means that in a separating structure, the independence of the outputs guarantees the separation of the sources. Hence, the (blind) separability of invertible bilinear mixing models remains as an open question.

- Repeat
  1. Repeat iterations (14.29) until convergence.
  2. Update the estimated values of the parameters l_1, l_2, q_1, q_2, based on a measure of statistical independence of y_1 and y_2.
- Until convergence.

For the second step (updating the values of the parameters), Hosseini and Deville have proposed two methods. In [50], they use an iteration which is very similar to what is used in the original Hérault-Jutten algorithm [48] and is based on nonlinear decorrelation as a measure of independence. Then in [51] they develop a maximum likelihood (ML) criterion, and use a steepest ascent iteration for maximizing it in the second step of the above general algorithm. Another method based on minimizing the mutual information of the outputs has been recently proposed by Mokhtari *et al.* [78].

14.3.7 A class of separable nonlinear mappings

Due to Darmois' results [33,34], which are very interesting for linear mixtures, a natural approach is to apply a transform so that nonlinear mixtures become linear ones.

14.3.7.1 Multiplicative mixtures

As an example, consider the multiplicative mixture

$$x_j(t) = \prod_{i=1}^{N} s_i^{\alpha_{ji}}(t), \quad j = 1, \ldots, N \tag{14.30}$$

where $s_i(t)$'s are independent sources with values in $\mathbb{R}^{*+}$. Taking the logarithm of (14.30) leads to

$$\log x_j(t) = \sum_{i=1}^{N} \alpha_{ji} \log s_i(t), \quad j = 1, \ldots, N \tag{14.31}$$

which is a linear mixture of the random variables $\log s_i(t)$.

This kind of mixture can model dependence between temperature and magnetic field in Hall effect sensors [9], or between incident light and object reflectance in an image [41]. Consider now the first example in more detail. The Hall voltage [88] is equal to

$$V_H = kBT^{\alpha}, \tag{14.32}$$

where α depends on the semi-conductor type (N or P), because temperature influence is related to mobility of majority carriers. Thus, using sensors of both N-type and P-type, one gets:

$$\begin{cases} V_{H_N}(t) = k_N B(t) T^{\alpha_N}(t), \\ V_{H_P}(t) = k_P B(t) T^{\alpha_P}(t). \end{cases} \tag{14.33}$$

For simplifying notations, in the following we forget the variable t. Since the temperature T is positive, but the magnetic field B can be both positive or negative, after taking the logarithm, one obtains the following equations:

$$\begin{cases} \log|V_{H_N}| = \log k_N + \log|B| + \alpha_N \log T, \\ \log|V_{H_p}| = \log k_P + \log|B| + \alpha_P \log T. \end{cases} \tag{14.34}$$

These equations are related to a linear mixture of two sources $\log|B|$ and $\log T$. They can be easily solved by simple decorrelation because B appears with the same power in both equations (14.33). It is then simpler to compute directly the ratio of the two equations (14.33):

$$R = \frac{V_{H_N}}{V_{H_p}} = \frac{k_N}{k_P} T^{\alpha_N - \alpha_P} \tag{14.35}$$

which only depends on the temperature T. For estimating the magnetic field, it is sufficient to estimate the parameter k such that $V_{H_N} R^k$ is uncorrelated with R. One can then deduce $B(t)$, up to a multiplicative gain. The final estimation of B and T requires a calibration step for restoring sign and scale.

This idea is also used for homomorphic filtering [95] in image processing, more precisely for removing the incident light contribution from the object reflectance. Assuming the incident light contribution is a low frequency band signal, a simple low-pass filtering applied on the logarithm of the image is a simple but efficient processing method.

14.3.7.2 Linearizable mappings

Extension of the Darmois-Skitovic theorem to nonlinear mappings has been addressed by Kagan *et al.* [62]. These results have been considered again in the framework of source separation of nonlinear mixtures by Eriksson and Koivunen [41]. The main idea is to consider particular mappings $\mathcal{F}$, satisfying an *addition theorem* in the sense of the theory of functional equations.

A simple example. Let us consider the nonlinear mapping $\mathcal{F}(s_1, s_2)$ of two independent random variables s_1 and s_2:

$$\begin{cases} x_1 = (s_1 + s_2)(1 + s_1 s_2)^{-1}, \\ x_2 = (s_1 - s_2)(1 - s_1 s_2)^{-1}. \end{cases} \tag{14.36}$$

By using the following change of variables $u_i = \tan^{-1}(s_i)$, Eq. (14.36) becomes:

$$\begin{cases} x_1 = \tan(u_1 + u_2), \\ x_2 = \tan(u_1 - u_2). \end{cases} \tag{14.37}$$

Applying again the transform $\tan^{-1}$ to the variables x_i, and denoting $v_i = \tan^{-1}(x_i)$, one finally gets:

$$\begin{cases} v_1 = \tan^{-1}(x_1) = u_1 + u_2, \\ v_2 = \tan^{-1}(x_2) = u_1 - u_2. \end{cases} \tag{14.38}$$

These equations are linear mixtures of the two independent variables u_1 and u_2. This result is simply due to the fact that $\tan(a+b)$ (and $\tan(a-b)$) is a function of $\tan a$ and $\tan b$, in other words because there exists a function $\mathcal{F}$ such that $\tan(a+b) = \mathcal{F}(\tan a, \tan b)$.

General result. More generally, Kagan *et al.* [62] show that this property appears provided that there exist a transform $\mathcal{F}$ and an invertible function f with values in the open set $\mathfrak{S}$ of $\mathbb{R}$ satisfying the following property (called the addition theorem)

$$f(s_1+s_2)=\mathcal{F}[f(s_1),f(s_2)]. \tag{14.39}$$

The properties required for the transform $\mathcal{F}$ (for two variables, but generalization to higher dimensions is straightforward) are the following:

- $\mathcal{F}$ is continuous, at least separately with respect to each variable.
- $\mathcal{F}$ is commutative, i.e. $\forall(u,v)\in\mathfrak{S}^2, \mathcal{F}(u,v)=\mathcal{F}(v,u)$.
- $\mathcal{F}$ is associative, i.e. $\forall(u,v,w)\in\mathfrak{S}^3$, $\mathcal{F}(\mathcal{F}(u,v),w)=\mathcal{F}(u,\mathcal{F}(v,w))$.
- There exists a neutral element $e\in\mathfrak{S}$ such that $\forall u\in\mathfrak{S}, \mathcal{F}(u,e)=\mathcal{F}(e,u)=u$.
- $\forall u \in \mathfrak{S}$, there exists an inverse element $u^{-1} \in \mathfrak{S}$ such that $\mathcal{F}(u,u^{-1}) = \mathcal{F}(u^{-1},u)=e$.

In other words, denoting $u \circ v = \mathcal{F}(u,v)$, these conditions imply that the set $\mathfrak{S}$ with the operation $\circ$ is a commutative group. Under this condition, Aczel [5] shows that there exists a monotonous and continuous function $f:\mathbb{R}\to\mathfrak{S}$ such that

$$f(x+y)=\mathcal{F}(f(x),f(y))=f(x)\circ f(y). \tag{14.40}$$

In fact, by applying f^{-1} (which exists since f is monotonous) to Eq. (14.40), one gets

$$x+y=f^{-1}(\mathcal{F}(f(x),f(y)))=f^{-1}(f(x)\circ f(y)). \tag{14.41}$$

By using the associative property of $\mathcal{F}$ and the relation (14.40), and posing $y=x$, one can define a multiplication by an integer c, denoted $\star$:

$$c\star f(x)=f(cx). \tag{14.42}$$

This multiplication can be extended to a multiplication by real variables α: $\alpha\star f(x)= f(\alpha x)$.

By computing the inverse f^{-1} and posing $f(x)=u$, one obtains:

$$cf^{-1}(u)=f^{-1}(c\star u). \tag{14.43}$$

Then, for any constant $c_1,\ldots,c_n$ and for any random variables $u_1,\ldots,u_n$, the following relation holds:

$$c_1f^{-1}(u_1)+\ldots c_nf^{-1}(u_n)=f^{-1}(c_1\star u_1\circ\ldots\circ c_n\star u_n). \tag{14.44}$$

Finally, Kagan *et al.* [62] present the following theorem:

THEOREM 14.6
Let $u_1, \ldots, u_n$ be independent random variables such that

$$\begin{cases} x_1 = a_1 \star u_1 \circ \ldots \circ a_n \star u_n \\ x_2 = b_1 \star u_1 \circ \ldots \circ b_n \star u_n \end{cases} \tag{14.45}$$

are independent, and such that the operators $\star$ and $\circ$ satisfy the above conditions. Denoting by f the function defined by the operator $\circ$, $f^{-1}(u_i)$ is Gaussian if $a_i b_i \neq 0$.

In fact, this theorem is nothing but the Darmois-Skitovich theorem with light modifications: by applying f^{-1} to equations (14.45), posing $f^{-1}(u_i) = s_i$ and taking into account the properties of the operators $\circ$ and $\star$, one gets exactly Darmois' equations [34].

Application to source separation. This theorem can be used for source separation. With such mixtures, a source separation algorithm consists of three main steps [41]:

1. Transform using f^{-1} the nonlinear observations x_i in order to obtain linear mixtures of "new" sources $s_i = f^{-1}(u_i)$.
2. Solve the linear mixtures of s_i using a method for linear source separation (e.g. ICA).
3. Restore the actual independent sources by applying on sources s_i the transform $u_i = f(s_i)$.

Unfortunately, this algorithm is not blind, since the mapping f must be known. If this condition is not satisfied, a possible architecture is a three stage cascade. The first stage consists of nonlinear blocks (for instance, multi-layer perceptrons) able to approximate f^{-1}. The second stage is a matrix $\mathbf{B}$ able to separate sources in linear mixtures. The structure of the third stage, which must approximate the mapping f, is similar to the structure of the first stage. One can remark that the first two stages are similar to the separating structure of PNL mixtures.[10] One can then compute the independent (distorted) sources s_i using a separation algorithm for PNL mixtures. Then, using the first stage (which provides an approximation of f^{-1}), one can identify the third stage (which must approximate f) and restore the initial independent sources.

PNL mixtures are similar to these nonlinear mappings. In fact, they are more general since the nonlinear functions f_i can be different and unknown. Consequently, the algorithms developed for source separation in PNL mixtures [99,14,2,45,46] can be used for separating these nonlinear mixtures blindly, avoiding the third stage. Other examples of nonlinear mappings satisfying the addition theorem are proposed in [62,41]. However, realistic mixtures belonging to this class of mappings do not seem to be commonplace, except for PNL mixtures (14.22) and multiplicative mixtures (14.30).

[10]In fact, the first stage is slightly simpler since the functions f_i are all the same, contrary to PNL mixtures.

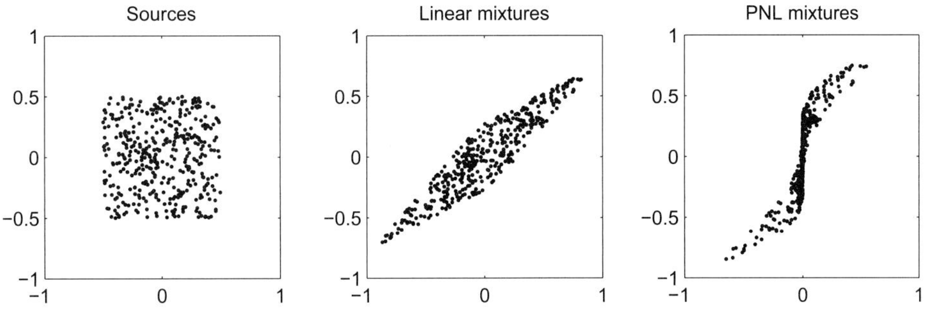

FIGURE 14.6

Joint distributions of sources **s** (left), of linear mixtures **e** (middle) and of PNL mixtures **x** (right).

14.4 PRIORS ON SOURCES

In this section, we show that priors on sources can relax indeterminacies or simplify algorithms. The first example takes into account the fact that the sources are bounded. The second one exploits the temporal correlation of the sources.

14.4.1 Bounded sources in PNL mixtures

Let us consider sources with a non-zero probability density function (pdf) on a bounded support, with non-zero values at the bounds of the support. As an example, sources having uniform distribution or the distribution associated to a sinewave signal with regular sampling satisfy this condition. For the sake of simplicity, we only consider PNL mixtures (Fig. 14.2) of two sources, but the results can be easily extended to mixtures with more than two sources. Using the independence condition, $p_{s_1 s_2}(s_1, s_2) = p_{s_1}(s_1) p_{s_2}(s_2)$, one can deduce that the joint distribution of the random vector $\mathbf{s}$ is bounded by a rectangle. After a linear mixture $\mathbf{A}$, the joint distribution $\mathbf{e} = \mathbf{As}$ is bounded in a parallelogram. Then, after componentwise mapping with nonlinear functions f_i, the joint distribution of the PNL observations $\mathbf{x}$ is bounded by a distorted parallelogram (see Fig. 14.6). One can prove [15] the following theorem:

THEOREM 14.7
Consider the mapping

$$\begin{cases} z_1 = g_1(e_1) \\ z_2 = g_2(e_2) \end{cases} \tag{14.46}$$

where g_1 and g_2 are two analytic functions.[11] *Assume that the borders of any parallelogram in the plane (e_1, e_2) are transformed to the borders of another parallelogram in the plane (z_1, z_2),*

[11] A function is called to be analytic on an interval, if one can expand it in Taylor series on this interval.

and that the borders of this parallelogram are not parallel to the coordinate axes. Then there exist real constants a_1, a_2, b_1, and b_2 such that

$$\begin{cases} g_1(u) = a_1 u + b_1 \\ g_2(u) = a_2 u + b_2 \end{cases} \tag{14.47}$$

Remarks

- This theorem provides another separability proof of PNL mixtures, for bounded sources.
- The existence of the constants b_1 and b_2 clearly shows another indeterminacy on the sources: in fact, sources are estimated up to an additive constant. This indeterminacy also exists in linear mixtures, but disappears because we assume that sources have zero mean. In other words, in PNL mixtures, the estimated sources can be written as $y_i(t) = \alpha_i s_{\sigma(i)}(t) + \beta_i$.
- One also finds that if the linear part $\mathbf{A}$ of the PNL model is not mixing, i.e. if $\mathbf{A}$ is diagonal, the joint distribution of the mixtures $\mathbf{e} = \mathbf{As}$ is bounded in a rectangle (a parallelogram whose borders are parallel to axes). Consequently the PNL observations $\mathbf{x} = \mathbf{f}(\mathbf{e})$ are bounded in a rectangle, too, and the estimation of nonlinear functions f_i is impossible.
- These results have recently been generalized for more than two sources with bounded densities in [102].

This theorem suggests a two step geometrical algorithm for separating bounded sources in PNL mixtures:

- Estimate the invertible functions g_1 and g_2 which transform the joint distribution of observations into a parallelogram. Using the above theorem, this step compensates the nonlinear distortion of the mixture.
- Separate the sources in the resulting linear mixture, by using an algorithm for linear ICA.

Details of the algorithm and a few experimental results are given in [15]. This method shows that the bounded source assumption provides very useful extra information, which simplifies source separation in PNL mixtures: linear and nonlinear parts of the separation structure can be optimized independently, according to two different criteria.

14.4.2 Temporally correlated sources in nonlinear mixtures

As we have explained in previous sections, independence is generally not a sufficient assumption for separating the sources in nonlinear mixtures. Without regularization, ICA can provide either *good solutions* corresponding to a σ-diagonal mapping of sources, or *bad solutions* for which estimated sources are still mixtures of original sources. The main question is then the following: how can we distinguish good solutions from the bad ones? Hosseini and Jutten [52] suggest that temporal correlation between successive samples of each source can be useful for this purpose.

14.4.2.1 A simple example

Let us return to the simple example presented in section 14.2.3.2, where we assume now that sources s_1, s_2, y_1 and y_2 are signals. If the signals $s_1(t)$ and $s_2(t)$ are temporally correlated and mutually independent, one can write:

$$\mathbb{E}\{s_1(t_1)s_2(t_2)\} = \mathbb{E}\{s_1(t_1)\}\mathbb{E}\{s_2(t_2)\}, \quad \forall t_1, t_2. \tag{14.48}$$

So, in addition to the decorrelation condition

$$\mathbb{E}\{y_1(t)y_2(t)\} = \mathbb{E}\{y_1(t)\}\mathbb{E}\{y_2(t)\}, \tag{14.49}$$

many supplementary equations, allowing us to reject the *bad solutions* $y_1(t)$ and $y_2(t)$, can be written:

$$\mathbb{E}\{y_1(t_1)y_2(t_2)\} = \mathbb{E}\{y_1(t_1)\}\mathbb{E}\{y_2(t_2)\} \quad \forall t_1 \neq t_2. \tag{14.50}$$

It is clear that if $y_1(t)$ and $y_2(t)$ are the original sources (up to any σ-diagonal transform), the above equality holds $\forall t_1, t_2$. Moreover, if the independent components are obtained from mapping (14.6) (this mapping is nonlinear, but preserves independence of random *variables*), the right hand side term of (14.50) is equal to zero because y_1 and y_2 are zero-mean Gaussian variables. The left hand side term of (14.50) is equal to

$$\begin{aligned}\mathbb{E}\{y_1(t_1)y_2(t_2)\} &= \mathbb{E}\{s_1(t_1)\cos(s_2(t_1))s_1(t_2)\sin(s_2(t_2))\} \\ &= \mathbb{E}\{s_1(t_1)s_1(t_2)\}\mathbb{E}\{\cos(s_2(t_1))\sin(s_2(t_2))\}.\end{aligned} \tag{14.51}$$

If $s_1(t)$ and $s_2(t)$ are temporally correlated, there probably exists a pair t_1, t_2 such that (14.51) is not equal to zero (of course, this depends on the nature of the temporal correlation between successive samples of each source), so that the equality (14.50) does not hold, and the solution can be rejected. In fact, the two stochastic *processes* $y_1(t)$ and $y_2(t)$ obtained by (14.6) are not independent although at each instant their samples (which are two independent random *variables*) are independent.

This simple example shows how by using temporal correlation of the sources, one can distinguish the σ-diagonal mappings, which preserve independence, or, at least reduce the set of non σ-diagonal mappings preserving independence. Here, we just use the cross-correlation of the signals (second order statistics), which is a first (but rough) step towards independence. We could also consider supplementary equations for improving the independence test, for example, by using cross-correlations of orders greater than two for $y_1(t_1)$ and $y_2(t_2)$, which must satisfy:

$$\mathbb{E}\{y_1^p(t_1)y_2^q(t_2)\} = \mathbb{E}\{y_1^p(t_1)\}\mathbb{E}\{y_2^q(t_2)\}, \quad \forall t_1, t_2, \forall p, q \neq 0. \tag{14.52}$$

14.4.2.2 Darmois decomposition with colored sources

As we presented in subsection 14.2.3, another classical example for showing the non-identifiability of nonlinear mixtures is the Darmois decomposition method. Let us consider two random signals $s_1(t)$ and $s_2(t)$, whose samples are independent and identically distributed (iid), and assume that $x_1(t)$ and $x_2(t)$ are nonlinear mixtures of them. By using the Darmois decomposition [33,55], one can derive new signals $y_1(t)$ and $y_2(t)$ which are statistically independent, although the related mapping is a mixing mapping (that is, not a σ-diagonal mapping):

$$\begin{cases} y_1(t) = F_{x_1}(x_1(t)), \\ y_2(t) = F_{x_2|x_1}(x_1(t), x_2(t)). \end{cases}$$

Let us denote by F_{x_1} and $F_{x_2|x_1}$ the cumulative probability density functions of the observations. If the sources are temporally correlated, one can show [52] that the independent components y_1 and y_2 obtained from the above decomposition generally do not satisfy the following equality for $t_1 \neq t_2$:

$$p_{y_1,y_2}(y_1(t_1)y_2(t_2)) = p_{y_1}(y_1(t_1))p_{y_2}(y_2(t_2)). \tag{14.53}$$

Conversely, σ-diagonal mappings of the actual sources, which can be written $y_1 = f_1(s_1)$ and $y_2 = f_2(s_2)$, clearly satisfy the above equality due to the assumption of independent sources. So the above equations can be used for rejecting (or at least for restricting) non σ-diagonal solutions provided by ICA using the Darmois decomposition.

Of course, this theoretical result does not constitute a proof of separability of nonlinear mixtures for temporally correlated sources. It simply shows that even weak prior information on the sources is able to reduce the typical indeterminacies of ICA in nonlinear mixtures. In fact, with this information, ICA provides many equations (constraints) which can be used for regularizing the solutions and achieving semi-blind source separation.[12]

14.5 INDEPENDENCE CRITERIA

14.5.1 Mutual information

ICA algorithms exploit the independence assumption. This assumption can be written using contrast functions, for which we shall seek the simplest expression. For linear mixtures, the simplest contrast functions are based on fourth-order cumulants. However, generally, one can consider that these particular contrasts are derived from the contrast $-I\{\mathbf{y}\}$, the opposite of mutual information, using various approximations of the pdf [10,31,53]. In this subsection, we shall mainly focus on the minimization of the mutual information, and then in less detail on a quadratic independence criterion.

Denoting by $H\{\mathbf{x}\} = -\mathbb{E}\{\log p_{\mathbf{x}}\}$ the differential entropy, one can write

$$I\{\mathbf{y}\} = \sum_i H\{y_i\} - H\{\mathbf{y}\}, \tag{14.54}$$

[12] *Semi-blind* since prior information is used.

where $H\{y_i\}$ is the marginal entropy of the estimated source y_i, and $H\{\mathbf{y}\}$ is the joint entropy of the estimated source vector.

14.5.1.1 Special case of nonlinear invertible mappings

For nonlinear invertible mappings $\mathcal{B}$, the estimated source vector is $\mathbf{y} = \mathcal{B}(\mathbf{x})$, and with a simple change of variable one can write [32]

$$I\{\mathbf{y}\} = \sum_i H\{y_i\} - H\{\mathbf{x}\} - \mathbb{E}\{\log|\det \mathbf{J}_{\mathcal{B}}|\}, \tag{14.55}$$

where $\mathbf{J}_{\mathcal{B}}$ is the Jacobian matrix of the mapping $\mathcal{B}$. For estimating the separation structure $\mathcal{B}$, minimizing $I\{\mathbf{y}\}$ is then equivalent to minimizing the simplified criterion

$$C(\mathbf{y}) = \sum_i H\{y_i\} - \mathbb{E}\{\log|\det \mathbf{J}_{\mathcal{B}}|\} \tag{14.56}$$

because $H\{\mathbf{x}\}$ does not depend on $\mathcal{B}$.

This criterion is simpler to use since its estimation and optimization only requires estimation of the mathematical expectation of a Jacobian and of the marginal entropies, and consequently only of marginal pdf (and not of joint multivariate pdf).

14.5.1.2 Mutual information for PNL mixture

For PNL mixtures, the separating mapping is $\mathcal{B}(\mathbf{x}) = \mathbf{B} \circ \mathbf{h}$, where $\mathbf{B}$ is a regular matrix, and $\mathbf{h}(\mathbf{x})$ is the vector $(h_1(x_1), \ldots, h_N(x_N))^T$. The simplified criterion then becomes

$$C(\mathbf{y}) = \sum_i H\{y_i\} - \log|\det \mathbf{B}| - \mathbb{E}\left\{\log\left|\prod_i h_i'(x_i)\right|\right\}. \tag{14.57}$$

Linear part. With respect to the linear part (matrix $\mathbf{B}$), the minimization of the criterion (14.57) leads to the same estimating equations as for linear mixtures:

$$\frac{\partial C(\mathbf{y})}{\partial \mathbf{B}} = \mathbb{E}\{\boldsymbol{\varphi}_{\mathbf{y}}(\mathbf{y})\mathbf{x}^T\} - \mathbf{B}^{-T} = 0. \tag{14.58}$$

In this relation, the components $\{\boldsymbol{\varphi}_{\mathbf{y}}(\mathbf{y})\}_i$ of the vector $\boldsymbol{\varphi}_{\mathbf{y}}$ are the score functions of the components y_i of the estimated source vector $\mathbf{y}$:

$$\{\boldsymbol{\varphi}_{\mathbf{y}}(\mathbf{y})\}_i = \varphi_{y_i}(y_i) = -\frac{d}{dy_i} \log p_{y_i}(y_i) = -\frac{p'_{y_i}(y_i)}{p_{y_i}(y_i)} \tag{14.59}$$

where $p_{y_i}(y_i)$ is the pdf of y_i, and $p'_{y_i}(y_i)$ its derivative. By multiplying from the right with $\mathbf{B}^T$, one obtains the estimating equations

$$\mathbb{E}\{\boldsymbol{\varphi}_{\mathbf{y}}\mathbf{y}^T\} - \mathbf{I} = 0. \tag{14.60}$$

In practice, one can derive from this equation an equivariant algorithm by computing the natural gradient or the relative gradient [30,27,10]. This provides equivariant performance, i.e. which does not depend on $\mathcal{A}$, for noiseless mixtures.

Nonlinear part. Concerning the nonlinear stage, by modeling the functions $h_i(\cdot)$ with parametric models $h_i(\boldsymbol{\theta}_i, \cdot)$, the gradient of the simplified criterion (14.57) can be written [99]

$$\frac{\partial C(\mathbf{y})}{\partial \boldsymbol{\theta}_k} = -\mathbb{E}\left\{\frac{\partial \log |h_k'(\boldsymbol{\theta}_k, x_k)|}{\partial \boldsymbol{\theta}_k}\right\} - \mathbb{E}\left\{\sum_{i=1}^{N} \varphi_{y_i}(y_i) b_{ik} \frac{\partial h_k(\boldsymbol{\theta}_k, x_k)}{\partial \boldsymbol{\theta}_k}\right\}, \tag{14.61}$$

where x_k is the k-th component of the observation vector, and b_{ik} is the element ik of the separation matrix $\mathbf{B}$. Of course, the exact equation depends on the parametric model. In [99], a multi-layer perceptron (MLP) was used for modeling each function $h_k(\boldsymbol{\theta}_k, .)$, $k = 1, \ldots, N$, but other models, parametric or not, can be used [2].

By considering a non-parametric model of the nonlinear functions h_i, i.e. by simply considering the random variables $z_k = h_k(x_k)$, one obtains the following estimating equations [98,2]:

$$\mathbb{E}\left\{\sum_{i=1}^{N} b_{ik} \varphi_{y_i}(y_i) \mid z_k\right\} = \varphi_{z_k}(z_k) \tag{14.62}$$

where $\mathbb{E}\{.|.\}$ denotes conditional expectation.

Influence of accuracy of score function estimation. Contrary to linear mixtures, separation performance for nonlinear mixtures strongly depends on the estimation accuracy of the score function (14.59) [99]. In fact, for the linear part it is clear that if one is close to the solution, i.e. if the estimated outputs y_j are mutually independent, the estimating equations (14.60) become:

$$\mathbb{E}\{\varphi_{y_i}(y_i) y_j\} = \mathbb{E}\{\varphi_{y_i}(y_i)\}\mathbb{E}\{y_j\} = 0, \quad \forall i = 1, \ldots, N \text{ and } i \neq j, \tag{14.63}$$

because the random variables y_j are zero mean. And the equality holds whatever the accuracy of the score function estimation is!

For the nonlinear part, the estimating equations (14.62) depend on score functions of both y_i and z_k. The equations are satisfied only if the score function estimation of both y_i and z_k is very accurate. This result explains the weak performance obtained (except if the mixtures are weakly nonlinear) by estimating the pdf using a fourth-order Gram-Charlier expansion [114,99], and then deriving the score functions by derivation. For harder nonlinearities, it is much better to compute a kernel estimate of the pdf. One can also estimate directly the score functions based on a least square minimization [87,97].

14.5.2 Differential of the mutual information

Minimization of the criterion (14.57) is simple since it only requires estimation of the marginal pdf's. However, this leads to biased estimates [3]. Moreover, this method is not applicable to convolutive mixtures, since there exist no simple relationships between pdf of $\mathbf{x}(t)$ and $\mathbf{y}(t) = [\mathbf{B}(z)]\mathbf{x}(t)$, where the filter matrix $\mathbf{B}(z)$ models (in discrete time) the convolutive mixture. In [17,13], Babaie-Zadeh and colleagues considered direct minimization of mutual information (14.54), and we present the main results in the following.

14.5.2.1 Definitions

For a random vector $\mathbf{x}$, marginal and joint score functions are defined as follows.

Definition 14.3 (MSF)
The marginal score function (MSF) of a random vector $\mathbf{x}$ *is the vector denoted by* $\boldsymbol{\varphi}_{\mathbf{x}}(\mathbf{x})$, *whose* i*-th component is equal to*

$$\{\boldsymbol{\varphi}_{\mathbf{x}}(\mathbf{x})\}_i = -\frac{d \log p_{x_i}(x_i)}{d x_i}. \tag{14.64}$$

Definition 14.4 (JSF)
The joint score function (JSF) of a random vector $\mathbf{x}$ *is the gradient of* $-\log p_{\mathbf{x}}(\mathbf{x})$. *It is denoted by* $\boldsymbol{\psi}_{\mathbf{x}}(\mathbf{x})$, *and its* i*-th component is equal to*

$$\{\boldsymbol{\psi}_{\mathbf{x}}(\mathbf{x})\}_i = -\frac{\partial \log p_{\mathbf{x}}(\mathbf{x})}{\partial x_i}. \tag{14.65}$$

One can now define the score function difference (SFD) of a random vector $\mathbf{x}$.

Definition 14.5 (SFD)
The score function difference (SFD) of $\mathbf{x}$ *is the difference between marginal and joint score functions:*

$$\boldsymbol{\beta}_{\mathbf{x}}(\mathbf{x}) \triangleq \boldsymbol{\varphi}_{\mathbf{x}}(\mathbf{x}) - \boldsymbol{\psi}_{\mathbf{x}}(\mathbf{x}). \tag{14.66}$$

14.5.2.2 Results

One can show [17] the following result concerning SFD:

Proposition 14.8
The components of the random vector $\mathbf{x} = (x_1, \ldots, x_N)^T$ *are independent if and only if* $\boldsymbol{\beta}_{\mathbf{x}}(\mathbf{x}) \equiv 0$, *i.e. if and only if:*

$$\boldsymbol{\psi}_{\mathbf{x}}(\mathbf{x}) = \boldsymbol{\varphi}_{\mathbf{x}}(\mathbf{x}). \tag{14.67}$$

More generally, SFD can be seen as a differential for mutual information (MI) by using the following theorem [17].

THEOREM 14.9 (Differential of MI)
Let $\mathbf{x}$ *be a random vector and* $\boldsymbol{\delta}$ *a "small" random vector with the same dimension. Then, one has:*

$$I\{\mathbf{x}+\boldsymbol{\delta}\}-I\{\mathbf{x}\}=\mathbb{E}\{\boldsymbol{\delta}^T\beta_{\mathbf{x}}(\mathbf{x})\}+o(\boldsymbol{\delta}) \tag{14.68}$$

where $o(\boldsymbol{\delta})$ *represents the higher order terms in* $\boldsymbol{\delta}$*, and* β_x *is the score function difference (SFD) of* $\mathbf{x}$.

Recall that, for a multivariate (differentiable) function $f(\mathbf{x})$, one has:

$$f(\mathbf{x}+\boldsymbol{\delta})-f(\mathbf{x})=\boldsymbol{\delta}^T\cdot(\nabla f(x))+o(\boldsymbol{\delta}). \tag{14.69}$$

By comparing the above equation with (14.68), one observes that the SFD can be interpreted as the "stochastic gradient" of mutual information. Finally, the following theorem [17] clearly shows that mutual information has no "local minima".

THEOREM 14.10
Let $\mathbf{x}_0$ *be a random vector whose pdf is continuously differentiable. If for any "small" random vector* $\boldsymbol{\delta}$*, the condition* $I\{\mathbf{x}_0\}\le I\{\mathbf{x}_0+\boldsymbol{\delta}\}$ *holds, then* $I\{\mathbf{x}_0\}=0$.

14.5.2.3 Practical consequences

Using the differential of MI requires one to estimate the marginal (MSF) and joint (JSF) score functions. Now, JSF is a multivariate function of N variables (its dimension is equal to the source number N). Its estimation is very costly as the number of sources increases. Fast algorithms for computing joint and conditional entropies and score functions have been proposed by Pham [86]. With these methods, the computational load increases with the factor 3^{n-1}, where n is the dimension.

It is also interesting to exploit the fact that MI has no local minima. At first glance, this result seems contradictory with other works [112]. In fact, it is not: observed local minima are related to the parametric model of the separation structure. Thus although the MI itself has no local minima, even for linear mixtures $I\{\mathbf{Bx}\}$ as a function of $\mathbf{B}$ can have local minima. Following this observation, Babaie-Zadeh and colleagues [16,13] proposed a new class of algorithms, called Minimization-Projection (MP) algorithms, which estimate $\mathbf{y}=\mathcal{B}(\mathbf{x})$ in two successive steps:

- minimization step of $I\{\mathbf{y}\}$, without constraint related to the model $\mathcal{B}$;
- projection step, where the model $\mathcal{B}$ is estimated by minimizing the mean-square error $\mathbb{E}\|\mathcal{B}(\mathbf{x})-\mathbf{y}\|^2$.

MP algorithms can be designed for any mixture model, especially for linear convolutive models, PNL mixtures with memory (convolutive PNL) and without memory.

14.5.3 Quadratic criterion

Following the works of Kankainen [63] and Eriksson *et al.* [40], Achard *et al.* [4] proposed a quadratic measure of dependence. The measure is based on a kernel $\mathcal{K}$, which has the property that its Fourier transform is almost everywhere non-zero. For N random variables $\mathbf{s} = (s_1, \ldots, s_N)^T$, one defines the quadratic criterion C_Q:

$$C_Q(s_1, \ldots, s_N) = \int D_{\mathrm{s}}^2(u_1, \ldots, u_N) d u_1, \ldots, d u_N, \tag{14.70}$$

where

$$D_{\mathrm{s}}(u_1, \ldots, u_N) = \mathbb{E}\left\{\prod_{i=1}^{N} \mathcal{K}\left(u_i - \frac{s_i}{\sigma_{s_i}}\right)\right\} - \prod_{i=1}^{N} \mathbb{E}\left\{\mathcal{K}\left(u_i - \frac{s_i}{\sigma_{s_i}}\right)\right\}, \tag{14.71}$$

in which σ_{s_i} is a scaling factor. From this definition, and using the *kernel trick* [79,18], the criterion (14.70) can be easily computed. One obtains a simple estimator, which does not suffer from the curse of dimensionality, and computes its asymptotic properties. One can also show that the measure is related to quadratic error between the first joint characteristic function and the product of the first marginal characteristic functions. The choice of the kernel $\mathcal{K}$ remains an open issue, although it is robust with respect to experimental results.

14.6 A BAYESIAN APPROACH FOR GENERAL MIXTURES

Bayesian inference methods are well-known for the quality of their results, the simplicity for taking into account priors in the model or in the sources, and their robustness. The main drawback is that their computational cost can be quite high, preventing sometimes their application to realistic unsupervised or blind learning problems where the number of unknown parameters and values to be estimated can grow very large.

Bayesian approaches have been used for source separation and ICA in linear mixtures and their various extensions [11,66,43,53,91,64]. Valpola *et al.* have developed variational Bayesian methods for various nonlinear mixture models in many papers, the main ones being [67,106–109]. Their research efforts have been summarized in [61,49]. In the following, we present briefly the principles of their basic approach to nonlinear BSS called Nonlinear Factor Analysis (NFA). For more details and various extensions, see the references mentioned above.

Variational Bayesian learning, also formerly called Bayesian ensemble learning [23,68], is in general based on approximation which is fitted to the posterior distribution of the parameter(s) to be estimated. The approximative distribution is often chosen to be Gaussian because of its simplicity and computational efficiency. In our setting, the method tries to estimate the sources $\mathbf{s}(t)$ and the mixing mapping $\mathcal{A}(\mathbf{s}(t))$ which have most probably generated the observed data $\mathbf{x}(t)$. Roughly speaking, this provides the regularization that is necessary for making the nonlinear BSS problem solvable and tractable.

14.6.1 The nonlinear factor analysis (NFA) method

14.6.1.1 The model and cost function

The NFA method assumes that the data are generated by a noisy nonlinear mixture model (14.1)

$$\mathbf{x}(t) = \mathcal{A}(\mathbf{s}(t)) + \mathbf{b}(t) \tag{14.72}$$

where $\mathbf{x}(t)$ and $\mathbf{s}(t)$ are respectively the observation and source vectors for the sample index or time t, $\mathbf{b}(t)$ is the additive noise term at time t, and $\mathcal{A}(\cdot)$ is a nonlinear mixing mapping. The dimensions P and N of the vectors $\mathbf{x}(t)$ and $\mathbf{s}(t)$, respectively, are usually different, and the components of the mapping $\mathcal{A}(\cdot)$ are smooth enough real functions of the source vector $\mathbf{s}(t)$.

The nonlinear factor analysis (NFA) algorithm approximates the observed data using an MLP network with one hidden layer:

$$\hat{\mathcal{A}}(\mathbf{s}(t)) = \mathbf{A}_2 \tanh(\mathbf{A}_1 \mathbf{s}(t) + \mathbf{a}_1) + \mathbf{a}_2, \tag{14.73}$$

where $(\mathbf{A}_1, \mathbf{a}_1)$ and $(\mathbf{A}_2, \mathbf{a}_2)$ are the weight matrices and bias vectors of the hidden layer and of the output layer, respectively. In this equation and in the following, the functions tanh and exp are applied componentwise to their argument vectors.

Let us denote by

- $\mathbf{X} = \{\mathbf{x}(1), \ldots, \mathbf{x}(T)\}$ the set of T observation vectors, $\mathbf{S} = \{\mathbf{s}(1), \ldots, \mathbf{s}(T)\}$ the set of T associated source vectors;
- $\boldsymbol{\theta}$ the vector containing all the unknown model parameters, including sources, noise and the parameters of the MLP network;
- $p(\mathbf{S}, \boldsymbol{\theta}|\mathbf{X})$ the theoretical *a posteriori* pdf and $q(\mathbf{S}, \boldsymbol{\theta})$ its parametric approximation.

In the variational Bayesian framework, one assigns a prior distribution for each parameter of the vector $\boldsymbol{\theta}$. For instance, assuming that the noise vector $\mathbf{b}(t)$ is jointly Gaussian and spatially and temporally white, one can write the likelihood of the observations

$$p(\mathbf{X}|\mathbf{S}, \boldsymbol{\theta}) = \prod_{i,t} p(x_i(t)|\mathbf{s}(t), \boldsymbol{\theta}) = \prod_{i,t} N(x_i(t); \mathcal{A}_i(\mathbf{s}(t)), \exp(2v_i)) \tag{14.74}$$

where $N(x; \mu, \sigma^2)$ represents a Gaussian distribution of variable x, with mean equal to μ and variance equal to σ^2, and $\mathcal{A}_i$ is the i-th component of the nonlinear mapping $\mathcal{A}$. Variances are parameterized according to exponential law $\exp(2v)$, where v is a parameter with a Gaussian *a priori*:

$$v \sim N(m_v, \sigma_v). \tag{14.75}$$

The goal of the Bayesian approach is to estimate the *a posteriori* pdf of all parameters of the vector $\boldsymbol{\theta}$. This is obtained by estimating a distribution $q(\mathbf{S}, \boldsymbol{\theta})$ which approximates the true *a posteriori* distribution $p(\mathbf{S},\boldsymbol{\theta}|\mathbf{X})$. The difference between the approximation $q(\mathbf{S},\boldsymbol{\theta})$ and the true pdf $p(\mathbf{S},\boldsymbol{\theta}|\mathbf{X})$ is measured using the Kullback-Leibler divergence:

$$K\{q,p\} = \int_{\mathbf{S}}\int_{\boldsymbol{\theta}} q(\mathbf{S},\boldsymbol{\theta})\ln\frac{q(\mathbf{S},\boldsymbol{\theta})}{p(\mathbf{S},\boldsymbol{\theta}|\mathbf{X})}d\boldsymbol{\theta}d\mathbf{S}. \tag{14.76}$$

The posterior distribution $p(\mathbf{S},\boldsymbol{\theta}|\mathbf{X})$ cannot usually be evaluated, and therefore the actual cost function used in variational Bayesian learning is

$$C = K\{q,p\} - \ln p(\mathbf{X}) = \int_{\mathbf{S}}\int_{\boldsymbol{\theta}} q(\mathbf{S},\boldsymbol{\theta})\ln\frac{q(\mathbf{S},\boldsymbol{\theta})}{p(\mathbf{S},\boldsymbol{\theta},\mathbf{X})}d\boldsymbol{\theta}d\mathbf{S}. \tag{14.77}$$

This can be split into two parts arising from the denominator and numerator of the logarithm:

$$C_q = \int_{\mathbf{S}}\int_{\boldsymbol{\theta}} q(\mathbf{S},\boldsymbol{\theta})\ln q(\mathbf{S},\boldsymbol{\theta})d\boldsymbol{\theta}d\mathbf{S}, \tag{14.78}$$

$$C_p = -\int_{\mathbf{S}}\int_{\boldsymbol{\theta}} q(\mathbf{S},\boldsymbol{\theta})\ln p(\mathbf{S},\boldsymbol{\theta},\mathbf{X})d\boldsymbol{\theta}d\mathbf{S}. \tag{14.79}$$

The cost function (14.77) can be used also for model selection as explained in [68]. In the nonlinear BSS problem, it provides the necessary regularization. For each source signal and parameter, its posterior pdf is estimated instead of some point estimate. In many cases, an appropriate point estimate is given by the mean of the posterior pdf of the desired quantity, and the respective variance provides at least a rough measure of the confidence of this estimate.

For evaluating the cost function $C = C_q + C_p$, we need two things: the exact formulation of the joint probability density $p(\mathbf{S},\boldsymbol{\theta},\mathbf{X})$, and its parametric approximation $q(\mathbf{S},\boldsymbol{\theta})$. Usually the joint pdf $p(\mathbf{S},\boldsymbol{\theta},\mathbf{X})$ is a product of simple terms due to the definition of the model. It can be written

$$p(\mathbf{S},\boldsymbol{\theta},\mathbf{X}) = p(\mathbf{X}|\mathbf{S},\boldsymbol{\theta})p(\mathbf{S}|\boldsymbol{\theta})p(\boldsymbol{\theta}). \tag{14.80}$$

The pdf $p(\mathbf{X}|\mathbf{S},\boldsymbol{\theta})$ has already been evaluated in (14.74), and the pdf's $p(\mathbf{S}|\boldsymbol{\theta})$ and $p(\boldsymbol{\theta})$ are also products of univariate Gaussian distributions. They can be obtained directly from the model structure [67,109].

The cost function can be minimized efficiently if one assumes that the parameters $\boldsymbol{\theta}$ and $\mathbf{S}$ are independent:

$$q(\mathbf{S},\boldsymbol{\theta}) = q(\mathbf{S})q(\boldsymbol{\theta}). \tag{14.81}$$

Finally, one assumes that parameters θ_i are Gaussian and independent, too:

$$q(\boldsymbol{\theta}) = \prod_i q(\theta_i) = \prod_i N(\bar{\theta}_i,\tilde{\theta}_i). \tag{14.82}$$

The distribution $q(\mathbf{S})$ follows a similar law. Estimation and evaluation of the cost function $C = C_q + C_p$ is discussed in detail in [67,106].

The NFA algorithm also assumes that the sources $\mathbf{s}(t)$ are Gaussian: $\mathbf{s}(t) \sim N(0, \exp(2\mathbf{v}_s))$. This assumption leads to a nonlinear PCA (principal component analysis) subspace only where the independent sources lie. The independent sources are then estimated by finding an appropriate rotation of this subspace using a standard linear ICA algorithm, such as FastICA [53].

14.6.1.2 The learning method

The parameters of the approximating distribution $q(\mathbf{S}, \boldsymbol{\theta})$ are optimized using gradient based iterative algorithms. During one sweep of the algorithm all the parameters are updated once, using all the available data. One sweep consists of two different phases. The order of computations in these two phases is the same as in the standard back-propagation algorithm for MLP networks [47] but otherwise the learning procedure is quite different. The most important differences are that in the NFA method learning is unsupervised, the cost function is different, and unknown variables are characterized by distributions instead of point estimates.

In the forward phase, the distributions of the outputs of the MLP networks are computed from the current values of the inputs. The value of the cost function is also evaluated as explained in the previous subsection. In the backward phase, the partial derivatives of the cost function with respect to all the parameters are fed back through the MLP and the parameters are updated using this information.

An update rule for the posterior variances $\tilde{\theta}_i$ is obtained by differentiating (14.77) with respect to $\tilde{\theta}_i$, yielding [67,109]

$$\frac{\partial C}{\partial \tilde{\theta}_i} = \frac{\partial C_p}{\partial \tilde{\theta}_i} + \frac{\partial C_q}{\partial \tilde{\theta}_i} = \frac{\partial C_p}{\partial \tilde{\theta}_i} - \frac{1}{2\tilde{\theta}_i}. \tag{14.83}$$

Equating this to zero yields a fixed-point iteration:

$$\tilde{\theta}_i = \left[2\frac{\partial C_p}{\partial \tilde{\theta}_i} \right]^{-1}. \tag{14.84}$$

The posterior means $\bar{\theta}_i$ can be estimated from the approximate Newton iteration [67,109]

$$\bar{\theta}_i \leftarrow \bar{\theta}_i - \frac{\partial C_p}{\partial \bar{\theta}_i} \left[\frac{\partial^2 C}{\partial \bar{\theta}_i^2} \right]^{-1} \approx \bar{\theta}_i - \frac{\partial C_p}{\partial \bar{\theta}_i} \tilde{\theta}_i. \tag{14.85}$$

In the NFA method, one tries to learn a nonlinear model (14.72), (14.73) with many parameters and unknowns in a completely blind manner from the data $\mathbf{X}$ only. Therefore care is required especially during the beginning phase of learning. Otherwise the whole learning method could converge to some false solution which is far from the correct one. The initialization process and tricks for avoiding false solutions are explained in [67,61].

14.6.2 Extensions and experimental results

The original NFA algorithm has been extended by modeling the sources as mixtures of Gaussians instead of plain Gaussians in [67,107,106]. This NIFA (NIFA means nonlinear independent factor analysis) method provides somewhat or slightly better estimation results than the NFA method followed by linear ICA at the expense of more complicated learning process and higher computational load. Experimental results with simulated data [67,107,53], for which the true sources are known, show that the NFA method followed by linear ICA and the NIFA method are able to approximate pretty well the true sources. These methods have been applied also to real-world data sets, including 30-dimensional pulp data [53,67,107] and speech data, but interpretation of the results is somewhat difficult, requiring problem specific expertise.

Furthermore, the NFA method has been extended for sources generated by a dynamic system [109], leading to a nonlinear dynamic factor analysis (NDFA) method. The NDFA method performs much better than the compared methods in blind estimation of the dynamic system and its source signals. It has been successfully applied also to the detection of changes in the states (sources) of the dynamic process [58]. There the NDFA method performed again much better than the compared state-of-the-art techniques of change detection.

The work on applying variational Bayesian learning to nonlinear BSS and related problems has been reported and summarized in more detail in [110,61,49], with more references and results. MATLAB codes of NFA and NDFA algorithms are available on the Web [108].

14.6.3 Comparisons on PNL mixtures

The NFA method has been applied on PNL mixtures (14.22) with additive noise $\mathbf{b}(t)$ and compared with the method based on mutual information minimization (MI) [56,57].

MI minimization is more efficient than NFA when the mixing model is a PNL model, even with additive noise. On the other hand, NFA can separate mixings which are locally non-invertible provided that they are globally invertible. This means that the nonlinear functions f_i are not invertible but the global mapping $\mathcal{A}$ is.

These experimental results show the importance of structural constraints on the achieved performance. Accurate modeling of the mixing and separating system is then an essential step. It is directly related to the existence, simplicity, performance, and robustness of separating methods, even in the Bayesian framework.

14.7 OTHER METHODS AND ALGORITHMS

14.7.1 Algorithms for PNL mixtures

A large number of works and algorithms have been devoted to PNL models. These methods mainly differ by the independence criterion [70], parameterization of the nonlinearity [85,2], or exploitation of temporal correlation [116] or of a Markovian model for the sources [69]. If certain conditions hold, geometrical approaches, which avoid statistical estimations but require a large number of samples, can be used. They have been proposed and studied in several papers [90,15,104,103,80,81].

Furthermore, a Bayesian method using MLP network structure has been introduced for blind separation of underdetermined post-nonlinear mixtures in [113]. For the same problem, a spectral clustering approach is proposed for sparse sources in [105]. A method based on flexible spline neural network structures and minimization of mutual information is proposed for convolutive PNL mixtures in [111].

A very simple and efficient idea has also been proposed for enhancing the convergence speed of source separation algorithms in PNL mixtures. From the central limit theorem, one can claim that the linear mixtures e_i (before application of the nonlinear functions f_i) are approximately Gaussian. One can then achieve a rough approximation of the inverse h_i of the function f_i, by enforcing $(h_i \circ f_i)(e_i)$ as being Gaussian, too. This idea, developed independently by Solé *et al.* [93] and Ziehe *et al.* [117], leads to very simple and fast estimation of h_i:

$$\hat{h}_i = \Phi^{-1} \circ F_{x_i}, \tag{14.86}$$

where F_{x_i} is the cumulative probability density function of the random variable x_i, and Φ is the cumulative probability density function of the Gaussian random variable. Of course, the Gaussian assumption is just an approximation, but the method is robust with respect to this assumption and provides an initial value for greatly h_i[13] which increases the convergence speed of algorithms [94]. Recently, this method has been extended to PNL mixtures in which the linear mixtures are close to Gaussian in [115].

14.7.2 Constrained MLP-like structures

Marques and Almeida [75,8] generalized the Infomax principle [21] to nonlinear mixtures. To this purpose, they propose a separation structure $\mathcal{B}$ realized using a multi-layer perceptron (MLP) neural network. Samples $y_i(t)$ at the output of the MLP network are then transformed by a mapping F_i in order to provide $z_i = F_i(y_i)$, whose distribution is uniform in $[0, 1]$. It is evident that F_i always exists: it is the cumulative probability density function of y_i. Since mutual information is invariant for any diagonal invertible

[13]Sometimes surprisingly accurate.

mapping, one can write:

$$I\{\mathbf{y}\} = I\{\mathbf{z}\} = \sum_i H\{z_i\} - H\{\mathbf{z}\}. \qquad (14.87)$$

Since the random variables z_i are always uniformly distributed in $[0, 1]$, their entropies are constants. Consequently, minimizing $I\{\mathbf{y}\}$ is equivalent to maximizing $H\{\mathbf{z}\}$. Under the generic name MISEP, Almeida proposed a few algorithms, which mainly differ according to the parameterization of F_i, in which the parameters of the MLP network are updated in order to maximize $H\{\mathbf{z}\}$ [7].

14.7.3 Other approaches

Tan, Wang, and Zurada [101] have proposed a radial basis function (RBF) neural network structure for approximating the separating mapping (14.2). Their contrast function consists of both the mutual information and partial moments of the estimated separated sources, which are used to provide the regularization needed in nonlinear BSS. Simulation results are presented for several artificially generated nonlinear mixture sets, confirming the validity of the method introduced in [101].

Levin has developed a nonlinear BSS method based on differential geometry and phase-space density in [71]. In [24], the authors claim that temporal slowness complements statistical independence well, and that a combination of these principles leads to unique solutions of the nonlinear BSS problem. The authors introduce an algorithm called independent slow feature analysis for nonlinear BSS. In [74], a noisy nonlinear version of ICA is proposed. Assuming that the pdf of sources is known, the authors derive a learning rule based on maximum likelihood estimation. A new method for solving nonlinear BSS problems is derived in [76] by exploiting second-order statistics in a kernel induced feature space. Experimental results are presented on realistic nonlinear mixtures of speech signals, gas multisensor data, and visual disparity data. A new nonlinear mixing model called by the authors "multinonlinearity constrained mixing model", and a separation method for it, are proposed in [42]. New nonlinear mixture models called by the authors "additive-target mixtures" and "extractable-target mixtures", and a separation method for them based on recurrent neural networks, are introduced in [35].

More references, especially to early works on nonlinear ICA and BSS, can be found in the reviews [7,61].

14.8 A FEW APPLICATIONS

Currently, a few real-world problems have been modeled by nonlinear mixtures. We present here three examples which seem quite interesting.

14.8.1 Chemical sensors

ISFET-like chemical sensors are designed with field effect transistors (MOSFET), whose metal gate is replaced by a polymer membrane which is sensitive to certain ions (see [22] and references inside). Each sensor typically provides a drain current

$$I_d = A + B \log\left(a_i + \sum_j k_{ij} a_j^{z_i/z_j}(t)\right), \tag{14.88}$$

where A and B are constants depending on technological and geometrical parameters of the transistor, k_{ij} measures the sensitivity of the sensor to secondary ions, and a_i and z_i are the activity and the valence of the ion i, respectively. Of course, activities of different ions act like sources and are assumed independent, which is a realistic assumption. Moreover, the unknown parameters A, B, and k_{ij} may vary between the sensors, and this spatial diversity allows the use of source separation methods.

In the general case, where the ions have different values, the quantity inside the log is nonlinear with respect to a_j. But if the different (or the main) ions have the same valence, the ratios z_i/z_j are equal to 1, and the responses provided by ISFET sensors can be modeled by PNL mixtures. The problem is, however, simpler that the generic PNL problem, since the nonlinearities f_i are known functions (log). PNL methods can then be applied in a semi-blind manner, leading to simpler and more efficient algorithms. In fact, from the physical model (14.88), one can see that the observation (the drain current) I_d can be linearized by applying $\exp[(I_d - A)/B]$. Concerning the nonlinear part of the separation structure, one then uses the parametric model $h_i(u) = \exp[(u - \alpha)/\beta]$, where α and β are the two parameters to be estimated [20]. In this way, one gets good estimates of the unknown activities a_i up to a constant, due to the scaling indeterminacy. This estimation process is then finalized with a calibration step.

For ions with different valences, the ratio $k = z_i/z_j$ is no longer equal to 1, and the mixture is much more intricate than a PNL mixture. This problem has been addressed assuming first that the log nonlinearity is canceled as suggested above. The remaining mixture is then still nonlinear:

$$x_i = a_i + \sum_j k_{ij} a_j^k(t). \tag{14.89}$$

This problem can be solved by extending the recursive structure suggested by Hosseini and Deville [50,51] for this particular mapping. Parameters of the separating structure can be estimated by considering either higher-order statistics [36], mutual information [37], sparsity of sources [38], or in a Bayesian framework [39], possibly also using temporal prior between successive values of ion activities.

14.8.2 Gas sensors

In a tin-oxide gas sensor, the oxygen concentration initially absorbed decreases if a combustible gas is present, which increases the sensor resistivity. If N different gases are present, one can model the variation of the conductance [72] of each sensor i by

$$G_i = G_0 + \sum_j^N \left(a_{ij} C_j^{-r_j} + \sum_{k=j}^N a_{ijk} C_j^{-r_j} C_k^{-r_k} \right) \tag{14.90}$$

where a_{ij}, a_{ijk}, r_j, and r_k are unknown parameters and C_j is the concentration of gas j. Denoting $C_j^{-r_j} = s_j$, the conductance varies according to a bilinear mixture of sources s_j. For two gases and two sensors, Bedoya [19] has shown that one can estimate the concentration of both gases by using methods developed by Hosseini and Deville [50] for source separation in bilinear mixtures.

14.8.3 Show-through removal

Let us consider a sheet of paper, of which both sides have been printed. If the paper sheet is thin, each side is in fact a mixture of the front and back images due to transparency. This phenomenon is called the show-through effect. Almeida [6] proposes to get two different mixtures (spatial diversity), by successively scanning the front and the back side of the paper sheet. After geometrical registration, he gets two nonlinear mixtures of the front and back images. He then uses his MISEP algorithm for separating the sources (front and back images) from their nonlinear mixtures. The separation performance is better than with a source separation algorithm for linear mixtures, but not very high. The separation results could be enhanced by employing a better model of the nonlinearities in the observations.

This problem has been then addressed by Merrikh-Bayat *et al.* in [77] using a refined model of the image mixtures and a suitable separation structure. First, their model takes into account the nonlinearity of the mixture, which is experimentally estimated by an exponential function. This function can be approximated by a first-order expansion, leading then to a bilinear model of mixtures. Merrikh-Bayat *et al.* [77] apply then the algorithm of Hosseini and Deville [50,51] (the recurrent structure of Fig. 14.5) to these mixtures to obtain the result shown in the middle column of Fig. 14.8. This result shows that although the bilinear mixing model may be effective for removing show-through, it leaves some traces of the borders of the texts on the other side.

Merrikh-Bayat *et al.* [77] then suggest that this problem arises because the bilinear mixing model and the separating structure of Fig. 14.5 are too simple to model the show-through effect in the scanned documents. In effect, when the light passes through the paper, it is scattered in different directions, and hence what we see from a side is a blurred version of the actual image of the other side. In other words, what the scanner sees is not "the image of recto mixed with the image of verso" but is "the image of recto mixed with a blurred version of the image of verso". But the model (14.27) and the separating structure of Fig. 14.5 do not take into account this blurring phenomenon. To remedy this problem, Merrikh-Bayat *et al.* propose the separating structure of Fig. 14.7. In this structure, near the convergence, y_2 is close to the image of the verso, and before subtracting it from x_1 (the scanned image of recto), it passes through a blurring filter $W(z)$. This corresponds

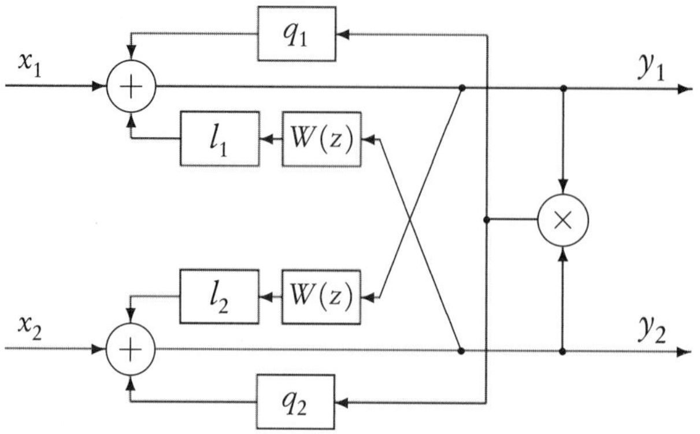

FIGURE 14.7

The structure proposed by Merrikh-Bayat *et al.* [77] for solving the show-though problem in scanned documents.

to writing the first equation of (14.27) as

$$x_1 = s_1 - l_1[W(z)]s_2 - q_1 s_1 s_2. \tag{14.91}$$

A similar operation is done to y_1, too. Note that the authors did not use $W(z)$ in the bilinear term, because the original equation is itself an approximate equation, and also because q_1 is usually very small. Then, as a first-order approximation, they choose the blurring filters $W(z)$ as fixed low-pass filters (3×3 point filters with all coefficients equal to 1). They then apply the same algorithm of Hosseini and Deville [51] which had been originally designed for the structure of Fig. 14.5 to this new structure, and they obtain the result shown in right column of Fig. 14.8. Applying Hosseini and Deville's algorithm on a different structure is just a first order approximation, justified by the fact that although the image of recto and its blurred version are different, they are not too different, and the same algorithm may give a first order approximation on the new separating structure. The results of Fig. 14.8 show that the approach using the structure of Fig. 14.7 may be very promising for solving the show-through effect in scanned documents.

We finish this section by emphasizing that in this application (removing the show-through effect) an important problem is registration of the images of both sides, that is, ensuring that the corresponding pixels of the images at hand correspond to same points on the paper. In effect, while scanning the other side of the paper, there is usually some amount of shift and rotation, which highly affect the quality of the results given by different algorithms. In the results of this section, it had been assumed that the scanned images of recto and verso have already been registered.

14.9 CONCLUSION

Blind source separation in nonlinear mixtures is generally not possible using merely ICA. In fact, for any random vector **s** with independent components, one can design a

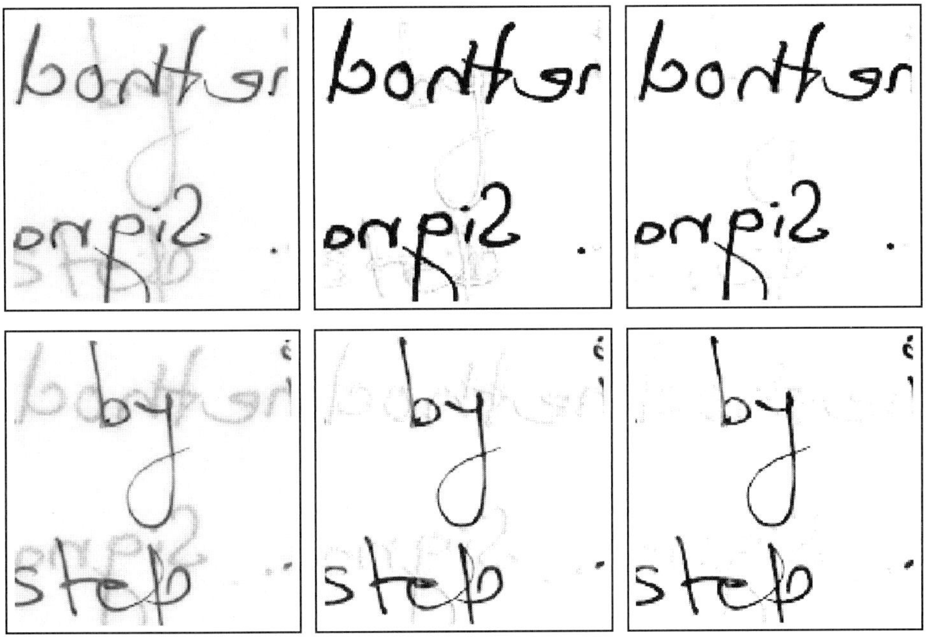

FIGURE 14.8

Results of removing show-through effect in scanned documents. Left column: Two registered images distorted by show-through. Middle column: Separation result using the structure of Fig. 14.5. Right column: Separation result using the structure of Fig. 14.7 for fixed $W(z)$'s.

mapping $\mathcal{H}$ which provides independent components $\mathbf{y}$ although the mapping is still mixing (not σ-diagonal).

In this chapter, we explored three main ideas for separating sources in nonlinear mixtures:

- The first one is based on structural constraints (on mixing and separating structures), which restrict the solution space. Post-nonlinear mixtures constitute a particularly interesting example, due to their realism, the theoretical separability characterized by weak indeterminacies, and the algorithmic performance.
- The second one consists in using prior information on sources, like the coloredness of temporal sources, or boundedness of sources. This idea is more *ad hoc* since each prior leads to particular processing and algorithm. It can lead to simpler algorithms, or simply restrict the solution space due to additional constraints introduced by extra equations.
- The third approach, which is quite promising although computationally demanding, is the Bayesian framework, which allows to take into account in a unified probabilistic formalism prior information and constraints on sources and mixtures.

In the framework of ICA for post-nonlinear mixtures, we used mutual information (MI) and showed that optimization of the nonlinear part of the model requires accurate estimation of the score functions, contrary to the estimation of the linear part estimation which allows very rough estimation. For overcoming the MI estimation problems, other criteria can be introduced, e.g. the quadratic criterion, whose estimation based on kernel trick is both simple and efficient.

In the future, the most efficient methods will be able to jointly take into account structural constraints and priors on sources. Finally, interest in nonlinear source separation methods will be relevant only if they can be applied to actual problems and applications. The most promising application domains currently include show-through removal and the design of smart sensor arrays, especially for chemicals and gases which can be classified in structurally constrained models.

Acknowledgments

This chapter is based on results obtained since 1996, in the framework of PhD theses and post-doctoral works, especially by A. Taleb, M. Babaie-Zadeh, S. Hosseini, S. Achard, and L. Duarte and in cooperation with D.T. Pham and other partners of the European project BLISS (BLInd Sources Separation and applications, IST-1999-14190), especially H. Valpola, J. Karhunen, A. Honkela and A. Ilin (Helsinki University of Technology, Finland), as well as K.-R. Müller, A. Ziehe and S. Harmeling (Fraunhofer Institute, Berlin, Germany).

Software

Bayesian Matlab software is available in [108]. Software (*applets* Java and Matlab sources) for PNL algorithms are available on the Web pages of GIPSA-lab:
http://www.lis.inpg.fr/demos/sep_sourc/ICAdemo/index.html
Other software (Matlab sources for PNL algorithms and NFA, especially) are available on the Web pages of the European project BLISS:
http://www.lis.inpg.fr/bliss/deliverables.php.

References

[1] S. Achard, C. Jutten, Identifiability of post nonlinear mixtures, IEEE Signal Processing Letters 12 (2005) 423–426.

[2] S. Achard, D. Pham, C. Jutten, Blind source separation in post nonlinear mixtures, in: Proc. of the 3rd Workshop on Independent Component Analysis and Signal Separation, ICA2001, San Diego (California, USA), 2001, pp. 295–300.

[3] S. Achard, D. Pham, C. Jutten, Criteria based on mutual information minimization for blind source separation in post-nonlinear mixtures, Signal Processing 85 (2004) 965–974.

[4] S. Achard, D. Pham, C. Jutten, Quadratic dependence measure for nonlinear blind source separation, in: S.-I. Amari, A. Cichocki, S. Makino, and N. Murata, (Eds.), Proc. of 4th Int. Symp. on Independent Component Analysis and Blind Source Separation, ICA2003, Nara, Japan, April 2003, pp. 263–268.

[5] J. Aczel, Lectures on Functional Equations and Their Applications, Academic Press, New York, 1966.

[6] L. Almeida, Separating a real-life nonlinear image mixture, Journal of Machine Learning Research 6 (2005) 1199–1229.

[7] L. Almeida, Nonlinear Source Separation, in: Synthesis Lectures on Signal Processing, vol. 2, Morgan & Claypool Publishers, 2006.

[8] L. Almeida, Linear and nonlinear ICA based on mutual information, in: Proc. IEEE 2000 Adaptive Systems for Signal Processing, Communications, and Control Symposium, AS-SPCC, Lake Louise, Canada, October 2000, pp. 117–122.

[9] L. Almeida, C. Jutten, H. Valpola, Realistic models of nonlinear mixtures, EU project BLISS (IST1999-14190) Report D5, May 2001. Available at http://www.lis.inpg.fr/bliss/deliverables.php.

[10] S. Amari, A. Cichocki, H. Yang, A new learning algorithm for blind signal separation, in: M. Mozer et al., (Eds.), Advances in Neural Information Processing Systems 1996, vol. 8, Cambridge, MA, USA, MIT Press, 1996, pp. 757–763.

[11] H. Attias, Independent factor analysis, Neural Computation 11 (1999) 803–851.

[12] M. Babaie-Zadeh, On blind source separation in convolutive and nonlinear mixtures, PhD Thesis, INPG, Grenoble, France, September 2002.

[13] M. Babaie-Zadeh, C. Jutten, A general approach for mutual information minimization and its application to blind source separation, Signal Processing 85 (2005) 975–995.

[14] M. Babaie-Zadeh, C. Jutten, K. Nayebi, Separating convolutive post non-linear mixtures, in: Proc. of the 3rd Workshop on Independent Component Analysis and Signal Separation, ICA2001, San Diego (California, USA), 2001, pp. 138–143.

[15] M. Babaie-Zadeh, C. Jutten, K. Nayebi, A geometric approach for separating post nonlinear mixtures, in: Proc. of the XI European Signal Processing Conf., EUSIPCO 2002, vol. II, Toulouse, France, September 2002, pp. 11–14.

[16] M. Babaie-Zadeh, C. Jutten, K. Nayebi, Minimization-projection (MP) approach for blind source separation in different mixing models, in: Proc. of the 4th Int. Symp. on Independent Component Analysis and Blind Source Separation, ICA2003, Nara, Japan, April 2003, pp. 1083–1088.

[17] M. Babaie-Zadeh, C. Jutten, K. Nayebi, Differential of mutual information, IEEE Signal Processing Letters 11 (2004) 48–51.

[18] F. Bach, M. Jordan, Kernel independent component analysis, Journal of Machine Learning Research 3 (2002) 1–48.

[19] G. Bedoya, Non-linear Blind Signal Separation for Chemical Solid-State Sensor Arrays, PhD Thesis, Technical Univ. of Catalonia, Dept. of Electrical Eng., Barcelona, Spain, 2006.

[20] G. Bedoya, C. Jutten, S. Bermejo, J. Cabestany, Improving semiconductor-based chemical sensor arrays using advanced algorithms for blind source separation, in: Proc. of the ISA/IEEE Sensors for Industry Conference, SIcon 04, New Orleans, USA, January 2004, pp. 149–154.

[21] A. Bell, T. Sejnowski, An information-maximization approach to blind separation and blind deconvolution, Neural Computation 7 (1995) 1129–1159.

[22] S. Bermejo, C. Jutten, J. Cabestany, ISFET source separation: Foundations and techniques, Sensors and Actuators B: Chemical, B (2006) 222–233.

[23] C. Bishop, Pattern Recognition and Machine Learning, Springer, 2006.

[24] T. Blaschke, T. Zito, L. Wiskott, Independent slow feature analysis and nonlinear blind separation, Neural Computation 19 (2007) 994–1021.

[25] G. Burel, Blind separation of sources: A nonlinear neural algorithm, Neural Networks 5 (1992) 937–947.

[26] J.-F. Cardoso, Blind signal separation: Statistical principles, Proceedings of the IEEE 9 (1998) 2009–2025.

[27] J.-F. Cardoso, B. Laheld, Equivariant adaptive source separation, IEEE Transactions on Signal Processing 44 (1996) 3017–3030.

[28] J.-F. Cardoso, A. Souloumiac, Blind beamforming for non gaussian signals, IEE Proceedings-F 140 (1993) 362–370.
[29] A. Cichocki, S.-I. Amari, Adaptive Blind Signal and Image Processing – Learning Algorithms and Applications, J. Wiley, 2002.
[30] A. Cichocki, R. Unbehauen, E. Rummert, Robust learning algorithm for blind separation of signals, Electronics Letters 30 (1994) 1386–1387.
[31] P. Comon, Independent component analysis, a new concept? Signal Processing 36 (1994) 287–314.
[32] T. Cover, J. Thomas, Elements of Information Theory, Wiley Series in Telecommunications, 1991.
[33] G. Darmois, Analyse des liaisons de probabilité, in: Proc. of Int. Statistics Conferences 1947, vol. III A, Washington (D.C.), 1951, p. 231.
[34] G. Darmois, Analyse générale des liaisons stochastiques, Revue de l' Institut International de Statistique 21 (1953) 2–8.
[35] Y. Deville, S. Hosseini, Recurrent networks for separating extractable-target nonlinear mixtures, part I: Non-blind configurations, Signal Processing 89 (2009) 378–393.
[36] L. Duarte, C. Jutten, Blind source separation of a class of nonlinear mixtures, in: Proc. of the 7th Int. Conf. on Independent Component Analysis and Signal Separation, ICA2007, London, United Kingdom, in: Lecture Notes in Computer Science, vol. 4666, Springer-Verlag, September 2007, pp. 41–48.
[37] L. Duarte, C. Jutten, A mutual information minimization approach for a class of nonlinear recurrent separating systems, in: Proc. of the 2007 IEEE Int. Workshop on Machine Learning for Signal Processing, MLSP2007, Thessaloniki, Greece, August 2007.
[38] L. Duarte, C. Jutten, A nonlinear source separation approach for the Nicolsky-Eisenman model, in: Proc. of the 16th European Signal Processing Conf., EUSIPCO2008, Lausanne, Switzerland, August 2008.
[39] L. Duarte, C. Jutten, S. Moussaoui, Ion-selective electrode array based on a Bayesian nonlinear source separation method, in: Proc. of the 8th Int. Conf. on Independent Component Analysis and Signal Separation, ICA2009, Paraty, Brasil, in: T. Adali, et al. (Eds.), Lecture Notes in Computer Science, vol. 5441, Springer, March 2009, pp. 662–669.
[40] J. Eriksson, A. Kankainen, V. Koivunen, Novel characteristic function based criteria for ICA, in: Proc. of the 3rd Int. Conf. on Independent Component Analysis and Signal Separation, ICA2001, San Diego, CA, USA, December 2001, pp. 108–113.
[41] J. Eriksson, V. Koivunen, Blind identifiability of class of nonlinear instantaneous ICA models, in: Proc. of the XI European Signal Proc. Conf., EUSIPCO2002, vol. 2, Toulouse, France, September 2002, pp. 7–10.
[42] P. Gao, W. Woo, S. Dlay, Nonlinear signal separation for multinonlinearity constrained mixing model, IEEE Transactions on Neural Networks 17 (2006) 796–802.
[43] M. Girolami (Ed.), Advances in Independent Component Analysis, Springer-Verlag, 2000.
[44] M. Haritopoulos, H. Yin, N. Allison, Image denoising using self-organizing map-based non-linear independent component analysis, Neural Networks 15 (2002) 1085–1098.
[45] S. Harmeling, A. Ziehe, B. Blankertz, K.-R. Müller, Non-linear blind source separation using kernel feature bases, in: Proc. of the 3rd Int. Conf. on Independent Component Analysis and Signal Separation, ICA2001, San Diego, CA, USA, December 2001, pp. 102–107.
[46] S. Harmeling, A. Ziehe, B. Blankertz, K.-R. Müller, Kernel-based nonlinear blind source separation, Neural Computation 15 (2003) 1089–1124.
[47] S. Haykin, Neural Networks – A Comprehensive Foundation, 2nd ed., Prentice Hall, 1998.
[48] J. Hérault, C. Jutten, B. Ans, Détection de grandeurs primitives dans un message composite par une architecture de calcul neuromimétique en apprentissage non supervisé, in: Actes du X ème colloque GRETSI, Nice, France, May 1985, pp. 1017–1022.
[49] A. Honkela, H. Valpola, A. Ilin, J. Karhunen, Blind separation of nonlinear mixtures by variational Bayesian learning, Digital Signal Processing 17 (2007) 914–934.

[50] S. Hosseini, Y. Deville, Blind separation of linear-quadratic mixtures of real sources using a recurrent structure, in: Proc. of the 7th Int. Work-Conference on Artificial and Natural Neural Networks, IWANN2003, Menorca, Spain, in: Lecture Notes in Computer Science, vol. 2686, Springer-Verlag, June 2003, pp. 241–248.

[51] S. Hosseini, Y. Deville, Blind maximum likelihood separation of a linear-quadratic mixture, in: C. Puntonet, A. Prieto (Eds.), Proc. of the 5th Int. Conf. on Independent Component Analysis and Blind Signal Separation, ICA2004, Granada, Spain, in: Lecture Notes in Computer Science, vol. 3195, Springer-Verlag, September 2004, pp. 694–701.

[52] S. Hosseini, C. Jutten, On the separability of nonlinear mixtures of temporally correlated sources, IEEE Signal Processing Letters 10 (2003) 43–46.

[53] A. Hyvärinen, J. Karhunen, E. Oja, Independent Component Analysis, J. Wiley, 2001.

[54] A. Hyvärinen, E. Oja, A fast fixed-point algorithm for independent component analysis, Neural Computation 9 (1997) 1483–1492.

[55] A. Hyvärinen, P. Pajunen, Nonlinear independent component analysis: Existence and uniqueness results, Neural Networks 12 (1999) 429–439.

[56] A. Ilin, S. Achard, C. Jutten, Bayesian versus constrained structure approaches for source separation in post-nonlinear mixtures, in: Proc. of the 2004 Int. J. Conf. on Neural Networks, IJCNN2004, Budapest, Hungary, July 2004, pp. 2181–2186.

[57] A. Ilin, A. Honkela, Post-nonlinear independent component analysis by variational Bayesian learning, in: C. Puntonet, A. Prieto (Eds.), Proc. of the 5th Int. Conf. on Independent Component Analysis and Blind Signal Separation, ICA2004, Granada, Spain, in: Lecture Notes in Computer Science, vol. 3195, Springer-Verlag, September 2004, pp. 766–773.

[58] A. Ilin, H. Valpola, E. Oja, Nonlinear dynamical factor analysis for state change detection, IEEE Transactions on Neural Networks 15 (2004) 559–575.

[59] C. Jutten, Calcul Neuromimétique et Traitement du Signal, Analyse en Composantes Indépendantes, PhD Thesis, INPG, Univ. Grenoble, France, 1987 (in French).

[60] C. Jutten, J. Hérault, Blind separation of sources, Part I: An adaptive algorithm based on a neuromimetic architecture, Signal Processing 24 (1991) 1–10.

[61] C. Jutten, J. Karhunen, Advances in blind source separation (BSS) and independent component analysis (ICA) for nonlinear mixtures, International Journal of Neural Systems 14 (2004) 267–292.

[62] A. Kagan, Y. Linnik, C. Rao, Extension of Darmois-Skitovic theorem to functions of random variables satisfying an addition theorem, Communications in Statistics 1 (1973) 471–474.

[63] A. Kankainen, Consistent testing of total independence based on empirical characteristic functions, PhD Thesis, University of Jyväskylä, Jyväskylä, Finland, 1995.

[64] K. Knuth, E. Kuruoglu (Eds.), Digital Signal Processing 17 (2007), Special issue on Bayesian source separation.

[65] M. Korenberg, I. Hunter, The identification of nonlinear biological systems: LNL cascade models, Biological Cybernetics 43 (1995) 125–134.

[66] H. Lappalainen, Ensemble learning for independent component analysis, in: Proc. Int. Workshop on Independent Component Analysis and Signal Separation, ICA'99, Aussois, France, 1999, pp. 7–12.

[67] H. Lappalainen, A. Honkela, Bayesian nonlinear independent component analysis by multi-layer perceptrons, in: M. Girolami (Ed.), Advances in Independent Component Analysis, Springer-Verlag, 2000, pp. 93–121.

[68] H. Lappalainen, J. Miskin, Ensemble learning, in: M. Girolami (Ed.), Advances in Independent Component Analysis, Springer-Verlag, Berlin, 2000, pp. 75–92.

[69] A. Larue, C. Jutten, S. Hosseini, Markovian source separation in non-linear mixtures, in: C. Puntonet, A. Prieto (Eds.), Proc. of the 5th Int. Conf. on Independent Component Analysis and Blind Signal Separation, ICA2004, Granada, Spain, in: Lecture Notes in Computer Science, vol. 3195, Springer-Verlag, September 2004, pp. 702–709.

[70] T.-W. Lee, B. Koehler, R. Orglmeister, Blind source separation of nonlinear mixing models, in: Neural Networks for Signal Processing VII, Proc. of the 1997 IEEE Signal Processing Society Workshop, IEEE Press, 1997, pp. 406–415.
[71] D. Levin, Using state space differential geometry for nonlinear blind source separation, Journal of Applied Physics 103 (2008) Article ID 044906, 12 pages.
[72] E. Llobet, X. Vilanova, J. Brezmes, X. Correig, Electrical equivalent models of semiconductor gas sensors using PSICE, Sensors and Actuators B 77 (2001) 275–280.
[73] E. Lukacs, A characterization of the Gamma distribution, Annals of Mathematical Statistics 26 (1955) 319–324.
[74] S. Maeda, W.-J. Song, S. Ishii, Nonlinear and noisy extension of independent component analysis: Theory and its application to a pitch sensation model, Neural Computation 17 (2005) 115–144.
[75] G. Marques, L. Almeida, Separation of nonlinear mixtures using pattern repulsion, in: Proc. Int. Workshop on Independent Component Analysis and Signal Separation, ICA'99, Aussois, France, 1999, pp. 277–282.
[76] D. Martinez, A. Bray, Nonlinear blind source separation using kernels, IEEE Transactions on Neural Networks 14 (2003) 228–235.
[77] F. Merrikh-Bayat, M. Babaie-Zadeh, C. Jutten, A nonlinear blind source separation solution for removing the show-through effect in the scanned documents, in: Proc. of the 16th European Signal Processing Conf., EUSIPCO2008, Lausanne, Switzerland, August 2008.
[78] F. Mokhtari, M. Babaie-Zadeh, C. Jutten, Blind separation of bilinear mixtures using mutual information minimization, in: 2009 IEEE Int. Conf. on Machine Learning for Signal Processing, MLSP2009, Grenoble, France, September 2009.
[79] K.-R. Müller, S. Mika, G. Rätsch, K. Tsuda, B. Schölkopf, An introduction to kernel-based learning algorithms, IEEE Transactions on Neural Networks 12 (2001) 181–201.
[80] T. Nguyen, J. Patra, A. Das, A post nonlinear geometric algorithm for independent component analysis, Digital Signal Processing 15 (2005) 276–294.
[81] T. Nguyen, J. Patra, S. Emmanuel, gpICA: A novel nonlinear ICA algorithm using geometric linearization, EURASIP Journal on Advances in Signal Processing (2007) (Special issue on Advances in Blind Source Separation, Article ID 31951, 12 pages).
[82] A. Parashiv-Ionescu, C. Jutten, G. Bouvier, Source separation based processing for integrated Hall sensor arrays, IEEE Sensors Journal 2 (2002) 663–673.
[83] L. Parra, Symplectic nonlinear component analysis, in: M. Mozer et al., (Eds.), Advances in Neural Information Processing Systems 1995, vol. 8, MIT Press, Cambridge, MA, USA, 1996, pp. 437–443.
[84] L. Parra, G. Deco, S. Miesbach, Statistical independence and novelty detection with information-preserving nonlinear maps, Neural Computation 8 (1996) 260–269.
[85] H. Peng, Z. Chi, W. Siu, A semi-parametric hybrid neural model for nonlinear blind signal separation, International Journal of Neural Systems 10 (2000) 79–94.
[86] D. Pham, Fast algorithms for estimating mutual information, entropies and score functions, in: A. Cichocki, S. Makino, N. Murata, (Eds.), Proc. 4th Int. Symp. on Independent Component Analysis and Blind Signal Separation, ICA2003, S.-I. Amari, Nara, Japan, April 2003, pp. 17–22.
[87] D. Pham, P. Garat, C. Jutten, Separation of mixtures of independent sources through a maximum likelihood approach, in: J. Vandewalle, R. Boite, M. Moonen, A. Oosterlinck (Eds.), Signal Processing VI, Theories and Applications. vol. 2, Elsevier, Brussels, Belgium, August 1992, pp. 771–774.
[88] R. Popovic, Hall-Effect Devices, Adam Hilger, Bristol, 1991.
[89] S. Prakriya, D. Hatzinakos, Blind identification of LTI-ZMNL-LTI nonlinear channel models, IEEE Transactions on Signal Processing 43 (1995) 3007–3013.
[90] C. Puntonet, M. Alvarez, A. Prieto, B. Prieto, Separation of sources in a class of post-nonlinear mixtures, in: Proc. of the 6th European Symp. on Artificial Neural Networks, ESANN'98, Bruges, Belgium, April 1998, pp. 321–326.

[91] S. Roberts, R. Everson (Eds.), Independent Component Analysis: Principles and Practice, Cambridge Univ. Press, 2001.

[92] S. Senecal, P.-O. Amblard, L. Cavazzana, Particle filtering equalization method for a satellite communication channel, Journal of Applied Signal Processing 15 (2004) 2317–2327.

[93] J. Solé, M. Babaie-Zadeh, C. Jutten, D. Pham, Improving algorithm speed in PNL mixture separation and Wiener system inversion, in: S.-I. Amari, A. Cichocki, S. Makino, N. Murata, (Eds.), Proc. of 4th Int. Symp. on Independent Component Analysis and Blind Source Separation, ICA2003, Nara, Japan, April 2003, pp. 639–644.

[94] J. Solé, C. Jutten, D. Pham, Fast approximation of nonlinearities for improving inversion algorithms of PNL mixtures and Wiener systems, Signal Processing 85 (2005) 1780–1786.

[95] T. Stockham, T. Cannon, R. Ingerbretsen, Blind deconvolution through digital signal processing, Proceedings of the IEEE 63 (1975) 678–692.

[96] A. Taleb, A generic framework for blind source separation in structured nonlinear models, IEEE Transactions on Signal Processing 50 (2002) 1819–1830.

[97] A. Taleb, C. Jutten, Nonlinear source separation: The postlinear mixtures, in: Proc. of the 5th Europ. Symp. on Artificial Neural Networks, ESANN'97, Bruges, Belgium, April 1997, pp. 279–284.

[98] A. Taleb, C. Jutten, Batch algorithm for source separation in post-nonlinear mixtures, in: Proc. First Int. Workshop on Independent Component Analysis and Signal Separation, ICA'99, Aussois, France, 1999, pp. 155–160.

[99] A. Taleb, C. Jutten, Source separation in post-nonlinear mixtures, IEEE Transactions on Signal Processing 47 (1999) 2807–2820.

[100] A. Taleb, J. Sole, C. Jutten, Quasi-nonparametric blind inversion of Wiener systems, IEEE Transactions on Signal Processing 49 (2001) 917–924.

[101] Y. Tan, J. Wang, J. Zurada, Nonlinear blind source separation using a radial basis function network, IEEE Transactions on Neural Networks 12 (2001) 124–134.

[102] F. Theis, P. Gruber, On model identifiability in analytic postnonlinear ICA, Neurocomputing 64 (2005) 223–234.

[103] F. Theis, E. Lang, Postnonlinear blind source separation via linearization identification, in: Proc. of the 2004 Int. J. Conf. on Neural Networks, IJCNN2004, Budapest, Hungary, July 2004, pp. 2199–2204.

[104] F. Theis, C. Puntonet, E. Lang, Nonlinear geometrical ICA, in: S.-I. Amari, A. Cichocki, S. Makino, and N. Murata, (Eds.), Proc. 4th Int. Symp. on Independent Component Analysis and Blind Signal Separation, ICA2003, Nara, Japan, April 2003, pp. 275–280.

[105] S. V. Vaerenbergh, I. Santamaria, A spectral clustering approach to underdetermined postnonlinear blind source separation of sparse sources, IEEE Transactions on Neural Networks 17 (2006) 811–814.

[106] H. Valpola, Nonlinear independent component analysis using ensemble learning: Theory, in: Proc. of the 2nd Int. Workshop on Independent Component Analysis and Blind Signal Separation, ICA2000, Helsinki, Finland, 2000, pp. 251–256.

[107] H. Valpola, X. Giannakopoulos, A. Honkela, J. Karhunen, Nonlinear independent component analysis using ensemble learning: Experiments and discussion, in: Proc. of the 2nd Int. Workshop on Independent Component Analysis and Blind Signal Separation, ICA2000, Helsinki, Finland, 2000, pp. 351–356.

[108] H. Valpola, A. Honkela, X. Giannakopoulos, MATLAB codes for the NFA and NDFA algorithms, Available at http://www.cis.hut.fi/projects/bayes/software/, 2002.

[109] H. Valpola, J. Karhunen, An unsupervised ensemble learning method for nonlinear dynamic state-space models, Neural Computation 14 (2002) 2647–2692.

[110] H. Valpola, E. Oja, A. Ilin, A. Honkela, J. Karhunen, Nonlinear blind source separation by variational Bayesian learning, IEICE Transactions (Japan) E86-A (2003) 532–541.

[111] D. Vigliano, R. Parisi, A. Uncini, An information theoretic approach to a novel nonlinear independent component analysis paradigm, Signal Processing 85 (2005) 997–1028.

[112] F. Vrins, M. Verleysen, On the entropy minimization of a linear mixture of variables for source separation, Signal Processing 85 (2005) 1029–1044.

[113] C. Wei, W. Woo, S. Dlay, Nonlinear underdetermined blind signal separation using Bayesian neural network approach, Digital Signal Processing 17 (2007) 50–68.

[114] H. Yang, S.-I. Amari, A. Cichocki, Information-theoretic approach to blind separation of sources in non-linear mixture, Signal Processing 64 (1998) 291–300.

[115] K. Zhang, L.-W. Chan, Extended gaussianization method for blind separation of post-nonlinear mixtures, Neural Computation 17 (2005) 425–452.

[116] A. Ziehe, M. Kawanabe, S. Harmeling, K.-R. Müller, Separation of post-nonlinear mixtures using ACE and temporal decorrelation, in: Proc. of the 3rd Int. Conf. on Independent Component Analysis and Signal Separation, ICA2001, San Diego, CA, USA, December 2001, pp. 433–438.

[117] A. Ziehe, M. Kawanabe, S. Harmeling, K.-R. Müller, Blind separation of post-nonlinear mixtures using gaussianizing transformations and temporal decorrelation, in: S.-I. Amari, A. Cichocki, S. Makino, N. Murata, (Eds.), Proc. of 4th Int. Symp. on Independent Component Analysis and Blind Source Separation, ICA2003, Nara, Japan, April 2003, pp. 269–274.

CHAPTER

Semi-blind methods for communications

15

V. Zarzoso, P. Comon, and D. Slock

15.1 INTRODUCTION

15.1.1 Blind source separation and channel equalization

The problem of BSS arises in a wide variety of real-life applications, which helps explain the intense interest this research area has attracted over the last years. A typical example is encountered in multi-user communication systems, where several mobile users sharing the transmission medium over the same time-frequency-code slot cause *co-channel interference (CCI)* with each other. CCI is due to signals from different spatial origins interfering with the signal of interest, giving rise to *spatial mixtures* observed at the receiving end. Hence, CCI cancellation can naturally be set out as a problem of blind source separation (BSS) in instantaneous linear mixtures, where the signal transmitted by each co-channel user represents one of the sources. These transmission scenarios are also known as instantaneous or static *multi-input multi-output (MIMO)* channels.

In digital communication systems, transmission effects such as multipath propagation and limited bandwidth produce linear distortion over the transmitted signals, causing *intersymbol interference (ISI)* at the receiver output, even if the channel is excited by a single input. Such distortions become more significant as the transmission rate and the user mobility (in the context of wireless communications) increase. ISI arises when a transmitted signal gets corrupted by time-delayed versions of itself, thus generating at the receiver end what could be described as *temporal mixtures*. The problem of ISI suppression is referred to as *deconvolution* or *channel equalization*. Classically, equalization is based on *training* or *pilot symbols* known by the receiver, which leads to a reduced bandwidth utilization. *Blind* equalization methods spare the use of training information [31,65,69,78], with the subsequent gain in bandwidth efficiency. If the transmitted symbols are temporally statistically independent (e.g. iid sequences), channel equalization can be formulated as a BSS problem of independent sources in instantaneous linear mixtures, that is, it accepts an ICA model [7,89,99]. Note that this model still holds if each symbol sequence is a linear process instead of an iid process. In the ICA formulation, the mixing matrix exhibits a Toeplitz structure fully characterized by the channel impulse response. In time-dispersive multi-input channels typically associated with high data rate multi-user wireless communication systems, both ISI and CCI need to be tackled simultaneously, which calls for *spatio-temporal equalization* techniques or, in the BSS/ICA jargon, for BSS methods in convolutive mixtures. These transmission scenarios are also referred to as convolutive or dynamic MIMO systems.

On account of the above connections, generic methods for BSS used in other applications could also be employed to perform digital channel equalization. However, digital communication channels present particular features that can be capitalized on to improve the source recovery. Firstly, digital modulations have finite support or, in other words, they contain only a small number of possible complex amplitudes. Criteria such as the *constant modulus (CM)* or the *constant power (CP)* are specifically adapted to the blind estimation of signals with such modulations and, as shown in Chapter 3, constitute valid contrasts for the separation and extraction of these signals in linear mixtures, either instantaneous or convolutive. The CM has long been used in blind equalization [31,65,78], whereas the CP criterion has been recently proposed for inputs with *q-ary phase shift keying* (*q-PSK*) modulation, for an arbitrary integer $q \geqslant 2$ [14]. These principles can be considered as *quasi-deterministic* rather than statistical criteria, in the sense that signals with adapted modulations cancel exactly (in the absence of noise) the sample version of the contrasts for any data length. As a result, these contrasts offer the potential of achieving good performance even for short sample size. A related benefit is that constellation-adapted criteria spare the input statistical independence assumption [14]. Secondly, training symbols known by the receiver can be incorporated into the transmitted signal to assist the equalization process, thus giving rise to *semi-blind methods*. By appropriately combining pilot information with blind criteria, semi-blind methods can outperform traditional training-based techniques at a fraction of the bandwidth utilization and with just a moderate increase in computational cost.

15.1.2 Goals and organization of the chapter

The present chapter studies a number of strategies for semi-blind equalization of digital communication channels. Only direct equalization, i.e. without previous channel identification, is addressed. For completeness, semi-blind channel estimation will also be treated, though briefly, at the end of the chapter. Our focus is on digital-modulation based contrasts like the CM and CP and, more particularly, semi-blind criteria that can be derived from them. In addition, we develop two other strategies aiming at improving the deficiencies of blind equalizers, which can also be employed in conjunction with semi-blind criteria: equalizer initialization by means of algebraic solutions and iterative search based on optimal step size computation. Algebraic methods are associated with challenging matrix and tensor decomposition problems analogous to those found in BSS based on statistical independence (ICA problem). Efficient iterative equalization techniques can be developed by the optimal step size approach presented at the end of Chapter 6. As shown there for the kurtosis contrast, the CM and CP criteria also admit algebraic solutions for the step size *globally* optimizing the contrast function along the search direction at each iteration. As demonstrated throughout the chapter, the combination of these three strategies, namely semi-blind contrasts, algebraic initialization and optimal step-size iterative search, leads to equalizers with increased robustness, high convergence speed and modest complexity. For reasons of space, our attention is restrained to the basic *single-input single-output (SISO)* system characterized by a single user without space-time diversity. However, these results are readily extended to the multi-channel case, including *single-input multi-output (SIMO)* systems [95] and MIMO systems typical of multi-user environments [96–98], more directly related to the BSS problem.

We begin the exposition by briefly reviewing the basic concepts of training-based and blind equalization in section 15.2; some strategies for improving their limitations are then summarized in section 15.3. The signal model and notational conventions that will be employed throughout the rest of the chapter are presented in section 15.4. Semi-blind equalization criteria are put forward in section 15.5. Algebraic solutions to the corresponding contrast optimization problems are then studied in section 15.6, whereas iterative methods based on algebraic optimal step-size optimization (introduced in the kurtosis contrast in the fully blind case in Chapter 6) are the topic of section 15.7. A thorough experimental study aiming to illustrate the performance of the presented techniques is reported in section 15.8. Finally, section 15.9 comments on the related problem of semi-blind channel identification. For the sake of clarity, the reader is referred to references [92–95] for details, proofs and other mathematical derivations.

15.2 TRAINING-BASED AND BLIND EQUALIZATION

15.2.1 Training-based or supervised equalization

Traditional equalization techniques are based on a sequence of symbols known by the receiver, the so-called pilot or training sequence, incorporated into the transmitted signal frame. The supervised equalizer is simply obtained by *optimal Wiener filtering* of the received signal using the pilot sequence as desired output. In this context, Wiener filters are also known as *minimum mean square error (MMSE)* equalizers and, in practical settings involving finite data, are usually obtained as the solution to the associated *least squares (LS)* problem. Due to their simplicity and robustness, supervised methods are employed by most of the current wireless communication systems; for instance, the second-generation GSM standard dedicates 20% of each burst to training [73]. The periodic transmission of pilot sequences reduces the useful data rate, and proves particularly ineffective in *broadcast*, *multicast* or non-cooperative environments, where synchronization is difficult [80,84]. Bandwidth utilization can be improved by reducing the pilot sequence length, but the sequence may become too short relative to the channel delay spread and additional information is thus necessary to render the supervised approach more robust in these conditions. The use of additional information in this context can sometimes be considered as a regularization [43].

15.2.2 Blind equalization

In the late 1970s, the drawbacks of training-based methods spurred a rising interest in blind techniques. The papers by Sato [65], Godard [31] and Treichler and Agee [78] are the pioneering contributions to the blind approach. By sparing the pilot sequence, blind equalization techniques increase the effective transmission rate and alleviate the need for synchronization.

In the fundamental SISO case, *non-minimum phase* systems cannot be identified using circular second-order statistics (SOS) only [44]. It is thus necessary to resort, explicitly or otherwise, to *higher-order statistics (HOS)*, or to non-circular SOS [36]. Essentially, most blind methods aim to restore at the equalizer output a known property of the input signal, such as a modulation alphabet with a finite number of symbols or with constant modulus. Under certain conditions, the use of these properties through the minimization of an appropriate cost function, or the maximization of a contrast function, guarantees the extraction of the signal of interest.

Despite their improved bandwidth utilization and versatility, blind methods present a number of important shortcomings:

- indeterminacy of the amplitude and/or the phase of the equalized signal;
- multi-modality, that is, existence of local extrema in the cost function to be optimized;
- increased computational complexity;
- larger data volume (block size) than supervised techniques required for the same equalization quality;
- in certain situations, slow convergence or tracking of the variations of the system parameters.

The last two drawbacks are mainly due to HOS estimation errors, which are typically more important than those of SOS for the same sample size.

15.2.3 A classical blind criterion: the constant modulus

The CM criterion [78] – a particular member of the more general family of Godard methods [31] – is perhaps the most widespread blind equalization principle, probably due to its simplicity and flexibility. Indeed, the CM criterion is easy to implement in an iterative fashion and can also deal with non-CM modulations at the expense of an increased estimation error due to constellation mismatch. The CP criterion, also studied in this chapter, can be considered as a modification of Godard's family, with a power parameter adapted to the number of symbols in the constellation; hence the name "constant power". Although Godard methods are globally convergent in the combined channel-equalizer space, they present suboptimal equilibrium points in the equalizer space [21,22]. Such points correspond to stable local extrema associated with filters unable to open sufficiently the eye diagram at the equalizer output, so that the detection device cannot extract the transmitted symbols with a reasonably low probability of error. Suboptimal equilibria are often called *spurious solutions* in the literature. Yet these solutions are typically very close to Wiener equalizers, which may question the term "spurious". In any case, the presence of suboptimal attractors renders the performance of gradient-type iterative algorithms based on Godard criteria very dependent on the initial value of the equalizer impulse response.

15.3 OVERCOMING THE LIMITATIONS OF BLIND METHODS

As discussed in [21,22], among other works, the convergence problems of iterative blind SISO equalizers require *ad hoc* strategies for suitable filter tap *initialization*, and even for maintaining the tap trajectories far from spurious attractor basins. Three such strategies are algebraic solutions, multi-channel systems and semi-blind approaches.

15.3.1 Algebraic solutions

Algebraic methods (sometimes called analytic) provide an equalization solution in a finite number of operations, and can always be employed as judicious initializations to iterative equalizers. An algebraic CM solution is obtained in [25], where the CM criterion is formulated as a nonlinear *least squares (LS)* problem. Through an appropriate transformation of the equalizer parameter space, the nonlinear system becomes a linear LS problem subject to certain constraints on the solution structure. Recovering of the correct structure is particularly important when multiple *zero forcing (ZF)* solutions exist; for instance, in all-pole channels with over-parameterized *finite impulse response (FIR)* equalizers, several ZF equalization delays are possible. From a matrix algebra perspective, enforcing this structure can be considered as a matrix diagonalization problem, where the transformation matrix is composed of the equalizer vectors. Once a non-structured solution has been obtained via pseudo-inversion, the minimum-length equalizer can be extracted by a subspace-based approach or other simple procedures for structure restoration.

The blind equalization method of [25] has strong connections with the *analytical CM algorithm (ACMA)* of [82] for BSS. ACMA yields, in the noiseless case, exact algebraic solutions for the spatial filters extracting the sources from observed instantaneous linear mixtures. It is interesting to note that the recovery of separating spatial filters from a basis of the solution space is equivalent to the joint diagonalization of the corresponding matrices. This joint diagonalization can be performed by the generalized *Schur decomposition* [32] of several (more than two) matrices, for which a convergence proof has not yet been found. Either for source separation or channel equalization, ACMA requires special modifications to treat signals with one-dimensional alphabets (e.g. binary) [25,81, 82]. Such modifications give rise to the *real ACMA (RACMA)* method [81].

Other solutions aiming at estimating algebraically the best SISO equalizer, or to identify the SISO channel, when the input belongs to a known alphabet, have been proposed in [1,18,19,28,33,36,42,46,48,76,83,86,90]. The *discrete alphabet* hypothesis is then crucial, and replaces the assumption of statistical independence between symbols [14], which is no longer necessary. The alphabet-based CP criterion also admits algebraic solutions, which, as reviewed in section 15.6, can be considered as a generalization of the algebraic CM solutions. Algebraic CP solutions are linked to challenging tensor decomposition problems. For a q-symbol constellation, the minimum-length equalizer can be determined from the joint decomposition of qth-order tensors, which, in turn, is linked to the rank-1 linear combination problem in the tensor case. To surmount the lack of effective tools for performing this task, approximate solutions can be proposed in the form of a subspace method exploiting the particular structure of the tensors associated with satisfactory equalization solutions. As opposed to [25], the subspace method proposed here takes into account a complete basis of the solution space. The use of this additional information allows one to increase the robustness of the algorithm with respect to the structure of the minimum-length equalizer. Moreover, the proposed blind algebraic solution deals naturally with binary inputs (BPSK, MSK) without any modifications.

15.3.2 Multi-channel systems

Multi-channel implementations, enabled by time oversampling or the use of multiple sensors, can avoid some of the deficiencies of blind SISO equalizers. Indeed, SIMO channels can be identified blindly by using SOS, regardless of their phase (minimal or non-minimal). Moreover, FIR SIMO channels can be perfectly equalized in the absence of noise by FIR filters [55,71,77]. However, the channel must verify strict diversity conditions, and a good number of these methods do not work when the channel length is overestimated [16]. In any case, the indeterminacy problems remarked in section 15.2.2 remain in the blind context.

Similarly, Godard SIMO equalizers do not present suboptimal minima for noiseless channels satisfying certain length and zero conditions [49]. All minima are indeed global, and coincide with MMSE solutions associated with achievable equalization delays. This feature results in iterative blind equalizers with improved performance. In the presence of noise, however, some of these minima become local and the respective equalizers provide different MSE performance [40]. Depending on its performance, a local minimum can also lead to a suboptimal solution. Consequently, the need for strategies to avoid local extrema remains pertinent in the multi-channel context. In some practical scenarios, it is not possible to attain the degree of spatio-temporal diversity required for a SIMO formulation, due to an insufficient *excess bandwidth* or hardware constraints limiting the number of receiving sensors (consider, for instance, the reduced spatial diversity available in a mobile phone). These are among the reasons for which this chapter is mainly focused on the SISO scenario, even if principles extend beyond that case.

15.3.3 Semi-blind approach

The combination of a training-based and a blind criterion can avoid their respective drawbacks while preserving their advantages. Indeed, it has been shown that any channel (SISO or SIMO) is identifiable from a small number of known symbols. Thanks to the use of a blind criterion, the pilot sequence necessary to estimate a channel of given length can become shorter relative to the training-only solution; spectral efficiency can thus be increased for a fixed estimation quality. As a result, semi-blind techniques often outperform supervised and blind techniques. When they fail, their semi-blind association can succeed [16]. On the one hand, the semi-blind approach can be interpreted as the regularization of the conventional supervised approach, avoiding its performance degradation for insufficient pilot length. On the other hand, the incorporation of a few pilot symbols "smoothens" the cost function, suppressing local minima, accelerating convergence and eliminating the indeterminacies of fully blind criteria. These features will be illustrated throughout the chapter.

The performance and robustness of the semi-blind approach justify the interest in this kind of technique. The fact that many of the current as well as future communication systems include pilot sequences in their definition standards (in particular to assist synchronization) provides a strong additional motivation for semi-blind equalization techniques. However, their use in currently available commercial products is rather limited.

In the context of algebraic methods, it has been recalled above that ACMA requires a joint diagonalization stage (a costly QZ iteration) in the general case where multiple solutions exist [82], although its complexity can be alleviated if the different solutions are delayed versions of each other [25]. The *semi-blind ACMA (SB-ACMA)* proposed in [74] avoids the costly joint diagonalization step of its blind version by constraining the spatial filter or *beamformer* to lie in a certain subspace associated with the pilot symbol vector. Nevertheless, the uniqueness of this semi-blind solution as well as its performance in the presence of noise remains to be ascertained in more detail.

15.4 MATHEMATICAL FORMULATION

15.4.1 Signal model

A digital signal $s(t) = \sum_n s_n \delta(t - nT)$ is transmitted at a known symbol rate $1/T$ through a time dispersive channel with impulse response $h(t)$. The channel is linear and time invariant (at least over the observation window), and has a stable causal possibly non-minimum phase transfer function. The baseband signal at the receiver output is given by $x(t) = r(t) + v(t)$, where $r(t) = h(t) \star s(t)$ denotes the noiseless observation and $v(t)$ an additive noise independent of $s(t)$. Assuming perfect synchronization and carrier residual elimination, symbol-rate sampling produces the discrete-time output:

$$x_n = r_n + v_n = \sum_k h_k s_{n-k} + v_n \tag{15.1}$$

where $x_n = x(nT)$ while h_k, s_n and v_n can be defined similarly. Each observed sample consists of a noisy linear mixture of time-delay versions of the original data, a phenomenon known as ISI. The goal of channel equalization or deconvolution is to recover the original data from the signal corrupted by convolutive channel effects (ISI) and noise. To this end, we seek a baud-rate FIR discrete-time equalizer with coefficients $\mathbf{f} = [f_1, \ldots, f_L]^{\mathrm{T}} \in \mathbb{C}^L$. The equalizer vector is sought so that the equalizer output

$$y_n = \mathbf{f}^{\mathrm{H}} \mathbf{x}_n$$

is an accurate estimate of the source symbols s_n, where:

$$\mathbf{x}_n = [x_n, x_{n-1}, \ldots, x_{n-L+1}]^{\mathrm{T}}.$$

Signal blocks composed of N_d symbol periods are observed at the channel output. These samples can be stored in a Toeplitz matrix:

$$\mathbf{X} = [\mathbf{x}_{L-1}, \mathbf{x}_L, \ldots, \mathbf{x}_{N_d-1}] \tag{15.2}$$

with dimensions $L \times N$, where $N = (N_d - L + 1)$. A similar signal model holds if the channel output is sampled at an integer multiple of the symbol rate (fractional sampling), if there exist multiple spatially separated sensors at reception (spatial oversampling), or if several signal sources transmit simultaneously, giving rise to additional CCI (multi-input system).

To enable the semi-blind mode of operation, we further assume that the transmitted block includes a pilot or training sequence composed of N_t symbols, denoted by $\check{\mathbf{s}} = [\check{s}_0, \check{s}_1, \ldots, \check{s}_{N_t-1}]^{\mathrm{H}}$. For the sake of simplicity, the training symbols are assumed to appear, without loss of generality, at the beginning of each block. It has been proven that, as far as channel estimation is concerned, the location of the pilot sequence at the beginning of the block is generally suboptimal [16]; this result probably applies as well to the direct equalization problem under study here. Nevertheless, the following results can be easily extended to an arbitrary location of the pilot sequence, including the optimal placement analyzed in [16].

15.4.2 Notations

Scalars, vectors and tensors (of which matrices are considered as particular cases) are denoted by lowercase (a), bold lowercase ($\mathbf{a}$) and bold uppercase ($\mathbf{A}$) symbols, respectively, except structures derived from Kronecker tensorial products, as detailed below. As will be explained in section 15.6, tensor structures will be employed in the derivation of algebraic solutions to the CP contrast. $\mathbf{I}_n$ refers to the identity matrix with dimensions $(n \times n)$, whereas $\mathtt{0}_n$ represents the vector with n zeroes; $\|\cdot\|$ is the conventional L^2 norm. $(\mathbf{A})_{i_1 i_2 \ldots i_q}$ stands for the $(i_1, i_2, \ldots, i_q)$-element of qth-order tensor $\mathbf{A}$. $\mathbb{R}\mathrm{e}(\cdot)$ and $\mathbb{I}\mathrm{m}(\cdot)$ represent the real and imaginary parts, respectively, of their complex argument; $j = \sqrt{-1}$ is the imaginary unit. Symbols $\otimes$, $\odot$ and $\circledcirc$ denote, respectively, the Kronecker product, the element-wise product and the outer product. Given a vector $\mathbf{a} \in \mathbb{C}^L$, we define its qth-order tensor product as the following rank-1 tensor: $\mathbf{a}^{\circledcirc q} = \underbrace{\mathbf{a} \circledcirc \cdots \circledcirc \mathbf{a}}_{q}$. For instance, matrix $\mathbf{a}^{\circledcirc 2} = \mathbf{a} \circledcirc \mathbf{a}$ can also be written as $\mathbf{a}\mathbf{a}^{\mathrm{T}}$. Note that on an appropriate basis, tensor $\mathbf{a}^{\circledcirc q}$ will have vector $\mathbf{a} \otimes \mathbf{a} \otimes \cdots \otimes \mathbf{a}$ as coordinates[1]; but this representation does not take into account the reduced dimension of the associated space due to symmetries. Indeed, a symmetric tensor of order q and dimension L can be stored in a vector $\mathbf{vecs}\{\mathbf{A}\}$ that contains only the $L_q = \binom{L+q-1}{q}$ different components of $\mathbf{A}$. Moreover, they can be normalized by the number of times they appear, so as to preserve the Frobenius norm [13]. In particular, we will write $\mathbf{a}^{\oslash q} = \mathbf{vecs}\{\mathbf{a}^{\circledcirc q}\}$. Similarly,

[1] Recall that $\otimes$ denotes the Kronecker product, whereas $\circledcirc$ denotes the tensor (outer) product.

unvecs$_q\{\mathbf{b}\}$ denotes the symmetric qth-order tensor made up of the elements of vector $\mathbf{b}$ with dimension L_q.

15.5 CHANNEL EQUALIZATION CRITERIA

15.5.1 Supervised, blind and semi-blind criteria

The supervised MMSE criterion aims at the minimization of the cost function:

$$\Upsilon_{\text{MMSE}}(\mathbf{f}) = \mathbb{E}\{|y_n - \check{s}_{n-\tau}|^2\}. \tag{15.3}$$

Symbol τ represents the equalization delay, on which performance strongly depends, and $\check{s}$ denotes the pilot sequence, as introduced before.

Concerning blind approaches, the CM criterion is defined by:

$$\Upsilon_{\text{CM}}(\mathbf{f}) = \mathbb{E}\{(|y_n|^2 - \gamma)^2\} \tag{15.4}$$

where $\gamma = \mathbb{E}\{|s_n|^4\}/\mathbb{E}\{|s_n|^2\}$ is an alphabet-dependent constant.

Another widely used criterion is the standardized cumulant [44], whose most elegant introduction is due to Donoho [26]. At fourth order, the standardized cumulant is called *kurtosis* (see also Chapters 3 and 6), and is given by

$$\Upsilon_{\text{KM}}(\mathbf{f}) = \frac{\text{cum}_4\{y_n\}}{\text{cum}_2^2\{y_n\}} \tag{15.5}$$

where $\text{cum}_2\{y_n\}$ is the variance of y_n and $\text{cum}_4\{y_n\} = \text{cum}\{y_n, y_n^*, y_n, y_n^*\}$ can be defined as:

$$\text{cum}_4\{y_n\} = \mathbb{E}\{|y_n|^4\} - 2\,\mathbb{E}\{|y_n|^2\}^2 - |\mathbb{E}\{y_n^2\}|^2.$$

When the source sequence s_n is white, the channel performs a mixture of independent random variables, so that the observation x_n is more Gaussian than s_n. The maximization of this criterion, or its square modulus when its sign is unknown, renders the equalizer output as non-Gaussian as possible. This *kurtosis maximization (KM) criterion* has widely been used for SISO and MIMO channel equalization, as well as for source separation in instantaneous linear mixtures (static MIMO channels). More generally, it is shown by Proposition 3.11, page 84, with slightly different notations that if the rth-order source cumulant $\text{cum}_r\{s_n\}$ is not null, one can maximize the normalized rth-order cumulant of the equalizer output, $\text{cum}_r\{y_n\}/\text{cum}_2^{r/2}\{y_n\}$.

For the application of the blind *constant power (CP)* approach, the transmitted symbols are assumed to belong to a q-PSK digital modulation, represented by the finite alphabet $\mathscr{A}_q = \{a^k\}_{k=0}^{q-1}$, where $a^q = d$ depends on the constellation; for instance, $(q, d) = (2, 1)$ for BPSK and $(q, d) = (4, 1)$ for QPSK sources. In addition, allowing a time-varying d, the above definitions are directly extended to modulations other than PSK, such as

MSK [33], which can be described by $(q, d_n) = (2, (-1)^n)$. As $s_n \in \mathcal{A}_q$, it follows that $s_n^q = d_n$. Consequently, a rather natural way to measure the proximity of the equalizer output to the original symbols is through the criterion:

$$\Upsilon_{\mathrm{CP}}(\mathbf{f}) = \mathbb{E}\{|y_n^q - d_n|^2\}. \tag{15.6}$$

This function is a particular case of the more general class of *alphabet polynomial fitting (APF)* criteria, where the equalizer output constellation is matched to that of the source, characterized by the complex roots of a specific polynomial [14,62]. In the context of BSS, the criterion is equivalent, for a sufficiently low noise level, to the *maximum a posteriori (MAP)* principle [10,34]. Moreover, it has been proved in [14] that, when the global channel-equalizer impulse response is of finite length and the input signal is sufficiently exciting, the global minima of the sample average of (15.6) in the combined channel-equalizer space correspond to ZF solutions. Nevertheless, this result does not guarantee that the desired solutions can always be attained. Indeed, spurious extrema can appear when the cost function is observed from the equalizer parameter space, due to the finite equalizer length, as remarked in [21,22] for Godard criteria. The existence of suboptimal extrema in the CP criterion will be illustrated by some simple experiments in section 15.8.

The linear combination of the above cost functions provides in a quite natural fashion the *semi-blind CM-MMSE (SB-CM-MMSE) and CP-MMSE (SB-CP-MMSE) criteria*:

$$\Upsilon_{\mathrm{SB-CM}}(\mathbf{f}) = \lambda\Upsilon_{\mathrm{MMSE}}(\mathbf{f}) + (1-\lambda)\Upsilon_{\mathrm{CM}}(\mathbf{f}) \tag{15.7}$$

$$\Upsilon_{\mathrm{SB-CP}}(\mathbf{f}) = \lambda\Upsilon_{\mathrm{MMSE}}(\mathbf{f}) + (1-\lambda)\Upsilon_{\mathrm{CP}}(\mathbf{f}). \tag{15.8}$$

Parameter λ is a real-valued constant in the interval [0, 1]. It can be considered as the relative degree of confidence in the blind and pilot-based parts of the criterion.

Note that, in practice, mathematical expectations are replaced by sample averages over the data available in the observed signal block.

15.5.2 Relationships between equalization criteria

The CP criterion (15.6) bears close resemblance to the Godard class [31], which in the PSK case becomes:

$$\Upsilon_{\mathrm{G}}^{(q,2)}(\mathbf{f}) = \mathbb{E}\{(|y_n|^q - |s_n|^q)^2\} = \mathbb{E}\{(|y_n|^q - |d_n|)^2\}. \tag{15.9}$$

For $q = 2$, this function corresponds to the CM criterion [31,78]. For BPSK sources and real-valued channels and equalizers, the CP and CM criteria are identical; in this case, we anticipate that the algebraic treatment of CP minimization (section 15.6) is also equivalent to that of ACMA for binary modulations (RACMA) [25,81]. The parallelism between the CM and CP cost functions suggests the existence of local extrema for the latter, even in the case $q > 2$.

The phase insensitivity of the CM criterion is one of its main interests, as it allows the independent operation of the equalization and carrier recovery stages [31,78]. However, for the same reason the carrier residual cannot be detected or identified by using this criterion. By contrast, the CP criterion can incorporate an appropriate carrier residual compensation mechanism into the algorithm or, otherwise, it requires the previous suppression of the residual before applying the algorithm. On the other hand, all PSK constellations being CM, the CM criterion does not make the difference between PSK modulations; similarly, the more general criterion (15.9) cannot privilege a particular PSK modulation. In contrast, criterion (15.6) explicitly takes into account the discrete nature of PSK alphabets, so that it may exhibit better discriminating properties among CM constellations.

If d_n is replaced by the available pilot symbols $\check{s}_n$, the CP cost function (15.6) reduces, with $q = 1$, to the MMSE supervised equalization principle (15.3). This fact will be exploited when designing semi-blind iterative methods in section 15.7.

We have seen that the KM criterion (15.5) maximizes non-Gaussianity at the equalizer output. This is also what the CM criterion does by forcing the equalizer output to approach the unit circle. To realize this connection, it is interesting to compare the CM and KM criteria. First, note that the KM criterion, as opposed to the CM, is insensitive to scale. Hence, we can write $y_n = \rho \bar{y}_n$, where ρ is a positive scale factor, and also $\mathbf{f} = \rho \bar{\mathbf{f}}$, where $\bar{\mathbf{f}}$ has fixed norm. This normalization is equivalent to fixing the variance of $\bar{y}_n$. Let us minimize $\Upsilon_{\mathrm{CM}}(\rho \bar{\mathbf{f}})$ with respect to ρ. We obtain $\rho_0^2 = \gamma^2 \mu_2 / \mu_4$, denoting $\mu_r = \mathbb{E}\{\bar{y}_n^r\}$. For this optimal value of ρ, we have:

$$\frac{1}{\gamma^2} \Upsilon_{\mathrm{CM}}(\rho_0 \bar{\mathbf{f}}) = 1 - \frac{\mu_2^2}{\mu_4}.$$

The CM criterion can then be linked to the KM criterion, provided that the source has a distribution with second-order circular symmetry, that is, $\mathbb{E}\{s_n^2\} = 0$. Under this assumption, $\Upsilon_{\mathrm{KM}}(\bar{\mathbf{f}}) + 2 = \mu_4/\mu_2^2$ and we have the following simple relationship:

$$\frac{1}{\gamma^2} \Upsilon_{\mathrm{CM}}(\rho_0 \bar{\mathbf{f}}) = 1 - \frac{1}{\Upsilon_{\mathrm{KM}}(\bar{\mathbf{f}}) + 2} \tag{15.10}$$

which shows that both criteria have the same stationary points in $\bar{\mathbf{f}}$. Hence, this equivalence, already established in [23, Chapter 4] [59–61], applies to most complex alphabets. However, the equivalence does not hold any more when the channel is complex-valued and the source real-valued (e.g. PAM modulations); indeed, in such a case the observation is no longer second-order circular.

Semi-blind CM-MMSE criterion (15.7) was initially proposed in [43], but using the so-called "CMA 1-2" cost instead of the "CMA 2-2" cost (15.4). The originality was to surmount the deficiencies of the LS solution to (15.3) (see the next section) when the pilot sequence is not long enough, an improvement known by its regularization capabilities. In addition, it has been proven that the incorporation of pilot symbols is capable of reducing the probability of converging towards spurious solutions due to the non-convexity of the CM cost function. The techniques presented in the following sections further reduce the impact of local extrema on equalization performance while accelerating convergence. As far as iterative techniques are concerned (section 15.7), we propose to minimize the hybrid criteria CM-MMSE (15.7) and CP-MMSE (15.8) through an efficient gradient algorithm where the step size is determined algebraically at each iteration by computing exhaustively all roots of a low degree polynomial. As demonstrated in the numerical experiments of section 15.8, this optimal step size accelerates convergence and makes performance closely approach the MMSE bound, even from a reduced number of pilot symbols. In addition, these optimal step-size iterative techniques can be judiciously initialized with the aid of the algebraic solutions presented next.

15.6 ALGEBRAIC EQUALIZERS

Perfect ZF equalization of a SISO channel is possible when both of the following conditions hold:

C1. The channel admits a noiseless Mth-order auto-regressive (AR) model.
C2. The FIR equalizer length is sufficient, $L \geqslant L_0$, with $L_0 = (M+1)$.

Indeed, a channel satisfying C1 can be equalized by an FIR filter $\mathbf{f}_0$ with minimum length L_0. If the equalizer filter is over-parameterized, i.e. its length verifies $L > L_0$, there exist $P = (L - L_0 + 1)$ exact ZF solutions, each one corresponding to a different equalization delay:

$$\mathbf{f}_p = [\mathbf{0}_{p-1}^{\mathrm{T}}, \mathbf{f}_0^{\mathrm{T}}, \mathbf{0}_{P-p}^{\mathrm{T}}]^{\mathrm{T}}, \qquad 1 \leqslant p \leqslant P. \tag{15.11}$$

As will be seen in the following, under these conditions the MMSE, CM and CP criteria can be perfectly minimized (even canceled if the sources verify the conditions of each criterion), and the global minimum can be computed algebraically, that is, without iterative optimization. The algebraic solution to the CP criterion (section 15.6.2) can be considered as an extension of the ACMA algorithm [82] to the CP principle; consequently, it can be referred to as *algebraic constant power algorithm (ACPA)*. The algebraic solutions to the supervised and blind criteria are later combined (section 15.6.3), giving rise to algebraic semi-blind equalizers. In practice, even if conditions C1–C2 are not satisfied, algebraic solutions can be used as judicious initializations for iterative equalizers (section 15.7).

15.6.1 Algebraic MMSE equalizer

It is well known that the MMSE criterion (15.3) is minimized by the Wiener-Hopf solution:

$$\mathbf{f}_{\tau\mathrm{MMSE}} = \mathbf{R}_x^{-1}\mathbf{p}_\tau, \quad \text{with } \mathbf{R}_x = \mathbb{E}\{\mathbf{x}_n\mathbf{x}_n^{\mathrm{H}}\} \text{ and } \mathbf{p}_\tau = \mathbb{E}\{\mathbf{x}_n\check{s}_{n-\tau}^*\}.$$

Assuming that the source signal is normalized (i.e. it has zero mean and unit variance), the mean square error (MSE) of the MMSE solution with delay τ is given by:

$$\text{MSE}_\tau = 1 - \mathbf{p}_\tau^{\text{H}} \mathbf{R}_x^{-1} \mathbf{p}_\tau.$$

Let the observed block associated with the pilot symbols at delay τ be denoted by:

$$\check{\mathbf{X}}_\tau = [\mathbf{x}_\tau, \mathbf{x}_{\tau+1}, \ldots \mathbf{x}_{\tau+N_t-1}]$$

with $N_t \geqslant L$. Canceling the criterion (15.3) is tantamount to solving the linear system:

$$\check{\mathbf{X}}_\tau^{\text{H}} \mathbf{f} = \check{\mathbf{s}}. \qquad (15.12)$$

However, such a system does not generally have an exact solution, as it consists of more equations than unknowns. Its LS solution is given by:

$$\mathbf{f}_{\tau \text{LS}} = (\check{\mathbf{X}}_\tau \check{\mathbf{X}}_\tau^{\text{H}})^{-1} \check{\mathbf{X}}_\tau \check{\mathbf{s}}, \qquad (15.13)$$

which we consider here as the algebraic solution to the MMSE criterion (15.3). This solution exists and is unique as long as matrix $\check{\mathbf{X}}_\tau$ is full rank, which is the case in the presence of noise. In the noiseless case, the whole observation matrix $\mathbf{X}$ given in (15.2) has rank L_0, so that it exists an infinite number of solutions to system (15.12) as soon as $L > L_0$. Under conditions C1–C2, the minimum-norm solution is given by $\mathbf{f}_{\tau \text{LS}} = \check{\mathbf{X}}_\tau^{\dagger} \check{\mathbf{s}}$, where $(\cdot)^\dagger$ denotes the Moore-Penrose pseudo-inverse. This solution corresponds to one of the exact ZF equalizers (15.11), which are identical up to a delay. In the presence of noise, the impact of delay on equalization performance may become important. The optimal delay in the MMSE sense, τ_{opt}, can be determined by comparing the MSE of the different equalization delays according to the MSE expression (see top of this page):

$$\tau_{\text{opt}} = \arg\min_\tau \text{MSE}_\tau = \arg\max_\tau \mathbf{p}_\tau^{\text{H}} \mathbf{R}_x^{-1} \mathbf{p}_\tau.$$

15.6.2 Algebraic blind equalizers

The algebraic solution to the CM criterion has been developed at length in [25,81,82]. Hence, we only describe in this section the solution to the CP criterion that we naturally refer to as ACPA. We will see that the search for such solutions can be associated with interesting tensor decomposition problems.

15.6.2.1 Determining a basis of the solution space

The exact minimizers of (15.6) are the solutions to the system of equations:

$$(\mathbf{f}^{\text{H}} \mathbf{x}_{n+L-1})^q = d_n, \quad n = 0, 1, \ldots, N-1. \qquad (15.14)$$

This nonlinear system can be linearized by taking into account that $(\mathbf{f}^{\mathrm{H}}\mathbf{x}_n)^q = \mathbf{f}^{\oslash q \mathrm{H}}\mathbf{x}_n^{\oslash q}$, and can be compactly expressed as:

$$\mathbf{X}^{q\mathrm{H}}\mathbf{w} = \mathbf{d} \tag{15.15}$$

where $\mathbf{X}^q = [\mathbf{x}_{L-1}^{\oslash q}, \mathbf{x}_L^{\oslash q}, \ldots, \mathbf{x}_{N_d-1}^{\oslash q}]$ and $\mathbf{d} = [d_0, d_1, \ldots, d_{N-1}]^{\mathrm{H}}$. Equation (15.15) must be solved under the following structural constraint: $\mathbf{w} \in \mathbb{C}^{L_q}$ must be of the form $\mathbf{w} = \mathbf{f}^{\oslash q}$, for certain vector $\mathbf{f} \in \mathbb{C}^L$.

Under conditions C1–C2, there must be P linearly independent solutions. Consequently, the dimension of the null space of $\mathbf{X}^{q\mathrm{H}}$, denoted $\ker(\mathbf{X}^{q\mathrm{H}})$, is $(P-1)$, and the solutions of (15.15) can be expressed as an affine space of the form $\mathbf{w} = \mathbf{w}_0 + \sum_{p=1}^{P-1}\alpha_p\mathbf{w}_p$, where $\mathbf{w}_0$ is a particular solution to the non-homogeneous system (15.15) and $\mathbf{w}_p \in \ker(\mathbf{X}^{q\mathrm{H}})$, for $1 \leqslant p \leqslant (P-1)$.

As in [82], we find it more convenient to work in a vector space, obtained through a unitary transformation $\mathbf{Q}$ with dimensions $(N \times N)$, such that $\mathbf{Qd} = [\sqrt{N}, \mathbf{0}_{N-1}^{\mathrm{T}}]^{\mathrm{T}}$. For instance, $\mathbf{Q}$ can be a Householder transformation [32] or, if $\mathbf{d}$ is composed of N equal values (as is the case for PSK sources), an N-point DFT matrix. Then, denoting:

$$\mathbf{QX}^{q\mathrm{H}} = \begin{bmatrix} \mathbf{r}^{\mathrm{H}} \\ \mathbf{R} \end{bmatrix},$$

system (15.15) reduces to:

$$\begin{cases} \mathbf{r}^{\mathrm{H}}\mathbf{w} = \sqrt{N} \\ \mathbf{Rw} = \mathbf{0}_{N-1} \end{cases}$$

under the constraint $\mathbf{w} = \mathbf{f}^{\oslash q}$. Similarly to [82, Lemma 4], it is possible to prove that this problem is equivalent to the solution of:

$$\begin{cases} \mathbf{Rw} = \mathbf{0}_{N-1} \\ \mathbf{w} = \mathbf{f}^{\oslash q} \end{cases}$$

followed by a scaling factor to enforce:

$$\mathbf{c}^{\mathrm{H}}\mathbf{w} = 1, \quad \text{with } \mathbf{c} = \frac{1}{\|\mathbf{d}\|^2}\sum_{n=0}^{N-1} d_n \mathbf{x}_n^{\oslash q} \tag{15.16}$$

or, equivalently:

$$\frac{1}{\|\mathbf{d}\|^2}\sum_{n=0}^{N-1} d_n (\mathbf{f}^{\mathrm{H}}\mathbf{x}_n)^q = 1. \tag{15.17}$$

If $\dim\ker(\mathbf{X}^{q\mathrm{H}}) = (P-1)$ and

$$N_d \geqslant L_q + L_0 - 1 \tag{15.18}$$

(or $N > L_q - P$), then $\dim\ker(\mathbf{R}) = P$, where L_0 is defined in condition C2 above. Hence, all the solutions of $\mathbf{Rw} = \mathbf{0}$ are linearly spanned by a basis $\{\mathbf{w}_k\}_{k=1}^{P}$ of $\ker(\mathbf{R})$. Such a basis can be computed from the *singular value decomposition (SVD)* of matrix $\mathbf{R}$ by taking its P least significant right singular vectors. The structured solutions $\{\mathbf{f}_p^{\oslash q}\}_{p=1}^{P}$ are also a basis of this subspace and, as a result, there exists a set of scalars $\{\alpha_{pk}\}_{p,k=1}^{P}$ such that:

$$\mathbf{f}_p^{\oslash q} = \sum_{k=1}^{P} \alpha_{pk}\mathbf{w}_k, \quad 1 \leqslant p \leqslant P \tag{15.19}$$

where matrix $(\mathbf{A})_{kp} = \alpha_{pk}$ is full rank. The problem of finding structured solutions to the linearized problem (15.15) is hence a particular *subspace fitting* problem with structural constraints. In terms of qth-order tensors, Eq. (15.19) can be rewritten as:

$$\mathbf{f}_p^{\otimes q} = \sum_{k=1}^{P} \alpha_{pk}\mathbf{W}_k, \qquad 1 \leqslant p \leqslant P \tag{15.20}$$

where $\mathbf{W}_k = \mathbf{unvecs}_q\{\mathbf{w}_k\}$. This is the tensorial *rank-1 linear combination problem*, which can be stated as follows: given a set of qth-order tensors $\{\mathbf{W}_k\}$, find the scalars $\{\alpha_{pk}\}$ in Eq. (15.20) yielding rank-1 tensors. The obtained rank-1 tensors correspond to $\{\mathbf{f}_p^{\otimes q}\}$. This tensor decomposition is generally a non-trivial task [13,15].

Before resuming our search for algebraic solutions to the CP contrast, it is interesting to remark that satisfactory algebraic equalization can be achieved in practice with observation windows shorter than the sample size bound (15.18), as illustrated by the numerical analysis of section 15.8.

15.6.2.2 Structuring the solutions: a subspace approach

A subspace method reminiscent of [55] can be used to recover the minimum-length equalizer $\mathbf{f}_0$ from a basis of (generally) unstructured solutions $\{\mathbf{w}_k\}_{k=1}^{P}$. The subspace fitting problem (15.19) can be written in compact form as $\mathbf{WA} = \mathbf{F}$, with $\mathbf{W} = [\mathbf{w}_1, \ldots, \mathbf{w}_P]$ and $\mathbf{F} = [\mathbf{f}_1^{\oslash q}, \ldots, \mathbf{f}_P^{\oslash q}]$. Since $\mathbf{A}$ is full rank, matrices $\mathbf{W}$ and $\mathbf{F}$ span the same column space, denoted by $\mathrm{range}(\mathbf{W}) = \mathrm{range}(\mathbf{F})$. In particular, $\forall \mathbf{u}_i \in \ker(\mathbf{W}^{\mathrm{H}})$, $\mathbf{u}_i^{\mathrm{H}}\mathbf{F} = \mathbf{0}_P^{\mathrm{T}}$. There are $\dim\ker(\mathbf{W}^{\mathrm{H}}) = (L_q - P)$ such linearly independent vectors.

Since equalization solutions have the form (15.11), the corresponding columns of $\mathbf{F}$ have a particular structure whereby the elements non associated with the minimum-length equalizer $\mathbf{f}_0$ are all zero. The remaining $L_{0q} = \binom{L_0+q-1}{q}$ elements form $\mathbf{f}_0^{\oslash q}$. Let σ_p describe the set of L_{0q} positions of $\mathbf{f}_0^{\oslash q}$ in $\mathbf{f}_p^{\oslash q}$, that is, $\sigma_p = \{j_1 + L(j_2-1) + \cdots + L^{q-1}(j_q-1)\}$, with $j_k \in [p, p+L_0-1]$, $k = 1, \ldots, q$, and $j_1 \geqslant j_2 \geqslant \ldots \geqslant j_q$. Similarly, $(\mathbf{u}_i)_{\sigma_p} \in \mathbb{C}^{L_{0q}}$ is the sub-vector composed of the elements of $\mathbf{u}_i$ in positions σ_p. Denote $\mathbf{U}_i = [(\mathbf{u}_i)_{\sigma_1}, \ldots, (\mathbf{u}_i)_{\sigma_P}] \in \mathbb{C}^{L_{0q} \times P}$. Hence:

$$\mathbf{u}_i^{\mathrm{H}}\mathbf{F} = \mathbf{0}_P^{\mathrm{T}} \quad \Leftrightarrow \quad \mathbf{U}_i^{\mathrm{H}}\mathbf{f}_0^{\oslash q} = \mathbf{0}_P.$$

The above equalities define a set of $P(L_q - P)$ linear equations, characterized by matrix $\mathbf{U} = [\mathbf{U}_1, \ldots, \mathbf{U}_{L_q-P}] \in \mathbb{C}^{L_{0q} \times P(L_q-P)}$, where the unknowns are the components of $\mathbf{f}_0^{\oslash q}$. As long as $L > L_0$, this linear system determines, up to a scale factor, the well-structured vector $\mathbf{f}_0^{\oslash q}$; its amplitude can then be set via (15.17) from a zero-padded version of the estimate of $\mathbf{f}_0$, yielding one of the solutions $\mathbf{f}_p$ in Eq. (15.11). In practice, we minimize the quadratic form $\|\mathbf{U}^{\mathrm{H}}\mathbf{f}_0^{\oslash q}\|^2 = \mathbf{f}_0^{\oslash q\mathrm{H}}\mathbf{U}\mathbf{U}^{\mathrm{H}}\mathbf{f}_0^{\oslash q}$, which leads to the estimation of $\mathbf{f}_0^{\oslash q}$ as the least significant left singular vector of matrix $\mathbf{U}$.

Once matrix $\mathbf{F}$ has been reconstructed, an LS estimate of coefficients $\{\alpha_{kp}\}$ can be obtained as $\hat{\mathbf{A}}_{\mathrm{LS}} = (\mathbf{W}^{\mathrm{H}}\mathbf{W})^{-1}\mathbf{W}^{\mathrm{H}}\mathbf{F} = \mathbf{W}^{\dagger}\mathbf{F}$. The elements of $\hat{\mathbf{A}}_{\mathrm{LS}}$ provide a solution to the rank-1 linear combination problem.

15.6.2.3 Recovering the equalizer vector from its symmetric tensor

In order to recover the equalizer impulse response $\mathbf{f}_0$ from its symmetric vectorization $\mathbf{f}_0^{\oslash q}$, it is possible to perform the SVD of a matrix unfolding of $\mathbf{f}_0^{\otimes q} = \mathbf{unvecs}_q\{\mathbf{f}_0^{\oslash q}\}$ [11, 35]. Denote by $\mathbf{F}_0 \in \mathbb{C}^{L_0 \times L_0^{q-1}}$ the matrix with elements $(\mathbf{F}_0)_{i_1, i_2+L_0(i_3-1)+\ldots L_0^{q-2}(i_q-1)} = (\mathbf{f}_0^{\otimes q})_{i_1 i_2 i_3 \ldots i_q}$. Then, $\mathbf{F}_0 = \mathbf{f}_0\bar{\mathbf{f}}_0^{\mathrm{T}}$, with $(\bar{\mathbf{f}}_0)_{i_2+L_0(i_3-1)+\ldots L_0^{q-2}(i_q-1)} = (\mathbf{f}_0)_{i_2}(\mathbf{f}_0)_{i_3}\ldots(\mathbf{f}_0)_{i_q}$. Hence, $\mathbf{f}_0$ can be estimated (up to a scale factor) as the dominant left singular vector of the matrix unfolding of $\mathbf{F}_0$. This matrix has rank one in the absence of noise.

In the presence of noise, it is generally no longer possible to express the estimated vector $\hat{\mathbf{f}}_0^{\oslash q}$ as the symmetric vectorization of a rank-1 tensor. In other words, no vector $\mathbf{f}_0$ exists such that $\hat{\mathbf{f}}_0^{\oslash q} = \mathbf{vecs}_q\{\mathbf{f}_0^{\otimes q}\}$ is verified exactly. Consequently, the matrix unfolding will not be of rank one, and the SVD-based procedure explained above will only yield an approximate solution. One is actually facing the problem of the *rank-1 approximation* to the symmetric tensor $\hat{\mathbf{f}}_0^{\otimes q}$. To date, only iterative solutions, e.g. inspired on the *iterative power method* [17,41], have been proposed to solve this problem. However, our experiments reveal that the solution previously described for the noiseless case is a good initialization.

15.6.2.4 Other structuring methods

In the context of the CM criterion, a subspace method similar to that of section 15.6.2.2 was proposed in [25, section III.C], operating on a single non-structured (LS) solution (see also [24]). This structure forcing procedure can be interpreted as the diagonalization of the matrix associated with the non-structured solution. By contrast, our approach takes advantage of a whole basis of the solution subspace, which should lead to an improved robustness, particularly for large values of P. The method of [35] and [25, section III.B] is based on the observation that the first L components of a solution $\mathbf{w}_k$ are equal to $\tilde{\alpha}_{k1} f_1^{q-1}[f_1, \sqrt{q}f_2, \ldots, \sqrt{q}f_{L_0-1}, \sqrt{q}f_{L_0}, \mathbf{0}_{P-1}^{\mathrm{T}}]^{\mathrm{T}}$, $\tilde{\alpha}_{kp} = (\mathbf{A}^{-1})_{pk}$, from which $\mathbf{f}_0$ can be extracted. This method is simple and ingenious, but inaccurate when coefficient $\tilde{\alpha}_{k1}$ or the first term f_1 of the equalizer are small relative to the noise level.

To surmount this limitation, we can note that the last components of $\mathbf{w}_k$ are equal to [25, section III.B]:

$$\tilde{\alpha}_{kP} f_{L_0}^{q-1} [\mathbf{0}_{P-1}^{\mathrm{T}}, \ldots, \sqrt{q} f_1, \sqrt{q} f_2, \ldots, \sqrt{q} f_{L_0-1}, f_{L_0}]^{\mathrm{T}}.$$

Appropriately combined with the estimation carried out from the first L components, this second option can provide an improved estimation of $\mathbf{f}_0$. In the simulation study of section 15.8, we employ the following heuristic (suboptimal) linear combination. Let us suppose that the filters estimated from the first and the last non-overlapping components of a non-structured solution are, respectively, $\hat{\mathbf{f}}_1 = \beta_1 \tilde{\mathbf{f}}_0$ and $\hat{\mathbf{f}}_2 = \beta_2 \tilde{\mathbf{f}}_0$, with $\tilde{\mathbf{f}}_0 = \mathbf{f}_0 / ||\mathbf{f}_0||$. Then, the unit-norm minimum-length equalizer LS estimate is given by $\hat{\hat{\mathbf{f}}}_0 = [\hat{\mathbf{f}}_1, \hat{\mathbf{f}}_2]\boldsymbol{\gamma}$, with $\boldsymbol{\gamma} = \boldsymbol{\beta}^* / ||\boldsymbol{\beta}||^2$, $\boldsymbol{\beta} = [\beta_1, \beta_2]^{\mathrm{T}}$. The coefficients of $\boldsymbol{\beta}$ can simply be estimated from the equation $\beta_i = ||\hat{\mathbf{f}}_i||$, $i = 1, 2$. This type of linear *maximal ratio combining* is reminiscent of the *RAKE* receiver and the matching filter [58]. Robustness can still be improved by exploiting a whole set $\{\mathbf{w}_k\}$ instead of a single solution, as explained above.

15.6.2.5 Approximate solution in the presence of noise

In the presence of additive noise at the sensor output, conditions C1–C2 are no longer satisfied, and an exact solution of (15.14) may not exist. An approximate solution in the LS sense can be obtained by minimizing $||\mathbf{X}^{q\mathrm{H}}\mathbf{w} - \mathbf{d}||^2$, under the structural constraint $\mathbf{w} = \mathbf{f}^{\oslash q}$. This minimization generally requires an iterative method, as detailed in section 15.7.

Nevertheless, the guidelines for determining an exact solution in the noiseless case can still provide a sensible initialization to an iterative equalizer in the noisy case. After applying the transformation $\mathbf{Q}$, the LS problem proves equivalent to the minimization of the quadratic form:

$$|\mathbf{c}^{\mathrm{H}}\mathbf{w} - 1|^2 + ||\mathbf{R}\mathbf{w}||^2. \tag{15.21}$$

To find a basis of the solution space, we seek a set of vectors minimizing $||\mathbf{R}\mathbf{w}||^2$ (for instance, the least significant P right singular vectors of $\mathbf{R}$), then structure them as in section 15.6.2.2 and finally normalize the solution to satisfy $\mathbf{c}^{\mathrm{H}}\mathbf{w} = 1$ [cf. Eqs (15.16)–(15.17)]. Although suboptimal, this solution will be tested in the experimental study of section 15.8.

A more accurate solution can be determined by realizing that expression (15.21) represents a non-negative quadratic form in vector $[\mathbf{w}^{\mathrm{T}}, 1]^{\mathrm{T}}$. Formulating the problem in the projective space, we can look for the least significant eigenvector $\mathbf{v}_{\mathrm{m}}$ of matrix:

$$\begin{bmatrix} \mathbf{R}\mathbf{R}^{\mathrm{H}} + \mathbf{c}\mathbf{c}^{\mathrm{H}} & -\mathbf{c} \\ -\mathbf{c}^{\mathrm{H}} & 1 \end{bmatrix}$$

and take as an approximate estimation of $\mathbf{w}$ the first $\dim(\mathbf{w})$ components of $\mathbf{v}_{\mathrm{m}}$ normalized by the first one.

15.6.3 Algebraic semi-blind equalizers

By extending the above ideas, we can also develop algebraic solutions to the semi-blind criterion CP-MMSE (15.8), the solutions to criterion CM-MMSE (15.7) being obtained in a totally analogous manner. To minimize algebraically the CP-MMSE criterion, we seek the simultaneous solution of systems (15.12) and (15.15):

$$\breve{\mathbf{X}}_{\tau}^{\mathrm{H}}\mathbf{f} = \breve{\mathbf{s}} \tag{15.22}$$

$$\mathbf{X}^{q\mathrm{H}}\mathbf{w} = \mathbf{d} \tag{15.23}$$

under the structural constraint $\mathbf{w} = \mathbf{f}^{\oslash q}$, where now

$$\mathbf{X}^{q} = [\mathbf{x}_{\tau+N_t}^{\oslash q}, \mathbf{x}_{\tau+N_t+1}^{\oslash q}, \ldots, \mathbf{x}_{N_d-1}^{\oslash q}]$$

and $\mathbf{d} = [d_{N_t}, d_{N_t+1}, \ldots, d_{N_d-\tau-1}]^{\mathrm{H}}$. Note that only the symbols not employed in the supervised part contribute now to the blind part of the criterion.

The case where conditions C1–C2 are verified is trivial, since both solutions of the composite system are exact and identical. Hence, let us first consider the case of a noisy AR channel with a sufficiently long equalizer. A suboptimal solution can be obtained by combining the solutions computed separately for the two sub-systems [11,35]. Let $\hat{\mathbf{f}}_{\mathrm{MMSE}}$ denote the solution of (15.22) and $\hat{\mathbf{f}}_{\mathrm{CP}}^{\oslash q}$ that of (15.23) associated with the same equalization delay τ; these solutions are computed as explained in sections 15.6.1 and 15.6.2, respectively. Let us unfold $\mathbf{unvecs}_q\{\hat{\mathbf{f}}_{\mathrm{CP}}^{\oslash q}\}$ into a matrix $\mathbf{F}_{\mathrm{CP}}$ with dimensions $(L \times L^{q-1})$, as described in section 15.6.2.3. Then, the joint solution to (15.22)–(15.23) can be approximated by the dominant left singular vector of matrix $\mathbf{F}_{\mathrm{SB}} = [\lambda\hat{\mathbf{f}}_{\mathrm{MMSE}}, (1-\lambda)\mathbf{F}_{\mathrm{CP}}]$. In the noiseless case, solutions $\hat{\mathbf{f}}_{\mathrm{MMSE}}$ and $\hat{\mathbf{f}}_{\mathrm{CP}}$ coincide with the dominant left singular vector of the rank-1 matrix $\mathbf{F}_{\mathrm{SB}}$; an iterative search is not necessary.

In the case of an FIR channel, no exact solution to system (15.22)–(15.23) exists, even in the absence of noise. However, the two sub-systems can be solved separately in the LS sense and the respective solutions can then be combined according to the above SVD-based procedure. We refer to this method as *semi-blind algebraic constant power algorithm (SB-ACPA)*.

The combined solution just described can initialize an iterative minimization algorithm aiming to refine this algebraic approximate solution.

15.7 ITERATIVE EQUALIZERS

15.7.1 Conventional gradient-descent algorithms

In practice, exact ZF equalization may not be feasible, due to noise or just to an insufficient equalizer length. In such cases, the cost function must be minimized iteratively, for instance, via a gradient-descent or a Newton algorithm. We describe here gradient-descent methods; these results can easily be extended to Newton implementations. If a good initialization has been obtained, only a few iterations will usually be necessary for convergence.

We define the complex gradient of a generic real-valued function $\Upsilon(\mathbf{f})$ with respect to complex variable $\mathbf{f}$ as:

$$\nabla\Upsilon(\mathbf{f}) = \nabla_{\mathbf{f}_r}\Upsilon(\mathbf{f}) + j\nabla_{\mathbf{f}_i}\Upsilon(\mathbf{f})$$

where $\mathbf{f}_r = \mathbb{R}\text{e}(\mathbf{f})$ and $\mathbf{f}_i = \mathbb{I}\text{m}(\mathbf{f})$ represent the real and imaginary parts, respectively, of vector $\mathbf{f}$. Up to an inconsequential scale factor, this definition corresponds to Brandwood's complex gradient [3]. Accordingly, the gradients of the CM (15.4) and CP (15.6) criteria can be expressed as:

$$\nabla\Upsilon_{\text{CM}}(\mathbf{f}) = 4\mathbb{E}\left\{(\mathbf{f}^{\text{H}}\mathbf{x}_n)^*[|\mathbf{f}^{\text{H}}\mathbf{x}_n|^2 - \gamma]\mathbf{x}_n\right\} \tag{15.24}$$

$$\nabla\Upsilon_{\text{CP}}(\mathbf{f}) = 2q\mathbb{E}\left\{(\mathbf{f}^{\text{H}}\mathbf{x}_n)^{q-1}[(\mathbf{f}^{\text{H}}\mathbf{x}_n)^q - d_n]^*\mathbf{x}_n\right\}. \tag{15.25}$$

From the relationships remarked in section 15.5.2, the gradient of MMSE criterion (15.3) can be computed from that of the CP criterion by setting $q = 1$ and replacing $\check{s}_n$ by d_n in expression (15.25). This yields:

$$\nabla\Upsilon_{\text{MMSE}}(\mathbf{f}) = 2\mathbb{E}\{[(\mathbf{f}^{\text{H}}\mathbf{x}_n) - \check{s}_n]^*\mathbf{x}_n\}. \tag{15.26}$$

The gradients of the semi-blind CM-MMSE and CP-MMSE criteria are simply obtained by linear combination of (15.24)–(15.26) according to (15.7) and (15.8). We refer to the resulting iterative methods as *constant modulus algorithm (CMA)* and *constant power algorithm (CPA)*; and their semi-blind versions as *semi-blind constant modulus algorithm (SB-CMA)* and *semi-blind constant power algorithm (SB-CPA)*.

As a judicious initialization in the blind case, we can employ the equalizer vector provided by an algebraic method, such as the direct LS (generally non-structured) solution of the linearized problem (15.15), $\hat{\mathbf{f}}_{\text{LS}} = (\mathbf{X}^{q\text{H}})^{\dagger}\mathbf{d}$, or the structured solution described in section 15.6.2. In the semi-blind case, the algebraic solution of section 15.6.3 becomes applicable as initialization. At each iteration, the equalizer vector can be adjusted by means of a gradient-based update:

$$\mathbf{f}^{+} = \mathbf{f} - \mu\nabla\Upsilon(\mathbf{f}). \tag{15.27}$$

Iterations are stopped when

$$\frac{||\mathbf{f}^{+} - \mathbf{f}||}{||\mathbf{f}||} < \eta/N \tag{15.28}$$

where η is a small positive constant.

We advocate the use of *block* or *batch* implementations [10], also known as fixed-window methods [61], rather than stochastic algorithms. The latter approximate the gradient by a single-sample estimate, which may be seen as dropping the expectation operator in the gradient expression. This simplification, which in the case of the CM criterion gives rise to the stochastic-gradient CMA, generally leads to slow convergence and poor final accuracy. Indeed, a single parameter, μ, must control at the same time the step size in the search trajectory and the implicit statistical average; this is a difficult balance. Stochastic algorithms found justification when the available computer power was rather limited. Nowadays, computer power is no longer the limiting factor of equalization performance, but the algorithms that are implemented, or the operating conditions (e.g. channel non-stationarity).

By contrast, batch methods estimate the gradient from a whole block of channel output samples, using the same data block at each iteration. This gradient estimate is more accurate and thus improves the convergence speed and equalization quality of the resulting algorithm [10,61]. Moreover, tracking capabilities are not necessarily sacrified, since good performance can be achieved from small data blocks; it suffices that the channel be stationary over the (short) observation window. Block methods are particularly suited to burst transmission systems (e.g. TDMA). The possibility of combining batch and stochastic operations in iterative optimization methods is discussed in section 6.4.3.

It is well known that gradient-based algorithms for blind equalization, despite their simplicity, present numerous drawbacks such as lack of robustness to local extrema, dependence on initialization and slow convergence [21,22,40]. These problems persist in block implementations, even though convergence is often accelerated. When the function to be optimized is convex in the unknowns, this problem can be alleviated with more elaborate approaches such as the conjugate gradient [57]. Nevertheless, the blind and semi-blind functions based on the CM and CP criteria (section 15.5) are not convex. This leads us to consider alternative optimization strategies without compromising the simplicity and numerical convenience of the implementation.

15.7.2 Algorithms based on algebraic optimal step size

15.7.2.1 Step-size polynomials

Exact *global* line search aims at finding the step size minimizing the cost function along the search direction:

$$\mu_{\text{opt}} = \arg\min_{\mu} \Upsilon(\mathbf{f} - \mu\mathbf{g}).$$

A possible search direction is simply the gradient, $\mathbf{g} = \nabla\Upsilon(\mathbf{f})$. These algorithms are generally unattractive due to their complexity, because the one-dimensional minimization must typically be carried out by costly numerical techniques. Another drawback is the orthogonality between successive gradient vectors (see section 15.7.2.3), which, depending on the initialization and the shape of the cost-function surface, can slow down convergence [57].

However, it has been observed in [14,34] that, for a number of equalization criteria including the CM, the CP and their semi-blind versions studied herein, functional $\Upsilon(\mathbf{f}-\mu\mathbf{g})$ is a rational function in the step size μ. This allows us to find μ_{opt} algebraically, so that it is possible to *globally* minimize the cost function in the descent direction while reducing complexity. Indeed, for the CM criterion (15.4), some algebraic manipulations show that the derivative of $\Upsilon_{\text{CM}}(\mathbf{f}-\mu\mathbf{g})$ with respect to μ is the following cubic:

$$p(\mu) = b_3\mu^3 + b_2\mu^2 + b_1\mu + b_0. \tag{15.29}$$

Its real coefficients are given by [93,95]:

$$\begin{aligned} b_3 &= 2\mathbb{E}\{a_n^2\}, & b_2 &= 3\mathbb{E}\{a_n b_n\} \\ b_1 &= \mathbb{E}\{2a_n c_n + b_n^2\}, & b_0 &= \mathbb{E}\{b_n c_n\} \end{aligned}$$

where $a_n = |g_n|^2$, $b_n = -2\mathbb{R}\mathrm{e}(y_n g_n^*)$, and $c_n = (|y_n|^2 - \gamma)$, with $g_n = \mathbf{g}^{\mathrm{H}}\mathbf{x}_n$. Similarly, for the CP criterion (15.6), the optimal step size μ_{opt} is found among the roots of the $(2q-1)$th-degree polynomial [92]:

$$p_{\text{CP}}(\mu) = \sum_{m=0}^{2q-1} \mathbb{R}\mathrm{e}(b_m)\mu^m \tag{15.30}$$

where

$$b_m = \begin{cases} \displaystyle\sum_{p=0}^{m} (m+1-p)\mathbb{E}\{a_{m+1-p}^* a_p\} - (m+1)\mathbb{E}\{a_{m+1}^* d_n\}, & 0 \leqslant m \leqslant q-1 \\ \displaystyle\sum_{p=m+1-q}^{q} (m+1-p)\mathbb{E}\{a_{m+1-p}^* a_p\}, & q \leqslant m \leqslant 2q-1 \end{cases}$$

with

$$a_p = (-1)^p \binom{q}{p} g_n^p y_n^{q-p}, \quad 0 \leqslant p \leqslant q.$$

The step-size polynomial of the MMSE criterion is easily determined by taking into account the link between the MMSE and CP criteria observed in section 15.5.2, which leads to:

$$p_{\text{MMSE}}(\mu) = b_1\mu + b_0 \tag{15.31}$$

$$b_1 = \mathbb{E}\{|g_n|^2\}, \qquad b_0 = -\mathbb{R}\mathrm{e}\left(\mathbb{E}\{g_n^*(y_n - \check{s}_n)\}\right). \tag{15.32}$$

The polynomials defining the optimal step size for the semi-blind CM-MMSE and CP-MMSE criteria are made up of polynomials (15.29), (15.30) and (15.31) according to the linear combinations of the respective cost functions (15.7)–(15.8). A similar polynomial (a quartic) is obtained for the kurtosis contrast, leading to the *RobustICA* algorithm described in section 6.11.2.

Once the coefficients have been determined, the roots of the optimal step-size polynomial can be obtained as explained in section 15.7.2.2 below. The optimal step size corresponds to the root attaining the minimal value of the cost function, thus leading to the *global* minimization of $\Upsilon(\cdot)$ in the descent direction. After determining μ_{opt}, the equalizer vector coefficients are updated as in (15.27), and the process is repeated with the new equalizer and gradient vectors, until convergence, which is tested with (15.28). We refer to this technique as *optimal step-size (OS) algorithm*, which gives rise, in particular, to the blind *OS-CMA* and *OS-CPA* algorithms and to the semi-blind *OS-SB-CMA* and *OS-SB-CPA* algorithms.

To improve numerical conditioning in the determination of μ_{opt}, it is useful to normalize the gradient vector $\mathbf{g}$ beforehand. Since the pertinent parameter is the search direction $\tilde{\mathbf{g}} = \mathbf{g}/\|\mathbf{g}\|$, this normalization does not cause any inconvenience. As a consequence, vector $\mathbf{g}$ is replaced by $\tilde{\mathbf{g}}$ when computing the optimal step-size polynomial coefficients, as well as in the update rule (15.27).

15.7.2.2 Root extraction

Standard procedures such as *Cardan's formula*, or often less costly iterative methods [27, 45], are available to extract the roots of cubics (15.29) and (15.30) with $q = 2$; an efficient Matlab™ implementation, valid for polynomials with real or complex coefficients, is given in [57] (see also [95]). For solving quartics, elementary algebra textbooks present the method developed by *Ferrari*, Cardan's student, in the 16th century. From the end of the 18th century, we have known that polynomials of degree higher than four cannot be solved by radicals; one thus needs to resort to iterative methods.

Concerning the roots of cubics (15.29) and (15.30) with $q = 2$, two options are possible: either all three roots are real-valued, or one is real and the two other form a complex-conjugate pair. In the first case, one just needs to verify which one provides the smallest value of $\Upsilon(\mathbf{f} - \mu\mathbf{g})$. In our computer experiments, when a complex-conjugate pair exists, it is the real root that typically minimizes the cost function. Even when the real root does not produce the minimal value of $\Upsilon(\cdot)$, it often provides lower MSE at the equalizer output than the complex roots. Real roots are thus preferred. This observation is also applicable to polynomials of higher degree - for instance, (15.30) with $q > 2$. Another possibility, employed in the RobustICA method based on the kurtosis contrast (section 6.11.2 of this book) is to consider only the real parts of the roots.

15.7.2.3 Convergence of optimal step-size algorithms

By construction of exact line search algorithms, gradient vectors of consecutive iterations are orthogonal, which, depending on initialization and the shape of the cost-function surface, can slow down convergence [57]. Gradient orthogonality is mathematically expressed as $\mathbb{R}\mathrm{e}(\mathbf{g}^{\mathrm{H}}\mathbf{g}^{+}) = 0$, with $\mathbf{g}^{+} = \nabla\Upsilon(\mathbf{f}^{+})$. This relationship can easily be derived by taking into account that

$$\frac{\partial \Upsilon(\mathbf{f} - \mu\mathbf{g})}{\partial \mu} = -\mathbb{R}\mathrm{e}\left(\mathbf{g}^{\mathrm{H}}\nabla\Upsilon(\mathbf{f} - \mu\mathbf{g})\right) = 0.$$

In our numerical experiments, the optimal step-size algorithms have always converged in fewer iterations than its fixed step-size counterparts [95]. Fast convergence and improved stability have also been reported in [87]. Moreover, the probability of converging to local extrema is decreased with the optimal step-size strategy, as shown empirically in [95] and section 15.8.

15.7.2.4 Variants

The coefficients of OS-CMA (15.29) and OS-CPA (15.30) cubics, the latter with $q = 2$, can also be determined from the sensor-output statistics, computed before starting the iterations [92,93,95]. This alternative requires the previous computation of the covariance matrix and the whole fourth-order cumulant tensor of the channel output. Indeed, the equalizer-output statistics can be deduced by multi-linearity; this way of computing the cumulants is called *deductive estimation* in [9]. Both alternatives are equivalent regarding equalization performance and convergence speed measured in terms of iterations. The only difference lies in their computational cost in terms of number of operations (section 15.7.2.5).

The algebraic optimal step-size technique can also be applied to other equalization criteria. For instance, the *kurtosis maximizaton (KM)*, likewise called Shalvi-Weinstein criterion [69] in the context of blind SISO equalization, can also be globally optimized along a given direction by rooting a fourth-degree polynomial in μ; all stationary points can be computed by Ferrari's formula for quartics. This naturally gives rise to the OS-KMA, developed in the context of BSS and referred to as RobustICA algorithm in [96] (see also section 6.11.2 of this book), with a complexity per iteration similar to OS-CMA's. On the other hand, the optimal step-size technique remains applicable if data are prewhitened, for instance through a QR decomposition of the observation matrix, as in the QR-CMA method of [61]. Prewhitening improves conditioning and can accelerate convergence under the hypothesis of iid inputs. Finally, by using the Hessian of the cost function, the optimal step-size technique can easily be combined with Newton optimization, as well as with any other method constructing successive search directions $\{\mathbf{g}_k\}$.

15.7.2.5 Computational complexity

The complexity of the optimal step-size technique is dominated by the computation of polynomial coefficients [Eqs (15.29), (15.30), etc.]. In practice, mathematical expectations are replaced by sample averages over the observed signal block. The cost of these averages for (15.29) is of order $O(LN)$ per iteration, for data blocks composed of N vectors $\mathbf{x}_n$. For the alternative procedure based on the previous computation of the second- and fourth-order moments of the sensor output (section 15.7.2.4), the cost per iteration is approximately of the order of $O(L^4)$, with an additional initial cost of $O(L^4N)$ operations. Depending on the number of iterations needed for convergence and the relative values of N and L, this initial burden can render the second method (that we refer to as OS-CMA-2) more costly than the first one (OS-CMA-1) [93,95]. Similar alternatives are possible for the OS-CPA and OS-KMA algorithms.

Table 15.1 sums up the computational cost of different optimal step-size techniques in terms of number of real-valued *floating point operations (flops)*; a flop represents a

Table 15.1 Computational cost in terms of number of flops for different iterative equalization algorithms in the case of real-valued signals and filters. L: number of equalizer filter coefficients; N: number of data vectors in the observed data burst

	Initialization	Per iteration
OS-CMA-1	–	$(3L+10)N$
OS-CMA-2	$\left[\binom{L+3}{4}+\binom{L+1}{2}\right]N$	$6L^4+3L^2+2L$
OS-CPA	–	$[3L+q(q+4)]N$
OS-KMA	–	$(5L+12)N$
SG-CMA	–	$2(L+1)$
CMA	–	$2(L+1)N$
QR-CMA [61]	L^2N	$(2L+3)N$
RLS-CMA [5]	–	$2L(2L+3)$
AAF-CMA [70]	–	$6L$

product or a division followed by an addition, and typically corresponds to a multiply-and-accumulate (MAC) cycle in a digital signal processor (DSP). Also considered are other representative equalization techniques, especially those based on the CM criterion: the stochastic CMA (SG-CMA), the QR-CMA of [61], the recursive least squares CMA (RLS-CMA) of [5] and the accelerating adaptive filtering CMA (AAF-CMA) of [70]. Only dominant terms in the pertinent parameters (L,N) are retained in the flop counting, under the hypothesis of real-valued signals and filters. In the complex case, the cost is around four times that of the real case with the same parameters. Note that the cost of the optimal step-size polynomial root extraction is independent of (L,N) and can thus be considered as negligible (see section 15.7.2.2).

The complexity per iteration of the OS-CPA and OS-CMA-1 is of the same order of magnitude, for moderate alphabet size relative to the equalizer length; both algorithms present practically the same cost for BPSK sources ($q = 2$). Finally, the complexity per iteration of the semi-blind techniques is essentially the same as that of their blind counterparts.

15.8 PERFORMANCE ANALYSIS

By means of a detailed empirical analysis, this section evaluates the performance of the different methods studied in this chapter.

15.8.1 Performance of algebraic blind equalizers

We begin by comparing the performance of the algebraic blind equalizers based on the CP criterion developed in section 15.6. The methods considered are: the direct non-structured LS solution of (15.15) ("LS, no struct"); the structuring method of [35] from the first non-overlapping components of the LS solution ("LS, top"); *idem*, from the last components ("LS, bottom"); the maximal ratio combination of the first and last components ("LS, top+bottom"); *idem*, from a whole basis of solutions ("basis, top+bottom"); and the subspace method of section 15.6.2.2 ("basis, subspace"). The "LS, top", "LS, bottom", "LS, top+bottom" and "basis, top+bottom" solutions are explained in section 15.6.2.4. After estimating the symmetric Kronecker vectorization in the direct LS and subspace solutions, the respective equalizer vectors are obtained from the SVD-based rank-1 tensor approximation described in section 15.6.2.3. The performance of the supervised MMSE receiver (15.13) is also computed as a reference. In the first simulation example, a QPSK signal ($q = 4$) excites the AR-1 channel:

$$H_1(z) = \frac{1}{1 - 0.5z^{-1}}, \quad |z| > 0.5$$

with a pole located at $z_p = 0.5$. The impulse response of this channel is well approximated by an order-50 FIR filter. ISI is perfectly canceled for the equalizer $\mathbf{f}_0 = [1, -0.5]^{\mathrm{T}}$, with a dominant first coefficient. The minimal equalizer length is thus $L_0 = 2$, but we suppose its length has been overestimated as $L = 5$, generating $P = 4$ possible ZF solutions, which are just delayed versions of each other [as in (15.11)]. Complex circular additive white Gaussian noise corrupts the channel output, with a *signal-to-noise ratio (SNR)* given by $\mathbb{E}\{|r|^2\}/\mathbb{E}\{|v|^2\}$ where r and v are defined in (15.1). Blocks of size $N_d = 100$ symbol periods are observed, and performance indices are averaged over ν independent *Monte Carlo (MC)* iterations, with $\nu N_d \geqslant 10^5$. Figure 15.1(a) shows the *symbol error rate (SER)* obtained by the algebraic equalizers as a function of the SNR. The performance of the direct LS solution stresses the need for structuring. Yet structuring from only the last components of the LS solution ("LS, bottom") also offers poor results. By contrast, the other methods present a superior performance, just 2–4 dB over the MMSE bound. It is interesting to note that the first components of the LS solution provide the best results for moderate SNR in this scenario. This superiority, however, depends on the optimal equalizer configuration, as shown by the next example.

We repeat the experiment, but moving the AR channel pole to $z_p = 2$, and taking a causal implementation of the channel transfer function:

$$H_2(z) = \frac{1}{1 - 2z^{-1}}, \quad |z| < 2$$

by delaying the truncated impulse response. The minimum-length equalizer is now $\mathbf{f}_0 = [1, -2]^{\mathrm{T}}$, with a dominant last coefficient. Figure 15.1(b) shows the algebraic equalization results. The performance of the"LS-top" method degrades considerably, and becomes similar to that of the"LS-bottom" method in the previous experiment. The performance of the subspace-based structuring method remains practically unchanged compared to the simulation of Fig. 15.1(a), thus showing its robustness to the relative weight of the equalizer coefficients.

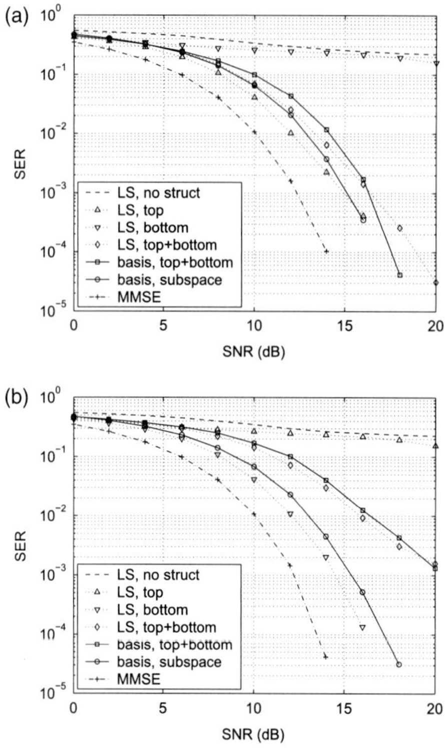

FIGURE 15.1

Algebraic blind equalization based on the CP criterion for different structuring methods, with a QPSK input ($q = 4$), $N_d = 100$ symbol periods, $L = 5$ ($L_0 = 2$), 1000 MC iterations: (a) channel $H_1(z)$; (b) channel $H_2(z)$.

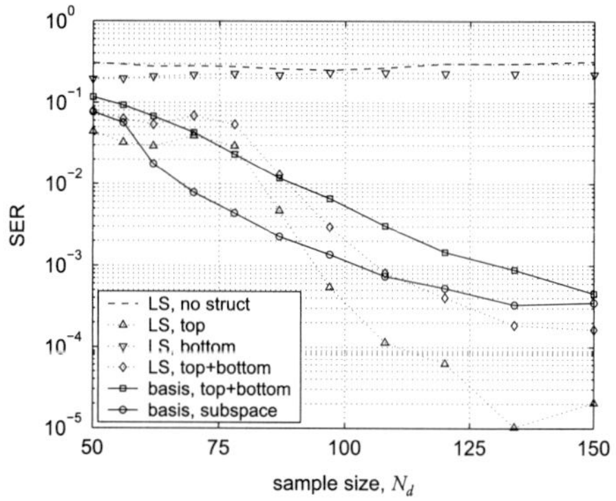

FIGURE 15.2

Algebraic blind equalization based on the CP criterion. Channel $H_1(z)$, QPSK input ($q = 4$), $L = 5$ ($L_0 = 2$), SNR = 15 dB, ν MC iterations, with $\nu N_d \geqslant 10^5$.

Figure 15.2 assesses the *sample size* needs of algebraic solutions, under the general conditions of the first experiment and with SNR = 15 dB. Satisfactory equalization from a basis of the solution space is obtained even under the bound imposed by (15.18) for this simulation example, $N_d \geqslant 71$. The subspace approach provides better results for short observation windows. However, the simplified procedure combining the first and last components of a single non-structured (LS) solution seems to yield good results for a sufficient sample size.

The algebraic semi-blind solutions will be evaluated from section 15.8.4.

15.8.2 Attraction basins of blind and semi-blind CP equalizers

The following experiment evaluates the iterative methods based on the CP criterion, in blind as well as semi-blind operation (section 15.7). In particular, we aim at illustrating

Table 15.2 Average number of iterations for convergence in the experiments of Figs 15.3 and 15.4

Step size	Blind	Semi-blind
Fixed	422	363
Optimal	11	9

the ability of the optimal step-size technique in escaping from *spurious solutions* and that of the training sequence in eliminating them.

We observe a burst of $N_d = 200$ symbols with SNR = 10 dB at the output of channel $H_1(z)$ excited by a BPSK input. Figure 15.3(a) shows the contour lines (in the equalizer parameter space) of the logarithm of the CP criterion (15.6) for $L = L_0 = 2$, computed from the data. Solid lines represent the trajectories of the equalizer coefficients updated by the CPA (section 15.7.1) from 16 different initial configurations (marked by "+") and $\eta = 10^{-5}$ in termination criterion (15.28); convergence points are marked by "×". A fixed step size $\mu = 10^{-2}$ is chosen to obtain the fastest convergence without compromising stability. The plot also shows the MMSE solutions with delays zero and one, $\mathbf{f}_{\text{MMSE},0} = [0.85, -0.38]^{\text{T}}$ and $\mathbf{f}_{\text{MMSE},1} = [0, 0.70]^{\text{T}}$, yielding an output MSE of −8.66 and −4.98 dB, respectively. From most starting points, the algorithm converges to the desired solutions, near the optimal-delay MMSE equalizer. However, the trajectories get trapped in stable extrema located at $\pm[0.01, 0.58]$, near the suboptimal-delay MMSE equalizer. The attraction basins of these spurious solutions are not negligible and can have a significant negative impact on equalization performance. CPA requires, on average, over 400 iterations to converge (Table 15.2).

Under identical conditions, and operating on the same observed data, the trajectories of the OS-CPA equalizer (section 15.7.2) are plotted in Fig. 15.3(b). Not only are undesired solutions avoided, but also convergence is considerably accelerated relative to the previous case: just about 10 iterations suffice (Table 15.2).

Using $N_t = 10$ pilot symbols and a confidence parameter $\lambda = 0.5$, the contour lines of semi-blind CP-MMSE criterion (15.8) have the shape shown in Fig. 15.4(a). The introduction of training data modifies the CP cost function by stressing the global minimum near the optimal MMSE solution while suppressing the suboptimal equilibria and the previously admissible equilibrium point symmetrically located across the origin. The optimal step size still leads to good equalization solutions (Fig. 15.4(b)) and, again, notably accelerates convergence (Table 15.2).

Similar results for the CM criterion are reported in [95].

15.8.3 Robustness of optimal step-size CM equalizers to local extrema

The following experiment demonstrates the faster convergence speed of the OS-CMA compared with the fixed step-size CMA and the RLS-CMA of [5], as well as its ability to escape the attraction basins of undesired equilibria in the CM cost surface.

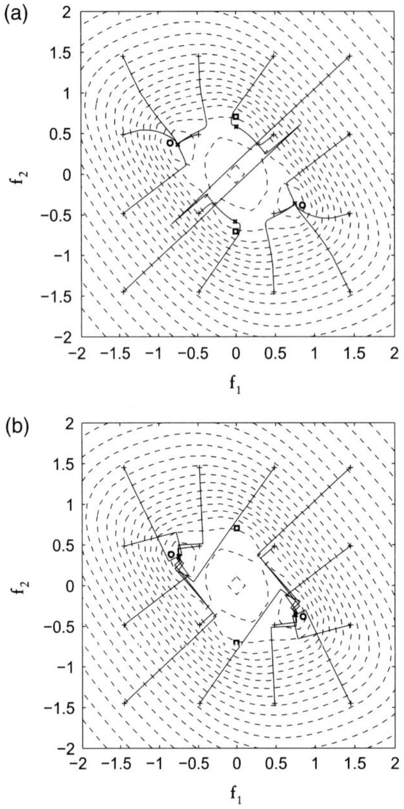

FIGURE 15.3

(Dashed lines) Blind CP criterion contour lines. (Solid lines) Iterative equalizer trajectories: (a) CPA with $\mu = 10^{-2}$; (b) OS-CPA. Channel $H_1(z)$, BPSK input ($q = 2$), $N_d = 200$ symbol periods, $L = L_0 = 2$, SNR = 10 dB. "+": initial point; "×": final point; "o": optimal-delay MMSE solution; "□": suboptimal-delay MMSE solution.

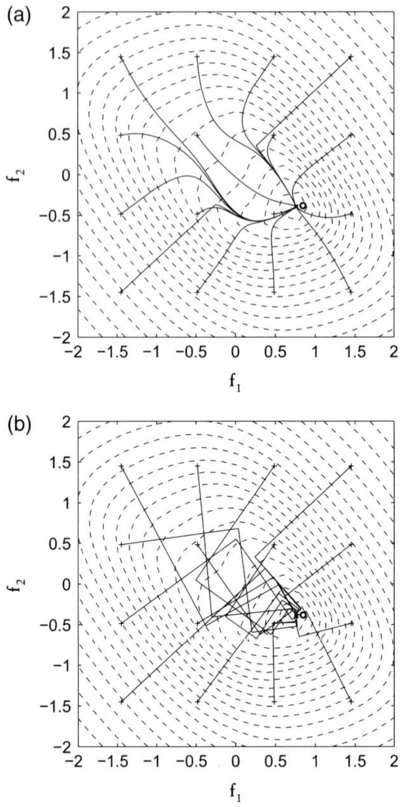

FIGURE 15.4

(Dashed lines) Semi-blind CP-MMSE criterion contour lines. (Solid lines) Iterative equalizer trajectories: (a) SB-CPA with $\mu = 10^{-2}$, (b) OS-SB-CPA. Same conditions as in Fig. 15.3, with $N_t = 10$ pilot symbols and $\lambda = 0.5$. "+": initial point; "×": final point; "o": optimal-delay MMSE solution.

Bursts of $N_d = 200$ baud periods are observed at the output of a channel oversampled at twice the symbol rate (fractionally-spaced SIMO system) excited by a BPSK source ($\gamma = 1$) and corrupted by additive white Gaussian noise with 10-dB SNR. We choose the channel with impulse response:

$$\{0.7571, -0.2175, 0.1010, 0.4185, 0.4038, 0.1762\}$$

corresponding to the second example of [40, section 2.4, pp. 82–83], and a 4-tap equalizer. This system presents the theoretical output MMSE against equalization delay profile of

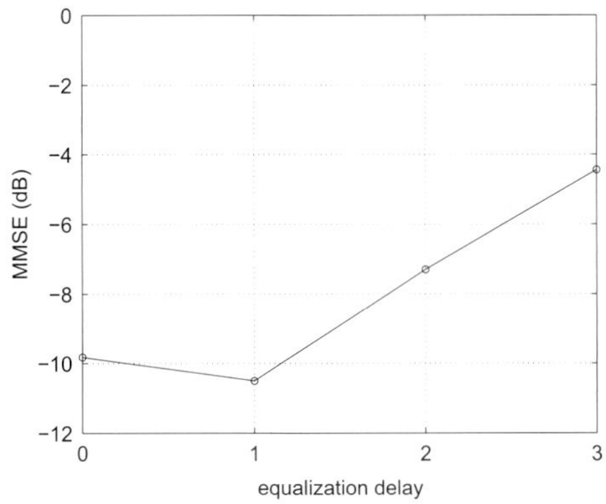

FIGURE 15.5

Theoretical output MMSE as a function of the equalization delay, for the experiment of section 15.8.3.

Fig. 15.5: delay 1 provides the best MMSE performance, closely followed by delay 0; the worst performance is obtained by delay 3. The initial equalizer coefficients are drawn randomly from a normalized Gaussian distribution before processing each signal block. The same initialization is used for all methods. A fixed step size $\mu = 0.025$ is found to prevent the divergence of the CMA. Following the guidelines given for the RLS-CMA in [5], we use the typical forgetting factor $\lambda_{\text{RLS}} = 0.99$ and an inverse covariance matrix initialized at the identity ($\delta = 1$). The samples of the observed signal block are re-used as many times as required. Iterations are stopped as soon as Eq. (15.28) is satisfied, with $\eta = 0.1\mu = 0.0025$. An upper bound of 1000 iterations is also set.

The evolution of the CM cost and the equalizer output MSE, averaged over 1000 independent signal blocks, is plotted in Fig. 15.6(a & b). The normalized histogram of equalization delays obtained by the three methods appears in Fig. 15.6(c), while Table 15.3 summarizes their computational cost. The CMA and the RLS-CMA often achieve the same suboptimal equalization delays. In contrast, the OS-CMA converges more frequently near the optimal-delay MMSE equalizer setting, and requires around an order of magnitude fewer iterations. Indeed, the CMA, the OS-CMA and the RLS-CMA converge to one of the two best equalization delays (0 or 1) with a probability of 67.8%, 86.6% and 73.1%, respectively. The OS-CMA obtains the best performance with an affordable complexity, which, thanks to its fast convergence, is always below that of the classical fixed step-size CMA.

15.8.4 CP equalizers for a non-minimum phase channel

We evaluate now the performance of algebraic and iterative solutions for the blind CP and semi-blind CP-MMSE criteria in the non-minimum phase channel of [25, section V],

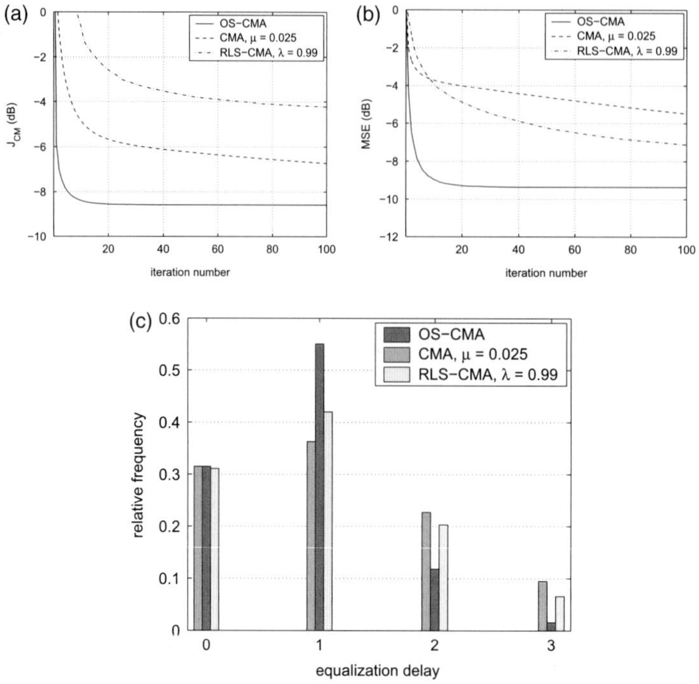

FIGURE 15.6

Iterative CM equalizers. Performance of the classical CMA ($\mu = 0.025$), the OS-CMA and the RLS-CMA ($\lambda_{\rm RLS} = 0.99$) for the SIMO system of section 15.8.3 and a 4-tap equalizer with random Gaussian initialization. (a) Evolution of CM cost function; (b) evolution of equalizer output MSE; (c) normalized histogram of equalization delay. Results are averaged over 1000 signal realizations.

Table 15.3 Average computational cost for convergence of the CM-based iterative equalizers in the experiment of Figs 15.5 and 15.6

COST	CMA	OS-CMA		RLS-CMA
		OS-CMA-1	OS-CMA-2	
Iterations	565	38		286
Total flops ($\times 10^3$)	1124.4	166.4	69.5	26.3

given by:

$$H_3(z) = (-0.033 + 0.014j) + (0.085 - 0.039j)z^{-1} - (0.232 - 0.136j)z^{-2}$$
$$+ (0.634 - 0.445j)z^{-3} + (0.070 - 0.233j)z^{-4}$$
$$- (0.027 + 0.071j)z^{-5} - (0.023 + 0.012j)z^{-6}. \tag{15.33}$$

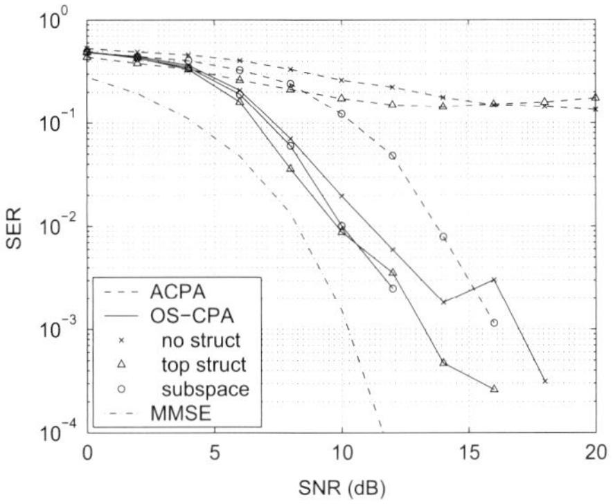

FIGURE 15.7

Blind CP equalization. The OS-CPA is initialized with different ACPA solutions. Channel $H_3(z)$, QPSK input ($q = 4$), $N_d = 100$ symbol periods, $L = 5$ ($L_0 = 3$), 200 MC iterations.

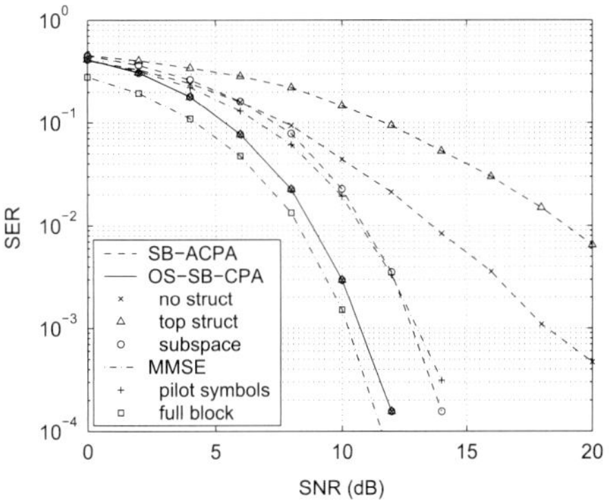

FIGURE 15.8

Semi-blind CP-MMSE equalization, in the same conditions as Fig. 15.7, with $N_t = 10$ pilot symbols and $\lambda = 0.5$. The OS-SB-CPA is initialized with different SB-ACPA solutions.

This 6th-order FIR channel can be perfectly equalized by an FIR filter with $L_0 = 3$ coefficients, but we choose $L = 5$. From a data block of $N_d = 100$ symbols and using several structuring procedures, the algebraic solutions to the blind CP criterion (section 15.6) yield the dotted-line curves shown in Fig. 15.7. These algebraic solutions are then employed to initialize the OS-CPA described in section 15.7.2, generating the results displayed by the solid lines in Fig. 15.7. The gradient-descent iterations refine the algebraic estimates, approaching the MMSE bound.

The performance of the semi-blind CP-MMSE methods is summarized in Fig. 15.8, for the same scenario with $N_t = 10$ pilot symbols and $\lambda = 0.5$. Algebraic estimates are first determined by combining the blind and supervised solutions as explained in section 15.6.3 (dotted lines), and then used to initialize the OS-SB-CPA of section 15.7.2 (solid lines). MMSE performance bounds (dashed lines) are determined by computing the LS solution (15.13) to the MMSE criterion. Two MMSE curves are obtained: using only the pilot sequence, as in a conventional receiver, and using the whole data block (MMSE bound); this bound is obviously unreachable in practice since the whole bandwidth would be used for training.

The advantages of the semi-blind approach are remarkable. In the first place, the performance of algebraic solutions is improved compared to the purely blind case. In the second place, the OS-SB-CPA exhibits identical performance regardless of initialization and nearly reaches the MMSE bound. The exploitation of "blind symbols" in addition to the training sequence improves the conventional receiver and almost attains the MMSE bound. Moreover, convergence speed is increased relative to the fully blind case, especially for low SNR, as shown in Fig. 15.9.

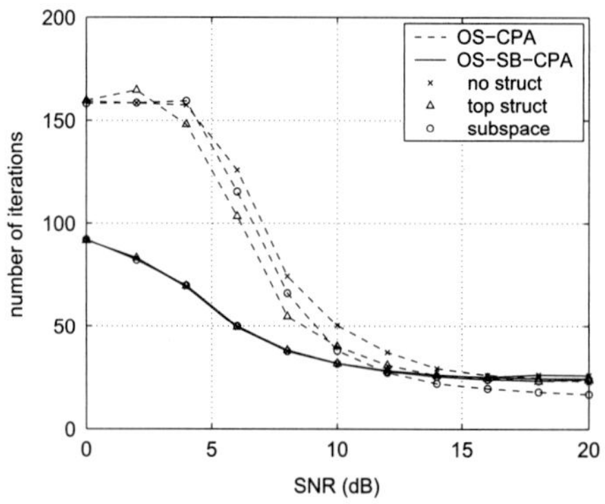

FIGURE 15.9

Average number of iterations for the three initializations of the OS-CPA (blind) and OS-SB-CPA (semi-blind) in the experiment of Figs 15.7 and 15.8.

15.8.5 Blind CM and semi-blind CM-MMSE equalizers

A zero-mean unit-variance QPSK-modulated input excites the same non-minimum phase channel $H_3(z)$ [Eq. (15.33)] of [25, section V], whose output is corrupted by complex circular additive white Gaussian noise. An FIR filter of length $L = 5$ is used to equalize the channel, aiming at the optimal-MMSE delay ($\tau_{\text{opt}} = 6$ at 20-dB SNR). Bursts of $N_d = 100$ symbols are observed at the channel output, generating a total of $N = 96$ channel-output vectors. We choose $\lambda = 0.5$ and $\mu = 10^{-3}$ for the fixed step-size algorithms. Iterations are stopped when (15.28) is verified, with $\eta = 0.1\mu$. Equalization quality is again measured in terms of SER, which is estimated by averaging over 500 independent bursts.

We first compare several fully blind criteria ($N_t = 0$). The algebraic solution of [25, section II-B] is called "DK-top", and corresponds to the structuring method described in section 15.6.2.4 based on the first elements of the non-structured LS solution to the CM criterion. Iterative solutions are obtained by the fixed step-size CMA (section 15.7.1) with three different initializations: *first-tap* filter, *center-tap* filter and DK-top solution. MMSE receiver and bound curves are also plotted for reference. Figure 15.10 shows that the algebraic solution is only useful as an initial point for the iterative blind receiver, whose performance depends strongly on the initialization used.

In the same scenario, the performance of the fixed step-size SB-CMA (section 15.7.1) is summarized in Fig. 15.11. The algebraic SB-ACMA solution of [74] is also considered; the semi-blind approach to the DK-top method (SB-DK-top) is enabled by the SVD-based procedure described in section 15.6.3. Although the inclusion of training information improves the DK-top method compared with the blind case (Fig. 15.10), the SB-ACMA proves superior and outperforms the conventional receiver for sufficient SNR. Nevertheless, the SB-ACMA can still be improved if used as initialization for the iterative SB-CMA, whose performance becomes practically independent of initialization for a low to moderate SNR. A performance flooring effect is observed for higher SNR. As

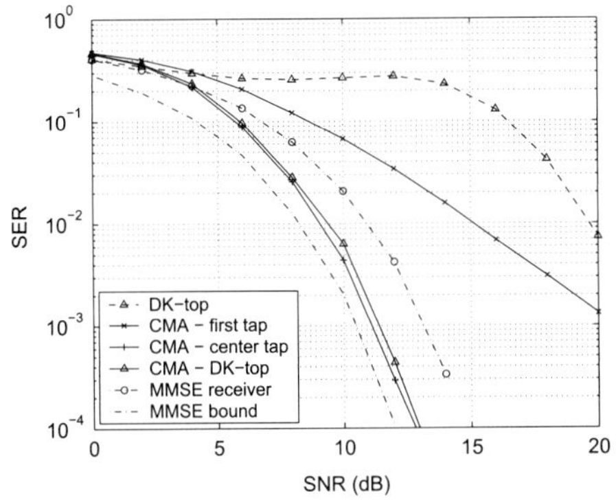

FIGURE 15.10

Blind CM equalization. Channel $H_3(z)$, QPSK input, $N_d = 100$ symbol periods, $L = 5$ ($L_0 = 3$), 500 MC iterations. Solid lines: CMA with fixed step size $\mu = 10^{-3}$ and different initializations.

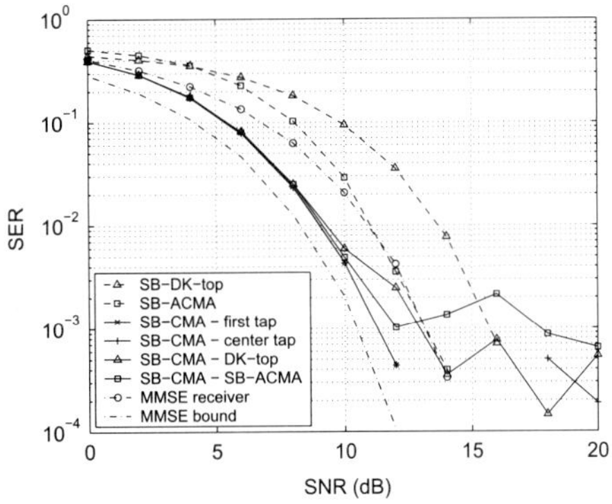

FIGURE 15.11

Semi-blind CM-MMSE equalization, under the conditions of Fig. 15.10 and 10% of pilot symbols. Solid lines: fixed step-size SB-CMA with different initializations.

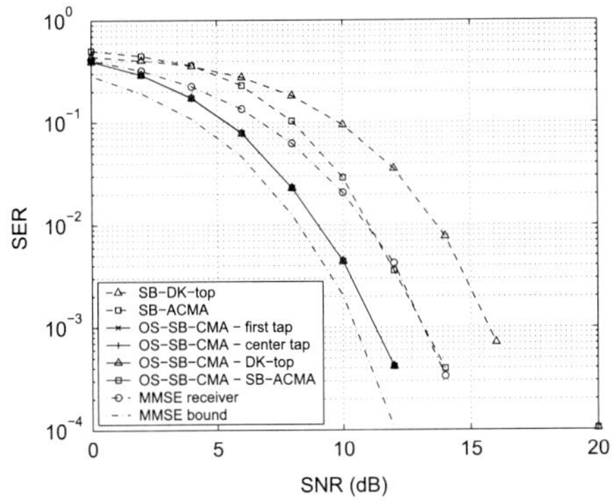

FIGURE 15.12

Semi-blind CM-MMSE equalization, under the conditions of Figs 15.10 and 15.11. Solid lines: OS-SB-CMA with different initializations.

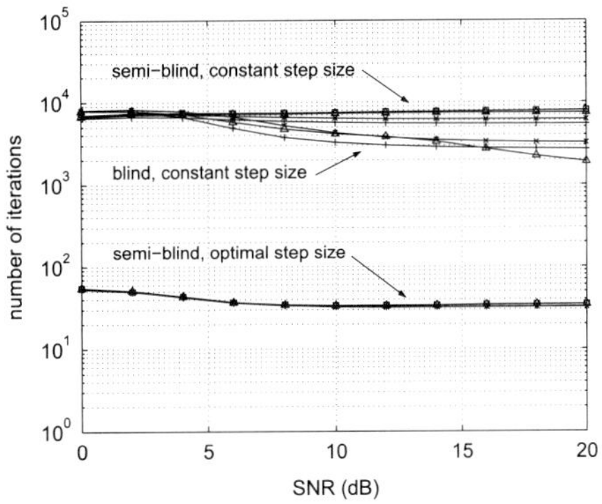

FIGURE 15.13

Average number of iterations for the iterative equalizers in the experiments of Figs 15.10–15.12.

highlighted by Fig. 15.13, the number of iterations required for convergence increases relative to the blind scenario. This increase is probably due to the flattening of the CM cost function when incorporating training data. A similar effect has been observed for the CP criterion in section 15.8.2.

Figures 15.12 and 15.13 show that the performance of the OS-SB-CMA (section 15.7.2) is virtually independent of initialization, and its iteration count is reduced by around two orders of magnitude relative to the constant step-size techniques. Furthermore,

(a)

(b)

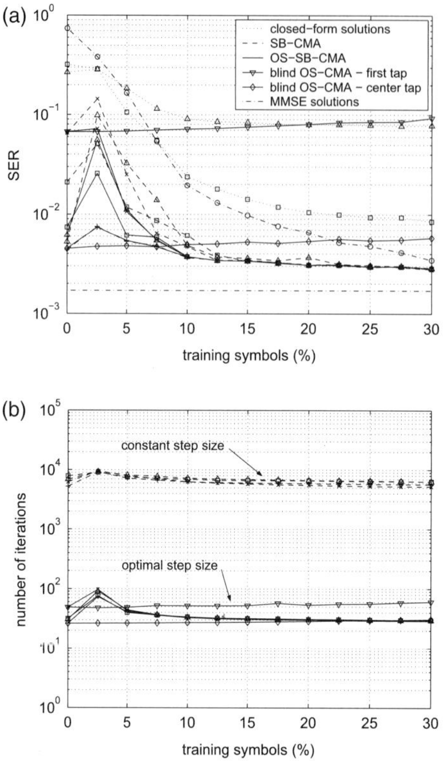

FIGURE 15.14

CM equalization with a variable number of pilot symbols in the transmitted burst, under the conditions of Figs 15.10–15.12. MMSE solutions (MMSE receiver and bound) are as in such figures. Iterative equalizers' initializations: "×": first tap; "+": center tap; "Δ": SB-DK-top; "□": SB-ACMA. (a) SER performance. (b) Average number of iterations for the iterative equalizers.

the performance flooring observed for the fixed step-size SB-CMA at high SNR now disappears.

15.8.6 Influence of pilot-sequence length

Under the same previous conditions, Fig. 15.14 illustrates the performance of semi-blind techniques as a function of the percentage of symbols in the transmitted block used for training, computed as $N_t/N \times 100\%$, for a 10-dB SNR. The OS-CMA using only the "blind symbols" is also tested for two different initializations. The SB-ACMA equalizer only outperforms the conventional receiver for short pilot sequences, and always benefits from gradient-descent iterations. The OS-SB-CMA slightly improves the SB-CMA for short training and all initializations, while maintaining its reduced complexity over the whole training-length range. For reasonable pilot-sequence sizes, semi-blind methods are capable of attaining the conventional MMSE receiver performance while improving spectral efficiency (decreasing the pilot-sequence length), thus increasing the effective transmission rate. With the appropriate initialization, fully blind processing outperforms the semi-blind methods for short pilot sequences, as if the use of too few training symbols could somehow confuse the blind receiver; a similar effect is observed for sufficient training, where "blind symbols" appear to divert the conventional receiver from its satisfactory solution. Yet the performance of the blind OS-CMA in this scenario depends strongly on initialization, although, as shown in sections 15.8.2 and 15.8.3, the optimal step-size approach provides certain immunity to local extrema.

A very similar behavior of CP-based equalizers against the pilot-sequence length has been reported in [92].

15.8.7 Influence of the relative weight between blind and supervised criteria

The performance of the semi-blind CP-MMSE methods as a function of confidence parameter λ is illustrated in Fig. 15.15, obtained in the same scenario as in section 15.8.4 with $N_t = 10$ pilot symbols. Equalization results are gradually improved as more weight is laid on the known data. Performance then deteriorates as the blind part of the criterion is neglected and the equalization is left to entirely depend on just a few training symbols; hence the SER increase up to the conventional MMSE receiver level when λ approaches 1. Consequently, this severe increase is not observed with longer training windows. Over a wider range of λ (roughly, in the interval $[0.3, 0.9]$), the influence of initialization on equalization quality and convergence speed of the OS-SB-CPA does not seem to be significant and, for practically all $\lambda \in]0, 1[$, the iterative semi-blind methods outperform the conventional equalizer.

Figure 15.15(b) also shows that, for certain value of confidence parameter ($\lambda \approx 0.7$), the cost-function surface seems best adapted to the operation of the optimal step-size algorithm, so that convergence is obtained in the minimum number of iterations. This optimal value of λ will generally depend on the specific system conditions, the sample size and the SNR.

15.8.8 Comparison between the CM and CP criteria

A final experiment makes a brief illustrative comparison between the CP and CM criteria in semi-blind operation (10% training). A co-channel interferer with the same modulation as the desired signal (QPSK) and a given signal-to-interference ratio (SIR) is added at the output of channel $H_3(z)$. The respective top-structuring analytic solutions are first obtained, and then used as initial points for the optimal-step size iterations. Figure 15.16 shows that, although the SB-ACPA solution is poorer than SB-ACMA's in this particular scenario, the OS-SB-CPA improves its CM counterpart with half the number of iterations.

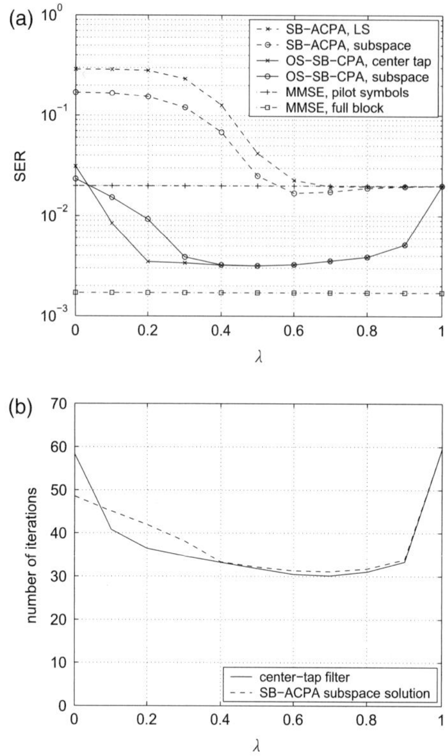

FIGURE 15.15

Impact of confidence parameter λ on the performance of semi-blind CP-MMSE methods. Channel $H_3(z)$, QPSK input ($q = 4$), $N_d = 100$ symbol periods, $N_t = 10$ pilot symbols, $L = 5$ ($L_0 = 3$), $\text{SNR} = 10$ dB, 500 MC runs. (a) SER performance. (b) Average number of iterations for the two initializations of the OS-SB-CPA.

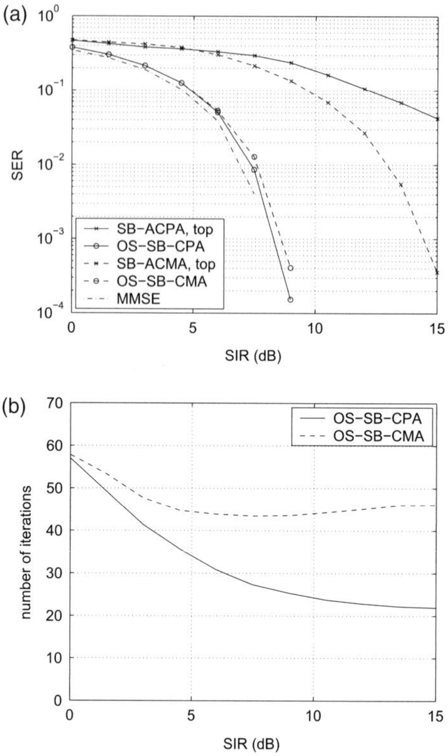

FIGURE 15.16

Semi-blind equalization with the CP and CM criteria. The analytic solutions are obtained using the top structuring method. Channel $H_3(z)$, QPSK input ($q = 4$), QPSK co-channel interferer, $N_d = 200$ symbol periods, $N_t = 20$ pilot symbols, 100 MC runs. (a) SER performance. (b) Average number of iterations for the iterative equalizers.

15.9 SEMI-BLIND CHANNEL ESTIMATION

We conclude this chapter by discussing the indirect approach to channel equalization. This approach consists of two stages: the channel is estimated in the first stage; equalization is then performed in a second stage. Whereas the direct approach is usually limited to linear equalizers as described in the preceding sections, the indirect approach allows the exploitation nonlinear equalization techniques such as the Viterbi algorithm.

As explained throughout the chapter, the basic semi-blind approach refers to the simultaneous exploitation of known pilot (or training) sequences and blind information.

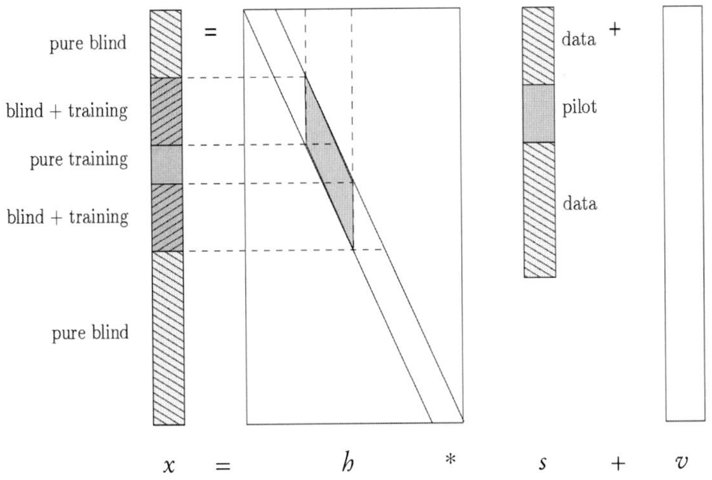

FIGURE 15.17

Received signal structure for a frequency-selective channel.

The blind information can arise from multichannel structures (SIMO, MIMO), possibly obtained after oversampling (e.g. as in, CDMA, where the spreading factor can be viewed as an oversampling factor), or from non-Gaussianity of the transmitted signals (constant modulus, finite alphabet, etc.). The channel can be time-invariant or time-varying, in which case it can be modeled with a Basis Expansion Model (BEM). Due to various forms of memory (delay spread in the time domain, Doppler spread in the frequency domain), the received signal components often contain a mixture of known and unknown symbols. An example of a signal in the time domain passing through a channel with memory is illustrated in Fig. 15.17 [cf. Eq. (15.1)]. Due to the channel memory, some received signal samples generally contain pure training symbols, others contain purely unknown data symbols and still others contain a mixture of both.

In this scenario, the optimal approach is to jointly estimate all unknown quantities (symbols and channel) with an optimal criterion such as maximum likelihood, if possible by incorporating also prior information (if available). Various forms of simpler suboptimal approaches have been proposed in the literature. A class of suboptimal approaches aim to optimize a weighted sum of a training-based and a blind channel estimation criterion; see [16] for an extensive discussion of these techniques for the SIMO case and [53] for the MIMO case.

With perfect Channel State Information at the Receiver (CSIR), no CSI at the Transmitter (CSIT) and iid channel elements, the optimal input signal is a zero-mean spatio-temporally white Gaussian noise. Any deviation from this (side information) will lower the perfect CSIR channel capacity. However, there is usually no CSIR, so that any such deviation may allow channel estimation, leading to an increase in actual channel capacity (see [100] for optimal input distributions in the absence of CSIR). Possible forms of side information are enumerated and briefly discussed below:

- Higher-order statistics of data symbols [4,8].
- Finite alphabet (FA) of unknown symbols (see also the algebraic methods presented earlier in the chapter). Note that the use of FA symbols instead of Gaussian inputs constitutes an introduction of side information and thus a reduction of full CSIR channel capacity. The FA can be exploited through iterative channel estimation and data detection; e.g. see [20,50,67,75,101], or [91] with two-level Kalman filtering. In [63], it is shown that when constraints on the input symbols such as those based on the FA property only leave a discrete ambiguity, then the Cramer-Rao Bound (CRB), which is a local bound, for channel estimation is the same as if the unknown symbols were known.
- Channel coding in unknown symbols, exploited, for example, through turbo detection and estimation. In [68], a channel estimation CRB is provided when data symbol channel coding is exploited, involving the minimum distance amplification introduced by the channel code. As the SNR increases from low to high values, this CRB moves from the case of the data symbols being unknown and Gaussian to the case of known pilot symbols.
- Partial FA knowledge, such as constant modulus (e.g. the 8-PSK modulation used in the EDGE standard) [37,51,64].
- Some training/pilot symbols, only enough to allow iterative joint data detection/channel estimation to converge.
- Symbol modulus variation pattern, which is a particular form of transmitter-induced cyclostationarity. Some of the techniques proposed to exploit this property lead to wide-sense cyclostationarity [79], without consistency in SNR. The technique proposed in [47], though, is deterministic.
- Space-time coding redundancies through reduced rate linear precoding, introducing subspaces in the transmitted signal covariance, e.g. Alamouti or other orthogonal space-time coding schemes [6,51,56].
- Guard intervals in time or frequency, as in [66,88], or cyclic prefix structure.
- Symbol stream color. In [39], it is shown that colored inputs can be separated if their spectra are linearly independent. Correlation can be introduced by linear convolutive precoding, e.g. in the form of MIMO prefiltering. In [47], an example of low rate precoding appears since the same symbol sequence gets distributed over all transmit antennas; see [53] for a more detailed discussion.
- Known pulse shape, which can be exploited when the received signal is oversampled with respect to the symbol rate (temporal oversampling).
- The spreading codes in CDMA. Direct sequence spectrum spreading (DS-CDMA) is a special case in which the oversampling factor corresponds to the spreading factor, a sample is called a chip, and a memoryless SIMO prefilter corresponds to an instantaneous multiplication with the spreading code (which can be time-varying in the case of long/aperiodic/pseudo-random codes or time-invariant as in the case of short/periodic/deterministic codes). Of course, CDMA can be combined with oversampling with respect to the chip rate and exploitation of a chip pulse shape.

The use of different spreading codes for different inputs allows for fairly robust blind source separation and channel estimation; see [29,30,38,51].

- Transmitter-induced non-zero mean, also known as superimposed training. Besides the use of time multiplexed (TM) pilots as in the semi-blind techniques presented in this chapter, a recent twist (which is actually not so recent) on the training paradigm is the appearance of superimposed (SI, also called embedded) pilots. SI pilots are actually classical in CDMA standards, which use a pilot signal, sometimes combined with TM pilots. In [101], SI pilot based channel estimates are used to initialize an iterative receiver. In [2], optimization of a mixture of TM and SI pilots is considered. The continuous SI pilots form actually a pilot signal and their large duration leads to quasi-orthogonality with the data. It is found that for large enough and equivalent pilot power, both pilot forms lead to similar performance. Only the channel estimation (CRB) is considered though, as performance indicator. In [85], the effect of both types of pilots on the throughput is considered and TM pilots appear to be favored. Indeed, pilots not only allow channel estimation but also influence the data detection. The presence of TM pilots leads to reduced ISI in frequency-selective channels with time-domain transmission. Semi-blind channel estimation and detection with SI pilots is considered in [54]. An important question here is: is orthogonality of pilots and data desirable? The answer may depend on how mixed information (pilot/data) is used and combined.
- Spatial multiplexing schemes that achieve the optimal rate-diversity trade-off do not typically introduce any blind information (other than that provided by Gaussian white inputs) for the channel estimation. In [52], for instance, a previously introduced linear prefiltering scheme was shown to attain this optimal trade-off. Since the prefilter is a MIMO all-pass filter, it leaves the white vector input white. However, perturbations of optimal trade-off achieving schemes can be derived that introduce side information.

Hence, some questions that so far have only been very partially answered are: what is the optimal amount of side information to maximize capacity, as more side information reduces capacity with CSIR but also reduces channel estimation error and hence increases capacity? More importantly, what is the optimal distribution of side information over the various forms? Note that, strictly speaking, blind approaches are based on just exploiting second-order information and/or subspaces. The exploitation of any form of side information mentioned above should be called a semi-blind approach.

Some other research avenues include:

- Multiuser case. In this scenario, the number of unknowns per received sample increases further. Whereas spatial multiplexing is the cooperative case of multi-input, the multi-user case corresponds to the non-cooperative version. Differentiation of users at the level of SOS can be obtained through coloring (e.g. CDMA) as mentioned earlier. In [88], a semi-blind multiuser scenario is considered.
- Non-coherent approaches.
- Channel estimation for the transmitter. Questions that arise here involve not only channel estimation but also its possible quantization and (digital or analog)

retransmission. A key issue here also is the degree of reciprocity of the channel or, for example, its pathwise parameters (direction, delay, Doppler shift, power). Another issue is the effect of sensor array design on channel estimation and reciprocity, e.g. beamspace (beam selection should be reciprocal).

- Description of channel variation in terms of user mobility.
- Bayesian (semi-)blind: from deterministic unknown channels to fading random channels [72].

15.10 SUMMARY, CONCLUSIONS AND OUTLOOK

The present chapter has addressed the problem of channel equalization, which consists in recovering the information emitted through a time-dispersive propagation medium. Source separation and channel equalization can be considered as dual problems whose goal is to unravel, respectively, spatial and temporal mixtures of the source(s). Yet the particularities of digital communication systems allow the design of more specific source recovery techniques, some of which have been presented in this chapter.

Our focus has been on SISO channel equalization. Several semi-blind criteria have easily been defined by combining purely blind criteria based on the finite alphabet of digital signals, such as the CM and CP principles, and the conventional training-based MMSE equalizer. Under certain conditions (essentially, the existence of exact ZF equalizers and input signals adapted to the blind part of the criteria), the global minima of such semi-blind objectives can be attained algebraically. These non-iterative solutions are unaffected by the presence of local extrema on the cost-function surface. The algebraic treatment of the CP criterion, resulting in the ACPA equalizer, is similar to that of ACMA, but does not require special modifications to treat binary modulations. Algebraically, the proposed subspace method provides a particular solution to the challenging rank-1 tensor linear combination problem. In our numerical study, this subspace approach proves more robust than other structuring methods, but the blind algebraic solutions offer a low tolerance to noise, particularly for long equalizers. This tolerance can be slightly improved by semi-blind techniques from just a few pilot symbols. The key point limiting performance is probably the SVD-based rank-1 tensor approximation employed to extract the equalizer vector from the estimated symmetric tensor, as described in sections 15.6.2.3 and 15.6.3. A refinement of this rank-1 tensor approximation, such as that obtained by the power method [17,41], could alleviate this limitation.

In general, algebraic solutions can only approximate a good equalization setting, and iterative techniques are generally necessary to find the global minimum of the criterion; an iterative method can also be used to refine an algebraic solution. An exact *global* line search technique based on block iterations has been proposed, allowing an optimal algebraic adaptation of the step size at each iteration; this adaptation only involves the roots of a polynomial that can be solved by radicals. The optimal step-size iterative algorithm offers a very fast convergence and, in semi-blind mode, yields equalization results very close to the MMSE bound while increasing the useful transmission rate and the robustness to the equalizer vector initialization. These benefits have been demonstrated by the numerical experiments presented in the last part of the chapter.

In summary, the impact of local minima and slow convergence typical of blind equalizers can be limited with the incorporation of training sequences, giving rise to semi-blind criteria. To further alleviate these drawbacks, the present chapter has endowed semi-blind criteria with a number of additional strategies:

- judicious initialization with *algebraic solutions*;
- iterative updates operating on signal *blocks* (or bursts);
- *one-dimensional global minimization* (exact line search) with an optimal step size computed algebraically.

These strategies are not exclusive to the equalization principles considered in this chapter (CM, CP), but can also profit other criteria such as the KM [96] (see also Chapter 6) or those of [34].

Avenues of further research could include the following aspects, which have been left aside in our analysis: a theoretical study of the local extrema of the CP criterion; the improvement of the SVD-based technique to recover the equalizer vector; the robust automatic detection of the number of ZF solutions and the optimal equalization delay [97]; the theoretical optimal choice of confidence parameter λ (e.g. based on an asymptotic variance analysis); the evaluation and reduction of carrier-residual effects on CP equalizers [11,12] (although the inclusion of pilot information may already play an important role in their compensation); and an exhaustive comparison, both theoretical and experimental, of other equalization principles with those presented in this chapter. Other challenging open issues in the related topic of semi-blind channel estimation have also been discussed in section 15.9.

References

[1] A. Belouchrani, J.-F. Cardoso, Maximum likelihood source separation for discrete sources, in: Proc. EUSIPCO-94, VII European Signal Processing Conference, Edinburgh, UK, September 13–16, 1994, pp. 768–771.

[2] L. Berriche, K. Abed-Meraim, J.-C. Belfiore, Cramer-Rao bounds for MIMO channel estimation, in: Proc. IEEE ICASSP, Montreal, Canada, May 17–21, 2004.

[3] D.H. Brandwood, A complex gradient operator and its application in adaptive array theory, IEE Proceedings F: Communications Radar and Signal Processing 130 (1983) 11–16.

[4] J.-F. Cardoso, Blind signal separation: Statistical principles, Proceedings of the IEEE 86 (1998) 2009–2025.

[5] Y. Chen, T. Le-Ngoc, B. Champagne, C. Xu, Recursive least squares constant modulus algorithm for blind adaptive array, IEEE Transactions on Signal Processing 52 (2004) 1452–1456.

[6] J. Choi, Equalization and semi-blind channel estimation for space-time block coded signals over a frequency-selective fading channel, IEEE Transactions on Signal Processing 52 (2004) 774–785.

[7] S. Choi, A. Cichocki, Blind equalisation using approximate maximum likelihood source separation, Electronics Letters 37 (2001) 61–62.

[8] A. Cichocki, S. Amari, Adaptive Blind Signal and Image Processing: Learning Algorithms and Applications, John Wiley & Sons, Inc., Chichester, UK, 2002.

[9] P. Comon, Analyse en Composantes Indépendantes et identification aveugle, Traitement du Signal 7 (1990) 435–450. Numero special non lineaire et non gaussien.
[10] P. Comon, Block methods for channel identification and source separation, in: Proc. IEEE Symposium on Adaptive Systems for Signal Processing, Communications and Control, Lake Louise, Alberta, Canada, October 1–4, 2000, pp. 87–92.
[11] P. Comon, Blind equalization with discrete inputs in the presence of carrier residual, in: Proc. 2nd IEEE International Symposium on Signal Processing and Information Theory, Marrakech, Morocco, December 2002.
[12] P. Comon, Independent component analysis, contrasts, and convolutive mixtures, in: Proc. 2nd IMA International Conference on Mathematics in Communications, University of Lancaster, UK, December 16–18, 2002.
[13] P. Comon, Tensor decompositions: State of the art and applications, in: J.G. McWhirter, I.K. Proudler (Eds.), Mathematics in Signal Processing V, Clarendon Press, Oxford, UK, 2002, pp. 1–24.
[14] P. Comon, Contrasts, independent component analysis, and blind deconvolution, International Journal of Adaptive Control and Signal Processing 18 (2004) 225–243 (Special issue on Blind Signal Separation).
[15] P. Comon, B. Mourrain, Decomposition of quantics in sums of powers of linear forms, Signal Processing 53 (1996) 93–107 (Special issue on Higher-Order Statistics).
[16] E. De Carvalho, D. Slock, Semi-blind methods for FIR multichannel estimation, in: G.B. Giannakis, Y. Hua, P. Stoica, L. Tong (Eds.), Signal Processing Advances in Wireless and Mobile Communications. Vol. 1: Trends in Channel Estimation and Equalization, Prentice Hall, Upper Saddle River, NJ, 2001, pp. 211–254. (Ch. 7).
[17] L. De Lathauwer, P. Comon, et al. Higher-order power method, application in Independent Component Analysis, in: NOLTA Conference, vol. 1, Las Vegas, 10–14 December 1995, pp. 91–96.
[18] J. P. Delmas, P. Comon, Y. Meurisse, Identifiability of BPSK, MSK, and QPSK FIR SISO channels from modified second-order statistics, in: IEEE Spawc'06, Cannes, July 2–5 2006.
[19] J.P. Delmas, Y. Meurisse, P. Comon, Performance limits of alphabet diversities for FIR SISO channel identification, IEEE Transactions on Signal Processing 57 (2008) 73–82.
[20] R. Demo Souza, J. Garcia-Frias, A.M. Haimovich, A semi-blind receiver for iterative data detection and decoding of space-time coded data, in: Proc. IEEE-COM WCNC Conf., 2004.
[21] Z. Ding, C.R. Johnson, R.A. Kennedy, On the (non)existence of undesirable equilibria of Godard blind equalizers, IEEE Transactions on Signal Processing 40 (1992) 2425–2432.
[22] Z. Ding, R.A. Kennedy, B.D.O. Anderson, C.R. Johnson, Ill-convergence of Godard blind equalizers in data communication systems, IEEE Transactions on Communications 39 (1991) 1313–1327.
[23] Z. Ding, Y. Li, Blind Equalization and Identification, Dekker, New York, 2001.
[24] K. Dogançay, K. Abed-Meraim, Y. Hua, Convex optimization for blind equalization, in: Proc. ICOTA-98, 4th International Conference on Optimization: Techniques and Applications, Perth, Australia, July 3–5, 1998, pp. 1017–1023.
[25] K. Dogançay, R.A. Kennedy, Least squares approach to blind channel equalization, IEEE Transactions on Signal Processing 47 (1999) 1678–1687.
[26] D. Donoho, On minimum entropy deconvolution, in: Applied Time-Series Analysis II, Academic Press, 1981, pp. 565–609.
[27] E. Durand, Solutions numériques des équations algébriques, vol. I, Masson, Paris, 1960.
[28] E. Gassiat, F. Gamboa, Source separation when the input sources are discrete or have constant modulus, IEEE Transactions on Signal Processing 45 (1997) 3062–3072.
[29] I. Ghauri, D.T.M. Slock, Blind and semi-blind single user receiver techniques for asynchronous CDMA in multipath channels, in: Proc. IEEE Globecom, Sydney, Australia, Nov. 1998.
[30] I. Ghauri, D.T.M. Slock, MMSE-ZF receiver and blind adaptation for multirate CDMA, in: Proc. IEEE Vehicular Technology Conference, Amsterdam, the Netherlands, September 1999.

[31] D.N. Godard, Self-recovering equalization and carrier tracking in two-dimensional data communication systems, IEEE Transactions on Communications 28 (1980) 1867–1875.
[32] G.H. Golub, C.F. Van Loan, Matrix Computations, 3rd ed., The Johns Hopkins University Press, Baltimore, MD, 1996.
[33] O. Grellier, P. Comon, Blind equalization and source separation with MSK inputs, in: Proc. SPIE Conference on Advances in Signal Processing, San Diego, CA, July 19–24, 1998, pp. 26–34.
[34] O. Grellier, P. Comon, Blind separation of discrete sources, IEEE Signal Processing Letters 5 (1998) 212–214.
[35] O. Grellier, P. Comon, Closed-form equalization, in: Proc. SPAWC-99, 2nd IEEE Workshop on Signal Processing Advances in Wireless Communications, Annapolis, MD, May 9–12, 1999, pp. 219–222.
[36] O. Grellier, P. Comon, B. Mourrain, P. Trebuchet, Analytical blind channel identification, IEEE Transactions on Signal Processing 50 (2002) 2196–2207.
[37] M.A.S. Hassan, B.S. Sharif, W. Woo, S. Jimaa, Semiblind estimation of time varying STFBC-OFDM channels using Kalman filter and CMA, in: Proc. IEEE ISCC, 2004.
[38] B. Hochwald, T.L. Marzetta, C.B. Papadias, A transmitter diversity scheme for wideband CDMA systems based on space-time spreading, IEEE Journal of Selected Areas in Communications 19 (2001).
[39] Y. Hua, J.K. Tugnait, Blind identifiability of FIR-MIMO systems with colored input using second-order statistics, IEEE Signal Processing Letters (2000).
[40] C.R. Johnson, P. Schniter, I. Fijalkow, L. Tong, et al., The core of FSE-CMA behavior theory, in: S.S. Haykin (Ed.), Unsupervised Adaptive Filtering, Vol. II: Blind Deconvolution, John Wiley & Sons, New York, 2000, pp. 13–112. (Ch. 2).
[41] E. Kofidis, P.A. Regalia, On the best rank-1 approximation of higher-order supersymmetric tensors, SIAM Journal on Matrix Analysis and Applications 23 (2002) 863–884.
[42] M. Kristensson, B. Ottersten, D.T.M. Slock, Blind subspace identification of a BPSK communication channel, in: Proc. 30th Asilomar Conf. Sig. Syst. Comp., Pacific Grove, CA, 1996.
[43] A.M. Kuzminskiy, L. Féty, P. Foster, S. Mayrargue, Regularized semi-blind estimation of spatio-temporal filter coefficients for mobile radio communications, in: Proc. XVIème Colloque GRETSI, Grenoble, France, September 15–19, 1997, pp. 127–130.
[44] J.L. Lacoume, P.O. Amblard, P. Comon, Statistiques d'ordre supérieur pour le traitement du signal, Collection Sciences de l'Ingénieur, Masson, 1997. téléchargeable gratuitement à www.i3s.unice.fr/~comon/livreSOS.html.
[45] C. Lanczos, Applied Analysis, Dover, New York, 1988.
[46] J. Lebrun, P. Comon, A linear algebra approach to systems of polynomial equations with application to digital communications, in: Proc. EUSIPCO-2004, XII European Signal Processing Conference, Vienna, Austria, September 6–10, 2004.
[47] G. Leus, P. Vandaele, M. Moonen, Deterministic blind modulation-induced source separation for digital wireless communications, IEEE Transactions on Signal Processing 49 (2001) 219–227.
[48] T.H. Li, K. Mbarek, A blind equalizer for nonstationary discrete-valued signals, IEEE Transactions on Signal Processing 45 (1997) 247–254. Special issue on communications.
[49] Y. Li, Z. Ding, Global convergence of fractionally spaced Godard (CMA) adaptive equalizers, IEEE Transactions on Signal Processing 44 (1996) 818–826.
[50] Y. Li, L. Yang, Semi-blind MIMO channel identification based on error adjustment, in: Proc. IEEE Conf. Neural Netw. Sig. Proc., Nanjing, December 2003.
[51] Z. Liu, G.B. Giannakis, S. Barbarossa, A. Scaglione, Transmit-antennae space-time block coding for generalized OFDM in the presence of unknown multipath, IEEE Journal of Selected Areas in Communications 19 (2001) 1352–1364.
[52] A. Medles, D. Slock, Achieving the optimal diversity-vs-multiplexing tradeoff for MIMO flat channels with QAM space-time spreading and DFE equalization, IEEE Transactions on Information Theory 52 (2006).

[53] A. Medles, D.T.M. Slock, Blind and semiblind mimo channel estimation, in: H. Bölcskei, D. Gesbert, C. Papadias, A.-J. van der Veen (Eds.), Space-Time Wireless Systems, From Array Processing to MIMO Communications, Cambridge University Press, 2006.
[54] X. Meng, J.K. Tugnait, Semi-blind channel estimation and detection using superimposed training, in: Proc. IEEE ICASSP, Montreal, Canada, May 17–21, 2004.
[55] E. Moulines, P. Duhamel, J.-F. Cardoso, S. Mayrargue, Subspace methods for the blind identification of multichannel FIR filters, IEEE Transactions on Signal Processing 43 (1995) 516–525.
[56] H.J. Pérez-Iglesias, J.A. García-Naya, A. Dapena, L. Castedo, V. Zarzoso, Blind channel identification in Alamouti coded systems: A comparative study of eigendecomposition methods in indoor transmissions at 2.4 GHz, European Transactions on Telecommunications 19 (2008) 751–759.
[57] W.H. Press, S.A. Teukolsky, W.T. Vetterling, B.P. Flannery, Numerical Recipes in C. The Art of Scientific Computing, 2nd ed., Cambridge University Press, Cambridge, UK, 1992.
[58] J.G. Proakis, Digital Communications, 4th ed., McGraw-Hill, New York, 2000.
[59] P.A. Regalia, On the equivalence between the Godard and Shalvi-Weinstein schemes of blind equalization, Signal Processing 73 (1999) 185–190.
[60] P.A. Regalia, Blind deconvolution and source separation, in: J.G. McWhirter, I.K. Proudler (Eds.), Mathematics in Signal Processing V, Clarendon Press, Oxford, UK, 2002, pp. 25–35.
[61] P.A. Regalia, A finite-interval constant modulus algorithm, in: Proc. ICASSP, 27th International Conference on Acoustics, Speech and Signal Processing, vol. III, Orlando, FL, May 13–17, 2002, pp. 2285–2288.
[62] L. Rota, P. Comon, Blind equalizers based on polynomial criteria, in: Proc. ICASSP-2004, 29th International Conference on Acoustics, Speech and Signal Processing, vol. IV, Montreal, Canada, May 17–21, 2004, pp. 441–444.
[63] B. Sadler, M. Kozik, T. Moore, Bounds on bearing and symbol estimation with side information, IEEE Transactions on Signal Processing 49 (2001) 822–834.
[64] A. Safavi, K. Abed-Meraim, Blind channel identification robust to order overestimation: A constant modulus approach, in: Proc. IEEE ICASSP, 2003.
[65] Y. Sato, A method of self-recovering equalization for multi-level amplitude modulation, IEEE Transactions on Communications 23 (1975) 679–682.
[66] A. Scaglione, G. Giannakis, S. Barbarossa, Redundant filterbank precoders and equalizers, part II: Blind channel estimation, synchronization, and direct equalization, IEEE Transactions on Signal Processing 47 (1999) 2007–2022.
[67] A. Scaglione, A. Vosoughi, Turbo estimation of channel and symbols in precoded MIMO systems, in: Proc. IEEE ICASSP, Montreal, Canada, May 17–21, 2004.
[68] A. Scherb, V. Kühn, K.-D. Kammeyer, Cramer-Rao lower bound for semiblind channel estimation with respect to coded and uncoded finite-alphabet signals, in: Proc. 38th Asilomar Conf. Signals, Systems & Computers, Pacific Grove, CA, November 2004.
[69] O. Shalvi, E. Weinstein, New criteria for blind deconvolution of nonminimum phase systems (channels), IEEE Transactions on Information Theory 36 (1990) 312–321.
[70] M.T.M. Silva, M. Gerken, M.D. Miranda, An accelerated constant modulus algorithm for space-time blind equalization, in: Proc. EUSIPCO-2004, XII European Signal Processing Conference, Vienna, Austria, September 6–10, 2004, pp. 1853–1856.
[71] D.T.M. Slock, Blind fractionally-spaced equalization, perfect-reconstruction filter banks and multichannel linear prediction, in: Proc. ICASSP-94, 19th International Conference on Acoustics, Speech and Signal Processing, vol. IV, Adelaide, Australia, Apr. 19–22, 1994, pp.585–588.
[72] D.T.M. Slock, Bayesian blind and semiblind channel estimation, in: Proc. IEEE-SP Workshop on Sensor Array and Multichannel Signal Processing, Spain, July 2004.
[73] R. Steele, L. Hanzo (Eds.), Mobile Radio Communications, 2nd ed., John Wiley & Sons, New York, 1999.
[74] A.L. Swindlehurst, A semi-blind algebraic constant modulus algorithm, in: Proc. ICASSP-2004, 29th International Conference on Acoustics, Speech and Signal Processing, vol. IV, Montreal, Canada, May 17–21, 2004, pp. 445–448.

[75] S. Talwar, A. Paulraj, Blind separation of synchronous co-channel digital signals using an antenna array – Part II: Performance analysis, IEEE Transactions on Signal Processing 45 (1997) 706–718.
[76] S. Talwar, M. Viberg, A. Paulraj, Blind estimation of multiple co-channel digital signals arriving at an antenna array: Part I, algorithms, IEEE Transactions on Signal Processing (1996) 1184–1197.
[77] L. Tong, G. Xu, T. Kailath, Blind identification and equalization based on second-order statistics: A time domain approach, IEEE Transactions on Information Theory 40 (1994) 340–349.
[78] J.R. Treichler, B.G. Agee, A new approach to multipath correction of constant modulus signals, IEEE Transactions on Acoustics, Speech and Signal Processing 31 (1983) 459–472.
[79] M. Tsatsanis, G. Giannakis, Transmitter induced cyclostationarity for blind channel estimation, IEEE Transactions on Signal Processing 45 (1997) 1785–1794.
[80] J.K. Tugnait, L. Tong, Z. Ding, Single-user channel estimation and equalization, IEEE Signal Processing Magazine 17 (2000) 16–28.
[81] A.-J. van der Veen, Analytical method for blind binary signal separation, IEEE Transactions on Signal Processing 45 (1997) 1078–1082.
[82] A.-J. van der Veen, A. Paulraj, An analytical constant modulus algorithm, IEEE Transactions on Signal Processing 44 (1996) 1136–1155.
[83] A.-J. van der Veen, S. Talwar, A. Paulraj, Blind estimation of multiple digital signals transmitted over FIR channels, IEEE Signal Processing Letters 2 (1995) 99–102.
[84] A.-J. van der Veen, S. Talwar, A. Paulraj, A subspace approach to blind space-time signal processing for wireless communication systems, IEEE Transactions on Signal Processing 45 (1997) 173–190.
[85] A. Vosoughi, A. Scaglione, On the effect of channel estimation error with superimposed training upon information rates, in: Proc. IEEE International Symposium on Information Theory, Montreal, Canada, June–July 2004.
[86] T. Wigren, Avoiding ill-convergence of finite dimensional blind adaptation schemes excited by discrete symbol sequences, Signal Processing 62 (1997) 121–162. Elsevier.
[87] C. Xu, J. Li, A batch processing constant modulus algorithm, IEEE Communications Letters 8 (2004) 582–584.
[88] Y. Zeng, T.S. Ng, A semi-blind channel estimation method for multiuser multiantenna OFDM systems, IEEE Transactions on Signal Processing 52 (2004).
[89] H. Yang, On-line blind equalization via on-line blind separation, Signal Processing 68 (1998) 271–281.
[90] D. Yellin, B. Porat, Blind identification of FIR systems excited by discrete-alphabet inputs, IEEE Transactions on Signal Processing 41 (1993) 1331–1339.
[91] J. Yue, K.J. Kim, T. Reid, J. Gibson, Joint semi-blind channel estimation and data detection for MIMO-OFDM systems, in: Proc. IEEE-CAS Symposium on Emerging Technologies: Mobile and Wireless Communications, Shanghai, China, May 2004.
[92] V. Zarzoso, P. Comon, Blind and semi-blind equalization based on the constant power criterion, IEEE Transactions on Signal Processing 53 (2005) 4363–4375.
[93] V. Zarzoso, P. Comon, Blind channel equalization with algebraic optimal step size, in: EUSIPCO-2005, XIII European Signal Processing Conference, Antalya, Turkey, September 4–8, 2005.
[94] V. Zarzoso, P. Comon, Semi-blind constant modulus equalization with optimal step size, in: ICASSP-2005, 30th International Conference on Acoustics, Speech and Signal Processing, vol. III, Philadelphia, PA, March 18–23, 2005, pp. 577–580.
[95] V. Zarzoso, P. Comon, Optimal step-size constant modulus algorithm, IEEE Transactions on Communications 56 (2008) 10–13.
[96] V. Zarzoso, P. Comon, Robust independent component analysis by iterative maximization of the kurtosis contrast with algebraic optimal step size, IEEE Transactions on Neural Networks 21 (2010).
[97] V. Zarzoso, A.K. Nandi, Blind MIMO equalization with optimum delay using independent component analysis, International Journal of Adaptive Control and Signal Processing 18 (2004) 245–263. Special issue on Blind Signal Separation.

[98] V. Zarzoso, A.K. Nandi, Exploiting non-Gaussianity in blind identification and equalization of MIMO FIR channels, IEE Proceedings – Vision, Image and Signal Processing 151 (2004) 69–75. Special issue on Non-Linear and Non-Gaussian Signal Processing.

[99] Y. Zhang, S.A. Kassam, Blind separation and equalization using fractional sampling of digital communications signals, Signal Processing 81 (2001) 2591–2608.

[100] L. Zheng, D. Tse, Communication on the Grassmann manifold: A geometric approach to the non-coherent multiple antenna channel, IEEE Transactions on Information Theory 48 (2002) 359–383.

[101] H. Zhu, B. Farhang-Boroujeny, C. Schlegel, Pilot embedding for joint channel estimation and data detection in MIMO communication systems, IEEE Transactions on Signal Processing 7 (2003) 30–32.

CHAPTER

Overview of source separation applications

16

Y. Deville, C. Jutten, and R. Vigario

16.1 INTRODUCTION

16.1.1 Context

Whereas the previous chapters of this book mainly concern Blind Source Separation (BSS) methods, based in particular on Independent Component Analysis (ICA), in this chapter and the subsequent ones, the emphasis is put on applications of such techniques. In each application domain (see also the next three chapters), priors are generally available and BSS shifts to semi-blind source separation (SBSS): exploiting priors is very important for designing simpler and more efficient algorithms. Finally, results are relevant only if assumptions (used in the separation criterion, or contrast function) and priors are satisfied by extracted sources.

Due to the current fast growth of the (blind or semi-blind) source separation field, the objectives of this chapter are restricted to a historical introduction of source separation applications, a strategical discussion on how to apply ICA, and descriptions of a few original applications, which are not addressed in the following three applications chapters and which allow us to present here various features and issues of BSS applications.

Moreover, BSS is a recent field, so that even when various methods are applied to similar problems in different papers, each of these papers often corresponds to specific conditions and associated databases. Consequently, it is not easy to compare the performances obtained in such investigations. A recent approach to this end is to propose BSS problems in international competitions, e.g. for audio source separation, and public databases, such as [73] for acoustic signals.

16.1.2 Historical survey

A good insight into the trends in source separation applications may be obtained by considering the first events that brought the BSS community together. The first special session ever devoted to this topic in an international conference is presumably the one which took place at the "1995 International Symposium on Nonlinear Theory and its Applications" (NOLTA'95). This session mainly consisted of theoretical contributions, some of which were illustrated by simulations performed with synthetic data. However, it also included a single[1] applications-driven paper [27]. That paper deals with data transmission, specifically with a multi-antenna BSS-based system, which aims at identifying several simultaneously present people or objects carrying radio-frequency tags. That system was developed by an industrial company, by extending a commercially available product which only applied to the situation when a single tag is present at a particular time. We will come back to this application in more detail in section 16.4.

Anticipating the subsequent developments reported below, it is worth mentioning that between the above-mentioned extremes, i.e. theoretical contributions and an industrial application, the NOLTA'95 session also included two papers in which the proposed algorithms were applied to already partly realistic signals. More precisely, the considered sources themselves were real signals. However, the mixed signals processed by BSS algorithms were artificially derived from these sources, by numerically computing linear instantaneous mixtures of these sources. These mixed signals were therefore clearly unrealistic (i.e. instantaneous mixtures of audio signals [6] instead of the convolutive mixtures which are provided by actual microphones, as discussed in Chapter 19), or at least they were not provided by the sensors of an actual setup or justified by the description of the operation of a specific system (see the artificial instantaneous mixtures of images in [20]). Nevertheless, these investigations were a first step of importance towards the more refined works mentioned below.

Applications to electromagnetic transmission, which were considered at NOLTA'95, then became even more prominent in the "Industrial applications" session organized in 1996 in Grenoble (France), in the framework of a summer school on higher-order statistics, with the main emphasis on BSS [17,18],[2] [25]. This session, which was probably the first of this kind, also included a paper [34] reporting BSS analyses of signals measured in mechanical structures. These analyses concerned: (i) the monitoring of movements of dams, and (ii) non-destructive control of the generators of nuclear power plants. Whereas the above papers emanated from industrial companies, other BSS applications developed by universities were presented in a poster session of this summer school. They covered the following three fields:

- the separation of speech and noise signals [10][3]: although the considered mixtures of these sources were still artificially created, they now had a convolutive nature;
- rotating machine monitoring [14], using artificial mixed signals derived from an experimental source;
- and, to a lesser extent because this paper did not contain experimental results, seismic data analysis [77].

A few months later, a one-day workshop mainly devoted to BSS was organized at the "1996 Conference on Neural Information Processing Systems" (NIPS'96).

[1]A second paper [42] in that session also involved real mixed signals, namely speech signals recorded in an anechoic chamber or a normal room. However, these signals were not processed by means of BSS algorithms in a "traditional sense". This paper is therefore not considered in the applications of BSS methods hereafter.

[2]For descriptions in English of the activities of this team during that period, the reader may refer to [19] and references therein.

[3]For a related paper in English, the reader may refer to [61].

It included various presentations concerning the application of BSS methods to two fields which are of increasing importance, i.e. (i) speech and audio signals, some of which were already recorded in completely realistic conditions, and (ii) biomedical data, i.e. electroencephalograms (EEG) and event-related potentials (ERP).

Other conference sessions focused on BSS and including application papers were subsequently organized. Moreover, many applications were described outside such focused meetings.[4] Therefore, a detailed account of all the application papers thus published cannot be provided here. However, a more complete review of these publications would confirm the major trends that appeared above. This is reflected in the last event in the BSS community mentioned here because of its importance, i.e. the first one-week workshop (ICA'99) completely dedicated to BSS, which took place in 1999 and which corresponds to a further step in the development of this domain, as compared to isolated conference sessions. This workshop included three sessions devoted to applications, which especially dealt with the three major application fields that had emerged at this stage, i.e. electromagnetic transmission systems, biomedical applications and audio.

The above-defined three application fields are still of high importance today. They are therefore respectively addressed in detail in Chapters 17 to 19. In addition, the current chapter not only presents several examples of applications in these and other fields, but also various general guidelines for using BSS in practice, and considerations about the issues related to applications, such as the robustness of the signals estimated from real-world data. This chapter is therefore organized as explained below.

16.1.3 Organization of this chapter

The remainder of this chapter consists of three main parts and a conclusion.

Section 16.2 first defines general challenging problems to be considered for successfully addressing source separation applications. Among them, section 16.3 focuses on concerns with the robust use of ICA in practice, and provides some suggestions on how to overcome these concerns.

Beyond the above general considerations, we then provide a detailed description of two types of applications, in order to clearly illustrate various aspects of the parameters to be taken into account and issues related to practical BSS applications. We thus, for example, discuss the relevance of extracted sources and limitations of ICA methods, and we provide comparisons of the performance of various BSS methods when applied to real-world signals and systems. The first two examples in this part of the chapter concern electromagnetic transmission systems (see section 16.4), which belong to a traditional application field for BSS, as explained above. The last example deals with astrophysics (see section 16.5), which is a domain where BSS applications appeared more recently.

In sections 16.6 and 16.7, we then discuss in more detail some aspects of BSS applications: this part deals with the properties of the sources and mixtures, such as their

[4]It should also be noted that the early paper [5] published in 1982, before the BSS field explicitly emerged, already concerned satellite communications.

dimensions (e.g. time series analysis or image processing), or the availability or lack of a physical mixture model, depending on the chosen application. These considerations are illustrated by means of actual examples, both by referring to the applications previously detailed in sections 16.4 and 16.5, and to some brief additional examples. The latter applications concern various problems, such as electrocardiogram analysis, optical imaging in the biomedical field or astrophysics.

Some major conclusions drawn from this chapter are provided in section 16.8, together with references about software and data that the reader may freely obtain on the Internet to start applying BSS methods.

16.2 HOW TO SOLVE AN ACTUAL SOURCE SEPARATION PROBLEM

16.2.1 Blind or semi-blind

A "fully blind" approach to source separation would assume strictly no prior knowledge about either source properties or type of mixture. However, the source separation problem cannot be solved in such conditions. Therefore, the most classical source separation approaches, which are called "blind methods", are based on generic priors. Three types of such priors have, in particular, been used in the literature, thus yielding three main classes of source separation methods, i.e.

- statistical independence of the sources, which leads to Independent Component Analysis (ICA);[5]
- positivity, which, in particular, leads to Non-negative Matrix Factorization (NMF);
- sparsity, which results in Sparse Component Analysis (SCA) and allows one to solve the source separation problem for more sources than observations (underdetermined case).

ICA may therefore be used when the only available prior concerning the sources is their statistical independence (or when more specific priors are also available but one prefers not to use them). ICA methods are powerful, but have a few drawbacks:

- The sources to be processed in practice are perhaps actually not independent.

[5]ICA methods based on the independence and non-Gaussianity of source signals were historically the first proposed methods. Source independence was therefore considered as the default hypothesis at that time, and BSS methods based on this hypothesis were called "blind methods". Some authors therefore called "semi-blind methods" subsequently proposed approaches which were based on different hypotheses. However, it should be clear that these other methods are not necessarily "less blind", i.e. more restrictive, than the original methods based on independence and non-Gaussianity. For instance, the BSS methods which use the temporal correlation of the source signals and are based on second-order statistics only request the signals to be uncorrelated, which is less constraining than independence (but these methods request temporal correlation, on the other hand). Similarly, SCA methods are less restrictive than ICA in the sense that they are applicable to dependent sources, and even correlated ones (but, on the contrary, they request source sparsity). Therefore, as opposed to the above historical definition of semi-blind BSS methods, one may now have the following point of view. Three main classes of BSS methods currently exist, i.e. ICA, NMF and SCA, as explained hereafter. All these methods are based on generic priors and are therefore called "blind methods" in this chapter. On the contrary, other methods use more specific priors and are called "semi-blind methods" hereafter.

- Since ICA algorithms are based on statistical independence, they require one to take into account higher (than 2) order statistics.
- ICA algorithms cannot separate Gaussian sources.

As opposed to the above generic priors, or in addition to them, some *specific* priors on sources or on mixtures are available in many applications. It is then strongly recommended to exploit these specific priors for designing simpler and more efficient source separation algorithms. These specific algorithms may then be called "semi-blind" methods. For instance, if sources are colored with different spectra, or non-stationary, one can design efficient algorithms based on second-order statistics, and which have the advantage of being able to separate Gaussian sources. Generally speaking, it is actually important to use all the available information we have concerning sources and observations: adding priors is a simple way to restrict the possible sets of solutions (it is a kind of regularization), and hence to achieve a unique solution, more simply.

Although the above priors were stated to be specific (as opposed to the more widely used "generic priors"), it should be clear that they are very weak in the sense that they are related to a wide (and qualitative) property which can apply to a large class of signals. Some specific priors are not sufficient for solving source separation alone. They may then be combined with independence (or uncorrelation) or other generic priors in a joint source separation criterion. For instance, one can use both source independence and positivity properties.

16.2.2 ICA for BSS

If we choose to use ICA, a few important questions are: can one apply ICA like a simple blackbox? Can one use any ICA algorithms? In other words, can one be confident in the results provided by ICA?

In fact, such questions are very usual for any estimation method, and ICA is nothing but a particular one. Basically, any estimation method is based on three ingredients: a parametric model, a criterion and an optimization algorithm. The parametric model provides a simple representation and restricts the solution to a particular space. The criterion, i.e. statistical independence for ICA methods, is a measure of the quality of the solution. The optimization algorithm is the way for optimizing (usually minimizing or maximizing) the criterion in the parametric model space. If we assume the optimization algorithm does not stop in spurious local extrema, ICA algorithms always converge to a solution which will be *optimal with respect to the chosen criterion in the parametric model space*.

When applying ICA, we consider P-dimensional observation vectors, $\mathbf{x}(t)$, which are assumed to be mixtures – through an unknown mapping $\mathscr{A}$ from $\mathbb{R}^N$ to $\mathbb{R}^P$ – of N-dimensional unobserved sources, $\mathbf{s}(t)$, assumed to be statistically independent:

$$\mathbf{x}(t) = \mathscr{A}(\mathbf{s}(t)). \tag{16.1}$$

In the following, we assume there are more observations than sources ($P \geq N$) and $\mathscr{A}$ is invertible. Then, ICA methods result in estimating a separating transform $\mathscr{B}$ such that:

$$\mathbf{y}(t) = \mathscr{B}(\mathbf{x}(t)) = (\mathscr{B} \circ \mathscr{A})(\mathbf{s}(t)), \tag{16.2}$$

with statistically independent components. As detailed in previous chapters, theoretical results prove that independence is sufficient for estimating the unknown sources, especially for linear mixtures, instantaneous [22] or convolutive [84,61,72], and even for particular nonlinear mixtures [74,3,36,2,48].

Of course, the estimated transform $\mathscr{B}$, and sources $\mathbf{y}(t)$, even optimal in the sense of the criterion, only leads to relevant solutions if:

- the nature of the separating transform $\mathscr{B}$ is suited to the mixing model $\mathscr{A}$;
- the independence assumption is actually satisfied by the unknown sources, $\mathbf{s}(t)$.

On the contrary, i.e. when the assumed mixing model (and then the selected separating model) is wrong or the source independence assumption is not satisfied, ICA can lead to irrelevant results.

16.2.3 Practical use of BSS and associated issues

In order to successfully apply BSS methods (e.g. ICA) to pratical problems, one first has (i) to define the considered relationship between the observations and the sources, which is obtained by modeling the physics of the system when this is possible, and (ii) to check if the assumed source properties (e.g. independence) are realistic. This is illustrated by means of various examples in this chapter, especially in section 16.4 and 16.5 where we proceed from "natural", i.e. simple, configurations to more complex ones. These examples also show various issues encountered in practical applications of BSS.

In addition to the above considerations, one often needs to make sure that a suitable amount of data exists, in order to correctly estimate the total amount of unknown parameters of the separation. With the increase of high-dimensional recordings, if the number of available samples is not sufficiently large, the algorithms will overfit or overlearn the data. In practice, most ICA approaches based on higher-order statistics will then produce sources presenting a single spike or bump, with values close to zero everywhere else, whereas temporal decorrelation methods will result in rather sinusoidal estimates [67].

ICA algorithms are usually derived under strict assumptions, such as statistical independence, stationarity and noiseless data. Yet, these assumptions may not always be fulfiled in practice. One often finds that the estimated sources change slightly every time the analysis is performed. Reasons for this fact may lie in the inherently stochastic nature of most ICA algorithmic implementations, the presence of noise or the deviation from that strict independence assumption.

These issues are discussed in more detail in the following section.

16.3 OVERFITTING AND ROBUSTNESS

We will now focus on two crucial questions regarding estimation and optimization methods. One addresses the balance between the information content in the data and that in the independence optimizing criterion, and how prone data is to overfit the contrast function. The second question deals with the robustness of certain ICA algorithms. Understanding how to identify, measure and control such phenomena is of paramount importance for a good use of the methodology considered in this chapter.

16.3.1 Overfitting

When the number of free unknown parameters in ICA is too high, as compared to the available sample size, the ICA model is likely to overfit or overlearn the data. A thorough study of overlearning phenomena in higher-order statistics ICA can be found in [67]. There, we see that overlearned components have a single spike or bump, and are practically zero everywhere else. The spike solution is typically due to a poor estimate of the ICA model, either because the contrast function is not appropriate or because it is inaccurately estimated. Note that true underlying sources can coexist with overfitting bumps or spikes. Figure 16.1 illustrates such phenomena. There, we clearly see a set of five physiological artifacts, present in magnetoencephalographic measurements, comprising cardiac, muscle and ocular activity. Similarly, we can observe a set of overfits that occupy the time domain rather randomly.

Solutions to the overfitting problem include, in addition to the acquisition of more data, a reduction of the dimensions of the data. Such a process can be performed during whitening, which typically precedes most ICA algorithms. One may need to take into account that a too drastic dimensionality reduction may result in a poor separation, as the resulting space will have a reduced degree of freedom, possibly insufficient for the required subsequent rotation. One may not be able to avoid that estimates retain most of the mixing nature of the observed data.

The overlearning problem is especially severe and cannot be solved properly in practice either by acquiring more samples or by dimension reduction, when the data have strong time dependencies, such as a $1/f$ power spectrum. This spectral characteristic is typical of many natural data [4]. In EEG and MEG, this problem greatly reduces the potential of ICA in analyzing the weak but complex ongoing brain signals, making the problem important to solve. Note as well that often a bump has a structure rather close to that of the desired source estimates. Such is the case, for example, in the analysis of averaged event-related recordings, where each signal deflection may be easily confused with an overfit (c.f. [82]). Similar considerations could be made for the analysis of fMRI, electrocardiogram or any sparse bumpy data type. On the other hand, in overfitting components there is no clear way to determine where the bump will occur. This variability can be used as an indication of a bad estimate.

To avoid the overlearning bumps, in addition to the suggestions above, filtering strategies were proposed in [67], with particular emphasis on high-pass filtering and data-driven singular spectrum analysis [85]. Another option is to use the innovation process,

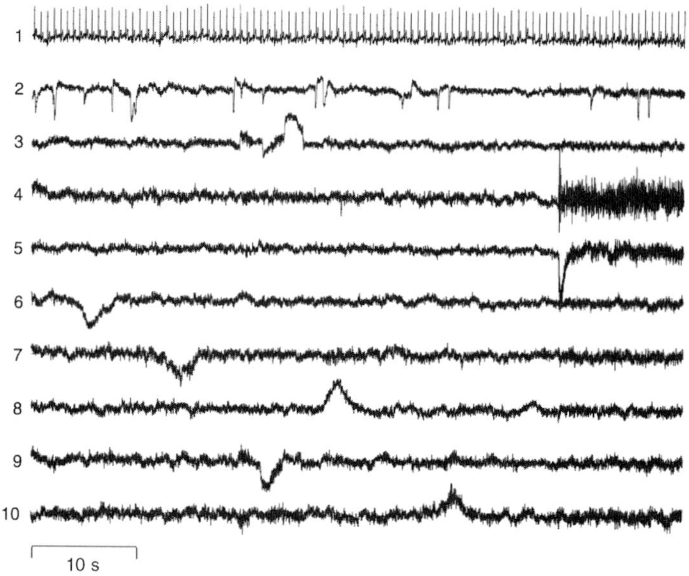

FIGURE 16.1

Example of the coexistence of artifactual independent components (top five) and bump-like overfits (bottom five).
Source: From [67].

rather than the recordings themselves. The innovation process is defined as the error of a stochastic process and the best prediction of it given its past [39]. It has been shown [43] that the solutions of ICA from the original and the innovation data sets should be identical. Furthermore, the innovations are usually less temporally-dependent and more non-Gaussian than the original processes.

Temporal decorrelation methods, such as SOBI [7] and TDSEP [89] have their own forms of overlearning as well, resulting in various periodic solutions, mostly sinusoidal. Dealing with such overfits should benefit as well, from the solutions presented above. Yet, one should make sure that the frequency ranges which are filtered out do not overlap those of the desired sources.

16.3.2 Robustness

ICA algorithms are usually derived under strict assumptions, such as source statistical independence and stationarity, and noiseless recordings. Independence and sparseness are strongly connected [52], which can result in a natural tendency of any ICA algorithm to bias towards sparse, rather than strictly independent, solutions.

One concern for a wider adoption of ICA as a routine processing tool, is that of the consistency of the estimated sources, when the algorithm parameters are freely changed. The estimated sources, or independent components, may change each time the analysis is performed. Even if sometimes small, such variability is often unacceptable, and reduces the trust in the found components. There are a number of possible factors influencing the estimation variability. One possibility is that the strict theoretical assumption of statistical independence may not hold for the data. Arguably, the good performance of ICA in practice, even under ill-conditioned situations, can be in part attributed to the fact that the sparse components often form a very natural and easily interpretable decomposition of the set [55,82]. The algorithmic implementation of ICA may also be inherently stochastic. Furthermore, additive noise or other features of the data can cause variations in the solutions, since the algorithms are typically derived for the noise-free case.

If an expert supervision exists, such behavior is usually overcome by comparing the estimated components with the known sources, or with references obtained by other methods (c.f. [13]). On the other hand, if the evaluation is based directly on the experimental set-ups, one can, at most, validate the expectations encoded within that design. This is still the default configuration in, for example, brain event-related studies, whether using EEG/MEG or fMRI recordings. One then overlooks crucial phenomena that deviate somewhat from the strictly imposed stimuli.

Another approach, which preserves the data-driven nature of ICA and BSS, is based on resampling and multiple-runs, in a bootstrapping [35] manner, to identify consistently reproducible components [33]. Most commonly the goal has been to analyze the stability of an algorithm, or to select the best performing model (c.f. [66]). Connections with bagging (c.f. [11]), as well as Bayesian theory (c.f. [21]) exist.

The approach proposed in [56] relies on running an ICA algorithm twice, with the inputs of the second run consisting of resampled copies of the first run's estimates. Consistent components will have a permutation vector in the second run, whereas poorly estimated and overfitted components will not be represented at the output of the second run. Himberg *et al.* [40] proposed instead a visualization method [45], based on clustering components from multiple runs of ICA with different starting conditions.

In [86], the resampled and multiple-run approach is revisited, but with a more intuitive and interpretable clustering strategy, based on a multi-path correlation analysis of the estimated components. The method can identify consistent components and neglect overfits. The latter would correspond to estimates appearing only once within the multiple runs. Additionally, it can reliably identify components that contain some variability, or are difficult to separate based on statistical independence alone, regardless of whether the component represents stimulus-related activation or not.

In the following example (Fig. 16.2), we see a set of robust independent components found from an fMRI experiment, together with their respective time courses. The gray-level bands around the temporal mean correspond to the spread of the cluster estimates for each time point. During the recordings, the subject listened to spoken safety instructions in 30 s intervals, interleaved with 30 s resting periods (for further details, see [86]). Notice that the component that follows best the given onset-offset pattern has very little variance, whereas the noisy component exhibits the highest degree of variability. The third frame shows an intermediate condition.

Other aspects of practical use of BSS are detailed by considering different types of applications in the following sections.

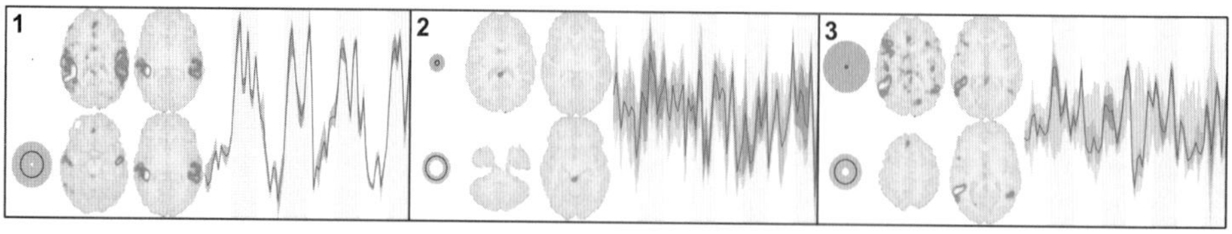

FIGURE 16.2

Three examples of consistent independent components, and their respective time courses. Gray levels around the mean time course correspond to quantiles of spread of the respective cluster. *Source*: Adapted from [86].

16.4 ILLUSTRATION WITH ELECTROMAGNETIC TRANSMISSION SYSTEMS

16.4.1 A variety of source natures and mixture configurations

As shown in the above historical survey, a significant number of investigations reported in the literature concern systems which aim at transmitting information (speech, data ...) by means of electromagnetic signals. Although this class of applications includes quite different systems, they lead to the same generic configuration, defined as follows. The emission part of the considered communication system contains one or several source transducers and its reception part includes a set of sensors. Both these source transducers and reception sensors here consist of electromagnetic antennas. Each sensor thus simultaneously receives various "source signals", as detailed below, together with "noise signals", which correspond to all the undesired electromagnetic interferences received by the antennas with non-negligible levels. These sensors thus provide a set of mixed signals, from which one aims at extracting the source signal(s) of interest in the considered application, by using BSS methods. This common framework gives rise to a wide range of situations, depending on the encountered sources and mixture:

- In the most classical applications, the signals to be separated correspond to physically distinct electromagnetic sources, i.e. to emitters situated in different locations. However, BSS methods have also been used in the case when a single original electromagnetic signal propagates through multiple paths to the reception antennas. Each antenna then receives a superposition of signals, which are respectively associated with all these propagation paths and which are the signals to be separated. The most general situation includes these two aspects, thus leading to multi-emitter and multi-path separation.
- The nature of the mixture which occurs between the source signals results from various source properties and/or propagation conditions, such as the narrow-band nature of the sources, the considered modulation/demodulation scheme, the use of low-frequency modulating signals, and the negligible propagation delays of electromagnetic signals over short distances. Part of the electromagnetic transmission applications thus leads to linear instantaneous mixtures, which are more easily handled than more general convolutive mixtures (while convolutive mixtures are, for example, encountered in almost all acoustic applications).

BSS methods have therefore been tested in quite different speech or data transmission applications. The most compact ones include single-board identification systems, where radio-frequency (RF) signals typically propagate over a few tens of centimeters [27,28,30]. On the contrary, applications related to telecommunication networks, military high-frequency transmission [19] or satellite communications [5], involve more complex systems, whose reception antennas are situated up to thousands of kilometers away from emission locations. Intermediate propagation distances are also encountered in such applications as secondary radar garbling processing for air traffic control [17]. The above references concern investigations which include experimental tests carried out with actual systems, whereas various other works were only performed with artificial data, numerically created by using a model of the considered configuration (especially of the modulation process). In the remainder of this section, we detail two of these experimental investigations, which illustrate several above-defined phenomena.

16.4.2 A case study on radio-frequency identification (RFID)

16.4.2.1 Overall mixture and separation configuration

We first consider an application which involves short-range propagation and standard separation, i.e. the separation of signals emitted by sources situated in different locations. This application concerns identification problems. The term "identification" here refers to the recognition of people, animals or objects that must be performed in many everyday-life situations, such as owner identification, before starting car engines, access control for restricted areas, cattle identification or control of the flow of manufactured products in factories. In the past, the approaches used to perform such identifications were mainly based on mechanical devices (such as keys for starting car engines), or human operators (e.g. visual inspection of people, cattle or products in the above examples). These approaches are progressively being replaced by various types of electronic systems, especially by systems based on RF communication, thus leading to the currently booming RFID (Radio-Frequency IDentification) field.

The investigation reported here started from a commercially available RF identification system [64,65]. This system is shown in Fig. 16.3. It consists of a base station inductively coupled to portable identifiers or "tags". These tags contain an LC resonator, a controller and non-volatile programmable memory (EEPROM). The memory contents are specific to each tag and allow one to identify the tag-bearer (person, object ...). The basic mode of operation of this system may be modeled as follows. The base station emits an RF sine wave, which is received by a single tag. This tag is thus powered and answers by emitting a sine wave at the same frequency (due to inductive coupling), modulated by its encoded memory contents. The base station receives this signal, demodulates it, and decodes it so as to determine the memory contents [30]. The overall identification system then checks these data and controls its actuators accordingly.

This type of system yields various advantages (see details, e.g. in [30]). However, when two tags are simultaneously placed in the RF field of the base station, both tags

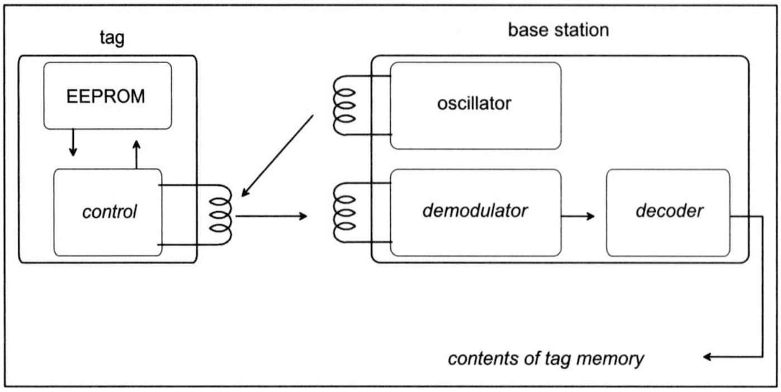

FIGURE 16.3

Single-tag RF identification system [30].

answer this station at the same time. The demodulated signal derived by this station is then a mixture of two components, and cannot be decoded by this basic station. This system is therefore unable to identify two simultaneously present tag-bearers. A few attempts to solve this problem have been presented in the literature, but they have various drawbacks (see details, e.g. in [30]). These drawbacks are avoided thanks to the modified version of this system, based on BSS techniques, which was introduced by Deville and Andry in [27–29] and subsequently extended in [30]. The prototype thus developed aims at simultaneously handling two tag signals. It relies on a base station containing two reception antennas and two demodulators, which yield two different mixed signals (see Fig. 16.4). These baseband mixed signals are processed by a BSS unit, which extracts the two components corresponding to the two tags. Then, by decoding these separated signals, the memory contents of the two tags are independently obtained.

The exact separation of a higher number n of tag signals may be achieved by the proposed approach, by using n reception antennas and demodulators. Anyway, it should be noted that the number of tags to be simultaneously handled remains low in most practical applications, thanks to a natural limitation which occurs in the considered version of the system: only the tags that are close to the base station (i.e. within a few cm or tens of cm) are activated and therefore participate in the mixture.

16.4.2.2 Source properties and mixing model

The sources to be restored consist of the encoded tag memory contents and may therefore be considered as statistically independent signals in most cases, i.e. if the tag memory bitstreams of interest are “quite different”. More details about the properties of these types of source signals are provided in [30].

Moreover, in an ideal model of the system described above, the mixed signals provided by the demodulators and to be processed by the BSS unit are restricted to their simplest possible form, i.e. they are linear instantaneous mixtures of the sources corresponding to the two tags. This characteristic of the mixtures results from the fact that (i) this system uses an amplitude modulation/demodulation scheme in which the modulating signal

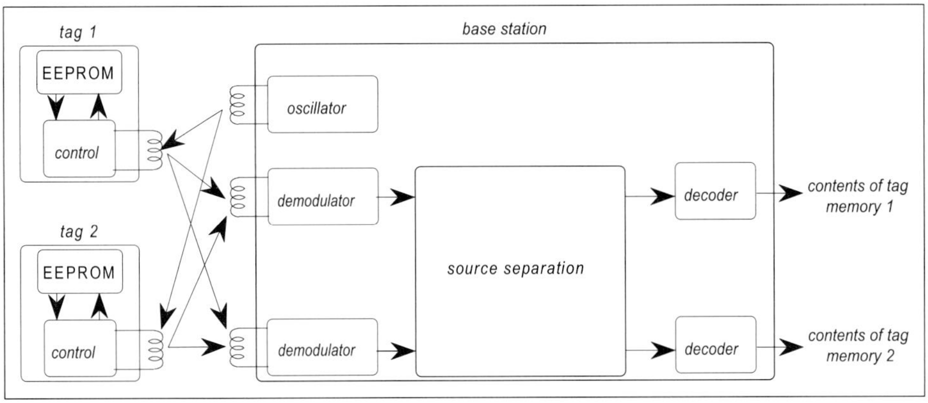

FIGURE 16.4

Multi-tag RF identification system [30].

and carrier have moderate frequencies and (ii) the source signals propagate over short distances. This may be shown as follows.

Let us consider the successive steps of the propagation and processing of the considered signals. Initially, a carrier signal $V sin\omega t$ is emitted by the base station. It is received by each tag j with a delay τ_j and a scale factor α_j (related to the attenuation of this signal which occurs during propagation). The signal received by tag j at time t is thus $\alpha_j V sin[\omega(t-\tau_j)]$. Assuming that this tag performs an ideal amplitude modulation of this carrier by its encoded tag data signal $s_j(t)$, the signal emitted by tag j at time t is $s_j(t)\alpha_j V sin[\omega(t-\tau_j)]$.

Now consider one of the reception antennas of the base station (the same principle applies to each of these antennas). It receives each emitted tag signal with a delay τ'_j and a scale factor α'_j. Each tag signal received by this antenna at time t is thus $a_j s_j(t-\tau'_j)V sin[\omega(t-\tau_j-\tau'_j)]$, with $a_j = \alpha_j\alpha'_j$. This antenna ideally performs a linear superposition of the contributions received from the two tags. The overall signal received by this antenna at time t is thus

$$a_1 s_1(t-\tau'_1)V sin[\omega(t-\tau_1-\tau'_1)] + a_2 s_2(t-\tau'_2)V sin[\omega(t-\tau_2-\tau'_2)]. \quad (16.3)$$

This overall received signal may be rewritten as

$$\begin{aligned} &sin\omega t[a_1 V s_1(t-\tau'_1)cos\phi_1 + a_2 V s_2(t-\tau'_2)cos\phi_2] \\ &+cos\omega t[a_1 V s_1(t-\tau'_1)sin\phi_1 + a_2 V s_2(t-\tau'_2)sin\phi_2] \end{aligned} \quad (16.4)$$

with $\phi_j = -\omega(\tau_j + \tau'_j) = -2\pi f L_j/c$, where f is the carrier frequency, L_j is the overall propagation distance (i.e. from the emitting antenna to tag j and then back to the reception antenna) and with $\tau'_j = l'_j/c$, where l'_j is the propagation distance from the tag to the reception antenna only.

In the considered system, these parameters have the following typical numerical values: $f = 125\ kHz$ and $L_j < 10\ cm$, so that $|\phi_j| < 2.6 * 10^{-4} \ll 1$; in addition, $l'_j < 5\ cm$, so that $\tau'_j < 1.7 * 10^{-10}s \ll T \simeq 10^{-3}s$, where T is the period of the modulating signal. Due to these values, the overall signal received by an antenna may be approximated by $sin\omega t[a_1 V s_1(t){+}a_2 V s_2(t)]$. The resulting output of an ideal amplitude demodulator is therefore $a_1 V s_1(t){+}a_2 V s_2(t)$. This signal, which is provided to the source separation unit, is thus indeed a linear instantaneous mixture of the modulating signals $s_1(t)$ and $s_2(t)$ of the two tags.

It should be noted that this result also applies to the case when each tag signal propagates through several paths to the base station: each path then yields an individual contribution; however, when neglecting all propagation delays as explained above, all these contributions are merged into a single overall signal which follows the above model.

16.4.2.3 Performance of on-line BSS methods, convergence speed/accuracy trade-off

Based on the above modeling of the mixing phenomenon, BSS methods suited to linear instantaneous mixtures are used in the considered system. Five such methods were selected in this investigation, due to their features, as explained in [30]. These methods are:

- the Hérault-Jutten recurrent neural network [46,47];
- its counterpart based on a direct structure, i.e. the Moreau-Macchi network [53,58];
- the modified version of the latter network based on a self-normalization of the network outputs, which was proposed by Cichocki *et al.* [20];
- the two networks defined by Deville *et al.* in this investigation [30]. These networks are respectively derived from the Hérault-Jutten recurrent and Moreau-Macchi direct networks by introducing a Normalized Weight Updating rule for their weights and are therefore respectively called the NWUr and NWUd networks hereafter.[6]

The first set of experiments performed with all these versions of the identification system aimed at comparing the convergence speed/accuracy trade-offs achieved by all these BSS methods in various situations, so as to select the best suited approaches. The convergence speed is measured by the number of samples T_c required for all network weights for converging to their equilibrium values.[7] Convergence accuracy is here assessed by means of the Signal-to-Interference Ratio Improvement (SIRI), i.e. the improvement of the Signal-to-Interference Ratio measured between the inputs and outputs of the considered BSS unit. SIRI is here always measured after convergence

[6] In this investigation, the separating functions f and g of all these networks were set to $f(x) = x^3$ and $g(x) = x$, as the considered sources were shown to be globally sub-Gaussian [30].

[7] T_c is here estimated from the plots representing the evolution of the network weights vs time.

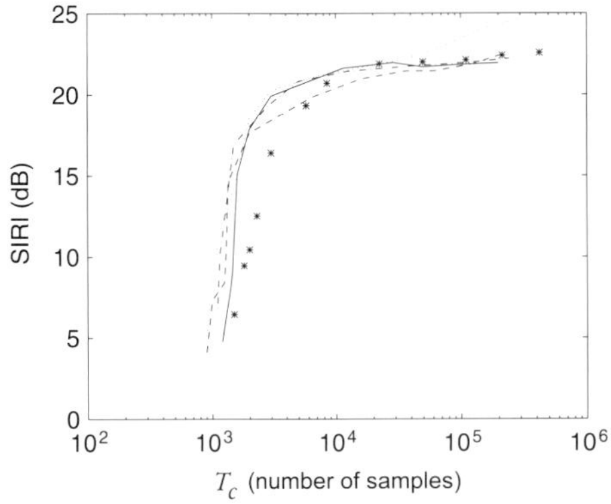

FIGURE 16.5

SIRI vs convergence time T_c, for moderately mixed RF sources. Each plot corresponds to a neural network: Hérault-Jutten: -.-. Moreau-Macchi: .... Cichocki: * * NWUr: --- NWUd: - -.
Source: Adapted from [30].

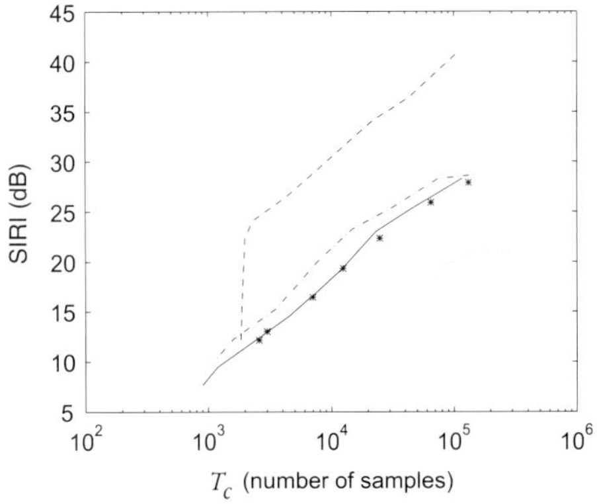

FIGURE 16.6

SIRI vs convergence time T_c, for highly mixed RF sources. Each plot corresponds to a neural network: Hérault-Jutten: -.-. Moreau-Macchi: .... Cichocki: * * NWUr: --- NWUd: - - (for the Moreau-Macchi and Cichocki networks, lower values of T_c than those provided in this figure cannot be reached, as T_c and *SIRI* then become very sensitive to an increase of the adaptation gain, and these networks eventually diverge when this gain is further increased).
Source: Adapted from [30].

has been reached, because constraints on the convergence time will be introduced in a different way further in this investigation. These types of performance parameters and the overall procedure for determining the performance of a BSS method are discussed in more detail in Chapter 19.

These first experiments were performed with artificial mixtures of real sources (files corresponding to these types of sources are available on the Internet at [26]). More precisely, each individual source signal was independently obtained by placing a single tag in the RF field of the base station and consists of the sampled output of a demodulator of the considered system. These source signals were then numerically mixed according to a linear instantaneous model. It should be noted that, although artificial, the mixed signals thus obtained correspond to the type of mixture actually encountered in the experimental setup defined above. The convergence speed/accuracy trade-off achieved by each network is determined by performing tests for various values of the network adaptation gain, deriving the values of T_c and $SIRI$ obtained in these conditions and plotting the resulting variations of $SIRI$ vs T_c. Figures 16.5 and 16.6 show the results obtained in two cases, i.e. respectively by mixing the source signals with the matrices

$$\begin{bmatrix} 1 & 0.4 \\ 0.3 & 1 \end{bmatrix} \text{ and } \begin{bmatrix} 1 & 0.98 \\ 0.98 & 1 \end{bmatrix}. \tag{16.5}$$

The first matrix yields moderately mixed sources. It is similar to the mixture coefficient values actually found in the considered experimental setup. The second matrix corresponds to a much more difficult situation, as the two mixed signals provided to the BSS networks are then very similar and the sources are therefore expected to be quite hard to separate for the considered BSS methods.[8] This second matrix therefore makes it possible to analyze the robustness of the considered approaches to ill-conditioned mixtures. Such highly mixed sources could, in particular, be encountered in long-range extensions of this system, when the two tags are very close to one another as compared to their distances with respect to the reception antennas.

The main conclusions that may be drawn from these figures result from the range of convergence times T_c required in practical applications, i.e. preferably about 2000 samples, but possibly up to 10,000 samples as explained in [30]. Figures 16.5 and 16.6 then show that the Moreau-Macchi network should preferably not be used in the considered application, as it cannot achieve the desired T_c for highly mixed sources. The Cichocki network is also not attractive here because: (i) it cannot reach $T_c \simeq 2000$ samples, or its $SIRI$ is then rather low and (ii) for any T_c in the considered range, its $SIRI$ is lower than or equal to those of the remaining three networks, i.e. Hérault-Jutten, NWUr and NWUd. Among the latter three networks, the preferred ones depend as follows on the main parameter of interest in the considered application. All three networks can reach $T_c \simeq 2000$ samples with an acceptable $SIRI$, but this is almost the limit achievable by the NWUd network. Therefore, if minimizing T_c is of utmost importance in the considered application, the Hérault-Jutten and NWUr networks are the best suited solutions. On the contrary, if the emphasis is laid on the SIRI of the network for highly mixed sources, while the value of T_c is not critical (within the considered range), the NWUd network should be preferred.

Up to this point of the discussion, we only studied the performance, in terms of T_c and $SIRI$, of the considered networks. But another feature of these networks should also be taken into account in the considered application, i.e. their ability to operate in a self-normalized (i.e. "automated") way, so that they are insensitive to the unknown signal magnitudes [30]. Then, the Hérault-Jutten network is to be avoided. In other words, the preferred networks in the considered application are NWUr and NWUd, and the eventual selection between these two networks depends on whether the emphasis is laid on a low T_c or on $SIRI$ for highly mixed sources, as explained above.

Only these two networks were therefore considered in the second step of this investigation. The corresponding experiments were performed with real mixed signals, recorded at the outputs of the two demodulators of the actual system, while two tags were simultaneously placed in the RF field of the base station and therefore emitted source signals. The method used in that case to assess performance is different from the approach considered above, mainly because the source signals are not available anymore here, but also because, in this final stage of the evaluation, one should preferably use

[8]We here restrict ourselves to the BSS methods which were considered in [27–30]. Since then, equivariant BSS methods have been proposed. Their performance is independent from the mixing matrix. The second matrix considered here therefore does not correspond to a difficult situation for these improved methods.

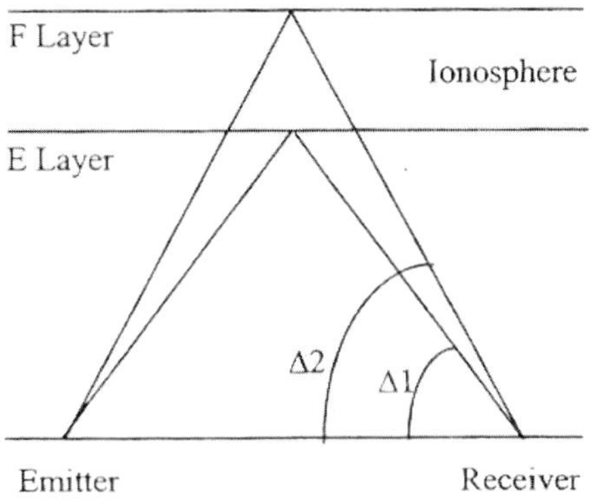

FIGURE 16.7

High-frequency propagation paths involving reflections on several layers of the ionosphere.
Source: From [19].

a performance criterion which is as close as possible to the parameter which is of importance in the considered application. The alternative method used here therefore consists in applying the BSS network outputs to the decoders of the system. These decoders then provide the estimated tag bitstreams. Comparing these data with those actually stored in the tags (which are known in these tests but unknown during normal operation, of course) here shows that these data are exactly the same. In other words, this system achieves a perfect restoration of the sources from an application point of view, in the sense that it restores the bitstreams of the tags without any errors.

16.4.3 A system with multi-path ionospheric propagation

16.4.3.1 Introducing several source signals from a single emitter

The second application that will now be described was reported by Chevalier *et al.* in [19]. It involves propagation over much longer distances than in the above identification system. It uses an experimental digital 8-PSK signal situated in a part of the High-Frequency (HF) band, i.e. [6 MHz, 11 MHz]. This signal is emitted from a town and received by a 5-sensor array situated in another town, which is 340 km away from the emission location. This experiment does not involve direct propagation between the emission and reception antennas. Instead, the emitted signal is reflected on several layers of the ionosphere, as shown in Fig. 16.7. Each reception antenna therefore receives a superposition of several signals, which correspond to the different propagation paths of the emitted signal.

These real antenna signals are first converted to an intermediate frequency and digitized. The associated baseband analytical signals are then derived. The authors show that the latter signals are linear instantaneous mixtures of the contributions respectively

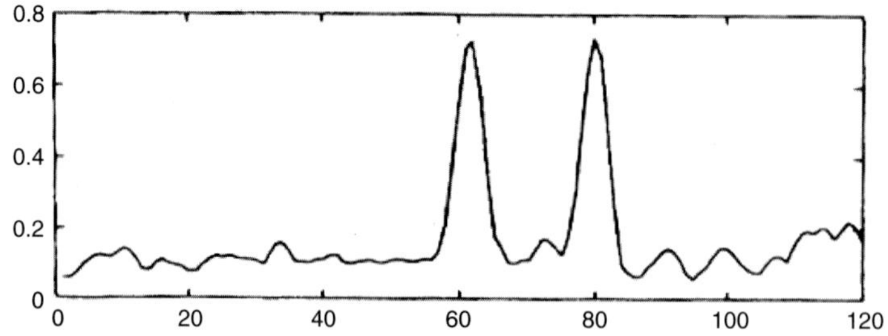

FIGURE 16.8

Synchronization function of the mixed signals.
Source: From [19].

associated with each propagation path (with superposed noise). The initial configuration is thus reformulated in terms of a BSS problem. BSS methods are therefore applied to these mixed signals in order to separate their components. BSS is thus used for multi-path separation and each of the outputs of the BSS unit aims at extracting the signal associated with an individual propagation path.

16.4.3.2 Source properties and performance criterion

Based on the above problem formulation, performance is here measured in terms of the quality of multi-path separation. The parameter used to characterize the presence of multiple paths in the baseband signals available before applying BSS is the synchronization function of these signals. This discrete-time function $C(n)$ is essentially the cross-correlation of a temporal window of the sampled received signals and of the time-limited, known, training sequence that they contain. This function has correlation peaks at times n such that the training sequence is synchronized to one of the paths contained in the received signals.

Figure 16.8 shows such an experimental synchronization function in a configuration involving two-path propagation. These paths yield two peaks in this function $C(n)$, which respectively correspond to samples $n \simeq 62$ and $n = 80$. The delay between these two paths is therefore about 18 samples, which corresponds to about 5 symbols, since the signals are processed so as to obtain 4 samples per transmitted symbol. As a first approach, the contributions of the two considered sources in the mixed signals available at any time may therefore be associated with different (and hence independent) symbols in the sequence of symbols which corresponds to the emitted signal. To be more precise, one should take into account that the actual emitted signal is a filtered version of this symbol sequence. This emission filter introduces some temporal dependence in the emitted signal. However, this dependence is only significant in the short term. The sources associated with different (and moreover non-adjacent) symbols may therefore be considered as almost independent. This condition is a key issue for many BSS methods, i.e. for the approaches based on ICA. It was not initially guaranteed to be fulfiled in this

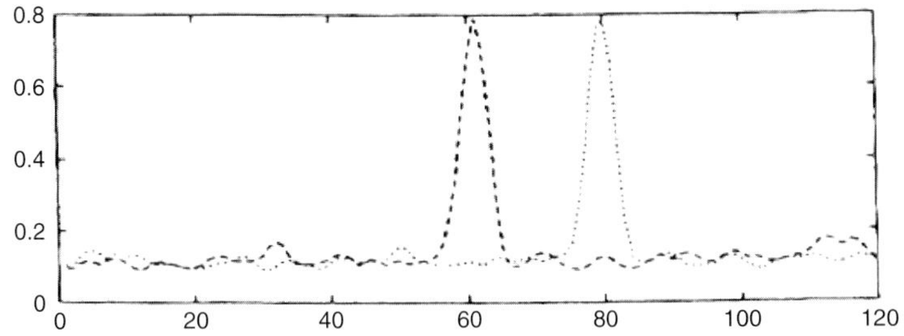

FIGURE 16.9

Synchronization functions of the outputs of the BSS unit. Each of the two plots corresponds to one output.
Source: From [19].

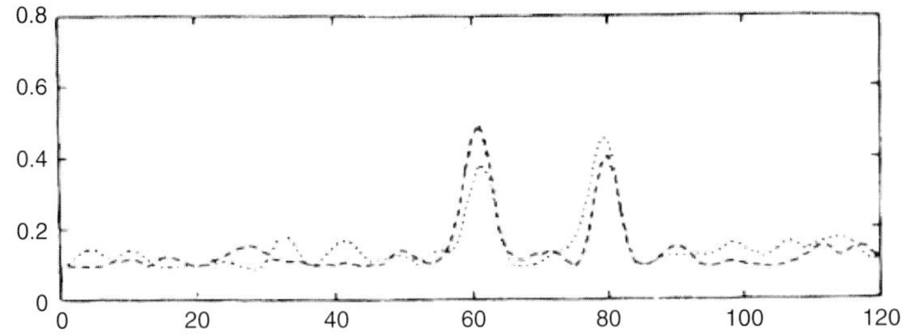

FIGURE 16.10

Synchronization functions of the outputs of the BSS unit, for a shorter observation period than in the previous figure.
Source: From [19].

application, as all the source signals introduced in the BSS formulation of this problem are derived from the same initial physical emitter.

16.4.3.3 Performance of some BSS methods

Synchronization functions $C(n)$ may then be computed for the outputs of the BSS unit, in order to analyze their ability to separate the propagation paths. This is illustrated in Fig. 16.9: each function $C(n)$ associated with a single output of the BSS unit contains only one peak, which shows that this output succeeds in extracting the corresponding path. This result was obtained by processing 20-ms signal windows (which correspond to about 50 samples) with the JADE algorithm.

A more detailed analysis then consists in determining if multi-path separation may also be achieved with shorter signal windows. Figure 16.10 represents the functions $C(n)$ obtained with 5-ms windows, i.e. about 12 samples. Each of them still contains two major peaks, which proves that the propagation paths are not separated in the corresponding output of the BSS unit.

Other results including more difficult configurations are reported in [19]. This shows that HF ionospheric multi-paths may be separated in less than 100 ms by the JADE method in most situations of practical interest, provided these paths have low spatial and temporal dependence. Comon's method described in [22] yields similar results. In addition, it is more robust with respect to the over-estimation of the number of sources. These experimental tests confirm the results previously obtained by means of simulations with artificial signals.

16.4.4 Using other signal properties

Other mixture identification and/or source separation approaches have also been applied to communication signals. Part of these methods take into account specific properties of these signals, such as their cyclostationarity [37,68] or their discrete nature [23], which in addition allows one to handle under-determined mixtures [23]. For instance, the tag identification application that we described above was recently revisited with a method intended for cyclo-stationary sources [68], which allows one to reduce the required signal size to about 300 samples, i.e. 19 symbols.[9] Applications of BSS methods to communications, especially using cyclo-stationarity properties, are described in more detail in Chapter 17. Another application field is considered hereafter.

16.5 EXAMPLE: ANALYSIS OF MARS HYPERSPECTRAL IMAGES

In this section, we illustrate the general ideas introduced at the beginning of this chapter using the analysis of Mars hyperspectral images, whose goal is to classify the planet Mars surface (see [60] for details). We first derive a mixing model of observations (hyperspectral images) based on physics; then, after choosing a spatial ICA decomposition, we present the ICA results and discuss relevance of extracted "independent" components (IC). Finally, we show how one can improve results using extra information, i.e. following a semi-blind approach, which is here based on positivity.

16.5.1 Physical model of hyperspectral images

16.5.1.1 The Omega spectrometer

The OMEGA spectrometer, carried by Mars Express spacecraft on an elliptical orbit, has a spatial resolution range from 300 m to 4 km. Here, we focus on a data set consisting of a single hyperspectral data cube of the South Polar Cap of Mars in the local summer where CO_2 ice, water ice and dust[10] were previously detected [9,71]. This data cube consists of 256 spectral planes from 0.93 μm to 5.11 μm with a resolution of about 0.020 μm. The reflectance[11] depends on the different compounds present at the pixel

[9]The numbers of samples and symbols required in all the tag identification methods studied above could be further reduced. This results from the fact that, in the considered experimental system, the sampling frequency was fixed, thus leading to 16 samples per symbol. However, good performance could also be achieved in a modified system which would sample the signals at a lower frequency.

[10]These three chemical constituents are called "endmembers" in the following, as usual in astrophysics.

[11]The reflectance is the ratio between the irradiance leaving each pixel toward the sensor and the solar irradiance at the ground.

location. Here, we just consider the simplest "geographic" mixture [49], where each pixel is a patchy area made of several pure compounds. This type of mixture happens when the spatial resolution is not large enough to observe the complex geological combination pattern. The total reflectance in this case will be a weighted sum of the pure constituent reflectances. The weights (abundance fractions) associated with each pure constituent are surface proportions inside the pixel and the mixing $\mathcal{A}$, since linear, reduces to a mixing matrix **A**.

In the experiments, we just consider $N_f = 174$ channels collected in the infrared region, removing noisy, hot and dead spectral planes. The spatial size of the data sets varies from $323 \leq N_x \leq 751$ and $N_y \in \{64, 128\}$, for a total pixel number: $41,344 \leq N_x \cdot N_y \leq 56,832$. Each data set (i.e. hyperspectral image) is then a data cube of size $N_x \times N_y \times N_f$ which contains between 7,193,856 and 9,888,768 pixels depending on the image size.

16.5.1.2 Observation model

Under geographic mixture model and assumptions concerning atmospheric contributions [75], the radiance factor at location (x,y) and wavelength λ is modeled as:

$$L(x,y,\lambda) = \left(\rho_a(\lambda) + \Phi(\lambda) \sum_{p=1}^{P} \alpha_p(x,y)\rho_p(\lambda) \right) \cos\left[\theta(x,y)\right] \tag{16.6}$$

where $\Phi(\lambda)$ is the spectral atmospheric transmission, $\theta(x,y)$ the angle between the solar direction and the surface normal (solar incidence angle), P the number of endmembers in the region of coordinates (x,y), $\rho_p(\lambda)$ the spectrum of the p-th endmember, $\alpha_p(x,y)$ its weight in the mixture and $\rho_a(\lambda)$ the radiation that does not arrive directly from the area under view. With simple algebra, this model can also be written as:

$$L(x,y,\lambda) = \sum_{p=1}^{P} \alpha'_p(x,y) \cdot \rho'_p(\lambda) + E(x,y,\lambda) \tag{16.7}$$

where

$$\begin{cases} \alpha'_p(x,y) &= \alpha_p(x,y)\cos\left[\theta(x,y)\right], \\ \rho'_p(\lambda) &= \Phi(\lambda)\rho_p(\lambda), \\ E(x,y,\lambda) &= \rho_a(\lambda)\cos\left[\theta(x,y)\right]. \end{cases} \tag{16.8}$$

As it can be seen in Eq. (16.8), the true endmember spectra are corrupted by the atmospheric attenuation, $\Phi(\lambda)$, and the abundance fractions are corrupted by the solar angle effect, $\cos\left[\theta(x,y)\right]$ (geometrical effect). However, since the abundance fraction is proportional to the quantity of each constituent in the geographical mixture, from the mixture model (16.7) and Eq. (16.8) one can see the abundance fractions are not altered by the geometrical effect since:

$$c_p(x,y) = \frac{\alpha'_p(x,y)}{\sum_{j=1}^{P} \alpha'_j(x,y)} = \frac{\alpha_p(x,y)}{\sum_{j=1}^{P} \alpha_j(x,y)}. \tag{16.9}$$

16.5.2 Decomposition models based on ICA

We now consider a hyperpectral data cube with N_f spectral images of $N_z = (N_x \times N_y)$ pixels. For simplicity, we assume raw vectorized images $I(n, \lambda_k)$, where

$$n = (i-1)N_y + j \tag{16.10}$$

is the *spatial index* (i and j are the initial row and column image indices) and k, $k = 1, \ldots, N_f$, is the *spectral index* for wavelength λ_k. We can then propose two decompositions of the hyperspectral images, i.e. spectral and spatial decompositions.

16.5.2.1 Spectral mixture decomposition

In the spectral model, we approximate the spectrum of each pixel as a weighted sum of N_c basis spectra:

$$I_n(\lambda_k) \approx \sum_{p=1}^{N_c} a_{(n,p)} \psi_p(\lambda_k), \quad \forall n = 1, \ldots, N_z, \tag{16.11}$$

where $\psi_p(\lambda_k)$, for $p = 1, \ldots, N_c$, are the constituent reflectance spectra, the number N_c is chosen for achieving a desired approximation accuracy, and $a_{(n,p)}$ are the mixing coefficients, which constitute a $N_z \times N_c$ matrix.

If ICA is used for the estimation, then the N_c basis spectra ψ_p, $p = 1, \ldots, N_c$, should be statistically independent. Moreover, the p-th column of the matrix $\mathbf{A}$ is the vectorized image associated to the basis spectrum ψ_p. With this model, we only have $N_f = 174$ samples for estimating the huge matrix $\mathbf{A}$ which has $N_z \times N_c \approx 250{,}000$ parameters (taking $N_c = 5$). The estimation problem is intractable except if one can select a small number of pixels (with respect to the sample number) among the N_z.

16.5.2.2 Spatial mixture decomposition

In the spatial model, we approximate the whole image at each frequency as a weighted sum of N_c basis images. Thus, at each wavelength λ_k, the observed image $I_{\lambda_k}(n)$ reads:

$$I_{\lambda_k}(n) \approx \sum_{p=1}^{N_c} a_{(\lambda_k,p)} Im_p(n), \quad \forall k = 1, \ldots, N_f. \tag{16.12}$$

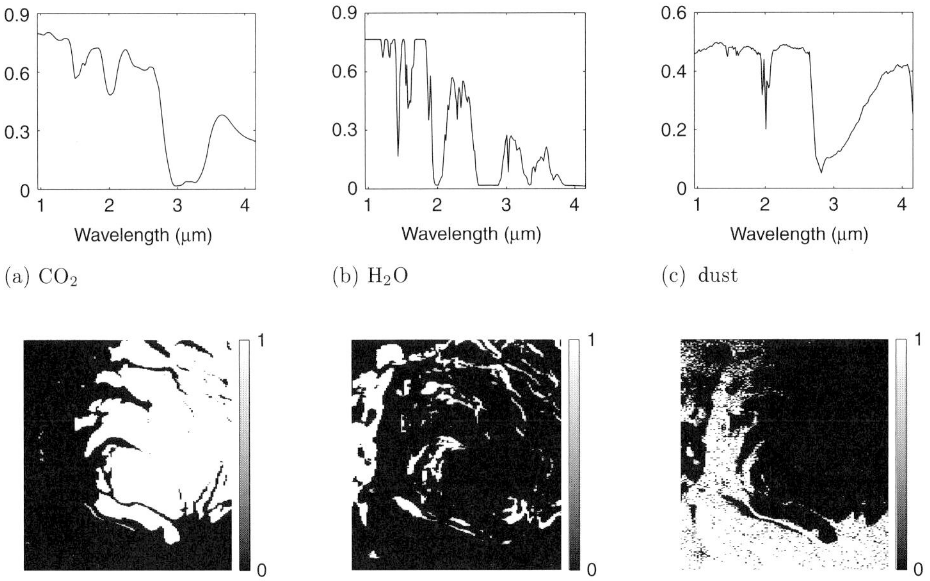

(a) CO_2 (b) H_2O (c) dust

(d) CO_2 (e) H_2O (f) dust

FIGURE 16.11

Reference spectra (top) and classification masks (bottom).

If ICA is used for the estimation, then the N_c basis images $Im_p(n)$, $p = 1,\ldots,N_c$, should be statistically independent. Moreover, the k-th column of the matrix $\mathbf{A}$[12] is the spectrum associated with the basis image Im_k. According to this model, we have a very large number of samples $N_z \approx 50{,}000$ for estimating the matrix $\mathbf{A}$ which has $N_f \times N_c < 900$ parameters (taking $N_c = 5$).

16.5.3 Reference data and classification

Since no ground truth is available on Mars, we need some reference information about the three main endmembers: dust, and CO_2 and H_2O ices. Two kinds of reference data are available (Fig. 16.11):

- reference classification masks obtained with the wavanglet classification method [71], which produces classification masks (Fig. 16.11, (d) to (f)) which are neither unique nor complete, i.e. pixels can belong to more than one class and some pixels are not classified;
- reference spectra: spectra of CO_2 and H_2O ices are simulations based on a radiative transfer model in realistic physical conditions of Mars [32], and, thus, they

[12] Although the mixing matrices are different in spatial and spectral models, we use the same notation $\mathbf{A}$.

are atmosphere-free simulations; conversely, the dust reference spectrum is derived from an OMEGA's observation, and, consequently, it is corrupted by the atmospheric transmission.

16.5.4 ICA results on hyperspectral images

Following the mixture model (16.7), one considers the data can be viewed as a linear mixture of sources, according to the ICA spatial decomposition (16.12).

16.5.4.1 Spatial ICA

In this experiment, we use two data sets. The first one is the original data set (RDS, for raw data set), while the second one is a preprocessed (done by astrophysicists) data set (PDS), obtained from the original data set by canceling the geometrical effect, atmospheric attenuation and a few known defects of the sensors.

Choosing the number of ICs. The first practical issue is to choose the number, N_c, of independent components (ICs), which must be at least equal to the number of sources present in the mixtures. If N_c gets larger, the accuracy of the approximation (16.12) increases, but extra ICs can be difficult to interpret. The choice is made using principal component analysis: on RDS, with 7 principal components, 98.58% of the variance of the initial image is preserved. In Fig. 16.12, we show the 7 ICs extracted with JADE [15].

Relevance of the ICs. A second step consists in evaluating the relevance of each component IC_k in the approximation. This is done by measuring the relative quadratic loss ϵ_k, obtained by canceling the IC_k, i.e. obtained when replacing the N_c-order approximation $\hat{I}_{N_c}$ with the (N_c-1)-order, denoted $\hat{I}_{N_c|N_k}$:

$$\epsilon_k = -10\log_{10}\left(\frac{P_{\hat{I}_{N_c}} - P_{\hat{I}_{N_c|N_k}}}{P_{\hat{I}_{N_c}}}\right), \tag{16.13}$$

where the energies of the approximated images are computed as:

$$P_{\hat{I}_{N_c}} = \sum_{m=1}^{N_f}\sum_{n=1}^{N_z}\left(\sum_{p=1}^{N_c} a_{(\lambda_m,p)} Im_p(n)\right)^2 \tag{16.14}$$

$$P_{\hat{I}_{N_c|N_k}} = \sum_{m=1}^{N_f}\sum_{n=1}^{N_z}\left(\sum_{p=1,p\neq k}^{N_c} a_{(\lambda_m,p)} Im_p(n)\right)^2. \tag{16.15}$$

These values, computed for the ICs estimated with the two data sets, are given in Table 16.1.

Finally, one has to wonder if the ICs are relevant and especially if they are robust with respect to the number of ICs and to the ICA algorithm. For this purpose, we did two sets of experiments:

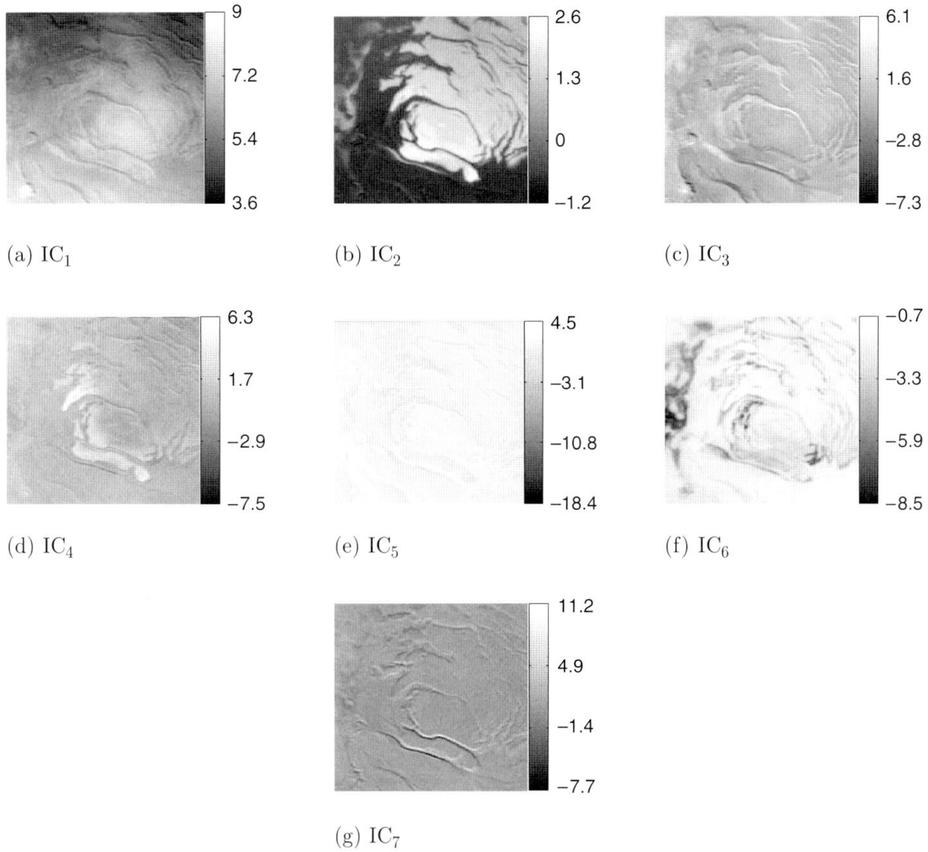

FIGURE 16.12

Independent components of the RDS hyperspectral image of the Martian south polar cap computed with JADE with 7 ICs.

1. We compared results obtained with the two data sets (RDS and PDS). We checked that a high-quality reconstruction is achieved with 4 ICs for PDS and with 7 ICs for RDS. If more than 4 (or 7) ICs are used, one always gets the same main 4 (or 7) ICs, while the others have very small contributions on the image reconstruction and cannot be interpreted. For this reason, we chose 4 ICs for PDS and 7 ICs with RDS.

2. We ran three ICA algorithms: FastICA [44] with various non-linearities in the symmetric or deflation versions and JADE [15]. In all these experiments, the results (ICs and reconstruction performance) are very similar and we preferred JADE since it has a low computational load and does not require parameters, except a stopping criterion.

Table 16.1 Independent components estimated with JADE. First column indicates the index of the IC. IC interpretation (see text for details) is given in column 2. The third column refers to the figure number (from 16.12(a) to 16.12(g)). The fourth and fifth columns are the losses in dB (ϵ_k) obtained if IC_k is not used in the approximation, for raw or preprocessed data, respectively. For this comparison, we did one experiment with 7 ICs for both raw and preprocessed data sets

k	Identification	Figure	RDS data	PDS data
			ϵ_k [dB]	ϵ_k [dB]
1	Solar angle effect	16.12(a)	32.6	1.3
2	CO_2 ice	16.12(b)	16.3	10.7
3	Atmospheric effect	16.12(c)	12.2	0.88
4	Intimate mixture?	16.12(d)	6.8	6.6
5	Corrupted line	16.12(e)	6.2	–
6	H_2O ice	16.12(f)	7.1	5.9
7	Channel shift	16.12(g)	2.0	0.1

16.5.4.2 ICs interpretation

In the spatial approximation, each IC_k can be viewed as an image, while column k of the estimated mixing matrix is the spectrum related to IC_k. So, IC interpretation can be done comparing the IC image or spectra to the reference classification masks or reference spectra (Fig. 16.11). Computing correlation with classification masks, the components IC_2 and IC_6 can be easily identified (Table 16.1) to, respectively, CO_2 ice and H_2O ice. In addition, correlations of columns 2 and 6 of $\mathbf{A}$ with reference spectra confirm this interpretation. Conversely, the spectrum associated with IC_4 has typical bands of dust, and CO_2 and H_2O ices: this could be due to a complex mixture (perhaps nonlinear, called intimate mixture by physicists) or to the dependence of those components.

We cannot interpret the other components (IC_1, IC_3, IC_5 and IC_7) with spectral as well as mask references. However, we remark that, on the preprocessed data, the energies of these ICs are very small (Table 16.1, last column). This means that these ICs are strongly reduced by the preprocessing, and thus must be related to phenomena canceled by the preprocessing. In fact, IC1 (Fig. 16.12(a)) has a luminance gradient which is characteristic of the solar angle effect ($E(x,y,\lambda)$ term in Equation (16.8)). Other ICs can be interpreted with the help of astrophysicists: IC_7 (Fig. 16.12(g)) is due to a misalignment between two parts of the detector; IC_3 can be associated with the transmission in the atmosphere effect; IC_5 (Fig. 16.12(e); the first line in this image has a very low response) corresponds to a corrupted line in the data set, due to a known sensor failure.

16.5.4.3 Independence of spatial sources

Spatial ICA decomposition assumes that the source images are independent. However, it is clear that this assumption is wrong.

This can be showed by computing the covariance matrix of the reference masks:

$$R_s\{I_{\mathrm{dust}}, I_{\mathrm{CO_2}}, I_{\mathrm{H_2O}}\} = \begin{pmatrix} 1 & -0.61 & -0.24 \\ -0.61 & 1 & -0.25 \\ -0.24 & -0.25 & 1 \end{pmatrix}.$$

From this covariance matrix, it is clear that CO_2 and H_2O ices have weak correlation, while CO_2 ice and dust are strongly correlated. Consequently, when using spatial ICA, components of CO_2 and H_2O ices can be retrieved while dust is not retrieved as a separate IC, since it is very similar to the negative of CO_2 ice.

16.5.5 Discussion

First, the physical model of observations shows that the extracted sources cannot be exactly the desired sources: in this example, endmember spectra cannot be dissociated from spectral atmospheric transmission.

Secondly, the source statistical independence, the hypothesis on which ICA is based, is not satisfied. In particular, in the endmember classification, it appears that dust and CO_2 ice are strongly correlated. Thus, the reliability of ICA is not certain, and the relevance of the extracted ICs is poor. Consequently, other methods, based on priors satisfied by the data, must be investigated.

Finally, the spatial ICA decomposition provides ICs which can be interpreted as artifacts or endmembers. However – and it is a consequence of the independence violation – the decomposition done by spatial ICA leads to a matrix **A** which has some non-positive columns. These columns cannot therefore be considered as spectra.

16.5.6 Beyond ICA: semi-blind source separation

The main constraint in data decomposition of hyperspectral mixtures is the positivity of both the mixing coefficients and the source signals. However, the positivity constraint alone does not lead to a unique solution except under some particular conditions [31,59]. Thus, additional assumptions are required to select a particular solution among the admissible ones. This can be performed using either constrained least square [76,51,50,12] or penalized least square [62,41,1] estimations. One can also address the problem in the more general Bayesian framework (see Chapter 12 for details concerning the Bayesian approach) and detailed results can be found in the literature [60]. In this work, Moussaoui *et al.* considered a spectral decomposition, using a restricted number of pixels, selected by spatial ICA: 50 pixels among the most significant ones are chosen in each IC.

Then, applying Bayesian Positive[13] Source Separation (BPSS) to the RDS hyperspectral images presented in section 16.5.1, spectral ICA on this restricted number of points leads

[13]Spectrum and mixing coefficients positivity is insured by using Gamma priors on probability density function of these variables.

to the results shown in Fig. 16.13. The results of the separation using BPSS with the mixture spectra provided by the selected pixels are post-processed to correct scale and ordering ambiguities and deduce abundance fractions. The identification of the spectra is straightforward from the correlation with the reference spectra.

Finally, the quality of the approximation can be measured in the spatial domain by the square difference in signal to noise ratio.

After scaling and permutation, the identified spectra are plotted in Fig. 16.13, together with the reference spectra. One can note the similarity between the estimated spectra and the references ones. This similarity is lower for both CO_2 ice and H_2O ice in the spectral region near 2 μm because of the presence of a deep atmospheric band. On the contrary, the extracted dust source is relatively in better agreement with the associated reference spectrum (see Eq. (16.8)) which does not contain the atmospheric transmission.

The results show that spatial as well as spectral components extracted using positivity constraints are relevant, while the ICs extracted by ICA are spurious, as suspected since the independence assumption was wrong!

Finally, instead of considering an average performance (approximation accuracy), one can represent the spatial accuracy of the BPSS reconstruction by computing the signal error (recontruction error) ratio (SER) in dB. The spatial SER (Fig. 16.14) shows that there exist a few places (with darkest pixels) where the approximation is poor. At these places, one can wonder if either the linear model (16.7) is locally relevant, or other endmembers are present, i.e. N_c must be increased.

16.5.7 Conclusion

In this section, we explained that ICA should be applied carefully for providing relevant results.

First, a physical modeling of the observations should be used whenever possible for selecting the assumed nature (linear instantaneous or convolutive, or nonlinear) of the mixing system and for choosing a suitable separating system. Then, the determination of the IC number is often a useful step for reducing the computational load. After running an ICA algorithm, interpretation of ICs is usually not so simple. It requires priors, interactive discussions and complementary experimentation by experts of the application domain.

In particular, it is important to check if ICs (i.e. estimated sources) satisfy (i) the independence assumption, and (ii) physical constraints (like spectrum positivity in this example). In the above example, independence of spatial endmember abundances is not satisfied and it leads to wrong results: for instance, the spectrum related to IC4 has negative values! For avoiding such irrelevant results, it can be much more efficient to use all the available extra information, like the positivity of spectra and of mixture entries in this example.

The overall approximation quality is measured as the average quadratic error between the ICA approximated data and the observed data. Even when this average is good, the approximation quality can be poor locally, i.e. at particular points (x, y, λ). One may

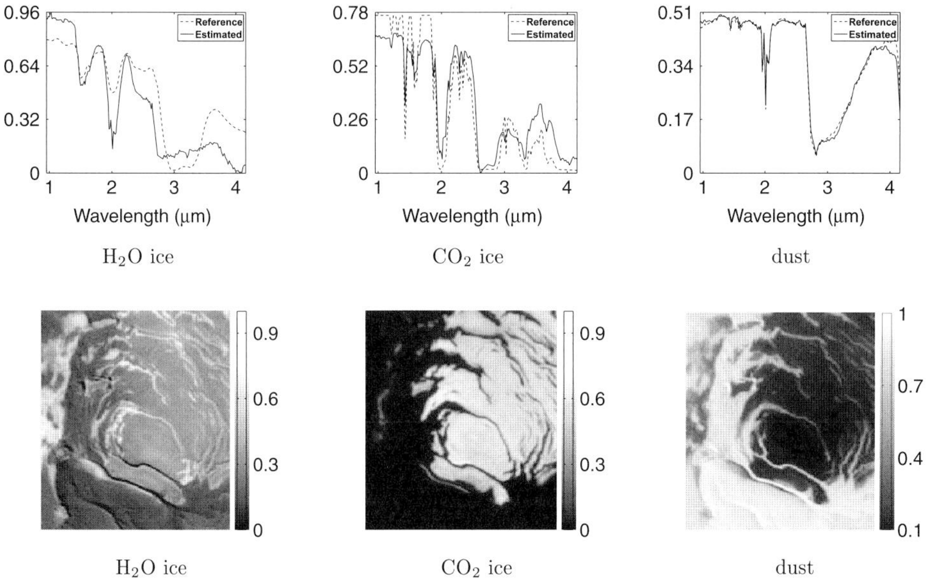

FIGURE 16.13

Estimated and reference spectra (top) and abundance fractions (bottom) for the three endmembers.

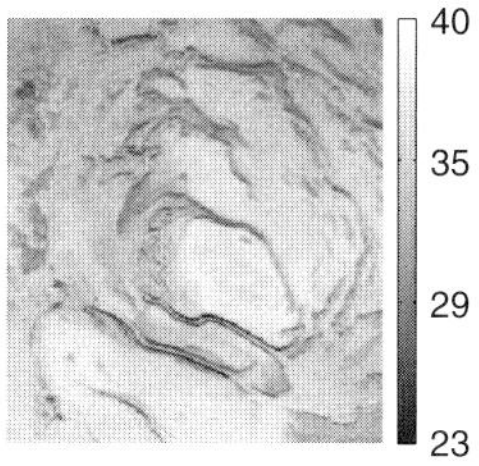

FIGURE 16.14

The Spatial Signal Error Ratio (SER) in dB provides a spatial representation of the local approximation quality.

then suspect the linear model (16.7) no longer holds, and further investigations assuming other endmembers (i.e. other sources) or a nonlinear model could be done.

Finally, if possible, we recommend consideration of semi-blind source separation methods. These approaches exploit specific priors, even very weak ones like temporal correlation [78,7] or non-stationarity [54,63]. They often lead to more efficient – sometimes simpler, too – algorithms. General frameworks exploiting various priors are currently being intensively explored. Methods assuming coloration and/or non-stationarity can be used in the time domain or in the frequency domain (after short-time Fourier transform) and have two main advantages: (i) they lead to very efficient algorithms based on joint diagonalization of matrices, (ii) they use second-order statistics and are able to separate Gaussian sources. Although it has a high computational cost, a recent approach is very attractive, too: Bayesian source separation methods [57] are able to manage any prior knowledge provided that it can be stated in probabilistic terms.

16.6 MONO- VS MULTI-DIMENSIONAL SOURCES AND MIXTURES

16.6.1 Time, space and wavelength coordinates

As shown by the above examples, the sources and mixtures may be either mono- or multi-dimensional, depending not only on the data available in the considered application, but also on the way one decides to process them. From that point of view, we have already introduced three types of configurations up to this point in the chapter:

1. The radio-frequency identification system described in section 16.4.2 is typical of many BSS applications, where a set of source objects simultaneously emit time-varying signals, which propagate to a set of sensors, so that these sensors provide mixtures of the emitted source signals. In such applications, each source signal is one-dimensional since it is a time series, whereas the set of sensors samples, at different locations, constitutes an overall field which may, for example, be electromagnetic (see section 16.4.2) or acoustic (see Chapter 19). The mixture is then three-dimensional in the sense that each observed mixture corresponds to a specific sampling point in the three-dimensional physical space. Note that this three-dimensional physical structure of the sensors does not prevent one from then re-indexing all measured signals with a single index. This commonly used approach consists in gathering all observed values, at a given time, in a one-dimensional vector.
2. On the contrary, in the first analysis of Mars data described in section 16.5.4, each observed mixture corresponds to a specific point along a one-dimensional axis, associated with wavelength, whereas each source is a two-dimensional pattern with spatial coordinates, i.e. an image. An overall Mars data set thus consists of a spatio-spectral cube, with two spatial coordinates (x, y) and one wavelength coordinate. Again, each source or observed image may then be rearranged as a one-dimensional vector, as was done in section 16.5.2 by replacing the pixel indices i and j by the single index n defined in (16.10).
3. Moreover, the nature[14] of the above data cube does not impose the way to analyze it; i.e. this cube may also be seen as a set of one-dimensional sources, where each

[14]The nature of the data axes discussed here should not be confused with the amount of data available along each axis. The latter aspect may yield additional concerns as discussed in section 16.5.2.1. However, such concerns may be handled as explained in section 16.5.6.

of these sources depends on wavelength, i.e. is a spectrum. Each observed mixture of these sources then corresponds to a specific point in the above-defined two-dimensional spatial coordinate system (x, y). Here too, all these observed values for a given wavelength may then be rearranged as a one-dimensional vector.

This alternative approach based on spectral sources, which was more briefly discussed in section 16.5.6, has also been used in various other investigations, e.g. still in astrophysics, although possibly with different prior knowledge. For instance, in [8] Berné *et al.* process observed mixed infrared spectra from Photo-Dissociation Regions (PDRs), in order to first retrieve the corresponding source spectra. In various PDRs, they thus obtain two source spectra, which are interpreted to respectively correspond to Very Small carbonaceous Grains (VSGs) and Polycyclic Aromatic Hydrocarbons (PAHs). In addition to these spectra, the authors then derive the "contribution" of each of the two estimated sources in each pixel of the observed data cube associated with the considered PDR. This then makes it possible to derive a spatial map of the ratio of these two source contributions, as shown in Fig. 16.15. This map is of high importance, because it provides an additional validation of the results obtained with the considered BSS methods:

- in this investigation, the source spectra are not known in detail beforehand, since they correspond to at least partly unknown chemical species. The estimated source spectra are therefore a first result of this approach. Astrophysics experts can then check that these extracted spectra seem to be relevant, e.g. in terms of positiveness, locations of bands. However, any other additional means to confirm the relevance of the blind signal processing methods used to derive these spectra is of interest to increase confidence in these results.
- The above-mentioned spatial map then provides such a means: when computing the ratio of the contributions of the two sources in each pixel and then deriving the resulting spatial map of this ratio, no assumption is made about the spatial properties of the processed data (since the spectra observed for all pixels are arranged in an arbitrary order to form a vector which is then processed by BSS methods). Yet, it turns out that the map thus obtained (see Fig. 16.15) does have a spatial structure, which is physically relevant. More precisely, the location of the illuminating star corresponds to the bottom-right corner of Fig. 16.15. This figure shows that, when moving away from this star, the luminance in this figure monotonically increases, i.e. the ratio of the contributions of the two sources increases. In other words, this figure proves that one of the sources is particularly close to the illuminating star, while the other source is prominent far from that star. The ability of BSS methods to extract such physical structure, although this information is not used in these methods, is an argument in favor of the relevance of the results there of from an astrophysicist point of view.

The above discussion shows how estimated sources (i.e. spectra, in this example) may be interpreted when little prior knowledge about them is available. This example should be contrasted with the investigation of Mars data described in section 16.5, where more prior knowledge was available, thanks to reference spectra for dust, CO_2 and H_2O.

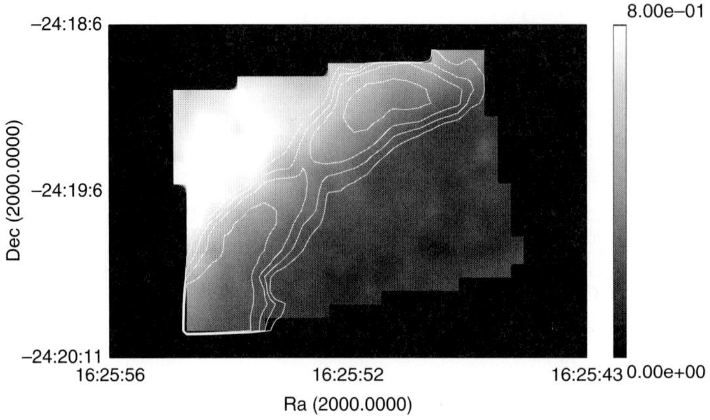

FIGURE 16.15

Spatial map of the ratio of the contributions of the two estimated sources in the ρ-Ophiuchi filament photo-dissociation region [8].

The above examples show that different situations may be encountered in BSS applications, from the point of view of the dimensions of the source data and of the space where mixtures of these sources are observed. However, these examples do not cover all possible cases. In particular, various other applications also involve data cubes, but with spatio-temporal variations instead of the above spatio-spectral ones. We describe two such applications hereafter.

16.6.2 Analyzing video frames from cortical tissues

Our first illustration of spatio-temporal BSS concerns a biomedical application, which was reported by Schiessl and colleagues in [70] (and [69]). It deals with the analysis of the changes of light reflectance over time which occur in cortical tissues as a response to stimulation. The measured data consist of a series of video frames, e.g. recorded by a CCD camera. This leads to a different formulation of the BSS problem, as compared to the framework of time-series analysis which was the first type of configuration considered in section 16.6.1. More precisely, whereas each measured signal sample was indexed by its time position and originating sensor in the above-mentioned configuration, these two variables are here respectively replaced by the spatial pixel position (seen as a one-dimensional series) in a frame and by the frame index within the frame series. Each source is thus an image, while each observed mixture of all sources corresponds to a specific time.

These data are analyzed as follows in the considered investigation. They are sphered (i.e. whitened and normalized) and processed by BSS methods. Three such algorithms are benchmarked, i.e. "infomax" extended with natural gradient, FastICA, and 2-lag decorrelation (in the space domain instead of the usual time domain, due to the above-mentioned variables). Tests performed with artificial sources and mixing show that the

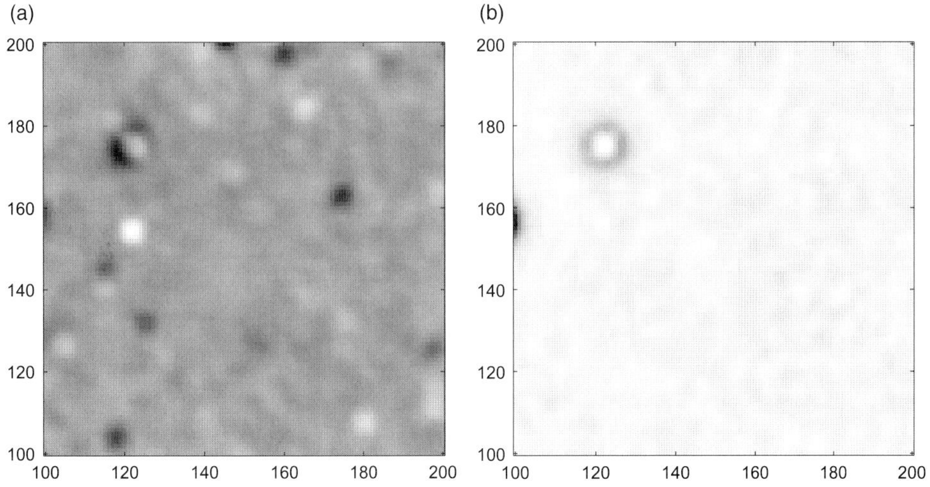

FIGURE 16.16

Application of BSS to astrophysical images: (a) a recorded image corresponding to the Andromeda Galaxy (M31), (b) an extracted image: resolved stars [38] (courtesy of M. Funaro).

decorrelation algorithm outperforms the other two, especially for noisy data. Actual recordings, taken from the primary visual cortex of a macaque monkey, are then used. The corresponding tests confirm the ability of the decorrelation algorithm to separate the signal specific to the visual stimulus from biological noise and artifacts.

16.6.3 Extracting components from a time series of astrophysical luminance images

Various other application fields give rise to the same type of image processing configuration as in the above example, i.e. they require one to extract source images from a temporal series of recorded images which are mixtures of these sources. For instance, in the astrophysics domain, Funaro *et al.* present in [38] the analysis of a series of 35 luminance images corresponding to the Andromeda galaxy (M31), recorded over 35 nights by the same telescope. An example of these recorded images is provided in Fig. 16.16(a).

These data are processed as follows. The authors first apply a Principal Component Analysis (PCA) to these images, in order to reduce their noise by projecting them on a 10-dimensional subspace, which keeps more than 90% of their energy. They then process the data thus obtained by means of the FastICA method. They thus extract various artifacts, such as cosmic rays, that they can then subtract from the observed images in order to denoise them. Besides, this reveals such phenomena as resolved stars (see Fig. 16.16(b)), which were much less visible in the initial data.

In this investigation, two types of results are derived from BSS methods. As explained above, the first output consists of spatial functions, each composed of an extracted source image. For each extracted source, the second output is the temporal function[15] composed of the mixture coefficients associated to this source in the observations recorded over these 35 nights. The latter function defines the temporal variations of the luminance of the considered source. It includes specific structures for some types of sources, e.g. a very high activity in a single recorded image if the source is a cosmic ray, large-magnitude variations around a constant value for resolved stars...

Spatio-temporal data cubes are encountered in other application fields, e.g. in fMRI. These data are more easily handled by considering them as composed of two-dimensional sources, with one observed mixture associated with each recording time, as in the above two examples. One might think of using the dual approach, by considering that each observation consists of the time series associated with each pixel or voxel. However, that dual approach results in ill-conditioned data and poor parameter estimation, unless some pixels/voxels are selected for processing, in the same way as for the spatio-spectral data considered in section 16.5.2.

16.7 USING PHYSICAL MIXTURE MODELS OR NOT

In addition to the dimension of the mixing model that we discussed in the previous section, a parameter of utmost importance when developing a BSS application is the nature (linear instantaneous, convolutive, nonlinear) of this model, because one should select accordingly the model used in the separating system. As already noted above, one should therefore derive a physical model of the relationship between the source signals and the observations, whenever this is possible. We have illustrated this approach in detail in this chapter, by means of two examples dealing with communications (see section 16.4.2.2) and astrophysics hyperspectral images (see section 16.5.1). BSS investigations not based on such physical models have also been presented in the literature, however. Even for a single class of signals, different approaches may be used from that point of view, as shown below by considering two types of investigations that have been reported for electrocardiogram (ECG) analysis.

16.7.1 Mother vs fetus heartbeat separation from multi-channel ECG recordings

One of the most classical biomedical applications of BSS methods is related to ECG analysis. It concerns the extraction of fetal heartbeats from a set of ECG signals recorded by means of cutaneous leads placed on the mother's skin. The signals thus obtained also include mother's heartbeats and noise components. Moreover, physical considerations

[15]This again corresponds to the above-mentioned duality with respect to the case when one analyzes time series by means of observations recorded at different spatial positions, e.g. by using RF antennas or microphones: in that case, the mixture coefficients associated with a given source define a spatial "signature" of the source, instead of the temporal signature involved in the application described here.

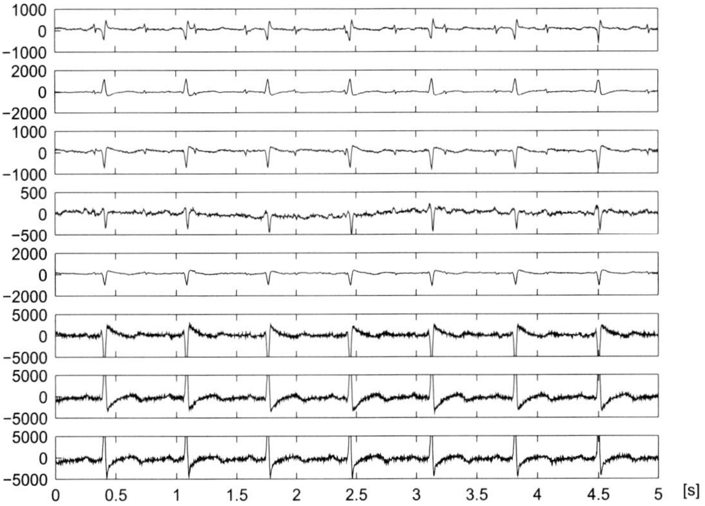

FIGURE 16.17

Eight-channel ECG recording [24] (courtesy of L. De Lathauwer).

show that the recorded signals are linear instantaneous mixtures of fetal heartbeats, mother's heartbeats and noise components [24].

This fetal heartbeat extraction problem was already tackled three decades ago by means of adaptive noise canceling techniques, which are now conventional [83]. The performance thus achievable was especially limited by the need for such techniques to use reference signals, i.e. unmixed sources. BSS algorithms now provide a more general framework which avoids this restriction and makes it possible to reconsider such applications while expecting higher performance. Results obtained in such applications are reported for example in [24,87,88]. A comparison of BSS methods and of more classical approaches based on Singular Value Decomposition (SVD) is also provided in [24].

A typical investigation of fetal ECG extraction, reported by De Lathauwer *et al.* [24], is illustrated in Figs 16.17 and 16.18. The measured signals are shown in Fig. 16.17 (such signals are available on the Internet: see section 16.8). All recorded channels include an almost periodic component, corresponding to the mothers's heartbeats. Each beat yields a major peak[16] in these signals. Seven such peaks appear over the recorded period (i.e. 5 s). In addition, the top three channels of Fig. 16.17 may be expected to contain a lower-magnitude, almost periodic, component. However, the associated waveform cannot be easily interpreted from these signals.

Applying the linear instantaneous BSS method considered in [22] to these recordings yields the signals shown in Fig. 16.18 (two other methods led to similar results in [24]). The mother's 7-peak periodic signals clearly appear in the top three output channels (an

[16]This peak corresponds to the R wave of the cardiac cycle. This wave is part of the QRS complex, which reflects ventricular depolarization. This depolarization entails ventricular contraction.

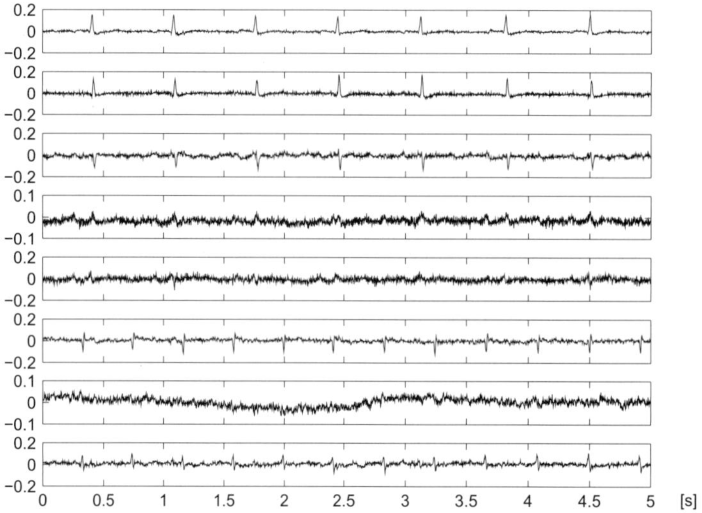

FIGURE 16.18

Source estimates derived by BSS method from ECG recording [24] (courtesy of L. De Lathauwer).

interpretation of the dimension of the associated subspace is provided in [24]). Moreover, another periodic signal is extracted, especially in the sixth channel. The considered recording period contains 12 peaks, i.e. 12 cycles of this signal. This periodic signal, which has a higher frequency than the mother's heartbeats, corresponds to the fetal ECG.

16.7.2 Analysis of heart control from single-channel ECG

ECG signals have also been analyzed with other goals. In particular, Vetter *et al.* [81] presented an investigation where the BSS aspect of the problem appears in a much less natural way than above to non-specialists. This investigation concerns the analysis of the control of the heart by the autonomic nervous system, whose alterations have been shown to play an important role in many pathophysiological situations. This heart control system contains two antagonistic parts, corresponding to the cardiac sympathetic (CSNA) and parasympathetic (CPNA) nervous activities. Variations in these activities influence heart behavior and yield modifications in the ECG. The reported investigation aims at extracting the original CSNA and CPNA signals only from a single-channel observed ECG, or more precisely from two parameters derived from it. These parameters are the successive so-called RR and QT intervals, which respectively correspond to the time interval between adjacent heartbeats and to the duration of a specific portion of the ECG cycle.

This separation of CSNA and CPNA signals from their RR and QT mixtures is performed by means of a BSS method suited to linear instantaneous mixtures. Unlike in most applications, this linear instantaneous mixing structure is not selected as a result of detailed modeling of the considered physical system, which would show that the measured signals provide this type of mixture. Instead, the approach used here contains two aspects. Qualitative physiological knowledge corresponding to the above-mentioned heart control model is first used. It shows that each of the RR and QT parameters depends on the CSNA and CPNA signals, i.e. is a mixture (of unspecified type at this stage) of these signals (and possibly of others signals, which is confirmed below). Then, a linear instantaneous mixture model is selected, based on two motivations. On the one hand, this investigation is focused on a "small-signal" approximation and only aims at extracting the most salient features of the variations of the CSNA and CPNA signals, which leads the authors to use linear modeling. On the other hand, they want to develop a simple analysis tool and they therefore only consider an instantaneous mixture model.

The other major issue of the BSS problem thus introduced concerns the independence of the sources to be restored (or at least their uncorrelation). It has been shown that the CSNA and CPNA signals are not independent. However, previous work of the authors leads them to assume that two independent components, respectively sensitive to the CSNA and CPNA signals, may be derived from the observed ECG parameters.

The BSS method applied to the considered ECG parameters is a classical approach intended for temporally correlated sources, often referred to as SOBI (Second-Order Blind Identification) [7]. Moreover, it is preceded by a noise reduction stage based on PCA. This stage aims at reducing the influence of all the "noise" signals contained by the considered ECG parameters, in addition to the mixed contributions of CNSA and CPNA. These noise sources correspond to the influence of respiration and unknown stochastic phenomena on ECG parameters, together with measurement and quantization noise.

In biomedical applications, BSS methods often provide estimates of hidden variables, such as CSNA and CPNA here, which are not accessible in humans. These estimated variables cannot be compared to the lacking original sources and are therefore hardly interpreted. This makes it difficult to validate the operation of BSS methods in such applications. The authors here solve this problem by using specific experimental protocols. These protocols are selected because they elicit or inhibit sympathetic or parasympathetic responses and therefore make it possible to check if the proposed approach is able to highlight changes in the levels of CPNA and CNSA. This shows the effectiveness of linear instantaneous BSS methods in this application and the need to use a denoising stage. This approach outperforms the traditional indicator based on Fast Fourier Transform (FFT).[17]

It should be mentioned that the same authors previously reported a related approach [79,80]. However, the latter method requires non-causal convolutive BSS algorithms and simultaneous recordings of ECG and arterial blood pressure, which may be cumbersome in clinical applications. The above-described approach therefore has the advantage of requiring only instantaneous BSS methods and recording of one ECG channel.

[17] The permutation and amplitude indeterminacies are also solved in [81] by using prior knowledge.

16.7.3 Additional comments about performance evaluation

The application that we just described highlighted the problem of performance evaluation in applications where source signals are not accessible. A general solution to this problem however, exists in cases when, even if the source signals to be restored are not known, prior information about their properties is available. Indeed, one may indirectly verify that the considered BSS methods are successful, by checking to what extent the source estimates that they provide actually have the expected properties.

This approach is, for example, used in [16], still in the framework of cardiac signal analysis. This paper concerns the extraction of atrial activity during atrial fibrillation episodes. The authors take advantage of the fact that atrial activity has a narrowband spectrum, which contains a major peak at a frequency situated between 3 and 9 Hz, unlike the other signals involved in this application. The approach proposed for approximately evaluating the quality of the extraction of atrial activity by means of a BSS method then consists in measuring the spectral concentration of the estimated source. A BSS method is considered to yield good performance in this application if the spectrum of the signal that it provides is concentrated around its peak situated in the frequency band ranging from 3 to 9 Hz. On the contrary, this spectrum is more spread out if the restored signal contains undesired contributions resulting from other sources.

16.8 SOME CONCLUSIONS AND AVAILABLE TOOLS

In this chapter, we discussed various general aspects of BSS applications and we illustrated them by means of examples corresponding to different domains. Various other applications are also presented in the subsequent chapters of this book. We hope that this overall description will convince the reader involved in signal, image or data processing in general that BSS methods may be applied to quite different fields. From that point of view, we would like to stress again that various problems may be *reformulated* in terms of BSS configurations, although they do not initially exhibit a clear set of mixtures of the same source signals: e.g. see the extraction of sympathetic and parasympathetic activities from a single ECG recording that we described in this chapter.

If we aim at proposing a general procedure for applying BSS methods, we should therefore include a "step zero" in it. This step consists, for the user of this procedure, in always keeping an eye on BSS methods and wondering whether the practical problems that he/she wants to address may be (re-)expressed in terms of mixtures of unknown sources, from which he wants to restore these sources. If this is the case, the user should then precisely define these sources and the mixture which occurs between them, especially in order to determine the properties of these sources (statistical independence, spatial uncorrelation and temporal coloration, sparsity ...) and the nature of the mixture (linear or not and, if it is linear, instantaneous or convolutive). He will then be able to select accordingly the BSS methods that he will apply to these data. Furthermore, it is crucial that the user knows the overfitting nature of the methods he intends to use. Robust uses of the methods typically reduce this overfit phenomenon; they give a hint on possible departures from the assumptions made and evidence some possible subspace relations in the data.

To help the reader make first practical steps in the BSS field, we here provide a few addresses of Internet websites which contain free software packages and/or data that may be used to perform initial tests:

- The website which is to be the central point for storing software and data, or links towards such items, is:
  http://www.tsi.enst.fr/icacentral/.
- A significant number of ICA-based BSS methods are also available in the ICALAB toolbox, which may be downloaded from:
  http://www.bsp.brain.riken.go.jp/ICALAB/.
- Toolboxes for NMF (Non-negative Matrix Factorization) and NTF (Non-negative Tensor Factorization) are available at:
  http://www.bsp.brain.riken.jp/ICALAB/nmflab.html.
- The reader interested in data corresponding to the applications detailed in this chapter[18] may get part of them from:
  http://www.ast.obs-mip.fr/deville
  and, concerning ECG signals, from:
  http://homes.esat.kuleuven.be/~smc/daisy/daisydata.html.

References

[1] S.A. Abdallah, M.D. Plumbley, Polyphonic transcription by non-negative sparse coding of power spectra, in: Proc. International Conference on Music Information Retrieval, ISMIR'2004, Barcelona, Spain, October 2004, pp. 318–325.

[2] S. Achard, C. Jutten, Identifiability of post nonlinear mixtures, IEEE Signal Processing Letters 12 (2005) 423–426.

[3] M. Babaie-Zadeh, C. Jutten, K. Nayebi, A geometric approach for separating post nonlinear mixtures, in: Proc. of the XI European Signal Processing Conf., EUSIPCO 2002, vol. II, Toulouse, France, 2002, pp. 11–14.

[4] P. Bak, C. Tang, K. Wiesenfeld, Self-organized criticality, Physical Review A 38 (1988) 364–374.

[5] Y. Bar-Ness, J. Carlin, M. Steinberger, Bootstrapping adaptive cross pol cancelers for satellite communications, in: Proceedings of the IEEE International Conference on Communication, ICC'82, Miami, USA, June 1982, pp. 4F.5.1–4F.5.5.

[6] A.J. Bell, T.J. Sejnowski, Fast blind separation based on information theory, in: Proceedings of the 1995 International Symposium on Nonlinear Theory and Its Applications, NOLTA'95, vol. 1, Las Vegas, U.S.A, December 1995, pp. 43–47.

[7] A. Belouchrani, K. Abed-Meraim, J.-F. Cardoso, E. Moulines, A blind source separation technique using second-order statistics, IEEE Transactions on Signal Processing 45 (1997) 434–444.

[18] The webpage http://www.ast.obs-mip.fr/deville mentioned here also concerns the subsequent chapter about audio applications (Chapter 19). Larger datasets for audio applications are, for example, also available at [73], as stated at the beginning of the current chapter.

[8] O. Berné, C. Joblin, Y. Deville, J. Smith, M. Rapacioli, J. Bernard, J. Thomas, W. Reach, A. Abergel, Analysis of the emission of very small dust particles from spitzer spectro-imagery data using blind signal separation methods, Astronomy & Astrophysics 469 (2007) 575–586.

[9] J.-P. Bibring, et al., Perennial water ice identified in the south polar cap of Mars, Nature 428 (2004) 627–630.

[10] S. Bozinoski, H.-L.N. Thi, Séparation de sources à bande large dans un mélange convolutif, in: Proceedings of Ecole des Techniques Avancées en Signal Image Parole, Grenoble, France, September 1996, pp. 303–310.

[11] M. Breakspear, M. Brammer, E. Bullmore, P. Das, L. Williams, Spatiotemporal wavelet resampling for functional neuroimaging data, Human Brain Mapping 23 (2004) 1–25.

[12] R. Bro, S. De Jong, A fast non-negativity constrained least squares algorithm, Journal of Chemometrics 11 (1997) 393–401.

[13] V. Calhoun, T. Adali, V. McGinty, J. Pekar, T. Watson, G. Pearlson, fMRI activation in a visual-perception task: network of areas detected using the general linear model and independent component analysis, NeuroImage 14 (2001) 1080–1088.

[14] V. Capdevielle, C. Servière, Séparation de signaux de machines tournantes, in: Proceedings of Ecole des Techniques Avancées en Signal Image Parole, Grenoble, France, September 1996, pp. 311–318.

[15] J.-F. Cardoso, A. Souloumiac, Blind beamforming for non Gaussian signals, IEE Proceedings-F 140 (1993) 362–370.

[16] F. Castells, J.J. Rieta, J. Millet, V. Zarzoso, Spatiotemporal blind source separation approach to atrial activity estimation in atrial tachyarrhythmias, IEEE Transactions on Biomedical Engineering 52 (2005) 258–267.

[17] E. Chaumette, D. Muller, Séparation de sources aux ordres supérieurs: application au garbling en radar secondaire, in: Proceedings of Ecole des Techniques Avancées en Signal Image Parole, Grenoble, France, September 1996, pp. 245–253.

[18] P. Chevalier, Les statistiques d'ordre supérieur en traitement d'antenne, in: Proceedings of Ecole des Techniques Avancées en Signal Image Parole, Grenoble, France, September 1996, pp. 237–243.

[19] P. Chevalier, V. Capdevielle, P. Comon, Performance of ho blind source separation methods: experimental results on ionospheric HF links, in: Proceedings of the International Workshop on Independent Component Analysis and Blind Signal Separation, ICA'99, Aussois, France, January 1999, pp. 443–448.

[20] A. Cichocki, W. Kasprzak, S. Amari, Multi-layer neural networks with a local adaptive learning rule for blind separation of source signals, in: Proceedings of the 1995 International Symposium on Nonlinear Theory and Its Applications, NOLTA'95, Las Vegas, U.S.A, December 95, pp. 61–65.

[21] M. Clyde, H. Lee, Bagging and the bayesian bootstrap, in: Proc. 8th Int. Workshop on Artificial Intelligence and Statistics, AISTATS 2001, Key West, Florida, January 2001.

[22] P. Comon, Independent component analysis, a new concept?, Signal Processing 36 (1994) 287–314.

[23] P. Comon, Blind identification and source separation in 2×3 under-determined mixtures, IEEE Transactions on Signal Processing 52 (2004) 11–22.

[24] L. De Lathauwer, B. D. Moor, J. Vanderwalle, Fetal electrocardiogram extraction by blind source subspace separation, IEEE Transactions on Biomedical Engineering 47 (2000) 567–572.

[25] Y. Deville, Application of the Hérault-Jutten source separation neural network to multi-tag radio-frequency identification systems, in: Proceedings of Ecole des Techniques Avancées en Signal Image Parole, Grenoble, France, September 1996, pp. 265–272.

[26] Y. Deville, http://www.ast.obs-mip.fr/deville, See page dealing with data, (2009).

[27] Y. Deville, L. Andry, Application of blind source separation techniques to multi-tag contactless identification systems, in: Proceedings of the 1995 International Symposium on Nonlinear Theory and Its Applications, NOLTA'95, vol. 1, Las Vegas, U.S.A, December 1995, pp. 73–78.

[28] Y. Deville, L. Andry, Système d'échange de données comportant une pluralité de porteurs de données, French Patent no. 95 10444, (1995) (this patent was initially filed in France and subsequently extended to several countries), filed on September 6, 1995.

[29] Y. Deville, L. Andry, Application of blind source separation techniques to multi-tag contactless identification systems, IEICE Transactions on Fundamentals of Electronics, Communications and Computer Sciences E79-A (1996) 1694–1699.
[30] Y. Deville, J. Damour, N. Charkani, Multi-tag radio-frequency identification systems based on new blind source separation neural networks, Neurocomputing 49 (2002) 369–388.
[31] D. Donoho, V. Stodden, When does non-negative matrix factorization give a correct decomposition into parts? in: Advances in Neural Information Processing Systems, NIPS'2003, Cambridge, United States, 2003.
[32] S. Douté, B. Schmitt, Y. Langevin, J.-P. Bibring, F. Altieri, G. Bellucci, B. Gondet, F. Poulet, the MEX OMEGA team, South Pole of Mars: Nature and composition of the icy terrains from Mars Express OMEGA observations, Planetary and Space Science 55 (2007) 113–133.
[33] J. Duann, T. Jung, S. Makeig, T. Sejnowski, Consistency of infomax ICA decomposition of functional brain imaging data, in: Proc. 4th Int. Symposium on Independent Component Analysis and Blind Signal Separation, ICA 2003, Nara, Japan, April 2003, pp. 289–294.
[34] G. D'Urso, Les techniques de séparation de sources appliquées à la surveillance des instalations EDF, in: Proceedings of Ecole des Techniques Avancées en Signal Image Parole, Grenoble, France, September 1996, pp. 255–263.
[35] B. Efron, R. Tibshirani, An Introduction to the Bootstrap, Routledge/Taylor Francis Group, Oxford, UK, 1994.
[36] J. Eriksson, V. Koivunen, Blind identifiability of class of nonlinear instantaneous ICA models, in: Proc. of the XI European Signal Proc. Conf., EUSIPCO 2002, vol. 2, Toulouse, France, September 2002, pp. 7–10.
[37] A. Ferréol, P. Chevalier, L. Albera, Second-order blind separation of first- and second-order cyclostationary sources – application to AM, FSK, CPFSK, and deterministic sources, IEEE Transactions on Signal Processing 52 (2004) 845–861.
[38] M. Funaro, E. Oja, H. Valpola, Independent component analysis for artefact separation in astrophysical images, Neural Networks 16 (2003) 469–478.
[39] S. Haykin, Adaptive Filter Theory, 3rd ed., Prentice-Hall International, Inc., 1996.
[40] J. Himberg, A. Hyvärinen, F. Esposito, Validating the independent components of neuroimaging time series via clustering and visualization, NeuroImage 22 (2004) 1214–1222.
[41] P.O. Hoyer, Non-negative sparse coding, in: Proceedings of IEEE Workshop on Neural Networks for Signal Processing, NNSP'2002, 2002, pp. 557–565.
[42] J. Huang, N. Ohnishi, N. Sugie, Sound separation based on perceptual grouping of sound segments, in: Proceedings of the 1995 International Symposium on Nonlinear Theory and Its Applications, NOLTA'95, vol. 1, Las Vegas, U.S.A, December 1995, pp. 67–72.
[43] A. Hyvärinen, Complexity pursuit: separating interesting components from time series, Neural Computation 13 (2001) 883–898.
[44] A. Hyvärinen, J. Karhunen, E. Oja, Independent component analysis, in: Adaptive and Learning Systems for Signal Processing, Communications, and Control, John Wiley, New York, 2001.
[45] Icasso, MATLAB package, http://www.cis.hut.fi/jhimberg/icasso, (2003).
[46] C. Jutten, J. Hérault, Une solution neuromimétique au problème de séparation de sources, Traitement du Signal 5 (1988) 389–403.
[47] C. Jutten, J. Hérault, Blind separation of sources, Part I: An adaptive algorithm based on neuromimetic architecture, Signal Processing 24 (1991) 1–10.
[48] C. Jutten, J. Karhunen, Advances in blind source separation (BSS) and independent component analysis (ICA) for nonlinear mixtures, International Journal of Neural Systems 14 (2004) 1–26.
[49] N. Keshava, J.F. Mustard, Spectral unmixing, IEEE Signal Processing Magazine 19 (2002) 14–57.
[50] C. Lawson, R. Hanson, Solving Least-Squares Problems, Prentice-Hall, 1974.
[51] D. Lee, H. Seung, Learning the parts of objects by non-negative matrix factorization, Nature 401 (1999) 788–791.
[52] Y. Li, A. Cichocki, S. Amari, Analysis of sparse representation and blind source separation, Neural Computation 16 (2004) 1193–1204.

[53] O. Macchi, E. Moreau, Self-adaptive source separation, Part I: convergence analysis of a direct linear network controlled by the Hérault-Jutten algorithm, IEEE Transactions on Signal Processing 45 (1997) 918–926.
[54] K. Matsuoka, M. Ohya, M. Kawamoto, A neural net for blind separation of nonstationary signals, Neural Networks 8 (1995) 411–419.
[55] M. McKeown, T. Sejnowski, Independent component analysis of FMRI data: examining the assumptions, Human Brain Mapping 6 (1998) 368–372.
[56] F. Meinecke, A. Ziehe, M. Kawanabe, K. Müller, A resampling approach to estimate the stability of one-dimensional or multidimensional independent components, IEEE Transactions on Biomedical Engineering 49 (2002) 1514–1525.
[57] A. Mohammad-Djafari, A Bayesian approach to source separation, in: Bayesian Inference and Maximum Entropy Methods in Science and Engineering (Proceedings 19th International Workshop on Maximum Entropy and Bayesian Methods (MaxEnt 99), in: J.T. Rychert, G.J. Erickson, C.R. Smith (Eds.), AIP Conference Proceedings, vol. 567, Boise, Idao, 1999, pp. 221–244.
[58] E. Moreau, O. Macchi, Self-adaptive source separation, Part II: comparison of the direct, feedback and mixed linear network, IEEE Transactions on Signal Processing 46 (1998) 39–50.
[59] S. Moussaoui, D. Brie, J. Idier, Non-negative source separation: Range of admissible solutions and conditions for the uniqueness of the solution, in: Proceedings of IEEE International Conference on Acoustics, Speech, and Signal Processing, ICASSP'2005, Philadelphia, USA, March 2005, pp. 289–292.
[60] S. Moussaoui, H. Hauksdóttir, F. Schmidt, C. Jutten, J. Chanussot, D. Brie, S. Douté, J.A. Benediksson, On the decomposition of Mars hyperspectral data by ICA and Bayesian positive source separation, Neurocomputing 71 (2008) 2194–2208.
[61] H.-L. Nguyen Thi, C. Jutten, Blind source separation for convolutive mixtures, Signal Processing 45 (1995) 209–229.
[62] P. Paatero, U. Tapper, Positive matrix factorization: a nonnegative factor model with optimal utilization of error estimates of data values, Environmetrics 5 (1994) 111–126.
[63] D.T. Pham, J.-F. Cardoso, Blind separation of instantaneous mixtures of nonstationary sources, IEEE Transactions on Signal Processing 49 (2001) 1837–1848.
[64] Philips Semiconductors Data Sheet, Om 4282 RF-identification, hardware description and tutorial. ID-No: 8962D26CEA20068F.
[65] Philips Semiconductors Data Sheet, Om 4282 RF-identification, software command reference and RS 232 transmission protocol. ID-No: 7F3C1206084C995E.
[66] J. Rao, R. Tibshirani, The out-of-bootstrap method for model averaging and selection, tech. rep., University of Toronto, Canada, 1997.
[67] J. Särelä, R. Vigário, Overlearning in marginal distribution-based ICA: analysis and solutions, Journal of Machine Learning Research 4 (2003) 1447–1469.
[68] H. Saylani, Y. Deville, S. Hosseini, M. Habibi, A multi-tag radio-frequency identification system using a new blind source separation method based on spectral decorrelation, in: Second International Symposium on Communications, Control and Signal Processing, ISCCSP 2006, Marrakech, Morocco, March 2006.
[69] I. Schiessl, M. Stetter, J. Mayhew, S. Askew, N. McLoughlin, J. Levitt, J. Lund, K. Obermayer, Blind separation of spatial signal patterns from optical imaging records, in: Proceedings of the International Workshop on Independent Component Analysis and Blind Signal Separation, ICA'99, Aussois, France, January 1999, pp. 179–184.
[70] I. Schiessl, M. Stetter, J. Mayhew, N. McLoughlin, J. Lund, K. Obermayer, Blind signal separation from optical imaging recordings with extended spatial decorrelation, IEEE Transactions on Biomedical Engineering 47 (2000) 573–577.
[71] F. Schmidt, S. Douté, B. Schmitt, Wavanglet: an efficient supervised classifier for hyperspectral images, IEEE Transactions on Geoscience and Remote Sensing 45 (5) (2007) 1374–1385. Part 2.
[72] C. Simon, P. Loubaton, C. Jutten, Separation of a class of convolutive mixtures: a contrast function approach, Signal Processing 81 (2001) 883–887.

[73] S.S.E.C. (SiSEC), http://sisec.wiki.irisa.fr/tiki-index.php, See pages dealing with datasets, (2008).
[74] A. Taleb, C. Jutten, Source separation in post-nonlinear mixtures, IEEE Transactions on Signal Processing 47 (1999) 2807–2820.
[75] D. Tanre, M. Herman, P.Y. Deschamps, A. de Leffe, Atmospheric modeling for space measurements of ground reflectances, including bidirectional properties, Applied Optics 18 (1979) 3587–3594.
[76] R. Tauler, B. Kowalski, S. Fleming, Multivariate curve resolution applied to spectral data from multiple runs of an industrial process, Analytical Chemistry 65 (1993) 2040–2047.
[77] N. Thirion, J. Mars, F. Glangeaud, Séparation d'ondes dans le cas des signaux de prospection sismique, in: Proceedings of Ecole des Techniques Avancées en Signal Image Parole, Grenoble, France, September 1996, pp. 357–366.
[78] L. Tong, V.C. Soon, Y.-F. Huang, R. Liu, Amuse: A new blind identification algorithm, in: Proc. IEEE ISCAS, New Orleans, LA, May 1990, pp. 1784–1787.
[79] R. Vetter, J.-M. Vesin, P. Celka, U. Scherrer, Observer of the autonomic cardiac outflow in humans using non-causal blind source separation, in: Proceedings of the International Workshop on Independent Component Analysis and Blind Signal Separation, ICA'99, Aussois, France, January 1999, pp. 161–166.
[80] R. Vetter, J.-M. Vesin, P. Celka, U. Scherrer, Observer of the human cardiac sympathetic nerve activity using noncausal blind source separation, IEEE Transactions on Biomedical Engineering 46 (1999) 322–330.
[81] R. Vetter, N. Virag, J.-M. Vesin, P. Celka, U. Scherrer, Observer of autonomic cardiac outflow based on blind source separation of ECG parameters, IEEE Transactions on Biomedical Engineering 47 (2000) 578–582.
[82] R. Vigário, J. Särelä, V. Jousmäki, M. Hämäläinen, E. Oja, Independent component approach to the analysis of EEG and MEG recordings, IEEE Transactions on Biomedical Engineering 47 (2000) 589–593.
[83] B. Widrow, J. Glover, J. McCool, J. Kaunitz, C. Williams, R. Hearn, J. Zeidler, E. Dong, R. Goodlin, Adaptive noise cancelling: principles and applications, Proceedings of the IEEE 63 (1975) 1692–1716.
[84] D. Yellin, E. Weinstein, Criteria for multichannel signal separation, IEEE Transactions on Signal Processing 42 (1994) 2158–2168.
[85] P. Yiou, D. Sornette, M. Ghil, Data-adaptive wavelets and multi-scale singular spectrum analysis, Physica D 142 (2000) 254–290.
[86] J. Ylipaavalniemi, R. Vigário, Analyzing consistency of independent components: An fMRI illustration, NeuroImage 39 (2008) 169–180.
[87] V. Zarzoso, A. Nandi, Non-invasive fetal electrocardiogram extraction: blind separation vs. adaptive noise cancellation, IEEE Transactions on Biomedical Engineering 48 (2001) 12–18.
[88] V. Zarzoso, A. Nandi, E. Bacharakis, Maternal and foetal ECG separation using blind source separation methods, IMA Journal of Mathematics Applied in Medicine and Biology 14 (1997) 207–225.
[89] A. Ziehe, K. Müller, TDSEP – an effective algorithm for blind separation using time structure, in: Proc. Int. Conf. on Artificial Neural Networks, ICANN'98, Skövde, Sweden, 1998, pp. 675–680.

CHAPTER

Application to telecommunications

17

P. Chevalier and A. Chevreuil

17.1 INTRODUCTION

Operational context. For several decades, we have observed a big rise of radio communications, or wireless communications, in both the civilian and the military areas, from the HF band (ionospheric links) to the EHF band (satellite communications), by way of cellular networks and digital broadcasting developments in the UHF band. The increasing need of high bit rates, the necessity of sharing the limited spectral resource, the frequency reuse concept of cellular networks and the potential for intentional or non-intentional jamming are some of the reasons for the high spectral congestion in most of the available frequency bands. This increasing development of radio communications generates an increasing need for both spectrum monitoring by civilians under state control agencies and radio surveillance by military administrations. More precisely, the spectrum control problem consists in blindly analyzing all the sources which are detected in the frequency band of the receiver. For some applications, the number of sources may be high. This is, in particular, the case for HF radio surveillance or VUHF airborne spectrum monitoring over dense urban areas. In such conditions, the spectrum control requires the estimation of many parameters belonging to spectrally overlapping sources. This requires a pre-processing of the data consisting in isolating the different components of the observed signals, i.e. *blind source separation* (BSS). Moreover, in most cases, the sources propagate through multi-path propagation channels due to multiple reflections on ionospheric layers (HF), on buildings (UHF in urban areas), on natural scatterers such as vegetations, mountains, etc. Spectrum control systems have then generally to blindly separate convolutive mixtures of sources. These may be either analogical or digital, either zero-mean or not, either deterministic (carriers) or random (modulated sources), with potentially different modulations, carrier residues and bit rates. Besides, most of the received sources share two particular properties: cyclostationarity [42,43] and non-Gaussianity. Cyclostationarity means that the sources statistics are periodic or poly-periodic functions of time, with variations depending on the modulation, baud rate (for digital modulations) and frequency offset [44,40]. Non-Gaussianity concerns all the radio communication sources except *orthogonal frequency division multiplex* (OFDM) ones. In this respect, two BSS approaches are possible.

The mixture as generally convolutive. The first approach consists in implementing blind separators of convolutive mixtures of sources, either indirectly, i.e. from the blind identification of all the propagation channels, or directly from the data (with no prior identification of the channels).

Second-order (SO) indirect techniques [111,1,2,49,48,14] assume *finite impulse response multi-input and multi-output* (FIR-MIMO) channels and mainly correspond to MIMO extensions of the subspace [111,1,48] or linear prediction [2,49] methods introduced for *single input and multi-output* (SIMO) systems in [73,99,79] and [91] respectively. These methods require very specific assumptions that limit their use in operational contexts. Indeed, the *independent and identically distributed* (iid) assumption on the sources is crucial for linear prediction methods [2,49]. As far as the subspace approaches [1,48] are concerned, the original methods rely on the wide-sense stationarity of the sources; nevertheless, the extension to cyclostationary sources does not have an impact on the algorithms: this is legitimate, after considering the *time-average* auto-correlation function in lieu of the ordinary auto-correlation function. Though theoretically appealing, the field of applications of the subspace method is actually very *narrow*. Indeed, the SO indirect methods require strong assumptions on the channels. The first one is related to the notion of diversity between the channels; as far as the second one is concerned, the delay spread of all the channels (as a function of the number of symbols) has to be known to the receiver. If the first assumption may not be so restrictive for multipath channels (it is required for the sources to be jointly estimated from the mixture via a linear time-invariant system), the second condition is prohibitive. Indeed, the notion of delay spread (or channel order for digital sources) is not well defined in a real environment and the estimation of a so-to-say order is difficult. To conclude on indirect SO methods, it has been noticed that it is possible to benefit from the non-stationarity (specifically the cyclostationarity) of the data and identify the unknown channels from their cyclic-statistics as was remarked in [3]. The method relies on the separability of the different contributions in the domain of the cyclic frequencies. This means that the sources must have distinct cyclic frequencies (assumption made in [14]); besides, the method requires the prior knowledge of these cyclic frequencies: these two reasons make too restrictive the field of application. Concerning this latter point, it is well reported that the estimation of these cyclic frequencies is not so easy from a limited time observation, due to the bandwidth limitation of the signals which makes the involved cyclic SO statistics be very weak [30]. As far as *higher-order* (HO) indirect techniques [53,98,92,117,103, 104] are concerned, they mainly assume MA, AR or ARMA propagation channels and suffer from lack of global convergence and poor estimation accuracy. Moreover, most of these techniques also assume stationary sources and either iid sources [98,92,117,103, 104] or colored sources [53]. For these reasons, the available indirect blind separators of convolutive mixtures of sources are not appropriate in the general context of spectrum monitoring of radio communications.

On the other hand, the direct methods, which aim at providing independent outputs, each involving a single source, seem to be more promising in the context of spectrum monitoring of radio communications signals. Most of the available direct blind source separators of convolutive mixtures of sources exploit the HO statistics of the data and have been developed under very specific conditions on the sources, such as iid [97,117, 105,106,70,28] or stationary continuous phase modulation (CPM) sources [17] having the same baud rate (see Chapter 8 of this book for a complete survey). Due to these strong restrictions on the sources, the above methods cannot cope with the complex operational contexts previously described. Most of the available blind separators allowing one to process convolutive mixtures of cyclostationary sources require the prior estimation of the cyclic frequency of the sources [113,8,87] (see also Chapter 8, section 8.6 of this book), which is difficult and costly (see [30,78] for the simple single-user case). The scarce blind separators of this family which do not require the prior estimation of the cyclic frequencies of the sources have been developed very recently [51,59,60,38]. However, [51] considers the case of complex circular linear modulation, whereas in [59, 60] the sources are either linear modulations or CPM, but the assumption of circularity is central. The case of non-circular linearly modulated sources is considered in [38]; nonetheless, the available blind separators of convolutive mixtures of sources cannot be implemented in all the operational contexts encountered in spectrum monitoring of radio communications; however, the approach given in [60] is a good reference for a certain number of scenarios.

Exploiting the specularity of the channels. The second approach relies on the specificity that most of the propagation channels encountered in practice are temporally specular. A channel is said to be specular in the time domain if it is composed of a (small) finite number of paths with different delays and complex attenuations, i.e. if its impulse response is a finite linear combination of Dirac functions. This property corresponds to a sparsity property in the time domain. This allows one to see the model (*a priori* convolutive) as an instantaneous mixture. In other words, this approach considers that two given paths associated with a given transmitter can be seen as two sources. Of course these two paths are not stricto sensu statistically independent but this assumption is all the more valid as the paths are well separated (with respect to the symbol period for digital sources). After achieving the BSS, a post-processing of the blindly separated paths may be implemented in order to associate and judiciously combine the separated paths of a given source and, hence, to optimize the reception of the latter (or at least of some of them). Available blind separators of instantaneous mixtures of radio communications sources assume statistical independent sources and take advantage of the specific properties of the latter (or at least of some of them), such as the *constant modulus* (CM) [102,101, 100,47,64,89,90,109,77,76,63,72,10,19,112], the discrete alphabet [95,110,7,93,96,94,107, 62,39,50,85,74,108,69,68,118], the cyclostationarity [4,88,16,119,115,65,18,75,66,81,45, 71,3,12,116,56,84,55,54,33] or the non-Gaussianity [32,46,61,15,27,31,82,6,67,11,83,5]. Some blind techniques which rely on the CM properties of the sources aim at extracting only a particular CM source corrupted by interferences [102,100,101,47,64,72,10,19], with the risk of a capture effect in the presence of CM interferences. Some others aim at separating all the sources from a multistage CM array [90,89,76,77,63,112] or from an analytical CM algorithm [107]. The first ones suffer from an error propagation effect from the first stage to the others and from a slow and irregular convergence especially for weak sources and short data sets. The second ones are much more complex to implement than the iterative techniques. Besides, sources encountered in practice are not necessarily

CM, which may *a priori* limit the pertinence of these methods. Techniques which exploit the discrete alphabet property of the sources generally assume that the latter share the same alphabet and the same symbol period, and are sampled at the symbol rate, which is not realistic for the operational context of spectrum monitoring. Techniques exploiting the cyclostationarity of the sources aim at blindly extracting either a particular cyclostationary source among interference [4,88,16,119,115,65,18,75,66,81,45] or several cyclostationary sources [71,3,12,116,56,84,55,54,33]. They all require the prior estimation of the first-order [56], the SO [4,88,16,119,115,65,18,66,81,71,3,12,116,84,55,54,33] or the HO cyclic frequencies [75,45] of the sources of interest, which is a difficult and complex task to implement. The above discussion has proved that these methods rely on too specific characteristics and thus cannot cope with the very general mixtures encountered in many contexts of spectrum monitoring of radio communications signals. However, some methods solely exploit a common feature of most of radio communications signals: the statistical independence and the non-Gaussian character of the sources, which make them very appropriate. Among these techniques, *fourth-order* (FO) cumulant-based techniques [32,46,15,27,31,11,5] are the most powerful for complex-valued signals. One of the main advantages of such techniques is that they can cope with all kinds of sources waveforms, which is a very important point for spectrum monitoring of radio communications signals. Techniques developed in [32,46] aim at extracting only one non-Gaussian sources corrupted by Gaussian interferences whereas techniques presented in [15,27,31, 11,5] aim at separating several non-Gaussian sources.

However, these techniques have been developed for stationary sources and one may wonder how they behave in the presence of cyclostationary sources with arbitrary cyclic frequencies. This question has been investigated for the first time in [34] for arbitrary zero-mean cyclostationary non-Gaussian sources, where an analytical performance analysis of the JADE algorithm [15] has been presented. The conditions under which the JADE algorithm is robust to the cyclostationary property of the sources have been given, showing off the good behavior of the latter in most cases of practical interest. This analysis has been later extended in [57] for the COM2 method [27] in the presence of zero-mean circular cyclostationary sources with similar conclusions. A similar analysis for non-zero mean cyclostationary sources can also be found in [36] and [35] for SO and HO blind source separation methods respectively. From these results, a general methodology of BSS of arbitrary cyclostationary sources is presented in [25].

This survey shows that the FO cumulant-based techniques developed for the blind separation of instantaneous mixtures of statistically independent and non-Gaussian sources seem very appropriate for the spectrum monitoring of radio-communication sources. In the sequel, we provide precise arguments for this claim.

Content of the chapter. In this context, the aim of the chapter, which is composed of 7 parts, is mainly to describe the implementation and the performance of such a technique for the spectrum control of radio communications. The observation model and statistics jointly with the problem formulation are presented in section 17.2. The two possible approaches of blind source separation in radio communications are synthesized in section 17.3. Section 17.4 introduces the ultimate separators of instantaneous mixtures of sources. Section 17.5 analyzes the behavior of FO cumulant-based blind source separators, such as JADE, in the presence of cyclostationary sources when the cyclic frequencies of the observations are unknown. A performance comparison of the convolutive and the instantaneous approaches is then presented in section 17.6. Finally section 17.7 concludes the chapter.

17.2 DATA MODEL, STATISTICS AND PROBLEM FORMULATION

17.2.1 Observation model

17.2.1.1 Convolutive model

We consider an array of P *narrow-band* (NB) sensors and we call $\mathbf{x}(t)$ the vector of complex amplitudes of the signals at the output of these sensors. Each sensor is assumed to receive the contribution of N (with $N \leq P$) zero-mean, NB, statistically independent cyclostationary sources corrupted by additive noise. Under these assumptions, the observation vector can approximately be written as follows:

$$\begin{aligned}\forall t \in \mathbb{R}, \quad \mathbf{x}(t) &= \sum_{i=1}^{N} \int_{\mathbb{R}} \mathbf{a}_i(t,\tau) s_i(t-\tau) d\tau + \mathbf{v}(t) \\ &\triangleq \int_{\mathbb{R}} \mathbf{A}(t,\tau)\mathbf{s}(t-\tau) d\tau + \mathbf{v}(t) \end{aligned} \tag{17.1}$$

where $\mathbf{v}(t)$ is the noise vector, assumed zero-mean, stationary, circular, Gaussian and spatially white in order to ease the analysis; $\mathbf{s}(t)$, independent of the noise, is the vector whose components $s_i(t)$ $(1 \leq i \leq N)$, assumed to be zero-mean and cyclostationary, are the complex amplitudes of the sources; $\tau \mapsto \mathbf{a}_i(t,\tau)$ is the vector of impulse responses, at time t, of the propagation channels between the source i and the P sensors; $\mathbf{A}(t,\tau)$ is the $P \times N$ mixing matrix whose columns are the vectors $\mathbf{a}_i(t,\tau)$.

17.2.1.2 Specular model

In most operational contexts encountered in practice, the propagation channels are specular in the time domain, which means that they are composed of a finite number of paths with different delays and complex attenuations. This property of the channels corresponds to a sparsity property in the time domain. Under this assumption, the vector $\mathbf{a}_i(t,\tau)$ can be written as

$$\mathbf{a}_i(t,\tau) = \sum_{m=1}^{M_i} \alpha_{im}(t) e^{i2\pi \nu_{im} t} \delta(\tau - \tau_{im}(t)) \mathbf{a}_{im}(t) \tag{17.2}$$

where M_i is the number of propagation multi-paths for the source i, $\alpha_{im}(t)$, ν_{im} and $\tau_{im}(t)$ are respectively the complex amplitude, the frequency offset[1] and the delay of path m of

[1]Note that the frequency offset comprises both the Doppler shift of the path and the potential clock jitter, and the carrier residue of the associated source.

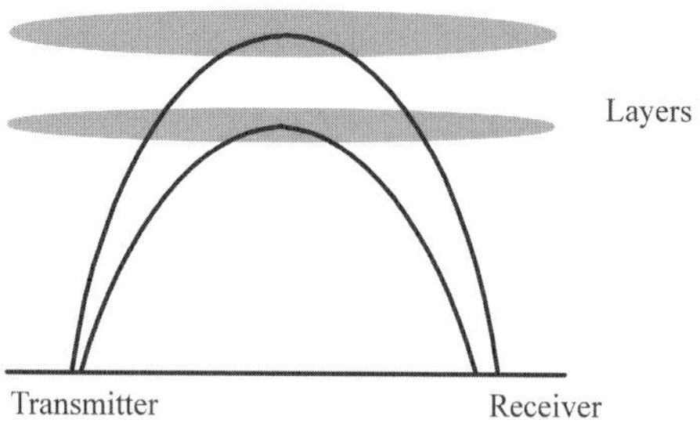

FIGURE 17.1

An HF ionospheric link.

source i at time t, called in the following path im; $\mathbf{a}_{im}(t)$ is the equivalent steering vector (or spatial signature) of path m_i at time t, which depends on the array, the *direction of arrival* (DOA), the polarization and the angular spread of this path.

The propagation vector $\mathbf{a}_i(t, \tau)$ is said to be time-variant if it is a function of t and it is time-invariant otherwise. Time variations of $\mathbf{a}_i(t, \tau)$ occur when $\nu_{im} \neq 0$ and/or when $\alpha_{im}(t)$, $\tau_{im}(t)$ or $\mathbf{a}_{im}(t)$ are not a constant function of time.

It is for example the case for long range ionospheric propagation in the HF band [3–30 MHz]. In this context, the multiple paths are due to different reflexions of waves on the different ionospheric layers of atmosphere (see Fig. 17.1) and the angular spread of paths may be often neglected with respect to the 3 dB beamwidth of the array. The time variations of the channels are due to the periodic movements of the ionospheric layers. In this case, the time variations of $\tau_{im}(t)$ and $\mathbf{a}_{im}(t)$ are much slower than the time variations of $\alpha_{im}(t)$, which is often modelized [114] as a stationary process corresponding to the output of a filter of band B whose input is a white noise. In this model, the band B corresponds to the Doppler spread of the channel, whose typical value is such that $0.1 \text{ Hz} \leq B \leq 1 \text{ Hz}$. Thus, despite its time variation over a long observation duration, the channel may be assumed to be time invariant, to within the frequency offset, over a duration of several tenths of milliseconds. Under this assumption and for a reception bandwidth which is less than 10 kHz, M_i may be such that $1 \leq M_i \leq 5$ and the delay spread of the channel may reach a few ms. Another example corresponds to propagation in an urban area in the UHF band. In this case, the multiple separated paths (i.e. the different delays τ_{im}) are due to reflexions of waves on buildings (see Fig. 17.2) while scatters in the neighborhood of the receiver generate flat fading and angular spread. This angular spread may be very important for the downlink (base station toward mobile for cellular radio communications) but is much slower for the uplink. Again, the time variations of $\tau_{im}(t)$ and $\mathbf{a}_{im}(t)$ are much slower than the time variations of $\alpha_{im}(t)$, which are mainly due to the speed of vehicles. In this case, for a vehicle with a speed of 150 km/h, the Doppler spread is around $B = 100$ Hz at 900 MHz. Thus the propagation channels may be assumed to be time invariant over a duration of 500 μs, which approximately corresponds to the duration of a GSM burst. Under this assumption and for reception bandwidth of 270 KHz (GSM receiver bandwidth), M_i may be such that $1 \leq M_i \leq 6$, with a

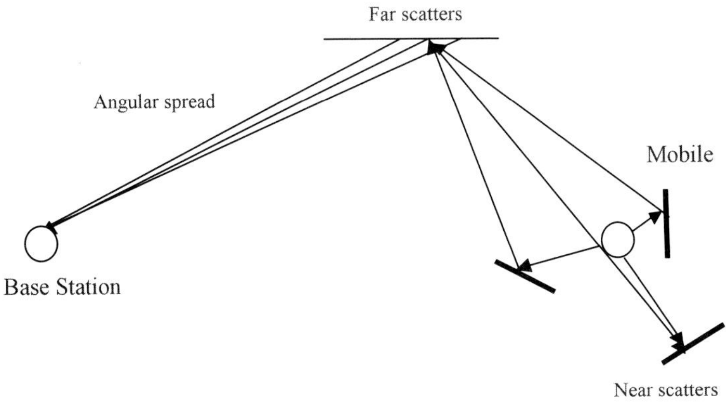

FIGURE 17.2

Multipath propagation in urban area.

strong power gradient between the paths, and the delay spread of the channel may reach a few μs.

Substituting (17.2) into (17.1), we obtain, for specular propagation channels in the time domain

$$\forall t \in \mathbb{R} \quad \mathbf{x}(t) = \sum_{i=1}^{N} \sum_{m=1}^{M_i} \alpha_{im}(t) e^{\imath 2\pi \nu_{im} t} s_i(t - \tau_{im}(t)) \mathbf{a}_{im}(t) + \mathbf{v}(t) \tag{17.3}$$

and for the observation duration over which the channels may be assumed to be time invariant to within the frequency offset, expression (17.3) becomes

$$\forall t \in \mathbb{R} \quad \mathbf{x}(t) = \sum_{i=1}^{N} \sum_{m=1}^{M_i} \alpha_{im} e^{\imath 2\pi \nu_{im} t} s_i(t - \tau_{im}) \mathbf{a}_{im} + \mathbf{v}(t). \tag{17.4}$$

17.2.1.3 Instantaneous model

The time specularity of propagation channels allows one to consider the convolutive mixture of N statistically independent sources as depicted in (17.1) as an instantaneous mixture of both the sources and their delayed versions, which are no longer necessarily statistically independent. Assuming time invariant propagation channels over the observation interval, model (17.4) may be written as

$$\begin{aligned} \forall t \in \mathbb{R} \quad \mathbf{x}(t) &= \sum_{i=1}^{N} \sum_{m=1}^{M_i} e^{\imath 2\pi \nu_{im} t} s_{im}(t) \mathbf{a}_{im} + \mathbf{v}(t) \\ &= \sum_{i=1}^{N} \sum_{m=1}^{M_i} \tilde{s}_{im}(t) \mathbf{a}_{im} + \mathbf{v}(t) \end{aligned}$$

where $s_{im}(t) \triangleq \alpha_{im} s_i(t - \tau_{im})$ and $\tilde{s}_{im}(t) \triangleq e^{\imath 2\pi \nu_{im} t} s_{im}(t)$. More compactly, we may factorize $\mathbf{x}(t)$ as

$$\mathbf{x}(t) = \sum_{i=1}^{N} \tilde{\mathbf{A}}_i \tilde{\mathbf{s}}_i(t) + \mathbf{v}(t) = \tilde{\mathbf{A}}\tilde{\mathbf{s}}(t) + \mathbf{v}(t). \tag{17.5}$$

$\tilde{\mathbf{A}}_i$ is the $P \times M_i$ matrix whose columns are the vectors $\mathbf{a}_{im}$ and $\tilde{\mathbf{s}}_i(t)$ is the $M_i \times 1$ vector of components $\tilde{s}_{im}(t)$; $\tilde{\mathbf{A}}$ is the $P \times \tilde{N}$ matrix defined by $\tilde{\mathbf{A}} = [\tilde{\mathbf{A}}_1, \ldots, \tilde{\mathbf{A}}_N]$ and $\tilde{\mathbf{s}}(t)$ is the $\tilde{N} \times 1$ vector defined by $\tilde{\mathbf{s}}(t) = [\tilde{\mathbf{s}}_1(t)^T, \ldots, \tilde{\mathbf{s}}_N(t)^T]^T$ where

$$\tilde{N} = \sum_{i=1}^{N} M_i. \tag{17.6}$$

Expression (17.5) describes a noisy instantaneous mixture of $\tilde{N}$ zero-mean cyclostationary NB paths, not necessarily statistically independent, with complex envelope and spatial signature respectively equal to $\tilde{s}_{im}(t)$ and $\mathbf{a}_{im}$ $(1 \le m \le M_i, 1 \le i \le N)$. In the following, to simplify the notations, the columns of $\tilde{\mathbf{A}}$ and the components of $\tilde{\mathbf{s}}(t)$ are respectively denoted by $\tilde{\mathbf{a}}_i$ and $\tilde{s}_i(t)$ $(1 \le i \le \tilde{N})$.

17.2.2 Data statistics

17.2.2.1 SO statistics

Presentation. Under the previous assumptions, the SO statistics of the data are characterized by the two correlation matrices $\mathbf{R}_{\mathbf{x}}(t,\tau)$ and $\mathbf{C}_{\mathbf{x}}(t,\tau)$, which are time dependent and defined by

$$\begin{aligned}
\mathbf{R}_{\mathbf{x}}(t,\tau) &\triangleq \mathbb{E}\{\mathbf{x}(t)\mathbf{x}(t-\tau)^H\} \\
&= \sum_{i=1}^{N} \tilde{\mathbf{A}}_i \mathbf{R}_{\tilde{\mathbf{s}}_i}(t,\tau)\tilde{\mathbf{A}}_i^H + \eta_2(\tau)\mathbf{I} \\
&= \tilde{\mathbf{A}}\mathbf{R}_{\tilde{\mathbf{s}}}(t,\tau)\tilde{\mathbf{A}}^H + \eta_2(\tau)\mathbf{I}
\end{aligned} \tag{17.7}$$

$$\begin{aligned}
\mathbf{C}_{\mathbf{x}}(t,\tau) &\triangleq \mathbb{E}\{\mathbf{x}(t)\mathbf{x}(t-\tau)^T\} \\
&= \sum_{i=1}^{N} \tilde{\mathbf{A}}_i \mathbf{C}_{\tilde{\mathbf{s}}_i}(t,\tau)\tilde{\mathbf{A}}_i^T \\
&= \tilde{\mathbf{A}}\mathbf{C}_{\tilde{\mathbf{s}}}(t,\tau)\tilde{\mathbf{A}}^T
\end{aligned} \tag{17.8}$$

where $\eta_2(\tau)$ is the noise correlation at lag τ on each sensor and $\mathbf{I}$ the identity matrix of order P. The matrices $\mathbf{R}_{\tilde{\mathbf{s}}_i}(t,\tau) \triangleq \mathbb{E}\{\tilde{\mathbf{s}}_i(t)\tilde{\mathbf{s}}_i(t-\tau)^H\}$ and $\mathbf{C}_{\tilde{\mathbf{s}}_i}(t,\tau) \triangleq \mathbb{E}\{\tilde{\mathbf{s}}_i(t)\tilde{\mathbf{s}}_i(t-\tau)^T\}$, diagonal if all the paths of source i are uncorrelated, are the first and second correlation matrix of the vector $\tilde{\mathbf{s}}_i(t)$. The matrices $\mathbf{R}_{\tilde{\mathbf{s}}}(t,\tau) \triangleq \mathbb{E}\{\tilde{\mathbf{s}}(t)\tilde{\mathbf{s}}(t-\tau)^H\}$ and

$\mathbf{C}_{\tilde{\mathbf{s}}}(t,\tau) \triangleq \mathbb{E}\{\tilde{\mathbf{s}}(t)\tilde{\mathbf{s}}(t-\tau)^T\}$, diagonal if all the paths of all the sources are uncorrelated, are the first and second correlation matrix of the vector $\tilde{\mathbf{s}}(t)$.

For a cyclostationary path j $(1 \le j \le \tilde{N})$, the SO statistics of $\tilde{s}_j(t)$ are periodic or *poly-periodic* (PP) functions of t and thus have Fourier serial expansions showing off the SO cyclic statistics of $\tilde{s}_j(t)$. These SO cyclic statistics are directly related to the baud-rate and the frequency offset of the path j. As a consequence of this result, matrices $\mathbf{R}_{\tilde{\mathbf{s}}_i}(t,\tau)$ and $\mathbf{C}_{\tilde{\mathbf{s}}_i}(t,\tau)$, as functions of t, are also PP. The same property hence holds for $\mathbf{R}_{\mathbf{x}}(t,\tau)$ and $\mathbf{C}_{\mathbf{x}}(t,\tau)$. As a consequence, these matrix-valued functions have a Fourier serial expansion and we can write

$$\mathbf{R}_{\tilde{\mathbf{s}}_i}(t,\tau) = \sum_{\alpha\in\mathcal{A}_i} \mathbf{R}_{\tilde{\mathbf{s}}_i}^{(\alpha)}(\tau)e^{\imath 2\pi\alpha t} \tag{17.9}$$

$$\mathbf{C}_{\tilde{\mathbf{s}}_i}(t,\tau) = \sum_{\beta\in\mathcal{B}_i} \mathbf{C}_{\tilde{\mathbf{s}}_i}^{(\beta)}(\tau)e^{\imath 2\pi\beta t} \tag{17.10}$$

$$\begin{aligned}\mathbf{R}_{\mathbf{x}}(t,\tau) &= \sum_{i=1}^{N}\sum_{\alpha\in\mathcal{A}_i} \tilde{\mathbf{A}}_i\mathbf{R}_{\tilde{\mathbf{s}}_i}^{(\alpha)}(\tau)\tilde{\mathbf{A}}_i^H e^{\imath 2\pi\alpha t} + \eta_2(\tau)\mathbf{I} \\ &= \sum_{\alpha\in\mathcal{A}} \mathbf{R}_{\mathbf{x}}^{(\alpha)}(\tau)e^{\imath 2\pi\alpha t}\end{aligned} \tag{17.11}$$

$$\begin{aligned}\mathbf{C}_{\mathbf{x}}(t,\tau) &= \sum_{i=1}^{N}\sum_{\beta\in\mathcal{B}_i} \tilde{\mathbf{A}}_i\mathbf{C}_{\tilde{\mathbf{s}}_i}^{(\beta)}(\tau)\tilde{\mathbf{A}}_i^T e^{\imath 2\pi\beta t} \\ &= \sum_{\beta\in\mathcal{B}} \mathbf{C}_{\mathbf{x}}^{(\beta)}(\tau)e^{\imath 2\pi\beta t}\end{aligned} \tag{17.12}$$

where we have set $\mathcal{A} = \bigcup_{i=1}^{N}\mathcal{A}_i$ and $\mathcal{B} = \bigcup_{i=1}^{N}\mathcal{B}_i$. $\mathcal{A}_i$ is the set of the cyclic frequencies of $\mathbf{R}_{\tilde{\mathbf{s}}_i}(t,\tau)$, also called SO cyclic frequencies of source i. $\mathcal{B}_i$ is the set of the cyclic frequencies of $\mathbf{C}_{\tilde{\mathbf{s}}_i}(t,\tau)$, also called SO conjugate cyclic frequency of source i. These SO cyclic and conjugate cyclic frequencies are directly related to the baud rate and the frequency offsets of the multiple paths of source i. If $\alpha\in\mathcal{A}$ and $\beta\in\mathcal{B}$, then α and β are the SO cyclic frequencies of $\mathbf{R}_{\mathbf{x}}(t,\tau)$ and $\mathbf{C}_{\mathbf{x}}(t,\tau)$ respectively. The quantities $\mathbf{R}_{\tilde{\mathbf{s}}_i}^{(\alpha)}(\tau)$ and $\mathbf{C}_{\tilde{\mathbf{s}}_i}^{(\beta)}(\tau)$ are respectively the first and the second cyclic correlation function of $\tilde{\mathbf{s}}_i$. The same holds for $\mathbf{R}_{\mathbf{x}}^{(\alpha)}(\tau)$ and $\mathbf{C}_{\mathbf{x}}^{(\beta)}(\tau)$ with regard to $\mathbf{x}(t)$. These quantities can be expressed as

$$\mathbf{R}_{\tilde{\mathbf{s}}_i}^{(\alpha)}(\tau) = \langle \mathbf{R}_{\tilde{\mathbf{s}}_i}(t,\tau)e^{-\imath 2\pi\alpha t}\rangle_c \tag{17.13}$$

$$\mathbf{C}_{\tilde{\mathbf{s}}_i}^{(\beta)}(\tau) = \langle \mathbf{C}_{\tilde{\mathbf{s}}_i}(t,\tau)e^{-\imath 2\pi\beta t}\rangle_c \tag{17.14}$$

$$\begin{aligned}\mathbf{R}_{\mathbf{x}}^{(\alpha)}(\tau) &= \langle \mathbf{R}_{\mathbf{x}}(t,\tau)e^{-\imath 2\pi\alpha t}\rangle_c \\ &= \tilde{\mathbf{A}}\mathbf{R}_{\tilde{\mathbf{s}}}^{(\alpha)}(\tau)\tilde{\mathbf{A}}^H + \eta_2(\tau)\delta(\alpha)\mathbf{I}\end{aligned} \tag{17.15}$$

$$\begin{aligned} \mathbf{C}_{\mathbf{x}}^{(\beta)}(\tau) &= \langle \mathbf{C}_{\mathbf{x}}(t,\tau) e^{-\imath 2\pi\beta t}\rangle_c \\ &= \tilde{\mathbf{A}}\mathbf{C}_{\tilde{\mathbf{s}}}^{(\beta)}(\tau)\tilde{\mathbf{A}}^T. \end{aligned} \tag{17.16}$$

In these equations, $\langle f(t)\rangle_c$ corresponds, for a continuous-time function $f(t)$, to the temporal mean operation of $f(t)$ over an infinite observation interval; $\delta(.)$ is the Kronecker series; $\mathbf{R}_{\tilde{\mathbf{s}}}^{(\alpha)}(\tau)$ and $\mathbf{C}_{\tilde{\mathbf{s}}}^{(\beta)}(\tau)$ stand respectively for the first and second cyclic correlation matrix of $\tilde{\mathbf{s}}(t)$. For instance, SO BSS methods such as SOBI [9] exploit the information contained in several matrices

$$\mathbf{R}_{\mathbf{x}}^{(0)}(\tau) = \langle \mathbf{R}_{\mathbf{x}}(t,\tau)\rangle_c \triangleq \mathbf{R}_{\mathbf{x}}(\tau)$$

for several values of τ. Methods such as JADE [15] or COM2 [27] exploit the information contained both in $\mathbf{R}_{\mathbf{x}}(0) \triangleq \mathbf{R}_{\mathbf{x}}$ and in the temporal mean of *fourth-order* (FO) statistics of the data (see section 17.2.2.2). One should notice that for continuous-time stationary sources, the quantities defined by (17.13) to (17.16) are zero except for the null ($\alpha = 0$) cyclic frequency.

In practice, the observations are sampled at a sample period denoted by T_e. The second-order statistics $\mathbf{R}_{\mathbf{x}}(t,\tau)$ and $\mathbf{C}_{\mathbf{x}}(t,\tau)$ are defined only at the time-instants $t = kT_e$ and $\tau = \ell T_e$ multiple of the sampling period. Of course, the expansions (17.11) and (17.12) still hold. As far as the computation of $\mathbf{R}_{\mathbf{x}}^{(\alpha)}(\tau)$ and $\mathbf{C}_{\mathbf{x}}^{(\beta)}(\tau)$ is concerned, it is shown in [80] that for band-limited sources, $\langle . \rangle_c$ in the expressions (17.13) to (17.16) can be replaced by the discrete-time temporal mean $\langle . \rangle_d$ provided that the data are sufficiently oversampled. In particular, a sample period such that $T_e < \frac{1}{2B}$, where B is the observation bandwidth, ensures the equivalence between $\langle . \rangle_c$ and $\langle . \rangle_d$.

Estimation. In situations of practical interest, the SO statistics of the data are not known *a priori* and have to be estimated from T sampled data,

$$\mathbf{x}(m) \triangleq \mathbf{x}(mT_e) \quad (1 \leq m \leq T),$$

by temporal averaging operations, using the cyclo-ergodicity property [13]. More precisely, the matrices $\mathbf{R}_{\mathbf{x}}(t,\tau)$ and $\mathbf{C}_{\mathbf{x}}(t,\tau)$ defined by (17.11) and (17.12) respectively have to be estimated, for $(t,\tau) = (kT_e, \ell T_e)$, from the cyclic statistics $\mathbf{R}_{\mathbf{x}}^{(\alpha)}(\ell T_e)$ and $\mathbf{C}_{\mathbf{x}}^{(\beta)}(\ell T_e)$ respectively and from the associated cyclic frequencies. Empirical estimates of $\mathbf{R}_{\mathbf{x}}^{(\alpha)}(\ell T_e)$ and $\mathbf{C}_{\mathbf{x}}^{(\beta)}(\ell T_e)$ are respectively given by

$$\hat{\mathbf{R}}_{\mathbf{x}}^{(\alpha)}(\ell T_e) \triangleq \frac{1}{T}\sum_{m=1}^{T} \mathbf{x}(m)\mathbf{x}(m-\ell)^H e^{-\imath 2\pi\alpha m T_e} \tag{17.17}$$

$$\hat{\mathbf{C}}_{\mathbf{x}}^{(\beta)}(\ell T_e) \triangleq \frac{1}{T}\sum_{m=1}^{T} \mathbf{x}(m)\mathbf{x}(m-\ell)^T e^{-\imath 2\pi\beta m T_e}. \tag{17.18}$$

For zero-mean, cyclostationary and band-limited vectors $\mathbf{x}(t)$ having a cyclo-ergodicity property [13] and for sufficiently oversampled data, the above empirical estimators $\hat{\mathbf{R}}_{\mathbf{x}}^{(\alpha)}(\ell T_e)$ and $\hat{\mathbf{C}}_{\mathbf{x}}^{(\beta)}(\ell T_e)$ are asymptotically unbiased and consistent estimates of $\mathbf{R}_{\mathbf{x}}^{(\alpha)}(\ell T_e)$

and $\mathbf{C}_{\mathbf{x}}^{(\beta)}(\ell T_e)$ respectively [29]. In particular, $\hat{\mathbf{R}}_{\mathbf{x}}^{(0)}(0)$ and $\hat{\mathbf{C}}_{\mathbf{x}}^{(0)}(0)$ are respectively asymptotically unbiased and consistent estimates of $\mathbf{R}_{\mathbf{x}} = \mathbf{R}_{\mathbf{x}}(0)$ and $\mathbf{C}_{\mathbf{x}} = \mathbf{C}_{\mathbf{x}}(0) \triangleq \mathbf{C}_{\mathbf{x}}^{(0)}(0)$.

17.2.2.2 FO statistics

Presentation. The FO statistics of the data which are considered in the following correspond to the instantaneous quadricovariance matrix $\mathbf{Q}_{\mathbf{x}}(t)$, whose element $[i,j,k,\ell]$, denoted by $\mathbf{Q}_{\mathbf{x}}[i,j,k,\ell](t)$, corresponds to the cumulant

$$\mathrm{cum}\{x_i(t), x_j(t)^*, x_k(t)^*, x_\ell(t)\}$$

defined by

$$\begin{aligned}\mathrm{cum}\{x_i(t), x_j(t)^*, x_k(t)^*, x_\ell(t)\} = {} & \mathbf{M}_{\mathbf{x}}[i,j,k,\ell](t) \\ & - \mathbf{R}_{\mathbf{x}}[i,j](t)\mathbf{R}_{\mathbf{x}}[\ell,k](t) - \mathbf{R}_{\mathbf{x}}[i,k](t)\mathbf{R}_{\mathbf{x}}[\ell,j](t) \\ & - \mathbf{C}_{\mathbf{x}}[i,\ell](t)\mathbf{C}_{\mathbf{x}}[j,k](t)^* \end{aligned} \tag{17.19}$$

In (17.19), $\mathbf{M}_{\mathbf{x}}[i,j,k,\ell](t) \triangleq \mathbb{E}\{x_i(t)x_j(t)^* x_k(t)^* x_\ell(t)\}$, $\mathbf{R}_{\mathbf{x}}[i,j](t) \triangleq \mathbb{E}\{x_i(t)x_j(t)^*\}$ and $\mathbf{C}_{\mathbf{x}}[i,j](t) \triangleq \mathbb{E}\{x_i(t)x_j(t)\}$ are time-dependent. Arranging these terms in a natural way in $\mathbf{Q}_{\mathbf{x}}(t)$ [26], we deduce from (17.5) that for a Gaussian noise, $\mathbf{Q}_{\mathbf{x}}(t)$ can be written as

$$\begin{aligned}\mathbf{Q}_{\mathbf{x}}(t) &= \sum_{i=1}^{N} \left[\tilde{\mathbf{A}}_i \otimes \tilde{\mathbf{A}}_i^*\right] \mathbf{Q}_{\tilde{\mathbf{s}}_i}(t) \left[\tilde{\mathbf{A}}_i \otimes \tilde{\mathbf{A}}_i^*\right]^H \\ &= \left[\tilde{\mathbf{A}} \otimes \tilde{\mathbf{A}}^*\right] \mathbf{Q}_{\tilde{\mathbf{s}}}(t) \left[\tilde{\mathbf{A}} \otimes \tilde{\mathbf{A}}^*\right]^H \end{aligned} \tag{17.20}$$

where $\otimes$ is the Kronecker product and $\mathbf{Q}_{\tilde{\mathbf{s}}_i}(t)$ and $\mathbf{Q}_{\tilde{\mathbf{s}}}(t)$ are the quadricovariance of $\tilde{\mathbf{s}}_i(t)$ and $\tilde{\mathbf{s}}(t)$ respectively.

FO BSS methods such as JADE [15] or COM2 [27] exploit the information contained in the temporal mean $\mathbf{Q}_{\mathbf{x}}[i,j,k,\ell] \triangleq \langle \mathbf{Q}_{\mathbf{x}}[i,j,k,\ell](t)\rangle_c$ of the data quadricovariance elements. From (17.19), we have:

$$\begin{aligned}\mathbf{Q}_{\mathbf{x}}[i,j,k,\ell] = {} & \mathbf{M}_{\mathbf{x}}[i,j,k,\ell] \\ & - \langle \mathbf{R}_{\mathbf{x}}[i,j](t)\mathbf{R}_{\mathbf{x}}[\ell,k](t)\rangle_c - \langle \mathbf{R}_{\mathbf{x}}[i,k](t)\mathbf{R}_{\mathbf{x}}[\ell,j](t)\rangle_c \\ & - \langle \mathbf{C}_{\mathbf{x}}[i,\ell](t)\mathbf{C}_{\mathbf{x}}[j,k](t)^*\rangle_c \end{aligned} \tag{17.21}$$

where $\mathbf{M}_{\mathbf{x}}[i,j,k,\ell] \triangleq \langle \mathbf{M}_{\mathbf{x}}[i,j,k,\ell](t)\rangle_c$. Then, using (17.11) and (17.12) for $\tau = 0$ and using the fact that $\langle e^{i2\pi\alpha t}\rangle_c = \delta(\alpha)$, we obtain

$$\begin{aligned}\mathbf{Q}_{\mathbf{x}}[i,j,k,\ell] = {} & \mathbf{M}_{\mathbf{x}}[i,j,k,\ell] \\ & - \sum_{\alpha \in \mathscr{A}} \mathbf{R}_{\mathbf{x}}^{(\alpha)}[i,j]\mathbf{R}_{\mathbf{x}}^{(-\alpha)}[\ell,k] - \sum_{\alpha \in \mathscr{A}} \mathbf{R}_{\mathbf{x}}^{(\alpha)}[i,k]\mathbf{R}_{\mathbf{x}}^{(-\alpha)}[\ell,j] \\ & - \sum_{\beta \in \mathscr{B}} \mathbf{C}_{\mathbf{x}}^{(\beta)}[i,\ell]\mathbf{C}_{\mathbf{x}}^{(\beta)}[j,k]^* \end{aligned} \tag{17.22}$$

where $\mathbf{R}_{\mathrm{x}}^{(\alpha)}[i,j]$ and $\mathbf{C}_{\mathrm{x}}^{(\beta)}[i,j]$ are the $[i,j]$ coefficients of the matrices $\mathbf{R}_{\mathrm{x}}^{(\alpha)} \triangleq \mathbf{R}_{\mathrm{x}}^{(\alpha)}(\tau = 0)$ and $\mathbf{C}_{\mathrm{x}}^{(\beta)} \triangleq \mathbf{C}_{\mathrm{x}}^{(\beta)}(\tau = 0)$ respectively. The expression (17.22) shows that $\mathbf{Q}_{\mathrm{x}}[i,j,k,\ell]$ depends on all the SO cyclic statistics of the observations. For stationary observations, the SO cyclic statistics of the observations are zero for the non-zero cyclic frequencies; $\mathbf{Q}_{\mathrm{x}}[i,j,k,\ell]$ is then denoted by $\mathbf{Q}_{\mathrm{x},a}[i,j,k,\ell]$ and, thanks to (17.22), we have

$$\begin{aligned} \mathbf{Q}_{\mathrm{x},a}[i,j,k,\ell] \triangleq\ & \mathbf{M}_{\mathrm{x}}[i,j,k,\ell] \\ & - \mathbf{R}_{\mathrm{x}}[i,j]\mathbf{R}_{\mathrm{x}}[\ell,k] - \mathbf{R}_{\mathrm{x}}[i,k]\mathbf{R}_{\mathrm{x}}[\ell,j] \\ & - \mathbf{C}_{\mathrm{x}}[i,\ell]\mathbf{C}_{\mathrm{x}}[j,k]^* \end{aligned} \tag{17.23}$$

Substituting (17.23) into (17.22), we finally obtain

$$\begin{aligned} \mathbf{Q}_{\mathrm{x}}[i,j,k,\ell] =\ & \mathbf{Q}_{\mathrm{x},a}[i,j,k,\ell] \\ & - \sum_{\alpha \neq 0} \mathbf{R}_{\mathrm{x}}^{(\alpha)}[i,j]\mathbf{R}_{\mathrm{x}}^{(-\alpha)}[\ell,k] - \sum_{\alpha \neq 0} \mathbf{R}_{\mathrm{x}}^{(\alpha)}[i,k]\mathbf{R}_{\mathrm{x}}^{(-\alpha)}[\ell,j] \\ & - \sum_{\beta \neq 0} \mathbf{C}_{\mathrm{x}}^{(\beta)}[i,\ell]\mathbf{C}_{\mathrm{x}}^{(\beta)}[j,k]^*. \end{aligned} \tag{17.24}$$

Estimation. In situations of practical interest, the FO statistics of the data are not known *a priori* and have to be estimated from the sampled data, $\mathbf{x}(m) = \mathbf{x}(mT_e)$ $(1 \leq m \leq T)$, by temporal averaging operations, using the cyclo-ergodicity property [13]. In practice, T snapshots are available in order to estimate $\mathbf{Q}_{\mathrm{x}}[i,j,k,\ell]$ defined by (17.21). Considering the Eq. (17.23), the empirical estimator $\hat{\mathbf{Q}}_{\mathrm{x}}[i,j,k,\ell]$ is defined as:

$$\begin{aligned} \hat{\mathbf{Q}}_{\mathrm{x}}[i,j,k,\ell] \triangleq\ & \hat{\mathbf{M}}_{\mathrm{x}}[i,j,k,\ell] \\ & - \hat{\mathbf{R}}_{\mathrm{x}}[i,j]\hat{\mathbf{R}}_{\mathrm{x}}[\ell,k] - \hat{\mathbf{R}}_{\mathrm{x}}[i,k]\hat{\mathbf{R}}_{\mathrm{x}}[\ell,j] \\ & - \hat{\mathbf{C}}_{\mathrm{x}}[i,\ell]\hat{\mathbf{C}}_{\mathrm{x}}[j,k]^* \end{aligned} \tag{17.25}$$

where

$$\hat{\mathbf{M}}_{\mathrm{x}}[i,j,k,\ell] \triangleq \frac{1}{T}\sum_{m=1}^{T} x_i(m)x_j(m)^* x_k(m)^* x_\ell(m) \tag{17.26}$$

$$\hat{\mathbf{R}}_{\mathrm{x}}[i,j] \triangleq \frac{1}{T}\sum_{m=1}^{T} x_i(m)x_j(m)^* \tag{17.27}$$

$$\hat{\mathbf{C}}_{\mathrm{x}}[i,\ell] \triangleq \frac{1}{T}\sum_{m=1}^{T} x_i(m)x_\ell(m). \tag{17.28}$$

Under the assumption of zero-mean cyclostationary, cyclo-ergodic and band-limited observations, the estimators $\hat{\mathbf{M}}_{\mathrm{x}}[i,j,k,\ell]$, $\hat{\mathbf{R}}_{\mathrm{x}}[i,j]$ and $\hat{\mathbf{C}}_{\mathrm{x}}[i,\ell]$ respectively defined by (17.26)–(17.28) are, for sufficiently oversampled data, unbiased and consistent estimates of $\mathbf{M}_{\mathrm{x}}[i,j,k,\ell]$, $\mathbf{R}_{\mathrm{x}}[i,j]$ and $\mathbf{C}_{\mathrm{x}}[i,\ell]$ respectively. As a consequence, as T becomes infinite, the empirical estimator $\hat{\mathbf{Q}}_{\mathrm{x}}[i,j,k,\ell]$, defined by (17.25), converges towards $\mathbf{Q}_{\mathrm{x},a}[i,j,k,\ell]$ defined by (17.23), and not towards $\mathbf{Q}_{\mathrm{x}}[i,j,k,\ell]$ defined by (17.24). Thus, for observations having no energy at the non-zero cyclic frequencies, which is the case for stationary observations, the quantities $\mathbf{Q}_{\mathrm{x},a}[i,j,k,\ell]$ and $\mathbf{Q}_{\mathrm{x}}[i,j,k,\ell]$ are equal and $\hat{\mathbf{Q}}_{\mathrm{x}}[i,j,k,\ell]$ is an asymptotically unbiased and consistent estimate of $\mathbf{Q}_{\mathrm{x}}[i,j,k,\ell]$. This explains the suffix a in $\mathbf{Q}_{\mathrm{x},a}[i,j,k,\ell]$, since we may call the empirical estimate $\hat{\mathbf{Q}}_{\mathrm{x}}[i,j,k,\ell]$ an *apparent* estimate of the temporal mean of $\mathbf{Q}_{\mathrm{x}}[i,j,k,\ell](t)$. However, in the general case of sufficiently over-sampled cyclostationary observations, $\mathbf{Q}_{\mathrm{x},a}[i,j,k,\ell] \neq \mathbf{Q}_{\mathrm{x}}[i,j,k,\ell]$ and the empirical FO statistics estimator is an asymptotically *biased* estimate, of $\mathbf{Q}_{\mathrm{x}}$. This was shown for the first time in [34]. We must then wonder what may be the consequences of this bias, in the estimation of $\mathbf{Q}_{\mathrm{x}}$, on the performances of the current blind source separators which exploit the FO statistics of the data, such as JADE [15] or COM2 [27]. This question is analyzed in section 17.5 of this chapter.

We have shown that the empirical estimator of the FO data statistics does not require any information about the cyclic frequencies of the observations but has the drawback of being asymptotically biased for zero-mean cyclostationary observations. On the other hand, it is possible to generate an unbiased and consistent FO statistics estimator of the data provided the SO cyclic frequencies and cyclic statistics of the data have been estimated. Indeed, thanks to (17.24), we may introduce the estimate

$$\begin{aligned}\hat{\mathbf{Q}}_{\mathrm{x},2}[i,j,k,\ell] = \hat{\mathbf{Q}}_{\mathrm{x}}[i,j,k,\ell] &- \sum_{\alpha\neq 0}\hat{\mathbf{R}}_{\mathrm{x}}^{(\alpha)}[i,j]\hat{\mathbf{R}}_{\mathrm{x}}^{(-\alpha)}[\ell,k] \\ &- \sum_{\alpha\neq}\hat{\mathbf{R}}_{\mathrm{x}}^{(\alpha)}[i,k]\hat{\mathbf{R}}_{\mathrm{x}}^{(-\alpha)}[\ell,j] \\ &- \sum_{\beta\neq 0}\hat{\mathbf{C}}_{\mathrm{x}}^{(\beta)}[i,\ell]\hat{\mathbf{C}}_{\mathrm{x}}^{(\beta)}[j,k]^{*}\end{aligned} \tag{17.29}$$

where $\hat{\mathbf{R}}_{\mathrm{x}}^{(\alpha)}[i,j]$ and $\hat{\mathbf{C}}_{\mathrm{x}}^{(\beta)}[i,j]$ are the (i,j) entries of the matrices respectively defined by (17.17) and (17.18) (by setting $\ell = 0$). Under the assumption of zero-mean cyclostationary and cyclo-ergodic observations, the estimator (17.29) corresponds in the steady state to the true matrix element $\mathbf{Q}_{\mathrm{x}}[i,j,k,\ell]$ defined by (17.22) [29]. As the implementation of $\hat{\mathbf{Q}}_{\mathrm{x},2}[i,j,k,\ell]$ requires the knowledge of all the cyclic frequencies of the observations, the first task to do is to detect all (or at least the principal) active SO cyclic frequencies of the observations by implementing one of the cyclic detectors developed these last years [41,30,78]. Then, knowing the SO active cyclic frequencies of the observations and provided the data are sufficiently oversampled, the element $\mathbf{Q}_{\mathrm{x}}[i,j,k,\ell]$ can be estimated through (17.29), from T data snapshots.

17.2.3 Formulation of the problem

The problem addressed in this chapter is to blindly extract, without knowing or estimating the SO cyclic frequencies of the data, the information contained in each of the received sources from a linear filtering of the data, resorting to SO and FO statistics of these latter. For each analogical source i, the goal is to generate the best possible estimate of a scaled, potentially Dopplerized and delayed version, $\alpha_i s_i(t - \tau_i)e^{i2\pi\nu_i t}$ of $s_i(t)$. For each digital source i, the goal is to restitute, in the best conditions of reception, the symbols that are conveyed by this source. This requires the prior estimation of the symbol duration and the frequency offset of each source. This may also be done after a preliminary step of estimation of a scaled, potentially Dopplerized and delayed version of $s_i(t)$, called BSS.

17.3 POSSIBLE METHODS

17.3.1 Treating the mixture as a convolutive one

17.3.1.1 Problem

As was specified in the introduction, the convolutive approaches are not designed to cope with analogical modulations; we hence restrict the field of applications to mixtures of digital modulations. We hence face the problem of estimating all the unknown parameters of several digital modulations (symbol period, frequency offsets, channels, type of modulation, etc.) and the associated symbols when these sources are observed through a (stationary) unknown non-flat channel. This is a very ambitious goal. For this reason, it seems reasonable, and even mandatory, to treat the problem in *two steps*. The first consists in blindly separating, to within a temporal filtering, the mixed sources from the data by a well-chosen spatio-temporal processing, compatible with sources of arbitrary nature and baud rate. The second step consists, for each separated source, in achieving the *single-input single-output (SISO)* blind channel deconvolution or equalization: this step implicitly assumes estimating such parameters as the symbol period, the frequency offset of the source to be restored (this latter estimation is considerably easier once the BSS is achieved).

It has been explained in the introduction that among the available BSS methods of convolutive mixtures, few of them could be used to solve (and in a partial manner), the previous problem of BSS of convolutive mixture of arbitrary cyclostationary sources. In the following, we focus on the approach introduced in [60,58] which does not require the prior SO cyclic frequency estimation of the data. In addition to the restrictions on the use of convolutive BSS methods (see the conditions in section 17.3.1.3), we may *a priori* understand why these methods are likely to show poor performances in practical scenarios as compared to the instantaneous methods described in section 17.5. Indeed, the so-called convolutive approaches do not exploit the specularity of the physical channels. On the one hand, relaxing the constraint on specularity provides, if anything, *general* BSS methods (robust to the distribution of the delays). On the other hand, the lack of information on the channel structure makes the separation much more difficult and suboptimal. In other words, we clearly have a trade-off between generality (convolutive mixtures) and performance (convolutive mixtures with specular constraints on the channels). This will be further illustrated.

17.3.1.2 Assumptions

The model of the data (17.4) is briefly recast in our framework. As is discussed in the introduction, the channel may be considered as time-invariant during the data acquisition, to within the frequency offsets of the paths. Denote by $h(t)$ a square-summable function such that its Fourier Transform $\hat{h}$ verifies: $\hat{h}(\mathsf{f}) = 0$ outside the sampling frequency band $(-\frac{1}{2T_e}, \frac{1}{2T_e})$ and $\hat{h}(\mathsf{f}) = 1$ for any f in the support of the power spectral density of $\mathbf{x}(t)$. Hence $\mathbf{x}(t) = \int h(u)\mathbf{x}(t-u)du$. Looking at (17.4), it yields[2]

$$\mathbf{x}_i(t) = \int \sum_{m=1}^{M_i} (h(u-\tau_{im})\alpha_{im}\mathbf{a}_{im}e^{\imath 2\pi \nu_{im}\tau_{im}})e^{\imath 2\pi \nu_{im}(t-u)}s_i(t-u)du$$

where $\mathbf{x}_i(t)$ is the contribution of the source i in the mixture. We deduce that $\mathbf{x}_i(t)$ can be seen as a convolution of the Dopplerized source $\tilde{s}_i(t) \triangleq e^{\imath 2\pi \nu_{im} t}s_i(t)$ when the frequency offsets associated with the source i are the same for all the associated paths, i.e. if $\nu_{im} = \nu_i$ $(1 \leq m \leq M_i)$. This condition is assumed to hold for all the sources. The further processings require digital data; in this respect, we consider a sampling period T_e. Suppose that T_e fullfills the Shannon sampling condition w.r.t. the signal $\mathbf{x}(t)$. We deduce that

$$\mathbf{x}(nT_e) = \sum_{i=1}^{N}\sum_{\ell \in \mathbb{Z}} \mathbf{a}_i(\ell)\tilde{s}_i((n-\ell)T_e)$$

where we have defined $\mathbf{a}_i(\ell) = \sum_{m=1}^{M_i} \alpha_{im}\mathbf{a}_{im}e^{\imath 2\pi \nu_i \tau_{im}} h(\ell T_e - \tau_{im})$. The definitions $\mathbf{x}(n) \triangleq \mathbf{x}(nT_e)$ and $s_i(\ell) \triangleq s_i(\ell T_e)$ are clearly abuses of notation, but no confusion is possible in the following since the indices n, ℓ are integers. This means that the discrete time data $\mathbf{x}(n)$ follow the convolutive model

$$\mathbf{x}(n) = \sum_{\ell \in \mathbb{Z}} \mathbf{A}(\ell) \begin{pmatrix} \tilde{s}_1(n-\ell) \\ \vdots \\ \tilde{s}_N(n-\ell) \end{pmatrix} \tag{17.30}$$

where $\mathbf{A}(\ell)$ is the $P \times N$ matrix whose i-th column is $\mathbf{a}_i(\ell)$.

We want to illustrate here the sparsity of the channels evoked in the introduction. We have considered the context of ionospheric transmission whose parameters are given in section 17.6. We have plotted in Fig. 17.3 the impulse response of an equivalent discrete

[2] Unless specified, we consider in this section a noiseless scenario.

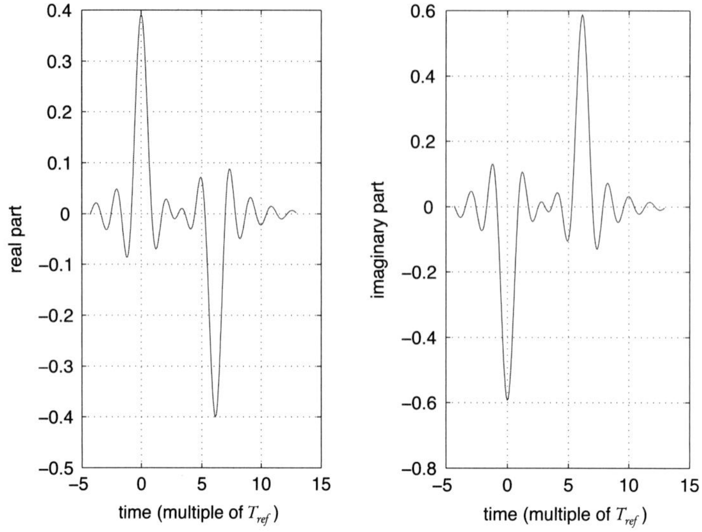

FIGURE 17.3

Impulse response of the channel source 1 → antenna 1.

time channel between the first source and the first antenna. As expected, the presence of the two paths may be easily seen.

The model (17.30) formally shows the convolutive structure of the mixture. It is very general and is not dependent on the distribution of the multi-paths. In particular, the specularity of the model is *hidden* in this equation: in the sequel of this section, we do not mention this structure; i.e. no assumption on the sparsity of the impulse response $(\mathbf{A}(k))_{k\in\mathbb{Z}}$ is done.

As was explained in the introduction perhaps the most general blind separation method is the one described in [60,58]. However, it does not make sense to consider such a method for analogical sources. We particularize the analysis for *digital modulations*, and more specifically, the *linear* ones. This means that the complex envelope of the k-th transmitted signal is:

$$\forall t \in \mathbb{R}, \quad s_k(t) = \sigma_k \sum_{n\in\mathbb{Z}} d_k(n) c_k(t - nT_k) \tag{17.31}$$

where T_k is the symbol period, $(d_k(n))_{n\in\mathbb{Z}}$ is the sequence of symbols, assumed to be iid zero mean, with normalized variance. $c_k(t)$ is a square-root raised-cosine filter with roll-off γ_k ($0 \le \gamma_k \le 1$) such that the spectral occupancy for this signal is

$$\left[-\frac{1+\gamma_k}{2T_k}, \frac{1+\gamma_k}{2T_k}\right]. \tag{17.32}$$

This shaping filter is such that

$$\frac{1}{T_k}\int c_k(t)^2 dt = 1.$$

The parameter $\sigma_k^2 = \langle \mathbb{E}\{|s_k(t)|^2\}\rangle_c$ is the power of the transmitted signal. In [60,58], the fact that the signals are complex circular plays a central role. In what follows, we face the general case; however, a condition on the non-circular sources is shown to be needed in order to achieve the separation.

17.3.1.3 A CMA-based BSS of convolutive mixture

We consider a particular BSS algorithm, based on the *constant modulus algorithm* (CMA). We will shortly discuss the theoretical conditions under which this algorithm is expected to converge. As will be understood, these conditions are not generically fulfilled: the method might fail in separating the sources. The practical considerations will follow. We will attempt to put emphasis on the limitations of the above separation method in an operational scenario.

Blind criterion. As was justified in Chapter 8, sections 8.5 and 8.6 of this book, a variable $(1 \times P)$ filter $\mathbf{b}[z] = \sum_k \mathbf{b}(k)z^{-k}$ is searched such that

$$y(n) = \sum_{k\in\mathbb{Z}} \mathbf{b}(k)\mathbf{x}(n-k)$$

only involves one of the sources. Thanks to (17.30), we may choose to write $y(n)$ as

$$y(n) = \sum_{i=1}^{N}\sum_{k\in\mathbb{Z}} g_i(k)\tilde{s}_i(n-k) \tag{17.33}$$

where the $(1 \times N)$ coefficients $\mathbf{g}(k) \triangleq (g_1(k), \ldots, g_N(k))$ are given[3] through the equation $\mathbf{g}[z] = \mathbf{b}[z]\mathbf{A}[z]$.

A function of the statistics of $(y(n))_{n\in\mathbb{Z}}$ is considered whose argument minima (or maxima) achieve the separation. The problem is the same as the one set previously in this book: see Chapter 8, section 8.6. As the data are cyclostationary, we recall that any function specifically based on averaged cumulants – say for instance $\mathrm{cum}\{y(n), y(n)^*, y(n), y(n)^*\}$ – cannot be implemented in our *blind* context since it requires knowledge of the cyclic frequencies of the mixtures (see section 17.2.2.2). As an example (see Chapter 8, section 8.6), the function

$$f\left(\frac{\langle \mathrm{cum}\{y(n), y(n)^*, y(n), y(n)^*\}\rangle_d}{\langle \mathbb{E}\{|y(n)|^2\}\rangle_d^2}\right)$$

[3]We recall that $\mathbf{g}[z] = (g_1[z], \ldots, g_N[z])$ is called the *equivalent filter*, which takes into account both the channel matrix and the separator.

where $f(.)$ is a convex function on $\mathbb{R}$ such that $f(0) = 0$, is a contrast function, but no blind algorithm can be derived from this result. On the other hand, the following function was considered:

$$\Upsilon_f(\mathbf{g}) = f\left(\frac{\langle \mathbb{E}\{|y(n)|^4\}\rangle_d - 2}{\langle \mathbb{E}\{|y(n)|^2\}\rangle_d^2}\right).$$

In this respect, it was shown that the contrast property requires conditions. However, in favorable contexts, this directly leads to a blind algorithm, since this function only depends on averaged moments, whose asymptotically unbiased and consistent estimation is achieved by the empirical moment estimates.

In this chapter, we rather consider the function

$$\Upsilon_{\mathrm{CMA}}(\mathbf{g}) = \left\langle \mathbb{E}\{(|y(n)|^2 - 1)^2\}\right\rangle_d. \tag{17.34}$$

We briefly justify this choice. First of all, one easily notices that, in the simple case when the sources are all complex-circular, the minimization of $\Upsilon_{\mathrm{CMA}}(\mathbf{g})$ and $\Upsilon_f(\mathbf{g})$ are closely related, at least for the choice $f = -\mathrm{Id}$; see [86]. So why do we wish to focus on Υ_{CMA} rather than on $\Upsilon_{-\mathrm{Id}}$? The minimization of Υ_{CMA} is numerically easier since the function is a fourth-order polynomial of the parameters. Besides, one notices that the function $\Upsilon_f(.)$ is homogeneous of order 0: a minimizer $\mathbf{g}_\sharp[z]$ is never an isolated point since all the vectors $\lambda\mathbf{g}_\sharp[z]$ achieve the minimum for any $\lambda \in \mathbb{C}$ which may be associated with numerical difficulties.

In the sequel, we do not wish to provide *all* the details on the theoretical conditions in a general scenario under which the minimizers of the function $\Upsilon_{\mathrm{CMA}}(\mathbf{g})$ are separating filters. We rather stick to the point of providing practical information.

First step: separation. As a preliminary step, we notice that the minimization of $\Upsilon_{\mathrm{CMA}}(\mathbf{g})$ given by (17.34) is equivalent to finding filtered versions of the observation such that the output of the filter has the smallest fluctuations in modulus. In the general operational context of this chapter, the (continuous-time) transmitted sources do not have a constant modulus. Hence, it is not intuitively easy to conclude that the minimization of $\Upsilon_{\mathrm{CMA}}(\mathbf{g})$ leads to a separating filter: the expression of $\Upsilon_{\mathrm{CMA}}(\mathbf{g})$ has hence to be worked out.

Criterion optimization. We suggest expanding $\Upsilon_{\mathrm{CMA}}(\mathbf{g})$ in order to draw some conclusions. We invoke the general definition of the cumulant and we write

$$\mathbb{E}\{|y(n)|^4\} = \mathrm{cum}\{y(n), y(n)^*, y(n), y(n)^*\} + 2(\mathbb{E}\{|y(n)|^2\})^2 + |\mathbb{E}\{y(n)^2\}|^2.$$

We may write $y(n) = \sum_{i=1}^N y_i(n)$ where $y_i(n)$ is the contribution of the source i in $y(n)$. This contribution may be normalized and we define

$$\tilde{y}_i = \frac{1}{||g_i||} y_i(n)$$

where

$$\|g_i\| = \sqrt{\langle \mathbb{E}\{|y_i(n)|^2\}\rangle_d}.$$

We have: $y(n) = \sum_{i=1}^{N} \|g_i\|\tilde{y}_i$. As a consequence, this allows us to claim that a filter $\mathbf{g}[z]$ is a separating one if and only if the numbers $\|g_i\|$ are all null except one.

As $\Upsilon_{\text{CMA}}(\mathbf{g}) = \langle \mathbb{E}\{|y(n)|^4\}\rangle_d - 2\langle \mathbb{E}\{|y(n)|^2\}\rangle_c + 1$, the expansion of this function is carried on after expanding $\mathbb{E}\{|y(n)|^4\}$ as

$$\mathbb{E}\{|y(n)|^4\} = 2\mathbb{E}\{|y(n)|^2\} + |\mathbb{E}\{y(n)^2\}|^2 + \text{cum}\{y(n), y(n)^*, y(n), y(n)^*\}.$$

We deduce, after some algebra (see Chapter 8, section 8.6) that

$$\begin{aligned} \Upsilon_{\text{CMA}}(\mathbf{g}) = &\sum_{i=1}^{N} \|g_i\|^4 \beta(\tilde{y}_i) \\ &+ \sum_{i \neq j} \|g_i\|^2 \|g_j\|^2 \lambda(\tilde{y}_i, \tilde{y}_j) \\ &+ 1 - 2\sum_{i=1}^{N} \|g_i\|^2 \end{aligned} \tag{17.35}$$

where we have defined

$$\beta(\tilde{y}_i) = \langle \mathbb{E}\{|\tilde{y}_i|^4\}\rangle_d \tag{17.36}$$

and

$$\lambda(\tilde{y}_i, \tilde{y}_j) = 2 + 4\sum_{\alpha \in \mathscr{A}_+^*} K\left(R_{\tilde{y}_i}^{(\alpha)} R_{\tilde{y}_j}^{(\alpha)*}\right) + \sum_{\beta \in \mathscr{B}} C_{\tilde{y}_i}^{(\beta)} C_{\tilde{y}_j}^{(\beta)*} \tag{17.37}$$

where $\mathscr{A}_+^*$ is the set of the SO non-null positive cyclic frequencies of the data. Consistently with the previous notation, we have set

$$\begin{aligned} R_{\tilde{y}_i}^{(\alpha)} &\triangleq R_{\tilde{y}_i}^{(\alpha)}(0) \\ &= \langle \mathbb{E}\{|\tilde{y}_i(n)|^2\} e^{-\imath 2\pi \alpha n}\rangle_d \\ C_{\tilde{y}_i}^{(\beta)} &\triangleq C_{\tilde{y}_i}^{(\beta)}(0) \\ &= \langle \mathbb{E}\{\tilde{y}_i(n)^2\} e^{-\imath 2\pi \beta n}\rangle_d. \end{aligned}$$

The minimization of this function is perilous [58]; as far as we know, the general solution has not been given yet. We aim rather at providing the simplest sufficient conditions on the sources ensuring that the minimization in question is achieved when the filter $\mathbf{g}[z]$ is separating.

In this respect, we recall a classical result in BSS.

LEMMA 17.1

Given parameters β_i $(1 \le i \le N)$ *such that* $1 \le \beta_i < 2$ $(1 \le i \le N)$, *the minimization of the functional in* $\rho_1, \ldots, \rho_N$

$$\Phi_{\beta_1,\ldots,\beta_N}(\rho_1,\ldots,\rho_N) = \sum_{i=1}^{N} \rho_i^4 \beta_i + 2\sum_{i \neq j} \rho_i^2 \rho_j^2 + 1 - 2\sum_{i=1}^{N} \rho_i^2$$

is achieved if and only if all the $\rho_i = \pm\frac{1}{\sqrt{\beta_{i^\sharp}}}\delta(i - i^\sharp)$, *where* $i^\sharp$ *is an index such that* $\beta_{i^\sharp} = \min_{i=1,\ldots,N} \beta_i$.

This result may be recast into the framework of BSS: it is equivalent to the fact that the arguments minima of the CMA are separating in the context of a mixture of N iid zero-mean complex-circular sources [103]. Now, it is possible to connect this elementary result to the minimization of (17.35). In this respect, we provide sufficient conditions on the sources which allow one to link directly our minimization problem with the result of Lemma 17.1:

CONDITION 1

The symbol periods of the sources are all different.

CONDITION 2

The sources are complex circular or the conjuguate cyclic frequencies are all different.

A direct consequence of Condition 1 is that the strictly positive cyclic frequencies (if any) of the sources are different; in other words, if $\alpha \in \mathcal{A}_+^*$, then α cannot be the cyclic frequency of two sources. This says that for any (i, j), $i \neq j$, we have: $R_{\tilde{y}_i}^{(\alpha)} R_{\tilde{y}_j}^{(\alpha)*} = 0$. In the same manner, Condition 2 implies that $C_{\tilde{y}_i}^{(\beta)} C_{\tilde{y}_j}^{(\beta)*} = 0$. In view of (17.37), Conditions 1 and 2 together imply that

$$\forall i \neq j, \quad \lambda(\tilde{y}_i, \tilde{y}_j) = 2.$$

It yields

$$\Upsilon_{\text{CMA}}(\mathbf{g}) = \Phi_{\beta(\tilde{y}_1),\ldots,\beta(\tilde{y}_N)}(||g||_1, \ldots, ||g||_N).$$

The minimization is very simple to handle.

CONDITION 3

For any source i, there exists a filter $g_i[z]$ such that $\beta(\tilde{y}_i) < 2$.

Under this condition, we may invoke Lemma 17.1. And we conclude that $\Upsilon_{\text{CMA}}(\mathbf{g})$ is minimized only when the filter $\mathbf{g}[z]$ has a single non-null component, i.e. when the separation is achieved.

Conversely, one has to investigate the conditions under which there exists a filter $\mathbf{b}[z]$ such that $\mathbf{g}[z]$ is separating. A sufficient condition is that the matrix $\mathbf{A}[z]$ be left invertible, which is generically true when $P \geq N$ (see first section of Chapter 8).

REMARK R17.1 (applicability of the approach)
Condition 1 is apparently the most limiting one, since it imposes the condition that all symbol periods are different. Relaxing this condition leads to a tougher minimization problem but, at least if the number of sources sharing a given symbol period does not exceed 4, the arguments minima of $\Upsilon_{\text{CMA}}(\mathbf{g})$ can be shown to be separating [58]. Condition 2 is not so restrictive, since the frequency offsets of the (non-circular) sources, in the context of spectum monitoring, are likely to be different. Nonetheless, if ever two frequency offsets associated with two non-circular sources are equal, this may strongly impact the method. Indeed, it can be shown in this case that the minimizers of $\Upsilon_{\text{CMA}}(\mathbf{g})$ are not always separating filters [38]. Condition 3 is true for linear modulations and *CPM modulations* (see [60]) and does not seem to be a limiting point.

Deflation. The procedure of deflation is rather standard: see section 8.5 of Chapter 8. It allows one to extract another source, once the first one has been extracted according to the process described above (CMA). Denote by $y(n)$ the scalar-valued signal involving only one of the sources (computed by means of the CMA). We define the "new" mixture as

$$\mathbf{x}^{(2)}(n) = \mathbf{x}(n) - \sum_k \mathbf{t}(k) y(n-k), \quad n \in \mathbb{Z} \tag{17.38}$$

where $(\mathbf{t}(k))_{k\in\mathbb{Z}}$ is the impulse response of a filter with 1 input and P outputs.This filter is searched as a solution of the least-squares problem:

$$\text{minimize} \quad \epsilon(\mathbf{t}) = \langle \mathbb{E}\{\mathbf{x}^{(2)}(n)^H \mathbf{x}^{(2)}(n)\}\rangle_d. \tag{17.39}$$

It hence can be seen that the cyclo-stationarity of the data does not impact the deflation procedure as compared to the stationary case. The deflation approach implies that, after extracting the first source, the CMA is run on the deflated mixture $\mathbf{x}^{(2)}(n)$. This theoretically prevents the CMA from extracting the source previously estimated. The method is iterated until all the sources have been extracted. Due to accumlations of errors (the deflation step cannot exactly cancel the contribution of the extracted sources) the second extraction is in general affected by strong interferences. This is enough to fail the procedure in an operational context. However, it is possible to alleviate this main drawback thanks to a re-initialization scheme introduced in [106].

Re-initialization. Suppose that the first source is extracted. After running the CMA on the deflated mixture, one obtains an estimate $y^{(2)}(n)$ (supposedly not accurate, following the previous discussion) of another source (say number 2). Thanks to a basic least square procedure, it is possible to compute a $1 \times P$ filter $\hat{\mathbf{b}}[z]$ which achieves the minimum quadratic distance between $[\hat{\mathbf{b}}[z]]\mathbf{x}(n)$ and $y^{(2)}(n)$. As $y^{(2)}(n)$ is not accurate, $\hat{\mathbf{b}}[z]$ is also a bad estimate of an extractor of the second source. However bad it might be, it lessens the contributions of all the sources except the number 2. One may think of running the CMA on the *initial* mixture $\mathbf{x}(n)$ while imposing $\hat{\mathbf{b}}[z]$ as the initial filter. In simulations, we have noticed tremendous improvements over the standard deflation.

Second-step: SISO blind equalization. After convergence of the method at any of the N steps, a SIMO ($1 \times P$) filter of order L $\mathbf{b}[z] = \sum_{k=0}^{L} \mathbf{b}(k)z^{-k}$ is computed. It has supposedly the ability of extracting one of the sources. Assume that $\mathbf{b}[z]$ removes all the sources except the first. It means that the signal $y(n) = [\mathbf{b}[z]]\mathbf{x}(n)$ involves the first source, i.e. the series of symbols $(d_1(n))_{n\in\mathbb{Z}}$. A major question arises: how may one face the estimation of the transmitted symbols? As a preliminary step, we may focus on the mathematical model of the signal $y(n)$. In the following lemma, $c(t)$ is a continuous-time impulse response and T_1 is the duration of a symbol $d_1(n)$.

LEMMA 17.2

We consider $x(t) = \sum_{k\in\mathbb{Z}} d_1(k)c(t - kT_1)$. *If* T_e *is a sampling period in accordance with the Shannon sampling condition, and* $x(n) \triangleq x(nT_e)$, *then the signal* $y(n) = [b(z)]x(n)$, *where* $b[z]$ *is any digital filter can be written as*

$$y(n) = \sum_{n\in\mathbb{Z}} d_1(n)f(t - kT_1)\Big|_{t=nT_e} \tag{17.40}$$

where $(f(t))_{t\in\mathbb{R}}$ *is a certain function depending on* $b[z]$ *and* $(c(t))_{t\in\mathbb{R}}$ *which has the same bandwidth as* $(c(t))_{t\in\mathbb{R}}$.

The proof can be found in [52]. This means that the output $y(n)$ of the filter computed by minimizing the constant modulus function is, up to the additive noise and interference terms, of the form (17.40). It is important to notice that the "shaping filter" $(f(t))_{t\in\mathbb{R}}$ is unknown to the receiver. Of course, this function is not of the Nyquist type. In other words, the estimation of the symbols reduces to the equalization of a signal of the type (17.40). This is a single-input/single output (SISO) equalization. A possible approach is the following:

1. Re-sample the signal at a fraction of the period symbol T_1.
2. Then run a fractional equalizer on this re-sampled signal.

Of course, the estimation of the symbols requires the knowledge (or at least an accurate estimate) of the symbol period. In the simulations, and *only for this step*, the symbol period is assumed known. We did not compute an estimate of this quantity, since we wish to measure the performance of the separation step, not that of conventional estimates. Notice, however, that the estimation of the symbol period is, after the extraction is processed, to a degree simpler than before the extraction; indeed, there is a single cyclic frequency to detect in $y(n)$ whereas there are as many cyclic frequencies to detect in the observed data as the number of sources.

Once the re-sampling at a fraction of the symbol period is done, a SISO equalizer is computed. We have chosen to run a fractional CMA (see [37] and the references therein). One should notice that this equalization step also achieves the symbol synchronization.

17.3.1.4 Application: limitations of the convolutive method

We focus on the separation step, and we provide practical information on the (non-)ability of the convolutive method described above to separate the sources. The context is the following. We consider that $N = 2$ sources co-exist in the HF band. In this case, the complex envelope (up to the $\sqrt{2}$ multiplicative factor) of the k-th ($k = 1, 2$) transmitted source can be written according to (17.31). The frequency offsets of the two sources are supposed to be null. We have chosen the two following modulations:

- Source 1: 8-PSK with $\gamma_1 = 0.1$, and $T_1 = T_{\text{ref}}$
- Source 2: QPSK with $\gamma_2 = 0.16$, and $T_2 = T_{\text{ref}}$

where we set $T_{\text{ref}} = 0.4$ ms. We have fixed a sampling period $T_e = T_{\text{ref}}/1.3$ which meets the Shannon sampling condition. The considered algorithms are fed with the data sampled at the sampling period T_e and the number of available samples is T. We consider the stationary model depicted in (17.4). In this equation, we have set $\alpha_{im} = \lambda_{im} e^{-\imath 2\pi f_0 \tau_{im}}$ where λ_{im} is a positive attenuation and f_0 the carrier, set to 10 MHz. The frequency offset is assumed to be negligible during the data acquisition. In addition, the antenna array has the following topology: it is a uniform linear array (ULA). The distance between two successive sensors is fixed to 15 m, which corresponds to half the wave length of the signals. We have supposed that the number of sensors P varies. The sensors are disposed along the (Ox) axis. The steering vector $\mathbf{a}_{im}$ in (17.4) depends on the angles of arrival of the considered path, i.e. the elevation ϕ_{im} and the azimuth θ_{im} angles. The exact expression of vector $\mathbf{a}_{im}$ can be found in [26]. We have considered a typical case for which each source is associated with 2 multi-paths, i.e.

$$M_1 = M_2 = 2.$$

The noise is assumed to be complex Gaussian with power spectral density equal to $N_0/2$ in the sampling band of frequencies. We have supposed that $\sigma_1^2 = 2\sigma_2^2$ (the QPSK source is 3 dB below the 8-PSK source. We set $\mathcal{E}_1 = \sigma_1^2 T_1$ and we consider the parameter of interest

$$\frac{\mathcal{E}_1}{N_0} = 10 \text{ dB}.$$

The separation filter $\mathbf{b}[z]$ varies in the set of finite impulse response filters; let L denote its order, i.e. $\mathbf{b}[z] = \sum_{k=1}^{L} \mathbf{b}_k z^{-k}$. We wish to illustrate how crucial the choice of this parameter L is. In this respect, we consider the channel given in Table 17.1.

As a measure of performance, we have computed the ultimate symbol error rate (SER); in this respect, we considered, for a given set of parameters, 100 independent trials, and for each trial, a long frame of duration $100000 T_{\text{ref}}$ is synthesized.

The considered numbers of antennas are $P = 4$, $P = 7$ and $P = 10$ and the performance associated with each source can respectively be seen in Figs 17.4–17.6 as a function of the number of taps L of the separating filter $\mathbf{b}[z]$. The first case ($P = 4$) corresponds to a saturated sensor array. When the number of snapshots is too small (e.g. corresponding to

Table 17.1 Physical paramaters of the channels

	Source1/path1	Source1/path2	Source2/path1	Source2/path2
Attenuation	$\lambda_{11} = 0$ dB	$\lambda_{12} = 0$ dB	$\lambda_{121} = 0$ dB	$\lambda_{22} = 0$ dB
Delay	$\tau_{11} = 0.025 T_{\text{ref}}$	$\tau_{12} = 2.12 T_{\text{ref}}$	$\tau_{21} = 0.012 T_{\text{ref}}$	$\tau_{22} = 2.17 T_{\text{ref}}$
Azimuth	$\theta_{11} = 10°$	$\theta_{12} = 10°$	$\theta_{21} = 20°$	$\theta_{22} = 20°$
Elevation	$\phi_{11} = 30°$	$\phi_{11} = 50°$	$\phi_{11} = 10°$	$\phi_{11} = 30°$

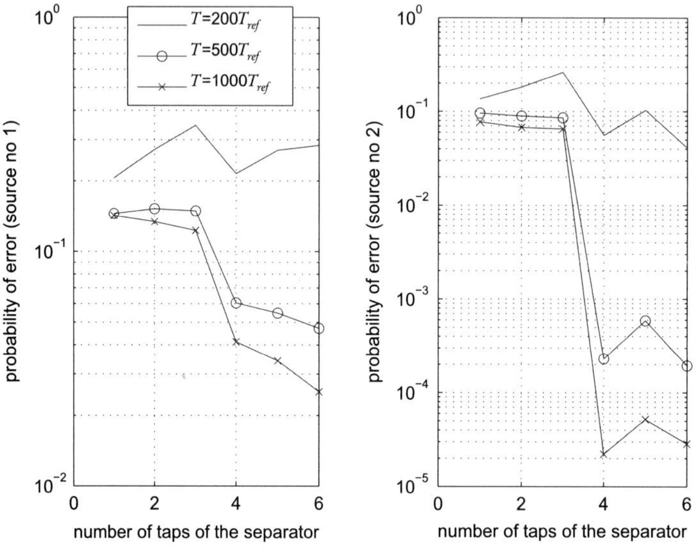

FIGURE 17.4

Performance of the convolutive approach for $P = 4$.

the transmission of 200 symbols), the algorithm fails to compute a good separator: on the one hand, if $L = 1$, the sensor array being saturated, there does not exist any $1 \times P$ vector $\mathbf{b}$ which is able to separate a source with a good SINR; on the other hand, if $L \geq 4$ (this is the condition required if we want $\mathbf{b}[z]$ to combine the different paths), the number of parameters to estimate is too big in relation to the number of available data. However, this latter point is all the less valid as the number of data increases. Indeed, we notice the tremendous gain in performance for $L \geq 4$ when the number of transmitted symbols is 500 or 1000. When the number of antennas is sufficient ($P = 7$ or $P = 10$), there exist vectors $\mathbf{b}$ able to separate the paths with a good SINR. In this case, we may conclude that

1. for $P = 7, 10$ and 200 transmitted symbols, the best strategy is to choose $L = 1$, i.e. the smallest number of taps to estimate;
2. for $P = 7$ and for 500, 1000 transmitted symbols, the best strategy consists in choosing $L = 4$, i.e. the minimum length that allows the separation filter to

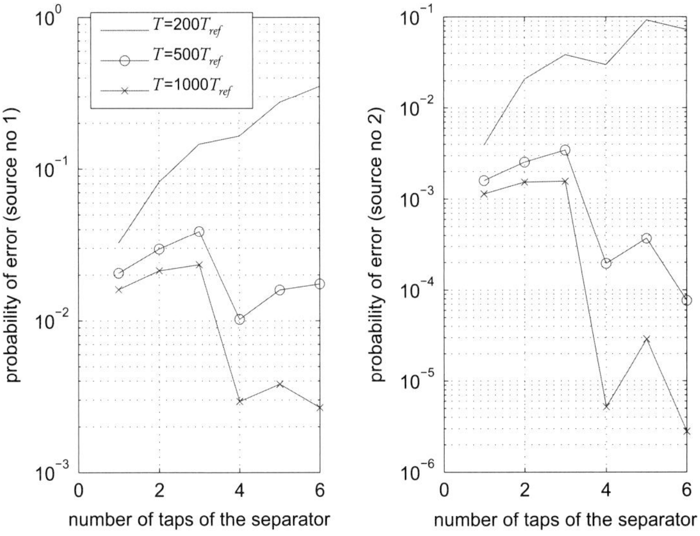

FIGURE 17.5

Performance of the convolutive approach for $P = 7$.

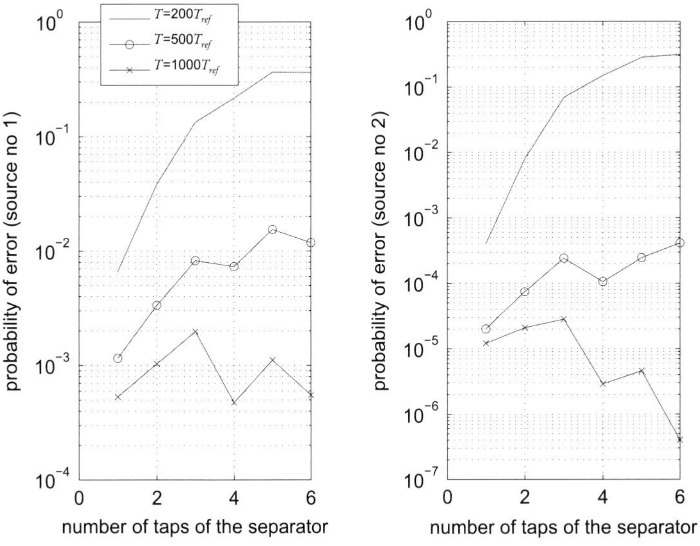

FIGURE 17.6

Performance of the convolutive approach for $P = 10$.

combine two paths; nevertheless the "convolutive gap" noticed when $P = 4$ is not as clear;

3. for $P = 10$ and 500, 1000 transmitted symbols, the best strategy is not clear since the performance depends on the considered source. This is a case when the number

of samples is too big to systematically consider $L = 1$ and not big enough to consider $L = 4$.

As far as the first step of the algorithm is concerned, i.e. the extraction of a source, one may consider the strategies:

1. Choose L big enough. By this statement, we mean that the action of $\mathbf{b}[z]$ is optimal: the filter may take into account *all* the paths and hence may combine them optimally in order to achieve the minimization of the Godard criterion. This means that the parameter L should verify $(L+1)T_e \geq \max_{i,m}(\tau_{im})$. The number of complex-valued coefficients to estimate when $\mathbf{b}[z]$ varies is PL.
2. Choose $L = 1$. This reduces the number of coefficients to compute.

The advantages of the first strategy are its optimality and the fact that it may face the case of difficult channels ($N \leq P \leq \tilde{N}$), i.e. when the sensor array is saturated. Its drawback is the number of coefficients to compute: it fits any situation where the number of data is big (the order of $20PL$); this means situations associated with channels whose variations are slow.

For rapidly varying channels, the second strategy is clearly to be preferred: it is useless to use a spatio-temporal filter if the memory of this latter is too small in order to combine two paths. It requires that $P > \tilde{N}$.

17.3.1.5 Advantages and drawbacks

The advantages of the convolutive approach as far as the BSS is concerned are summarized hereafter.

- It requires the minimum number of antennas ($P \geq N$) to separate the sources, whatever the number of paths may be.
- It does not require a time specularity property of the channels. In particular, the prior estimation of the number of paths associated with a given source is not required.
- It implicitly takes into account all the paths (it processes an automatic recombination of the paths) if the number of taps of the separating filter is sufficient.
- It is a blind approach in the sense that it does not require the prior estimation of the SO cyclic frequencies of the observations.

The drawbacks of the convolutive approach are summarized hereafter.

- No algorithm seems to be currently available to process, in all cases, convolutive mixtures of arbitrary cyclostationary sources, without strong assumptions about the sources.
- For propagation channels which have a long delay spread and which are time invariant over a short interval duration, such as in the HF band, purely spatial filters ($L = 1$) should be used for the BSS from a given number of antennas. This may prevent the algorithm from taking into account the energy of all the paths of a given source.
- The frequency offsets of the paths associated with a given source have to be equal.

17.3.2 Treating the mixture as an instantaneous one

17.3.2.1 Problem

When the mixture is treated as an instantaneous one, i.e. when the time specularity of the propagation channel is exploited, a blind spatial processing of the data has to be implemented to restore, to within potential scales, delays and frequency offsets, the complex envelope of the different paths of the sources, with no prior estimation of the SO cyclic frequencies of the observations. In other words, such an approach aims at restoring, to within a diagonal and a permutation matrix, the components $\tilde{s}_i(t)$ $(1 \leq i \leq \tilde{N})$ of the vector $\tilde{\mathbf{s}}(t)$ from the observation vector described by (17.5). The problem is hence to implement a $(P \times \tilde{N})$ separator $\mathbf{B}$, whose output vector, $\mathbf{y}(t) \triangleq \mathbf{B}^H\mathbf{x}(t)$, corresponds, to within a diagonal matrix $\boldsymbol{\Lambda}$ and a permutation matrix $\boldsymbol{\Pi}$, to an estimate $\hat{\tilde{\mathbf{s}}}(t)$ of $\tilde{\mathbf{s}}(t)$. This problem is synthesized by the following equation

$$\begin{aligned}\mathbf{y}(t) &= \mathbf{B}^H\mathbf{x}(t)\\ &= \boldsymbol{\Lambda}\boldsymbol{\Pi}\hat{\tilde{\mathbf{s}}}(t).\end{aligned} \tag{17.41}$$

This requires of course a prior blind estimation of the number of sources to be processed, from the statistics of the data. As explained in the introduction, among the available BSS of instantaneous mixtures, those exploiting the statistical independence and the non-Gaussianity of the sources are the most appropriate for this problem. This is the choice we make despite the fact that the different paths of a source may be correlated to each other. The behavior of such a BSS in the presence of zero-mean cyclostationary sources is investigated in section 17.5.

17.3.2.2 Post-processing of the separated paths

Paths association (optional). After the previous BSS step, the uncorrelated paths of the sources are potentially separated whereas the correlated ones remain mixed together. In this context, in order to maximize the received power for each source, one may first wish to associate all the separated paths of a given source. This association is justified only if some paths have similar power. Otherwise, the selection of the main path is sufficient. Ideally, two outputs of the separator $\mathbf{B}$ which are associated with the same source correspond to two scaled, delayed and Dopplerized versions of the same signal. The detection of two paths of a given source then requires the estimation of the correlation level between delayed and Dopplerized versions of the outputs of the separators. This operation is similar to the Range-Doppler ambiguity function computation in radar applications. The detection of the correlation maxima of this ambiguity function makes possible the association of the outputs associated with the same source and the estimation of the delays between the paths. The functional scheme of this step is depicted in Fig. 17.7.

Paths recombination (optional) and equalization/demodulation. Once the different paths of the same source have been associated, the problem is to judiciously combine these outputs to both optimize the received energy for the considered source and

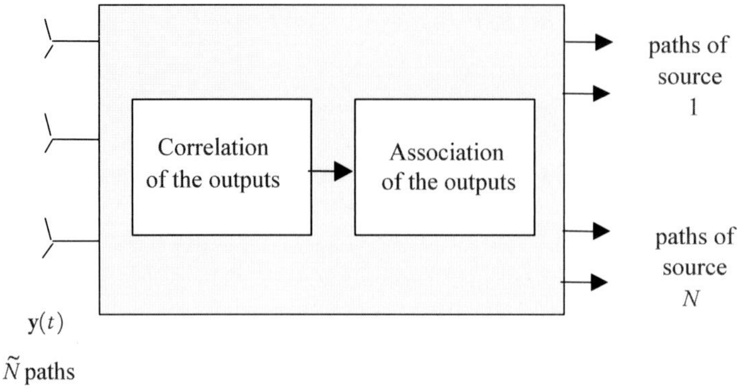

FIGURE 17.7

Association of separated paths.

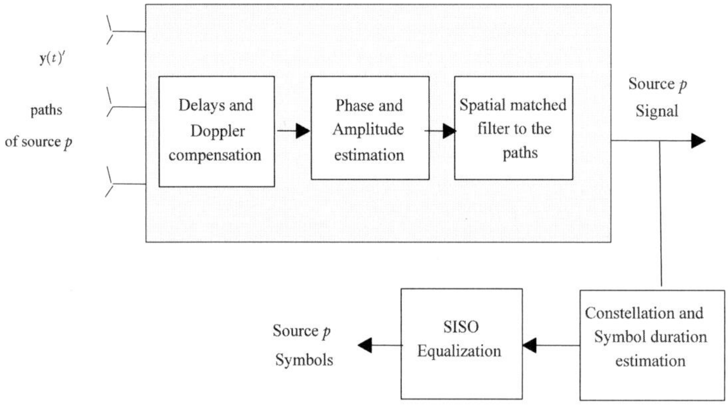

FIGURE 17.8

Path recombination (first method).

to remove the potential residual inter-symbol interference present in the associated separator outputs. For this purpose two methods are possible.

The first method consists in a first step in compensating the frequency offsets and the delays of the selected outputs of the separators (after the association process). In a second step, an estimation of the differential amplitude and phase of the compensated outputs is implemented. This may be done for example from the estimation of the principal eigenvector of an estimate of the correlation matrix of the compensated outputs. In a third step, a spatial matched filter built from the second step is implemented on the compensated outputs. Such a processing is valid whatever the kind of source. Finally, to recover the symbols of the source when the latter is digital, a SISO blind equalization/demodulation may finally be implemented from the output of the third step and from an estimation of the modulation and the symbol duration of the processed source. The functional scheme of such a method is illustrated in Fig. 17.8.

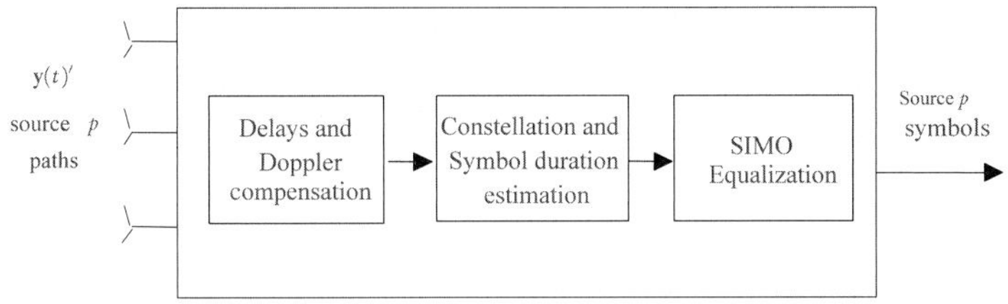

FIGURE 17.9

Path recombination (second method).

The second method is limited to digital sources. The first step is the same as the one depicted above. In a second step, a SIMO blind equalization is implemented from the associated outputs and from a prior estimation of both the modulation and the symbol period of the considered source. Let us note that in the presence of only one source, the proposed method (BSS of the paths + Blind SIMO equalization of the outputs of the separator) allows one to reduce the number of coefficients of the equalizer. The functional scheme of such a method is illustrated in Fig. 17.9.

17.3.2.3 Advantages and drawbacks

The advantages of the instantaneous approach are summarized hereafter.

- It makes possible the use of simple, powerful BSS methods with quick convergence, whatever the kind of non-Gaussian signals.
- It does not require the estimation of the SO cyclic frequencies of the observations.

The drawbacks of the instantaneous approach are summarized hereafter.

- It requires propagation channels with a time specularity property.
- It increases the risk of array saturation if the total number of paths becomes greater than the number of sensors, i.e. if $\tilde{N} > P$. In such situations, only the most powerful and the least Gaussian paths are separated in the separation step. However, this risk decreases when P increases.
- It requires the prior estimation of the number of sources and the number of paths associated with each source.

As a consequence, the use of instantaneous methods for applications such as the spectrum monitoring of radiocommunications sources in the HF band seems very appropriate.

17.4 ULTIMATE SEPARATORS OF INSTANTANEOUS MIXTURES

Before analyzing the behavior of blind methods exploiting the non-Gaussianity and the statistical independence of sources, for the separation of instantaneous mixtures of cyclostationary paths or sources, we briefly describe the ultimate source separators for this problem, described in [21], i.e. the separators that BSS methods aim at implementing asymptotically, when the statistics of the observations are perfectly estimated. In this respect, we introduce the concept and the performance of the optimal separator on the one hand, and the performance of the ultimate separator on the other hand. The description is first made for uncorrelated paths. The case of correlated paths is discussed for a second time.

17.4.1 Source separator performance

Assuming, for the first time, that the $\tilde{N}$ paths are uncorrelated, the concept of source separator performance should verify some properties in order to be pertinent. In particular it should:

- be able to characterize quantitatively the restitution quality of each of the $\tilde{N}$ paths by an arbitrary source separator;
- make possible the comparison of two arbitrary separators for the restitution of each path;
- be related, under some hypotheses, to an ideal performance criterion, such as *reception operational characteristic* (ROC) curves for detection problems or *symbol error rate* (SER) for demodulation of digital signals, depending on the application;
- be invariant by a diagonal and a permutation matrix.

Such a criterion has been introduced in [21] and is based on the *signal to interference-plus-noise ratio* (SINR) of each path at the output of the separator. In order to introduce it, let us call interference for the path k, all the paths j with $j \neq k$. Then we define the SINR of path k at output i of separator $\mathbf{B}$ by the following expression:

$$\mathrm{SINR}_k(\mathbf{b}_i) = \sigma^2_{\tilde{s}_k} \frac{|\mathbf{b}_i^H \tilde{\mathbf{a}}_k|^2}{\mathbf{b}_i^H \mathbf{R}_{\backslash k} \mathbf{b}_i} \tag{17.42}$$

where $\sigma^2_{\tilde{s}_k} = \langle \mathbb{E}\{|\tilde{s}_k(t)|^2\}\rangle_c$; $\tilde{s}_k(t)$ and $\tilde{\mathbf{a}}_k$ are respectively the complex envelope and the steering vector of the k-th path; $\mathbf{b}_i$ is the i-th column of $\mathbf{B}$; $\mathbf{R}_{\backslash k}$ is the total noise correlation matrix temporal mean for the path k, defined by $\mathbf{R}_{\backslash k} \triangleq \mathbf{R}_{\mathbf{x}} - \sigma^2_{\tilde{s}_k} \tilde{\mathbf{a}}_k \tilde{\mathbf{a}}_k^H$.

With these definitions, we define the maximum SINR for each path k at the outputs of $\mathbf{B}$, denoted by $\mathrm{SINRM}_k(\mathbf{B})$, by the maximum value of the quantities $\mathrm{SINR}_k(\mathbf{b}_i)$, when i varies from 1 to $\tilde{N}$. Finally, the performance index of the separator $\mathbf{B}$ is defined by the $1 \times \tilde{N}$ vector

$$C(\mathbf{B}) = (\mathrm{SINRM}_1(\mathbf{B}), \ldots, \mathrm{SINRM}_{\tilde{N}}(\mathbf{B})). \tag{17.43}$$

The restitution quality of the path k by the separator $\mathbf{B}$ can be evaluated by the quantity $\mathrm{SINRM}_k(\mathbf{B})$. We verify that $\mathbf{B}$ and $\mathbf{B\Lambda\Pi}$ have the same performance, which is meaningful, since they belong to the same equivalence class. Moreover, we say that a separator $\mathbf{B}_1$ is better than a separator $\mathbf{B}_2$ for the restitution of the path k in a given context if $\mathrm{SINRM}_k(\mathbf{B}_1) > \mathrm{SINRM}_k(\mathbf{B}_2)$. For correlated paths, the concept of SINRM is more difficult to define. In this case, a good separator for a path k_1 which is correlated with a path k_2 should generate a mixture of paths k_1 and k_2 but with a good ratio between the power of the latter mixture and the power of noise plus other uncorrelated paths $j \neq k_1$ and $j \neq k_2$.

17.4.2 Ultimate separator

17.4.2.1 Optimal separator

The optimal source separator, for uncorrelated paths, is the separator that gives the best performance for the restitution of each path. For each path k, it maximizes, over all the possible separators $\mathbf{B}$, the quantities $\mathrm{SINRM}_k(\mathbf{B})$ with respect to the k-th column of $\mathbf{B}$. It implements a *spatial matched filter* (SMF) to each path, and is defined by the following expression:

$$\mathbf{B}_{\mathrm{SMF}} = \mathbf{R}_{\mathrm{x}}^{-1}\tilde{\mathbf{A}}\mathbf{\Lambda\Pi} \tag{17.44}$$

Its output vector is given by

$$\begin{aligned} \mathbf{y}_{\mathrm{smf}}(t) &\triangleq \mathbf{B}_{\mathrm{SMF}}^{H}\mathbf{x}(t) \\ &= \mathbf{\Lambda}^{H}\mathbf{\Pi}^{H}\tilde{\mathbf{A}}^{H}\mathbf{R}_{\mathrm{x}}^{-1}\mathbf{x}(t) \end{aligned} \tag{17.45}$$

which gives, by using model (17.5)

$$\mathbf{y}_{\mathrm{smf}}(t) = \mathbf{\Lambda}^{H}\mathbf{\Pi}^{H}(\tilde{\mathbf{A}}^{H}\mathbf{R}_{\mathrm{x}}^{-1}\tilde{\mathbf{A}}\tilde{\mathbf{s}}(t) + \tilde{\mathbf{A}}^{H}\mathbf{R}_{\mathrm{x}}^{-1}\mathbf{v}(t)). \tag{17.46}$$

We deduce from (17.46) that in the presence of noise, the exact restitution of the paths is not possible. Moreover, there is no reason in a noisy context why the matrix $\tilde{\mathbf{A}}^{H}\mathbf{R}_{\mathrm{x}}^{-1}\tilde{\mathbf{A}}$ should be diagonal; hence, in most situations, an interference residue exists at each output of the optimal separator. As a consequence, interference is not completely canceled, except when the matrix $\tilde{\mathbf{A}}^{H}\mathbf{R}_{\mathrm{x}}^{-1}\tilde{\mathbf{A}}$ is diagonal or when the noise is absent [21]. In fact the optimal separator makes a trade-off between the noise minimization and the interference rejection, hence the result. Moreover, in the presence of noise, Gaussian or not, the outputs of the optimal separator show no specific properties of statistical independence (even at the fourth-order) [20].

Nevertheless, in the absence of noise, the matrix $\tilde{\mathbf{A}}^{H}\mathbf{R}_{\mathrm{x}}^{-1}\tilde{\mathbf{A}}$ is always diagonal [21]; notice that in this context, $\mathbf{R}_{\mathrm{x}}$ is not invertible when $P > \tilde{N}$, hence $\mathbf{R}_{\mathrm{x}}^{-1}$ is to be understood as $\mathbf{R}_{\mathrm{x}}^{\dagger}$, the pseudo-inverse of $\mathbf{R}_{\mathrm{x}}$. In this case, the interference is completely nulled by the optimal separator and the exact restitution of the paths is possible. In the absence of noise, the optimal separator has a very specific structure, described in [20]: when the sources are statistically independent (and non-Gaussian), the outputs of the optimal separator are indeed also independent. These properties are at the origin of the development of cumulant-based BSS methods as explained in [20].

17.4.2.2 Weighted least-square separator

For uncorrelated paths, the source separator, optimal with respect to the criterion (17.43), that completely nulls the interferences at each output, corresponds to the *weighted least-square* (WLS) source separator [21], denoted by $\mathbf{B}_{\text{WLS}}$. This separator generates the WLS estimate of $\tilde{\mathbf{s}}(t)$, i.e. the estimate minimizing

$$C_{\text{WLS}}(\tilde{s}(t)) \triangleq \left(\mathbf{x}(t) - \tilde{\mathbf{A}}\tilde{\mathbf{s}}(t)\right)^H \mathbf{R}_{\text{v}}^{-1} \left(\mathbf{x}(t) - \tilde{\mathbf{A}}\tilde{\mathbf{s}}(t)\right) \tag{17.47}$$

where $\mathbf{R}_{\text{v}} \triangleq \mathbb{E}\{\mathbf{v}(n)\mathbf{v}(n)^H\}$ is the auto-correlation of the noise. The WLS equivalence class contains the separators defined by

$$\mathbf{B}_{\text{WLS}} = \mathbf{R}_{\text{v}}^{-1}\tilde{\mathbf{A}}\left(\tilde{\mathbf{A}}^H\mathbf{R}_{\text{v}}^{-1}\tilde{\mathbf{A}}\right)^{-1}\mathbf{\Lambda\Pi} \tag{17.48}$$

whose output vector $\mathbf{y}_{\text{wls}}(t) \triangleq \mathbf{B}_{\text{WLS}}^H\mathbf{x}(t)$ is given by

$$\mathbf{y}_{\text{wls}}(t) = \mathbf{\Lambda}^H\mathbf{\Pi}^H\left(\tilde{\mathbf{A}}^H\mathbf{R}_{\text{v}}^{-1}\tilde{\mathbf{A}}\right)^{-1}\tilde{\mathbf{A}}^H\mathbf{R}_{\text{v}}^{-1}\mathbf{x}(t). \tag{17.49}$$

which can be written, according to model (17.5):

$$\mathbf{y}_{\text{wls}}(t) = \mathbf{\Lambda}^H\mathbf{\Pi}^H\left(\tilde{\mathbf{s}}(t) + \tilde{\mathbf{v}}(t)\right) \tag{17.50}$$

where

$$\tilde{\mathbf{v}}(t) \triangleq \left(\tilde{\mathbf{A}}^H\mathbf{R}_{\text{v}}^{-1}\tilde{\mathbf{A}}\right)^{-1}\tilde{\mathbf{A}}^H\mathbf{R}_{\text{v}}^{-1}\mathbf{v}(t).$$

The WLS corresponds to the optimal one whenever the latter completely nulls the interferences, i.e. when $\tilde{\mathbf{A}}^H\mathbf{R}_{\text{v}}^{-1}\tilde{\mathbf{A}}$ is diagonal or in the absence of noise. It reduces to the LS separator when $\mathbf{R}_{\text{v}}$ is proportional to the identity matrix, i.e. when the noise is spatially white. The properties of this separator are described in [21] and [20].

17.4.3 Ultimate performance

The computation of the optimal separator performance index allows us to understand the way in which physical parameters such as the *signal-to-noise ratio* (SNR), the angular separation of the sources, the noise spatial coherence, the number of paths and sensors and the array geometry, influence the output performance. Besides, these optimal performance indices represent an upper bound to the performance achievable by any linear spatial and time-invariant blind separator of NB sources.

17.4.3.1 The two-path case

Assuming two temporally uncorrelated paths, the SINRM_1 at the output of the optimal separator is given by

$$\text{SINRM}_1(\mathbf{B}_{\text{SMF}}) = \epsilon_1 \left(1 - \frac{\epsilon_2}{1+\epsilon_2}|\alpha_{12}|^2\right) \tag{17.51}$$

where

$$\epsilon_k \triangleq \sigma^2_{\tilde{s}_k} \tilde{\mathbf{a}}_k^H \mathbf{R}_{\text{v}}^{-1} \tilde{\mathbf{a}}_k \tag{17.52}$$

and where α_{12} is the spatial correlation coefficient between the paths 1 and 2 in the metric of $\mathbf{R}_{\text{v}}^{-1}$, defined by

$$\alpha_{12} \triangleq \frac{\tilde{\mathbf{a}}_1^H \mathbf{R}_{\text{v}}^{-1} \tilde{\mathbf{a}}_2}{\left(\tilde{\mathbf{a}}_1^H \mathbf{R}_{\text{v}}^{-1} \tilde{\mathbf{a}}_1\right)^{1/2} \left(\tilde{\mathbf{a}}_2^H \mathbf{R}_{\text{v}}^{-1} \tilde{\mathbf{a}}_2\right)^{1/2}}. \tag{17.53}$$

This coefficient is a function of the noise spatial coherence matrix, the angular separation between the sources, the array geometry, the number and the kind of antennas, the polarization of the sources and antennas. Notice that $\text{SINRM}_2(\mathbf{B}_{\text{SMF}})$ is obtained from (17.51) by exchanging indices 1 and 2. Eq. (17.51) shows off all the physical parameters that control the optimal source separator performance. It shows in particular that $\text{SINRM}_1(\mathbf{B}_{\text{SMF}})$ is always proportional to the input SNR of path 1. Besides, $\text{SINRM}_1(\mathbf{B}_{\text{SMF}})$ is maximum and equal to ϵ_1 in the absence of interference ($\epsilon_2 = 0$) or when the two paths are orthogonal in the $\mathbf{R}_{\text{v}}^{-1}$ metric ($\alpha_{12} = 0$). In these latter situations, $\text{SINRM}_1(\mathbf{B}_{\text{SMF}})$ is, for spatially white noise and identical antennas, proportional to P. Nevertheless, in the general case, the presence of a second path degrades the performance and $\text{SINRM}_1(\mathbf{B}_{\text{SMF}})$ becomes a decreasing function of ϵ_2 and $|\alpha_{12}|^2$. As $|\alpha_{12}|^2$ increases toward 1, which occurs when the angular separation between the paths shrinks, the optimal source separator becomes more and more inoperative. For a strong interference ($\epsilon_2 \gg 1$), as long as $|\alpha_{12}|^2$ is not too close to unity, expression (17.51) becomes

$$\text{SINRM}_1(\mathbf{B}_{\text{SMF}}) \approx \epsilon_1(1 - |\alpha_{12}|^2). \tag{17.54}$$

This expression no longer depends upon the input SNR of the interference path since the latter is rejected under the background noise level. In this case $\text{SINRM}_1(\mathbf{B}_{\text{SMF}})$ is only controlled by ϵ_1 and $|\alpha_{12}|^2$. Notice that (17.54) exactly corresponds to $\text{SINRM}_1(\mathbf{B}_{\text{WLS}})$ whatever the power and the angular separation between the paths. This confirms that in all cases, $\text{SINRM}_1(\mathbf{B}_{\text{WLS}}) \leq \text{SINRM}_1(\mathbf{B}_{\text{SMF}})$, except when $|\alpha_{12}| = 0$ (in this case, the two quantities coincide).

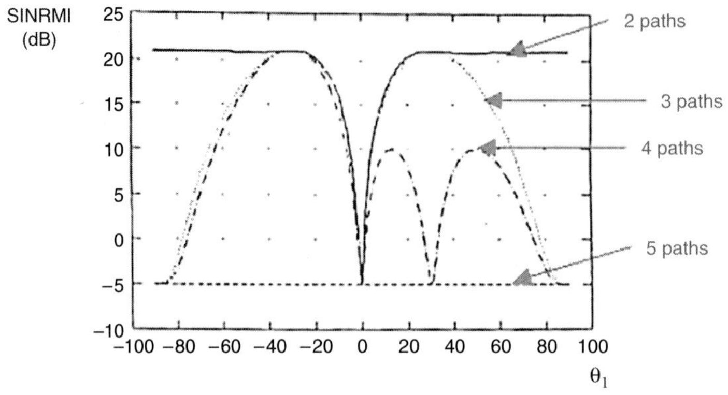

FIGURE 17.10

Performance of the optimal separator.

17.4.3.2 General case

The previous results can be generalized to an arbitrary number of uncorrelated paths, provided that $\tilde{N} \leq P$, as shown in [21]. In particular, it is shown in [21] that an increase of the number of paths necessarily decreases the output performance, as illustrated in Fig. 17.10. This figure shows the variations of $\text{SINRM}_1(\mathbf{B}_{\text{SMF}})$ as a function of θ_1, the direction of arrival of the first path, for several values of $2 \leq \tilde{N} \leq 5$ and for a uniform linear array of $P = 4$ antennas, assuming that the input SNR of all the paths is equal to 20 dB. The DOAs of paths 2, 3, 4 and 5 are such that $\theta_2 = 0°$, $\theta_3 = 90°$, $\theta_4 = 30°$ and $\theta_5 = -30°$. Note the degradation in performance as the number of paths increases and as the angular separation between paths 1 and the others decreases.

17.5 BLIND SEPARATORS OF INSTANTANEOUS MIXTURES

We analyze in this section the behavior of blind separation methods exploiting the non-Gaussianity and the statistical independence of sources, in the case of instantaneous mixtures of cyclostationary paths. We limit the analysis to the JADE algorithm, introduced in [15], which involves SO and FO statistics of the data.We firstly recall the main steps of the JADE algorithm, initially developed for stationary and statistically independent non-Gaussian paths. Then we analyze the behavior of the JADE algorithm in the presence of cyclostationary uncorrelated and correlated paths respectively, when the empirical cumulant estimators of the data are used for its implementation. Finally we illustrate the performance of JADE in a cyclostationary context.

17.5.1 JADE for stationary uncorrelated paths

We briefly recall in this section the main steps of the JADE algorithm, originally developed for stationary and statistically independent non-Gaussian paths. This algorithm aims at estimating the steering vectors of all the paths. Once identified, the steering vectors are used to build, for each path, a well suited spatial filter such as the spatial matched filter described in the previous section. The blind identification procedure requires the *prewhitening* of the data, by the pseudo-inverse, denoted by $\mathbf{W}$, of a square root of the matrix $\mathbf{R}_m = \tilde{\mathbf{A}}\mathbf{R}_{\tilde{\mathbf{s}}}\tilde{\mathbf{A}}^H$, computed from the $\mathbf{R}_{\mathbf{x}}$ matrix, where $\mathbf{R}_{\tilde{\mathbf{s}}} = \langle \mathbf{R}_{\tilde{\mathbf{s}}}(t,0)\rangle_c$. This prewhitening operation aims at orthonormalizing the paths steering vectors and these latter are searched as columns of a unitary matrix $\mathbf{U}$. If we note $\mathbf{z}(t) \triangleq \mathbf{W}\mathbf{x}(t)$ the whitened observation vector, the matrix $\mathbf{U}$ is chosen in the JADE algorithm so as to optimize a FO criterion or contrast function, depending on the elements of the temporal mean, $\mathbf{Q}_{\mathbf{z}}$, of the quadricovariance matrix of vector $\mathbf{z}(t)$. Using the multilinearity property of the cumulants, this matrix $\mathbf{Q}_{\mathbf{z}}$ can be easily computed from (17.5) and is given, following (17.20), for FO uncorrelated paths, by

$$\mathbf{Q}_{\mathbf{z}} = \sum_{i=1}^{\tilde{N}} \tilde{c}'_i \left[\tilde{\mathbf{a}}'_i \otimes \tilde{\mathbf{a}}'^*_i\right]\left[\tilde{\mathbf{a}}'_i \otimes \tilde{\mathbf{a}}'^*_i\right]^H \tag{17.55}$$

$$= \left[\tilde{\mathbf{A}}' \otimes \tilde{\mathbf{A}}'^*\right] \mathbf{Q}_{\tilde{\mathbf{s}}'} \left[\tilde{\mathbf{A}}' \otimes \tilde{\mathbf{A}}'^*\right]^H \tag{17.56}$$

where $\tilde{\mathbf{A}}'$ is the $\tilde{N} \times \tilde{N}$ unitary matrix of the whitened paths steering vectors $\tilde{\mathbf{a}}'_i$ $(1 \le i \le \tilde{N})$, $\mathbf{Q}_{\tilde{\mathbf{s}}'}$ corresponds to the temporal mean of the quadricovariance of $\tilde{\mathbf{s}}'(t)$, the normalized version[4] of $\tilde{\mathbf{s}}'(t)$, and $\tilde{c}'_i \triangleq \mathbf{Q}_{\tilde{\mathbf{s}}'}[i,i,i,i]$ is the autocumulant of the (normalized) path i.

For example, assuming no Gaussian paths for simplicity, the JADE method [15] is based on the fact that the $\tilde{N}$ orthonormalized vectors $[\tilde{\mathbf{a}}'_i \otimes \tilde{\mathbf{a}}'^*_i]$ $(1 \le i \le \tilde{N})$ are eigenvectors of the $\mathbf{Q}_{\mathbf{z}}$ matrix associated with its $\tilde{N}$ non-zero eigenvalues $\tilde{c}'_i$, which also correspond to the $\tilde{N}$ eigenvalues having the greatest absolute value. Then, an arbitrary eigenvector, $\mathbf{r}$, of $\mathbf{Q}_{\mathbf{z}}$ associated with a non-zero eigenvalue is necessarily a linear combination of the $\tilde{N}$ vectors $[\tilde{\mathbf{a}}'_i \otimes \tilde{\mathbf{a}}'^*_i]$. Mapping the $\tilde{N}^2$ components of $\mathbf{r}$ into the $\tilde{N} \times \tilde{N}$ matrix $\mathbf{R}$ such that $\mathbf{R}[k,\ell] = \mathbf{r}[\tilde{N}(k-1)+\ell]$ $(1 \le k,\ell \le \tilde{N})$, the matrix $\mathbf{R}$ associated with the vector $\mathbf{r}$, becomes a linear combination of the matrices $\mathbf{M}_i \triangleq \tilde{\mathbf{a}}'_i\tilde{\mathbf{a}}'^H_i$ associated with the eigenvectors $[\tilde{\mathbf{a}}'_i \otimes \tilde{\mathbf{a}}'^*_i]$. In these conditions, it is easy to verify [15] that the unitary matrix $\tilde{\mathbf{A}}'$ is, to within a permutation and a unitary diagonal matrix, the only one which jointly diagonalizes the $\tilde{N}$ matrices $\mathbf{R}_k$ $(1 \le k \le \tilde{N})$ associated with the $\tilde{N}$ orthonormalized eigenvectors $\mathbf{r}_k$ of $\mathbf{Q}_{\mathbf{z}}$ related to a non-zero eigenvalue. In fact, it is suggested in [15] to jointly diagonalize the $\tilde{N}$ matrices $\mathbf{R}_k$ $(1 \le k \le \tilde{N})$ weighted by the associated eigenvalues λ_k to be able to process at most one Gaussian source. In other words, the unitary matrix $\tilde{\mathbf{A}}'$ maximizes, with respect to the unitary matrix variable $\mathbf{U} \triangleq (\mathbf{u}_1, \ldots, \mathbf{u}_{\tilde{N}})$, the

[4] Each component of $\tilde{\mathbf{s}}'(t)$ has a unit power.

following joint diagonalization criterion which is also a contrast function [15]

$$C(\mathbf{U}) = \sum_{k=1}^{\tilde{N}} \sum_{\ell=1}^{\tilde{N}} |\mathbf{u}_\ell^H \lambda_k \mathbf{R}_k \mathbf{u}_\ell|^2. \tag{17.57}$$

In situations of practical interest, the SO and FO statistics of the data are not known and have to be estimated from the sampled data, $\mathbf{x}(m)$ $(1 \leq m \leq T)$, by the empirical estimators presented in section 17.2.2.

17.5.2 JADE for cyclostationary uncorrelated paths

17.5.2.1 Problem formulation

In the presence of cyclostationary observations, it has been shown in section 17.2.2 that the empirical estimator of the FO statistics temporal mean of the data is asymptotically biased. In this context, one may wonder what may be the consequences of this bias on the performance of JADE. The analysis of this question for cyclostationary uncorrelated paths, first reported in [34], is precisely the purpose of this section.

17.5.2.2 Apparent time average quadricovariance of $\mathbf{z}(t)$

Taking into account the results of section 17.2.2.2, we deduce from (17.24) that in the presence of zero-mean cyclostationary and cyclo-ergodic observations, the FO contrast function (17.57) is not, in the steady state, a function of the $\mathbf{Q_z}$ matrix, as it should ideally be, but becomes a function of the matrix $\mathbf{Q}_{\mathbf{z},a}$, the apparent temporal mean of the quadricovariance of $\mathbf{z}(t)$ (see section 17.2.2.2). Using the multilinearity property of $\mathbf{Q}_{\mathbf{z},a}[i,j,k,l]$, we deduce from (17.5), (17.23) and (17.56) that, in the presence of cyclostationary and uncorrelated paths, the matrix $\mathbf{Q}_{\mathbf{z},a}$ can be written as

$$\mathbf{Q}_{\mathbf{z},a} = \left[\tilde{\mathbf{A}}' \otimes \tilde{\mathbf{A}}'^*\right] \mathbf{Q}_{\tilde{\mathbf{s}}',a} \left[\tilde{\mathbf{A}}' \otimes \tilde{\mathbf{A}}'^*\right]^H \tag{17.58}$$

where $\mathbf{Q}_{\tilde{\mathbf{s}}',a}$ is the apparent temporal mean of the quadricovariance of $\tilde{\mathbf{s}}'(t)$, whose element $[i,j,k,\ell]$, denoted by $\mathbf{Q}_{\tilde{\mathbf{s}}',a}[i,j,k,\ell]$ is defined by Eq. (17.23) with the indice $\tilde{\mathbf{s}}'$ instead of $\mathbf{x}$. Comparing (17.56) and (17.58), we must wonder whether some of the coefficients $\mathbf{Q}_{\tilde{\mathbf{s}}',a}[i,j,k,\ell]$ with $[i,j,k,\ell] \neq [i,i,i,i]$, may not be equal to zero and, in this case, how the presence of these terms in $\mathbf{Q}_{\mathbf{z},a}$ may modify the behavior of the JADE method. The first question is adressed in section 17.5.2.3 whereas the second one is adressed in sections 17.5.2.4 to 17.5.2.6.

17.5.2.3 Analysis of $\mathbf{Q}_{\tilde{\mathbf{s}}',a}$

General case. In the presence of $\tilde{N}$ uncorrelated paths, the matrices $\mathbf{Q}_{\tilde{\mathbf{s}}'}$ and $\mathbf{Q}_{\tilde{\mathbf{s}}',a}$ are $\tilde{N}^2 \times \tilde{N}^2$ matrices. However, while the only non-zero elements of the matrix $\mathbf{Q}_{\tilde{\mathbf{s}}'}$ are the $\tilde{N}$ elements $\mathbf{Q}_{\tilde{\mathbf{s}}'}[i,i,i,i]$ $(1 \leq i \leq \tilde{N})$, it is not necessarily the case for the matrix $\mathbf{Q}_{\tilde{\mathbf{s}}',a}$. Taking into account the symmetries of the matrix $\mathbf{Q}_{\tilde{\mathbf{s}}',a}$, we find that $\mathbf{Q}_{\tilde{\mathbf{s}}',a}[i,i,j,j] = \mathbf{Q}_{\tilde{\mathbf{s}}',a}[i,j,i,j]$ and the analysis of the structure of $\mathbf{Q}_{\tilde{\mathbf{s}}',a}$ can be deduced from the analysis of the elements $\mathbf{Q}_{\tilde{\mathbf{s}}',a}[i,i,i,i]$, $\mathbf{Q}_{\tilde{\mathbf{s}}',a}[i,i,j,j]$ and $\mathbf{Q}_{\tilde{\mathbf{s}}',a}[i,j,j,i]$ with $i \neq j$ $(1 \leq i,j \leq \tilde{N})$, given from (17.24) by

$$\mathbf{Q}_{\tilde{\mathbf{s}}',a}[i,i,i,i] = \mathbf{Q}_{\tilde{\mathbf{s}}'}[i,i,i,i] + 2\sum_{\alpha \in \mathscr{A}^*} \mathbf{R}_{\tilde{\mathbf{s}}'}^{(\alpha)}[i,i]\mathbf{R}_{\tilde{\mathbf{s}}'}^{(-\alpha)}[i,i] + \sum_{\beta \in \mathscr{B}^*} \mathbf{C}_{\tilde{\mathbf{s}}'}^{(\beta)}[i,i]\mathbf{C}_{\tilde{\mathbf{s}}'}^{(\beta)}[i,i]^* \tag{17.59}$$

$$\mathbf{Q}_{\tilde{\mathbf{s}}',a}[i,i,j,j] = \sum_{\alpha \in \mathscr{A}^*} \mathbf{R}_{\tilde{\mathbf{s}}'}^{(\alpha)}[i,i]\mathbf{R}_{\tilde{\mathbf{s}}'}^{(-\alpha)}[j,j] \tag{17.60}$$

$$\mathbf{Q}_{\tilde{\mathbf{s}}',a}[i,j,j,i] = \sum_{\beta \in \mathscr{B}^*} \mathbf{C}_{\tilde{\mathbf{s}}'}^{(\beta)}[i,i]\mathbf{C}_{\tilde{\mathbf{s}}'}^{(\beta)}[j,j]^* \tag{17.61}$$

where $\mathscr{A}^*$ and $\mathscr{B}^*$ are respectively the set of non-null cyclic frequencies of the data and the set of the conjugate non-null frequencies of the data, $\mathbf{R}_{\tilde{\mathbf{s}}'}^{(\alpha)}[i,i]$ and $\mathbf{C}_{\tilde{\mathbf{s}}'}^{(\beta)}[i,i]$ are the coefficients $[i,i]$ of the first and second cyclic correlation matrix of $\tilde{\mathbf{s}}'(t)$ at lag $\tau = 0$ and for the cyclic frequencies α and β respectively. In other words, $\mathbf{R}_{\tilde{\mathbf{s}}'}^{(\alpha)}[i,i]$ and $\mathbf{C}_{\tilde{\mathbf{s}}'}^{(\beta)}[i,i]$ correspond to the first and second cyclic correlation coefficient of the path i, for $\tau = 0$ and for the cyclic frequencies α and β respectively. Expression (17.59) shows that in the general case of zero mean uncorrelated cyclostationary paths, $\mathbf{Q}_{\tilde{\mathbf{s}}',a}[i,i,i,i] \neq \mathbf{Q}_{\tilde{\mathbf{s}}'}[i,i,i,i]$, which shows that the empirical estimator of FO data statistics modifies the FO normalized auto-cumulant of the paths. Moreover, the Eqs (17.60) and (17.61) show that if Condition 4, defined hereafter, is verified for two different paths i and j, then at least one of the two expressions (17.60) and (17.61) does not cancel, which means in some way that these two paths, although statistically uncorrelated, become apparently FO correlated to each other [34]. The condition in question is defined by

CONDITION 4
There exists at least a pair of different paths i and j sharing, for $\tau = 0$, at least one non-zero cyclic frequency on the same SO cyclic correlation function.

In this case, the algebraic structures of $\mathbf{Q}_{\mathbf{z},a}$ and $\mathbf{Q}_{\mathbf{z}}$ are no longer the same and the JADE method may be affected by the cyclostationarity of the paths. Note that for SO circular sources (for example the M-PSK sources with $M > 2$) the expression (17.61) cancels and Condition 4 concerns only the first SO cyclic correlation function.

Case of linearly modulated paths. In the particular case of linearly modulated paths i with symbol duration T_i, frequency offset ν_i and bandwidth B_i, it is shown in [34] that Condition 4 reduces to Condition 5 defined by the following

CONDITION 5
There exists at least a pair of different paths i and j such that one of the following conditions is verified:

1. Two band-limited paths i and j such that $1/T_i \leq B_i$ and $1/T_j \leq B_j$ have the same baud-rate or generate solution $(k_i, k_j) \neq (0,0)$ such that $|k_\ell| \leq \lfloor B_\ell T_\ell \rfloor, (\ell = i, j)$, to the equation $\frac{k_i}{T_i} = \frac{k_j}{T_j}$.
2. Two SO non-circular paths have the same non-zero frequency offset.
3. For two SO non-circular paths i and j, at least one of which is band-limited, the equation $2\nu_i + \frac{k_i}{T_i} = 2\nu_j + \frac{k_j}{T_j} \neq 0$ has at least one solution $(k_i, k_j) \neq (0,0)$, with $k_i = 0$ if the source i is not filtered, where $|k_\ell| \leq \lfloor B_\ell T_\ell \rfloor, (\ell = i, j)$.

17.5.2.4 Eigenstructure of the matrix $\mathbf{Q}_{\mathbf{z},a}$

Using the results of section 17.5.2.3 in (17.58), we deduce the general expression of $\mathbf{Q}_{\mathbf{z},a}$ for zero-mean cyclostationary uncorrelated paths, given by

$$\begin{aligned} \mathbf{Q}_{\mathbf{z},a} &= \sum_i \mathbf{Q}_{\mathbf{s}',a}[i,i,i,i] \left[\tilde{\mathbf{a}}'_i \otimes \tilde{\mathbf{a}}'^*_i\right] \left[\tilde{\mathbf{a}}'_i \otimes \tilde{\mathbf{a}}'^*_i\right]^H \\ &+ \sum_{i,j\ i\neq j} \mathbf{Q}_{\mathbf{s}',a}[i,i,j,j] \left[\tilde{\mathbf{a}}'_i \otimes \tilde{\mathbf{a}}'^*_i\right] \left[\tilde{\mathbf{a}}'_j \otimes \tilde{\mathbf{a}}'^*_j\right]^H \\ &+ \sum_{i,j\ i\neq j} \mathbf{Q}_{\mathbf{s}',a}[i,j,i,j] \left[\tilde{\mathbf{a}}'_i \otimes \tilde{\mathbf{a}}'^*_j\right] \left[\tilde{\mathbf{a}}'_i \otimes \tilde{\mathbf{a}}'^*_j\right]^H \\ &+ \sum_{i,j\ i\neq j} \mathbf{Q}_{\mathbf{s}',a}[i,j,j,i] \left[\tilde{\mathbf{a}}'_i \otimes \tilde{\mathbf{a}}'^*_j\right] \left[\tilde{\mathbf{a}}'_j \otimes \tilde{\mathbf{a}}'^*_i\right]^H \end{aligned} \qquad (17.62)$$

where it is easy to verify that

$$\mathbf{Q}_{\mathbf{s}',a}[i,i,j,j] = \mathbf{Q}_{\mathbf{s}',a}[i,j,i,j] = \mathbf{Q}_{\mathbf{s}',a}[j,j,i,i] = \mathbf{Q}_{\mathbf{s}',a}[j,i,j,i]$$

is a real quantity denoted by $c'_{ij,a}$ and that

$$\mathbf{Q}_{\mathbf{s}',a}[i,j,j,i] = \mathbf{Q}_{\mathbf{s}',a}[j,i,i,j]^*,$$

denoted by $\tilde{c}'_{ij,a}$.

To simplify the developments, we limit, in the following, the analysis to *the two paths case*. In these conditions, the $\mathbf{Q}_{\mathbf{z},a}$ matrix has, in the general case, a rank equal to four with the four potentially non-zero eigenvalues given, after some easy computations, by

$$\lambda_\pm = \frac{1}{2}\left(c'_{1,a} + c'_{2,a} \pm \left[(c'_{1,a} - c_{2,a})^2 + 4(c'_{12,a})^2\right]^{1/2}\right) \qquad (17.63)$$

$$\delta_\pm = c'_{12,a} \pm |\tilde{c}'_{12,a}| \qquad (17.64)$$

where $c'_{i,a} \triangleq \mathbf{Q}_{\tilde{\mathbf{s}}',a}[i,i,i,i]$ $(i = 1,2)$. If, as it has been done in section 17.5.1, to each eigenvector $\mathbf{r}$ of $\mathbf{Q}_{\mathbf{z},a}$, we associate a matrix $\mathbf{R}$ such that $\mathbf{R}[k,\ell] = \mathbf{r}[\tilde{N}(k-1)+\ell]$ $(1 \leq k,\ell \leq \tilde{N})$, the four matrices $\mathbf{R}_{\lambda_-}$, $\mathbf{R}_{\lambda_+}$, $\mathbf{R}_{\delta_-}$ and $\mathbf{R}_{\delta_+}$ associated with the four eigenvectors $\mathbf{r}_{\lambda_-}$, $\mathbf{r}_{\lambda_+}$, $\mathbf{r}_{\delta_-}$ and $\mathbf{r}_{\delta_+}$ respectively can be written as

$$\mathbf{R}_\lambda = x_\lambda \tilde{\mathbf{a}}'_1 \tilde{\mathbf{a}}'^H_1 + y_\lambda \tilde{\mathbf{a}}'_2 \tilde{\mathbf{a}}'^H_2 + z_\lambda \tilde{\mathbf{a}}'_1 \tilde{\mathbf{a}}'^H_2 + t_\lambda \tilde{\mathbf{a}}'_2 \tilde{\mathbf{a}}'^H_1 \tag{17.65}$$

where, for each $\lambda = \lambda_\pm$ or $\delta_\pm$, the complex numbers $x_\lambda, y_\lambda, z_\lambda, t_\lambda$ are solutions of the following system of equations:

$$(c'_{1,a} - \lambda)x + c'_{12,a}y = 0 \tag{17.66}$$

$$c'_{12,a}x + (c'_{2,a} - \lambda)y = 0 \tag{17.67}$$

$$(c'_{12,a} - \lambda)z + \tilde{c}'_{12,a}t = 0 \tag{17.68}$$

$$\tilde{c}'^*_{12,a}z + (c'_{12,a} - \lambda)t = 0. \tag{17.69}$$

It can easily be verified that if $\lambda_\pm \neq \delta_\pm$, the so-called eigen-matrices $\mathbf{R}_{\lambda_-}$ and $\mathbf{R}_{\lambda_+}$ are such that $(z_\lambda, t_\lambda) = (0,0)$ $(\lambda = \delta_\pm)$ whereas the eigen-matrices $\mathbf{R}_{\delta_-}$ and $\mathbf{R}_{\delta_+}$ are such that $(x_\lambda, y_\lambda) = (0,0)$ $(\lambda = \delta_\pm)$. In other words, $\mathbf{R}_{\lambda_-}$ and $\mathbf{R}_{\lambda_+}$ become linear combinations of the two matrices $\mathbf{M}_i \triangleq \tilde{\mathbf{a}}'_i \tilde{\mathbf{a}}'^H_i$ $(i = 1,2)$, whereas $\mathbf{R}_{\delta_-}$ and $\mathbf{R}_{\delta_+}$ become linear combinations of $\mathbf{M}_{12} \triangleq \tilde{\mathbf{a}}'_1 \tilde{\mathbf{a}}'^H_2$ and $\mathbf{M}_{21} \triangleq \tilde{\mathbf{a}}'_2 \tilde{\mathbf{a}}'^H_1$. Otherwise, i.e. if one $\lambda_\pm$ is equal to one $\delta_\pm$, the two eigenmatrices $\mathbf{R}_\lambda$ are linear combinations of the matrices $\mathbf{M}_1, \mathbf{M}_2, \mathbf{M}_{12}, \mathbf{M}_{21}$.

17.5.2.5 Blind identification from $\mathbf{Q}_{\mathbf{z},a}$ by the JADE method

Referring to the description of the JADE method given in section 17.5.1, in the two uncorrelated and cyclostationary paths case, the 2×2 unitary matrix $\mathbf{U}$ solution to the blind identification problem, maximizes the criterion (17.57) where the two matrices $\mathbf{R}_1$ and $\mathbf{R}_2$ appearing in the latter are the so-called eigenmatrices of $\mathbf{Q}_{\mathbf{z},a}$ associated with the two eigenvalues having the greatest absolute value. In these conditions, we deduce from the previous section that if $|\lambda_\pm| > |\delta_\pm|$ the matrices $\mathbf{R}_1$ and $\mathbf{R}_2$ correspond to $\mathbf{R}_{\lambda_-}$ and $\mathbf{R}_{\lambda_+}$, linear combinations of $\mathbf{M}_1$ and $\mathbf{M}_2$, which are jointly diagonalized, to within a unitary diagonal and permutation matrices, by the matrix $\mathbf{U} = \tilde{A}'$. In other words, if Condition 6 presented hereafter is verified, the blind identification of the whitened source steering vectors by the JADE method, and thus the JADE method itself, is, in the steady state, not affected by the apparent FO dependence ($c'_{12,a} \neq 0$ or $\tilde{c}'_{12,a} \neq 0$) of the two paths. Condition 6 is defined by the following:

CONDITION 6

$$\left|c'_{1,a} + c'_{2,a} \pm [(c'_{1,a} - c'_{2,a})^2 + 4(c'_{12,a})^2]^{1/2}\right| > 2\left|c'_{12,a} \pm |\tilde{c}'_{12,a}|\right|.$$

This result shows that the JADE method is able to accommodate itself to some configurations of apparent FO correlated paths. In particular, if we assume two synchronized and linearly modulated sources with the same half-Nyquist pulse function, associated with not a too high roll-off ($\gamma = 0.5$ for example) and the same baud-rate, the JADE method accommodates itself to two QPSK sources or to two BPSK sources either with the same zero frequency offset or with different frequency offsets such that $\nu_1 - \nu_2 \neq k/T_1$, despite their apparent FO dependence.

However, if Condition 6 is not verified, at least one of the matrices $\mathbf{R}_1$ and $\mathbf{R}_2$ to be jointly diagonalized corresponds to $\mathbf{R}_{\delta_-}$ or $\mathbf{R}_{\delta_+}$ which are not diagonalized by $\mathbf{U} = \tilde{\mathbf{A}}'$. In this case, it is easy to verify, from the process of the joint diagonalization of two potentially non-Hermitian matrices described in [15], that the matrix $\tilde{\mathbf{A}}'$ no longer maximizes the criterion (17.57) and that the two orthonormalized vectors $\mathbf{u}_1$ and $\mathbf{u}_2$, corresponding to the two columns of $\mathbf{U}$, become linear combinations of $\tilde{\mathbf{a}}'_1$ and $\tilde{\mathbf{a}}'_2$

$$i = 1,2: \quad \mathbf{u}_i = \alpha_i \tilde{\mathbf{a}}'_1 + \beta_i \tilde{\mathbf{a}}'_2 \tag{17.70}$$

where $\alpha_i^2 + \beta_i^2 = 1$ $(i = 1,2)$ and $|\alpha_1\alpha_2| = |\beta_1\beta_2|$. In these conditions, the blind identification stage of the JADE method is perturbed by the apparent FO dependence of the sources and the behavior of the JADE method is modified. This non-ideal behavior of the blind identification of the whitened source steering vectors generates a degradation of the source separation process but this does not necessarily mean that the blind separation of the two sources is no longer possible, as is shown in the following section. Thus, as a summary, the blind identification of the whitened source steering vectors and then the performances of the JADE method are modified in the steady state by the apparent FO dependence of the two uncorrelated paths if Condition 6 is not verified. This is for example the case in the presence of two synchronized BPSK sources having the same half-Nyquist pulse function and the same non-zero frequency offset, whatever their baud-rate. Indeed, in this case, it is shown in [34] that $\mathbf{R}_1$ and $\mathbf{R}_2$ correspond to $\mathbf{R}_{\delta_-}$ and $\mathbf{R}_{\delta_+}$ and we deduce from [15] that the process of the joint diagonalization of $\lambda_1\mathbf{R}_1$ and $\lambda_2\mathbf{R}_2$ gives (17.70) with $\alpha_1 = \frac{\sqrt{2}}{2}$, $\alpha_2 = -\frac{\sqrt{2}}{2}e^{\imath\Phi}$ where Φ is an angle depending on $\lambda_1 = \delta_-$ and $\lambda_2 = \delta_+$.

17.5.2.6 Blind source separation

Following the description of the JADE method in section 17.5.1, from the blindly identified whitened source steering vectors $\mathbf{u}_1$ and $\mathbf{u}_2$, it is possible to obtain, to within a scalar factor, an estimate of the true steering vectors of the sources, defined by $\hat{\mathbf{a}}_i = \mathbf{W}^\dagger \mathbf{u}_i$, where $\mathbf{W}^\dagger$ is the pseudo-inverse of a square root of the matrix $\mathbf{R}_m$ (see section 17.5.1). Using (17.70), these two vectors can be written as

$$i = 1,2: \quad \hat{\mathbf{a}}_i = \alpha_i \sigma_{\tilde{s}_1} \tilde{\mathbf{a}}_1 + \beta_i \sigma_{\tilde{s}_2} \tilde{\mathbf{a}}_2 \tag{17.71}$$

where $\sigma_{\tilde{s}_i}$ has been defined after (17.42). In these conditions, the optimal linear and time invariant spatial filter associated with the vector $\hat{\mathbf{a}}_i$ can be computed. It corresponds to the spatial matched filter $\mathbf{b}_i$ for the path i, defined by $\mathbf{b}_i \triangleq R_x^{-1}\hat{\mathbf{a}}_i$ [21]. Then, for each path k ($k = 1,2$), the performance of the JADE method can be evaluated in terms of output SINRM for each source. Assuming orthogonal (i.e. $\tilde{\mathbf{a}}_1^H\tilde{\mathbf{a}}_2 = 0$) and strong paths ($\epsilon_1, \epsilon_2 \gg 1$) to symplify the analysis, it is easy to show that, for each source k, the quantity SINRM_k is given by

$$\mathrm{SINRM}_k = \epsilon_k \left(1 - \frac{\epsilon_k}{1+\epsilon_k}\gamma_{\alpha,\beta}\right) \tag{17.72}$$

where $0 \leq \gamma_{\alpha,\beta} \leq 1$ is defined by

$$\gamma_{\alpha,\beta} \triangleq \min\left(\frac{|\alpha_1|}{|\beta_1|}, \frac{|\beta_1|}{|\alpha_1|}\right). \tag{17.73}$$

The expression (17.72) shows that SINRM_k does not depend on the input signal to noise ratio of the source difference of the source k and is a decreasing function of $\gamma_{\alpha,\beta}$, which means that the separation process by the JADE method degrades as $\gamma_{\alpha,\beta}$ increases. The performance of the JADE method is optimal and the SINRM_k is maximum and equal to ϵ_k when $\gamma_{\alpha,\beta} = 0$, i.e. when the blind identification of the whitened source steering vector is perfect. On the contrary, SINRM_k becomes minimum when $\gamma_{\alpha,\beta} = 1$, i.e. when $|\alpha_1| = |\beta_1|$, which occurs, for example, in the presence of two synchronized BPSK sources having the same half-Nyquist pulse function and the same non-zero frequency offset, whatever their baud-rate, as shown at the end of section 17.5.2.5. In this case, the quantity SINRM_k tends to be equal to 1 or 0 dB for each source k, which shows the absence of source separation. As a summary, building, for each received source, an SMF from the blind identification of the source steering vectors by the JADE method, implemented from empirical estimators of data statistics, generates, in the steady state, an optimal source separator in spatially Gaussian noise as long as Condition 6 is verified, which confirms the results of [20]. However, as soon as this condition is no longer verified, the latter separator becomes sub-optimal and its performance decreases as $\gamma_{\alpha,\beta}$ increases. This shows in particular that despite the apparent FO dependence of the sources, the source separation, although degraded, remains possible as long as $\gamma_{\alpha,\beta}$ is not too high. This condition is verified in most situations of practical interest. However, the separation is no longer possible whatever the relative values of the strong input SNR of the sources as $\gamma_{\alpha,\beta}$ approaches unity, which occurs in very specific situations described previously.

17.5.3 JADE for cyclostationary correlated paths

The behavior of the JADE algorithm, implemented from empirical estimators of data statistics, in the presence of correlated cyclostationary paths has been initialized in [23]. In such situations, the behavior of the JADE method is more controlled by the temporal correlation of the paths than by their cyclostationarity as described previously. It has been shown in [23] that an SO temporal correlation of paths implies an FO temporal correlation between them. In general, this FO dependence implies that the blind estimates of the steering vectors are actually linear combinations of steering vectors of correlated paths. In this case, correlated paths remain mixed at the output of the optimal separator. This is not really a problem when a post-processing of blind channel equalization is implemented from the outputs of the separator. Otherwise, some fading may occur at the outputs of the separator. More precisely, it is shown in [23] that SO correlated paths which are separated by more than 0.4 T, where T is the symbol duration, remain separated by the JADE algorithm. However, when the delay between the paths is lower than 0.4 T, the paths remain mixed at the output of the JADE method. Thus the JADE method accommodates itself from a low SO temporal correlation between paths.

17.5.4 Performance illustration

Hypotheses. The results presented in the previous sections are illustrated in Figs 17.11–17.13 where two linearly modulated and statistically independent NB sources having the same symbol period T are assumed to be received by a circular array of $P = 5$ uniformly spaced sensors with a radius 0.55λ (λ is the wavelength). The two sources are assumed to have the same input SNR of 10 dB, to be synchronized and to have a pulse function corresponding to a half-Nyquist filter with a roll-off equal to $\gamma = 0.3$ for Fig. 17.11 and to $\gamma = 0.5$ for Figs 17.12 and 17.13. Moreover, the two sources are assumed to be orthogonal to each other, which is in particular the case when their angle of arrival is such that $\theta_1 = 50°$ and $\theta_2 = 91°$. Besides, the source 1 has a non-zero carrier residue such that $\nu_1 T_e = 0.1$. Finally, the SINRM_k ($k = 1, 2$) at the output of the JADE method, computed in these figures, are averaged over 200 realizations.

BPSK sources. Under the previous assumptions, the Fig. 17.11 shows, for several values of ν_2, the variations of the SINRM_1 at the output of both the theoretical spatial matched filter (SMF) and the JADE separator, implemented from the *empirical* (EMP) SO and FO statistics estimator, as a function of the number of snapshots T. We considered two BPSK paths, having a symbol period such that $T_1 = T_2 = 6T_e$. This chosen sampling rate ensures the equivalence between the continuous-time and the discrete-time temporal mean operations. The carrier residue of the source 2 is such that $(\nu_1 - \nu_2)T_e \in \{0, 0.005\ 0.015, 0.1\}$. Note the non-separation of the two sources when $\nu_1 = \nu_2\ (\neq 0)$, even in the steady-state ($T \to \infty$) since $\text{SINRM}_1 = 0$ dB. Note also the decreasing convergence speed of the JADE separator as $\nu_1 - \nu_2$ decreases. This latter result shows that for BPSK sources having different carrier residue, while the steady-state output performance is not affected by the use of the empirical FO statistics estimator, since Condition 6 is verified, the output performance obtained from a short time observation is all the better as the difference between the carrier residues is big.

QPSK sources. Finally, Figs 17.12 and 17.13 show the variations of the SINRM_1 at the output of both the theoretical SMF and the JADE separator, implemented from both the empirical and the unbiased FO statistics estimator, as a function of the number of

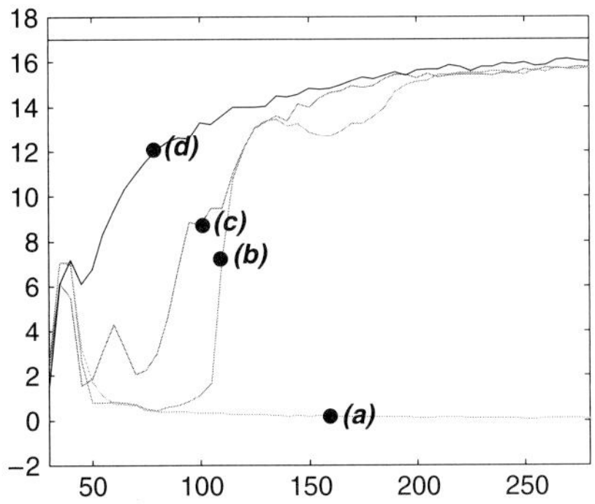

FIGURE 17.11

SINRM$_1$ at the output of the theoretical SMF and JADE as a function of T. $P = 5$, $\tilde{N} = 2$ half-Nyquist BPSK paths, $\gamma = 0.3$, $\theta_1 = 50°$, $\theta_2 = 91°$, SNR = 10 dB, EMP, $T_1 = T_2 = 6T_e$, $\nu_1 T_e = 0.1$, $(\nu_1 - \nu_2)T_e = 0$ (a) 0.005 (b), 0.015 (c) or 0.1 (d).

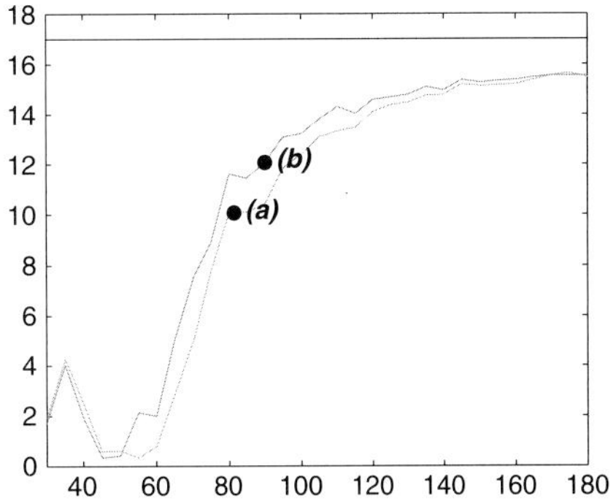

FIGURE 17.12

SINRM$_1$ at the output of the theoretical SMF and JADE as a function of T. $P = 5$, $\tilde{N} = 2$ half-Nyquist QPSK paths ($\gamma = 0.5$), $\theta_1 = 50°$, $\theta_2 = 91°$, SNR = 10 dB, $T_1 = T_2 = 6T_e$, $\nu_1 T_e = 0.1$, $(\nu_1 - \nu_2)T_e = 0$ EMP (a), EXH (b).

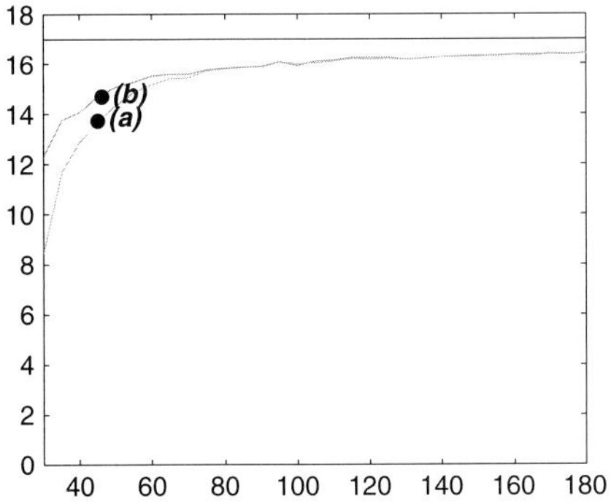

FIGURE 17.13

SINRM$_1$ at the output of the theoretical SMF and JADE as a function of T. $P = 5$, $\tilde{N} = 2$ half-Nyquist QPSK sources ($r = 0.5$), $\theta_1 = 50°$, $\theta_2 = 91°$, SNR = 10 dB, $T_1 = T_2 = 2T_e$, $\nu_1 T_e = 0.1$, $(\nu_1 - \nu_2)T_e = 0$ EMP (a), EXH (b).

snapshots T. Two QPSK paths are considered, having the same non-zero carrier residue. In Fig. 17.12, we have chosen $T_1 = T_2 = 6T_e$ whereas in Fig. 17.13, $T_1 = T_2 = 2T_e$. For the exhaustive estimator, the cyclic frequencies $\alpha_1 = -1/T_1$ and $\alpha_2 = 1/T_2$ have been taken into account in (17.29) in addition to the zero cyclic frequencies. In this case, the SO circularity of the two sources ($\tilde{c}'_{12,a} = 0$) together with the weak value of the coefficient $c_{12,a'}$ (indeed, $c_{12,a'} = 0.052$ for $T_1 = 6T_e$ and $c_{12,a'} = 0.198$ for $T_1 = 2T_e$) implies that

Condition 6 is verified ($c'_{1,a} = c'_{2,a} = -0.825$ for $T_1 = 6T_e$ and $c'_{1,a} = c'_{2,a} = -0.547$ for $T_1 = 2T_e$), which explains the good steady-state performances of the JADE method implemented from the empirical estimators. However, we note in Fig. 17.12 the slower convergence speed of the output SINRM_1 associated with the use of the empirical estimator compared to that associated with the unbiased one.

17.6 INSTANTANEOUS APPROACH VERSUS CONVOLUTIVE APPROACH: SIMULATION RESULTS

In this section, we aim at comparing, from a numerical point of view, the performances of an instantaneous BSS approach, on the one hand, and a convolutive BSS approach on the other hand. The context is that of a ionospheric propagation. We consider the case of two sources, each of which propagates through two paths. When the sensor array is far from being saturated, the instantaneous approach (which explicitly exploits the specular character of the distribution of the paths) shows better results than the convolutive approach (which does not specifically takes into account the specular structure of the channel). This is all the more spectacular as the number of available samples becomes smaller.

We have considered the parameters specfied in section 17.3.1.4.

17.6.1 BSS algorithms and measures of performance

As seen in the previous sections, we aim at comparing a CMA-based convolutive method, on the one hand, with an instantaneous method on the other hand.

Instantaneous method (JADE). Both the number of sources and the number of paths associated with each source are assumed to be known. The algorithm computes a certain $P \times (M_1 + M_2)$ matrix $\mathbf{B}$, any column of which – say $\mathbf{b}_n$ – is such that

$$y(k) = \mathbf{b}_n^H \mathbf{x}(k)$$

expectedly corresponds, up to a complex scaling and an additive noise term, to one of the paths, i.e. to $s_i(kT_e - \tau_{im})$ for certain indices (i, m). An optimal solution should consist in optimally combining all the paths associated with a given source in order to benefit from the energy of all the paths (see 17.3.2.2). Here, we adopt the sub-optimal approach consisting in picking only one of the paths. Moreover, we assume that the receiver is able to choose, among the $\tilde{N}$ columns of $\mathbf{B}$, the one associated with the best SINRM. We wish to access the performance of the approach in terms of probabilities of errors. In this respect, a demodulator is to be provided. To this latter, the following tasks are assigned:

- re-sampling;
- symbol-synchronization (in order to remove the unknown time-shift);
- rotation: in order to remove the complex scaling indeterminacy.

The re-sampling requires that the symbol-period of the extracted source be known: this assumption is made here. As far as the synchronization is concerned, we have run a fractional CMA (with an over-sampling factor of two): the order of the synchronization filter is fixed to 10. The aim of this section is to show the ability of JADE to accurately separate the sources; now, a strong bias involved by the synchronizer would not ease the understanding of the performance and it is desirable to alleviate the impact of the synchronizer. In this respect, we have computed the coefficients of the synchronization filter from more data (1000 transmitted symbols). The removing of the scaling indeterminacy is done by the (unrealistic) correlation with 1000 transmitted symbols of the extracted source. Once the parameters of these steps are computed, a long mixture of duration $100000T_{\text{ref}}$ with the same transmission parameters is generated. The symbols of the two sources are estimated and, after a hard-decision scheme, the number of incorrect symbols is computed.

Convolutive method. Basically, the three step post-processing described above regarding the instantaneous method is the same. In the convolutive case, however, the fractional works jointly as an equalizer (recombination of the contributions of the paths associated with the extracted source) and a synchronizer (see section 17.3.1.3).

17.6.2 Performances

We have measured the performances for two cases: when $T = 200T_e$ (which corresponds to the transmission of 200 symbols) and when $T = 500T_e$. Though quite unrealistic in a practical scenario, the second case is illuminating. As far as the convolutive method is concerned, and in view of the conclusions of section 17.3.1.4, the time-depth of the separator to be computed $\mathbf{b}[z]$ is chosen to be 0, i.e. $L = 1$, when the number of data snapshots is small ($T = 200T_e$); otherwise ($T = 500T_e$), we have chosen $L = 4$.

The sensor array is almost saturated. We indeed have considered the case of a ULA (the distance between two consecutive sensors is half the wavelength) with $P = 5$ antennas. As the factor $\frac{\mathcal{E}_1}{N_0}$ varies, we have measured the performances given in Fig. 17.14. It can be noticed that the performances are poor. This is instrinsically due to the weak number of antennas. In this situation, the convolutive method shows a gain (for $T = 500T_e$) in performance. This originates in the fact that the convolutive algorithm takes into account the two paths of a given source. However, if the delay between the paths of the two sources increases, the advantage shown previously collapses, since the recombination of the paths requires considering a bigger L.

More antennas. For $P = 7$ and $P = 10$, we have obtained the results respectively given in Figs 17.15 and 17.16. This clearly shows the superiority of the JADE approach over the convolutive one when the number of data snapshots is small ($T = 200T_e$).

Although the superiority of the JADE method over the spatial CMA ($L = 1$) is clear, we nevertheless notice that the performance of the latter is not poor. In this respect, we recall that the simulation is not completely fair since, in the JADE approach, the best paths (i.e. those associated with the most favorable SINRs) are considered for the

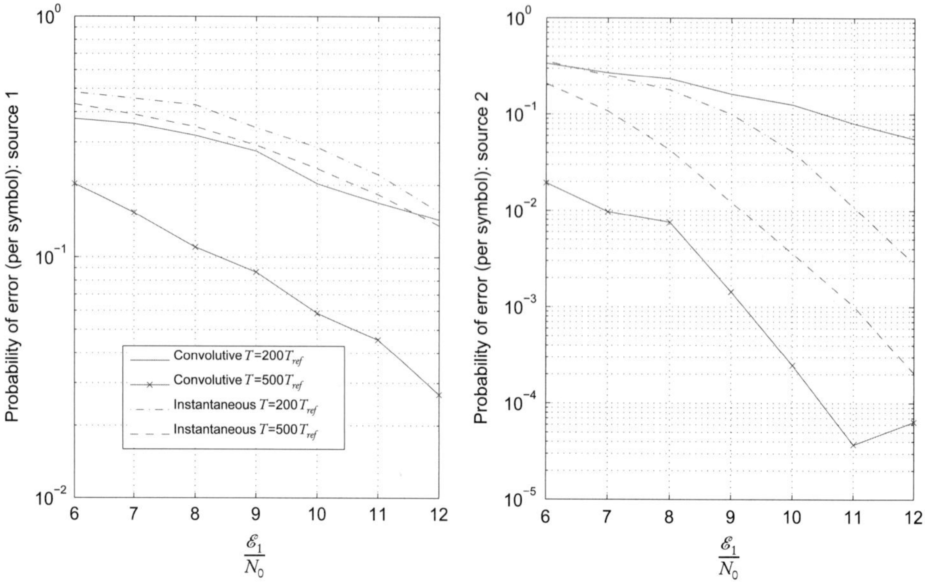

FIGURE 17.14

Performances for $P = 5$.

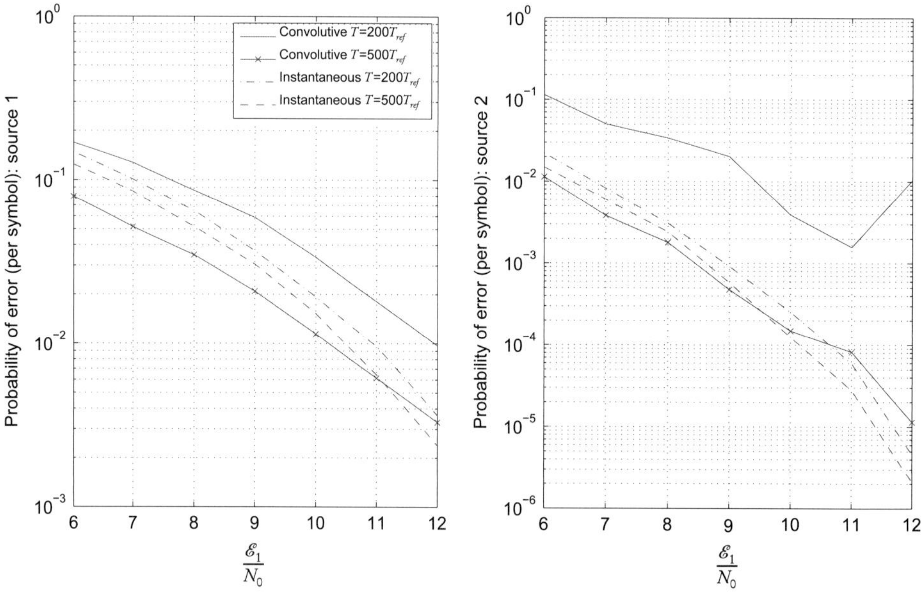

FIGURE 17.15

Performances for $P = 7$.

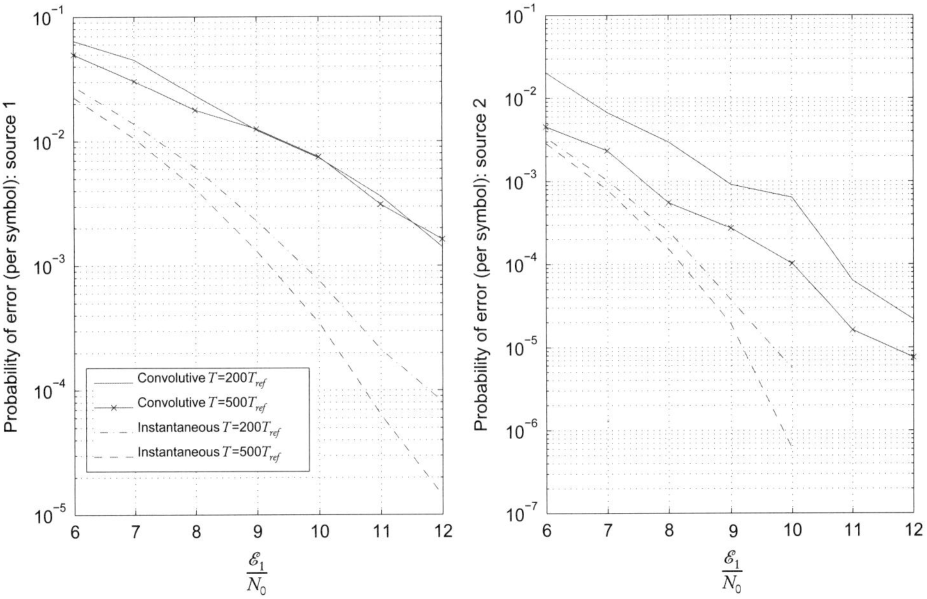

FIGURE 17.16

Performances for $P = 10$.

estimation of the symbols, which is not realistic. Moreover, the JADE method assumes the number of sources and the number of paths associated with each source are known.

17.7 CONCLUSION

In this chapter, BSS methods have been considered for the spectrum monitoring of radio communications. The problem is generally to blindly separate convolutive mixtures of N zero-mean cyclostationary non-Gaussian sources from an array of P antennas such that $N \leq P$. The number P of antennas depends on the frequency band and is typically equal to 5 in the VUHF band whereas it is around 10 for the HF band. The number of sources, mainly non-Gaussian for radio communications applications, also depends on the application and may be typically ranged between 1 and 3 for HF applications. The main problem posed by the cyclostationarity of the sources for BSS is that the empirical estimator of the FO statistics of the observations is asymptotically biased. On the other hand, the use of an asymptotically unbiased FO statistics estimator requires the prior estimation of the SO cyclic frequencies of the data, which is difficult and costly. The challenge is then to implement powerful BSS algorithms which do not require any estimation of the SO cyclic frequencies of the sources. In this respect two approaches are possible. The first one consists in directly processing the convolutive mixture of sources. However, among all the algorithms available to process convolutive mixtures of sources (see Chapter 8), few are able to process convolutive mixtures of arbitrary cyclostationary sources without exploiting the knowledge of the SO cyclic frequencies of the sources. Such methods have been developed very recently [60,58]. They are deflation-based methods relying on the optimization of a CMA temporal mean criterion. These methods are powerful but require some hypotheses on the sources. For this reason they cannot cope with all the situations encountered in practice. Moreover, for propagation channels which have a long delay spread and which can be considered as time invariant over a short duration only, as in the HF band, only filters which are very limited in size may be used for the BSS, which may not be sufficient to separate the sources. The second approach consists in exploiting the time specularity property shared by most propagation channels encountered in practice, by processing the mixture as an instantaneous one. The problem then consists in blindly separating all the paths of all the sources, no longer statistically independent. A post processing may then be potentially implemented to associate and to further combine all the paths of a given source to increase the reception performance. The most powerful BSS methods for the previous problem correspond to FO cumulant-based methods exploiting both the non-Gaussianity and the statistical independence of the sources. Among such methods the JADE method [15] is very attractive. In spite of the fact that such methods have been developed for stationary sources, it has been shown that they remain very powerful for zero-mean cyclostationary sources in most situations of practical interest. Such methods have many advantages over convolutive ones. They are able to process any kind of non-Gaussian sources, analog or digital, and can cope with propagation channels with long delay spread and short time invariance duration. Their main drawback lies in the risk of array saturation, i.e. that the total number of paths becomes greater than the number of antennas. However, this risk decreases when the number of antennas of the array increases. Such methods have already been successfully tested on both synthetic data [22] and real-data [24]: they are very promising for future operational use.

Acknowledgment

This work has been carried out in part in the frame of the research project "Decotes" ANR06-BLAN-0074.

References

[1] K. Abed-Meraim, P. Loubaton, E. Moulines, A subspace algorithm for certain blind identification problems, IEEE Trans. Inform. Theory 43 (1997) 499–511.

[2] K. Abed-Meraim, E. Moulines, P. Loubaton, Prediction error for second-order blind identification: Algorithms and statistical performance, IEEE Trans. Signal Process. 45 (1997) 694–705.

[3] K. Abed-Meraim, Y. Xiang, H. Manton, Y. Hua, Blind source separation using second-order cyclostationary statistics, IEEE Trans. Signal Process. 49 (2001) 694–701.

[4] B. Agee, S. Schell, W. Gardner, Spectral self-coherence restoral: A new approach to blind adaptive signal extraction using antenna arrays, Proc. IEEE 78 (1990) 753–767.

[5] L. Albera, A. Ferréol, P. Chevalier, P. Comon, ICAR: A tool for blind source separation using fourth order statistics only, IEEE Trans. Signal Process. 53 (2005) 3633–3643.
[6] S. Amari, A. Cichocki, Adaptive blind signal processing – neural network approaches, Proc. IEEE 86 (1998) 2026–2048.
[7] K. Anand, G. Mathew, V. Reddy, Blind separation of multiple co-channel BPSK signals arriving at an antenna array, IEEE Signal Process. Lett. 2 (1995) 176–178.
[8] J. Antoni, F. Guillet, M.E. Badaoui, F. Bonnardot, Blind separation of convolved cyclostationary processes, Signal Process. 85 (2005) 51–66.
[9] A. Belouchrani, K. Abed-Meraim, J.-F. Cardoso, E. Moulines, A blind source separation technique using second-order statistics, IEEE Trans. Signal Process. 45 (1997) 434–444.
[10] T. Biedka, W. Tanter, J. Reed, Convergence analysis of the least squares constant modulus algorithm in interference cancellation applications, IEEE Trans. Signal Process. 48 (2000) 491–501.
[11] E. Bingham, H. Hyvarinen, A fast fixed-point algorithm for independent component analysis of complex valued signals, Int. J. Neural Syst. 10 (2000) 1–8.
[12] N. Bouguerriou, M. Haritopoulos, C. Capdessus, L. Allam, Novel cyclostationary-based blind source separation algorithm using second order statistical properties: Theory and application to the bearing defect diagnosis, Mech. Syst. Signal Process. 19 (2005) 1260–1281.
[13] R. Boyles, W. Gardner, Cycloergodic properties of discrete-parameter non stationary stochastic processes, IEEE Trans. Inform. Theory 39 (1983) 105–114.
[14] I. Bradaric, A. Petropulu, K. Diamantaras, On the blind identifiability of FIR-MIMO systems with cyclostationnary inputs using second-order statistics, IEEE Trans. Signal Process. 51 (2003) 434–441.
[15] J. Cardoso, A. Souloumiac, Blind beamforming for non gaussian signals, IEE Proc-F 140 (1993) 362–370.
[16] L. Castedo, A. Figueiras-Vidal, An adaptive beamforming technique based on cyclostationary signal properties, IEEE Trans. Signal Process. 43 (1995) 1637–1650.
[17] M. Castella, P. Bianchi, A. Chevreuil, J.-C. Pesquet, A blind source separation framework for detecting CPM sources mixed by a convolutive mimo filter, Signal Process. 86 (2006) 1950–1967.
[18] Y. Chen, Z. He, T. Ng, P. Kwok, RLS adaptive blind beamforming algorithm for cyclostationary signals, Electron. Lett. 35 (1999) 1136–1138.
[19] Y. Chen, T. Le-Ngoc, B. Champagne, C. Xu, Recursive least squares constant modulus algorithm for blind adaptive array, IEEE Trans. Signal Process. 52 (2004) 1452–1456.
[20] P. Chevalier, On the blind implementation of the ultimate source separators for arbitrary noisy mixtures, Signal Process. 78 (1999) 277–287.
[21] P. Chevalier, Optimal separation of independent narrow-band sources – concept and performance, Signal Process. 73 (1999) 27–47.
[22] P. Chevalier, L. Albera, A. Ferréol, P. Comon, Comparative performance analysis of eight blind source separation methods on radio communications signals, in: Proc. IJCNN'04, Budapest, July 2004.
[23] P. Chevalier, V. Capdevielle, P. Comon, Behaviour of HO blind source separation methods in the presence of cyclostationary correlated multipaths, in: IEEE SP Workshop on HOS, Alberta (Canada), July 1997, pp. 363–367.
[24] P. Chevalier, V. Capdevielle, P. Comon, Performance of HO blind source separation methods: Experimental results on ionospheric HF links, in: Proc. Workshop on Independent Component Analysis, ICA 99, Aussois (France), January 1999, pp. 443–448.
[25] P. Chevalier, A. Ferréol, L. Albera, Méthodologie générale pour la séparation aveugle de sources cyclostationnaires arbitraires – application à l'écoute passive des radiocommunications, in: Proc. GRETSI, Paris (France), September 2003.
[26] P. Chevalier, A. Ferréol, L. Albera, High resolution direction finding from higher order statistics: The 2q-MUSIC algorithm, IEEE Trans. Signal Process. 54 (8) (2006) 2986–2997.
[27] P. Comon, Independent component analysis, a new concept? Signal Process. 36 (1994) 287–314. Special issue on higher-order statistics.

[28] P. Comon, L. Rota, Blind separation of independant sources from convolutive mixtures, IEICE Trans. Fundam Electron. Commun. Comput. Sci. E86-A (2003) 542–549.

[29] A.V. Dandawaté, G.B. Giannakis, Asymptotic theory of mixed time averages and kth-order cyclic-moment and cumulant statistics, IEEE Trans. Inform. Theory 41 (1995) 216–232.

[30] A. Dandawatté, G. Giannakis, Statistical tests for presence of cyclostationarity, IEEE Trans. Signal Process. 42 (1994) 2355–2369.

[31] N. Delfosse, P. Loubaton, Adaptive blind separation of independant sources: A deflation approach, Signal Process. 45 (1995) 59–83.

[32] M. Dogan, J. Mendel, Cumulant-based blind optimum beamforming, IEEE Trans. Aerosp. Electron. Syst. 30 (1994) 722–741.

[33] A. Ferréol, P. Chevalier, Higher order blind source separation using the cyclostationarity property of the signals, in: Proc. ICASSP, Munich (Germany), April 1997, pp. 4061–4064.

[34] A. Ferréol, P. Chevalier, On the behavior of current second and higher order blind source separation methods for cyclostationary sources, IEEE Trans. Signal Process. 48, April 2002; (2002), pp. 1712–1725.

[35] A. Ferréol, P. Chevalier, L. Albera, Higher order blind separation of non zero-mean cyclostationary sources, in: Proc. EUSIPCO, Toulouse (France), September 2002, pp. 103–106.

[36] A. Ferréol, P. Chevalier, L. Albera, Second order blind separation of first and second order cyclostationary sources – application to AM, FSK, CPFSK and deterministic sources, IEEE Trans. Signal Process. 52 (2004) 845–861.

[37] I. Fijalkow, A. Touzni, J. Treichler, Fractionally spaced equalization using CMA: Robustness to channel noise and lack of disparity, IEEE Trans. Signal Process. 46 (1998) 227–231.

[38] E. Florian, A. Chevreuil, P. Loubaton, Blind source separation of convolutive mixtures of non circular linearly modulated signals with unknown baud rates, in: Proc. European Signal Processing Conference, 2008.

[39] F. Gamboa, E. Gassiat, Source separation when the input sources are discrete or have constant modulus, IEEE Trans. Signal Process. 45 (1997) 3062–3072.

[40] W. Gardner, Spectral correlation of modulated signals: Part I – analog modulations, IEEE Trans. Commun. 35 (1987) 584–594.

[41] W. Gardner, Signal interception, a unifying theorical framework for feature detection, IEEE Trans. Commun. 36 (1988).

[42] W. Gardner, Exploitation of the spectral redundancy in cyclostationary signals, Signal Process. Mag. 8 (1991) 14–37.

[43] W. Gardner, Cyclostationarity in Communications and Signal Processing, IEEE Press, New York, 1994.

[44] W. Gardner, W. Brown, C.-K. Chen, Spectral correlation of modulated signals: Part II – digital modulation, IEEE Trans. Commun. 35 (1987) 595–601.

[45] G. Gelli, D. Mattera, L. Paura, Blind wideband spatio-temporal filtering based on higher-order cyclostationarity properties, IEEE Trans. Signal Process (53) (2005) 1282–1290.

[46] E. Gonen, J. Mendel, Applications of cumulants to array processing – part III: Blind beamforming for coherent signals, IEEE Trans. Signal Process. 45 (1997) 2252–2264.

[47] R. Gooch, J. Lundell, The CM array: An adaptive beamformer for constant modulus signals, in: Proc. ICASSP, Tokyo (Japan), April 1986, pp. 2523–2526.

[48] A. Gorokhov, P. Loubaton, Subspace-based techniques for blind separation of mixtures with temporally correlated sources, IEEE Trans. Circuit. Syst. 44 (1997) 813–820.

[49] A. Gorokhov, P. Loubaton, Blind identification of MIMO-FIR systems: A generalized linear prediction approach, Signal Process. 73 (1999) 104–124.

[50] O. Grellier, P. Comon, Blind separation of discrete sources, IEEE Signal Process. Lett. 5 (1998) 212–214.

[51] S. Houcke, A. Chevreuil, P. Loubaton, Blind source separation of a mixture of communication sources with various symbol periods, IEICE Trans. Fundam. Electron. Commun. Comput. Sci. 82 (1999) 1–9.

[52] S. Houcke, A. Chevreuil, P. Loubaton, Blind equalization: Case of an unknown symbol period, IEEE Trans. Signal Process. 51 (2003) 781–793.
[53] Y. Inouye, K. Hirano, Cumulant-based blind identification of linear multi-input-multi-output systems driven by colored inputs, IEEE Trans. Signal Process. 45 (1997) 1543–1552.
[54] M. Jafari, S. Alty, J. Chambers, New natural gradient algorithm for cyclostationary sources, IEE Proc.-Vis. Image Signal Process. 151 (2004) 62–68.
[55] M. Jafari, J. Chambers, D. Mandic, Natural gradient algorithm for cyclostationary sources, Electron. Lett. 38 (2002) 758–759.
[56] M. Jafari, W. Wang, J. Chambers, T. Hoya, A. Cichocki, Sequential blind source separation based exclusively on second-order statistics developed for a class of periodic signals, IEEE Trans. Signal Process. 54 (2006) 1028–1040.
[57] P. Jallon, A. Chevreuil, Separation of instantaneous mixtures of cyclostationary sources, Signal Process. 87 (2007) 2718–2732.
[58] P. Jallon, A. Chevreuil, P. Loubaton, Separation of digital communication mixtures with the CMA: Case of unknown symbol rates, Signal Process. (2009).
[59] P. Jallon, A. Chevreuil, P. Loubaton, P. Chevalier, Separation of convolutive mixtures of cyclostationary sources: A contrast function based approach, in: Proc. ICA, Grenade (Spain), 2004.
[60] P. Jallon, A. Chevreuil, P. Loubaton, P. Chevalier, Separation of convolutive mixtures of linear modulated signals using the constant modulus algorithm, ICASSP'05, Philadelphia, in: ICASSP, Philadelphia, 2005.
[61] C. Jutten, J. Hérault, Blind separation of sources, part I: An adaptive algorithm based on neuromimetic architecture, Signal Process. 24 (1991) 1–10.
[62] A. Kannan, V. Reddy, Maximum likelihood estimation of constellation vectors for blind separation of co-channel BPSK signals and its performance analysis, IEEE Trans. Signal Process. 45 (1997) 1736–1741.
[63] A. Keerthi, A. Mathur, J. Shynk, Misadjustment and tracking analysis of the constant modulus array, IEEE Trans. Signal Process. 46 (1998) 51–58.
[64] O. Kwon, C. Un, J. Lee, Performance of constant modulus adaptive digital filters for interference cancellation, Signal Process. 26 (1992) 185–196.
[65] J. Lee, Y. Lee, Robust adaptive beamforming for cyclostationary signals under cycle frequency error, IEEE Trans. Antennas Propag. 47 (1999) 233–241.
[66] J. Lee, Y. Lee, W. Shih, Efficient robust adaptive beamforming for cyclostationary signals, IEEE Trans. Signal Process. 48 (2000) 1893–1901.
[67] T. Lee, M. Girolmi, T. Sejnowski, Independent component analysis using an extended infomax algorithm for mixed subgaussian and supergaussian sources, Neural Comput. 11 (1999) 417–441.
[68] Q. Li, E. Bai, Z. Ding, Blind source separation of signals with known alphabets using ϵ-approximation algorithms, IEEE Trans. Signal Process. 51 (2003) 1–10.
[69] T. Li, N. Sidiropoulos, Blind digital signal separation using successive interference cancellation iterative least squares, IEEE Trans. Signal Process. 48 (2000) 3146–3152.
[70] Y. Li, K. Liu, Adaptive blind source separation and equalization for multiple-input/multiple-output systems, IEEE Trans. Inform. Theory 44 (1998) 2864–2876.
[71] Y. Liang, A. Leyman, B. Soong, Blind source separation using second-order cyclic statistics, in: Proc. SPAWC, Paris (France), 1997, pp. 57–60.
[72] D. Liu, L. Tong, An analysis of constant modulus algorithm for array signal processing, Signal Process. 73 (1999) 81–104.
[73] H. Liu, G. Xu, Closed-form blind symbol estimation in digital communications, IEEE Trans. Signal Process. 43 (1995) 2714–2723.
[74] O. Macchi, E. Moreau, Adaptive unsupervised separation of discrete sources, Signal Process. 73 (1999) 49–66.
[75] M. Martone, Adaptive multistage beamforming using cyclic higher order statistics (CHOS), IEEE Trans. Signal Process. 47 (1999) 2867–2873.

[76] A. Mathur, A. Keerthi, J. Shynk, A variable step-size CM array algorithm for fast fading channel, IEEE Trans. Signal Process. 45 (1997) 1083–1087.
[77] A. Mathur, A.V. Keerthi, J.J. Shynk, R.P. Gooch, Convergence properties of the multistage constant modulus array for correlated sources, IEEE Trans. Signal Process. 45 (1997) 280–296.
[78] L. Mazet, P. Loubaton, Cyclic correlation based symbol rate estimation, in: Proc. Asilomar Conference on Signals, Systems, and Computers, October 1999, pp. 1008–1012.
[79] E. Moulines, P. Duhamel, J.-F. Cardoso, S. Mayrargue, Subspace methods for the blind identification of multi-channel FIR filters, IEEE Trans. Signal Process. 43 (1995) 516–525.
[80] A. Napolitano, Cyclic higher-order statistics: Input/output relations for discrete- and continuous-time mimo linear almost-periodically time-variant systems, Signal Process. 42 (1995) 147–166.
[81] A. Orozco-Lugo, M. Lara, D. McLernon, H. Muro-Lemus, Multiple packet reception in wireless ad hoc networks using polynomial phase-modulating sequences, IEEE Trans. Signal Process. 51 (2003) 2093–2110.
[82] D. Pham, Blind separation of instantaneous mixture of sources via an independent component analysis, IEEE Trans. Signal Process. 44 (1996) 2768–2779.
[83] D. Pham, Mutual information approach to blind source separation of stationary sources, IEEE Trans. Inform. Theory 48 (2002) 1935–1946.
[84] D. Pham, Blind separation of cyclostationary sources using joint block approximate diagonalization, in: Proc. of Independent Component Analysis, London (UK), September 2007.
[85] A. Ranheim, A decoupled approach to adaptive signal separation using an antenna array, IEEE Trans. Veh. Technol. 48 (1999) 676–682.
[86] P. Regalia, On the equivalence between the Godard and Shalvi-Weinstein schemes of blind equalization, Signal Process. 73 (1999) 185–190.
[87] K. Sabri, M. Taoufiki, A. Adib, D. Aboutadjine, M.E. Badaoui, F. Guillet, Separation of convolutive mixtures of cyclostationary sources by reference contrasts, Int. J. Comput. Sci. Netw. Secur. 6 (2006) 224–229.
[88] S. Schell, W. Gardner, Blind adaptive spatiotemporal filtering for wide-band cyclostationary signal, IEEE Trans. Signal Process. 41 (1993) 1961–1964.
[89] J. Shynk, R. Gooch, The constant modulus array for cochannel signal copy and direction finding, IEEE Trans. Signal Process. 44 (1996) 652–660.
[90] J. Shynk, A. Keerthi, A. Mathur, Steady state analysis of the multistage constant modulus array, IEEE Trans. Signal Process. 44 (1996) 948–962.
[91] D. Slock, Blind fractionally-spaced equalization, perfect-reconstruction filter-banks and multichannel linear prediction, in: Proc. of ICASSP, vol. 4, 1994, pp. 585–588.
[92] A. Swami, G.B. Giannakis, S. Shamsunder, Multichannel ARMA processes, Trans. Signal Process. 42 (1994) 898–913.
[93] A. Swindelhurst, S. Daas, J. Yang, Analysis of a decision directed beamformer, IEEE Trans. Signal Process. 43 (1995) 2920–2927.
[94] S. Talwar, A. Paulraj, Blind separation of synchronous co-channel digital signals using an antenna array – Part II: Performance analysis, IEEE Trans. Signal Process. 45 (1997) 706–718.
[95] S. Talwar, M. Viberg, A. Paulraj, Blind estimation of multiple co-channel digital signals using an antenna array, Signal Process. Lett. 1 (1994) 29–31.
[96] S. Talwar, M. Viberg, A. Paulraj, Blind separation of synchronous co-channel digital signals using an antenna array – Part I: Algorithms, IEEE Trans. Signal Process. 44 (1996) 1184–1197.
[97] H. Thi, C. Jutten, Blind source separation for convolutive mixture, Signal Process. 45 (1995) 209–229.
[98] L. Tong, Y. Inouye, R. Liu, A finite step global convergence algorithm for parameter estimation of multichannel MA processes, IEEE Trans. Signal Process. 40 (1992) 2547–2558.
[99] L. Tong, G. Xu, T. Kailath, Blind identification and equalization based on second-order statistics: A time-domain approach, IEEE Trans. Inform. Theory 40 (1994) 340–349.
[100] J. Treichler, M. Larimore, New processing techniques based on constant modulus adaptive algorithm, IEEE Trans. Acoust. Speech Signal Process. 33 (1985) 420–431.

[101] J. Treichler, M. Larimore, The tone capture properties of CMA-based interference suppressors, IEEE Trans. Acoust. Speech Signal Process. 33 (1985) 946–958.
[102] J.R. Treichler, B.G. Agee, A new approach to multipath correction of constant modulus signals, IEEE Trans. Acoust. Speech Signal Process. 31 (1983) 459–472.
[103] J. Tugnait, Blind spatio-temporal equalization and impulse response estimation for MIMO channels using a Godard cost function, IEEE Trans. Signal Process. 45 (1997) 268–271.
[104] J. Tugnait, Identification and deconvolution of multi-channel non-gaussian processes using higher-order statistics and inverse filter criteria, IEEE Trans. Signal Process. 45 (1997) 658–672.
[105] J. Tugnait, On blind separation of convolutive mixtures of independent linear signals in unknown additive noise, IEEE Trans. Signal Process. 46 (1998) 3117–3123.
[106] J. Tugnait, Adaptive blind separation of convolutive mixtures of independent linear signals, Signal Process. 73 (1999) 139–152.
[107] A.V.D. Veen, Analytical method for blind binary signal separation, IEEE Trans. Signal Process. 45 (1997) 1078–1082.
[108] A.V.D. Veen, Blind separation of BPSK sources with residual carriers, Signal Process. 73 (1999) 67–79.
[109] A.V.D. Veen, A. Paulraj, An analytical constant modulus algorithm, IEEE Trans. Signal Process. 5 (1996) 1136–1155.
[110] A.V.D. Veen, S. Talwar, A. Paulraj, Blind estimation of multiple digital channels transmitted over FIR channels, IEEE Signal Process. Lett. 2 (1995) 99–102.
[111] A.V.D. Veen, S. Talwar, A. Paulraj, A subspace approach to blind space-time signal processing for wireless communication systems, IEEE Trans. Signal Process. 45 (1997) 173–190.
[112] V. Venkataraman, J. Shynk, A multistage hybrid constant modulus array with constrained adaptation for correlated sources, IEEE Trans. Signal Process. 55 (2007) 2509–2519.
[113] W. Wang, M. Jafari, S. Sanei, J. Chambers, Blind separation of convolutive mixtures of cyclostationary signals, Internat. J. Adapt. Control Signal Process. 18 (2004) 279–298.
[114] C. Watterson, J. Juroshek, W. Bensema, Experimental confirmation of an HF model, Collection Information, IEEE Trans. Commun. 18 (1970) 792–803.
[115] Q. Wu, K. Wong, Blind adaptive beamforming for cyclostationary signals, IEEE Trans. Signal Process. 44 (1996) 2757–2767.
[116] Y. Xiang, W. Yu, H. Zheng, S. Nahavandi, Blind separation of cyclostationary signals from instantaneous mixtures, in: Proc. 5th World Congress on Intelligent Control and Automation, Hangzhou (China), June 2004, pp. 309–312.
[117] D. Yellin, E. Weinstein, Multi-channel signal separation: Methods and analysis, Trans. Signal Process. 44 (1996) 106–118.
[118] J. Yu, Y. Cheng, Blind estimation of finite alphabet digital signals using eigenspace-based beamforming techniques, Signal Process. 84 (2004) 895–905.
[119] S. Yu, J. Lee, Adaptive array beamforming for cyclostationary signals, IEEE Trans. Antennas Propag. 44 (1996) 943–953.

CHAPTER

Biomedical applications

18

L. Albera, P. Comon, L.C. Parra, A. Karfoul,
A. Kachenoura, and L. Senhadji

In this chapter we focus on the use of Independent Component Analysis (ICA) in biomedical systems. Several studies dealing with ICA-based biomedical systems have been reported during the last decade. Nevertheless, most of these studies have only explored a limited number of ICA methods, namely SOBI [7], FastICA [39] and InfoMax [53]. In addition, the performance of ICA algorithms for arbitrary electro-physiological sources is still almost unknown. This prevents us from choosing the best method for a given application, and may limit the role of these methods in biomedical systems.

The purpose of this chapter is first to show the interest of ICA in biomedical applications such as the analysis of human electro-physiological signals. Next, we aim at studying twelve of the most widespread ICA techniques in the signal processing community and identify those that are most appropriate for biomedical signals.

18.1 INTRODUCTION

The previous chapters presented various categories of algorithms performing ICA. The main difference between the various methods lies in their approach for measuring statistical independence between random variables. One group of algorithms such as InfoMax [35], FastICA [38] or PICA [48] measure independence using Mutual Information (MI), which is directly related to the definition of independence via the Kullback-Leibler divergence [6]. Alternatively, some algorithms use the normalized Differential Entropy (DE), which is a special distance to normality also referred to as negentropy (see Chapter 3 page 81 or Chapter 6 page 181 for a brief description). More precisely, the InfoMax method solves the ICA problem by maximizing the DE of the output of an invertible nonlinear transform of the expected sources using the natural gradient algorithm [35] (see Chapter 6 for more details). The PICA algorithm uses the parametric Pearson model in the score function in order to minimize the MI. Using either a deflation or a simultaneous optimization scheme, FastICA extracts each component by maximizing an approximation of the DE of the expected source by means of an approximate Newton iteration (which actually often reduces to a fixed-step gradient, as shown in section 6.10.4).

Another group of algorithms measure independence indirectly via cumulants which are easier to compute [59,3,2,49] (see Chapter 3). These statistical measures are very useful to build a good optimization criterion, referred to as contrast [18,19, definition 5]. This comes essentially from two important properties: (i) if at least two components or groups of components are statistically independent, then all cross-cumulants involving these components are zero and (ii) if a variable is Gaussian, then all its High Order (HO) cumulants are zero. Note that cross-cumulants share two other useful properties. For real-valued random variables they are symmetric arrays since the value of their entries does not change with the permutation of their indices, and they satisfy the multi-linearity property [59] (see also Chapter 9 page 335).

Thus, numerous cumulant-based techniques were proposed such as JADE [10,11], COM2 [18,19], COM1 [21,20], SOBI in both conventional [7] and robust whitening [105] versions, STOTD [52], SOBIUM [51], TFBSS [36,30] and $\text{FOBIUM}_{\text{ACDC}}$. For more detailed description, one can refer to Chapter 5 for methods COM2, JADE and STOTD, to Chapter 7 for the SOBI algorithm, to Chapter 9 for the SOBIUM method and to Chapter 11 for the TFBSS approach. In the SOBI and TFBSS algorithms, ICA is solved by jointly diagonalizing several Second Order (SO) cumulant matrices, say covariance matrices, well-chosen in the time and the time-frequency planes, respectively. Also based on SO cumulants, the SOBIUM algorithm extends the SOBI concept to the case of under-determined mixtures (i.e. more sources than sensors). Regarding the $\text{FOBIUM}_{\text{ACDC}}$ method, it is based on a non-orthogonal joint diagonalization of several Fourth Order (FO) cumulant matrices, well-known as quadricovariance matrices, using the diagonalization ACDC scheme [100].

It is noteworthy that the $\text{FOBIUM}_{\text{ACDC}}$ method is a variant of FOBIUM [29], which has the advantage of processing both sub- and super-Gaussian sources. As far as the COM2 method is concerned, a maximization of a FO contrast function is used to extract independent components. In addition, the JADE algorithm [10] solves the ICA problem by jointly diagonalizing a set of eigenmatrices of the quadricovariance matrix of the whitened data. Following the same spirit as the JADE algorithm, STOTD [52] jointly diagonalizes third order slices extracted from the FO cumulant array of the whitened data.

We stress that a good performance of each technique is subject to some specific conditions or assumptions on the processed sources and the additive noise. For instance some methods require sources to be colored (i.e. temporarily correlated); others do not allow Gaussian noise with unknown spatial coherence, etc. Figure 18.1 sheds light on the essential hypotheses required for each method considered in this chapter, and for which a good behavior is guaranteed.

This specific chapter aims at giving some insights into the numerical complexity of many of these ICA algorithms. In addition, it gives a comparative performance analysis based on simulated signals in electro-encephalography. This chapter hopefully constitutes a useful reference for researchers from the biomedical community, especially for those who are not familiar with ICA techniques.

Characteristics / Algorithms	Stat. order	Whitening	Allowed Gaussian sources	Necessity of source cumulants of the same sign	Necessity of sources with different spectra	Utilization of the source coloration	Utilization of the source non-stationarity	Noise with unknown spatial coherence
SOBI	2	Yes	All	No	Yes at order 2	Yes	No	No
$SOBI_{rob}$	2	Yes (Robust)	All	No	Yes at order 2	Yes	No	No
TFBSS	2	Yes	All	No	Yes at order 2	Yes	Yes	No
SOBIUM	2	No	All	No	Yes at order 2	Yes	No	No
COM2	4	Yes	1	No	No	No	No	No
JADE	4	Yes	1	No	No	No	No	No
$FOBIUM_{ACDC}$	4	No	0	No	Yes at order 4	Yes	No	Yes if Gauss.
STOTD	4	Yes	1	No	No	No	No	No
$FastICA_{sym}$	4	Yes	1	No	No	No	No	No
$FastICA_{def}$	4	Yes	1	No	No	No	No	No
PICA	Up to 4	Yes	1	No	No	No	No	No
INFOMAX	4	Yes	1	No	No	No	No	No

FIGURE 18.1

Twelve ICA methods with their main characteristics.

18.2 ONE DECADE OF ICA-BASED BIOMEDICAL DATA PROCESSING

Advances in data recordings technologies and digital signal processing have enabled recordings and analysis of vast amounts of multidimensional biomedical data. The extraction of the essential features from the data becomes therefore paramount. The use of ICA in biomedical systems is now very widespread since the ICA concept is very easy to understand: it decomposes the data as a linear combination of statistically independent components. However, biomedical signals pose a challenge as it is often difficult to ascertain a ground truth that could be used to evaluate the accuracy of the ICA decomposition, or even its meaning. The following questions therefore deserve careful consideration: can the data be modeled as a linear combination of physically independent random processes? Are there a fixed and known number of such independent processes? Do they represent independent source signals or more generally independent source subspaces? The answers to these questions are not simple and thus the results of ICA for biomedical signals often requires a significant level of expertise. This section will survey biomedical applications where ICA plays a central role to provide an overview of the assumptions and motivations for its use.

18.2.1 Electromagnetic recordings for functional brain imaging

As explained in [4], functional brain imaging is a relatively new and multidisciplinary research field, which encompasses techniques devoted to a better understanding of the human brain through noninvasive imaging of the electro-physiological, hemodynamic, metabolic, and neurochemical processes underlying normal and pathological brain

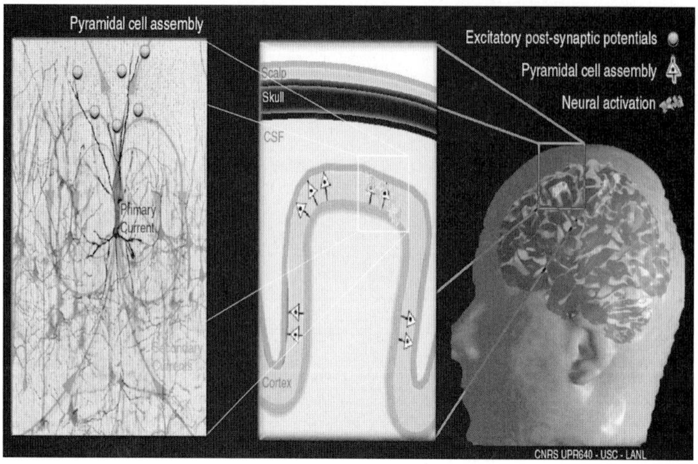

FIGURE 18.2

Left: Excitatory postsynaptic potentials (EPSPs) are generated at the apical dendritic tree of a cortical pyramidal cell and trigger the generation of a current that flows through the volume conductor from the non-excited membrane of the soma and basal dendrites to the apical dendritic tree sustaining the EPSPs. Some of the current takes the shortest route between the source and the sink by traveling within the dendritic trunk, while conservation of electric charges imposes that the current loop be closed with extracellular currents flowing even through the most distant part of the volume conductor. **Center**: Large cortical pyramidal nerve cells are organized in macro-assemblies with their dendrites normally oriented to the local cortical surface. This spatial arrangement and the simultaneous activation of a large population of these cells contribute to the spatio-temporal superposition of the elemental activity of every cell, resulting in a current flow that generates detectable EEG and MEG signals. **Right**: Functional networks made of these cortical cell assemblies and distributed at possibly multiple brain locations are thus the putative main generators of MEG and EEG signals. *Source*: From [4] with permission.

function. These imaging techniques are powerful tools for studying neural processes in the normal working brain. Clinical applications include improved understanding and treatment of serious neurological and neuropsychological disorders such as intractable epilepsy, schizophrenia, depression, Parkinson's disease and Alzheimer's disease. Due to the unsurpassed temporal resolution of electro-encephalography (EEG) and magneto-encephalography (MEG), and to their widespread use by clinicians and scientists, we decided to focus on both these functional brain imaging tools.

EEG and MEG are two complementary techniques that measure, respectively, the scalp electric potentials and the magnetic fields, outside the head produced by currents in neural cell assemblies (Fig. 18.2). They directly measure electrical brain activity and allow for studies of the dynamics of neural networks or cell assemblies, which occur

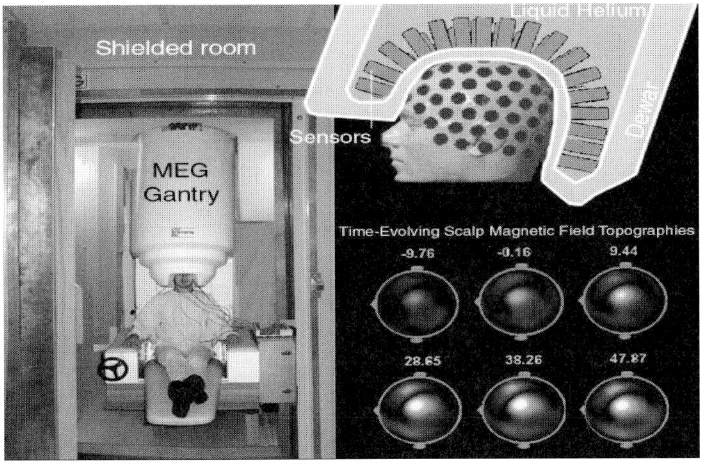

FIGURE 18.3

Typical scalp magnetic fields are on the order of a 10 billionth of the earth's magnetic field. MEG fields are measured inside a magnetically shielded room for protection against higher-frequency electromagnetic perturbations (left). MEG sensors use low-temperature electronics cooled by liquid helium (upper right) stored in a Dewar (left and upper right). Scalp magnetic fields are then recorded typically every millisecond. The resulting data can be visualized as time-evolving scalp magnetic field topographies (lower right). These plots display the time series of the recorded magnetic fields interpolated between sensor locations on the subject's scalp surface. This MEG recording was acquired as the subject moved his finger at time 0 (time relative to movement ($t = 0$) is indicated in ms above every topography). Data indicate early motor preparation prior to the movement onset before peaking at about 20 ms after movement onset.
Source: From [4] with permission.

at typical time scales on the order of tens of milliseconds as shown in Fig. 18.3. EEG was born in 1924 when the German physician Hans Berger first measured traces of brain electrical activity in humans. Although today's electronics and software for EEG analysis benefit from the most recent technological developments, the basic principle remains unchanged since Berger's time. EEG consists of measurements of a set of electric potential differences between pairs of scalp electrodes. The sensors may be either directly glued to the skin (for prolonged clinical observation) at selected locations directly above cortical regions of interest, or fitted in an elastic cap for rapid attachment with near uniform coverage of the entire scalp. Research protocols can use up to 256 electrodes. In many clinical and research applications, EEG data are analyzed using pattern analysis methods to associate characteristic differences in the data with differences in patient populations or experimental paradigm. Typical EEG scalp voltages are on the order of tens of microvolts and thus readily measured using relatively low-cost scalp electrodes and high-impedance high-gain amplifiers.

In contrast, characteristic magnetic fields produced by neural currents are extraordinarily weak, on the order of several tens of femtoTeslas, thus necessitating sophisticated sensing technology. MEG was developed in physics laboratories and especially in low-temperature and superconductivity research groups. In the late 1960s, Zimmerman co-invented the SQUID (Superconducting QUantum Interference Device) – a supremely sensitive amplifier that has since found applications, ranging from airborne submarine sensing to the detection of gravitational waves – and conducted the first human magnetocardiogram experiment using a SQUID sensor at MIT. SQUIDs can be used to detect and quantify minute changes in the magnetic flux through magnetometer coils in a superconducting environment. Cohen, also at MIT, made the first MEG recording a few years later [17]. Recent developments include whole-head sensor arrays for the monitoring of brain magnetic fields at typically 100 to 300 locations.

To our knowledge, Makeig *et al.* [56] and Vigario [89] were the first to apply ICA to EEG data. Now ICA is widely used in the EEG/MEG research community [56,89, 90,93,92,84,85,94,65,31,69,37,47,91,22,47,40,43,45,44,41,25,57,63]. Applications of ICA mainly include artifact detection and removal [85,65,69,37,40,43,45,44,92,41,89], analysis of event-related response averages [92,56,45,44,57], and single-trial EEG/MEG [84,45,25, 63]. In order to illustrate more precisely the use of ICA, two applications are discussed in the sequel. The latter are devoted to the denoising problem in EEG and MEG, respectively. Such a problem plays an important role in the analysis of electromagnetic recordings.

Indeed, the signals of interest are very often recorded in the presence of noise signals, which can be decomposed into the sum of internal and external noises. The external noise is the instrumentation noise. The internal "noise" comprises all normal physiological activities that may generate electrical currents but that are not of interest to a specific study. A common example of physiological noise is the activity linked to movements of the eyes and eye blinks [89]. Similarly, jaw-muscle activity caused by chewing or talking leads to large broad-band signals that overshadow the more subtle brain signals. Figure 18.4 displays a few channels of EEG signals contaminated by such artifacts – the naming convention for the different channels is based on the standardized 10-20 system (see Fig. 18.19). The cardiac cycle, as well as contamination coming from a digital watch (see [91, figure 3]), can also disturb the EEG/MEG signals of interest.

Choosing an inactive voltage reference for recording is one of the oldest technical problems in EEG. Since commonly used cephalic references contaminate EEG and can lead to misinterpretation, the elimination of the reference contribution is of fundamental interest. This is the purpose of the work presented in [37] by applying ICA to interictal recordings. In fact, the classical EEG recording techniques may complicate the extraction of accurate information. For instance, in coherence analysis, the presence of a common reference signal in EEG recordings results in a distortion of the synchrony values observed and may destroy the intended physical interpretation of phase synchrony [78,34].

Other assemblies such as bipolar EEG, average common reference EEG and Laplacian EEG can be used in order to obtain reference-free EEG. But all of them also present strong drawbacks. The bipolar EEG, obtained by subtracting the potentials of two

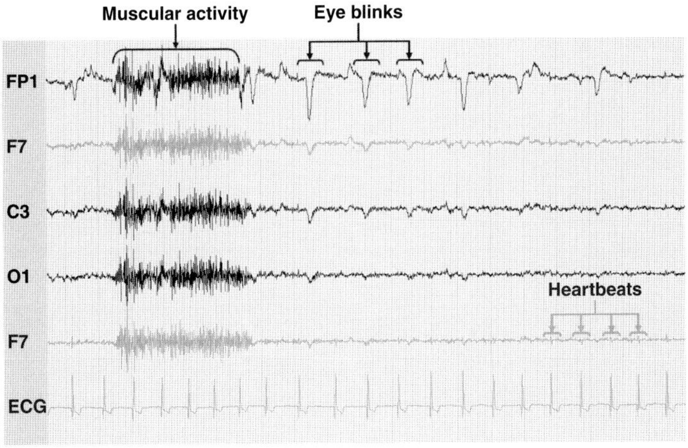

FIGURE 18.4

Some surface electrical activities recorded by a few EEG electrodes, derived from the standardized 10-20 system, and an ECG sensor.

nearby electrodes, will remove all signals common to the two channels, including the common reference but also information from dipoles with certain locations and tangential orientations. The average reference EEG, obtained by subtracting the average of all recording electrodes (i.e. the average reference) from each channel does not suit the classical 10-20 system coverage [46,34] and would need a denser coverage of the head surface [46] to be really efficient. Eventually, Laplacian maps, based on the second spatial derivative in an attempt to remove non-local contributions to the potential, eliminate all the distributed sources regardless whether or not they originate from the reference and are biased toward superficial, radial sources. In addition, the three latter techniques can also lead to misinterpretation of the synchronization results.

In order to overcome the drawbacks of the classical methods, Hu *et al.* proposed to use ICA for identification and removal of the reference signal contribution from intracranial EEG (iEEG) recorded with a scalp reference signal. All EEG potential measurements reflect the difference between two potentials, and can then be factorized as a static mixture of two source subspaces corresponding to the reference signal and signals of interest, respectively. The former subspace is 1-dimensional by definition, while the latter subspace may have a higher dimension. As far as the statistical independence between both subspaces is concerned, as mentioned by the authors, it should at least be approximately true because of the high resistivity of the skull between scalp electrode and intracranial electrodes. Then the FastICA method is used in order to automatically extract the scalp reference signal based on its independence from the bipolar iEEG data. When iEEG and scalp EEG (Fig. 18.5(a)) are recorded simultaneously using the same scalp reference, the scalp reference signal R_2 extracted by ICA from iEEG can also be used in order to clean the scalp EEG. The results of the latter procedure are displayed in Fig. 18.5(c) in comparison with other techniques such as bipolar scalp

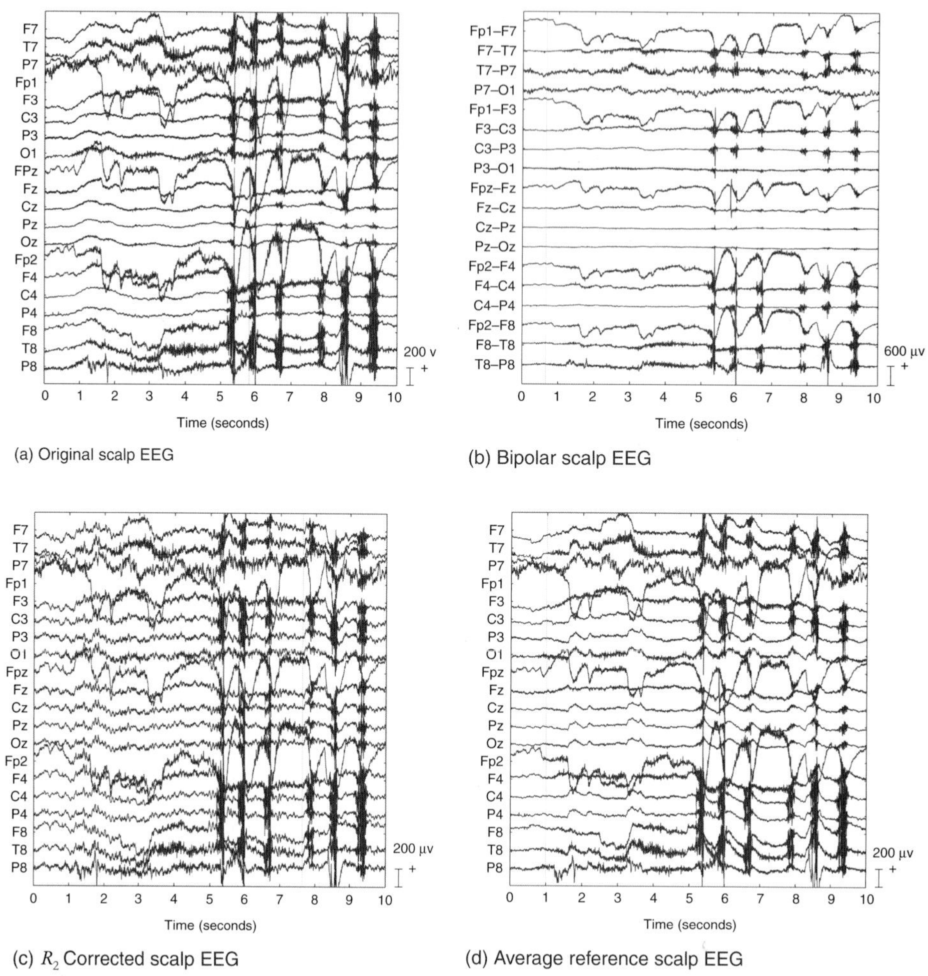

FIGURE 18.5

Removal of the reference contribution from scalp EEG recorded simultaneously with iEEG using the same scalp reference. (a) A 10-s sample of scalp EEG recorded simultaneously with iEEG using the same scalp reference. The segment is remarkable for the large muscle artifact due to the patient chewing between 5 s and 10 s. (b) Bipolar scalp EEG is created by subtracting the EEG signal from adjacent electrodes. This demonstrates that in this case the muscle artifact is not only from the common scalp reference. (c) The corrected scalp EEG using R_2 to remove the reference contribution. (d) The scalp EEG obtained using the average scalp reference. *Source*: From [37] with permission.

EEG (Fig. 18.5(b)) and average reference scalp EEG (Fig. 18.5(d)). The R_2 corrected EEG is clearly advantageous over the average reference corrected. The average reference corrected EEG in this case has removed the diffuse cerebral activity that remains evident in the R_2 corrected EEG. The R_2 corrected EEG is also advantageous over the bipolar EEG because the latter usually leads to smaller amplitudes (see for instance F7 and T7 in Fig. 18.5(c) and F7–T7 in Fig. 18.5(b)) and causes misinterpretation of EEG.

MEG is useful in preparation of epilepsy surgery in order to localize current dipoles from surface epileptic spikes allowing for cortex mapping and intracranial electrodes placement. As a result, several approaches for epileptic spike detection were proposed such as morphological analysis, template matching, predictive filtering and ICA analysis (see [65] and the references therein). In practice these techniques have various limitations. For instance, methods based on morphological analysis and template matching do not take into account the spatial structure of the measurements since they were developed for single channel data [65]. Predictive filtering techniques are less appropriate for MEG data since they were basically proposed for EEG recordings, well known to have a better SNR [65].

Now concerning spike detection techniques, the majority of them require visual inspection or interpretation of independent components, and manual cluster analysis to discard spurious sources [65]. In order to overcome these drawbacks, Ossadtchi *et al.* proposed the use of ICA in a fully automated way [65]. It is noteworthy that the latter authors jointly identify the epileptic spikes and the corresponding current dipole positions in an efficient way.

First, an ICA analysis via the InfoMax algorithm [35] is used to recover the spike-like components from MEG measurements assumed to correspond to a noisy static mixture of two statistically independent source subspaces, say, focal epileptic and background activity sources interfered with by a spatially independent instrumentation noise. The static mixture assumption is justified by the quasistatic electromagnetic properties of MEG data. In our opinion, the independence assumption between the three classes of signals holds true in practice due to (i) the macroscopic desynchronization between the neurons contributing to the epileptic spikes and those contributing to the background activity, and (ii) the physical independence between intracerebral activities and instrumentation noise.

Second, a spikyness index is used to select the components with spike-like characteristics among all those extracted by InfoMax and only the most spiky sources are retained in order to reconstruct a denoised epileptic surface of MEG data. Then, a focal neuronal source localization step is realized on these new surface data by means of the RAP-MUSIC algorithm [62], where the current dipole model is fitted in the vicinity of each previously detected spike (i.e. with a temporal window of size 16 ms). As a result, the retained spiky sources are those that fit a current dipole with more than 95%. Next, a clustering procedure is applied on all localized dipoles in order to reduce the number of detected spiky sources. The retained sources are those falling within one of the clusters automatically determined using a distance metric taking into account both the location and the time courses of the considered sources. Finally, a cluster significance step based on a statistical test is used to refine the results by excluding all non-statistically significant clusters among all those defined in the previous step. This elaborate process is a good example of the type of expert knowledge that is sometimes required when interpreting the results of ICA on biomedical signals.

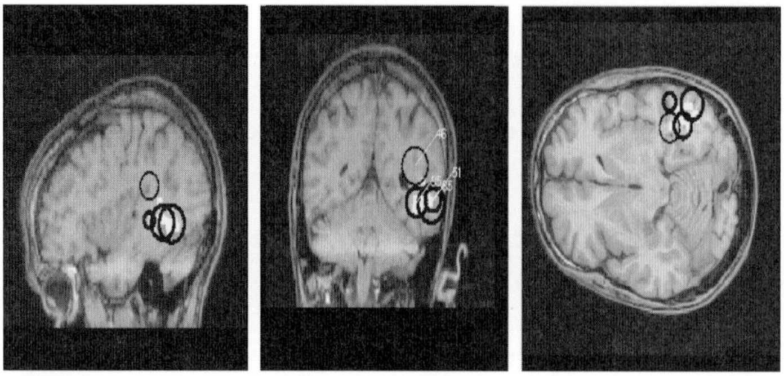

FIGURE 18.6

Manual (white square dots) and automated (circles) detection and clustering results for one subject.
Source: From [65] with permission.

The proposed method showed similar performance in terms of abnormal activity detection, compared to the one using the conventional four-phase clinical procedure, performed on four subjects requiring invasive electrode recordings for localization of the seizure origin for surgical planning. For one of the four considered subjects, Fig. 18.6 shows the performance of the proposed method (circles) for both abnormal spike activity detection and its spatial location determination. Reported results were compared to results of manual detection performed by a qualified examiner (yellow square dots) [65]. Probable epileptogenic clusters were indicated by circles with thicker lines while thinner lines indicate non-epileptogenic clusters as determined by their location and averaged time courses. On the other hand, in order to evaluate the necessity of spatial and temporal investigation of putative clusters of localized sources during the clustering step, two different clusters were studied in Figs 18.7 and 18.8, respectively. Figure 18.7 shows the average time course of dipoles localized in the temporal lobe. Due to its spike-like shape, the investigated cluster was retained as a result. However, the inspection of the average time course of the second cluster in Fig. 18.8 shows a less-descriptive spike-like characteristic, hence its elimination.

18.2.2 Electrocardiogram signal analysis

The electrocardiogram (ECG) reflects the electrical activity of the heart, which is usually recorded with surface electrodes placed on the chest, arms and limbs. A typical ECG signal of a normal heartbeat (Fig. 18.9) is composed of a series of waves: the P-wave describes the sequential depolarization of the right and the left atria, the QRS complex, related to the depolarization of the right and the left ventricles and the T-wave, generated by the ventricular repolarization [79]. In clinical practice, there exist many systems dedicated to ECG signal acquisition [97,79]. The standard 12-lead ECG seems to be the most widely used by the physician especially when waveform morphology is required. It is obtained by placing an electrode on each wrist (VR and VL), one electrode on the left foot (VF) and six precordial electrodes on the chest (V1 to V6). The positions

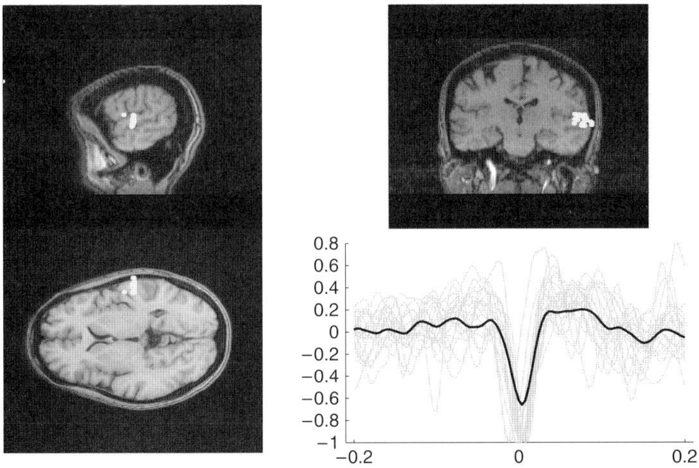

FIGURE 18.7

Location in three orthogonal views of an apparent dipole cluster (cluster 1; localized in the temporal lobe) found for one subject. This cluster was retained by the method since its average time course represents a potentially epileptogenic region.
Source: From [65] with permission.

of the nine electrodes are typically chosen to capture the electric activity of the heart from different angles reconstructing the spatial dynamics of the heart's electrical activity. Although ECG recording techniques are very effective, the distortions caused by noises and artifacts are still very significant. Indeed, in many practical situations, the ECG signal is contaminated by different types of noises and artifacts, such as sinusoidal 50/60 Hz power-line, electrode movements and broken wire contacts, but also interfering physiological signals such as those related to muscle movements and breathing. In addition, some arrhythmias may cause various disturbances in the regular rhythm of the heart and thus generate ECG waves, which are very different from those of the normal heartbeats. Hence, the main objective in ECG signal analysis is two-fold: (i) to denoise the ECG signal in order to enhance the SNR of the signals of interest and (ii) to separate the different bioelectric sources of the heart, such as ventricular activity (VA) and atrial activity (AA), in order to characterize some specific arrhythmias.

Several approaches to ECG analysis have been reported such as linear noise filtering [71], adaptive filtering [96,83,86], neural network [99,87] and wavelet transform [77]. The majority of these methods have various drawbacks. Cardiac signal often overlap with various sources of noises and artifacts in time and frequency domains. Thus extracting the signal of interest may be a difficult task. In addition, as noted in Chapter 16, some of these methods usually need the use of reference signals, which make their performance highly dependent on the reference electrode positions. Due to these limitations, reliable signal processing tools for signal enhancement and detection and noise reduction are crucial for cardiac diagnosis and therapy. In contrast to the methods mentioned above, ICA has the potential of extracting signal sources even if they are superimposed

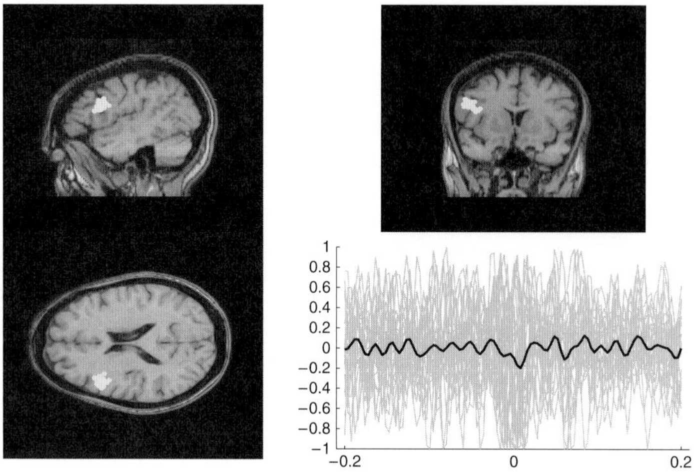

FIGURE 18.8

Location in three orthogonal views of an apparent dipole cluster (cluster 2) found for the same subject presented in Fig. 18.7. This cluster was discarded by the method since its average time course does not match with the epileptogenic region activity.
Source: From [65] with permission.

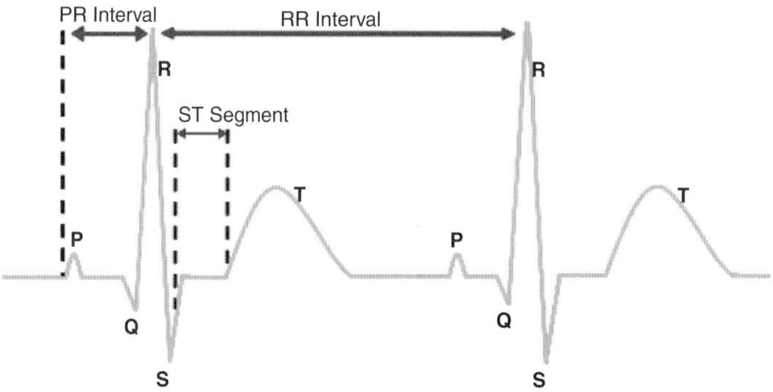

FIGURE 18.9

The ECG wave and its important features.

in the time-frequency plane. Moreover, ICA can estimate the sources by taking into account their spatio-temporal correlation, physiological prior information and mutual statistical independence. Therefore, it appears natural to consider ICA techniques as a potential tool for ECG analysis. ICA has been applied to ECG for various purposes. These include artifact and noise removal [98,5,82,14], analysis of the autonomic control of the heart [88], ventricular arrhythmia detection and classification [66], extraction of the Fetal ECG (FECG) from maternal recordings [23,9,24,104] and atrial activity extraction for Atrial Fibrillation (AF) [74,72,73,12,102]. To show how the ICA technique can be applied to ECG signal analysis, we discuss two of the most frequent applications in more detail, namely the extraction of fetal ECG and extraction of arterial activity.

The FECG can provide the clinician with valuable information on the well-being of the fetus and facilitates early diagnosis of fetal cardiac abnormalities and other pathologies. For instance, hypoxia may cause an alteration in the PR and the RR intervals [95], whereas a depression of the ST segment may be associated with acidosis [81]. Invasive recordings provide FECG with high SNR [67], but the procedure requires the use of intrauterine electrodes and is therefore only performed during labor when there is access to the fetal scalp. Non-invasive recordings with several electrodes placed on the mother's body can provide FECG. However, obtaining clean and reliable FECG is challenging due to the interfering ECG from the mother (MECG) and other interfering signals such as respiration, the electro-myogram (EMG), electrode movements and baseline drift. The mother's QRS wave occasionally overlaps with the fetal QRS wave making even visual detection of the individual beat of the fetus impossible, let alone any automated beat detection. Efforts to extract fetal ECG include adaptive filtering, linear decomposition of single and multi-channel recordings and nonlinear projection methods (see [76, chapter 2]) including ICA which we will discuss next.

De Lathauwer *et al.* [23] show that the separation of FECG from mother's skin electrodes can be approached as a linear instantaneous blind source separation problem. More precisely, the authors discuss two important aspects: (i) the nature of the occurring signal and (ii) the characteristics of the propagation from bioelectric sources to skin electrodes. Regarding the first aspect, the authors state in accordance with the work of [68] that the MECG-subspace is characterized by a three-dimensional vector signal, whereas the dimension of the FECG-subspace is subject to changes during the pregnancy period [68]. The transfer function between the bioelectric sources to body surface electrodes is assumed to be linear and resistive [64]. Finally, the high propagation velocity of the electrical signal in human tissues validates the instantaneous assumption of the model. De Lathauwer *et al.* [23] show that the application of the ICA method proposed in [19] to eight ECG observations (see Fig. 16.17 in Chapter 16) recorded from five skin electrodes located on the mother's abdominal region (electrodes placed near the fetus) and three electrodes positioned near the mother's heart (on the mother's thoracic region) provides very interesting preliminary results (see section 16.7.1 for more details). A complete and extended study of [23] is presented by the same team in [24]. They investigate the case of atypical fetal heart rate (FHR) and the case of fetal twins. To do so, the FHR is artificially obtained using the real observations depicted in Fig. 16.17. More precisely, a small piece of data around $t = 0.75$ s in Fig. 16.17 was copied to $t = 3.5$ s to simulate an extrasystolic fetal heartbeat and the QRS wave around $t = 2$ s was removed. Figure 18.11(a) illustrates that the application of ICA on the latter artificial observations succeeds in the extraction of FECG from multi-lead potential recordings on the mother's skin. Figure 18.10 shows artificial observations of the fetal twins. The data of this figure were obtained by shifting both the sixth and the eighth components of Fig. 16.18 to artificially generate a new independent ECG attributed to the second fetus. These two signals were then merged with the original observations (Fig. 16.17) after multiplication with random mixing vectors. Figure 18.11(b) shows that ICA is also able

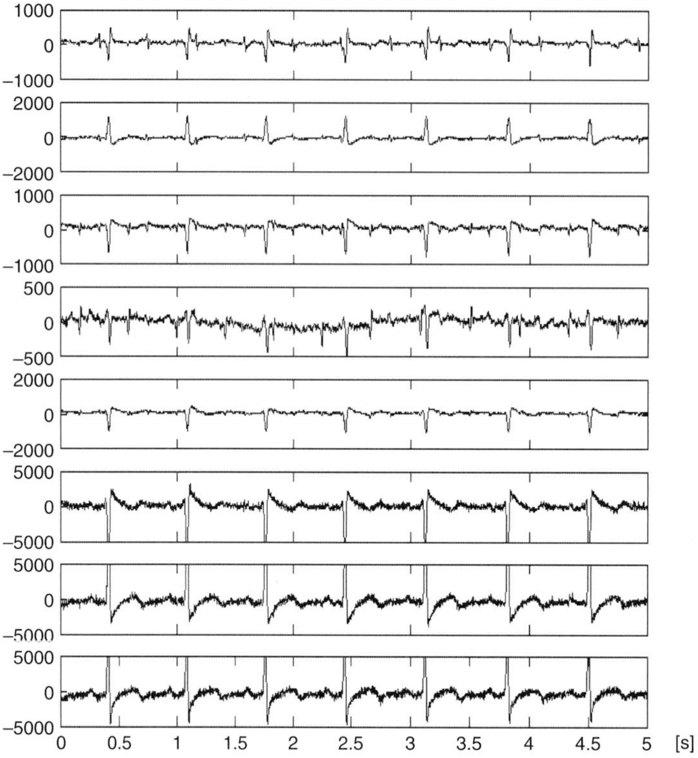

FIGURE 18.10

8-channel set of observations containing heartbeats of fetal twins.
Source: From [24] with permission.

to discriminate the two fetal twin ECG subspaces, say components 6 and 8 for the first fetus and component 7 for the second fetus.

In [104], Zarzoso *et al.* compare a standard adaptive filtering method, namely multi-reference adaptive noise canceling (MRANC) [96], with the ICA method proposed in [103]. Figure 18.12 shows one of the cutaneous electrode recordings used in the study. The first five channels correspond to the electrodes placed on the mother's abdominal region and the last three signals are related to the mother's chest electrodes. The results obtained with MRANC, using the three thoracic leads as a reference, are depicted in Fig. 18.13. Figure 18.14 displays the FECG contribution to the abdominal electrodes (first five signals) and to chest electrodes (last three signals) of the original recordings obtained with ICA. More precisely, the ICA method is first applied to the raw data to isolate the fetal subspace and then to reconstruct the surface FECG from the FECG-subspace only. Contrary to the MRANC method where the estimated components are still corrupted by the baseline drift (signal 4 of Fig. 18.13), the ICA method seems to be more effective for reconstructing the FECG contribution. Indeed, the baseline drift is eliminated in Fig. 18.14 and all fetal components are less noisy in comparison to those

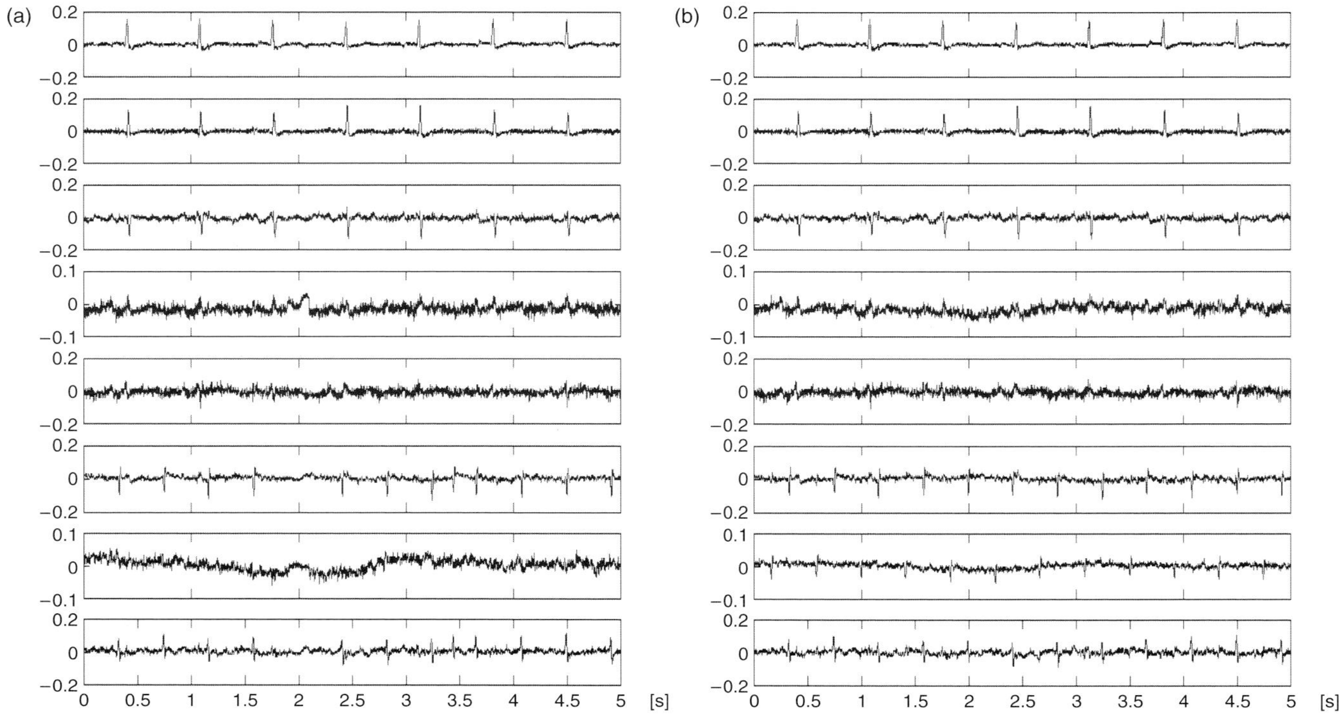

FIGURE 18.11

(a) Source estimates obtained from data containing an extrasystole around $t = 3.5$ s and missing a fetal heartbeat around $t = 2$ s; and (b) source estimates obtained from the data of Fig. 18.10.
Source: From [24] with permission.

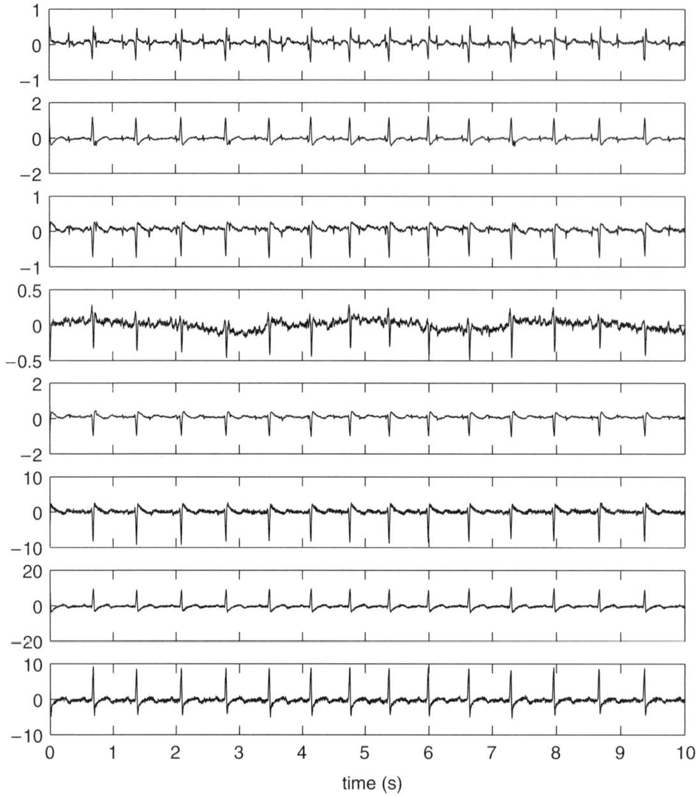

FIGURE 18.12

A cutaneous electrode recording from a pregnant woman.
Source: From [104] with permission.

obtained by the MRANC method. It is also important to point out that, because ICA does not need to use reference signals, Zarzoso *et al.* are able to reconstruct the FECG contribution to the eight electrodes, either thoracic or abdominal.

Atrial fibrillation (AF) is one of the most common arrhythmias managed in human cardiology. In the general population it has a prevalence of 0.4-1%, increasing to around 9% among those over 80 years of age [16,32]. AF is a supraventricular tachyarrhythmia characterized by uncoordinated atrial activation. The result is the replacement of the P waves by rapid oscillations or fibrillatory waves that vary in size, shape and timing [32]. AF is associated with increased mortality and hospitalization in the general population, and understanding of pathological mechanisms underlying AF using non-invasive diagnosis tools such as surface ECG is crucial to improve patient treatment strategies. However, due to the low SNR of AA on surface ECG the analysis of AF remains difficult. Some methods reported in the literature to enhance AA involve direct suppression of QRS-T by subtracting: (i) a fixed or adaptive template [86,80] or (ii) an estimated

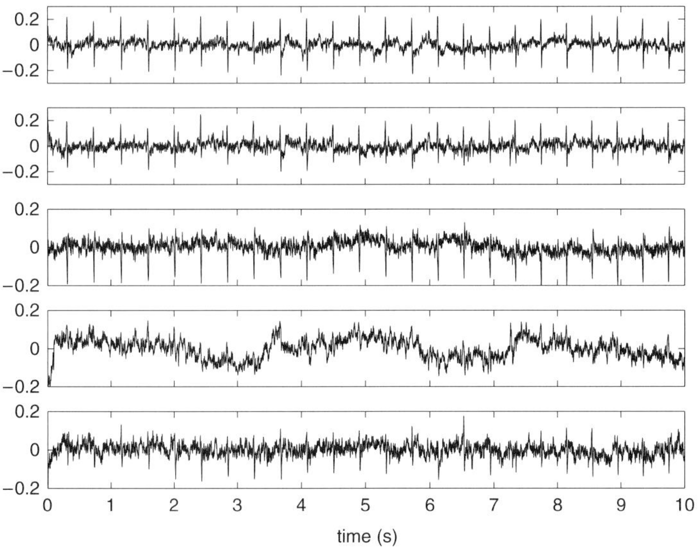

FIGURE 18.13

FECG contribution to the abdominal electrodes (first five signals) of Fig. 18.12 obtained by the MRANC method.
Source: From [104] with permission.

QRS-T complex [87]. Other approaches, based on Principal Component Analysis (PCA) [50,27,54] try to derive a relatively small number of uncorrelated linear combinations (principal components) of a set of random zero-mean variables while retaining as much of the information from the data as possible [19,26]. The major drawbacks of the QRS-T subtraction approaches are the use of a small number of QRS-T templates and the high sensitivity to QRS morphological changes. Regarding the PCA based methods, the orthogonality of the bioelectrical sources of the heart is not physiologically justified. This orthogonality condition can only be obtained through a correct orthogonal lead, known as the Frank lead system after its inventor [97,79].

The AF problem was tackled by means of ICA-based methods in [73]. The authors justify three basic considerations regarding AA and VA, and the way in which both activities are acquired from the surface electrodes: (i) the independence of VA and AA, (ii) their non-Gaussianity and (iii) the observations follow an instantaneous linear mixing model. First, due to the bioelectrical independence of the atrial and ventricular regions [55], the atrial and ventricular electrical sources can reasonably be considered as statistically independent. Second, the distributions of AA and VA are sub-Gaussian and super-Gaussian, respectively (see section II.C [73] for more details). Finally, the bioelectric theory has modeled the torso as an inhomogeneous volume conductor [68,58] which justifies that ECG surface recordings can be assumed to arise as a linear instantaneous transformation of cardiac bioelectric sources. Hence, the application of ICA-based methods to extract AA from 12-lead ECG seems to be well suited. The authors apply the FastICA algorithm

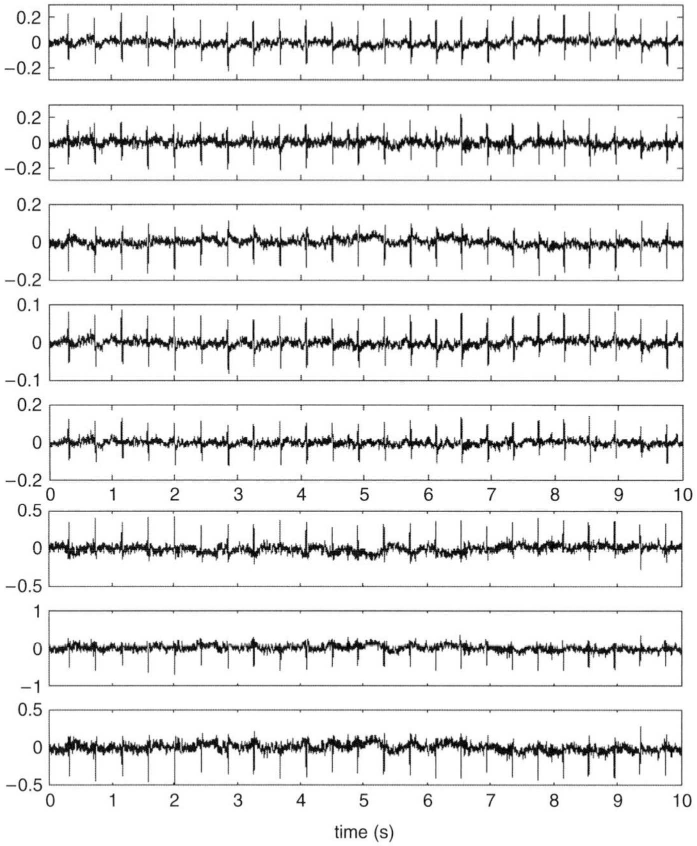

FIGURE 18.14

FECG contribution to the abdominal electrodes (first five signals) and to the thoracic electrodes (last three signals) of Fig. 18.12 obtained with ICA.
Source: From [104] with permission.

[38] on the 12-lead ECG of seven patients suffering from AF, where they consider that the source subspace is composed of AA, VA and other interferences. The use of FastICA is justified by the fact that this algorithm demonstrates a rather fast convergence. As other algorithms, it can operate in a deflation mode, which be stopped as soon as the AA sources have been extracted. Figure 18.15(a) displays a 12-lead ECG with an AF episode. Fibrillation waves are observed in several leads, especially leads II, III, aVF and V1. The results obtained after applying ICA are depicted on Fig. 18.15(b). The authors estimate the kurtosis of the extracted sources and show that the three first separated sources have a sub-Gaussian distribution and hence are candidates to be related to AA. The sources 4 to 7 are associated with Gaussian noise and artifact, whereas the components 8 to 12, which present a super-Gaussian distribution, contain a VA.

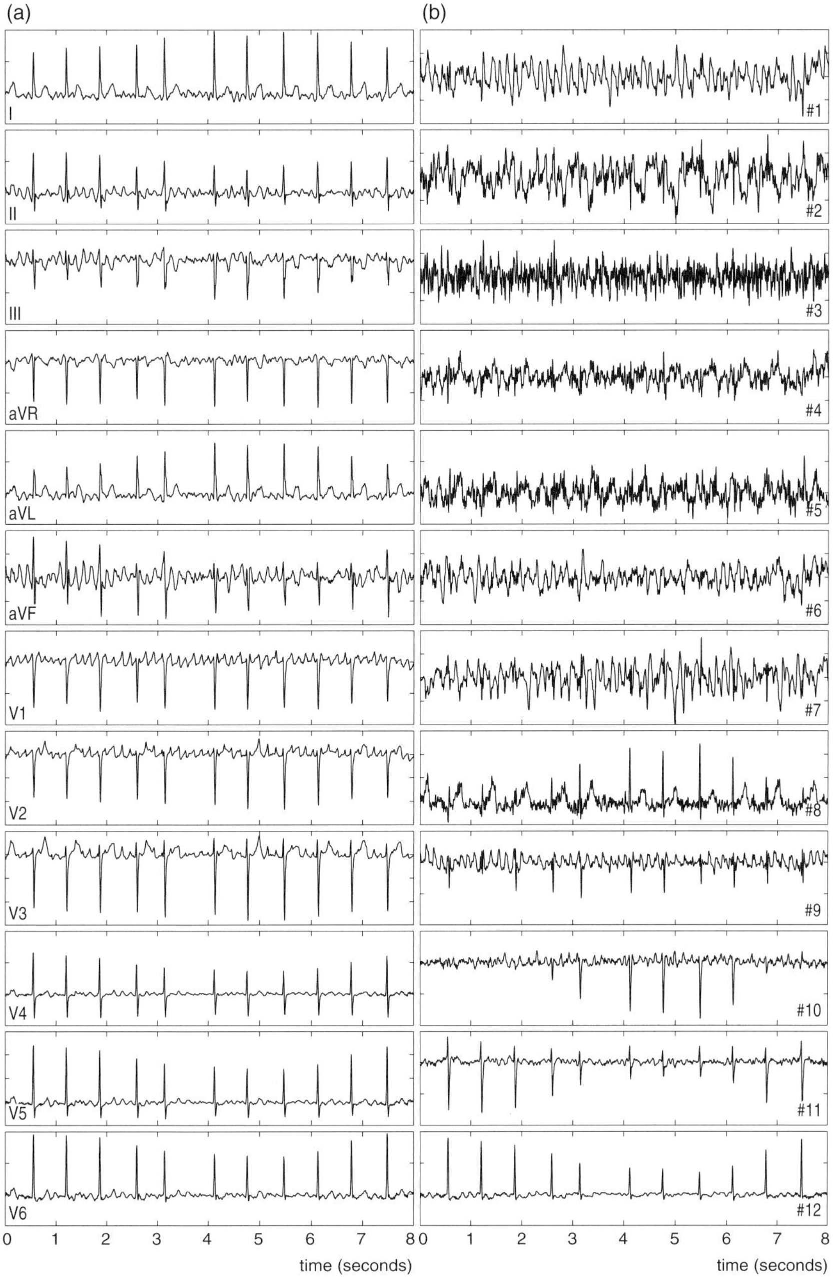

FIGURE 18.15

Inputs and results of the ICA separation process: (a) A 12-lead ECG from a patient in AF and (b) source estimates obtained by ICA and sorted from lower to higher kurtosis values. *Source*: From [73] with permission.

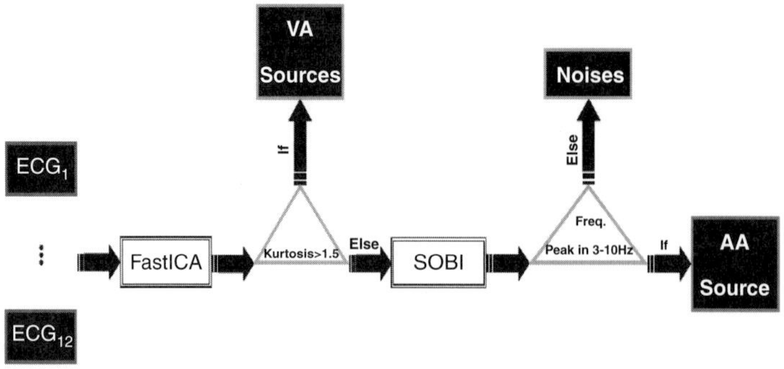

FIGURE 18.16

Block diagram of the spatio-temporal method proposed in [12].

One important problem that arises when ICA is used in a biomedical context is to automatically select and classify independent sources of interest, as addressed in Chapter 16 (Section 16.7.3). Typically, in the above example the question is: how to choose the most informative AA sources among the three first extracted components of Fig. 18.15(b)? The authors solve this problem by exploiting some prior information about spectral content of AA during an AF episode. Indeed, the AA signal exhibits a narrow-band spectrum with a main frequency peak, f_p, between 3.5-9 Hz [50]. Thus, they apply a spectral analysis over all the sources with sub-Gaussian distribution (kurtosis < 0) and choose the first component of Fig. 18.15(b) as the AA source because it presents a major peak at frequency $f_p = 6.31$ Hz.

The previous study only exploits the spatial diversity introduced by the different placement of the electrodes on the body. However, the exploitation of the temporal correlation of the AA source may improve the analysis of AF episode. Based on the narrow-band character of the AA, Castells *et al.* [12] propose a spatio-temporal blind source separation technique which takes advantage of both the spatial and the temporal information contained in the 12-lead ECG. This technique (Fig. 18.16) consists of an initial ICA [38], which aims at removing the non-Gaussian interferences such as VA sources. The second step, based on the SOBI algorithm [7], exploits the narrow-band nature of the AA source in order to improve its extraction from near-Gaussian source subspaces [102]. To evaluate the spatio-temporal ICA method, the authors simulate the synthesized ECG with a known AA source and then validate the results on 14 real-life ECGs AF data. The results obtained on the simulated database demonstrate the effectiveness of the spatio-temporal ICA method in comparison with spatial ICA [73]. Indeed, the correlation coefficients, between the real AA and the extracted AA, lie between 0.75 and 0.91, whereas the correlation coefficients of the spatial ICA method lie between 0.64 and 0.80. Regarding the real-life data, Fig. 18.17(a) shows that the estimated AA source provided by the spatio-temporal ICA method (bottom) seems to be less noisy in comparison to the AA source obtained by the spatial ICA technique (top). Figure 18.17(b) displays the spectral concentration around the main frequency peak, $f_p = 6.31$ Hz, for the whole real-life database. The spectral concentration is computed as the percentage of the signal power in the

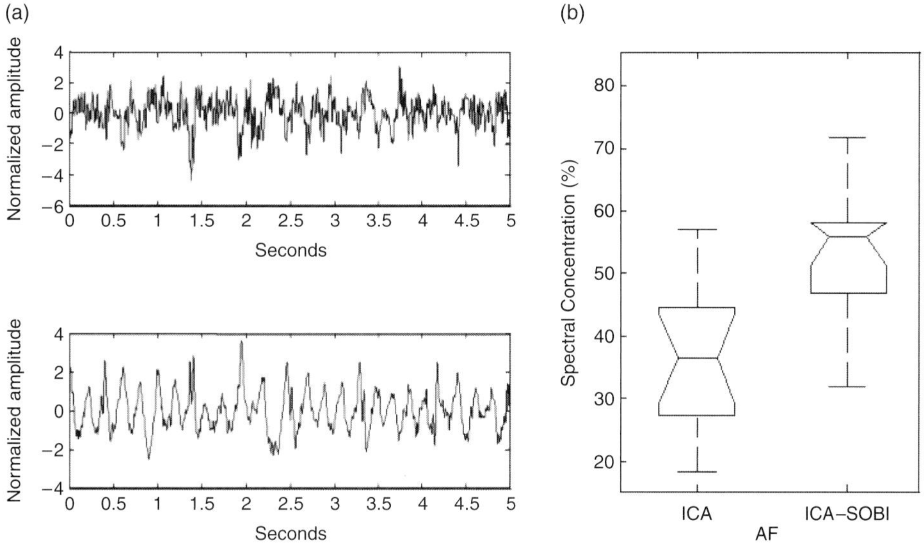

FIGURE 18.17

(a) An example where the spatio-temporal (ICA-SOBI) approach outperforms the spatial ICA and (b) spectral concentration of AA for AF.
Source: From [12] with permission.

frequency interval $[0.82f_p, 1.17f_p]$. According to this criterion, the spatio-temporal ICA method clearly outperforms the spatial-only ICA method.

18.2.3 Other application fields

It would be regrettable to close this bibliographical survey without citing other application fields such as functional Magnetic Resonance Imaging (fMRI) where ICA was used, in particular since fMRI was one of the first biomedical fields explored by ICA methods [60,61]. In fact, fMRI is a technique that provides the opportunity to study brain function non-invasively and has been utilized in both research and clinical areas since the early 1990s. The most popular technique utilizes Blood Oxygenation Level Dependent (BOLD) contrast, which is based on the differing magnetic properties of oxygenated (diamagnetic) and deoxygenated (paramagnetic) blood. When brain neurons are activated, there is a resultant localized change in blood flow and oxygenation which causes a change in the MR decay parameter. These blood flow and oxygenation (vascular or hemodynamic) changes are temporally delayed relative to the neural firing, a confounding factor known as hemodynamic lag. Since the hemodynamic lag varies in a complex way from tissue to tissue, and because the exact transfer mechanism between the electrical and hemodynamic processes is not known, it is not possible to completely recover the electrical process from the vascular process. Nevertheless, the vascular process remains an informative surrogate for electrical activity. However, relatively low image contrast-to-noise ratio of the BOLD effect, head movement, and undesired physiological sources of variability (cardiac, pulmonary) make detection of the activation-related signal changes difficult. ICA has shown to be useful for fMRI analysis for several reasons. ICA finds systematically non-overlapping, temporally coherent brain regions without constraining the temporal domain. The temporal dynamics of many fMRI experiments are difficult to study with functional magnetic resonance imaging (fMRI) due to the lack of a well-understood brain-activation model. ICA can reveal inter-subject and inter-event differences in the temporal dynamics. A strength of ICA is its ability to reveal dynamics for which a temporal model is not available. ICA also works well for fMRI as it is often the case that one is interested in spatially distributed brain networks. A more exhaustive study of ICA for fMRI is proposed by Calhoun *et al.* [8].

Another, but not the last, application of ICA in biomedical engineering is its use in Magnetic Resonance Spectroscopy (MRS) contexts [70]. MRS is a recent diagnosis method that was adopted into clinical practice. As explained in [75], it consists in measuring different biochemical markers by tuning to particular nuclear resonance frequencies, thus providing precise characterization of tissue and/or a means for optimizing the SNR. It allows for the non-invasive characterization and quantification of molecular markers with clinical utility for improving detection, identification, and treatment for a variety of diseases, most notably brain cancers. The interpretation of MRS data is quite challenging: a typical dataset consists of hundreds of spectra, typically having low SNR with numerous and overlapping peaks. More particularly, the observed spectra are a combination of different constituent spectra. ICA can therefore be used to identify the shapes of the underlying constituent source spectra and their contribution to each voxel in the MRS data. Its use aims at leading to a better quantification of each reasonance peak.

The third application of ICA concerns Electromyographic (EMG) signals. These EMG signals acquired from skin electrodes may be generated by different muscles. EMG may overlap in the time and frequency domains, which makes the classical filtering approaches less appropriate for EMG data separation. Farina *et al.* [28] show that, under certain assumptions, surface EMG may be considered as a linear instantaneous mixture. Experimental signals were collected by considering two muscles, the flexor carpi radialis and the pronator teres. The choice of these two muscles is motivated by the fact that it is possible to produce contractions in which only one muscle is active at a time. The ICA method, proposed in [30], is then applied on three surface observations. The obtained results show that the source separation is not perfectly reached. The authors explain this limitation by the fact that the linear instantaneous mixture is only appropriate in the case of small muscles placed close to each other. Nevertheless, the reported results are still very interesting and show that ICA is a promising approach for surface EMG signal separation: the correlation coefficients between the separated sources and the reference sources (obtained by the arrays located directly over the muscles) is higher than 0.9.

18.3 NUMERICAL COMPLEXITY OF ICA ALGORITHMS

Although the ultimate goal of a signal separation approach is the quality of such a separation, reflected on the estimated source signals, it is interesting to compare the various

ICA approaches from a numeral complexity viewpoint. Numerical complexity is defined here as the number of floating point operations required to execute an algorithm (flops). A flop corresponds to a multiplication followed by an addition. But, in practice, only the number of multiplications is considered since, most of the time, there are about as many (and slightly more) multiplications as additions. In order to simplify the expressions, the complexity is generally approximated by its asymptotic limit, as the size of the problem tends to infinity. We shall subsequently denote, with some small abuse of notation, the equivalence between two strictly positive functions f and g:

$$f(x) = \mathcal{O}[g(x)] \text{ or } g(x) = \mathcal{O}[f(x)] \tag{18.1}$$

if and only if the ratio $f(x)/g(x)$ tends to 1 as $x \to \infty$. In practice, knowing whether or not an algorithm is computationally costly becomes as important as knowing its performances in terms of SNR. Yet, despite its importance, the numerical complexity of the ICA algorithms is poorly addressed in the literature. This section first addresses the complexity of some elementary mathematical operations needed by ICA algorithms. Then, the numerical complexity of various ICA algorithms is reported with comparisons to each other as a function of the number of sources.

18.3.1 General tools

Many ICA algorithms use standard eigen-value decomposition (EVD) or singular value decomposition (SVD), for instance when a whitening step is required to reduce the dimensions of the space. In addition to the latter decompositions, many operations can be considered as elementary such as solving a linear system, matrix multiplication, joint diagonalization of several matrices and computation of cumulants, when cumulant-based algorithms are considered.

- Let $\mathbf{A}$ and $\mathbf{B}$ be two matrices of size $(P \times N)$ and $(N \times P)$, respectively. Then the numerical complexity of their product $\mathbf{G} = \mathbf{AB}$ is equal to P^2N flops, since each element of $\mathbf{G}$ requires N flops to be computed. The latter amount can be reduced to $(P^2 + P)N/2 = \mathcal{O}[P^2N/2]$ flops if $\mathbf{G}$ is symmetric.
- The solution of a $P \times P$ linear system via the LU decomposition requires approximately $\mathcal{O}[4P^3/3]$ flops.
- The numerical complexity of the SVD of $\mathbf{A} = \mathbf{U}\mathbf{\Lambda}\mathbf{V}^{\mathrm{T}}$ is given by $\mathcal{O}[2P^2N+4PN^2+14N^3/3]$ flops when it is computed using the Golub-Reinsch algorithm [33]. This amount can be considerably reduced to $\mathcal{O}[2P^2N]$ when $\mathbf{A}$ is tall (i.e. $P \gg N$) by resorting to Chan's algorithm [13], known to be suitable in such a case.
- The numerical complexity of the EVD $\mathbf{G} = \mathbf{L}\mathbf{\Sigma}\mathbf{L}^{\mathrm{T}}$ is $\mathcal{O}[4P^3/3]$ flops.

Based on the previous expressions, the numerical SO whitening procedure is equal to $\mathcal{O}[TP^2/2 + 4P^3/3 + NPT]$ flops when it is achieved using the EVD while it is equal to $\mathcal{O}[2TP^2]$ flops when the SVD is used.

Some ICA algorithms [7,105,10,51,30] are based on the joint approximate diagonalization of a set of M matrices $\mathbf{G}_m$ $(1 \leq m \leq M)$ of size $(P \times P)$. Recall that the joint diagonalization problem is defined as the search for a linear transformation that jointly diagonalizes the target matrices $\mathbf{G}_m$. Two main classes of joint diagonalization techniques can be distinguished: the orthogonal and the non-orthogonal methods. Orthogonal joint diagonalization is defined when the diagonalizing matrix is unitary while the non-orthogonal one only requires invertibility. A Jacobi-like algorithm such as the JAD algorithm [11] is commonly used for joint orthogonal diagonalization. Its numerical complexity is equal to $IP(P-1)(4PM+17M+4P+75)/2$ flops if the M matrices $\mathbf{G}_m$ are symmetric where I stands for the number of executed sweeps. On the other hand, the ACDC algorithm [100] is a good choice when non-orthogonal joint diagonalization is concerned. The numerical complexity of the latter is given, for a full sweep AC phase with a single interlacing with the DC one, by $MNP^2+2P^3/3+(3P-1)/3+4N^3/3+(M+1)N^2P+(M+1)N^2+MNP^2$ flops per iteration.

Finally, regarding cumulants estimation, the computation of the $2q$-th order cumulant of a P-dimensional random process requires $(2q-1)T$ flops where T stands for the data length. Consequently, the number of flops required to compute one $2q$-th order cumulant array utilizing all its symmetries is then given by $(2q-1)Tf_{2q}(P)$ flops where $f_{2q}(P)$ denotes the number of its free entries and is given as a function of P, for $q=1,2,3$, by:

$$f_2(P) = \frac{P^2+P}{2} = \mathcal{O}\left[\frac{P^2}{2}\right] \tag{18.2}$$

$$f_4(P) = \frac{1}{8}P(P+1)(P^2+P+2) = \mathcal{O}\left[\frac{P^4}{8}\right] \tag{18.3}$$

$$f_6(N) = \frac{P^6}{72}+\frac{P^5}{12}+\frac{13P^4}{72}+\frac{P^3}{4}+\frac{22P^2}{72}+\frac{P}{6} = \mathcal{O}\left[\frac{P^6}{72}\right] \tag{18.4}$$

Table 18.1 summarizes the numerical complexities of the elementary operations considered in this chapter.

18.3.2 Complexity of several ICA algorithms

This section aims at giving insights into the numerical complexity of twelve ICA-based algorithms evaluated in this chapter, as a function of the number N of sources, the number P of sensors and the data length T. As mentioned in the introduction, these algorithms are JADE, COM2, SOBI, SOBI_{rob}, TFBSS, InfoMax, PICA, $\text{FastICA}_{\text{sym}}$, $\text{FastICA}_{\text{def}}$, $\text{FOBIUM}_{\text{ACDC}}$, SOBIUM and STOTD. Their numerical complexity is given in Table 18.2.

For a given number N of sources, the minimal numerical complexity of the previous methods is obtained by minimizing the values of I,P,M and T provided a good extraction of the sources is guaranteed. A good rule of thumb is to use at least $I_{\text{min}} = 1+\text{floor}(N^{1/2})$ sweeps as in [19]. The minimum value of P is equal to $P_{\text{min}} = N$ for all the methods considered even if SOBIUM and $\text{FOBIUM}_{\text{ACDC}}$ are able to identify underdetermined (i.e. $N > P$) mixtures of sources. We refer to Chapter 9 for a treatment

Table 18.1 Numerical complexity of elementary operations generally used in the ICA methods. $\mathbf{A}$ and $\mathbf{B}$ are two matrices of size $(P \times N)$ and $(N \times P)$, respectively. I and M stand for the number of executed sweeps and the number of matrices to be jointly diagonalized, respectively. $f_{2q}(P)$ denotes the number of free entries in the $2q$-th order cumulant array. J_1 is the number of iterations required for the convergence of the ACDC algorithm

	Numerical complexity (flops)
$\mathbf{G} = \mathbf{AB}$	P^2N
Lin. system solving	$4P^3/3$
SVD of $\mathbf{A}$	$2P^2N + 4PN^2 + 14N^3/3$
EVD of $\mathbf{A}$	$4P^3/3$
JAD [11] (Symmetric case)	$IP(P-1)(4PM + 17M + 4P + 75)/2$
ACDC [100]	$(MNP^2 + 2P^3/3 + (3P-1)/3 + 4N^3/3 + (M+1)N^2P + (M+1)N^2 + MNP^2)J_1$
Estimation of the $2q$-th order cumulants array	$(2q-1)Tf_{2q}(P)$

of under-determined mixtures. The minimum value of M is chosen equal to $M_{\min} = 6$ for SOBIUM, SOBI, SOBI_{rob} and $\text{FOBIUM}_{\text{ACDC}}$. The minimum value of T depends on several parameters such as N, P, the FO marginal cumulants and the SNR of sources. Therefore, it is chosen to be the same for all methods.

In summary, it is difficult to compare computational complexity across algorithms because the input parameters are different. But it is still possible if a common source extraction quality is imposed, which yields an estimation of the latter parameters. Another possibility would have been to impose a common overall numerical complexity for all methods as in [101, Fig. 1], and to look at performances.

In our experiment, these parameters are empirically estimated from over 200 realizations in the context of the Mu-based BCI system described in section 18.4 when $P = 6$ sensors are used to recover $N = 2$ sources from $T = 10000$ data samples and for an SNR of 5 dB. Figure 18.18 shows the variations of the minimal numerical complexity of the twelve methods as a function of the number of sources N. In the sequel and for the sake of readability, all the methods considered are classified into three categories to be depicted in all figures: (i) SO statistics methods (TFBSS, SOBIUM, SOBI and SOBI_{rob}), (ii) HO cumulant-based methods ($\text{FOBIUM}_{\text{ACDC}}$, STOTD, COM2 and JADE) and (iii) iterative MI-based methods ($\text{FastICA}_{\text{def}}$, $\text{FastICA}_{\text{sym}}$, PICA and InfoMax). The average number of iterations used in the iterative methods is equal to $6, 3, 17, 476$ and 12 for $\text{FastICA}_{\text{def}}$, $\text{FastICA}_{\text{sym}}$, InfoMax, $\text{FOBIUM}_{\text{ACDC}}$ and PICA, respectively. Regarding the TFBSS method, the number of used matrices M_1 and M_2 is equal to 8446 and 139 respectively while both the used number of time N_t and frequency N_f bins are set to $T/2$ with a smoothing window of size $N_f/10$. As depicted in Fig. 18.18, the TFBSS method requires generally the largest number of calculations compared to the other methods followed by $\text{FastICA}_{\text{def}}$. Regarding the SOBIUM method, it shows an equivalent

Table 18.2 Numerical complexity of some ICA-based algorithms

	Computational complexity
	P: number of sensors, N: number of sources, $J_i, i \in \{1,\ldots,6\}$ is the number of iterations in the i-th iterative method. Q is the complexity required to compute the roots of a real 4th degree polynomial by Ferrari's technique in the COM2 algorithm. L_w, N_t, N_f, M_1 and M_2 are respectively the smoothing window's length, the number of time bins, the number of frequency bins, the number of matrices referred to the time-frequency point wherein sources are of significant energy, and the number of matrices among those M_1 ones with only one active source in the considered time-frequency point in the TFBSS algorithm. As far as the SOBIUM2 algorithm is concerned, m_1, n_1, m_2 and n_2 denote $\max(P^2,M), \min(P^2,M), \max(P^4,N(N+1)/2)$ and $\min(P^4,N(N+1)/2)$, respectively.
SOBI_{rob}	$MTP^2/2+5M^2P^3-M^3P^3/3+2MP^2N+MN^2P+MN^2+(MN^2+4N^3/3)J_1+MN+MP^2+2P^3/3+PN+(3P-N)N^2/3+IN(N-1)(17M+75+4N+4NM)/2$
SOBI	$MTP^2/2+4P^3/3+(M-1)P^3/2+IN(N-1)(17(M-1)+75+4N+4N(M-1))/2$
TFBSS	$\min(TP^2/2+4P^3/3+NPT+N,2TP^2)+2N\log_2 N+N+(T+L_w+\log_2(L_w))N_tN_fN(N+1)/2+2M_1N^3/3+3T_2+IN(N-1)(4NM_2+17M_2+4N+75)/2$
SOBIUM	$MTP^2/2+\min(7m_1n_1^2+11n_1^3/3,3m_1n_1^2)+2P^2N+P^2N^2+N(N-1)(4N^3/3+N^4(N-1))/2+2N(N+1)P^4+(2P^3/3+(3P-1)/3)N+\min(7m_2n_2^2+11n_2^3/3,3m_2n_2^2)$
COM2	$\min(TP^2/2+4P^3/3+NPT,2TP^2)+IN^2Q/2+\min(12If_4(N)N^2+2IN^3+3Tf_4(N)+TN^2,13ITN^2/2)$
JADE	$\min(TP^2/2+4P^3/3+NPT,2TP^2)+3Tf_4(N)+TN^2+\min(4N^6/3,8N^3(N^2+3))+IN(N-1)(75+21N+4N^2)/2$
STOTD	$\min(TP^2/2+4P^3/3+NPT,2TP^2)+3TN^4/8+12IN^2(N^2-1)$
$\text{FOBIUM}_{\text{ACDC}}$	$3TMf_4(P)+(2P^6/3+2MNP^4+(M+1)N^2P^2+MN^2+4N^3/3+N^2+P^2)J_2+NIP(P-1)(4P^2+21P+75)/2$
PICA	$\min(TP^2/2+4P^3/3+NPT,2TP^2)+(N^3+(T+1)N^2+3NT)J_3$
InfoMax	$\min(TP^2/2+4P^3/3+NPT,2TP^2)+(N^2+N^3+4N+5TN)J_4$
$\text{FastICA}_{\text{def}}$	$\min(TP^2/2+4P^3/3+NPT,2TP^2)+(2(N-1)(N+T)+5TN(N+1)/2)J_5$
$\text{FastICA}_{\text{sym}}$	$\min(TP^2/2+4P^3/3+NPT,2TP^2)+2P^3/2+(16N^3/3+N^2+3TN^2)J_6$

computational complexity compared to those of SOBI and SOBI_{rob} up to $N=6$, while an important increase in its complexity can be noted beyond $N=6$. $\text{FOBIUM}_{\text{ACDC}}$ seems to be more costly than STOTD, COM2 and JADE where they show a comparable number of calculations especially for $P<10$. Finally, InfoMax, PICA and $\text{FastICA}_{\text{sym}}$ show a comparable number of calculations.

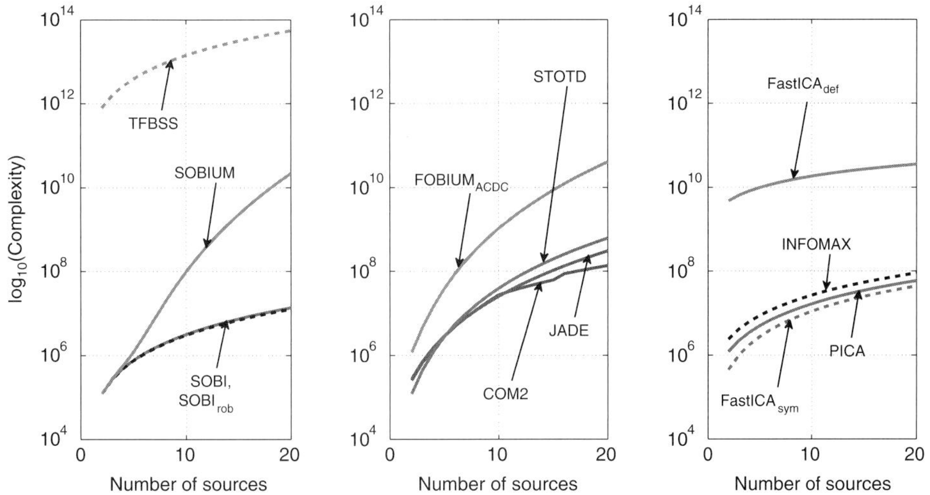

FIGURE 18.18

Minimal numerical complexity as a function of the number of sources for ten thousand data samples and SNR of 5 dB.

18.4 PERFORMANCE ANALYSIS FOR BIOMEDICAL SIGNALS

The goal of this section is to compare the performance of ICA methods w.r.t realistic biomedical signals. To do so, we will simulate EEG signals to be used for a brain-computer interface (BCI) system based on the Mu-rhythm with seven surface electrodes located around sensorimotor cortex (Fig. 18.19(a)). ICA-based BCI systems are now of great research interest thanks to the potential of interpreting brain signals in real-time. So far only a limited set of ICA algorithms has been explored for this application [47], namely FastICA and InfoMax, whereas many other ICA-based algorithms could be used. This point is investigated by comparing the twelve aforementioned ICA algorithms, namely SOBI, $SOBI_{rob}$, COM2, JADE, TFBSS, $FastICA_{def}$, $FastICA_{sym}$, InfoMax, PICA, the STOTD and $FOBIUM_{ACDC}$.

In such a context the surface observations can be considered as a noisy mixture of one source of interest, namely the Mu-rhythm, and artifact sources such as the ocular activity. The intracerebral Mu wave located in the motor cortex (Fig. 18.19(b)) is simulated using the parametric model of Jansen [42] whose parameters are selected to derive a Mu-like activity. The ocular signal is obtained from our database. As far as the additive noise is considered, it is modeled as the sum of the instrumental noise and the background EEG activity. A Gaussian vector process is used to simulate the instrumental noise while a brain volume conduction of 800 independent EEG sources is generated using the Jansen model [42] in order to simulate a surface background EEG activity. Finally, the mixing matrix is defined as the concatenation of two columns modeling the head volume conduction [1] of the Mu and the ocular activities.

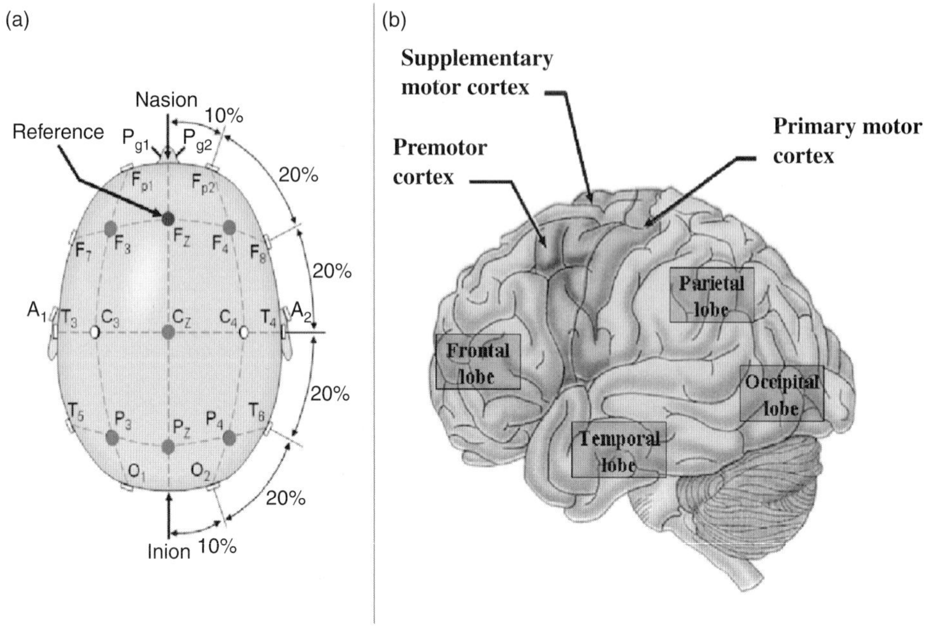

FIGURE 18.19

(a) The international 10-20 electrode placement system and (b) the surface of the left cerebral hemisphere, viewed from the side.

18.4.1 Comparative performance analysis for synthetic signals

Two studies are considered hereafter to evaluate the twelve ICA algorithms we have tested in the context of Mu-based BCI systems. In the first study, the quality of the source extraction is evaluated, as a function of the data samples for different SNR values. In the second study, the behavior of the ICA methods is examined in the case of an overestimation of the number of sources, for a fixed data length of $T = 10000$ and different SNR values. Moreover, the InfoMax algorithm is implemented with a prewhitening step. All reported results are obtained by averaging the performance criterion presented hereafter over 200 realizations. Note that a new trial of both sources and noise is generated at each realization.

18.4.1.1 Performance criterion

Two separators, $\mathbf{B}^{(1)}$ and $\mathbf{B}^{(2)}$ can be compared with the help of the criterion introduced by Chevalier [15]. The quality of the extracted component is directly related to its Signal to Interference-plus-Noise Ratio (SINR). More precisely, the SINR of the n-th source at the i-th output of the separator $\mathbf{B} = [\mathbf{b}_1, \ldots, \mathbf{b}_N]$ is defined by:

$$\mathrm{SINR}_n(\mathbf{b}_i) = \pi_n \frac{|\mathbf{b}_i{}^{\mathrm{T}} A_n|^2}{\mathbf{b}_i{}^{\mathrm{T}} \mathbf{R}_{\mathbf{x}}^{(n)} \mathbf{b}_i} \tag{18.5}$$

where π_n represents the power of the n-th source, $\mathbf{b}_i$ the i-th column of the separator $\mathbf{B}$ and $\mathbf{R}_{\mathbf{x}}^{(n)}$ is the total noise covariance matrix for the n-th source, corresponding to the estimated data covariance matrix $\mathbf{R}_{\mathbf{x}}$ in the absence of the n-th source. On the basis of these definitions, the reconstruction quality of the n-th source at the output of the separator $\mathbf{B}$ is evaluated by computing the maximum of $\text{SINR}_n(\mathbf{b}_i)$ with respect to i where $1 \leq i \leq N$. This quantity is denoted by $\text{SINRM}_n(\mathbf{B})$. The performance of a source separator $\mathbf{B}$ is then defined by the following line vector SINRM($\mathbf{B}$):

$$\text{SINRM}(\mathbf{B}) = (\text{SINRM}_1(\mathbf{B}), \ldots, \text{SINRM}_N(\mathbf{B})). \tag{18.6}$$

In a given context, a separator $\mathbf{B}^{(1)}$ is better than another $\mathbf{B}^{(2)}$ for retrieving the source n, provided that $\text{SINRM}_n(\mathbf{B}^{(1)}) > \text{SINRM}_n(\mathbf{B}^{(2)})$. The criterion given by Eq. (18.6) allows for a quantitative performance evaluation and comparison of various ICA algorithms. However, the use of this criterion requires knowledge of its upper bound, which is achieved by the optimal source separator, in order to completely evaluate the performance of a given ICA method. It is shown in [15] that the optimal source separator corresponds to the separator $\mathbf{B}$(SMF) whose columns are the Spatial Matching Filters (SMF) associated with the different sources. It is defined to within a diagonal matrix and a permutation by $\mathbf{B}(\text{SMF}) = \mathbf{R}_{\mathbf{x}}^{-1}\mathbf{A}$ where $\mathbf{A}$ is the true mixture.

18.4.1.2 Impact of both the SNR and the number of samples

Four experiments are realized in order to evaluate the behavior of the twelve ICA methods as a function of the data samples for different SNR values, i.e. SNR $= -5, 5, 15$ and 25 dB. As depicted in Figs 18.20–18.23, the SOBI_{rob} method exhibits, generally, quasi-optimal performance in extracting both sources (i.e. less than 1 dB from the optimal SINRM). Regarding the COM2, JADE, TFBSS, SOBI, $\text{FastICA}_{\text{def}}$, $\text{FastICA}_{\text{sym}}$, InfoMax, PICA and STOTD methods, their performance is comparable with a gap of 2 dB, approximately, from the optimal SINRM for SNR values less than or equal to 15 dB. This gap is reduced for SNR of 25 dB and for sufficient data samples as shown in Fig. 18.23. Sometimes, the PICA algorithm suffers from convergence problems especially for an SNR of 5 dB as depicted in Fig. 18.21, which is probably due to the insufficiently small stop criterion, i.e. 10^{-4}. As far as the SOBIUM algorithm is concerned, its behavior seems to be comparable with the one of the SOBI_{rob} method for ocular activity extraction (i.e. SINRM_1). However, it shows similar behavior to the ones of COM2, JADE, $\text{FastICA}_{\text{def}}$ and $\text{FastICA}_{\text{sym}}$ for the Mu wave extraction, except for an SNR of -5 dB, which seems to strongly affect it. Finally, the $\text{FOBIUM}_{\text{ACDC}}$ algorithm shows a very good performance in extracting the ocular activity, especially for low SNR values (Figs 18.20 and 18.21) and small data samples where, in such a case, it outperforms the classical ICA methods such as COM2, JADE, $\text{FastICA}_{\text{sym}}$ and $\text{FastICA}_{\text{def}}$. A comparable performance, but with a slower convergence speed, can be noted for the Mu wave extraction provided that the SNR is not too low because for an SNR of -5 dB, $\text{FOBIUM}_{\text{ACDC}}$ shows a poor performance as compared to SOBIUM.

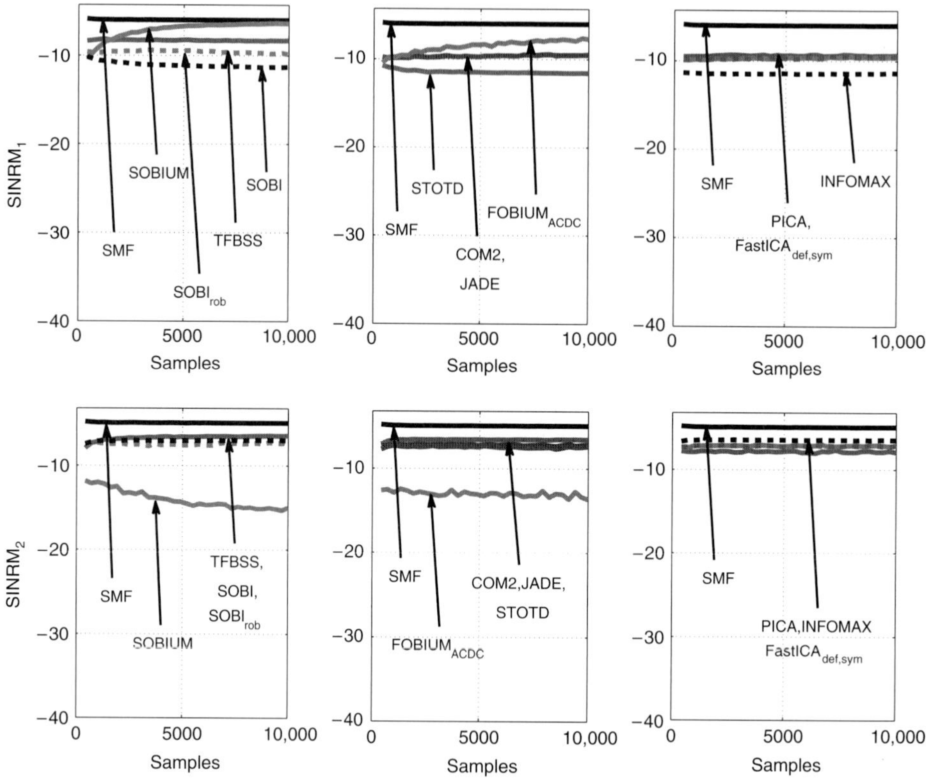

FIGURE 18.20

SINRM_n for $n \in \{1,2\}$ as a function of the data samples with an SNR of −5 dB, at the output of twelve ICA methods in the context of Mu-based BCI systems. SINRM_1 and SINRM_2 express respectively the extraction quality measure of the ocular and Mu activities compared to the optimal SMF filter. (i) SO-cumulant based methods (left column), (ii) HO cumulant-based methods (middle column) and (iii) iterative MI-based methods (right column).

18.4.1.3 Impact of the overestimation of the number of sources

Since the estimation of the number of sources, when it is unknown, is essential in ICA methods, it should be interesting to examine the behavior of the latter when an overestimation of the number of sources occurs. Therefore, two experiments are realized showing the performance of the twelve methods as a function of the estimated number of sources, named N_{est}, for a true number of sources equal to 2 and for different SNR values, i.e. 5 dB and 25 dB.

For an SNR of 5 dB, Fig. 18.24 shows that the methods $\mathrm{SOBI}_{\mathrm{rob}}$, SOBI, COM2, JADE, TFBSS, $\mathrm{FastICA}_{\mathrm{sym}}$, $\mathrm{FastICA}_{\mathrm{def}}$ and STOTD, are insensitive to the overestimation of the number of sources. As far as the InfoMax and the $\mathrm{FOBIUM}_{\mathrm{ACDC}}$ algorithms are concerned, their behaviors are comparable, for ocular activity extraction, to those of

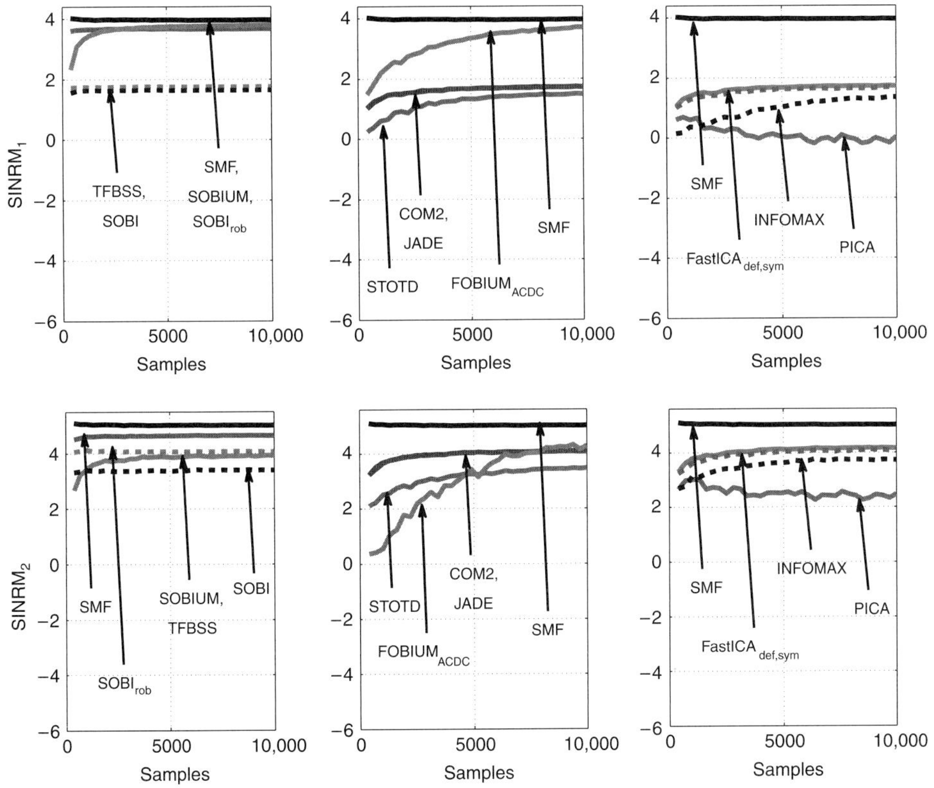

FIGURE 18.21

SINRM_n for $n \in \{1,2\}$ as a function of the data samples with an SNR of 5 dB, at the output of twelve ICA methods in the context of Mu-based BCI systems. SINRM_1 and SINRM_2 express respectively the extraction quality measure of the ocular and the Mu activities compared to the optimal SMF filter. (i) SO-cumulant based methods (left column), (ii) HO cumulant-based methods (middle column) and (iii) iterative MI-based methods (right column).

the latter methods. In spite of this, they seem to be less effective when the Mu activity extraction (i.e. SINRM_2) is concerned, especially for the InfoMax method, which shows a higher sensitivity. On the other hand, the performance of PICA can be considered as biased for an overestimation of the number of sources contrary to the latter methods, especially for the iterative ones such as $\mathrm{FastICA}_{\mathrm{sym}}$, $\mathrm{FastICA}_{\mathrm{def}}$, as shown in Fig. 18.24. Moreover, a comparable performance with the one corresponding to InfoMax can be noted for the PICA method, especially for Mu wave extraction. This behavior is probably due to the bias estimation of the moments required for the estimation of the Pearson model's parameters or to the high presence of noise which was probably extracted instead of the source of interest. Regarding the SOBIUM algorithm, it is the most affected by the overestimation of the number of sources, as shown in Fig. 18.24. Such behavior may be explained by the miss-estimation of the subspace spanned by the

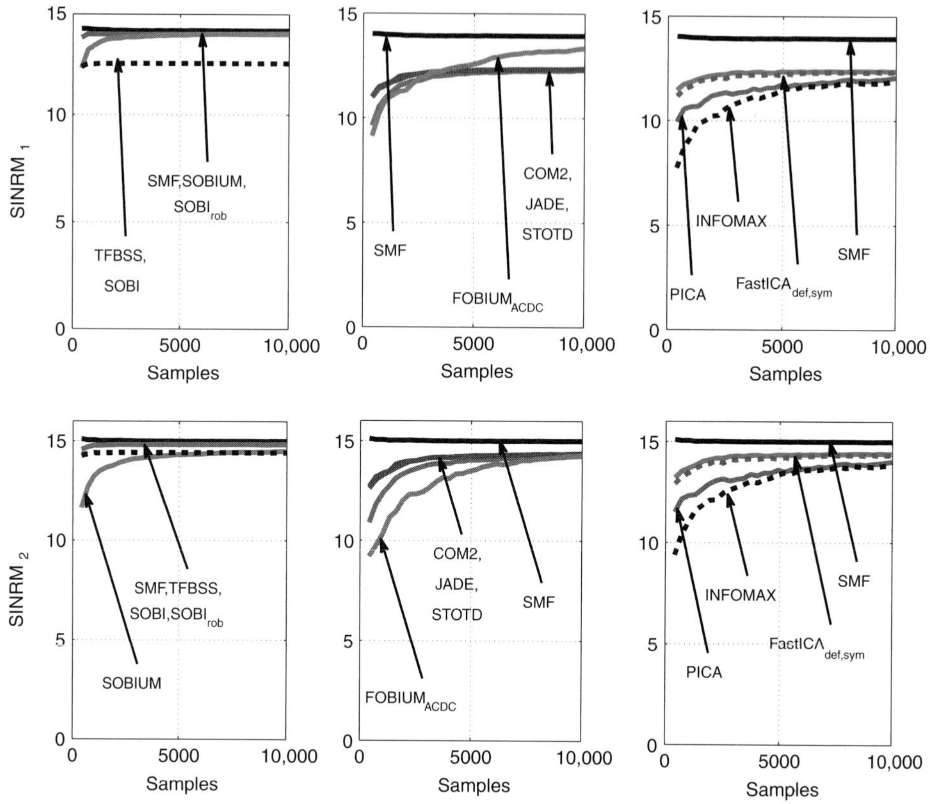

FIGURE 18.22

SINRM_n for $n \in \{1,2\}$ as a function of the data samples with an SNR of 15 dB, at the output of twelve ICA methods in the context of Mu-based BCI systems. SINRM_1 and SINRM_2 express respectively the extraction quality measure of the ocular and the Mu activities compared to the optimal SMF filter. (i) SO-cumulant based methods (left column), (ii) HO cumulant-based methods (middle column) and (iii) iterative MI-based methods (right column).

matrices required for the joint diagonalization or, as for the PICA algorithm, by the high presence of noise that probably hides the source of interest. Thus, a good way to circumvent such a problem for both the SOBIUM and the PICA algorithms, is to extract as many sources as sensors. Such a solution guarantees the extraction of the sources of interest and enhances the source extraction quality as depicted in Figs 18.24 and 18.25.

Finally, for an SNR of 25 dB, SOBI_{rob}, SOBI, PICA, SOBIUM, $\text{FOBIUM}_{\text{ACDC}}$, InfoMax show a quasi-similar behavior compared to the case of an SNR of 5 dB. But, contrary to the previous case, COM2, STOTD, $\text{FastICA}_{\text{sym}}$, $\text{FastICA}_{\text{def}}$, JADE and TFBSS methods seem to be slightly sensitive to the overestimation of the number of sources, as depicted in Fig. 18.25 but with some superior performances of the COM2 and STOTD algorithms with respect to both JADE and TFBSS.

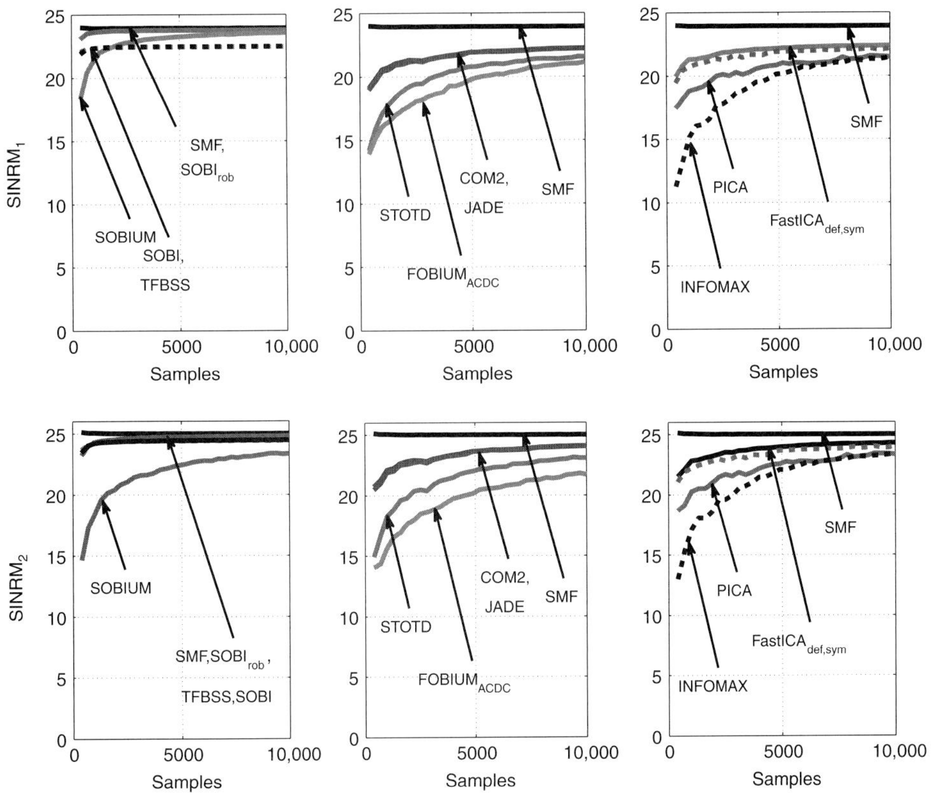

FIGURE 18.23

SINRM_n for $n \in \{1,2\}$ as a function of the data samples with an SNR of 25 dB, at the output of twelve ICA methods in the context of Mu-based BCI systems. SINRM_1 and SINRM_2 express respectively the extraction quality measure of the ocular and the Mu activities compared to the optimal SMF filter. (i) SO-based methods (left column), (ii) HO cumulant-based methods (middle column) and (iii) iterative MI-based methods (right column).

18.4.1.4 Summary of both studies

This section gives a summary of the computer results obtained from both previous studies. The main conclusions are as follows:

- SOBI_{rob} is the most powerful method compared to the other methods considered, especially those based on SO statistics. Its quasi-optimal performance (when its requested assumptions are satisfied) is not limited to source extraction with moderate numerical complexity, but also to its insensitivity to the overestimation of the number of sources.
- Regarding the HO cumulant-based methods, STOTD, COM2 and JADE show a good performance with moderate numerical complexity. One can notice some weakness of the JADE algorithm with respect to an overestimation of the number

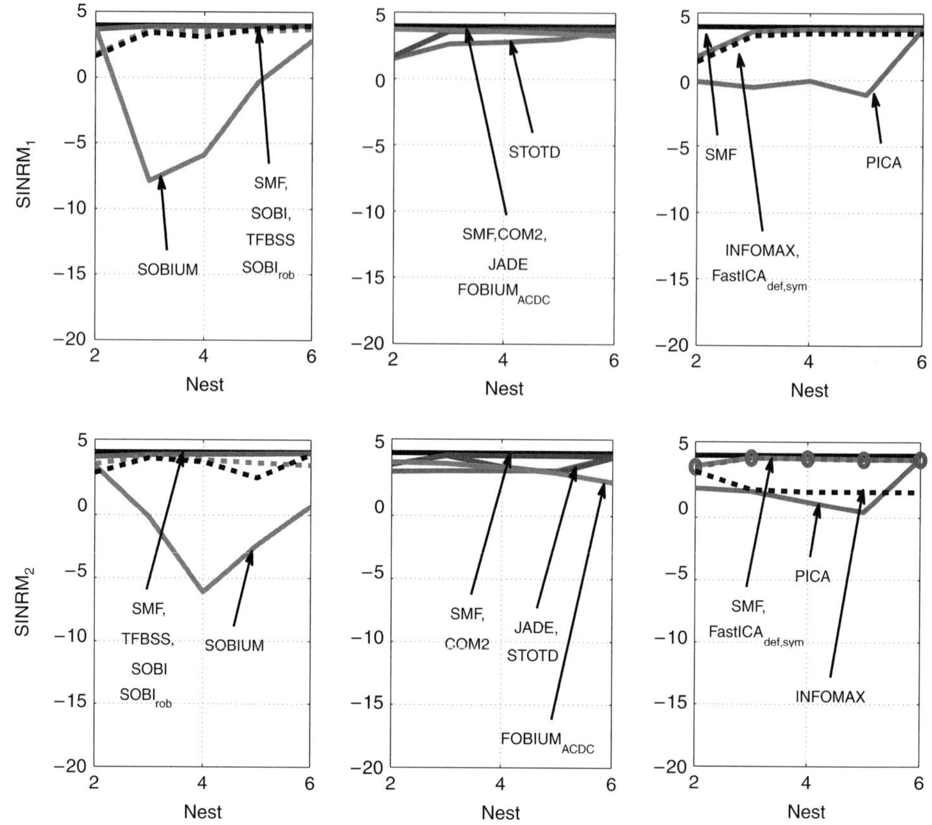

FIGURE 18.24

SINRM$_n$ for $n \in \{1,2\}$ as a function of the estimated number of sources for 10,000 data samples and with an SNR of 5 dB at the output of twelve ICA methods in the context of Mu-based BCI systems. SINRM$_1$ and SINRM$_2$ express respectively the extraction quality measure of the ocular and the Mu activities compared to the optimal SMF filter. (i) SO-cumulant based methods (left column), (ii) HO-cumulant based methods (middle column) and (iii) iterative methods (right column).

of sources. On the other hand, one could resort to artifact removal as a preprocessing step (provided reasonable SNR) using the FOBIUM$_{ACDC}$ method, and then apply COM2 or JADE or STOTD. It is worth noting that using reasonably small data samples for artifact removal would be a good trade-off between the performance and the computational complexity of the FOBIUM$_{ACDC}$ method.

- Both FastICA$_{def}$ and FastICA$_{sym}$ show generally a good performance for source extraction over the other iterative MI-based algorithm. But the FastICA$_{sym}$ algorithm should be preferred over the FastICA$_{def}$ when a constraint on the computational burden is imposed. However, iterative algorithms (such as FastICA) always yield a larger cumulated computational burden than SOBI or HO methods, when they are imposed to reach the same performance level.

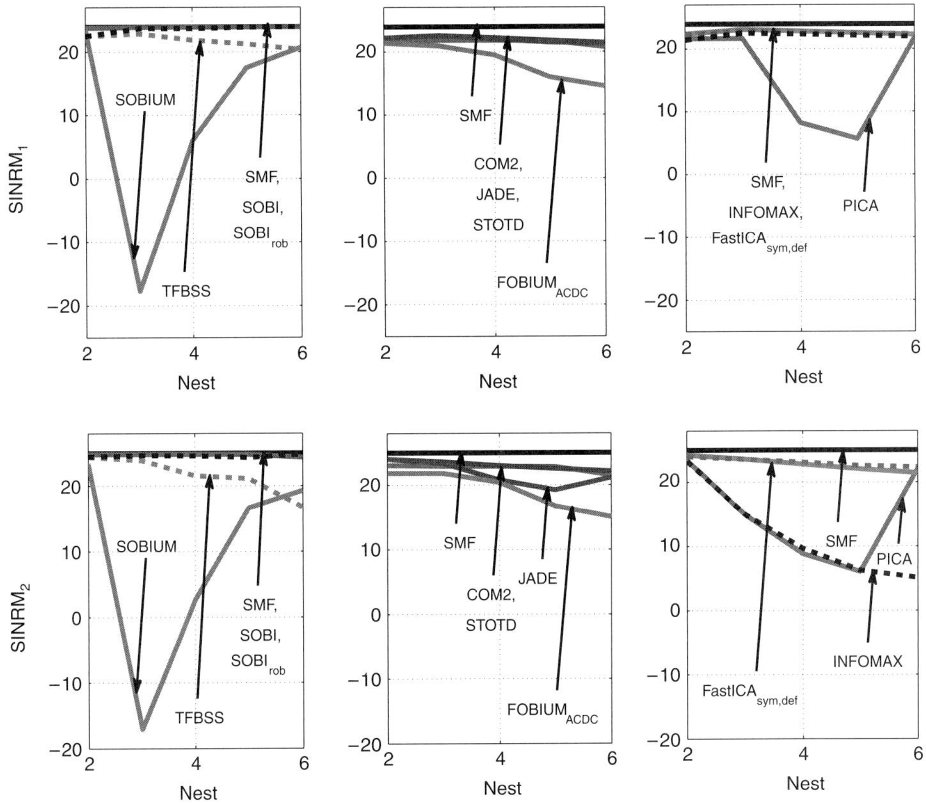

FIGURE 18.25

SINRM_n for $n \in \{1,2\}$ as a function of the estimated number of sources for 10,000 data samples and with an SNR of 25 dB at the output of twelve ICA methods in the context of Mu-based BCI systems. SINRM_1 and SINRM_2 express respectively the extraction quality measure of the ocular and the Mu activities compared to the optimal SMF filter. (i) SO-cumulant based methods (left column), (ii) HO-cumulant based methods (middle column) and (iii) iterative methods (right column).

18.4.2 ICA of real data

In this section, we present an example of using ICA in the real data world. As we have already seen in our experiments, the SOBI_{rob} algorithm seems to be the most powerful method for simulated data. Therefore, it is applied to the observations depicted in Fig. 18.4. These observations present some surface electrical activities recorded by a subset of 13 EEG electrodes, derived from the standardized 10-20 system during a twenty second period with a sampling rate of 256 Hz. Several artifact structures are evident, such as horizontal eye movements, eye blinks and muscle activity. In addition, other disturbances with weaker power, such as electrode movements and baseline drift, contaminate the data. The obtained results, depicted in Fig. 18.26, show that components S3 and S5 represent clearly an activation of eye blinks and horizontal eye movements, respectively. Regarding the components S2 and S4, they represent the muscle activity

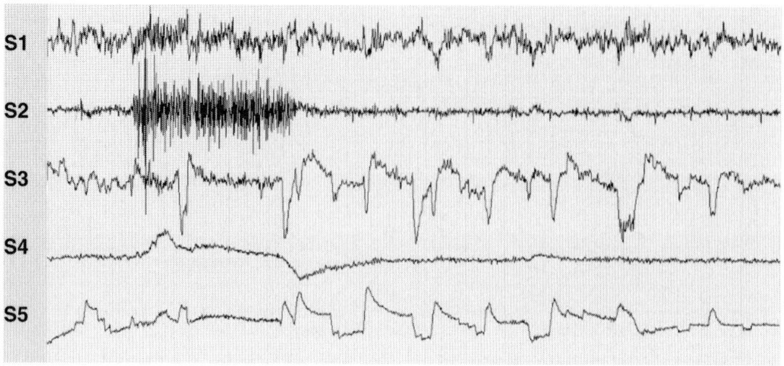

FIGURE 18.26

Result obtained by applying ICA to the signals depicted in Fig. 18.4.

and a slow movement which is probably related to the electrode movement. In addition, S1 displays the cerebral activity, which is considered in this experiment as the signal of interest. However, a slight alteration in S1, caused by eye blinks, still exists, which implies that the source extraction is not well-performed. This example shows that, even if the ICA can not perfectly separate the signal of interest from the artifacts, it guarantees the SNR enhancement.

18.5 CONCLUSION

The strength of ICA in the context of biomedical applications lies in its ability to extract signals despite the lack of reference signals or training labels. When this information is lacking, as is so often the case in biomedical signals, one can not use conventional regression methods to remove noise or extract signals of interest. ICA exploits the spatial diversity; infact independent sources contribute additively to multiple simultaneous recordings. In all applications of ICA it is important to verify that the sources of interest do indeed satisfy these assumptions: diversity, independence, and linearity. For biomedical signals in particular, the use of ICA also imposes on the user the extra burden to demonstrate that the extracted components are physiologically meaningful. This is typically done by comparing the extracted sources to the known properties of the signals of interest, such as their spatial distribution, spectral characteristics, or temporal regularities. This invariably requires the user to have a solid knowledge-based system in the biomedical domain. In summary, one could say that the power of ICA should be used responsibly.

References

[1] L. Albera, A. Ferreol, D. Cosandier-Rimele, I. Merlet, F. Wendling, Brain source localization using a fourth-order deflation scheme, IEEE Trans. Biomedical Eng. 55 (2008) 490–501.

[2] P.O. Amblard, M. Gaeta, J.L. Lacoume, Statistics for complex variables and signals – Part II: Signals, Signal Processing, Elsevier 53 (1996) 15–25.
[3] P.O. Amblard, M. Gaeta, J.L. Lacoume, Statistics for complex variables and signals – Parts I and II: Variables, Signal Processing, Elsevier 53 (1996) 1–13.
[4] S. Baillet, J.C. Mosher, R.M. Leahy, Electromagnetic brain mapping, IEEE Signal Processing Magazine 18 (2001) 14–30.
[5] A.K. Barros, A. Mansour, N. Ohnishi, Removing artefacts from electrocardiographic signals using independent component analysis, Neurocomputing 22 (1998) 173–186.
[6] M. Basseville, Distance measures for signal processing and pattern recognition, Signal Processing 18 (1989) 349–369.
[7] A. Belouchrani, K. Abed-Meraim, J.F. Cardoso, E. Moulines, A blind source separation technique using second-order statistics, IEEE Transactions on Signal Processing 45 (1997) 434–444.
[8] V.D. Calhoun, T. Adali, L.K. Hansen, J. Larsen, J.J. Pekar, ICA of functional MRI data: An overview, in: ICA 03, Fourth International Symposium on Independent Component Analysis and Blind Signal Separation, Nara, Japan, April 1–4 2003, pp. 909–914.
[9] J.F. Cardoso, Fetal electrocardiogram extraction by source subspace separation, in: IEEE International Conference on Acoustics Speech and Signal Processing, 1998, pp. 1941–1944.
[10] J.F. Cardoso, A. Souloumiac, Blind beamforming for non-gaussian signals, IEE Proceedings F 140 (6) (1993) 362–370.
[11] J.F. Cardoso, A. Souloumiac, Jacobi angles for simultaneous diagonalization, SIAM Journal Matrix Analysis and Applications 17 (1996) 161–164.
[12] F. Castells, J.J. Rieta, J. Millet, V. Zarzoso, Spatiotemporal blind source separation approach to atrial activity estimation in atrial tachyarrhythmias, IEEE Transactions on Biomedical Engineering 52 (2005) 258–267.
[13] T.F. Chan, An improved algorithm for computing the singular value decomposition, ACM Transactions on Mathematical Software 8 (1982) 72–83.
[14] M.P.S. Chawla, H.K. Verma, V. Kumar, Artifacts and noise removal in electrocardiograms using independent component analysis, International Journal of Cardiology 129 (2008) 278–281.
[15] P. Chevalier, Optimal separation of independent narrow-band sources: Concept and Performances, Signal Processing, Elsevier 73 (1999) 27–47.
[16] M.K. Chung, Current clinical issues in atrial fibrillation, Cleveland Clinic Journal of Medicine 70 (2003) 6–11.
[17] D. Cohen, Magnetoencephalography: Evidence of magnetic fields produced by alpha rhythm currents, Science 161 (1972) 664–666.
[18] P. Comon, Independent component analysis, in: J.-L. Lacoume (Ed.), Higher Order Statistics, Elsevier, Amsterdam, London, 1992, pp. 29–38.
[19] P. Comon, Independent Component Analysis, a new concept?, Signal Processing, Elsevier 36 (1994) 287–314.
[20] P. Comon, From source separation to blind equalization, contrast-based approaches, in: ICISP 01, International Conference on Image and Signal Processing, Agadir, Morocco, May 3–5 2001, pp. 20–32.
[21] P. Comon, E. Moreau, Improved contrast dedicated to blind separation in communications, in: ICASSP 97, 1997 IEEE International Conference on Acoustics Speech and Signal Processing, Munich, April 20–24 1997, pp. 3453–3456.
[22] M. Congedo, C. Gouy-Pailler, C. Jutten, On the blind source separation of human electroencephalogram by approximate joint diagonalization of second order statistics, Clinical Neurophysiology 119 (2008) 2677–2686.
[23] L. de Lathauwer, D. Callaerts, B. de Moor, J. Vandewalle, Fetal electrocardiogram extraction by source subspace separation, in: IEEE Workshop on Higher Order Statistics, Girona, Spain, June 12–14 1995, pp. 134–138.
[24] L. de Lathauwer, B. de Moor, J. Vandewalle, Fetal electrocardiogram extraction by blind source subspace separation, IEEE Transactions on Biomedical Engineering 47 (2000) 567–572.

[25] S. Debener, S. Makeig, A. Delorme, A.K. Engel, What is novel in the novelty oddball paradigm? functional significance of the novelty P3 event-related potential as revealed by independent component analysis, Cognitive Brain Research 22 (2005) 309–321.

[26] K.L. Diamantaras, S.Y. Kung, Principal Component Neural Networks: Theory and Applications, in: Adaptive and Learning Systems for Signal Processing, Communications and Control Series, John Wiley, New York, 1996.

[27] L. Faes, G. Nollo, E.O.M. Kirchner, F. Gaita, R. Riccardi, R. Antolini, Principal component analysis and cluster analysis for measuring the local organization of human atrial fibrillation, Medical and Biological Engineering and Computing 39 (2001) 656–663.

[28] D. Farina, C. Fevotte, C. Doncarli, R. Merletti, Blind separation of linear instantaneous mixtures of non-stationary surface myoelectric signals, IEEE Transactions on Biomedical Engineering 9 (2004) 1555–1567.

[29] A. Ferreol, L. Albera, P. Chevalier, Fourth order blind identification of underdetermined mixtures of sources (FOBIUM), IEEE Transactions On Signal Processing 53 (2005) 1254–1271.

[30] C. Fevotte, C. Doncarli, Two contributions to blind sources separation using time-frequency distributions, IEEE Signal processing Letters 11 (2004) 386–389.

[31] A. Flexer, H. Bauer, J. Pripfl, G. Dorffner, Using ICA for removal of ocular artifacts in EEG recorded from blind subjects, Neural Networks 18 (2005) 998–1005.

[32] V. Fuster, E. Ryden, W. Asinger, S. Cannom, J. Crijns, R.L. Frye, J.L. Halperin, G.N. Kay, W.W. Klein, S. Levy, R.L. Mcnamara, E.N. Prystowsky, L.S. Wann, D.G. Wyse, ACC/AHA/ESC guidelines for the management of patients with atrial fibrillation, Journal of the American College of Cardiology 38 (2001) 1266i–1266lxx.

[33] G.H. Golub, C. Reinsch, Singular value decomposition and least squares solutions, in: J.H. Wilkinson (Ed.), Handbook for Automatic Computation, in: Linear Algebra, vol. II, Springer-Verlag, New York, 1970.

[34] R. Guevara, J.L. Velazquez, V. Nenadovic, R. Wennberg, G. Senjanovic, L.G. Dominguez, Phase synchronization measurements using electroencephalographic recordings: What can we really say about neuronal synchrony? Neuroinformatics 3 (2005) 301–314.

[35] S. Haykin (Ed.), Source Separation: Models, Concepts, Algorithms and Performance in Unsupervised Adaptive Filtering, in: Blind Source Separation of Series in Adaptive and Learning Systems for Communications, Signal Processing, and Control, vol. I, Wiley, 2000.

[36] A. Holobar, C. Févotte, C. Doncarli, D. Zazula, Single autoterms separation based on bilinear time-frequency representations, in: EUSIPCO, Toulouse, France, Sept. 2002, pp. 565–568.

[37] S. Hu, M. Stead, G.A. Worrell, Automatic identification and removal of scalp reference signal for intracranial EEGs based on independent component analysis, IEEE Transactions on Biomedical Engineering 54 (2007) 1560–1572.

[38] A. Hyvarinen, Fast and robust fixed-point algorithms for independent component analysis, IEEE Transactions On Neural Networks 10 (3) (1999) 626–634.

[39] A. Hyvarinen, J. Karhunen, P. Oja, Independent Component Analysis, Wiley Interscience, Simon Haykin, 2001.

[40] J. Iriarte, E. Urrestarazu, M. Valencia, M. Alegre, A. Malanda, C. Viteri, J. Artieda, Independent component analysis as a tool to eliminate artifacts in EEG: A quantitative study, Journal of Clinical Neurophysiology 20 (2003) 249–257.

[41] C.J. James, O.J. Gibson, Temporally constrained ICA: An application to artifact rejection in electromagnetic brain signal analysis, IEEE Transactions on Biomedical Engineering 50 (2003) 1108–1116.

[42] B.H. Jansen, V.G. Rit, Electroencephalogram and visual evoked potential generation in a mathematical model of coupled cortical columns? Biological Cybernetics 73 (1995) 357–366.

[43] C.A. Joyce, I.F. Gorodnitsky, M. Kutas, Automatic removal of eye movement and blink artifacts from EEG data using blind component separation, Psychophysiology 41 (2004) 313–325.

[44] T.-P. Jung, S. Makeig, M. Westerfield, J. Townsend, E. Courchesne, T.J. Sejnowski, Removal of eye activity artifacts from visual event-related potentials in normal and clinical subjects, Clinical Neurophysiology 11 (2000) 1754–1758.

[45] T.-P. Jung, S. Makeig, M. Westerfield, J. Townsend, E. Courchesne, T.J. Sejnowski, Analysis and visualization of single-trial event-related potentials, Human Brain Mapping 14 (2001) 166–185.

[46] M. Junghofer, T. Elbert, D.M. Tucker, C. Braun, The polar average reference effect: A bias in estimating the head surface integral in EEG recording, Clinical Neurophysiology 110 (1999) 1149–1155.

[47] A. Kachenoura, L. Albera, L. Senhadji, P. Comon, ICA: a potential tool for BCI systems, IEEE Signal Processing Magazine, special issue on Brain-Computer Interfaces 25 (2008) 57–68.

[48] J. Karvanen, V. Koivunen, Blind separation methods based on Pearson system and its extensions, Signal Processing, Elsevier 82 (2002) 663–673.

[49] J.L. Lacoume, P.O. Amblard, P. Comon, Statistiques d'ordre supérieur pour le traitement du signal, in: Collection Sciences de l'Ingénieur, Masson, 1997.

[50] P. Langley, J.P. Bourke, A. Murray, Frequency analysis of atrial fibrillation, in: IEEE Computers in Cardiology, 2000, pp. 65–68.

[51] L.D. Lathauwer, J. Castaing, Blind identification of underdetermined mixtures by simultaneous matrix diagonalization, IEEE Transactions on Signal Processing 56 (2008).

[52] L.D. Lathauwer, B.D. Moor, J. Vandewalle, Independent component analysis and (simultaneous) third-order tensor diagonalisation, IEEE Transactions on Signal Processing 49 (2001) 2262–2271.

[53] T.W. Lee, M. Girolami, T.J. Sejnowski, Independent component analysis using an extended infomax algorithm for mixed sub-gaussian and super-gaussian sources, Neural Computation 11 (1999) 417–441.

[54] R. Legarreta, Component selection for PCA-based extraction of atrial fibrillation, in: IEEE Computers in Cardiology, 2006, pp. 137–140.

[55] S. Levy, G. Breithardt, R.W.F. Campbell, A.J. Camm, J.C. Daubert, M. Allessie, E. Aliot, A. Capucci, F. Cosio, H. Crijns, L. Jordaens, R.N.W. Hauer, F. Lombardi, B. Lüderitz, Atrial fibrillation: Current knowledge and recommendations for management, European Heart Journal 419 (1998) 1294–1320.

[56] S. Makeig, A.J. Bell, T.-P. Jung, T.J. Sejnowski, Independent component analysis of electroencephalographic data, Advances in Neural Information Processing Systems 8 (1996) 145–151.

[57] S. Makeig, T.-P. Jung, A.J. Bell, D. Ghahermani, T.J. Sejnowski, Blind separation of auditory event-related brain responses into independent components, Proceedings of National Academy of Sciences of the United States of America 94 (1997) 10979–10984.

[58] J. Malmivio, R. Plonsey, Bioelectromagnetism, in: Principles and Applications of Bioelectric and Biomagnetic Fields, Oxford University Press, New York, 1995.

[59] P. McCullagh, Tensor Methods in Statistics, in: Monographs on Statistics and Applied Probability, Chapman and Hall, 1987.

[60] M.J. McKeown, S. Makeig, G.B. Brown, T.B. Jung, S.S. Kindermann, A.J. Bell, T.J. Sejnowski, Analysis of fMRI data by blind separation into independent spatial components, Human Brain Mapping 6 (1998) 160–188.

[61] M.J. McKeown, T.J. Sejnowski, Independent component analysis of fMRI data: Examining the assumptions, Human Brain Mapping 6 (1998) 368–372.

[62] J.C. Mosher, R.M. Leahy, Source localization using Recursively Applied and Projected (RAP) music, IEEE Transactions On Signal Processing 47 (1999) 332–340.

[63] J.L. Onton, A. Delorme, S. Makeig, Frontal midline EEG dynamics during working memory, Neuroimage 27 (2005) 341–356.

[64] T. Oostendorp, Modeling the Fetal ECG, PhD thesis, K. U. Nijmegen, 1989.

[65] A. Ossadtchi, S. Baillet, J.C. Mosher, D. Thyerleid, W. Sutherling, R.M. Leahy, Automated interictal spike detection and source localization in magnetoencephalography using independent components analysis and spatio-temporal clustering, Clinical Neurophysiology, Elsevier 115 (2004) 508–522.

[66] M.I. Owis, A.B.M. Youssef, Y.M. Kadah, Characterization of ECG signals based on blind source separation, Medical and Biological Engineering and Computing 40 (2002) 557–564.
[67] M.I. Owis, A.B.M. Youssef, Y.M. Kadah, A successive cancellation algorithm for fetal heartrate estimation using an intrauterine ECG signal, IEEE Transactions on Biomedical Engineering 49 (2002) 943–954.
[68] R. Plonsey, Bioelectric Phenomena, McGraw-Hill, New York, 1969.
[69] F. Poree, A. Kachenoura, H. Gauvrit, C. Morvan, G. Carrault, L. Senhadji, Blind source separation for ambulatory sleep recording, Transactions on Information Technology in Biomedicine 10 (2006) 293–301.
[70] J. Pulkkinen, A.M. Häkkinen, N. Lundbom, A. Paetau, R.A. Kauppinen, Y. Hiltunen, Independent component analysis to proton spectroscopic imaging data to human brain tumours, European Journal of Radiology 56 (2005) 160–164.
[71] R.M. Rangayyan, Biomedical signal snalysis: A case study approach, in: IEEE Press Series on Biomedical Engineering, John Wiley, New York, 2002.
[72] J.J. Rieta, F. Castells, C. Sanchez, J. Igual, ICA applied to atrial fibrillation analysis, in: International Conference of Independent Component Analysis and Blind Signal Separation, 2003, pp. 59–64.
[73] J.J. Rieta, F. Castells, C. Sanchez, V. Zarzozo, J. Millet, Atrial activity extraction for atrial fibrillation analysis using blind source separation, IEEE Transactions on Biomedical Engineering 51 (2004) 1176–1186.
[74] J.J. Rieta, V. Zarzoso, J. Millet, R. Garcis, R. Ruiz, Atrial activity extraction based on blind source separation as an alternative to QRST cancellation for atrial fibrillation analysis, in: IEEE Computers in Cardiology, 2000, pp. 69–72.
[75] P. Sajda, S. Du, T. Brown, R. Stoyanova, D. Shungu, X. Mao, L. Parra, Nonnegative matrix factorization for rapid recovery of constituent spectra in magnetic resonance chemical shift imaging of the brain, IEEE Transactions on Medical Imaging 23 (2004) 1453–1465.
[76] R. Sameni, Extraction of fetal cardiac signals from an array of maternal abdominal recordings, PhD thesis, Institut Polytechnique de Grenoble, July 2008.
[77] C. Sanchez, J. Millet, J.J. Rieta, F. Castells, J. Rodenas, R. Ruiz, V. Ruiz, Packet wavelet decomposition: An approach for atrial activity extraction, in: IEEE Computers in Cardiology, 2002, pp. 33–36.
[78] S.J. Schiff, Dangerous phase, Neuroinformatics 3 (2006) 315–318.
[79] L. Sornmo, P. Laguna, Bioelectrical signal processing in cardiac and neurological applications, in: Biomedical Engineering, Elsevier, Academic Press, USA, 2005.
[80] M. Stridh, L. Sörnmo, Spatiotemporal QRST cancellation techniques for analysis of atrial fibrillation, IEEE Transactions on Biomedical Engineering 48 (2001) 105–111.
[81] E.M. Symonds, D. Sahota, A. Chang, Fetal electrocardiography, in: Cardiopulmonary medicine, Imperial College Press, London, 2001.
[82] H. Taigang, G. Clifford, L. Tarassenko, Application of independent component analysis in removing artefacts from the electrocardiogram, Neural Computing and Applications 15 (2006) 105–116.
[83] T. Talmon, J. Kors, J. Von, Adaptive gaussian filtering in routine ECG/VCG analysis, IEEE Transactions On Acoustics, Speech and Signal Processing 34 (1986) 527–534.
[84] A.C. Tang, B.A. Pearlmutter, N.A. Malaszenko, D.B. Phung, Independent components of magnetoencephalography: Single-trial response onset times, NeuroImage 17 (2002) 1773–1789.
[85] A.C. Tang, B.A. Pearlmutter, N.A. Malaszenko, D.B. Phung, B.C. Reeb, Independent components of magnetoencephalography: Localization, Neural Computation 14 (2002) 1827–1858.
[86] N. Thakor, Z. Yi-Sheng, Applications of adaptive filtering to ECG analysis: noise cancellation and arrhythmia detection, IEEE Transactions on Biomedical Engineering 38 (1991) 785–794.
[87] C. Vasquez, A.I. Hernandez, F. Mora, G. Carrault, G. Passariello, Atrial activity enhancement by Wiener filtering using an artificial neural network, IEEE Transactions On Biomedical Engineering 48 (2001) 940–944.

[88] R. Vetter, N. Virag, J.M. Vesin, P. Celka, U. Scherrer, Observer of autonomic cardiac outflow based on blind source separation of ECG parameters, IEEE Transactions On Biomedical Engineering 47 (2000) 578–582.

[89] R. Vigario, Extraction of ocular artefacts from EEG using independent component analysis, Electroencephalography Clinical Neurophysiology 103 (1997) 395–404.

[90] R. Vigario, V. Jousmaki, M. Hamalainen, R. Hari, E. Oja, Independent component analysis for identification of artifacts in magnetoencephalographic recordings, in: Proceedings of the 1997 Conference on Advances in Neural Information Processing System, 10, MIT Press Cambridge, MA, USA, Denver, USA, 1998, pp. 229–235.

[91] R. Vigario, E. Oja, BSS and ICA in neuroinformatics: From current practices to open challenges, IEEE Reviews in Biomedical Engineering 1 (2008) 50–61.

[92] R. Vigario, J. Sarela, V. Jousmaki, M. Hamalainen, E. Oja, Independent component approach to the analysis of EEG and MEG, IEEE Transactions On Biomedical Engineering 47 (2000) 589–593.

[93] R. Vigario, J. Särelä, V. Jousmäki, E. Oja, Independent component analysis in decomposition of auditory and somatosensory evoked fields, in: ICA'99, Second International Symposium on Independent Component Analysis and Blind Signal Separation, Aussois, France, Jan. 1999, pp. 167–172.

[94] R. Vigario, A. Ziehe, K. Muller, G. Wubbeler, L. Trahms, B.M. Mackert, G. Curio, V. Jousmaki, J. Sarela, E. Oja, Blind Decomposition of Multimodal Evoked Responses and DC Fields, MIT Press, Cambridge, MA, USA, 2003.

[95] J.A. Westgate, A.J. Gunn, L. Bennet, M.I. Gunning, H.H.D. Haan, P.D. Gluckman, Do fetal electrocardiogram PR-RR changes reflect progressive asphyxia after repeated umbilical cord occlusion in fetal sheep?, Pediatric Research 44 (1998) 419–461.

[96] B. Widrow, J. Glover, J. Mccool, J. Kaunitz, C. Williams, H. Hearn, J- Zeidler, E. Dong, R. Googlin, Adaptive noise cancelling: principles and applications, Proc. IEEE 63 (1975) 1692–1716.

[97] T. Winsor, Primer of Vectorcardiography, Lea and Febiger, Philadelphia, 1972.

[98] J.O. Wisbeck, A.K. Barros, R. Ojeda, Application of ICA in the separation of breathing artefacts in ECG signals, in: International Conference on Neural Information Processing, (ICONIP'98), 1998, pp. 211–214.

[99] J.O. Wisbeck, R.G. Ojeda, Application of neural networks to separate interferences and ECG signals, in: Proceedings of the 1998 Second IEEE International Caracas Conference, 1998, pp. 291–294.

[100] A. Yeredor, Non-orthogonal joint diagonalization in the least-squares sense with application in blind source separation, IEEE Transactions on Signal Processing 50 (2002) 1545–1553.

[101] V. Zarzoso, P. Comon, Comparative speed analysis of FastICA, in: M.E. Davies, C.J. James, S.A. Abdallah, M.D. Plumbley (Eds.), ICA'07, in: Lecture Notes in Computer Sciences, vol. 4666, Springer, London, UK, 2007, pp. 293–300.

[102] V. Zarzoso, P. Comon, Robust independent component analysis for blind source separation and extraction, in: International IEEE Engineering in Medicine and Biology Conference, 2008, pp. 3344–3347.

[103] V. Zarzoso, A.K. Nandi, Blind separation of independent sources for virtually any source probability density function, IEEE Transactions on Signal Processing 47 (1999) 2419–2432.

[104] V. Zarzoso, A.K. Nandi, Noninvasive fetal electrocardiogram extraction: blind separation versus adaptive noise cancellation, IEEE Transactions on Biomedical Engineering 48 (2001) 12–18.

[105] A. Ziehe, K.R. Muller, TDSEP – an efficient algorithm for blind separation using time structure, in: ICANN'98, Proceedings of the 8th International Conference on Artificial Neural Networks, 1998, pp. 675–680.

CHAPTER

Audio applications

19

E. Vincent and Y. Deville

Acoustic signal processing is one of the earliest fields in which the source separation problem was studied. Indeed, most available acoustic signals are mixtures of several sources. Although good separation may currently be obtained for some simple synthetic mixtures, the separation of real-world signals requires basic competence in this field and remains difficult to achieve. It should therefore be stressed that acoustics is among the most difficult application fields of source separation under investigation.

Early acoustic source separation systems relied on fixed or adaptive beamforming, which remains in use today [12]. These systems require some prior knowledge, such as the relative positions of the microphones and the target source or the time intervals during which the target source is inactive. A review of the performance of five such systems is provided in [34]. In practice, prior knowledge about the sources or the mixing system is rarely available, so that blind source separation (BSS) systems must be used instead.

In this chapter, we discuss the application of the BSS techniques reviewed in this book to the separation of audio signals, with main emphasis on convolutive independent component analysis (ICA) (see Chapter 8) and sparse component analysis (SCA) (see Chapter 10). In particular, we characterize real-world audio mixtures and summarize the performance of existing systems for such mixtures. A more detailed review of audio BSS systems is provided in [35]. Applications of BSS to underwater acoustics, medical acoustics, ultrasounds and infrasounds are not considered. Interested readers may refer to for example [23,25,49,72]. The structure of the chapter is as follows. In section 19.1, we describe the various types of audio mixtures and source separation objectives encountered in real-world applications. Then, we investigate the properties of audio mixtures exploited by BSS techniques in section 19.2 and discuss the application of ICA and SCA in sections 19.3 and 19.4 respectively. We conclude and provide references to alternative audio BSS systems in section 19.5.

19.1 AUDIO MIXTURES AND SEPARATION OBJECTIVES

The need for BSS arises with various real-world signals, including meeting recordings, hearing aid signals, music CDs and radio broadcasts. These signals are obtained via different techniques, which results in different signal properties. Also, the objective to be achieved with the help of BSS depends on the considered application.

Handbook of Blind Source Separation: Independent Component Analysis and Applications

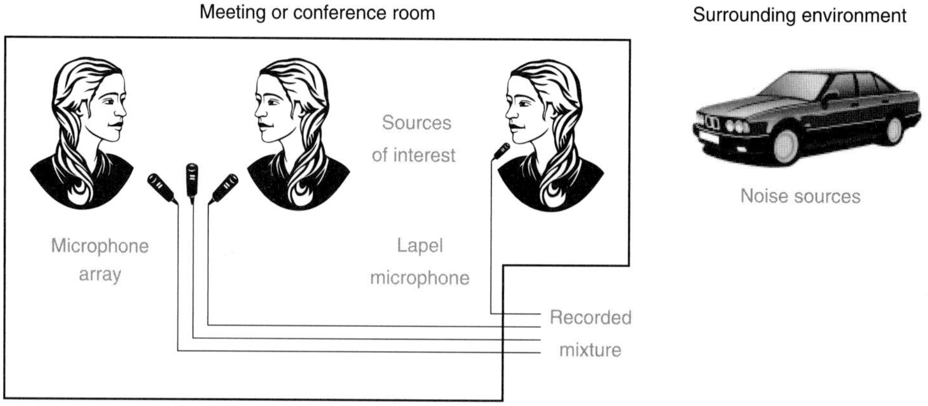

FIGURE 19.1

Examples of recording techniques used in a meeting context.

19.1.1 Recorded mixtures

The simplest way of converting a sound scene into a digital signal is to record all the sources simultaneously using one or several microphones, as shown in Fig. 19.1. Each microphone receives sound waves coming from all sources and provides an observed signal which is a mixture of the original source signals. This technique is employed, for example, for meeting recordings and live concert broadcasts. Common microphone arrangements may involve both near-field and far-field microphones with various spatial directivities and spacings [9,12].

The source mixing phenomenon results from the simultaneous propagation of sound waves through air from the sources to the microphones and the acoustic-electric conversion performed by the microphones. In normal situations, these phenomena are both linear. The effect of acoustic-electric conversion may therefore be conceptually merged with that of propagation and is not explicitly considered in the following.

The mixing process may be modeled by combining the following properties. Firstly, the output signal of each microphone is an additive superposition of the contributions associated with individual sources. Secondly, the contribution of each source to each microphone is obtained by linear filtering, i.e. convolution, of the original source signal by a source-microphone filter.[1] The simplest model for this filter is obtained by considering omnidirectional microphones and a single direct propagation path between each source and each microphone. This results in a time delay and an amplitude attenuation, which are respectively proportional and inversely proportional to the path length. This anechoic model may be used in specific environments, such as anechoic rooms or open air areas with dampening ground. However, in most practical situations, the propagation

[1]This property is meaningful only for point sources, e.g. speakers and small musical instruments, that produce sound from a single point in space. By contrast, some sources, e.g. piano or drums, can produce sound at different spatial positions at the same time. These extended sources cannot be defined by a single source signal and must be modeled as a collection of point sources associated with different source signals.

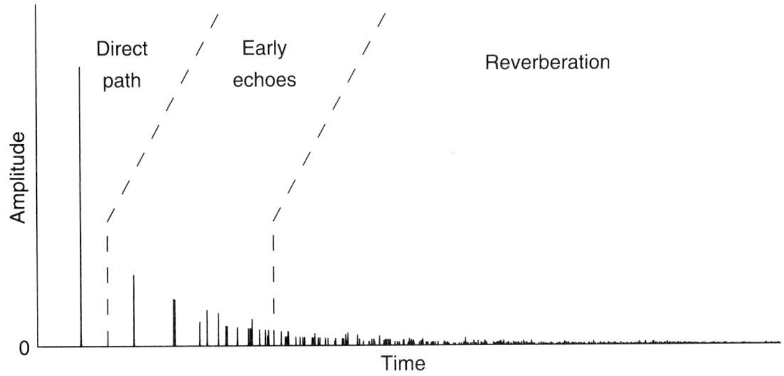

FIGURE 19.2

Theoretical shape of a source-microphone impulse response.

of the sound waves between each source and each microphone involves a large number of paths associated with successive *echoes* on the encountered obstacles, such as walls, furniture and people's bodies within a room [32]. Each echo introduces additional delay and attenuation. The superposition of the delays and attenuations associated with individual echoes can be modeled by an overall *source-microphone impulse response*. This response is quite sensitive to the location of the corresponding source and microphone and generally varies over time due to possibly uncontrolled source movements.

The theoretical shape of a source-microphone impulse response is illustrated in Fig. 19.2. It involves a main peak corresponding to the *direct path* between the source and the microphone, followed by some major peaks due to early echoes and an exponentially decreasing tail due to late echoes called reverberation. The main peak is more prominent for near-field microphones than for far-field ones. The positions of early echoes may be predicted to some extent given the geometry of the room, while reverberation occurs at diffuse near-random locations. The amount of reverberation is measured by the *reverberation time* RT_{60}, which is the delay after which the magnitude of the echoes becomes 60 dB smaller than that of the sound source. This quantity is on the order of 150-500 ms in office rooms and 1-2 s in concert halls. Therefore, the length of the finite impulse response (FIR) mixing filters required to accurately model these impulse responses ranges from a few hundred to a few ten thousand taps at standard sampling rates.

Actual source-microphone impulse responses may be acquired by playing some wide-band test signal, e.g. a chirp signal, through a loudspeaker placed at the location of the source, dividing the Fourier transform of the signal recorded at the microphone by that of the test signal and applying the inverse Fourier transform [32]. Figure 19.3 depicts this experimental setup in a car environment. Example source-microphone impulse responses measured in this car environment and in a large concert room are shown in Figs 19.4 and 19.5 respectively. These responses follow the overall shape of Fig. 19.2, with peaks being replaced by sine cardinal functions due to finite sampling rate. In particular, it can be seen that the peaks of the responses occur at different times for each microphone.

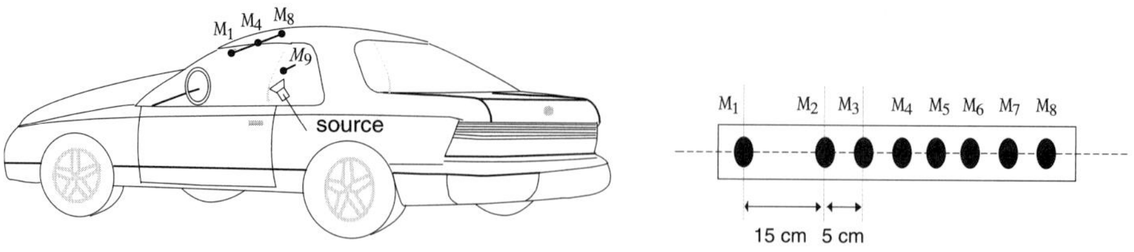

FIGURE 19.3

Experimental setup for in-car source-microphone impulse response measurements and BSS experiments [13] (left). Sound played through a loudspeaker placed at the driver's head position is recorded via an array of microphones numbered M_1-M_8 (right) placed at the top of the windshield and a lapel microphone numbered M_9.

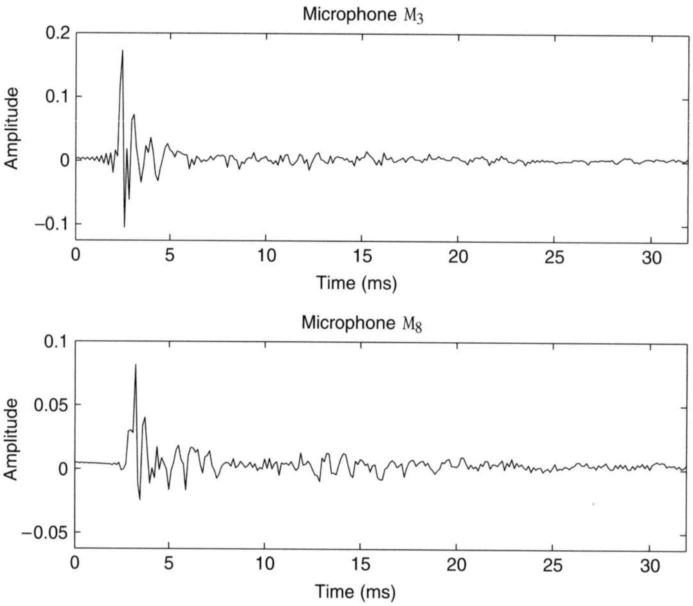

FIGURE 19.4

Measured in-car source-microphone impulse responses using the setup in Fig. 19.3 [13].

19.1.2 Synthesized mixtures

An alternative way of rendering a sound scene is to record individual sources or small groups of sources separately in a studio and mix the recorded signals together using a mixing desk or dedicated software, as illustrated in Fig. 19.6. This synthetic mixing technique is used, for example, for pop music and movie soundtracks, since it allows

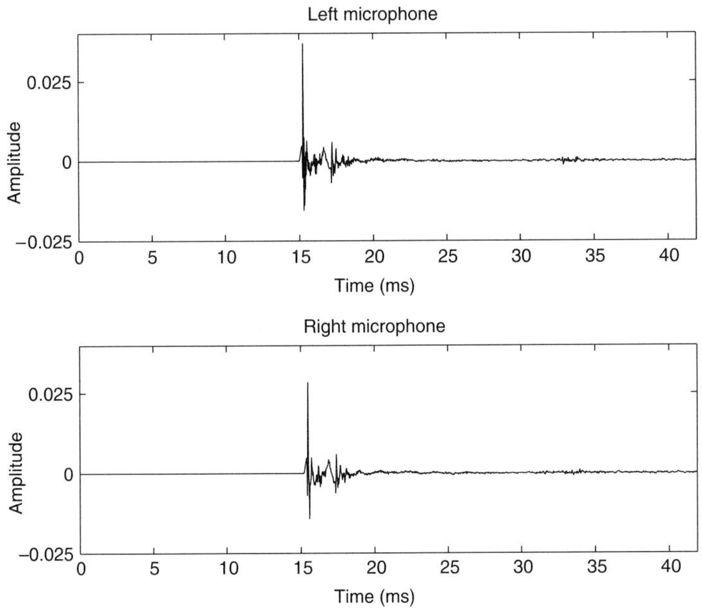

FIGURE 19.5

Measured concert room source-microphone impulse responses for a pair of omnidirectional microphones with 20 cm spacing at 4.5 m distance from the source [63].

the application of different audio processing effects to each source or each group of sources prior to mixing. The mixture signal is typically in stereo (two-channel) or some other multichannel format. The contribution of each source to each mixture channel is generated using one or more effects such as "pan", which scales the source signal by a positive mixing gain, and "reverb", which simulates reverberation by applying synthetic FIR and infinite impulse response (IIR) mixing filters. In the following, we use the term "mixing filter" to refer both to source-microphone impulse responses and to synthetic mixing filters.

19.1.3 Separation objectives and performance evaluation

In all mixing situations, the objective of BSS is to extract one or several source signals from the observed multichannel mixture signal, with other source signals being regarded as undesired noise. The signals of interest depend on the application. For instance, in the context of speech enhancement for mobile phones, the only source signal of interest is the user's speech. Undesired sources may then include speech signals from surrounding people and environmental noises produced by cars, wind or rain. On the contrary, the so-called "cocktail-party" application refers to the situation when the

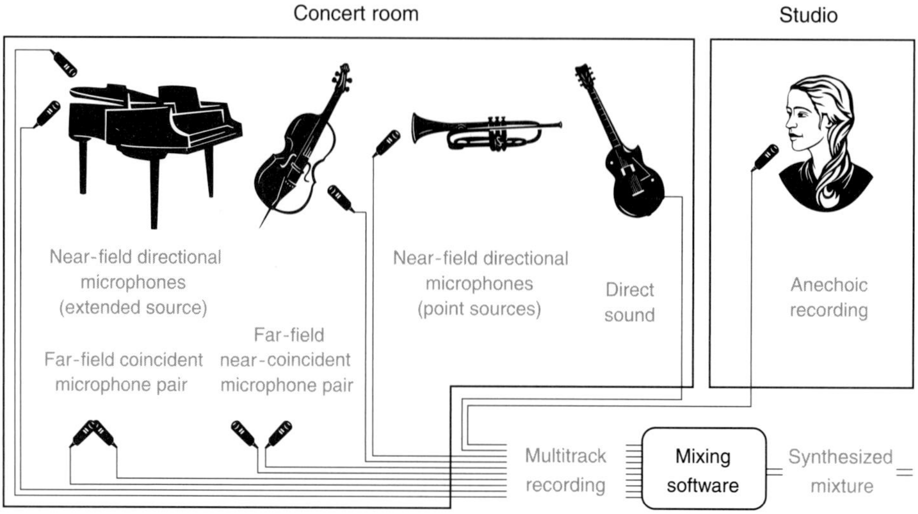

FIGURE 19.6

Example recording techniques underlying a music CD track.

observed mixture signal results from several people simultaneously speaking in a room and all speech signals are of interest. Noise may then originate from clinking glasses or footsteps.

In practice, perfect separation is rarely achieved, for example because the signal models and the assumptions behind BSS systems are not always met in real-world situations. The level of the source of interest is then typically increased within each estimated source signal, but distortions remain with respect to the original source signal. Such distortions may include filtering of the source of interest, residual sounds from other sources of interest known as *interference*, residual noise from undesired sources and additional "gurgling" sounds known as *artifacts*. Minimizing one type of distortion alone, e.g. interference, may result in increasing another type of distortion, e.g. residual noise. A suitable tradeoff is typically sought depending on the application. For instance, filtering distortion degrades automatic speech recognition abilities while "gurgling" artifacts are particularly annoying for hearing aid applications.

In order to evaluate established and novel BSS systems, a range of performance metrics have been developed, assuming that either the original source signals or their contribution to the mixture are available as references. Many investigations have been performed with artificial convolutive mixtures generated by digital filtering of individual source recordings by random FIR mixing filters or measured source-microphone impulse responses. More realistic mixtures may be obtained by successively activating each source, recording it by all microphones and digitally adding the contributions of all sources at each microphone [58]. These contributions are then available as references for evaluation, while the resulting mixture signal is close to the one that would have been recorded if all sources were active at the same time, provided that the positions of the sources are fixed and background noise is negligible. Moving sources are less easily handled because each source acts as an obstacle that affects the impulse responses for other sources. For a more detailed discussion of this issue, and, more generally speaking, of adequate recording situations, see [58].

Assuming that reference source signals are available, a natural way of evaluating an audio BSS system is to ask a panel of human subjects to listen to and rate the quality or the intelligibility of each estimated source signal as compared to the reference. This makes it possible to determine whether the distortions between the two signals are audible and how perceptually disturbing they are on average. This approach is relevant if the estimated source signals are to be eventually listened to in the considered application, e.g. when one aims at denoising the speech of a mobile phone user before transmitting it to the other party. In the case when the estimated source is to be used as the input of another signal processing system, the performance of the BSS system should instead be assessed by the performance of the overall system. For instance, when BSS is used as a speech enhancement front-end for automatic speech recognition, the *word error rate* (WER) of the overall system can be compared to that of the plain speech recognizer applied either to the reference source signal or to the mixture signal [20]. The *word error rate decrease* (WERD) due to source separation can then be derived.

In a development context, these application-specific performance metrics may be replaced by generic metrics which are more quickly computed and more easily interpreted. Such metrics are typically defined by decomposing each estimated source signal as the sum of a term corresponding to the reference signal and one or several terms representing distortions, as illustrated in Fig. 19.7. The relative importance of these terms is then measured by means of energy ratios expressed in decibels (dB). The most straightforward metric is the *signal-to-distortion ratio* (SDR) measuring overall distortion [58]. The *signal-to-interference ratio* (SIR) assessing rejection of the other sources of interest was also defined in [58] for determined or over-determined mixtures. In [65], these metrics were generalized to all types of mixtures and supplemented by the *signal-to-noise ratio* (SNR) assessing rejection of undesired sources and the *signal-to-artifacts ratio* (SAR) quantifying the remaining distortion including "gurgling" artifacts. Alternative metrics such as SDR improvement (SDRI) [14] and SIR improvement (SIRI) [45] have also been employed. Recent studies have shown that perceptual quality assessments and speech recognition rates can be approximated by different weightings of similar metrics [22,47].

A complete protocol for performance evaluation should not only include the definitions of the metrics, but also the specification of the time interval over which they are measured. Indeed, performance may vary over time due, for example, to a variable number of active sources or to the use of an online algorithm. In the latter case, performance is typically measured after a fixed period during which it is acceptable from an application point of view to let the algorithm converge. In order to assess the difficulty of separating a given mixture, it may also be interesting to compute the best possible performance or *oracle performance* theoretically achievable by various classes of algorithms, as explained in [66].

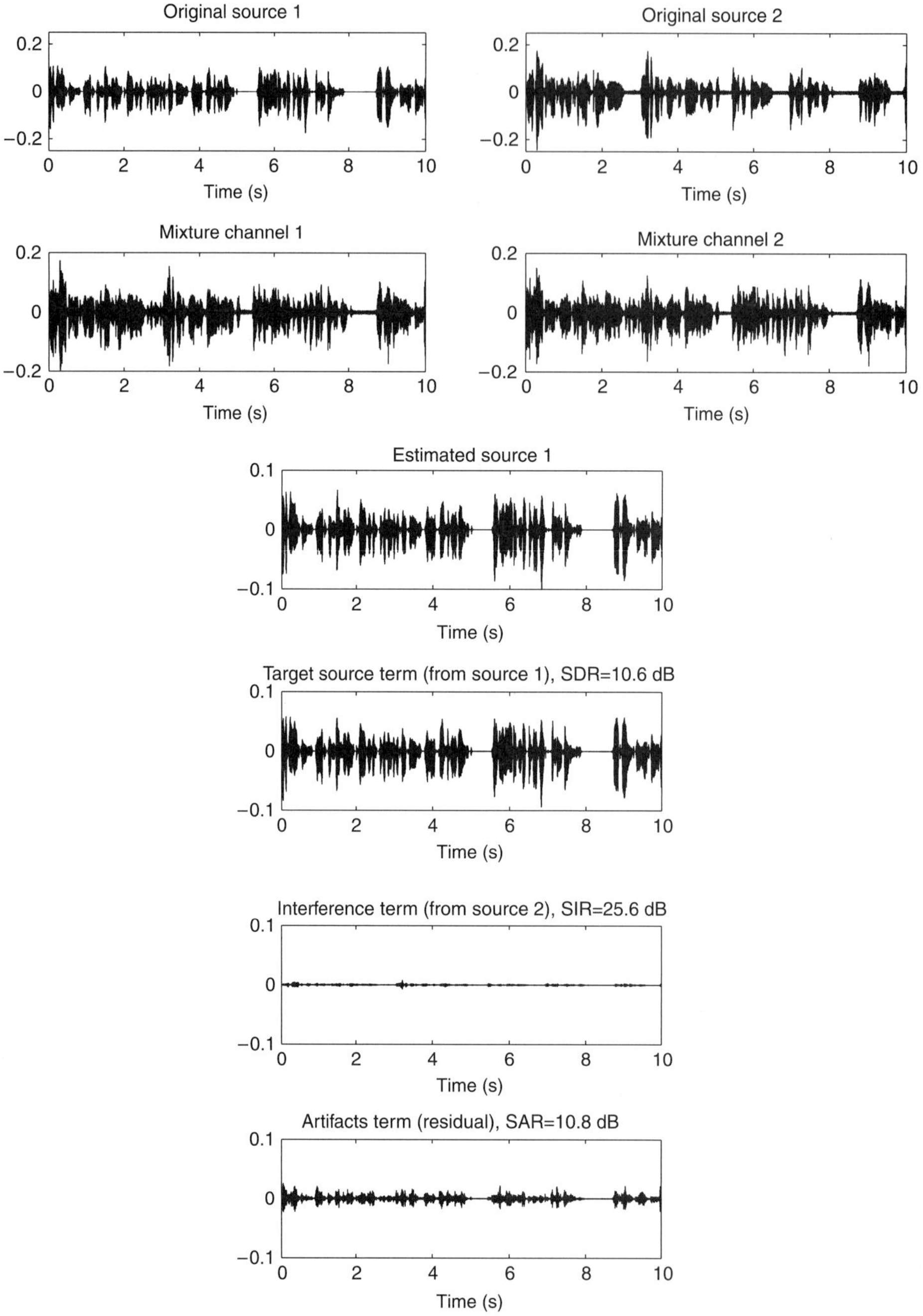

FIGURE 19.7

Example performance evaluation for the separation of an anechoic stereo speech mixture via the algorithm in [73] using the metrics in [65].

19.2 USABLE PROPERTIES OF AUDIO SOURCES

Most of the BSS techniques reviewed in this book rely on the independence of the source signals or on their sparsity in a suitable representation domain. Surprisingly, these assumptions have not all been validated for audio signals in the literature. In the following, we show that they are satisfied in most but not all situations, so that these BSS techniques can indeed be applied in these situations. Additional assumptions made by alternative techniques are reviewed in subsequent sections.

19.2.1 Independence

ICA and second-order techniques often achieve source separation by minimizing some dependency measure between the estimated source signals. Common measures include *mutual information* (see section 2.2) and *Gaussian mutual information* averaged over disjoint time frames (see section 4.6), respectively defined as

$$I\{s_1,\ldots,s_N\} = -\mathbb{E}\log\frac{p_{s_1}(s_1)\ldots p_{s_N}(s_N)}{p_{s_1,\ldots,s_N}(s_1,\ldots,s_N)} \tag{19.1}$$

$$GI\{s_1,\ldots,s_N\} = \frac{1}{Q}\sum_{q=1}^{Q}\frac{1}{2}\log\frac{\det\operatorname{diag}\widehat{\mathbf{R}}_s(q)}{\det\widehat{\mathbf{R}}_s(q)}. \tag{19.2}$$

The resulting separation performance is related to the value of these measures over the true source signals [51], which is assumed to be small. The validity of this assumption depends on the considered mixture [51]:

- Audio sources exhibit significant dependencies on average over short durations, which decrease with increasing signal duration in most but not all situations.
- Speech and music sources exhibit similar dependencies on average.

In order to illustrate these claims, we computed experimental dependency values over the audio BSS dataset in [66], which contains sixty pairs of speech or music sources sampled at 22.05 kHz, collected from English audio books by different speakers and from synchronized multitrack recordings. All pairs of signals were partitioned into disjoint segments of equal duration, from 2^7 samples (5.7 ms) to 2^{18} samples (11.9 s). The above dependency measures were computed for each segment and summarized by their median value over all segments for a given pair of sources. Mutual information was estimated via the software in [30] and Gaussian mutual information by (19.2) with $Q = 8$ time frames per segment. In order to estimate the variance of the estimators, this experiment was also conducted for 30 pairs of independent Gaussian white noise signals, resulting in dependency values below 10^{-2} for all durations.

The dependencies measured between audio sources plotted in Fig. 19.8 appear much larger than between independent noise signals, regardless of their duration. This is partly explained by the short-term periodicity of some speech and music sounds [60].

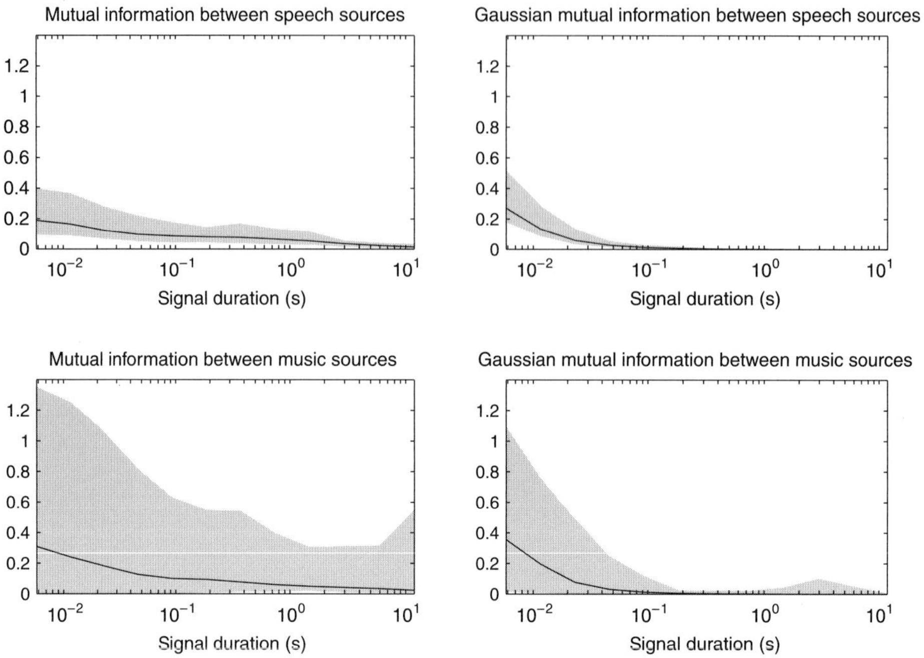

FIGURE 19.8

Mutual information and Gaussian mutual information between the source signals in [66]. The black line and the gray areas denote respectively the median and the range between minimum and maximum measured values.

Fortunately, both measures decrease with increasing duration on average. This phenomenon is observed for all pairs of sources, except one pair of electronically-generated music sources whose dependency remains high for durations above 2 s due to repetitions of the same note samples over time. Gaussian mutual information is much smaller than mutual information for durations above 100 ms, indicating that the non-stationary Gaussian source model is more appropriate for audio sources than the stationary non-Gaussian model. The results also show that both dependency measures span a larger range for music than for speech, but that they are similar for both types of sources on average, except for very short durations below 20 ms. Note that this contradicts earlier claims that the dependency between the source signals should be larger for music than for speech [1], due to the fact that musicians often play synchronous sounds as specified by the rules of musical harmony, while speakers tend to speak freely without attention to distant speakers.

19.2.2 Sparsity

In addition to independence, SCA and Bayesian separation techniques exploit prior distributions over the source signals, especially in the time domain or in the time-frequency domain.[2] Sparse distributions (see section 10.2) are generally employed, reflecting the assumption that the coefficients of the source signals in the chosen domain concentrate around zero and exhibit few significant values. A popular family of distributions is the *generalized Gaussian* family [14,64,71]

$$p_{s_n|\tau,\beta}(s_n(t)) \propto \exp[-\beta |s_n(t)|^\tau] \tag{19.3}$$

where $s_n(t)$ denotes the source coefficients in the time domain. This equation may also be applied in any other representation domain by replacing $s_n(t)$ with the source coefficients in that domain. The shape parameter τ governs the sparsity of the distribution: the smaller it is, the more coefficients concentrate around zero. The distributions with shape parameter $\tau < 2$ are generally considered as sparse and those with $\tau > 2$ as non-sparse with respect to the Gaussian distribution associated with $\tau = 2$. A common choice of the shape parameter is $\tau = 1$, resulting in the Laplacian distribution [14,70]. Again, the resulting performance depends on the actual sparsity of the true sources [73].

The sparsity assumption is suited to time-domain real-world speech recordings in, for example, meeting situations when people speak in turn and remain silent most of the time. Analyzing its validity in a more general context requires basic understanding of the physics of speech and music. Speech signals consist of a sequence of phonemes, which typically involve periodic, noisy or transient sounds. Periodic sounds are louder and composed of sinusoidal components at discrete frequencies. Successive phonemes build up into words separated by small silence intervals. Music signals also involve near-periodic or transient sounds termed notes, with the difference that some instruments may play several notes at a time. Successive notes generally have similar loudness, although music can exhibit a large dynamic range in the long term, and may be played with or without silence in between, depending on the style of the composition. Overall, due to silence intervals and/or periodicity,

- audio sources are generally sparse in the time-frequency domain;
- in the time domain, speech sources are generally sparse but only some music sources are sparse.

Experiments with audio data confirm these claims. The sixty speech and music sources from [66] were transformed into the time-frequency domain via the short-time Fourier transform (STFT) with a sine window. The length of the window was varied from 1 sample to the whole duration of the signals. A generalized Gaussian distribution was then fitted to the set of STFT coefficients of each source in the maximum likelihood sense and sparsity was quantified via the estimated shape parameter τ. The results are displayed in Fig. 19.9, where the leftmost part of each plot (1-sample i.e. 4.5×10^{-5} s STFT window) corresponds to measuring sparsity of the untransformed time-domain sources. All tested sources are sparse in the time-frequency domain. The median shape

[2] In this chapter, we consider linear time-frequency representations as in Chapter 10 as opposed to the quadratic time-frequency representations studied in Chapter 11.

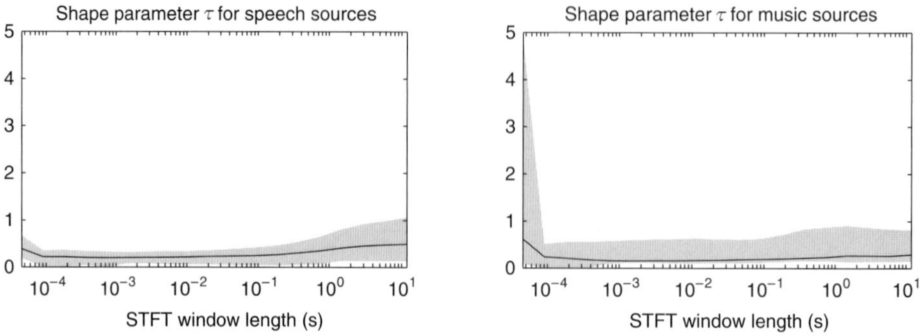

FIGURE 19.9

Generalized Gaussian shape parameters for the STFTs of the source signals in [66]. The black line and the gray areas denote respectively the median and the range between minimum and maximum measured values.

parameter τ then varies between 0.2 and 0.4, indicating that generalized Gaussian priors with $\tau < 1$ provide a better fit than Laplacian priors. Speech sources are also sparse in the time domain, with a median shape parameter of 0.6. However, not all music sources are sparse in the time domain: three distorted electric guitar sources are associated with a shape parameter $\tau > 2$.

19.3 AUDIO APPLICATIONS OF CONVOLUTIVE ICA

The study of the previous section showed that the source signals underlying most audio mixtures with long enough duration are measurably independent. This property allows the separation of determined or over-determined mixtures via ICA. Determined audio mixtures are typically of convolutive nature, hence standard ICA techniques for instantaneous mixtures cannot be applied. Convolutive ICA techniques (see Chapter 8) must be employed instead. Convolutive ICA separates the sources by convolving the mixture signals with multichannel FIR unmixing filters, often chosen so as to maximize some contrast function (see Chapter 3) of the resulting source estimates. These filters can be expressed either in the time domain or in the frequency domain, leading to two distinct classes of algorithms. In the following, we analyze the effect of multichannel filtering on audio mixtures and the resulting theoretical maximum separation performance. We then review some convolutive ICA algorithms and mention their actual performance when available.

19.3.1 Multichannel filtering

The unmixing filters estimated by convolutive ICA perform adaptive null *beamforming* [4]: they attenuate interfering sounds based on their spatial positions at each frequency. Each source generates a large number of sounds at different positions – due to echoes on the surface of the recording room – that may all be attenuated to different extents. Let $B_{np}(t)$ be the time-domain unmixing filters for source n. Denoting by $F_{p,\mathbf{z}}(t)$ the acoustic impulse response representing the direct path between some position $\mathbf{z}$ and microphone p, the response of the unmixing filters to a sound of frequency ν at position $\mathbf{z}$ is given by [12]:

$$\check{G}_n(\mathbf{z},\nu) = \sum_{p=1}^{P} \check{B}_{np}(\nu)\check{F}_{p,\mathbf{z}}(\nu) \tag{19.4}$$

where $\check{B}_{np}(\nu)$ and $\check{F}_{p,\mathbf{z}}(\nu)$ are the Fourier transforms of $B_{np}(t)$ and $F_{p,\mathbf{z}}(t)$.

Each acoustic impulse response is the combination of the microphone transfer function depending on the sound *direction of arrival* (DOA) and a delay and a gain depending on the traveled distance. In the particular case when the sources are in the far field, i.e. far from the microphones with respect to their wavelength, and the microphones have similar spatial directivities, $F_{p,\mathbf{z}}(t)$ can be approximated by a constant gain for all microphones and a distinct delay $\delta_p(\mathbf{z})$ for each microphone. The *inter-microphone time difference* (ITD) of arrival is then equal to [12]:

$$\text{ITD}_{pq}(\mathbf{z}) = \delta_q(\mathbf{z}) - \delta_p(\mathbf{z}) = \frac{d_{pq}}{c}\cos\theta_{pq}(\mathbf{z}) \tag{19.5}$$

where d_{pq} is the distance between the microphones, c the speed of sound, i.e. about 340 m/s, and $\theta_{pq}(\mathbf{z})$ the DOA relative to the microphone axis oriented from p to q. Therefore, for any reference microphone p, the magnitude response (also termed the *directivity pattern*) of the unmixing filters for source n is given up to a multiplicative constant by

$$|\check{G}_n(\mathbf{z},\nu)| \propto \left| \sum_{q=1}^{P} \check{B}_{nq}(\nu) \exp\left(-2i\pi\nu \frac{d_{pq}}{c}\cos\theta_{pq}(\mathbf{z})\right) \right|. \tag{19.6}$$

Fig. 19.10 represents possible unmixing filters for a target source at a DOA of 125° recorded in a concert room, in the presence of an interfering source at a DOA of 70° whose mixing filters are shown in Fig. 19.5. The unmixing filter system in this example was computed by inversion of the mixing filter system, resulting in maximum SDR [66]. The directivity pattern of the unmixing filters typically exhibits several notches at each frequency, among which one is close to the DOA of the interfering source. The exact position of that notch varies with frequency, so as to reject not only direct sound from that DOA but also echoes from other DOAs [39]. Other notches contribute to the attenuation of reverberation, although complete removal of reverberation is not feasible due to its diffuse DOA [39]. No notches appear at the frequencies ν where the target and interference DOAs result in the same ITD up to a multiple of $1/\nu$. Separation via multichannel filtering is hence not infeasible at these frequencies.

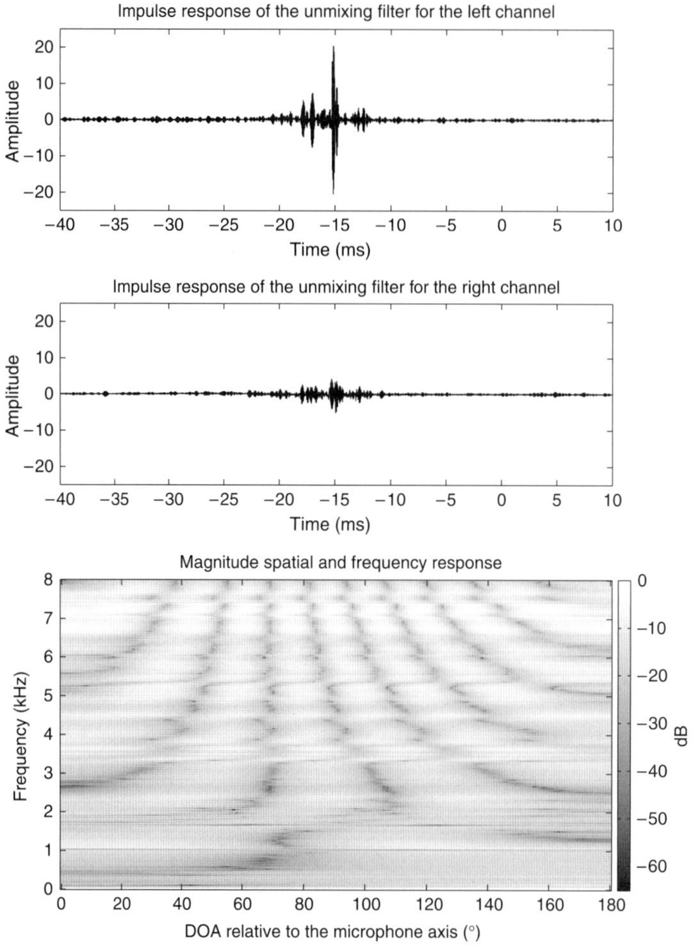

FIGURE 19.10

Example unmixing filters for a target source at a DOA of 125° recorded in a concert room in the presence of an interfering source at a DOA of 70°.

Interference cancellation may be improved in theory by increasing the number of notches or their spatial width. This requires increasing the number of microphones or the length of the unmixing filters [12]. Hence the separation performance achievable via multichannel filtering admits a theoretical upper bound for a given mixture depending on these quantities. Figure 19.11 displays the average upper bound measured in terms of SDR over the two-source convolutive mixtures of the audio BSS dataset in [66]. These mixtures were obtained by filtering sixty pairs of speech or music sources with two or more mixing filters of varying reverberation times. Theoretical upper bounds were computed via the software in [66]. The results show that

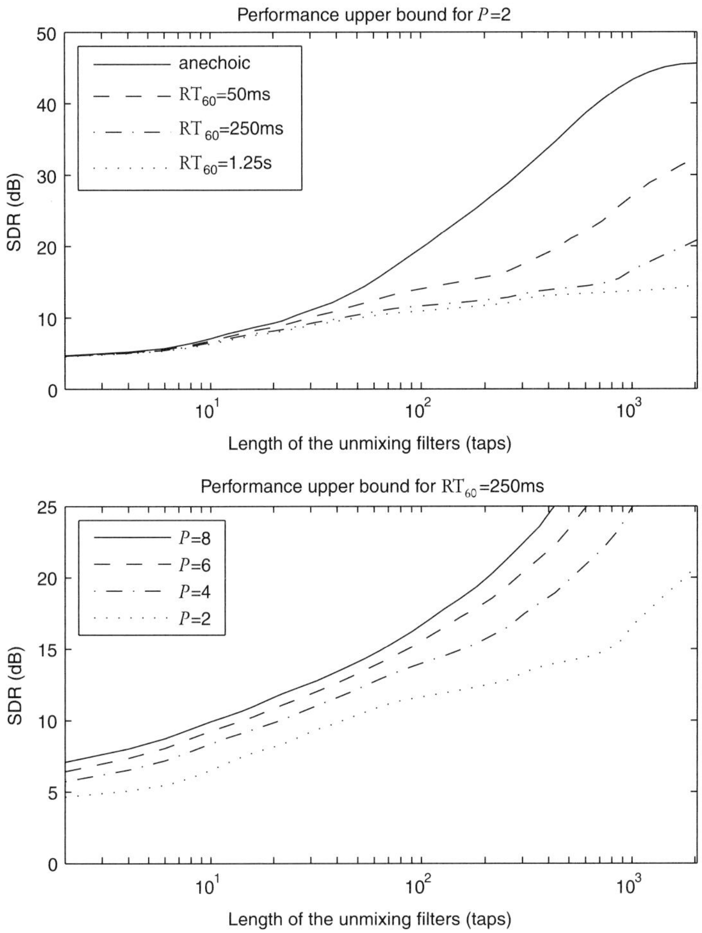

FIGURE 19.11

Average performance upper bound of multichannel filtering for the two-source convolutive audio mixtures in [66] sampled at 22.05 kHz, as a function of the number of mixture channels P and the reverberation time RT_{60}.

multichannel filtering can potentially provide an SDR of 20 dB or more for determined mixtures with short reverberation time up to $\mathrm{RT}_{60} = 50$ ms using unmixing filters of a few hundred taps. This upper bound decreases with longer reverberation times; nevertheless an SDR up to 15 dB may still be achieved with unmixing filters of a few hundred taps [39]. Increasing the number of microphones results in monotonous increase of the SDR. Additional experiments have shown that the upper bound further decreases by a few dB when the sources are moving [8,66]. Indeed interference must then be canceled over a range of positions even within a short duration.

19.3.2 Time-domain convolutive ICA

19.3.2.1 Example algorithms

Convolutive ICA computes suitable FIR unmixing filters by maximizing some contrast function (see Chapter 3) of the estimated sources. Time-domain convolutive ICA algorithms implement this approach by computing the impulse response of the unmixing filters according to some time-domain contrast function. Example contrast functions previously used for audio include:

- likelihood under an iid source model [14,15,18,33,61,71] (see section 4.5)
- likelihood under a Gaussian non-stationary source model [28] (see section 4.6)
- higher-order statistics-based contrasts [41] (see Chapters 3 and 5)
- time-delayed decorrelation [19,69] (see Chapter 7).

Likelihood-based contrast functions have been derived from different iid source priors, corresponding to a variety of score functions φ_n (see Chapter 4):

- linear: $\varphi_n(s_n) = s_n$ [14,15] (Gaussian prior)
- cubic: $\varphi_n(s_n) = s_n^3$ [14,18,61] (generalized Gaussian prior with $\tau = 4$)
- hyperbolic: $\varphi_n(s_n) = \tanh(3\, s_n)$ [15]
- negentropic: $\varphi_n(s_n) = s_n \exp(-s_n^2/2)$ [61]
- sign: $\varphi_n(s_n) = |s_n|^{-1} s_n$ [14,15,24,33] (Laplacian prior)
- root: $\varphi_n(s_n) = |s_n|^{\tau-2} s_n$ [71] (generalized Gaussian prior with $\tau < 2$)
- adaptive: $\varphi_n(s_n) = \omega_{n1} s_n + \omega_{n2} s_n^3 + \omega_{n3} |s_n|^{-1/2} s_n$, with weights ω_{n1}, ω_{n2} and ω_{n3} adaptively derived from the estimated sources [14].

Note that the use of the linear score function requires mixing filters to be strictly causal as justified in [14] and that the above adaptive score function is specific to [14] but that alternative adaptive functions could be similarly defined. The algorithm in [14] performs explicit whitening of the estimated sources, hence applying the above score functions to whitened sources instead of actual sources. Unmixing filters are computed by iterative online or batch gradient-based maximization of the contrast function. Some algorithms rely on prewhitening and DOA-based initialization [24] or initialization via shorter unmixing filters [69] to favor convergence to a relevant maximum of the contrast function. Few algorithms address the separation of mixtures of moving sources [28,69]. References to other time-domain convolutive ICA algorithms may be found in [35,46].

19.3.2.2 Performance evaluation

The above algorithms were each evaluated on different audio data by their authors. The results in [33,71] are unfortunately not comparable with others due to the lack of a specified performance measure or to the use of a non-standard measure. The performance of the algorithms in [14,15,18,19,24,28,61,69] is reported in Table 19.1, using the sign score function in [15] and the negentropic score function in [61]. A SIRI up to 15 dB is achieved by the latest algorithms for two-channel recordings of two speech sources at fixed locations, while a SIRI of 7 dB is obtained with abrupt microphone movements. The former figure compares favorably with the SDR upper bound plotted

Table 19.1 Performance evaluation of some time-domain convolutive ICA algorithms on P-channel mixtures of N speech sources

Algorithm	N	P	Mixing technique	Performance
Choi & Cichocki [15]	2	2	Convolution by synthetic FIR filters of order 21	SDRI = 5.0 dB
Ehlers & Schuster [19]	2	2	Microphone recording from loudspeaker playback (office room)	WERD = 80%
Charkani & Deville [14]; Albouy & Deville [2]	2	2	Convolution by synthetic FIR filters of order 13	WERD = 92%
Ito *et al.* [28]	2	2	Microphone recording from playback on moving loudspeakers ($RT_{60} = 810$ ms)	SDR = 8.9 dB
Douglas *et al.* [18]	2	2	Convolution by recorded room impulse responses ($RT_{60} = 130$ ms)	SIRI = 8.9 dB
Thomas *et al.* [61]	2	2	Convolution by recorded head related transfer functions of order 64	SIR = 8.0 dB
Gupta & Douglas [24]	2	2	Microphone recording from loudspeaker playback ($RT_{60} = 300$ ms)	SIRI = 15.3 dB
	3	3		SIRI = 17.1 dB
Wehr *et al.* [69]	2	2	Convolution by room impulse responses recorded for moving microphones ($RT_{60} = 50$ ms)	SIRI = 7.0 dB

in Fig. 19.11, which equals for instance 21 dB for 2048-tap filters with $RT_{60} = 250$ ms. When coupled with a small-vocabulary automatic speech recognizer, some algorithms achieve a dramatic decrease of the WER of 80% or more. Interestingly, it has been shown experimentally in [28] that, when recording the mixture via directional microphones, placing them at the same position improves performance compared to placing them apart.

The algorithm by Charkani and Deville [14] was studied more deeply by its authors than other algorithms. Five experiments were conducted over two-channel mixtures of speech or speech plus noise sampled at 8 kHz, considering several mixing situations so as to quantify their impact on the separation performance. Four contrast functions were also compared, using the linear, cubic, sign and adaptive score functions defined above. In all experiments, performance was assessed in terms of SIRI. Due to the online nature of the algorithm, this quantity was measured over a 10 s time interval after skipping the initial 10 s of the estimated sources. Although the measured SIRIs may not be up to the state of the art anymore, they provide an interesting view of the factors affecting separation performance that remains valid for more recent algorithms. We summarize below the main results of this investigation. A detailed description, including the values of mixing filter coefficients, may be found in [14].

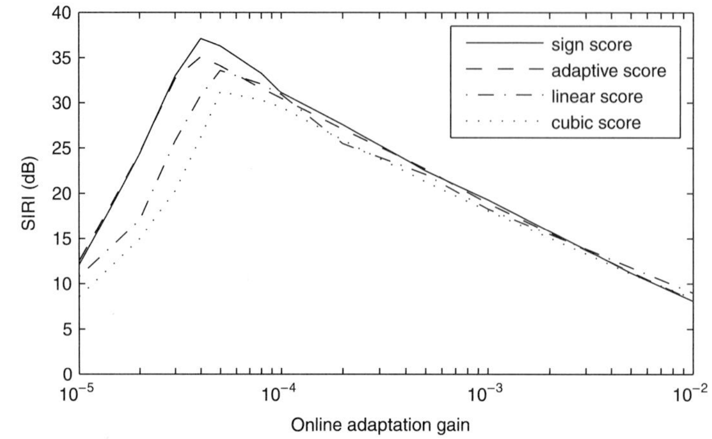

FIGURE 19.12

Performance of Charkani and Deville's time-domain convolutive ICA algorithm [14] over synthetic 4th-order convolutive two-channel mixtures of two speech sources, as a function of the online adaptation gain and the chosen score function.

The first experiment aims to highlight the relation of the achieved SIRI to the online adaptation gain of the algorithm. This relation is illustrated in Fig. 19.12 for synthetic convolutive mixtures obtained by convolving two speech signals with strictly causal FIR filters of order 4. For all contrast functions, the SIRI increases with increasing adaptation gain up to a critical gain beyond which it decreases. This phenomenon can be explained as follows. When the adaptation gain is very small, the algorithm converges slowly so that the time interval necessary to achieve convergence may overlap with the time interval over which the SIRI is measured. As a consequence, the measured SIRI is small because the estimated source signals are still partly mixed. Choosing a larger adaptation gain increases convergence speed so that convergence is reached before the beginning of the time interval over which the SIRI is measured. However, further increasing the adaptation gain yields less accurately separated output signals, due to larger fluctuations of the unmixing filter coefficients around equilibrium values once convergence has been reached. This in turn results in a reduced SIRI. The critical adaptation gain yielding maximum SIRI is slightly different for each contrast function. Therefore, in the following experiments, performance is always measured by selecting the best adaptation gain *a posteriori* for each contrast function.

The second experiment presented in Table 19.2 compares the performance of the algorithm over the synthetic mixtures considered in the first experiment to synthetic speech mixtures obtained with longer strictly causal FIR filters of order 13. These filters remain extremely short compared to real-world source-microphone impulse responses. Nevertheless, the range of SIRI drops from 31 to 37 dB for 4th-order mixtures to 16 to 26 dB for 13th order mixtures. For both types of mixtures, maximum performance is achieved with the sign score function, closely followed by the adaptive score function.

Table 19.2 Performance of Charkani and Deville's time-domain convolutive ICA algorithm [14] over synthetic two-channel mixtures of two speech sources, as a function of the order of the FIR mixing filters and the chosen score function. The best adaptation gain is assumed for each mixture

Mixing filter order (taps)	SIRI (dB)			
	Linear score	Cubic score	Sign score	Adaptive score
4	33.6	31.2	37.1	35.2
13	17.9	16.4	25.8	25.2

Table 19.3 Performance of Charkani and Deville's time-domain convolutive ICA algorithm [14] over mixtures of two speech sources played by loudspeakers and recorded in a room, as a function of the microphone spacing and the chosen score function. The best adaptation gain is assumed for each mixture

Microphone spacing (cm)	SIRI (dB)			
	Linear score	Cubic score	Sign score	Adaptive score
5	3.7	4.3	7.4	8.0
10	5.2	5.1	9.1	9.1
15	4.9	4.6	10.0	10.3
20	4.8	5.7	9.6	10.0
25	5.0	6.3	10.0	10.0
30	5.4	6.4	9.7	10.0

The linear and cubic score functions perform worse, particularly when the mixing filters are longer. This is consistent with the fact that whitened speech signals are sparse in the time domain, as demonstrated in the previous section. The Laplacian prior underlying the sign score function is sparse. The adaptive score function can also model a range of sparse priors, among which the sparsest is the generalized Gaussian prior with shape parameter $\tau = 3/2$ associated with $\varphi_n(s_n) = |s_n|^{-1/2} s_n$, which is less sparse than the Laplacian. By contrast, the linear and cubic score functions correspond to non-sparse priors.

The third experiment addresses the separation of recordings performed in a room without acoustic treatment. Two speech signals are emitted by loudspeakers and recorded by an array of microphones. Tests are successively carried out considering different couples of microphone signals corresponding to increasing microphone spacings. Several mixtures of the considered two speech sources are thus independently processed, which makes it possible to investigate the influence of microphone spacing on performance. The resulting SIRIs are shown in Table 19.3. Performance is clearly reduced compared to synthetic mixtures, with SIRIs in the range between 4 and 10 dB. The adaptive score function now provides slightly better performance than the sign score function. Also, the

Table 19.4 Performance of Charkani and Deville's time-domain convolutive ICA algorithm [14] over mixtures of two speech sources played by loudspeakers and recorded in a car, as a function of the microphone spacing and the chosen score function. The best adaptation gain is assumed for each mixture

Microphone spacing (cm)	SIRI (dB)			
	Linear score	Cubic score	Sign score	Adaptive score
5	3.0	1.8	3.4	3.2
10	3.9	2.4	5.8	6.2
15	4.0	4.0	7.8	7.8
20	4.9	5.2	7.6	7.6
25	5.2	5.6	7.2	7.2
30	6.9	7.5	9.5	9.5

SIRI can be seen to decrease when the microphones have close locations, thus illustrating the lack of robustness of the algorithm to ill-conditioned mixtures.

The fourth experiment involves microphone recordings again, except that they are performed in a steady car, with the engine off. The source loudspeakers are situated at the driver's and front passenger's head positions and the microphones are arranged as in Fig. 19.3. The resulting SIRIs listed in Table 19.4 yield the same comments as with room recordings, except that a larger performance degradation occurs here when the microphone spacing is smaller.

In the fifth and last experiment, the signals recorded when playing a speech source through the loudspeaker situated at the driver's head position are added to real noise recordings made in a moving car with the same microphone array. This situation is closer to the real-world issue of speech enhancement for in-car hands-free use of mobile phones, except that the target source does not move here. The algorithm by Charkani and Deville was applied to segregate the speech source. The sign score function was employed for that source and the linear score function for the noise source, in accordance with their statistical distributions. The influence of microphone spacing is crucial in this context because it results in conflicting requirements. The noise source is spatially diffuse, due, for example, to air flow around the car, vibrations of the car structure and engine noise, so that its recordings at the microphones are not coherent. Therefore multichannel filtering can achieve only limited attenuation of this source. This issue can be addressed using closer microphones. The noise signals recorded by the two microphones then become more coherent, hence more compatible with the assumed convolutive mixture model where the contributions of a source to two microphones are equal up to a filter. However, this approach conflicts with the performance degradation for close microphones discussed above. As a result, the best SIRI obtained for this mixture was about 2 dB, which does not yield the desired communication quality.

Overall, these experiments indicate that separation performance over stereo mixtures dramatically decreases with

- increasing reverberation time;
- decreasing microphone distance;
- diffuse interfering sources.

These limitations remain a concern for more recent time-domain convolutive ICA algorithms and can be mitigated to a certain extent by using more than two microphones, as in many modern beamforming-based hands-free devices.

19.3.3 Frequency-domain convolutive ICA

19.3.3.1 Example algorithms

As an alternative to time-domain convolutive ICA, frequency-domain convolutive ICA algorithms compute the frequency responses of the unmixing filters according to some frequency-domain contrast function. The mixture channels are first transformed into the time-frequency domain by means of the STFT. The STFT coefficients of the sources are then estimated by multiplication of the mixture STFT coefficients with a complex-valued unmixing matrix within each frequency bin. Time-domain source signals are eventually obtained by inversion of the STFT.

The unmixing matrices in distinct frequency bins may be constrained to represent the Fourier transform of some time-domain filters. While this ensures approximate equivalence between frequency-domain and time-domain filtering, this also implies joint estimation of all matrices [45,37]. Most algorithms bypass this constraint and estimate the unmixing matrix in each frequency bin independently instead. This is computationally more efficient but the estimated source signals are distorted due to the approximation of linear convolution by circular convolution. This issue can be partly addressed at low computation cost by increasing the overlap between successive STFT windows or by smoothing the estimated unmixing matrices over frequency by a carefully designed filter [57].

The problem of estimating a complex-valued unmixing matrix within each frequency bin is equivalent to the BSS of instantaneous mixtures. It can be addressed by an instantaneous ICA technique maximizing some contrast function (see Chapter 3) of the estimated source STFT coefficients. Example contrast functions employed in this context include:

- quadratic contrasts under a Gaussian non-stationary source model [27,44,45,52]
- likelihood under a Gaussian non-stationary source model [48] (see section 4.6)
- likelihood under a non-Gaussian iid source model [40,56,59] (see section 4.5)
- likelihood under a non-Gaussian non-stationary source model [37]
- time-delayed decorrelation [26] (see Chapter 7).

Note that alternative BSS techniques, including those based on quadratic time-frequency representations (see Chapter 11), could also possibly be used in this context, but have not been used so far. Non-gaussian likelihood-based contrasts were again derived from various nonlinear score functions φ_n (see Chapter 4), applied to the STFT coefficients $\check{s}_n$ of the sources:

- $\varphi_n(\check{s}_n) = |\check{s}_n|^{-1}\check{s}_n$ [37,56] (circular Laplacian prior)
- $\varphi_n(\check{s}_n) = \tanh(100|\check{s}_n|)|\check{s}_n|^{-1}\check{s}_n$ [40] (circular smoothed Laplacian prior)
- $\varphi_n(\check{s}_n) = \tanh(\Re(\check{s}_n)) + \tanh(\Im(\check{s}_n))$ [59] (non-circular prior).

The first two score functions are theoretically more appropriate than the third one, since the STFT coefficients of audio sources have circular distributions in the complex plane [37]. A few frequency-domain ICA techniques also consider the separation of mixtures of moving sources [40].

By exploiting the independence of the sources, frequency-domain convolutive ICA techniques can estimate the source signals within each frequency bin. However, due to the inherent permutation ambiguity of BSS (see section 1.2), additional information is needed to identify which source estimates correspond to the same source within different frequency bins. This so-called *permutation problem* may be addressed via two complementary approaches not detailed elsewhere in this book that we present below.

19.3.3.2 Approaches to the permutation problem

A first approach is to rely on the dependencies between the frequency responses of acoustic mixing filters at different frequencies. Assuming a near-anechoic recording environment and similar directivities for all microphones, the relative acoustic impulse response between the sound of a far-field source n as recorded on microphones p and q can be approximated as a simple delay filter, where the delay is equal to the ITD defined in (19.5) as a function of the DOA θ_{pqn}. The frequency response $\check{A}_{pn}(\nu)$ of the mixing filters at frequency ν can be deduced from the estimated unmixing matrix $\check{\mathbf{B}}(\nu)$ by matrix inversion $\check{\mathbf{A}}(\nu) = \check{\mathbf{B}}^{-1}(\nu)$, or by pseudo-inversion when there are fewer sources than microphones. Provided that the unmixing matrix was accurately estimated, we get

$$\frac{\check{A}_{qn}(\nu)}{\check{A}_{pn}(\nu)} \approx \exp\left(-2i\pi\nu\frac{d_{pq}}{c}\cos\theta_{pqn}\right). \tag{19.7}$$

This equation shows that the ratio between the frequency response of the estimated mixing filter for microphone q and that of any reference microphone p varies continuously over frequency under the above conditions. As a consequence, some authors have assumed that the ratio between the unmixing coefficients at these two microphones also varies continuously over frequency. This assumption has been employed to define a continuity constraint over frequency-domain unmixing coefficients in neighboring frequency bins in [59] or equivalently a constraint over the length of the corresponding time-domain unmixing filters in [45]. These constraints are implemented by modifying ICA iterative update rules. The above equation may also be exploited to estimate source permutations between non-neighboring frequency bins. This has been done either by adding a constraint to the ICA update rules ensuring proximity between the two sides of the equation [44] or by computing frequency-dependent estimates of the source DOAs via

$$\theta_{pqn}(\nu) = \arccos\left(-\frac{c}{2\pi\nu d_{pq}}\angle\left(\frac{\check{A}_{qn}(\nu)}{\check{A}_{pn}(\nu)}\right)\right), \tag{19.8}$$

with $\angle$ denoting the phase of a complex number in $(-\pi, \pi]$, deriving a single average DOA θ_{pqn} for each source by clustering and permuting the sources so that their frequency-dependent DOAs are aligned with the average DOAs [40]. In the latter case, standard ICA update rules are used and permutation is applied as a postprocessing step once ICA has converged. This helps in avoiding local maxima issues arising with the above constrained optimization-based techniques [37]. However, frequency-dependent DOAs may be computed only for frequencies ν below the critical frequency c/d_{pq}, since the ITD can be estimated only up to a multiple of $1/\nu$ above that frequency due to phase rolling. This critical frequency is equal to 6.8 kHz for a microphone spacing of 5 cm or 340 Hz for a spacing of 1 m. This phenomenon known as *spatial aliasing* may be circumvented by computing the average DOAs over frequencies below the critical frequency and considering all possible DOAs above that frequency, computed either from (19.7) or by locating the notches of the directivity patterns of the unmixing filters in (19.6) [27]. When directional microphones are used or when some sources are in the near field, the *inter-microphone intensity difference* (IID) defined as

$$\text{IID}_{pqn}(\nu) = 20\log_{10}\left|\frac{\check{A}_{qn}(\nu)}{\check{A}_{pn}(\nu)}\right| \tag{19.9}$$

may also be used to help identify the correct permutations [55].

Figures 19.13 and 19.14 depict the frequency-dependent ITD and IID computed from the pairs of source-microphone impulse responses shown in Figs 19.4 and 19.5. The ITD is approximately constant above 1.2 kHz or 1.7 kHz respectively. However, large deviations up to 1 ms from the average ITD can be observed at lower frequencies. These deviations due to the fact that echoes are stronger than direct sound at certain frequencies are particularly noticeable for in-car impulse responses, which include many strong echoes, whereas the concert room impulse responses include only a few significant echoes. Also, aliased ITDs corresponding to the same phase difference up to a multiple of 2π become closer to the true ITD when frequency increases. Therefore, in reverberant environments, the ITD is a reliable cue for the estimation of source permutations in the medium frequency range only [56]. The IID appears less reliable since it exhibits deviations up to 40 dB from the average IID over the whole frequency range.

An alternative approach to address the permutation problem is to exploit the dependency between the time series of time-frequency coefficients of a given source in different frequency bins. This dependency can be measured for instance by linear correlation

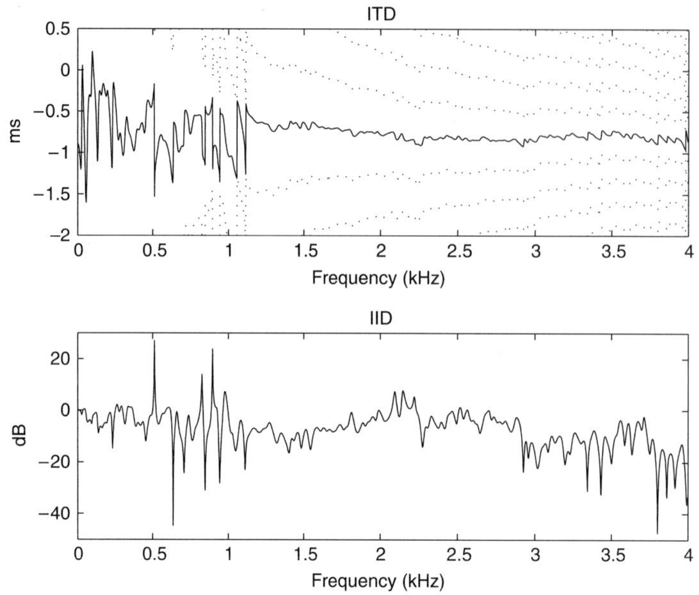

FIGURE 19.13

Measured frequency-dependent ITD and IID for the in-car source-microphone impulse responses of Fig. 19.4. Aliased ITDs corresponding to the same phase difference are represented as dotted curves.

between the centered magnitude STFT coefficients $v_n(\nu, t)$ defined as [56]:

$$v_n(\nu,t) = |\check{s}_n(\nu,t)| - \frac{1}{T}\sum_{t=1}^{T}|\check{s}_n(\nu,t)| \tag{19.10}$$

where $\check{s}_n(\nu, t)$ is the STFT coefficient of source n at frequency ν and time t. The correlation between the time series of centered magnitude STFT coefficients of source k at frequency ν and source l at frequency ν' is given by

$$MC\{s_k, s_l\}(\nu,\nu') = \frac{\frac{1}{T}\sum_{t=1}^{T} v_k(\nu,t)v_l(\nu',t)}{\sqrt{\left(\frac{1}{T}\sum_{t=1}^{T} v_k(\nu,t)^2\right)\left(\frac{1}{T}\sum_{t=1}^{T} v_l(\nu',t)^2\right)}}. \tag{19.11}$$

For any pair of frequency bins ν and ν', this quantity is assumed to be larger for the same source ($k = l$) than for different sources ($k \neq l$). Indeed, provided that the sources are independent, correlation should be near zero for different sources but positive for the same source due to the wideband character of speech and music sounds. Many

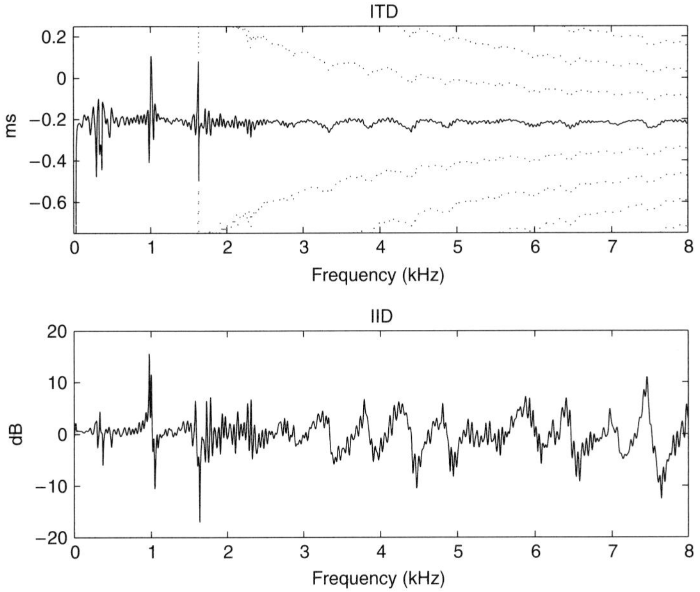

FIGURE 19.14

Measured frequency-dependent ITD and IID for the concert room source-microphone impulse responses of Fig. 19.5. Aliased ITDs corresponding to the same phase difference are represented as dotted curves.

alternative dependency measures have been considered, including the correlation between non-centered magnitude STFT coefficients after additional temporal smoothing [26], the correlation between non-centered power STFT coefficients [52], the Euclidean distance between centered log-magnitude STFT coefficients [48] and the likelihood of a non-stationary source model with common variance for all frequencies [37]. Several algorithms have also been proposed to derive source permutations from these measures, including deciding permutations for each frequency bin successively after sorting them in an appropriate order [26], deciding permutations for increasingly large sets of neighboring bins in a hierarchical clustering fashion [52] or interleaving permutations with ICA update rules [37].

We investigated the suitability of this approach over the speech and music sources from [66], arranged into sixty pairs of sources. The STFT was computed with sine windows of length 1024 (46 ms). On average, the correlation between the centered magnitude STFT coefficients of any two frequency bins was larger for the same source than for different sources in 97% of the cases for music but in 89% of the cases only for speech. This is illustrated by example correlation values between two pairs of speech and music sources given in Figs 19.15 and 19.16 respectively. The correlation between centered magnitude STFT coefficients of the same source is always large for neighboring bins, but may be smaller for bins further apart [56]. For instance, coefficients at

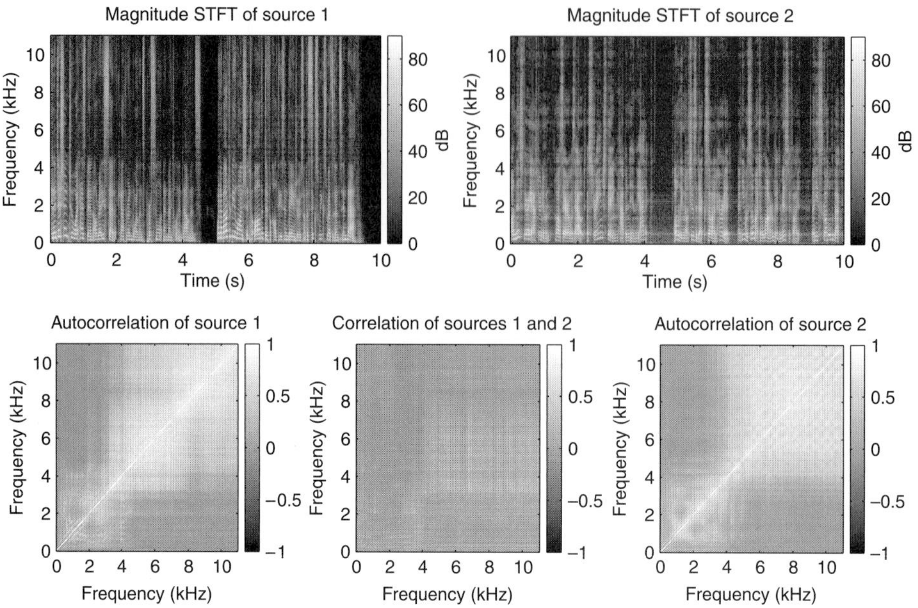

FIGURE 19.15

Across-frequency correlation $MC\{s_k, s_l\}(\nu, \nu')$ between the centered magnitude STFT coefficients of two female speech sources as a function of ν and ν' for $k = l = 1$ (lower left), $k = 1$ and $l = 2$ (lower center) and $k = l = 2$ (lower right).

low and high frequencies are almost uncorrelated in the case of speech. This can be explained by the fact that low frequency bins mostly contain periodic speech sounds, whose time evolution is distinct from that of transient speech sounds present at higher frequencies.

Overall, Figs 19.13 to 19.16 suggest that DOA and across-frequency magnitude correlation are complementary cues to address the permutation problem, as argued in [56]: magnitude correlation is reliable between neighboring frequency bins whatever their frequency, while DOA is reliable between frequency bins further apart but not over the whole frequency range. This observation has led to the design of a joint approach, consisting of fixing permutations in the frequency bins where the confidence in the DOA is sufficiently high, then deciding permutations in the remaining frequency bins based on magnitude correlation. DOAs are deemed to be reliable in a given frequency bin if the estimated frequency-dependent DOAs do not differ much from the average DOAs over the whole frequency range and the resulting inter-microphone phase differences are sufficiently distinct. Experimental comparison of all approaches showed that magnitude correlation-based permutation results in better separation performance than DOA-based permutation for speech sources, but that the joint exploitation of both cues performs even better [56]. Another joint approach relying on inter-frequency similarity of the unmixing matrices and on magnitude correlation has been proposed in [48]. References

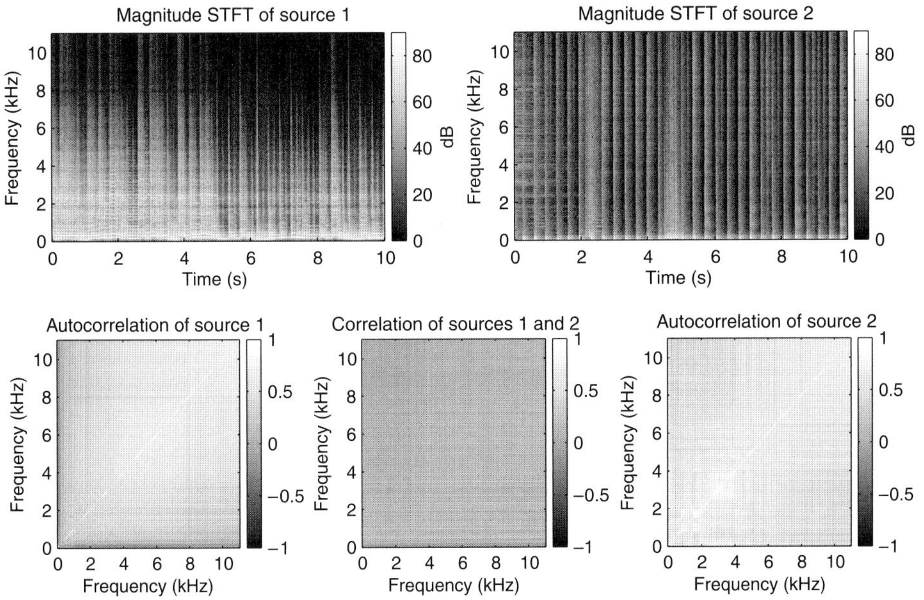

FIGURE 19.16

Across-frequency correlation $MC\{s_k, s_l\}(\nu, \nu')$ between the centered magnitude STFT coefficients of two synchronized music sources, namely an electric guitar and a drum kit, as a function of ν and ν' for $k = l = 1$ (lower left), $k = 1$ and $l = 2$ (lower center) and $k = l = 2$ (lower right).

to alternative frequency-domain convolutive ICA algorithms and solutions to the permutation problem can be found in [35,46].

19.3.3.3 Performance evaluation

The above frequency-domain convolutive ICA algorithms were each evaluated on different audio data by their authors. The results in [48,59] are unfortunately not comparable with others due to the lack of an objective performance measure or to the use of a non-standard measure. The performance of the algorithms in [26,27,37,40,44,45,52,56] is reported in Table 19.5, considering a few experimental situations per algorithm. A SIRI up to 16 dB is achieved by the latest algorithms for two-channel recordings of two speech sources at fixed positions in an office room. Additional SIRI on the order of a few dB can be achieved by using more microphones than sources. These figures are similar to those obtained with state-of-the-art time-domain convolutive ICA algorithms, as discussed in section 19.3.2.2, and compare favorably with the SDR upper bound in Fig. 19.11. The results of algorithm [45] further indicate that a limited SIRI of 5 dB is sufficient to provide a 30% decrease of WER in the context of large-vocabulary automatic speech recognition. Detailed experimental evaluations involving alternative room sizes, reverberation times, numbers of microphones or moving sources are provided in [27,40,45].

Table 19.5 Performance evaluation of some frequency-domain convolutive ICA algorithms on P-channel mixtures of N speech sources

Algorithm	N	P	Mixing technique	Performance
Ikeda & Murata [26]	2	2	Convolution by simulated room impulse responses ($RT_{60} = 140$ ms)	SIR = 10.3 dB
Parra & Spence [45]	2	2	Microphone recording of actual speakers (office room)	SIRI = 4.6 dB WERD = 30%
Parra & Alvino [44]	3	8	Microphone recording from loudspeaker playback ($RT_{60} = 100$ ms)	SIRI = 7.9 dB
Ikram & Morgan [27]	2	2	Convolution by simulated room impulse responses ($RT_{60} = 40$ ms)	SIR = 6.8 dB
			Convolution by simulated room impulse responses ($RT_{60} = 130$ ms)	SIR = 4.7 dB
Mitianoudis & Davies [37]	2	2	Convolution by synthetic filters (single echo)	SDRI = 5.9 dB
			Convolution by recorded impulse responses (conference room)	SDRI = 3.7 dB
Mukai *et al.* [40]	2	2	Microphone recording from loudspeaker playback ($RT_{60} = 130$ ms)	SIRI = 16.0 dB
Sawada *et al.* [56]	2	2	Convolution by recorded impulse responses ($RT_{60} = 300$ ms)	SIR = 16.8 dB
Rahbar & Reilly [52]	2	4	Microphone recording from loudspeaker playback (RT_{60} in the range from 120 ms to 880 ms)	SIR = 21.0 dB
			Microphone recording from loudspeaker playback (RT_{60} in the range from 490 ms to 1.25 s)	SIR = 16.0 dB

19.4 AUDIO APPLICATIONS OF SCA

While most of the chapters of this book deal with ICA and hence with determined or over-determined mixtures, audio mixtures are often under-determined. These mixtures can be separated by SCA (see Chapter 10), as justified by the sparsity properties of audio sources examined in section 19.2.2. The range of application of SCA is not limited to under-determined mixtures but extends to mixtures of any number of sources. Among the available SCA techniques, the most popular belong to the class of time-frequency masking (see section 10.6). After representing the mixture in the time-frequency domain, separation is achieved by finding a set of $M \leq P$ active sources in each time-frequency bin and estimating the coefficients of active and inactive sources via two different operations. The time-frequency coefficients of the M active sources are derived by multiplying the mixture coefficients with the pseudo-inverse of the estimated mixing matrix restricted to these sources, while those of the N-M inactive sources are set to zero. In the particular case when a single source is assumed to be active, the resulting technique is called *binary masking*. While the study in Chapter 10 focuses on instantaneous mixtures, these techniques are also applicable to convolutive mixtures by modeling the mixing process as a complex-valued mixing matrix in each frequency bin. In the following, we analyze the potential separation performance achievable via time-frequency masking on audio mixtures, then review some SCA algorithms and compare their actual performance.

19.4.1 Time-frequency masking

Time-frequency masking attenuates interfering sounds based on their time-frequency distribution and to a lesser extent on their spatial position. It is most often expressed using an STFT representation, although the modified discrete cosine transform (MDCT) [38] or perceptually motivated time-frequency representations [63] have also been employed. Interfering sounds in time-frequency bins where the target source is inactive are fully canceled by the zeroing operation, while those in time-frequency bins where the target source is active are attenuated by the pseudo-inversion operation which results in a form of beamforming [54]. Contrary to multichannel filtering, time-frequency masking typically results in little residual interference within the estimated sources [65]. However, it generates "gurgling" artifacts, due to time-frequency discontinuities induced by the zeroing operation. These discontinuities can be reduced by choosing a time-frequency transform with as much overlap as possible between neighboring bins and by computing the inverse time-frequency transform using a smooth synthesis window [5].

Figure 19.17 illustrates the separation of three male speech sources from a two-channel instantaneous mixture via binary time-frequency masking using the STFT with a 46 ms sine window. The time-frequency masks in this example were computed using knowledge of the true sources so as to assess the maximum achievable SDR [66]. The results indicate that the three sources have reasonably disjoint time-frequency supports, since the estimated sources are similar to the true sources. Nevertheless, missing regions can be found in the time-frequency representation of each estimated source, for instance at low frequencies around $t = 1.3$ s for source 1, which is dominated by another source in this region.

In theory, separation performance can be improved by estimating $M \geq 2$ active sources per time-frequency bin instead of $M = 1$ active source for binary masking or by choosing the length of the time-frequency analysis window so that the resolution of the time-frequency transform is appropriate and the sources are maximally disjoint. Figure 19.18 shows the performance upper bound achievable by time-frequency masking as a function of these quantities, averaged over the two-channel, three-source mixtures of the audio BSS dataset in [66]. Instantaneous mixtures and convolutive mixtures with reverberation time $RT_{60} = 250$ ms are both considered. Theoretical upper bounds were computed via the software in [66]. The results show that binary time-frequency masking can potentially provide an SDR of 13 dB over instantaneous and convolutive mixtures using a window length of the order of 100 ms. More precisely, the optimal window length equals 60 ms for speech [73] and 200 ms for music [66] on average. For instantaneous mixtures,

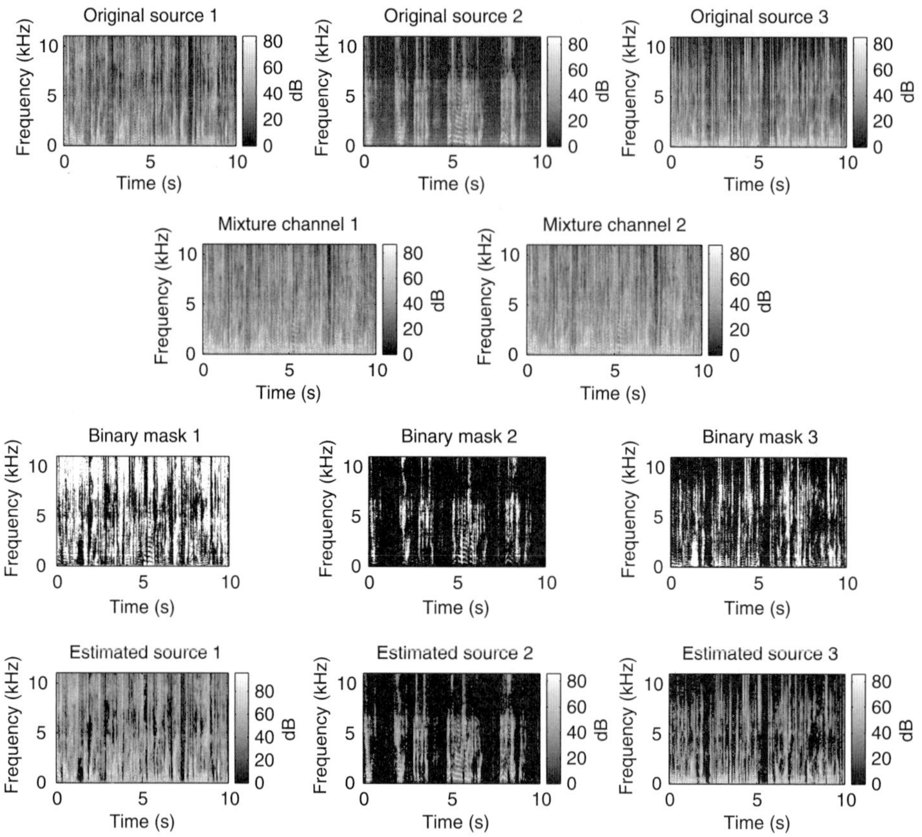

FIGURE 19.17

Example binary time-frequency masks for the separation of three male speech sources from a two-channel instantaneous mixture using the STFT. Time-frequency bins are colored in white whenever the source is considered to be active and in black otherwise.

estimating two active sources per time-frequency bin instead of one can improve the SDR by up to 10 dB for the same window length. For convolutive mixtures, this can also improve performance by up to 7 dB as compared to binary masking. However, the optimal window length becomes much longer in this case, since the approximation of the mixing process by a complex-valued mixing matrix in each frequency bin is valid for long window lengths only.

19.4.2 Instantaneous SCA

19.4.2.1 Example algorithms

SCA algorithms for instantaneous mixtures typically consist of two successive steps: the estimation of the mixing matrix via a geometrical approach (see section 10.4) and the separation of the source signals given the estimated matrix (see section 10.6).

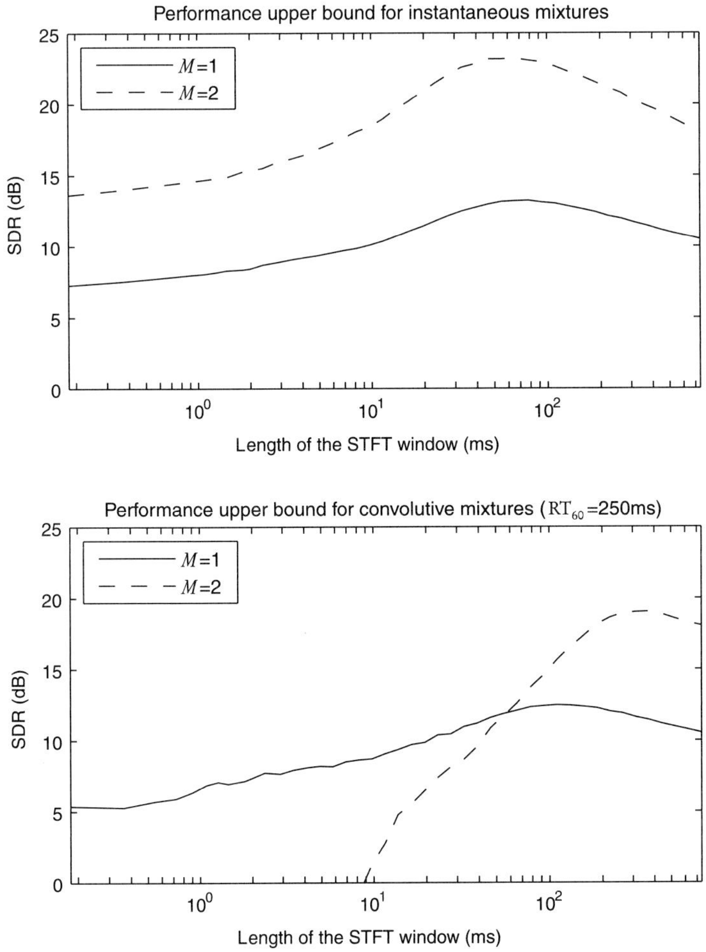

FIGURE 19.18

Average performance upper bound of time-frequency masking for the two-channel three-source audio mixtures in [66], as a function of the assumed number M of active sources per time-frequency bin.

Many approaches assume that the mixing matrix has non-negative entries and can be equivalently parameterized in terms of IID. Most approaches also rely on a *spatial coherence* measure quantifying the likelihood that a single source be active in each time-frequency bin (see sections 10.4.3 and 11.4). Example approaches previously used for mixing matrix estimation in the context of audio include:

- fuzzy clustering of the mixture MDCT coefficients [38]
- detection of local maxima of the smoothed magnitude-weighted histogram of the mixture STFT coefficients above a certain magnitude threshold [74]

- hard coherence-weighted clustering of the mixture STFT coefficients [6]
- hard magnitude-weighted clustering of the mixture STFT coefficients above a certain coherence threshold [10]
- hard coherence and power-weighted clustering of the mixture STFT coefficients [7]
- selection of the mixture STFT coefficients with highest coherence [1,16,17].

Various coherence measures have been defined, based either on the ratio between the largest eigenvalue and the sum of other eigenvalues of the covariance matrix of the mixture coefficients over some neighborhood of the considered time-frequency bin [6, 7,16,17] or on the variance of IID or similar direction-related quantities over such neighborhood [1,10,17].

Given the mixing matrix, several approaches have also been investigated to estimate audio source signals, including:

- binary masking of the mixture MDCT coefficients [38] (see section 10.6.2)
- soft masking of the mixture STFT coefficients over each channel [7]
- minimization of the ℓ^1-norm of the estimated source STFT coefficients, under the constraint that $M = P$ sources are active per time-frequency bin [74] (see section 10.6.4)
- minimization of the ℓ^1-norm of the estimated source STFT coefficients [11] (see section 10.6.4)
- minimization of the ℓ^τ-norm of the estimated source STFT coefficients with $\tau \ll 1$, resulting in $M = P$ active sources per time-frequency bin [64] (see section 10.6.4).

19.4.2.2 Performance evaluation

The algorithms in [6,10,11,38,64] were compared over a set of two-channel instantaneous mixtures of three to four audio sources in the context of the 2007 Stereo Audio Source Separation Evaluation Campaign (SASSEC) [67]. The results presented in Table 19.6 show that all algorithms provide reasonable interference rejection, as indicated by an average SIR above 13 dB, but that they introduce different amounts of artifacts as measured by the SAR. While binary masking provides a low average SAR of 5 dB, ℓ^τ-norm minimization achieves a SAR above 11 dB for a similar or slightly larger SIR. The average SDR of the algorithm in [64] over three-source mixtures equals 12 dB, which is more than 10 dB below the SDR upper bound for time-frequency masking with $M = 2$ active sources per time-frequency bin displayed in Fig. 19.18. Additional results for non-blind algorithms, non-SCA-based algorithms or unpublished algorithms over the same data can be found in [67].

19.4.3 Convolutive SCA

19.4.3.1 Example algorithms

SCA has also been used for the separation of anechoic or convolutive mixtures. The estimation of the mixing system for anechoic mixtures has been addressed in a similar way as for instantaneous mixtures by transforming the mixture signal into the STFT domain

Table 19.6 Average performance of some SCA algorithms on a common set of two-channel instantaneous mixtures of three to four audio sources, from [67]. The SDR achieved by algorithm [38] is not meaningful, since it estimated the contribution of the sources to the mixture up to an arbitrary scaling only, while this scaling was blindly estimated by the other algorithms

Sources	Average perf.	Algorithm		
		Bofill *et al.* [10]+[11]	Mitianoudis & Stathaki [38]	Arberet & Vincent [6]+[64]
Music $N=3$	SDR (dB)	4.5	N/S	12.1
	SIR (dB)	14.1	13.8	17.7
	SAR (dB)	12.7	7.0	15.0
Speech $N=4$	SDR (dB)	3.8	N/S	8.5
	SIR (dB)	11.6	12.7	14.2
	SAR (dB)	8.9	3.5	9.5

and parameterizing the mixing system by a common set of parameters for all frequency bins, namely ITDs and IIDs or related quantities. Due to the spatial aliasing phenomenon discussed in section 19.3.3.2, the ITDs can be derived from the STFT coefficients at a particular frequency ν above the critical frequency c/d_{pq} only up to some multiple of $1/\nu$. This phenomenon can be either circumvented by choosing small microphone spacings d_{pq} or addressed for any microphone spacing by combining the ITDs estimated at different frequencies. Example approaches previously used for audio include:

- hard weighted clustering of the mixture STFT coefficients, in the absence of spatial aliasing [73]
- selection of the mixture STFT regions with highest coherence spanning a number of successive frequency bins and derivation of the ITD by phase unwrapping [50].

The mixing system for convolutive mixtures can also be parameterized in terms of ITDs and IIDs. However, the estimation of these parameters is more difficult than with anechoic mixtures, since they vary with frequency. The problem of identifying the frequency-dependent mixing parameters corresponding to the same source is equivalent to the permutation problem of frequency-domain ICA (see section 19.3.3.2) and can be addressed using the same assumption, namely that the frequency-dependent ITDs and/or IIDs for a given source are close to the average ITD and IID associated with its DOA. This assumption has been employed for instance within the following approaches:

- hard clustering of the mixture STFT coefficients in each frequency bin and alignment of frequency-dependent ITDs with the average ITDs determined in the absence of spatial aliasing [70]

- joint hard clustering of the mixture STFT coefficients in all frequency bins under the constraint that cluster centers at each frequency correspond to the same ITD and IID [55]
- joint soft clustering of the mixture STFT coefficients in all frequency bins under the constraint that cluster centers at each frequency correspond to the same ITD and IID [29,36].

Some approaches have also been proposed relying on the assumption that the sources are often inactive over some time intervals. While this assumption is stronger than the time-frequency sparsity assumption employed by SCA, it is typically satisfied by speech. In [3], the mixing parameters for each source are directly obtained from the time frames when this source only is active, selected based on some coherence criterion. In [21], a single ITD is computed for each time frame and each microphone pair and soft clustering is performed to group ITDs corresponding to the same source. The latter is one of the rare approaches for the estimation of the mixing parameters of moving sources.

Once the mixing parameters have been inferred, the separation of the source signals can be addressed in a similar way as with instantaneous mixtures by modeling the mixing process as a complex-valued mixing matrix in each frequency bin. Approaches previously used for audio include:

- binary masking of the mixture STFT coefficients [55,73] (see section 10.6.2)
- masking of the mixture STFT coefficients with $M = P - 1$ active sources per bin [54] (see section 10.6.3)
- soft masking of the mixture STFT coefficients over each channel [29,36]
- minimization of the ℓ^1-norm of the estimated source STFT coefficients, under the constraint that $M = P$ sources are active per time-frequency bin [70] (see section 10.6.4)
- minimization of the ℓ^τ-norm of the estimated source STFT coefficients with $\tau \ll 1$, resulting in $M = P$ active sources per time-frequency bin [64] (see section 10.6.4)
- minimization of the sum of the ℓ^1-norm of the estimated source STFT coefficients and the ℓ^2-norm of the residual between the mixture STFT coefficients and the product of the mixing matrix and the estimated source coefficients [62] (see section 10.6.4).

19.4.3.2 Performance evaluation

Table 19.7 reports the average performance of the algorithms in [29,36,55] in the context of the 2007 SASSEC campaign mentioned above [67]. Evaluation was conducted on two-channel convolutive mixtures of three to four audio sources with revereberation time $RT_{60} = 250$ ms, obtained either by convolution with simulated room impulse responses or by microphone recording from loudspeaker playback with two different microphone spacings, namely 5 cm or 1 m. The average SDR varies between 0 and 4 dB depending on the algorithm and the mixing technique, that is more than 15 dB below the theoretical SDR upper bound for time-frequency masking with $M = 2$ active sources per time-frequency bin displayed in Fig. 19.18 and more than 8 dB below the theoretical upper

Table 19.7 Average performance of some SCA algorithms on a common set of two-channel convolutive mixtures of three to four audio sources with reverberation time $\mathrm{RT}_{60} = 250$ ms, from [67]. The average performance of algorithm [29] was computed over speech data in a single mixing situation only, since results for music data or for other mixing situations were not submitted

Mixing technique	Mic. spacing	Average perf.	Algorithm		
			Izumi *et al.* [29]	Mandel *et al.* [36]	Sawada *et al.* [55]
Synthetic	5 cm	SDR (dB)		0.9	0.2
		SIR (dB)		−2.7	4.4
		SAR (dB)		14.1	7.5
	1 m	SDR (dB)		0.7	0.6
		SIR (dB)		−0.4	4.2
		SAR (dB)		10.7	7.5
Recording	5 cm	SDR (dB)	1.6	1.2	1.8
		SIR (dB)	1.3	−1.9	4.2
		SAR (dB)	6.2	13.0	6.8
	1 m	SDR (dB)		2.1	3.6
		SIR (dB)		0.8	6.9
		SAR (dB)		8.0	6.8

bound for binary masking. This poor performance appears mostly due to low interference rejection. Indeed, while the range of SAR is comparable to that typically achieved on instantaneous mixtures in Table 19.6, the SIR is much lower and reaches 7 dB at most. Additional results over the same data can again be found in [67]. On average, performance is slightly higher with 1 m microphone spacing than with 5 cm spacing, which is in line with the effect of microphone spacing on the separation of over-determined recordings discussed in section 19.3.2.2. Typical WERs obtained when coupling SCA and automatic speech recognition are provided in [53].

19.4.3.3 Performance comparison of convolutive ICA and SCA

As explained earlier, the range of applications of SCA are not restricted to under-determined mixtures but also cover determined and over-determined mixtures that can be separated via convolutive ICA. Both approaches were compared in [55] on a set of determined or over-determined mixtures of two to three speech sources recorded by two to four microphones in a room with reverberation time $RT_{60} = 130$ ms. The average SIRI was equal to 14.4 dB for binary time-frequency masking and 16.7 dB for frequency-domain convolutive ICA. Time-frequency masking with $M \geq 2$ active sources per bin was unfortunately not considered, and hence a more general experimental comparison of SCA *vs.* ICA remains to be made.

19.5 CONCLUSION

In this chapter, we have experimentally shown that the general assumptions of BSS, namely source independence and sparsity, are valid for many audio signals. Some of the BSS algorithms reviewed in previous chapters, such as time-domain convolutive ICA or instantaneous SCA, can hence be readily applied to audio data. These assumptions appear insufficient, however, for the separation of under-determined convolutive mixtures. Specific properties of acoustic mixing filters are then employed in addition. These properties can also be exploited to speed up convolutive ICA by performing separation in the frequency domain. From a cognition perspective [68], all the algorithms presented in this chapter rely primarily on spatial cues, that is on the fact that the sources have different spatial positions and that the induced ITDs and IIDs over the mixture channels are somewhat constant over the frequency range.

While determined or over-determined mixtures with static sources and moderate reverberation can now be separated with satisfactory performance, few real-world audio signals meet these constraints. The separation of such signals remains a difficult problem, due to challenging mixing situations including a large number of sources, source movements and/or reverberation which decrease the amount of information carried by spatial cues. Theoretical performance upper bounds indicate that significant improvements could be achieved in the future, while remaining in the framework of time-frequency masking. For instance, suitable time-frequency masks may also be estimated by modeling dependencies between the magnitude STFT coefficients of each source via, for example, a hidden Markov model (HMM) [31] or via non-negative matrix factorization (NMF) (see Chapter 13). By exploiting spectro-temporal cues instead of spatial cues, this approach can potentially address the source separation problem for single-channel mixtures. A promising extension of this approach consists of building Bayesian models of audio signals jointly exploiting spatial and spectro-temporal cues, in the spirit of the auditory processing of complex sound scenes [68]. While early models were trained on separate data for each source [42,63], recent advances have made it possible to exploit such models in a blind framework by learning their parameters from a mixture [43]. We believe that research will ultimately result in high-quality BSS for a large range of real-world audio signals in the coming years and allow many applications of BSS to audio signal processing and information retrieval.

Acknowledgments

E. Vincent would like to thank Maria G. Jafari, Samer A. Abdallah, Mark D. Plumbley and Mike E. Davies for advice about parts of this chapter.

References

[1] F. Abrard, Y. Deville, A time-frequency blind signal separation method applicable to underdetermined mixtures of dependent sources, Signal Processing 85 (2005) 1389–1403.

[2] B. Albouy, Y. Deville, Improving noisy speech recognition with blind source separation methods: Validation with artificial mixtures, in: Proceedings of the 5th International Worshop on Electronics, Control, Modelling, Measurement and Signals, ECM2S01, 2001, p. 271.

[3] B. Albouy, Y. Deville, Alternative structures and power spectrum criteria for blind segmentation and separation of convolutive speech mixtures, in: Proceedings of the 4th International Symposium on Independent Component Analysis and Blind Signal Separation, ICA'03, 2003, pp. 361–366.

[4] S. Araki, S. Makino, Y. Hinamoto, R. Mukai, T. Nishikawa, H. Saruwatari, Equivalence between frequency-domain blind source separation and frequency-domain adaptive beamforming for convolutive mixtures, EURASIP Journal of Applied Signal Processing 11 (2003) 1157–1166.

[5] S. Araki, S. Makino, H. Sawada, R. Mukai, Reducing musical noise by a fine-shift overlap-add method applied to source separation using a time-frequency mask, in: Proceedings of the 2005 IEEE International Conference on Acoustics, Speech, and Signal Processing, ICASSP'05, 2005, pp. III–81 - III–84.

[6] S. Arberet, R. Gribonval, F. Bimbot, A robust method to count and locate audio sources in a stereophonic linear instantaneous mixture, in: Proceedings of the 6th International Conference on Independent Component Analysis and Blind Source Separation, ICA'06, 2006, pp. 536–543.

[7] C. Avendano, J.-M. Jot, Frequency domain techniques for stereo to multichannel upmix, in: Proceedings of the 22nd AES Conference on Vitual, Synthetic and Entertainment Audio, 2002, pp. 121–130.

[8] R.V. Balan, J.P. Rosca, S.T. Rickard, Robustness of parametric source demixing in echoic environments, in: Proceedings of the 3rd International Conference on Independent Component Analysis and Blind Signal Separation, ICA'01, 2001, pp. 144–148.

[9] B. Bartlett, J. Bartlett, Practical Recording Techniques: The Step-by-step Approach to Professional Recording, 4th ed., Focal Press, Boston, MA, 2005.

[10] P. Bofill, Identifying single source data for mixing matrix estimation in instantaneous blind source separation, in: Proceedings of the 18th International Conference on Artificial Neural Networks, ICANN'08, 2008, pp. 759–767.
[11] P. Bofill, E. Monte, Underdetermined convoluted source reconstruction using LP and SOCP, and a neural approximator of the optimizer, in: Proceedings of the 6th International Conference on Independent Component Analysis and Blind Source Separation, ICA'06, 2006, pp. 569–576.
[12] M.S. Brandstein, D.B. Ward (Eds.), Microphone Arrays: Signal Processing Techniques and Applications, Springer, New York, NY, 2001.
[13] N. Charkani, Séparation auto-adaptative de sources pour les mélanges convolutifs. Application à la téléphonie mains-libres dans les voitures, PhD thesis, INP Grenoble, France, 1996.
[14] N. Charkani, Y. Deville, Self-adaptive separation of convolutively mixed signals with a recursive structure. Part II: Theoretical extensions and application to synthetic and real signals, Signal Processing 75 (1999) 117–140.
[15] S. Choi, A. Cichocki, Adaptive blind separation of speech signals: Cocktail party problem, in: Proceedings of the International Conference on Speech Processing, ICSP'97, 1997, pp. 617–622.
[16] Y. Deville, M. Puigt, Temporal and time-frequency correlation-based blind source separation methods. Part I: Determined and Underdetermined Linear Instantaneous Mixtures, Signal Processing 87 (2007) 374–407.
[17] Y. Deville, M. Puigt, B. Albouy, Time-frequency blind signal separation: Extended methods, performance evaluation for speech sources, in: Proceedings of the International Joint Conference on Neural Networks, IJCNN'04, 2004, pp. 255–260.
[18] S.C. Douglas, H. Sawada, S. Makino, Natural gradient multichannel blind deconvolution and speech separation using causal FIR filters, IEEE Transactions on Speech and Audio Processing 13 (2005) 92–104.
[19] F. Ehlers, H.G. Schuster, Blind separation of convolutive mixtures and an application in automatic speech recognition in a noisy environment, IEEE Transactions on Signal Processing 45 (1997) 2608–2612.
[20] D.P.W. Ellis, Evaluating speech separation systems, in: P. Divenyi (Ed.), Speech Separation by Humans and Machines, Kluwer Academic Publishers, New York, NY, 2004, pp. 295–304.
[21] M. Fallon, S.J. Godsill, A. Blake, Joint acoustic source location and orientation estimation using sequential Monte Carlo, in: Proceedings of the 9th International Conference on Digital Audio Effects, DAFx'06, 2006, pp. 203–208.
[22] B. Fox, A. Sabin, B. Pardo, A. Zopf, Modeling perceptual similarity of audio signals for blind source separation evaluation, in: Proceedings of the 7th International Conference on Independent Component Analysis and Signal Separation, ICA'07, 2007, pp. 454–461.
[23] C.M. Gallippi, G.E. Trahey, Adaptive clutter filtering via blind source separation for two-dimensional ultrasonic blood velocity measurement, Ultrasonic imaging 24 (2002) 193–214.
[24] M. Gupta, S.C. Douglas, Beamforming initialization and data prewhitening in natural gradient convolutive blind source separation of speech mixtures, in: Proceedings of the 7th International Conference on Independent Component Analysis and Signal Separation, ICA'07, 2007, pp. 462–470.
[25] F.M. Ham, N.A. Faour, J.C. Wheeler, Infrasound signal separation using independent component analysis, in: Proceedings of the 21st Seismic Research Symposium, SRS'99: Technologies for Monitoring the Comprehensive Nuclear-Test-Ban Treaty, 1999, pp. II–133 – II–140.
[26] S. Ikeda, N. Murata, An approach to blind source separation of speech signals, in: Proceedings of the 8th International Conference on Artificial Neural Networks, ICANN'98, 1998, pp. 761–766.
[27] M.Z. Ikram, D.R. Morgan, A beamforming approach to permutation alignment for multichannel frequency-domain blind speech separation, in: Proceedings of the 2002 IEEE International Conference on Acoustics, Speech, and Signal Processing, ICASSP'02, 2002, pp. I–881 – I–884.
[28] M. Ito, Y. Takeuchi, T. Matsumoto, H. Kudo, M. Kawamoto, T. Mukai, N. Ohnishi, Moving-source separation using directional microphones, in: Proceedings of the 2nd IEEE International Symposium on Signal Processing and Information Technology, ISSPIT'02, 2002, pp. 523–526.

[29] Y. Izumi, N. Ono, S. Sagayama, Sparseness-based 2ch BSS using the EM algorithm in reverberant environment, in: Proceedings of the 2007 IEEE Workshop on Applications of Signal Processing to Audio and Acoustics, WASPAA'07, 2007, pp. 147–150.

[30] A. Kraskov, H. Stögbauer, P. Grassberger, Estimating mutual information, Physical Review E 69 (2004).

[31] T.T. Kristjánsson, J.R. Hershey, P.A. Olsen, S.J. Rennie, R.A. Gopinath, Super-human multi-talker speech recognition: The IBM 2006 speech separation challenge system, in: Proceedings of the 9th International Conference on Spoken Language Processing, Interspeech'06, 2006, pp. 97–100.

[32] H. Kuttruff, Room Acoustics, 4th ed., Spon Press, London, UK, 2000.

[33] R.H. Lambert, A.J. Bell, Blind separation of multiple speakers in a multipath environment, in: Proceedings of the 1997 IEEE International Conference on Acoustics, Speech, and Signal Processing, ICASSP'97, 1997, pp. I-423 - I-426.

[34] M.E. Lockwood, D.L. Jones, R.C. Bilger, C.R. Lansing, W.D. O'Brien, B.C. Wheeler, A.S. Feng, Performance of time- and frequency-domain binaural beamformers based on recorded signals from real rooms, Journal of the Acoustical Society of America 115 (2004) 379–391.

[35] S. Makino, T.-W. Lee, H. Sawada (Eds.), Blind Speech Separation, Springer, Dordrecht, the Netherlands, 2007.

[36] M.I. Mandel, D.P.W. Ellis, T. Jebara, An EM algorithm for localizing multiple sound sources in reverberant environments, in: Advances in Neural Information Processing Systems, NIPS 19, 2007, pp. 953–960.

[37] N. Mitianoudis, M.E. Davies, Audio source separation of convolutive mixtures, IEEE Transactions on Speech and Audio Processing 11 (2003) 489–497.

[38] N. Mitianoudis, T. Stathaki, Underdetermined source separation using mixtures of warped Laplacians, in: Proceedings of the 7th International Conference on Independent Component Analysis and Signal Separation, ICA'07, 2007, pp. 236–243.

[39] R. Mukai, S. Araki, H. Sawada, S. Makino, Evaluation of separation and dereverberation performance in frequency domain blind source separation, Acoustical Science and Technology 25 (2004) 119–126.

[40] R. Mukai, H. Sawada, S. Araki, S. Makino, Blind source separation for moving speech signals using blockwise ICA and residual crosstalk subtraction, IEICE Transactions on Fundamentals of Electronics, Communications and Computer Sciences E87-A (2004) 1941–1948.

[41] H.-L. Nguyen Thi, C. Jutten, Blind source separation for convolutive mixtures, Signal Processing 45 (1995) 209–229.

[42] J. Nix, V. Hohmann, Combined estimation of spectral envelopes and sound source direction of concurrent voices by multidimensional statistical filtering, IEEE Transactions on Audio, Speech and Language Processing 15 (2007) 995–1008.

[43] A. Ozerov, C. Févotte, Multichannel non-negative matrix factorization in convolutive mixtures with application to blind audio source separation, in: IEEE International Conference on Acoustics, Speech and Signal Processing, ICASSP'09, Taipei, Taiwan, April 19–24, 2009, pp. 3137–3140.

[44] L.C. Parra, C.V. Alvino, Geometric source separation: merging convolutive source separation with geometric beamforming, IEEE Transactions on Speech and Audio Processing 10 (2002) 352–362.

[45] L.C. Parra, C. Spence, Convolutive blind separation of non-stationary sources, IEEE Transactions on Speech and Audio Processing 8 (2000) 320–327.

[46] M.S. Pedersen, J. Larsen, U. Kjems, L.C. Parra, A survey of convolutive blind source separation methods, in: J. Benesty, M.M. Sondhi, Y. Huang (Eds.), Springer Handbook of Speech Processing, Springer, New York, NY, 2007, pp. 1065–1094.

[47] L.D. Persia, M. Yanagida, H.L. Rufiner, D. Milone, Objective quality evaluation in blind source separation for speech recognition in a real room, Signal Processing 87 (2007) 1951–1965.

[48] D.-T. Pham, C. Servière, H. Boumaraf, Blind separation of speech mixtures based on nonstationarity, in: Proceedings of the 7th International Symposium on Signal Processing and its Applications, ISSPA'03, 2003, pp. II–73 - II–76.

[49] A. Pietilä, M. El-Segaier, R. Vigário, E. Pesonen, Blind source separation of cardiac murmurs from heart recordings, in: Proceedings of the 6th International Conference on Independent Component Analysis and Blind Source Separation, ICA'06, 2006, pp. 470–477.

[50] M. Puigt, Y. Deville, Time-frequency ratio-based blind separation methods for attenuated and time-delayed sources, Mechanical Systems and Signal Processing 19 (2005) 1348–1379.

[51] M. Puigt, E. Vincent, Y. Deville, Validity of the independence assumption for the separation of instantaneous and convolutive mixtures of speech and music sources, in: Proceedings of the 8th International Conference on Independent Component Analysis and Signal Separation, ICA'09, 2009.

[52] K. Rahbar, J.P. Reilly, A frequency domain method for blind source separation of convolutive audio mixtures, IEEE Transactions on Speech and Audio Processing 13 (2005) 832–844.

[53] N. Roman, S. Srinivasan, D.L. Wang, Binaural segregation in multisource reverberant environments, Journal of the Acoustical Society of America 120 (2006) 4040–4051.

[54] J.P. Rosca, C. Borss, R.V. Balan, Generalized sparse signal mixing model and application to noisy blind source separation, in: Proceedings of the 2004 IEEE International Conference on Acoustics, Speech, and Signal Processing, ICASSP'04, 2004, pp. III–877 – III–880.

[55] H. Sawada, S. Araki, R. Mukai, S. Makino, Grouping separated frequency components with estimating propagation model parameters in frequency-domain blind source separation, IEEE Transactions on Audio, Speech and Language Processing 15 (2007) 1592–1604.

[56] H. Sawada, R. Mukai, S. Araki, S. Makino, A robust and precise method for solving the permutation problem of frequency-domain blind source separation, IEEE Transactions on Speech and Audio Processing 12 (2004) 530–538.

[57] H. Sawada, R. Mukai, S.F.G.M. de la Kethulle de Ryhove, S. Araki, S. Makino, Spectral smoothing for frequency domain blind source separation, in: Proceedings of the 2003 International Workshop on Acoustic Echo and Noise Control, IWAENC'03, 2003, pp. 311–314.

[58] D. Schobben, K. Torkkola, P. Smaragdis, Evaluation of blind signal separation methods, in: Proceedings of the 1st International Workshop on Independent Component Analysis and Blind Signal Separation, ICA'99, 1999, pp. 261–266.

[59] P. Smaragdis, Blind separation of convolved mixtures in the frequency domain, Neurocomputing 22 (1998) 21–34.

[60] D. Smith, J. Lukasiak, I.S. Burnett, An analysis of the limitations of blind signal separation applications with speech, Signal Processing 86 (2006) 353–359.

[61] J. Thomas, Y. Deville, S. Hosseini, Time-domain fast fixed-point algorithms for convolutive ICA, IEEE Signal Processing Letters 13 (2006) 228–231.

[62] M. Togami, T. Sumiyoshi, A. Amano, Sound source separation of overcomplete convolutive mixtures using generalized sparseness, in: Proceedings of the 2006 International Workshop on Acoustic Echo and Noise Control, IWAENC'06, 2006.

[63] E. Vincent, Musical source separation using time-frequency source priors, IEEE Transactions on Audio, Speech and Language Processing 14 (2006) 91–98.

[64] E. Vincent, Complex nonconvex l_p norm minimization for underdetermined source separation, in: Proceedings of the 7th International Conference on Independent Component Analysis and Signal Separation, ICA'07, 2007, pp. 430–437.

[65] E. Vincent, R. Gribonval, C. Févotte, Performance measurement in blind audio source separation, IEEE Transactions on Audio, Speech and Language Processing 14 (2006) 1462–1469.

[66] E. Vincent, R. Gribonval, M.D. Plumbley, Oracle estimators for the benchmarking of source separation algorithms, Signal Processing 87 (2007) 1933–1950.

[67] E. Vincent, H. Sawada, P. Bofill, S. Makino, J.P. Rosca, First stereo audio source separation evaluation campaign: Data, algorithms and results, in: Proceedings of the 7th International Conference on Independent Component Analysis and Signal Separation, ICA'07, 2007, pp. 552–559.

[68] D.L. Wang, G.J. Brown (Eds.), Computational Auditory Scene Analysis: Principles, Algorithms and Applications, Wiley, New York, NY, 2006.

[69] S. Wehr, A. Lombard, H. Buchner, W. Kellermann, "Shadow BSS" for blind source separation in rapidly time-varying acoustic scenes, in: Proceedings of the 7th International Conference on Independent Component Analysis and Signal Separation, ICA'07, 2007, pp. 560–568.

[70] S. Winter, H. Sawada, S. Araki, S. Makino, Overcomplete BSS for convolutive mixtures based on hierarchical clustering, in: Proceedings of the 5th International Conference on Independent Component Analysis and Blind Signal Separation, ICA'04, 2004, pp. 652–660.

[71] H.-C. Wu, J.C. Príncipe, Generalized anti-Hebbian learning for source separation, in: Proceedings of the 1999 IEEE International Conference on Acoustics, Speech, and Signal Processing, ICASSP'99, 1999, pp. II–1073 - II–1076.

[72] Z. Xinhua, Z. Anqing, F. Jianping, Y. Shaoqing, Study on blind separation of underwater acoustic signals, in: Proceedings of the 5th IEEE International Conference on Signal Processing, ICSP'00, 2000, pp. III–1802 - III–1805.

[73] O. Yılmaz, S.T. Rickard, Blind separation of speech mixtures via time-frequency masking, IEEE Transactions on Signal Processing 52 (2004) 1830–1847.

[74] M. Zibulevsky, B.A. Pearlmutter, P. Bofill, P. Kisilev, Blind source separation by sparse decomposition in a signal dictionary, in: S.J. Roberts, R.M. Everson (Eds.), Independent Components Analysis: Principles and Practice, Cambridge University Press, Cambridge, UK, 2001, pp. 181–208.

Glossary

$\mathbf{x}$	vector of components x_p, $1 \leq p \leq P$
s, x, y	sources, observations, separator outputs
N	number of sources
P	number of sensors
T	number of observed samples
$\star$	convolution
$\mathbf{A}$	matrix with components A_{ij}
$\mathbf{A}$, $\mathbf{B}$	mixing and separation matrices
$\mathbf{G}$, $\mathbf{W}$, $\mathbf{Q}$	global, whitening, and separating unitary matrices
$\check{g}$	Fourier transform of g
$\widehat{\mathbf{s}}$	estimate of quantity $\mathbf{s}$
$p_\mathbf{x}$	probability density of $\mathbf{x}$
ψ	joint score function
φ_i	marginal score function of source s_i
Φ	first characteristic function
Ψ	second characteristic function
$\mathbb{E}\mathbf{x}$, $\mathbb{E}\{\mathbf{x}\}$	mathematical expectation of $\mathbf{x}$
$I\{\mathbf{y}\}$ or $I(p_\mathbf{y})$	mutual information of $\mathbf{y}$
$K\{\mathbf{x};\mathbf{y}\}$ or $K(p_\mathbf{x}; p_\mathbf{y})$	Kullback divergence between $p_\mathbf{x}$ and $p_\mathbf{y}$
$H\{\mathbf{x}\}$ or $H(p_\mathbf{x})$	Shannon entropy $\mathbf{x}$
$\mathscr{L}$	likelihood
$\mathscr{A}$, $\mathscr{B}$	mixing, and separating (nonlinear) operators
$\mathrm{cum}\{x_1, \ldots, x_P\}$	joint cumulant of variables $\{x_1, \ldots, x_P\}$
$\mathrm{cum}_R\{y\}$	marginal cumulant of order R of variable y

$\mathbf{Q}^{\mathrm{T}}$	transposition of matrix $\mathbf{Q}$
$\mathbf{Q}^{\mathrm{H}}$	conjugate transposition of matrix $\mathbf{Q}$
$\mathbf{Q}^{*}$	complex conjugation of matrix $\mathbf{Q}$
$\mathbf{Q}^{\dagger}$	pseudo-inverse of matrix $\mathbf{Q}$
Υ	contrast function
$\mathbb{R}$	real field
$\mathbb{C}$	complex field
$\widehat{\mathbf{A}}$	estimator of mixing matrix
$\mathrm{diag}\,\mathbf{A}$	vector whose components are the diagonal of matrix $\mathbf{A}$
$\mathrm{Diag}\,\mathbf{a}$	diagonal matrix whose entries are those of vector $\mathbf{a}$
$\mathrm{trace}\,\mathbf{A}$	trace of matrix $\mathbf{A}$
$\det\mathbf{A}$	determinant of matrix $\mathbf{A}$
$\mathrm{mean}\,\mathbf{a}$	arithmetic average of component of vector $\mathbf{a}$
$\check{s}(\nu)$	Fourier transform of process $s(t)$
$\otimes$	Kronecker product between matrices
$\otimes$	tensor product
$\bullet_j$	contraction over index j
$\mathrm{krank}\{\mathbf{A}\}$	Kruskal's k-rank of matrix $\mathbf{A}$

Index

D

E

T